Evolution and Ecology of Zooplankton Communities

Evolution and Ecology of Zooplankton Communities

Special Symposium Volume 3
American Society of Limnology
and Oceanography

W. Charles Kerfoot, Editor

University Press of New England
Hanover, New Hampshire and
London, England   1980

Copyright © 1980 by Trustees of Dartmouth College
All rights reserved
Library of Congress Catalog Card Number 80-50491
International Standard Book Number 0-87451-180-1
Printed in the United States of America

Library of Congress Cataloging in Publication data
will be found on the last printed page of this book.

Proceedings of a symposium on the structure of zooplankton communities held at Dartmouth College on 20-25 August 1978, sponsored by The American Society of Limnology and Oceanography, The National Science Foundation, and Dartmouth College.

The University Press of New England

Sponsoring Institutions:

Brandeis University
Clark University
Dartmouth College
University of New Hampshire
University of Rhode Island
Tufts University
University of Vermont

Contents

Preface   xi
Opening Remarks   xv
Section Organizers   xvii
Authors   xix

I. Life at Low Reynolds Numbers, Swimming Behavior, Orientation in Time and Space   1

1. The Animal and Its Viscous Environment
   Robert E. Zaret   3
2. Visual Observations of Live Zooplankters: Evasion, Escape, and Chemical Defenses
   W. Charles Kerfoot, Dean L. Kellogg, Jr., and J. Rudi Strickler   10
3. Optical Orientation of Pelagic Crustaceans and Its Consequence in the Pelagic and Littoral Zones
   H. Otto Siebeck   28
4. Morphological, Physiological, and Behavioral Aspects of Mating in Calanoid Copepods
   Pamela I. Blades and Marsh J. Youngbluth   39
5. Adaptive Responses to Encounter Problems
   Jeroen Gerritsen   52

II. Directed Movements: Vertical Migration   63

6. Introductory Remarks: Causal and Teleological Aspects of Diurnal Vertical Migration
   J. Ringelberg   65
7. Vertical Migrations of Zooplankton in the Arctic: A Test of the Environmental Controls
   Claire Buchanan and James F. Haney   69
8. Diurnal Vertical Migration in Aquatic Microcrustacea: Light and Oxygen Responses of Littoral Zooplankton
   Dewey G. Meyers   80
9. Aspects of Red Pigmentation in Zooplankton, Especially Copepods
   J. Ringelberg   91
10. The Vertical Distribution of Diaptomid Copepods in Relation to Body Pigmentation
    Nelson G. Hairston, Jr.   98

Contents

11. Experimental Studies on Diel Vertical Migration
    Richard N. Bohrer 111
12. Seasonal Patterns of Vertical Migration: A Model for *Chaoborus trivittatus*
    Louis A. Giguère and Lawrence M. Dill 122
13. Diel Vertical Migration of Pelagic Water Mites
    Howard P. Riessen 129
14. Adaptive Value of Vertical Migration: A Simulation Model Argument for the Predation Hypothesis
    David Wright, W. John O'Brien, and Gary L. Vinyard 138

III. Rotifer Feeding and Population Dynamics 149

15. Behavioral Determinants of Diet Quantity and Diet Quality in *Brachionus calyciflorus*
    Peter L. Starkweather 151
16. Feeding in the Rotifer *Asplanchna:* Behavior, Cannibalism, Selectivity, Prey Defenses, and Impact on Rotifer Communities
    John J. Gilbert 158
17. A Chemostat System for the Study of Rotifer-Algal-Nitrate Interactions
    Martin E. Boraas 173

IV. The Copepod Filter-Feeding Controversy 183

18. Comparative Morphology and Functional Significance of Copepod Receptors and Oral Structures
    Marc M. Friedman 185
19. Chemosensory Feeding and Food-Gathering by Omnivorous Marine Copepods
    S. A. Poulet and P. Marsot 198
20. Grazing Interactions among Freshwater Calanoid Copepods
    Sumner Richman, Scott A. Bohon, and Stephen E. Robbins 219
21. Grazing Interactions in the Marine Environment
    Percy L. Donaghay 234
22. Catching the Algae: A First Account of Visual Observations on Filter-Feeding Calanoids
    Miguel Alcaraz, Gustav-Adolf Paffenhöfer, and J. Rudi Strickler 241

V. Plant-Herbivore Interactions with Emphasis upon Cladocerans 249

23. Nutrient Recycling as an Interface between Algae and Grazers in Freshwater Communities
    John T. Lehman 251
24. The Importance of "Threshold" Food Concentrations
    Winfried Lampert and Ursula Schober 264
25. Nutritional Adequacy, Manageability, and Toxicity as Factors that Determine the Food Quality of Green and Blue-Green Algae for *Daphnia*
    Karen Glaus Porter and John D. Orcutt, Jr. 268
26. Filtering Rates, Food Size Selection, and Feeding Rates in Cladocerans—Another Aspect of Interspecific Competition in Filter-Feeding Zooplankton
    Z. Maciej Gliwicz 282

27. Resource Characteristics Modifying Selective Grazing by Copepods
    Donald C. McNaught, David Griesmer, and Michele Kennedy ... 292
28. *Aphanizomenon* Blooms: Alternate Control and Cultivation by *Daphnia pulex*
    Michael Lynch ... 299
29. Zooplankton as Algal Chemostats: A Theoretical Perspective
    Robert A. Armstrong ... 305

VI. Genetics, Demographics, and Life Histories of Zooplankton ... 313

30. The Genetic Structure of Zooplankton Populations
    Charles E. King ... 315
31. The Genetics of Cladocera
    Paul D. N. Hebert ... 329
32. An Analysis of the Precision of Birth and Death Rate Estimates for Egg-Bearing Zooplankters
    William R. DeMott ... 337
33. Habitat Selection and Population Growth of Two Cladocerans in Seasonal Environments
    Stephen T. Threlkeld ... 346
34. Seasonal Variation in the Sizes at Birth and at First Reproduction in Cladocera
    David Culver ... 358
35. Predation, Enrichment, and the Evolution of Cladoceran Life Histories: A Theoretical Approach
    Michael Lynch ... 367
36. Size-Selective Predation on Zooplankton
    Barbara E. Taylor ... 377
37. Some Aspects of Reproductive Variation among Freshwater Zooplankton
    J. David Allan and Clyde E. Goulden ... 388
38. Evolutionary Aspects of Diapause in Freshwater Copepods
    Kåre Elgmork ... 411
39. When and How to Reproduce: A Dilemma for Limnetic Cyclopoid Copepods
    Jens Petter Nilssen ... 418

VII. Cyclomorphosis ... 427

40. Environmental Control of Cladoceran Cyclomorphosis via Target-Specific Growth Factors in the Animal
    Jürgen Jacobs ... 429
41. Seasonal Changes in Size at Maturity in Small Pond *Daphnia*
    Donald J. Brambilla ... 438
42. The Genetic Component of Cyclomorphosis in *Bosmina*
    Robert W. Black ... 456
43. Perspectives on Cyclomorphosis: Separation of Phenotypes and Genotypes
    W. Charles Kerfoot ... 470
44. Dimorphic *Daphnia longiremis*: Predation and Competitive Interactions between the Two Morphs
    W. John O'Brien, Dean Kettle, Howard Riessen, David Schmidt, and David Wright ... 497

VIII. Predation, Prey Vulnerability, Zooplankton Composition — 507

45. Variation among Zooplankton Predators: The Potential of *Asplanchna*, *Mesocyclops*, and *Cyclops* to Attack, Capture, and Eat Various Rotifer Prey
    Craig E. Williamson and John J. Gilbert — 509
46. The Predatory Feeding of Copepodid Stages III to Adult *Mesocyclops leuckarti* (Claus)
    C. D. Jamieson — 518
47. Selection of Prey by *Chaoborus* Larvae: A Review and New Evidence for Behavioral Flexibility
    Robert A. Pastorok — 538
48. The Effects of an Introduced Invertebrate Predator and Food Resource Variation on Zooplankton Dynamics in an Ultraoligotrophic Lake
    Stephen T. Threlkeld, James T. Rybock, Mark D. Morgan, Carol L. Folt, and Charles R. Goldman — 555
49. Odonate "Hide and Seek": Habitat-Specific Rules?
    Dan M. Johnson and Philip H. Crowley — 569
50. Alewives (*Alosa pseudoharengus*) and Ciscoes (*Coregonus artedii*) as Selective and Non-Selective Planktivores
    John Janssen — 580
51. The Roles of Zooplankter Escape Ability and Fish Size Selectivity in the Selective Feeding and Impact of Planktivorous Fish
    Ray W. Drenner and Steven R. McComas — 587
52. The Effect of Prey Motion on Planktivore Choice
    Thomas M. Zaret — 594
53. Selective Predation by Zooplankton and the Response of Cladoceran Eyes to Light
    John L. Confer, Gregory Applegate, and Christine A. Evanik — 604
    Commentary: Transparency, Body Size, and Prey Conspicuousness
    W. Charles Kerfoot — 609
54. Predation Pressure from Fish on Two *Chaoborus* Species as Related to Their Visibility
    Jan A. E. Stenson — 618

IX. Spatial and Temporal Aspects of Community Structure: Micro- and Macrogeographic Patterns — 623

55. Evidence for Stable Zooplankton Community Structure Gradients Maintained by Predation
    William M. Lewis, Jr. — 625
56. Relationships between Trout and Invertebrate Species as Predators and the Structure of the Crustacean and Rotiferan Plankton in Mountain Lakes
    R. Stewart Anderson — 635
57. Zoogeographic Patterns in the Size Structure of Zooplankton Communities with Possible Applications to Lake Ecosystem Modeling and Management
    W. Gary Sprules — 642
58. Chydorid Cladoceran Assemblages from Subtropical Florida
    Thomas L. Crisman — 657
59. Structure of Zooplankton Communities in the Peten Lake District, Guatemala
    Edward S. Deevey, Jr., Georgiana B. Deevey, and Mark Brenner — 669

60. Systematic Problems and Zoogeography in Cyclopoids
    U. Einsle 679
61. Zooplankton and the Science of Biogeography: The Example of Africa
    Henri Jean Dumont 685
62. Species Richness and Area in Galapagos and Andean Lakes: Equilibrium
    Phytoplankton Communities and a Paradox of the Zooplankton
    Paul Colinvaux and Miriam Steinitz 697

X. Community Structure: Experimental and Theoretical Approaches 713

63. Breaking the Bottleneck: Interactions of Invertebrate Predators and
    Nutrients in Oligotrophic Lakes
    William E. Neill and Adrienne Peacock 715
64. Foundations for Evaluating Community Interactions: The Use of
    Enclosures to Investigate Coexistence of *Daphnia* and *Bosmina*
    W. Charles Kerfoot and William R. DeMott 725
65. The Inadequacy of Body Size as an Indicator of Niches in the Zooplankton
    B. W. Frost 742
66. Dynamic Energy-Flow Model of the Particle Size Distribution in Pelagic
    Ecosystems
    William Silvert and Trevor Platt 754

Appendix. The Program 765
Index of Species 773
General Index 781

Preface

Early in 1975, sensing a general interest for a conference on the evolution and ecology of zooplankton communities and some dissatisfaction with the restrictive format of conventional meetings, Rudi Strickler (then at Johns Hopkins University) and I circulated a brief inquiry. The letter and attached questionnaire proposed an organized program but one to be conducted informally, if sufficient interest was voiced by aquatic ecologists. The response to the questionnaire was enthusiastic, endorsements coming from workers in thirteen countries (United States, Canada, Great Britain, Belgium, New Zealand, Spain, Czechoslavakia, West Germany, Sweden, The Netherlands, Norway, Poland, and Austria). Many of the respondents generously agreed to donate their time and effort to organizing the conference.

Because of potential financial and academic conflict with the 20th Congress of the International Society of Theoretical and Applied Limnology (SIL) in Copenhagen, Denmark, during the summer of 1977, the special symposium was scheduled for late in the summer of 1978. During the intervening years the initial excitement continued as more details on organisms emerged. Even within the general public there was a growing awareness of aquatic environments and a growing concern about deteriorating water quality.

Not only are there important economic considerations closely tied to lake management and eutrophication (since the zooplankton community serves as an important trophic link between phytoplankton and fish), but there are also strong academic reasons for investing in aquatic research. In aquatic ecology, the relatively short lifespan of planktonic organisms, the opportunity to use enclosure experiments to manipulate field populations, the structural simplicity of pelagic communities, and the clearly defined limits of lake ecosystems have all played central roles in the development of a vital and sturdy branch of science. The resulting theoretical development of experimentally testable hypotheses based upon biological interactions, and the often dramatic results of simple exclusion or perturbation experiments, have focused attention on aquatic communities. Subsequent research upon competitive and predatory interactions has spurred attempts to formulate models of community structure. The resulting controversies have also attracted a host of bright young students to the field.

It is to this spirit of active inquiry that the volume is dedicated. The symposium is the third in a series of special conferences, initially sanctioned in 1970, when the American Society of Limnology and Oceanography decided that it should take a leadership role in public affairs relating to aquatic resources. Additionally, within the membership there is increasing dissatisfaction with the inability of traditional conference formats to inspire either synthesis or creative thought. At most large conferences there is usually inadequate time alloted for discussion, research summaries are crammed into stiflingly short intervals, papers are haphazardly tossed together under loosely organized general topics, and concurrent sessions divide the time and attention of most participants. To circumvent these problems, in Special Symposium III the format would be simple. In place of the traditional schedule we would substitute a carefully organized program, with frequent breaks, aimed at creating an atmosphere for creative ferment, synthesis, and lively debate. Papers would be centered on fundamental subject areas, topics would build one upon another, and themes would progress from the level of individual behavior through population dynamics to community structure in such a way as to promote the exchange of ideas.

The symposium was held at Dartmouth College, 20–25 August 1978. Obviously many people contributed to the success of the meetings, and aided publication of its proceedings. I especially want to thank the section chairmen and chairwoman, discussion moderators, supporting staff, and participants for their continued effort and enthusiasm. Special credit goes to Gene E. Likens and George Saunders, past presidents of the American Society of Limnology and Oceanography (ASLO), for their advice and continued encouragement; to the National Science Foundation, for its support (NSF DEB78-19557) of foreign speakers; and to Lyn Cole for her efforts in copy editing the formidable number of manuscripts that resulted. I also want to thank several persons on the local staff, whose efforts made the meetings and correspondence go so smoothly: Lucie Zelazny, Mona Mort, Charles Levitan, Richard Aronson, Dean Kellogg, Mary Poulson, Carol Green, Eve Parnell, Robert MacMillen, and his staff at the Dartmouth Conference Center.

All manuscripts were reviewed, edited, and tailored somewhat toward the style of the ASLO journal, *Limnology and Oceanography* (*L&O*). I assumed an editorial prerogative in rearranging the order of certain "section" and "contributed" papers. For those who might be interested in the original sequence, the program is printed in the Appendix. I wish to extend my final thanks to Tom McFarland and the staff at the University Press of New England for their aid and assistance in publishing such a lengthy volume at a reasonable price.

Contributions focused upon several central themes: the importance of appreciating behavior, the manner by which organisms encounter and capture their food, and the clear distinction of the difference between processes which usually limit population growth (e.g., food reserves) and those which can heavily influence individual fitness (e.g., selective predation). New and exciting among the contributions were (1) the first clear demonstration (in talks and films) that "filter-feeding" in copepods is not filtering at all, but a high-speed, discriminative manipulation of currents and individual particles, (2) a reemphasis upon nutrients regenerated by zooplankton grazing, (3) the recognition that the seasonal population dynamics of rotifers and cladocerans involve successional events, for the first time termed "replacement cycles," and (4) the beginnings of a tendency to recognize that, for some as yet unknown reason, tropical freshwater zooplankton communities are impoverished. Common among participants was the desire for a fusion of newly emerging concepts of food-gathering and predation into a synthetic, predictive theory. Some of the more theoretical papers pointed toward this goal.

Regrettably, the papers in this volume show a bias toward freshwater environments. Part of the bias is intentional, a consequence of the decision by the principal organizers to emphasize biological interactions in easily manipulated communities. The physical hardships of investigating interactions among oceanic species and the bewildering numbers of marine invertebrates suggest that comparable insights will come harder for these systems. Just as clearly, work in both freshwater and marine systems is still in its infancy. I can think of no finer tribute to the participants of this symposium than that their contributions will inspire efforts toward a joint understanding of both freshwater and marine zooplankton communities.

My hope is that the student and professional will find the collection as exciting and interesting to read as we have found it challenging to assemble.

Hanover, N.H. *W. Charles Kerfoot*
August 1979

Opening Remarks

Nearly a decade ago the American Society of Limnology and Oceanography, Inc., instituted a series of special symposia. The notion was that the symposia would be held on an irregular basis and should treat special subjects of any conceivable topic for which there was a recognized need to analyze, digest, distill, and synthesize information. The first two symposia treated theoretical and applied aspects of water pollution. Although this third ASLO special symposium does not treat pollution directly, it provides insight into aquatic perturbations by placing into perspective the basic precepts of zooplankton ecology.

We have recognized for some time an underlying bubbling ferment within zooplankton ecology. In the past, largely because of the diverse nature of practitioners and publications, it was difficult to discern what in fact was the state of the science. The idea for a symposium on zooplankton was born in 1974, both as a stimulus for synthesis and as an opportunity for reflection on how zooplankton interact with both fishes and phytoplankton. The efforts came to fruition in August 1978, in an informal symposium that included brief individual presentations, panel discussions, and open constructively critical discussions from the floor. Participation was limited in the number of participants but diverse in their interest and geographical range, as the open nature of discussions required some restrictions on direct participation. The setting for the symposium was superb. Dartmouth College provided an environment that was both beautiful, peaceful, and without distracting influences. The enthusiasm of the participants for the subject matter was abundantly obvious: they attacked central questions with both optimism and healthy skepticism. The tenor of the symposium was one of unqualified success.

ASLO, of course, is pleased to have helped to organize the symposium. We are particularly indebted to Dr. W. C. Kerfoot for his dedication to the idea, the organization, the conduct, and the publication of the symposium. We hope that the enthusiasm of the participants will be conveyed as a landmark publication and as a document of special interest to many.

George W. Saunders, President
American Society of Limnology
and Oceanography, Inc.

Section Organizers*

J. D. Allan (genetics, life histories)
University of Maryland
College Park, Maryland

J. J. Gilbert (rotifer feeding)
Dartmouth College
Hanover, New Hampshire

W. C. Kerfoot (cyclomorphosis, invertebrate predation)
Dartmouth College
Hanover, New Hampshire

D. G. Meyers (littoral ecology)
University of Wisconsin
Madison, Wisconsin

W. J. O'Brien (cyclomorphosis, invertebrate and vertebrate predation)
University of Kansas
Lawrence, Kansas

K. G. Porter (plant-herbivore interactions, zoogeography)
University of Georgia
Athens, Georgia

S. Richman (filter-feeding dynamics)
Lawrence University
Appleton, Wisconsin

J. R. Strickler (behavior, vertical migration)
University of Ottawa
Ottawa, Canada

T. M. Zaret (vertical migration)
University of Washington
Seattle, Washington

*addresses are those at the time of the conference

Authors

Miguel Alcaraz   Instituto de Investigaciones Pesqueras, Barcelona, Spain

J. David Allan   Department of Zoology, University of Maryland, College Park, Maryland 20742, U.S.A.

R. Stewart Anderson   Canadian Wildlife Service, c/o Biology Department, University of Calgary, Calgary, Alberta T2N 1N4, Canada

Gregory Applegate   Biology Department, Ithaca College, Ithaca, New York 14850, U.S.A.

Robert A. Armstrong   Department of Ecology and Evolution, State University of New York, Stony Brook, New York 11794, U.S.A.

Robert W. Black   Department of Ecology and Evolution, State University of New York, Stony Brook, New York 11794, U.S.A.

Pamela I. Blades   Harbor Branch Foundation, Inc., RRI, Box 196, Fort Pierce, Florida 33450, U.S.A.

Scott A. Bohon   Department of Biology, Lawrence University, Appleton, Wisconsin 54911, U.S.A.

Richard N. Bohrer   Department of Oceanography, Dalhousie University, Halifax, Nova Scotia B3H 4J1, Canada

Martin E. Boraas   Department of Biology, The Pennsylvania State University, University Park, Pennsylvania 16802, U.S.A.

Donald J. Brambilla   Department of Biology, University of South Carolina, Columbia, South Carolina 29208, U.S.A.

Mark Brenner   Florida State Museum, University of Florida, Gainesville, Florida 32611, U.S.A.

Claire Buchanan   Department of Zoology, University of New Hampshire, Durham, New Hampshire 03824, U.S.A.

Paul Colinvaux   Department of Zoology, The Ohio State University, Columbus, Ohio 43210, U.S.A.

John L. Confer   Biology Department, Ithaca College, Ithaca, New York 14850, U.S.A.

Thomas L. Crisman   Department of Environmental Engineering Sciences, University of Florida, Gainesville, Florida 32611, U.S.A.

Philip H. Crowley T. H. Morgan School of Biological Sciences, University of Kentucky, Lexington, Kentucky 40506, U.S.A.

David Culver Department of Zoology, The Ohio State University, Columbus, Ohio 43210, U.S.A.

Edward S. Deevey, Jr. Florida State Museum, University of Florida, Gainesville, Florida 32611, U.S.A.

Georgiana B. Deevey Florida State Museum, University of Florida, Gainesville, Florida 32611, U.S.A.

William R. DeMott Department of Biological Sciences, Dartmouth College, Hanover, New Hampshire 03755, U.S.A.

Lawrence M. Dill Department of Biological Sciences, Simon Fraser University, Burnaby, British Columbia V5A 1S6, Canada

Percy L. Donaghay Oregon State University, School of Oceanography, Corvallis, Oregon 97331, U.S.A.

Ray W. Drenner Biology Department, Texas Christian University, Fort Worth, Texas 76129, U.S.A.

Henri Jean Dumont University of Ghent, Limnology Division, Ledeganckstraat 35, 9000, Gent, Belgium

U. Einsle Staatliches Institut für Seenforchung und Fischereiwesen, Abt. Max-Auerbach-Institut, Schiffstrasse 56, D-775, Konstanz, West Germany

Kåre Elgmork Zoological Institute, University of Oslo, P. Box 1050 Blindern, Oslo 3, Norway

Christine A. Evanik Biology Department, Ithaca College, Ithaca, New York 14850. U.S.A.

Carol L. Folt Institute of Ecology and Environmental Studies, University of California, Davis, California 95616, U.S.A.

Marc M. Friedman School of Veterinary Medicine, University of Pennsylvania, 3800 Spruce Street, Philadelphia, Pennsylvania 19104, U.S.A.

B. W. Frost Department of Oceanography, WB-10, University of Washington, Seattle, Washington 98195, U.S.A.

Jeroen Gerritsen Department of Zoology, University of Georgia, Athens, Georgia 30602, U.S.A.

Louis A. Giguére Department of Biological Sciences, Simon Fraser University, Burnaby, British Columbia V5A 1S6, Canada

John J. Gilbert Department of Biological Sciences, Dartmouth College, Hanover, New Hampshire 03755, U.S.A.

Z. Maciej Gliwicz Department of Hydrobiology, University of Warsaw, Nowy Swiat 67, Warsaw, Poland

Charles R. Goldman Institute of Ecology and Division of Environmental Studies, University of California, Davis, California 95616, U.S.A.

Clyde E. Goulden Academy of Natural Sciences, Philadelphia, Pennsylvania 19103, U.S.A.

David Griesmer State University of New York at Albany, Department of Biological Sciences, Albany, New York 12222, U.S.A.

Nelson G. Hairston, Jr. Department of Zoology, University of Rhode Island, Kingston, Rhode Island 02881, U.S.A.

James F. Haney Department of Zoology, University of New Hampshire, Durham, New Hampshire 03824, U.S.A.

Paul D. N. Hebert Department of Biology, University of Windsor, Windsor, Ontario N9B 3P4, Canada

Jürgen Jacobs Zoologisches Institut, Universität München, Luisenstrasse 14, D-8000, München 2, West Germany

C. D. Jamieson Department of Zoology, University of Otago, Dunedin, New Zealand

John Janssen Biology Department, Loyola University of Chicago, 6525 N. Sheridan, Chicago, Illinois 60626, U.S.A.

Dan M. Johnson Biology Department, East Tennessee State University, Johnson City, Tennessee 37601, U.S.A.

Dean L. Kellogg, Jr. Department of Zoology, University of Texas, Austin, Texas 78712, U.S.A.

Michele Kennedy State University of New York at Albany, Department of Biological Sciences, Albany, New York 12222, U.S.A.

W. Charles Kerfoot Department of Biological Sciences, Dartmouth College, Hanover, New Hampshire 03755, U.S.A.

Dean Kettle Department of Systematics and Ecology, University of Kansas, Lawrence, Kansas 66045, U.S.A.

Charles E. King Department of Zoology, Oregon State University, Corvallis, Oregon 97331, U.S.A.

Winfried Lampert Fachbereich Biologie (Zoologie) J. W. Goethe-Universität, Siesmayerstrasse 70, 6000 Frankfurt / a.M., West Germany

John T. Lehman Division of Biological Sciences, University of Michigan, Ann Arbor, Michigan 48109, U.S.A.

William M. Lewis, Jr. Department of Environmental, Population, and Organismic Biology, University of Colorado, Boulder, Colorado 80309, U.S.A.

Michael Lynch Department of Ecology, Ethology, and Evolution, University of Illinois, Urbana, Illinois 61801, U.S.A.

P. Marsot Institute national de la recherche scientifique, Oceanologie, 310 Ave. Ursulines, Rimouski, Quebec G5L 3A1, Canada

Steven R. McComas Biology Department, Texas Christian University, Fort Worth, Texas 76129, U.S.A.

Donald C. McNaught Department of Ecology and Behavioral Biology, University of Minnesota, Minneapolis, Minnesota 55455, U.S.A.

Dewey G. Meyers Academy of Natural Sciences, Department of Limnology and Ecology, Philadelphia, Pennsylvania 19103, U.S.A.

Mark D. Morgan University of Texas Marine Science Institute, Port Aransas, Texas 78373, U.S.A.

William E. Neill Institute of Animal Research Ecology, University of British Columbia, Vancouver, British Columbia V6T 1W5, Canada

Jens Petter Nilssen Zoological Institute, University of Oslo, P.O. Box 1050, Blindern, Oslo 3, Norway

W. John O'Brien Department of Systematics and Ecology, University of Kansas, Lawrence, Kansas 66045, U.S.A.

John D. Orcutt, Jr. Institute of Ecology, University of Georgia, Athens, Georgia 30602, U.S.A.

Gustav-Adolf Paffenhöfer Skidaway Institute of Oceanography, Savannah, Georgia 31406, U.S.A.

Robert A. Pastorok Biology Department, University of Puget Sound, Tacoma, Washington 90416, U.S.A.

Adrienne Peacock Department of Zoology, University of British Columbia, Vancouver, British Columbia V6T 1W5, Canada

Trevor Platt Marine Ecology Laboratory, Bedford Institute of Oceanography, Dartmouth, Nova Scotia B2Y 4A2, Canada

Karen Glaus Porter Department of Zoology, University of Georgia, Athens, Georgia 30602, U.S.A.

S. A. Poulet Institute national de la recherche scientifique, Oceanologie, 310 Ave. Ursulines, Rimouski, Quebec G5L 3A1, Canada

Sumner Richman Department of Biology, Lawrence University, Appleton, Wisconsin 54911, U.S.A.

Howard P. Riessen Osborn Memorial Laboratories, Department of Biology, Yale University, New Haven, Connecticut 06520, U.S.A.

J. Ringelberg Laboratory of Aquatic Ecology, University of Amsterdam, Kruislan 320, Amsterdam, Netherlands

Stephen E. Robbins Department of Biology, Lawrence University, Appleton, Wisconsin 54911, U.S.A.

James T. Rybock Fugro Incorporation, 444 NE. Ravenna Blvd., Seattle, Washington 98115, U.S.A.

David Schmidt Department of Systematics and Ecology, University of Kansas, Lawrence, Kansas 66045, U.S.A.

Ursula Schober Limnologisches Institut, Mainaustrasse 212, 7750 Konstanz-Egg, West Germany

H. Otto Seibeck Zoologisches Institut, Universität München, Luisenstrasse 14, D–8000 München 2, West Germany

William Silvert Marine Ecology Laboratory, Bedford Institute of Oceanography, Dartmouth, Nova Scotia B2Y 4A2, Canada

W. Gary Sprules   Department of Zoology, Erindale College, University of Toronto, Mississauga, Ontario L5L 1C6, Canada

Peter L. Starkweather   Department of Biological Sciences, University of Nevada, Las Vegas, Nevada 89154, U.S.A.

Miriam Steinitz   Department of Zoology, The Ohio State University, Columbus, Ohio 43210, U.S.A.

Jan A. E. Stenson   Department of Zoology, University of Göteborg, Fack, S-400 33 Göteborg 33, Sweden

J. Rudi Strickler   Australian Institute of Marine Science, PMB No. 3, Townsville MSO, Queensland 4810, Australia

Barbara E. Taylor   Department of Zoology, NJ-15, University of Washington, Seattle, Washington 98195, U.S.A.

Stephen T. Threlkeld   University of Oklahoma Biological Station, Kingston, Oklahoma 73439, U.S.A.

Gary L. Vinyard   Biology Department, University of Nevada, Reno, Nevada 00000, U.S.A.

Craig E. Williamson   Department of Biological Sciences, Dartmouth College, Hanover, New Hampshire 03755, U.S.A.

David Wright   Department of Systematics and Ecology, University of Kansas, Lawrence, Kansas 66045, U.S.A.

Marsh J. Youngbluth   Harbor Branch Foundation, Inc., RRI, Box 196, Fort Pierce, Florida 33450, U.S.A.

Robert E. Zaret   Zoology Department, Duke University, Durham, North Carolina 27706, U.S.A.

Thomas M. Zaret   Department of Zoology, Institute for Environmental Studies, FM-12, University of Washington, Seattle, Washington 98195, U.S.A.

I

Life at Low Reynolds
Numbers, Swimming Behavior,
Orientation in Time and Space

1. The Animal and Its Viscous Environment

Robert E. Zaret

Abstract

A derivation of a form of Newton's Second Law of Motion appropriate for fluid mechanics suggests the importance of two parameters, the Reynolds number, and the Strouhal number. The Reynolds number measures the importance of the fluid's inertia. As the Reynolds number associated with a moving animal decreases, some propulsion mechanisms, such as those used by fish, become ineffective. As the velocity of a fluid flowing past a flat plate decreases, the tendency of the fluid to slow down near the wall extends farther away from the plate. This effect may explain how planktonic organisms that rarely encounter plates or walls in nature can detect them in laboratories. The effect may also explain how zooplankters can detect changes in their own concentrations. The Strouhal number measures the importance of imposed oscillations. As the Strouhal number associated with a flow increases, the flow changes character. In particular, the flow produced by an oscillating object may have a steady component even when the Reynolds number associated with the flow is small. This effect should be considered when trying to understand the feeding mechanisms of zooplankters.

Owners of pets are known for their tendency to psychoanalyze their pets. Those who do not own pets are generally either bored or outraged by such discussions. In doing so they miss a game which is fun for its own sake, but which also has its practical aspects.

Here I discuss water as experienced by some of our favorite pets, small zooplankters. The properties of fundamental importance to small zooplankters are properties that humans tend to overlook. I shall try to demonstrate how understanding the differences between human and zooplankter interactions with water can further our understanding of zooplankton ecology.

I shall begin with a rather oversimplified description of some equations from theoretical fluid mechanics. This description (based on White 1974) demonstrates the mathematical derivation of two parameters that are useful in discussing differences between human and zooplankter interactions with water. The notation is not conventional, but has been chosen to enhance the intuitive value of the presentation.

The second part of this work demonstrates manifestations of changes in one particular parameter, the Reynolds number. The demonstration combines mathematics, physical experiments, and some experiments with animals. The physical experiments are demonstrated in the movie *Low Reynolds number flow* (featuring Sir G. I. Taylor and available through Educational Services Incorporated). I did the biological experiments with the cameras and assistance of J. R. Strickler.

My discussion of theoretical fluid mechanics begins with some differences between the mathematical description of particle mechanics and the mathematical description of fluid mechanics. In describing particle physics (which most of us studied in elementary physics), we can reasonably describe the properties of each particle in the system. We can describe the velocity, position, temperature, mass, etc. for each particle. In describing the physics of a fluid, such an approach is unreasonable. Instead, the usual approach is to designate positions in the fluid and to describe the properties of all fluid particles at each position. Thus velocity is given as a function of the position, rather than of the individual particles. This change in the method of describing physical properties makes mathematical analyses much more difficult for fluid mechanics than for particle physics.

The first step is to provide a method for specifiying location. I use a cartesian coordinate system and give location as

$$x = (x_1, x_2, x_3) \qquad (1)$$

where x_i is the distance parallel to the ith axis. Thus x_1 is another name for x, x_2 is another name for y, and x_3 is another name for z. Notation with subscripts make equations easier to write. Generally, the choice of origin for the coordinate system affects the interpretation of the solution to a problem in fluid mechanics. For an object moving through a fluid, the origin is usually attached either to the object or to the fluid infinitely far away from the object. As an analogy, we can imagine trying to take pictures of the water moving past a boat. We might stand on the boat or we might stand on shore and wait until the boat is in front of our camera. The equations I present are appropriate for any choice of origin.

I present velocities as

$$v = (v_1, v_2, v_3) \qquad (2)$$

where v_i is the velocity component parallel to the ith axis.

When a fluid particle moves from point A to point B, the velocity of the particle generally changes. The change may occur because the velocity of all particles at point B changes with time, and/or because the velocity of the particles at point B differs from the velocity of the particles at point A. If x is the position of a particle, the total change in its velocity is given as

$$\frac{\partial v}{\partial t} + \sum_{j=1}^{3} \left(\frac{\partial v}{\partial x_j} \frac{dx_j}{dt} \right) . \qquad (3)$$

This formula can also be derived using the chain rule.

We know that the velocity of the particle is given by

$$v_j = \frac{dx_j}{dt} ,$$

so that the total change in the particle's velocity can be given as

$$\frac{\partial v}{\partial t} + \sum_{j=1}^{3} \left(v_j \frac{\partial v}{\partial x_j} \right) . \qquad (4)$$

In more conventional notation

$$\frac{\partial v_i}{\partial t} + \sum_{j=1}^{3} v_j \left(\frac{\partial v_i}{\partial x_j} \right) \qquad (5)$$

where Eq. 5 represents changes in the ith velocity component for a fluid particle. Because the velocity of the fluid at any point is the same as the velocity of any fluid particle at that point, Eq. 5 also represents changes in the ith velocity component for the fluid at any point x. Note that Eq. 5 is derived using Lagrangian coordinates (which specify the positions of individual particles), but is evaluated using Eulerian coordinates (which specify locations in the fluid).

One set of equations basic to theoretical fluid mechanics is derived from Newton's Second Law of Mechanics. For a particle of fixed mass m, the law is given as

$$F = ma \qquad (6)$$

where F is the force vector and a is the acceleration vector. Rearranging, the law can be given as

$$a = F/m. \quad (7)$$

Equation 5 gives an expression for acceleration. One force exerted on the fluid is gravity, which directly affects only the vertical component of the velocity. Thus if we define the x_3 axis as vertical, the appropriate term for the gravitational force is

$$g\delta_{3i} \quad \text{(gravity)} \quad (8)$$

where g is the gravitational acceleration and

$$\delta_{3i} \equiv \begin{cases} 1 \text{ if } i = 3 \\ 0 \text{ if } i \neq 3 \end{cases}.$$

Another force arises from the pressure gradient in the fluid:

$$-\frac{1}{\rho}\frac{\partial P}{\partial x_i} \quad \text{(pressure)} \quad (9)$$

where ρ is the density of the fluid and P is the pressure. The negative sign arises from conventions about directions of the forces. The final force I consider arises from viscosity—the tendency of a fluid to resist shear. The viscous force has been found, experimentally, to fit a relatively simple formula (at least for air and water under most conditions):

$$\frac{\mu}{\rho}\sum_{j=1}^{3}\frac{\partial^2 v_i}{\partial x_j^2} \quad \text{(viscosity)} \quad (10)$$

where μ is the viscosity of the fluid. Combining all the terms gives

$$\frac{\partial v_i}{\partial t} + \sum_{j=1}^{3}\left(v_j\frac{\partial v_i}{\partial x_j}\right) = -\frac{1}{\rho}\frac{\partial P}{\partial x_i} + g\delta_{3i}$$

$$+ \frac{\mu}{\rho}\sum_{j=1}^{3}\left(\frac{\partial^2 v_i}{\partial x_j^2}\right). \quad (11)$$

Note that Eq. 11 represents three equations (one for each velocity component).

When density is constant, pressure is usually redefined to include the gravity effects:

$$p \equiv P - \rho g \delta_{3i}$$

so Eq. 11 becomes

$$\frac{\partial v_i}{\partial t} + \sum_{j=1}^{3}\left(v_j\frac{\partial v_i}{\partial x_j}\right) = -\frac{1}{\rho}\frac{\partial p}{\partial x_i} +$$

$$\frac{\mu}{\rho}\sum_{j=1}^{3}\left(\frac{\partial^2 v_i}{\partial x_j^2}\right). \quad (12)$$

The set of equations given by Eq. 12 is a set of three nonlinear, second-order partial differential equations which must be solved for the velocity components v_i. These equations are known as the Navier-Stokes equations. In even the simplest problem in fluid mechanics, Eq. 12 must be solved along with an equation representing conservation of mass and a set of boundary conditions which give the velocities on any surfaces. This simplest set of differential equations is applicable to a range of problems with a precision that would make most theoretical biologists envious. For a wide variety of flows, Eq. 12 and the equation for mass conservation remain unchanged, while the boundary conditions change. Unfortunately, the system has been solved exactly for very few practical flows.

Thus a generally necessary step in developing a solution is determining which terms in Eq. 12 can be neglected. The usual approach to this step is to scale the equation to known parameters.

Suppose we have an object whose speed is sinusoidal:

$$U \sin \omega t.$$

Then we may assume that (1) as U (the maximum speed of the object) increases, so does the fluid velocity at any point; (2) as ω increases, so does the rapidity with which the velocity at any point changes with time; and (3) as L (the length of the object) increases, so do the distances over which the velocity changes (i.e. spatial gradients become gentler). Thus we define a set of nondimensional variables:

$x'_i \equiv x_i/L$

$v'_i \equiv v_i/U$

$t' \equiv t\omega$.

After some straightforward manipulations, Eq. 12 becomes

$$\frac{\omega L}{U} \frac{\rho U L}{\mu} \frac{\partial v'_i}{\partial t'} + \frac{\rho U L}{\mu} \sum_{j=1}^{3} \left(v'_j \frac{\partial v'_i}{\partial x'_j} \right)$$

$$= -\frac{L^2}{\mu U} \frac{\partial p}{\partial x_i} + \sum_{j=1}^{3} \left(\frac{\partial^2 v'_i}{\partial x'_j{}^2} \right). \quad (13)$$

I have postponed introducing a nondimensional form of the pressure term until now because the logic is a little more involved. Remember that the purpose of scaling is to determine which terms can be ignored. Physical experiments and mathematical analyses have shown that the pressure term must always be included. Thus the nondimensional form of the pressure may vary, depending on the value of the other parameters. Actually, absolute pressure is not important for flows which are adequately described by Eq. 12. Only the pressure gradient is important, so it should be scaled directly. For the current discussion, an appropriate form of the nondimensional pressure gradient is

$$\frac{\partial p'}{\partial x'_i} \equiv \frac{L^2}{\mu U} \frac{\partial p}{\partial x_i}.$$

The final version of Newton's Second Law of Motion as appropriate for a theoretical analysis of the flow past zooplankters is

$$SR \frac{\partial v'_i}{\partial t'} + R \sum_{j=1}^{3} \left(v'_j \frac{\partial v'_i}{\partial x'_j} \right)$$

$$= -\frac{\partial p'}{\partial x'_i} + \sum_{j=1}^{3} \left(\frac{\partial^2 v'_i}{\partial x'_j{}^2} \right) \quad (14)$$

where

$S = \dfrac{\omega L}{U}$ Strouhal number,

and

$R = \dfrac{\rho U L}{\mu}$ Reynolds number.

From Eq. 14 we can see that two nondimensional parameters are important in determining which terms may be neglected. When the Reynolds number is small (<0.1), the nonlinear terms (which are multiplied by the Reynolds number) can be neglected. The resulting approximation is called the Stokes approximation. The Oseen approximation is often appropriate for flows with a Reynolds number about equal to 0.1. In this approximation, the sum of the nonlinear terms is replaced by

$$U \frac{\partial v_i}{\partial x_1}.$$

When the Strouhal number is small or zero (as in steady flow), the unsteady term (which is multiplied by the Strouhal number and the Reynolds number) can be neglected. For flows that are neither steady, nor periodic, the scaling for time is less direct. The swimming motion of most cladocerans is an example of such motion. These arguments about which terms can be neglected follow from the idea that, when the equations are properly scaled, the magnitude of each derivative will approximate unity.

To emphasize the importance of the Reynolds number, I temporarily ignore the unsteady term. In other words, I assume that the Strouhal number is very small or zero and further assume that the speed of the object is given by U. Newton's Second Law may thus be given as

$$R \sum_{j=1}^{3} \left(v'_j \frac{\partial v'_i}{\partial x''_j} \right) = -\frac{\partial p'}{\partial x'_i} + \sum_{j=1}^{3} \left(\frac{\partial^2 v'_i}{\partial x'_j{}^2} \right). \quad (15)$$

Notice that the sum on the left is nonlinear; that is, the sum involves a multiple of v_i and its derivatives. In particular, this sum is quadratic. Thus, when an object changes direction, the sign of the sum on the left does not necessarily change. The sum all the way on the right is linear, so its sign will change when the object changes direction. Therefore, when the sum on the left is important, the magnitude of the pressure term depends on both the speed and direction of the motion of an object. When the Reynolds number is small, the sum on the left is negligible and the magnitudes of the remaining terms become independent of the direction of an object's motion. This feature of low Reynolds number flows is called kinematic reversibility and is admirably demonstrated in the G. I. Taylor movie mentioned earlier.

The conditions within the movie demonstration were, of course, arranged so that the Reynolds number was very low, ≪1.0. For comparison, the Reynolds number associated with a 1.5 m human swimming 1.0 m·s^{-1} is 1.5×10^6 and the Reynolds number associated with a 5cm fish swimming 1.0 cm·s^{-1} is 5.0×10^3. The Reynolds numbers associated with zooplankters range from 0.1 for a sinking *Bosmina* to 3 for a swimming *Bosmina* to 500 for some copepods. Thus the flow past a zooplankter is generally not completely reversible. Movies of swimming *Leptodora* and *Bosmina* show the extent to which the flow may be reversible. As reversibility becomes more important, swimming mechanisms that depend on irreversibility become ineffective (Purcell 1977). The ineffectiveness of a fish's swimming mechanism at low Reynold's number is demonstrated in the G. I. Taylor film.

We can surmise another feature of low-Reynolds number flow by examining Eq. 15. The sum on the left is approximately proportional to U^2, while the sum on the right is approximately proportional to U. Thus, the sum on the left will be more affected by changes in fluid velocity. In particular, velocity gradients associated with a wall are likely to be steeper as the Reynolds number (and thus the importance of the sum on the left) increases. The flow past a semi-infinite flat plate is a classic problem and is known as the Blasius problem (see White 1974 for a discussion of the problem). The fluid infinitely far from the plate is assumed to move with speed U. The fluid next to the plate is assumed to move with speed zero. This latter assumption agrees very well with experiments. Most of the flow is virtually unaffected by the wall except in a small region near the wall. The size of this region decreases with the speed of the flow (*see* Fig. 1). White (1946)

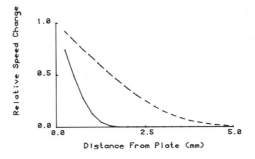

Fig. 1. Solution of the Blasius problem for flow past a flat plate. Solution calculated using Runge-Kutta algorithm given in appendix of White (1974). The figure shows velocity profiles along a line perpendicular to plate and crossing it 1 mm downstream from its leading edge. Dashed line—1 mm·s^{-1}; solid line—10 mm·s^{-1}.

experimented with wires that were dropped lengthwise in cylinders. The distance at which the cylinder walls affected the flow increased with decreasing Reynolds number. When the Reynolds number was 10^{-4}, walls several hundred wire-diameters from the wire decreased the speed of the wire. The G. I. Taylor movie provides a demonstration in which two identical balls are dropped in a liquid near a wall and the ball closer to the wall moves more slowly.

I conducted a set of experiments to test the hypothesis that zooplankters may sense the presence of a wall and may change their behavior as a result. I put a movable wall in a 5 cm × 5 cm tank in front of a movie camera. The positions of the wall and the tank could be adjusted so that the focal plane of the camera remained midway between the adjustable wall and the wall of the tank as the gap between the two walls varied. The gap was set at each of five distances once in each of three series, with the order within each series determined from a random-digit table. Thus, the order was [8, 6, 10, 2, 4], [6, 8, 10, 4, 2], [4, 10, 6, 2, 8], . . . (the numbers refer to the width of the gap in mm). Each time the gap was

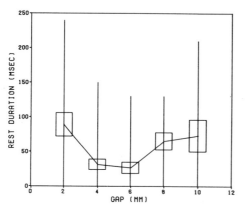

Fig. 2. Effect of crowding *Bosmina*. Horizontal line connects mean interval for each gap width. Vertical lines show ranges; boxes show 95% confidence levels for means (assuming data are normally distributed).

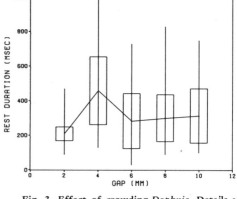

Fig. 3. Effect of crowding *Daphnia*. Details as in Fig. 2.

set, I used the camera to record the movements of an animal. *Bosmina* was tested between 1500 and 1700 hours one day and *Daphnia pulex* was tested between 1500 and 1600 hours another day. The camera operated at 100 frames·s^{-1}, so the temporal resolution was 10 ms. Since the camera did not start or stop instantly, the film analysis ignored any interval which began within 20 frames of the start of a sequence or ended within 10 frames of the end of a sequence.

The results shown in Figs. 2 and 3 represent the interval between strokes as a function of the gap between walls. These results can be compared with those of Zaikin and Rudyakov (1976), who showed that the average swimming speed of copepods and ostracods first increased and then decreased with increased crowding. An ANOVA (Sokal and Rohlf 1973) was significant at the 0.5% level for *Bosmina* [$F(4, 222) = 19.3$], but was not significant at the 5.5% level for *Daphnia* [$F(4, 51) = 2.14$]. Table 1 shows the results of several Mann-Whitney U-tests. The time taken for each stroke varied little. *Daphnia* took 91.9 ± 21.1 ms per stoke and *Bosmina* took 55.4 ± 9.8 ms.

These results have several important implications. Most directly, they support the notion that results of experiments conducted with confined animals should be interpreted with caution. A more positive implication is that measurements of interstroke intervals may provide a useful technique for measuring the effects of changes in other environmental conditions, such as temperature and light.

One possible objection to these experiments is that planktonic organisms rarely meet walls in nature. In fact, we should be somewhat surprised that planktonic organisms are able to detect and avoid walls in the laboratory. Schröder (1960) suggested that animals use echo location to detect walls. However, zooplankters have sensory hairs which probably detect changes in velocity (Kerfoot pers. comm.; Strickler and Bal 1973). Therefore, an animal could detect the presence of a wall by sensing a decrease in its own velocity.

I further speculate that zooplankters may be able to detect changes in the concentrations of objects (including other zooplankters and algae) in the water by sensing changes in their own velocity. The fact that the velocity of an object does depend on the concentration of objects in the water is demonstrated in the G. I. Taylor movie. Unfortunately, my experiments with walls suggest that the concentration of objects might have to be very high to have any effect.

Finally, I want to comment briefly on the importance of "unsteady" effects. Returning to Eq. 14, we can see that, as the Strouhal number becomes large, the term containing

Table 1. Mann-Whitney U-test statistics. First column lists gaps being compared; numbers refer to gap width in mm. Second column lists number of rest intervals analyzed for each group being compared. Third column lists U-test statistic using convention in Sokal and Rohlf (1973). Fourth column lists value of normal approximation calculated using formula of Sokal and Rohlf (1973, p. 219). Last column lists significance level.

Comparison	Numbers	U_s	t_s	P
Bosmina				
4 vs. 6	79, 56	2,413.5	0.917	>0.05
8 vs. 10	26, 31	407.5	0.072	>0.05
2 vs. 4+6	35, 135	4,051.0	6.302	<0.001
2 vs. 8+10	35, 57	1,233.0	1.898	<0.05
4+6 vs. 8+10	135, 57	5,713.0	5.365	<0.001
Daphnia				
2 vs. 4	18, 10	132.0		<0.025
2 vs. 6+8+10	18, 29	340.0	0.503	>0.05
4 vs. 6+8+10	10, 29	192.5	1.767	<0.05

the time derivative becomes important. The appearance of this additional term might be expected to correspond to a qualitative change in the flow. Riley (1967) presented an analysis which suggests that, when the Strouhal number associated with an oscillating sphere is $\geqslant 1$, the nonlinear terms may contribute a steady (i.e. nonoscillating) component to the flow even when the associated Reynolds number is $\ll 1$. Such a phenomenon may be associated with the feeding appendages of zooplankters.

References

PURCELL, E. M. 1977. Life at low Reynolds number. Am. J. Phys. **45**: 3-11.

RILEY, N. 1967. Oscillatory viscous flows. Review and extension. J. Inst. Math. Appl. **3**: 419-434.

SCHRÖDER, R. 1960. Echoorientierung bei *Mixodiaptomus laciniatus*. Naturwissenschaften **47** (Sonderdr.): 548-549.

SOKAL, R. R., and F. J. ROHLF. 1973. Introduction to biostatistics. Freeman.

STRICKLER, J. R., and A. K. BAL. 1973. Setae of the first antennae of the copepod *Cyclops scutifer* (Sars): Their structure and importance. Proc. Natl. Acad. Sci. **70**: 2656-2659.

WHITE, F. M. 1946. The drag of cylinders in fluids at slow speeds. Proc. Roy. Soc. A. **186**: 472-479.

WHITE, F. M. 1974. Viscous fluid flow. McGraw-Hill.

ZAIKIN, A. N., and Yu. A. RUDYAKOV. 1976. Rate of movement of planktonic crustaceans. Oceanology. **16**: 516-519.

2. Visual Observations of Live Zooplankters: Evasion, Escape, and Chemical Defenses

W. Charles Kerfoot, Dean L. Kellogg, Jr., and J. Rudi Strickler

Abstract

Direct visual observations of feeding, locomotion, and predator-prey interactions allow several insights into the evolution of zooplankton body shape and coloration. Under intermediate Reynolds Number flows, zooplankton body shape can be subject to molding by individual competitive or defensive compromises, and yet the variance of shape within the three dominant freshwater groups (rotifers, cladocerans, and copepods) is remarkably heterogeneous. Rotifers and cladocerans show much phenotypic flexibility with and between species, whereas copepods are surprisingly conservative. These differences seem to be a consequence of strong selection for streamlining in copepods and/or of defensive patterns.

The biased foraging of dominant predators, such as visually feeding fishes and grasping copepods, constitutes the force, in conjunction with competition, which drives prey toward certain adaptive modes. Of importance here is how the predator detects and recognizes the prey, and if the initial predisposition of escape or rejection involves sight, mechanoreception, or taste. Many small cladocerans have converged upon a secondary defense that involves playing dead (akinesis), which foils the detection and handling capabilities of predatory copepods, while water mites have developed distasteful compounds and conspicuous warning coloration which deter fishes.

All too often, in studies of zooplankton communities, one senses a peculiar absence of detail, a disquieting feeling of unfamiliarity with the life history details or the behavior of the species that are discussed. Since aquatic communities, and especially freshwater communities, contain no more dominants (common species of both plants and animals) than terrestrial communities, this unfamiliarity is based less upon complexity and more upon accessibility and ease of observation. Because zooplanktonic species are characteristically small (frequently bordering upon the microscopic), often transparent, and usually drift about in a turbulent medium, they are difficult to observe and study in their natural environment.

Though zooplankton watching may never approach the popularity of bird watching as a national pastime, direct visual observations of zooplankton feeding, locomotion, and predator-prey interactions permit valuable insights into the dynamics of species interactions. Unfortunately, careful study of these aspects often requires moderately high magnification, high-speed cinematographic techniques, and familiarity with fluid dynamics. Yet the rewards are great, for much of the fuzziness of zooplanktonic research seems centered on inadequate appreciation and documentation of be-

---

This research was sponsored through NSF grant DEB 76-20238.

havioral characteristics central to food-gathering and predator-prey modeling. Admittedly, dwelling upon the particulars of certain behaviors may err in the direction of too much detail, yet erring in the opposite direction by gathering too little information may lead to a general failure to recognize the evolutionary intricacies which ultimately mold both the interactions and morphology of community members.

In this paper, following a brief discussion of feeding, locomotion, and defenses in freshwater zooplankton species, we consider two kinds of predator-prey interactions. In one case, the predator is blind and perceives the prey with mechanoreceptors. In the other case, the cues are visual but the outcome is influenced greatly by taste. Of importance is the realization that the biased foraging of dominant predators, such as visually feeding fishes and grasping copepods, constitutes the force, in conjunction with competition, which drives prey and predator toward certain adaptive modes, hence which indirectly molds the structure of aquatic communities.

Zooplankton feeding, locomotion, and propensity towards certain prey defenses: rotifers, cladocerans, and copepods

The zooplankton of lakes and oceans consist mainly of small, transparent organisms whose bodies range between 0.01 and 0.3 cm in length. In freely swimming species the swimming movements carry the body at mean speeds which vary between 0.02 to 1.0 cm/sec, so by approximation ($R_e = L \cdot v \cdot \rho / \mu$; see R. Zaret, this volume) the Reynolds number ranges from about .2 to 100. These intermediate values of Reynolds number become important when we realize that extremely large ($R_e > 10^3$) or small ($R_e < 10^{-3}$) values describe fluid flow conditions where pressure forces (high R_e, turbulent drag) or surface forces (low R_e, viscous drag) drive swimming animals toward uniform morphology and behavior. For example, the high Reynolds number flows experienced by dolphins, whales, squids, seals, and fishes have forced body shapes toward a basic streamlined, tear-drop shape profile which greatly reduces the effects of turbulent drag. Likewise, under the low Reynolds number conditions experienced by microscopic motile forms, various organisms (unicellular algae, male gametes of both plants and animals, bacteria) have converged upon similar body shapes and flagellar movement as a basic way of moving through a relatively viscous medium. We are fortunate that within the intermediate range of Reynolds numbers experienced by zooplankton the consequences of ecological interactions can potentially influence the external morphology of the body, for some interesting questions arise out of patterns of variation and uniformity.

In one sense, the morphology of the three major freshwater zooplankton groups (rotifers, cladocerans, and copepods) reflects their phylogenetic predispositions toward food-gathering and locomotion. Although all three groups include species that move water in order to extract suspended particulate matter (the so-called "filter-feeders") and others that move their bodies in order to grasp either suspended or swimming objects ("encounter feeders," many of which are omnivorous or predatory), locomotion is performed in fundamentally different ways. Figure 1 illustrates the general body plans of typical rotifers, cladocerans, and copepods. Almost all rotifers feed and swim using modified cilia, usually by creating currents

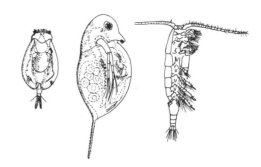

Fig. 1. General body form and appendages of rotifers (left), cladocerans (middle), and copepods (right; only appendages on left side drawn). Redrawn from original figures in H. B. Ward and G. C. Whipple (1918) *Fresh-water Biology*, Wiley and Sons.

with their coronal cilia. The swimming movement is generally smooth and regular, there being little distinction between average and instantaneous swimming speeds. With the exception of a few species (notably *Polyarthra*) which have developed secondary structures (paddles) especially designed for evasive escape, rotifers show limited ability to accelerate away from danger, although many possess flexible bodies which greatly aid maneuverability. In general, rotifers are slightly smaller than either cladocerans or copepods. These latter two groups, although distantly related, show considerable divergence in basic body plans. In freely swimming cladocerans, the body is propelled solely through strokes from a single appendage, the modified second antennae. The first antennae are usually greatly reduced, while the thoracic appendages are shielded within a tough chitinous carapace (in herbivorous species) and are modified for food-gathering. The chitinous exoskeleton (head shield and carapace) is generally inflexible, smooth, and adorned with few pores or sensory setae (DuMont and Van De Velde 1975; Halcrow 1976; Schultz 1977; Schultz and Kennedy 1977; for an example, see Fig. 2). In marked contrast, freely swimming copepods usually possess four pairs of thoracic appendages (the swimming legs) modified for locomotion, a flexible body, two pairs of extended and movable antennae, and an abdomen that is freely articulated with the thorax. The relatively thin chitinous exoskeleton is covered by extensive fields of mechano- and chemoreceptors and the antennae bristle with sensory setae (Strickler and Bal 1973; Strickler 1975a, b; Raymont et al. 1974; Gharagozlou-van Ginneken and Bouligand 1975; for an example, see Fig. 3). Calanoid copepods generally swim along a smooth path, usually propelled by the rapid stroking of their second antennae (Fig. 4), though cyclopoid copepods move in a "hop-and-sink" manner (Strickler 1970, 1975a). Cladocerans may hang motionless for brief moments and then jump up to a new position [e.g. *Diaphanosoma*, large *Daphnia* (Fig. 5)] or appear to move continuously forward (e.g. *Bosmina, Chydorus*). However, in both calanoids like *Diaptomus* and cladocerans like *Bosmina* the steady motion is only apparent, as movement and rest periods alternate at brief intervals (e.g. the movement phase lasts about 25 milliseconds in swimming *Bosmina*; Zaret and Kerfoot, 1980). Both calanoid and cyclopoid copepods are capable of using their four ventrally situated pairs of legs for a series of powerstrokes, by stroking in a 4-3-2-1 metachronical pattern and retrieving all legs in a single recovery stroke (Strickler 1975a). In cladocerans and copepods, the distinction between average and instantaneous swimming speeds is important for calculating Reynolds number flows (Strickler 1975a, Strickler and Twombly 1975) and when considering predator-prey dynamics (Kerfoot 1978).

Because of their reliance upon a single pair of appendages for locomotion, cladocerans generally have limited locomotory abilities (with the exception of a few species like *Diaphanosoma*, which possess enormously enlarged second antennae), while copepods are able to accelerate rapidly, sustain great speeds, and maneuver quickly (Kerfoot 1978). During the power strokes of cladocerans, the body accelerates forward only to maximum speeds around 0.5-2.0 cm/sec, while in normal cyclopoid hops it may attain a maximum of between 4.0-10.0 cm/sec. When predatory copepods attack, or when all copepods execute an avoidance response (escape reaction), successive power strokes of the swimming legs may combine with tail flapping to accelerate the body to speeds approaching 20-35 cm/sec (Strickler 1975a). For these reasons, during an encounter with attacking copepods, cladocerans and rotifers often play tortoise, relying upon secondary defenses (spines, hard coats or sheaths, gelatinous coverings) to deter the predators, while herbivorous copepods play hare, escaping with a burst of speed.

Of the three major groups, the two (rotifers and cladocerans) that are generally not capable of rapid evasive tactics are characterized by highly variable morphology between species, while copepods are uniformly conservative in body shape (Fig. 6). The lack of body shape variation within

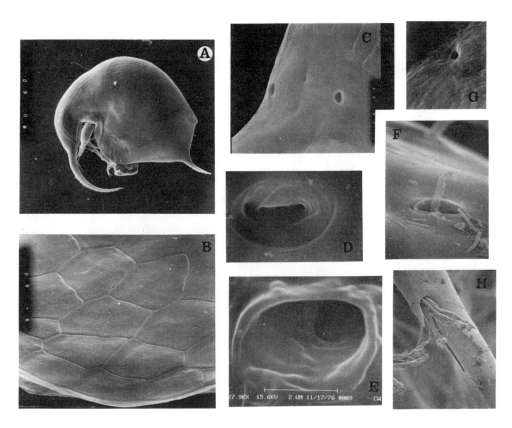

Fig. 2. Surface texture and sensory pores of a cladoceran, *Bosmina longirostris* (long-featured form): (A) side view of parthenogenetic female; (B) close-up of carapace surface, showing smooth texture; (C) example of the 6 frontal and 2 lateral pores found on parthenogenetic females; this view shows 3 pores situated on the forehead, immediately above the base of the antennules (includes left and right sensory setae with a single medial pit); (D) enlargement of medial pit; (E) enlargement of pore associated with seta, looking down into pore; (F) same type of pore as before, viewed from the side; (G) single medial sensory seta and pit located dorsally on head shield; and (H) sensory setae projecting from pore located midway down antennule (SEM photos by Y. Tsukada, M. Friedman, L. Zelazny, and W.C.K.).

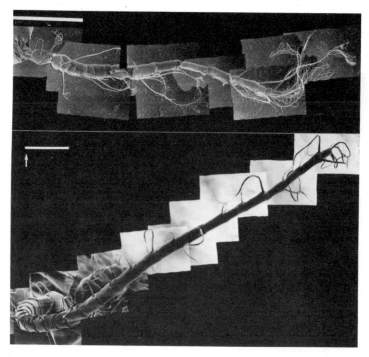

Fig. 3. Extensive fields of sensory hairs on the first antennae of two calanoid copepods (*Acartia tonsa* above, *Skistodiaptomus oregonensis* below). Arrow points to scale bar = 0.1 mm (SEM photomosaic by Y. de Avendano).

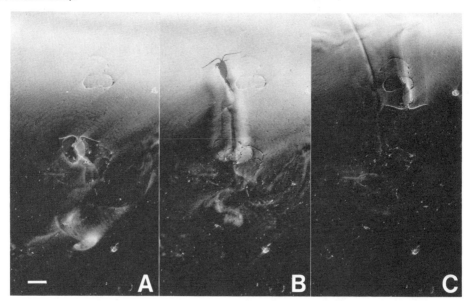

Fig. 4. Swimming movement of a predatory calanoid copepod, *Epischura nordenskioldi*: (A) directional change, illustrating flexibility of body and swimming motions; (B) straight-line foraging motion, propelled almost exclusively by strokes of the second antennae; and (C) reversal of path, showing region scanned by outstretched first antennae. Schlieren photos taken using configuration discussed in Strickler (1977). (scale bar = 1mm).

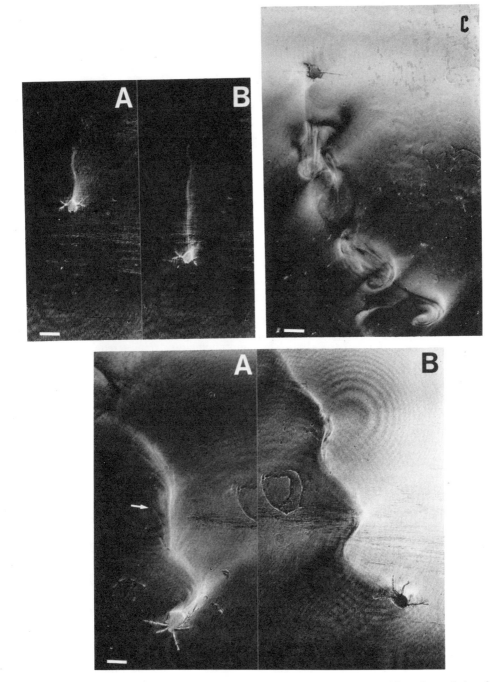

Fig. 5. Swimming wakes of *Daphnia*, *Epischura*, and water mites. Above: (A) juvenile *Daphnia pulex* creates an indistinct wake that quickly dissipates; pattern trailing a stroking juvenile; (B) pattern two seconds later; (C) wake created by *Epischura* as it changes direction; micro-eddies caused by kicking of swimming legs, flap of abdomen, and trailing caudal setae. Below: (A) helical swimming pattern of the water mite *Piona constricta*, arrow points to disturbance created by single leg; (B) change in optical path orientation initiates apparent brightness reversal and highlights swimming pattern (techniques as in Fig. 4, scale bar = 1 mm).

freely swimming copepods is even more astounding, when we consider that all three groups of zooplankton live in similar habitats and experience similar predators.

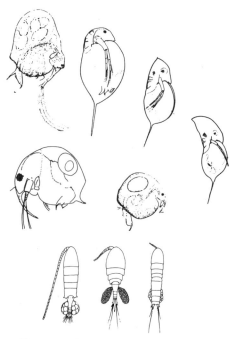

Fig. 6. Variable morphology of cladocerans contrasted with rather invariant teardrop or torpedo shape of free-living copepods (all except *Bosmina* taken from Edmondson, W. T. (ed.) 1957. Ward and Whipple's *Fresh-water Biology*. Wiley and Sons, New York; with permission).

Strickler (1975a) has hypothesized that the slight differences between cyclopoid and calanoid body shapes, i.e. the more teardrop shape of cyclopoids and the more torpedo shape of calanoids, reflect the importance of streamlining in reducing drag during "escape" reactions. Likewise, he has hypothesized that the uniformity within each order is a consequence of strong selection (via energetic losses associated with increased drag) against animals that depart from these basic body plans. Subsequent attempts to document flow dynamics and to calculate flow characteristics during escape reactions, using Schlieren optics and high-speed photography (Strickler 1975b, 1977), has provided inconclusive results. While the Reynolds numbers during escape reactions approached relatively large values (R_e = 500–1,000), both the visual evidence and the fluid flow calculations revealed laminar conditions around the moving body (Strickler 1975b, 1977; Lehman 1977). Although these findings do not eliminate streamlining as the likely cause of uniform body shape within copepods, they complicate this line of reasoning and suggest the necessity of scale model construction in future studies.

An alternative, or contributing, explanation for the lack of morphological variation in copepod body shape concerns their primary defensive response, i.e. the swift, evasive escape reaction. If these animals are able to utilize escape responses effectively to avoid approaching predators, there will be little selection for the kinds of secondary defenses characteristically found in rotifers and cladocerans, because copepods will generally not be captured and handled. Not only would the evolution of secondary morphological defenses require exposure to prolonged periods of handling, but the development of spines, sheaths, or thickened exoskeletons might seriously detract from the flexibility and neurophysiological equipment necessary for escape reactions.

Predatory copepods and akinesis

Through biased selection, predators may drive prey toward radically different end points, depending upon how the predator detects and recognizes the prey and how the prey perceives the predator. Since the dominant predators in open waters are often visually foraging fishes or invertebrates, the general transparency and small size of zooplankton is an interesting example of cryptic morphology against a background that continuously changes in brightness and color. Beyond this obvious risk, zooplankton additionally face exposure to a variety of predatory invertebrates, many of which are not restricted to the photic zone or any diurnal pattern of foraging. In these instances, the predators perceive their prey with mechanoreceptors which sense water disturbances or fire upon direct contact with prey.

Using the progression of events prior to ingestion, we can conveniently classify prey defenses against invertebrate predators into two broad categories: primary (precapture) and secondary (postcapture) responses. Primary responses may include swift evasive flight ("escape reactions," Strickler 1975a), retaliatory combat, or akinesis ("motionless," "dead-man," or "playing dead" reactions, Fryer 1968; Kerfoot 1977a, 1978; Smirnov 1978). Secondary responses may include the use of, or reliance upon, spines (Gilbert 1966; Kerfoot 1975), thick chitinous walls (Fryer 1968; Kerfoot 1977b, 1978), or gelatinous sheaths (Allen 1973).

Free-swimming cyclopoid copepods usually detect prey by sensing water disturbances via mechanoreceptors on their first and second antennae (Strickler and Bal 1973; Strickler 1975b), although olfactory setae can be used to distinguish motionless items (Kerfoot and Peterson, in press). Awareness of the approaching prey appears to take place in the sink phase of normal hop-and-sink swimming behavior (Strickler 1971, 1975) for novel stimuli will induce prolonged sinking periods (Gerritsen 1978). In the sink phase, the body sinks 10° off the vertical, the first antennae are fully extended, and the swimming legs are in rest position (Fig. 7). Detection, attack, or avoidance may occur in microseconds, often coupled with a variety of evasive or pursuit maneuvers. Swimming prey are usually attacked with a short leap, coordinated with tonglike extension and adduction of mouthparts, from a distance of .5–2 body lengths.

In the attack, the initial movements closely resemble normal hop behavior. In normal hop-and-sink swimming, the first two antennae stroke for initial acceleration and flatten back against the cephalothorax, while the body flexes for direction. Two milliseconds later, the swimming legs begin to stroke in a 4-3-2-1 metachronal sequence, each leg taking about 5 ms, while the cyclopoid flaps its abdomen for sustained high speed. In the power stroke, *Cyclops* accelerates much faster than a swimming *Bosmina* and can reach a much greater sustained speed (Figs. 8–9). Thrust comes mainly from the successive kicks of the swimming legs, executed within the

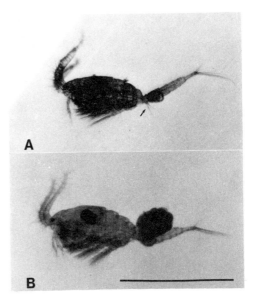

Figure 7. Swimming position of *Cyclops scutifer* in the sinking phase of hop-and-sink behavior: (A) male (arrow points to distinctive orientation of the fifth thoracic appendages); (B) female with egg sacs. In both sexes the first antennae are held above, and second antennae held below, the rostrum to maximize the field of perception (scale bar = 1 mm).

wake of the body, and a reextension of the antennae, all lasting about 10 ms. A single hop usually carries the *Cyclops* about 1 body length. In the attack, however, the *Cyclops* is capable of executing a series of multiple strokes with its swimming legs, of slightly elevating the head region, and of extending the mouthparts toward the prey. Prey are usually attacked in front of the body, within a cone-shaped region extending out between the first antennae (Kerfoot 1978).

Because the flexible body and swimming legs of *Cyclops* allow it great momentary acceleration and speed, and because it perceives prey through water disturbances, its biased predation has driven prey toward interesting extremes. Many herbivorous copepods, equally capable of moving at great speeds, simply flee an approaching *Cyclops*. Some larger-bodied cladocerans (*Daphnia*, Fig. 5) rely upon a joint strategy of escape

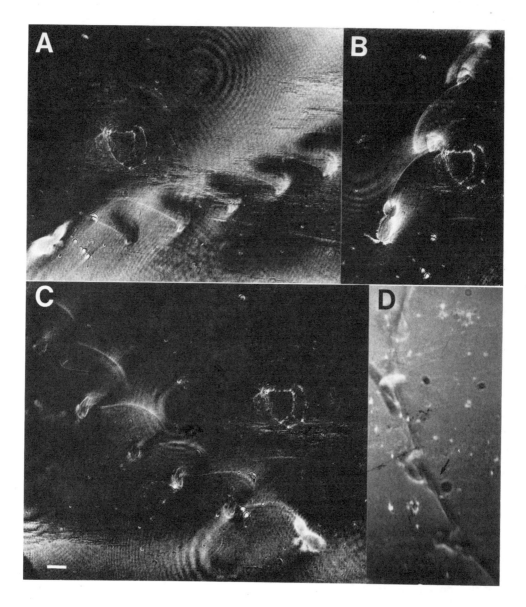

Figure 8. Swimming patterns of *Cyclops scutifer*: (A) typical hop-and-sink swimming pattern, movement from right to left (note how persistent pattern might aid in mate recognition); (B) extended distance between jumps evident with gravity-aided downward movement; (C) note consistent length of typical hops; and (D) contrasting, straight-line, escape response (arrow indicates position of abdominal flap). See Fig. 4 for details on photographic technique (scale bar = 1 mm).

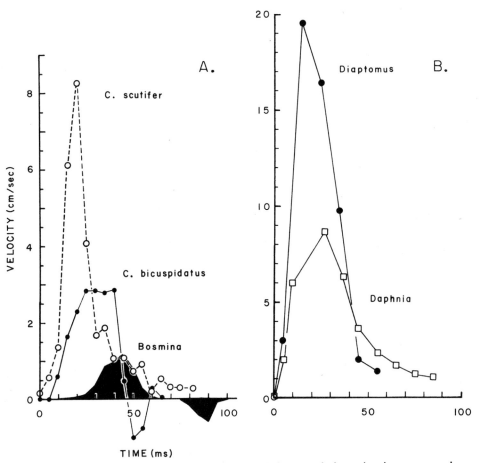

Figure 9. Instantaneous velocities of copepods and cladocerans during swimming power strokes and escape reactions: (A) velocities of *Cyclops scutifer* (Strickler 1975a); *C. bicuspidatus* and *Bosmina longirostris* (Kerfoot 1978) during power strokes; (B) escape velocities of *Diaptomus franciscanus* and *Daphnia pulex* (Lehman 1977).

and deterence (protected partially by a hard, chitinous carapace). When attacked by *Cyclops*, smaller bodied and more slowly moving cladocerans, such as free-swimming *Bosmina* and *Chydorus*, go into a peculiar motionless state termed akinesis (playing dead). In akinesis, the vulnerable second antennae are either folded into a recessed groove behind the tusklike antennules (*Bosmina*) or tucked completely under the head shield (*Chydorus*), and the ventral carapace margins are clamped securely together (Fig. 10). Thus from the point of impact with the predator, *Bosmina* and *Chydorus* will sink passively at about 0.6–$0.8 mm \cdot s^{-1}$. If, during a sequence of attacks, a *Cyclops* loses contact with a falling *Bosmina* or *Chydorus*, the predator appears to have considerable difficulty relocating its prey. Apparently a passively falling *Bosmina* or *Chydorus* produces either such a different signal, or so little disturbance in comparison with a swimming animal, that it becomes undetectable to the predator. However, if the predator presses its attack and remains in contact, the secondary defenses of both prey (thick chitinous exoskeletons, spines of *Bosmina*) serve as effective deterrents.

The importance of bad taste and predator learning: water mites and mimicry

As mentioned earlier, in freshwater zooplankton communities most species have evolved modifications in size, shape, trans-

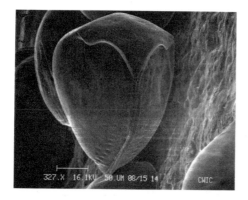

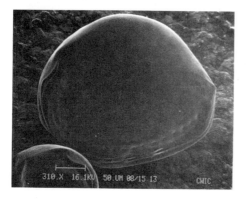

Figure 10. Akinesis in *Chydorus sphaericus* (front and lateral views of animal fixed during motionless posture).

parency, or behavior which aid concealment from visual predators. Yet, despite the myriad adaptations designed to reduce individual conspicuousness, there are certain organisms that are so gaudily colored that they appear to advertise their presence in open water. We refer to water mites (Order Hydracarina). Not only does this order constitute one of the few distinctively freshwater groups among the animal kingdom, but it is characterized by species that are relatively large and brightly colored in vivid reds, oranges, greens, blues, browns, and blacks (Pennak 1953). One might suppose, as a few have done (Elton 1923; Cloudsley-Thompson 1947), that these creatures are advertising their distastefulness, much as is commonly found among brightly colored aposematic lepidopterans or beetles (Ford 1964).

Certainly there is opportunity for the evolution of unpalatability in the zooplankton, for most aquatic predators taste prey before ingestion, and ones with highly developed nervous systems (e.g., fishes, salamanders) can learn to recognize foreign tasteful items and improve foraging effectiveness (Ware 1971). Potentially important avenues for the evolution of distasteful defenses include: (1) the ingestion of noxious algae by filter-feeders (this alternative requires sacrifice by individual prey, close sibling resemblance, and learning by predators, since prey must be either injured or eaten before the distasteful substance would act upon the predator); (2) attraction of distasteful symbionts or epibionts; or (3) the synthesis or sequestering of repugnant chemical compounds.

One obvious problem with the evolution of unpalatability in aquatic communities concerns the variety of potential predators. Often the predators in a given trophic level come from such diverse phylogenetic backgrounds that for an unpalatable substance or toxin to be effective it must either block a basic physiological process or be extremely potent to specific, dominant predators. As mite life histories are exceedingly complex, and development often prolonged (Pennak 1953; Uchida 1932), the ability of an individual to survive certainly might be enhanced by the evolution of unpalatable compounds against dominant visual predators. Yet commonly occurring visual predators include numerous fishes, amphibians (especially salamander larvae), larval and adult insects, and a few suspected crustaceans (*Polyphemus, Mysis, Neomysis*). To evolve substances that would protect all life stages against such a broad array of vertebrate and invertebrate visual predators seems a herculean feat, but certainly one worth investigating. Moreover, since water mites can form both a part of the open-water assemblage of large lakes (Gliwicz 1975; Riessen, this volume) and the littoral community of small lakes, ponds, and bogs, the targets and effectiveness of any evolved unpalatable compounds might give us valuable insights into the predator-prey nexus of aquatic communities.

Another intriguing, and related, question involves the predominance of certain colors among water mites. Of the more than 325 species of water mites described from North America (Pennak 1953; Newell 1959), the genera can be divided roughly into four groups according to color:

1. All species in the genus are bright red, e.g. *Limnochares, Diplodontus, Hydryphantes, Eylais*;
2. Most species in the genus are bright red, but a few are non-red, e.g. *Hydrachna*;
3. A few of the species in the genus are bright red, although most are non-red, e.g., *Arrenurus, Piona*;
4. All the species in the genus are non-red, e.g. *Hygrobates, Unionicola*.

The convergence of many distant lineages upon red coloration, first pointed out by Elton (1923), suggests an example of Müllerian mimicry, i.e. a number of species all of which are protected by unpalatable substances, yet which have come to resemble one another. Two possibilities could explain the convergence: red coloration is advertising an especially potent unpalatable substance and/or the color red is learned faster and remembered longer than alternative colors.

Results of preliminary investigations. To examine the general unpalatability of water mites to a variety of visual predators, during the spring of 1977 we obtained 29 species of water mites from five ponds and reservoirs around Hanover, New Hampshire (Mud Pond, Canaan, N.H.; an unnamed pond on the Hanover Golf Course, near Route 10; a roadside ditch in Wayne Johnson Bog, Norwich, Vermont; Storrs Pond, Hanover; and Hanover Reservoir, Hanover Center). We first tested these species on 19 local and exotic fishes. The following species were seined from local streams and ponds: golden shiner fingerlings (*Notemigonus*), white suckers (*Catostomus*), black bullhead adults and fry (*Ictalurus*), eastern banded killifish (*Fundulus*), largemouth bass fingerlings (*Micropterus*), pumpkinseed fingerlings (*Lepomis*), bluegill fingerlings (*Lepomis*), yellow perch fingerlings (*Perca*), black-nosed dace (*Rhinichthys*), and young carp (*Cyprinus*). The following were purchased from local pet stores: mosquitofish (*Gambusia*), stickleback (*Gasterosteus*), goldfish (*Carassius*), Siamese fighting fish (*Betta*), swordtails (*Xiphophorus*), rosy barbs (*Barbus*), neon tetra (*Hyphessobrycon*), and Australian rainbows (*Melanotaenia*).

When given the opportunity to ingest water mites, i.e. when water mites were released with *Daphnia* into 20-liter aquaria containing fish, few species ever purposely ingested water mites following initial mouthing experiences. While naive native fish would spot, approach, attack and mouth mites, they would usually not swallow the prey, and subsequently would learn to avoid or ignore mites. However, two exotic fishes (the goldfish, *C. auratus*, and the Australian rainbows, *M. nigrans*) would ingest large numbers of water mites. From these preliminary observations, we must conclude that water mites are generally unpalatable to a wide variety of native and non-native fishes.

The same result was not obtained from all other groups of visual predators. In these preliminary experiments, we placed several mites in 500ml beakers with the following visual predators: two species of adult salamanders (eastern newt, *Notophthalmus*, and the spotted salamander, *Ambystoma*), water scorpions (*Ranatra*), giant water bugs (*Lethocerus*), various dragonfly and damselfly larvae (*Aeschna*, three other unidentified genera), and larvae of the predaceous diving beetle (*Dytiscus*). In addition, we sampled a variety of water mites ourselves, by crushing an individual mite and placing it upon our tongues.

The responses by the predators were varied. All amphibians ate mites, but seemed to prefer other food (e.g. small crickets) if offered simultaneously with water mites. All insect adults and larvae attacked mites, but the responses to contact and ingestion differed considerably. Water scorpions and giant water bugs would capture, handle, and suck mites, but young individuals might die following such events. Predaceous diving beetle larvae would generally attack most species of water mite, but violently rejected mites after contact. Some dragonfly and damselfly larvae showed discrimination, eating certain mite species, but rejecting others (e.g. *Aeschnia* would reject *Hydrachna*). Our own taste buds found the water mite flavor bland but not disagreeable. Of course, in proportion to our body weight, we tasted only small amounts.

These findings are in general agreement with scattered accounts from the published literature. For example, Cloudsley-Thompson (1947) tasted water mites and found "no particularly unpleasant flavour." He also offered mites to newts (*Molge*) and observed them eating several species. McAtee (1932) claimed that practically all the frogs, toads, and salamanders he tested ate water mites. McAtee also found 535 water mites in the gizzard of a single green-winged teal and 594 in the stomach of a pied-billed grebe. Several authors mention the rejection of water mites by fishes (Elton 1923, sticklebacks and mudminnows; Cloudsley-Thompson 1946, 1947, although noting that sticklebacks ate mites). Piersig (1900) observed water scorpions seizing and sucking water mites without apparent hestiation, while Cloudsley-Thompson (1947) reported that mites placed in a small beaker with dragonfly nymphs (*Platetrum depressum*) and water-boatmen (*Noto-*

TABLE 1. Potency of "mite substance" upon a guppy (*Lebistes reticulatus*). Concentrations expressed as a fraction of dry weight.

Mite Type	Concentration of Mite in Powder	Concentration of Powder in Pellet	Mite Concentration in Pellet	Reactions
Control	0.0000	0.0000	0.0000	ingested
Control	1.0000	1.0000	1.0000	rejected
"V" Mud Pond (red)	.3561	.2048	.0729	rejected
		.1192	.0424	rejected
		.0603	.0215	ingested
"T" Mud Pond (red)	.2692	.0707	.0190	rejected
		.0301	.0081	rejected
		.0041	.0011	ingested
"D" Golf Course (yellow)	.1949	.2048	.0399	rejected
		.0760	.0148	rejected
		.0462	.0090	rejected
		.0364	.0071	rejected
		.0126	.0025	ingested
		.0112	.0022	ingested
"B" Golf Course (yellow)	.2787	.3759	.1048	rejected
		.0495	.0138	ingested
		.0213	.0059	ingested
	.1283	.2637	.0338	rejected
		.1633	.0209	ingested
		.1613	.0207	ingested
"M" Reservoir (yellow-brown)	.3459	.5271	.1823	rejected
		.0641	.0222	rejected
		.0264	.0091	ingested

necta glauca) were killed overnight but not eaten.

The foregoing accounts, although admittedly anecdotal, suggest a pattern of heterogeneous palatability. While the gaudy colors of water mites appear to advertise their general distastefulness to fishes, the defense appears ineffective against other vertebrates, and of questionable effectiveness against certain invertebrates.

Predator learning and Mullerian mimicry. Despite the apparent inedibility of water mites, they are frequently found in the stomachs of fish (Swynnerton and Worthington 1940; Frost 1943; Cloudsley-Thompson 1947; Kerfoot and Frost, manuscript). Are some mites cheaters (automimics) which produce no unpalatable substances, or are the mistakes a consequence of limited learning ability in fishes?

To evaluate the potency of the unpalatable substances produced by water mites, we ground up dried water mites, mixed a measured portion of the powdered material with edible food (Gold Medal flour), and fed the mixture to a test fish species (the common guppie, *Lebistes reticulatus*). As a control, the fish was offered pure flour balls alternately with "mite powder" balls. Since many control balls were fed between "mite powder" balls, the fish never built up an aversion to flour balls, nor could they discern any visual difference between experimental or control food balls. During the experiments, we varied the concentration of "mite powder" from variously colored mites, and recorded acceptance or rejection of the mixtures (Table 1). Although the results are again admittedly preliminary and subject to vagaries of fish choice, they demonstrated that balls containing between 4.0-0.7 percent mite powder (by weight) would cause rejection, and that the substance was an especially potent one in yellow and red mites (Kerfoot, unpubl.).

We next attempted to measure the learning response of fishes, in order to evaluate the second hypothesis, i.e. that fish learn faster on red items and remember this color longer than alternative drab colors. To quantify the fish's degree of "interest" in a food item, responses were scored on a scale of 0 to 5, where: 0, no interest; 1, slight interest (fish orients towards prey); 2, approach; 3, attack; 4, fish mouths prey repeatedly and rejects; 5. swallows prey. Fish were placed in separate 20 liter aquaria, fed standard laboratory food (TetraMin℗ and mixed copepods), and allowed to acclimate to their surroundings. Water mites and large, mature *Daphnia pulex* were then offered to fishes, following a standard routine. In this routine a single feeding "trial" consisted of presenting a fish with 10 successive *Daphnia* followed by a single water mite. Several trials could then be strung together, end to end, in a single day to form either a short (8-trial) or prolonged (≥ 30-trial) feeding "bout." For example, in a short bout a fish would be fed 80 *Daphnia* and presented with 8 water mites, all offered in a 10 *Daphnia*: 1 water mite sequence. If during a bout a fish showed any disinterest in *Daphnia*, the series was discontinued until the fish's appetite returned. In this manner we could ensure that the fish's interest in food items remained high throughout the duration of the bout. Although highly regimen-

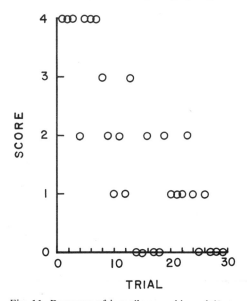

Fig. 11. Response of juvenile pumpkinseed (*Lepomis gibbosus* during prolonged feeding bout consisting of 29 trials (mite = *Hydrachna* sp.).

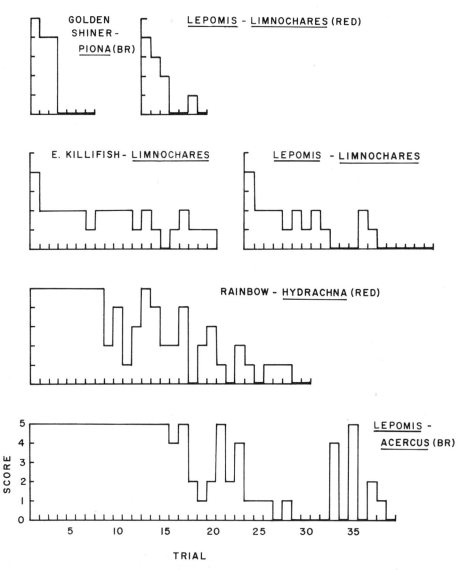

Fig. 12. Responses of a variety of fishes to brown and red water mites during short and prolonged feeding bouts. See text for meaning of score values.

ted, the routine also attempted to duplicate the natural situation, where fish generally encounter many edible items before chancing upon a water mite (Fig. 11).

The feeding bouts clearly demonstrated that fish learn to avoid water mites and that their memory involves both short-term and long-term components. For example, Figure 12 presents the responses of a variety of fishes that were exposed to red (*Limnochares, Hydrachna*) or brown mites (*Acercus*). In all cases, fish were deprived of food for 24 hours prior to feeding bouts. As the results demonstrate, during extended bouts all fish either learned to ignore water mites or greatly reduced their interest. Learning was generally more erratic upon brown mites than upon red mites.

If experienced fishes were starved for several days, or if they went for lengthy periods without being offered mites, their interest in mites would return to high levels. For example, the two fishes in Figure 12 which initially consumed several water mites (a 55mm standard length adult female Australian Rainbow and a 35mm standard length juvenile pumpkinseed) were kept from experiencing water mites for three months prior to the extended feeding bouts. Likewise, if initially naive fishes were frequently exposed to water mites, their average scores would decrease to low levels. For example, Figure 13 illustrates the results of five short-term feeding bouts (each 8 trials long) performed over an 11-day span. The species of fish were the same as before, and no individuals had laboratory experience with water mites prior to the experiment. Initially the pumpkinseed, which probably had encountered mites in nature before it was captured, ate one mite and subsequently showed strongly lowered interest scores. The rainbow initially consumed all water mites and showed no obvious signs of rejection. Over the span of eleven days, however, both fishes converged toward the same low score. The high variance around the mean bout scores, indicated by the large 95 percent confidence intervals, resulted from the short-term learning responses during each bout. From these results we can conclude that fish actively learn to avoid distasteful water mites, yet continued exposure seems necessary to maintain low interest levels. That recognition requires continual reinforcement probably accounts for the occasional incidence of water mites in fish's stomachs.

As mentioned earlier, in the presence of visually feeding fishes there is evidence that many taxa have converged upon red coloration. Around the town of Hanover often 3-5 species of red water mites are found in littoral margins of lakes where fishes abound. These species (typically *Limnochares americana*, *Diplodontus* sp., *Eylais* sp., *Hydryphantes* sp., *Hydrachna* sp., and *Piona* sp.) are large-bodied (2-5 mm), active crawlers or swimmers, and are exceedingly conspicuous. Generally bright reds are rare colors in pond waters, and are strikingly

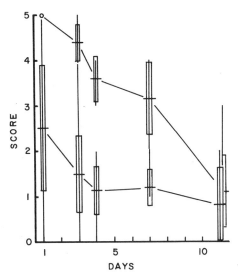

Figure 13. Evidence for long-term learning. Responses of an Australian Rainbow (upper line) and pumpkinseed (lower line) during five short-term feeding bouts performed over the course of 11 days. Horizontal bars give average scores, vertical bars = 95% confidence limits, vertical lines = ranges of scores. Notice the eventual convergence of responses toward a low mean score.

apparent against bottom sediments and green macrophytes. Since fish have long been known to recognize red faster than other colors (Brown 1937), we might conclude that a fish's prior experience with red mites confers protection upon similarly colored species. Evidence for this conclusion is presented in Figure 14, where after single-species bouts, different species of mites are substituted in a sequence of trials. In these experiments, a single pumpkinseed (45 mm juvenile) showed rapid learning on red mites and a more erratic pattern on brown mites. When offered both red and brown mites (red *Limnochares americanus*, brown *Acercus torris* and *Piona* sp.) in a sequence of trials, the fish consistently avoided the red mites while often attacking or consuming brown mites. If brightly colored red mites of the genus *Hydrachna* were placed in a sequence of *Limnochares* introductions, the second species appeared to benefit from the fish's experience with the first red-colored mite, whereas introduction of a second or third species of brown mite into a brown mite sequence merely resulted in the same erratic scoring pattern (Fig. 14). We must conclude

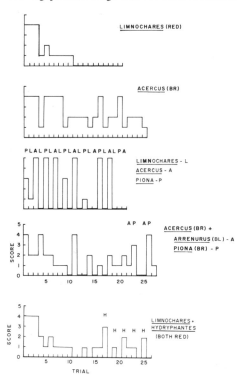

Figure 14. Responses of juvenile pumpkinseeds to different sequences of water mites. Top three diagrams show responses of a single pumpkinseed to sequences of red and brown mites alone, and then to a mixed sequence. Bottom two diagrams show responses of single pumpkinseeds to mixtures of distinct, but similarly colored, mites.

from these results that the fish is keying principally upon the color of water mites, and can associate red coloration with especially distasteful experiences.

While the source of the unpalatable substance(s) produced by water mites is not yet known, certain sites are suspected. All water mites possess large skin glands which open to the surface, and each opening is provided with sensory hairs (usually there are seven pairs of skin glands; Pennak 1953). Piersig (1900) suggested that these glands might secrete some unpleasant fluid.

Conclusions

In a sense the *Cyclops-Bosmina* and fish-mite predator-prey stories represent only two of the various interactions going on within aquatic communities. These intriguing duets emphasize that the ecological interaction, in this case predation, is also the machinery that ultimately directs the evolution of both behavior and morphology. Yet to understand the details of evolutionary adjustments and (eventually) how closely tailored natural communities are, we must appreciate the importance of directly observing food-gathering or predator-prey interactions. Only then can we comprehend why water mites provide the exception that proves the rule and why certain groups, like the rotifers and cladocerans, show amazing flexibility and variety in body shapes while other groups, such as the copepods, are rather invariant and monotonously uniform.

References

ALLAN, J. D. 1973. Competition and the relative abundances of two cladocerans. Ecology 54: 484–498.

BROWN, F. A. 1937. Responses of the large mouth black bass to colors. Nat. Hist. Bull. 21: 33–55.

CLOUDSLEY-THOMPSON, J. L. 1947. The edibility of Hydracarina. The Naturalist 822: 116–118.

DUMONT, H. J., and I. VAN DE VELDE. 1976. Some types of head-pores in the Cladocera as seen by scanning electron microscopy and their possible functions. Biol. Jb. Dodonaea 44: 135–142.

ELTON, C. S. 1923. On the colours of water-mites. Proc. Zool. Soc. London (1922) 82: 1231–1239.

FORD, E. B. 1964. *Ecological Genetics*. Methuen, London, 335 pp.

FROST, W. E. 1943. The natural history of the minnow, *Phoxinus phoxinus*. J. Anim. Ecol. 12: 139–162.

FRYER, G. 1968. Evolution and adaptive radiation in the Chydoridae (Crustacea; Cladocera): A study in comparative functional morphology and ecology. Phil. Trans. R. Soc. Lond. Ser. B 254: 221–385.

GERRITSEN, J. 1978. Instar-specific swimming patterns and predation of planktonic copepods. Int. Ver. Theor. Angew. Limnol. Verh. 20: 2531–2536.

GHARAGOZLOU-VAN GINNEKEN, I. D., and Y. BOULIGAND. 1975. Studies on the fine structure of the cuticle of *Porcellidium* Crustacea Copepoda. Cell Tiss. Res. **159**: 399-412.

GILBERT, J. J. 1966. Rotifer ecology and embryological induction. Science **151**: 1234-1237.

GLIWICZ, Z. M., and E. BIESIADKA. 1975. Pelagic water mites (Hydracarina) and their effect on the plankton community in a neotropical man-made lake. Arch. Hydrobiol. **76**: 65-88.

HALCROW, K. 1976. The fine structure of the carapace integument of *Daphnia magna* Straus (Crustacea, Branchiopods). Cell Tiss. Res. **169**: 267-276.

KERFOOT, W. C. 1975. The divergence of adjacent populations. Ecology **56**: 1298-1313.

——. 1977a. Implications of copepod predation. Limnol. Oceanogr. **22**: 316-325.

——. 1977b. Competition in cladoceran communities: The cost of evolving defenses against copepod predation. Ecology **58**: 303-313.

——. 1978. Combat between predatory copepods and their prey: *Cyclops*, *Epischura*, and *Bosmina*. Limnol. Oceanogr. **23**: 1089-1102.

——, and K. PETERSON. in press. Predatory copepods and *Bosmina*: Replacement cycles and further influences of predation upon prey reproduction. Ecology

LEHMAN, J. T. 1977. On calculating drag characteristics for decelerating zooplankton. Limnol. Oceanogr. **22**: 170-172.

McATEE, W. L. 1932. Smithson. Misc. Coll. **85**, No. 7.

NEWELL, I. M. 1959. Acari. pp. 1080-1116. In Ward & Whipple, *Freshwater Biology* [W. T. Edmondson, ed.], Wiley and Sons.

PENNAK, R. W. 1953. *Fresh-water invertebrates of the United States*. Ronald Press, New York.

PIERSIG, R. 1900. Deutschlands Hydrachniden. Zoologica **22**: 32.

RAYMONT, J., S. KRISHNASWAMY, M. A. WOODHOUSE, and R. L. GRIFFIN. 1974. Studies on the fine structure of Copepoda. Observations on *Calanus finmarchicus* (Gunnerus). Proc. R. Soc. Lond. B. **185**: 409-424.

SCHULTZ, T. W. 1977. Fine structure of the ephippium of *Daphnia pulex* (Crustacea: Cladocera). Trans. Amer. Micros. Soc. **96**: 313-321.

——, and J. R. KENNEDY. 1977. Analyses of the integument and muscle attachment in *Daphnia pulex*. J. Submicr. Cytol. **9**: 37-51.

SMIRNOV, N. N. 1977. Morphofunctional basis of life mode of Cladocera: VIII. Akinesis in Cladocera. Zool. Zh. **56**: 571-472.

STRICKLER, J. R. 1970. Ueber das Schwimmverhalten von Cyclopoiden bei Verminderungen der Bestrahlungsstarke. Schweiz. Z. Hydrol. **32**: 150-180.

——. 1975a. Swimming of planktonic *Cyclops* species (Copepoda, Crustacea): Pattern, movements and their control, pp. 599-613. In T. Y.-T. Wu et al. (eds.) Swimming and Flying in Nature, v. 2. Plenum.

——. 1975b. Intra- and interspecific information flow among planktonic copepods: Receptors. Int. Ver. Theor. Angew. Limnol. Verh. **19**: 2951-2958.

——. 1977. Observation of swimming performances of planktonic copepods. Limnol. Oceanogr. **22**: 165-170.

——, and A. K. Bal. 1973. Setae of the first antennae of the copepod *Cyclops scutifer* (Sars.): Their structure and importance. Proc. Nat. Acad. Sci. **70**: 2656-2659.

——, and S. TWOMBLY. 1975. Reynolds number, diapause, and predatory copepods. Int. Ver. Theor. Angew. Limnol. Verh. **19**: 2943-2950.

SWYNNERTON, G. H., and E. B. WORTHINGTON. 1940. Note on the food of fish in Haweswater (Westmorland). J. Anim. Ecol. **9**: 183-187.

UCHIDA, T. 1932. Some ecological observations on water mites. Jour. Fac. Sci. Hokkaido Imp. Univ. Ser. VI Zool. **1**: 143-165.

WARE, D. M. 1971. Predation by rainbow trout (*Salmo gairdneri*): the effect of experience. J. Fish. Res. Bd. Can. **28**: 1847-1852.

ZARET, R. E., and W. C. KERFOOT. 1980. The swimming technique of *Bosmina longirostris*. Limnol. Oceanogr. **25**: 126-133.

3. Optical Orientation of Pelagic Crustaceans and Its Consequence in the Pelagic and Littoral Zones

H. Otto Siebeck

Abstract

In the limnetic zone, pelagic crustaceans show a symmetrical optical orientation which often coincides with a negative geotactic orientation. In the littoral zone there is a discrepancy between this optical orientation and negative geotactic orientation. It is thought that in the littoral zone the symmetrical light orientation dominates and forces the animal's body to assume an inclination toward the limnetic zone. The result is a swimming movement away from shore. Therefore this behavior ("avoidance of shore") is a by-product of symmetrical optical orientation.

The distribution of pelagic zooplankters in lakes—if due to the behavior of the animals at all—is determined in particular by two prominent behavioral patterns: daily periodic vertical migration (e.g. Cushing 1950; Siebeck 1960), and "Uferflucht", i.e. avoidance of shore (Siebeck 1968; Preissler 1974).

There must be a correlation between these two behavioral patterns, for there is hardly an example of a pelagic zooplankter which both exhibits daily periodic vertical migration and which also is prevalent in the shore zone or vice versa, almost no example of a zooplankter which does not undergo vertical migration but which avoids the shore zone. In the past (Welch 1935), this connection between shore avoidance and vertical migration was the reason for the widely held opinion that shore avoidance is a direct consequence of vertical migration. It appeared to be a quite plausible idea that an animal which arrives prematurely at the bottom of the littoral zone during its migration into deeper water would change its course and descend along the slope, thereby moving away from shore by necessity.

A few years ago I proved that this assumption is incorrect. One can show that the animals leave the shore region along a substantially horizontal path even after the end of, or before the beginning of vertical migration (Siebeck 1964, 1968). Nonetheless there must be a relationship between these two behavioral patterns. To analyze this, one must remember that spatial orientation consists basically of two components: orientation of the body in space (primary orientation) and orientation of the direction of local movement (secondary orientation) (Lindauer 1963).

Of the many possible spatial positions of the body, there is always one preferred position which is usually reflected in body structure. In the case of aquatic animals, for instance, this position is influenced not only by the force of gravity, but also by the hydrodynamic properties of the animal's body. Dead *Mixodiaptomus laciniatus* and dead *Daphnia magna*, for example, sink in an upright position when the antennae are spread apart. They assume a similar position when encountered in the dark, and when certain brightness-distribution criteria are met in the light.

If a pelagic copepod, e.g. *M. laciniatus*,

is observed in an aquarium that is illuminated only from the top, it is striking that the animals gather beneath the lamp and assume the aforementioned approximately vertical body position there. Now and then they ascend, retaining this upright body position until they reach the surface. This activity is intensified markedly if we position the light source on the side of the aquarium. The animals now try to swim toward the light source almost without interruption, bumping against the glass violently. In particular, they do this at the beginning of the constellation. How can this difference in behavior be explained? I think there is only one possibility: When the lamp is located above the water level, the animal can assume its normal spatial position, and in this position it perceives the light coming from the same direction as always in the pelagic zone. If the light impinges laterally, there are two positions, both of which are unusual: either the animal chooses its normal position relative to the force of gravity, in which case the light will come from an unusual direction, or it chooses its normal position relative to the light, in which case it will assume a body position that will be unusual relative to the force of gravity. Although the eye of *M. laciniatus* is not immobile, it is, unlike *Daphnia*'s, apparently incapable of assuming almost any arbitrary position relative to the longitudinal axis of the body, but is fixed in position relative to the light source (Jander 1966). Moreover, there is apparently no compromise position between the light sources and the force of gravity as is the case, for example, with many fish (v. Holst 1935).

Littoral crustaceans, e.g. *Macrocyclops albidus* or *Eucyclops serrulatus*, react with considerably less variation under these two situations. They wander around everywhere in the aquarium, resting here and there, assuming different body positions on the substrate in each case.

Actually, this difference between pelagic and littoral inhabitants is not surprising if one considers that the littoral animals are subject again and again to new constellations as they swim around or climb on the aquatic plants and on the substrate. There is no doubt that the variability in time and space is considerably greater in this habitat than in the pelagic zone.

As a preliminary result of these considerations, let us note that, as far as spatial orientation is concerned, pelagic animals show a substantially closer correlation to the brightness pattern than littoral animals. Let us now look more closely at this fact (Fig. 1).

First of all, a few remarks are necessary concerning the brightness distribution under water. It is a well-known fact that the characteristics of brightness distribution are determined by the refraction of light at the surface. If we imagine an underwater light source, only those light rays which impinge on the surface at an angle $<48.6°$ will penetrate through it into the atmosphere. The respective rays are therefore all located within a cone whose apex coincides with the light source and whose base describes a circular disk on the surface of the water. This circular disk is the "window" through which the aquatic animals can look through the water into the atmosphere above. Compared to its optical environment, this window is especially bright. It is therefore a conspicuous optical marker for checking body position. This check can be improved if it is combined with gravitational orientation.

It is also important in this context to emphasize an essential difference which results when the brightness distribution is compared close to the shore and far from the shore. For the sake of simplicity, let us assume that the sky is uniformly bright. In this case the window on the surface of the water will also be uniformly bright. Far away from shore this applies to the entire extent of the window, but close to shore it applies only to the region not affected by the elevation of horizon on the shore.

Under these conditions *M. laciniatus*, for example, exhibits strikingly different behavior depending on the two brightness distributions. If the window is uniformly bright, we do not see any preferred direction of swimming. If the intensity increases, the animals migrate into deeper water. If the intensity decreases, they migrate upward.

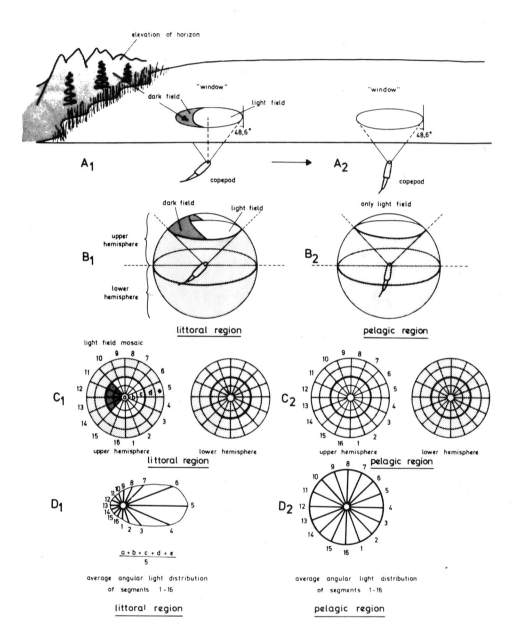

Fig. 1. A_1 —Influence of elevation of horizon on brightness distribution in window: a dark field is produced. A_2 —Dark field disappears when distance from shore is adequate. $B_1 B_2$ —If one imagines copepod eye to be located at center of a sphere, brightness pattern can be projected onto surface of sphere. Noticeable difference in window is indicated close to and far away from shore. $C_1 C_2$ —Upper and lower hemispheres are projected onto a plane. Values from 1 to 16 can be entered in fields a, b, c, d, e to describe mosaic. $D_1 D_2$ —This drawing illustrates mean brightness distribution of upper hemisphere for littoral and pelagic zones (vector diagrams).

If part of the window is darkened because of elevation of the horizon, we observe horizontal migration. Like all of the other pelagic crustaceans we have tested up to now, *M. laciniatus* prefers a direction that leads it away from the elevation of horizon. This means that, in a lake, the animals will move away from shore.

In this context it must be stressed in particular that the elevation of horizon virtually influences only the brightness distribution within the window, i.e. the brightness area above the animal, and not the area which lies in the direction of its migratory destination.

It has been shown by experimental means that this horizontal migration always occurs in the plane of symmetry of the brightness pattern (Siebeck 1968). This means that the animal is oriented in such a way that it receives the same optical information from the top left that it does from the top right. The animal is thus in a situation characterized by symmetrical stimulation (Fig. 2). That observation, however, is not sufficient to determine that this and no other direction of migration is preferred, for there are many potential migratory directions within the symmetry plane of the brightness pattern. Furthermore, the animal would be in a situation with symmetrical stimulation in all directions. There must therefore be additional information responsible for causing the animal to choose only one direction of swimming. What information is this?

If we look at the brightness distribution, in which the animals do not prefer any horizontal migratory directions, we will find evidence that will enable us to answer the question. As we have already seen, this is the case when a punctiform light source is located above the position of the animal or when the window is uniformly illuminated. The brightness distribution in both cases can

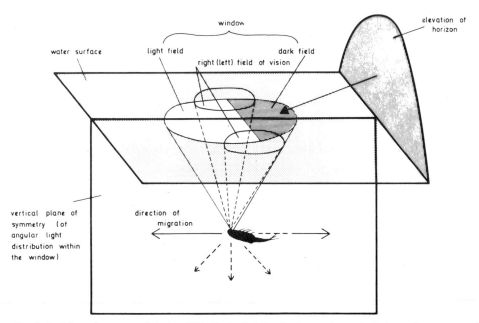

Partially darkened window in the littoral zone

Fig. 2. Position of a copepod during "Uferflucht." Animal migrates in vertical plane of symmetry of brightness pattern. It is located in a position characterized by symmetrical stimulation of opposing fields of vision. This applies to any arbitrary direction within this plane of symmetry.

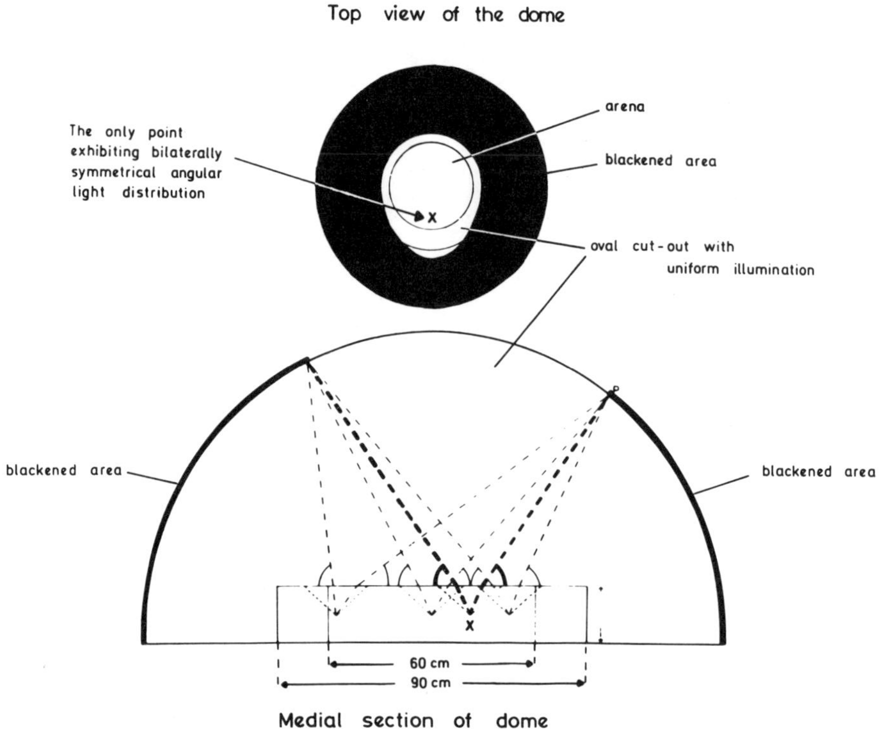

Fig. 3. Experimental setup. Top is a top view; bottom is a longitudinal section. Plexiglass dome has been blackened except for an oval segment. Every location in arena has a different brightness distribution. Only one location is bilaterally symmetrical (indicated by heavy dashed line).

be subdivided by an arbitrary number of planes of symmetry whose sectional plane coincides with the vertical. Such brightness patterns are radially symmetrical. This is a rare event under natural conditions. However, it can be shown by experimentation that a bilaterally symmetrical brightness distribution is sufficient to stop or prevent substantial horizontal migration. This statement will be proved with the aid of the following laboratory experiment. A white plexiglass dome is placed over a circular arena. The dome has been blackened except for an oval segment, which is uniformly bright (Fig. 3). The consequence is that every location inside the arena has a different brightness distribution within the window (Fig. 4). A few of these brightness-distribution patterns are illustrated by the vector diagram in Figure 4.

The following situation is especially important. There is only one single location in the entire arena at which the brightness pattern can be subdivided by two vertical, perpendicularly oriented planes of symmetry. The brightness pattern at this location is therefore bilaterally symmetrical. It is precisely at this location that the animals gather, i.e. the animals prefer a vertical body position at this one location.

If we follow the paths of a few animals from start to finish, we find that animals released at the side of the arena do not head directly for their ultimate destination. Instead, they choose the migratory direction which coincides with the vertical plane of symmetry of the brightness pattern at that location. Some animals are very sensitive to location-dependent changes in the position of the plane of symmetry, while others initially retain the direction in which they once started and make an obvious correction later. The animals released in the plane of symmetry of the oval segment head directly for their destination. This migration, how-

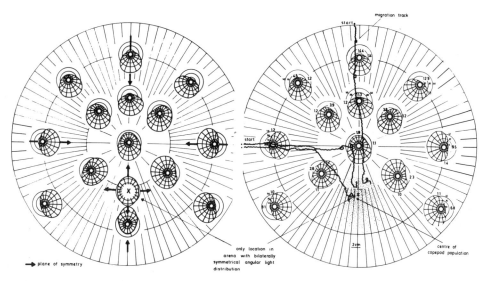

Fig. 4. Vector diagrams indicate mean brightness distribution for various locations within arena (*see Fig. 1*). Only at X is brightness distribution bilaterally symmetrical. Three swimming tracks and positions of all copepods at end of experiment (small dots) are shown on right side of drawing.

ever, as shown before, cannot be compared with the purposeful movement of a predator that has caught sight of its prey. The only essential aspect in each case is that the plane of symmetry of the brightness pattern is always directed along this line to the ultimate destination (Fig. 4).

How can this behavior be interpreted? Imagine a model copepod that has a pigment cup ocellus, located at its cranial end symmetrical to the longitudinal axis of the body, which consists of four pigment cups with divergent optical axes. Two of the divergent axes should be located in a common plane, the respective pigment cups thus forming a pair.

Our model copepod should be capable of aligning the position of its body in such a way that all pigment cups are illuminated as much as possible. If there is a punctiform light source positioned above the water surface or if the window is uniformly illuminated, this copepod will orient itself vertically, for only in this position can it ensure that all pigment cups will be illuminated in whole or in part by the window (Fig. 5).

We now darken a portion of the window by elevating the horizon. If the copepod retains its position, one of the pigment cups will be darkened in whole or in part. Symmetry thus exists for only one pair of the four pigment cups. By turning the longitudinal axis of its body in the plane of the two unequally illuminated pigment cups, in particular toward the more strongly illuminated one, the animal is able to find a new position in which these two pigment cups are once again illuminated equally. In so doing, however, the longitudinal axis of the animal's body has been rotated away from the normal position to a greater or lesser extent. The animal is thus forced to move away from the side of the dark field. In a lake, this behavior means the animal will leave the shore region.

There are other possible interpretations, which I do not want to discuss here. Instead, I should like to discuss briefly a very obvious objection to a special feature of the model copepod, namely the four pigment cups, because a true copepod has only three. Hence, only two pigment cups are arranged to form a pair. These are likely used to compare the brightness at the top left with the brightness at the top right. The respective pair of pigment cups required to compare the brightness at the top front with the top rear, however, is missing. This group of three

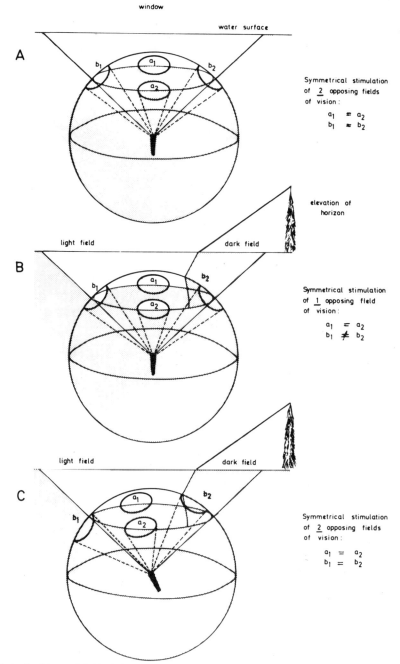

Fig. 5. A—Positions of field of view when model copepod is in upright vertical position. When window is uniformly illuminated, animal experiences symmetrical stimulation of two opposing fields of vision. B—As in A, but window has been partially darkened by elevating horizon. C—Positions of field of view when animal is in inclined position. Symmetrical stimulation has been re-established by inclination in plane $b_1 - b_2$.

cups nevertheless fulfills the prerequisites to function completely analogously to the group of four cups. It can be shown that there are sensory cells in the unpaired pigment cup which are positioned in a common plane with sensory cells from one of the paired pigment cups. It is therefore possible to compare the brightness in two planes that are perpendicular to each other. Incidentally, this is also shown by the observations of nauplii as described by Jander (1966).

There are also other important differences between our model copepod and a true copepod. For instance, the pigment cup in a true copepod is not positioned exactly at the top of the anterior pole, but is displaced somewhat ventrally. In order to direct the pigment cups to the window, the copepod must turn slightly dorsally. Finally, it must also be mentioned that the pigment cup ocellus in many copepods is not rigidly connected to the body, but is mobile. It can thus orient itself toward the window within certain limits irrespective of body position. This opens up a wide field of new and interesting questions, to be analyzed, at least in part, by utilizing the new techniques developed by Strickler (1977).

I should like to leave the realm of interpretation at this point and return to the last experimental result I discussed. We now know that the body position of *M. laciniatus* under natural lighting conditions is to some degree dependent on the brightness distribution within the window. It is also certain that *M. laciniatus* attempts to assume a position of symmetrical stimulation of two opposing fields of vision in this brightness pattern. Because of the brightness distributions prevailing in the pelagic zone, the result must be that the vertical body position is dominant in this habitat. If we retain these criteria of orientation, a more horizontal body position directed away from the shore must be prevalent in the shore vicinity. "Uferflucht" is thus the consequence of an orientation that results from the peculiarities of the brightness pattern within the window in the shore region. This means that Uferflucht is a by-product of behavior that ensures a vertical position in the pelagic zone (Siebeck 1973).

This position is especially important in the pelagic zone, because pelagic animals make daily periodic vertical migrations, which are not a luxury but are vital to them. If such vertical migration is to take place without wasting energy, it must be oriented precisely. Hence, the capacity of an organism to orient itself vertically is of utmost importance.

The relationship between Uferflucht and vertical migration, raised at the outset of this paper, is therefore that daily periodic vertical migration presupposes vertical orientation. Negative geotactic spatial orientation is one method of achieving this, and symmetrical stimulation of two opposing fields of vision relative to the window—which, like gravity, constitutes an invariant—is another. It is possible that higher precision is attained because of the synergism of both methods. If this orientation relative to the window occurs as a stereotype reaction, it must produce the above-mentioned spatial orientation in the littoral zone, with Uferflucht as the consequence.

There are, of course, brightness patterns within the window even in the pelagic zone, e.g. when the sun sets, which cause an inclined position in the animals and thus a corresponding horizontal migration. Nevertheless, large mass displacements of the pelagic crustaceans are normally not to be expected because the sun changes its position and because the horizontal drift is considerable and varies at different times and at different depths. The elevation of the horizon is a stable factor affecting the brightness pattern in the vicinity of shore. The current near the shore is almost parallel to the shore, so that the animals can migrate transversely to the current.

The behavior of pelagic crustaceans is so stereotyped under the conditions of Uferflucht that it is in no way influenced by currents. Only if the current direction coincides with the plane of symmetry of the littoral brightness pattern and the animals are drifting in the direction of the elevation

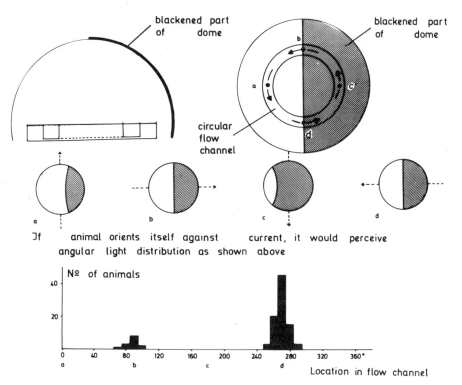

Fig. 6. Top shows experimental setup. Circular flow channel is located beneath plexiglass dome, half of which has been darkened. Arrow indicates direction of flow. Brightness distribution in window is illustrated schematically for locations a, b, c, d. Animals compensate for current only at b and d as shown by lower graph.

of horizon is there any increase in the swimming activity (Siebeck unpub.). Under these conditions, there is more likely to be compensation or even overcompensation for the current flow rate. In smaller lakes this behavior plays an important role near the lake outlet (Regnauer 1975). It has been found that under a snow-covered ice layer, during thick fog, or during a very dark night large numbers of animals drift into the outlet. However, if the elevation of horizon—irrespective of whether it is interrupted by the lake outlet—is effective in a segment of the lake in front of the outlet, and the drift is not too strong there, only here do the animals still have a chance of escaping from the drift. This is shown by the following experiment employing a circular flow channel beneath the plexiglass dome, half of which is blackened in this case (Fig. 6). Compensation for the current is found only in the channel where the vertical plane of symmetry of the brightness pattern within the window coincides with the direction of current flow. At locations where these two directions intersect at an angle of 90°, the drift becomes fully effective, because here the animals migrate transversely to the direction of current flow.

The stereotyped behavior of the various pelagic crustaceans is even more conspicuous if we examine the behavior of littoral crustaceans. Pelagic crustaceans leave the shore region until the dark field has disappeared almost entirely. Littoral crustaceans react in a more differentiated manner (Link 1975).

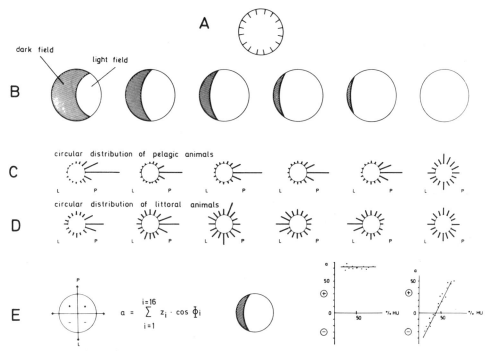

Fig. 7. A—Top view of arena for field experiments. B—Brightness distribution within window at different distances from shore. C—Distribution of animals in arena. Migratory directions toward pelagic zone (P) are preferred. D—Distribution of animals in arena. Migratory directions are dependent on size of dark field. A large dark field area induces animals to migrate toward pelagic zone, whereas a small dark field area induces animals to migrate toward littoral zone. E—Evaluation of circular distributions (see Link 1975 for further details). HU is elevation of horizon.

As long as the area of the dark field exceeds about 40 percent, they behave in principle like pelagic crustaceans: they prefer to migrate in directions leading away from shore. If the proportion of the dark field declines below about 30 percent they prefer directions leading toward shore. There is also a percentage of the dark field which lies between these two limits where the animals prefer neither the one nor the other but several directions. This behavior is sufficient to keep the littoral crustaceans in the littoral zone solely by virtue of optical criteria (Fig. 7).

I do not have the space here to discuss these findings in detail. I have mentioned them to show that the following expectation is correct: the optically more complex littoral habitat requires the organisms to orient themselves in a more differentiated manner than is demanded of them, in particular the filter-feeding animals, in the optically simpler pelagic zone. Here the problem of meaningful orientation for vertical migration comes to the fore. *Mixodiaptomus laciniatus* achieves almost vertical orientation because it tries to establish a symmetrical stimulation of two opposing fields of vision relative to the window. This positioning in the pelagic zone frequently means coincidence with a negative geotactic orientation. Such coincidence is normally not possible in the littoral zone. In this case, preference is given to optical orientation. It forces the animal to assume a position in which it must leave the littoral zone. Therefore, there is indeed a relationship between Uferflucht and vertical migration.

References

CUSHING, D. H.   1950.   The vertical migration of planktonic Crustacea. Biol. Rev. **26**: 158-192.

HOLST, E. v., 1935. Über den Lichtdruck-enreflex bei Fischen. Pubbl. Staz. Zool. Napoli. **25**: 143-158.

JANDER, R. 1966. Die Phylogenie von Orientierungsmechanismen der Arthropoden. Verh. Dtsch. Zool. Ges. **29**: 266-305.

LINDAUER, M. 1963. Allgemeine Sinnesphysiologie, Orientierung im Raum. Fortschr. Zool. **16**: 58-140.

LINK, G. 1975. Freilandexperimente zur Bedeutung der Horizontüberhöhung für das Verhalten verschiedener Litoralcrustaceen. Inauguraldiss., Univ. München. 114 p.

PREISSLER, K. 1974. Vergleichende ökologisch-physiologische Untersuchungen an Pelagial- und Litoralcrustaceen unter Berücksichtigung der räumlichen Helligkeitsverteilung. Inauguraldiss., Univ. München. 145 p.

REGNAUER, A. 1975. Untersuchungen zum Verhalten pelagischer Crustaceen im Seeausfluss. Inauguraldiss., Univ. München. 131 p.

SIEBECK, O. 1960. Untersuchungen über die Vertikalwanderung planktischer Crustaceen unter besonderer Berücksichtigung der Strahlungsverhältnisse. Int. Rev. Gesamten Hydrobiol. **45**: 381-454.

—, 1964. Ist die Uferflucht pelagischer Crustaceen eine Folge der Vertikalwanderung? Arch. Hydrobiol. **64**: 419-427.

—, 1968. Uferflucht und optische Orientierung pelagischer Crustaceen. Arch. Hydrobiol. Suppl. 35(1), p. 1-118.

—, 1973. Untersuchungen zur Biotopbindung einheimischer Pelagial-Crustaceen. Verh. Ges. Oekolog. **2**:11-24.

STRICKLER, J. R. 1977. Observation of swimming performances of planktonic copepods. Limnol. Oceanogr. **22**: 165-170.

WELCH, P.S. 1935. Limnology. McGraw-Hill.

4. Morphological, Physiological, and Behavioral Aspects of Mating in Calanoid Copepods

Pamela I. Blades and Marsh J. Youngbluth

Abstract

The mating behavior of calanoid copepods is precise, ritualized and species-specific. Observations of living copepods indicate that the ritual begins with a chemically induced, mate-seeking behavior, proceeds through various morphologically and physiologically directed movements and orientations of the male with respect to the female, and terminates with the attachment of a spermatophore to a precise location on the female urosome. The elaborate mating behavior of the calanoid copepod *Labidocera aestiva* Wheeler provides an example of how chemoreception, mechanoreception, certain modified appendages, sensory and non-sensory integumental structures, and spermatophore complexes have evolved to ensure that the female is fertilized.

Reproductive behavior is an important, but neglected aspect of zooplankton ecology. The ability to attract and recognize a specific mate and to synchronize physical contact and copulatory movements is vital for propagation. With regard to copepods, J. Gerritsen (this volume) discusses the variety of behavioral responses these animals may exhibit when encountering prey items, predators, and mates, while J. R. Strickler and Bal (1973), using elaborate cinematographic techniques to record the swimming behavior of copepods, have learned that copepods can distinguish potential mates, predators, or prey by mechanoreception and probably chemoreception.

It is apparent from these studies that specific behavioral and sensory processes have evolved to ensure that male and female copepods meet. Once the sexes are coupled, what other physiological events and behavioral movements guarantee that a male will fertilize a female? Mating behavior that is precise, ritualized, and species-specific has been suggested (Fleminger 1967; Lee 1972; Hopkins and Machin 1977). This theory is based on inspections of preserved copepods, including observations of specially modified appendages, unique spermatophore structures, and a specific attachment site for the spermatophore complex on the female urosome. It has also been proposed that sensory structures—e.g. spines, hairs, and pores present on the integument of copepods—may play an important role in the mating encounter (Fleminger 1973; Fleminger and Hulsemann 1977; Mauchline 1977; Mauchline and Nemoto 1977).

Observations of mating between living calanoid copepods are as yet rare (Katona 1975; Blades 1977; Blades and Youngbluth 1979) and usually incomplete (Gauld 1957; Jacobs 1961). These studies, however, do provide examples of the specificity and variability of reproductive behavior. Our purpose here is twofold: to discuss how certain modified appendages, integumental structures, and spermatophore complexes may be used by copepods to achieve fertilization, and to summarize what has been

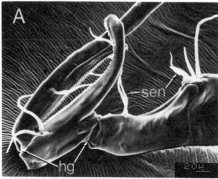
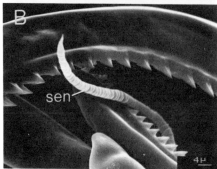
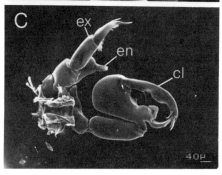

Fig. 1. Modified appendages of male *Labidocera aestiva*. A—Prehensile right antennule in grasping position. Note hinges (hg) and sensory hairs (sen). B—Sensory hair on right antennule. C—Fifth pair of swimming legs (ex, exopod; en, endopod; cl, chela).

learned from investigations of reproductive morphology and behavior of copepods. The mating behavior of the pontellid copepod, *Labidocera aestiva* Wheeler, is used as an example.

We thank Julianne Piraino of the Smithsonian Institution, Fort Pierce Bureau, for her skillful operation of the scanning electron microscope. We also thank Kevin J. Eckelbarger of the Harbor Branch Foundation, Inc., for the information obtained from his transmission electron microscopic examination of our specimens and for his suggestions and corrections to the manuscript.

Morphology

To understand the events that occur throughout the mating encounter of *L. aestiva*, we give a brief review of the relevant morphological structures of the male and female.

The right antennule of the male is geniculate and prehensile (Fig. 1A). This appendage is able to fold back upon itself, because of the presence of two hinges, one between segments 17 and 18, the other adjoining segment 18 and fusion segment 19–21. The antennule is also equipped with various sensory hairs (Fig. 1A, B). The fifth pair of swimming legs of the male is asymmetrical (Fig. 1C). The right portion is uniramous and chelate, whereas the left leg is biramous, consisting of a unisegmented endopod and a double-segmented exopod. The distal segment of the exopod terminates in a bifurcation of one short and one long spine, and two thick patches of long hairs occur along the inner margin (Fig. 2A). The distal section of the endopod (hereafter called the corrugated endopod) is partially encircled by a series of parallel ridges, providing the endopod with a rough external surface (Fig. 2B, C, and D).

The spermatophore of *L. aestiva* is representative of calanoids which possess chitinous coupling devices designed to attach at a precise location on the female. The spermatophore coupler of *L. aestiva* consists of an anterior shield and a posterior coupling plate, connected by two lateral bridges (Fig. 3). A cement-like substance is present on the external surface of the anterior shield, the right lateral bridge, and the posterior coupling plate, in addition to the inner surface of the posterior coupling

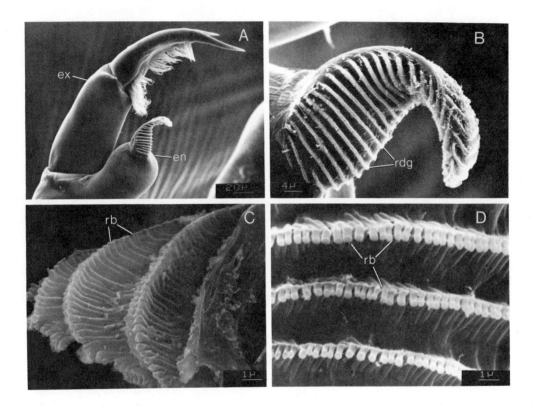

Fig. 2. Left fifth leg of male *Labidocera aestiva*. A—Double-segmented exopod (ex) and unisegmented endopod (en). B—Corrugated endopod showing parallel ridges (rdg). C, D—Higher magnification of ridges showing double, distally flattened ribs (rb).

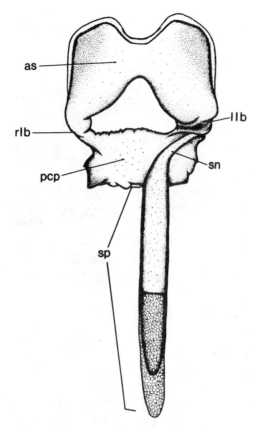

Fig. 3. Diagrammatic illustration of spermatophore complex of *Labidocera aestiva* (as, anterior shield; rlb, right lateral bridge; llb, left lateral bridge; pcp, posterior coupling plate; sp, spermatophore proper; sn, spermatophore neck). Composite from photographs, not drawn to scale).

plate. When the spermatophore complex is correctly attached to the female, the anterior shield fits loosely, without cement, over the ventral and ventrolateral surfaces of the genital segment, covering the gonopore (Fig. 4A). The posterior coupling plate, however, is firmly cemented to the right ventrolateral surface of the abdominal segment (Fig. 4C, D). The tubelike spermatophore proper lies over the midventral axis of the female urosome and is free from the coupling device except where the tube narrows to become the spermatophore neck. This latter section is firmly held within a groove sculptured into the posterior coupling plate. Anterior to the groove, the spermatophore neck narrows to become the spermatophore stalk which passes under the left lateral bridge and continues forward beneath the anterior shield to terminate at the gonopore (Fig. 4B).

The urosome of the female is composed of a symmetrical genital segment, a slightly narrower abdominal segment, and an anal segment fused with the caudal furcae. The gonopore, located midventrally on the genital segment, is shielded by a genital plate that is hinged along the anterior margin of the pore (Fig. 5A).

The ventral and lateral surfaces of the female genital segment are sites for groups of nonsensory hairs and spines of varying sizes and shapes (Fig. 5B, C, D). Transmission electron microscopy (TEM) has revealed these projections to be epicuticular structures.

A cluster of 20 to 23 unusual pores, referred to as pit-pores, are present on the right ventrolateral surface of the female abdominal segment (Fig. 6). Their morphology is that of a shallow crater, about 6–10 μm in diameter. At the base of the crater is a small, round pore within which is a conical structure, bearing a small opening at its tip. TEM has revealed that each pit-pore extends through the cuticle and opens into a lumen. The lumen is lined with an extracellular matrix produced by what is believed to be a complex, secretory-like cell. When attached to the female, the posterior coupling plate of the spermatophore is cemented to the surface of this pit-pore area.

Mating Behavior

The mating ritual appears to be a sequence of precise movements directed by chemical and mechanical signals. In describing the mating behavior of *L. aestiva*, therefore, we shall combine our behavioral observations of live animals with speculations on events that may occur at the sensory level.

During the initial physical contact, the male uses his geniculate right antennule to hold the female's caudal setae (Fig. 7A). We have speculated that specific hairs on the antennule are excited when it closes. This stimulus may signal the male to move to the next position, i.e. to secure the caudal furcae

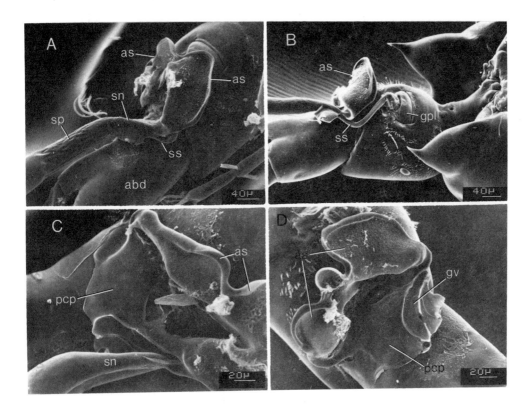

Fig. 4. Spermatophore attachment to urosome of female *Labidocera aestiva*. A—Left ventrolateral view. Anterior shield (as) covers ventral and ventrolateral surfaces of genital segment including gonopore (considerable shrinkage occurs during SEM preparation) (sp, spermatophore proper; sn, spermatophore neck; ss, spermatophore stalk; abd, abdomen). B—Anterior shield (as) has been pulled back to expose spermatophore stalk (ss) enroute to gonopore (gpl, genital plate). C, D—Posterior coupling plate (pcp) cemented over surface of pit-pore field on abdomen (C—right lateral view, D—right ventrolateral view). Spermatophore tube removed exposing supporting groove (gv).

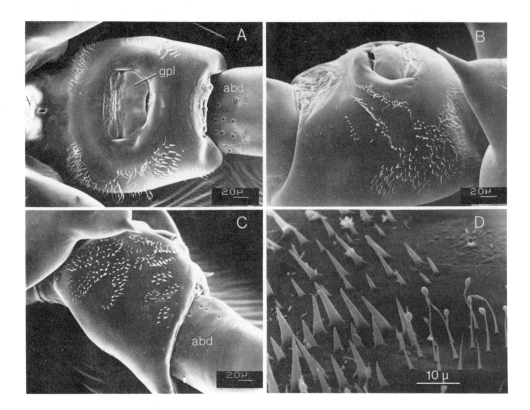

Fig. 5. Genital segment of female *Labidocera aestiva*. A—Ventral view showing gonopore shielded by genital plate (gpl) (abd, abdomen). B—Left lateral view showing groups of epicuticular spines and hairs. C—Right lateral view of same. D—Higher magnification of spines and hairs. Bulbous tips on hairs are artifacts.

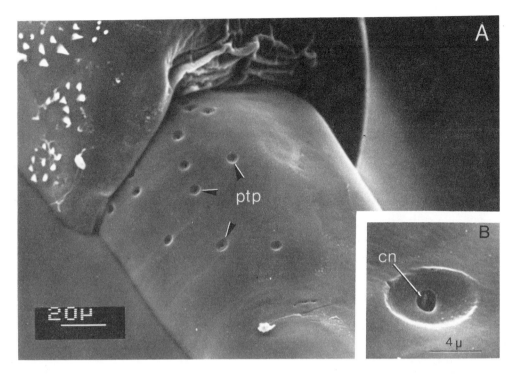

Fig. 6. Pit-pores (ptp) on abdomen of female *Labidocera aestiva*. A—Right ventrolateral view. B—Higher magnification showing crater, pore, and conical structure (cn).

of the female with the chelate fifth leg. Having accomplished this, the male releases his antennule and lies in the same plane as the female but is oriented posteriorly with respect to her longitudinal axis. Their ventral surfaces face in the same direction (Fig. 7B).

Our behavioral observations in the laboratory have shown that the spines on the male chela may also function as mechanoreceptors, and it is likely that these sensilla are stimulated when the male closes the chela on the female's caudal furcae. Signals from the sensilla consequently elicit a precopulatory "stroking" behavior. Using the chela to hold the female, the male rhythmically draws himself near to, then drifts away from, her urosome. While in the close position, the exopod and endopod of the male's left fifth leg can be seen moving rapidly back and forth against the surface of the female's urosome (Fig. 7B, 8). Further examination has revealed that the male uses the long hairs on the exopod of the left leg to stroke rapidly the right ventrolateral field of the female's genital segment. If the hairs and spines on this surface are nonsensory, this behavior may aid the male in mate identification and in obtaining a specific orientation with respect to the female. As the brushing of the exopod continues, the long terminal spine on the same limb probes around the edges of the female gonopore. This tactile stimulation may signal her genital plate to open to accept the spermatophore stalk. Concurrent with the exopodal movements just described, the male's corrugated endopod also vigorously strokes the surface of the pit-pore field on the female's abdomen. Information obtained from SEM, TEM, and behavioral observations suggests that during this precopulatory stroking the pit-pores may simultaneously secrete a solvent that is spread over the abdominal surface by the movements of the male's endopod. The combined effect of these mechanical and chemical processes may be to remove any cement remaining from a previously attached spermatophore, and thus prepare the ab-

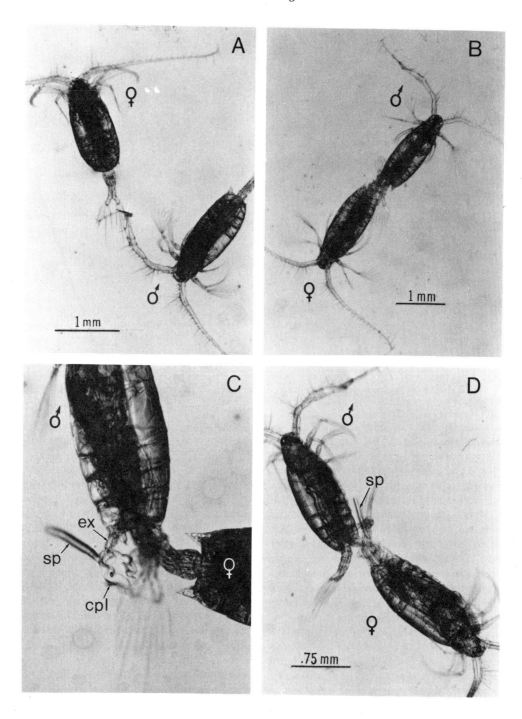

Fig. 7. Light micrographs of *Labidocera aestiva* mating behavior. A—Initial capture of female by male. B—Precopulatory "stroking" position (*see also Fig. 8*). C—Spermatophore extrusion. Dorsal view of male showing left fifth leg exopod (ex) holding coupling plate (cpl) of fully extruded spermatophore (sp, spermatophore proper). D—Copulatory position and spermatophore attachment to the female.

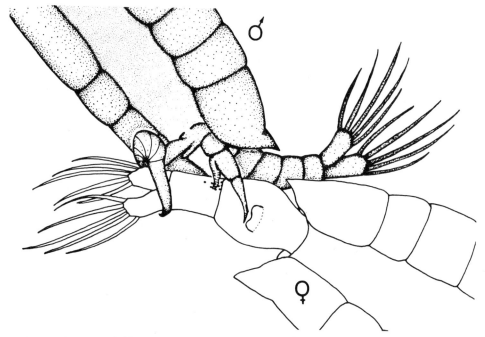

Fig. 8. Diagrammatic illustration of precopulatory "stroking" of *Labidocera aestiva*. Long hairs on terminal segment of exopod not shown. Composite from photographs, not drawn to scale.

dominal surface for placement of a new spermatophore.

This precopulatory behavior normally continues for 2–5 minutes, or until a spermatophore is extruded. As the spermatophore emerges, the male lowers his left fifth leg to his gonopore, where the long hairs on the exopod adhere to the cement on the outside of the posterior coupling plate (Fig. 7C). This allows the male to quickly and accurately guide the spermatophore to the female. He holds the spermatophore against the female's urosome (Fig. 7D) until it is securely fastened (ca. 30 seconds) and then releases her.

The contents of the spermatophore are discharged into the female within about 30 minutes. During this process the pit-pores may again secrete a solvent that dissolves the cement and loosens the posterior coupling plate. When the spermatophore is empty, the female quickly positions the long, symmetrical exopods of her fifth legs under the lateral edges of the anterior shield and pulls the loosened spermatophore complex off the urosome.

Discussion

The preceding remarks described morphological, physiological, and behavioral adaptations relating to the mating behavior of *L. aestiva*. We now review the interaction of these factors.

The mating ritual presumably begins when the male detects pheromones released by a female. Previous investigations have shown that male copepods exhibit a distinctly different swimming behavior when in close proximity to a female or to chemical substances released from females (Katona 1973; Kittredge et al. 1974; Griffiths and Frost 1976). This behavior consists of a series of loops, tight turns, and somersaults in *L. aestiva* (Parker 1902), *Eurytemora affinis* (Katona 1973), *Calanus pacificus*, and *Pseudocalanus* sp. (Griffiths and Frost 1976), and most likely allows the male to search a small area intensively. Griffiths and Frost (1976) presented autoradiographic evidence that aesthetase hairs on the antennules of male *C. pacificus* are sites for chemoreception and may function to preceive chemicals released by a female. Furthermore, Katona (1973) and Kittredge et al. (1974) suggested that the male might be able to detect a chemical gradient. Following such a gradient may lead the male into the swimming wake of a female. Mate-specific swimming vibrations could be detected at this point and enhance the male's initial capture response. Strickler and Bal (1973) provided TEM data indicating that setae on the antennules of male *Cyclops scutifer* are likely to be mechanoreceptors. Thus both chemical and mechanical stimuli may promote contact between the sexes.

Our observations of *L. aestiva* and those of *Centropages typicus* (Blades 1977) imply that sensory information may also be conveyed through setae located on the male's geniculate antennule and fifth legs. When the antennule of these copepods is closed around the caudal setae of a female, tactile stimulation of hairs along the inner margin of the hinge (Fig. 1A) may signal the male to proceed with the mating ritual and secure the female's caudal furcae with his chelate fifth leg. Precopulatory stroking movements by the left fifth leg of *L. aestiva* can be artificially induced when the male closes the chela around a small needle. This response suggests that setae on the opposing edges of the chela (Fig. 1C) are mechanoreceptors and are excited when the chela grasps the female. It is also possible that the tufts of long hair on the left leg exopod of the male *L. aestiva* provide sensory input. The precopulatory stroking of these hairs across the spines of the female's genital field could inform the male that he has captured the correct mate, that he is properly oriented with respect to her urosome, and that a spermatophore is or is not attached to her urosome.

In some species the mating ritual may be regulated primarily by pheromones; in others a combination of chemical cues and tactile stimulation is needed. Hopkins and Machin (1977) suggested that chemicals may be used to advertise the receptive condition of the female, i.e. the presence or absence of a spermatophore. Fertilization, resulting from a spermatophore correctly fastened to the female genital opening, might inhibit the production and release of attracting pheromones. The male would be informed of the mating status of the female before physical contact and thus conserve the time and energy spent in capturing her. Likewise, to extrude a spermatophore before knowing the receptive state of the female could result in multiple spermatophore attachments and a consequent waste of gametes. A reduction in pheromone production below a threshold level might cause the male to retain his spermatophore.

Evidence to support these ideas may be found in the mating behaviors of *E. affinis* (Katona 1973) and *C. typicus* (Blades 1977). Males of *E. affinis* extrude a spermatophore before physical contact, and males of *C. typicus* do so while using the antennule to hold the female (Fig. 9). Neither species exhibits a precopulatory tactile inspection of the female. *Labidocera aestiva* males, on the other hand, extrude a spermatophore only after seizing the caudal furcae of the female, stroking her urosome, and inspecting the genital area. These observations suggest that this species has evolved a less efficient system of chemical communication and must also rely on tactile inspection of the female.

Behavior that may illustrate communication feedback between mates has been observed. For example, certain movements of the female may provide stimuli to reinforce the male's behavior. Blades reported that a male *C. typicus* extruded a spermatophore but failed to attach it to a dead female he had seized with his antennule. Male *L. aestiva* will cease precopulatory stroking when a female is artificially subdued with forceps. If she is released and begins swimming again, the male will resume the stroking movements. When adults of *L. aestiva* were crowded together in culture dishes, the males were occasionally noted to couple with other males. When one male seized another male, they often wrestled vigorously for a few minutes before separating. A spermatophore is rarely extruded or attached to one of the males during such an encounter. Females, however, are much less active when

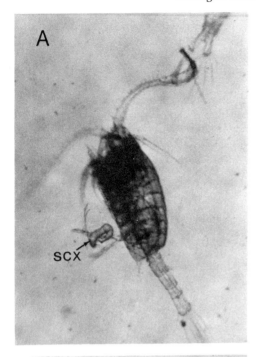

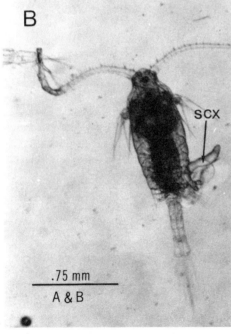

Fig. 9. Light micrographs of spermatophore extrusion by male *Centropages typicus*. A—Left lateral view. B—Ventral view (scx, spermatophore complex).

seized by a male. It thus appears that the difference in male and female swimming activity may be a determining factor in mate recognition and, perhaps, spermatophore extrusion.

Copepods in many other families have appendages morphologically similar to those found on pontellids, suggesting that analogous movements may exist in their mating behaviors. For example, a modified antennule and fifth leg are present on the male in the following calanoid families: Pontellidae, Centropagidae, Temoridae, Metridiidae, Pseudodiaptomidae, Lucicutiidae, Heterorhabdidae, Augaptilidae, Arietillidae, Candaciidae, Bathypontiidae, Acartiidae, and Tortanidae (Brodskii 1967). A prehensile, geniculate antennule and a chelate fifth leg are structures shared by *L. aestiva*, *C. typicus*, and *E. affinis*. These three species exhibit similar movements in the mating encounter. As mentioned previously, the geniculate antennule of the male is constructed for the initial capture of the female. It provides the male with an extended reach and has one or more hinges which enable certain modified segments to close around the female. In the calanoid families mentioned above, males that lack a distinctly chelate fifth leg nevertheless have some form of asymmetry and modification in that limb which most likely functions to hold the female. Since *L. aestiva*, *C. typicus*, and *E. affinis* males transfer the spermatophore to the female urosome with the left fifth leg, we assume that it is used by other calanoids in a similar fashion. This same appendage may have additional functions, as demonstrated by the precopulatory stroking of *L. aestiva*.

Although similar movements among copepods may form a basic pattern for mating behavior, additional, unique characteristics ultimately define a mating ritual that is specific to each species. These individualized mating behaviors therefore function to prevent species hybridization.

Morphological specializations might include differences in the length and structure of the male's fifth pair of swimming legs and location of the gonopore on the female, in addition to the presence or absence of integumental organs on or near the genital area and spermatophore shape and attachment to the female. For example, spermatophore structures vary from simple tubes affixed to the female by an adhesive-like quality of secretions discharged from the spermatophore, e.g. *Temora turbinata* (Blades unpub.), to tubes supported by complicated coupling devices, e.g. *Labidocera* spp. (Fleminger 1967). The shape of the coupling plates corresponds to the morphology of the female urosome and designates precise attachment points.

The following are some examples of species-specific behavioral characteristics: (1) The moment of spermatophore extrusion, e.g. as previously described for *L. aestiva*, *C. typicus*, and *E. affinis*. (2) Orientation of the male with respect to the female. (3) The duration of certain movements in the ritual [e. g. *C. typicus* males may remain in the initial antennule hold for several hours; *L. aestiva* moves quickly from this position to the chelate hold. The male *L. aestiva* releases the female soon after a spermatophore is attached; male *Pontellopsis villosa* have been observed (Blades unpub.,) holding the spermatophore to the female as it discharges]. (4) The length of time a spermatophore, or part of the coupler, stays on the female. The whole spermatophore coupler is removed from the female *L. aestiva* soon after it has discharged. In *C. typicus*, however, the posterior coupling plate remains on the female urosome after the rest of the spermatophore device has been discarded (Lee 1972). Fleminger (1975) reported that females of the *Labidocera trispinosa* species subgroup retain the spermatophore until eggs are shed. The relative importance of these and other behavioral characteristics will become evident when more is known about the mating rituals of other copepods.

Many questions remain to be answered. How do the males of those copepods that have little or no modifications to their appendages, e.g. *Calanus* spp., capture and fertilize females? Is mechanoreception more important than chemoreception for some copepod species? Can the unique patterns of integumental pores seen on copeods (Fleminger and Hulsemann 1977) be related to behavioral control systems? Are specific pheromones produced by each species? We believe that further studies of live copepods will be the most suitable way to relate structure and physiology to functional, reproductive behavior.

References

BLADES, P. I. 1977. Mating behavior of *Centropages typicus* (Copepoda: Calanoida). Mar. Biol. 40: 47-64.

——, and M. J. YOUNGBLUTH. 1979. Mating behavior of *Labidocera aestiva* (Copepoda: Calanoida). Mar. Biol. 51: 339-355.

BRODSKII, K. A. 1967. Calanoida of the far eastern seas and polar basin of the USSR. Israel Program Sci. Transl. Jerusalem.

FLEMINGER, A. 1967. Taxonomy, distribution, and polymorphism in the *Labidocera jollae* group with remarks on evolution within the group (Copepoda: Calanoida). Proc. U. S. Natl. Mus. 120: 1-61.

——. 1973. Pattern, number, variability, and taxonomic significance of integumental organs (sensilla and glandular pores) in the genus *Eucalanus* (Copepoda, Calanoida). Fish. Bull. 71: 965-1010.

——. 1975. Geographical distribution and morphological divergence in American coastal-zone planktonic copepods of the genus *Labidocera*, p. 392-419. In L. E. Cronin [ed.], Estuarine research, v. 1. Academic.

——, and K. HULSEMANN. 1977. Geographical range and taxonomic divergence in North Atlantic *Calanus* (*C. helgolandicus, C. finmarchicus,* and *C. glacialis*). Mar. Biol. 40: 233-238.

GAULD, D. T. 1957. Copulation in calanoid copepods. Nature 180: 510.

GRIFFITHS, A. M., and B. W. FROST. 1976. Chemical communication in the marine planktonic copepods *Calanus pacificus* and *Pseudocalanus* sp. Crustaceana 30: 1-8.

HOPKINS, C. C. E., and D. MACHIN. 1977. Patterns of spermatophore distribution and placement in *Euchaeta norvegica* (Copepoda: Calanoida). J. Mar. Biol. Ass. U. K. 57: 113-131.

JACOBS, J. 1961. Laboratory cultivation of the marine copepod (*Pseudodiaptomus coronatus* Williams. Limnol. Oceanogr. 6: 443-446.

KATONA, S. K. 1973. Evidence for sex pheromones in planktonic copepods. Limnol. Oceanogr. 18: 574-583.

——. 1975. Copulation in the copepod *Eurytemora affinis* (Poppe, 1880). Crustaceana 28: 89-95.

KITTREDGE, J. S., F. T. TAKAHASHI, J. LINDSEY, and R. LASKER. 1974. Chemical signals in the sea: marine allelochemics and evolution. Fish. Bull. 72: 1-11.

LEE, C. M. 1972. Structure and function of the spermatophore and its coupling device in the Centropagidae (Copepoda: Calanoida). Bull. Mar. Ecol. 8: 1-20.

MAUCHLINE, J. 1977. The integumental sensilla and glands of pelagic Crustacea. J. Mar. Biol. Assoc. U. K. 57: 973-994.

——, and T. NEMOTO. 1977. The occurrence of integumental organs in copepodid stages of calanoid copepods. Bull. Plankton Soc. Jpn. 24: 32-38.

PARKER, G. H. 1902. The reactions of copepods to various stimuli and the bearing of this on daily depth migrations. Bull. U. S. Fish Comm. 21: 103-123.

——, and A. K. BAL. 1973. Setae of the first antennae of the copepod *Cyclops scutifer* (Sars): their structure and importance. Proc. Natl. Acad. Sci. 70: 2656-2659.

5. Adaptive Responses to Encounter Problems

Jeroen Gerritsen

Abstract

A model of random encounters among zooplankton shows that encounter probability depends on the encounter radius, density of prey, and swimming speeds of both predators and prey. The model predicts two optimal tactics for efficient predators: ambush and cruising. Swimming direction influences encounter probability; cruising predators can maximize their encounter rates by swimming orthogonally to the predominant prey direction, while prey can minimize their encounter rates by swimming parallel to predators. Investigation of the swimming behavior of certain zooplankters indicates that predatory *Cyclops scutifer* adults swim with a strong horizontal component, while some of their potential prey swim primarily vertically. *Cyclops* males, on the other hand, swim in random directions at much lower average speeds. It is proposed that they use an ambush strategy to encounter females and to balance their greater risk of predation, because they may make mistakes in identifying females.

Encounter, or the meeting of individual organisms in space, is the central process governing direct interactions between animals and their food, mates, and predators. During its lifetime, a free-living animal may encounter predators, it must encounter food if it is to grow, and, if the animal is sexual, it must encounter mates if it is to reproduce. The rates and probabilities of these encounters influence biological interactions and can therefore play a large part in the evolution and distribution of organisms.

Human beings are terrestrial animals, and, as such, our perception of encounters with other organisms is primarily two-dimensional. Terrestrial animals have recognizable places to which they can return, many have home ranges, and the substrate is immobile. Planktonic animals have none of these: there are no recognizable places, other organisms are encountered in three rather than in two dimensions, and circulation and turbulence result in continual movement.

Yet these animals do not move about passively, allowing currents to carry them wherever; they have definite, directed behavior and able sensory systems which lead them toward and away from stimuli and enable them to detect, avoid, or pursue each other.

We here inquire into the process of encounter in three dimensions and consider different behavioral alternatives available to such animals as they encounter predators, food, and mates. These processes cannot be easily separated for zooplankton because encounters with food, predators, and mates may occur at any time. There are circadian variations in behavior, particularly in feeding, but we may expect all three to be active in the same short time interval.

I thank J. P. Gilman, J. Kou, and J. R.

Strickler for help and constructive comments. Parts of this research were funded by NSF grant DEB76-02096.

Encounter Model

The basis of many encounter models is an analogy of animals to gas molecules in the kinetic theory of gasses. In these models, small objects (i.e. molecules or animals) collide randomly with each other in space, and the goal of the model is to find the probability of such a collision. Since animals do not follow the Maxwell-Boltzmann velocity distribution, the exact solutions from gas theory are inapplicable to biological problems, yet the general approach may yield valuable insights.

The first application of the analogy to biology was by Yapp (1956) for estimating the density of animals observed on line transects. The two-dimensional solution of random encounters was solved independently by Koopman (1956) for antisubmarine warfare and by Skellam (1958), also for line transects. The importance of the three-dimensional case for zooplankton was recognized by Katona (1973) in an investigation of sex pheromones, and he briefly considered an encounter model. The three-dimensional encounter model was more fully developed by Gerritsen and Strickler (1977) and is the basis of the strategies and swimming behaviors discussed here.

The model assumes that space is homogeneous and three-dimensional, that animals are randomly distributed in the space, that the animals swim in random directions with a uniform probability distribution in space, that predators have a constant encounter radius, R, within which they can detect other animals, and that the animals are considered dimensionless points in space, which is reasonable if encounter distances are measured center-to-center. An expression for the rate that a predator encounters prey is derived by integrating over all directions the number of moving prey that would enter the encounter sphere of a moving predator

$$Z_p = \frac{\pi R^2 N_b}{3} \left(\frac{u^2 + 3v^2}{v}\right) \text{ for } v \geq u$$

$$= \frac{\pi R^2 N_b}{3} \left(\frac{v^2 + 3u^2}{u}\right) \text{ for } u \geq v \quad (1)$$

where Z_p is the encounter rate of a predator with its prey, N_b is the density of prey, and v and u are the mean speeds of predator and prey. The probability of encounter in a time interval t is Poisson-distributed, and the Poisson parameter is $Z_p t$.

From Eq. 1 we see that the parameters affecting encounter are encounter radius, swimming speeds of the animals, and density. Encounter radius depends on the size, nature, and sensitivity of an animal's sensory systems, on conditions in the environment and on the kind of organisms encountered. I consider here how swimming behavior affects encounter probability, in that swimming speed, as a component of behavior, influences encounter probability and that swimming direction also influences encounter probability.

Swimming Speed

From Eq. 1 we see that swimming speeds positively affect encounter rate. This means that a predator can increase its speed to increase its encounter rate and, conversely, that a prey animal can slow down to reduce its encounter rate. It should be noted, however, that the greater of the two speeds has the stronger effect in determining encounter rates. This is important for slow animals in that a further decrease in speed of a slow-moving prey would result in a negligible decrease of its encounter probability with a fast-moving predator. Likewise, a slow predator cannot appreciably increase encounter rates with prey until it swims at least as fast as the prey.

When we consider the energetic cost of swimming versus the energetic rewards of food capture, we find that the encounter model predicted two optimal swimming strategies for predators: ambush predators that specialize on fast-moving prey, and cruising predators that prey on slow-moving prey. Energy spent swimming increases with

Table 1. Trophic Status and Swimming Speeds, in mm/s, of Various Zooplankton Species
P—Predator; 0—Omnivore; F—Filter-Feeder

Species	Trophic Status	Speed	Source
Rotifera			
Asplanchna sieboldi	P-O	0.7	Gerritsen unpubl.
Cladocera			
Bosmina longirostris	F	1.2	Gerritsen and Kerfoot in prep.
Daphnia galeata mendotae	F	1.2	Gerritsen unpubl.
D. pulex	F	1.5	Gerritsen unpubl.
D. magna	F	0.6-1.2	Ringelberg 1964
Diaphanosoma leuchtenbergianum	F	0.7	Gerritsen unpubl.
Diaphanosoma	F	0.3	Swift and Fedorenko 1975
Polyphemus	P	7.6	Swift and Fedorenko 1975
Copepoda			
Cyclops scutifer nauplii	F	0.3	Gerritsen 1978
Cyclops scutifer cop. 1-4, adult ♂	F-O?	1.4	Gerritsen 1978
Cyclops scutifer cop. 5, adult ♀ stages	F-O	3.2	Gerritsen 1978
C. bicuspidatus thomasi	P-O	1.7	Gerritsen and Kerfoot in prep.
C. abyssorum prealpinus	P-O	3.6	Strickler 1970
C. vicinus lobosus	P-O	3.8	Strickler 1970
Diaptomus ashlandi	F	0.3	Gerritsen and Kerfoot in prep.
D. tyrelli	F	0.5	Swift and Fedorenko 1975
D. kenai adults	P	5.5	Swift and Fedorenko 1975
Acanthodiaptomus denticornis	F	0.9-1.2	Gerritsen unpubl.
Epischura nevadensis	P	4.4	Gerritsen and Kerfoot in prep.
E. lacustris	P	2.0	Gerritsen unpubl.
Insecta			
Chaoborus sp. (while foraging)	P	<0.5	Gerritsen unpubl.

the square of the swimming speed, but energy gained from feeding does not, resulting in an optimum cruising speed for predators that consume slow-moving prey. A similar conclusion had been reached by Ware (1975) in considering optimal foraging speed for planktivorous fish.

Within the zooplankton, invertebrate predators fall into one of two classes: ambush or cruising. Most ambush predators cling to a substrate (e.g. *Hydra*, damselfly larvae), but the best known free-swimming ambush predators in fresh water are the larvae of the phantom midge *Chaoborus*. Predatory crustaceans all seem to be cruising predators in fresh water. Data on average swimming speeds of some zooplankton are summarized in Table 1. These are average speeds over all directions; yet it is not the complete picture (see below), since many species show directional preferences. The average speed of predatory crustaceans is significantly higher than the average speed of filter-feeders (t_{16} = 4.991, p = 0.002, Table 2). The predatory rotifer *Asplanchna* was not included among predators, because it eats smaller rotifers and protozoans rather than crustacean filter-feeders.

Table 2. Comparison of Mean Swimming Speeds of Crustacean Filter-Feeders and Predators

	n	Mean Speed	2 SE	t	p
Predators	8	3.98	1.35	4.99	<0.002
Filter-feeders	10	0.89	0.287		

The filter-feeders swim at a moderate rate, fast enough to encounter their small (10-50 μm), passively moving, and dispersed food particles, and yet slow enough to reduce the probability of encountering predators. The crustacean predators swim faster, presumably at an optimal speed to exploit their prey, which are filter-feeders and rotifers (Table 1). The presence of faster animals now allows the existence of such ambush predators as *Chaoborus* in the same environment, which are better able to exploit fast-moving prey (Gerritsen and Strickler 1977). Field and laboratory studies have shown that these predators do consume a greater relative proportion of fast-moving prey than slow-moving prey. *Chaoborus* larvae remove a greater proportion of the fast-swimming *Diaptomus kenai* than of other zooplankters (Fedorenko 1975; Swift and Fedorenko 1975), and they consume *Cyclops* copepodites at a faster rate than *Cyclops* nauplii, which swim more slowly (Lewis 1977; Gerritsen 1978).

Swimming Direction

In the original encounter model, swimming direction was not taken into account, and the probability distribution of swimming directions was assumed to be uniform. However, many zooplankters are known to have oriented swimming behavior (e.g. Smith and Baylor 1953; Ringelberg 1964), which may affect encounter probabilities. The assumption of a uniform distribution of swimming directions was relaxed to see if direction was important.

Directional swimming was assumed to be characterized by either of two motions: vertical (up or down) swimming or horizontal swimming (with no preference for any direction in the horizontal plane). Horizontal, directional swimming occurs in the avoidance of shore response ("Uferflucht": Siebeck 1964, 1969), when pelagic zooplankton find themselves near a shore, but this kind of swimming is not known in open-water zones.

In considering optimal swimming direction, the important quantity is the mean relative speed between a predator and its prey population (or between a prey animal and its predator population). In the uniform case (Eq. 1), i.e. where the probability distribution of swimming directions is uniform, the mean relative speed is

$$\bar{w} = \frac{(u^2 + 3v^2)}{3v} \quad (2)$$

where $\bar{w}$ is the mean relative speed. The model is most sensitive when $v = u$, so that Eq. 2 reduces to

$$\bar{w} = \frac{4}{3} v. \quad (3)$$

If all animals (predator and prey) are confined to a single plane and speeds are equal, the mean relative speed becomes (Koopman 1956)

$$\bar{w} = \frac{4}{\pi} v. \quad (4)$$

If predators now swim vertically and prey are horizontal, all animals meet at right angles and the relative speed is

$$\bar{w} = \sqrt{2}\, v. \quad (5)$$

Eqs. 4 and 5 represent the extremes of the options available to a predator confronted with horizontally swimming prey. The highest encounter rate for a predator results from its swimming vertically (Eq. 5), which gives an 11 percent advantage over swimming horizontally. Conversely, the lowest encounter rate for prey results when they swim in the same plane as their predators.

The second alternative occurs when all prey swim vertically up or down, with equal probability. A predator swimming vertically will then encounter half the prey at a relative speed of $2v$, but it will not encounter the other half (those swimming in the same direction), which results in a mean relative speed of v. Encounter rates of horizontally swimming predators with vertical prey are then proportional to Eq. 5, giving horizontal predators a 41 percent advantage over vertical ones, when their prey swim vertically up and down.

We can conclude from this exercise that the best way for a predator to maximize encounters with prey is to swim at right angles to prey movement, and the best way for prey to minimize encounters is to swim in the same direction as predators. This conclusion is sensitive to the ratio of the two speeds as shown by Eq. 1; the larger of the two speeds effectively determines encounter rate, and if the speeds of predator and prey differ considerably, the advantages of moving orthogonally are severely reduced (Fig. 1).

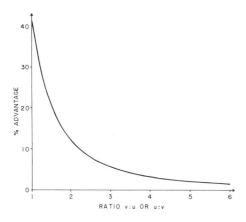

Fig. 1. Advantage of swimming horizontally vs. swimming vertically for a predator searching for prey that are swimming vertically, as the ratio of the two speeds deviates from 1.

If we consider now a prey population divided in halves so that half of the individuals swim in the horizontal plane and half swim vertically, then predators that swim in the horizontal plane have an 11 percent higher encounter rate than predators that swim vertically. Similarly, if the prey animals swim faster along certain linear dimensions than along other dimensions, then predators that swim orthogonal to that direction experience a faster encounter rate. The relative advantage to predators of swimming orthogonally to prey depends on the frequency and speed of the prey that move in the favored (orthogonal) direction. This advantage reaches a maximum when all prey are moving in the favored direction at the same swimming speed as the predator (Fig. 1).

A linear component in zooplankton swimming is present when the animals are engaged in vertical migration, in continual up-and-down swimming (Bainbridge 1952), or when the swimming motion involves up-and-down bobbing. If a proportion of prey are swimming vertically, we should expect foraging predators to swim primarily horizontally. However, only if a majority of prey swim horizontally should we expect predators to swim vertically.

If we look at an animal's movement in a vertical plane, we can measure both the angle from the vertical at which the animal is swimming and the average speed along that angle.

Measuring the directions of all animals in a population gives a circular frequency distribution of swimming angles about the vertical plane. These circular distributions ($0°$ = up; $180°$ = down; between $\pm\ 0-180°$, movement to left or right of vertical) and direction-specific swimming speeds of some zooplankton are drawn in Figs. 2–10. These data are n-observations of 20–40 animals each, over periods of 10–30s, where n represents animal-seconds of observations. The data were taken from a series of 16-mm films of the swimming animals (Gerritsen 1978). Three patterns emerge from these circular distributions: (1) uniform distribution, or no discernible pattern (*Cyclops* nauplii, C3 at 2155 hours and males: Figs. 2, 4, 7, 8); (2) frequent horizontal swimming (*Cyclops* C3 at 1300 hours and females: Figs. 3, 5, 6); and (3) frequent vertical swimming (*Asplanchna* and *Daphnia:* Figs. 9 and 10).

Adult females of *Cyclops scutifer*, which are predaceous, swim more frequently and swim faster in the horizontal plane than in other directions (Figs. 5 and 6). This behavior pattern is absent in nauplii (Fig. 2) and early copepodites (C1 and C2) and can be seen in stage 3 copepodites during the day but not at night. Horizontal swimming is more pronounced in stage 4 copepodites and becomes fully developed in stage 5, when males are indistinguishable from adult females in their swimming behavior. The behavior of the males is discussed more fully below.

Encounter Problems 57

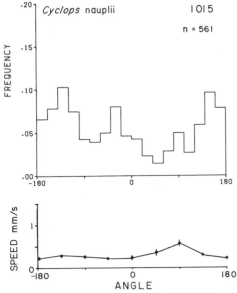

Fig. 2. Circular swimming distribution of *Cyclops scutifer* nauplii at 1015 hours. Angle of 0 is vertically up, ± 180 is down; 1-s observations. Histogram on the top is the circular frequency distribution in a vertical plane; graph on the bottom is the mean swimming speed as a function of direction.

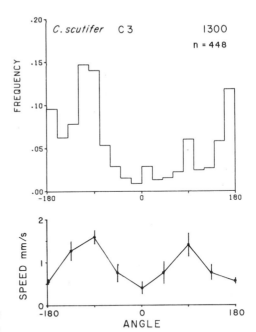

Fig. 3. *Cyclops scutifer* copepodite 3 at 1300 hours. Details as in Fig. 2.

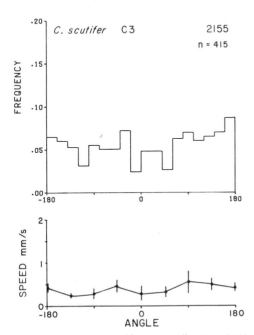

Fig. 4. *Cyclops scutifer* copepodite 3 at 2155 hours. Details as in Fig. 2.

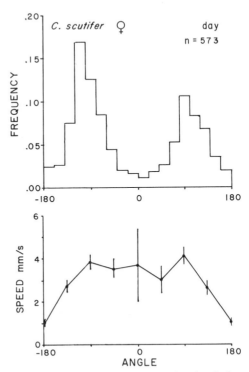

Fig. 5. *Cyclops scutifer* adult females during day. Details as in Fig. 2.

58 J. Gerritsen

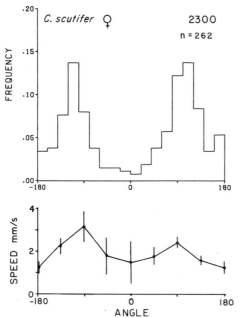

Fig. 6. *Cyclops scutifer* adult females at night. Details as in Fig. 2.

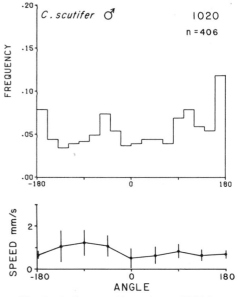

Fig. 7. *Cyclops scutifer* males at 1020 hours. Details as in Fig. 2.

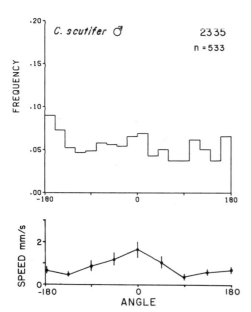

Fig. 8. *Cyclops scutifer* males at 2335 hours. Details as in Fig. 2.

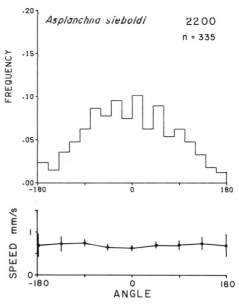

Fig. 9. *Asplanchna sieboldi* at 2200 hours. Details as in Fig. 2.

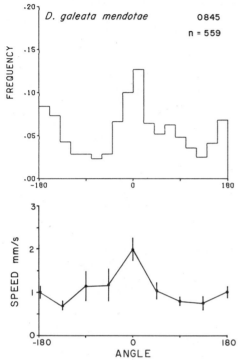

Fig. 10. *Daphnia galeata mendotae* at 0845 hours. Details as in Fig. 2.

If some of the prey of *Cyclops* have distribution modes in the vertical direction, it may be advantageous, in terms of encounter probability, for *Cyclops* to swim horizontally. These animals prey on smaller crustaceans and rotifers, and Figs. 9 and 10 show that some species do swim vertically, at least some of the time. *Asplanchna* is a suitable prey for *C. scutifer*, but *Daphnia* is generally too large for the copepods to grasp. Other predators, however, also swim horizontally. *Cyclops abyssorum* and *C. vicinus* swim faster in the horizontal, though not more frequently (5.5 and 5.8 mm/s, respectively: Strickler 1970), and unquantified observations of *Epischura nevadensis* indicate that this predator also swims primarily horizontally. The fact that horizontal swimming arises at the same time that *C. scutifer* becomes more predaceous also supports the optimal encounter hypothesis.

This is not to say that there are no other reasons for animals to swim in preferred planes or lines. Vertical migration and the "avoidance of shore" response each represents large-scale patterns of directed swimming and may occur frequently enough for predators to exploit them. Horizontal swimming may be a way for an animal to maintain a preferred depth while keeping up a constant swimming speed. Observations of filter-feeding diaptomid copepods (*D. ashlandi* and *Acanthodiaptomus denticornis*) show that these animals also swim in the horizontal. Up-and-down bobbing, or hop and sink, such as shown by *Daphnia* (Fig. 10), may be an energetically efficient way of maintaining depth while feeding (Haury and Weihs 1976). Hop and sink behavior has been described for many species of zooplankters, occurs in *C. scutifer* as well, and is more pronounced in juvenile stages and at night when the animals are less active (Figs. 3, 4, and 8). Encounter probability may not be the only reason why predatory *C. scutifer* swim horizontally, but the model predicts that the animals get an advantage from such behavior when their prey swim with vertical hop and sink motions.

Interacting Encounter Rates

If we consider encounter rate alone, it is apparent that the optimal swimming strategy of an animal depends very much on the swimming behavior of the encountered or avoided target (summarized in Table 3). When there is more than one interaction (i.e. with predators, prey, and mates), optimal swimming behavior is difficult to determine, because of different "currencies" of optimal behavior (Cody 1974). Energetic

Table 3. Summary of Swimming Behaviors for Optimal Search or Avoidance

| | Optimal | |
Object Quality	Search	Avoidance
Speed		
Slow	Fast	Slow
Fast	Slow*	Slow
Swimming direction		
Vertical	Horizontal	Vertical
Horizontal plane	Vertical	Horizontal
Random	—	—

*Energetically optimal.

efficiency is commonly used to predict optimal foraging behavior, but this is difficult to reconcile with the fitness of avoiding predators and finding mates.

As an example, we can observe the behavior of *C. scutifer*, particularly with regard to mating. Male *Cyclops* actively pursue females, usually as follows. When a female passes near a male, the male will jump into the wake of the female, chase her, and attempt to capture her with his geniculate antennules. Females generally try to escape a pursuing male, but they usually do not attempt escape if the males have already grasped their abdomen. Males are not very selective in pursuit; films indicate that they frequently give chase to other males and already gravid females, as well as virgin females. If they capture an inappropriate mate, however, they quickly release it.

Since *Cyclops* males actively pursue females, the burden of mate detection and recognition is on the males. Pheromones play a part in this, but the extent is unclear (Katona 1973; Griffiths and Frost 1976). Exact detection and location of mates in water via a chemical gradient seems unlikely because of slow diffusion rates in water (Wilson 1970) and the fact that females are not stationary, i.e. they swim away from any gradient that would be set up by diffusion. It would be possible for a female to lay down a chemical trail in her wake, enabling males to follow more accurately once they have found the trail, but predators can also take advantage of such a trail. The experimental evidence of Griffiths and Frost (1976) indicates that male copepods increase their mate-searching activity in the presence of pheromones, thus increasing their encounter probability with females when it is likely that females are nearby. Lastly, chemicals could indicate mates on contact, resulting in the rejection of inappropriate mates, as shown by *Cyclops* males.

Cyclops males begin to pursue females when the females pass them, usually jumping into their wake and catching them from the rear. The initial response is too quick to be chemically induced, and must be a reaction to mechanical disturbances caused by the swimming female. The result is that males and females respond differently to mechanical stimuli: moderate vibrations from a needle in the water elicit escape responses from females, but males stop and frequently approach the source of the vibration. If the strength of the stimulus is increased, there is a uniform escape response from both sexes (Gerritsen 1978).

Males have difficulty in distinguishing between signals of females and predators, both of which are larger than male *Cyclops*. With the moderate stimulus, there is insufficient information for a male to tell whether it comes from a female or from something else, and it frequently responds by stopping ("stop and listen" behavior), and sometimes by approaching the stimulus as well. Males may then be exposed to a predator for a longer time, and they may even swim toward a waiting predator.

Cyclops scutifer males, however, swim more slowly than the females (Figs. 7 and 8), and in this way they reduce their encounter rate with predators, particularly with such ambush predators as *Chaoborus*. Because male *Cyclops* are less likely to avoid a *Chaoborus* before it strikes, but have a slower speed than females, predation by *Chaoborus* on male and female *Cyclops* is about the same (Gerritsen 1978). The males essentially follow an ambush strategy to find mates, waiting for females to pass and then giving chase. They have no preferred direction of swimming, usually sinking passively until a stimulus is detected and then swimming very rapidly in chase. Although the average swimming speed of males is slow, their maximum attainable speeds are at least as fast, and perhaps faster, than that of the females. Males also swim upward faster at night (Fig. 8), in contrast to females, which swim primarily in the horizontal.

The cost, for males, of an ambush strategy in finding mates is a reduced encounter rate with their own food—smaller and slower crustaceans and rotifers. Female *Cyclops* are apparently sexually receptive for only a short time after reaching maturity, and males are short-lived, so there is no necessity for males to eat a great deal. The most fit males are those which mate the most, not those which live longest. A male that is eaten before it has a chance to mate, however, is obviously unfit.

If females are slow-moving for other reasons (e.g. filter-feeders), males do not have the luxury of following an ambush tactic to find mates, and they must swim fast. Male rotifers are small, they do not eat, and they swim much faster than the slow-moving females (Wesenberg-Lund 1923).

Conclusion

I have identified here some of the benefits and disadvantages of certain kinds of swimming behavior. Of the two components of swimming behavior discussed, swimming direction is secondary to swimming speed in determining encounter probabilities, so we would expect selection to be stronger for speed than for direction. Density distribution (i.e. patchiness) also affects encounter probability and may further influence the evolution of various components of swimming behavior (Murdie and Hassell 1973). As noted above, there may be other reasons for certain swimming behavior, and swimming behavior may therefore be an important preadaptation for a species invading a new habitat. We must note for zooplankton encountering predators, prey, and mates that there is no generally optimal swimming behavior which satisfies all three requirements. Encounter probability and other selective forces acting on swimming behavior indicate that the tactics of other organisms largely dictate the optimal behavior for an animal (Maynard Smith 1976), so that the behavior of an animal is intricately linked to the kind of community in which the animal occurs.

References

BAINBRIDGE, R. 1952. Underwater observations on the swimming of marine zooplankton. J. Mar. Biol. Assoc. U.K. 31: 107–112.

CODY, M. L. 1974. Optimization in ecology. Science 183: 1156–1164.

FEDORENKO, A. Y. 1975. Feeding characteristics and predation impact of *Chaoborus* (Diptera, Chaoboridae) larvae in a small lake. Limnol. Oceanogr. 20: 250–258.

GERRITSEN, J. 1978. Instar-specific swimming patterns and predation of planktonic copepods. Int. Ver. Theor. Angew. Limnol. Verh. 20: 2531–2536.

——, and J. R. STRICKLER. 1977. Encounter probabilities and community structure in zooplankton: a mathematical model. J. Fish. Res. Bd. Can. 34: 73–82.

GRIFFITHS, A. M., and B. W. FROST. 1976. Chemical communication in the marine planktonic copepods *Calanus pacificus* and *Pseudocalanus* sp. Crustaceana 30: 1–8.

HAURY, L., and D. WEIHS. 1976. Energetically efficient swimming behavior of negatively bouyant zooplankton. Limnol. Oceanogr. 21: 797–803.

KATONA, S. K. 1973. Evidence for sex pheromones in planktonic copepods. Limnol. Oceanogr. 18: 574–583.

KOOPMAN, B. O. 1956. The theory of search. I. Kinematic bases. Oper. Res. 4: 324–346.

LEWIS, W. M. 1977. Feeding selectivity of a tropical *Chaoborus* population. Freshwater Biol. 7: 311–325.

MAYNARD SMITH, J. 1976. Evolution and the theory of games. Am. Sci. 64: 41–55.

MURDIE, G., and M. P. HASSELL. 1973. Food distribution, searching success and predator-prey models, p. 87–101. *In* M. S. Bartlett and R. W. Hiorns [eds.], The mathematical theory of the dynamics of biological populations. Academic Press.

RINGELBERG, J. 1964. The positively phototactic reaction of *Daphnia magna* Straus: a contribution to the understanding of vertical migration. Neth. J. Sea Res. 2: 319–406.

SIEBECK, O. 1964. Researches on the behavior of planktonic crustaceans in the littoral. Int. Ver. Theor. Angew. Limnol. Verh. 15: 746–751.

——. 1969. Spatial orientation of planktonic crustaceans. 1. The swimming behavior in a horizontal plane. Int. Ver. Theor. Angew. Limnol. Verh. 17: 831–847.

SKELLAM, J. G. 1958. The mathematical foundations underlying the use of line transects in animal ecology. Biometrics 14: 385–400.

SMITH, F. E., and E. R. BAYLOR. 1953. Color responses in the Cladocera and their ecological significance. Am. Nat. 87: 49–55.

STRICKLER, J. R. 1970. Über das Schwimmverhalten von Cyclopoiden bei Verminderungen der Bestrahlungsstärke. Schweiz. Z. Hydrol. 32: 150–180.

SWIFT, M. C., and A. Y. FEDORENKO. 1975. Some aspects of prey capture by *Chaoborus* larvae. Limnol. Oceanogr. 20: 418–425.

WARE, D. M. 1975. Growth, metabolism and optimal swimming speed in a pelagic fish. J. Fish Res. Bd. Can. 32: 33–41.

WESENBERG-LUND, C. 1923. Contributions to the biology of the Rotifera. I. The males of the rotifera. Kgl. Dan. Vidensk. Selsk. Skr. Nat. Math. Afd. 8 4: 189–346.

WILSON, E. O. 1970. Chemical communication within animal species, p. 133–156 *In* E. Sondheimer and J. B. Simeone, eds. Chemical Ecology. Academic Press, N.Y.

YAPP, W. B. 1956. The theory of line transects. Bird Study 3: 93–104.

II

Directed Movements: Vertical Migration

6. Introductory Remarks: Causal and Teleological Aspects of Diurnal Vertical Migration

J. Ringelberg

Of late, research in diurnal vertical migration (d.v.m.) has increased again. Interest is especially directed toward the so-called "ultimate causes." It seems to me these causes are easily confused with causes as used, for instance, in physics. There is a long tradition to do this in the literature on vertical migration. The animals following an optimal or preferred light intensity have been used by many workers as an explanation of swimming movement. Of course, teleological explanations have a function in allowing us to understand biological phenomena. The problem, however, is to explain how the goal comes about. In many cases teleological aspects are evolutionary adaptations, and the origin of goal-directed behavior must be looked for in a mechanism of evolution at work in the history of the species.

Teleological explanations as well as causal explanations have a place in biology. Both types are complementary in biological understanding, and one can never replace the other. Therefore, in the scheme of aspects and mechanisms of diurnal vertical migration that I offer in Fig. 1, both domains are placed next to each other. In one case, i.e. the avoidance of predation, two possible mechanisms to achieve this are presented. In this case the interrelation between the two aspects of behavior is shown.

In my opinion there is no reason to replace the traditional names of causal and teleological explanations by such terms as proximate and ultimate causes, respectively.

The danger is that after formulating the ultimate cause of a particular phenomenon, research will be stopped, as if the phenomenon were completely explained.

Diurnal vertical migration is a movement. Therefore, in Fig. 1, it is divided into two components: velocity and direction. At the physiological base, two mechanisms must be present: first a kinetic mechanism, and second an orientation mechanism. The stimuli initiating these mechanisms can be called releasing stimuli and directing stimuli, respectively. These names are used in ethology (Tinbergen 1951). At least in *Daphnia*, the first stimulus can be defined as a change in light intensity relative to the prevailing absolute intensity (Ringelberg 1964; Ringelberg et al. 1967). The second mechanism, also in *Daphnia*, is thought to be an optical beacon found in the contrast present underwater at about $49°$ with the vertical axis (Ringelberg 1964). The orientation mechanism is used to bring the body into an upward or downward position during migration (in passive sinking no orientation seems to be necessary). Both the releasing, or movement-initiating stimulus, and the directing stimulus must be present to evoke the behavior of migration. It is well known, that in a particular species the extension of d.v.m. can vary with age, sex, reproductive state, etc. Therefore, the physiological state of the animal must influence the internal motivation to migrate. An endogenous rhythm in a physiological state variable might be present, thereby enhancing the motivation to respond to

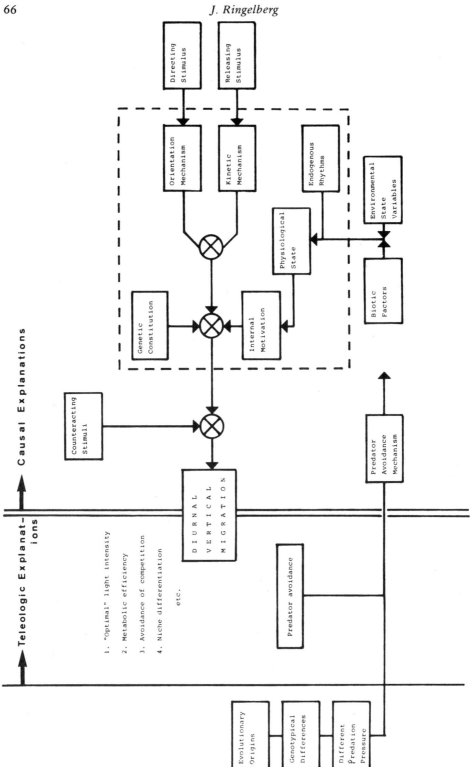

Fig. 1. Scheme of relations of factors and mechanisms involved in diurnal vertical migration.

the necessary external stimuli. Abiotic environmental variables such as daylength, temperature, and oxygen content influence the physiological state. The same holds for biotic factors such as the presence of food, crowding, and predators. The variable expression of d.v.m. justifies systematic research into the influence of abiotic and biotic environmental state variables.

Very little information is available on the influence of the genetic constitution of zooplankton. In teleological explanations, however, a change in the genotypical constitution of the population often plays a role. If all factors inside the animal (indicated by the broken line boundary) and outside the animal are beneficial, a d.v.m. can take place. Nevertheless, what I call counteracting stimuli might be present to decrease the velocity or distance of swimming. These counteracting stimuli are, for instance, met by a migrating animal approaching a thermocline. The change in temperature per time can cancel the effect of the relative change in light intensity; there is thus a question of balance between the strength-effect of one compared to the other. In some cases migration through the thermocline is possible, while in other cases swimming will stop. Experimental research into these balances must be done.

Several teleological explanations (ultimate causes) were offered at this conference. As examples, some are listed again. The basic thought in common among/with these teleological functions is that they give the individual (population?) a higher survival value. Several of these functions must be at work at the same time. It is even possible to think of opposing strategies. Research could be directed to validate the bonus of a particular strategy in relation to another operating at the same time. In this way these teleological functions are brought within the limits of science. Studies along these lines might contribute to evolution theory in general.

It is quite possible that more than one mechanism (or complex of mechanisms) leads to the same goal. Diurnal vertical migration is thought to result from predator avoidance (Zaret and Suffern 1976; Zaret unpubl.). How is this achieved? Two possibilities are presented in the scheme. One mechanism operates by a perception of the predator resulting in a change in physiological state—(An analogous example is presented by the spine induction in *Brachionus* caused by *Asplanchna*; Gilbert 1967.). Here the teleological explanation is directly connected with the causal. The other possibility is that predators change the gene frequency distribution in the population. Predation during the daytime in surface waters would select against nonmigrating individuals. In this case, genotypical differences in behavior must be a prerequisite established in the evolutionary history of the species.

The presented considerations must be looked upon as a contribution to clarify and conduct research. It is stressed that they are only partly based on facts. To what extent this is the case can easily be found by reading reviews (Cushing 1951; Ringelberg 1964; Segal 1970; Hutchinson 1967).

What is needed includes, among others:

1. Precise descriptions of the vertical movements of migrating populations must be produced. To facilitate the calculation of time of migrations and the overall displacement velocity, observations at very short time intervals around sunrise and sunset (± 2 hours) must be done. Simultaneously a continuous record of the light intensity must be made in order to calculate stimulus values. This measurement can be done at a fixed depth. Knowledge of the extinction coefficient is necessary to calculate the underwater light climate. Abiotic state variables as mentioned in the scheme can easily be measured.

2. Laboratory experiments must be designed to find out the mechanism involved in different species. Threshold stimuli, swimming speed, etc. have to be systematically determined.

3. Hypotheses must be generated that can be tested by experiments or validated by new observations. If experimental evidence is contrary, then the hypothesis must be rejected. The formulation of a secondary

hypothesis to rescue a beloved primary one must be avoided.

4. It would be worthwhile to design an organized research program. This can be begun by working according to an agreed-upon scheme.

References

CUSHING, D. H. 1951. The vertical migration of planktonic Crustacea. Biol. Rev. 26: 158–192.
GILBERT, J. J. 1967. *Asplanchna* and posterolateral spine production in *Brachionus calyciflorus*. Arch. Hydrobiol. 64: 1–62.
HUTCHINSON, G. E. 1967. A treatise on limnology, v. 2. Wiley.
RINGELBERG, J. 1964. The positively phototactic reaction of *Daphnia magna* Straus, a contribution to the understanding of diurnal vertical migration. Neth. J. Sea Res. 2: 319–406.
———, J. VAN KASTEEL, and H. SERVAAS. 1967. The sensitivity of *Daphnia magna* Straus to changes in light intensity at various adaptation levels and its implication in diurnal vertical migration. Z. Vergl. Physiol. 56: 397–407.
SEGAL, E. 1970. Light, invertebrates, vertical distribution, p. 194–206. *In* Kinne, O. [ed.], Marine ecology, v. 1: Environmental factors, Part 1. Wiley-Interscience.
TINBERGEN, N. 1951. Study of instinct. Clarendon.
ZARET, T. M., and J. S. SUFFERN. 1976. Vertical migration in zooplankton as a predator avoidance mechanism. Limnol. Oceanogr. 21: 804–813.

7. Vertical Migrations of Zooplankton in the Arctic: A Test of the Environmental Controls

Claire Buchanan and James F. Haney

Abstract

Diel vertical migration studies of several zooplankton species were done in the arctic in experimental columns and in situ under a range of light cycles, including continuous sunlight. Populations did not migrate under continuous light in midsummer, but began to migrate when photoperiods shifted to <20h. The migrations occurred when the relative rate of change in light intensity was most rapid around sunset and sunrise. A diel cycle of unoriented swimming activity was observed under all light cycles in some of the species investigated. The 24-h-day depth of a population could be altered by temperature, food concentration, age, and diel cycles of activity. The association of zooplankton behavior with morphological and physiological features is discussed in terms of the selective forces for vertical migration.

The arctic provides a unique area in which to study the environmental controls of diel vertical migrations in zooplankton populations. It is generally accepted that the diel light-dark cycles in temperate and tropical regions, and especially the rapid changes in the light environment around sunset and sunrise, play an important role in regulating the diel vertical migrations. Since these light-dark cycles are never absent in temperate and tropical regions, hypotheses concerning the stimuli and mechanisms of vertical movement and the controls of depth distribution have necessarily been tested in the laboratory under artificial light conditions and light cycles. As the complexity of zooplankton light responses (e.g. Heberday 1949; Scheffer et al. 1958; Hazen and Baylor 1962; Ringelberg 1964; Waterman 1960; Siebeck 1968; McNaught 1971) and the effects of environmental conditions on zooplankton light responses (e.g. Clarke 1930, 1932; Baylor and Smith 1957; Itoh 1970; Stavn 1970; Kikuchi 1930, 1936) have become more apparent, laboratory tests of hypotheses concerning diel vertical migrations have faced the necessity of simulating natural light environments in order to directly compare laboratory results with data from the field. In the arctic, experimental tests of the hypotheses can be done in situ and in experimental apparatus under naturally occurring extreme light cycles.

In the arctic during midsummer, the sun is continuously above the horizon. The rapid changes in light which are usually associated with the evening and morning and the lack of light which is associated with the night in temperate and tropical regions do not occur. Except when altered by clouds, diel changes in light intensity are slow and light intensities remain relatively high throughout the day. When the sun begins to go below the horizon for a few hours each day, night con-

---

This research was supported by NSF grant DPP76-80605 and Water Resource Research Center grant A-044-NH.

ditions approach those found at lower latitudes; however, changes in the light environment at sunset and sunrise are still slow. Finally, several weeks before the autumn equinox, the diel pattern of light fluctuations in the arctic is directly comparable to those in the temperate and tropical regions (Fig. 1). Diel studies of vertical

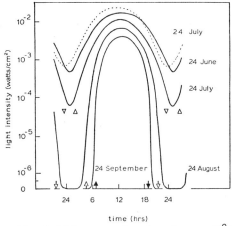

Fig. 1. Light curves for clear days at 68°N (solid line) and 74°N (dashed line). Arrows indicate sunset and sunrise on 24 July (▽ △), 24 August (◊◊), and 24 September (↓↑). Light curves for 68°N based on measurements made with an International Light model 700 radiometer (June–August) at Toolik, Alaska, and measurements made by the Smithsonian Institution (September) (Smithson. Meteorol. Tables 1918). Light curve for 74°N based on measurements made with a pyranometer (Eppley model 2) at Resolute Bay, N.W.T., 1973.

distributions and movements of zooplankton populations done in midsummer, late summer, and autumn in the arctic consequently compare the behaviors of zooplankton under continuously high light intensities, alternate light-dark conditions with a slow rate of change at sunset and sunrise, and alternate light-dark conditions with a rapid rate of change at sunset and sunrise.

We here present the first results of some diel studies done in the Alaskan and Canadian arctic under a wide range of light cycles from midsummer to autumn. The experiments were designed to test several current hypotheses concerning the stimuli and other environmental controls of diel vertical migrations. The studies of the proximal causes of migration also allowed us to raise questions about the effectiveness of selective forces acting on migration in zooplankton. In the second part of this paper we discuss the variability in diel vertical migrations of zooplankton and speculate briefly on the interactions of zooplankton behavior and selective forces. The methods and results of the diel studies are described in more detail elsewhere (in prep.).

Methods

In situ diel studies of the vertical distributions of zooplankton populations were done in shallow tundra thaw ponds at Barrow, Alaska, and in lakes at Resolute Bay, N.W.T., and Toolik, Alaska. Small plankton nets towed through the water on the end of a stick were used to sample zooplankton in shallow systems; plankton closing nets and Schindler traps were used in the deep systems. Diel experiments were done with zooplankton populations placed in narrow plexiglass columns filled with water and surrounded by black plastic which blocked light from the sides and created a vertical light gradient in the chamber. The vertical distributions and swimming activities of zooplankton in the columns were viewed through slits in the black plastic chamber. Observations were recorded on a tape recorder and later transcribed. Incident light intensity during most of the diel studies was measured with a radiometer and recorded on a continuous chart recorder.

Diel Vertical Migrations

Zooplankton populations did not undergo diel vertical migrations when exposed to continuous daylight in the arctic. Populations tended to remain at a constant mean depth and to maintain the same vertical distributions throughout the day (Fig. 2). This population behavior indicates that the preferendum hypothesis of diel vertical migration does not correctly describe the mechanism of the diel movements of zooplankton. The preferendum hypothesis has several versions, but generally states that diel vertical migrations occur as zooplankton popula-

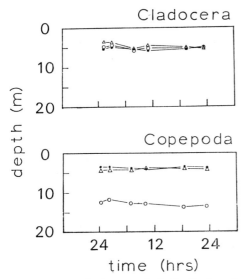

Fig. 2. Zooplankton vertical distributions in continuous daylight; Toolik Lake, Alaska, 21 July 1976. Cladocera: ●—*Bosmina longirostris*, △—*Holopedium gibberum*; ○—*Daphnia longiremis* forma *cephala*. Copepoda: ●—*Heterocope septentrionalis*; ○—*Cyclops scuifer*; △—*Diaptomus*.

tions stay with an optimum or preferred zone of light intensity which moves down and up in the water column during the daylight hours. If the preferred zone of light intensity moves more rapidly than zooplankton are able to travel (for example, at sunset and sunrise at lower latitudes), the photoreceptors of the zooplankton are forced to adapt to lower or higher light levels as the zooplankton fall behind their preferred zone. As a result of this adaptation, zooplankton acquire a new preferred zone of light intensity as they follow their old one. If the preferendum hypothesis of diel vertical migration is correct, one would expect arctic zooplankton to maintain an unchanged preferred zone of light intensity and undergo diel vertical migrations during the midsummer period. The migrations would be uninterrupted in deep lakes such as Toolik where the physical boundaries of the lake would not prevent the animals from following a preferred zone (Fig. 3). Since zooplankton populations did not undergo diel vertical migrations, it would seem that absolute light intensity in the form of a preferred or optimal zone is not the proximal cause of diel vertical migrations. More specifically, absolute light intensity does not stimulate the rapid vertical movements which occur at sunset and sunrise or regulate vertical distributions of the populations during midday.

Most of the zooplankton species investigated began to undergo classical diel vertical

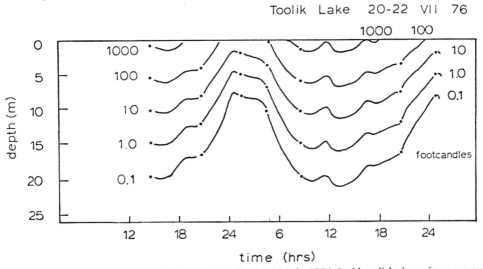

Fig. 3. Isopleths of light intensity in Toolik Lake, 20-22 July 1976. Incident light intensity measured continuously with IL700 radiometer and light extinction in lake measured at seven times with a Lambda light meter. Isopleths calculated from incident light intensity and extinction coefficient.

Table 1. Characteristics of some light cycles at Toolik, Alaska, 1976 and 1977. 1–10 ergs/cm^2/sec is the intensity threshold of the optical orientation mechanism of *Daphnia magna* (Ringelberg 1964). Stimulus, or relative rate of change in light intensity, equals $I_0 - I_t / I_0 \cdot t_{(sec)}$. Averages of the maximum stimuli at dawn and dusk are given.

	Photoperiod h	< 1–10 ergs/cm^2/sec*	Stimulus (sec^{-1}) Dusk	Dawn	
24–25 July 77	21.5	0 h	−0.0003	+0.0004	clear
27–29 July 76	20.5	0 h	−0.0003	+0.0004	partly cloudy
10–13 Aug 77	18.2	3.5 h	−0.0004	+0.0005	overcast
17–18 Aug 76	17.0	5.2 h	−0.0015	+0.0017	sun through clouds
18–19 Aug 77	16.9	5.3 h	−0.0008	+0.0009	overcast
24–25 Aug 77	16.2	6.5 h	−0.0009	+0.0015	overcast → clearing

*Ringelberg 1964.

migrations in the arctic when the photoperiod shifted from continuous daylight to <20 h of sunlight. Individual species started to migrate at different photoperiods, with cladocerans migrating earliest and copepods migrating later. Photoperiods <20 h are characterized by dark nights and by increasingly rapid changes in light intensity at dawn and dusk (Table 1). Migrations tended to occur when the relative rates of change in light intensities were most rapid at dawn and dusk. These diel studies in the late summer and autumn at Toolik, Alaska, strongly support the hypothesis proposed by Ringelberg (1964) and McNaught and Hasler (1964) which states that rapid changes in light intensity at dawn and dusk stimulate the rapid vertical movements of zooplankton populations at those times. Rapidly increasing light stimulates a downward movement; rapidly decreasing light stimulates an upward movement. In the absence of these stimuli (e.g. in continuous daylight) zooplankton would not be expected to undergo rapid vertical movements in the water column as the midsummer diel studies in the arctic demonstrate.

Some of the zooplankton species investigated tended to undergo reverse diel migrations before changing to the classical patterns of diel vertical migrations. It is possible that, initially, dawn and dusk stimuli were too weak to motivate rapid vertical movements in zooplankton populations; however, nights may have been dark enough to stop the use of an optical mechanism for orientation if zooplankton were using such a mechanism to orient themselves in the bright light gradients during the daytime. Ringelberg (1964) suggests that 1–10 ergs/cm^2/sec is the intensity threshold of the optical orientation mechanism for *Daphnia magna*. Below this threshold, *D. magna* is not able to use its compound eye to orient its body axis or swimming direction in a light gradient. As the photoperiod shortens in the arctic during late summer, nights with light intensities <1–10 ergs/cm^2/sec are experienced before threshold stimuli are experienced at dawn and dusk (Table 1). If species other than *D. magna* also have intensity thresholds of 1–10 ergs/cm^2/sec for photic orientation, they would be unable to use the optical mechanism during nights. At night they would use other environmental cues and orientation mechanisms to orient their body axes and swimming directions, and consequently could redistribute their populations upward or downward. Parker (1902), Clarke (1930, 1932), and Kikuchi (1938) found differences in the vertical distributions of several marine and freshwater species under bright and dim or dark conditions in the laboratory; although no absolute light intensities were measured, their observations suggest that many zooplankton species do use an optical mechanism for orientation when sufficient light gradients occur. Investigations by Siebeck (1968), Siebeck and Ringelberg (1969), Ringelberg

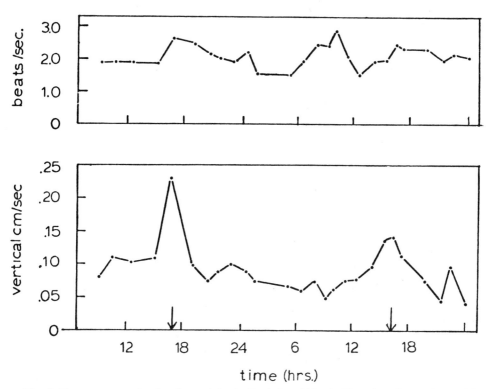

Fig. 4. Two measures of swimming activity in *Daphnia middendorffiana* juveniles, 3–4 June 1977, Toolik, Alaska. Data points represent average vertical centimeters traveled per second and antennal beats per second for 8 to 12 individuals. Arrows point out peak activities.

(1964), and others indicate that the optical mechanism for orientation probably involves orientation of the body axis and swimming direction to contrasts in the 3-dimensional distribution of underwater light.

A diel cycle of swimming activity was observed in most of the zooplankton species studied under continuous daylight in the arctic and under alternate dark-light in temperate regions. Peak activities typically occurred between 1500–2100 hours in the arctic. Cycles of swimming activity in zooplankton have been observed enough in the laboratory to indicate that the activity is based on an endogenous rhythm and timed by a zeitgeber (Stearns 1975; Hart and Allanson 1976). In the arctic the cladoceran activity pattern was most clearly expressed as temporal changes in the rate of vertical movement (vertical distance traveled per unit time) of individuals randomly observed in a population (Fig. 4). Despite the increased vertical movement of individuals during periods of peak activity, population mean depth usually remained unchanged, suggesting that the behavior is more kinetic (unoriented) than tactic (oriented, e.g. oriented to light). In unusual situations when zooplankton populations were close to the bottom of the experimental column, diel increases in unoriented swimming activity could force the populations away from the bottom and result in a small vertical migration. Esterly (1917), Harris (1963), and Enright and Hamner (1967) have observed circadian rhythms of what first appears to be oriented swimming activity (vertical migrations) under constant light or dark conditions. It is possible that the boundaries of their experimental apparatus may also have translated some of the unoriented (kinetic) swimming activity of their experimental animals into directional movements of the population and were partially responsible

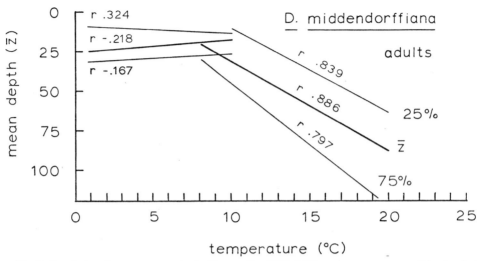

Fig. 5. Correlations between mean depth ($\bar{z}$) quartiles (25%, 75%), and temperatures at $\bar{z}$ for *Daphnia middendorffiana* adults, June 1977, Toolik, Alaska. Regressions for temperatures between 1° and 10°C are insignificant. Regressions for temperatures between 8° and 20°C are significant ($P \leq 0.01$).

for the observed vertical migrations.

Environmental and physiological factors appear to have important roles in controlling the vertical distributions of zooplankton populations during the midday phase of diel vertical migrations. In temperate regions the rapid changes in light intensity at sunrise initially direct zooplankton populations into deeper waters. The midday depth of a population does not remain constant at the depth where the morning descent ended, however, and upward and downward drifts during the daytime are often observed. Since maximum light penetration and the maximum distance of prey detection by visual predators occurs at midday, daytime depth of a zooplankton population will influence its susceptibility to predators. When arctic zooplankton are not stimulated by light changes to undergo diel vertical migrations, other factors which control the vertical distributions of populations during daylight hours assume major importance. To examine the effects of a few of these factors, we did several diel studies in the arctic, in situ, in the experimental columns, and in the laboratory.

Extreme water temperatures usually depressed the mean depth of a population. Cold temperatures depressed the mean depths of imported temperate species, e.g. *D. magna* (see Fig. 8), and warm temperatures depressed the mean depth of arctic species, e.g. *Daphnia middendorffiana* (Fig. 5). Extreme water temperatures are sometimes experienced by zooplankton in shallow arctic ponds (and in the experimental columns) because these heated and cooled rapidly. Since the temperature cycle closely follows the light cycle in small bodies of water, diel vertical migrations sometimes occurred. Columns were experimentally cooled and heated in several diel studies in the arctic in order to confirm that correlations between temperature and mean depth were caused by temperature. Lake zooplankton in the arctic rarely experience extreme water temperatures during midsummer since they do not undergo diel migrations and hence do not travel through the thermocline.

Starved populations of a predator (*Heterocope septentrionalis*) and a grazer (*D. magna*) tended to disperse in the water column and lower their mean depths (Figs. 6 and 7). The vertical distribution of prey species (*Daphnia longiremis* forma *typica*, *Daphnia pulex*) did not appear to influence the vertical distributions of the predator *H. septentrionalis*.

Older individuals of cladoceran species tended to be lower in the water column than younger individuals (Fig. 8). The segregation

Vertical Migrations in Arctic

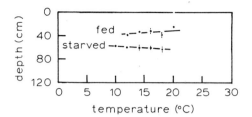

Fig. 6. Mean depth ($\bar{z}$) of *Heterocope septentrionalis*, starved and fed, during midsummer period of continuous daylight, Toolik, Alaska. Forty measurements of mean depth were made; data points represent average mean depths for $2°C$ intervals, ±SE. *Heterocope* were fed *Daphnia pulex* which maintained an average mean depth of 66.4 cm in the columns or *Daphnia longiremis* forma *typica* which maintained an average mean depth of 110 cm. *Heterocope* distribution was not significantly affected by prey distribution.

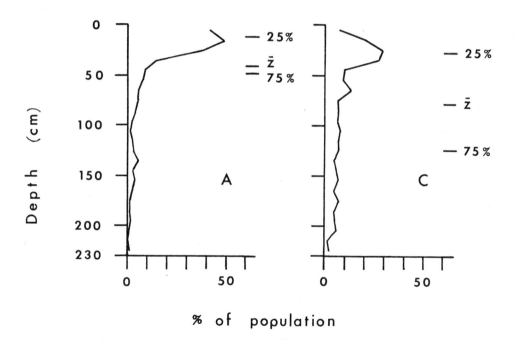

Fig. 7. Average daytime vertical distributions of *Daphnia magna* in 220-cm columns, maintained at two concentrations of *Scenedesmus* spp., 27 January-3 February 1977. Columns exposed to diffuse natural light from 1600–0800 hours and to constant, artificial light from 0800–1600 hours (incident light intensity = 1.35×10^{-5} W/cm^2). —89,600 cells/ml (8.96×10^{-5} cc/ml); ·····9,900 cells/ml (9.9×10^{-6} cc/ml).

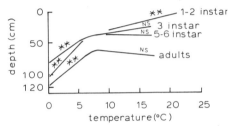

Fig. 8. Correlations between mean depths ($\bar{z}$) of four age groups of *Daphnia magna* exposed to continuous daylight (June 1976 and 1977) and temperature, Toolik, Alaska. Regression coefficients (r) are given for linear portions of curves. Asterisks—significant at $P \leq 0.01$; NS—no significance.

was observed in well fed animals over a wide range of water temperatures and light cycles; however, no observations were made of zooplankton populations held at different food concentrations.

Variations in Vertical Migrations

It was apparent in these arctic diel studies that environmental (e.g. temperature, food density) and physiological (e.g. age, diel cycles of activity) factors can influence the daytime vertical distributions of zooplankton populations. These results agree with vertical migration studies done in temperate regions under less extreme light cycles. Kikuchi (1930), Langford (1938), Plew and Pennak (1949), Herman (1963), Healey (1967), and others have observed seasonal variation in daytime vertical distributions of zooplankton in temperate lakes and oceans and suggest that changes in various environmental and physiological factors caused the seasonal differences. Pennak (1944), Kikuchi (1938), Hutchinson (1967), and others have noted dissimilar daytime vertical distributions of the same species in separate lakes and attribute the dissimilarities to different environmental conditions. The phototactic reaction, expressed as rapid vertical movements in the water column, can also be modified by environmental and physiological factors. Rapid changes in light initially stimulate the reactions; however, environmental parameters like carbon dioxide, oxygen, ph, temperature, and light intensity, and physiological parameters like age, sex, and reproductive condition will modify and even reverse the reactions (*see* Cushing 1951; Baylor and Smith 1957; Hutchinson 1967; Vinogradov 1970). Besides altering the behavior of individuals, environmental parameters may also modify the features that mediate behavior. For example, gross morphological features of neurons in the visual systems of clones of *D. magna* are reproduced identically; finer details, however, are not identical (Macagno et al. 1973). The relative concentrations of the four visual pigments in *Daphnia* photoreceptors are determined by the photic environment of the animal rather than by genetic instruction (McNaught 1971). Overall, environmental and physiological factors seem to be responsible for much of the variation observed in zooplankton behavior.

Intraspecific variation is essential to the selection processes of evolution. Speculations on the evolutionary cause(s) of a trait assume that heritable variations (alleles) were partly responsible for the expressed variation in the ancestral trait, and the selection forces being proposed acted upon the genome of the ancestral population as well as its array of phenotypes. The assumption seems valid, since alleles that produce variant phenotypes in modern species can be either removed or established in population gene pools by selective forces. Behavior is one component of an organism's phenotype, but unlike structures it has no physical form. The result of interactions between an organism's sensory, nervous, effector, and skeletal organs, behavior is strongly influenced by the physiological and environmental factors that modify the structures and functions of the organs. Consequently, relationships between the organism's genotypes and its behaviors are rarely straightforward. We would like to pose the following questions with respect to zooplankton behavior: Can genotypic differences within a species population produce variation in the vertical distributions and movements of zooplankton? Can selective forces act on populations to change the average behavior if little of the variation in behavior is due to heritable differences?

Several investigations of zooplankton, and primarily of cyclomorphic species, have indicated that some of the variation in zooplankton vertical distributions and movements is directly related to genotypic differences in the population. Natural populations of cyclomorphic species appear to contain polymorphic genotypes for helmet formation. Although helmet induction and growth rate are governed by environmental factors (i.e. temperature, turbulence, light intensity, nutrition), individuals from different clones will begin helmet formation at different threshold stimuli (Brooks 1946) and produce different size helmets under identical environmental conditions (Jacobs 1961). Helmets are known to alter *Daphnia* locomotion and orientation (Jacobs 1964, 1965) and appear to affect the vertical distributions and movements of *Daphnia* in natural systems (Brooks 1964; Worthington 1931; Woltereck [*cited in* Hutchinson 1967]) so some of the variation found in the vertical distributions of cyclomorphic species may be directly related to genetic differences. Studies of other types of polymorphism show that some polymorphisms will not affect zooplankton behavior; for example, diel vertical migrations of horned and unhorned individuals of the polymorphic *Ceriodaphnia cornuta* are indistinguishable (Zaret 1972). Differences in diel vertical migrations of females and males have been observed (*see* Hutchinson 1967), suggesting that secondary sex characteristics affect the light responses and locomotion of zooplankton; however, no controlled studies have been done. Endogenous cycles, thought to be controlled by gene action (*see* Brown 1973), are known to affect light responses and swimming behaviors of zooplankton. Cycles of 1.8 min in swimming movements (Daan and Ringelberg 1969), 8 min and 24 h in the strength of the phototactic reaction to light stimuli (Clarke 1932), 24 h in zooplankton swimming activity (Hart and Allanson 1976; Buchanan in prep.), and long-term in swimming behavior and light responses, related to ecdysis (molt) cycles (Clarke 1932; Haney et al, unpub.) have been observed. Allelic differences in the endogenous cycles could create consistent variation in zooplankton behavior; however, these differences have not been looked for or studied. The evidence to date that relates genetic differences to behavioral variation is mostly inferential, but it does suggest that such relationships occur within species. Behavioral variation caused by genetic differences has not been demonstrated in natural populations of species.

Selective forces can shift the behavior of a population if the population contains two or more alleles which produce different behaviors (e.g. Dobzhansky and Spassky 1969). However, if behaviors are continuously modified by changes in environmental factors, it is possible that allelic expression may be masked and cannot be acted upon by selective forces. One attempt has been made under semicontrolled conditions to examine the effect of heavy selection pressure on zooplankton behavior, and the results suggest that heritable changes in zooplankton behavior are slow. Banta (1921) attempted to select artificially for and against a phototactic response (i.e. the rate at which individuals move toward a stationary, horizontal light source) in clones of parthenogenic *D. pulex* and *Daphnia longispina* raised under variable conditions. Banta assumed that mutations in these organisms which affected their behavior would be immediately evident and could be selectively established. However, he found that no significant differences due to selection had occurred after intensive selection for up to 203 generations (about 5 years) in 11 clones, although environmental factors did modify the phototactic reactions of both species during the experiment. Banta also studied five clones of parthenogenic *Simocephalus exspinosus*, a semibenthic species, and observed an apparent heritable change due to selection in at least one strain of one clone. *Simocephalus exspinosus*, however, is known to produce polymorphic intersexes (Banta 1939) and variant behaviors may have been produced by pleiotropic effects of the intersex genes (or by other associated polygenes or alleles) if the genes were activated by environmental stimuli.

Arctic lakes and ponds seem to offer many uncontrolled experiments of natural selection on the light responses and swimming behaviors of zooplankton, and the phototactic reaction in particular. For most of their active life, zooplankton in the arctic are exposed to continuous daylight and do not undergo rapid vertical movements. During these 2–4 months, random mutation and selection could lead to changes in the phototactic sensitivity of arctic populations. Many zooplankton species found in the arctic, however, began to show phototactic reactions at roughly the same threshold stimuli that evoke vertical movements in a temperate zooplankton, *D. magna*, suggesting that phototactic sensitivity has not changed drastically in arctic populations. Changes in arctic populations may be slow because of a masking effect of environmental variation or possibly because selection pressures are ineffective or counteract each other. In addition, the phototaxis mechanisms (ocelli, compound eyes) could be conservative traits. Several ecologically important behaviors are mediated by the rhabdomeric eyes of zooplankton, including the dorsal light reaction, shore avoidance, daytime spatial orientation, somersaulting (fright) reaction, and the phototactic reaction (to rapid changes in light); hence many eye mutations may not be readily incorporated in the gene pool of a species.

Despite the multiphyletic origins of zooplankton and their range of sensory receptors, nervous systems, skeletal systems, and swimming behaviors, representatives of most phyla show phototactic behaviors and undergo diel vertical migrations, suggesting that selective advantages favor the development of diel vertical migrations (Hutchinson 1967). Population studies have pointed out the selective advantages of vertical migration in modern communities and have stimulated much speculation on the evolutionary causes of vertical migrations (*see* Vinogradov 1970). Such hypotheses are stimulating and heuristically useful, but are limited, since they cannot be tested (Peters 1976). Experimental research associating behaviors with morphological and physiological features should improve our perception of the mechanisms of vertical migration. Investigations of selection on allelic frequencies of these features could help identify the selective forces that modify migration behavior. With this kind of information we can speculate more knowledgeably about the present evolution of diel vertical migrations and perhaps stimulate testable ideas about the mechanisms and ecological importance of this behavior.

References

BANTA, A. M. 1921. Selection in Cladocera on the basis of a physiological character. Carnegie Instit. Wash. Publ. 305. 170p.

———. 1939. Controls of male and sexual-egg production. Carnegie Instit. Wash. Publ. 513, p. 106–130.

BAYLOR, E. R., and F. E. SMITH. 1957. Diurnal migration of plankton crustaceans. Rec. Adv. Invertebr. Physiol., pp. 21–35.

BROOKS, J. L. 1946. Cyclomorphosis in *Daphnia*. Ecol. Monogr. 16:409–447.

———. 1964. The relationship between the vertical distribution and seasonal variation of limnetic species of *Daphnia*. Int. Ver. Theor. Angew. Limnol. Verh. 15:684–694.

BROWN, F. A., Jr. 1973. Biological rhythms, p. 429–456. *In* C. L. Prosser [ed.], Comparative animal physiology. Saunders.

CLARKE, G. L. 1930. Change of phototropic and geotropic signs in *Daphnia* induced by changes of light intensity. J. Exp. Biol. 7: 109–131.

———. 1932. Quantitative aspects of the change of phototropic signs in *Daphnia*. J. Exp. Biol. 9:180–211.

CUSHING, D. H. 1951. The vertical migration of planktonic Crustacea. Biol. Rev. 26:158–192.

DAAN, N., and J. RINGELBERG. 1969. Further studies on the positive and negative phototactic reaction of *Daphnia magna* Straus. Neth. J. Zool. 19:525–540.

DOBZHANSKY, T., and B. SPASSKY. 1969. Artificial and natural selection for two behavioral traits in *Drosophila pseudoobscura*. Proc. Nat. Acad. Sci. 62:75–80.

ENRIGHT, J. T., and W. M. HAMNER. 1967. Vertical diurnal migration and endogenous rhythmicity. Science 157:937–941.

ESTERLY, C. O. 1917. The occurrence of a rhythm in the geotropism of two species of plankton copepods when certain recurring external conditions are absent. Univ. Calif. Publ. Zool. 16:393–400.

HARRIS, J. E. 1963. The role of endogenous

rhythms in vertical migration. J. Mar. Biol. Assoc. U.K. 43:153-166.
HART, R. C., and B. R. ALLANSON. 1976. The distribution and diel vertical migration of *Pseudodiaptomus hessei* (Mrazek) (Calanoida: Copepoda) in a subtropical lake in southern Africa. Freshwater Biol. 6:183-198.
HAZEN, W. E., and E. R. BAYLOR. 1962. Behavior of *Daphnia* in polarized light. Biol. Bull. 123:243-252.
HEALEY, M. C. 1967. The seasonal and diel changes in distribution of *Diaptomus leptopus* in a small eutrophic lake. Limnol. Oceanogr. 12:34-38.
HEBERDEY, R. F. 1949. Das Unterscheidungsvermögen von *Daphnia* für Helligkeiten farbiger Lichter. Z. Vgl. Physiol. 31:89-111.
HERMAN, S. S. 1963. Vertical migration of the opossum shrimp, *Neomysis americana* Smith. Limnol. Oceanogr. 8:228-238.
HUTCHINSON, G. E. 1967. A treatise on Limnology, v. 2. Wiley.
ITOH, K. 1970. Studies on the vertical migration of zooplankton in relation to the conditions of underwater illumination. Sci. Bull. Fac. Agr. Jpn. 25(1):71-96.
JACOBS, J. 1961. Cyclomorphosis in *Daphnia galeata mendotae* Birge, a case of environmentally controlled allometry. Arch. Hydrobiol. 58:7-71.
———. 1964. Hat der höhe Sommerhelm zyklomorpher Daphnien einen Anpassungswert? Int. Ver. Theor. Angew. Limnol. Verh. 15:676-683.
———. 1965. Significance of morphology and physiology of *Daphnia* for its survival in predator-prey experiments. Naturwissenschaften 52 (6):141-142.
KIKUCHI, K. 1930. Diurnal migration of plankton Crustacea. Q. Rev. Biol. 5:189-206.
———. 1936. Studies on the vertical distribution of the plankton Crustacea I. A comparison of the vertical distribution of the plankton Crustacea in six lakes of middle Japan in relation to the underwater illumination and the water temperature. Rec. Oceanogr. Works Jpn. 9:61-85.
———. 1938. Studies on the vertical distribution of plankton Crustacea II. The reversal of phototropic and geotropic signs of the plankton Crustacea with reference to the vertical movement. Rec. Oceanogr. Works Jpn. 10:17-41.
LANGFORD, R. R. 1938. Diurnal and seasonal changes in the distribution of limnetic Crustacea of Lake Nipissing, Ontario. Univ. Toronto Stud. Biol. Ser. 45, p. 1-42.
MACAGNO, E. R., V. LOPRESTI, and C. LEVINTHAL. 1973. Structure and development of neuronal connections in isogenic organisms: variations and similarities in the optic system of *Daphnia magna*. Proc. Natl. Acad. Sci. 70:57-61.
McNAUGHT, D. 1971. Plasticity of cladoceran visual systems to environmental changes. Trans. Am. Microsc. Soc. 90:113-114.
———, and A. D. HASLER. 1964. Rate of movement of populations of *Daphnia* in relation to changes in light intensity. J. Fish Res. Bd. Can. 21:291-318.
PARKER, G. H. 1902. The reactions of copepods to various stimuli and the bearing of this on daily depth-migrations. Bull. U.S. Fish Comm. 21:103-123.
PENNAK, R. W. 1944. Diurnal movements of zooplankton organisms in some Colorado mountain lakes. Ecology 25:387-403.
PETERS, R. H. 1976. Tautology in evolution and ecology. Am. Nat. 110:1-12.
PLEW, W. F., and R. W. PENNAK. 1949. A seasonal investigation of the vertical movements of zooplankters in an Indiana lake. Ecology 30:93-100.
RINGELBERG, J. 1964. The positively phototactic reaction of *Daphnia magna* Straus: a contribution to the understanding of diurnal vertical migration. Neth. J. Sea Res. 2:319-406.
SCHEFFER, E., P. ROBERT, and J. MEDIONI. 1958. Reactions oculomotrices de la Daphnie (*Daphnia pulex* De Geer) en réponse à des lumières monochromatiques d'égale énergie. Sensibilité visuelle et sensibilité dermatoptique. C. R. Soc. Biol. 151:1000-1003.
SIEBECK, O. 1968. "Uferflucht" und optische Orientierung pelagischer Crustaceen. Arch. Hydrobiol. Suppl. 35(1), p. 1-118.
———, and J. RINGELBERG. 1969. Spatial orientation of planktonic crustaceans. Int. Ver. Theor. Angew. Limnol. Verh. 17:831-947.
STAVN, R. H. 1970. The application of the dorsal light reaction for orientation in water currents by *Daphnia magna* Straus. Z. Vgl. Physiol. 70:349-362.
STEARNS, S. C. 1975. Light responses of *Daphnia pulex*. Limnol. Oceanogr. 20:564-570.
VINOGRADOV, M. E. 1970. Vertical distribution of the oceanic zooplankton. Akad. Nauk SSSR.
WATERMAN, T. H. 1960. Interaction of polarized light and turbidity in the orientation of *Daphnia* and *Mysidium*. Z. Vgl. Physiol. 43:149-172.
WORTHINGTON, E. B. 1931. Vertical movements of fresh-water macroplankton. Int. Rev. Gesamten Hydrobiol. Hydrogr. 25:394-436.
ZARET, T. M. 1972. Predators, invisible prey, and the nature of polymorphism in the Cladocera (Class Crustacea). Limnol. Oceanogr. 17:171-184.

8. Diurnal Vertical Migration in Aquatic Microcrustacea: Light and Oxygen Responses of Littoral Zooplankton

Dewey G. Meyers

Abstract

The upward swimming reaction of littoral zooplankton was measured in response to reductions of illumination and oxygen concentrations. Decreased light intensities that simulated sunset conditions elicited rapid and pronounced upward movement in obligate plankters (*Daphnia magna* and *Ceriodaphnia reticulata*). Facultative plankters (*Chydorus sphaericus* and *Pseudochydorus globosus*) exhibited a delayed and comparatively depressed ascent. When attached to a substrate, they failed to respond at all. Lowered oxygen conditions, similar to those which occur diurnally in littoral macrophyte beds, did, however, elicit distinct upward swimming. *Chydorus sphaericus* reacted to higher oxygen levels (2.5 ppm) than *P. globosus* (1.25 ppm). It is suggested that the regulatory mechanism of diurnal vertical migration for littoral zooplankton differs between species—obligate forms cueing on light and facultative species responding primarily to diel oxygen fluctuations.

Daily variation in the self-directed, vertical positioning of zooplankton has evoked a relatively large number of investigations into its causes and adaptive significance. Initiation, control, and direction appear to be primarily regulated by exogenous changes in the intensity and angular distribution of light (Ringelberg 1964; McNaught and Hasler 1964) apparently based upon an endogenous rhythm (Harris 1963; Enright and Hamner 1967; Ringelberg and Servaas 1971). A profusion of speculations exist on the adaptive value of diurnal migration (Hutchinson 1967; McLaren 1963), often implicating competition (Dumont 1972) and predation (Zaret and Suffern 1976) as selective forces. Most studies have been conducted on pelagic or open-water zooplankton; no similar experimental research exists on aquatic microcrustacea of littoral freshwater habitats.

Littoral microcrustaceans of the Order Cladocera may be categorized by their primary mode of locomotion—"swimmers" (obligate zooplankters), e.g. Daphnidae and Bosminidae, and "crawlers," e.g. Chydoridae and Macrothricidae. Swimmers occupy the open-water areas within and between macrophyte beds and are known to vertically migrate on a diurnal basis (e.g. *Daphnia magna* Straus: Ringelberg 1964; *Ceriodaphnia quadrangula* O.F.M.: Szlauer 1963). Crawlers commonly inhabit the surfaces of hydrophytes and benthic muds and can be subdivided into two groups—substrate dwellers and facultative zooplankters. Substrate dwellers almost exclusively maintain contact with a surface and express diel periodicity in traversing the vertical extent of their macrophyte habitat, seldom venturing beyond its upper border (e.g. *Alonella excisa* Fischer: Whiteside 1974). Facultative zooplankters are those crawlers known to be less faithful to surface attachment and those that diurnally migrate into the open water above hydrophyte stands (e.g. *Chydorus sphaericus* O.F.M. and *Pseudochydorus globosus* Baird: Whiteside 1974). With only these few studies documenting the occurrence of diel vertical movements in littoral cladocerans, it is not surprising that little is known about the causes or selective advantage of this cyclic behavioral phenomenon.

I here attempt to clarify the causal agents of vertical migration in both obligate and facultative zooplankters of the littoral. Two series of experiments were conducted to test the effects of reduced light intensity and lowered oxygen concentrations on upward movement. The results of these tests provide insight into the possible adaptive significance and origin of diurnal migration of littoral and pelagic microcrustaceans.

I thank J. R. Strickler for valuable contributions to the development of this paper and for the use of laboratory space and equipment while I was a guest at Osborn Memorial Laboratories, Yale University. I also thank G. E. Hutchinson and L. Provasoli for valuable discussions; J. Neess, S. I. Dodson, C. Goulden, A. Tessier, K. Sellner, and F. Burchsted for constructive criticism; and B. Wilson for statistical advice.

Light experiments

Changes in light intensity have long been recognized as the primary environmental cue that initiates and controls diel vertical movements in aquatic organisms. Two outstanding reviews of the phototaxis hypothesis are contained in Ringelberg (1964) and Hutchinson (1967). Although phototactic migration has been widely accepted, quantitative experimental evidence on the controlling mechanism was not available until recently, when Ringelberg (1964) determined that a relative decrease in light intensity triggered the upward movements of *D. magna*. The phototaxis threshold values, above which upward swimming began, ranged from 11.5 to 12.2% for an instantaneous decrease and 0.17% sec^{-1} for a continuous illumination reduction. After laboratory determination, these results were verified in field studies (Ringelberg 1964; McNaught and Hasler 1964). Application of these findings to other vertically migrating zooplankters, although untested, has been assumed.

To investigate the response of various littoral microcrustacea to decreasing light intensity, I used the basic methodology of Ringelberg (1964). Experimental animals were individually tested in a glass cylinder (30 cm long, 2.5 cm diameter) containing filtered pond water and a sparse amount of algae (axenic mixture of *Scenedesmus obliquus* and *Chlamydomonas reinhardi*). The cylinder was placed in a rectangular plexiglass container (32 cm high), open at the top (20 cm long x 12 cm wide), darkened at the bottom (with a piece of black "construction" paper), and filled to the same 30-cm height with distilled water. Double containers enhanced visual observation by bringing the entire inner surface of the cylinder into focus.

Illumination was provided by a halogen lamp (quartz iodine, 100 W, 12 V). Its spectral composition closely approximated that of sunlight. The lamp was mounted above the experimental vessels and focused to produce a circular collimated beam of light directly over the cylinder with a diameter just the width of the top opening of the plexiglass container (12 cm). Translucent, white "tracing" paper (Canson, Vidalon No. 90) covered the top of the plexiglass vessel, thus eliminating horizontal markers and encouraging random horizontal movement (Siebeck 1969).

Initial light intensity was reduced by a series of 10 "window pane" glass filters that were individually inserted between the top of the experimental vessels and the light source. Spectral distribution of the light was not appreciably altered by the glass filters (92.70 ± 0.21%, mean transmittance ± 95% confidence limits for the visible spectrum, Beckman spectrophotometer model 25). Filters were contained in a specially constructed wooden box with 10 tracks that allowed each filter to be individually inserted into the light path. The exterior and interior of the box were painted with nonreflective black paint. The experimental vessels were surrounded with black material to further eliminate reflected light and to reduce external visual distractions to the test animals. A 3-cm vertical opening in the material allowed observation. All experiments were conducted in a darkroom at 20° C. Light intensities were recorded at the water surface with a LI-COR photometer model LI-185.

Twenty-five mature female animals from each of four species were tested: *D. magna* and *Ceriodaphnia reticulata* Jurine (obligate zooplankton) plus *C. sphaericus* and *P. globosus* (facultative plankters). Only *P. globosus* was collected from the field and acclimated to laboratory conditions before experimentation; the other three species were obtained from established laboratory populations maintained by J. R. Strickler.

Each obligate plankter was individually tested by being placed into the glass cylinder and allowed to acclimate to the initial light intensity (660 lux) for at least 30 min, while establishing its own horizontal level of hop-and-sink locomotion. Facultative plankters were only allowed a 40-sec acclimation period to avoid their settling on the sides of the cylinder. After acclimation, the animal was exposed to a series of 10 subthreshold light-intensity reductions (7.27 ± 0.28%, mean ± SE), one every 30 sec. The experiment concluded with an additional 10-min exposure to the final light intensity. Positional observations were recorded every 10 sec with the use of a cassette recorder and later transcribed to graphic form. Tests were conducted for the 2 hours during and immediately after sunset to coincide temporally with the upward swimming migration and endogenous rhythms of field populations.

The swimming reaction data of obligate zooplankton species to illumination reductions are shown in Fig. 1. Time lag between initiation of stimuli and upward movement was negligible; *D. magna* reacted immediately (Fig. 1a) and *C. reticulata* had a slight latency period of 40 sec (Fig. 1b). Ringelberg (1964), using 3% reductions at 30-sec intervals, observed a 231–286 sec lag period with *D. magna*. Absence of latency in the *D. magna* population I used may be due to

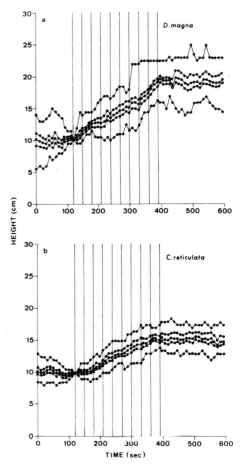

Fig. 1. Upward swimming response of littoral obligate zooplankters (a. *Daphnia magna*; b. *Ceriodaphnia reticulata*) to presentation of subthreshold reductions in illumination. Vertical lines designate stimuli presentation (every 30 sec). Central dotted line is mean, adjacent two lines are 95% confidence limits, and outer two lines are range; $n = 25$. (Response data for each animal shifted vertically to achieve a comparable height of 10 cm at presentation of first stimulus.)

the greater illumination reductions of 7.28% which, on a continuous basis, equal 0.24% sec^{-1}, well above the 0.17% sec^{-1} threshold demonstrated by Ringelberg.

Vertical magnitude, duration, and velocity of response were also greater for *D. magna* than *C. reticulata* (Table 1, Fig. 1). During the "pre-" and "post-" stimuli periods, the slopes of the lines generated by a linear regression analysis did not

Table 1. Comparison of magnitude, duration (and range), and velocity of upward swimming response of four littoral microcrustaceans exposed to 10 subthreshold illumination reductions ($\bar{x}$ = 7.28%).

Exptl sp.	Upward swimming Distance (cm)	Reaction Duration (sec)	Range	Swimming Velocity (cm/sec)
D. magna	9.5	330	120–450	0.029
C. reticulata	5.5	240	160–400	0.023
C. sphaericus	4.0	90	330–420	0.044
P. globosus	4.3	140	220–360	0.031

*Range of reactions interpreted as lowest mean position of a positive slope, after presentation of first stimulus, and terminating with highest mean position of that slope.

significantly differ from zero (horizontal) with the exception of the prestimuli negative slope of *C. reticulata* (Table 2). *Daphnia magna*, however, had a significantly steeper upward inclination, during the stimuli period (Fig. 2), reflecting its greater ascension velocity (Table 1).

The responses of the facultative zooplankton species also differed from one another (Fig. 3). Latency of reaction was pronounced, with *C. sphaericus* exhibiting the greater lag time of 210 sec (Fig. 3a). The *P. globosus* population demonstrated greater distance and duration of upward swimming; *C. sphaericus* had a high ascension velocity (Table 1), and consequently a steeper inclination (Fig. 2). The post stimuli period was marked by a downward

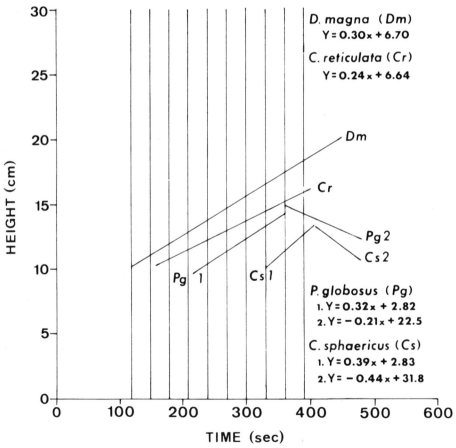

Fig. 2. Linear regressions and their corresponding equations for the "Mid" and "Post activity periods of the four microcrustaceans.

Table 2. Significance of linear regression slopes and r values for movements of four littoral microcrustaceans, during four activity periods of illumination reduction experiments, compared to zero (horizontal movement). Abbreviations: Pre = before stimuli presentation; Lag = after stimuli, before crustacean reaponse; Mid = reaction period; Post = after presentation of last stimulus.

Activity Period	D. magna		C. reticulata		C. sphaericus		P. globosus	
	Slope	r	Slope	r	Slope	r	Slope	r
Pre	0.240	0.06	∓0.005*	0.15				
Lag			0.432	0.07	0.214	0.05	0.854	0.01
Mid	0.0001*	0.84	0.0001*	0.78	0.0001*	0.28	0.0001*	0.38
Post	0.616	0.03	0.116	0.07	∓0.001*	0.24	∓0.0001*	0.22

*Highly significant, ∓negative slope.

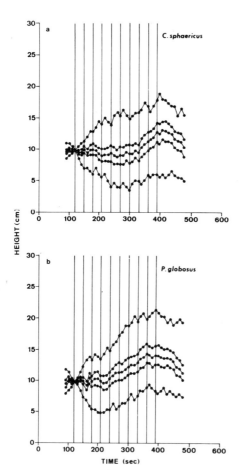

Fig. 3. As Fig. 1, except for facultative zooplankters (a. *Chydorus sphaericus*; b. *Pseudochydorus globosus*).

movement by both species (Fig. 2), and *P. globosus* population beginning its descent before presentation of the last stimulus (Fig. 3b).

In comparison to obligate plankters, the facultative species exhibited a slower response, smaller magnitude of vertical movement, reduced duration of reaction, and a rapid loss of upward gain before or during the post-stimuli phase. The duration of the lag period and magnitude of response are directly related to stimuli schedule and intensity (Ringelberg 1964). With stimuli presentation temporally constant, it would, therefore, follow that the further the experimental light reductions are below the threshold stimulus, the longer is the lag time and the shorter is the upward swimming response. The upward swimming behavior of facultative zooplankters, then, is less responsive to illumination reductions that simulate natural field conditions at sunset.

These crawler species, apparently, require a higher threshold stimulus than obligate plankters (about 50% for *C. sphaericus*). In addition, they were allowed a negligible acclimation period to avoid their clinging to the sides of the cylinder, because, once attached, light intensity reductions of as much as 100% failed to evoke either dislodgement or upward movement. Thus, facultative zooplankters are less reactive to light cues in initiating and maintaining upward vertical migration, and once having settled on a substrate, they seem to not respond at all.

Oxygen experiments

Since the experimental results of the light experiments indicate that facultative zooplankters (Chydoridae), if they react at all, do not readily respond to light cues, it is probable that another environmental variable is primarily responsible for eliciting diel migrations.

Oxygen concentrations in the littoral may be more temporally variable than in the pelagic (Hutchinson 1967; Hall et al. 1970; Loffler 1974) especially within macrophyte beds where diurnal fluctuations may range from as low as 0 ppm at 0500 hours to 12 ppm at 1500 hours (Roach and Wickliff 1934). Chydorids commonly inhabit the surfaces of macrophytes and benthic muds and, consequently, are exposed to low nocturnal oxygen concentrations due to the combined respiration of the animals and plants. During the day, however, photosynthesis by dense hydrophyte stands and their associated periphyton can supersaturate the water with oxygen, although overshadowing may cause a small area near the benthic substratum to remain anoxic (Buscemi 1958).

To test the effects of lowered oxygen conditions on the vertical movements and distribution of *C. sphaericus* and *P. globosus*, I exposed mature females to a stepdown series of four concentrations—8, 5, 2.5, 1.25 ppm at 20°C (87, 54, 27, and 14% saturation). The experimental apparatus was composed of a 40-cm length of glass tubing (1-cm diameter), covered at one end with nylon screening (104-μm mesh) and a loose rubber cap (1 cm high), and positioned within a larger glass cylinder (2.5-cm diameter, 35 cm long) containing a small amount of algae and filled with filtered pond water to a height of 30 cm. Oxygen reductions were accomplished with little temperature change (<0.5C°, 8-1.25 ppm) by bubbling nitrogen gas through the water and were measured before and after experimentation at each of the four concentrations with an oxygen meter (Y.S.I model 54), later verified by the standard Winkler method (Am. Public Health Assoc. 1971) to be accurate within ±0.05 ppm.

Ten animals from one of the two species were added to the glass tubing. The tubing was removed from the cylinder and drained through the nylon mesh to a 1-cm depth (maintained by the rubber collar). Draining allowed the chydorids to be concentrated but avoided exposing them to air. The water was adjusted to the next lower oxygen level, and the tubing was slowly lowered back into the cylinder so that the inrush of deoxygenated water would approximate an even distribution of animals throughout the final 30 cm of the water column. This technique of submersion and removal allowed a gradual, stepwise exposure to increasingly lower oxygen concentrations by maintaining the concentrated chydorids at their previously tested oxygen tension before exposure to the next lower level. Positional observations were taken every minute for 15 min, recording total numbers of animals located and swimming in the upper 5 cm. All experiments were conducted in a darkroom using a very low level of illumination (3 lux) provided by a red safelight (Kodak model A).

The data from the lowered oxygen experiments clearly indicate that an increased abundance and swimming activity occur near the water surface at an oxygen tension of 1.25 ppm (Figs. 4 and 5)—interspecific (Table 3) and intraspecific (Table 4) differences being highly significant. Although interspecific differences also differ at 2.5 ppm O_2 (Table 3), *C. sphaericus* seems to be the more responsive species, with a greater abundance and activity level in the upper 5 cm of the water column (Figs. 4 and 5), and a highly significant reaction difference between the 2.5 and 5 ppm O_2 levels (Table 4). After an initial period of 1 to 3 min of relatively high activity, the number of swimming chydorids declined, except for *P. globosus* at 1.25 ppm (Fig. 5b) and *C. sphaericus* at 1.25 and 2.5 ppm (Fig. 5a). Thus, it appears that the threshold below which upward swimming is initiated is species-dependent but may occur at oxygen

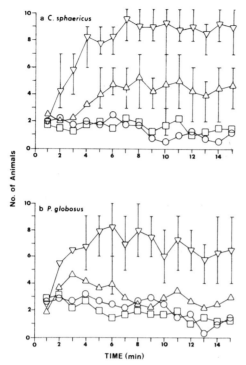

Fig. 4. Total abundance of facultative zooplankters (a. *C. sphaericus*; b. *P. globosus*) in upper 5 cm of a 30-cm experimental vessel at four oxygen concentrations (○ = 8, □ = 5, △ = 2.5, ▽ = 1.25 ppm). Symbols designate means; vertical lines represent ranges (only nonoverlapping ranges are graphed).

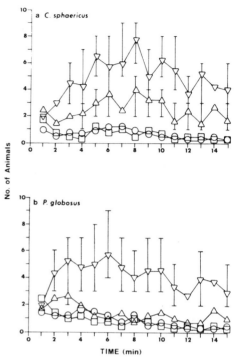

Fig. 5. As Fig. 4, except for swimming facultative zooplankters (a. *C. sphaericus*; b. *P. globosus*).

Table 3. Interspecific comparison using analysis of variance for profile data (Morrison 1967) on abundance and swimming activity of facultative zooplankters (*C. sphaericus* and *P. globosus*) in upper 5 cm of a 30-cm cylinder at four oxygen concentrations. Symbols: C = total number of animals present, S = number of swimming animals. First 3 min of observation excluded from analysis to allow immigration to equilibrate.

	Oxygen Concentration (ppm)			
Behavior	8	5	2.5	1.25
C	0.02*	0.76	0.01†	<0.01†
S	0.75	0.94	<0.01†	<0.01†

*Significant.
†Highly significant.

Table 4. Intraspecific comparison using analysis of variance for profile data on abundance and swimming activity of facultative zooplankters (Cs = *C. sphaericus*, Pg = *P. globosus*) in upper 5 cm of a 30-cm cylinder at four oxygen concentrations. Symbols: C = total number of animals present, S = number of swimming animals. First 3 min of observation excluded from analysis to allow immigration to equilibrate.

Test Species	Behavior	Compared Oxygen Concentration (ppm O_2)		
		8-5	5-2.5	2.5-1.5
Cs	C	0.208	<0.001*	<0.001*
	S	0.804	<0.001*	<0.001*
Pg	C	0.127	0.002†	<0.001*
	S	0.559	0.031	<0.001*

*Highly significant, α level of 0.004 given by Bonferroni technique.
†Significant.

levels of 2.5 ppm and higher. The abundance of actively swimming animals increases below this threshold, and, consequently, fewer animals settle on a substrate.

Lowered oxygen concentrations are known to affect the behavior and physiology of many different taxa of aquatic animals, including crustaceans (von Brand 1944; Prosser 1973; Davis 1975). Littoral or pond daphnids (*D. magna*: Heisey and Porter 1977; *D. pulex*: Kring and O'Brien 1976) exposed to decreased oxygen tensions maintain constant filtration rates until they reach a critical point ($\sim$3 mg liter^{-1}) where rates decrease; pelagic species (*D. galeata mendotae*: Heisey and Porter 1977), however, exhibit a linear reduction in filtration throughout the experimental oxygen range. These distinct, interhabitat (littoral to pelagic) responses may occur on a smaller scale as littoral, intrahabitat, behavioral differences. *Chydorus sphaericus* displays definite upward swimming at higher oxygen concentrations (2.5 ppm) than its more benthic counterpart, *P. globosus* (1.25 ppm), a species that may be routinely exposed to low oxygen levels (1-2 ppm: Buscemi 1958). Such response heterogeneity might be expected, since cladocerans that inhabit oxygen-deficient waters have been shown to be less reactive, physiologically and behaviorly, to oxygen reductions and, consequently, are believed to be better adapted to tolerate these conditions (Herbert 1954; Heisey and Porter 1977).

Many cladoceran species are known to produce hemoglobin when subjected to lowered oxygen conditions for an extended period (12 hr to 4 days) (Fox 1957; Smirnov 1971; Kring and O'Brien 1976). However, if presented with a choice, those littoral species tested were observed to swim away from lower and into higher oxygen levels. *Daphnia magna* selected 100% over 15% oxygen-saturated water (Ganning and Wulff 1966) and swam upward in response to oxygen levels of 2.7 cc liter^{-1} (Langford 1938). *Gammarus pulex* L. (Amphipoda, Crustacea) exposed to 1.4 mg O_2 liter^{-1} immediately swam to areas of higher oxygen concentration; 2.7 to 5.2 mg O_2 liter^{-1} elicted slower responses, and 7.2 mg. O_2 liter^{-1} none (Costa 1967). These results are in close agreement with the upward swimming reactions of facultative zooplankters exposed to similar oxygen tensions. Upward swimming, therefore, in the littoral crustaceans tested, has been initiated by lowered oxygen conditions and is likely to be the basis for diurnal migration in those species that are less responsive to light cues.

Other ancillary factors that may augment the effect of decreased oxygen concentration on vertical migration are carbon dioxide and temperature. Low carbon dioxide levels have been shown to elicit upward movement in *D. magna* (Langford 1938) and reverse negative phototaxis in *D. magna* (Loeb 1904) and in *C. sphaericus* (Smirnov 1971). Langford (1938) also suggested that, in the field, oxygen and carbon dioxide levels may act synergistically to raise the oxygen threshold from 2.7 to 3.5 cc liter^{-1}. Similarly, low temperatures are known to enhance upward migration (Loeb 1894; Langford 1938). In a more recent study, temperature change itself failed to induce vertical migratory behavior, and low oxygen tension was identified as the major regulatory factor (LaRow 1970: *Chaoborus punctipennis* Say larvae).

Discussion

Classical diurnal vertical migration (into the open water above the tops of the macrophyte beds) and diel vertical crawling movements on the surface of macrophytes have been confirmed within the littoral crawlers (Szlauer 1963; Whiteside 1974). *Chydorus sphaericus*, the most ubiquitous and abundant cladoceran crawler and the only chydorid that commonly inhabits the pelagic region (Hutchinson 1967), is most often cited as diurnally migrating and displays a variety of migration patterns. In general, these animals remain near the bottom or the lower portions of the hydrophyte beds during the day (Lang 1970; Smirnov 1971; Whiteside 1974; Shireman and Martin 1978); after sunset and during the night, *C. sphaericus* increases in density near the water surface and throughout the water column, dispersing downward at dawn (Berg and Nygaard 1929; Whiteside 1974; Buchanan 1975). A similar behavioral pattern was observed in the day-to-night activity patterns of laboratory populations (Meyers and Strickler 1978).

Based on the experimental data, it seems likely that oxygen is the major regulatory mechanism of diurnal migrations of crawlers, augmented by carbon dioxide, temperature, and, to a lesser degree, light. No similar experimental data exist on the adaptive significance of this cyclic, behavioral phenomenon within the littoral; competition and predation, however, are often suggested as selective forces in the pelagic.

Chydoridae, being the most species-diverse family of Cladocera, have evolved highly specialized feeding structures, and yet, the composition of their diets is basically uniform (Fryer 1968). There is distributional evidence to support the hypothesis that resource partitioning, then, may be dependent upon habitat selection both vertically (Lang 1970) and horizontally (Rybak et al. 1964). Until recently it was believed that grazing epiphytes off substrates was the exclusive feeding mode of chydorids. Mounting evidence, however, supports an alternate feeding behavior: filter-feeding suspended particulate matter directly from the water column by at least one species, *C. sphaericus* (Meyers and Strickler 1978; Haertel 1977). Although untested in the field, other species are known to filter feed in the laboratory (Fryer 1968).

One possible selective advantage to upward migration may then be that certain chydorids are able to use filtration to further partition their available resources by exploiting an alternate food source—phytoplankton. The selective pressures that contribute to the diel migration pattern of *P. globosus* are less apparent, but, as a scavenger, this species could be migrating in search of carrion or fatally weakened prey. Those chydorids that inhabit dense macrophyte beds and that move up the plant surface (substrate dwellers) may be migrating into an area of higher photosynthetic activity and the potentiality of a fresher, more nutritious food source (Whiteside 1974).

If the adaptive advantages of feeding higher in the water column are as great during the night as they seem, even stronger selection pressures must be causing the daylight descent. Based on the Brooks and Dodson (1965) hypothesis of size-selective predation, only those zooplankton larger than 1 mm were thought to be heavily preyed upon by planktivorous fish. Recently, however, through experimental investigations of littoral environments, it has been demonstrated that fish fry will feed upon prey of 0.5 mm and less—the size of most littoral zooplankters (Hall et al. 1970; Werner and Hall 1976; Phoenix 1976). Visual invertebrate predators of the littoral are also known to feed on this size range of prey (Johnson 1973). In the presence of predation pressure, it is likely that the facultative zooplankters that remain as cryptic and as low as possible within the camouflage of their macrophyte habitats would have a selective advantage during the day. Freed from visual predators during all but the brightest moonlit nights (Zaret and Suffern 1976), these animals would then be able to move up into an area of alternate food resources with relatively less danger.

Once active swimming and filter feeding are evolutionarily incorporated into the behavioral repertoire of a littoral, facultative zooplankter, the likelihood of preadaptation to an obligate swimming life-style is enhanced. Substantial morphological and behavioral evidence suggests this is the more likely hypothesis with littoral, benthic forms evolving into free-swimming pelagic species (Fryer 1968, 1974; Goulden 1968; Hutchinson 1967; Wesenberg-Lund 1926). Littoral, obligate zooplankters appear to express this transition from littoral benthic to pelagic existence in their feeding behaviors. As well as filter feeding, some species have been shown to graze or browse detrital particles off substrates (*D. magna*: Horton et al. 1979). Another transitional component could be the change in regulatory mechanism of diurnal vertical migration from oxygen to light. In this way, the animals could predetermine decreasing oxygen conditions without actually experiencing them. Total dependence on suspended food and successful foraging in the pelagic may have been the final steps by littoral, obligate zooplankton which preceded their actual colonization of the open waters.

Recent studies have implicated fish predation as an important component for the selection of diurnal vertical migration in pelagic zooplankton (Zaret and Suffern 1976). It would therefore seem possible that the initial or subsequent waves of pelagic colonizers, when exposed to planktivorous fish, maintained their diurnal migration patterns, now cued to light, as an adaptive advantage against the potent selective pressures wrought by visually hunting fish predators.

References

AMERICAN PUBLIC HEALTH ASSOCIATION. 1971. Standard methods for the examination of water and wastewater, 13th ed.
BERG, K., and G. HYGAARD. 1929. Studies of the plankton in the lake of Frederiksborg Castle. K. Dan. Vidensk. Selsk. Skr. 2. Nat. Mat. Afd. 9 1: 227-316.
BROOKS, J. L., and S. I. DODSON. 1965. Predation, body size, and composition of plankton. Science 150: 28-35.
BUCHANAN, C. 1975. A biological study of the Lakes Of The Clouds (with emphasis on zooplankton). M.S. Rep. Dep. Zool. Univ. New Hampshire.
BUSCEMI, P. S. 1958. Littoral oxygen depletion produced by a cover of *Elodea canadensis*. Oikos 9: 239-245.
COSTA, H. H. 1967. Responses of *Gammaris pulex* (L.) to modified environment. III. Reaction to low oxygen tensions. Crustaceana 13: 175-189.
DAVIS, J. C. 1975. Minimal dissolved oxygen requirements of aquatic life with emphasis on Canadian species: A review. J. Fish. Res Bd. Can. 32: 2295-2332.
DUMONT, H. J. 1972. A competition-based approach of the reverse migration in zooplankton and its implications, chiefly based on a study of the interactions of the rotifer *Asplanchna priodonta* (Gosse) with several Crustacea Entomostraca. Int. Rev. Gesamten Hydrobiol. 57: 1-38.
ENRIGHT, J. T., and W. M. HAMNER. 1967. Vertical diurnal migration and endogenous rhythmicity. Science 157: 937-941.
FOX, H. M. 1957. Hemoglobin in Crustacea. Nature 179: 148.
FRYER, G. 1968. Evolution and adaptive radiation in the Chydoridae (Crustacea: Cladocera): A study in comparative functional morphology and ecology. Trans. R. Soc. Lond. Ser. B 254: 221-385.
———. 1974. Evolution and adaptive radiation in the Macrothricidae (Crustacea: Cladocera): A study in comparative functional morphology and ecology. Trans. R. Soc. Lond. Ser. B 269: 137-274.
GANNING, B., and F. WULFF. 1966. A chamber for offering alternative conditions to small motile aquatic animals. Ophelia 3: 151-160.
GOULDEN, C. E. 1968. The systematics and evolution of the Moinidae. Trans. Am. Phil. Soc. 58: 1-101.
HAERTEL, L. 1977. Effects of zooplankton grazing on nuisance algae blooms. U.S. Dep. Interior, Water Res, Technol., Proj. A-047-SDAK.
HALL, D. J., W. E. COOPER, and E. E. WERNER. 1970. An experimental approach to the production dynamics and structure of freshwater animal communities. Limnol. Oceanogr. 15: 839-928.
HARRIS, J. E. 1963. The role of endogenous rhythms in vertical migration. J. Mar. Biol. Assoc. U.K. 43: 153-166.
HEISEY, D., and K. G. PORTER. 1977. The effect of ambient oxygen concentration on filtering and respiration rates of *Daphnia galeata mendota* and *Daphnia magna*. Limnol. Oceanogr. 22: 839-845.
HERBERT, M. R. 1954. The tolerance of oxygen deficiency in the water by certain Cladocera. Mem. Ist. Ital. Idrobiol. 8: 97-107.
HORTON, P. A., M. ROWAN, K. E. WEBSTER, and R. H. PETERS. 1979. Browsing and grazing by cladoceran filter feeders. Can. J. Zoo. 57: 206-212.
HUTCHINSON, G. E. 1967. A treatise on limnology, v. 2. Wiley.
JOHNSON, D. M. 1973. Predation by damselfly naiads on cladoceran populations: fluctuating intensity. Ecology 54: 251-268.
KRING, R. L., and W. J. O'BRIEN. 1976. Effect of varying oxygen concentrations on the filtering rate of *Daphnia pulex*. Ecology 57: 808-814.
LANG, K. L. 1970. Distribution and dispersion of the Cladocera of Lake West Okoboji, Iowa. Ph.D. thesis, Univ. Iowa.
LANGFORD, R. R. 1938. Diurnal and seasonal changes in the distribution of the limnetic Crustacea of Lake Nipissing, Ontario. Univ. Toronto Stud. Biol. Ser. 45: 1-142.
LAROW, E. J. 1970. The effect of oxygen tension on the vertical migration of *Chaoborus* larvae. Limnol. Oceanogr. 15: 357-362.
LOEB, J. 1894. Über die kunstliche umwandlung

positiv heliotropischer Tiere in negativ heliotropischer und umgekehrt. Pfluegers Archiv. Gesamte Physiol. Menschen Tiere **54**: 81–107.

———. 1904. The control of heliotropic reactions in freshwater Crustacea by chemicals, especially CO_2. Univ. Calif. Pub. Physiol. **2**: 1–3.

LOFFLER, H. 1974. Der Neusiedlersee. Molden Wein, Munich.

McLAREN, I. A. 1963. Effects of temperatures on growth of zooplankton and the adaptive value of vertical migration. J. Fish. Res. Bd. Can. **20**: 685–727.

McNAUGHT, D. C., and A. D. HASLER. 1964. Rate of movement of populations of *Daphnia* in relation to changes in light intensity. J. Fish. Res. Bd. Can. **21**: 291–316.

MEYERS, D. G., and J. R. STRICKLER. 1978. Morphological and behavioral interactions between a carnivorous aquatic plant (*Utricularia vulgaris* L.) and a chydorid cladoceran (*Chydorus sphaericus* O.F.M.). Int. Ver. Theor. Angew. Limnol. Verh. **20**: 2490–2495.

MORRISON, D. F. 1967. Multivariate statistical methods. McGraw-Hill.

PHOENIX, D. 1976. Temporal dynamics of a natural multipredator-multiprey system. Ph.D. thesis, Univ. Pennsylvania.

PROSSER, C. L. 1973. Comparative animal physiology. Saunders.

RINGELBERG, J. 1964. The positively phototactic reaction of *Daphnia magna* Straus. A contribution to the understanding of vertical migration. Neth. J. Sea. Res. **2**: 319–406.

———. and H. SERVAAS. 1971. A circadian rhythm in *Daphnia magna*. Oecologia **6**: 289–292.

ROACH, L. S., and E. L. WICKLIFF. 1934. Relationship of aquatic plants to oxygen supply, and their bearing on fish life. Trans. Am. J. Fish. **64**: 370–378.

RYBAK, M., J. J. RYBAK, and K. TARWID. 1964. Differences in Crustacea plankton based on the morphological character of the littoral of the lakes. Ekol. Pol. Ser. A **12**: 159–172.

SHIREMAN, J. V., and R. G. MARTIN. 1978. Seasonal and diurnal zooplankton investigations of a south-central Florida lake. Florida Sci. **41**: 193–201.

SIEBECK, O. 1969. Spatial orientation of planktonic crustaceans. 1. The swimming behavior in a horizontal plane. Int. Ver. Theor. Angew. Verh. Limnol. **17**: 831–840.

SMIRNOV, N. N. 1971. Fauna of the U.S.S.R., Crustacea. Vol. 1, No. 2. Chydoridae. [Isreal Program Sci. Transl., Jerusalem.]

SZLAUER, L. 1963. Diurnal migrations of minute invertebrates inhabiting the zone of submerged hydrophytes in a lake. Schweiz. Z. Hydrol. **25**: 56–64.

VON BRAND, T. 1944. Occurrence of anaerobiosis among invertebrates. A review. Biodynamica **4**: 183–228.

WERNER, E. E., and D. J. HALL. 1976. Niche shifts in sunfishes: experimental evidence and significance. Science **191**: 404–406.

WESENBERG-LUND, C. 1926. Contributions to the biology and morphology of the genus *Daphnia*, with some remarks on heredity. K. Dan. Vidensk. Selsk. Skr. 2. Nat. Mat. Afd. 8 **11**: 89–252.

WHITESIDE, M. C. 1974. Chydorid (Cladocera) ecology: seasonal patterns and abundance of populations in Elk Lake, Minnesota. Ecology **55**: 538–550.

ZARET, T. M., and J. S. SUFFERN. 1976. Vertical migration in zooplankton as a predator avoidance mechanism. Limnol. Oceanogr. **21**: 804–813.

9. Aspects of Red Pigmentation in Zooplankton, Especially Copepods

J. Ringelberg

Abstract

Red coloration in copepods is caused by carotenoids, such as β-cryptoxanthin, hydroxyechinenone and astaxanthin, synthesized from β-carotene present in the algal diet. A diurnal rhythm in carotenoid content seems to be present in some species. A possible function of the pigment might be photoprotection. Also, the carotenoids might function as energy storage. Attention is drawn to the existence of highly pigmented types and bluish, translucent types either as different populations in nearby localities, as simultaneous in one population, or as succeeding one another in time.

The striking red coloration of some planktonic crustaceans has been described by early hydrobiologists like Elster (1896), Blaas (1923), and Brehm (1938). Diaptomids especially can be very heavily pigmented. The pigment is found in fat globules along the gut wall and in the carapace epidermis, the ovaries, the eggs, and the body fluid. Interest in this topic has recently been revived, and to stimulate a critical discussion some research is reviewed here.

Analysis by paper chromatography (Ringelberg and Hallegraeff 1976) and high-pressure liquid chromatography (Paanakker and Hallegraeff 1978) of pigments of highly red-colored *Acanthodiaptomus denticornis* from Lac Pavin (France) has shown the pigment to consist of several carotenoids. Since cladocerans and copepods cannot synthesize these carotenoids de novo, the animals depend on those present in the algal diet. Some of the carotenoids have been identified to a certain extent. In *A. denticornis* β-cryptoxanthin and hydroxyechinenone-like fractions were found. A very important fraction (33.9%) was positively identified as astaxanthin (Paanakker and Hallegraeff 1978). As could be expected, β-carotene from the algae was present in the extracts of the animals. It is highly probable the carotenoids are metabolized from β-carotene only, following an oxidative sequence with β-cryptoxanthin and hydroxyechinenone as intermediates, thus leading to astaxanthin (Hallegraeff pers. comm.) (Fig. 1). This is the case in *A. denticornis*. In *Daphnia longispina* from Lake Maarsseveen (Netherlands) other intermediates are involved, suggesting species-specific enzymatic processes (Hallegraeff et al. 1978).

The fact that metabolic intermediates were found suggests the red pigments are synthesized daily and certainly metabolized further into colorless low-energy products. Therefore, it is not surprising that a daily change in carotenoid content was found in *A. denticornis* of Lac Pavin (Hallegraeff et al. 1978) (Fig. 2). Minima in carotenoid content were observed around noon on the three succeeding days of observation in August 1976. During the night, the pigments were built up again. This rhythm seems to correspond with nighttime grazing as described by Haney and Hall (1975), Stark-

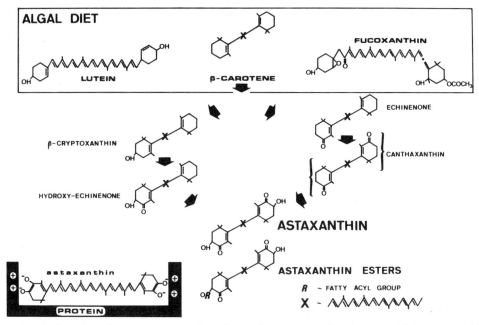

Fig. 1. Tentative scheme of biosynthetic pathway of metabolism of algal carotenoids into astaxanthin and astaxanthin esters, as major constituents of pigmentation in zooplankton. For theoretical reasons, it is unlikely that lutein and fucoxanthin are converted into astaxanthin. Most probable biosynthetic pathway of carotenoid metabolism in Crustacea would be an oxidative sequence, starting with β-carotene of dietary origin, involving hydroxy-carotenes (left, e.g., *Acanthodiaptomus*) or keto-carotenes (right, e.g. *Daphnia*) as intermediates, and ending in astaxanthin and astaxanthin esters, eventually as prosthetic groups of carotenoproteins. (G. M. Hallegraeff pers. comm.)

weather (1975), Duval and Geen (1975), and others. However, it is not as simple as that since the year before in Lac Pavin we found a minimum around midnight. In that case, the increase in pigmentation started after 0400-0500 hours (Fig. 3) (Ringelberg and Hallegraeff 1976). If diurnal feeding interacts with diurnal vertical migration as is supposed by Haney and Hall (1975) and Mackas and Bohrer (1976), it is not surprising to find the synchronizing of a carotenoid content rhythm changing, derived as it is of complicated rhythms in itself.

Now I would like to turn to another aspect of pigmentation in planktonic crustaceans. Brehm (1938), in a review on red-colored zooplankton, stressed its distribution in alpine mountain lakes and polar regions. In this context, one of the eight hypotheses on the finalistic aspects of red coloration that he mentioned is protection against ultraviolet light. Nearly 40 years later, experiments to test this hypothesis were started by, amongst others, Siebeck and Hairston. Siebeck (1978) using UV radiation found a higher lethal dose for the more pigmented *Daphnia pulex obtusa* than for the transparent *D. galeata* (Fig. 4). Moreover, he found the lethal dose is increased in both species if the animals are brought into daylight after UV exposure. This photoreactivation brings two photochemical mechanisms for UV protection to mind: an absorption mechanism transforming harmful radiation into harmless thermal energy, and a mechanism to absorb those radiation components essential for the reactivation process. Extensive series of experiments with visual light have been performed by Hairston (1976, 1978, 1979). He exposed a red and a more transparent morph of *Diaptomus nevadensis*, originating from different lakes, to blue

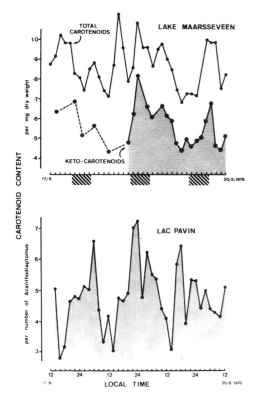

Fig. 2. Diurnal variation in carotenoid content of zooplankton of Lake Maarsseveen (Netherlands) (above) and Lac Pavin (France) (below). (From Hallegraeff et al. 1978.)

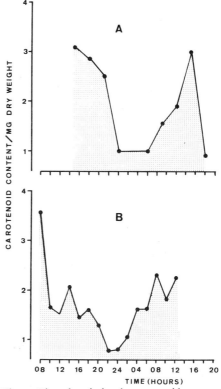

Fig. 3. Diurnal variation in carotenoid content of *Acanthodiaptomus* of Lac Pavin. A—18-19 August 1975; B—23-24 August 1975 (From Ringelberg and Hallegraeff 1976.)

fluorescent light. The results together with the dark control series are presented in Fig. 5. Especially at a high light intensity, survival is much lower than in the dark. Also, the red ones show a better survival than the translucent morphs. At low intensities, when mortality takes more time, the difference between the dark control and the light is less. Since no food was given, Hairston also mentioned starvation as an additional cause of mortality. The same kind of experiment was done with two different pigmented morphs of *Diaptomus kenai* to determine the mortality at a high intensity blue fluorescent light and at darkness (Hairston 1978). Again the red copepods have a better survival in the light; no significant difference between the two types was present in darkness. Starvation is mentioned as a possible additional factor of mortality in the light. Using *Diaptomus nevadensis* again, Hairston (1979) determined survival at two different temperatures. From these results (Fig. 6) we observe that mortality is higher in the light than in darkness, that it is higher at higher temperatures, and that it is higher in pale-colored copepods. Not much difference is present in the dark, neither between the two morphs nor at the two temperatures (with one exception). From all these clearcut results, Hairston concluded that the red carotenoid gives these copepods protection against high radiation near the surface.

This wealth of results certainly indicates that red pigmented copepods have a better survival in the light than translucent animals have. But do they also *prove* that the carotenoids have a light protection function? Let us assume starvation is indeed a factor of

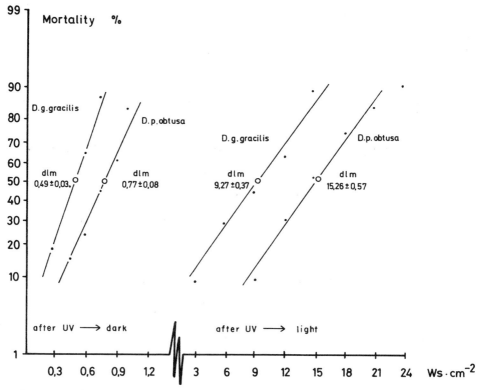

Fig. 4. Mortality imposed by ultraviolet radiation on *Daphnia galeata gracilis* and *D. pulex obtusa*. Horizontal axis: dose of UV radiation; vertical axis: percentage mortality. Results of two experiments are presented. Left—animals kept in dark after UV exposure; right—exposed to daylight after UV exposure (Siebeck 1978).

mortality and let us assume the keto-carotenoids act as a readily available source of energy (a hypothesis already mentioned: Brehm 1938). Furthermore, let us assume continuous light provides a stress situation (the higher the intensity, the greater the stress), leading to a higher metabolic activity as higher temperature also does. Then the red-colored animals can be expected to have a higher survival in the light, and a higher survival at high temperatures. Also both morphs must have a higher survival at low temperatures than at high temperatures. Nearly all results described by Hairston can also be interpreted in this way. Of course, new experiments must be devised to refute one or the other hypothesis.

Radiation protection is thought of as an advantage because it makes eating in the algal-rich surface layers possible. Indeed, Hairston (1978) found *D. nevadensis* and *D. sicilis* with the highest concentration of pigment near the surface. Is it right to assume these red animals are there because they can withstand high light intensities better than the pale ones found in deeper layers? Or were these red animals more pigmented because they recently had eaten? In many lakes of high transparency algal abundance is greatest at several meters depth. For instance, during summer the highest chlorophyll *a* concentration and the highest algal volume in Lac Pavin is found around 15 m. In August primary production was highest at 5 m (Flik et al. 1973; Hallegraeff et al. 1976). Also, as was the case in Lac Pavin, the most heavily pigmented copepods were found below 30 m.

To emphasize the possible function of

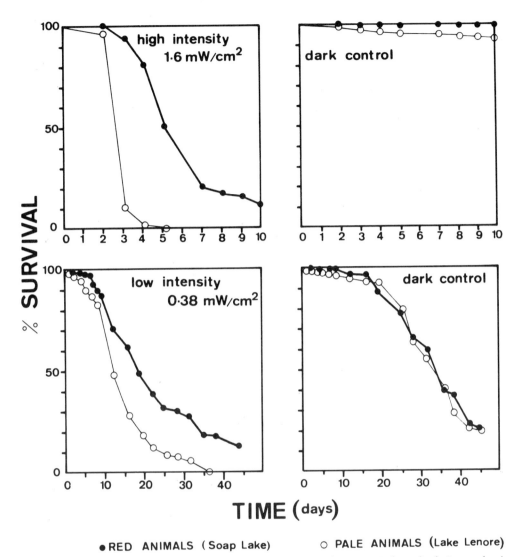

Fig. 5. Effect of continuous light on red *Diaptomus nevadensis* from Soap Lake and pale *D. nevadensis* from Lake Lenore. (From Hairston 1976.)

carotenoids as an energy storage product, the preliminary results of a temperature experiment are presented here. If the keto-carotenoids are readily metabolized, temperature must influence discoloration. Plastic bags with freshly caught *A. denticornis* were placed at 10°, 15°, and 25°C. After 12, 24, 48, and 72 hours, animals from a bag at each temperature were used for pigment analysis. As Fig. 7 shows, carotenoid content decreases more rapidly at 25°C than at the lower temperatures. Though I believe this experiment must be repeated, the results do not refute the hypothesis that the red keto-carotenoids are metabolized.

The presence of two morphs, a red one and a transparent bluish one of the same species, presents a problem in itself. Populations of the two types have been described living separately in different lakes but within the same region (Hairston 1976). Elster (1896) mentioned that the population of

96   J. Ringelberg

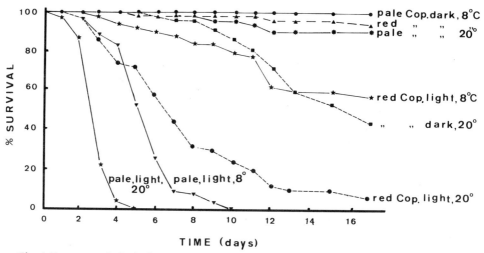

Fig. 6. Percent survival of *Diaptomus nevadensis* with large and small amounts of carotenoid pigment when exposed to blue fluorescent light at two temperatures. (Redrawn from Hairston 1979.)

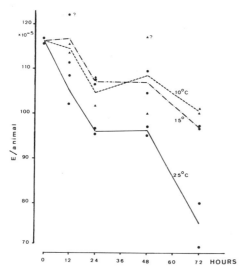

Fig. 7. Decrease at three temperatures of carotenoid content in *Acanthodiaptomus denticornis* from Lac Pavin. Ordinate—concentration in arbitrary extinction units per animal.

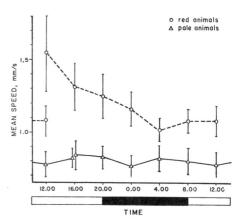

Fig. 8. Mean swimming speed of two morphs of *Acanthodiaptomus denticornis* from Lac Pavin. Vertical bars give ± 2 SE. (Gerritsen unpubl.)

red-colored *Diaptomus superbus* changes into a population of animals with a bluish tint during May. *Acanthodiaptomus denticornis* from Lac Pavin, on the other hand, is more pigmented in summer (Berthon 1975), although in August about 20% of the animals are of a transparent, bluish color (Ringelberg unpubl.). What is the exact status of these morphs in all these cases?

Probably the metabolism is different in the two morphs and, for certain, the behavior is different, as is shown by the observations of Gerritsen (unpubl.) with animals from Lac Pavin (Fig. 8).

I hope this short review stimulates a critical attitude toward the phenomenon of pigmentation in zooplankton.

References

BERTHON, J. L. 1975. Repartition spatio-temporelle, nutrition, composition chimique du zooplankton pelagique de Lac Pavin. Thesis, Univ. Clermont-Ferrand.

BLAAS, E. 1923. Über die Rotfärbung von *Diaptomus vulgaris*. Arb. Zool. Inst. Univ. Innsbruck **1**: 1-19.

BREHM, V. 1938. Die Rotfärbung von Hochgebirgsee-organismen. Biol. Rev. **13**: 307-318.

DUVAL, W. S., and G. H. GEEN. 1975. Diel rhythms in the feeding and respiration of zooplankton. Int. Ver. Theor. Angew. Limnol. Verh. **19**: 518-524.

ELSTER, A. 1896. Über einen Fundort von *Diaptomus superbus* nebst einigen Bemerkungen über Farben der Copepoden. Zool. Anz. 31.

FLIK, B. J. G., G. M. HALLEGRAEFF, and R. LINGEMAN. 1973. Limnological notes on Lac Pavin (Auvergne, France). Ann. Sta. Biol. Besse-en-Chandesse **8**: 119-147.

HAIRSTON, H. G., JR. 1976. Photoprotection by carotenoid pigments in the copepod *Diaptomus nevadensis*. Proc. Natl. Acad. Sci. **73**: 971-974.

———; 1978. Carotenoid photoprotection in *Diaptomus Kenai*. Int. Ver. Theor. Angew. Limnol. Verh. **20**: 2541-2545.

———. 1979. The effect of temperature on carotenoid photoprotection in the copepod *Diaptomus nevadensis*. Comp. Biochem. Physiol. **62A**: 445-448.

HALLEGRAEFF, G. M., R. LINGEMAN, and B. J. G. FLIK. 1976. Physical characteristics, phytoplankton standing crop and primary productivity at the end of summer in four French crater lakes. Ann. Sta. Biol. Besse-en-Chandesse **10**: 251-265.

———, I. J. MOUS, R. VEEGER, B. J. G. FLIK, and J. RINGELBERG. 1978. A comparative study on the carotenoid pigmentation of the zooplankton of Lake Maarsseveen (Netherlands) and of Lac Pavin (Auvergne, France). II. Diurnal variations in carotenoid content. Comp. Biochem. Physiol. **60B**: 59-62.

HANEY, J. F., and D. J. HALL. 1975. Diel vertical migration and filter-feeding activities of *Daphnia*. Arch. Hydrobiol. **75**: 413-441.

MACKAS, D., and R. BOHRER. 1976. Fluorescence analysis of zooplankton gut contents and an investigation of diel feeding patterns. J. Exp. Mar. Biol. Ecol. **25**: 77-85.

PAANAKKER, J. E., and G. M. HALLEGRAEFF. 1978. A comparative study on the carotenoid pigmentation of the zooplankton of Lake Maarsseveen (Netherlands) and of Lac Pavin (Auvergne, France). I. Chromatographic characterization of carotenoid pigments. Comp. Biochem. Physiol. **60B**: 51-58.

RINGELBERG, J., and G. M. HALLEGRAEFF. 1976. Evidence for a diurnal variation in carotenoid content of *Acanthodiaptomus denticornis* (Crustacea, Copepoda) in Lac Pavin (Auvergne, France), Hydrobiologia **51**: 113-118.

SIEBECK, O. 1978. UV-Toleranz und Photoreaktivierung bei Daphnien ans Biotopen verschiedener Höhrenregionen. Naturwissenschaften **65**: 390-391.

STARKWEATHER, P. L. 1975. Diel patterns of grazing in *Daphnia pulex* Leydig. Int. Ver. Theor. Angew. Limnol. Verh. **19**: 2851-2857.

10. The Vertical Distribution of Diaptomid Copepods in Relation to Body Pigmentation

Nelson G. Hairston, Jr.

Abstract

Carotenoid pigmentation has previously been shown to protect diaptomid copepods from injury by low intensity visible light. Predicted relationships between the vertical distribution of *Diaptomus sicilis* and *Diaptomus nevadensis*, their pigment contents, and the intensity of sunlight were studied in two lakes. As expected, on individual dates the most pigmented copepods of both species were found closest to the lakes' surfaces, while annually the average depth of *D. nevadensis* adults varied directly with solar intensity and inversely with their pigment content. The depth distributions of *D. sicilis* did not follow the predicted annual pattern. Since *D. nevadensis* is predatory, it is proposed that *D. sicilis* migrated oppositely to them as an avoidance strategy. Data on pigmented copepods from four other lakes support this interpretation. Laboratory and field observations of red diaptomid copepods suggest that photodamage may be a more important selective force than previously supposed.

A primary function of the red carotenoid pigment found in some populations of *Diaptomus nevadensis* Light and *Diaptomus kenai* Wilson is photoprotection. I have shown that under a variety of light and temperature conditions, darkly pigmented copepods survived significantly longer than pale copepods when both were exposed to natural intensities of visible blue light (Hairston 1976, 1978, 1979c). Supporting a photoprotective function is the observation that copepods fed a carotenoid-rich food gained pigment at a significantly higher rate when exposed to light than when kept in the dark, even though feeding rates in the two treatments were comparable (Hairston 1979a), and the observation that pale copepods showed a greater increase of avoidance behavior (accelerated swimming speed, sinking) than red copepods when both were exposed to photodamaging wavelengths of light (Hairston 1976).

These laboratory results were used to support the hypothesis that diaptomid copepods were pigmented red in lakes with few or no visually oriented predators as a result of natural selection by solar radiation, while in lakes with many predators such pigmentation made the copepods particularly susceptible to being consumed and was selected against (Hairston 1979a).

Carotenoid photoprotection is well documented in studies of bacteria and plants, and the physical and chemical mechanisms by which damage occurs and against which the pigment protects are known (Krinsky 1971). In photodynamic action, light is absorbed by a sensitizer molecule within the organism which passes some of the energy to oxygen. Excited oxygen binds with many different

---

This study was supported by ERDA contract AT(45-1)-2225-T23 to W. T. Edmondson.

intracellular compounds from water to DNA, rendering them functionless or even toxic. The wavelengths of light that cause the most damage are those absorbed most strongly by the sensitizer molecule. Known sensitizers such as porphyrins, flavins, and quinones have absorption maxima in the blue region of the spectrum. Carotenoids are active in photoprotection as quenchers of singlet-excited oxygen and are effective at protecting against photodamage by wavelengths from blue through near ultraviolet (Thomas 1977).

Several investigators have suggested a photoprotective function for copepod carotenoid pigments. Bright red diaptomid copepods occur in many high mountain lakes (Zschokke 1900) and both Hutchinson (1939) and Fairbridge (1943) proposed that pigmentation might serve as a defense against the intense sunlight of that region. Elster (1931) and Hutchinson (1937) suggested that pigmented copepods were found closer to the surfaces of the lakes they studied than were pale copepods, because the former were protected from damage by light. Berthon (1978) found that *Acanthodiaptomus denticornis* Wierzejski were most pigmented in midsummer when solar radiation was greatest and suggested that the carotenoid was photoprotective. The carotenoid content of adult *Diaptomus* often changes seasonally. The usual pattern is for the animals to be most pigmented in the winter and spring and least pigmented in the summer and fall (Brehm 1938; Siefken and Armitage 1968). I found for *D. nevadensis* and *D. sicilis* Forbes that the period of greatest pigmentation coincided with the period of greatest egg production and that the eggs and naupliar larvae contained high carotenoid concentrations (Hairston 1979b). Since the nauplii reside at the lake surface during the day, it is adaptive for the female to provide them with pigment as a protection from photodamage.

The function of carotenoids in protecting copepods from injury by solar radiation, as demonstrated in the laboratory, leads to some simple predictions for the field about the relationship between the pigmentation of copepods and their depth in the water column. It is expected that when sunlight is intense, individual copepods near the lake surface should have more pigment than those at greater depth. This should be true because individuals containing little pigment would be expected to avoid light more than those with much pigment. Furthermore, it might be expected that on a seasonal basis, there should be an inverse relationship between the average pigmentation, P, of individuals in a population and their average depth, D, in the water column ($P \propto 1/D$); a direct relationship between average depth and solar intensity, I, at the lake surface ($D \propto I$); and a direct relationship between average pigment content and solar intensity ($P \propto I$); Further consideration of these relationships shows, however, that all three cannot be true simultaneously. Accepting any two necessitates that the third be false. For example, $P \propto 1/D$ and $D \propto I$ implies $P \propto 1/I$, not the predicted $P \propto I$. For *D. nevadensis* and *D. sicilis* I have already shown that the relationship $P \propto I$ must be false since, as a result of their reproductive cycle, these copepods are most pigmented in the winter when solar intensity is least. Given this imposed seasonal pigmentation cycle it is expected that the other two predictions will be true. Copepod population depth will be inversely proportional to pigmentation (animals will occur at the greatest depth when they are least pigmented) and average copepod depth will be directly proportional to solar intensity (animals will occur at the greatest depth when sunlight is most intense). My purpose here is to report the results of a study of the pigmentation and vertical distributions of *D. nevadensis* and *D. sicilis* and to relate these to information available for several other species.

I thank W. T. Edmondson and M. Griffiths for advice and the use of their laboratories. R. A. Pastorok and B. E. Taylor provided data and ideas about the zooplankton of Lake McDonald and Hall Lake. I also benefited from discussions with J. T. Lehman and B. W. Frost.

Methods

The vertical distributions of *D. nevadensis* and *D. sicilis* were studied in Soap Lake and Lake Lenore, central Washington, USA, as a part of a larger investigation of copepod pigmentation (Hairston 1979a). Soap Lake is meromictic with a mixolimnion 20 m deep (salinity 17 g · liter^{-1}) and a monomolimnion 6 m thick. The lake is 3.6 km long, 1.3 km wide, has a mean depth of 7.5 m, and during the years of this study stratified thermally between June and September. Lake Lenore is holomictic, with a salinity of 1.7 g · liter^{-1}. It is 9.2 km long, 0.9 km wide, has a mean depth of 6.5 m, and only stratifies for brief periods during the summer. Both Soap Lake and Lake Lenore are described in more detail by Walker (1975).

Field trips were made on 22 dates between September 1973 and October 1974. Quantitative diagonal hauls were taken with a Clarke-Bumpus plankton sampler (135-μm-mesh nylon net) to collect copepods between 20 m and the surface of Soap Lake at 5-m intervals, and between 7 m and the surface of Lake Lenore at the intervals 7–5 m, 5–2.5 m, and 2.5–0 m. Additional plankton hauls were taken in each lake and kept alive in an ice chest for later pigment analysis. Solar intensity was recorded using a pyrheliometer. For each lake on each date pigment was extracted from about 100 adult *D. nevadensis* and about 200 adult *D. sicilis*. The extractions were made in 100% ethanol according to the method described by Hairston (1979a). Dry weights were found for 50 *D. sicilis* and 10 *D. nevadensis* on each date and pigment content of the copepods was determined in units of optical density (absorbance at 474 nm) per mg dry weight.

On 23 and 24 April 1976 and on 26 and 27 August 1976, animals were collected at 0, 1, 2, 4, and 7 m in Lake Lenore and at 0, 1, 2, 4, 8, and 16 m in Soap Lake. Four replicate pigment extractions were made on each species from each depth. Forty to 50 adult *D. sicilis* or 10 to 20 adult *D. nevadensis* were homogenized in 100% ethanol in each extraction. At depths where animals did not occur, no extraction was made. On 24 April in Lake Lenore, *D. nevadensis* was rare and samples from 1 and 2 m and from 4 and 7 m had to be mixed together to get enough animals for replicate extractions. On 27 August, in this lake, *D. nevadensis* was too rare to make any extractions. The 1-m sample of *D. sicilis* and *D. nevadensis* in Soap Lake collected on 26 August was partially decomposed when returned to the lab. The *D. sicilis* was discarded and the *D. nevadensis* values may be low due to loss of pigment. Quantitative plankton hauls were made at each depth to determine the vertical distributions of the copepods. In both lakes the samples were taken at midday in an order alternating between deep and shallow so that no sampling bias with time could occur.

Diel sampling series were made on several dates in both lakes. Plankton hauls were taken at noon, sunset, midnight, and sunrise from the same depths as those in the regular sampling scheme.

Results and Discussion

Pigmentation, solar intensity, and vertical distribution—Figures 1 and 2 show the relationships between the pigmentation and depth of *D. sicilis* and *D. nevadensis* in April and August 1976. For each species and date, the pigment content of copepods from the upper three depths was compared with that for the lower two depths using the Mann-Whitney U-test. In all cases but one, the animals nearer to the lake surface tended to be more pigmented than those from greater depth. In April there was no apparent relationship between depth and pigmentation of Soap Lake *D. sicilis* (U-test, $P = 0.34$), while in August it was pronounced (U-test, $P = 0.001$). On both dates the relationship was significant for Lake Lenore *D. sicilis* (U-test, April $P = 0.007$, August $P = 0.002$) as it was for Soap Lake *D. nevadensis* (U-test, April $P = 0.01$, August $P = 0.03$). In Lake Lenore the *D. nevadensis* collected in April appeared to be most pigmented near to the surface, but the difference was not significant, probably due to small sample size (U-test, $P = 0.06$). Thus in five of seven cases

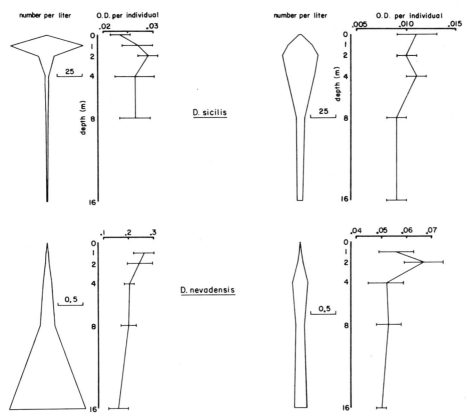

Fig. 1. Carotenoid content (± 1 SD) of *D. sicilis* and *D. nevadensis* adults from a series of depths in Soap Lake on two dates in 1976. *Diaptomus nevadensis* from 1 m on 26 August partially decomposed before analysis, so value may be too low.

the results support the hypothesis that copepods found near the lake surface should be those with the most pigment and therefore the greatest amount of photoprotection. The lack of a relationship in April for *D. sicilis* in Soap Lake was probably due to the large amount of pigment it contained at this time of year compared with August (Fig. 1) as well as to the relatively low transparency and surface light intensity on this date (Table 1).

The two species were distributed in April and August 1976 within the water columns of Soap Lake and Lake Lenore as if their carotenoid content was important in determining their sensitivity to sunlight. The second two hypotheses proposed in the introduction are that, over the course of a year, the average depths of the copepods should change in inverse proportion to their pigmentation and in direct proportion to the

Table 1. Solar intensities and Secchi disk transparencies at time of stratified hauls for copepod carotenoid content in Soap Lake and Lake Lenore. On 24 April 1976 the pyrheliometer pen failed.

	$cal \cdot cm^{-2} \cdot min^{-1}$	Secchi depth (m)
Soap Lake		
23 Apr 76	0.45	1.4
26 Aug 76	1.20	7.3
Lake Lenore		
24 Apr 76	—	5.1
27 Aug 76	0.80	2.5

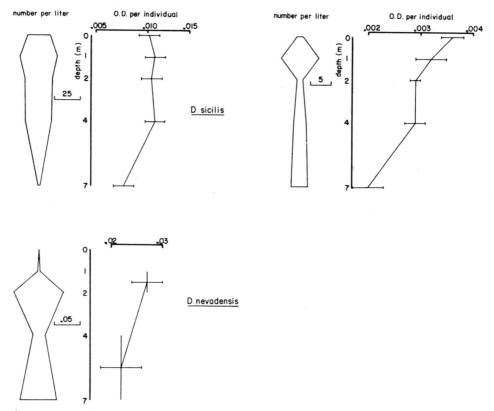

Fig. 2. Carotenoid content (± 1 SD) of *D. sicilis* and *D. nevadensis* adults from a series of depths in Lake Lenore on two dates in 1976.

intensity of sunlight. These predictions are supported by the data on *D. nevadensis* in both lakes. The vertical distribution changed seasonally in a way consistent with the hypothesis that the animals were avoiding strong summer sunlight (Fig. 3A, B). The average depth of the population in Soap Lake was greatest between July and September when light intensity and Secchi disk transparency were maximum (Figs. 4 and 5). In Lake Lenore, the *D. nevadensis* population depth was greatest between May and August, and Secchi disk transparency varied little during this period. Although the summer solstice occurs in June, the prevalence of cloud cover in the spring decreased sunlight during this period relative to summer.

In Soap Lake, which is deep and in which the copepods contained more pigment, there was a tendency for the *D. nevadensis* mean population not to continue to descend in higher light intensities beyond about 12 m. No such trend was observed in shallow Lake Lenore, and in midsummer *D. nevadensis* was abundant only near the lake bottom.

In winter, the average depths of the *D. nevadensis* populations in the two lakes moved upward. This rise was mainly the result of a tendency toward an even distribution more than a synchronized population ascent. Both Soap Lake and Lake Lenore copepods were most pigmented during the winter (Hairston 1979b, 1979c). Thus when copepod carotenoid content was greatest and winter solar intensity was least, the *D. nevadensis* appeared to be restricted to no

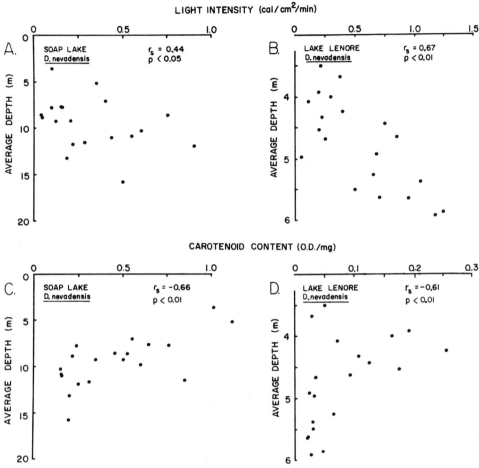

Fig. 3. Average depth of *D. nevadensis* adults in Soap Lake and Lake Lenore as a function of body pigmentation and of light intensity at lake surface at time of sampling.

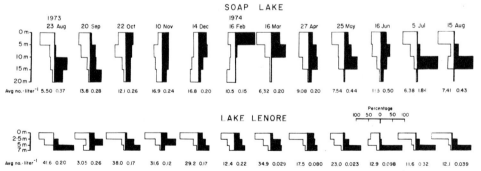

Fig. 4. Daytime vertical distributions of *D. nevadensis* (darkened) and *D. sicilis* (open) adults during August 1973 to August 1974.

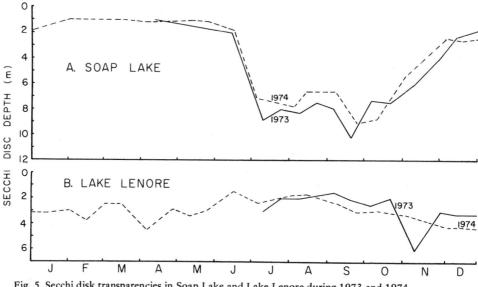

Fig. 5. Secchi disk transparencies in Soap Lake and Lake Lenore during 1973 and 1974.

particular depth. This tendency was most pronounced in Soap Lake where pigmentation was greatest (Fig. 3C, D).

While the vertical distributions of the *D. nevadensis* populations conformed to the predicted relationships with solar intensity and body pigmentation, the average depths of *D. sicilis* in the two lakes did not (Fig. 6). The population in Soap Lake was nearest to the surface when sunlight was most intense and descended during the winter (Fig. 4). In Lake Lenore, the depth of *D. sicilis* changed little throughout the year and no correlation with sunlight was found. The pigmentation of Soap Lake *D. sicilis* did not vary seasonally and was not correlated with the copepod's average depth in the lake, while in Lake Lenore the copepods were found deepest when pigmentation was greatest. Thus, although *D. sicilis* was distributed on single days in April and August 1976 as though pigmentation and sunlight were related, the populations did not behave as expected when studied over the course of a year.

The predator-prey relationship between D. nevadensis *and* D. sicilis.—The lack of expected correlations between depth, light intensity, and pigmentation for *D. sicilis* does not necessitate the conclusion that *Diaptomus* in lakes are insensitive to light or that carotenoid pigmentation does not moderate the reaction. Photosensitivity is only one of many factors influencing the vertical distribution of copepods (e.g. Hutchinson 1967). *Diaptomus nevadensis* behaves as if damage by sunlight were an important selective force. The average population depth of *D. sicilis* is apparently influenced principally by factors other than sunlight, but within the populations sampled in April and August 1976, pigmented animals were still distributed as if light were important.

Diaptomus sicilis is a significant prey item in the diet of *D. nevadensis* and its behavior may be strongly influenced as a result. In laboratory experiments adult *D. sicilis* were consumed by adult *D. nevadensis* at a rate of 0.5 prey $\cdot$ predator^{-1} $\cdot$ day^{-1} at natural prey densities, and at rates as high as 5 prey $\cdot$ predator^{-1} $\cdot$ day^{-1} at elevated prey densities (unpublished data; Anderson 1970 obtained similar results). In Soap Lake and Lake Lenore the importance of *D. sicilis* in the diet of *D. nevadensis* varied depending upon its abundance relative to other prey species. Over the period of this study (22 dates in 1973 and 1974), *D. sicilis* was found in 41% of the guts of *D.*

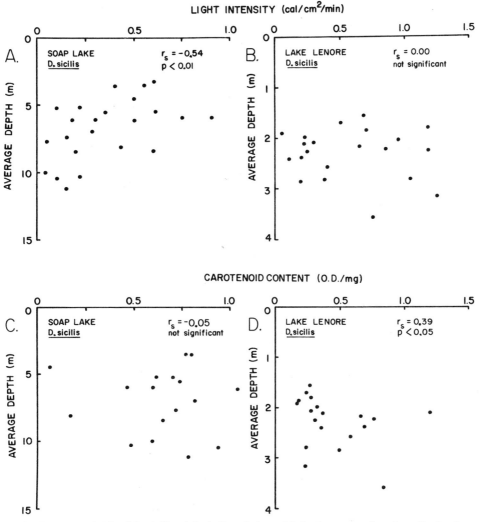

Fig. 6. Average depth of *D. sicilis* adults in Soap Lake and Lake Lenore as a function of body pigmentation and of light intensity at lake surface at time of sampling.

nevadensis collected from Soap Lake and 56% of the guts of *D. nevadensis* from Lake Lenore (Hairston 1979a).

If predation by *D. nevadensis* were a more important source of mortality than photodamage, the vertical distribution of *D. sicilis* would be expected to reflect this. An individual copepod may avoid being consumed by staying at a depth where the predators are not found. When predators are constrained to reside deep in the lake, the prey should stay close to the surface. As the predator population ascends, the prey should descend. Prey would be exposed to predation as the two populations pass, but only during the period of overlap. If the predators are not restricted to any depth, the prey have no refuge, but the encounter of prey by predators would be minimized if the prey population were evenly distributed. Under these conditions the predator population should also distribute evenly. This argument may attract an explanation based on group selection, since an even distribution might require that every individual in the population know the location of all others, but individual selection may be postulated if each individual is assumed to swim randomly throughout the range of the population. In Soap Lake and Lake Lenore, the vertical distributions of *D. sicilis* and *D. nevadensis* followed the expected pattern (Fig. 4). In summer, *D. nevadensis* was found deep in lakes while *D. sicilis* was close to the surface. In winter when the *D. nevadensis* population ascended and became more evenly distributed, *D. sicilis* also distributed more evenly. The winter ascent of *D. nevadensis* was much less marked in Lake Lenore than in Soap Lake, and accordingly the descent of *D. sicilis* in Lake Lenore was less marked than in Soap Lake.

During summer, *D. sicilis* was exposed in both lakes to more intense sunlight than *D. nevadensis*. In Lake Lenore the average summer pigment content of *D. sicilis* was 9 times that of *D. nevadensis*. In Soap Lake it was 2 times greater. During winter when predator and prey experienced more similar light regimes this difference was reduced to 3-fold in Lake Lenore and eliminated in Soap Lake. Note that at this time of year in Lake Lenore, *D. sicilis* was still located closer to the surface than *D. nevadensis*. It must be borne in mind, however, that other factors are also important in influencing pigmentation, including the presence of visually oriented predators, the timing of reproduction, and the nutritional state of the copepods (Hairston 1979a, b; Ringelberg and Hallegraeff 1976).

Vertical diel migration—Since *D. sicilis* and *D. nevadensis* showed contrasting patterns of vertical distribution seasonally, their diel migration patterns are naturally of interest. The predictions made for seasonal changes in distribution should also hold diurnally. Studies were carried out on six dates at Soap Lake and on four dates at Lake Lenore. Representative examples of the data obtained are given in Fig. 7 (all data are presented in Hairston 1977).

Although diel migrations were not extensive in either lake, the trends were consistent with the predictions. At Soap Lake in July neither predator nor prey migrated. *Diaptomus nevadensis* remained deep while *D. sicilis* stayed near the surface. On both the December and June dates at Soap Lake and the August and December dates at Lake Lenore, slight upward migrations of *D. nevadensis* were accompanied by slight reverse migrations by *D. sicilis*. In 1976 *D. nevadensis* became rare in Lake Lenore while the larvae of *Chaoborus flavicans* (Meigen) became relatively abundant. On 10 July these predators migrated up to an even distribution at midnight while their potential prey, *D. sicilis*, migrated down, also to an even midnight distribution.

Pigmented species of *Diaptomus* are found in many lakes (e.g. Brehm 1938) and it may be that migrating invertebrate predators partially specify the niches of many of them. In order to examine the generality of the behavior pattern exhibited by *D. sicilis*, data were obtained from studies of four additional lakes that described the vertical distributions of red copepods and invertebrate predators (Fig. 8). In Hall Lake, Washington, third and fourth instars of *Chaoborus trivittatus* Saether underwent a distinct diel migration, moving from deep water during the day to the surface at night. Over the same period, the red copepod *D. hesperus* Wilson and Light did a reverse migration (B. E. Taylor pers. comm.). A very darkly pigmented population of *D. kenai* in Findley Lake, Washington, was studied by Pederson (1974). *Chaoborus trivittatus* was abundant in the lake and individuals examined had obviously red guts. Although diel changes in the distribution of these predators were not studied, the prey showed a reverse migration. In Lake McDonald, Washington, third and fourth instars of *Chaoborus* migrated normally while the prey species, *D. franciscanus* Lilljeborg, moved down at night, though not as dramatically as in Hall Lake (R. A. Pastorok pers. comm.). Fedorenko (1975a, b) studied the vertical distributions and diets of *C. trivittatus* and *C. americanus* (Joh.) as well as the vertical distributions of their prey *D. kenai* and *D. tyrelli* Poppe in Eunice Lake, British Columbia. Fourth instars of *C. trivittatus* migrated normally while *D. kenai* underwent

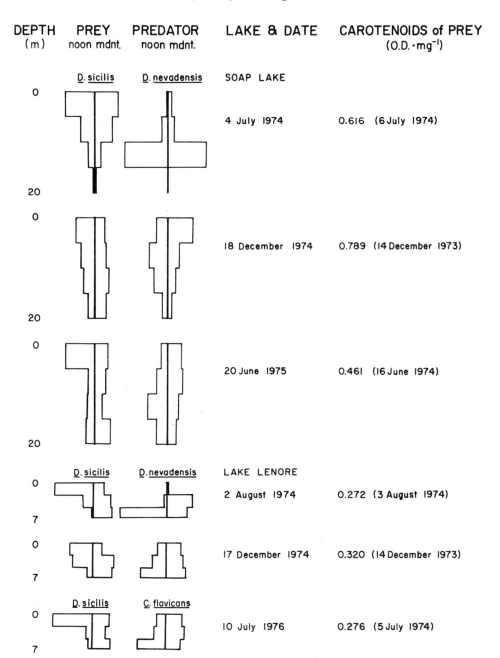

Fig. 7. Summary of diel migrations of *D. sicilis*, *D. nevadensis*, and *C. flavicans* in Soap Lake and Lake Lenore. Carotenoid content of prey determined using animals collected by a vertical plankton haul from entire water column.

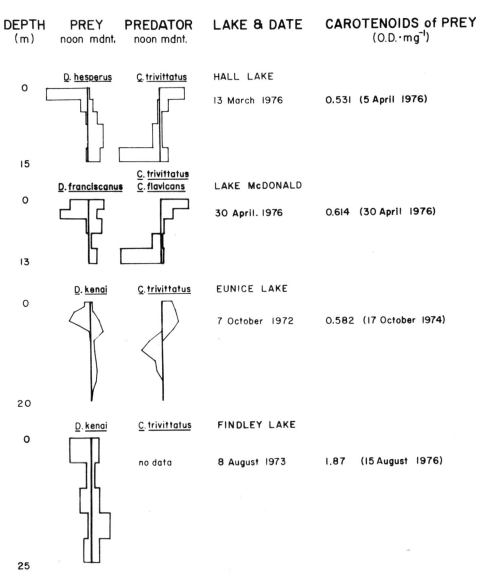

Fig. 8. Summary of diel migrations of pigmented copepods in Hall Lake (B. E. Taylor pers. comm.), Lake McDonald (R. A. Pastorok pers. comm.), Eunice Lake (Fedorenko 1975a), and Findley Lake (Pederson 1974). I determined carotenoid content of prey using animals collected from entire water column.

a distinct reverse migration. The smaller copepod, *D. tyrelli*, and the less abundant predator, *C. americanus*, did not migrate. Fedorenko's studies point out what may be an essential factor in the strategy of reverse migration to escape predation. If a prey population is to avoid the predator population by being at a different depth most of the time, it may be necessary that there be other prey species in the water column that do not migrate. Otherwise, at night the predators would follow the migrating prey, leading to an even distribution of both as seen in Soap Lake and Lake Lenore. In Eunice Lake the migrating prey species, *D. kenai*, made up 38% of the diet of *C. trivittatus*, but the predators accounted seasonally for a large amount of mortality in the

D. kenai population. *Diaptomus tyrelli* and other nonmigrating prey made up 62% of the predator's diet, but suffered relatively low mortality. In Lake McDonald, *D. franciscanus* comprised an average 8% of the diet of *Chaoborus*, while the nonmigrating *Daphnia pulicaria* Forbes made up 88%, but the relative effects on mortality of the prey are not known. In these lakes, selection pressure from *Chaoborus* predation produced reverse migration in populations of pigmented copepods. At the same time, *Chaoborus* obtained most of its diet from nonmigrating prey, but apparently did not produce sufficient mortality to have a selective effect on the vertical distributions of the prey.

Conclusions

Diel change in solar intensity is widely accepted as the proximate cue for the vertical movement of zooplankton (Hutchinson 1967), but has been rejected in the past as its ultimate cause (Marshall and Orr 1955), except in relation to visual predators such as fish. The information presented here and in other studies suggest that sunlight may be more important than previously supposed. Low intensity visible light damages a variety of crustaceans including copepods (Harvey 1930; Marshall and Orr 1955), cladocerans (Merker 1925), mysids (Smith 1970), and ostracods (Maguire 1960), and the quantity of light known to be lethal to diaptomids may penetrate more than 20 m in clear lakes (Hairston 1976). The observation that at least some species contain pigment that protects them from photodamage (Hairston 1976; Siebeck 1978) further attests to the importance of light as a selective force, particularly since predators such as fish should select strongly against such a visible characteristic. In this study pigmented copepods occurred in lakes as if solar radiation were an important determinant of their distribution. Individuals containing the most carotenoid were found nearest the surface; those with less pigment were located deeper. Populations residing near the surface during the day, apparently in response to invertebrate predation, were predictably highly pigmented. In a study of *Daphnia* populations in high altitude Himalayan lakes, Löffler (1969) found darkly colored *D. tibetana* in shallow ponds and clear lakes and unpigmented species of *Daphnia* in turbid lakes. If photodamage had been unimportant, no relationship would have been expected between water clarity and zooplankton pigmentation, and if fish predation had been significant, the least pigmented animals should have been found in the clearest water.

Direct injury by sunlight is attractive as an explanation for the vertical distributions of zooplankton because it links the proximate cue to the ultimate cause. Experimental support for this hypothesis is now slight because few studies of the damaging effects of sunlight have been carried out. Unambiguous field data will be difficult to collect, since nearly all evidence showing light to be the ultimate cause of migration would also support it as only the proximate cue. Carotenoid pigmentation adds a useful independent variable with which to formulate and test hypotheses.

References

ANDERSON, R. S. 1970. Predator-prey relationships and predation rates for crustacean zooplankters from some lakes in western Canada. Can. J. Zool. 48: 1229-1240.

BERTHON, J. L. 1978. Variations saisonnières en pigments caroténoides chez le Copépode *Acanthodiaptomus denticornis* dans un lac oligothrophe du Massif Central (Le Pavin). C. R. Acad. Sci. Paris 286: 899-900.

BREHM, V. 1938. Die Rotfärbung von Hochgebirgseeorganismen. Biol. Rev. 13: 307-318.

ELSTER, H. J. 1931. Über einen Fundort von *Diaptomus superbus* Schmeil, nebst einigen Bemerkunger über die Farben der Copepoden. Zool. Anz. 96: 245-251.

FAIRBRIDGE, W. S. 1943. West Australian fresh water calanoids (Copepoda). J. R. Soc. W. Aust. 29: 25.

FEDORENKO, A. Y. 1975a. Instar and species-specific diets in two species of *Chaoborus*. Limnol. Oceanogr. 20: 238-249.

———. 1975b. Feeding characteristics and predation impact of *Chaoborus* (Diptera, Chaoboridae) larvae in a small lake. Limnol. Oceanogr. 20: 250-258.

HAIRSTON, N. G., JR. 1976. Photoprotection by carotenoid pigments in the copepod *Diaptomus nevadensis*. Proc. Natl. Acad. Sci. 73: 971-974.

———. 1977. The adaptive significance of carotenoid pigmentation in *Diaptomus* (Copepoda). Ph.D. thesis, Univ. Wash., Seattle. 230 p.

———. 1978. Carotenoid photoprotection in *Diaptomus kenai*. Int. Ver. Theor. Angew. Limnol. Verh. 20: 2541-2545.

———. 1979a. The adaptive significance of color polymorphism in two species of *Diaptomus* (Copepoda). Limnol. Oceanogr. 24: 15-37.

———. 1979b. The relationship between pigmentation and reproduction in two species of *Diaptomus* (Copepoda). Limnol. Oceanogr. 24: 38-44.

———. 1979c. The effect of temperature on carotenoid photoprotection in the copepod *Diaptomus nevadensis*. Comp. Biochem. Physiol. 62A: 445-448.

HARVEY, J. M. 1930. The action of light on *Calanus finmarchicus* (Gunner.) as determined by its effect on the heart rate. Contrib. Can. Biol. N. S. 5: 83-93.

HUTCHINSON, G. E. 1937. Limnological studies in Indian Tibet. Int. Rev. Gesamten Hydrobiol. 35: 134-177.

———. 1939. Addendum. In F. Kiefer, Scientific results of the Yale North India Expedition. Mem. Indian Mus. 13: 83-203.

———. 1967. A treatise on limnology, v. 2. Wiley.

KRINSKY, N. I. 1971. Function, p. 669-716. *In* O. Isler [ed.], Carotenoids. Birkhauser.

LÖFFLER, H. 1969. High altitude lakes in Mt. Everest region. Int. Ver. Theor. Angew. Limnol. Verh. 17: 373-385.

MAGUIRE, B., JR. 1960. Lethal effect of visible light on cavernicolous ostracods. Science 132: 226-227.

MARSHALL, S. M., and A. P. ORR. 1955. The biology of a marine copepod. Oliver and Boyd.

MERKER, E. 1925. Die Empfindlichkeit feuchthäutiger Tiere im Lichte. Zool. Jahrb. Abt. Allg. Zool. Physiol. Tiere 42: 217-222.

PEDERSON, G. L. 1974. Plankton secondary production and biomass; seasonality and relation to trophic state in three lakes. Ph.D. thesis, Univ. Wash., Seattle.

RINGELBERG, J., and G. M. HALLEGRAEFF. 1976. Evidence for a diurnal variation in the carotenoid content of *Acanthodiaptomus denticornis* (Crustacea, Copepoda) in Lac Pavin (Auvergne, France). Hydrobiologia 51: 113-118.

SIEBECK, O. 1978. UV Toleranz und Photoreaktiviergun bei Daphnien aus Biotopen verschiedener Höhenregionen. Naturwissenschaften 65: 390-391.

SIEFKEN, M., and K. B. ARMITAGE. 1968. Seasonal variation in metabolism and organic nutrients in three *Diaptomus* (Crustacea: Copepoda). Comp. Biochem. Physiol. 24: 501-609.

SMITH, W. E. 1970. Tolerance of *Mysis relicta* to thermal shock and light. Trans. Am. Fish. Soc. 99: 418-422.

THOMAS, G. 1977. Effects of near ultraviolet light on microorganisms. Photochem. Photobiol. 26: 669-673.

WALKER, K. F. 1975. The seasonal phytoplankton cycles of two saline lakes in central Washington. Limnol. Oceanogr. 20: 40-53.

ZSCHOKKE, F. 1900. Die Tierwelt der Hochgebirgsseen. Denkschr. Schweiz. Naturforsch. Ges. 37.

11. Experimental Studies on Diel Vertical Migration

Richard N. Bohrer

Abstract

Experimental manipulations of factors potentially influencing vertical migration were performed in the Dalhousie Aquatron tower tank (depth, 10 m; volume, 100 m^3). The effects of food distribution, food concentration, and the presence of other species were investigated. Species used were *Pseudocalanus minutus, Temora longicornis, Calanus finmarchicus,* and *Tortanus discaudatus. Thalassiosira fluviatilis* was used as the food source. The presence of a subsurface food peak was found to decrease the ascent of the animals, so that nighttime depth corresponded to the depth of the food peak. Food concentration affected both the magnitude and the timing of the migrations with *Calanus* and *Pseudocalanus* being more affected than *Temora* and *Tortanus*, but the results at high food concentrations concerning a possible satiation effect were inconclusive. In experiments involving more than one species, migratory patterns differed in such a way as to result in the ecological isolation of the species via habitat selection. These results demonstrate the ability, which may be adaptive, of copepods to modify their migration behavior in response to food and other species. In addition, the variation among individuals in these experiments suggests that vertical migration should be considered a population phenomenon; the sources of this variation deserve further study.

Many zooplankton, some smaller than a few millimeters, daily migrate tens or hundreds of meters (Cushing 1951; Vinogradov 1968). Interest in the factors influencing these migrations originated early, and some information was obtained by correlating migratory patterns with environmental variables in the field (Russell 1926; Clarke and Backus 1956). An experimental approach is preferable for investigating the proximal causes (*see* Baker [*cited in* Lack 1954]) of biological phenomena, and several experimental studies on diel vertical migration have been made (Esterly 1917, 1919; Clarke 1930; Harris and Wolfe 1955; Ringelberg 1964; Enright and Hamner 1967).

One potential disadvantage of all of these investigations, with the exception of that by Enright and Hamner, is that they were conducted on a few individual animals in small laboratory vessels. These conditions may affect the behavior of the animals. Here I present some of the large-scale experimental studies I have performed using the Aquatron facilities at Dalhousie University (Balch et al. 1978). I chose to investigate the effects of food distribution, food concentration, and the presence of potential competitors on vertical migration. Since the ultimate cause of vertical migration may be related to resource exploitation (Kerfoot 1970; Enright 1977), all three factors might be expected to exert proximal effects.

I thank C. M. Boyd, D. Mackas, S. Smith, and S. Wilson for help with sampling. Comments on the manuscript by P. A. Lane, R. J. Conover, and C. M. Boyd were greatly appreciated, if not always followed.

This paper is part of a dissertation submitted in partial fulfillment of the requirements of the Ph.D. degree at Dalhousie University.

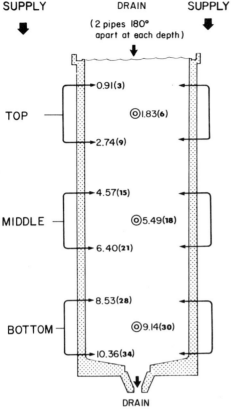

Fig. 1. Sectional view of tower tank with depths of supply and drain pipes. Depths in meters (feet).

used in these experiments differed in size and showed some vertical separation during the day, collections dominated by single species could be obtained by towing nets with the proper size mesh at the proper depth. Experimental animals were then transported to the lab in insulated containers and added to the tank.

Animals were allowed to acclimate for 1 day and then sampled periodically using a pump (capacity: 36 liters/min. sampling duration 2–3 min). Samples were taken at 1-m intervals during the experiment in March 1977 when an electronic counting technique (Boyd 1973; Mackas 1977) was used; samples from 1-m intervals were pooled every 2 m in the remaining experiments when microscope counts were made. Sampling times are given below. Temperature profiles were taken with a Beckman RS-5 unit and chlorophyll distributions were determined either by fluorescence in vivo or by spectrophotometric analysis of filtered water samples (Strickland and Parsons 1972: SCOR-UNESCO equations). In order to manipulate phytoplankton levels, *Thalassiosira fluviatilis* was cultured in 20-liter glass carboys and added at the experimental depth.

Methods

Experimental procedure—The experiments were performed in the Aquatron tower tank using zooplankton collected in the field. The tank is an indoor facility about 10 m deep with a volume of 100 m^3 and has ports for seawater supply and drainage at various depths (Fig. 1). Four 1,000-W phosphor-coated metal halide lamps and two 400-W mercury lamps give a maximum illumination of 0.16 cal/cm^2/min which is about 40% of midday sunlight at this latitude. At the beginning of each experiment the tank was filled with seawater filtered through four Graver pressure sand filters and the lights were arranged with timers to simulate the natural cycle of day, dusk, night, and dawn. Because the zooplankton

Data analysis—The vertical migration patterns were presented in three ways. In the first, abundance at each depth was expressed as a percentage of the total abundance and the deviation of this value from its daily mean ($p_i-\bar{p}$) plotted versus depth. This method was used when it appeared that many individuals were not migrating and served to emphasize those that were. The second method of presentation was a graph showing the depth of the 20, 40, 60, and 80 percentiles at each time period (Pennak 1943). The depth distribution of the animals was expressed in terms of percentage composition, and the depths of the different percentiles were calculated by interpolation. This method reduced the complete pattern to one graph but was subject to the inaccuracies of interpolation. The third method was to graph the numbers of animals occurring at the surface. It showed the periods of maximum ascent most clearly.

Statistical analysis of migratory patterns is difficult because no suitable methods exist for comparing whole patterns, the data points are not independent, and usually samples are not replicated. P. Lane and I are presently working on an approach that separates migratory patterns into various characteristics such as amplitude (or range), modality, timing, dispersion, depth distribution, and velocity, each of which can be described statistically. Migratory patterns can then be compared with regard to each characteristic.

Only the timing of the migration on different days is compared statistically here. Correlation coefficients were calculated between days by using the depth of the 40th percentile at each time period. These coefficients were calculated between days by using the depth of the 40th percentile at each time period. These coefficients were summarized in a correlation matrix and subjected to cluster analysis (Williams 1971). I used a clustering program which forms hierarchical groups using group average for the amalgamation rule (Dixon 1975). The result was a dendrogram showing which days were most similar.

Results

As there was only one tower tank, no simultaneous controls on the experiments were possible. Instead, migrations preceding manipulations were used as controls and assumed to be repeatable in the absence of manipulation. To test this assumption a 3-day experiment was performed in March 1977. The animals in the tank were mostly adult females of *Pseudocalanus minutus*. Although the phytoplankton distribution and temperature structure changed slightly due to phytoplankton sinking and surface warming (Fig. 2), the general features of the migrations were repeated. On each day the animals began to come up as early as 1400 hours, peaked in the early evening, began to go down or disperse during the night, and had their deepest distribution in late night or early morning (Fig. 3). The other experiments generally did not allow for repetition of migrations, but I believe the results of this experiment validate the conclusion that differences between days are due to experimental manipulations.

The effect of the vertical distribution of food on diel migration was tested in an experiment with *Temora longicornis*, mostly adult females. On the first day, in filtered

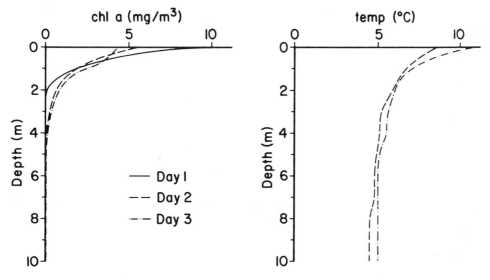

Fig. 2. Vertical distribution of chlorophyll *a* and temperature for 3-day experiment, March 1977.

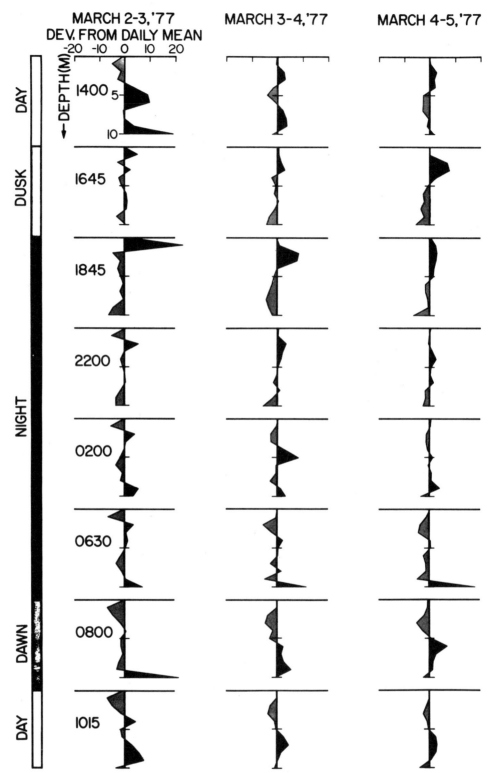

Fig. 3. Vertical distributions of *Pseudocalanus minutus* for 3-day experiment, March 1977. Abundance at each sampling depth expressed as a percentage of total abundance and deviation of this value from its daily mean plotted vs. depth.

water, the animals migrated to the surface after dusk and descended at dawn. On the second day, after the addition of phytoplankton at 3 m, the animals again began to ascend after dusk, but most of the population did not go to the surface. Instead, they remained at depth, centered on the phytoplankton layer (Fig. 4). Apparently the surface chlorophyll peak caused the animals to stop migrating, and their greater dispersion on the second day may have been related to the chlorophyll dispersion.

The effect of food concentration was tested in two experiments. In the first, a *Calanus finmarchicus* population of mixed late stages, mostly stage V, represented about 90% of the animals present with the remainder being adult *Tortanus discaudatus*. On the first day the animals were in filtered water; on the second day two carboys of phytoplankton were added; on the third day one carboy was added; and on the fourth day eleven carboys were added. Temperature structure showed only a general warming of the water with no qualitative changes (Fig. 5d).

The migration of *Calanus* changed every day (Fig. 5a) while that of *Tortanus* showed no change (Fig. 5c). On the first day *Calanus* migrated up at about midnight. After food was added on the second day, *Calanus* had a more pronounced migration with the peak ascent still occurring near midnight. However, on the third day, with similar food conditions, the animals were up before dusk and showed less overall migration. On the fourth day, with the highest food concentration, there was again a marked migration with the peak ascent this time occurring immediately after dusk. The migrations on the second and fourth days were similar in magnitude, but the time spent at the surface was longer on the second day when the food concentration was lower (Fig. 5b).

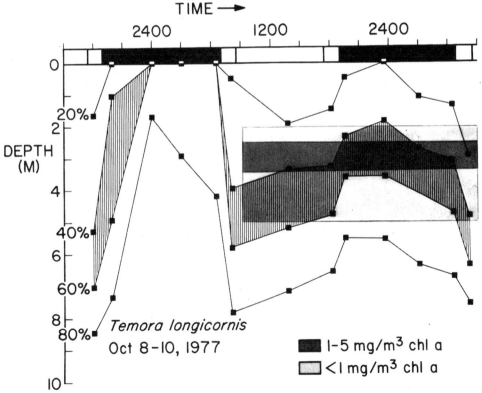

Fig. 4. Vertical migration pattern of *Temora longicornis* in experiment with phytoplankton added at 3 m on second day.

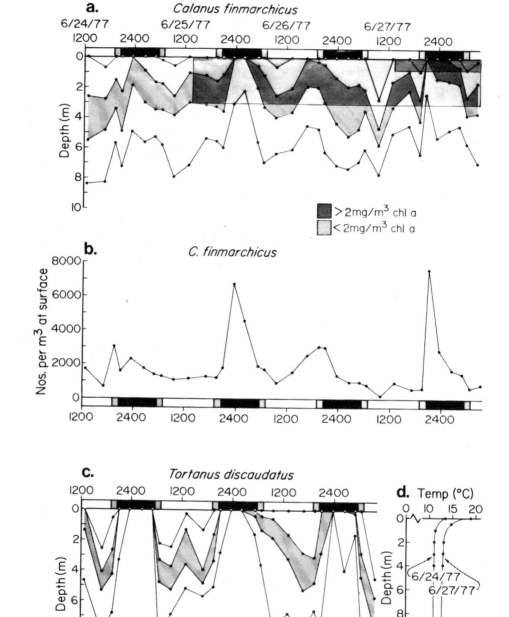

Fig. 5. Experiment with different food concentrations, June 1977. a, b—Vertical migration of *Calanus finmarchicus*; c—vertical migration of *Tortanus discaudatus*; d—vertical distribution of temperature.

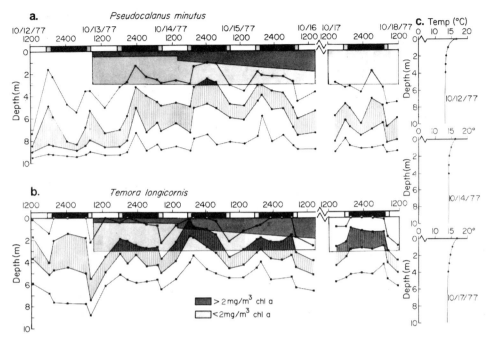

Fig. 6. Experiment with different food concentrations, October 1977. a—Vertical migration of *Pseudocalanus minutus*; b—vertical migration of *Temora longicornis*; c—vertical distribution of temperature.

Tortanus migrated up after dusk and remained up until dawn on each day. The fourth day was similar to the first three, but is not shown because the decline in abundance due to sampling and mortality made the data unreliable.

In the second experiment designed to study the effects of food concentration, adult females of *P. minutus* and late stages, mostly adult females, of *T. longicornis* were present with the latter being about ten times more abundant than the former. Two carboys of phytoplankton were added on the second day and nine carboys on the third. After the fourth day the top 2 m of the tank were washed out in an effort to return to the initial food conditions and to test the repeatability of the migrations observed at the beginning of the experiment. This procedure reduced the phytoplankton concentration to 1–2 mg/m^3 Chl *a*. Temperature structure changed little throughout the experiment (Fig. 6c).

The migration of *Pseudocalanus* varied with the experimental conditions (Fig 6a).

The animals had a bimodal migration on the first day with peaks at dusk and dawn. This bimodality was repeated on the last day, although the timing of the peaks was not exactly the same. The animals also had their deepest mean distributions on the first and last day. After food was added, the migratory pattern became unimodal with its peak occurring in the first half of the night. There was little difference among the 3 days with food, although on the 2 days with the higher food concentration the animals migrated to the surface earlier and remained there longer.

Temora exhibited a different migratory pattern (Fig. 6b). On the first and last days the animals had a unimodal pattern, coming up after dusk and remaining up throughout the night. When phytoplankton was present, the animals were up during dusk and slowly descended throughout the night. This pattern was most pronounced on the third day and least pronounced on the fourth day when there was some suggestion of a bimodal pattern.

Both species showed a change in migration following the addition of the phytoplankton and a return to their initial behavior after the reduction of the phytoplankton. There was little change in the migrations on the 3 days with food. The similarity of the first and last days, despite the fact that some phytoplankton was still present on the last day, suggested that the decrease in phytoplankton concentration was more important in producing this pattern than the absolute concentration. In order to determine if these conclusions could be demonstrated objectively, a cluster analysis was performed on the timing of the migrations. The analysis included the experiments depicted in Figs. 4 and 6 and another experiment with *Temora* and *Pseudocalanus* not shown here. All of these experiments were performed within a period of 1 month. Days were classified as either no food, low food, or high food. The results of the cluster analysis supported the previous interpretation of the data. Each species had two main groups corresponding to days without food and days with food (Fig. 7). The lower correlation among the days for *Pseudocalanus* meant that this species exhibited a greater heterogeneity in its migratory behavior.

Although both *Temora* and *Pseudocalanus* had relatively consistent patterns of migration for days with and without food, these patterns were different for each species and may indicate that the species were reacting to each other's presence. On each day, the timing of the migration in one species was the opposite of the other. *Temora* in the absence of food was highest in the water column throughout the night while *Pseudocalanus* was highest at dusk and dawn. Conversely, *Temora* in the presence of food was highest at dusk while *Pseudocalanus* was highest at night, except on the fourth day when the reverse was true (cf. Fig. 6a and b). The species also differed in vertical distribution. *Pseudocalanus* was deeper than *Temora* and this depth segregation was maintained during migration (Fig. 8). The maintenance of depth segre-

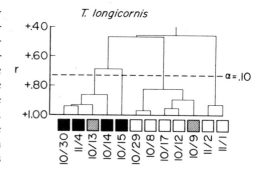

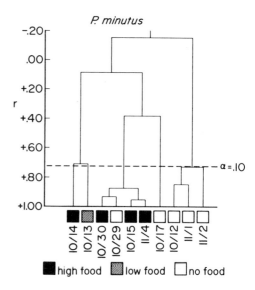

Fig. 7. Dendrogram for hierarchical classification of migratory patterns based on timing of migrations. Included are experiments shown in Figs. 4 and 6 and an additional experiment not shown.

gation and differences in timing, even when both species were changing both depths and timing, suggests that the differences between the species were not simply the result of fixed behavioral preferences.

Discussion

Many investigators have hypothesized that food conditions could influence the

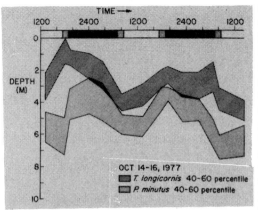

Fig. 8. Vertical migration of *Temora longicornis* and *Pseudocalanus minutus* from Figs. 6a and b showing centers of abundance defined by depth of 40th and 60th percentiles.

vertical distribution and vertical migration of zooplankton. Associations of animals with subsurface chlorophyll maxima have been found by Anderson et al. (1972) and Mullin and Brooks (1972). Hardy and Gunther (1936) suggested that the time spent at the surface by migrating animals might be inversely proportional to the phytoplankton concentration because full, or satiated, animals would descend. Pearre (1973), Mackas and Bohrer (1976), and Arashkevich (1977) have all described such behavior for zooplankton in the field.

My experiment with a subsurface phytoplankton layer clearly shows that such a layer can determine the depth to which the animals ascend. However, results with different food concentrations are inconclusive with respect to a satiation effect. *Calanus* did spend a longer time at the surface when the food concentration was lower, but the difference was not great. The experiments with *Temora* and *Pseudocalanus* do not even show this slight difference. The variation in migrations observed may have more to do with the change in food concentration from day to day than with the concentration itself. For example, in the *Calanus* experiment, migrations on day 2 and day 4, associated with increasing food concentrations, are more similar than migrations on day 2 and day 3, associated with the same concentration (Fig. 6a). This is also true in the experiment with *Temora* and *Pseudocalanus* where the migrations beginning on the 13 and 14 October, associated with increasing food concentrations, are more similar than the migrations beginning on 14 and 15 October, associated with the same concentration (Fig. 7). The influence of the previous food concentration may indicate some kind of conditioning aspect in the migratory behavior.

The mechanisms responsible for these shifts in behavior with food conditions are unknown. According to Ringelberg (1964), the major stimulus for migration is the relative change in light intensity; but in these experiments animals sometimes migrated before, during, or after the changes in intensity. This variation could be due to changes in subsurface light caused by the phytoplankton or to changes in the phototactic response caused by the recent feeding history of the animal (Singarajah et al. 1967). Whatever the mechanisms, the ability of zooplankton to modify migratory behavior under different food conditions does support the hypothesis that the adaptive value of vertical migration is related to the dynamics of resource utilization. Because of the patchiness of phytoplankton and the vertical current shear, migrating zooplankton are unlikely to experience the same conditions of food abundance and distribution from day to day. The interaction of these factors with the effects of phytoplankton on vertical migration can determine the spatial correlation between phytoplankton and zooplankton with important consequences for zooplankton feeding and hence production (Evans et al. 1977). Migratory behavior was more variable in *Calanus* and *Pseudocalanus*, both classified as herbivores on the basis of mouthpart morphology, than in *Tortanus*, a carnivore, and *Temora*, an omnivore. Either the ability to modify migratory behavior is more important for herbivores, due maybe to a greater variation in their resources, or the herbivores just respond more readily to changes in phytoplankton. Enright and Honegger (1977) observed a similar behavior in which *Calanus helgolandicus (Pacificus)*, a herbivore, varied in its time of ascent while *Metridia pacifica*, an omnivore, did not.

The evidence for an effect due to the presence of other species on vertical migration patterns is indirect. The experiment with *Temora* and *Pseudocalanus* shows that the species do have differences in depth distribution and in the timing of the migration. Even when each species changed its migratory pattern the change in both species was such that differences were still maintained. Lane (1975) and Tsalkina (1971) found similar patterns of species segregation in field studies of vertical migration. Such patterns in the field and in my experiments may be due either to differences in the response of each species to environmental stimuli or to a direct interaction between species similar to interference competition. The latter possibility was documented by Dumont (1972) as the cause of a reversal in migration by *Asplanchna priodonta*. A choice between these two explanations could be made by examining the migration of one species in the presence and absence of the other, but I was unable to do this because the species could not be obtained separately. However, whatever the explanation, the existence of the species segregation provides a partial ecological isolation for these species.

One aspect of these experiments which is independent of experimental manipulations is the existence of individual variation. It is this variation that makes vertical migration a population phenomenon. In Figs. 4-6 the dispersion is one aspect of this individual variation. Asynchrony and nonmigration, two other aspects, are impossible to tell apart in migration data such as mine. Individuals near the surface during the day or down deep at night may be either nonmigratory or asynchronous (Pearre 1979). The behavior for *Calanus* on days 2 and 4 shown in Fig. 5a,b could be due either to more individuals migrating to the surface or to the accumulation of asynchronous individuals at the surface. While individual variation is difficult to study in my experimental design, its existence in the response of zooplankton to light has been well documented (Esterly 1917; Spooner 1933; Daan and Ringelberg 1969). Esterly's results are especially interesting because he found that *Acartia clausi* collected at the surface had the opposite phototactic response from *A. clausi* collected at depth. It would be worth knowing whether this variation exists only at the level of the phenotype or also at that of the genotype.

Conclusions

I have shown that food distribution and food concentration affect vertical migration behavior and that these effects may be adaptive for animals living in a heterogeneous food environment. Spatial and temporal segregation of species has also been demonstrated and represents an ecological isolation mechanism for these species which may be important in their resource partitioning dynamics. Other characteristics of experimental migrations reveal that vertical migration should be considered a population phenomenon. The variation among individuals is real and probably important in considerations of the adaptive significance of diel vertical migration.

References

ANDERSON, G. C., B. W. FROST, and W. K. PETERSON. 1972. On the vertical distribution of zooplankton in relation to chlorophyll concentration, p. 341-345. *In* A. Y. Takenouti [ed.], Biological oceanography of the northern North Pacific Ocean and Bering Sea. Idemitsu Shoten.

ARASHKEVICH, Ye. C. 1977. Relationship between the feeding rhythm and the vertical migrations of *Cypridina sinuosa* (Ostracoda, Crustacea) in the western part of the equatorial Pacific. Oceanology 17: 466-469.

BALCH, N., C. M. BOYD, and M. M. MULLIN. 1978. Large-scale tower tank systems. Rapp. P.-V. Reun. Cons. Int. Explor. Mer 173: 13-21.

BOYD, C. M. 1973. Small scale spatial patterns of marine zooplankton examined by an electronic in situ zooplankton detecting device. Neth. J. Sea Res. 7: 103-111.

CLARKE, G. L. 1930. Change of phototropic and geotropic signs in *Daphnia* induced by changes of light intensity. J. Exp. Biol. 7: 109-131.

——, and R. H. BACKUS. 1956. Measurements of light penetration in relation to vertical migration and records of luminescence of deep-sea animals. Deep-Sea Res. 4: 1-14.

CUSHING, C. H. 1951. The vertical migration of planktonic organisms. Biol. Rev. 26: 158-192.

DAAN, H., and J. RINGELBERG. 1969. Further studies on the positive and negative phototactic reaction of *Daphnia magna* Straus. Neth. J. Sea Res. 4: 525-540.

DIXON, W. J. 1975. Biomedical computer programs. Univ. Calif. Press.

DUMONT, H. J. 1972. A competition-based approach of the reverse vertical migration in zooplankton and its implications, chiefly based on a study of the rotifer *Asplanchna priodonta* (Gosse) with several Crustacea Entomostraca. Int. Rev. Gesamten Hydrobiol. 57: 1-38.

ENRIGHT, J. T. 1977. Diurnal vertical migration: Adaptive significance and timing. Part 1. Selective advantage: A metabolic model. Limnol. Oceanogr. 22: 856-872.

——, and W. M. HAMNER. 1967. Vertical diurnal migration and endogenous rhythmicity. Science 157: 937-941.

——, and H. W. HONEGGER. 1977. Diurnal vertical migration: Adaptive significance and timing. Part 2. Test of the model: Details of timing. Limnol. Oceanogr. 22: 873-886.

ESTERLY, C. O. 1917. Specificity in behavior and the relation between habits in nature and reactions in the laboratory. Univ. Calif. Publ. Zool. Berkeley 16: 381-392.

——. 1919. Reactions of various plankton animals with reference to their diurnal migrations. Univ. Calif. Publ. Zool. Berkeley 19: 1-83.

EVANS, G. T., J. H. STEELE, and G. E. B. KULLENBERG. 1977. A Preliminary model of shear diffusion and plankton populations. Scot. Fish. Res. Rep. 9.

HARDY, A. C., and E. R. GUNTHER. 1936. The plankton of the South Georgia whaling grounds and adjacent water 1926-27. Discovery Rep. 11.

HARRIS, J. E., and U. K. WOLFE. 1955. A laboratory study of vertical migration. Proc. R. Soc. Lond. Ser. B 144: 329-354.

KERFOOT, W. B. 1970. Bioenergetics of vertical migration. Am. Nat. 104: 529-546.

LACK, D. 1954. The natural regulation of animal numbers. Oxford Univ. Press.

LANE, P. A. 1975. The dynamics of aquatic systems: A comparative study of the structure of four zooplankton communities. Ecol. Monogr. 45: 307-336.

MACKAS, D. L. 1977. Horizontal spatial variability and covariability of marine phytoplankton and zooplankton. Ph.D. thesis, Dalhousie Univ.

——, and R. BOHRER. 1976. Fluorescence analysis of zooplankton gut contents and an investigation of diel feeding patterns. J. Exp. Mar. Biol. Ecol. 25: 77-85.

MULLIN, M. M., and E. R. BROOKS. 1972. The vertical distribution of juvenile *Calanus* (Copepoda) and phytoplankton within the upper 50 m of water at La Jolla, California, p. 347-354. *In* A. Y. Takenouti [ed.], Biological oceanography of the northern North Pacific Ocean and Bering Sea. Idemitsu Shoten.

PEARRE, S., JR. 1973. Vertical migration and feeding in *Sagitta elegans* Verrill. Ecology 54: 300-314.

——. 1979. Problems of detection and interpretation of vertical migration. J. Plankton Res. In Press.

PENNAK, R. W. 1943. An effective method of diagramming diurnal movements of zooplankton organisms. Ecology 24: 405-407.

RINGELBERG, J. 1964. The positively phototactic reaction of *Daphnia magna* Straus. Neth. J. Sea Res. 2: 319-406.

RUSSELL, F. S. 1926. The vertical distribution of marine macroplankton. IV. The apparent importance of light intensity as a controlling factor in the behavior of certain species in the Plymouth area. J. Mar. Biol. Assoc. U.K. 14: 415-440.

SINGARAJAH, K. V., J. MOYSE, and E. W. KNIGHT-JONES. 1967. The effect of feeding upon the phototactic behavior of cirripede nauplii. J. Exp. Mar. Biol. Ecol. 1: 144-153.

SPOONER, G. M. 1933. Observations on the reactions of marine plankton to light. J. Mar. Biol. Assoc. U.K. 19: 385-438.

STRICKLAND, J. D. H., and T. R. PARSONS. 1972. A practical handbook of seawater analysis, 2nd ed. Bull. Fish. Res. Bd. Can. 167.

TSALKINA, A. V. 1971. Vertical distribution and diurnal migrations of mass species of Cyclopoida (Copepoda) in the western Equatorial Pacific, p. 229-239. *In* M. E. Vinogradov [ed.], Life activity of pelagic communities in the ocean tropics. [Transl. U.S. Dep. Commerce 1973.]

VINOGRADOV, M. E. 1968. Vertical distribution of the oceanic zooplankton. [Transl. U.S. Dep. Commerce 1970.]

WILLIAMS, W. T. 1971. Principles of clustering. Annu. Rev. Ecol. Syst. 2: 303-326.

12. Seasonal Patterns of Vertical Migration: A Model for *Chaoborus trivittatus*

Louis A. Giguère and Lawrence M. Dill

Abstract

Energy budgets were computed for 4th instar *Chaoborus trivittatus* living in a simplified hypothetical lake and utilizing two alternative migratory tactics. All larvae were allowed to spend 8 hours feeding near the surface (obtaining a fixed amount of food) and then allowed to either migrate to cool (5°C) deep water strata for a 16-h "resting" period or continue feeding in the warm (16°C) food-rich surface layers. The model is based on an earlier thorough energetic study of *Chaoborus* that includes temperature-dependent respiration rates, temperature and meal-size-dependent assimilation efficiencies, costs of swimming and attacking and the probability of prey capture at varying hunger levels. The larvae were found to derive maximal energy benefits by migrating when the density of *Diaptomus kenai*, its principal prey, is low (<0.2 individuals/liter) and by feeding near the surface when conditions are otherwise. A comparison of our findings to field data from a fishless lake suggests that some seasonal changes in patterns of vertical migration in *Chaoborus* can be accounted for on an energetic basis.

In fishless lakes near Vancouver, B.C., the predatory 4th instar larvae of the phantom midge *Chaoborus trivittatus* demonstrate striking changes in seasonal patterns of vertical migration. In June most larvae remain in the food-rich layers near the surface throughout the day and night and exhibit migrations of small amplitude (based on depth distribution profiles: Fedorenko 1975). Later on, in the summer and the fall, however, the larvae undergo migrations of increasing amplitude. Our purpose here was to investigate whether this pattern could be the result of individuals attempting to maximize their net rate of energy intake (ENET). We assume that individuals base their decision on whether to migrate on a comparison of their net energy intake using the two contrasting tactics (migratory and nonmigratory).

The data required to calculate daily energy budgets under various temperature and prey-size conditions were collected in the laboratory, using larvae from Gwendoline Lake, B.C. These data are presently being incorporated into a detailed simulation model of predation, migration, and growth strategies. In this paper we will generate a prediction about a migratory tactic based on a shortened version of the simulation model and a number of simplifying assumptions. This prediction is then compared to field data collected in another fishless lake (Eunice Lake, B.C.) by another researcher (Fedorenko 1975). Both lakes are situated in the U.B.C. Research Forest and contain the same set of species. Fourth instar *C. trivittatus* feed mostly on copepods of the genus *Diaptomus* (about 60–80% by volume), primarily the large *Diaptomus kenai* (Fedorenko 1975; Giguère unpubl.).

We thank R. L. Dunbrack and M. C. Swift for reviewing an earlier version of this paper.

Scenario for the model

Let us consider a hypothetical fishless lake with a simplified temperature stratification in two layers (16°C for 0-5 m and 6°C from 5 m down). Moreover, let us assume that most prey are found in the top 5 m of the lake during the daytime. *Diaptomus kenai*, for example, is usually rare below that depth during the day (<0.05 individuals/liter). We can imagine the following scenario, based on the timing and usual migratory patterns of the 4th instar: *Chaoborus* larvae come up at dusk to feed in the upper layers for a period of 8 hours. If we grant the larvae a composite meal of 0.8 J for this period (based on data in Fedorenko 1973), what is the "best" migratory tactic, i.e. the one permitting maximization of ENET over the next 16 hours? At dawn, an animal can either cease feeding and return toward the bottom of the lake or forgo vertical migration and continue feeding for the remainder of the day (16 hours). In the former case, greater metabolic efficiency (McLaren 1963) may be possible (at the cost of vertical migration and a potentially reduced assimilation efficiency). Non-migratory larvae, on the other hand, can count on greater opportunities to capture prey, provided that they are not fully satiated and that there are a sufficient number of easily handled energy-rich items available to compensate for higher metabolic rates at high temperature. For several reasons, we expect that this depends on the availability of the prey *D. kenai*. First, copepods are considered the preferred food of *Chaoborus* (Deonier 1943; Main 1953). Swüste et al. (1973) found a preference for copepods over daphnids in simple choice experiments, and Pastorok (1978) showed that diaptomid copepods are preferred to daphnids of equivalent body length at high total prey density (in a 1:1 ratio). Copepods are also more easily handled than daphnids due to their body shape, as shown by Swift and Fedorenko (1975) (*see also* Swüste et al. 1973). *Chaoborus* ingestion time (Swift and Fedorenko 1975) and digestion time (Fedorenko 1973; Giguère pers. obs.) for *Daphnia* are greater than for the prey *Diaptomus*. Moreover, the % energy assimilated by *Chaoborus* is greater for copepod prey than for *Daphnia* prey (Swift 1976). Parma (1969) has published data showing greater growth for *C. crystallinus* larvae fed with a copepod diet than those fed mainly *Daphnia pulex*. Finally, *D. kenai* ranks first or second in terms of its relative contribution to the total biomass of crustacean zooplankton in these lakes. Therefore we consider that *D. kenai* can make an energy boost possible for non-migratory 4th instar larvae. The following model attempts to compare the relative magnitude of this energy boost, given various densities of *D. kenai*, to the boost accrued from vertical migration.

The model

Data were collected in the laboratory, and the details of the methods and results will be presented elsewhere (Giguère in prep.). The following data were included in the model—

1. Capture rate: The probability of capturing a relatively large copepod (1.8 mm long, 0.3 μl volume) in a given time period (PRCAP) was obtained from laboratory predation experiments. This particular prey size was chosen because it is the average size of *D. kenai* in Gwendoline Lake. At a given density the capture rate for a larva having food in its crop is lower than for larvae starved for 1-2 days (crop volume = 0) (Fig. 1). This is caused by fewer attacks being triggered under partly satiated conditions and by a lower probability that an attack will be successful once launched. These data allow calculation of feeding rates, if it is further assumed that PRCAP increases linearly as a function of *D. kenai* density. This is not an unrealistic assumption considering the low prey density found in these oligotrophic lakes and the relatively short handling time of copepods by *Chaoborus* (103.6 sec for 1.8 mm length prey: Swift and Fedorenko 1975).

2. Metabolic costs: The energetic costs of three behaviors of the 4th instar larvae are required in the model. These are the

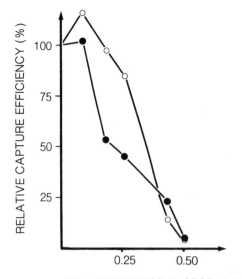

Fig. 1. Relative capture efficiency as a function of hunger for 4th instar *Chaoborus trivittatus* larvae. Capture rates expressed as a percentage of capture rate of 1-2-day starved larvae at each temperature. ○—13°C; ●—22°C.

the crop was studied at two different temperatures (13° and 22°C) but did not differ significantly with respect to this variable in one-year-old larvae. The following regression was used to express crop volume (μl) as a function of the hours since ingestion (TF), given an initial volume of 0.42 μl (equivalent to 0.8 J):

Crop volume (ml) = 0.42 − 0.04 (TF).

The prediction

What is the migratory tactic that will result in a maximum amount of net energy being secured by a 4th instar larva in a day? When we consider the case of a larva choosing vertical migration, we obtain the value 0.33 for ENET/day (Table 1 gives a

metabolic rate at rest (0.008 J/h at 5°C and 0.020 J/h at 16°C: Swift 1976), the metabolic cost while swimming (0.00014 J per swimming motion) and while attacking (0.00044 J per attack). The cost of vertical migration can then be estimated since vertical displacement is generally of the order of 5 m in our lakes and a larva covers roughly 2.5 cm vertically per swimming motion (based on observations in an artificial column in the laboratory). If we assume active swimming, an average round-trip therefore represents an expenditure of about 0.056 J.

3. Assimilation efficiency: Using a radiotracer technique, we evaluated the % assimilation of copepods by *Chaoborus* under four temperature conditions and five different meal volumes (Fig. 2). It was nearly constant over the complete range of possible meal volumes (up to 0.8 μl) at 22°C. At lower temperatures, it was constant up to a meal volume of 0.5 μl but decreased for larger meals (two large *D. kenai* ingested), particularly at low temperature.

4. Digestion rate: The rate of emptying of

Table 1. Energy budget for *Chaoborus* larvae undergoing vertical migration after 8 hours.

	J/day
Meal assimilation	0.70
Vertical migration	−0.06
Respiration	−0.34
Possibility of further capture	0.03
Total	0.33

complete breakdown). Some feeding is allowed on the low *D. kenai* density found below 5 m. In the case of no vertical migration, we computed the probable number of further captures by simulating predation hourly over a period of 16 hours. A random number table was used to determine whether prey capture occurred, by comparing the number picked ($0 < n < 1$) to the probability estimate for capture (PRCAP) at a given prey density and hunger level (updated each hour). Ten trials were run at each density; the results suggest that vertical migration should cease to occur at *D. kenai* densities above 0.20/liter if individuals wish to maximize ENET/day (Table 2).

Moreover, we found that larvae cannot increase ENET/day by staying near the surface until one more prey is captured and then migrating. This can be attributed to two factors: The PRCAP value in the early part of the 16-hour period is low and is combined with a lowered assimilation efficiency for all prey, including those

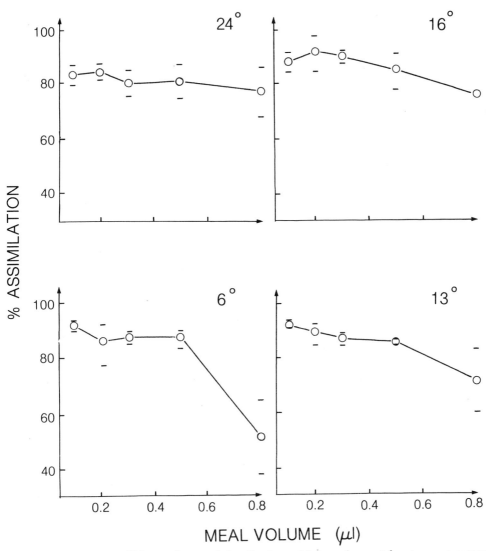

Fig. 2. Assimilation efficiency of copepods by *Chaoborus* 4th instar larvae at four temperature treatments and five meal volumes (with 95% confidence intervals).

Table 2. Energy budget (J/day) for *Chaoborus* larvae staying near the surface for 24 hours, at various *D. kenai* densities.

	Prey Density (No./liter)				
	0.1	0.2	0.3	0.4	0.5
Meal assimilation	0.77	0.85	0.90	0.94	0.99
Respiration	-0.51	-0.51	-0.51	-0.51	-0.51
Cost of attacks	-0.01	-0.01	-0.01	-0.01	-0.01
Total	0.25	0.33	0.38	0.42	0.47

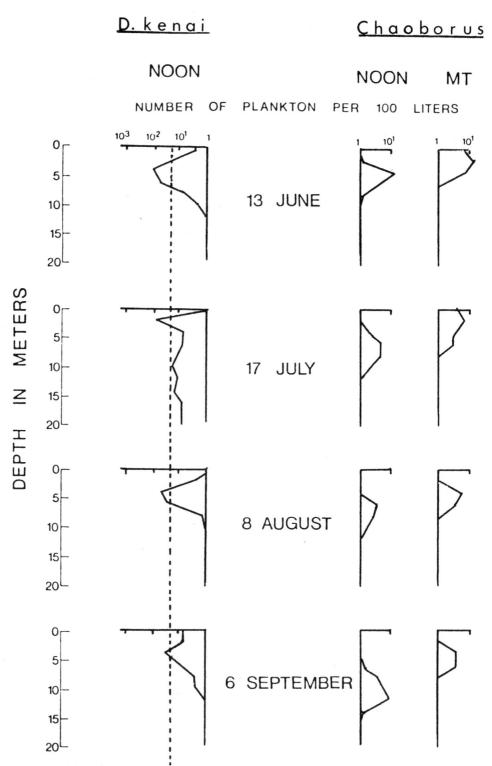

Fig. 3 Vertical distribution of zooplankton in Eunice Lake in 1972 (from Fedorenko 1973). Dotted line represents prey density above which simulation model predicts "no migration" for 4th instar *Chaoborus* larvae. MT—Midnight.

captured previously; and any potential savings due to a lowered respiration rate following capture late in the period will be outweighed by the cost of vertical migration.

Comparison with field data

Data on vertical distribution of zooplankton in Eunice Lake are available from Fedorenko (1973). We reproduce here the information on the vertical distribution of one-year-old 4th instar larvae of *C. trivittatus* at noon and midnight, as well as the noon distribution of *D. kenai* on four dates in 1972 (Fig. 3). The dotted line represents the density above which we predict cessation of vertical migration. Based on our probability model, we expect more larvae to reach the "correct" decision of not migrating as the density of *D. kenai* increases above this threshold. This is because a greater number of larvae will capture copepods as the density of the latter increases, up to a point where presumably most *Chaoborus* larvae can secure one or two more prey items and the larvae will remain near the surface "en masse." *Chaoborus* daytime and nighttime distribution overlap should then be maximal, i.e. little or no vertical migration should occur. To examine this relationship quantitatively, we plotted the area under the *D. kenai* curve above the critical threshold against the % overlap in *Chaoborus* density curves at noon and at midnight (all data were first converted from a logarithmic to a linear scale). The results confirm the predicted pattern of vertical migration (Fig. 4).

Discussion

The agreement between the field data and the model prediction is encouraging considering the simplifying assumptions under which the latter was derived. The next step will consist of testing the robustness of the model by relaxing the assumption of a fixed copepod size and using a more realistic temperature profile that matches field data on each date. Capture efficiency is known to vary with temperature, prey size, and prey type, and all prey species should be included to obtain a complete picture of *Chaoborus* predation.

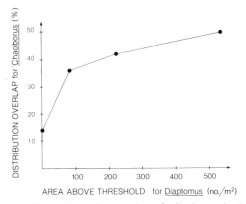

Fig. 4. Plot of % overlap of *Chaoborus* depth distribution at noon and at midnight versus area under curve of noon depth distribution of *D. kenai* lying above critical density threshold predicted by simulation model (*see Fig. 3*).

Our energy-efficiency hypothesis is an expansion of the metabolic efficiency hypothesis of McLaren (1963), who suggested that increasingly extensive migrations will be valuable as surface temperatures increase, since an animal can gain an energy bonus by feeding efficiently at high temperature (near the surface) and resting at low temperature (greater depth) while maintaining a low metabolic rate. Swift (1976) studied the energetics of *C. trivittatus* and his simulations demonstrated that a larva would maximize growth either by feeding near the surface or by migrating according to its own physiological rhythm. According to Swift, his conclusion differs from McLaren's because of different assumptions about the effect of temperature on digestion rate. However, both Swift and McLaren assume that assimilation efficiency is constant. Our data suggest that assimilation efficiency is temperature-dependent and that the effect is very much dependent on meal size (Fig. 5). Over a 24-hour period, *Chaoborus* assimilation decreases by 0.037 $J/°C$ for larger meals (0.8 μl) compared to a 0.025 $J/°C$ decrease in respiration. There is actually an increase of 0.003 $J/°C$ in assimilation for smaller meals (0.5 μl). Our next step will consist

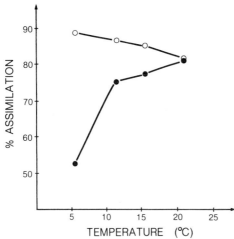

Fig. 5. Summary of assimilation efficiency of copepods by *Chaoborus* 4th instar larvae. ○—Small meal (0.5 µl); ●—large meal (0.8 µl).

of testing the effect of meal volume on migratory behavior of *Chaoborus* by prefeeding known amounts of food to larvae before introducing them into thermally stratified water columns in the laboratory. If our model is confirmed, we can predict vertical migration of low amplitude for *Chaoborus* in the summer since one would tend to associate high crustacean zooplankton production (and hence larger meals for *Chaoborus*) with higher surface temperature in temperate waters. This prediction is contrary to McLaren's.

References

DEONIER, C. C. 1943. Biology of the immature stages of the clear lake gnat (Diptera, Culicidae). Ann. Entomol. Soc. Am. **36**: 383–388.

FEDORENKO, A. Y. 1973. Predation interactions between zooplankton and two species of *Chaoborus* (Diptera, Chaoboridae) in a small coastal lake. M.S. thesis, Univ. British Columbia, Vancouver. 123 p.

———. 1975. Instar and species-specific diets in two species of *Chaoborus*. Limnol. Oceanogr. **20**: 238–249.

McLAREN, I. A. 1963. Effects of temperature on growth of zooplankton and the adaptive value of vertical migration. J. Fish. Res. Bd. Can. **20**: 685–727.

MAIN, R. A. 1953. A limnological study of *Chaoborus* (Diptera) in Hall lake, Washington. M.S. thesis, Univ. Wash., Seattle. 106 p.

PARMA, S. 1969. The life cycle of *Chaoborus crystallinus* (DeGeer) (Diptera, Chaoboridae) in a dutch pond. Int. Ver. Theor. Angew. Limnol. Verh. **17**: 888–894.

PASTOROK, R. A. 1978. Preferential feeding by *Chaoborus* larvae and its impact on the zooplankton community. Ph.D. thesis, Univ. Wash., Seattle. 238 p.

SWIFT, M. C. 1976. Energetics of vertical migration in *Chaoborus trivittatus* larvae. Ecology **57**: 900–914.

———, and A. Y. FEDORENKO. 1975. Some aspects of prey capture by *Chaoborus* larvae. Limnol. Oceanogr. **20**: 418–425.

SWÜSTE, H. F. J., R. CREMER, and S. PARMA. 1973. Selective predation by larvae of *Chaoborus flavicans* (Diptera, Chaoboridae). Int. Ver. Theor. Angew. Limnol. Verh. **18**: 1559–1563.

13. Diel Vertical Migration of Pelagic Water Mites

Howard P. Riessen

Abstract

Although water mites are not generally regarded as planktonic organisms they occur in abundance in the pelagic zone of Heney Lake, Québec. These mites exhibit a typical nocturnal vertical migration—they ascend from the bottom waters around sunset, peak at the surface about 2200 hours, and then undergo a long period of gradual sinking throughout the night and following day until sunset. The potential environmental cues of light, temperature, and oxygen are examined in relation to the migratory pattern and the adaptive value of this behavior is discussed in terms of predator avoidance, energetic efficiency, increased fecundity, and responses to food distribution.

Water mites are not typically regarded as planktonic organisms—they are known chiefly from ponds, streams, and the littoral zone of lakes, usually in close association with the bottom or some other substrate. Hutchinson (1967), however, reviewed a few cases in which water mites (especially members of the genera *Piona* and *Unionicola*) were collected in plankton samples, and Gliwicz and Biesiadka (1975) found concentrations of *Piona limnetica* reaching 150/m^3 in the pelagic zone of Madden Lake, Panama. In Heney Lake, Québec, I have found even greater concentrations of mites in the open water plankton, in excess of 250/m^3 during periods of population maxima. Little is known of the ecological relationships of water mites in general (*see* Pieczyński 1976) and even less as inhabitants of the pelagic zone of lakes. The life cycle of water mites involves seven distinct developmental stages (Böttger 1977), three of which (larva, nymph, and adult) are active. The nymph and adult are usually carnivorous while the larva is typically an ectoparasite on some aquatic insect (Böttger 1976).

This study investigates the vertical migration of pelagic water mites from Heney Lake; the possibility of this behavior in planktonic mites was first suggested by Viets (*cited in* Hutchinson 1967) for *Piona rotunda* from the Grosser Plöner See. The pattern of vertical migration in Heney Lake is documented, related to the environmental factors of light, temperature, and oxygen, and discussed in terms of its adaptive significance to the organism. I thank D. Smith, D. Safar, and M.-E. Coupal for help in the field, I. M. Smith for identification of the mites, and J. R. Strickler, R. Zaret, and E. McCauley for valuable discussion. This paper is based on portions of a thesis submitted to fulfill in part the requirements for the degree of Doctor of Philosophy, Yale University.

Methods

Heney Lake is about 100 km north of Ottawa, Ontario (46°02'N, 75°55'W); it is 9.3 km long and 2.6 km wide in its maximum dimensions and has a maximum depth of about 25 m. The lake is mesotrophic and alkaline (pH 8.0-8.8), with oxygen rarely, if ever, significantly depleted from the hypolimnion during the summer development of an unstable thermal stratification. The exact site for this study was a bay on the north end of the lake at which three stations were located, each at a depth of exactly 10.5 m and about 25 m from the other two.

Sampling was done on 5-6 June 1978 at 4-hour (0000-0400, 0600-1800) or 2-hour intervals (0400-0600, 1800-0000). Samples were taken at each station at depths of 1, 3, 5, 7, 9, and 10 m with a 40.6-liter Schindler-Patalas zooplankton trap (Schindler 1969); samples were preserved immediately in 6% formalin and all mites (adults and nymphs) in each sample were later examined and counted under 15x magnification with a binocular microscope. Water temperature was measured at 1-m intervals during each sampling period and incident light on the surface of the lake was determined every ½-2 hours using a Kahlsico radiometric submarine photometer (radiation measured in mW/cm^2). Sunrise occurred at 0516 and sunset at 2045 hours; there was a new moon on 5 June.

The water mites found in the plankton of Heney Lake during this period were dominated by *Piona constricta*[1] (89% of the individuals), with the remainder being *Unionicola crassipes*. Since too few *Unionicola* were found for a separate analysis based on species, all mites (excluding larvae) are treated together. The general pattern of vertical migration described below, however, appears to be the same for both *Piona* and *Unionicola*. In order to test the relative degree and significance of shifts in the vertical distribution of water mites, chi-square contingency tests were applied consecutively to the depth-frequency data for each sampling time and the one following it. This statistic tests only the relative distribution of the mites in the water column and does not depend on the total numbers found. High chi-square values indicate a major distributional shift between two consecutive sampling periods and critical values for this statistic indicate the level of significance of the change; the directional nature of the shift is not revealed by this method but can be easily determined by inspection of the depth-frequency data.

Results and discussion

Pattern of vertical migration. The depth-frequency data for water mites is shown in Fig. 1. During daylight hours most of the mites are located in the lower levels of the water column—there is also a tendency, subject to some variation, for the main body of the community to sink deeper as the afternoon progresses. The most pronounced shift in distribution occurs around sunset as the mites ascend in the water column, becoming at first more uniformly distributed and then concentrating in the upper waters. This evening ascent begins between 1800 and 2000 on 5 June, a day marked by overcast conditions, and peaks at 2200. On 6 June, a completely sunny day, the ascent is delayed, beginning between 2000 and 2200; the difference in distributions at 2200 can be attributed to the earlier start of the ascent on 5 June. After the evening surface maximum there is a definite nocturnal sinking of the mite populations—no morning ascent at sunrise takes place and the process of gradual sinking begun between 2200 and midnight continues throughout the day until sunset. The general pattern of movement can be classified as typical nocturnal migration (Hutchinson 1967), in which there is a single maximum near the surface between sunset and sunrise; the process is more precisely characterized by the continual nocturnal sinking—the third type of nocturnal migration mentioned by Hutchinson (1967).

Proximal causes of vertical migration. Changes in light intensity serve as the primary environmental cues for vertically migrating zooplankton (Hutchinson 1967; Forward 1976); other external factors such as temperature (Cushing 1951; Forward 1976), pH (Bayly 1963), and oxygen (LaRow 1970) may also play a role in deter-

---

[1] The genus *Piona* is currently undergoing revision; these individuals fall into the nodata species group and are provisionally identified as *Piona constricta* (Wolcott).

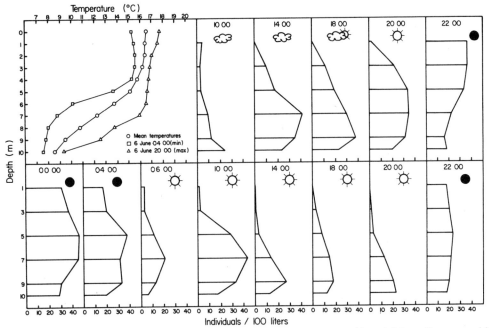

Fig. 1. Depth-frequency pattern of water mites for 5 June (upper panels) and 6 June (lower panels) (N=1,820). Temperature profile of maximum, minimum, and mean values (5 June 1000 to 6 June 2200) shown in upper left panel.

mining the position of a population in the water column. The pattern of incident radiation, both in terms of absolute light and its rate of change, and the degree of change between sampling periods in the vertical distribution of water mites (indicated by the chi-square statistic) are shown in Fig. 2. The greatest change in vertical distribution between any two time periods occurs between 2000 and 2200 (around sunset) on both days; the evening ascent actually begins earlier on 5 June (between 1800 and 2000: Fig. 1) but does not represent a significant shift until later. Either absolute changes or rates of change of incident light may account for the earlier ascent on 5 June. Light reaching the lake surface drops to levels of 2-3 mW/cm^2 by 1800 on 5 June, but not until about 1930 on 6 June; this is due to generally overcast conditions on 5 June. If this 2-3 mW/cm^2 level is a threshold for initiating the evening ascent, then the 1½ hour time difference between the two days readily explains the difference in timing of the migration. From another point of view, the maximum rate of decrease in light may be the cue for upward movement and, since it also occurs earlier on 5 June, it is hard to choose between the two alternatives.

A second significant shift occurs between 2200 and midnight, corresponding to the beginning stage of nocturnal sinking. The period immediately following the evening surface maximum shows the greatest degree of downward movement during the night—a pattern also found in vertically migrating populations of *Daphnia* by McNaught and Hasler (1964).

Significant distributional changes also occurred at midday on both 5 and 6 June. During the first day there was a shift upward in the water column during a period marked by a rapid increase in incident light followed by a slight decrease; the exact cause for this change is unclear and is further complicated by a lack of knowledge of the mite distribution before the change. On 6 June there is a movement downward in the water column, occurring during a period of generally in-

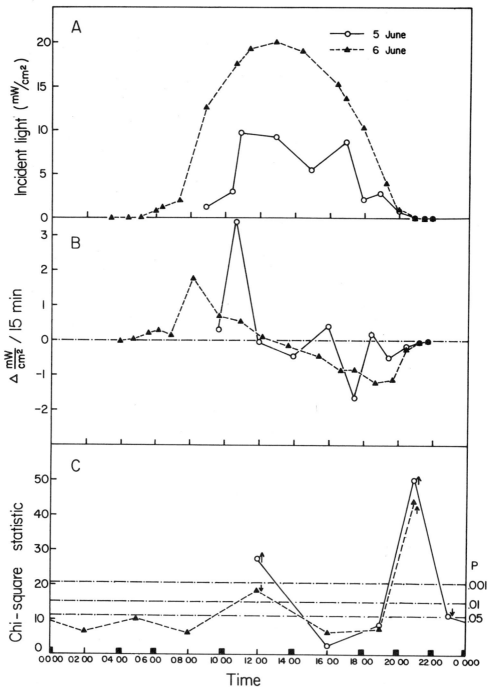

Fig. 2. A—Pattern of absolute incident light on Heney Lake, 5-6 June 1978. B—Rate of change of incident light (calculated from data of A). C—Chi-square statistic indicating degree of distributional shift in water mite population from one sampling period to next. Arrows signify direction of distributional shift, horizontal dashed lines mark critical values of the chi-square test for $P=0.05$, 0.01, 0.001 (5 df), and solid blocks show sampling times.

creasing light but well after the maximal rates of increase; this is an accentuation of the general pattern of downward movement during the period from just before midnight to sunset. The major distributional shifts at midday represent 4-hour intervals—twice as long as those around sunset and after dark; the water mites have more time to migrate before the next sampling period in these longer intervals and thus chi-square values for their vertical movement are overestimated in relation to the 2-hour sampling intervals, which makes the striking changes around sunset even more impressive.

There has been little observation or experimentation on the reactions of water mites to differences in light conditions. Uchida (1932) reported that water mites are active only in the daytime, resting on the bottom or on plants during the night or in cloudy weather. Pieczyński (1964, 1976) found that mites in the littoral and sublittoral of Mikolajskie Lake are also generally more mobile during the day than at night; the difference in daily mobility is less in the sublittoral than in the littoral. *Unionicola crassipes*, however, has a slightly higher mobility at night than during the day. The pelagic mites in Heney Lake (including *U. crassipes*) are at least as active during the night as during the day; thus it seems that the daily activity of water mites varies considerably in different regions of a lake.

Baylor and Smith (1953) observed that water mites reacted to a vertical beam of polarized light in the same way as cladocerans, "by swimming back and forth in the light at right angles to the plane of vibration" (p. 97); they also suggested a mechanism by which a water mite can orient under such conditions. It is not known whether this similarity in behavior toward light between water mites and cladocerans also exists for unpolarized sunlight or for light of different wavelengths.

The temperature structure of the water column (Fig. 1) appears not to be a factor in regulating the vertical migration of water mites. A thermocline between 4–7 m develops in the afternoon of 5 June, as the weather shifts from cloudy and windy to sunny and calm, and is maintained throughout the night—windy conditions returned late the next morning, progressively mixing more and more of the water column and breaking down the distinct thermocline that previously existed. The migration from deeper, cooler waters during the day into the warmer epilimnion at sunset involves crossing a sharp temperature gradient of about $2°C$ per meter depth (between 4–7 m on 5 June and 7–10 m on 6 June); nocturnal sinking brought the main body of water mites back down into the deeper waters by dawn and involved recrossing the same thermocline. None of the major shifts in vertical distribution (Fig. 2C) are marked by significant changes in the temperature structure of the lake and temperature thus appears not to have a discernible effect on the pattern of vertical migration of water mites.

Gliwicz and Biesiadka (1975) found that planktonic water mites in Madden Lake were uniformly distributed in the water column from January to May, a period marked by deep surface mixing, but disappeared from the hypolimnion of the lake from July to November when a strong thermal stratification and oxygen deficits in the metalimnion (negative heterograde oxygen distribution) became established. They proposed depleted oxygen as the proximal cause for the characteristic vertical distribution of mites in the latter half of the year, although the oxygen-deficit layer was also the thermal-gradient layer. While in Madden Lake the seasonal vertical distribution was apparently controlled by oxygen and/or temperature, diel vertical migration in Heney Lake is a clear response to changes in light intensity. Temperature gradients do not affect migrating mites in Heney Lake; neither is oxygen a determining factor. Readings taken from 1976–1977 showed it was never seriously depleted from bottom waters: values were consistently greater than 5 ppm.

Diel differences in population densities. The total number of mites found in the water column varied considerably over the study period (Fig. 3), from a maximum of

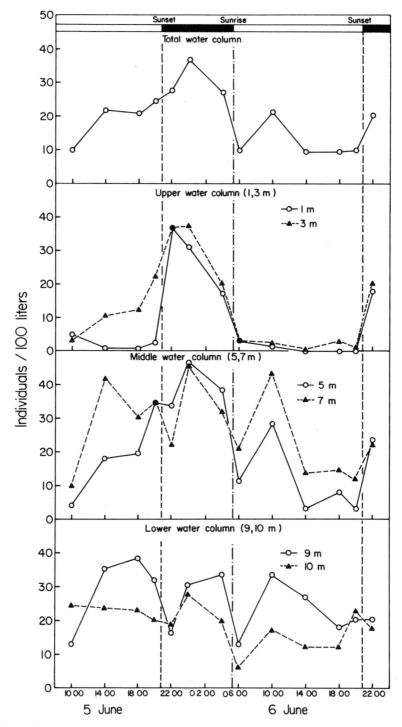

Fig. 3. Densities of water mites in entire water column (top panel) and at various depths throughout experimental period.

37/100 liters to a minimum of 10/100 liters; this variation was due mainly to changes in the upper layers of water with the deeper regions having considerably fewer differences in numbers over time. The main factor affecting these differences seems to be the presence or absence of daylight—differences are significant between numbers found during the day and those found at night (Mann-Whitney U-test, $P=0.05$). Densities increased distinctly around sunset and decreased sharply at sunrise. These changes might result from migration of a large portion of the mite population into or near the sediments during the day with subsequent emergence at night—a behavior similar to that seen in *Chaoborus* (Hutchinson 1967). Uchida (1932) reported that *Piona carnea* frequently buried itself in the mud; however, this behavior usually occurred during the evening or in cloudy weather.

Large-scale horizontal shifts due to wind conditions could be a second factor resulting in variation in numbers at different times. The lowest densities of mites generally occurred early on 5 June and throughout the afternoon and evening of 6 June—periods characterized by windy conditions—the calm weather between noon on 5 June and noon on 6 June generally showed high numbers of water mites.

Evolutionary causes and consequences. There are several theories to explain the adaptive advantages of vertical migration to zooplankton. This behavior is often viewed as a mechanism to avoid predation from visually orienting planktivores (Hutchinson 1967; Zaret and Suffern 1976). Since water mites are relatively large, darkly pigmented zooplankters (*Piona* and *Unionicola* adults exceed 1 mm long and are nearly as wide) and do not have the efficient escape reactions of copepods, they would seem to be easy targets for planktivorous fish. Their avoidance of surface waters during the day may thus be seen as an attempt to stay in the dimly lit areas of the lake where fish predation is less efficient. However certain aspects of the predator-prey interaction make this hypothesis less attractive.

Elton (1922) first documented the distasteful nature of water mites, which led sticklebacks to reject them after capture and to avoid them in the future. Water mites are not distasteful to all fish, however, as experiments at Heney Lake (pers. obs.) reveal that while the planktivorous banded killifish (*Fundulus diaphanus*) rejects adult *Piona constricta* after capture, the benthic-feeding brown bullhead (*Ictalurus nebulosus*) readily devours them. The fact that distasteful water mites are unharmed after capture and rejection by fish (Elton 1922; pers. obs.) makes this chemical adaptation an efficient anti-predator mechanism. Although water mites are often found in the alimentary tracts of some fish, their contribution to the food of these fish is small and the effect of fish predation on mite populations is negligible (Pieczyński 1976; Pieczyński and Prejs 1970). Among water mites, those species least affected by fish predation include *U. crassipes* and members of the genus *Piona* (Preczyński and Prejs 1970). The necessity of vertical migration for planktonic water mites to avoid fish predation and the ability of such predation to select for this behavior are thus in grave doubt.

A second class of arguments dealing with the adaptive advantages of vertical migration concern metabolic efficiency and increased fecundity. McLaren (1963) hypothesized that vertical migration is an evolutionary response to more efficient feeding at higher water temperatures and more efficient growth at lower water temperatures, which leads to larger size and hence to increased fecundity. In a thermally stratified lake, therefore, zooplankton migrate up to feed in the warm epilimnion and down to assimilate this food in the cooler hypolimnion. McLaren (1974) emphasized the "realized rates of increase involving natural mortality" (p. 92) and concluded that "vertical migration in thermally stratified waters may offer important demographic advantages to migrants whose fecundity is increased by development at low temperatures" (p. 100). These arguments are appealing for the vertical migration of water mites in Heney Lake, firstly because of the distinct thermal stratification of the water column and secondly because the development of the eggs and other inactive stages in the life history of water mites (*see* Böttger 1977) must take place on the bottom of the lake. A disadvantage of vertical migration is the slower rate of development of the individual when in the cooler hypolimnion (McLaren 1963, 1974), but since the inactive stages of water mites must develop there regardless of whether the animal migrates, there is no additional disadvantage in aquiring the behavior. Slower developmental rates for the active nymphs and adults will occur in the hypolimnion, however, and the necessary knowledge to build a life table modeling the effects of vertical migration (cf. McLaren 1974) on a population of water mites is lacking.

Vertical migration may also be viewed as a response to the food distribution of an organism (*see* Bainbridge 1961; McLaren 1963). Water mites are carnivores, showing food preference for cladocerans over copepods and rotifers (Gliwicz and Biesiadka 1975; Riessen in prep.); as such they may be expected to follow the movements of their prey. Since there is ample evidence that many herbivorous zooplankton, cladocerans as well as copepods, undergo distinct vertical migrations and that these are usually of the nocturnal or twilight form (Hutchinson 1967), water mites may be occupying the lower water column during the day and ascending at dusk to follow the movements of one or more of the cladoceran populations in the lake. It is not known at present, however, whether the cladoceran zooplankton of Heney Lake show marked vertical migrations, and if so, how close the pattern and timing is to that seen for the water mites.

The water mite vertical migration may have serious consequences for the herbivorous cladoceran community, whether or not it is a direct response to cladoceran vertical migration. The predatory effect of planktonic water mites on cladoceran populations can be highly significant (Gliwicz and Biesiadka 1975, Riessen in prep.) and may provide a strong selective pressure resulting in an altered pattern of vertical migration for their prey. At the very least, vertical migration of mites, coupled with their tactile predatory habits, results in an effect on the vertical migration of herbivorous zooplankton directly opposite to that imposed by visually orienting planktivorous fish.

While the adaptive significance of vertical migration to water mites is not entirely clear, it is more likely that the primary ultimate causes concern the attainment of food and its efficient utilization rather than the avoidance of predators.

References

BAINBRIDGE, R. 1961. Migrations, p. 431–463. *In* T. H. Waterman [ed;] , The physiology of Crustacea, v. 2. Academic.

BAYLOR, E. R., and R. E. SMITH. 1953. The orientation of Cladocera to polarized light. Am. Nat. **87**: 97–101.

BAYLY, I. A. E. 1963. Reversed diurnal vertical migration of planktonic Crustacea in inland waters of low hydrogen ion concentration. Nature **200**: 704–705.

BÖTTGER, K. 1976. Types of parasitism by larvae of water mites (Acari: Hydrachnellae). Freshwater Biol. **6**: 497–500.

———. 1977. The general life cycle of fresh water mites (Hydrachnellae, Acari). Acarologia **18**: 496–502.

CUSHING, D. H. 1951. The vertical migration of planktonic Crustacea. Biol. Rev. **26**: 158–192.

ELTON, C. S. 1922. On the colours of water-mites. Proc. Zool. Soc. Lond. (1922): 1231–1239.

FORWARD, R. B., JR. 1976. Light and diurnal vertical migration: photobehavior and photophysiology of plankton. Photochem. Photobiol. Rev. **1**: 157–209.

GLIWICZ, Z. M., and E. BIESIADKA. 1975. Pelagic water mites (Hydracarina) and their effect on the plankton community in a neotropical man-made lake. Arch. Hydrobiol. **76**: 65–88.

HUTCHINSON, G. E. 1967. A treatise on limnology, v. 2. Wiley.

LAROW, E. J. 1970. The effect of oxygen tension on the vertical migration of *Chaoborus* larvae. Limnol. Oceanogr. **15**: 357–362.

McLAREN, I. A. 1963. Effects of temperature on growth of zooplankton and the adaptive value of vertical migration. J. Fish. Res. Bd. Can. **20**: 685–727.

———. 1974. Demographic strategy of vertical migration by a marine copepod. Am. Nat. **108**: 91–102.

McNAUGHT, D. C., and A. D. HASLER. 1964. Rate of movement of populations of *Daphnia* in relation to changes in light intensity. J. Fish. Res. Bd. Can. **21**: 291–318.

PIECZYNSKI, E. 1964. Analysis of numbers, activity, and distribution of water mites (Hydracarina), and some other aquatic invertebrates in the lake littoral and sublittoral. Ekol. Pol. Ser. A **12**: 691–735.

———. 1976. Ecology of water mites (Hydracarnia) in lakes. Pol. Ecol. Stud. **2**(3): 5–54.

———, and A. PREJS. 1970. The share of water mites (Hydracarina) in the food of three species of fish in Lake Warniak. Ekol. Pol. Ser. A **18**: 445–452.

SCHINDLER, D. W. 1969. Two useful devices for vertical plankton and water sampling. J. Fish. Res. Bd. Can. **26**: 1948–1955.

UCHIDA, T. 1932. Some ecological observations on water mites. J. Fac. Sci. Hokkaido Imp. Univ. Ser. 6 Zool. **1**: 143–165.

ZARET, T. M., and J. S. SUFFERN. 1976. Vertical migration in zooplankton as a predator avoidance mechanism. Limnol. Oceanogr. **21**: 804–813.

14. Adaptive Value of Vertical Migration: A Simulation Model Argument for the Predation Hypothesis

David Wright, W. John O'Brien, and Gary L. Vinyard

Abstract

The predation hypothesis for the selective advantage of zooplankton vertical migration was tested by means of a model that simulated zooplankton population growth and fish predation rates under a variety of conditions. Non-migrating zooplankton were considered to remain at the surface where the light was bright and the water warm, theoretically increasing their susceptibility to fish predation but improving their growth and reproductive rates. Migrating zooplankton were considered to spend the days in darker, cooler, less productive waters where predation would be reduced but growth rates would be slowed. The simulation showed that, with one exception, migrating species always had higher rates of survivorship, indicating that the relative freedom from predation in dark waters offsets any disadvantage of reduced temperature and slowed growth rate. With *Bosmina longirostris*, a species that is too small to be heavily preyed upon by fish under any conditions, migration was not advantageous. Small life stages of copepods also were at an advantage if they remained near the surface. It therefore appears that fish predation is a major force maintaining migratory behavior in zooplankton.

The diel vertical migration of zooplankton—swimming upward during or around sunset and swimming downward during or around sunrise—is a universal behavioral characteristic of zooplankton (Hutchinson 1967). Vertical migration occurs in many different types of lakes and in tropical, temperate, and polar oceans. The behavior has been observed not only in crustacean zooplankton but also in a wide array of different marine zooplankton groups (Russell 1927; McLaren 1963). Thus there is no doubt that vertical migration is a widespread behavior of polyphyletic origin. Such similar behavior occurring in so many different animals in such diverse aquatic habitats seems certain to have potent selective advantages. However, there is still much debate over the nature of these selective advantages (Enright 1977). Zaret (pers. comm.) has listed eight published hypotheses for the selective force behind diel vertical migration; of these two have most frequently been set forth. First and most prevalent is what may be termed a resource hypothesis: that migrating populations achieve some energetic benefit either by regularly leaving and returning to their food source or by entering a cool hypolimnion (Enright 1977; Kerfoot 1970; McLaren 1963, 1974). This is a growth rate-dictated hypothesis. The second major hypothesis contends that the migrants lessen predation by moving where light intensities are too low for visual planktivores to find and catch them efficiently (Zaret and Suffern 1976). This, then, is a predation or death rate-dictated selective advantage.

As is well recognized, there is no way that either of these hypotheses can be proved correct and fully sufficient. What we would like to show is that the predation

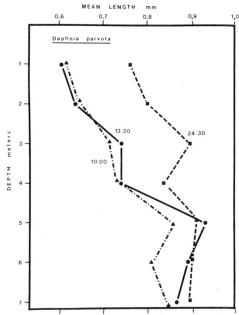

Fig. 1. Vertical size distribution of *Daphnia parvula* in Lone Star Lake on 1 and 2 July. Notice that there is a consistent strong size gradient during daylight hours which is much less pronounced during evening.

hypothesis cannot be negated in most cases and is sufficient to explain most observations of vertical migration. Therefore, although in the current climate the growth rate hypothesis seems to dominate, we believe the predation hypothesis is the one that should be accepted until proved false or inadequate.

We thank N. Slade for help and encouragement in the initial development of the planktivorous fish model. We also thank M. O'Brien for editorial help and B. R. Archinal for typing and secretarial assistance.

The case for predation

What is the major evidence for the predation hypothesis?

1. Under conditions of relatively high light intensity, planktivorous fish are extraordinarily rapacious. Werner and Hall (1974) found that in small pools, 10 small bluegill sunfish could consume hundreds of daphnid prey in a matter of minutes. In a more long term experiment, Hall et al. (1970) found that bluegills, when added in reasonable densities to quarter-acre ponds, completely eliminated all crustacean zooplankton in one summer. The well known ability of planktivorous fish to eliminate large zooplankton when introduced into virgin lakes and ponds speaks of the power of this type of predation.

2. Because planktivorous fish have been consistently found to be size-selective in their feeding, larger zooplankton within an assemblage are often far more susceptible to fish predation than are smaller ones. If vertical migration is related to predation, then larger zooplankton should migrate more deeply into lakes than smaller ones. This is commonly so, as first reported by Zaret and Suffern (1976) and also shown in Fig. 1, in which the mean length of *Daphnia parvula* sampled at seven depths is graphed. Additionally, Hutchinson (1967) reports that in several cases where daphnid congeners occur within the same lake the largest species is found deepest and the smallest migrates least.

3. If visual predation is important, then the light intensities to which zooplankton migrate should be related to visual attributes of planktivorous fish. Again, evidence exists for this interaction (Zaret and Suffern 1976; McNaught and Hassler 1964, 1966). For example, the amplitude of migration in turbid lakes is fairly slight, often only a few meters; whereas in oligotrophic lakes and seas, where the light intensity gradient is less, the amplitude of migration of the same species is quite marked. The predation hypothesis would claim that in both cases the animals migrate to a depth were the light intensity is such that planktivorous fish may have difficulty locating the migrant. Fig. 2 shows data indicating that this is so. By measuring the reactive distance of white crappie (an important planktivore in midwestern lakes of the U.S.) at different light intensities, we determined that the fish have increasing difficulty locating prey at light intensities below 10 lux. Maximum reduction of reactive distance occurs at 1 lux. This effect is far more pronounced for large prey than for small prey, suggesting that large prey benefit much more by migrating

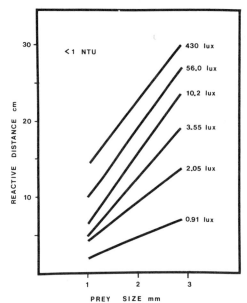

Fig. 2. Reactive distance of white crappie (*Pomoxis annularis*) as a function of prey sizes at different light intensities. Each line was developed from at least 20 points with lowest correlation coefficient of 0.70.

than do small species, which are difficult for visual predators to find under any conditions. Thus if larger zooplankton are migrating to reduce predation from fish, then they should migrate to depths where the light intensity is below 10 lux. Shown in Fig. 3 is the depth of migration of two common lake zooplankton, *D. parvula* and *Mesocyclops edax*, in a midwestern reservoir where white crappie are very common. As predicted, the mean population density lies at or below the 10-lux isophot.

Interestingly, the mean population density of two species common in this lake, *Bosmina longirostris* and *Diaptomus pallidus*, is not found below the 10-lux isophot; in fact, neither species changes its mean depth distribution much at all during the diel cycle (Fig. 4). However, even these exceptions support the predation hypothesis. *Bosmina longirostris* is quite small and would receive only modest benefit by reducing an already short reactive distance just a bit more, whereas *D. pallidus* is very

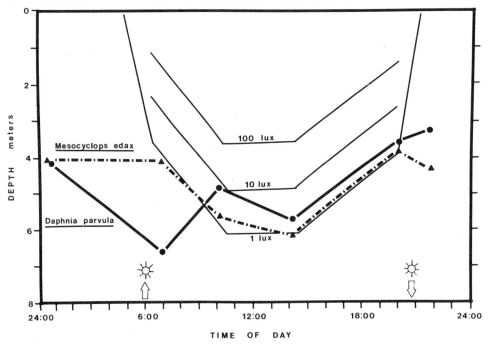

Fig. 3. Vertical diel distribution of mean population depth of *Daphnia parvula* and *Mesocyclops edax* in Lone Star Lake on 1 July. Notice that mean depth of each population is always below 10-lux isophot and often quite near 1-lux isophot.

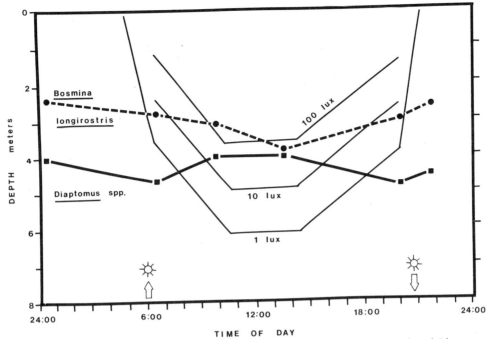

Fig. 4. Vertical diel distribution of mean population depth of *Bosmina longirostris* and *Diaptomus* spp. (certainly *D. pallidus* but perhaps other species as well). Mean depth of both populations shows little diel vertical variation and is not related to light intensity.

adept at evading fish suction attack (Drenner et al. 1978) and thus would gain little selective advantage by reducing the likelihood of being located.

The relationship between visual acuity and light intensity varies widely between fishes. The bluegill is more highly adapted to locating zooplankton at low light intensity than is the white crappie (Vinyard and O'Brien 1976). Both centrarchids have better visual acuity than the salmonids tested (Confer et al. 1978; Schmidt and O'Brien in prep.). The predation hypothesis predicts that the light intensity to which zooplankton migrate should vary as a function of the visual capabilities of the dominant planktivore.

Thus the predation hypothesis can account for the timing of migration (down at sunrise and up at sunset), for the depth of migration (to a light intensity of 10 lux or less), and even for the deeper migration of large animals (which require lower light conditions for reduced reactive distance). It can also provide an explanation for the absence of migration in small zooplankton and those that are relatively difficult for planktivorous fish to capture.

But what about the arguments put forth by those who espouse growth rate-dictated hypotheses? Does migration have energetic or resource-conserving advantages? What are the relative tradeoffs if a species migrates in terms of population growth versus death due to predation? The biggest problem with answering these questions is, of course, that there are no species or even similar species in which some migrate and some do not. Experimentally restraining a population from migrating would be so artificial as to prove little. Thus the only way to create a comparative situation is to mathematically simulate zooplankton population growth and predation mortality under the conditions of not migrating (high light intensities during the day) and migrating (low light intensities both at night and during the day). We here report the results of such a simulation using bluegill sunfish feeding on four species of zooplankton prey.

The simulation model

The feeding simulation was drawn directly from the published model of O'Brien et al. (1976) which is based on the phenomenon that when fish choose among several prey they choose the one that, by proximity or actual size, appears largest; at low prey densities this model becomes identical with that of Confer and Blades (1975) and very similar to that of Werner and Hall (1974). For the present study, certain changes were made to adapt the model to lake conditions. The visual field of the fishes was assumed to be a hemisphere, the shape that, in the original model, gave the best fit between predicted and observed feeding ratios (O'Brien et al. 1976). A typical fish was restricted to a given volume of the lake (either 6, 25, or 100 m^3) and located within that volume so that it could always survey its maximum perceptual volume. The probability of each prey size class appearing largest was not reassessed after every prey capture but once per hour or whenever a 10% change in the density of any prey class occurred since the last assessment. During each simulation sequence, the fish was considered to consume multiple prey (5 to 10) rather than a single prey; this procedure considerably reduced computer time and was found to produce results that were not significantly different from the lengthier method.

We simulated three densities of 8-cm bluegill sunfish: the high density was equal to 1 fish per 6 m^3; the medium, 1 fish per 25 m^3; and the low, 1 fish per 100 m^3. The high density is equivalent to distributing 50 kg of fish this size over 10 m of depth per ha, which is the same as the overall stocking rate of Hall et al. (1970). The medium density is 12 kg/ha distributed through 10 m depth, and the low is 3 kg/ha distributed through 10 m, which is very close to a value believed by Gliwicz and Prejs (1975) to be a low level of pelagic fish biomass.

The ability of the bluegill to locate prey, or its reactive distance, was treated as a function of prey body size and light intensity. In nature, daylight intensities near the surface greatly exceed 10 lux, while surface light intensities on moonless nights approach 0.48 lux (Vinyard 1977). Light intensities <1 lux are found in many lakes during the day (Wetzel 1975; O'Brien 1975). Thus the daytime reactive distances to each size class of a species considered to be nonmigrating were taken directly from the high light (10 lux) results of Vinyard and O'Brien (1976), while the low light reactive distances were taken from the 0.7-lux results (Vinyard and O'Brien 1976). These reactive distances were reduced by an additional third as suggested by Confer et al. (1978) and Hester (1968) to more accurately describe visual acuity within a three-dimensional field.

The fish was allowed to feed continuously, and the prey volume eaten per hour was dependent on a variety of rate functions. The rate at which a bluegill pursues zooplankton is temperature-dependent and can be defined by the following equation (Vinyard 1977):

$$PT \text{ (sec)} = [1.2579 + (-0.0366 \times \text{Temp})] \\ + [0.1106 + (-0.0028 \times \text{Temp})] \times \text{Distance}. \quad (1)$$

Search rates were set at 3 cm/sec (Vinyard 1977). Although search rates also decrease as a function of temperature, this relationship is not well quantified for bluegill. Maintaining a constant search speed of 3 cm/sec imposes a disadvantage on the migrating population in that the bluegill should be swimming slower where the water temperature is lower.

Temperature in well lighted areas was considered to be constant at 20°C, and to drop 10°C where light intensities were 0.7 lux.

The time required for bluegill to handle a prey was estimated from Werner (1974) and is a function of the size of both the bluegill and the prey:

$$HT \text{ (sec)} = 1.0 + 0.0045 \times \text{Exp} \langle 9.469 \times \\ \{\text{Prey size }(l)/[0.217 + (0.093 \\ \times \text{Fish length})]\}\rangle. \quad (2)$$

Bluegill digestion rates vary with the Q_{10} effect, decreasing by half with a 10°C drop

Table 1. Parameters used in model simulation for various zooplankton growth rates, egg number and development rates, and escape success per size class. References given in text.

Size Classes (mm)	Growth Rate (days)		Eggs	Egg Development Rate (days)		Escape % Successful
Daphnia galeata mendotae	10°C	20°C		10°C	20°C	
0.4			0			0
0.8	22.0	7.0	2			0
1.2	30.0	8.0	5	9.1	2.6	0
1.6	30.0	29.0	10			0
Daphnia magna	15°C	25°C		15°C	25°C	
1.0			0			0
1.75	9.0	4.1	0			0
7.5	9.5	4.3	5	5.5	2.5	0
3.25	22.0	10.0	14.5			0
4.0	63.0	28.7	30			0
Bosmina longirostris	10°C	20°C		10°C	20°C	
0.2			0			0
0.3	16.0	5.0	3			0
0.4	16.0	5.0	6	6.5	2.0	0
0.5	16.0	5.0	9			0
Diaptomus	10°C	20°C		10°C	20°C	
0.2			0			0
0.4	14.0	7.0	0			30
0.8	18.0	9.0	0	12.4	6.2	50
0.9	31.0	15.5	0			50
1.0			12			50

in temperature (Hoar and Randall 1969). At 20°C an 8-cm bluegill was considered to be able to eat a prey volume equivalent to 500 1.6-mm–long zooplankton per hour (Vinyard 1977; Confer et al. 1978). In most simulations, however, the prey volume eaten per hour did not approach this maximum.

The continuous range of prey sizes observed in nature was simplified to 4 or 5 size classes in each species (Table 1). Total body length, from the most anterior portion of the head to the base of the tail spine, was used for cladoceran body size, while cephalothorax length, or cephalothorax plus egg sac length, was used for the copepod body size. These sizes have been shown to stimulate comparable visibility reductions in small lake trout (Kettle and O'Brien 1978). Zooplankton volume, which was used to set limits on the size of ration consumed per hour, was considered to be a cubic function of length. The zooplankton species growth rates between size classes, reproductive values for each size class, and egg development times were taken from the literature: for *Daphnia magna* from Anderson (1932), Ryther (1954), Anderson and Jenkins (1942), and Green (1956); for *Daphnia galeata mendotae* from Hall (1964); for *B. longirostris* from Kerfoot (1974). Data for several diaptomid species were pooled from Confer and Cooley (1977), Ratzlaff (1966), and Ewers (1936) (*see* Table 1).

Zooplankton growth was treated as noncontinuous: each hour a given percentage of each size class was transformed to the next larger size as a function of food density and temperature. The birth rate of zooplankton was also treated as noncontinuous: each hour a given percentage of each size class was considered to give birth to individuals of the smallest size class. The number of eggs per individual in each size class was kept constant. Egg development times, and thus

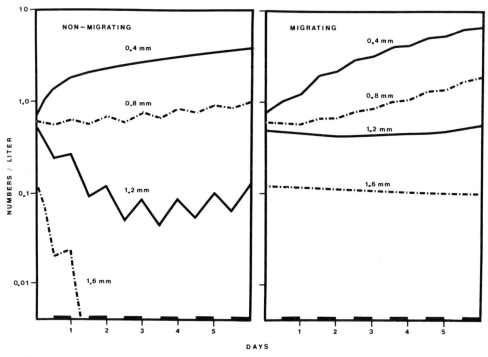

Fig. 5. Typical simulation run with *Daphnia galeata mendotae* exposed to fish densities of 1 per 25 m^3. Details given in text.

birth rates, were varied with temperature (Hall 1964). The rates of zooplankton reproduction we used (*see Table 1*) are probably a little higher than typical of field conditions.

The only death rate imposed on zooplankton was bluegill predation. The actual metabolic cost of vertical migration is thought to be minor (Vlymen 1970) and was not incorporated. The ability of some zooplankton, especially copepods, to evade the suction attack of fish is well known (Szlauer 1964, 1965; Confer and Blades 1975; Drenner et al. 1978) and was incorporated. Generation of a random number as described in O'Brien et al. (1976) was used to determine whether the escape probability had been exceeded.

Results and discussion

The typical behavior of the simulation is shown in Fig. 5 which represents a non-migrating and a migrating population of *D. galeata mendotae* at fish densities of 1 per 25 m^3. When the zooplankton do not migrate, the largest size class is depleted rapidly. Irregular growth patterns are particularly evident in the 1.2-mm size class and show the population increasing at night and decreasing during the day. However, when the zooplankton do migrate all four size classes are maintained, and the growth patterns, while still irregular in the two smallest size classes, are reversed: the increases occur at night and growth levels off during the day when the population is in the cool hypolimnion.

This contrast illustrates the most consistent finding of the simulation: that non-migrating populations, because they remain near the surface where light intensities and thus visibility to fish predators are high, are subject to high death rates due to predation, whereas migrating populations, because they move into a cold hypolimnion, must contend with slower egg development times and lower growth rates. Thus both strategies have disadvantages, and vertical migration

Table 2. Results of simulation model for different densities of planktivorous fish under conditions of migrating vs. nonmigrating zooplankton populations. Migrating population leaves surface during daylight hours for strata with low light intensity (0.7 luz) and 10°C colder water temperature. Figures given represent percent of initial adult biomass remaining after 5 days (under condition stated). Populations marked as extinct (Ext) indicate that all reproductive individuals have been eliminated and that the population will collapse.

	$100 \ m^3 / fish$ (%)	$25 \ m^3/fish$ (%)	$6 \ m^3/fish$ (%)
Daphnia magna			
Migrating	236	83	Ext
Nonmigrating	6	Ext	Ext
Daphnia galeata mendotae			
Migrating	163	103	50
Nonmigrating	124	21	Ext
Bosmina longirostris			
Migrating	441	384	105
Nonmigrating	1,158	802	304
Diaptomus spp.*			
Migrating	120	104	62
Nonmigrating	77	17	Ext

* 50% evasion success.

does not appear to be beneficial to a species' growth rate as has been suggested by McLaren (1963, 1974). In our simulations, and presumably also in nature, reducing predation during the day lowers the death rate enough to outweigh the negative effects on growth, and migrators are usually better able to survive and grow than are nonmigrators.

Table 2 shows the results of the simulations in terms of the percentage change in adult biomass after 5 days. For the larger species that cannot evade fish attack—*D. magna* and *D. galeata mendotae*—migration is always advantageous. However, the nonmigrating *B. longirostris* survive better than those that migrate. Because their small size makes the likelihood of visual predation on *Bosmina* very low whether they migrate or not, the negative effects of vertical migration become more evident. What little reduction in predation migrating *Bosmina* may receive is more than offset by their reduced rate of reproduction. This result is supported by field data (Fig. 4) showing that *Bosmina* generally migrate very little under most conditions. The adult diaptomid copepods, which have some ability to evade fish predation (Drenner et al. 1978), still gained some benefit from migrating. However, as Table 3 shows, the diaptomids were most successful when only the adults migrated while the nauplii and first size class of copepodites stayed in warmer conditions near the surface. This is a situation that is commonly observed in nature (Hutchinson 1967;

Table 3. Results of simulation model for *Diaptomus* spp. with fish density of 1/6 m³ and with varying size classes of a population undergoing vertical migration. Numbers within table represent percent of each initial size class remaining after 5 days under the various migration regimes. Notice that negative effects of predation are minimized for most size classes when only size classes 3 through 5 migrate.

Size Class	None	5	4-5	3-5	2-5	All
1	124	247	231	304	176	258
2	48	99	89	97	123	101
3	6	17	12	96	82	64
4	0.2	0.6	51	65	50	40
5	0.3	49	49	61	42	35

Zaret and Suffern 1976; Hart and Allanson 1976). Although Enright (1977) concluded that the failure of all life stages of copepods to migrate was an argument against the predation hypothesis, Table 3 suggests just the opposite. It also seems likely that the larger sizes of copepods which do migrate may be able to reduce the negative effects on their growth rate by staying higher in the water column than do cladocerans which are unable to evade fish attack. If copepods can stay closer to the euphotic zone where temperatures are higher and food is available all day, this may help to explain the apparent paradox that copepods persist in the presence of cladoceran competition even though their filtering rates are significantly lower.

Conclusions

In short, evidence from nature and this model supports predation pressure from visual planktivores as the major factor governing diel vertical migration. Species that are not heavily preyed upon by fish gain an advantage by not migrating into colder, less productive waters; for typical prey species, migration is always beneficial despite the accompanying reduction in growth rate. Until and unless further data can prove predation to be less potent a selective force than indicated by these studies, fish predation must be considered to be the primary influence on migratory behavior.

References

ANDERSON B. G. 1932. The number of preadult instars, growth, relative growth, and variation in *Daphnia magna*. Biol. Bull. **63**: 81-98.

———, and J. C. JENKINS. 1942. A time study of events in the lifespan of *Daphnia magna*. Biol. Bull. **83**: 260-272.

CONFER, J. L., and P. I. BLADES. 1975. Omnivorous zooplankton and planktivorous fish. Limnol. Oceanogr. **20**: 571-579.

———, and J. M. COOLEY. 1977. Copepod instar survival and predation by zooplankton. J. Fish. Res. Bd. Can. **34**: 703-706.

———, G. L. HOWICK, M. H. CORZETTE, S. L. KRAMER, S. FITZGIBBON, and R. LANDESBERG. 1978. Visual predation by planktivores. Oikos **31**: 27-37.

DRENNER, R. W., J. R. STRICKLER, and W. J. O'BRIEN. 1978. Capture probability: The role of zooplankter escape in the selective feeding of planktivorous fish. J. Fish Res. Bd. Canada, **35**: 1370-1373.

ENRIGHT, J. T. 1977. Diurnal vertical migration: Adaptive significance and timing. Part 1. Selective advantage: a metabolic model. Limnol. Oceanogr. **22**: 856-872.

EWERS, L. A. 1936. Propagation and rate of reproduction of some freshwater copepods. Trans. Am. Microsc. Soc. **55**: 230-238.

GLIWICZ, Z. M., and A. PREJS. 1975. Can planktivorous fish keep in check planktonic crustacean populations? A test of size-efficiency hypothesis in typical Polish lakes. Ekol. Pol. **25**: 567-591.

GREEN, J. 1956. Growth, size and reproduction in *Daphnia* (Crustacea: Cladocera). Proc. Zool. Soc. Lond. **126**: 173-203.

HALL, D. J. 1964. An experimental approach to the dynamics of a natural population of *Daphnia galeata mendotae*. Ecology **45**: 94-112.

———, W. E. COOPER, and E. E. WERNER. 1970. An experimental approach to the population dynamics and structure of freshwater animal communities. Limnol. Oceanogr. **15**: 839-928.

HART, R. C. and B. R. ALLANSON. 1976. The distribution and diel vertical migration of *Pseudodiaptomus hessei* (Mrazek) (Calanoida: Copepoda) in a subtropical lake in southern Africa. Freshwater Biol. **6**: 183-198.

HESTER, F. J. 1968. Visual contrast thresholds of the goldfish (*Carassive auratus*). Vision Res. **8**: 1315-1335.

HOAR, W. S., and D. J. RANDALL. 1969. Fish physiology, v. 1. Academic.

HUTCHINSON, G. E. 1967. A treatise on limnology, v. 2. Wiley.

KERFOOT, W. B. 1970. Bioenergetics of vertical migration. Am. Nat. **104**: 529-546.

KERFOOT, W. C. 1974. Egg-size cycle of a cladoceran. Ecology **55**: 1259-1270.

KETTLE, D., and W. J. O'BRIEN. 1978. Vulnerability of arctic zooplankton species to predation by small lake trout. J. Fish. Res. Bd. Can. **35**: 1495-1500.

McLAREN, I. 1963. Effects of temperature on the growth of zooplankton, and the adaptive value of vertical migration. J. Fish. Res. Bd. Can. **20**: 685-727.

———. 1974. Demographic strategy of vertical migration by a marine copepod. Am. Nat. **108**: 91-102.

McNAUGHT, D. C., and A. D. HASLER. 1964. Rate of movement of populations of *Daphnia* in relation to changes in light intensity. J. Fish. Res. Bd. Can. **21**: 291-318.

———, and ———. 1966. Photoenvironments of planktonic Crustacia in Lake Midriga. Int. Ver. Theor. Angew. Limnol. Verh. **16**: 194-203.

O'BRIEN, W. J. 1975. Nutrient limiting factors in turbid Kansas reservoirs. Completion Rep. Kansas Water Resour. Res. Inst. 41 p.

——, N. A. SLADE, and G. L. VINYARD. 1976. Apparent size as the determinant of prey selection by bluegill sunfish (*Lepomis macrochirus*). Ecology 57: 1304-1310.

RATZLAFF, W. 1976. Some aspects of the biology of *Diaptomus siciloides*. Ph. D. thesis, Univ. Kansas, Lawrence.

RUSSELL, E. S. 1927. The vertical distribution of plankton in the sea. Biol. Rev. 2: 213-263.

RYTHER, J. H. 1954. Inhibitory effects of phytoplankton upon the feeding of *Daphnia magna* with reference to growth, reproduction, and survival. Ecology 35: 522-533.

SZLAUER, L. 1964. Reaction of *Daphnia pulex* de Geer to the approach of different objects. Pol. Arch. Hydrobiol. 12: 6-16.

——. 1965. The refuge ability of plankton animals before models of plankton-eating animals. Pol. Arch. Hydrobiol. 13: 89-95.

VINYARD, G. L. 1977. Preference and accessibility as determinants of prey choice by the bluegill sunfish (*Lepomis macrochirus*). Ph. D. thesis, Univ. Kansas, Lawrence.

——, and W. J. O'BRIEN. 1976. Effect of light and turbidity on the reactive distance of bluegill (*Lepomis macrochirus*). J. Fish. Res. Bd. Can. 33: 2845-2849.

VLYMEN, W. J. 1970. Energy expenditure of swimming copepods. Limnol. Oceanogr. 15: 348-356.

WERNER, E. E. 1974. The fish size, prey size, handling time relation in several sunfishes and some implications. J. Fish. Res. Bd. Can. 31: 1531-1536.

——, and D. J. HALL. 1974. Optimal foraging and the size selection of prey by bluegill sunfish (*Lepomis macrochirus*). Ecology 55: 1042-1052.

WETZEL, R. G. 1975. Limnology. Saunders.

ZARET, T. M., and J. S. SUFFERN. 1976. Vertical migration in zooplankton as a predator avoidance mechanism. Limnol. Oceanogr. 21: 804-813.

III

Rotifer Feeding and Population Dynamics

15. Behavioral Determinants of Diet Quantity and Diet Quality in *Brachionus calyciflorus*

Peter L. Starkweather

Abstract

The feeding behavior of the monogonont rotifer *Brachionus calyciflorus* varies, depending on the type of food cell available in suspension. Three different food types, a bacterium *Aerobacter aerogenes*, a yeast *Rhodotorula glutinis*, and an alga *Euglena gracilis*, produce unique relationships between ingestion rate and biomass density. These patterns result from food-specific behaviors which the rotifers adopt when exposed to monospecific suspensions. When foods are mixed, *B. calyciflorus* feeding activity may be modified to produce complex relationships between ingestion rate and food density, relationships which are dissimilar from those obtained when the rotifers are fed on single food types. These or related behaviors may play a large part in determining the dietary interactions between the rotifers and their more heterogeneous food supplies in the natural planktonic community.

The rotifer *Brachionus calyciflorus* is a common planktonic inhabitant of mesotrophic and eutrophic lakes and ponds. While truly a cosmopolitan species, its occurrence, like that of ecologically similar zooplankters, is often seasonally restricted. The species can, however, reproduce rapidly and attain extremely high population numbers during those transient maxima. Because of their large populations, ubiquity, and rapid metabolic turnover, *B. calyciflorus* and its congeners are potentially important functional members of the suspension-feeding zooplankton community.

Of central importance in assessing the role of these rotifers in zooplankton communities is an understanding of their feeding biology. Of particular interest are estimations of the absolute rates at which the animals feed and the behavioral capability of individuals to regulate the quantity and quality of food in their diet. This paper summarizes the recent work on *B. calyciflorus* feeding behavior performed by J. J. Gilbert, T. M. Frost, and myself in our laboratory at Dartmouth College. In addition to the above-mentioned colleagues I thank M. Bean, who provided many hours of supportive technical assistance, and F. Bowsher, who prepared the manuscript.

Methods and materials

Our concern with the behavioral mechanisms used by *B. calyciflorus* while feeding and the quantitative estimation of feeding rates has led us to two distinct

---

Supported by NSF grants DEB 76-09768 and DEB 77-07541 (J. J. Gilbert) and DEB 78-02882 (J. J. Gilbert and P. L. Starkweather).

sets of methodologies. First, we have directly observed free-swimming and gently-restrained animals under differing feeding conditions and noted their variety of feeding behaviors (and often the relative frequency of behaviors). The details of our procedures are given elsewhere (Gilbert and Starkweather 1977, 1978). Secondly, we have made quantitative measurements of the feeding activity of *B. calyciflorus* when the rotifer was placed in pure suspensions and mixtures of the same foods used in the direct observations. The techniques used in these studies are summarized in our earlier work (Starkweather and Gilbert 1977a, b, 1978; Starkweather et al. in prep.).

In brief, we made direct observations of rotifers feeding in a variety of pure and mixed suspensions of several foods: bacteria, yeast, and green algae. We particularly noted the regulatory behaviors of food rejection and the formation of pseudotrochal "screens" which may occlude the entrance to the rotifers' buccal funnels. We quantified the latter activity by recording the proportion of *B. calyciflorus* which adopted the screening attitude while they swam through an optical field (90 magnifications, D.I.C. optics) in an observation chamber of about 22 × 22 × 1.0 mm.

For the quantitative feeding measurements we used radioisotopes (^{32}P and ^{33}P as H_3PO_4) to label food cells. We then introduced aliquots of radioactive suspensions to similar, but unlabeled, food preparations on which the rotifers had acclimated. We permitted the animals to ingest radioactive food for periods less than the gut passage time (*see* Starkweather and Gilbert 1977b), removed them from suspension, washed them free of residual radioactive cells, and prepared each group of 20–60 individuals for scintillation counting. Samples of the radioactive food suspensions were treated in a similar way to determine isotopic levels per cell and per microliter of cell suspension. A flow diagram of our procedure is shown in Fig. 1.

In both our direct observations and feed-

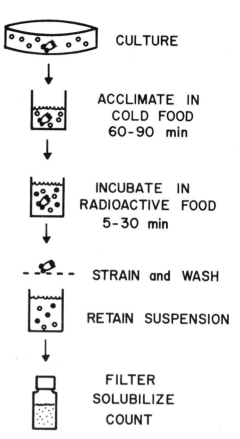

Fig. 1. Flow diagram of quantitative feeding estimate procedure using radioisotope-labeled food cells as tracers for consumption of *Euglena gracilis*, *Rhodotorula glutinis*, or *Aerobacter aerogenes*.

ing rate determinations, we principally used three food types, the bacterium *Aerobacter aerogenes*, the yeast *Rhodotorula glutinis*, and the flagellate alga *Euglena gracilis*.

Results and discussion

Our initial experiments compared the feeding activities of *B. calyciflorus* on pure suspensions of *E. gracilis* and *R. glutinis* (Starkweather and Gilbert 1977a). When feeding on the latter food, the rotifers show a continuous and roughly exponential decrease in clearance rate (μl animal^{-1} h^{-1}) with increased food density between 0.01 and 1,000 μg dry wt ml^{-1}. This decrease, while substantial (40 μl animal^{-1} h^{-1} to <0.5 μl animal

$^{-1}$ h^{-1}), still permits ingestion rate to rise in a density-dependent fashion throughout the tested range. Ingestion rates at yeast densities <0.1 μg ml^{-1} usually fall between 0.1 and 1.0 ng animal $^{-1}$h^{-1} (<100 cells h^{-1}) while those at the highest levels (>100 μg ml^{-1}) may exceed 100 ng animal $^{-1}$h^{-1} or between 5 and 10×10^3 cells h^{-1}.

As in the yeast suspensions, *B. calyciflorus* clearance rates for *E. gracilis* are strongly density-dependent, with maximal rates (45–50 μl animal $^{-1}$h^{-1}) found at low food density (0.1 μg ml^{-1}) and minimal values (ca. 0.1 μl animal $^{-1}$h^{-1}) in abundant food (100 μg ml^{-1}). This decline in clearance rate is more abrupt than that seen for yeast and the resulting ingestion rates are very nearly constant at *Euglena* biomass densities >5 μg ml^{-1}. The average ingestion rate for this broad region of density-independence is ca. 25 ng animal^{-1}h^{-1} or 40–50 *E. gracilis* cells consumed per rotifer per hour.

For a third food, the bacterium *Aerobacter aerogenes*, we have found another distinct pattern of feeding activity versus food density (Starkweather et al. in prep.). Clearance rates based on bacterial consumption are generally lower than those for either *Rhodotorula* or *Euglena*, except at higher food densities when bacterial clearance rates may slightly exceed those for the alga. The bacterial clearance rate values do not change systematically with variable food availability, average values falling between 0.1 and 1.0 μl animal $^{-1}$ h^{-1} for a biomass density range of 0.01–100.0 μg ml^{-1}. This pattern is reflected in ingestion rates for *B. calyciflorus* feeding on *A. aerogenes* that are strongly density-dependent, with a tenfold gain in biomass ingested per unit time for every order of magnitude increase in cell number.

Figure 2 shows a summary of these relationships for the three foods. The ingestion rate patterns have some similarities; each shows a region of density-dependence and each could be fit to common rectilinear or hyperbolic functions

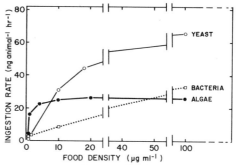

Fig. 2. *Brachionus calyciflorus* ingestion rate relationships to variable biomass densities of three foods: yeast, *Rhodotorula glutinis*; alga, *Euglena gracilis*; bacterium, *Aerobacter aerogenes*. Values shown are means (*n* = 3–5) of estimates reported elsewhere (*see text*); the lines are visually fitted.

using different constants. For this reason, it is possible to consider the differences among the feeding relationships as derived from the passive effects of different cell encounter rates or handling times, for instance, and not to active behavioral differences in the way the rotifers collect and ingest (or reject) the various foods. We have found, however, that *B. calyciflorus* does behave differently when feeding on the three foods and that these fundamental behavioral dissimilarities are better able to explain the diversity of observed patterns.

As noted in review elsewhere (Gilbert and Starkweather 1977), the morphology and function of the feeding apparatus of *Brachionus* are extremely complex. The coronal cilia produce water currents which are responsible for locomotion as well as feeding, with a circumapical band of cilia (the cingulum) contributing the major food-transport currents and the buccal funnel ciliary fields directing captured particles toward the mouth.

At least three distinct behaviors are used by *B. calyciflorus* to regulate the quantity of food that enters the gut. Particles that have entered the buccal funnel may be rejected before ingestion, presumably by a modification in ciliary action, and be passed from the coronal region through a ventral cleft in the pseudotrochus. Particles that have been accumulated in the distensible oral cavity may be forced back into the buccal region by the jaws and be subsequently rejected anteriorly from the funnel or through the ventral cleft. A third regulatory mechanism is the formation of "screens" over the buccal funnel, as mentioned earlier. The screens are formed by medially bent pseudotrochal cilia which prevent food particles (at least those in the size range between and including R. glutinis and E. gracilis) from being collected by the rotifers. When the pseudotrochal cilia are extended forward or laterally, there is no such screening effect and particles freely enter the buccal region.

There are other possible regulatory mechanisms which have been suggested before by ourselves and others, and which await further study. These include less obvious modification of ciliary actions involved in both feeding and locomotion, physical interference of high particle densities, intermittency of feeding effort, and behavioral variation in collection efficiency.

The incidence of pseudotrochal screening is affected both by food type and, for those foods which elicit substantial screening behavior, by cell density. *Rhodotorula glutinis* only rarely induces *Brachionus* to form pseudotrochal screens and the proportion of rotifers observed with screens at yeast densities of 1.0 $\mu g\ ml^{-1}$ is not significantly different from the proportion with screens at 100 $\mu g\ ml^{-1}$. At both low and high food densities *E. gracilis* elicits higher proportions of screening than that found with yeast. Table 1 summarizes these results as well as those for three other foods of intermediate size. These data suggest that food particle size may be an important determinant in eliciting pseudotrochal screening, with the smallest cells having the lowest values as well as the least density-dependence. The exception is found with *A. aerogenes*; the bacterium induces a low proportion (0.09 ± 0.03, n = 4) of *B. calyciflorus* to form screens at 1.0 $\mu g\ ml^{-1}$, but at 100 $\mu g\ ml^{-1}$ the bacterium induces a significantly greater proportion (0.41 ± 0.13, n = 9).

The frequent adoption of pseudotrochal screens by *B. calyciflorus* in moderate to high densities of *E. gracilis* helps explain the broad region of ingestion rate density-independence seen with that food (Fig. 2). The screening behavior effectively excludes excess cells from the buccal region and results in a strong regulation of feeding. For yeast (*R. glutinis*), screening regulation is seldom observed and ingestion rate is relatively unrestricted even at very high cell densities. Pseudotrochal screens are seemingly ineffective in restricting *Aerobacter* consumption, as evidenced by

Table 1. Proportions of *Brachionus calyciflorus* observed with pseudotrochal screens in 1 and 100 $\mu g\ ml^{-1}$ suspensions of five food types.

Food Type	Cell or Colony Mean Dry Wt ($\mu g \times 10^{-5}$)	Proportion with Screens			
		Particle Density ($\mu g\ ml^{-1}$)			
		1		100	
		mean ± SD	n	mean ± SD	n
Rhodotorula	1.97	0.04 ± 0.03	5	0.07 ± 0.06	12
Chlamydomonas	4.86	0.02 ± 0.01	5	0.14 ± 0.02	6
Ankistrodesmus	7.59	0.08 ± 0.03	8	0.34 ± 0.03	8
Scenedesmus	10.86	0.09 ± 0.06	8	0.45 ± 0.07	8
Euglena	50.10	0.20 ± 0.08	7	0.55 ± 0.13	25

a constant clearance rate coincident with a 3- to 5-fold change in screening. This observation makes the significance of the behavior in bacteria-fed rotifers obscure. Interestingly, and as described elsewhere (Starkweather et al. in prep.), bacteria-elicited screening may affect the consumption of other suspended foods.

We have utilized three classes of experiments to evaluate the quantitative feeding performance of *B. calyciflorus* when the rotifer is exposed to mixtures of the above foods. First, we have held the proportion of two foods in a mixture constant and varied the total biomass available. Secondly, we have varied the proportions of two cell types available in mixtures while holding the total biomass constant. Lastly, we have made fixed additions of one food type to variable cell densities of alternative foods.

In the first class of experiments we mixed equal quantities (by dry weight biomass) of *R. glutinis* and *E. gracilis* to obtain food suspensions with total densities between 0.1 and 200 $\mu g\ ml^{-1}$ (Starkweather and Gilbert in prep.). In these and the other experiments using mixed food suspensions we labeled the constituent cell types with different radioisotopes, allowing us to measure the clearance and ingestion rates for the two foods simultaneously.

The relationship between ingestion rate and food density for *R. glutinis* consumption was very similar to that observed in pure suspensions. In those experiments we found a direct proportionality between rate and density throughout the tested range. The series of ingestion rates obtained for *Euglena*, on the other hand, differed from both the yeast curve and the characteristic pure-suspension *Euglena* pattern. Below 10 $\mu g\ ml^{-1}$ total food density, the uptake of *Euglena* by *B. calyciflorus* resembled that found in pure suspensions, even including an interval (1–10 $\mu g\ ml^{-1}$) of density-independence. At higher densities, algal ingestion increased sharply with food density to levels comparable to those of yeast, a response very different from that seen when *Euglena* was in pure suspension.

The explanation for the change in ingestion pattern for *Euglena* mixed with yeast versus *Euglena* in pure suspension relies upon our direct observation of the rotifers' behavior in the two feeding conditions. The presence of as little as 1% yeast in a *Euglena* suspension reduces pseudotrochal screening by *B. calyciflorus* by a factor of 2–5 times. Without pseudotrochal screening, the algal cells are continuously accepted into the buccal funnel. At 100 $\mu g\ ml^{-1}$, *B. calyciflorus* consumes about 10 times as many *Euglena* when yeast is present in the suspension than when yeast is absent.

In the second group of *Euglena-Rhodotorula* experiments, we kept total food density constant (20 $\mu g\ ml^{-1}$) and varied the relative quantity of the two foods so that suspensions were composed of 20, 40, 60, or 80% of either cell type. For both foods, *B. calyciflorus* had ingestion rates proportional to the relative availability of the food types. *Euglena* ingestion rates did not increase with *Euglena* density in direct proportion, however, since ingestion about doubled as density changed from 4 (20%) to 16 (80%) $\mu g\ ml^{-1}$. This partial regulation, not found with yeast, resulted in an overall drop in total ingestion rate as *Euglena* density increased in the mixtures.

In the third set of experiments we combined fixed quantities of *Euglena* (10 or 100 $\mu g\ ml^{-1}$) and variable densities of *Rhodotorula* (Fig. 3). In all cases, *Euglena* consumption was uniform for a given algal density, irrespective of the number of yeast cells in suspension. However, the reduction of regulatory pseudotrochal screening caused by the presence of yeast with dense *Euglena* permitted ingestion rates for the algae that were higher at 100 $\mu g\ ml^{-1}$ than at 10 $\mu g\ ml^{-1}$. For the yeast, both the density-dependent pattern of ingestion rate and the absolute consumption values were equivalent at 10 and 100 $\mu g\ ml^{-1}$ of added *Euglena*. This result is consistent with those described

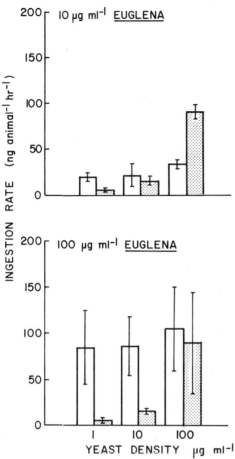

Fig. 3. Ingestion rates of *Brachionus calyciflorus* feeding on variable densities of *Rhodotorula glutinis* (yeast) with *Euglena gracilis* present at fixed densities of either 10 (upper panel) or 100 (lower panel) $\mu g \ ml^{-1}$. *Rhodotorula glutinis* ingestion rates represented by stippled bars, *E. gracilis* rates by open bars.

for the first two experimental schemes, i.e. *Euglena* consumption is strongly affected by the presence of yeast, while yeast ingestion is relatively insensitive to *Euglena* even when the latter is present at very high levels.

Conclusions

Brachionus calyciflorus has a variable capacity to regulate the quantity of food included in its diet depending upon the nature of the food cells available in suspension. These differences are effected by several active behavioral mechanisms which the rotifers exhibit when feeding on a variety of algal, fungal, and bacterial foods. Among the most important regulatory behaviors are rejection of collected food materials from the buccal funnel and oral cavity, as well as the formation of pseudotrochal screens which exclude at least some cell types from the buccal region. This last behavior seems to be an energetically favorable one since it diminishes the necessity of processing and rejecting excess or undesirable foods.

The behavior of *Brachionus* when fed one food type may change when that food is mixed with other particles. The resultant feeding rates are hybrids of the pure suspension patterns. This type of modification produces feeding rates for foods (e.g. *Euglena*) which may vary considerably depending upon the qualitative nature of the cell mixture. The consumption of other food types (e.g. *Rhodotorula*), however, is relatively unaffected by the presence or absence of alternative foods.

The types of feeding behaviors exhibited by *B. calyciflorus* are varied and their effects are substantial in regulating the quantity and quality of food in the rotifers' diet. The importance of these patterns in laboratory-cultured animals leads us to believe that similar behaviors almost certainly play a part in interactions between these rotifers and natural, heterogeneous food supplies.

References

GILBERT, J. J., and P. L. STARKWEATHER. 1977. Feeding in the rotifer *Brachionus calyciflorus* I. Regulatory mechanisms. Oecologia (Berl.) **28**:125-131.

—— and ——. 1978. Feeding in the rotifer *Brachionus calyciflorus* III. Direct observations on the effects of food type, food density, change in food type, and starvation on the incidence of pseudotrochal screening. Int. Ver. Theor. Angew. Limnol. Verh. **20**: 2382-2388.

STARKWEATHER, P. L., and J. J. GILBERT. 1977a. Feeding in the rotifer *Brachionus calyciflorus* II. Effect of food density on feeding rates using *Euglena gracilis* and *Rhodotorula glutinis*. Oecologia (Berl.) **28**:133-139.

—— and J. J. GILBERT. 1977b. Radiotracer

determination of feeding in *Brachionus calyciflorus*: the importance of gut passage times. Arch. Hydrobiol. Ergeb. Limnol. 8:261–263.

——— and J. J. GILBERT. 1978. Feeding in the rotifer *Brachionus calyciflorus* IV. Selective feeding on tracer particles as a factor in trophic ecology and in situ technique. Int. Ver. Theor. Angew. Limnol. Verh. 20:2389-2394.

16. Feeding in the Rotifer *Asplanchna:* Behavior, Cannibalism, Selectivity, Prey Defenses, and Impact on Rotifer Communities

John J. Gilbert

Abstract

The recent literature is reviewed, and some new, experimental results on cannibalism and the defenses of various rotifers against *Asplanchna* predation are provided. The bases of active and passive selective feeding in *Asplanchna* are considered, and some ideas on the evolution of cannibalism in *Asplanchna* and the egg-carrying habit in *Brachionus* are presented. *Asplanchna* has the capacity for both functional and rapid, numerical responses to prey density and, thus, can be expected to influence the survivorship of certain prey organisms and the species structure of zooplankton communities.

Asplanchna is a large, ovoviviparous, planktonic rotifer which is common in ponds and lakes throughout the world. Although omnivorous, it is an important predator and may exert a considerable influence on the survivorship of a variety of prey organisms, especially smaller rotifers. Owing to its potential for rapid, parthenogenetic reproduction, this predator can be expected to exhibit a rapid numerical as well as a functional response to prey density. In this respect, *Asplanchna* differs from planktonic, crustacean predators, whose life spans and generation times are very much longer than those of *Asplanchna* and which, therefore, exhibit primarily functional responses.

Here I review available information on various aspects of feeding in *Asplanchna*. Most of the relevant studies have been published within the last several years. In addition, some of my previously unpublished data and thoughts are presented where appropriate. The methods used to obtain these data are those of Gilbert (1976a, 1977b).

I thank J. R. Litton, Jr. and R. E. Magnien for improving the manuscript.

Feeding behavior

The feeding behavior of *Asplanchna intermedia* and *A. sieboldi* has been described and analyzed (Gilbert 1976a, 1977a, 1978). These predators swim randomly in the presence of prey and only respond to them when their coronae physically contact them. The *Asplanchna* do not seem to orient to prey at a distance and depend upon chance encounters for prey location. A similar conclusion was drawn by Pourriot (1965) for *A. brightwelli*. It is possible that *Asplanchna's* swimming speed or rate of change of swimming direction may be influenced by the presence of prey, but as yet there is no evidence for such kinetic responses.

This research was supported by National Science Foundation research grant DEB 76-09768 and Public Health Service research career development award K04-GM-70557 from the National Institute of General Medical Sciences.

When *Asplanchna* contacts a prey with its corona, it may or may not attack the prey. Attack or feeding behavior consists of at least four clearly recognizable responses which serve to first capture and then ingest the prey. The corona moves in response to the location of the prey so that the mouth is appressed to the prey. The mouth opens to admit the prey into the large pharynx. Muscles of the corona contract to narrow the diameter of the corona and entrap the prey into the pharynx. The pincerlike trophi or jaws manipulate, position, and then push the prey down the esophagus toward the blind stomach.

Asplanchna does not exhibit any of these responses when prey are contacted on regions of its body other than the corona. The receptors responsible for prey recognition and for initiating feeding responses, therefore, seem to be limited to the corona. Numerous sensory organs and structures on the corona of *Asplanchna* have been described (Nachtwey 1925; Waniczek 1930), but the ones involved in feeding behavior have not been identified.

The feeding responses of *Asplanchna* can be induced by chemical stimulation alone (Gilbert 1977a). Starved campanulate females of *A. intermedia* (clone 12C1) regularly attack and eat saccate-female clonemates. Aqueous homogenates of the saccate females and also filtrates of such homogenates induce characteristic feeding responses in these predators. The corona-movement, mouth-opening, corona-contraction, and jaw-movement responses observed in normal feeding behavior are all clearly elicited and closely coupled regarding their occurrence and intensity in different trials.

The activity of these filtrates is reduced by about 50% when they are heated at 100°C for 5 minutes (Gilbert unpubl.). This loss in activity may be due to denaturation of proteins.

Prey recognition and feeding behavior in *A. intermedia*, then, seems to be controlled by coronal contact chemoreceptors. Selective feeding behavior (*see below*) is probably mediated to a large extent by chemical differences among the surfaces of potential food types.

The role of the physical properties of particulate material in controlling feeding behavior in *Asplanchna* is not well known. In some species of *Asplanchna*, potential food items must be of a certain minimum size before they can elicit an attack response, be captured, or be ingested. Naumann (1923) indicated that the minimum particle size that could be eaten by an unidentified species was about 15 μm, and Tribush (1960) and Sorokin and Mordukhai-Boltovskaya (1962) found that *A. herricki* and *A. priodonta* could only poorly utilize small algae. Gilbert (unpubl.) found that *A. brightwelli, A. intermedia,* and *A. sieboldi* inefficiently capture and cannot subsist on small prey organisms, such as *Euglena gracilis*. Also, saccate females of the latter two species readily consume *Paramecium aurelia*, but the much larger, campanulate females of these species rarely respond to them (Gilbert unpubl.).

The basis for the inability of some species of *Asplanchna* to eat small food organisms has not been extensively investigated. For example, information is unavailable on the degree to which certain small organisms elicit feeding responses, have surfaces with chemical properties capable of eliciting feeding responses, can be captured if attacked, can be ingested if captured, and can be assimilated if ingested. Since saccate *A. sieboldi* readily attack *P. aurelia*, this prey must contain stimulatory chemicals on its surface. The failure of the much larger campanulates of this species to attack this organism, therefore, is probably due to its relatively small size in proportion to the corona of these females.

Resvoj (1926) believed that only relatively large particles elicited feeding in *A. sieboldi* because only these provided a threshold mechanical stimulus. The explanation is probably not so simple, since feeding responses can be induced with particle-free filtrates of homogenized prey. Perhaps particles with large surface areas may be needed to provide above-threshold chemical stimulation or to stimulate a requisite number of coronal chemoreceptors. It is also possible that mechanical stimulation may depress the threshold for chemical stimulation.

The upper size limit for prey that can be eaten by *Asplanchna* is a function of the size of the corona of the *Asplanchna*. Larger individuals can eat larger prey. The large, campanulate females of *A. intermedia* and *A. sieboldi*, for example, eat much larger prey than the smaller, saccate females of these species (Gilbert 1973a, b).

Although *Asplanchna* is raptorial, some species may also be suspension feeders. Gossler (1950) reported that in *A. priodonta* the ciliary covering of the mouth can withdraw small particles like nannoplankton from suspension and then transport them down into the pharynx. Gossler concluded that a small part of the diet of this species could be taken in through such filter feeding. Similarly, Tribush (1960) observed that *A. herricki* could ingest carmine particles and concluded that the species was both a predator and a suspension feeder.

As expected, the proclivity of *Asplanchna* to attack a potential prey is a function of its state of hunger. The data presented in Table 1 (from Gilbert unpubl.) show that, with increasing starvation time, two species of *Asplanchna* both attack a greater proportion of prey that contact their coronae and also capture a greater proportion of the attacked prey. Accordingly, all experiments conducted to test or compare the feeding responses of *Asplanchna* must control for this factor.

Sated animals do not exhibit feeding responses, even to preferred prey (Gilbert unpubl.). Accordingly, there probably is a feedback between some part of the digestive system and either the coronal chemoreceptors themselves or neurons between the receptors and the various muscles which affect the feeding behavior. The basis of this feedback is unknown but may be in the form of stretch receptors in the stomach which reflexly inhibit the positive response to chemostimulation.

Asplanchna may be very discriminating in its diet. Decisions on whether to try to eat a potential prey organism appear to be made primarily at the time of initial encounter, rather than after the prey has been attacked or captured (Gilbert 1978). The predator decides whether or not to attack a prey item that has contacted its corona. Once an attack is triggered, *Asplanchna* will try to capture and then ingest the prey, even if the prey-type is only infrequently attacked. For example, adult campanulate females of *A. intermedia* attack *Brachionus calyciflorus* and *Volvox aureus* much less frequently than *A. brightwelli*; however, if any of these three prey-types is attacked, there is a very high probability that the prey

Table 1. Effect of starvation time (ST) on the feeding behavior of two species of adult *Asplanchna* presented with newborn, female prey of their own clone and morphotype. Pairs of proportions analyzed using G-statistic.

Species, Clone, Morphotype	No. Predators Tested	ST (min)	No. Contacts with Prey	Proportion Contacted Prey Attacked		Proportion Attacked Prey Captured		Proportion Captured Prey Ingested
				G	P	G	P	
sieboldi, 10C6, saccate	6	195	54	0.83		0.09		0
				1.60	0.206	6.43	0.011	
	9	315	76	0.91		0.28		0
brightwelli 4B61, α-form	4	120	90	0.89		0.28		0
				20.69	5.4×10^{-6}	22.52	2.08×10^{-6}	
	7	255	154	1.00		0.60		0
				0		27.82	1.34×10^{-7}	
	3	345	59	1.00		0.76		0

will be captured and then ingested. This type of mechanism for selective feeding is very efficient, for objects can be recognized rapidly without having to be handled.

Asplanchna girodi has occasionally been observed to reject captured food. The alga *Microcystis* is not usually attacked by this species; however, when it is, it is often rejected after being taken into the pharynx (Gilbert unpubl.).

The subject of selective feeding in *Asplanchna* is considered in detail separately below.

Cannibalism

The potential for cannibalism in *Asplanchna* varies greatly both among species and also within species, according to the morphotype of the female. Laboratory investigations of cannibalism and its effects on female polymorphism have been conducted by Gilbert (1973a, b, 1975, 1976a,b,c, 1977a,b,c) on *A. intermedia*, *A. sieboldi*, *A. brightwelli*, and *A. girodi*. The major results of these and additional, unpublished studies are summarized below and in Tables 2 and 3.

Asplanchna intermedia. Adult females of the large, campanulate morphotype of this species are voracious cannibals on conspecific females of the smaller, saccate morphotype. About 85% of the adult saccate females contacted are attacked, and about 5% of the attacked individuals are subsequently captured and ingested (Gilbert 1976a,b, 1977b). These campanulate-female predators, however, only rarely attack newborn females belonging to the cruciform and campanulate morphotypes, the attack probabilities being 0.18 and 0.08, respectively (Gilbert 1976a). Furthermore, these campanulate predators almost never attack the males of their own species, responding to only 5% of those encountered (Gilbert 1976b, 1977a,b). The failure of these cannibalistic, campanulate females to attack cruciform and campanulate females, as well as males, is limited to individuals from their own species, for they readily attack both these female morphotypes and males belonging to *A. sieboldi* (Gilbert 1976a) (Table 2). Thus, intense cannibalism by the campanulate females of *A. intermedia* is restricted to saccate-female prey. The other female morphotypes and the males of this species are relatively immune from attack by these cannibals. All of the experiments with this species were conducted with one clone, and so the extent to which these findings may pertain to interactions among different clones or races of this species is not known. The subject of defenses against

Table 2. Summary of some predator-prey interactions within and between *Asplanchna intermedia* and *A. sieboldi*. Feeding responses of starved, adult predators after contact with prey are given as attack probabilities. The (+) sign indicates an unquantified but very high attack probability. (From Gilbert 1977d.)

Predator	Prey							
	A. intermedia				*A. sieboldi*			
	Saccate	Cruciform	Campanulate		Saccate	Cruciform	Campanulate	
	♀ Adults	♀ Young	♀ Young	♂	♀ Adults	♀ Young	♀ Young	♂
A. intermedia campanulate♀	0.84	0.18	0.08	0.05	+	0.94	0.81	0.77
A. sieboldi cruciform♀	0.86	+	0.96	0.58	0.20	0.02		0.51
A. sieboldi campanulate♀	+	+	+	+	+	1.00	0.97	0.82

Table 3. Feeding behavior of three species of adult *Asplanchna* when presented with newborn female prey of their own clone and morphotype.

Species, Clone, Morphotype	Replicate	No. Predators Tested	Predator Starvation Time (min)	No. Contacts with prey	Proportion Contacted Prey Attacked	Proportion Attacked Prey Captured	Proportion Captured Prey Ingested
sieboldi, 10C6, saccate	1	15	195–315	130	0.88	0.20	0
	2	11	360	123	0.87	0.13	0
	1–2	26	195–360	253	0.87	0.17	0
brightwelli, 4B61, α–form	1	17	180	233	0.94	0.16	0
	2	14	120–345	303	0.97	0.54	0
	3	7	210	59	0.58	0.06	0
	4	11	285	99	0.73	0.17	0
	1–4	49	120–345	694	0.89	0.34	0
girodi, 5A1	1	10	240	> 50	0	0	0
	2	10	150	> 50	0	0	0
	3	16	210	> 80	0	0	0
	4	17	195–320	> 85	0	0	0
	1–4	53	150–320	> 260	0	0	0

cannibalism is considered in detail separately below.

Although saccate females of *A. intermedia* have not been tested individually for their potential as cannibals, some results show that starved females from populations cultured on *Paramecium* may become cannibals. (Gilbert 1973a).

Asplanchna sieboldi. Adult campanulate females of *A. sieboldi* are more voracious than those of *A. intermedia* and readily attack conspecific females of all three morphotypes and males (Gilbert 1976a,b, 1977b). Attack probabilities for all of these prey-types are >0.77 (Table 2).

Although campanulate females of *A. sieboldi* can readily ingest conspecific saccate females, they have considerable difficulty capturing and ingesting males and even newborn campanulate and cruciform females of their own species (Gilbert 1976a, 1977b). They will, however, sometimes manage to eat these prey (Powers 1912; Gilbert 1976a). Thus, although not immune to attack, males and cruciform and campanulate females of this species are morphologically well protected against cannibalism (*see separate section below*).

Cruciform females of *A. sieboldi* will attack conspecific, saccate females but are much more inclined to attack saccate females of *A. intermedia* (Gilbert 1976a) (Table 2). Similarly, they only rarely attack conspecific, newborn, cruciform females (Gilbert 1976a) (Table 2). Thus, saccate and cruciform females of *A. sieboldi* are relatively immune to attack by cannibalistic, cruciform predators. Saccate prey that are attacked are readily captured and ingested (Gilbert unpubl.), but cruciform prey are morphologically protected from capture. *Asplanchna sieboldi* cruciforms readily attack conspecific males, but these males are also morphologically well-protected from being captured. The defenses of this species against cannibalism are considered in a separate section below.

Saccate females of *A. sieboldi* will try to eat newborn, conspecific females of their own morphotype, the probability of attack being 0.87; however, they are usually unable to capture these relatively large prey and cannot ingest the ones they do capture (Gilbert unpubl.) (Table 3).

Asplanchna brightwelli. The potential for cannibalism in the relatively small, α-form females of *A. brightwelli* is similar to that in the saccate females of *A. sieboldi*. Adult

Table 4. Feeding behavior of adult *Asplanchna girodi* (clone 5A1) when presented with four types of prey: newborn, female clonemates; newborn, α–form, female *Asplanchna brightwelli* (clone 4B61); short-spined *Brachionus calyciflorus* with no eggs attached to their loricae; and *Paramecium aurelia*.

Replicate	No. Predators Tested	Prey Tested	No. Contacts with Prey	No. Contacted Prey Attacked	Proportion Attacked Prey Captured	Proportion Captured Prey Ingested
1	10	*A. girodi*	> 50	0	0	0
	10	*A. brightwelli*	> 50	0	0	0
	10	*Paramecium*	10	10	1.00	1.00
2	10	*A. girodi*	> 50	0	0	0
	10	*A. brightwelli*	> 50	0	0	0
	10	*Brachionus*	> 51	16	0.19	1.00
	7	*Paramecium*	7	7	1.00	1.00
3	16	*A. girodi*	> 80	0	0	0
	16	*A. brightwelli*	> 80	0	0	0
	16	*Brachionus*	> 82	17	0.18	1.00
	13	*Paramecium*	13	13	1.00	1.00
4	17	*A. girodi*	> 85	0	0	0
	16	*A. brightwelli*	> 75	0	0	0
	16	*Brachionus*	> 70	35	0.26	1.00
	7	*Paramecium*	7	7	1.00	1.00
1–4	53	*A. girodi*	> 265	0	0	0
	52	*A. brightwelli*	> 255	0	0	0
	42	*Brachionus*	> 203	68	0.22	1.00
	37	*Paramecium*	37	37	1.00	1.00

females readily attack newborn individuals, the attack probability being 0.89; however, they capture these prey with some difficulty and are unable to swallow any of those that are captured (Table 3).

Females of the much larger, β-form of this species have not been tested individually in a similar manner, but they are known to be cannibalistic in cultures and to further increase in size as a result (Gilbert 1975).

Asplanchna girodi. The females of *A. girodi*, which exhibit no polymorphism in response to dietary tocopherol (Gilbert and Litton 1978), have little or no potential for cannibalism. Starved adult females showed no responses to newborn conspecifics after more than 265 encounters but readily attack *Brachionus* and, especially, *Paramecium* (Gilbert unpubl.; Tables 3 and 4).

Laboratory studies on cannibalism in *Asplanchna*, then, have shown that *A. sieboldi*, *A. intermedia*, and *A. brightwelli* all have the potential to attack, capture, and ingest conspecific prey. *Asplanchna sieboldi* is the most voracious cannibal, and *A. brightwelli* is the least. Within each species, the largest, female phenotypes have the greatest potential for cannibalism. In *A. sieboldi* and *A. intermedia*, males and certain female morphotypes are well protected from being attacked or captured by cannibalistic females (see separate section below).

A number of field studies have shown that cannibalism may also occur in natural populations of *Asplanchna*. Stomach content analyses have verified this for *A. sieboldi* (Wesenberg-Lund 1923; Hurlbert et al. 1972), *A. brightwelli* (Guiset 1977), and *A. girodi* (Guiset 1977). An extensive analysis by Salt et al. (1979) on the stomach contents of five species of *Asplanchna* over a one-year period showed that *A. sieboldi* would eat *A. brightwelli* and *A.*

priodonta but that none of the species seemed to be cannibalistic. The ecological significance of the potential for cannibalism, therefore, seems to be variable.

In conclusion, cannibalism in *Asplanchna* appears to be limited to those species with a horseshoe-shaped yolk gland. These species tend to be more voracious predators than those with a spherical yolk gland, such as *A. priodonta*, and, with the exception of *A. girodi*, also exhibit female polymorphism (Gilbert 1973b). The species with the largest and most polymorphic females have the greatest potential for cannibalism. Thus, *A. sieboldi*, *A. intermedia*, and *A. brightwelli* exhibit decreasing tendencies both for female size and polymorphism and for cannibalism. *Asplanchna girodi* appears to be monomorphic and to rarely, if ever, exhibit cannibalism.

It is difficult to assess the ecological significance of cannibalism in *Asplanchna*. In addition to being able to eat conspecific prey, cannibalistic females can also eat congeneric prey and relatively large prey of other taxa. Thus, I suggest that the selective pressure for large-sized females in polymorphic species may not have been cannibalism but the ability to eat congeneric competitors or, more likely, to take advantage of a variety of large-sized prey unavailable to smaller-sized females. Some indirect evidence for this view is the fact that several adaptations have evolved in species with the potential for cannibalism to limit mortality from cannibalism (*see separate section below*).

Defenses against cannibalism

Species of *Asplanchna* that have the potential to produce cannibalistic females also have adaptations to minimize mortality from cannibalism. Three types of mechanisms seem to have evolved for this purpose: body-wall outgrowths, large birth size, and failure of cannibalistic females to attack conspecific individuals (Gilbert 1973b, 1976a, 1977b).

Body-wall outgrowths are outpocketings of the body wall which are normally withdrawn by muscles. However, when the rotifer retracts its corona, as when it is attacked by another *Asplanchna*, the hydraulic pressure created in the body cavity counteracts the contractions of these muscles and causes the extension of the outgrowths. Individuals with extended outgrowths are effectively much larger than those without them and are correspondingly more difficult for *Asplanchna* to capture and ingest.

Body-wall outgrowths evolved in both females and males. They are largest in those species with the most cannibalistic or voracious females and absent in species which characteristically eat only relatively small prey (Gilbert 1973b). Cruciform females of *A. sieboldi*, for example, have very large body-wall outgrowths which prevent even newborn individuals from being captured by conspecific, adult campanulates (Gilbert 1976a). Similarly, the males of this species have very large outgrowths which protect them from these same predators; only 14% of the males that are attacked are captured (Gilbert 1977b).

The body-wall outgrowths of cruciform females in *A. sieboldi* are proportionately largest in young individuals (Powers 1912; Gilbert 1975), making them especially valuable for the smaller and hence more vulnerable individuals. The negative allometric growth of the body-wall outgrowths provides further evidence for their role as devices to restrict mortality through cannibalism.

The giant, campanulate females of *A. sieboldi* and *A. intermedia* have no body-wall outgrowths but are disproportionately larger at birth than the relatively small, saccate females of these species (Powers 1912; Gilbert 1976c). This large birth size is very effective in preventing the capture of young campanulates by adult campanulates. For example, only 3 of the 83 (4%) campanulate young of *A. sieboldi* attacked by conspecific, adult campanulates were subsequently captured (Gilbert 1976a).

In both *A. sieboldi* and *A. intermedia*, females of certain morphotypes generally fail to attack certain types of individuals from their own clones (Table 2). It is not known, though, whether these clonal recognition responses also apply more generally to different genotypes within each species. In *A. intermedia*, campanulate females readily attack only saccate-female clonemates and rarely attack males, cruciform females, or campanulate females (Gilbert 1976a). The potential susceptibility of saccate females to cannibalism in this species is probably not ecologically important, because saccate females are unlikely to be present under those conditions which induce cruciform and campanulate females (Gilbert 1976a).

The immunity of males, cruciform females, and campanulate females to attack by campanulate females in *A. intermedia* may be related to the fact that the body-wall outgrowths of the males and cruciform females in this species are smaller than those in the male and cruciform females of *A. sieboldi*, which are not immune to attack by campanulate clonemates (Gilbert 1976a,b, 1977b). This relationship suggests, perhaps, that in *A. intermedia* morphological defenses against capture by conspecific campanulates became less essential when susceptible individuals became immune to attack (Gilbert 1976a,b, 1977b).

Similarly, it is interesting to note that, perhaps for the same reason, male body size in *A. intermedia* is smaller than that in *A. sieboldi* (Gilbert 1976b, 1977b). In general, it is probably adaptive for the nonfeeding males of monogonont rotifers to be as small as possible, so that parental females can produce as many male offspring as possible (Gilbert 1977b). In *Asplanchna*, there seems to be a conflicting selective pressure for increased male size to prevent mortality from cannibalism (Gilbert 1977b). Thus, male size in different species of *Asplanchna* probably represents a balance between these two selective pressures and is determined to some extent by the tendency of the male to be cannibalized.

In *A. sieboldi*, it is the cruciform, and not the campanulate, females which often fail to attack certain types of clonemates. Other cruciform females and, to some extent, saccate females are immune to attack by these predators (Gilbert 1976a). In contrast, the campanulate females of this species attack all conspecific individuals (Gilbert 1976a).

The difference in the behavior of the campanulate females of *A. sieboldi* and *A. intermedia* may be related to the rarity of this morphotype in the former species. Only a small percentage of the individuals in populations of *A. sieboldi* seem to have the potential to produce campanulate offspring (Powers 1912; Gilbert 1975), while almost all individuals of *A. intermedia* have this potential (Gilbert 1975). Accordingly, the predominant, cannibalistic, female morphotype in *A. sieboldi* and *A. intermedia* may be the cruciform and campanulate, respectively. If this is the case, selective pressure may have favored the inhibition of cannibalistic attacks in the cruciform, and not the campanulate, morphotype of *A. sieboldi* (Gilbert unpubl.) The potential for relatively infrequent induction of campanulate females in *A. sieboldi* may have evolved, in part, to allow cannibalism when alternate prey are unavailable.

The basis for the failure of certain types of cannibalistic females to attack certain types of clonemates is not known. In *A. intermedia*, adult campanulate females respond similarly to filtrates from both saccate and campanulate clonemates, even though they rarely attack the campanulates (Gilbert 1977a). This result suggests that the tissue chemistries of the two morphotypes are similar and that the detectable differences between the two morphotypes are limited to the surfaces of the body wall (Gilbert 1977a). The cuticle of the campanulate females might contain an inhibitor for the feeding response or might prevent the chemical activity of the tissues from reaching the outside surface of the body (Gilbert 1977a).

Selective feeding

I suggest the following guidelines for considering questions of selective feeding, especially those involving differentiation of active and passive selection.

A predator may fail to eat an available organism for several reasons. First, the predator may not attack the organism. This may be because the organism fails to stimulate or inhibits attack responses or because it manages to escape before the predator can respond to it. Second, a predator may not capture an organism it has attacked. The organism may escape from the predator through a behavioral adaptation, may be of a size, shape, or texture which the predator mechanically cannot handle, or may possess physical or chemical characteristics which inhibit the feeding behavior of the predator. Third, for these same three types of reasons, the predator may reject or be unable to ingest a captured organism.

Active selective feeding by a predator involves choices by the predator on whether to attack, capture, or ingest organisms which can elicit feeding responses and which can be eaten. Passive selective feeding results from the failure of a predator to eat certain organisms, either because these organisms cannot elicit feeding responses in the predator or because they cannot be captured or ingested if attacked by the predator. The critical difference between active and passive selective feeding is that decisions by the predator to avoid certain potential prey organisms occur only in the former.

Adult campanulate females of *A. intermedia* attack *A. brightwelli* and *A. girodi* two to six times more readily than either *Volvox aureus* or *Brachionus calyciflorus* (Gilbert 1978). As indicated earlier, once any of these four types of prey are attacked, the probabilities that they will be captured and then ingested are very high for all of the prey-types (Gilbert 1978).

As another example, adult *A. girodi* do not respond to contact with either newborn conspecifics or *A. brightwelli* but do attack *B. calyciflorus* and, especially, *P. aurelia* (Gilbert unpubl.) (Table 4). These *A. girodi* predators have difficulty capturing the relatively large *Brachionus* but ingest all those that are captured. The paramecia are easily captured and ingested. I suspect that the failure of *A. girodi* to attack other *Asplanchna* may be due either to the absence of some stimulatory chemical in these rotifers or, more likely, to inhibition of feeding responses by large mechanical stimuli.

Further evidence for active selection in *Asplanchna* was provided by Pourriot (1965). He found that *A. brightwelli* would reject the rotifers *Lepadella* and *Lecane*, as well as the empty loricae of *Brachionus*. He did not mention, however, whether the *Asplanchna* responded to these objects and then released them or whether they failed to attack them.

Although *Asplanchna* probably generally eats only living prey, some species at least will eat dead organisms. For example, campanulate females of *A. intermedia* will readily consume freeze-killed *A. brightwelli* and campanulate clonemates, and saccate females of *A. intermedia* will eat feeeze-killed *A. brightwelli* (Gilbert unpubl.).

Investigations of the stomach contents of *Asplanchna* from natural populations have shown that the diets of various species in this genus are extremely diverse. Such studies have been performed for *A. sieboldi* (Resvoj 1926; Hurlbert et al. 1972; Salt et al. 1979), *A. brightwelli* (Erman 1962; Guiset 1977; Salt et al. 1979), *A. girodi* (Pejler 1961; Pourriot 1965; Guiset 1977; Salt et al. 1979), *A silvestri* (Salt et al. 1979), *A. priodonta* (Pejler 1957, 1961, 1965; Erman 1962; Nauwerck 1963; Gliwicz 1969; Ejsmont-Karabin 1974; Guiset 1977; Salt et al. 1979), and *A. herricki* (Kosova 1960; Tribush 1960). All of these species are omnivorous to some extent, but, depending on the species, utilize algal material to greater or lesser degrees.

Several of these investigations compared the relative abundances of organisms in the stomach contents with those in the natural environment and showed that *Asplanchna* feeds preferentially. However, this type of study cannot discriminate between active and passive selection.

Erman (1962) and Ejsmont-Karabin (1974) found that *A. brightwelli* and *A. priodonta*, respectively, both select animal over algal food. Some of Ejsmont-Karabin's data are summarized in Table 5.

Guiset (1977) compared the relative abundances of rotifer prey in the stomachs

Table 5. Food preferences of Asplanchna priodonta. Ivlev's electivity index, S, $= g-e/g+e$, where g and e are percentages of a food item in stomach of predator and in lake, respectively. (Data from Ejsmont-Karabin 1974.)

Food Item	S	
	Mikolajskie Lake	Lake Taltowisko
Keratella cochlearis	0.94	0.79
K. quadrata		0.81
Chromogaster sp.	0.92	
Trichocerca sp.	0.84	
Polyarthra sp.	0.84	0.40
Pompholyx sulcata		0.44
Codonella	0.98	0.62
Ceratium hirundinella	0.59	0.36
Cyclotella	0.09	− 0.73
Fragillaria	− 0.64	− 0.05
Melosira	− 0.96	− 0.81
Synedra	− 0.88	
Tabellaria	− 0.93	− 0.30

of *A. priodonta*, *A. girodi*, and *A. brightwelli* with those in their environments. The information he tabulated represents average values from many different localities. His results show that, out of a total of about 20 possible species of rotifer prey, *A. priodonta* positively selects only the smallest species (*Polyarthra remata, Anuraeopsis fissa, Pompholyx sulcata, Trichocerca pusilla/similis*), *A. brightwelli* positively selects numerous species (*Polyarthra* spp., *Synchaeta pectinata/grandis, A. fissa, P. sulcata, Filinia* spp., *Trichocerca* spp., *Hexarthra mira*), and *A. girodi* positively selects several species (*Polyarthra vulgaris/dolichoptera, S. pectinata/grandis, S. oblonga/kitina, T. pusilla/similis*). As mentioned previously, though, it is not possible to tell from this type of study which prey the *Asplanchna* are attacking and which they are unable to capture and ingest after an attack.

The extent to which the food preferences of *Asplanchna* may be altered by previous feeding history or by the frequency distribution of available prey-types has not been adequately investigated. It is known, however, that *Asplanchna intermedia* campanulates previously fed on *Brachionus calyciflorus* attack this prey no more or less readily than those previously fed on *A. brightwelli* (Gilbert 1978).

Defenses of planktonic rotifers against Asplanchna predation

Many rotifers which are readily attacked by *Asplanchna* have structural or behavioral adaptations that protect them to various degrees from being captured or ingested by this predator. In some cases, these adaptations definitely evolved in response to *Asplanchna* predation. The developmental responses of several species to a substance produced by *Asplanchna* fit into this category. In other cases, the selective pressure for the defensive adaptation is not clear. For example, the hard lorica and spines of *Keratella* and the escape behavior of *Polyarthra* are effective against a variety of predators and may have evolved in response to any one or more of them.

Asplanchna-induced developmental changes. The embryological development of several species of rotifer is influenced by a substance released into the environment by *Asplanchna*. In each case, the altered phenotype is characterized by an increase in the relative length of various body processes, and, in most cases, the *Asplanchna*-induced morph is extremely adaptive in that it is physically more difficult for *Asplanchna* to capture and ingest. Thus, the proximate factor responsible for this type of polymorphism is the same as the ultimate factor.

I suggest that this type of direct, developmental response to predation evolved only when the defensive structure is primarily effective against a single type of predator and when the prey organism has a short generation time. If the defensive adaptation were effective against a variety of predators, a direct response to just one predator might result in general asynchrony between the response and other predators, while direct responses to all predator-types might be mechanistically very complex if not impossible. If the prey had a long generation time, there might be a long lag time between the presence of the predator and the ability of the prey to produce offspring with the modified phenotype.

The following examples of *Asplanchna*-

induced polymorphism have been described and investigated.

Brachionus calyciflorus may have no posterolateral spines or ones varying in length from very short protrusions to structures as long as their loricae. The only known factor which can regularly induce very long spines is the *Asplanchna*-substance. The relationship between *B. calyciflorus* and *Asplanchna* was first discovered by Beauchamp (1952) and then explored in detail by Gilbert (1966, 1967) and Halbach (1970).

The *Asplanchna*-substance appears to be a peptide (Gilbert 1967), is fairly stable in aqueous media (Gilbert 1967; Halbach 1970), and is produced at inductive levels in natural environments (Gilbert 1967; Gilbert and Waage 1967; Halbach 1970). There is frequently a positive correlation in nature between the length of the posterolateral spines of *B. calyciflorus* and the population density of *Asplanchna* (Gilbert 1967; Gilbert and Waage 1967; Halbach 1969, 1970; Green and Lan 1974).

Brachionus calyciflorus with no or short posterolateral spines are very susceptible to *Asplanchna* predation, but those with long spines are usually well protected from being captured or ingested (Beauchamp 1952; Gilbert 1966; 1967; Halbach 1971). The posterolateral spines, which articulate with the body, normally extend straight behind the animal, but are spread out laterally by hydraulic pressure when the *Brachionus* retracts its corona after being attacked by an *Asplanchna* (Gilbert 1967).

Brachionus bidentata exhibits a similar type of polymorphism (Pourriot 1974). Posterolateral spines may be absent or up to about half the lorica length. In the presence of *Asplanchna*-substance, the posterolateral spines are long; otherwise, they are absent or relatively short. Although these spines are not movable, as they are in *B. calyciflorus*, long-spined individuals of all ages are well protected from being captured and eaten by *A. brightwelli*.

Brachionus urceolaris sericus develops a posterodorsal process in response to *Asplanchna*-substance (Pourriot 1964). This process, however, does not appreciably protect the individual from being ingested by *Asplanchna*.

Finally, *Filinia mystacina* exhibits a pronounced and adaptive polymorphic response to *Asplanchna* (Pourriot 1964). In the absence of *Asplanchna*-substance, the two movable, anterior, setiform appendages are about 350 μm in length or about twice the body length, and the nonmovable, posterior, setiform appendage is about 200-240 μm. In the presence of this substance, these appendages develop to lengths of 500-600 μm and 300-400 μm, respectively. Individuals with long setae are more difficult for *Asplanchna* to ingest than those with short setae. The long setae are especially effective against predation by *A. girodi* and permit the coexistence of the two species in laboratory cultures. However, even forms with long setae are eventually eliminated when cultured with the more voracious *A. brightwelli*.

Keratella. Observations on predator-prey behavior have shown that *Keratella cochlearis* with posterior spine lengths of about 50 μm are partially protected from capture by *Asplanchna*. In several experiments, the probability of *K. cochlearis* being attacked by adult *A. girodi* was 0.68, and the probabilities of the attacked *Keratella* being captured and then ingested by this predator were 0.89 and 0.65, respectively (Gilbert and Williamson 1978). In another experiment, four adult, saccate *A. sieboldi* captured 12 *K. cochlearis* but managed to ingest only one of these (Gilbert unpubl.).

The *Keratella* in these experiments were primarily protected from *Asplanchna* because captured individuals were difficult for this predator to ingest. *Keratella cochlearis* has a thick lorica with rigid spines and can be easily ingested only when it is positioned in the pharynx with its anterior end facing toward the esophagus; otherwise, the posterior spine catches the walls of the pharynx or the esophagus, preventing the animal from being manipulated into and down the esophagus (Gilbert and Williamson 1979). *Keratella* that are not swallowed are released unharmed (Gilbert and Williamson 1979).

Although *Keratella* may often be difficult for *Asplanchna* to ingest, this genus is a common food of *Asplanchna* (Pejler 1957, 1961, 1965; Kosova 1960; Erman 1962; Nauwerck 1963; Gliwicz 1969; Ejsmont-Karabin 1974; Guiset 1977; Salt et al 1979) and may be positively selected (Ejsmont-Karabin 1974).

Polyarthra. Observations on predator-prey behavior show that *Polyarthra* is practically completely protected from predation by *Asplanchna*. In several experiments, the probability of *P. vulgaris* being captured after contact with adult *A. girodi* was 0.01 (Gilbert and Williamson 1979). In a second series of experiments, the probability of *P. vulgaris* being captured after 52 contacts with 10 adult, saccate *A. sieboldi* was 0.02. In a third series of experiments, 12 young and adult cruciform-campanulate *A. sieboldi* contacted 112 *P. euryptera*, a species considerably larger than *P. vulgaris*, and captured only 4 or 4% of these (Gilbert unpubl.). In each of the three series of experiments, the *Asplanchna* were subsequently tested with *Brachionus* or congeneric prey and found to be responsive.

Polyarthra is a soft-bodied rotifer but possesses movable appendages controlled by well developed striated muscles. When *Polyarthra* is contacted by *Asplanchna*, it immediately raises its paddle-shaped appendages and "jumps" away a distance of up to about ten times its own body length (Gilbert and Williamson 1979). This escape response occurs so quickly that *Asplanchna* rarely initiates an attack (Gilbert and Williamson 1979). *Polyarthra* which are unhealthy or trapped on the air–water interface in laboratory vessels are readily attacked (Gilbert unpubl.), and all *Polyarthra* that are captured are quickly ingested (Gilbert and Williamson 1979; Gilbert unpubl.). Thus *Polyarthra* is an attractive prey for *Asplanchna* but can only rarely be captured.

Because of these results, it is surprising that, in natural populations of *Asplanchna*, *Polyarthra* may comprise an important part of the diet and may even be positively selected (Ejsmont-Karabin 1974; Guiset 1977). This paradox may be due to relatively high encounter rates between *Polyarthra* and *Asplanchna* and perhaps to the existence of relatively large proportions of unhealthy or old *Polyarthra* which cannot exhibit normal escape responses.

Effect of egg-carrying in Brachionus on susceptibility to Asplanchna predation

The eggs of *B. calyciflorus* are attached to the cloaca of the parent by very thin filaments (Beauchamp 1965; *see* photographs in Gilbert 1970). Parthenogenetic eggs remain attached to their mothers until they hatch. Depending on their birth rate, amictic females may carry one, two, and sometimes even three or four eggs. Amictic eggs are large, their length being about 0.60–0.68 that of the body length, and therefore they considerably increase the effective size of the parent. Measurements of the total lengths and widths [body with egg(s), if present] of adult *B. calyciflorus* with no, one, and two attached eggs are shown in Table 6. These data demonstrate that the

Table 6. Sizes (μm) of preserved adult *Brachionus calyciflorus* carrying 0, 1, and 2 eggs.

No. eggs carried	Total length		Max Width	
	Mean ± SE	N	Mean ± SE	N
0	201.7 ± 2.8	50	132.9 ± 1.3	50
1	284.6 ± 2.5	50	144.2 ± 0.8	50
2	312.1 ± 2.8	50	175.6 ± 1.7	50

total length and width of adults carrying one amictic egg are increased by about 40 and 9%, respectively, and that these same dimensions of adults carrying two amictic eggs are increased by about 55 and 33%, respectively.

The additional size contributed by these eggs makes *B. calyciflorus* more difficult for *Asplanchna* to capture and eat (Table 7). As the number of attached eggs increases from zero to two, both *A. sieboldi* and *A. brightwelli* have to encounter more *Brachionus* before they capture one, are less likely to ingest the captured *Brachionus*, and take more time to handle the captured *Brachionus* they do ingest.

It is also noteworthy that the attached

Table 7. Effect of eggs attached to the cloaca of adult, short-spined *Brachionus calyciflorus* (B. c.) on susceptibility of these rotifers to predation by *Asplanchna*.

Asplanchna Predator	Morphotype, Age of Predator	No. Eggs on B.C.	No. B. c. Encountered before a Capture		No. B. c. Captured	Proportion B. c. Released	Time (sec) between Capture and Ingestion of B. c.		No. Eggs Released from Captured B. c.
			Mean ± SE	N			Mean ± SE	N	
sieboldi	saccate, young	0	4.1 ± 0.6	14	23	0	48.1 ± 1.2	23	–
		1	6.1 ± 0.7	11	28	0.18	81.9 ± 2.6	23	1
		2	6.8 ± 1.1	11	29	0.24	125.5 ± 2.8	22	6
brightwelli	α-form, adult	0	7.6 ± 0.6	24	28	0.17	41.6 ± 1.1	24	–
		1	14.6 ± 1.3	19	23	0.21	56.1 ± 1.1	19	1
		2	16.0 ± 1.6	7	10	0.43	129.0 ± 4.7	7	5
brightwelli	α-form, young	0			27	0	38.6 ± 0.7	27	–
		1			0	–			–
		2			0	–			–

eggs of *B. calyciflorus* are sometimes liberated before the captured parent is ingested; this is especially true when the parent is carrying two eggs (Table 7). The fate of such separated eggs in a natural environment is not known, but in the laboratory such eggs develop normally and at the same rate as attached eggs (Haluszka and Gilbert unpubl.). Thus, even if an egg-carrying adult is captured by an *Asplanchna*, one or more of the eggs may be freed and so protected from predation.

Since carrying eggs must involve an energy demand on the parents, there must be some selective pressure for certain rotifers, such as *B. calyciflorus*, to carry their eggs. One type of selective pressure involves the survivorship of the embryos, and another type involves that of the parents. Although separated eggs survive well in the laboratory, they may not do so in nature. I believe that the results presented above suggest that the egg-carrying habit in *B. calyciflorus* may have evolved, at least in part, to reduce parental mortality from predation by *Asplanchna*. Protection from predation may have provided some selective pressure for the egg-carrying habit in other rotifers as well.

Effect of Asplanchna predation on rotifer community structure

Experiments with simple laboratory communities, with *A. girodi* as predator and *P. vulgaris* and *K. cochlearis* as prey, have shown that the probability of *Polyarthra* survival is 3.25 times that of *Keratella* survival (Gilbert and Williamson 1979). This result is in accordance with the behavioral observations described above, showing that *Polyarthra* almost always escapes from *A. girodi* while *Keratella* is ingested by this predator with a relatively high probability after an encounter (0.44) (Gilbert and Williamson 1979). It is likely that the relative survivals of *Polyarthra* and *Keratella* in the face of *Asplanchna* predation will be similar in natural populations, but this conjecture needs to be tested.

To date, there is very little information on the effect of *Asplanchna* on the structure of rotifer communities in nature. Edmondson (1960) and Zimmerman (1974) both found that the death rates of *K. cochlearis* were positively correlated with the density of *A. priodonta*. Also, Lewkowicz (1971) found that the productivity of *A. priodonta* and *A. brightwelli* together was positively associated with depletion of the biomass of other rotifers.

It is clear that much remains to be done regarding behavioral interactions between *Asplanchna* and its potential prey organisms. Furthermore, studies need to be conducted to assess the effects of these interactions on the species structure of natural communities.

References

BEAUCHAMP, P. de. 1952. Un facteur de la variabilité chez les rotifères du genre *Brachionus*. C. R. Hebd. Seances Acad. Sci. **234**: 573-575.

———. 1965. Classe des rotifères, p. 1225-1379. *In* P. P. Grassé [ed.], Traité de Zoologie, v. 4, fasc. 3. Masson & Cie.

EDMONDSON, W. T. 1960. Reproductive rates of rotifers in natural populations. Mem. Ist. Ital. Idrobiol. **12**: 21-77.

EJSMONT-KARABIN, J. 1974. Studies on the feeding of planktonic polyphage *Asplanchna priodonta* Gosse (Rotatoria). Ekol. Pol. **22**: 311-317.

ERMAN, L. A. 1962. On the utilization of the reservoirs trophic resources by plankton rotifers [in Russian]. Byull. Mosk. O-Va. Ispyt. Prir. Otd. Biol. **67**: 32-47.

GILBERT, J. J. 1966. Rotifer ecology and embryological induction. Science **151**: 1234-1237.

———. 1967. *Asplanchna* and posterolateral spine production in *Brachionus calyciflorus*. Arch. Hydrobiol. **64**: 1-62.

———. 1970. Monoxenic cultivation of the rotifer *Brachionus calyciflorus* in a defined medium. Oecologia **4**: 89-101.

———. 1973*a* The induction and ecological significance of gigantism in the rotifer *Asplanchna sieboldi*. Science **181**:63-66.

———. 1973*b* The significance of polymorphism in the rotifer *Asplanchna*. Humps in males and females. Oecologia **13**: 135-146.

———. 1975. Polymorphism and sexuality in the rotifer *Asplanchna*, with special reference to the effects of prey-type and clonal variation. Arch. Hydrobiol. **75**: 442-483.

———. 1976*a* Selective cannibalism in the rotifer *Asplanchna sieboldi*: contact recognition of morphotype and clone. Proc. Natl. Acad. Sci. **73**: 3233-3237.

———. 1976*b* Sex-specific cannibalism in the rotifer *Asplanchna sieboldi*. Science **194**: 730-732.

———. 1976*c*. Polymorphism in the rotifer *Asplanchna sieboldi*: biomass, growth and reproductive rate of the saccate and campanulate morphotypes. Ecology **57**:542-551.

———. 1977*a*. Control of feeding behavior and selective cannibalism in the rotifer *Asplanchna*. Freshwater Biol. **7**: 337-341.

———. 1977*b*. Defenses of males against cannibalism in the rotifer *Asplanchna*: size, shape, and failure to elicit tactile feeding responses. Ecology **58**:1128-1135.

———. 1977*c*. Effect of the non-tocopherol component of the diet on polymorphism, sexuality, biomass, and reproductive rate of the rotifer *Asplanchna sieboldi*. Arch. Hydrobiol. **80**: 375-397.

———. 1977*d*. Selective cannibalism in the rotifer *Asplanchna sieboldi*. Arch. Hydrobiol. Ergeb. Limnol. **8**: 267-269.

———. 1978. Selective feeding and its effect on polymorphism and sexuality in the rotifer *Asplanchna sieboldi*. Freshwater Biol. **8**:43-50.

———, and J. R. LITTON, JR. 1978. Sexual reproduction in the rotifer *Asplanchna girodi*: Effects of tocopherol and population density. J. Exp. Zool. **204**: 113-122.

———, and J. K. WAAGE. 1967. *Asplanchna*, *Asplanchna*-substance, and posterolateral spine length variation of the rotifer *Brachionus calyciflorus* in a natural environment. Ecology **48**: 1027-1031.

———, and C. E. WILLIAMSON. 1979. Predator-prey behavior and its effect on rotifer survival in associations of *Mesocyclops edax*, *Asplanchna girodi*, *Polyarthra vulgaris*, and *Keratella cochlearis*. Oecologia in press.

GLIWICZ, Z. M. 1969. Studies on the feeding of pelagic zooplankton in lakes with varying trophy. Ekol. Pol. **17**: 663-708.

GOSSLER, O. 1950. Funktionanalysen am Räderorgan von Rotatorien durch optische Verlangsamung. Oesterr. Zool. Z. **2**: 568-584.

GREEN, J., and O. B. LAN. 1974. *Asplanchna* and the spines of *Brachionus calyciflorus* in two Javanese sewage ponds. Freshwater Biol. **4**: 223-226.

GUISET, A. 1977. Stomach contents in *Asplanchna* and *Poloesoma*. Arch. Hydrobiol. Ergeb. Limnol. **8**: 126-129.

HALBACH, U. 1969. Räuber und ihre Beute: der Anpassungswert von Dornen bei Rädertieren. Naturwissenschaften **56**: 142-143.

———. 1970. Die Ursachen der Temporal variation von *Brachionus calyciflorus* Pallas (Rotatoria). Oecologia **4**: 262-318.

———. 1971. Zum Adaptivwert der zyklomorphen Dornenbildung von *Brachionus calyciflorus* Pallas (Rotatoria). I. Räuber-Beute-Beziehung in Kurzzeit-Versuchen. Oecologia **6**: 267-288.

HURLBERT, S. H., M. S. MULLA, and H. R. WILLSON. 1972. Effects of an organophosphorus insecticide on the phytoplankton, zooplankton, and insect populations of freshwater ponds. Ecol. Monogr. **42**: 269-299.

KOSOVA, A. A. 1960. Seasonal changes in plankton and benthos in the poloys of the lower zone of the Volga delta [in Russian]. Tr. Vses. Gidrobiol. O-Va. **10**: 102-134.

LEWKOWICZ, M. 1971. Biomass of zooplankton and production of some species of Rotatoria and *Daphnia longispina* in carp ponds. Pol. Arch. Hydrobiol. **18**: 215-223.

NACHTWEY, R. 1925. Untersuchungen über die Keimbahn, Organogenese und Anatomie von *Asplanchna priodonta* Gosse. Z. Wiss. Zool. **126**: 239-492.

NAUMANN, E. 1923. Spezielle Untersuchungen über die Ernährungsbiologie des tierischen Limnoplanktons. II. Über den Nahrungserwerb und die natürliche Nahrung der Copepoden und der Rotiferen des Limnoplanktons. Lunds Univ. Arsskr. N.F. Avd. 19(6): 1-17.

NAUWERCK, A. 1963. Die Beziehungen zwischen Zooplankton und Phytoplankton im See Erken. Symb. Bot. Ups. 17(5). 163 p.

PEJLER, B. 1957. Taxonomical and ecological studies on planktonic Rotatoria from central Sweden. K. Sven. Vetenskaps akad. Handl. 6(7): 1-52.

———. 1961. The zooplankton of Ösbysjön, Djursholm. I. Seasonal and vertical distribution of the species. Oikos 12: 225-248.

———. 1965. Regional-ecological studies of Swedish fresh-water zooplankton. Zool. Bidr. Upps. 36: 407-515.

POURRIOT, R. 1964. Étude experimentale de variations morphologiques chez certaines espèces de rotifères. Bull. Soc. Zool. Fr. 89: 555-561.

———. 1965. Recherches sur l'écologie des rotifères. Vie Milieu 21 (Suppl.) 224 p.

———. 1974. Relations prédateur-proie chez les rotifères: influence du prédateur (*Asplanchna brightwelli*) sur la morphologie de la proie (*Brachionus bidentata*). Ann. Hydrobiol. 5: 43-55.

POWERS, J. H. 1912. A case of polymorphism in *Asplanchna* simulating a mutation. Am. Nat. 46: 441-462, 526-552.

RESVOJ. P. 1926. Observations on the feeding of rotifers [in Russian]. Trav. Soc. Nat. Leningr. 56: 73-89.

SALT, G. W., G. F. SABBADINI, and M. L. COMMINS. 1979. Trophi morphology relative to food habits in six species of rotifers (Asplanchnidae). Trans. Am. Microsc. Soc. 97:469-485.

SOROKIN, Ju. I., and E. D. MORDUKHAI-BOLTOVSKAYA. 1962. A study of the nutrition of the rotifer *Asplanchna*, with the help of ^{14}C [in Russian]. Byull. Inst. Biol. Vodokhran. 12: 17-20.

TRIBUSH, T. M. 1960. Certain observations on rotifers of the family Asplanchnidae in the Rybinsk Reservior [in Russian]. Byull. Inst. Biol. Vodokhran. 6: 18-19.

WANICZEK, H. 1930. Untersuchungen über einige Arten der Gattung *Asplanchna* Gosse (*A. girodi* de Guerne, *A. brightwelli* Gosse, *A. priodonta* Gosse). Ann. Mus. Zool. Pol. 8: 109-321.

WESENBERG-LUND, C. 1923. Contributions to the biology of the Rotifera. I. The males of the Rotifera. Kgl. Dan. Vidensk. Selsk. 8 2(3): 191-345.

ZIMMERMAN, C. 1974. Die pelagischen Rotatorien des Sempachersees mit spezieller Berücksichtigung der Brachioniden und der Ernährungsfrage. Schweiz. Z. Hydrol. 36: 205-300.

17. A Chemostat System for the Study of Rotifer-Algal-Nitrate Interactions

Martin E. Boraas

Abstract

A method is described for long-term continuous culture of rotifers *(Brachionus)* on a green alga *(Chlorella)* in a defined medium, both in two-stage chemostats and in mixed culture. At constant environmental conditions, the cultures are stable and reproducible. Steady states are independent of initial conditions. In general, the results agree with previous work on bacteria and protozoa.

In this paper I present a method for the continuous culture of the rotifer *Brachionus calyciflorus* on the alga *Chlorella pyrenoidosa*. Continuous cultures have been used ever since Monod (1950) and Novick and Szilard (1950) independently published continuous culture techniques as a method for growing bacteria in steady states. Since that time, these techniques have been applied to many problems (Kubitschek 1968). In the past decade, studies of substrate-bacterial-protozoan systems have increased in frequency (see reviews by Curds and Bazin 1977 and Fredrickson 1977). However, the method only rarely has been used with metazoa (Droop and Scott 1978; and Lampert 1976).

Here I present the details of my procedures for growing *B. calyciflorus* and *C. pyrenoidosa* in chemostats, and point out some advantages and limitations of the technique. Both two-state and single-stage rotifer-algal chemostats are described. In addition, I briefly examine the usefulness of the method for testing hypotheses of predator-prey dynamics with respect to plant-herbivore-nutrient interactions.

Methods and materials

Continuous cultures of three types were maintained: single-stage algal, two-stage algal-rotifer, and mixed algal-rotifer chemostats. The single-stage algal cultures consisted of a glass vessel (Fig. 1) which contained unispecific nitrogen-limited *C. pyrenoidosa* growing at a controlled turnover rate at constant light, temperature, and nutrient concentration. An algal chemostat in steady-state which supplied algae to a second chemostat, containing *B. calyciflorus* held in the dark, comprised the two-stage cultures. Mixed cultures containing both algae and rotifer growing in a single chemostat vessel were held in the light under constant environmental conditions.

Chemostat design. The culture vessel was designed to meet the following criteria: maximum volume to minimize sampling effect, optimum geometry to allow for rapid mixing of input media and for light saturation under all growth conditions, and minimum internal horizontal surfaces where cells and debris might settle. The design in Fig. 1 represents an empirical compromise follow-

---

This study was supported by NSF grants BMS75-18749 and DEB77-29404 to D. B. Seale and F. M. Williams. Thanks go to F. M. Williams for valuable advice and to D. B. Seale and J. R. Barrett for critically reviewing the manuscript.

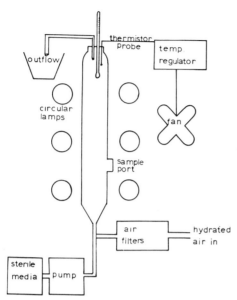

Fig. 1. Schematic diagram of continuous culture apparatus. Two growth tubes were coupled in series for two-stage cultures with modifications as described in text.

ing these criteria for the single-stage algal cultures. The vessel was cylindrical, 4.3 by 40 cm, with a conical bottom and a rounded top. Light was supplied by standard 8-inch "cool-white" circular lamps. The volume was about 570 ml. Air and fresh media were put in through a 0.5 mm capillary "T" joint. Sterile fresh media and air entered through the T joint at the bottom of the vessel. Spent media and cells were blown out the top by air pressure. Two ports, consisting of septa cut from silicone rubber stoppers, allowed penetration with a 25 gauge needle for samples and inoculations. The ports remained watertight after >1,000 punctures.

Air supplied by an air pump (Silent Giant: Aquarium Pump Supply, Inc.) was hydrated in distilled water. The flow rate was regulated with a flowmeter and bleed valve to 800 ml/min. The hydrated air was sterilized by passing it through two filters, a 8 × 16 cm cotton filter in series with a 47mm by 0.45μm cellulose acetate membrane filter. Both filters were warmed with a heat tape to prevent condensation. The sterile, hydrated air was passed to the cultures through high-quality gum rubber or silicone tubing. The temperature of the culture was regulated to 25°±0.05°C by a fan controlled by a relay driven by an amplified thermistor bridge. A 1 to 2-sec delay was inserted between the amplifier and the relay to prevent chatter.

The medium was modified Chu–10 (Williams 1971) with 50 μg/liter cyanocobalin (vitamin B_{12}) added. *Brachionus calyciflorus* would not reproduce in continuous culture without vitamin B_{12}.

With the vitamin, no evidence of inhibitory or dietary insufficiency of rotifers feeding on algae in continuous or batch cultures was observed. The vitamin did not affect algal growth. The medium was aerated for at least 24 hr before use and was passed to the culture through silicone rubber tubing. The empty culture vessels, tubing, air filters, and the media reservoir were connected and autoclaved before medium was added. Medium was sterile-filtered into the reservoir through a 0.22 μm cellulose acetate or polycarbonate membrane filter. After having been assembled and autoclaved, the culture vessel and tubing were flushed with at least 0.5 liters of medium to remove contaminants released during sterilization.

The algal culture vessel was modified for rotifer culture by adding an extra exit port, a three-way solenoid valve and timer, and a small magnetic stirring bar inside the vessel. With only one exit port, the rotifers became selectively concentrated in the culture. Apparently they avoided the bubbling action that forced the medium out of the exit port. This problem was solved by adding a second stepped exit port. The upper port (A) was normally open through the solenoid valve; this allowed only air to escape. The lower port (B) was normally closed. At one hour intervals, port A was closed and port B opened for about 30 sec. This action caused air pressure to force out accumulated medium and rotifers rapidly. Thus the rotifer cultures had constant input, but the output was a sawtooth wave with an amplitude <3% that of the culture volume. The culture volume could be adjusted by the length of the tube connected to port B. The magnetic stirring bar was included to scrape wall growth from the culture. Although every effort was made to keep the cultures bacteria-free, all cultures became contaminated sooner or later; wall growth ensued. Even in bacteria-free cultures, the algae tended to clump and to adhere to culture walls, perhaps because the rotifers released some substance into the media. All samples were taken from the sample ports with a 25-gauge needle after the silicone stopper was flamed. Algal and rotifer samples were collected independently.

Both algal and rotifer population sizes were measured with a Coulter Counter model B interfaced with a model C-100 Channelyzer. These instruments allowed the total particle number, a 100-channel size distribution, and a biovolume to be taken for both species with each sample. The biovolume was computed as the sum of the products of the numbers in each channel and the size in cubic micrometers corresponding to that channel. Later a model ZB counter replaced the model B; a teletype interface and a microcomputer (Ohio Sci. Instr.) also were added.

Algae were counted with a 50 μm aperture in a medium modified from Williams (1971): 35 g polyvinylpyrollidone-40 (Sigma Chemical), 4 g NaCl, 1 g sodium azide, and a phosphate buffer. The PVP–40 stabilized the cell volume, reduced settling of dense *C. pyrenoidosa* cells, and increased the viscosity (and thus slowed the movement of the cells through the aperture). The model B Counter has a slow rise time, which leads to distortion in the size distribution (H. Kubitschek pers. comm.). The viscous PVP-40 medium increases the time each cell spends in the aperture, which helps to compensate for the slow rise time of the model B. There is no significant difference between estimates of biovolume obtained with the tube-type model B and and solid state ZB when the PVP-40 medium was used. The counting medium was passed through a 0.22 μm membrane filter before use. Cells were diluted to 20–40 $\times$ 10^3 per ml. At least two counts from one dilution were taken for each sample.

Rotifer eggs were stripped from the females with a 25-gauge needle and a 30–50 ml plastic syringe while they were sampled. With practice >95% of the eggs could be stripped with no damage to adults or eggs. About 15 ml from the chemostat were weighed into a preweighed beaker and then diluted to 150 ml with an electrolyte (0.075% NaCl). Filtration was unnecessary. The beaker and remaining sample were reweighed after sampling; the difference in weight was the volume sampled. Two independent samples were collected and counted.

A primary difficulty with this Coulter Counter system is that the Coulter Channelyzer is rated to about ±1 channel horizontally. This translates to about a ±3% uncertainty in the biovolume computation. As this may be the major source of sampling error, I attempted to minimize it by frequent calibrations with standard latex particles and by matching the Counter to the Channelyzer daily. Another difficulty with this method is that rather large amounts of data are generated. Until recently more than 50% of the manhours for a study were spent manually collecting and analyzing numbers. With the addition of the microcomputer, however, the time and the human error associated with this aspect of data collection and reduction have been significantly reduced.

Algal biomass (dry weight) was measured by filtering samples from unispecific cultures onto 0.2 or 1.0 μm polycarbonate (Nucleopore) filters. Rotifers from two-stage or predator-prey cultures were filtered onto 8.0 or 12.0 μm polycarbonate filters 13-mm in diameter. The filters were dried at 60°C and weighed to 0.5 μg on a Cahn electrobalance. Percent N was determined with a Coleman nitrogen analyzer on a 2–3 mg dry weight sample accumulated from outflow from algal (*C. pyrenoidosa*) or two-stage (*B. calyciflorus*) cultures. Ammonia was measured by the method of Solórzano (1969) using a boric acid buffer at pH 10.0. Nitrate was measured as in Wood et al. (1967). Nitrate samples had to be diluted 3-5 fold to prevent inhibition of the cadmium column.

Chlorella pyrenoidosa Chick was obtained from the Culture Collection of Algae, University of Texas at Austin (catalog No. 26). Algal stock cultures were maintained on Chu-10 agar slants. The *B. calyciflorus* were generously provided by Dr. John Gilbert. The rotifer cultures were routinely kept in continuous or semi-continuous culture. In addition, resting eggs (frozen at $-10°C$ for up to three years) were hatched. Rotifers were isolated from bacteria and the eggs stripped from females as previously described; the stripped eggs were washed in bacteria-free algal culture and filtered onto a 10 μm sterile Nitex screen in a 13 mm stainless steel filter holder. While they were held on the screen, the eggs were rinsed vigorously with 0.22 μm-filtered 0.025% sodium hypochlorite and then rinsed with sterile algal suspension. After the rinses, the eggs were transferred to a sterilized tube containing algal suspension and allowed to hatch. As noted by Gilbert (1970), eggs in later developmental stages appeared to have a greater hatching success than younger eggs. Soon, after hatching, the juveniles were transferred to continuous culture vessels, else the algal cells were rapidly diminished by the growing rotifer population.

Results and discussion

The data here are representative of the algal and rotifer population structure and dynamics observed in the continuous cultures. A more complete analysis of the data will be presented elsewhere (Boraas in prep.).

Characteristics of rotifer and algal populations. Algal size distributions from unispecific single-stage cultures have a shoulder characteristic of *C. pyrenoidosa* (Fig. 2). Williams (1971) accounted for this shoulder by assuming that 75% of the mature cells divide into four neonatal cells and that 25% divide into two daughter

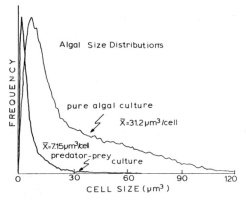

Fig. 2. Algal size distributions from steady state unispecific cultures.

cells. He presented a quantitative model for the distribution.

Size distributions of algae from the second stage of two-stage cultures and from the mixed cultures lack the characteristic shoulder. In the two-stage cultures the algae are in the dark; they have ceased to grow in biomass and have simply divided down to their minimum cell size for the prevailing environmental conditions. This explanation cannot hold for algae growing in mixed cultures; these cells often grow at or near their maximum rate and, hence, should be at or near their maximum cell size (Williams 1971). The possibility that this qualitative difference in size distribution is due to size-selective grazing by the rotifers (with subsequent effects on the age structure of the algae) is under investigation.

At 25°C, *B. calyciflorus* eggs hatch in 8-9 hr; about 6-7 hr are spent as juveniles and about 8-9 hr are necessary to produce the first egg. The sizes of the animals in these cohorts were followed through time to establish an age-size relationship. The egg-to-egg time is 24±1 hr at high food levels. The survivorship curve is rectangular with a lifespan of about 200 hr (see Halbach, 1970), during which time 23 ±3 eggs are produced by each amictic female. Size distributions of populations that are growing exponentially (Fig. 3a), that are declining (Fig. 3b), and that are in a steady state (Fig. 3c) demonstrate

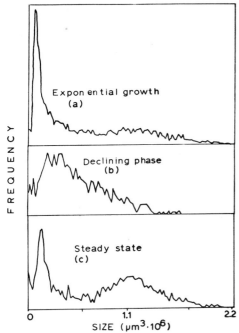

Fig. 3. Rotifer size distributions from mixed cultures at different rotifer growth rates.

the effects of age structure changes on the size distributions. In Figs. 3a and 3c, the left-hand peaks are eggs, the troughs are juveniles, and the second (right-hand) peaks are adults. During exponential growth, eggs are abundant. Adult rotifers are large, with full guts and developing eggs. When a steady state is reached, egg frequency declines in proportion to adults. During a population crash, usually due to a lack of food, debris (e. g. dead individuals and empty loricas) occupy the lower size classes and eggs are absent (Fig. 3b). Thus once the eggs are stripped from the adults, the condition and age structure of the population can be estimated directly from the size structure.

Conversion factors (biovolume, biomass and nitrogen). An important consideration in mixed species populations is the selection of appropriate units (Curds and Bazin 1977). Nitrogen concentration controlled the total attainable biomass of the system because it was ultimately limiting to the algae. However, it was not necessarily limiting to the animals, either in mixed or in two-stage culture.

The algae were much more dense (dry weight/wet volume) than the rotifers. The algal dry weight conversion factor was 0.57 g dry wt./cm^3 wet volume. The rotifer conversion factor ranged from 0.10 g/cm^3 with empty guts to 0.21 g/cm^3 with packed guts. Animals in steady-state chemostats ranged from 0.12 to 0.17 g/cm^3. In addition, algal N/biomass was variable. Consequently, with hindsight, biovolume *may* have been the single most unbiased population measure for this study. However, it is also clear that measuring as many different units as possible is desirable at the present state of the art.

Although I made no attempt to determine an accurate nitrogen balance, an analysis of the available data may help in guiding future work. Algae from pure cultures were about 7.0% N dry wt, while starved rotifers were 11.6% N; therefore conversions were 0.040 g N/cm^3 biovolume for algae and 0.012 g N/cm^3 for rotifers. For a culture (presented later in Fig. 5) which had a steady state rotifer biovolume of 2.20 10^8 μm^3/ml and 3.19 10^7 μm^3/ml algal biovolume, an assumption of constant chemical composition for both species gives 2.64 μg N/ml and 1.27 μg N/ml for the rotifers and algae respectively. Given the input nitrogen concentration of 7.0 μg N/ml, 3.1 μg N are not accounted for by the organisms. Ammonia, the rotifer excretory product, was virtually undetectable both in the culture itself and in exhaust gas bubbled through boric acid. A single measurement of dissolved organic N was less than 0.5 μg N/ml. Therefore it seems unlikely that the algae had a constant nitrogen content; e.g. they probably accumulated ammonia excreted from the rotifers. I have since estimated the rotifer gut volume to be about 9% of the total rotifer biovolume in growing populations (unpub.). If this were all algae, there was a total of 5.2 10^7 μm^3 as algae. If these algae had a two-fold increase in N/volume due to NH_4^+ accumulation, there could have been a total of 4.14 μg N/ml in suspended algae and in rotifer guts. The total N/ml computed in this manner, including suspended algae, algae inside rotifer guts, rotifers, and maximum dissolved N, is more than enough to account for the missing N (4.14 + 2.64 + 0.5 = 7.27 μg N/ml). The point of this exercise is that, although N is not necessarily the limiting factor for the rotifers, nitrogen rather than number, biovolume, or biomass, is conserved in this system.

Population dynamics in continuous cultures. For the most part, the results from unispecific algal cultures were very stable and repeatable. The culture conditions were similar to those described by Williams (1971). One perturbation not performed by Williams is shown in Fig. 4. Algae were inoculated into a culture vessel and allowed to reach a steady-state with a turnover rate of 0.020/hr. At about 0320 hr, the incident light was decreased by 40% from a crude measurement of 550 ft.-c. The cell number then increased abruptly, but biovolume did not change detectably. The cell number continued to rise slowly with a consequent drop in mean cell size. Thus algal cell size is a function of a variety of environmental and biological conditions. Biomass, as measured by biovolume, is much more closely tied to the chemical inputs to the population—the input nitrate concentration.

Both rotifer and algal populations entered steady states in two-stage chemostats. The culture shown in Fig. 5 demonstrated most of the characteristic features of these systems. The approach to the steady state was essentially critically damped. The rotifer steady-state was maintained with a coefficient of variation of about 10–15%; single-species *C. pyrenoidosa* cultures had a CV of less than 5%. At steady-state the rotifer culture contained an average of 97 adults, 70 juveniles, and 129 eggs per ml. The turnover rate was 0.0385/hr. At 1100 hr the turnover rate was increased to 0.070 /hr and the rotifers washed out, indicating that this turnover rate exceeded their maximum growth rate. Based on the approach to the steady state and the washout curve respectively, the maximum population growth rate was 0.055 and 0.056/hr, or a minimum doubling time of 12.5 hr (ln2/ 0.0555hr).

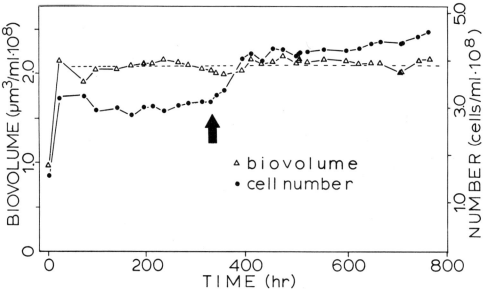

Fig. 4. Number and biovolume of algae in unispecific chemostat culture. Light intensity was decreased by 40% at arrow.

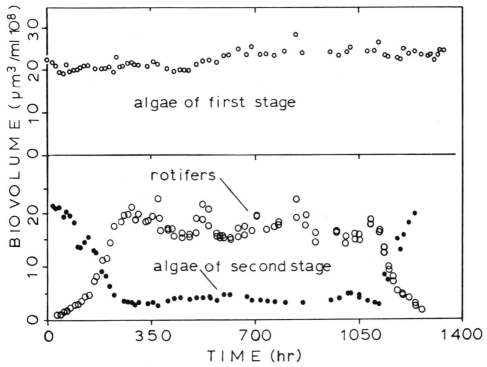

Fig. 5. Biovolume of rotifers and algae in the first stage and the second stage of a two stage chemostat system.

The predator-prey cultures showed either damped oscillations into a steady-state (Fig. 6) or were critically damped in mixed species chemostats (Fig. 7), depending on initial conditions. The overall patterns suggest nutrient limitation for each component of the system. Total algal biomass was bounded by the input nitrate concentration, and algal growth rate was a function of nitrate concentration within the culture. Rotifer biomass was bounded by the maximum algal biomass, and rotifer growth rate was a function of algal concentration and turnover rate. However, the limiting chemical constituent in the algae to the rotifers is not known at present.

Peaks and troughs of algae, rotifers, and nitrate appeared in a regular progression (Fig. 6). As the rotifer population increased, the algal concentration fell. Free nitrate appeared when the algae fell to about 0.5 10^8 μm^3 /ml. Algal levels continued to drop as the rotifers first peaked and then began to fall. The declining rotifer populations became severely food limited. Their egg production stopped (Fig. 3b); the adults became transparent with virtually no evidence of internal organs. Large numbers of dead animals appeared during the population crash. Normally there were <0.2 dead animals per ml; up to 6 dead per ml were present during the declining phases. When the rotifer concentration had declined to about 0.75 E8 μm^3 /ml, the algal population began to grow at specific growth rates near their maximum (0.125 /hr), causing a simultaneous drop in nitrate concentration. These oscillations eventually damped out, and the system approached a steady state with low algal and nitrate levels and relatively high rotifer concentrations.

The steady state was independent of initial conditions. At 1100 hr (double vertical arrow in Fig. 7), this culture suf-

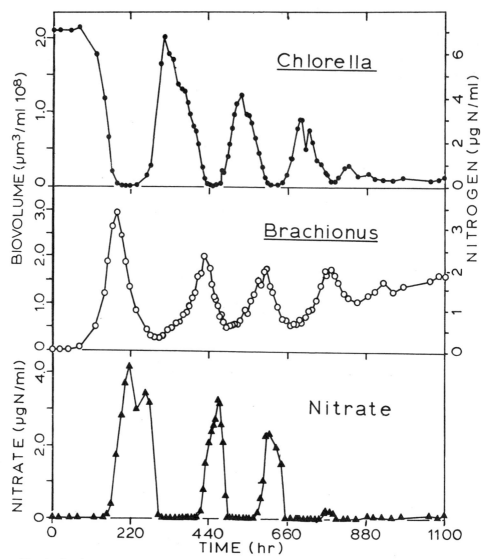

Fig. 6. Biovolume and calculated nitrogen of rotifers and algae and nitrate nitrogen in mixed continuous culture. Constant environmental conditions were obtained throughout culture.

fered a temperature shock of less than 30°C (5°C rise) for 5–10 h. The first two points of Fig. 7 are identical with the last two points of Fig. 6. After the temperature shock, although the adults remained active and continued to feed, they ceased egg production. The algal population began to increase as the rotifers washed out; free nitrate became and remained undetectable.

About 50 hr after the temperature shock, the rotifers resumed egg production. In this time, the algae had risen to 0.5 10^8 μm^3/ml. These events produced a new set of initial conditions: low levels of algae and rotifers and undetectable nitrate. The system rapidly returned to the previous steady state, with little or no overshoot.

A simple experiment indicated that the wall growth probably had little effect on the culture dynamics. Cultures rou-

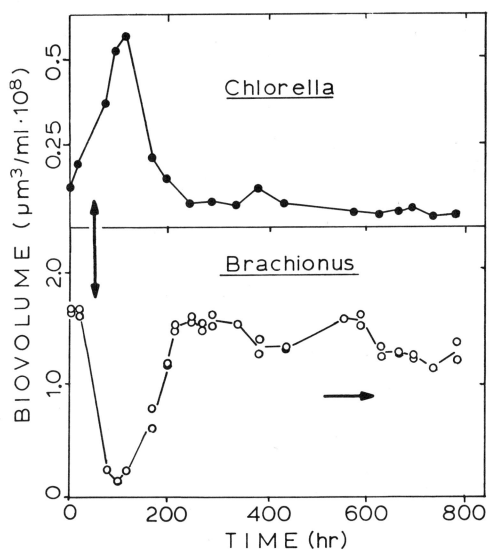

Fig. 7. Transient response and steady state of mixed culture after a temperature perturbation at double vertical arrow. After horizontal arrow, wall growth was allowed to accumulate.

tinely were scraped with a magnetic stirring bar at least once daily to remove wall growth. I allowed wall growth to accumulate in the culture after the time indicated by the horizontal arrow in Fig. 7. There was no apparent change in the predator-prey steady-state dynamics.

Theoretical implications

The behavior of single species algal cultures and of rotifers in two-stage cultures was consistent with classical chemostat theory (Herbert et al. 1956), particularly when respiration was included in the models (as in Williams 1972; Droop and Scott 1978). However, a preliminary analysis suggests that the mixed culture system does not behave as predicted by a simple coupling of two rectangular hyperbolae, one for algae on nitrate and one for rotifers on algae (Boraas unpub.). Oscillations in the model system are undamped and of a longer period than in the experimental system. Inclusion of internal (ammonia) cycling increases the frequency of oscillations, but does not produce damping. These observations are consistent with other studies (Curds and Bazin 1977; Frederickson 1977). The theory of single species growth in continuous culture is reasonably well developed, but the theory of interactions in mixed culture, particularly predator-prey interactions, is only beginning.

In general, logistic-based models usually have not been useful for simulating or for interpreting predator-prey data (Canale 1970; Jost et al. 1973; Dent et al. 1976), primarily because the mechanism of prey growth, uptake of nutrient, has not been explicitly incorporated into the models (see Curds and Bazin 1977).

References

CANALE, R. P. 1970. An analysis of models describing predator-prey interaction. Biotechnol. Bioeng. **12**: 353-378.

CURDS, C. R., and M. J. BAZIN. 1977. Protozoan predation in batch and continuous culture. Adv. Aquatic Microbiol. **1**:115-176.

DENT, J. E., M. J. BAZIN, and P. T. SAUNDERS. 1976. Behavior of *Dictyostelum discoideum* amoebae and *Escherichia coli* grown together in chemostat culture. Arch. Microbiol. **109**: 187-194.

DROOP, M. R., and J. M. SCOTT. 1978. Steady-state energetics of a planktonic herbivore. J. Mar. Biol. Ass. U. K. **58**: 749-772.

FREDRICKSON, A. G. 1977. Behavior of mixed cultures of organisms. Ann. Rev. Microbiol. **31**: 63-87.

GILBERT, J. J. 1970. Monoxenic cultivation of the rotifer *Brachionus calyciflorus* in a defined medium. Oecologia **4**: 89-101.

HALBACH, U. 1970. Einfluss der temperatur auf die populationsdynamik des planktischen Radertiers *Brachionus calyciflorus* Pallas. Oecologia **4**: 176-207.

HERBERT, D., R. ELSWORTH, and R. C. TELLING. 1956. The continuous culture of bacteria; a theoretical and experimental study. J. Gen. Microbiol. **14**: 601-622.

JOST, J. L., J. F. DRAKE, A. G. FREDRICKSON, and H. M. TSUCHIYA. 1973. Interactions of *Tetrahymena pyriformis, Escherichia coli, Azotobacter vinelandii* and glucose in a minimal medium. J. Bacteriol. **113**: 834-840.

KUBITSCHEK, H. E. 1970. Introduction to Research with Continuous Cultures. Prentice-Hall, N. J.

LAMPERT, W. 1976. A directly coupled, artificial two-step food chain for long-term experiments with filter-feeders at constant food conditions. Mar. Biol. **37**: 349-355.

MONOD, J. 1950. La technique de culture continue; theorie et applications. Ann. Inst. Pasteur **79**: 390-410.

NOVICK, A. and L. SZILARD. 1950. Description of the chemostat. Science. **112**: 715.

SOLÓRZANO, L. 1969. Determination of ammonia in natural waters by the phenolhypochlorite method. Limnol. Oceanogr. **14**: 799-801.

WILLIAMS, F. M. 1971. Dynamics of microbial populations. p. 197-267. *In* Patten [ed.], Systems analysis and simulation in ecology. vol. 1. Academic Press, N. Y.

WILLIAMS, F. M. 1972. Mathematics of microbial populations, with emphasis on open systems. Trans. Conn. Acad. Arts. Sci. **44**: 397-426.

WOOD, E. D., F. A. J. ARMSTRONG, and F. A. RICHARDS. 1967. Determination of nitrate in sea water by cadmium-copper reduction to nitrite. J. Mar. Biol. Assoc. U. K. **47**: 23-31.

IV

The Copepod Filter-Feeding Controversy

18. Comparative Morphology and Functional Significance of Copepod Receptors and Oral Structures

Marc M. Friedman

Abstract

Scanning and transmission electron microscopy are used to describe the structure of antennal mechanoreceptors and the chemoreceptors of various copepod mouthparts. Certain appendages and structures that are expected to participate in food handling, ingestion, and mastication are also described. These include the feeding chamber and its volume, glandular and other pores on the labrum (upper lip), the masticatory mandibular endite teeth, the width of the mouth entrance, and hair groups that trap food particles in the mouth. These characters are compared among four species of *Diaptomus* and among other genera.

The classical model of copepod filter feeding does not account for the ability of *Diaptomus* and *Acartia* to capture and ingest particles 0.5 μm in diameter (such abilities are revealed by electron micrographs of animals fixed while feeding) or for calculations indicating that filtering rates exhibited by *Diaptomus* are almost an order of magnitude larger than the maximum volume of water that can be pumped through the feeding chamber. Calculation of expected filtering rates using an expanded feeding envelope that includes the volume enclosed by outlying appendages suggests that these appendages participate in food capture. This hypothesis is supported by the presence of 40–80 contact chemoreceptors in the setae of these outlying appendages.

The structural complexity of the copepod mouth region indicates that key morphologic characters may be used to delineate ecological and evolutionary relationships.

The filter-feeding mechanism of calanoid copepods has been described as a combination water pump and filter screen (Storch and Pfisterer 1925; Cannon 1928; Lowndes 1935; Gauld 1966). In these classical studies, particle selection was regarded as being passive, caused by enhanced retention of larger particles on a filter screen and resulting from the distribution of the filter's pore sizes (Boyd 1976; Nival and Nival 1976). The discovery by Friedman and Strickler (1975) of numerous contact chemoreceptors in the mouthpart setae prompted them to suggest that many observed feeding behaviors are not passive, but dynamic, and are mediated by chemical cues. Recently, evidence from laboratory feeding experiments has been presented to support this hypothesis and to further indicate that filter feeding is indeed selective and also responsive to varying food conditions (Poulet and Marsot 1978; Donaghay at this conf.).

Due to the newness and extent of these data, the current picture of filter feeding is confusing. Forthcoming observations will undoubtedly expand the known range of copepod feeding behaviors into areas not yet clearly delimited. Physiological and behavioral limits, however, do exist. Copepod behavior, like that of other arthropods, is constrained by sensory structure and morphologic capability, both of which can be studied and described. Although copepod receptors are too small to be studied by electrophysiology, their function may be inferred from ultrastructural comparison with arthropod receptors whose functions have been determined (Friedman and Strickler 1975), and from laboratory experiments, observations, and high-speed motion pictures which verify the copepods' abilities to perform the predicted or requisite behavior (Strickler 1977; Poulet and Marsot 1978). The morphology of copepod mouthparts can be described by electron microscopy (EM), their respective functions determined by high-speed motion photography and receptor placement, and details provided by EM examination of copepods fixed while feeding (P. Blades at this conf.). Data obtained via these techniques, combined with the results of laboratory feeding experiments (Richman et al. 1977; S. Richman at this conf.) will help to frame a more complete understanding of copepod behavior.

The antennal mechanoreceptors of cyclopoid copepods have been implicated in predator avoidance, e.g. escape reactions (Strickler 1975). Calanoid copepods exhibit similar escape behavior. Their antennal mechanoreceptors are described below and compared with those of cyclopoid copepods. Although the role of mouthpart contact chemoreceptors in food selection has been discussed elsewhere (Friedman and Strickler 1975; Friedman 1977), their structures and distributions are presented here: I describe and compare specific oral appendages and structures among several species of *Diaptomus* and among other genera and evaluate the functional significance of these appendages using available data. I also consider the concept of the feeding chamber and construct a new view of an expanded feeding envelope (Fig. 1).

The structure of antennal mechanoreceptors

Strickler and Bal (1973) determined that certain hairs on the first antenna of *Cyclops scutifer* were mechanoreceptors, and that these were probably derived from two sheathed, ciliary neurons that exhibited a 9+0 basal body structure. Distal to the

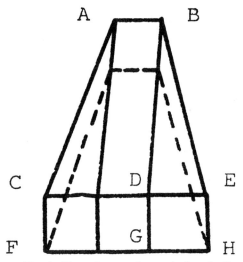

Fig. 1. Dimensions used to calculate feeding chamber volume.

basal body the number of neurotubules increased to form a bundle which passed from the antenna into the external mechanoreceptor hair. Because of difficulty in positioning the antenna for sectioning, they were unable to verify the neural continuity of the receptors or to describe their complete structure.

Apparently nearly identical mechanoreceptors are located on the first antennae of *Diaptomus pallidus* (Fig. 2). They are innervated by paired neurons that leave the antennal ganglion and continue parallel to it beneath the exoskeleton. Each neuron of a pair forms a 9+0 basal body, at which point it is enclosed by a dense, microtubular sheath. As the ensheathed receptor dendrites approach the surface of the first antenna, each dendrite elaborates 500–700 neurotubules that are bound into separate bundles. The paired bundles leave the neural sheath and the antenna and enter the external chitinous hair precisely apposed: One bundle of about 700 neurotubules is flattened along the outer edge of the combined bundle and attaches to the wall of the hair; the inner bundle of about 500 neurotubules remains ovoid and is cradled within the outer, attached bundle (Figs. 6,7). This conformation supports the suggestion made

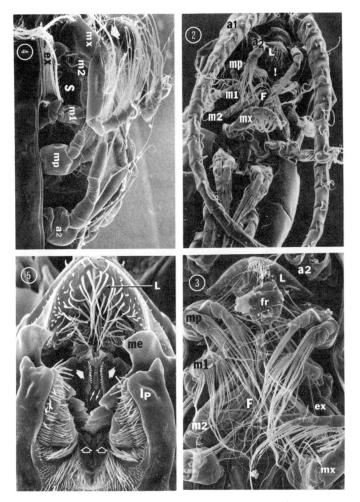

Fig. 2. Note that viewing orientation of each following figure is the same as that of its number. Abbreviations: a1, A1—first antenna; a2—second antenna; en, m1en—endites of first maxilla; ex, m1ex—exite of first maxilla; F—feeding chamber (filter chamber); L—labrum; lp—labial palp (lower lip); M—mechanoreceptor hair; m1—first maxilla; m2—second maxilla; me—mandibular endite; mp—mandibular palp; mx—maxilliped; nt—neurotubules; S—suction chamber of Cannon (1928); !—mouth.

Diaptomus pallidus, × 190. Anterio-ventral view, showing head and mouth appendages and first two pairs of swimming legs. Long first antennae, which in life are laterally extended, are folded close to body in a common fixation artifact.

Fig. 3. *Calanus finmarchicus* × 110. Overhead (ventral) view of feeding chamber, after removal of one maxilliped and both second antennae; Chitinous fragment (fr) positioned over mouth is probably an artifact generated during microdissection. Symbols as in Fig. 2.

Fig. 4. *Calanus finmarchicus*, × 120. View of feeding appendages from left side (anterior of animal is to right). Large arrow (extreme left) indicates setae of second antenna which overlie feeding chamber and extend posteriorly past maxillipeds. Symbols as in Fig. 2.

Fig. 5. *Diaptomus pallidus*, × 700. Looking through labial-palp (lower lip) cleft, through mouth to labrum. Mandibular endites (me) are inside of mouth, and all other appendages have been removed. Note prominent central hair channel of labrum, which is matched with a similar and opposite hair channel below setal brushes (unfilled arrows) which groom mandibular endite teeth. Two opposing hair channels form oral channel, which retains and concentrates food particles during feeding. Large solid arrows indicate open, glandular pores with elevated pore structures just above them. Small arrows indicate dome-shaped structures with terminal pores (*see also Fig. 11*). Symbols as in Fig. 2.

by Strickler and Bal (1973) that these receptors may function by the piezoelectric principle, with neurotubule shear forces generating an electric potential.

Mechanoreceptor hairs are often flattened along one edge (Fig. 7) and attached by external chitinous projections at their base, or on the antenna (Fig. 6), which prevent bending in certain directions. The superimposed asymmetry of neurotubule bundles and hair structure suggests that individual receptors may respond to vibrations from only one, or a few, directions. The multidirected array of mechanoreceptor hairs characteristic of the first antennae of copepods may thus be used to receive and analyze signals from several directions independently.

Structure of mouthpart receptors

Contact chemoreceptors are located on all of the mouthparts of *Diaptomus*, within setae of the mandibular palps, the second maxillae, and the first maxillary endites and palps. Receptors of undetermined function are located within the mouth in the mandibular endites and the labrum and also on the maxilliped setae.

The mouthpart setal receptors are housed entirely within the long chitinous setae used for food capture and handling. Each receptor contains one to five neurons, one or two of which are ciliary and possess 9+0 basal bodies. Distally, the number of neurotubules may increase to about 50 per ciliary neuron. The accompanying neurons lack a fully developed basal body and contain only a few singlet or doublet neurotubules (Fig. 8). These unsheathed receptors are protected by the chitinous wall of the seta, and communicate with the external environment through pores in the chitinous wall (Friedman and Strickler 1975).

Receptors in the labrum are similar to some setal receptors in that they consist of four neurons, one of which is ciliary. Distally each neuron may contain 150 neurotubules. The receptor is enclosed within an accessory cell that contains numerous support microtubules, thus representing a condition intermediate between the dense microtubular sheath of antennal mechanoreceptors and the unsheathed setal chemoreceptors.

The mandibular endite receptors consist of one or two ciliary neurons (Fig. 9) which may be accompanied by a single, nonciliary neuron. As in setal chemoreceptors, the ciliary neurons may contain 30–50 neurotubules in their distal regions. The mandibular receptors are enclosed in a chitinous, cuticular sheath and reportedly terminate in a pore at the edge of the mandible (Ong 1969); although I have been unable to confirm this using scanning microscopy. The dense chitinous sheath protects the receptors from buffeting as the mouthparts masticate food particles.

Diaptomus thus possesses several appendaged-mounted receptor groups: mechanoreceptive hairs on the first antennae; contact chemoreceptors in the numerous setae responsible for food capture and ingestion; and receptors of unknown function within the mouth. The chemoreceptor groups have been implicated in chemical (taste) selection (Friedman and Strickler 1975; Poulet and Marsot 1978; Donaghay at this conf.).

Mechanism of filter feeding

Filter feeding has been viewed as a process in which feeding currents, generated by the outlying setae of the mouthparts and the second antennae, transport food-laden water along the sides of the animal and into the rear of the midventral feeding chamber. This feeding current is then filtered through the second maxillary setae, which bound the feeding chamber (filter chamber) and act as a filtering mesh that retains captured particles (Cannon 1928; Lowndes 1935; Gauld 1966). Particles smaller than the smallest spaces between setae and setules (the minimum pore size of the filter mesh) cannot be retained and are lost in the excurrent water (Boyd 1976; Nival and Nival 1976).

In another paper (Friedman in prep.) I have presented data that contradict this last aspect of the "fixed-sieve" feeding hypothesis. Electron micrographs reveal that *Diaptomus* and *Acartia* capture and ingest bacteria and algae as small as 0.5 μm in

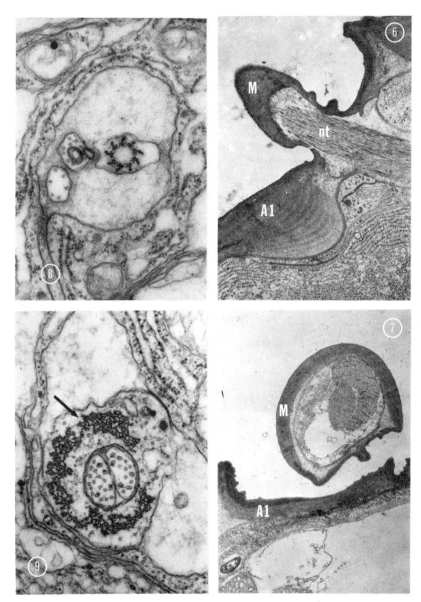

Fig. 6. *Diaptomus pallidus*, × 19,500. An oblique, longitudinal section through a part of first antenna and one mechanoreceptor hair, showing entry of neurotubule bundles. Symbols as in Fig. 2.

Fig. 7. *Diaptomus pallidus*, × 6,700. Cross section of a mechanoreceptor hair just distal to its point of origin on first antenna. The neurotubule bundles can be differentiated: outer bundle contains about 700 neurotubules and is flattened and attached to chitinous wall; inner bundle contains about 500 neurotubules and is partially enveloped by first bundle. Note that one side of hair is flattened, which may indicate directionality of bending. Symbols as in Fig. 2.

Fig. 8. *Diaptomus pallidus*, × 49,000. Cross section of a contact chemoreceptor in an endite seta of first maxilla (*see Fig. 14*). Only one neuron is ciliary and possesses a 9+0 basal body. Distally, this ciliary neuron may contain 30-50 neurotubules.

Fig. 9. *Diaptomus pallidus*, × 58,000. Cross section of a mandibular receptor containing two ciliary neurons, just distal to 9+0 basal bodies. Chitin sheath (arrow) is beginning to form. Distally, each neuron may contain 30-50 neurotubules.

Table 1. Feeding appendage measurements (from SEM micrographs) and calculated feeding chamber volumes.

Diaptomus And Other Genera	Cephalothorax Length Adult ♀ (mm)	Feeding Chamber Dimensions* (μm)				Second Maxillary Setule Spacing (μm)	Second Maxillary Setule Thickness (μm)	Calculated Feeding Chamber Volume ($\times 10^{-7} cm^3$)
		L_1	L_2	L_3	L_4			
D. siciloides	0.73	45	20	90	25	2	1.1	1.8
D. ashlandi	0.80	50	25	100	40	2	1.7	3.25
D. oregonensis	0.93	60	25	100	35	2	1.6	3.6
D. pallidus	0.84	70	30	100	25	3	2.2	3.85
Acartia tonsa	1.0	—	65	—	—	4.4	6.0	23.
Eurytemora affinis	0.73	60	15	120	55	—	—	5.0

*See Fig. 1., L_1 = CF, L_2 = AB, L_3 = BD, L_4 = DE.

diameter, even though their respective filters have minimum pore sizes of 2.0 μm and 4.4 μm (Table 1; see also Fig. 4). The fixed-sieve hypothesis also fails to account for the observed feeding behavior of *Acartia* and *Eurytemora* (Richman et al. 1977), *Diaptomus*, and other genera (summarized in Friedman and Strickler 1975).

Statics and dynamics

The complex dynamics of the calanoid filter-feeding mechanism are not discernible using standard techniques (Strickler 1977). The literature illustrates this point well: the three investigators who first published estimates of mouthpart oscillation rates gave significantly different figures (Storch and Pfisterer 1925; Cannon 1928; Lowndes 1935). Analysis of high-speed motion pictures taken by J. R. Strickler (1977, this volume) revealed that the true rates may exceed 60 Hz, a rate significantly higher than the 10–45 Hz previously reported. The techniques of scanning and transmission electron microscopy (SEM and TEM) which I have relied upon cannot directly provide such dynamic information on the filter-feeding mechanism. They do, however, permit accurate and detailed morphologic analysis of appendages and receptor structures, and allow examination of portions of the food-handling and ingestion processes which can be "frozen" by using animals fixed while feeding.

Receptors, filtering rates, and the classical model

The greatest number of contact chemoreceptors are located within the long setae that project ventrally over the mouth region and the feeding chamber (Figs. 2, 3). In the classical model of filter feeding proposed by Cannon (1928; see also Gauld 1966), oscillations of these setae aid in generating the water currents used for feeding. However, the presence in *Diaptomus* of 40–80 contact chemoreceptors in these setae argues for their more direct participation in food capture, even in these small filter-feeding genera.

Expected filtering rates

The classical filtering mechanism stipulates that food particles can be extracted from the feeding current only as they pass through the feeding ("filter") chamber. Thus, the maximum filtering (clearance) rate of an animal is solely dependent on the volume of water which is drawn through the feeding chamber. The classical model may therefore be tested by calculating the feeding chamber volume (V) from SEM measurements, by calculating an expected filtering rate (F) using the above feeding chamber volume and appendage oscillation rates measured from high-speed motion pictures (Strickler 1977, this volume), and by comparing the expected filtering rate with rates measured in laboratory experiments.

The feeding chamber volume is defined as that volume enclosed by the second maxillae (Figs. 3, 14) and can be represented by a trapezoidal solid (Fig. 1). I have measured its dimensions from scanning micrographs taken from various angles, after the removal of obscuring appendages by microdissection. The feeding chamber volume of *Diaptomus* or *Calanus* can be calculated from Fig. 1 using the formula

$$V = (DG)(AB)(BD) + (DG)(BD)(DE). \quad (1)$$

For example, for *Diaptomus oregonensis* (from Table 1),

$$V = 3.6 \times 10^{-7} \text{cm}^3. \quad (2)$$

The classical model of filter feeding further requires that all water which enters the feeding chamber in the feeding current must be drawn out via the lateral suction chambers (Figs. 4, 14), which are powered by a pumping action of the long exite setae of the first maxillae (Cannon 1928; Gauld 1966). In both *Diaptomus* and *Calanus* these suction chambers have a combined volume no larger than that of the feeding chamber. The exites are thus supposed to be capable of pumping twice the volume of the suction chambers on each appendage cycle, or twice the feeding chamber volume. We may therefore calculate the maximum expected filtering rate from the equation

$$F \text{ (ml animal}^{-1} \text{day}^{-1}) = (2)(V)(\text{No. of appendage cycles day}^{-1}).$$

The maximum reported oscillation rate for any calanoid is about 60 Hz (for *Diaptomus* and *Acartia*: Strickler 1977). Assuming the animals feed continuously at 60 Hz with a handling efficiency of 100%, for *D. oregonensis*, we would calculate

$$F_{(max)} = (2)(3.6 \times 10^{-7} \text{cm}^3 \text{cycle}^{-1})$$
$$(60 \text{ cycles s}^{-1})(8.64 \times 10^4 \text{s day}^{-1}) = 3.7 \text{cm}^3 \text{animal}^{-1} \text{day}^{-1}.$$

$$F = (4.2 \times 10^{-6} \text{cm}^3 \text{cycle}^{-1})(60 \text{ cycles s}^{-1})$$
$$(8.64 \times 10^4 \text{s day}^{-1})$$
$$= 22 \text{ cm}^3 \text{animal}^{-1} \text{day}^{-1}. \quad (3)$$

This rate is from 3–6 times lower than feeding rates measured for *D. oregonensis* by S. Richman (at this conf.) and 3.5 times lower than rates reported by McQueen (1970); both workers used natural food assemblages. I therefore conclude that observed filtering rates cannot be accounted for within the classical model of filter feeding.

The feeding envelope

Conover (1966) reported that when *Calanus hyperboreus* fed on very large particles, which it seized individually, it was able to capture them within a volume much larger than that enclosed by the second maxillae. This "expanded" feeding volume was that enclosed by outlying setae of the mouthparts. Extending this concept to our present discussion of filter feeding, we can calculate this "feeding envelope" volume in the same manner as we calculated feeding chamber volume. For *D. oregonensis* the expanded volume is $4.2 \times 10^{-6} \text{cm}^3$. Revising the equation for expected filtering rate, and substituting, we obtain

$$F = (4.2 \times 10^{-6} \text{cm}^3 \text{cycle}^{-1})(60 \text{ cycles s}^{-1})(8.64 \times 10^4 \text{s day}^{-1})$$
$$= 22 \text{ cm}^3 \text{animal}^{-1} \text{day}^{-1}. \quad (4)$$

This calculated feeding rate is much closer to the maximum value of 21.6 cm³ animal⁻¹ day⁻¹ reported by S. Richman (at this conf.). The results of this calculation and the presence of contact chemoreceptors in the outlying mouthpart setae suggest that they are not restricted to generating water currents, but somehow take part in food capture.

We can satisfy ourselves that the animals are able to process food particles at the rate indicated by the filtering volumes by calculating the average number of particles that *D. oregonensis* encounters per unit time. In a food particle concentration of 10^5 cells cm^{-3}, an animal would encounter

Fig. 10. *Diaptomus ashlandi*, × 1,300. Overhead view of mouth, showing web of fine hairs (arrow) stretching from the labial palps (lower lips) to labrum (upper lip). Symbols as in Fig. 2.

Fig. 11. *Diaptomus oregonensis*, × 4,300. Close-up of labrum, as shown in center of Fig. 5. Large arrows indicate open, glandular pores (0.9 µm diam) shown in TEM cross section in inset at upper left. Small arrow points to a triplet of dome structures with terminal pores. Note close similarity of these labral pores and those of *D. pallidus* (Fig. 5). Symbols as in Fig. 2.

Fig. 12. *Acartia tonsa*, × 1000. Overhead view of mouth after removal of most right-side appendages (left side of micrograph). Note cone- and dome-shaped structures (arrows) with terminal pores. Symbols as in Fig. 2.

Fig. 13. *Eurytemora affinis*, × 1,500. View of mouth and labrum. Right mandibular endite blade has been raised above its normal position, to expose teeth. Between teeth are small chitinous protuberances (arrow). Open pores and dome-shaped structures are quite similar to those of *Diaptomus*. Symbols as in Fig. 2.

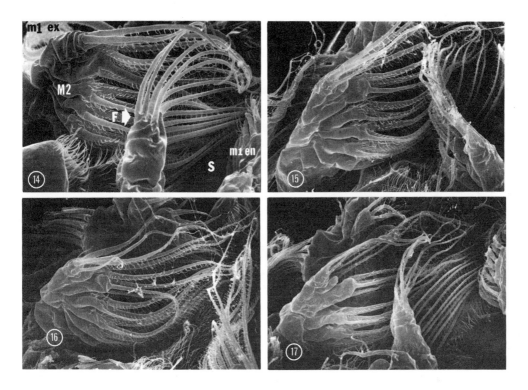

Figs. 14–17. Second maxillae of *D. pallidus* (× 900), *D. ashlandi* (× 1,000), *D. oregonensis* (× 975), and *D. siciloides* (× 1,100). Large arrow next to F in Fig. 14 denotes posterior-to-anterior direction of current flow reported by previous investigators (*see text*). Lateral suction chambers (S) of Cannon (1928) also shown. Symbols as in Fig. 2.

$$(4.2 \times 10^{-6} \text{ cm}^3 \text{ cycle}^{-1})(10^5 \text{ cells cm}^{-3}) = 0.42 \text{ cells cycle}^{-1}. \quad (5)$$

or only one cell each 2-3 appendage cycles.

Comparative morphology

In examining the mouthpart morphology of several calanoid genera, I noted similarities and differences in certain morphologic characters which are believed to have functional significance (Anraku and Omori 1963; Gauld 1966; Itoh 1970; Sullivan et al. 1975; Friedman 1977). While their full significance is not yet known, I will present and briefly discuss these characters with the view that we may soon be able to evaluate them functionally.

The detailed mouthpart morphology of *D. pallidus* has been described elsewhere (Friedman in prep.). *Diaptomus siciloides*, *D. ashlandi*, *D. oregonensis*, and *Eurytemora affinis* were supplied from Lake Michigan by S. Richman. *Acartia tonsa* was supplied by D. Heinle and R. Huff. The measurements listed in Table 1 are taken from micrographs of relatively few animals. A larger number of animals, sampled over the entire year, will be required to establish seasonal differences and statistically significant values.

Data from electron micrographs have provided substantial information concerning food handling and processing. The localization of ultra-small, ingested food particles in the mouth, esophagus, and gut of copepods is one example. The ingestion and mastication process of *Diaptomus* can be outlined as follows: captured food particles are transported anteriorly in the feeding chamber until they reach the labial palps (Fig 5); very small particles may pass directly into the mouth; larger particles are retained by the first maxillary endite setae which lie against the labial palps and guard the entrance to the mouth (Figs. 14-17); these endite setae contain contact chemoreceptors, and the setae may subsequently reject particles or pass them into the mouth; once in the mouth, food particles are retained within a central "oral channel" formed between two opposing rows of stout hairs (the hair channels) below the labrum and the labial palps (Figs. 5, 11, 12); particles are masticated by the toothed mandibular endites (Fig. 5), pushed down into the esophagus, and transported through the gut via peristalsis.

Comparative morphology of four Diaptomus *species*

Of the three Lake Michigan species, *D. siciloides* has by far the smallest feeding chamber volume, the narrowest labial palp width, the thinnest second maxillary setae, and the finest mandibular endite teeth (Table 1). However, all three species have an oral web of fine, chitinous hairs that stretch between the labrum and the labial palps (Fig. 10). These hairs may represent an adaptation for retaining small particles. Atop the labrum is a field of long chitinous hairs that can trap small particles during feeding.

The upper (ventral) portion of each mandibular endite blade contains two rows of teeth; a posterior row of stouter teeth and an anterior row of much finer teeth (Figs. 18-21). The stout posterior teeth result from the fusion of at least two finer teeth. The tooth arrangement allows the cradling of small particles during mastication (Fig. 21).

Diaptomus ashlandi and *D. oregonensis* have similar feeding chamber dimensions, setal thickness, and setule spacing, with the latter species having a longer cephalothorax. The feeding chamber volume of *D. oregonensis* is calculated to be only 10% larger than that of *D. ashlandi*, and its posterior teeth are 20% stouter (although similarly shaped: (Figs. 18, 20). These small apparent differences in related characters may be an artifact of small sample size or specimen preparation. If real, their significance is not clear. The mouthparts of these two species appear to be so similar that observed differences in feeding (S. Richman at this conf.) may be due to behavioral rather than morphologic specialization.

Diaptomus siciloides has a 9% shorter cephalothorax than that of *D. ashlandi*,

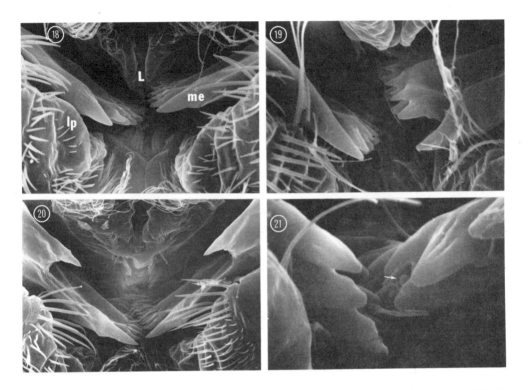

Figs. 18-20. Overhead close-up of mouth showing mandibular teeth of *D. ashlandi* (× 3,000), *D. siciloides* (x 5,000), and *D. oregonensis* (× 2,300). For comparison, note thickness of mandibular teeth of *D. pallidus* in Fig. 5. Symbols as in Fig. 2.

Fig. 21. *Diaptomus siciloides*, × 13,000. Arrow indicates a 0.4 μm diameter algal fragment cradled in mandibular teeth.

while its feeding chamber volume, setal thickness, and labial palp widths are 45%, 35%, and 20% smaller, respectively. Its significantly smaller mouthparts suggest that *D. siciloides* is highly adapted for feeding on very small and perhaps soft-bodied food particles (cf. Fig. 21). S. Richman (at this conf.) found that *D. siciloides* fed on the smallest size range of particles and had the lowest filtering rate.

Diaptomus pallidus lies between *D. ashlandi* and *D. oregonensis* in cephalothorax length, but its mouthparts are significantly stouter. It has the widest labial palp spacing, the largest feeding chamber volume, and the thickest second maxillary setae and mandibular teeth. The latter are almost three times wider than those of *D. oregonensis*. While *D. pallidus* does capture and ingest bacteria as small as $0.5 \times 2.0 \mu m$, electron micrographs of the gut reveal that these particles are not digested. The stouter teeth and appendages of *D. pallidus* may be adaptations for feeding on larger or harder particles.

Three characters are morphologically indistinguishable in all four *Diaptomus* species: 1) the field of chitinous hairs which traps food particles on the labrum (Fig. 5); 2) the opposing hair channels, below the labrum and the labial palps, which form the oral channel (Figs. 5, 11); and 3) the suite of pore structures on the labrum. These latter structures include: three singlet dome-shaped pore structures, followed by a triplet of similar structures, which run from top to bottom (ventral to dorsal) within the labral hair channel (Fig. 11); three or four pairs of open, glandular pores on either side of the labral hair channel (also found within the hair field atop the labrum); and a pore structure elevated above only one pair of the open pores. These consistent characters in the mouth region suggest that food processing is similar in all four species.

On the basis of labial palp width and feeding chamber volume (which is proportional to the volume of the feeding envelope) of the Lake Michigan species, *D. siciloides* would be expected to feed on the smallest particles and have the lowest filtering rate, while *D. oregonensis* should occupy the opposite position. The results of S. Richman (at this conf.) show that this is indeed the case. He also has found that *Eurytemora affinis* from Lake Michigan does not ingest particles below about 2 μm in diameter, but can ingest particles up to about 30 μm. These results are consistent with *Eurytemora*'s mouthpart morphology (Fig. 13). This species lacks the various hair structures of the labrum which retain small food particles. The top of the labrum is recessed to permit large particles to be passed over the labial palps and into the mouth. Also, the feeding chamber of *Eurytemora* is larger than that of *D. oregonensis*. Interestingly, the labral pore structures of *Eurytemora* are strikingly similar to those of *Diaptomus*, while its wing-shaped mandibular teeth are much stouter and resemble those of *Eucalanus*.

Acartia tonsa is another calanoid capable of feeding on very small food particles (Richman et al. 1977; Friedman in prep.). It has the oral channel and labral hair field which aid in small particle capture, and its labral pore structures are not unlike those of *Diaptomus*. Wilson (1973) found that *Acartia* was able to ingest whole plastic beads up to about 50 μm in diameter. Figure 12 reveals that the very wide labial palp spacing of about 65 μm is the reason for this unusual ability in a small copepod.

I have briefly discussed a number of morphologic characters that influence food processing and appear to correlate with observed feeding behaviors of certain calanoid species. Further analysis using electron microscopy, high-speed motion pictures, comparative morphology, and the results of feeding experiments will possibly complete our understanding of the functions of these characters and permit their use in determining ecological and evolutionary relationships.

References

ANRAKU, M., and M. OMORI. 1963. Preliminary survey of the relationship between the feeding habit and the structure of the mouthparts of marine copepods. Limnol. Oceanogr. 8: 116–126.

BOYD, C. M. 1976. Selection of particle sizes by

filter-feeding copepods: A plea for reason. Limnol. Oceanogr. **21**: 175–179.

CANNON, H. G. 1928. On the feeding mechanism of copepods, *Calanus finmarchicus* and *Diaptomus gracilis*. J. Exp. Biol. **6**: 131–144.

CONOVER, R. J. 1966. Feeding on large particles by *Calanus hyperboreus* (Kroyer), p. 187–194. *In* H. Barnes [ed.], Some contemporary studies in marine science. Allen & Unwin.

FRIEDMAN, M. M. 1977. Electron microscopic studies of the filter-feeding mechanism of calanoid copepods. Ph.D. thesis, The Johns Hopkins Univ., Baltimore. 100 pp.

———, and J. R. STRICKLER. 1975. Chemoreceptors and feeding in calanoid copepods (Arthropoda: Crustacea). Proc. Natl. Acad. Sci. **72**: 4185–4188.

GAULD, D. T. 1966. The swimming and feeding of planktonic copepods, p. 313–334. *In* H. Barnes [ed.], Some contemporary studies in marine science. Allen & Unwin.

ITOH, K. 1970. A consideration of feeding habits of planktonic copepods in relation to the structure of their oral parts. Bull. Plankton Soc. Jpn. **17**: 1–10.

LOWNDES, A. G. 1935. The swimming and feeding of certain calanoid copepods. Proc. Zool. Soc. Lond. **3**: 687–715.

McQUEEN, D. J. 1970. Grazing and food selection in *Diaptomus oregonensis* (Copepoda), Marion Lake, British Columbia. J. Fish. Res. Bd. Can. **27**: 13–20.

NIVAL, P. and S. NIVAL. 1976. Particle retention efficiencies of an herbivorous copepod, *Acartia clausi* (adult and copepodid stages): Effects on grazing. Limnol. Oceanogr. **21**: 24–38.

ONG, J. E. 1969. The fine structure of the mandibular sensory receptors in the brackish water calanoid copepod *Gladioferens pectinatus* (Brady). Z. Zellforsch. **97**: 178–195.

POULET, S., and P. MARSOT. 1978. Chemosensory grazing by marine calanoid copepods (Arthropoda: Crustacea). Science **200**: 1403–1405.

RICHMAN, S., D. R. HEINLE, and R. HUFF. 1977. Grazing by adult estuarine calanoid copepods of the Chesapeake Bay. Mar. Biol. **42**: 69–84.

STORCH, O. and O. PFISTERER. 1925. Der Fangaparat von *Diaptomus*. Z. Vergl. Physiol. **3**: 330–376.

STRICKLER, J. R. 1975. Intra- and interspecific information flow among planktonic copepods: Receptors. Int. Ver. Theor. Angew. Limnol. Verh. **19**: 2951–2958.

———. 1977. Observation of swimming performances of planktonic copepods. Limnol. Oceanogr. **22**: 165–169.

———, and A. K. BAL. 1973. Setae of the first antennae of the copepod *Cyclops scutifer* (Sars): their structure and importance. Proc. Natl. Acad. Sci. **70**: 2656–2659.

SULLIVAN, B. K., C. B. MILLER, W. T. PETERSON, and A. H. SOELDNER. 1975. A scanning electron microscope study of the mandibular morphology of boreal copepods. Mar. Biol. **30**: 175–182.

WILSON, D. S. 1973. Food size selection among copepods. Ecology **54**: 909–914.

19. Chemosensory Feeding and Food-Gathering by Omnivorous Marine Copepods

S. A. Poulet and P. Marsot

Abstract

Feeding processes of copepods are neither passive nor automatic phenomena. Raptorial feeding and impaction feeding are likely the two basic modes of food capture existing among copepods. The intensity of capture depends on the size of the particles and both on the anatomical and dynamic properties of the feeding appendages. The flexibility noticed in the regime of particles consumed over the entire size spectrum is related to the capability as well as to the length of time with which copepods can operate one mode of capture and can switch from one mode to the other. Small and large size particles can be retained either with the same or with variable efficiencies, depending on the abundance of each particle category in the water and also on the capture ability of the copepod species. Selectivity for food results from the stimulatory effects of physical and chemical agents on receptors presumably located in the filtering chamber and on the feeding appendages. Identification of food through sensory mechanisms seems to be more important than selectivity for sizes alone, as far as the dynamic equilibrium of the pelagic food web is concerned. Sensory selection should allow copepods to obtain their energy requirements with the least effort from the heterogenous mixtures of particles, a fraction of which constitutes potential food resources.

The distribution of particle numbers versus particle sizes shows that in the pelagic food web, particulate prey of small size are generally more abundant than large ones (Mullin 1965; Sheldon et al. 1967). This pattern seems to be well understood by copepods which have developed and adapted their feeding patterns to this structure (Parsons et al. 1967, 1969; Poulet 1973; Allan et al. 1977). We may also observe that "In the pelagic environment the maximum pressure for survival is exerted between the organisms themselves. It is further apparent that the medium in which they live allows for continual contact between predator and prey with little refuge" (Parsons 1976, p. 81-82). We can extend this statement to admit that the two most important parameters governing life in the sea are the growth rate and the size of the organisms within the ecosystem. For copepods the best tactic for their survival is to be small and numerous. Copepods are particle-feeders, whose survival depends on the utilization of the organic material available within the scattered and heterogenous assortment of particles which constitute their food resources. Growth and reproduction are two major functions which have to be achieved within the shortest period of time, before copepods become prey for other consumers. These functions are achieved by most species within a few days or weeks and they rely basically on ambient biophysical parameters as well as on the standing stock of food and feeding abilities. The feeding of copepods is different, since the group includes carnivores, omnivores, and herbivores (Anraku and Omori 1963; Arashkevich 1969; Itoh 1970).

If we accept the close relationship existing between food, feeding, and reproduction (Marshall and Orr 1955; Nassogne 1970; Harris and Paffenhöfer 1976; Paffenhöfer and Harris 1976; Zurlini et al. 1978), we can assume that the most abundant copepods should be those that are the most successful feeders. One proof of the ecological success of herbivorous and omnivorous species is reflected by their abundance within the plankton community compared to the scarcity of carnivores (Arashkevich 1969). The strategy for growth can be achieved only if the available stock of food is detected, tracked, and consumed with the minimum energy expenditure and at a maximum rate, whatever the changing conditions that are prevailing in the particulate food resources of the sea. The literature suggests that feeding patterns such as "switching mechanism" (Poulet 1973, 1974; Richman et al. 1977; Allan et al. 1977), "raptorial feeding" (Conover 1966), and "chemosensory grazing" (Poulet and Marsot 1978) are the basic tools used by marine herbivorous and omnivorous copepods to obtain their energy requirements. Re-examination of the "particle size selectivity" hypothesis as well as the "particle feeding efficiency" hypothesis indicates that these recent hypotheses, which may be satisfactory for some models, are simplifications of the feeding realities.

We thank D. Cossa and G. Drapeau for reviewing the manuscript, R. Couture, M. Morissette, M. Leclerc, and G. Bérubé for technical assistance, and E. Goulet for typing. This work was supported by grants from CNRC (A 9667; A 6491), from FCAC-DGES, and from FEC.

Raptorial feeding and impaction feeding: An alternative feeding theory

There are several recent hypotheses that consider the mechanisms of feeding by copepods (Nival and Nival 1973, 1976; Boyd 1976; Lehman 1976; Lam and Frost 1976; Frost 1977). Many workers generally suggest that copepods are indiscriminate filterers whose retention efficiency for different particle sizes depends on their filter structures which retain larger particles more efficiently than smaller ones. These workers also admit that larger particles are preferentially selected by copepods (Mullin 1963; Richman and Rogers 1969; Wilson 1973; Frost 1972, 1977). In contrast, experiments performed with natural particles show that the highest rates of ingestion occur at or near the biomass peaks observed in the size spectrum (Parsons et al. 1967, 1969; Poulet 1973, 1974, 1977; Conover unpubl.; Richman et al. 1977).

Poulet and Chanut (1975) have shown that the feeding of *Pseudocalanus minutus* on peaks in the particle spectrum is not a selective process based on sizes, but rather an active opportunistic mode of feeding based on the most concentrated stock of particles within a given size range (Fig. 1). Using the same statistical techniques as Poulet and Chanut (1975), Frost (1977) showed and confirmed that the size-frequency distribution of cells in controls and for the rations obtained by *Calanus pacificus* can be quite similar, even though computations of clearance rates indicate strong differential filtration of cells of different sizes—results that show increasing rates with increasing food sizes. Following this interpretation, it would appear that the efficiency curves computed for several species would look like those drawn by Nival and Nival (1973) or Boyd (1976). Similar curves computed for five sympatric copepod species, fed simultaneously on the same stocks of naturally occurring particles, vary both with time and biomass peak distributions over the particle spectrum (Poulet 1978; Fig. 2). If these efficiency curves describing particle retention were strictly related to the mesh aperture size of the filtering apparatus (Nival and Nival 1973, 1976; Boyd 1976; Frost 1977), then they should be constant independently of time and particle size distribution. Obviously, this is not the case judging from several experimental results (Conover unpubl.; Poulet 1977, 1978; Richman et al. 1977; Allan et al. 1977). When feeding on particles, copepods generally feed on peaks in the particle spectrum,

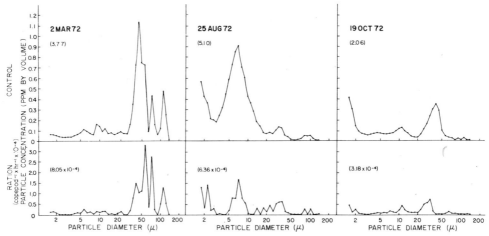

Fig. 1. Size frequency in control and ration. Total particle concentration and food uptake by *Pseudocalanus minutus* in parentheses (from Poulet and Chanut 1975).

but shift grazing pressure to other parts of the spectrum according to the occurrence and succession with time of biomass peaks (Fig. 1). Such results clearly demonstrate considerable flexibility in copepod feeding and strongly suggest that this shifting process (Poulet 1973; Richman et al. 1977) depends on behavioral phenomena.

The concept of a fixed sieve deduced from the theoretical performance of the filtering apparatus (Boyd 1976) does not seem to correlate with experimental results (Richman et al. 1977; Poulet 1977; Allan et al. 1977). In addition, this concept logically extends the classical hypothesis according to which the mouth parts of copepods (second maxilla) act as filters or sieves (reviewed by Marshall 1973). During the past decades, most workers dealing with feeding processes have tried to demonstrate this hypothesis and claimed that copepods preferentially select larger cells when fed with either single or mixed cultures of phytoplankton of different sizes. To verify this effect, experiments should have been done using mixtures of cells of different sizes but of equal concentrations, that is, strictly speaking, equal biomass per unit volume in order to allow choice by copepods. This experimental condition has been met only rarely (Table 1). The unit generally used for concentration has been set most often as the number of cells per volume. As a consequence, workers often claimed that at "equal" concentrations the largest cells were grazed at the highest rates. In reality, the food concentrations (part. vol. per fluid vol.; $\mu g\ C\ ml^{-1}$) of the larger cells were generally greater than those of the smaller species (as shown in Table 1), even though the larger ones were less numerous or equal in number. In our interpretation, under such conditions copepods always consumed preferentially cells having the most abundant biomass independent of size, because in that case these particles occupied the largest volume in water, and their chance to be encountered and captured by copepods was the highest. This interpretation agrees with the behavior currently observed for copepods feeding on naturally occurring particulate matter (Parsons et al. 1967; Poulet 1973; Conover unpubl.; Richman et al. 1977). Therefore, we think that the confusion introduced in the past, due to the fact that food offered was not equally available to the copepods, has led to some incorrect views on the copepod's feeding habits, even though experimental results were right and agreed with the modern concept of "preferential utilization of biomass peak" demonstrated for *P. minutus* (Poulet 1973, 1974). Following a different approach, Marshall and Orr (1955) fed

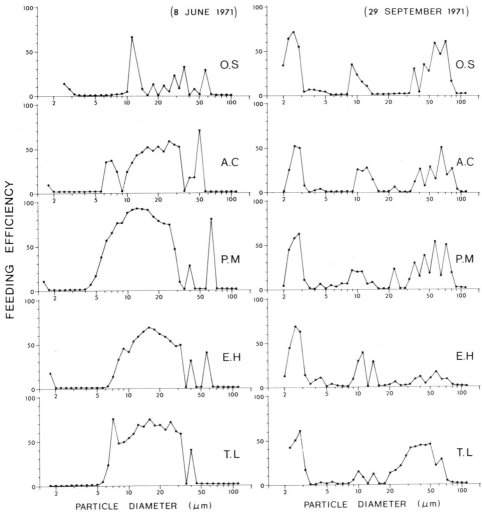

Fig. 2. Feeding efficiency curves (%) of five species of copepod fed simultaneously on particles at two different times of year. O.S—*Oithona similis*; A. C—*Acartia clausi*; P. M—*Pseudocalanus minutus*: E. H—*Eurytemora herdmani*; T. L—*Temora longicornis* (from Poulet 1978).

C. finmarchicus with cell cultures labeled with ^{32}P. In their experiments, an attempt was made to equate the biomass of the two foods offered. In a first set of experiments, radioactive cells of *Ditylum* (large size) were used in a mixture with nonradioactive *Chaetoceros* (small size). In a second set, the small diatoms were radioactive, and the large ones were not. If *C. finmarchicus* had selected large cells preferentially to small ones, the radioactivity of the copepods should have been higher in the first set than in the second one. In reality, the radioactivity was much the same in each set (Marshall and Orr 1955). It appears that copepods do not show much discrimination versus sizes, but take most of what predominates in biomass.

In nature, feeding strategy tends to maximize energy input per unit of energy expenditure (Richman et al. 1977). To do so, copepods must adapt their feeding behavior to the changing trophic environment which is characterized by successions of particle

Table 1. Results from feeding experiments performed with several cultivated species of phytoplankton showing preferential grazing by copepods on the most abundant biomass regardless of cell size. A—*Calanus pacificus* fed with nonmixed cells in separated containers (Frost 1972, from his table 1 and fig. 4); B—*Calanus finmarchicus* fed with mixtures of two cell categories (Mullin 1963, from his table 2); C—*Calanus pacificus* fed with mixtures of two cell categories (Frost 1977, from his table 2); D—*Calanus hyperboreus* fed on a mixture of various diatoms (Mullin 1963, from his fig. 5).

	cell vol. (μm^3)	cell C ($\mu g\ C \times 10^{-6}\ cell^{-1}$)	Food Concentration		Ingestion ($\mu g\ C\ h^{-1}\ cop^{-1}$)
			No. cells (cells ml^{-1})	cell biomass ($\mu g\ C \times 10^{-6}\ ml^{-1}$)	
A. *Calanus pacificus*:					
Coscinodiscus angstii	26,000	840	25	0.021	145
			50	0.044	250
			75	0.063	500
Coscinodiscus eccentricus	63,000	1,644	25	0.041	170
			50	0.082	380
			75	0.123	550
				(ppm)	(%)
B. *Calanus finmarchicus*:					
Ditylum brightwellii	1.2×10^5		50	1.2	74
Rhizosolenia setigera	7.6×10^4		50	0.76	26
Ditylum brightwellii	1.2×10^5		50	1.2	75
Gymnodinium nelsoni	7.4×10^4		50	0.75	25
Ditylum brightwellii	1.2×10^5		50	1.2	80
Striatella unipunctata	6.9×10^4		50	0.69	20
					(Ivlev's index, E)
C. *Calanus pacificus*:					
Coscinodiscus angstii					
small	40,160		310	12.44	−0.05
large	904,320		78	70.53	+0.23
small	40,160		227	9.11	−0.04
large	904,320		57	51.54	+0.23
					($ml\ d^{-1}\ cop^{-1}$)
D. *Calanus hyperboreus*:					
Coscinodiscus concinnus	6.9×10^6		0.6	4.1	390
Coscinodiscus sp. 2	1.5×10^6		0.9	1.04	110
Ditylum brightwellii	1.2×10^5		47.4	5.6	230
Rhizosolenia setigera	7.6×10^4		59.8	4.1	240
Striatella unipunctata	6.9×10^4		51.0	3.5	140
Thalassiosira fluviatilis	2.3×10^3		460	1.05	105
Cyclotella nana	2.1×10^2		1,467	0.3	75

biomass peaks with changing size modes (Poulet 1974). Theoretically, the same maximum ration (0.98-1.13 μg C h^{-1} copepod^{-1}) can be obtained from many cell categories, Calanus pacificus fed on four different species of algae ranging in size from 11 to 87 μm (Frost 1972; Table 2). For small Thalassiosira with a size 8 times smaller than large centric species, the number of ingested particles needs to be about 70 times higher, whereas the concentration must be 3 times higher (Table 2). Similarly, theoretical rations computed for P. minutus could be constant over five different categories of particle size, provided that the corresponding number of ingested particles is 10^4 times higher in the small size group (Table 2). On the average, the ration of P. minutus computed on a seasonal basis in each size group (Poulet 1974) leads to the same conclusions (Table 2). When copepods feed on particles of small sizes, they may obtain a substantial ration only if particle number increases to compensate for size reduction and to balance biomass. At the same time, copepods must adapt their feeding behavior to these changing conditions. Results from Table 2 and Fig. 1 demonstrate that some species obviously do so. The shifting process allows copepods to pick up particles from the entire size spectrum (3-50 μm) (Poulet 1974, 1978; Fig. 2). The lower and upper size limits are still unclear because of the technical limitations of the methods used. The upper limit ($\sim$ 300-500 μm) likely corresponds to the range of the microzooplankton and it corresponds rather to a carnivorous regime. The lower size limit of the food is estimated to be roughly 3 μm and corresponds to the minimum apperture size of the filtering appendages. Whatever the precise limits are, size of food has a 100-fold range between the smallest and largest item usually captured by copepods. It is unlikely that the properties of a sievelike apparatus generally ascribed to the filtering appendages could alone account for the feeding performances shown by most copepods over such a broad particle size spectrum. Gauld (1964) and Jϕrgensen (1966) have suggested that prey below 30-50 μm are filtered, whereas large ones are seized raptorially.

Filter feeding has been classically ascribed to the function of the second maxillae, which act as filters or sieves (Cannon 1928; Lowndes 1935). According to Pich (1966) the sieve theory implies 100% capture of particles larger than the pore size and 0% of those smaller. To a first approximation, this is consistent with the decreasing percentage capture of smaller particles found for copepods (Nival and Nival 1973, 1976; Boyd 1976). This is also apparently consistent with Wilson's (1973) theory, which states that Acartia could track its food, changing the size of particles it captures only by changing the pore of its filters. The "filter" of Acartia mainly consists of a basal part with four endites and a 5 segment endopod, all having long setae (see Marshall 1973). The setae are all setulate and the distance apart of this network gives some idea of the size of particles this filter can retain (Nival and Nival 1973). The aperture of the mesh is variable, being least near the proximal parts and greatest at the distal parts of both the limb and the setae, respectively. In most species the pore size varies from 2 to 30 μm (Jϕrgensen 1966; Vyshkvartseva and Gutel'makher 1971; Nival and Nival 1973; Marshall 1973). Marshall (1973) pointed out, that in the smaller copepods the distance is, surprisingly, much the same as in Calanus. Despite all the considerations listed above, some other observations on feeding mechanisms reveal several contradictions between theory and experimental facts. First, the percentage of small particles captured does not decrease with decreasing sizes below 30 μm (Fig. 2; Conover unpubl.; Poulet 1977, 1978; Richman et al. 1977). In addition, Wilson's theory does not make sense because muscles are absent in the setae of the second maxillae (Friedman and Strickler 1975). Without muscles, copepods cannot change the distance between consecutive setulae. The shifting process of biomass feeding demonstrated for several copepods (Poulet 1974, 1978; Richman et al. 1977) also provides further arguments against the sieve

Table 2. Relationship between size and concentration of particles in water and ingestion rates by copepods. A—*Calanus pacificus* (Frost 1972, from his figs. 2 and 4); B—*Pseudocalanus minutus* (Poulet and Chanut 1975, from their table 1).

	Phytoplankton species			
	Thalassiosira fluviatilis	*Coscinodiscus angstii*	*Coscinodiscus eccentricus*	*Centric sp.*
	A. *Calanus pacificus*			
Size (µm)	11	35	75	87
Food concn (μg C liter^{-1})	300	160	160	140
Max ration (μg C h^{-1} cop^{-1})	1.03	1.04	0.98	1.13
No. of cells ingested (cells h^{-1} cop^{-1})	3,200	200	100	45
	B. *Pseudocalanus minutus*			
Size range (µm)	1.58–3.57	4–8.98	10.1–22.6	25.4–57	64–144
Theoretical ration					
Volume*	1.5	1.5	1.5	1.5	1.5
Number†	19,894	716	86	2	~1
Average ration					
Volume*	0.35	0.89	1.57	4.43	2.03
Number†	4,761	426	90	6	1
Available food concn (ppm)	~0.88	~0.88	~1	2.05	1.28

* mg liter^{-1} 10^{-4} h^{-1} cop^{-1}.
† liter^{-1} h^{-1} cop^{-1}.

theory. The mechanisms by which feeding takes place are still poorly understood by most marine biologists, from Cannon (1928) and Lowndes (1935) onward. In brief, it was generally felt that the particles filtered and retained by the setulae and setules of the second maxillae are "scraped off" or "combed off" by the first maxillae endites and are then passed on by the maxillary endites and setae on the bases of the maxillipeds to the mandibles at the mouth.

This feeding pattern is very similar to the behavior described for the megalopa larvae of the crab *Pachycheles pubescens* (Gonor and Gonor, 1973). The same sequence of movements applied to copepods will present extreme difficulty with regard to coordination of the limbs. Limbs involved in this process are extremely active except the second maxillae, and vibrate at a high frequency (Cannon 1928; Lowndes 1935; Gauld 1966; Poulet 1977). Contrasting with the megalopa larvae, where limb movements are slow enough to permit good coordination of the maxillipeds and maxillae, the high speed of the copepods' limbs does not enable such coordination. In addition, the length of the setae carried by the second maxillae is either too short or too long with regard to the mouth aperture in most copepods, so that the particles "combed off" the maxillae would "drop out" before reaching the mandibles and would "fall off" before or after the mouth. As a matter of fact, "filtering", "combing", and "passing" remain mysterious processes which have been always postulated but never really described for either herbivorous or omnivorous copepods. In reality, the process of filter feeding can hardly be seen because it is too fast a process and involves too small a size of particles. When using a strobelight (Lowndes 1935; Poulet 1977) it is rather easy to slow down the speed of the limbs artificially. However, under such experimental conditions, we have never been able to observe "filtering," "combing," and "passing" operations enumerated for the mouthparts of copepods. In our opinion, filter feeding is likely not a sieving process, and it still deserves some new theoretical and further experimental considerations.

Rubenstein and Koehl (1977) have provided some theoretical views on the general mechanism of filter feeding which can serve as the foundation of a new alternative theory for the mechanism of feeding by copepods. By definition, filtration is the separation of particles from fluids by use of fibrous or porous media. Sieving is only one of the several mechanisms by which filters remove particles (Rubenstein and Koehl 1977). "Three elements are involved in any filtration system: the dispersed particles, the fluid medium, and the filter. By characterizing each of these elements, it is possible to predict the manner by which particles are captured" (Rubenstein and Koehl 1977, p. 982). Surprisingly, characteristics of the fluid and particles in the sea are more or less well known, whereas those of the filtering apparatus of copepods are still obscure. Rubenstein and Koehl have listed five mechanisms which logically could apply to copepods as well, by which a filter can remove suspended particles: (1) sieving, (2) interception, (3) impaction, (4) deposition, and (5) diffusion. According to the equations given by Ranz and Wang (1952), and by Pich (1966) (referred to by Rubenstein and Koehl 1977), it is possible to compare the relative magnitude of each mechanism through a dimensionless index N, provided that the physical parameters which characterize the particles, the fluid, and the filter are known. The relationship existing between velocity (V_0) and the intensity of capture (N) shows that feeding efficiency will vary with both the mode of capture and the velocity of the fluid (Fig. 3A). Rubenstein and Koehl (1977) have pointed out that at velocities lower than V_1 deposition and diffusion are the main modes of particle capture. At velocities between V_1 and V_2 interception predominates, and at velocities greater than V_2 impaction has the largest index of capture (Fig. 3A). Among the several equations they gave, only three seem to apply to the feeding mechanisms of copepods (Fig. 3B). At low velocity ($<V_1$) corresponding to section 1 and part of section 2 of the curve, modes of capture

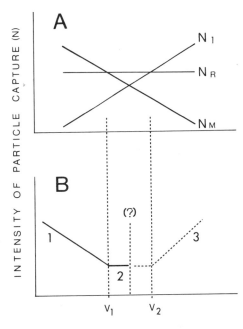

Fig. 3. Theoretical relationship existing between fluid velocity and particle capture applying to filter-feeding mechanisms. A. Where $N_I = [(\rho_p - \rho_m) d^2 p V_0] / 18\mu d_f$ corresponds to impaction; where $N_R = d_p/d_f$ corresponds to interception; and where $N_M = (KT/dp)/(1/3\pi \mu V_0 d_f)$ corresponds to motile deposition (after Rubenstein and Koehl 1977). ρ_p and ρ_m are densities of both particle and fluid; d_p and d_f are diameter of both particle and fiber; V_0 is velocity of the fluid; μ is viscosity of the fluid; K is Boltzmann's constant and T is absolute temperature. B. Sections of theoretical function applying to copepods. Section 1 is for raptorial feeding; section 3 is for impaction feeding and section 2 is a "transition" between the two modes of capture.

such as interception and deposition apply also as velocity decreases. This would lead to a raptorial mode of feeding. As velocity of the fluid increases and becomes higher than V_2, then, impaction prevails following section 3 of the curve in Fig. 3B. This would lead to an impaction mode of feeding. The transition between a raptorial and an impaction-feeding mode, or the reverse, should occur at a velocity falling somewhere between V_1 and V_2. Thus, food capture should be first of all dependent on velocity. The mode of feeding utilized by copepods could switch from raptorial to impaction and vice-versa only by changing the speed of the fluid. Does reality support this alternative theory? Results reported in the literature and our own observations seem to do so.

Several papers have been published on the raptorial feeding behavior of marine copepods. This mechanism can easily be seen because it deals with large particles and occurs at low speed. Descriptions have been written for marine carnivores such as *Candacia bradyi*, *Labidocera jollae*, *Sapphirina angusta*, and *Tortanus discaudatus* (Wickstead 1959; Lillelund and Lasker 1971; Heron 1973; Ambler and Frost 1974). Raptorial feeding behavior exists also among omnivorous and herbivorous marine copepods. It has been mentioned for *Acartia clausi* (Conover 1956; Petipa 1959) and for *Calanus helgolandicus* (Petipa 1965) as well as for *Calanus hyperboreus* (Conover 1966). When swimming in slow gliding movements (presumably when $V_0 < V_1$ in Fig. 3B), copepods may frequently contact large particles. As Conover (1966) mentioned, when encounter is successful, particles which have entered an area bounded dorsally by the ventral body wall, laterally by the exopods of the first maxillae, and ventrally by the maxillipeds are suddenly grasped by the second maxillae or maxillipeds and then manipulated by coordinated efforts of the mouthparts. Particles are then driven forward to the mandibles and to the mouth before consumption starts. Our observations have revealed similar behavior for *Pseudocalanus*, *Eurytemora*, and *Temora*. Conover (1966) mentioned that raptorial feeding and filter feeding are exclusive processes. In reality, in several copepod species these two patterns are combined, as animals alternatively switch from one mode to the other within short periods of time (Poulet 1974). Before and during raptorial feeding, the mouth-appendages may be almost motionless (before capture), yet move slowly with coordinated movements when driving particles toward the mouth (after capture); but under no circumstances do they create feeding currents. Moveover, during filtration, fast feeding currents drive large size particles toward the area close to the maxillae and maxillipeds. A sudden contact will stop filtration, which is instantaneously followed by a raptorial mode of feeding. Under such conditions, the velocity of the fluid is either below V_1 or close to zero, and section 1 of the curve in Fig. 3B should apply to the raptorial mode of capture generally described for copepods feeding on large particles (>30-50 μm).

Our alternative concept for filter feeding is more difficult to assess. It is known that small particles are driven toward the filter chamber by means of feeding currents. These currents are created by the vibrations of the second antennae, palps of the mandibles, first maxillae, and endopods of the maxillipeds (Cannon 1928; Lowndes 1935; Gauld 1966; Marshall 1973). The speed of these appendages can be rather high, i.e. can vary from a few hundred to several thousand oscillations per minute, depending on the species and on the developmental stages of the copepods (Lowndes 1935; Poulet 1977). The appendages create two vortices which have been described in detail in the past (Cannon 1928; Lowndes 1935), the smallest of the two being located within the filtering chamber (Cannon 1928). As the speed of the mouthparts increases, the velocity of the feeding currents increases as well. Copepods start filtration and become successful filterers when velocity increases, presumably changing from V_1 to V_2 and greater (Fig. 3B). Above V_2, impaction feeding becomes the most efficient mode of capture according to the theory (Rubenstein and Koehl 1977; Section N_1 in

Fig. 3A). The long and numerous setae (covered with setules) which are borne by the second maxillae, mandibles, first maxillae, and maxillipeds are used as whips and create the feeding currents (Lowndes 1935; Gauld 1966). The second maxillae also bear setae which are setulate (Gauld 1966). But, we do not believe that they act as the fibers of a sieve. Judging from the fact that the limbs show little motion and that their position looks like a funnel during the feeding process, we postulate that their major and basic function is to channel the feeding current within the filtering chamber toward the mouth. The feeding current acts as a jet stream knocking against the region of the mouth formed by the labrum and lower lip, which is covered with oriented setae and where are located the four apertures of the labral glands (Lowe 1935; Park 1965). Small particles (<30 μm) carried by this water flow will reach the fibrous region of the mouth and will likely adhere to the mucus produced and presumably spread around the mouth orifice. After capture, particles are likely pushed toward the inside of the mouth by the mandibles and then ingested. At this stage, our description is speculative because all the processes involved are almost impossible to observe. However, it seems to us that the concept of impaction feeding is one of the best ways to explain the capture efficiency demonstrated by copepods when feeding on small particles, and it agrees with the theoretical ideas recently discussed by Rubenstein and Koehl (1977). However, we disagree slightly with them. We believe that impaction is an important mechanism for most organisms (including copepods) feeding at high velocity on small and dense particles. But we do not think that the "filters" of copepods and their particulate food fall into the intermediate size and velocity range where direct interception is the main mode of capture. Interception is described by $N_R = d_p/d_f$ (Pich 1966; Rubenstein and Koehl 1977; Fig. 3A) and as described by this equation, it does not involve velocity. The same reasoning applies to sieving mechanisms. Herbivorous as well as omnivorous copepods have evolved (Lowndes 1935; Gauld 1966; Marshall 1973) very sophisticated tools whose main functions are to create feeding currents. We do not understand why copepods should capture food through interception or sieving alone without also using impaction, provided that they have developed the ability to regulate (increase or decrease) the velocity of their feeding appendages, and thus adjust fluid velocity, which, according to theory, tends to maximize food capture when impaction is operating. When impaction is used, the intensity of particle capture is directly proportional to velocity (Fig. 3A and legend of Fig. 3). Therefore, we assume that the whole process of food capture by copepods can be simply described by a curve deducted from Fig. 3B which would have a parabolic shape such as $y = (x - A)^2 + B$; where y is the intensity of capture, x is velocity of the feeding currents or vibration of the mouth appendages, and where A and B are constants (Fig. 4). In Fig. 4, sections a, b, and c are parts of the parabolic function and they correspond to large, small, and mixed size particle feeders. Depending on the relative abundance of plant and animal prey in the stock of available particles, as well as on the velocity of the appendages of a given copepod, a, b, and c could apply to carnivores, herbivores, and omnivores.

This alternative theory of feeding (Figs. 3 and 4) is based on two specific parameters: the density of the plumosus setae and the velocity of the feeding appendages. We must also assume that particles are channeled toward the mouth and adhere to the labrum and mouth regions where the setae and mucus produced by the labral glands act as collectors. There is no direct evidence for these last assumptions. Nevertheless, recent results have shown that *Aetideus divergens* is a less effective grazer on small size particles (< 49 μm) than *Pseudocalanus* and *Calanus* (Robertson and Frost 1977). This is perfectly coherent with the theory illustrated in Fig. 4. In addition, comparisons between *Aetideus* (Robertson and Frost 1977, their fig. 1) and *C. pacificus* (Frost 1972, his fig. 4) feeding on almost similar size categories of cells

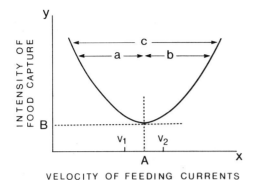

Fig. 4. Theoretical parabolic function $y=(x-A)^2+B$, relating intensity of food capture (y) to the velocity of feeding currents (x). A and B are constants. Section a applies to large size particle feeders (raptorial feeding); section b applies to small size particle feeders (impaction feeding); section c applies to mixed size particle feeders (switching from raptorial to impaction and reverse).

show that the former genus is less efficient than the latter one at all size categories used. According to Robertson and Frost (1977), *Aetideus* has not only fewer and shorter setae than *Pseudocalanus* and *Calanus*, but also the setules are more widely spaced (2–3 μm for *Pseudocalanus*, 4–7 μm for *Calanus*, and 8–9 μm for *Aetideus*). Unfortunately, the velocity of the vibrations of the feeding appendages have not been measured for these three copepods. We believe that the morphological differences between these copepods, in addition to differences likely to exist in their vortex velocities (Fig. 5), should be sufficient to explain the reduced capture efficiency on any particles demonstrated by a "slow feeder" (*Aetideus*) compared to some "fast feeders" (*Pseudocalanus* or *Calanus*). In order to illustrate the fundamental relationship existing between velocity and capture efficiency we have plotted ingestion rates versus copepod dry weight and versus the velocity measured for the feeding appendages of 3 species of copepods (Fig. 5). As was demonstrated by Paffenhöfer (1971), there is a negative relationship between body dry weight and ingestion rate per body dry weight, so that the ingestion per unit weight is less for *Calanus* than

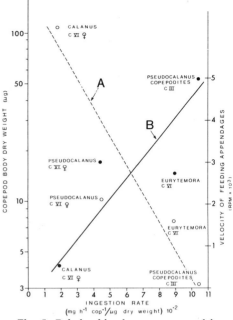

Fig. 5. Relationships between copepod ingestion rates per body dry weight and copepod body dry weight (A), and relative vibration of the mouth appendages (B). (Data from Cannon 1928; Lowndes 1935; Corner 1961; Poulet 1977, 1978).

for the three other small copepods (Fig. 5, curve A). When the same data on ingestion rates are plotted versus velocity, it is clear that ingestion rates per body dry weight increase from adult females of *Calanus* to copepodites of *Pseudocalanus* and it appears that the increase is directly proportional to velocity (Fig. 5, curve B). Direct measurements of food consumption by copepodites and adult females of *P. minutus* fed simultaneously on the same stocks of naturally occurring particles have revealed that impaction feeding by copepodites is more efficient than by adults for particles < 30 μm, whereas the amount of food obtained through raptorial feeding is similar for both age categories of copepods (Poulet 1977). The discrepancy noticed in the utilization of small particles between the copepod developmental stages has been related both to the anatomy and velocity of the feeding appendages. It has been demonstrated that the relative mesh aperture size of the network of the setae on the feeding appendages is smaller for copepodites, whereas their relative velocities are higher than those of adults (Poulet 1977, his table 2). Among copepods, high velocity of the feeding currents due to the fast vibrations of the feeding appendages combined with a dense network of plumose setae should favor impaction feeding and should enhance capture of small particles (Fig. 3B, 4; Poulet 1977). Reduction in the density of setulae associated with slower motion of the feeding appendages should reduce fluid velocity and in turn it should favor raptorial feeding with enhancement of the capture or large particles.

Chemosensory grazing

It has been claimed in a number of recent hypotheses that copepods indiscriminately filter suspensions of food particles following a passive and mechanical pattern, and further that apparent selectivity for food of different sizes is not behaviorally determined. Conover (unpubl.) has discussed some grazing realities, and he has pointed out that some mechanisms must still be elucidated to clarify the details of the food gathering and sorting process. It is clear from our previous results (Poulet 1974, 1978; and Figs. 1, 2) that copepods belonging to different species are able to select particles within the peaks among the size spectrum. These results have been confirmed and verified by other experiments carried out on similar genera (Conover unpubl.; Allan et al. 1977; Richman et al. 1977). Poulet and Chanut (1975, p. 710 and 712) have stated that "if selectivity really occurs, it should be based on characteristics other than size or total particle abundance," and "it is likely that selective feeding, when it occurs, is based on chemical sense."

Chemoreceptors described for several planktonic crustaceans (Ong 1969; Elofsson 1971; Fleminger 1973; Friedman and Strickler 1975) are located either on the antennae, the feeding appendages, or on the anterior tip or the body of the organisms. They are involved in the feeding and mating process (Griffiths and Frost 1976; Hamner and Hamner 1977). They are responsible for chemosensory tracking of food by planktonic shrimps (Hamner and Hamner 1977). Kittredge et al. (1974) already mentioned that copepods search for food. It is likely that switching mechanisms, as well as searching and tracking of biomass peaks already demonstrated for marine copepods (Poulet 1974; Conover unpubl.; Richman et al. 1977) are probably dependent on some receptors. It is clear that the switching responses demonstrated for both adult and young copepods (Poulet 1977; Allan et al. 1977) argue for a sensory mechanism, either mechanical reception (Wilson 1973; Skiver 1978) or chemical reception, in the copepod's feeding process. However, chemical perception seems more likely, but not exclusive, provided that the copepods are able to first grasp and then reject particles that are held in the mouthparts and moved about as if being tasted (Conover 1966, among others). Direct evidence for chemosensory feeding has been recently provided for some marine copepods (Poulet and Marsot 1978). According to Poulet's (1978) point of view, particles responsible for biomass peaks over the size spectrum have the most abundant biomass and should have a chemical attraction for copepods by means of their major chemical components or scent acting as feeding stimulants. The fact that plankton search for food and females (Kittredge et al. 1974; Katona 1975; Griffiths and Frost 1976; Poulet 1977; Allan et al. 1977) strongly suggests that chemical perception is a basic sensory mechanism that is found in many free-swimming organisms. This property has been successfully investigated for *A. clausi* and *Eurytemora herdmani* (Poulet and Marsot 1978), as well as for *C. finmarchicus* feeding on microcapsules. Our grazing experiments were conducted with copepods collected at the surface in the St. Lawrence estuary near Rimouski, Québec. A natural population of mixed adult and young copepods was used in each series of experiments. The animals were allowed to feed on microcapsules (particles consisting of thin semipermeable polymer membrane around aqueous microdroplets of cell homogenate, proteins, or amino-acids), with nylon protein walls permeable to small molecules (Chang 1972; Chang et al. 1966). Artificial food particles are known to be acceptable for a wide range of filter-feeders (Wilson 1973; Jones et al. 1974; Frost 1977; Skiver 1978). We produced microcapsules with peak concentrations in the sizes $\sim$ 8 and $\sim$ 50 μm by stirring protein emulsions at speeds of 5,000 and 3,000 rpm, respectively (Poulet and Marsot 1978).

In the first series of experiments, "enriched" capsules were produced with an encapsulated homogenate which consisted of a freeze-dried water-soluble fraction of concentrate (400 mg ml^{-1}) of naturally occurring particles, collected during a phytoplankton bloom, mixed with albumin (400 mg ml^{-1} in water). We measured ingestion rates at the end of each experiment by comparing Coulter counts (Coulter model TA, using a 280 μm aperture tube) of the microcapsule suspensions in bottles containing animals with control bottles which contained no animals. As the microcapsules were permeable to small molecules (pore size $\sim$ 18 Å: Chang 1972) we assumed that the enriched capsules would let small molecules originating from the homogenate diffuse out and thus stimulate the feeding. "Nonenriched" capsules were produced from albumin, individual molecules of which are too large ($\sim$ 100 Å) to escape. Depending on the type of experiments, we provided microcapsules and beads (Sephadex G-75) to the copepods. When fed simultaneously in separated bottles containing "enriched" and "nonenriched" microcapsules of small size, copepods preferentially grazed the "enriched" ones with ingestion rates three times higher than for "nonenriched" capsules (Fig. 6 A-1, A-2; Table 3). Production of particles below 15 μm, resulting from the breakdown of microcapsules through feeding activity, was almost double in bottles containing the "nonenriched" particles (Fig. 6 A-2; Table 3). When fed with large microcapsules, feeding by copepods was observed with maximum intensity occurring at the right side of the peaks (Fig. 6 B-1, B-2). Net ingestion could be measured only for "enriched" capsules (Fig. 6 B-1; Table 3) even though there was production of particles of smaller size. There was little or no net ingestion with unenriched particles, but very intense

Table 3. Feeding by *Acartia clausi* and *Eurytemora herdmani* on enriched microcapsules (containing homogenate of naturally occurring phytoplankton) and nonenriched microcapsules. Mean value ± SD; N is number of experiments and n is number of replicate measurements. Experiment A-1 and A-2, and B-1 and B-2 conducted simultaneously (from Poulet and Marsot 1978).

Exp type	Size mode (μm)	No. copepods in exp	Exp time (h)	Initial concn (particles liter^{-1})	Ingestion rate (particles liter^{-1} h^{-1} cop^{-1})	Particle debris production (%)	Consumption range (%) <15 μm	>40 μm
A-1	8*	646±52 (N=3)	4	11.6×10^6±3.1×10^5 (n=9)	344±12 (n=9)	3.10	100	—
A-2	8†	671±32 (N=3)	4	10.3×10^6±1.4×10^5 (n=9)	107±20 (n=9)	5.40	100	0
B-1	50*	640±54 (N=5)	4	12.1×10^5±2.7×10^5 (n=15)	59±19 (n=15)	12.95	—	100
B-2	50†	642±141 (N=5)	4	12.1×19^5±3.1×10^5 (n=15)	0 (n=15)	28.15	—	0
C	8†-50*	329±22 (N=2)	3	17.2×10^6±3.4×10^5 (n=6)	467±26 (n=6)	2.80	20	80
D	8*-50†	266±34 (N=6)	3	5.4×10^6±9.2×10^4 (n=18)	229±10 (n=18)	4.42	100	0

*Enriched.
†Non-enriched.

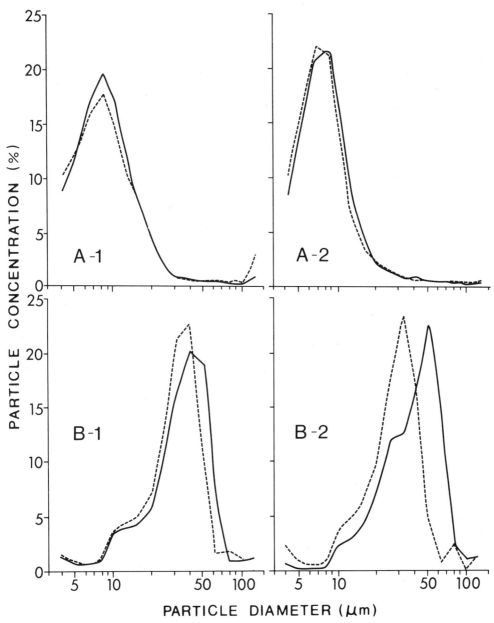

Fig. 6. Feeding activity of mixed populations of *Acartia clausi* and *Eurytemora herdmani* measured by comparing unimodal particle size distributions of microcapsules in control (———) and in experimental (with copepods, ----) bottles. Experiments A-1 and B-1 were performed with enriched microcapsules, experiments A-2 and B-2 with nonenriched microcapsules (from Poulet and Marsot 1978).

production of particle debris (Fig. 6 B-2; Table 3). In this case, particle size distributions were considerably modified toward the left side of the spectrum, presumably as a result of mastication followed by rejection (Conover 1966; Friedman and Strickler 1975; O'Connors et al. 1976). Ingestion rates were significantly different between experiments B–1 and B–2 (*t*-test, $p = 0.05$).

In nature, particle size distributions are not as simple as those shown in Fig. 6. Generally particle spectra in seawater are bi- or polymodal, permitting copepods to choose particles from among several peaks (Poulet 1974). This condition was artificially created in a second series of experiments in which bimodal spectra with peaks of approximately equal concentration alternately contained the "enriched" particles either in the large size range (Fig. 7C) or in the small size range (Fig. 7D). In experiment C ingestion occurred on both particle peaks, but it was four times higher for large particles containing the homogenate. Moreover, particle production was negligible (Table 3). Because of large production of small-sized particle debris in type D experiments, which often masked feeding responses below 20 μm, net ingestion rates of small capsules could hardly be measured. Then large-sized microcapsules were replaced by beads of similar sizes (Sephadex G-75) in order to reduce the production of debris. We assumed that the substitution did not affect the feeding mechanisms, providing that copepods can ingest hard plastic beads within this size range (Wilson 1973). Production of debris was then negligible (Table 3) and preferential feeding could be observed in the size range corresponding to the small "enriched" particles (Fig. 7D; Table 3). In each category of experiments A, B, C, and D, *Acartia* and *Eurytemora* always preferentially consumed microcapsules that were "enriched" with homogenates of naturally occurring phytoplankton. The results in Fig. 6 and 7 and Table 3 suggest that small copepods can discriminate between two types of particles on the basis of their chemical "scent," independent of size (Poulet and Marsot 1978). In a third series of similar experiments we fed *C. finmarchicus* with enriched and nonenriched capsules of ∿ 50 μm in size. We found that *Calanus* could distinguish the taste of its food. In each case, this copepod preferentially selected the enriched microcapsules (Table 4). Therefore, feeding by copepods is a sensory-determined behavioral process.

Production of capsule debris was noticed in each set of experiments. This production has always been higher for nonenriched capsules than for enriched ones when copepods were offered both categories simultaneously and it was often associated with very low or no apparent ingestion (Tables 3

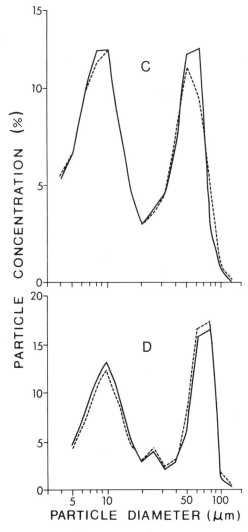

Fig. 7. Feeding activity of mixed populations of *Acartia clausi* and *Eurytemora herdmani* measured by comparing bimodal particle size distributions of microcapsules in control (——) and in experiments (with copepods, ----). C—Enriched microcapsules in large size range; D—enriched microcapsules in small size range (from Poulet and Marsot 1978).

Table 4. Feeding by *Calanus finmarchicus* on enriched microcapsules (containing a homogenate of naturally occurring phytoplankton) and nonenriched microcapsules. Experiments E-1 and E-2, and F-1 and F-2 conducted simultaneously. Details and units as in Table 3.

Exp	Size mode	No. copepods in exp	Exp time	Initial concn	Ingestion rate	Particle debris production
E-1	50*	70±0 (N=2)	4	$32.71 \times 10^3 \pm 344$ (n=6)	430±115 (n=6)	1.20
E-2	50†	70±0 (N=2)	4	29.02×10^3 465 (n=6)	0 ± 0 (n=6)	2.20
F-1	50*	80±0 (N=2)	5	28.99×10^3 327 (n=6)	603± 69 (n=6)	2.90
F-2	50†	80±0 (N=2)	5	23.47×10^3 151 (n=6)	50±120 (n=6)	2.90

*Enriched.
†Non-enriched.

and 4). Production was also higher for large capsules than for small ones. These results indirectly reflect some real behavioral criteria which are a part of the chemosensory grazing itself. Large capsules (∽ 50 μm) are likely to be captured raptorially whereas small ones (∽ 8 μm) are collected through impaction feeding, but in both cases copepods can taste either the outside (before mastication) or the inside (after mastication), depending probably on the duration of the handling process. When capsules have a satisfactory chemical taste they are recognized immediately and are readily ingested with minimum production of fragments. "Untasty" capsules are "manipulated" longer, "handled" and "chewed" by the mouthparts, so that this additional time allows copepods to search for any external or internal chemical "scent," before the capsules are torn apart, fragmented, and discarded. Very often, the handling of large particles by copepods is a wasting process, as has been reported in the past (Conover 1966). However, we do not think that the content of large nonenriched capsules could be sucked out before copepods discard unwanted fragments of the membranes, thus masking ingestion rates (O'Connors et al. 1976). If experiments A1-A2, B1-B2, E1-E2, and F1-F2 (Figs. 6, 7; Tables 3, 4) were always performed simultaneously with groups of copepods coming from identical samples to ensure identical experimental conditions, then the ingestion rates in experiments A1, B1, E1, and F1 would have been masked as well. The feeding responses measured for each of the three copepod species tested demonstrate, for the first time, chemical control over feeding behavior, as well as the induction of selectivity by stimulatory chemical compounds.

The stimulants responsible for the sensory stimuli permitting copepods to locate, identify, and select their prey are still unknown. Several workers (e.g. McLeese 1970; Shelton and Mackie 1971; Laverack 1974; Fuzessery and Childress 1975) have provided evidence that feeding stimulation in marine crustaceans requires several compounds, among which amino acids seem to be the most active substances. Others have shown that the planktonic shrimp *Acetes sibogae australis* (Hamner and Hamner 1977) has a feeding behavioral response that could vary according to the type of amino acids tested, while urea and glucose are nonstimulatory. The problem of chemical sensitivity and food recognition by copepods is under examination in our laboratory. Preliminary results (Table 5) demonstrate that marine copepods, like other crustaceans, also have variable feeding responses for several chemical compounds. A set of grazing experiments was carried out with *A. clausi* fed simultaneously in separated bottles containing nonenriched micro capsules which act as controls, and with enriched ones containing either amino acid (*L*-leucine or *L*-methionine) or glycolic acid; all capsules had a mean spherical diameter of ∽ 40 μm. Initial concentration of each chemical inside the microcapsules

Table 5. Feeding of *Acartia clausi* on microcapsules acting as controls (nonenriched) and enriched with three chemical compounds acting as potential feeding stimulants. Details and units as in Table 3.

Exp	Size mode	No. copepods in exp	Exp time	Initial concn	Ingestion rate	Particle debris production
Nonenriched	40	98±20 ($N=2$)	4	$13 \times 10^5 \pm 4.2 \times 10^4$ ($n=6$)	37±113 ($n=6$)	6.06
L-methionine	40	82±11 ($N=2$)	4	$12.9 \times 10^5 \pm 3.4 \times 10^4$ ($n=6$)	0± 0 ($n=6$)	4.18
L-leucine	40	86±17 ($N=2$)	4	$12.1 \times 10^5 \pm 2.9 \times 10^4$ ($n=6$)	257±105 ($n=6$)	5.72
Nonenriched	40	95± 3 ($N=3$)	4	$12.5 \times 10^5 \pm 3.3 \times 10^4$ ($n=6$)	0± 0 ($n=9$)	8.67
Glycolic acid	40	86± 5 ($N=5$)	4	$10.9 \times 10^5 \pm 1.9 \times 10^4$ ($n=9$)	0± 0 ($n=9$)	7.47

was 10^{-1} M before the experiments began. Ingestion rates measured in each set of experiments were not all the same, but seemed to change depending on the nature of the encapsulated compound (Table 5). Results show that L-leucine had a stimulant effect on feeding, whereas L-methionine and glycolic acid had an undetectable effect, judging from the ingestion rates (which were as low as those measured for nonenriched microcapsules).

Discussion

The first stage of the feeding process could start with a "detection phase," defined as the degree with which copepods can locate large algal patches or can locate micro-patches of particles belonging to biomass peaks over the spectrum. The problem of chemotropotaxis is rather controversial as far as zooplankton is concerned. Results from some workers (Bainbridge 1949, 1953) suggest that copepods might locate their food via chemotaxis; whereas others (Conover 1966; Friedman and Strickler 1975) assumed that chemotropotaxis is not normally used by zooplankters. Our own preliminary observations revealed that chemotropotaxis occurs for *A. clausi* and *E. herdmani* within a short distance (a few centimeters) (Fig. 8). However we do not know yet at what maximum distance this process allows copepods to detect chemical gradients originating from the particulate food. This first stage is followed by an "encounter phase," which is, to a first approximation, a random process. Let us assume that a given volume of water is occupied by two nonmobile suspended particles (one large, one small) and by one copepod swimming randomly and searching for food. The simplest way to express the probability that the large particle will be encountered by the copepod is $p_1 = f(n_1 x_1)$, while the probability of encountering the small one is $p_2 = f(n_2 x_2)$ (where n is the number and x is the diameter of each particle). Under such schematic conditions, it is obvious that the probability of encounter p_1 is higher than that of p_2, knowing that the largest particle occupies more space than the small one. The frequency with which the copepod encounters particles of different sizes will vary according to the frequency distribution of the diameter or volume of each particle category (Table 1). In the marine environment small particles generally predominate so that they are more frequently encountered by copepods than are the large ones. Growth of any particles within a given size range will tend to modify the simple linear relationship existing between number and size of particles (Sheldon et al. 1967), and thus the occupation of space, and will modify in turn the frequency of encounter itself. Encounter is followed by either avoidance or a "capture phase" defined as the ability of copepods to efficiently retain suspended particles of different sizes after encounter and contact are achieved. Depending on the mode of capture (Fig. 4), large and small particles, or both categories alternately, will be more or less efficiently captured (Robertson and Frost 1977;

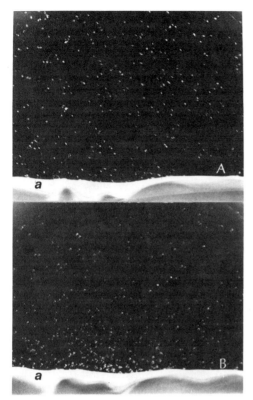

Fig. 8. Chemiotropotaxis among copepods. Natural population of mixed species of copepods (white spots) (*Acartia clausi*, and *Eurytemora herdmani*) placed in a petri dish (1 liter organic matter free seawater, 19 cm diameter) and separated by a Nitex curtain (a; 10 μm mesh size) from a natural mixture of phytoplankton species artificially concentrated (cell size <75 μm). Water and plankton samples were collected at the same location in the St. Lawrence Estuary. Experiments were achieved in a dark, cold room (4°C) during 6 hours. Pictures were taken every hour with a ¼ sec. flash bulb and ordinary camera: A—Random distribution of copepods at time $t = 0$ hour. B—Patchy distribution of copepods at time $t = 3$ hours.

Poulet 1977). This reflects the ability of retention, but not which particle will be selected. The fourth phase co-occurs with the third one and corresponds to an "identification phase". When contact is achieved between the physical or chemical stimulants of the particles and receptors of the feeding appendages (barro- or chemoreceptors), then copepods can evaluate the quality of their food (Conover 1966; Poulet and Marsot 1978; Tables 3, 4, 5). This is probably the only occasion when selectivity for food may really take place. The length of time, during which particles suitable as food are handled or are within the filtering chamber, should enhance feeding stimulation and should increase the process of capture. This phase can be followed by either an "ingestion phase" or "rejection phase," depending on the suitability of the food. Digestion and defecation should start soon after ingestion has begun. At that stage many details of the various phases deserve further research in order to verify the extent to which our alternative theory corresponds to some realistic feeding patterns. Nevertheless, direct observations of the swimming and feeding of copepods (*Pseudocalanus, Acartia, Eurytemora, Temora*, and *Calanus*), in addition to the results summarized in Figs. 1, 2, 6, and 7 and from Tables 3, 4, and 5, provide evidence that the six phases prevailing for the feeding process do exist to some extent.

It is obvious that the capture intensity will vary according to copepod species, following the anatomical properties of their feeding appendages (Anraku and Omori 1963; Arashkevich 1969; Robertson and Frost 1977; Poulet 1978) coupled with their relative vibrations (Poulet 1977; Fig. 5). Its variability over the particle size spectrum seems to be related to the mode of capture preferentially utilized by copepods. Also, raptorial feeding and impaction feeding are likely not achieved with equal success among copepods. Variable capture of particles over the spectrum or temporary utilization of particles of constant size (Poulet 1973, 1978; Conover unpubl.; Richman et al. 1977; Allan et al. 1977; Skiver 1978) seem to depend on the length of time during which each feeding mode lasts or alternates with other feeding modes. The capture of particles should be related to a specific parabolic function (Fig. 4) prevailing for a given species rather than to a differential selectivity for sizes. At this stage, food gathering is more a problem of behavioral capability than a matter of selectivity. Selectivity does not seem to be related to the anatomical characteristics of the feeding appendages (intersetal spacing, mainly) as postulated by the passive size selection hypothesis (e.g. Boyd 1976; Frost 1977). Selectivity logically starts only during the identification phase, after capture is achieved. It likely occurs when receptors located in the filtering chamber and on the feeding appendages (Cannon 1928; Ong 1969; Friedman and Stricker 1975) are activated either by physical or chemical stimulants (Poulet and Marsot 1978). According to Laverack (1974), there must be some contact between the stimulant and the membrane of the receptor. Therefore, it appears that switching, raptorial, and impaction feeding are mechanisms allowing copepods to capture any particles within a broad size spectrum, with variable intensity depending on the relative concentration of particles over the spectrum and on the dynamic properties of the mouth appendages. Selectivity appears to be a behavioral process initiated by stimulatory effects of chemical or physical compounds on specific receptors.

Baroreception has not been investigated in our study. However, Friedman and Strickler (1975) have postulated that simultaneous activations of several receptors would inform copepods about the dimension of particles. This property should not be exclusive of chemoreception, and it might account for some of the selectivity patterns for particle sizes demonstrated in *Pseudocalanus, Acartia,* and *Temora* (Skiver 1978). Chemosensory feeding (Poulet and Marsot 1978) should also be responsible for the active sensory feeding behavior of copepods. Chemosensory feeding seems to be a powerful tool used by most marine organisms to detect (Laverack 1974; Mackie and Grant 1974) and to select their food (Poulet and Marsot 1978). An understanding of these sensory functions, as well as knowledge of the origin and nature of the stimulants that underlie physical or chemical reception, is essential for comprehending the regulation and dynamics of the pelagic food web. Feeding is one of the most important activities of copepods. Strategy for growth depends upon the yield of sufficient food in correct proportions. Energetics is of some ecological importance if the demand for energy can be met with the least energy expenditure. Economy is achieved if suitable food can be successively identified, located, and obtained with the least effort from the heterogeneous mixtures of particles among which nonliving and unsuitable ones predominate (Sutcliffe 1972; Heinle and Flemer 1975; Poulet 1976; Roman 1977). Feeding mechanisms based on a sensory-determined behavioral process may allow copepods to do just this.

References

ALLAN, J. D., S. RICHMAN, D. R. HEINLE, and R. HULL. 1977. Grazing in juvenile stages of some estuarine calanoid copepods. Mar. Biol. 43: 317–331.

AMBLER, J. W., and B. W. FROST. 1974. The feeding behavior of a predatory planktonic copepod, *Tortanus discaudatus.* Limnol. Oceanogr. 19: 446–451.

ANRAKU, M., and M. OMORI. 1963. Preliminary survey of the relationship between the feeding habit and the structure of the mouth-parts of marine copepods. Limnol. Oceanogr. 8: 116–126.

ARASHKEVICH, Y. G. 1969. The food and feeding of copepods in the northwestern Pacific. Oceanology 9: 695–709.

BAINBRIDGE, V. 1949. Movement of zooplankton in diatom gradients. Nature 163: 910–911.

———. 1953. Studies on the interrelationships of zooplankton and phytoplankton. J. Mar. Biol. Assoc. U. K. 32: 385–447.

BOYD, C. 1976. Selection of particle sizes by filter-feeding copepods: A plea for reason. Limnol. Oceanogr. 21: 175–180.

CANNON, H. G. 1928. On the feeding mechanism of the copepods, *Calanus finmarchicus* and *Diaptomus gracilis.* Brit. J. Exp. Biol. 6: 131–144.

CHANG, T. M. S. 1972. Artificial cells. Thomas.

———, F. E. MACINTOSH, and S. G. MASON. 1966. Semipermeable aqueous microcapsules. 1. Preparation and properties. Can. J. Physiol. Pharmacol. 44: 115–128.

CONOVER, R. J. 1956. Biology of *Acartia clausi* and *A. tonsa.* Bull. Bingham Oceanogr. Collect. 15: 156–233.

———. 1966. Feeding on large particles by *Calanus hyperboreus* Kröyer), p. 187–194. *In* H. Barnes [ed.], Some contemporary studies in marine science. Allen and Unwin.

CORNER, E. D. S. 1961. On the nutrition and metabolism of zooplankton. 1. Preliminary observations on the feeding of the marine copepod *Calanus helgolandicus* (Claus). J. Mar. Biol. Assoc. U. K. **41**: 5-16.

ELOFSSON, R. 1971. The ultrastructure of chemoreceptors organ in the head of copepod crustaceans. Acta Zool. **52**: 299-315.

FLEMINGER, A. 1973. Pattern, number, variability and taxonomic significance of integumental organs (sensilla and glandular pores) in the genus *Eucalanus* (Copepoda, Calanoida). Fish. Bull. **71**: 965-1010.

FRIEDMAN, M. M., and J. R. STRICKLER. 1975. Chemoreceptors and feeding in calanoid copepods (Arthropoda: Crustacea). Proc. Natl. Acad. Sci. **72**: 4185-4188.

FROST, B. W. 1972. Effects of size and concentration of food particles on the feeding behavior of marine planktonic copepod *Calanus pacificus*. Limnol. Oceanogr. **17**: 805-815.

——. 1977. Feeding behavior of *Calanus pacificus* in mixtures of food particles. Limnol. Oceanogr. **22**: 472-491.

FUZESSERY, Z. M., and J. J. CHILDRESS. 1975. Comparative chemosensitivity to amino acids and their role in the feeding activity of bathypelagic and littoral crustaceans. Biol. Bull. **149**: 522-538.

GAULD, D. T. 1964. Feeding in planktonic copepods, p. 239-245. *In* D. J. Crisp [ed.], Grazing in terrestrial and marine environments. Blackwell.

——. 1966. The swimming and feeding of planktonic copepods, p. 313-334. *In* H. Barnes [ed.], Some contemporary studies in marine science. Allen and Unwin.

GONOR, S. L., and J. J. GONOR. 1973. Feeding, cleaning, and swimming behavior in larval stages of porcellanid crabs (Crustacea: Anomura). Fish. Bull. **71**: 225-234.

GRIFFITHS, A. M., and B. W. FROST. 1976. Chemical communication in the marine planktonic copepods *Calanus pacificus* and *Pseudocalanus* sp. Crustaceana **30**: 1-8.

HAMNER, P., and W. M. HAMNER. 1977. Chemosensory tracking of scent trails by the planktonic shrimp *Acetes sibogae australis*. Science **195**: 886-888.

HARRIS, R. P., and G. A. PAFFENHÖFER. 1976. Feeding, growth and reproduction of the marine planktonic copepod *Temora longicornis*. Müller. J. Mar. Biol. Assoc. U. K. **56**: 675-690.

HEINLE, D. R., and D. A. FLEMER. 1975. Carbon requirements of a population of the estuarine copepod, *Eurytemora affinis*. Mar. Biol. **31**: 235-247.

HERON, A. C. 1973. A specialized predatory prey relationship between the copepod *Saphirina angusta* and the pelagic tunicate *Thalia democratica*. J. Mar. Biol. Assoc. U. K. **53**: 429-435.

ITOH, K. 1970. A consideration on feeding habits of planktonic copepods in relation to the structure of their oral parts. Bull. Plank. Soc. Jpn. **17**: 1-10.

JONES, D. A., J. G. MANFORD, and P. A. GABBOTT. 1974. Microcapsules as artificial food particles for aquatic filter feeders. Nature **247**: 233-235.

JØRGENSEN, C. B. 1966. Biology of suspension feeding. Int. Ser. Monogr. Pure Appl. Biol. **27**: Pergamon.

KATONA, S. K. 1975. Copulation in the copepod *Eurytemora affinis* (Poppe, 1880). Crustaceana **28**: 89-95.

KITTREDGE, J. S., F. T. TAKAHASHI, J. LINDSEY, and R. LASKEI. 1974. Chemical signals in the sea: Marine allelochemics and evolution. Fish. Bull. **72**: 1-11.

LAM, R. K., and B. W. FROST. 1976. Model of copepod filtering response to changes in size and concentration of food. Limnol. Oceanogr. **21**: 490-500.

LAVERACK, M. S. 1974. The structure and function of chemoreceptor cells, p. 1-48. *In* P. T. Grant and A. M. Mackie [eds.], Chemoreception in marine organisms. Academic.

LEHMAN, J. T. 1976. The filter feeder as an optimal forager, and the predicted shapes of feeding curves. Limnol. Oceanogr. **21**: 501-516.

LILLELUND, K., and R. LASKER. 1971. Laboratory studies of predation by marine copepods on fish larvae. Fish. Bull. **69**: 655-667.

LOWE, E. 1935. On the anatomy of a marine copepod, *Calanus finmarchicus* (Gunnerus). Trans. R. Soc. Edinburgh **58**: 561-603.

LOWNDES, A. G. 1935. The swimming and feeding of certain calanoid copepods. Proc. Zool. Soc. Lond. 1935(2), 687-715.

MACKIE, A. M., and P. T. GRANT. 1974. Interspecies and intraspecies chemoreception by marine invertebrates, p. 105-141 *In* P. T. Grant and A. M. Mackie [eds.], Chemoreception in marine organisms. Academic.

McLEESE, D. W. 1970. Detection of dissolved substances by the American lobster (*Homarus americanus*) and olfactory attraction between lobsters. J. Fish. Res. Bd. Can. **27**: 1371-1378.

MARSHALL, S. M. 1973. Respiration and feeding in copepods. Adv. Mar. Biol. **11**: 57-120.

——, and A. O. ORR. 1955. On the biology of *Calanus finmarchicus*. 8. Food uptake, assimilation and excretion in adult and stage V *Calanus*. J. Mar. Biol. Assoc. U. K. **34**: 495-529.

MULLIN, M. M. 1963. Some factors affecting the feeding of marine copepods of the genus *Calanus*. Limnol. Oceanogr. **8**: 239-250.

——. 1965. Size fractionation of particulate organic carbon in the surface waters of the western Indian Ocean. Limnol. Oceanogr. **10**: 459-462.

NASSOGNE, A. 1970. Influence of food organisms on the development and culture of pelagic copepods. Helgol. Wiss. Meeresunters. **20**: 333-345.

NIVAL, P., and S. NIVAL. 1973. Efficacité de

filtration des copépodes pélagiques. Ann. Inst. Oceanogr. Paris **49**: 135-144.

——, and ——. 1976. Particle retention efficiencies of an herbivorous copepod *Acartia clausi* (adult and copepodite stages): Effects on grazing. Limnol. Oceanogr. **21**: 24-38.

O'CONNORS, H. B., L. B. SMALL, and P. L. DONAGHAY. 1976. Particle size modification by two size classes of the estuarine copepod *Acartia clausi*. Limnol. Oceanogr. **21**: 300-308.

ONG, J. E. 1969. The fine structure of the mandibular sensory receptors in the brackish water calanoid copepod *Gladioferens pectinatus* (Brady). Z. Zellforsch. **97**: 178-195.

PAFFENHÖFER, G. A. 1971. Grazing and ingestion rates of nauplii, copepodids and adults of the marine planktonic copepod *Calanus helgolandicus*. Mar. Biol. **11**: 266-298.

——, and R. P. HARRIS. 1976. Feeding, growth and reproduction of the marine planktonic copepod *Pseudocalanus elongatus* Boeck. J. Mar. Biol. Assoc. U. K. **56**: 327-344.

PARK, T. S. 1965. The biology of a calanoid copepod: *Epilabidocera amphitrites* McMurrich. Ph.D. thesis, Univ. Washington. 249 p.

PARSONS, T. R. 1976. The structure of life in the sea, p. 81-97. *In* D. H. Cushing and J. J. Walsh [eds.], The ecology of the seas. Saunders.

——, R. J. LeBRASSEUR, and J. D. FULTON. 1967. Some observations on the dependence of zooplankton grazing on the cell size and concentration of phytoplankton blooms. J. Oceanogr. Soc. Jpn. **23**: 10-17.

——, ——, ——, and O. D. KENNEDY. 1969. Production studies in the strait of Georgia. Part 2. Secondary production under the Fraser River plume, February to May, 1967. J. Exp. Mar. Biol. Ecol. **3**: 39-50.

PETIPA, T. S. 1959. Feeding of the copepod, *Acartia clausi* Giesbr. Tr. Biol. Sta. Sebastopol **11**: 72-100.

——. 1965. The food selectivity of *Calanus helgolandicus* (Claus), p. 100-110. *In* Investigation of the plankton in the Black Sea and Sea of Azov. Akad. Sci. Ukr. SSR. [Min. agr. Fish. Food, Great Britain Transl. N. S. 72.]

PICH, J. 1966. Theory of aerosol filtration by fibrous and membrane filters, p. 223-285. *In* C. N. Davies [ed.], Aerosol science. Academic.

POULET, S. A. 1973. Grazing of *Pseudocalanus minutus* on naturally occurring particulate matter. Limnol. Oceanogr. **18**: 564-573.

——. 1974. Seasonal grazing of *Pseudocalanus minutus* on particles. Mar. Biol. **25**: 109-123.

——; 1976. Feeding of *Pseudocalanus minutus* on living and non-living particles. Mar. Biol. **34**: 117-125.

——. 1977. Grazing of marine copepod developmental stages on naturally occurring particles. J. Fish. Res. Bd. Can. **34**: 2381-2387.

——. 1978. Comparison between five coexisting species of marine copepods feeding on naturally occurring particulate matter. Limnol. Oceanogr. **23**: 1126-1143.

——, and J. P. CHANUT. 1975. Non-selective feeding of *Pseudocalanus minutus*. J. Fish. Res. Bd. Can. **32**: 706-713.

——, and P. MARSOT. 1978. Chemosensory grazing by marine calanoid copepods (Arthropoda: Crustacea). Science **200**: 1403-1405.

RANZ, W. E., and J. B. WANG. 1952. Impaction of dust and smoke particles on surface and body collectors. Ind. Eng. Chem. **44**: 1371-1381.

RICHMAN, S., D. R. HEINLE, and R. HUFF. 1977. Grazing by adult estuarine calanoid copepods of the Chesapeake Bay. Mar. Biol. **42**: 69-84.

——, and J. N. ROGERS. 1969. The feeding of *Calanus helgolandicus* on synchronously growing populations of the marine diatom *Ditylum brightwellii*. Limnol. Oceanogr. **14**: 701-709.

ROBERTSON, S. B., and B. W. FROST. 1977. Feeding by an omnivorous planktonic copepod *Aetideus divergens* Bradford. J. Exp. Mar. Biol. Ecol. **29**: 231-244.

ROMAN, M. R. 1977. Feeding of the copepod *Acartia clausi* on the diatom *Nitzschia closterium* and brown algae (*Fucus vesiculosus*) detritus. Mar. Biol. **42**: 149-155.

RUBENSTEIN, D. I., and M. A. R. KOEHL. 1977. The mechanisms of filter feeding: Some theoretical considerations. Am. Nat. **111**: 981-994.

SHELDON, R. W., T. R. T. EVELYN, and T. R. PARSONS. 1967. On the occurrence and formation of small particles in seawater. Limnol. Oceanogr. **12**: 367-375.

SHELTON, R. G. J., and A. M. MACKIE. 1971. Studies on the chemical preferences of the shore crab, *Carcinus maenas* (L.). J. Exp. Mar. Biol. Ecol. **7**: 41-49.

SKIVER, J. 1978. Resource partitioning patterns of calanoid copepods. M.S. thesis, Dalhousie Univ.

SUTCLIFFE, W. H., JR. 1972. Some relations of land drainage, nutrients, particulate material and fish catch in two eastern Canadian bays. J. Fish. Res. Bd Can. **29**: 357-362.

VYSHKVARTSEVA, N. V., and B. L. GUTEL'MAKHER. 1971. Trapping ability of the filtering apparatus of some calanidae. Hydrobiol. J. **7**: 58-63.

WICKSTEAD, J. H. 1959. A predatory copepod. J. Anim. Ecol. **28**: 69-72.

WILSON, D. S. 1973. Food size selection among copepods. Ecology **54**: 909-914.

ZURLINI, G., I. FERRARI, and A. NASSOGNE. 1978. Reproduction and growth of *Euterpina acutifrons* (Copepods: Harpacticoida) under experimental conditions. Mar. Biol. **46**: 59-64.

20. Grazing Interactions among Freshwater Calanoid Copepods

Sumner Richman, Scott A. Bohon, and Stephen E. Robbins

Abstract

A comparative study was carried out on the grazing interactions of four calanoid copepod species feeding on natural phytoplankton assemblages in southern Green Bay and Lake Winnebago and on algal species used singly and in mixtures. The species studied were *Diaptomus siciloides, Diaptomus oregonensis, Diaptomus ashlandi*, and *Eurytemora affinis*. Particle distributions before and after feeding were analyzed using a model B Coulter Counter interfaced to a PDP 11/20 laboratory computer, which allowed for automated analyses of particles in 128 size categories and a size range of 6-20,000 μm^3 (2-33 μm in spherical equivalent diameter).

Comparisons of the four calanoid copepods revealed both similarities and differences in their feeding behavior. The three *Diaptomus* species could all feed over a broad range of particle sizes with strong evidence for selective feeding on biovolume peaks > 12 μm and on larger particles beyond the biovolume peak as well. Feeding on particles < 12 μm was nonselective as indicated by a constant filtering rate over the size range from 2.5-12 μm. Feeding on particles < 12 μm increased abruptly, indicating an active selection for those particles probably through a raptorial mode of feeding.

Diaptomus oregonensis had a broader feeding range than either *D. siciloides* or *D. ashlandi*; it included particles in the 2.5-30 μm range. *D. oregonensis* could also handle species of large platelike shape such as *Cosmarium* sp. and *Stephanodiscus* sp., whereas the other two species could not. *D. oregonensis* was a more active feeding, not only filtering large particles at higher rates (up to 0.9 ml animal^{-1} h^{-1}), but also producing more particles through fragmentation.

Eurytemora affinis had the most restricted feeing range. Significant filtering rates always occurred on particles > 12 μm. Feeding was primarily focused on the largest particles within the particle spectrum.

It is common in eutrophic ecosystems to find several species of copepods and cladocerans grazing on a wide variety of phytoplankton species (Hutchinson 1967; Porter 1973, 1977). Competition between zooplankton species may be reduced by the seasonal abundance of many food species and the seasonal cycles of the herbivores (Hutchinson 1967). In addition, freshwater ecosystems are not homogeneous, allowing herbivore species to be segregated in space and time (Makarewicz and Likens 1975; Bowers 1977; Bowers and Grossnickle 1978). The feeding niches of closely relat-

---

This investigation was supported in part by the University of Wisconsin Sea Grant College Program under an institutional grant from the National Oceanic and Atmospheric Administration, U. S. Department of Commerce.

ed herbivorous copepods could be separated by differences in their feeding behavior, although there is no convincing evidence for such separation. This problem is further complicated by the lack of comparative studies of zooplankton feeding on the complex array of particles that occur in freshwater.

Nevertheless, it has been demonstrated that filter feeders can discriminate among particles on the basis of size, shape, and taste. The subject has been reviewed recently by Porter (1977). Calanoid copepods may select for size classes in greatest abundance (Poulet 1973; Wilson 1973; Allan et al. 1977; Richman et al. 1977). Moreover, these animals may reject algae or distinguish between enriched and nonenriched food according to taste (Friedman and Strickler 1975; Poulet and Marsot 1978), may avoid certain shapes (Harvey 1937), or may be capable of raptorial feeding processes that involve the capture and breakage of large cells (Conover 1966; Parsons and Seki 1970; O'Conners et al. 1976). It has also been shown that copepods can switch feeding modes, feeding raptorially on large particles and passively filtering smaller sizes (Mullin 1963; Richman and Rogers 1969). Although most of these studies have used estuarine and marine copepods, this latter observation has been shown for the freshwater copepod *Diaptomus oregonensis* when fed natural phytoplankton assemblages (McQueen 1970).

There has been considerable interest in the mechanism underlying these feeding interactions. One viewpoint considers only the sieving mechanism of particle capture and regards filtration as a mechanical process similar to the operation of a "leaky sieve" which retains particles with an efficiency in direct relation to particle size (Boyd 1976; Lam and Frost 1976; Lehman 1976; Nival and Nival 1976; Frost 1977). However, by changing the velocity of water passing through the filters or by altering the diameter or adhesiveness of the filtering fibers, the range of particle sizes that can be captured most efficiently can be shifted (Rubenstein and Koehl 1977). Such shifts may explain the "tracking" of biomass peaks and selection for large particles frequently described (Wilson 1973; Poulet 1973; Richman et al. 1977; Allan et al. 1977). Further elaboration of the hydrodynamic principles that relate to mouthpart morphology and a better understanding of chemical perception is needed before we can fully understand the role of sensory-determined behavioral processes in filter feeding.

This study was undertaken to better understand the trophic interactions between phytoplankton and zooplankton in southern Green Bay. Since this plankton community is strongly influenced by species in Lake Winnebago that are transported through the Fox River (Gannon 1974), grazing interactions in this lake were also examined.

We acknowledge the work of S. Woods and C. Langdon, two former Lawrence students, who developed the electronic interface and computer software which automated our model B Coulter Counter and made possible the detailed analysis of the feeding data. We also thank P. Sager, University of Wisconsin, Green Bay, and his students for their help with the Green Bay collections.

Methods

Feeding experiments were performed using adult females of four copepod species: *Diaptomus siciloides, Diaptomus oregonensis, Diaptomus ashlandi,* and *Eurytemora affinis*. The study was conducted at four sampling sites located along the east and west shores of southern Green Bay. Samples were also collected along the northern shore of Lake Winnebago, near High Cliff State Park.

At each sampling station, animals and water samples were collected from an outboard motor boat 0.8–1.6 km offshore. Animals were captured with a 363 µm plankton net towed horizontally 1.5–3.1m below the surface. Individual tows lasted about 5 min. The contents of the nets were emptied into 1-liter plastic jugs and transported to the laboratory. Winter sampling required cutting holes in the ice and collecting animals via vertical tows extending from about 3.1 m deep to the surface.

Raw water samples were obtained by collecting water with 19-liter plastic carboys from just below the surface. These samples provided the experimental feeding media for most of our experiments.

After making the field collections, the zooplankton were sorted with eyedroppers to species under dissecting microscopes. All of the species were collected from Green Bay, but only *D. siciloides* and *D. oregonensis* were present in Lake Winnebago. The adult females were temporarily kept in 0.45 μm millipore-filtered lake water at temperatures close to those of the water mass from which the animals were collected. The animals were kept under these conditions for 12–24 h before the start of each feeding experiment.

To set up each feeding experiment, representative adult females of the desired species were resorted and placed in Syracuse dishes filled with millipore-filtered lake water. Pasteur pipettes were used to rinse the animals with additional millipore-filtered water to remove any adhering particles. This "washing" procedure was repeated 2–3 times before the animals were used. The contents of the Syracuse dishes (millipore-filtered water and animals) were poured into 25-ml beakers. Total water volumes therein were reduced to about 5 ml with pipettes.

In experiments using natural particle distributions, raw water was passed through a 153 μm mesh to remove zooplankton This water was then thoroughly mixed and 135 ml poured into experimental and control bottles. Hence, initially the feeding medium in all experimental and control bottles was identical. The animals were then added to the experimental bottle, filling it to the 140-ml level. The control bottle was filled with an equivalent (5 ml) volume of "animal water," water in which the animals had been placed before being used in the experiment. This procedure ensured that the only entities added to the experimental bottles but not to the controls were the animals themselves.

Several experiments were also conducted using laboratory-cultured algae. The species used were *Chlamydomonas reinhardi*, *Scenedesmus quadricauda*, *Pediastrum* sp., and *Cosmarium* sp. Unialgal cultures were grown in 1,000-ml Erlenmeyer flasks containing 250 ml of nutrient media. The media used were either Bristol's or Beyerick's solution enriched with soil-water extract. The culture flasks were exposed to continuous light and grown at temperatures ranging from 20°–25°C.

In lab-culture experiments, feeding media were prepared by adding aliquots of algal cultures to particle-free millipore-filtered lake water to produce the desired particle distribution. After thorough mixing, 135-ml samples of the medium were poured into experimental and control bottles. The bottles were "loaded" with either animals or "animal water" as described above.

In all experiments bottles were "run" in pairs. One bottle was experimental, to which zooplankton were added. The second was a control to which nothing but the feeding medium was added. Hence, comparison of experimental and control bottles after specified feeding durations allowed us to evaluate both "feeding" and "production." The former was found when the number of particles within particular size categories in the experimental bottle was significantly lower than the number of particles in the same size categories in the control bottle. Such a result indicates that particles were removed by the animals, an indication of feeding. Production was the opposite. This was suggested when the number of particles within particular size categories was significantly higher than in the control bottle. What we observe, then, is the net effect of production and utilization in these experiments. The t-test was used to evaluate statistically significant differences between the control and experimental bottles.

Grazing experiments were conducted in the dark to minimize algal growth. They were run at the temperature of the water from which the animals were collected. The experiments were run either for one time period of 8–10 h or involved a series of control and experimental bottle-pairs in which the animals were allowed to feed at selected time intervals between 2–24 h. The latter approach allowed us to analyze both where along the particle spectrum the animals began to feed and how feeding and particle production progressed with time. The number of animals used per bottle ranged from 35 to 50 depending on the species. After the bottles were loaded, they were placed on a plankton wheel and rotated end-over-end at 2 rpm. After predetermined time intervals, experimental bottles and their corresponding controls were removed from the wheel. The animals were removed from the experimental bottles with a 153 μm mesh as the contents of the bottles were poured into 150-ml beakers for Coulter Counter sampling. The control samples were filtered through the same net into 150-ml beakers as well, to ensure equal treatment.

After the contents of the experimental and control bottles were transferred to 150-ml beakers, 7.5 ml of 20% NaCl solution was added to each beaker by pipette, creating a 1% salt solution. The addition of salt was necessary as the particle counting system requires an electrolytic solution to measure particle volumes. No changes in particle distribution were observed due to the salt addition during the 15–20 min sampling time.

The particle distributions from experimental and control bottles were analyzed by sampling 10 replicate 0.5-ml aliquots from each bottle using a model B Coulter Counter interfaced to a PDP 11/20 laboratory computer. This automated particle counting system was equipped with a logarithmic amplifier, which allowed analysis of particles over a wide range of sizes. The system permitted the measurement of particles in 128 size categories. The actual particle sizes were measured in terms of particle volume. These measurements were then converted to spherical equivalent diameters.

The particle-counting equipment was calibrated to determine the particle volumes associated with each of the 128 channels by analyzing several solutions, each of which contained polystyrene beads of single known size and volume. These single-peak distributions were fit via a least-squares analysis to a Gaussian function. This established the channel location of each particle. The particle volumes and their peak channels were then fit, using another least-squares analysis, to a logarithmic relationship. This produced volume parameters which were used to calculate the particle volume for each channel for a given sensitivity setting. Two sensitivities were used in our experiments, depending on the distribution under study, covering size ranges of 6–8,000 μm^3 in volume (2–25 μm in spherical equivalent diameter) and 6–20,000 μm^3 in volume (2–33 μm in spherical equivalent diameter).

The system featured rapid data analysis after acquisition on the PDP 11/20. The analyses performed included calculation of mean counts (number of particles) from the 10 replicate experimental and the 10 replicate control samples in each size category, the conversion of counts per milliliter of water to particle volume per milliliter of water in each size category, and calculation of filtering rates in each size category using the equation of Gauld (1951). This parameter in these experiments represents the volume of water in each size category from which particles of that size were removed, produced, or both and can therefore be positive, zero, or negative depending on the net effect of particle production and utilization. Each of the three parameters was graphed as a function of spherical equivalent diameter with a Hewlett-Packard X-Y recorder.

Further data analysis was performed by transferring stores raw data on DEC tape to a PDP 11/45 computer. In addition to a print-out of the data described above, calculations were also made of the ingestion rate, calculated as total volume of food ingested per copepod per hour for each size category, the standard error of the experimental and control means for each size category, and a *t*-test calculation to evaluate the statistical significance of mean differences between control and experimental distributions.

Results

Comparative results of grazing experiments performed with the five zooplankton species fed on naturally occurring particle distributions from southern Green Bay and Lake Winnebago, Wisconsin, and unialgal and mixed algal cultures are shown in Figs. 1-11. Since *D. siciloides* was the most prevalent calanoid copepod species in the sampling locations, more data were obtained for this species than for the others.

Figure 1A depicts a typical particle distribution for the eastern shore of southern Green Bay in midsummer 1977. It is characterized by a broad biovolume "peak" composed of particles in the 12-27-μm spherical equivalent diameter range and a peak concentration of about 0.14 $\mu m^3 \times 10^6 ml^{-1}$. Changes in the distribution due to feeding by *D. siciloides* are shown at 8, 12, and 18 h.

At 8 h, feeding was restricted to the biovolume peak, from 13-19 μm. The filtering rate was greatest on the largest particles in this range (18-19 μm) at 0.11 ml animal $^{-1}h^{-1}$ (Fig. 1B). Spreading to larger particles was found at 12 h as the feeding range expanded to 12-26 μm with a maximum filtering rate of 0.12 ml animal^{-1} h^{-1} on 24-μm particles. Feeding emphasis upon even larger particles continued at 18 h. At this time the maximum filtering rate was 0.11 ml animal $^{-1}h^{-1}$ on 27-μm particles. In addition, feeding at lower rates spread to encompass smaller particles after feeding on larger ones for many hours, a shift which probably occurs as a result of the depletion of food in the preferred larger-size range. There is evidence of a net gain of particles in the 12-μm size category, but it is probably too slight to affect the major features described here.

In a somewhat different distribution taken from Lake Winnebago in February 1978, a similar feeding pattern was observed for *D. siciloides* (Fig. 2). This particle assemblage was unimodal, featuring a narrow biovolume peak at 13-14 μm and a maximum particle concentration of 0.015 $\mu m^3 \times 10^6 ml^{-1}$. At 4 h significant feeding occurred on the peak of the distribution and beyond, from 12-17 μm with maximum filtering rates of 0.5 ml animal $^{-1}h^{-1}$. After 6 h, feeding spread to smaller size categories, and by 10 h, after larger particles were consumed during earlier time periods, significant feeding occurred from 2.8-14 μm. Feeding on the larger particles was as much as 5 times greater than on those below the biovolume peak. Measured filtering rates were 0.5 ml animal $^{-1}h^{-1}$ as compared to 0.1 on the smaller particles.

The preference of *D. siciloides* for larger particles was further emphasized when copepods were fed the double-peaked distribution collected from Lake Winnebago in October 1977 (Fig. 3). Even though the smaller peak was greater in concentration, the animals fed on the larger particle biovolume peak between 12-19 μm. Particle fragmentation resulted in significant production between 8-10 μm which was subsequently fed upon in the later time interval. By 22.75 h, feeding spread to particles in the 2.5-8 μm range as particles in the larger sizes were depleted. Particle production was essentially balanced by particle utilization in the 8-10-μm range. Maximum filtering rates were 0.4 ml animal $^{-1}h^{-1}$ on 18-20 μm particles, whereas average filtering rates were about 0.15 ml animal $^{-1}h^{-1}$ on the 2.5-8 μm portion of the distribution. It is important to note that in all of the time series experiments with *D. siciloides*, there was low and uniform feeding on particles between about 2.5-12 μm and a sharp increase in feeding on larger particles on the 12-22 μm size categories. Furthermore, the larger particles are selected earlier in time with shifting to smaller food particles as the larger ones are consumed (Figs. 1-3).

Figure 4 shows the results of *D. oregonensis* feeding on the same Lake Winnebago particle distribution described earlier for

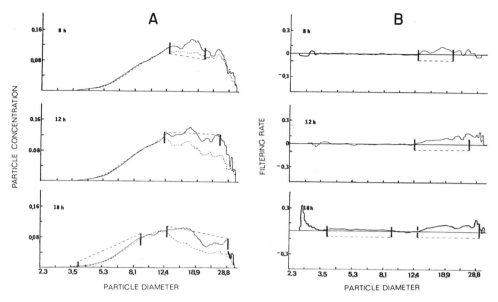

Fig. 1.A. Particle size-biovolume distributions for southeastern Green Bay water, July 1977, without feeding (solid lines) and after feeding (broken lines) for *Diaptomus siciloides* for time periods indicated. Bracketed sections show size ranges with significant differences between control and experimental bottles ($P=0.05$). B. Filtering rates as a function of particle size calculated from differences in control and experimental distributions shown in panel A. Bracketed sections same as panel A. On this and later figures particle concentration is in $\mu m^3 \times 10^6\, ml^{-1}$, particle diameter in μm, and filtering rate in ml animal^{-1} h^{-1}.

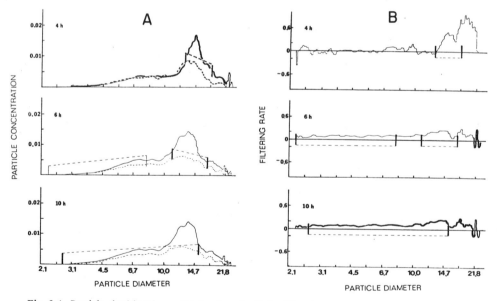

Fig. 2.A. Particle size-biovolume distributions for Lake Winnebago water, February 1978, *Diaptomus siciloides* (see Fig. 1A for explanation). B. Filtering rates calculated from distributions shown in panel A (see Fig. 1B for explanation).

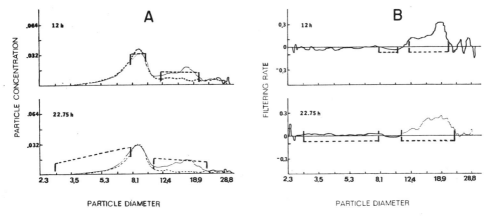

Fig. 3.A. Particle size-biovolume distributions for Lake Winnebago water, October 1977, *Diaptomus siciloides* (see Fig. 1A for explanation). B. Filtering rates calculated from distributions shown in panel A (see Fig. 1B for explanation).

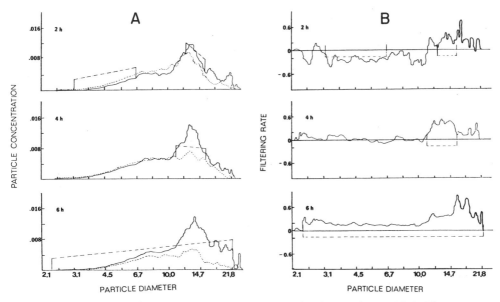

Fig. 4.A. Particle size-biovolume distributions for Lake Winnebago, February 1978, *Diaptomus oregonensis* (see Fig. 1A for explanation). B. Filtering rates calculated from distributions shown in panel A (see Fig. 1B for explanation).

D. siciloides in Fig. 2. These data show many of the same major features observed with *D. siciloides*. However, significant particle production was clearly evident in the small particle range of 3–6 μm after only 2 h, as feeding took place upon larger particles, those comprising the biovolume peak from 12–15 μm. By 4 h, the produced particles were consumed while selection for the larger particles continued with maximum filtering of particles in the 12–15-μm range at rates of about 0.5 ml animal^{-1}h^{-1}. By 6 h, feeding extended over the entire particle spectrum, ranging from 2.5–22 μm. Filtering rates from 0.4–0.9 ml animal^{-1} h^{-1} on particles >12 μm in the peak

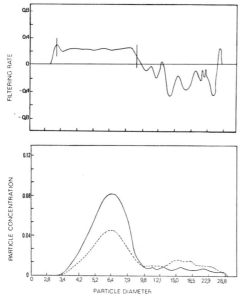

Fig. 5. Upper panel: filtering rates calculated from differences in control and experimental distributions shown in lower panel. Vertical bars bracket size range where significant differences occurred in the two distributions ($P=0.05$). Lower panel: particle size-biovolume distributions for *Chlamydomonas reinhardi* without feeding (solid line) and after feeding (broken line) by *Diaptomus siciloides*.

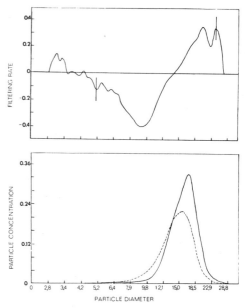

Fig. 6. Particle size-biovolume distributions for *Scenedesmus quadricauda*, with and without feeding by *Diaptomus siciloides* (see Fig. 5 for explanation).

of the distribution and beyond were much higher than the rest of the particle spectrum where rates of 0.2–0.3 ml animal^{-1} h^{-1} were measured.

The feeding behavior of *Diaptomus* species was further investigated by using laboratory algal species singly and in mixtures (Figs. 5–9). When the results of *D. siciloides* feeding on *C. reinhardi* are compared to those that used *S. quadricauda* as food, constant feeding over the entire *Chlamydomonas* distribution was found from 3–9 μm with a filtering rate of 0.2 ml animal^{-1} h^{-1}, whereas the filtering rate on the larger *Scenedesmus* peak in the 15–22 μm range was twice as great at 0.4 ml animal^{-1} h^{-1} (Figs. 5 and 6). Particle production in larger size categories suggested in the *Chlamydomonas* experiment was not significant and was probably due to variability in counting small numbers of particles in this size range. However, significant particle production was evident in the 5–12 μm sizes of the *Scenedesmus* population and was apparently due to the breaking of four-celled units into smaller fragments (Fig. 6). When *D. siciloides* were fed a mixture of these two algal species to produce two biovolume peaks of equal concentration, the feeding on each peak was the same as on each food species separately and demonstrated a clear preference for the larger cells of *Scenedesmus* (Fig. 7). Furthermore, this experiment shows that particle production did not obscure the main features of the observed feeding behavior, i.e. the feeding at a low and constant rate on the smaller-sized *Chlamydomonas* population and higher selection for the larger *Scenedesmus* cells.

Diaptomus siciloides were also fed on a trimodal biovolume distribution with the addition of *Cosmarium* sp. to the *Chlamydomonas* and *Scenedesmus* mixture to produce a third peak at about 28 μm. After 8 h, particle feeding and production

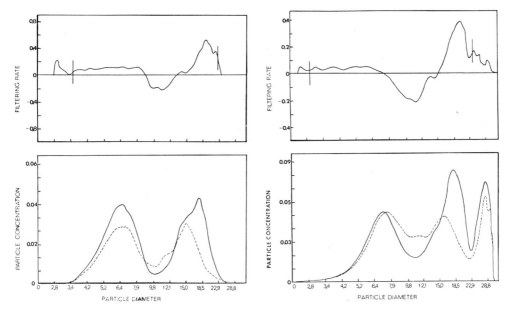

Fig. 7. Particle size-biovolume distributions for a mixture of *Chlamydomonas reinhardi* (left peak) and *Scenedesmus quadricauda* (right peak) with and without feeding by *Diaptomus siciloides* (see Fig. 5 for explanation).

Fig. 8. Particle size-biovolume distributions for a mixture of *Chlamydomonas reinhardi* (left peak), *Scenedesmus quadricauda* (middle peak) and *Cosmarium* sp. (right peak) with and without feeding by *Diaptomus siciloides* (see Fig. 5 for explanation).

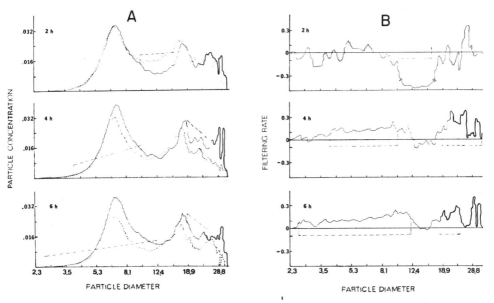

Fig. 9.A. Particle size-biovolume distributions for *C. reinhardi* (left peak), *Pediastrum* sp. (middle peak) and *Cosmarium* sp. (right peak) with and without feeding by *Diaptomus oregonensis* (see Fig. 1 for explanation). B. Filtering rates calculated from distributions shown in panel A. Minimum filtering rate is off scale at -0.83 ml animal^{-1} h^{-1} at 13 μm (see Fig. 1B for explanation).

was almost identical to that observed with the mixture of the latter two algal species (Figs. 7 and 8). Feeding on the *Cosmarium* sp., however, was not statistically significant, indicating a cell size and/or shape that this copepod was unable to consume or a taste that was avoided. Work to be described later adds further support to this contention. As in the previous experiments, the animals showed evidence for selective feeding on the larger cell and constant passive feeding on the smaller one.

Diaptomus oregonensis was also fed on a three biovolume peak distribution of algal species consisting of *C. reinhardi* (6 μm), *Pediastrum* sp. (18 μm) and *Cosmarium* sp. (28 μm). While initial feeding is partially obscured by a large amount of fragmentation of the two larger species, the time series provides further insights into the feeding behavior of this species. By 4 h, both the *Cosmarium* and *Pediastrum* have been significantly fed upon, indicating the ability of this species to feed more readily on larger cells (up to 30 μm) than *D. siciloides* (Fig. 9). The time series further shows this species feeding on the whole distribution, including the fragmented particles it initially produced. As in an earlier experiment (Fig. 4), feeding on large particles is greater than on small ones.

Further comparisons of feeding were made possible when the four calanoid copepod species were simultaneously collected in Little Sturgeon Bay, southern Green Bay, Wisconsin, in July 1977 (Fig. 10). During this time the concentration of particles was primarily due to the large diatom *Stephanodiscus* which produced a sharp peak in the 25–31 μm range. However, in one of the experiments (Fig. 10E) the distribution had more particles in the middle range, between 7–19 μm. These experiments show that neither *D. siciloides* nor *D. ashlandi* can feed on the *Stephanodiscus* peak, probably because of its size, shape, or taste. We observed a similar result when *D. siciloides* was provided with *Cosmarium*, another platelike cell with the same spherical equivalent diameter (Fig. 8).

The filtering rate values for *D. siciloides* and *D. ashlandi* (Fig. 10A,B) show that these two species have very similar feeding characteristics although more work needs to be done to confirm this contention. Significant feeding for *D. ashlandi* ranged from 3–11 μm and 15–19 μm with the gap between these two ranges probably due to particle production obscuring feeding. *D. ashlandi* was rarely encountered at the sampling sites, so only limited data were obtained to evaluate its feeding behavior.

The results of *D. oregonensis* feeding on the *Stephanodiscus* peaked distribution indicates that this species does select these large particles at high filtering rates (0.5–0.6 ml animal^{-1}h^{-1}) and also feeds on the rest of the distribution between 4–20 μm at lower rates (0.15–0.20 ml animal^{-1} h^{-1} (Fig. 10C). These results are consistent with previously described experiments for this species and show that its feeding behavior is basically similar to the other *Diaptomus* species. However, its feeding range is extended beyond the limits of *D. siciloides* and *D. ashlandi* to at least 30 μm in spherical equivalent diameter as compared to 20–22 μm for the other two species (Figs. 9, 10C).

In marked contrast to the *Diaptomus* species, *E. affinis* fed almost exclusively on the *Stephanodiscus* peak, showing a clear selection for these cells in the Little Sturgeon Bay distribution. After 10 h of feeding, these copepods nearly eliminated the biovolume peak between 19–33 μm with virtually no feeding on any other part of the distribution (Fig. 10D). In addition, particle production can be noted from 3.5 to 8 μm. When this particle distribution included greater particle concentrations below the *Stephanodiscus* peak, *E. affinis* expanded its feeding. The significant feeding range encompassed particles in the 12–31 μm size range and, as in the previous experiment, there was clear preference for the larger particles in the 20–30 μm biovolume peak with particle production in the 4–8 μm range (Fig. 10E).

Size selection as well as biovolume peak

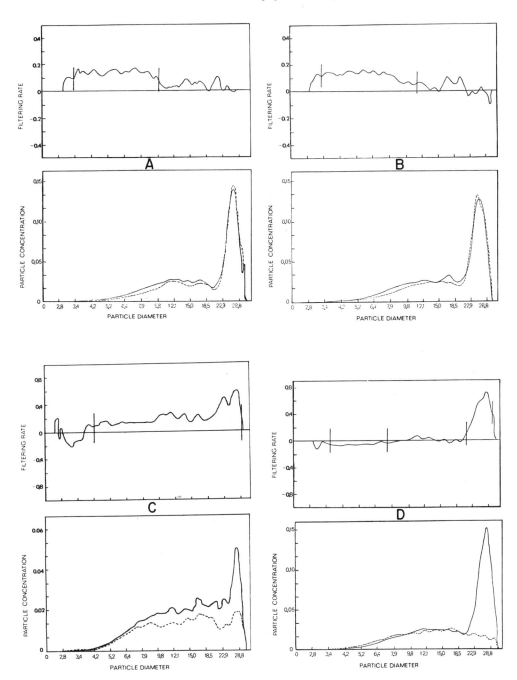

Fig. 10. Feeding on particle size-biovolume distributions from Little Sturgeon Bay, Green Bay, July 1977 by species below (*see* Fig. 5 for explanation). A. *Diaptomus siciloides*. B. *Diaptomus ashlandi*. C. *Diaptomus oregonensis*. D. *Eurytemora affinis*.

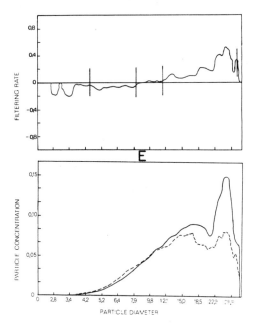

Fig. 10. E. *Eurytemora affinis*.

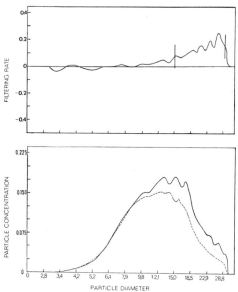

Fig. 11. *Eurytemora affinis* feeding on particle size-biovolume distribution from southeastern Green Bay, July 1977 (see Fig. 5 for explanation).

selection was further demonstrated for *E. affinis* from southeastern Green Bay when found in water that had a rounded unimodal biovolume distribution with a maximum at 12–18 μm (Fig. 11). The significant feeding range covered particles 13–29 μm in size. The filtering rate was skewed slightly toward large particles. This feeding reflects selection for large particles and those particles comprising the biovolume peak. In all of the experiments with *E. affinis*, feeding clearly favored the large end of the size spectrum between 12 and 33 μm in spherical equivalent diameter.

Discussion

Comparisons of the four calanoid copepods studied reveal both similarities and differences in their feeding behavior. The three *Diaptomus* species were all capable of feeding over a broad range of particle sizes with strong evidence for selective feeding on biovolume peaks >12 μm in size and on larger particles beyond the biovolume peak. Feeding on particles <12 μm was nonselective as indicated by a constant filtering rate over the size range from 2.5–12 μm in spherical equivalent diameter and 100–1,000 μm³ in particle volume. Feeding on particles >12 μm increased abruptly, indicating active selection for these particles. There is considerable evidence that marine copepods selectively remove larger particles (Mullin 1963; Conover 1966; Richman and Rogers 1969; Paffenhöfer 1971; Wilson 1973; Allan et al. 1977; Richman et al. 1977) but there is little in the literature concerning freshwater forms (McQueen 1970; Bowers 1977).

The major portion of the study investigated the feeding behavior of *D. siciloides*, the dominant herbivorous copepod taken in our collections from both southern Green Bay and Lake Winnebago. Short time-interval studies revealed that these animals start feeding on the biovolume peak and then extend their feeding range to both smaller and larger particles as material within the initial feeding ranges is decreased (Figs. 1, 2). When provided with two biovolume peaks, they selected the larger-sized peak even when the two peaks were of equal concentration or when the smaller-sized peak was greater in concentration (Figs. 3, 7). Furthermore, feeding was always passive, as indicated by constant filtering rates, on particles <12 µm in spherical equivalent diameter. This finding held even when there were distinct biovolume peaks below 12 µm (Figs. 3, 5, 7, 8)—a result that is different from similar studies conducted on estuarine forms (Richman et al. 1977).

Diaptomus siciloides could not handle large platelike cells such as *Cosmarium* (Fig. 8) or *Stephanodiscus* (Fig. 10A), although chemosensory avoidance may have been involved (Poulet and Marsot 1978). *D. siciloides* displayed an effective feeding range of from 2.5–22 µm in spherical equivalent diameter. A comparable result of feeding on particles <22 µm was shown for *D. minutus* by Bogdan and McNaught (1975) and a very similar feeding behavior was observed for *D. ashlandi* (Fig. 10B). Although more work is needed for the latter species, it did not feed on the large *Stephanodiscus* peak. However, it was able to feed on particles larger than the limit of 10 µm suggested for this species by Bowers (1977) and Bowers and Grossnickle (1978).

Diaptomus oregonensis had a broader feeding range than the other two diaptomid species. Its range included particles between 2.5–30 µm (Figs. 9, 12). Feeding behavior of this species is in close agreement with the results of McQueen (1970) for *D. oregonensis* feeding on natural phytoplankton. He found filtering rates of 0.3–0.5 ml animal^{-1}h^{-1} for large particles and 0.06–0.08 for particles <5.8 µm. Feeding on particles larger than about 8 µm resulted in almost 10 times higher filtering rates than on the smaller particles. This species was able to consume both *Cosmarium* and *Stephanodiscus* at higher filtering rates than smaller particles, showing the ability to feed on cells not consumed by the other species. *D. oregonensis* seems to be a more active and superfluous feeder.

It not only filtered particles at higher rates but its feeding was characterized by greater production of particles. The short time-interval feeding studies clearly showed an initial feeding at the highest rates on larger particles at or beyond the biovolume peak and a subsequent feeding on smaller particles, which were present in the distribution or produced through fragmentation, at lower rates (Figs. 4, 9).

Significant particle production also was observed for *D. siciloides*, although at a lesser magnitude than *D. oregonensis* (Figs. 3, 6, 7, 8). This form of particle modification, largely due to breakage of cells and fragmentation of colonial algae and chain-formers, is similar to that reported by O'Connors et al. (1976), who examined particle modification of the chain-forming diatom *Thalassiosira* and natural phytoplankton assemblages by *Acartia clausi*. Several other investigators observed the production of particles during copepod feeding, including Poulet (1974), Allan et al. (1977), Richman et al. (1977), and Poulet and Marsot (1978). Recently, Deason (unpubl.), through computer simulations, showed that such breakage may lead to either over- or underestimations of filtering rates and erroneous interpretations of complex selection for size and particle abundance. In her models, the extent of the error is related to breakage rates, length of the experiment, size of the largest colony, experimental volume, number of animals and their actual filtration rates. Even though we observed the net effect of production and utilization of particles in our study, the short time-interval feeding sequences helped to distinguish the two processes so that the major features of passive feeding and active selection were evident. Further evidence for particle production comes from the algal-culture feeding experiments where *D. siciloides* produced the expected results of passive feeding on *Chlamydomonas*, selection for *Scenedesmus*, and production between these two peaks when the two algae were mixed. Thus, the production of particles did not change the feeding patterns observed on the individual algal species (Figs. 5–8).

Eurytemora affinis had the most restricted feeding rate of the four copepods studied. In our experiments, significant filtering rates always occurred on particles >12 μm (Figs. 10D, E, 11). Feeding was primarily focused on the largest particles within the particle spectrum and this feeding produced particles in the small-sized range. Selection for particles in the biovolume peak, and on larger particles as well, was similar to that found by Richman et al. (1977) for this species in the Chesapeake Bay, suggesting a common feeding behavior for this copepod which lives in a wide salinity range. There was a greater emphasis on large particle feeding in this study, but this result may be due to the nature of the particle distributions provided. It may also be due to the separation of feeding niches in Little Sturgeon Bay where *D. siciloides* and *D. ashlandi*, *D. oregonensis* and *E. affinis* displayed three overlapping but, nevertheless, different size-dependent feeding behaviors on the same particle distribution (Steele and Frost 1977).

Similar passive feeding on particles <12 μm was evident for all three *Diaptomus* species. Friedman (1977) studied in detail the mouthpart morphology of the species used in our collections. He found that the mouthparts were adapted for small-particle filter feeding. Scanning electron microscopy revealed that all three species had virtually identical mouthpart structures. The extent of particle selection on larger particles and the overall feeding range did vary for the diaptomid species. *D. siciloides* and *D. ashlandi* showed smaller feeding ranges and large particle selection than did *D. oregonensis*. These results may be related to the volume of the feeding chamber of the three species, which were 1.8×10^{-7} cm^3 for *D. siciloides*, 3.25×10^{-7} cm^3 for *D. ashlandi* and 3.6×10^{-7} cm^3 for *D. oregonensis* (M. Friedman at this conf.). On the other hand, *E. affinis* has second maxillae and mandibular teeth that are quite different when compared to the *Diaptomus* species and its feeding chamber volume is considerably greater (5.0×10^{-7} cm^3: Friedman 1977). These observations are consistent with the different pattern of feeding observed for this species. Friedman further suggests that larger feeding chamber volumes and the increased rate of flow of incurrent, food laden water will result in increased filtering rates and selective retention of larger particles.

There are several means by which an organism can capture certain sizes of particles more effectively than others. Some of the mechanisms of filter feeding have been recently detailed by Rubenstein and Koehl (1977) and include direct interception, inertial impaction, gravitational deposition, motile particle deposition, and electrostatic attraction. An animal can actively change the size of the food particles it retains, not only by changing its pore size, but also by altering the velocity of water passing through the filter or the adhesive diameter of the filtering fibers. Obviously, retention of particles by an animal's filter and the transport of particles to the mouth are also important. The movement of particles to the mouth through a raptorial mode of feeding by grasping large particles one at a time was clearly demonstrated for calanoid copepods in the elegant high-speed motion picture films that Strickler has shown at this symposium. Further understanding of the hydrodynamic principles relating to the morphology and ecology of these organisms is necessary before we can understand the complexities of the filter-feeding behaviors described for these freshwater calanoid copepods.

References

ALLAN, J. D., S. RICHMAN, D. R. HEINLE, and R. HUFF. 1977. Grazing in juvenile stages of some estuarine calanoid copepods. Mar. Biol. 43:317–331.

BOGDAN, K. D., and D. C. McNAUGHT. 1975. Selective feeding by *Diaptomus* and *Daphnia*. Int. Ver. Theor. Angew. Limnol. Verh. 19: 2935–2942.

BOYD, C. M. 1976. Selection of particle sizes in filter-feeding copepods: A plea for reason. Limnol. Oceanogr. 21:175–180.

BOWERS, J. A. 1977. The feeding habits of *Diaptomus ashlandi* and *Diaptomus sicilis* in Lake Michigan and the seasonal vertical distribution of chlorophyll at a nearshore station. Ph.D. thesis, Univ. Wisconsin-Madison. 138 p.

——, and N. E. GROSSNICKLE. 1978. The herbivorous habits of *Mysis relicta* in Lake Michigan. Limnol. Oceanogr. 23:767–776.

CONOVER, R. J. 1966. Feeding on large particles by *Calanus hyperboreous* (Kroyer), p. 187–194. *In* H. Barnes [ed.] Some contemporary studies in marine science. Allen and Unwin.

FRIEDMAN, M. M. 1977. Electron microscopic studies of the filter-feeding mechanism of calanoid copepods. Ph.D. thesis, Johns Hopkins Univ.

——, and J. R. STRICKLER. 1975. Chemoreceptors and feeding in calanoid copepods (Arthropoda:Crustacea). Proc. Natl. Acad. Sci. 72:4185–4188.

FROST, B. W. 1977. Feeding behavior of *Calanus pacificus* in mixtures of food particles. Limnol. Oceanogr. 22:472–491.

GANNON, J. E. 1974. The crustacean zooplankton of Green Bay, Lake Michigan. Proc. l7th Conf. Great Lakes Res. 1974:28–51.

GAULD, D. T. 1951. The grazing rate of planktonic copepods. J. Mar. Biol. Assoc. U. K. 29:295–706.

HARVEY, W. H. 1937. Note on selective feeding by *Calanus*. J. Mar. Biol. Assoc. U. K. 22:97–100.

HUTCHINSON, G. E. 1967. Treatise on limnology, vol. 2. Wiley.

LAM, R. K., and B. W. FROST. 1976. Model of copepod filtering response to changes in size and concentration of food. Limnol. Oceanogr. 21:490–500.

LEHMAN, J. T. 1976. The filter feeder as an optimal forager, and the predicted shapes of feeding curves. Limnol. Oceanogr. 21:501–516.

McQUEEN, D. J. 1970. Grazing rates and food selection in *Diaptomus oregonensis* (Copepoda), Marion Lake, British Columbia. J. Fish. Res. Bd. Can. 27:13–20.

MAKAREWICZ, J. C., and G. E. LIKENS. 1975. Niche analysis of a zooplankton community. Science 190:1000–1003.

MULLIN, M. M. 1963. Some factors affecting the feeding of marine copepods of the genus *Calanus*. Limnol. Oceanogr. 7:239–249.

NIVAL, P., and S. NIVAL. 1976. Particle retention efficiencies of an herbivorous copepod, *Acartia clausi* (adult and copepodite stages): Effects on grazing. Limnol. Oceanogr. 21:24–38.

O'CONNERS, H., L. F. SMALL, and P. L. DONAGHAY. 1976. Particle size modification by two size classes of the estuarine copepod *Acartia clausi*. Limnol. Oceanogr. 21:300–308.

PAFFENHOFER, G. A. 1971. Grazing and ingestion rates of nauplii, copepodids, and adults of the marine planktonic copepod *Calanus helgolandicus*. Mar. Biol. 11:286–298.

PARSONS, T. R., and H. SEKI. 1970. Importance and general implications of organic matter in aquatic environments, p. 1–27. *In* D. W. Hood [ed.] Organic matter in natural waters. Inst. Mar. Sci. (Alaska) Occas. Publ. 1.

PORTER, K. G. 1973. Selective grazing and differential digestion of algae by zooplankton. Nature 244:179–180.

——. 1977. The plant-animal interface in freshwater ecosystems. Am. Sci. 65:159–170.

POULET, S. A. 1973. Grazing of *Pseudocalanus minutus* on naturally occurring particulate matter. Limnol. Oceanogr. 18:564–573.

——. 1974. Seasonal grazing of *Pseudocalanus minutus* on particles. Mar. Biol. 25:109–123.

——, and P. MARSOT. 1978. Chemosensory grazing by marine calanoid copepods (Arthropoda:Crustacea). Science 200:1403–1405.

RICHMAN, S., D. R. HEINLE, and R. HUFF. 1977. Grazing by adult estuarine calanoid copepods of the Chesapeake Bay. Mar. Biol. 42:69–84.

——, and J. N. ROGERS. 1969. The feeding of *Calanus helgolandicus* on synchronously growing populations of the marine diatom *Ditylum brightwelli*. Limnol. Oceanogr. 14:701–709.

RUBENSTEIN, D. I., and M. A. R. KOEHL. 1977. The mechanisms of filter feeding: Some theoretical considerations. Am. Nat. 111:981–993.

STEELE, J. H., and B. W. FROST. 1977. The structure of plankton communities. Phil. Trans. R. Soc. Lond. Ser. B. 280: 485–534.

WILSON, D. S. 1973. Food size selection among copepods. Ecology 54:909–914.

21. Grazing Interactions in the Marine Environment

Percy L. Donaghay

Abstract

The physical properties of copepod filtering appendages suggest an increasing filtering capacity for large versus small particles. This prediction agrees with laboratory data for some species, but does not agree with field experiments or laboratory experiments for other species. The expected results for *Acartia clausi* are shown to be modified by previous feeding history, presence of nonfood particles, and food quality. Important interactions exist between each of these factors. Alteration of patterns of filter movement, combing rejection, and postcombing rejection are considered as possible mechanisms for modification of expected filtering patterns. An experiment with food and nonfood particles of identical size is used to demonstrate that *A. clausi* can postcomb reject unwanted particles.

Careful study of the morphology of the feeding appendages of copepods has shown that pore size formed by the setae and setules are approximately normally distributed (Nival and Nival 1973, 1976). Filtering appendages can be used to directly capture particles (as in the raptorial feeding behavior for *Acartia*: Conover 1956) or to act as a concentrating device without direct contact (as for the cladoceran *Daphnia*: Porter unpubl.). Regardless of whether particles are trapped on the filter or only concentrated by it, a filtering efficiency curve can be calculated from the variance in setule spacing (Fig. 1a) (Nival and Nival 1973, 1976). These filtering-efficiency curves can be used to predict relative filtering rates as a function of particle size.

Such expected filtering rate-particle size relationships may also be derived experimentally, as e.g. by Frost for *Calanus pacificus* (Frost 1972). These curves are developed by determining filtering rates over a variety of concentrations and for a variety of different-sized particles. The maximum filtering rates (observed at low concentrations) provide an estimate of the relative efficiency of filtering for those sized particles. These maximum filtering rates can then be plotted versus particle size to form a filtration efficiency curve (Fig. 2). Such experimentally derived relationships most accurately reflect only the mechanical properties of the filter when the different-sized particles used are of equal food quality and identical particle shape (nearly spheroid in shape). These conditions can be met by using size clones of a single species of algae as was done by Frost (1972).

The strictly mechanical properties of the filter will result in higher filtering and in-

---

Supported by NSF grants OCE 76–10347 and OCE 78–08635. I thank S. Schnack, R. D. Leatham, and B. L. Dexter for technical assistance and discussions.

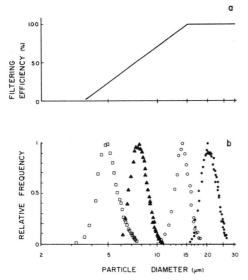

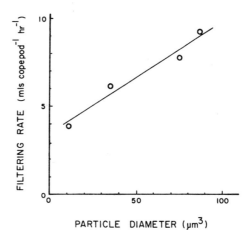

Fig. 2. Experimentally derived filtering vs. particle size for *Calanus pacificus* (redrawn from Frost 1972).

Fig. 1 a. Particle selectivity curve for *Acartia clausi* (redrawn from Nival and Nival 1976). b. Size distribution of the algae *T. pseudonana* (□), 8-μm spheres (▲ ranging in size from 6 to 11 μm), *T. fluviatilis* (○), and 20-μm spheres (● 95% >15 μ) showing relative position of particle-size peaks relative to theoretical Nival function. Size distribution peaks are relative volume distributions of particles used in experiments. Note separation between peaks (redrawn from Donaghay and Small, in press).

gestion rates for larger particles, providing the particles are of equal quality. It should be noted that this is a purely passive selection and requires no behavioral response by the copepod. Hereafter I shall term this "passive selection" to distinguish it from "active selection." Any deviation of the observed filtering responses from expectations based on the physical properties of the filter requires some behavioral response on the part of the animal and is thus "active selection." As has been pointed out by Boyd (1976), Lam and Frost (1976), and Lehman (1976), passive selection can be used to explain apparent selection for larger particles over smaller ones. This is true for inert spheres (Wilson 1973) or for foods of equal value (Frost 1977). It can also be shown to be true for *Acartia clausi* when fed two different-sized species of *Thalassiosira* (Fig. 3).

Recent experiments in our laboratory

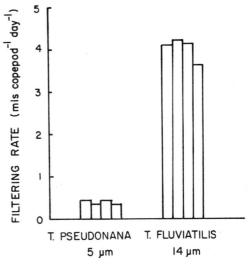

Fig. 3. Filtering rates for *Acartia clausi* fed a mixture of 14-μm *Thalassiosira fluviatilis* and 5-μm *T. pseudonana*. Bars represent separate replicate experiments.

(Donaghay and Small, in press) and field experiments (Richman et al. 1977) clearly demonstrate that the results expected from the mechanical properties of the filter can be strongly modified by the type of food spectra, the animals' previous feeding history, and by presence of nonfood particles. Richman et al. (1977) have demonstrated that filtering curves can be altered by the types of natural particle spectra offered. The responses reported by them range from generally increasing filtration rates with increasing size (as expected from passive selection) to feeding in only very narrow sections of the particle spectra. It must be assumed that these patterns are the result of differing food quality associated with individual sections of the particle spectra and of the resultant active selection by copepods for certain food types.

Previous feeding history can also be shown to modify filtering responses for *A. clausi*. Donaghay and Small (in press) fed *Acartia* females on equal concentrations of *Thalassiosira pseudonana* (5-μm) and *Thalassiosira fluviatilis* (15-μm) (see Fig. 1b for sizes relative to the filtering function). Both species are small, single-celled centric diatoms. The animals were preconditioned on each food for 4 days to ensure that the copepods would be fully acclimated to those foods and so that any alteration of setal spacing (as suggested by the model of Wilson 1973) would have occurred. The ingestion response was then examined on both foods separately and together (Fig. 4). Figure 4a shows that ingestion of the small cell occurred only with *T. pseudonana* preconditioned animals, but animals with both preconditionings ingested the larger cell, *T. fluviatilis*. These results demonstrate that previous feeding history can modify the expected responses of copepods based on filter structure alone. These results are also consistent with the setal spacing alteration mechanism suggested by Wilson (1973).

The presence of nonfood particles was also shown by Donaghay and Small (in press) to modify expected mechanical filtering responses. After performing the above experiment, inert latex spheres of 8-μm size were added to the food suspension of 5- and 15-μm *Thalassiosira* cells (Fig. 1b). This experiment tested the hypothesis that alteration of setal spacing was the animals' only mechanism for rejecting unwanted particles. In all cases spheres were avoided.

However, the feeding patterns showed a clear interaction with previous feeding history (Fig. 4b). Animals preconditioned on the large food avoided spheres by not ingesting either spheres or *T. pseudonana*. Those preconditioned on small cells ingested both food species but no spheres. Since altered setal spacing cannot explain the latter behavior, the *T. pseudonana* preconditioned animals must have rejected the spheres after capture on the filter using a postcapture rejection mechanism. This experiment demonstrates that *A. clausi* has a postcapture rejection mechanism and that presence of nonfood particles strongly interacts with previous feeding history to modify responses based on mechanical properties of the filter.

More recently Poulet and Marsot (1978) have demonstrated that food quality can alter the expected filtering pattern. They offered microencapsulated particles containing either phytoplankton- or a nonphytoplankton-derived material. The animals repeatedly selected the phytoplankton-containing particles regardless of whether they were the larger or smaller sized particles offered.

Both the results of Poulet and Marsot (1978) and Donaghay and Small (in press) can be interpreted as evidence for either of two types of postcapture rejection. Donaghay and Small (in press) suggested that their data could be explained either by the selective removal of particles of a given size by the multiple entry and exit of the combing appendage along the filter or by particle-by-particle postcombing rejection (after removal of those particles from the maxillae). The first type of rejection, termed "combing selection," is possible because the variance in pore size is oriented along the filter axis and allows the animal to select a certain sized particle by combing at the appropriate place along the filter (Donaghay in prep.).

In order to test these two hypotheses and to further investigate how expected filtering curves could be modified, a rejection experiment was run. Both hypothesized mechanisms for particle rejection will have identical results except when

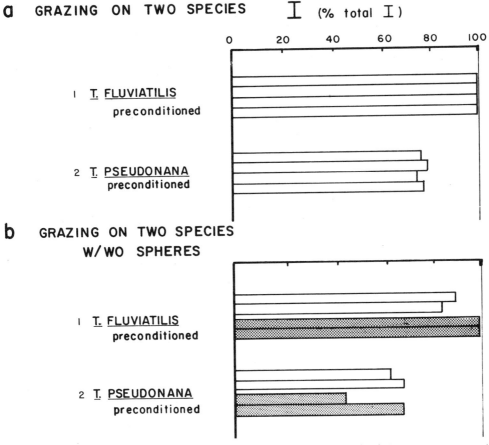

Fig. 4 a. Effects of preconditioning on ingestion behavior of *Acartia clausi*. Bars represent percent of total ingestion represented by ingestion of *T. fluviatilis* calculated as

$$I_{T.f.} \% = \frac{\text{Ingestion rate on } T.f.}{\text{Ingestion rate on } T.p. + \text{ingestion rate on } T.f.} \times 100.$$

a1—*T.f.* preconditioned animals grazing on *T. pseudonana* plus *T. fluviatilis*; a2—*T.p.* preconditioned animals grazing on *T. pseudonana* plus *T. fluviatilis*.

b. Grazing on two food peaks with or without spheres present. Two replicate flasks without spheres are open bars; two replicate flasks with spheres represented by stippled bars. b1—*T.f.* preconditioned animals; b2—*T.p.* preconditioned animals. (Redrawn from Donaghay and Small in press).

food and nonfood particles of identical size are offered together. Under these conditions, both foods and nonfoods will occur at similar positions along the filter and therefore size-based rejection will not be possible. Thus, rejection of the inert spheres in this situation will be a clear demonstration of a postcombing rejection mechanism. To test the hypothesis that combing rejection is the only mechanism, six replicate groups of 20 *A. clausi* females each were fed a mix of 15-μm *T. fluviatilis* and 15-μm spheres. The relative sizes and abundances of these particles (Fig. 5) were such that the sphere distribution completely overlapped the food distribution. As a result, spheres were equal to or greater in number than the food in all size classes. After 1, 2, and 24 hours, two groups of animals were removed from the grazing chambers. The animals were preserved and prepared for gut content analysis using the methods of Schnack (1975). The feces were collected on 35-μm Nitex and preserved, and the number of spheres per feces were counted on an inverted microscope. At the end of 24 hours, grazed and control suspensions were counted and sized with a Coulter Counter (after Donaghay and Small in press) and filtering rate curves were calculated using the equations of Frost (1972). When the resulting filtering curves are compared to those obtained with pure *T. fluviatilis* suspensions (Fig. 5b and d), it is clear that the presence of nonfood particles altered the filtering function from an increasing function with size to a uniform rate at all sizes. In a similar experiment with food particles also present at a smaller size (in addition to the food-sphere mix at 15 μm), the filtering curve became maximal at the smallest sizes (Fig. 5e and f). In this case, the observed filtering rate is a decreasing function of particle size. Thus, the presence of nonfood particles in the same size range as a food will result in significant modification of the mechanistic filtering function.

The above filtering curves tell us only that the observed filtering pattern was modified. They tell us nothing about whether the nonfood particles were ingested, and nothing about the mechanism involved. In order to answer these questions, the numbers of spheres in the animal guts and in the feces were counted and the total number of spheres ingested per flask was calculated. These totals were divided by the number of animals per flask to determine the number of spheres ingested per animal per day (Table 1). Based on the 24-hour data, each animal ingested an average of 101 spheres per day. This number is very small when compared to the total number of particles ingested per day (22,000 per animal per day). Also, recall that the food and spheres were equally abundant, and thus approximately equal numbers of

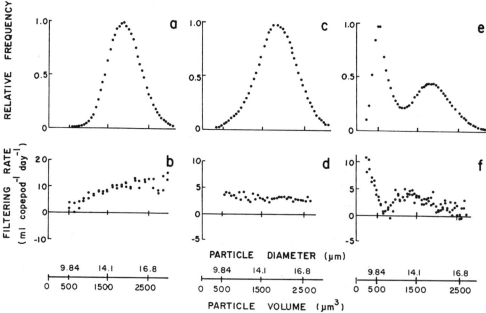

Fig. 5. Varying types of filtration curves (b,d,f) derived from *Acartia clausi* grazing on particle distribution composed of pure 15-μm *Thalassiosira fluviatilis* (b); approximately equal numbers of 15-μm inert plastic spheres (d) and a mix of 11-μm *T. fluviatilis*, and the same equal number mix of 15-μ *Thalassiosira* and 15-μ spheres as in d (f). Relative frequency (relative particle number distributions) for each of these experiments is shown directly above filtering curves of a, c, and e respectively. Circles represent means of four replicate experiments for d and individual pairs of replicates for b and f.

Table 1. *Acartia clausi* grazing on particle mixes of *Thalassiosira fluviatilis* and spheres.

Replicate	Particle ingestion (per copepod/day)	Sphere ingestion (per copepod/day)
Particle mix 1*		
1	19,560	143
2	22,089	94
3	29,149	95
4	21,755	73
Mean	23,138	101

Error rate: $\dfrac{\text{of spheres ingested}}{\text{of spheres handled}} = \dfrac{101}{24,063} = 0.42\%$

(1 sphere ingested per 238 handled)

Particle mix 2†		
1	10,357	79
2	9,328	47
3	11,021	82
4	11,631	59
Mean	10,584	67

Error rate: $\dfrac{\text{of spheres ingested}}{\text{of spheres handled}} = \dfrac{67}{11,132} = 0.6\%$

(1 error in 166 encounters)

*Particle mix 1: 11-μm cells, 0; 15-μm cells, 2,456; 15-μm spheres, 2,629.
†Particle mix 2: 11-μm cells, 2,358; 15-μm cells, 1,302; 15-μm spheres, 1,410.

both food and spheres should have been eaten. The difference between the observed and expected results shows a clear postcombing rejection capability for *Acartia* with an error rate of only 1 in 200. This is a rather impressive capability. The filtering rate data, particularly from the two peak experiments, indicates that although *Acartia* can reject unwanted particles on a one-by-one basis, it also responds to the presence of a nonfood by shifting its filtering activity to other parts of the particle spectrum.

The above demonstration of a particle-by-particle postcapture rejection capability does not disprove the use of alteration of setal spacing, combing rejection, or other behavioral actions as mechanisms for modification of feeding patterns. However, since postcombing rejection is a more sophisticated mechanism, visual observations will be required to confirm these other mechanisms, if they exist. Visual observations of a variety of copepods feeding in different types of particle suspensions suggest that the way in which the filter is used to remove particles from the water may be altered in response to changes in that particle spectra. For example, after feeding on a suspension containing both spheres and food, *A. clausi* stops using a seining motion (Conover 1956) and switches to very brief bursts of feeding activity. These changes in filter motion must almost certainly affect the way particles are captured and handled and thereby must affect the observed filtering curves.

In conclusion, it appears that by using a variety of behavioral tools *A. clausi* can radically modify the filtering behavior expected from the morphology of the filter. The observed feeding patterns on complex particle mixtures are strongly controlled by food quality and preconditioning. In our efforts to understand the capabilities of *A. clausi*, it has become apparent that the Coulter Counter is but one of several powerful tools needed to solve grazing problems. Careful study of feeding appendage morphology, gut content, fecal pellet analysis, and behavioral observation all play important roles. Their combined use, where appropriate, is often more powerful than any one alone.

Care should be taken in extrapolating these results to all species. Work in our laboratory with *Eurytemora* and *Calanus marshallae* has shown that while *Eurytemora* has many of the capabilities that *Acartia* has, *Calanus* appears to be much more limited in the extent to which behavior may be used to modify filtering patterns. As a result, much more work is needed before we can attempt to develop a universal feeding model. The differences observed between species may be an important element in controlling population dynamics of these species and in generating observed community structures.

References

BOYD, C. M. 1976. Selection of particle sizes

by filter feeding copepods: A plea for reason. Limnol. Oceanogr. 21:175–180.
CONOVER, R. J. 1956. Oceanography of Long Island Sound, 1952–1954. VI. Biology of *Acartia clausi* and *A. tonsa*. Bull. Bingham Oceanogr. Collect. 15:156–233.
DONAGHAY, P. L., and L. F. SMALL. In press. Food selection capabilities of the estuarine copepod *Acartia clausi*. Marine Biology.
FROST, B. 1972. Effects of size and concentration of food particles on the feeding behavior of the marine planktonic copepod *Calanus pacificus*. Limnol. Oceanogr. 17:805–815.
———. 1977. Feeding behavior of *Calanus pacificus* in mixtures of food particles. Limnol. Oceanogr. 22:472–491.
LAM, R. K., and B. W. FROST. 1976. Model of copepod filtering response to changes in size and concentration of food. Limnol. Oceanogr. 21:490–500.
LEHMAN, J. T. 1976. The filter-feeder as an optimal forager, and the predicted shapes of feeding curves. Limnol. Oceanogr. 21:501–516.
NIVAL, P., and S. NIVAL. 1973. Efficacité de filtration des copépodes planctoniques. Ann. Inst. Oceanogr. 49:135–144.
———, and ———. 1976. Particle retention efficiencies of an herbivorous copepod, *Acartia clausi* (adult and copepodite stages): Effects on grazing. Limnol. Oceanogr. 21:300–308.
POULET, S. A., and P. MARSOT. 1978. Chemosensory grazing by marine calanoid copepods (Arthropoda: Crustacea). Science 200:403–405.
RICHMAN, S., D. R. HEINLE, and R. HUFF. 1977. Grazing by adult estuarine copepods of the Chesapeake Bay. Mar. Biol. 42:69–84.
SCHNACK, S. 1975. Untersuchungen zur Nahrungsbiologie der Copepoden (Crustacea) in der Kieler Bucht. Ph.D. thesis, Univ. Kiel. 144 p.
WILSON, D. S. 1973. Food size selection among copepods. Ecology 54:909–914.

22. Catching the Algae: A First Account of Visual Observations on Filter-Feeding Calanoids

Miguel Alcaraz, Gustav-Adolf Paffenhöfer, and
J. Rudi Strickler

Abstract

The phenomenon of reinventing the wheel occurs regularly when one does research in a manner contrary to the rules of the National Science Foundation. About 80 years ago researchers started a very lively debate on the nutrition and mechanics of the food gathering of zooplankters. Neither the views of Cannon (1928) nor the opposing ones of Storch (1929) could solve the problem of how copepods filter algae from the water and what constitutes a filter in the first place. Storch concluded, "So setzen wir unsere Hoffnungen auf Mikro-Zeit-Lupenaufnahmen." However, his camera could only film at 70 frames per second and had only a "5 Ampere-Bogenlampe." Why couldn't this early work tell us how copepods filterfeed? I will address this question only with a few remarks and leave its solution to some multilingual scholar in the history of biology.

On Tuesday evening, 22 August 1978, there was a special session of this conference devoted to movies. First R. Zaret showed a film made by the National Committee for Fluid Mechanics Films on "Low Reynolds Number Flows," starring Sir Geoffrey Taylor. After some discussion and a short break I presented films on the filter feeding of *Eucalanus crassus*. These films were made at the University of Ottawa in July 1978 with the aid of three grants from the National Research Council of Canada and the help of M. Alcaraz and G. Paffenhofer. I concentrated on the technical aspects of filming, while Alcaraz specialized in gluing the animal on a doghair and Paffenhofer cultured the algae and animals.

This report is a very personal account of the thoughts and experiences we had; detailed research reports of the findings will be published elsewhere. I will also outline the optical techniques we applied and comment on various publications on the topic of filter-feeding mechanisms (without necessarily claiming that my thoughts are correct or complete).

The Problems

The problems of visually observing the processes involved in filter feeding can be broken into two subsets. Figure 1 illustrates the point in a caricature by D. Irvine, a fluid dynamicist from The Johns Hopkins University. On one side there is the scientist who tries to play with the animal—he uses different approaches—which should, however, not be too far offtrack. The imaginary batter represents all the odd facts and circumstances which try to interrupt the game. I will first elucidate a few problems the animal has and then deal with the ones of the researcher. It is clear that an animal that is in an entirely strange environment cannot and will not perform the same way as in the depths of an ocean. The question is, therefore, what is the smallest ocean the scientist has to provide for the filter-feeding copepod without telling the animal that it is really contained in a vessel within a laboratory.

In trying to answer this question we begin with evaluating a few results from

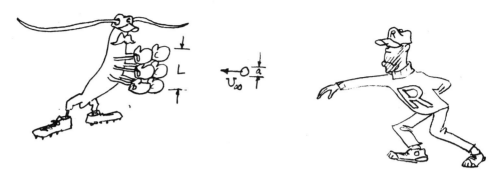

Fig. 1. Caricature of a researcher in the topic of "filter feeding": a modeler's approach (by D. Irvine, The Johns Hopkins University, 1975).

research done on the filtering rate of *Calanus helgolandicus* on *Prorocentrum micans*. Paffenhöfer (1976) found that at a food concentration of 14 cells/ml, the copepod swept clear 1,600 ml per day. The astonishing fact is that the copepod could use only 1.6×10^{-7} of the volume that was moved by the mouthparts. The animal's problem seems to be how to get rid of all the useless water rather than how to find the little part of value. If the water is recirculated without enhancing the chance of capturing more algae it would obviously be of no value. Two components help the copepod find this small part in the water volume: first, a sensory system that can detect living organic material of small size in water within reach of the animal; and second, a mechanical system that allows the animal to scan a large volume of water. The characteristics of this latter system are dependent on the sensory mode, even if sensing is lacking. In the latter case the best system would be a sieve where all the water gets sucked, or pushed, through. If there is a sensory system, however, there should be scanning movements rather than sieving ones, and the better the sensory system the more the movements will resemble a searching pattern. Sieving water costs energy and the resistance of the water passing through the sieve depends on the mesh size and on the relative speed of the water molecules with respect to the sieve. A vacuum pump is needed to filter lake water through a 5 µm membrane filter (it takes less time if the filter does not clog due to the particles retained). Because the animals are very small, so too are their appendages and their food. Water is, for them, a quite viscous environment and it therefore costs a considerable amount of energy to press a large amount of water through a small rigid filter. On the other hand, the animals have several appendages, including ones which groom the filter and transport the particles retained therein to the mouth. It would be interesting to study the conditions of having a sensory input only at the level of grooming and thereafter—this probably depends on the encounter rate of the scanning system with the algae. For example, the above-mentioned animal encounters an alga about every 4 seconds, a rate which is very different than for *Daphnia*, where 20 cells per second is a reasonable approximation. It seems to me that to enclose the pumping and sieving system in a housing to avoid damage done by predators also has its drawbacks. It must cost the *Daphnia* more energy to scan the same volume of water than it costs copepods, as there is the additional viscous drag from the carapace. In an environment with high concentrations of food this might not be a serious problem when compared with the advantage of a better filtering system. In oceans, however, with low algal concentrations, this could be a limiting factor for the dispersion of the cladocerans.

The above remark about recirculation is of some interest. In all descriptions of the filtering movements, strong currents are mentioned. The water comes from the space between the maxillipeds and the first pair of legs, flows between the first and second maxillae toward the labrum, and leaves this space given by the filtering walls of the maxillae through the gaps in the mesh. Most accounts indicate that the same water will now be recirculated to where it started. These eddies—one on each side of the animal—are supplemented with two larger eddies which move near the animal as they go from the second antennae toward the maxillipeds and outward in circles back to the second antennae. Cannon (1928) first described these currents (p. 137–138 show the results of many observations). Storch and Pfisterer (1926), Storch (1929), Lowndes (1935), and Marshall and Orr (1972) agreed in general with this description. However, if the animals fed using this pattern they would not be able to get enough food! They would get only the odd alga entering the eddies and all of the algae growing in these small water masses. Cannon himself (p. 141) points toward a possible solution. The maxillipeds and the second antennae move rhythmically apart and together, forming a pumping mechanism similar to the wings of a butterfly. The above-mentioned researchers all observed their animal in a small drop of water (e.g. in a depression slide), and, given the pumping mechanism, the water had to be recirculated. Obviously that ocean was too small. An eddy then only makes sense when there are algae in it that the animal "wants" to catch. I will come back to this point in discussing our results.

I have clearly established (Strickler 1975) that copepods sense the displacement of water by other zooplankters, fish, and the pipettes of researchers. It is a vibration if I consider one phase as sufficient—an encounter with another zooplankter would thus "sound" like a click. A copepod can in this way perceive not only swimming particles but also nonswimming ones like the walls of a vessel. Because of the wall effect, water near the wall is less easily displaced by the swimming animal than water far from the wall (R. Zaret at this conf.), and thus copepods can sense even the slowing of the filtering current by a boundary. There is a threshold for every sensory mode. Viscosity slows the currents and thus transfers the kinematic energy to heat. Therefore, at a certain distance from a feeding animal a wall will not interfere with that animal.

I first set the distance between the animal and the nearest wall arbitrarily at 2 cm, and later at 1 cm (about 8 body lengths). In both cases (for one animal) I used 120 ml of fresh unfiltered seawater and the animal was at least 5 cm from the bottom of the vessel—the water was not stirred in order to prevent fecal pellets from entering the filtering currents.

For some zooplankters this ocean would still be too small. *Eucalanus crassus* can spend a long time in 120 ml of water until it bumps into a wall, as this species does not cruise like *Labidocera* spp. or *Epischura* spp. However, one might wait a long time for the animal to swim into focus.

We "crazy-glued" the end of a fine doghair to the fourth thoracic segment of a copepod. The hair was about 1 cm long and ended in locking forceps which were mounted on a micromanipulator. If the animals made a swimming movement with their legs, they could jump a normal distance and then slowly be put back in focus; even while feeding they moved a little against the pull of the hair. The bond between the hair and the dry chitin occurred quickly and in order to keep the animal from drying out, it was placed on wet filter paper. All tethered animals used for our studies survived more than a week, and some even for months. Daphnids molt too often to be kept for such a long time on the same hair. All animals had to be checked for debris and algae sticking to their skeleton and cleaned by blowing water over their bodies.

This method cannot be used to accurately study the animal's filtering currents however, since the animal normally moves against a current. An alternative pattern is shown by *E. crassus* which "flicks" with its abdomen and remains for a long time within the same area. We were most interested in the food handling of algal chains by the mouthparts and in this case the disadvantages of the tethering are not too serious.

Films done on free-swimming animals show the same frequencies and rhythms as in animals tethered to the doghair.

The frequency of the movements of the mouthparts reaches 60 Hz, and to observe the paths they describe, the camera had to be run at 500 frames per second. The exposure time is given by the dimensions of the objects which are to be observed. If a seta with a diameter of 5 μm moves during the exposure more than ¾ of its diameter it will be indistinguishable against the background. Dark-field illumination does not give enough energy to film at such high speeds. As a result of these limitations, I chose an exposure time of 111 μsec and thus could resolve about 3 μm for slow-moving appendages and 5 μm for moving algae.

With a distance of 1 cm between the animal and the nearest wall I had to use a Luminar lens from Carl Zeiss (25 mm) with a numerical aperture of 0.15. This would give me a theoretical resolution of a little better than 2 μm. The animal was illuminated by a long working distance condenser with a numerical aperture of 0.32 (Carl Zeiss, Phako IV Z7). For a light source, I used an Osram ZBO 75 W/2 Xenon high pressure burner in a Carl Zeiss housing with a highly stabilized power supply (many thanks to D. Brown for its use). Two interference filters (670 nm and N. D. = 1.00) kept the animal from overheating. The films were done at the full energy of the lamp with a light green absorption filter in order to enhance the structures of the mouthparts. One of two oculars from Carl Zeiss (Mipro 125 mm and 63 mm Kpl) formed the final magnification. A Locam high-speed camera (Redlake) with built-in beam splitter was mounted on a Cambo stand, and an additional beam splitter with a photocell (Carl Zeiss) was used to control the light level and allowed a second observer. Films were taken only when both observers agreed that the situation was perfect. I used Eastman Ektachrome video news film No. 7250 with a speed of 400 ASA, and in this way the turnover time was less than 2 hours, given the news-generating environment of a capital city. Lowndes (1934) remarked that high-speed filming is expensive (3 pence a foot), a fact which has not changed over the years—100 feet of film resulted in about 3 sec of good film. The algae were distributed in the water at natural densities just before the filming. The whole optical setup had a horizontal optical axis and was mounted on a Newport Research optical table (Fig. 2) floating on a nitrogen cushion. Between the ocular and the front lens of the Zeiss beam splitter there was a 0.1–cm air gap. In this way there were no vibrations from the camera nor from other sources that could have interfered with the animal and the filming.

Studying filter feeding that uses intermittent illumination does not work well. First, only oscillatory movements can be slowed down, and second, Cannon (1928, p. 132) has already pointed out that the "limb movement is never absolutely regular." Since moving algae do not oscillate, a chopped light source does not help in observing them. These are some of the problems and the means by which I solved them. Needless to say, I perceive this solution as no more than the most convenient and not necessarily as the best.

We were interested in the handling of algal chains by *E. crassus*, and we also wanted to see if the second maxillae formed a rigid wall in order to function as a sieve. For example, consider these questions: Why do second maxillae have so many joints (*see* scanning micrographs: Friedman 1977)? Why doesn't the water go outside the filter chamber to the second antennae, as it would in this way encounter less resistance? How does a chain of algae get broken into smaller parts, and how do cells get between the mandibles through a suction or a stuffing mechanism?

The sensory mode

If a female calanoid copepod ingests more food, it will produce more offspring. Thus there will be a selective pressure favoring a copepod that can capture more algae and, consequently, those that can scan a larger volume of water. It can forage in two ways: first, it can increase the relative speed

Fig. 2. Optical set up. Xe—Xenon light source; I—interference filter; Mi—micro-positioner; Co—condenser; V—vessel with tethered animal; Ob—objective; Oc—ocular; 1B—first beam-splitter; 2B—second beam-splitter; L—Locam camera; St—camera stand; N-C—nitrogen cushion.

between the algae and itself, and second, it can scan a larger cross section per time. Pumping more water through the capture area (e.g. a sieve) costs increasingly more energy, and fluid mechanical and structural reasons will set an upper limit to this process. On the other hand, the animal could enlarge the cross section. This will still mean an increase in energy expenditure, but at larger volumes this solution would be cheaper in terms of energy and structural modifications. Additionally, there would be a much higher theoretical upper limit to the process. For copepods it would be impossible to structurally enlarge the cross section beyond half a body length, as the animal would then become quite vulnerable to predation. Salps and appendicularia did, however, develop in this direction. The introduction of a sensory system serves just as well to increase the cross section, except that the animal now needs structural and/or behavioral modifications to capture a "sensed" alga. The situation resembles that of a predator-prey interaction, including a chain of events with their conditional probabilities (Gerritsen and Strickler 1977).

What is the sensory mode and what are the structural and behavioral modifications? Of the three possible sensory modes (visual, chemical, and mechanical), only the chemical can perform this function. To perceive an alga visually at a distance of 500 μm necessitates two eyes with high spatial and temporal resolution. Furthermore, the animals would have to feed in an illuminated environment, which would also lead to disadvantageous interactions with fish. An alga could be detected as a disturbance in the water, mechanoreceptors sensing the boundary layer around it moving with the current (wall-effect). In the case where the alga is capable of movement, the displacement of the water could be perceived; however such a sensory system would have to be very sensitive and all other water movements would have to be compensated for. This mode is not impossible, yet the use of chemical cues is easier and has an additional advantage—not only can the animal sense the presence of a particle, but also it can distinguish between particles of value and debris.

Any alga will leak some chemicals, and there will be an active space (Wilson 1970) around it which is elongated in the direction of the current bringing the alga closer to the copepod. Friedman and Strickler (1975) described chemoreceptor sites at the setae of the mouthparts of calanoid copepods. The setae themselves are the sensillae, with pores of about 10-nm diameter. This multiple receptor system of the copepod allows it to detect the different algae according to their size, shape, and the amount of chemicals they leak. Additionally, algae of different species and age will "smell" differently. Chemoreception here is analogous to hearing in humans—the difference between Brahms and Haydn does not depend on the presence of a certain frequency.

Research on the filtering rate of copepods on algae has shown that chemoreception is indeed used. Early in this century Esterly (1916) found that sometimes the animals ingest carmine particles immediately and sometimes not at all. Donaghay (at this conf.) showed a behavioral response of the animals toward a change of the food quality. Poulet and Marsot (at this conf.) used artificial particles with different olfactory characteristics, and the animals fed accordingly. Does this system enhance the capture of an object of interest, or does it serve only to reject particles of no interest?

The capture of an alga

Friedman (1977) concentrated on the structures of copepod mouthparts and without such a detailed survey it would be impossible to evaluate a film on filter feeding. His micrographs show clearly that the mouthparts have many joints. Different pictures of the same appendage reveal the possible range of the movements and the dynamic situation can be described as follows.

In a plane perpendicular to the axis of the animal at the height of the first maxillae we could draw contour lines of capture probabilities—the area between the maxillae and the body itself having the highest value. If an alga is on the symmetry axis or not far from it, the chances of its being captured at a certain distance are higher than if it is at the same distance but more to the left or right. If the animal now senses the presence of an alga, it should alter the contour lines in such a way that the alga lies within an area of high capture probability. What the animal has to do is "shovel" the water between the maxillae and the alga in order to get the alga closer to the maxillae. The second maxillae should touch each other, and when the alga is close enough, they should open like the wings of a butterfly and suck the alga into the space between the setae. Closing this space carefully ensures that the alga does not get washed out again.

Our films show exactly this pattern. Every mouthpart can move independently of the other. If there are no algae within sensing range (= encounter radius), the animal maintains a flow of water passing through this space. If it perceives the presence of an alga, it changes the movements and uses the viscous forces to bring the alga within reach of an appendage. Sometimes this happens so perfectly that this appendage (it could be any from the second antenna to the maxillipeds) can get hold of the alga with its setae and bring it chopstick style in front of the mouth. Most of the time, the algae had to be shoveled close to the mouth in between the maxillae. Once there, they were invariably at the mercy of the copepod. One question arising is whether the recognition of a nearby alga by the setae of one mouthpart is signaled through the nervous system to the other mouthparts, or if all appendages work independently, each relying on its own chemoreception to perform well.

The handling of chains

Without diminishing the value of a further account of our findings, I would like to describe a few observations on the way the copepods handle algae once they are between the second maxillae.

An animal can spend some time (up to a second) in turning the alga to a good position so that it can be pushed and sucked between the mandibles. The second maxillae can open and close and consequently swirl the alga to a better position. The endites of the first maxillae help in stuffing the alga onto the tips of the setae of the second maxillae closest to the mandibles. The upper labrum with all its fine setae flips down and puts "the lid on the machine" (=working mandibles).

A colony of algae, like *Rhizosolenia indica* or *Lauderia borealis*, will break in the process of being moved to the mouth opening, only a few cells being ingested in the first round. The mouthparts will immediatly start to recircuiate the water, which contains the remainders of the chain and try to capture all of the cells still at large. We have observed that from an average chain of 12 cells only 4 are successfully captured, the other 8 escaping through the setae of the second maxillae, to be swept away with the filtering current. We have also observed rejections of particles and algae which were already between the setae of the second maxillae in front of the mouth. The second maxillae opened and pushed the alga away, again using the viscous water bond between the mouthpart and the alga.

Most of the algae which were not captured just passed close by the setae of the moving mouthparts, sometimes even through the spaces between them. At each incident, the animal already had some algae in front of its mouth. If, at this moment, the filtering current was set up in such a way that it recirculated the algae, we could not observe this with our methods. Recirculation in this context would make sense, although most researchers who have described such eddies have also given the animals an ample supply of algae.

Rhythms and viscosity

One should keep in mind that for a copepod, water is a viscous environment. Asimov (1966) made one error in his story *Fantastic voyage*. A body of very small size cannot move by using a propeller—it needs a flagella! The molecular structure of the submarine and its passengers had been miniturized, but not that of the surrounding water (otherwise the Brownian motion would not have been a danger). In our case the films prove clearly that there are only viscous forces involved and hardly any inertia (*see* Purcell 1977). When the mouthparts stop, so does the movement of the alga—the latter does not coast. This is an advantage to the animal, as it can stop the filtering movements and in this way stop the current without losing the alga. The smaller the animal, the more energy, relative to its size, will be used to start the movements again. Small copepods, like *Pseudodiaptomus* spp., freshwater *Diaptomus* spp., and *Temora* spp., do not stop their filtering movements. *Eucalanus crassus*, however, makes about four movements and then stops—this pattern could save energy and better enhance the probability of perceiving the chemical stimuli. The animal does not stop filtering when an alga has been sensed; it will, instead, alter the pattern in order to capture the alga. *Acartia clausi* also stops its movements; however, it uses its first pair of legs to generate the filtering currents.

Remarks

The above ideas and explanantions may not be entirely correct, and further research will undoubtedly increase our knowledge about food gathering. It is not astonishing, however, that copepods might feed in such a way. Most crustaceans have highly diverse behavioral patterns of food handling, and one additional interesting question would be why cladocerans do not exhibit such a fine system. It could be that they gave it up to find more protection from invertebrate predation. It could also be that their filtering system is even more advanced than that of copepods.

The questions arising from our approach to the topic of "filter feeding" are hundredfold. We could not observe very many animals in very many situations, and how the animals ingest small particles, like

Chlamydomonas spp., will be our next target. In addition, exact knowledge of the nature of the propulsion system and of the flow around an untethered animal will allow us to understand better the animal's evolution and the selective pressures in the past and in our polluting times.

We thank all the colleagues who helped and/or challenged us, among them K. G. Porter, R. J. Conover, G. E. Hutchinson, R. Margalef, L. Provasoli, and H. Riessen. Also many thanks to the administration of the University of Ottawa for providing a stimulating atmosphere.

References

ASIMOV, I. 1966. Fantastic voyage. Bantam Press.

CANNON, H. G. 1928. On the feeding mechanism of the copepods, *Calanus finmarchicus* and *Diaptomus gracilis*. Brit. J. Exp. Biol. 6: 131–144.

ESTERLY, C. O. 1916. The feeding habits and food of pelagic copepods and the question of nutrition by organic substances in solution in the water. Univ. Calif. Publ. Zool. 16: 171–184.

FRIEDMAN, M. M. 1977. Electron microscopic studies of the filter-feeding mechanism of calanoid copepods. Ph. D. thesis, The Johns Hopkins Univ., Baltimore.

——, and J. R. STRICKLER. 1975. Chemoreceptors and feeding in calanoid copepods (Arthropoda: Crustacea). Proc. Natl. Acad. Sci. 72: 4185–4188.

GERRITSEN, J., and J. R. STRICKLER. 1977. Encounter probabilities and community structure in zooplankton: a mathematical model. J. Fish. Res. Bd. Can. 34: 73–82.

LOWNDES, A. G. 1934. The polygraphic process. Discovery Rep. 15: 341–344.

——. 1935. The swimming and feeding of certain calanoid copepods. Proc. Zool. Soc. Lond. 3: 687–715.

MARSHALL, S. M., and A. P. ORR. 1972. The biology of a marine copepod. Springer-Verlag.

PAFFENHÖFFER, G. -A. 1976. Continuous and nocturnal feeding of the marine planktonic copepod *Calanus helgolandicus*. Bull. Mar. Sci. 26: 49–58.

PURCELL, E. M. 1977. Life at low Reynolds number. Am. J. Phys. 45: 3–11.

STORCH, O. 1929. Analyse der Fangapparate niederer Krebse auf Grund von Mikro-Zeitlupenaufnahmen. Biol. Gen. 5: 39–59.

——, and O. PFISTERER. 1926. Der Fangapparat von *Diaptomus*. Z. Vergl. Physiol. 3: 330–376.

STRICKLER, J. R. 1975. Intra- and interspecific information flow among planktonic copepods: receptors. Int. Ver. Theor. Angew. Limnol. Verh. 19: 2951–2958.

WILSON, E. O. 1970. Chemical communication within animal species, p. 133–155. *In* E. Sondheimer and J. G. Simeone [eds.], Chemical ecology. Academic Press.

V

Plant-Herbivore Interactions with
Emphasis upon Cladocerans

23. Nutrient Recycling as an Interface between Algae and Grazers in Freshwater Communities

John T. Lehman

Abstract

Based on nutrient budgets and in situ measurements, epilimnetic zooplankton in Lake Washington supply 10 times more P and 3 times more N to the surface-mixed layer during the summer months (June to September) than enters from all external sources combined. The remineralization of nutrients that is mediated by crustacean zooplankton is sufficient to supply a sizable fraction of the daily P and N requirements of the phytoplankton, based on areal rates of primary production. Nutrients that are egested and excreted into the water by the animals are rapidly reabsorbed by the algae; thus ambient pools of dissolved nutrients remain small, whereas turnover rates of the pools are rapid.

Ever since Cooper (1935) demonstrated the rapidity with which nutrients were liberated from enclosed plankton, and Gardiner (1937) showed that much of the release was mediated by the zooplankton, many other workers have tried to estimate nutrient regeneration from zooplankton (e.g. Marshall and Orr 1961; Pomeroy et al. 1963; Martin 1968; Butler et al. 1969; Jacobsen and Comita 1976). Some of these investigators and others have further tried to relate the derived quantities to overall nutrient budgets or to the estimated demands of primary producers (e.g. Harris 1959; Barlow and Bishop 1965; Ganf and Blažka 1974). The studies all agree that contributions by the animals can be substantial and immediate (Cushing and Vucetic 1963; Cushing and Nicholson 1963; Dugdale 1976).

These studies and others amply demonstrated the likelihood that nutrient regeneration by the grazers might represent an important source of inorganic nutrients to freshwater and marine phytoplankton. A complete analysis of the relative contributions made by zooplankton to overall nutrient budgets is not feasible in most lakes, however, because hydrologic budgets and the associated fluvial nutrient fluxes are insufficiently known. Furthermore, the rapid cycling of nutrients, especially P, between grazers and algae complicates attempts to quantify internal fluxes in any basin. In order to identify and measure the fluxes of minor nutrients which may guide the productivity and species composition of lacustrine plankton communities, a rigorous evaluation of nutrient budgets in at least one basin seems essential. This paper reports efforts to quantify inputs of P and N to the surface mixed layer in

---

This paper is part of a Ph.D. thesis submitted to the University of Washington. W. T. Edmondson provided financial support for the study and made available much of his unpublished data on Lake Washington.

Lake Washington (Seattle, Washington) from external sources, from vertical flux across the thermocline, and from remineralization by zooplankton, and then compares those inputs with rates of primary production.

Methods

Rates of input of P and N (PO_4-P, total-P, NO_3-N, total-N) to Lake Washington from fluvial and aerial sources and the standard errors of the estimated fluxes have been computed by Lehman (1978), and those values are used here.

Vertical mass transport across the thermocline was computed from measured concentration gradients and from estimates of vertical eddy diffusion based on the flux gradient method of Jassby and Powell (1975). Following Powell and Jassby (1974), I established an empirical relation between eddy diffusivity and the Brunt-Väisälä frequency calculated from vertical temperature profiles. That empirical relation was subsequently used to estimate diffusivities from the temperature profiles directly (Lehman 1978). Samples for determination of dissolved nutrients were collected at intervals of 5 m (or less) in the region of the thermocline during the summer months (June to September), making it possible to compute concentration gradients across the thermocline (mg m^{-4}). Values for eddy diffusivity (m^2 day^{-1}) were computed as described above for depth intervals corresponding to those of the concentration gradients. By multiplying the gradients by the diffusivities, vertical mass flux (mg m^{-2} day^{-1}) into the epilimnion could be computed for any chemical species measured.

Remineralization of dissolved nutrients as a result of egestion and excretion by pelagic zooplankton was assessed by a series of enclosure experiments, placed in situ in Lake Washington during September 1977. Lake water was collected 100–200 m north of the Evergreen Point Floating Bridge, over 60 m of water. Samples were taken from a depth of 1.5 m by successive casts of a 10-liter Van Dorn sampler, and 20 liter volumes were enclosed in polyethylene bags. Zooplankton assemblages in some of the bags were either decreased in abundance by coarse filtration of the water (130-μm aperture) or were augmented by the addition of a freshly collected sample of animals (130-μm aperture net haul). In some cases, initial concentrations of dissolved inorganic nutrients inside the enclosures were increased by addition of standard solutions of Na_2HPO_4 and $(NH_4)_2SO_4$ (addition of 0.5 μM PO_4-P, 7.5 μM NH_3-N). The bags were then tied tightly closed at the mouth, weighted, and suspended at 1.5 m by mooring them from the open frame of the Webster Point lighthouse at the mouth of Union Bay.

At intervals ranging from 2 to 24 h, the bags were opened, the contents were mixed, and a 500-ml sample was removed for nutrient analysis. Filtration was done in the field as soon as the samples were removed from the bags, and filtered samples were returned to the lab for nutrient analysis. Analyses were usually started within 1 h of collection for both PO_4-P and NH_3-N (methods followed Edmondson 1977). At termination of each experiment, after final samples were collected for nutrient analysis, the entire remaining contents of the bags were filtered through a plankton net of 70-μm aperture, and the retained material was saved for complete enumeration of the zooplankton. When the preserved plankton samples were counted, size distributions of the animals were determined. Only crustacean zooplankton were included in these exercises, because preliminary examinations had revealed that they constituted more than 95% of the animal biomass retained in the 70-μm net samples at this time of year (autumn). To obtain a single index of plankton abundance, I converted the counts to units of dry weight of zooplankton per unit volume of water. The conversion made use of length-weight regressions that have been determined for the *Daphnia* spp., *Diaptomus* spp., and *Epischura nevadensis* in Lake Washington; published regressions were used for *Bosmina longirostris* and *Cyclops* spp. (Bottrell et al. 1976).

Results

Inputs of nutrients to Lake Washington vary seasonally, with the maximal fluxes occurring during the winter, at the time of peak fluvial discharge. The minimum nutrient input to the lake occurs during the summer, when total fluvial discharge is minimal. Mean daily loading rates during the summer periods (June to September) of 1970 to 1977 are listed in Table 1 for several nutrient categories. Before evaluating the probable contributions of these nutrient additions to planktonic production, however, the absolute magnitudes of these inputs must first be compared with other sources of P and N.

Diffusive flux. A second source of dissolved nutrients to the surface mixed layer is mass flux vertically across the thermocline. Eddy diffusivity (K) was calculated for all dates during June–September of 1970–1977 for which vertical profiles of nutrients and temperature had been measured. Concentration gradients across the thermocline were combined with mean values for K in the same interval in order to estimate mass flux on each date. Once firmly established, the thermocline remains at relatively constant depth during the entire summer, so changes in thermocline area need not be considered here. Mean summer values for vertical flux of inorganic N and PO_4-P are shown in Table 2. The apparent decrease in the flux of N from 1970 to 1977 may be partly artificial: before 1974, water samples were collected at intervals of 2 or 3 m near the thermocline; thereafter the interval was increased to 5 m. The net result is that concentration gradients at the thermocline may be somewhat underestimated in recent years. Nonetheless, it is clear from Table 2 that vertical flux of inorganic N (chiefly NO_3-N) is of comparable magnitude to the measured input of NO_3-N from the watershed and airshed during the summer months (cf. Table 1). For PO_4-P, on the other hand, vertical mass flux is negligible compared with loading from sources external to the lake.

Regeneration of P and N by zooplankton. Results from an experiment designed to measure rates of nutrient remineralization by zooplankton in situ are shown in Fig. 1. Abundances of the animals in several of the enclosures were artificially increased, and the accumulations of dissolved nutrients were followed for 30 h. Analyses showed an almost linear accumulation of both NH_3-N and PO_4-P during the entire incubation. The slopes of the individual lines are the observed rates of nutrient regeneration, and they are summarized in Table 3, after being normalized to the abundances of zooplankton in the different enclosures. Thus expressed, the ratio

Table 1. Addition of nutrients (mg m^{-2} day^{-1}) to Lake Washington from all sources external to lake for June to September of years listed.

	NO_3-N		PO_4-P		Total P		Total N	
	Mean	(SE)	Mean	(SE)	Mean	(SE)	Mean	(SE)
1970	3.27*	(0.26)	0.268	(0.013)	0.779	(0.044)	5.78*	(0.32)
1971	4.36	(0.24)	0.303	(0.013)	0.747	(0.036)	9.25	(0.30)
1972	5.69	(0.36)	0.345	(0.017)	0.925	(0.046)	11.65	(0.43)
1973	3.28	(0.48)	0.292	(0.026)	0.675	(0.065)	7.04	(0.62)
1974	5.89	(0.67)	0.300	(0.023)	1.044	(0.072)	10.37	(0.78)
1975	4.13	(0.49)	0.316	(0.022)	0.732	(0.057)	10.09	(0.61)
1976	5.64	(0.59)	0.340	(0.021)	0.832	(0.057)	10.87	(0.69)
1977	5.10	(0.51)	0.406	(0.028)	1.099	(0.082)	11.75	(0.69)

* August to September only.

Table 2. Vertical transport of dissolved inorganic N and P (mg m^{-2} day^{-1}) across the thermocline of Lake Washington by eddy diffusion for June to September of years listed.

	(NO_3 -N + NO_2 -N + NH_3 -N)			PO_4 -P		
	Flux	(SE)	n	Flux	(SE)	n
1970	3.84	(0.68)	6	0.0067	(0.0040)	6
1971	3.25	(0.55)	7	0.0194	(0.0073)	7
1972	2.78	(0.33)	9	0.0016	(0.0011)	9
1973	2.97	(0.66)	8	0.0154	(0.0078)	8
1974	2.22	(0.21)	7	0.0067	(0.0021)	7
1975	2.06	(0.21)	9	0.0147	(0.0058)	9
1976	1.57	(0.21)	8	0.0134	(0.0077)	8
1977	1.56	(0.16)	8	0.0399	(0.0215)	8
1970–1977	2.48	(0.16)	62	0.0148	(0.0035)	62

Table 3. Measured rates of regeneration of NH_3 -N, soluble reactive P (SRP), and dissolved P (DP) for series of enclosures shown in Fig. 1 (Lake Washington, 15–16 September 1977).

Zooplankton mg dry wt liter^{-1}	Regeneration rates µg (mg dry wt)$^{-1}$ day^{-1}			Ratios by weight	
	NH_3-N	SRP	DP	N:SRP	N:DP
1.22	21.8	1.5	3.2	14.4	6.8
2.02	19.9	1.7	2.7	11.6	7.3
3.27	17.3	2.1	3.1	8.4	5.6
4.42	17.8	2.7	3.4	6.6	5.2

of apparent regenerated NH_3-N to PO_4-P decreased markedly with increased abundance of animals in the enclosures. This is principally because rates of accumulation of PO_4-P per unit mass of zooplankton seem to increase with the total abundance of animals.

Any simple interpretation of the results from an experiment like this is complicated by the fact that measured concentrations of nutrients inside the enclosures represent a balance between rates of nutrient remineralization by the zooplankton and the simultaneous rates of uptake of those same nutrients by phytoplankton. The well-documented ability of phytoplankton for rapid uptake and storage of PO_4-P provides a special complication here. Not only will uptake interfere with measurement of absolute rates of nutrient remineralization when rates of the latter are low, but rates of uptake can be expected to increase in magnitude as nutrients begin to accumulate in the water, because uptake is substrate-dependent. If one adds zooplankton to water filtered free of algae, one again invites bias in estimates of remineralization, because the animals process and recycle nutrients more rapidly when they are actively feeding than when they are starving (Conover and Corner 1968; Taguchi and Ishii 1972; Takahashi and Ikeda 1975; Mayzaud 1976; Ikeda 1977). Similarly, drastically increasing abundances of zooplankton above natural levels, as was done in the experiment shown in Fig. 1, carries with it the danger that food supplies may be rapidly exhausted and that the behavior of the animals may change. Furthermore, initial abundances of some of the algae may be changed in the process. Note, for instance, that in

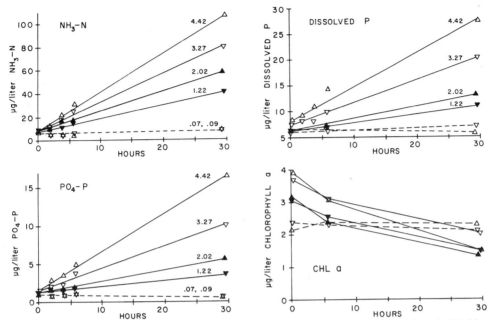

Fig. 1. Concentrations of several substances inside plastic bag enclosures incubated in Lake Washington, 15–16 September 1977. Zooplankton abundances (mg dry wt liter^{-1}).

Fig. 1, the initial quantities of Chl a were elevated in the bags that received zooplankton additions. This was because some large colonies of *Botryococcus* (>150-μm diameter) were collected with the animals in the field. Although the *Botryococcus* almost certainly did not contribute to the food supply available to the zooplankton, owing to its large size, it may have absorbed some of the dissolved nutrients. What one seeks, therefore, is an alternative method that permits an assessment of remineralization rates, corrected for uptake, and which does not require large manipulations of the abundances of either phytoplankton or zooplankton.

One approach to this problem is to make use of the nonlinear response of uptake rates to nutrient concentrations. Whether or not nutrient uptake by phytoplankton is best represented by a rectangular hyperbola (e. g. *see* Brown et al. 1978), it is clear that rates of uptake become saturated when ambient concentrations are sufficiently elevated. In addition, a concentration change of 1 μg liter^{-1} can be anticipated to have much less of a relative effect when it occurs against a background of 10 μg liter^{-1} or more than when background concentrations are of the same magnitude as the change itself. By adding inorganic N and P to all the enclosures initially, the unknown uptake by the phytoplankton can be held relatively constant and independent of nutrient fluxes from the animals during short experiments. Rates of regeneration can then be assessed by comparing net rates of change of nutrient concentrations with or without added zooplankton. Figure 2 shows results from one such experiment in which dissolved nutrients are removed more slowly when animals are present in abundance than when they are comparatively rare.

Because of the experimental design, the results shown in Fig. 2 can be used to estimate uptake and regeneration in situ by using the following simple model: assume, first, that the phytoplankton population (A) changes exponentially during short time intervals t, such that

$$A = A_0 e^{rt} \qquad (1)$$

where r, the *net* rate of growth of the

population, may be either positive or negative and can be determined empirically. Next, assume that the initial nutrient additions have effectively saturated the uptake rates of the algae, so that

$$dS/dt = -uA + cZ \quad (2)$$

where S is nutrient concentration (μg liter^{-1}), Z is mg dry weight of zooplankton per liter, and A is algal abundance, expressed here as μg Chl a liter^{-1}. The coefficient u is the saturated rate of nutrient uptake in μg of N or P (μg Chl a)$^{-1}$ day^{-1}; c is the specific regeneration rate of the zooplankton assemblage in μg of N or P (mg dry wt)$^{-1}$ day^{-1}. By integrating this equation over a time period of 1 day, and by assuming that zooplankton abundance inside the enclosures does not change significantly in that time, one computes

$$\Delta S = -u(A - A_0)/r + cZ. \quad (3)$$

Since $(A - A_0)/r$ is the mean abundance of phytoplankton during the 1-day period ($\bar{A}$), the equation can be rewritten

$$-\Delta S/\bar{A} = -cZ/\bar{A} + u \quad (4)$$

and both c and u can be computed by linear regression on the empirical data, i. e. by plotting $\Delta S/\bar{A}$ versus $Z/\bar{A}$. Results calculated for three 24-h intervals are shown in Table 4. The data for both 29–30 September and 30 September–1 October were taken from the experiment shown in Fig. 2. In all cases, predicted values for uptake and regeneration were significantly greater than zero. Furthermore, as one might anticipate from the similarities of the zooplankton assemblages among those periods, there are no statistically valid differences between calculated regeneration rates of either PO$_4$-P or NH$_3$-N. Mean values computed for the estimates and their errors are 2.66 (SE=0.51) μg PO$_4$-P (mg dry wt zoop.)$^{-1}$ day^{-1} and 13.68 (SE=1.49) μg NH$_3$-N (mg dry wt zoop.)$^{-1}$ day^{-1}. Both of these values compare favorably with estimates computed from the simpler experiment shown in Fig. 1, although the value for regeneration of NH$_3$-N is a bit lower than the prior estimates (cf. Table 3).

In order to compare the quantities of N and P released to the water by zooplankton with the amounts of those nutrients that are supplied from other sources, consider that during September 1977, the zooplankton, mostly *Daphnia*, *Epischura*, and *Diaptomus*, constituted a mass of about 100 mg dry wt m^{-3} in the epilimnion of Lake Washington. The depth of the mixed layer was about 15 m, and the zooplankton were relatively uniformly distributed there (A. Litt and W. T. Edmondson pers. comm.). On the basis of the regeneration rates just calculated, the zooplankton would be expected to contribute ca. 4.0 mg PO$_4$-P m^{-2} day^{-1} and 20.5 mg NH$_3$-N m^{-2} day^{-1} directly to the epi-

Table 4. Rates of simultaneous uptake and remineralization of dissolved N and P by epilimnetic plankton of Lake Washington, autumn 1977.

	Uptake μg (μg Chl a)$^{-1}$ d^{-1}	(SE)	Regeneration μg (mg dry wt zoop.)$^{-1}$ d^{-1}	(SE)
PO$_4$-P				
27–28 Sep	0.54	(0.09)	1.73	(0.50)
29–30 Sep	0.88	(0.15)	1.95	(0.44)
30 Sep–1 Oct	2.04	(0.20)	4.30	(1.38)
NH$_3$-N				
27–28 Sep	2.25	(0.38)	11.93	(2.23)
29–30 Sep	3.91	(0.40)	12.41	(2.85)
30 Sep–1 Oct	2.78	(0.38)	16.69	(2.62)

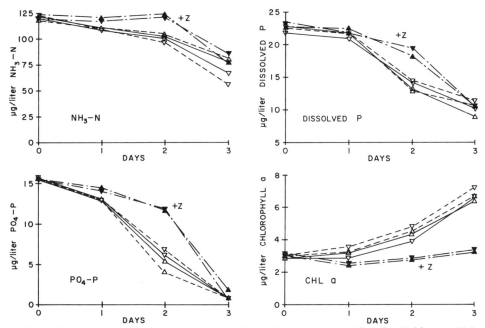

Fig. 2. Concentrations of several substances inside enclosures suspended in Lake Washington, 29 September–2 October 1977. Concentrations of inorganic P and N were artificially elevated in all bags at start of experiment. +Z denotes bags which received added quantities of zooplankton (0.62 and 0.69 mg dry wt liter^{-1}). Abundances in other enclosures were 0.08 and 0.16 (———), and 0.10 and 0.04 (– – –) mg dry wt liter^{-1}, different symbols distinguish replicates.

limnion, because the animals do not show any strong diel migratory behavior in the lake.

Mean rates of carbon fixation in Lake Washington from June to September of several years are summarized in Table 5 from data supplied by D. Allison and W. T. Edmondson (pers. comm.). Requirements of the phytoplankton for N and P have been computed from the summer means for C fixation, by assuming a ratio of 40:7:1 by weight for C:N:P in the algal cells (Vallentyne 1974).

Discussion

The figures presented for areal rates of C fixation in Lake Washington (Table 5) are based on 24-h incubations and are corrected for dark bottle uptake of ^{14}C. They are assumed to represent net production over a 24-h period. In fact, the values compare very well with independent determinations of productivity made in 1974 (Devol 1979) with shorter incubations and ETS estimates of dark respiration. The 362.2 (SE=82.5) mg C m^{-2} day^{-1}

Table 5. Areal rates of C fixation in Lake Washington from June to September of years listed. Phytoplankton "demands" for N and P have been calculated from C fixation by assuming C:N:P = 40:7:1 by weight.

	FIXATION			DEMANDS	
	mg C m^{-2} day^{-1}			mg m^{-2} day^{-1}	
	Mean	(SE)	n	N	P
1972	253.0	(44.9)	7	44.3	6.3
1973	293.6	(38.4)	8	51.4	7.3
1974	362.2	(82.5)	8	63.4	9.1
1975	492.8	(54.4)	9	86.2	12.3
1976	225.6	(36.0)	9	39.5	5.6
1977	484.9	(155.2)	9	84.9	12.1

listed here for summer 1974, is not significantly different from the 374 mg C m^{-2} day^{-1} measured by Devol for 11 July–2 October 1974. Estimated "demands" for N and P based on these production rates are of course suspect for the reasons discussed by Banse (1974), but they nonetheless provide a rough index for comparison.

The value computed for remineralization of PO_4-P by epilimnetic zooplankton during September 1977, 4 mg P m^{-2} day^{-1}, is an order of magnitude higher than the rate of supply of PO_4-P to the euphotic zone of Lake Washington from all external sources during the summer (Tables 1 and 2). Likewise, the remineralization of NH_3-N by the animals is larger by a factor of three than combined vertical flux and allochthonous inputs of inorganic N. Abundances of zooplankton were just as great, or greater, during the rest of the summer, so if the values are biased, they are probably underestimates of summertime regeneration.

Not only are the recycling fluxes large in magnitude, but the N and P are supplied to the water in chemical forms that are immediately available to phytoplankton. Because most of the external input of nutrients to Lake Washington occurs during the winter, and because in situ productivity is low at that time, concentrations of dissolved nutrients reach a maximum by February or March. Ambient concentrations of PO_4-P and NO_3-N are both rapidly exhausted in the spring bloom, however, and values remain low and almost constant during the summer months. Because fluvial and diffusive inputs are low in magnitude, and ambient pools are small, recycling appears to supply almost all of the P that is available for biological production during the summer. The estimates made here for rates of remineralization by the zooplankton suggest that the animals do a large part of the recycling.

The analysis followed here intentionally ignores leakage directly from living algal cells and rapid cycling of P among dissolved, colloidal, and particulate fractions (Lean 1973; Rigler 1973). These fluxes may be large in magnitude, but they cannot be included in the same category as the other inputs identified and measured in this study. Directional flux of a nutrient across a cell's membrane can be very large, but if it is balanced by an equal flux in the opposite direction, its significance at the level of the whole cell, or of populations of cells, is completely neutral. As Pomeroy et al. (1963) have pointed out, the sources of nutrients which make possible de novo production of phytoplankton biomass are of foremost concern here.

More important, perhaps, is the suggestion by Golterman (1973) that autolysis of algal cells can release sizable quantities of nutrients to the water, and that microbial processes may represent an appreciable source of P and N. Zooplankton, too, rapidly release P to the water after death, probably by autolysis (Marshall and Orr (1961). It may be, however, that incompletely digested material that is egested by the zooplankton represents an important part of the substrate for the microbes, and thus even those rates depend on the grazers. In the field experiments described, no efforts were made to distinguish whether all the P and N released to the water was an immediate result of zooplankton excretion or whether some of the nutrients may have been rapidly liberated from egested cell fragments. In either case, the zooplankton played an important role in the regeneration process.

Some of the values which have been reported in the literature for regeneration of nutrients by freshwater and marine zooplankton are listed in Table 6. The list is not extensive, particularly for marine species, but it does include many small copepods of the same size range as individuals found in Lake Washington, as well as a few of the larger *Calanus* species. The marine species were included primarily to take advantage of the wider array of measurements available for nitrogen excretion.

Most of the studies report rates of regeneration of dissolved inorganic P and N. There is some disagreement among workers regarding the fractions of total P and N

Table 6. Measured rates of regeneration of NH_3-N and PO_4-P by some freshwater and marine zooplankton. Water for the incubations was either filtered (F) or unfiltered (U) to remove phytoplankton.

Species	Temp °C	µg dry wt animal^{-1}	N µg (mg dry wt)$^{-1}$ day^{-1}	P µg (mg dry wt)$^{-1}$ day^{-1}	
Freshwater					
Daphnia magna, F	20–22	250.0		0.8	Rigler 1961
Daphnia rosea, F	20–22	13.4		2.0*	Peters and Lean 1973
Daphnia pulex, F	15,20,25	26.0	5.1		Jacobsen and Comita 1966
mostly Cladocera, U	5	1.0		2.0	
	20	0.9		4.6	Barlow and Bishop 1965
mostly Copepoda, U	5	3.6		1.9	
	20	2.9		6.0	
Thermocyclops hyalinus, U	27.3	0.25–1.6	42.1	11.3	Ganf and Blažka 1974
Brackish water and marine					
Asplanchna sp., U	18	0.5		1.5	
Acartia tonsa, U naup.	18	3.0		1.4	Hargrave and Geen
CII-CIV	18	5.0		1.3	1968
CV-CVI	18	10.0		1.0	
Acartia clausi + *A. tonsa*, U		5–10	12.2	4.0	Martin 1968
Acartia clausi, F	15	4.1	52.0		Mayzaud 1973
Acartia clausi, F	5	5.8	8.0		Mayzaud 1976 (day 0)
78% *A. clausi*, U	4.5–5.5	8.4	44.8	5.4	Harris 1959
80% *A. clausi*, U	12.5–18.0	2.6	42.4	12.9	
Calanus hyperboreus, U	4–6	450–3,250	0.1–1.1		Conover and Corner 1968
C. finmarchicus and *C. helgolandicus*, F	10–14	116–226	5.0–14.9*	1.0–2.3*	Butler et al. 1969
C. finmarchicus, F	5	260	0.9		Mayzaud 1976 (day 0)
mixed marine zooplankton, U	19–29			6.0*	Satomi and Pomeroy 1965
	12–15		2.2–8.4		Jawed 1973
	17–22	2	19.8		Smith and Whitledge 1977
		3	32.3		

*Values are for total P or total N, including dissolved organics.

that are supplied in organic form. Johannes and Webb (1965) indicated that release of organic N, particularly amino N, might be substantial for natural plankton, and others working with concentrated natural collections have sometimes found the same (Le Borgne 1973). Corner and Newell (1967), on the other hand, reported that *Calanus helgolandicus* would consistently release about 75% of its N as NH_3-N and that release of amino N was significant only when animals were severely crowded. Mullin et al. (1975) have cautioned that concentrated and unsorted net samples of zooplankton like those used by Johannes and Webb (1965) may overestimate release rates, because of contributions that leak from dead and injured animals. In general, the workers who have examined the release products of single species conclude that most of the N appears as NH_3-N (e.g. Butler et al. 1969; Jawed 1969), but that a measurable fraction, usually <25%, consists of other compounds (e.g. Mayzaud 1973; Mayzaud and Dallot 1973). A similar dichotomy of opinion and evidence exists for relative rates of release of inorganic and organic P. Some workers, again dealing with net plankton, find as much as half of the P in organic form (Pomeroy et al. 1963; Satomi and Pomeroy 1965; Le Borgne 1973). Studies with single species find that most of the release in short term assays is PO_4-P (Rigler 1961; Butler et al. 1969; Peters and Lean 1973).

In studies which have examined release products of freshwater *Daphnia*, PO_4-P and NH_3-N seem to dominate (Peters and Lean 1973; Jacobsen and Comita 1976). The substances are reabsorbed swiftly by the phytoplankton (Ganf and Blažka 1974; Porter 1976) or even by the epizoic communities of the animals (Rigler 1961). Peters and Lean (1973) have, in fact, postulated that rapid cycling of PO_4-P, more so than of any organic P released by the zooplankton, is one of the factors responsible for fractionally large net accumulation of organic P in the course of some experiments. Some support for this notion can be seen in Table 3, where the lowest specific rate of accumulation of PO_4-P occurs at low animal densities and where there is also the greatest opportunity for phytoplankton to have an impact on final measured concentrations. Important also in this regard is the necessary distinction between excretion and egestion. Rates of nutrient release decline quickly when zooplankton are separated from their food supply (Mayzaud 1976); the character of the released substances may change as well. Methods designed to assess both mechanisms seem to be the most applicable to the issue addressed here: overall reutilization of the elements N and P.

Rates measured in this study (Tables 3 and 4) were well within the range of reported values (Table 6), which in many cases are measures of excretion alone. The values I computed may be underestimates by several percentages because organic contributions were neglected, but I reasoned that the organic compounds may not all be immediately available to algal cells. In any case, my best estimates carry sufficiently large standard errors (19% for P and 11% for N) that small corrections will not alter them significantly.

It proved impossible to analyze the data for simultaneous uptake and release of total P (Fig. 2) in the same manner as PO_4-P and NH_3-N, basically because the total P analysis had a higher background value and lower inherent accuracy (shorter spectrophotometric path length), thus making the values for fractionally small ΔS comparatively imprecise. Nonetheless, changes of total dissolved P were of the same magnitude as changes in PO_4-P, making it seem likely that uptake and release of inorganic P were the dominant processes of interest. It is furthermore noteworthy that although separate estimates of rates of regeneration (Table 4) were not significantly different for either N or P, differences did exist among calculated values for saturated uptake rates on different days. The variability probably comes from two sources. First, the rates were normalized to chlorophyll concentrations even though the rates are in fact species specific.

Under the influence of selective grazing the character of the plankton may have changed enough in some of the bags to have made a difference. Secondly, effects of local weather, particularly changes in day-to-day cloudiness, etc. may have been important to the changes registered over a 24-h period.

Whether physiological death of algal cells is an important process in the epilimnia of many lakes (Jassby and Goldman 1974) and whether nutrient release from such cells represents important nutrient sources (Golterman 1973) cannot be resolved by this work. What is already clear, however, is that allochthonous sources of nutrients supply only a small fraction of the P and N used for daily production in Lake Washington during the summer. Other sources, not mentioned, include the microzooplankton (Johannes 1964, 1968) and release from littoral sediments. Of these, the sediments are likely to be inconsequential as a source because mass budgets of the water column of Lake Washington show that throughout the summer net fluxes of P and N are *into* the sediments (W. T. Edmondson and J. T. Lehman unpubl.). Although the microzooplankton constitute only a small fraction of the total volume of plankton in Lake Washington, their greater intrinsic metabolic rates may still cause their contribution to be sizable. Because only macrozooplankton were affected when animal abundances were altered, the contribution from microzooplankton was not assessed here. However, because the measured rates of regeneration are large enough to support a sizable fraction of the phytoplankton requirements (cf. Table 5), and contributions from rotifers and protozoa could only increase that fraction further, it is unlikely that recycling from other sources seriously outweighs that by zooplankton. Based on ETS estimates of zooplankton respiration and O:P conversion ratios, Devol (1979) similarly concluded that most of the calculated algal "demand" for P could be supplied by zooplankton during the summer in Lake Washington.

The issue raised by the importance of nutrient regeneration by zooplankton to the availability of P and N is an exciting one. On the one hand, the remineralization might be regarded as having a general fertilizing effect on algal growth and production, which is offset by the perpetual harvesting activities of the grazers, but its biological significance increases when we consider its species specific effects. Because not all species of phytoplankton are grazed uniformly, although they do compete for inorganic nutrients, it is very likely that herbivorous zooplankton may be removing some species through their feeding efforts, while at the same time supporting a substantial part of the production of potential competitors of those species. In Lake Washington, remineralization of P by the zooplankton appears to be more important relative to external sources than is regeneration of N. The observation is doubly significant when coupled with the fact that planktonic production during the spring appears to be limited by P (Edmondson 1970). The availability of the limiting nutrient during the summer, therefore, may be governed by the activities of the grazers. After the lake receives substantial nutrient inputs from the high fluvial discharges of the winter, production effectively runs on ambient nutrients to the extent that the population dynamics of the phytoplankton, and species succession, may be dependent just as much on the remineralization of nutrients as on the mortality caused by grazing zooplankton.

References

BANSE, K. 1974. The nitrogen-to-phosphorus ratio in the photic zone of the sea and the elemental composition of the plankton. Deep-Sea Res. **21**:767–771.

BARLOW, J. P., and J. W. BISHOP. 1965. Phosphate regeneration by zooplankton in Cayuga Lake. Limnol. Oceanogr. **10**(suppl.):R15–R25.

BOTTRELL, H. H., and others. 1976. A review of some problems in zooplankton production studies. Norw. J. Zool. **24**:419–456.

BROWN, E.J., R.F. HARRIS, and J.K. KOONCE. 1978. Kinetics of phosphate uptake by aquatic microorganisms: deviations from a simple

Michaelis-Menten equation. Limnol. Oceanogr. 23:26–34.
BUTLER, E. I., E. D. S. CORNER, and S. M. MARSHALL. 1969. On the nutrition and metabolism of zooplankton. 6. Feeding efficiency of *Calanus* in terms of nitrogen and phosphorus. J. Mar. Biol. Assoc. U. K. 49: 977–1001.
CONOVER, R. J., and E. D. S. CORNER. 1968. Respiration and nitrogen excretion by some marine zooplankton in relation to their life cycles. J. Mar. Biol. Assoc. U. K. 48:49–75.
COOPER, L. H. N. 1935. Rate of liberation of phosphate in sea water by the breakdown of plankton. J. Mar. Biol. Assoc. U. K. 20:197–200.
CORNER, E. D. S., and B. S. NEWELL. 1967. On the nutrition and metabolism of zooplankton. 4. The forms of nitrogen excreted by *Calanus*. J. Mar. Biol. Assoc. U. K. 43:373–386.
CUSHING, D. H., and H. F. NICHOLSON. 1963. Studies on a *Calanus* patch. 4. Nutrient salts off the north-east coast of England in the spring of 1954. J. Mar. Biol. Assoc. U. K. 43:373–386.
———, and T. VUCETIC. 1963. Studies on a *Calanus* patch. 3. The quantity of food eaten by *Calanus finmarchicus*. J. Mar. Biol. Assoc. U. K. 43:349–371.
DEVOL, A. H. 1979. Respiratory enzyme activity of net zooplankton from two lakes of contrasting trophic state. Limnol. Oceanogr. 24(5): 893–905.
DUGDALE, R. C. 1976. Nutrient cycles, p. 141–172. *In* D. H. Cushing and J. J. Walsh [eds.], The ecology of the seas. Blackwell.
EDMONDSON, W. T. 1970. Phosphorus, nitrogen and algae in Lake Washington after diversion of sewage. Science 169:690–691.
———. 1977. Trophic equilibrium in Lake Washington. Rep. EPA-600/3-77-087. U. S. EPA, Washington, D. C.
GANF, G. G., and P. BLAŽKA. 1974. Oxygen uptake, ammonia and phosphate excretion by zooplankton of a shallow equatorial lake (Lake George, Uganda). Limnol. Oceanogr. 19:313–325.
GARDINER, A. C. 1937. Phosphate production of planktonic animals. J. Cons. Cons. Int. Explor. Mer 12:144–146.
GOLTERMAN, H. L. 1973. Vertical movement of phosphate in freshwater, p. 509–538. *In* E. J. Griffith et al. [eds.], Environmental phosphorus handbook. Wiley.
HARGRAVE, B. T., and G. H. GEEN. 1968. Phosphorus excretion by zooplankton. Limnol. Oceanogr. 13:332–343.
HARRIS, E. 1959. The nitrogen cycle in Long Island Sound. Bull. Bingham Oceanogr. Collect. 17:31–65.
IKEDA, T. 1977. The effect of laboratory conditions on the extrapolation of experimental measurements to the ecology of marine zooplankton. 4. Changes in respiration and excretion rates of boreal zooplankton species maintained under fed and starved conditions. Mar. Biol. 41:241–252.
JACOBSEN, T. R., and G. W. COMITA. 1976. Ammonia-nitrogen excretion in *Daphnia pulex*. Hydrobiologia 51:195–200.
JASSBY, A. D., and C. R. GOLDMAN. 1974. Loss rates from a lake phytoplankton community. Limnol. Oceanogr. 19:618–627.
———, and T. POWELL. 1975. Vertical patterns of eddy diffusion during stratification in Castle Lake, California. Limnol. Oceanogr. 20:530–543.
JAWED, M. 1969. Body nitrogen and nitrogen excretion in *Neomysis rayii* Murdoch and *Euphausia pacifica* Hansen. Limnol. Oceanogr. 14:748–754.
———. 1973. Ammonia excretion by zooplankton and its significance to primary productivity during summer. Mar. Biol. 23:115–120.
JOHANNES, R. E. 1964. Phosphorus excretion and body size in marine animals: microzooplankton and nutrient regeneration. Science 146:923–924.
———. 1968. Nutrient regeneration in lakes and oceans. Adv. Microbiol. Sea 1:203–213.
———, and K. L. WEBB. 1965. Release of dissolved amino acids by marine zooplankton. Science 150:76–77.
LEAN, D. R. S. 1973. Phosphorus dynamics in lakewater. Science. 179:678–680.
LeBORGNE, R. P. 1973. Étude de la respiration et de l'excretion d'azote et de phosphore des populations zooplanktoniques de l'upwelling mauritanien (mars-avril, 1972). Mar. Biol. 19:249–257.
LEHMAN, J. T. 1978. Aspects of nutrient dynamics in freshwater communities. Ph.D. thesis, Univ. Washington, Seattle. 180 p.
MARSHALL, S. M., and A. P. ORR. 1961. On the biology of *Calanus finmarchicus*. 12. The phosphorus cycle: excretion, egg production, autolysis. With an addendum "The turnover of phosphorus by *Calanus finmarchicus*" by R. J. Conover. J. Mar. Biol. Assoc. U. K. 41:463–488.
MARTIN, J. H. 1968. Phytoplankton-zooplankton relationships in Narragansett Bay. 3. Seasonal changes in zooplankton excretion rates in relation to phytoplankton abundance. Limnol. Oceanogr. 13:63–71.
MAYZAUD, P. 1973. Respiration and nitrogen excretion of zooplankton. 2. Studies of the metabolic characteristics of starved animals. Mar. Biol. 21:19–28.
———. 1976. Respiration and nitrogen excretion of zooplankton. 4. The influence of starvation on the metabolism and the biochemical composition of some species. Mar. Biol. 37: 47–58.

——, and S. DALLOT. 1973. Respiration et excretion azotéc du zooplankton. 1. Etude des niveaux métaboliques de quelques espèces de Mediterranée occidentale. Mar. Biol. 19:307-314.

MULLIN, M. M., M. J. PERRY, E. H. RENGER, and P. M. EVANS. 1975. Nutrient regeneration by some oceanic zooplankton: a comparison of methods. Mar. Sci. Commun. 1: 1-13.

PETERS, R., and D. LEAN. 1973. The characterization of soluble phosphorus released by limnetic zooplankton. Limnol. Oceanogr. 18: 270-279.

POMEROY, L. R., H. M. MATHEWS, and H. S. MIN. 1963. Excretion of phosphate and soluble organic phosphorus compounds by zooplankton. Limnol. Oceanogr. 8:50-55.

PORTER, K. G. 1976. Enhancement of algal growth and productivity by grazing zooplankton. Science 192:1332-1334.

POWELL, T., and A. JASSBY. 1974. The estimation of vertical eddy diffusivities below the thermocline in lakes. Water Resour. Res. 10:191-198.

RIGLER, F. H. 1961. The uptake and release of inorganic phosphorus by *Daphnia magna* Straus. Limnol. Oceanogr. 6:165-174.

——. 1973. A dynamic view of the phosphorus cycle in lakes, p. 539-572. *In* E. J. Griffith et al. [eds.], Environmental phosphorus handbook. Wiley.

SATOMI, M., and L. R. POMEROY. 1965. Respiration and phosphorus excretion in some marine populations. Ecology 46:877-881.

SMITH, S. L., and T. E. WHITLEDGE. 1977. The role of zooplankton in the regeneration of nitrogen in a coastal upwelling system off northwest Africa. Deep-Sea Res. 24:49-56.

TAGUCHI, S., and H. ISHII. 1972. Shipboard experiments on respiration, excretion, and grazing of *Calanus cristatus* and *C. plumchrus* (Copepoda) in the northern North Pacific, p. 419-431. *In* A. Y. Takenouti [ed.], Biological oceanography of the northern North Pacific Ocean. Idemitsu Shoten.

TAKAHASHI, M., and T. IKEDA. 1975. Excretion of ammonia and inorganic phosphorus by *Euphausia pacifica* and *Metridia pacifica* at different concentrations of phytoplankton. J. Fish. Res. Bd. Can. 32:2189-2195.

VALLENTYNE, J. R. 1974. The algal bowl: lakes and man. Misc. Spec. Publ. 22. Environ. Can. 185 p.

24. The Importance of "Threshold" Food Concentrations

Winfried Lampert and Ursula Schober

Abstract

Food shortage may be a structuring factor for zooplankton communities. In the laboratory "threshold concentrations" of food were determined which were sufficient for the maintenance of the body weight of *Daphnia pulex* and for egg production. The same general trend as in tracer experiments and long term growth studies could be demonstrated in the field, if the number of eggs per adult was related to the concentration of ingestible carbon. Threshold concentrations obtained in the field come close to the laboratory results.

Lakes of different trophy have different zooplankton communities and it has been shown numerous times that the community of the same lake is altered when the lake is enriched with nutrients. One excellent example, Lake Washington, has been discussed by Edmondson at this conference. Under similar circumstances in Lake Constance not only has the maximum number of *Daphnia* quintupled during 10 years, but also its seasonal period of abundance has expanded (Elster and Schwoerbel 1970). Although the number of *Eudiaptomus* has also increased, the more rapid increase of *Daphnia* has led to a shift in the *Daphnia-Eudiaptomus* ratio from 1 to 4. Similar observations are reviewed by McNaught (1975).

The most probable reasons for these alterations are changes in the nutritional conditions associated with the composition and concentration of food. A species which shows a more pronounced response to better food conditions will be favored by eutrophication as long as there are no other interactions (e.g. inhibition by blue-greens).

Food conditions in the field fluctuate very widely, so we can assume that there are periods when the animals starve and other periods when they have sufficient food. Therefore an enhancement of the food conditions can mean either a general increase in the food concentration or the reduction of periods of starvation. An analysis of this problem requires a knowledge of the minimum amount of food which is needed by an individual or a population to survive. Since in filter feeders the amount of food ingested is dependent on the food concentration, we must know the critical concentration of food below which the animal will starve. This "threshold concentration" should be different under various environmental conditions. We here demonstrate how such thresholds can be determined in the laboratory and discuss whether it is possible to transfer the laboratory data to the field.

This paper was supported by Deutsche Forschungsgemeinschaft.

Laboratory measurements

On an individual basis, we can define the threshold food concentration at which an animal is just able to equalize its metabolic losses so that it does not grow, yet does not lose weight either. Under these conditions production is zero. If we have data on the dependency of the assimilation rate and the respiration rate on the food concentration, we can calculate the production rate from the basic equation:

$$P = A - T$$

where P is production rate, A is assimilation rate, and T is metabolic losses.

Two examples of such calculations using data of radiotracer experiments (Lampert 1977a) are presented in Fig. 1. The produc-

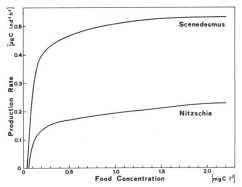

Fig. 1. Production rates of *Daphnia pulex* (2 mm, 25 µg C) fed different concentrations of algae at 20°C. Curves have been calculated from data of radiotracer experiments.

tion curves are calculated for *Daphnia pulex* 2 mm in size (25 µg C) at 20°C. The lower assimilation rates obtained for the diatom *Nitzschia* result in a shift of the threshold to a higher food concentration. Nevertheless, the threshold concentrations are very low (around 50 µg C liter^{-1}). The thresholds calculated in this way, for several environmental conditions, are presented in Lampert (1977b). Between 10° and 20°C, temperature has only a slight effect, but at 25°C the minimum food concentration required is increased. If *Scenedesmus acutus* is used as food, the threshold concentration ranges from 0.04 to 0.12 mg C liter^{-1}. Slightly higher values have been found for other algal species and in the case of *Staurastrum* the threshold is never reached because the assimilation rate for this unsuitable food is too low to sustain the metabolic losses.

For the survival of a population the metabolic losses must be more than equalized by production (e.g. to compensate for mortality, predation). So the energy input (i.e. food concentration) must be higher than in the former threshold case. This "threshold food concentration for the population" is not as easily definable, because the losses are quite variable depending on the external conditions. However, the minimum requirement to sustain a population is that the animals produce eggs. So we can define a second threshold by determining experimentally the dependency of the egg production on the food concentration.

Although several investigators have studied this problem, the results cannot be related to exact food levels as the experiments have been carried out in small standing containers where the food concentration changes due to sedimentation and grazing during the feeding period. Some data on *Daphnia rosea* can be found in Peters (1972), who used a plankton wheel for the experiments.

In Fig. 2, Peters' results at 20°C are compared with the egg production of *D. pulex*. These data have been obtained from

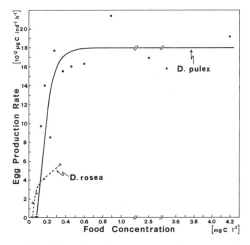

Fig. 2. Results of long term experiments on the dependence of egg production of *Daphnia* on food concentration. Food for *D. pulex* (1st brood): *Scenedesmus*; food for *D. rosea* (Peters 1972) *Rhodotorula*, temperature 20°C. Egg production is zero at "threshold concentration." Note that thresholds are similar to those in Fig. 1.

long term growth experiments under constant food conditions in a through-flow device as described in Lampert (1976). The food for *D. rosea* was *Rhodotorula*, for *D. pulex* it was *Scenedesmus*. The absolute rates in *D. pulex* are higher because the animals are larger. It can be clearly seen that the threshold concentration for the egg production obtained by totally different methods comes close to the results of the tracer experiments (Fig. 1), considering that we should expect values which are a little higher. Moreover, the threshold for *D. rosea* seems to be lower than for *D. pulex*.

Field studies

In field studies the effect of a single factor is obscured by the simultaneous changing of several other environmental parameters, so it is difficult to see the effect of food concentration from field correlation analysis. We present an attempt to isolate the food factor in the field in Fig. 3.

These are preliminary results of zooplankton studies in Lake Constance relating the average number of eggs carried by an adult *Daphnia* (*longispina* group) to food concentration. The food basis is the particulate carbon of the fraction of cells small enough to be ingested by the daphnids. The timing of egg numbers has been adjusted to correspond to a date equal to half the developmental time of the eggs before sampling. (For reasons and details on methods *see* Lampert 1978.) The data of example A (Fig. 3) came from in a nearshore station at the Limnological Institute, Konstanz-Egg, during 3 months in spring 1976. Panels B and C (Fig. 3) cover a whole-year cycle from the end of 1976 to the end of 1977. B gives results of a sampling station in the more eutrophicated eastern part of the lake, while C presents data of the deepest part of the lake, combined with a few points from a similar station in the western part. Although there is a remarkable scattering in the points, the results are principally the same as in the laboratory experiments. In each case, food concentrations can be detected where no

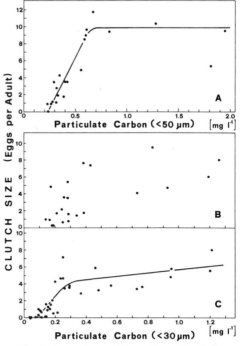

Fig. 3. Dependence of fecundity of *Daphnia* sp. in Lake Constance on availability of ingestible carbon. Three examples from different sites and time periods (*see text*). Note "threshold food concentrations" below which no eggs are produced.

eggs are present, i.e. which lie below the threshold concentration. These low concentrations are not only correlated with low temperatures, they also appear in summer during times of extremely clear water.

Since the food in the lake is a mixture of various algae, detritus, and bacteria, we cannot expect the same threshold concentrations as in the laboratory with unialgal food. Nevertheless, the threshold shown in panel C comes very close to the laboratory results. It is a little higher in panel A, but this may be due to the fact that 50 μm as the upper limit of ingestible food is too high. There are several reasons for the scattering of the points. One of the reasons for the clearer results shown in panel A is the relatively short period of time (although there is a temperature rise from 6°C to 17°C during that time) and the narrow sampling interval (up to three times a week). As the food conditions can change very rapidly, the sampling interval must be short to get reliable results. A period of 2 weeks may often be too long.

Discussion

Adequate studies to explain the role of food quantity in structuring zooplankton communities are lacking. Peters (pers. comm.) found the minimum food concentrations of egg production for several Cladocera to be similar, but the resolution of the available data was not good enough to clearly define the thresholds. However, in a fluctuating environment food concentrations often fall below a critical value for short times. In this case slight differences in the ability of zooplankton to withstand starvation may be important for the success of a species. This was demonstrated for marine copepods by Dagg (1977).

As we were able to show, there is a possibility of applying data from simplified laboratory experiments to complex field conditions. Needed now are more precise experiments comparing different zooplankton species at very low food levels, although the experimental work at low food concentrations requires more effort and care than work above the incipient limiting level.

References

DAGG, M. 1977. Some effects of patchy food environments on copepods. Limnol. Oceanogr. **22**: 99–107.

ELSTER, H. -J., and J. SCHWOERBEL. 1970. Beiträge zur Biologie und Populationsdynamik der Daphnien im Bodensee. Arch. Hydrobiol. Suppl. 38, p. 18–72.

LAMPERT, W. 1976. A directly coupled, artificial two-step food chain for long-term experiments with filter-feeders at constant food concentrations. Mar. Biol. **37**: 349–355.

———. 1977a, b. Studies on the carbon balance of *Daphnia pulex* de Geer as related to environmental conditions. III. Production and production efficiency. IV. Determination of the "threshold" concentration as a factor controlling the abundance of zooplankton species. Arch. Hydrobiol. Suppl. 48, p. 336–360, 361–368.

———. 1978. A field study on the dependence of the fecundity of *Daphnia* spec. on food concentration. Oecologia **36**: 363–369.

McNAUGHT, D. C. 1975. A hypothesis to explain the succession from calanoids to cladocerans during eutrophication. Int. Ver. Theor. Angew. Limnol. Verh. **19**: 724–731.

PETERS, R. H. 1972. Phosphorus regeneration by zooplankton. Ph. D. thesis, Univ. Toronto. 205 p.

25. Nutritional Adequacy, Manageability, and Toxicity as Factors that Determine the Food Quality of Green and Blue-Green Algae for *Daphnia*

Karen Glaus Porter and John D. Orcutt, Jr.

Abstract

The handling properties, toxicity, and nutritional adequacy of two species of algae with equivalent cell sizes were examined for their effects on zooplankton growth and reproduction. Cell concentrations of 10^3, 10^4, and 10^5 cells cc^{-1} of the green alga *Chlamydomonas reinhardi* (GMS$^-$) and 10^3 and 10^5 cells cc^{-1} of the toxin-producing strain of *Anabaena flos-aquae* (NRC-44-1), in both cellular and filamentous forms were fed to cohorts of *Daphnia magna*. The growth parameters—body size at death, percent growth, and number of instars—and the life history parameters—average lifespan, LT 50, age of initial reproduction, instar at initial reproduction, mean generation time (T), net reproductive rate (R_0), and intrinsic rate of increase (r)—were determined using life tables. Ingestion and filtering rates were measured using radiotracer techniques. Rejection rates were measured by observation.

Growth, survivorship, ingestion, and filtration rates of *Daphnia* were lower, and rejection rates were higher, on *A. flos-aquae* than on *C. reinhardi*. The toxic effect of *A. flos-aquae* was independent of nutritional adequacy. Although *Anabaena* promoted growth, it reduced survivorship below that of animals fed no food. Growth rates of *Daphnia* were enhanced and survivorship reduced when the alga was supplied in cellular form. These effects were less pronounced when the filamentous form was supplied as food. Feeding rate measurements showed that cells are ingested more rapidly than filaments. *Chlamydomonas reinhardi* cells were ingested more rapidly and rejected less frequently than *A. flos-aquae* cells, indicating that the blue-greens can be discriminated against on the basis of taste, not size or shape. *Anabaena flos-aquae* (NRC-44-1), therefore, has nutritional, handling, and toxic properties that render it a poor food source for *Daphnia*. The effect of toxic blue-greens in nature will be related to their growth conditions, the probability of their encounter with *Daphnia*, and their tolerance of zooplankton species and strains.

The quality of different algae as food for zooplankton is of interest because algae are at the base of aquatic food chains and their abundance and composition can be altered by management practices and human disturbance. Nutritional adequacy, as determined by chemical or caloric content, has been the major factor considered in evaluating algal food quality (e.g. Richman 1958; Schindler 1968). However, handling properties and toxicity may also determine whether an alga can support continued zooplankton growth. Blue-green algae are of particular interest because their abundance in eutrophic waters suggests that they do not readily enter aquatic food chains.

Conflicting evidence exists for the suitability of blue-green algae as food for grazing zooplankton. Blue-greens produce toxins lethal to farm animals and wildlife (*see* Gorham 1964; Gentile 1971) and inhibit the growth of bacteria (*see* Gentile 1971)

This research was supported by NSF grant DEB77-25354.

and other algae (Keating 1978). Accumulated toxins may produce ciguatera in marine fishes (Moikeha and Chu 1971). Yet there is some question as to their toxic effect on zooplankton, their natural consumers (Lefevre 1950; Fryer 1957; Gentile and Maloney 1969; Schindler 1971; Arnold 1971; Crowley 1973; O'Brien and deNoyelles 1973; Porter 1977). Blue-greens are rarely found in the guts of zooplankton (see Porter 1975, 1977), yet some food chains are based on them (Wiegert and Mitchell 1973; Moriarity et al. 1973; Seale et al. 1975). Some field studies report a reduction in zooplankton populations during times of the year and in environments in which blue-green algae are abundant (Burns 1968; Porter 1973, 1977) while others do not (Haney 1973; Haertel 1976).

These conflicting results may be due to a variety of causes. Zooplankton species and strains may differ in their abilities to utilize, detoxify, or discriminate against blue-greens. Zooplankton have been found to ingest particles selectively on the basis of taste as well as abundance (Poulet and Marsot 1978). The relative and absolute abundances of blue-greens, as compared with other suitable food items in natural plankton, will determine the probability of their encounter by zooplankton and if nutrients and energy are available from other food sources to supplement or detoxify them. Growth medium can influence algal chemical composition and food quality for zooplankton (Provasoli et al. 1970). It may also influence toxin production. Toxicity (Carmichael and Gorham 1977, 1978) and nutritional quality can vary among genetically different strains and among different growth phases of the same strain. Also, bacteria growing in association with blue-greens can reduce their toxicity (Carmichael and Gorham 1977). Laboratory studies that use unispecific algae may involve similar problems especially due to uncontrolled changes in algal concentration, nonaxenic cultures, and bacterial growth. However, these studies show more consistently that zooplankton survivorship, growth, and reproduction are lower when these animals are fed blue-greens than when they are fed other algae such as greens (Arnold 1971; Lampert 1977).

The inadequacy of blue-greens as food for zooplankton may be due to their inadequacy as sources of nutrients for zooplankton growth and reproduction. They may do so by lacking essential nutrients or by containing an imbalanced amount of essential nutrients. Also, their nutrients and energy may be in unassimilable forms. They may also contain lower total concentrations of nutrients essential for zooplankton growth than equivalent volumes of other algae. Gas vacuoles present in some blue-greens reduce the volume of nutritive protoplasm in each cell. Some blue-green cells with thick walls and gelatinous sheaths can pass intact through the guts of grazers (Gibor 1956; see Porter 1975, 1977) and, therefore, require extensive trituration and digestion to release nutrients. A range of assimilation rates has been measured for blue-greens using radioactively labeled carbon as a tracer (Arnold 1971; Lampert 1977). These results suggest that most blue-greens are digested and assimilated as easily as green algae of high food quality. However, these rates do not necessarily indicate the quality of the nutrients assimilated.

Blue-greens that form large colonies and filaments may be inaccessible to some zooplankton and therefore not available as food. Colonies and filaments clog the filtering apparatus of *Daphnia* and are rejected with the abdominal claw (McMahon and Rigler 1963; Burns 1968; Porter 1975, 1977). Rejection is the major mechanism for discrimination and selective feeding in *Daphnia* (Porter 1977; unpublished observations and film analyses). Blue-greens may also be unmanageable and require extensive manipulation and modification before they can be broken up and ingested. This reduction in handling efficiency can result in low apparent filtering rates of zooplankton fed colonial blue-greens (Arnold 1971).

Blue-greens also may be toxic when ingested by their natural herbivores, although no study has clearly shown this. Extracts of *Microcystis aeruginosa* and

Aphanizomenon flos-aquae were lethal when added to *Daphnia* cultures (Stangenberg 1968; Gentile and Maloney 1969). Under nonaxenic conditions, Arnold (1971) found reduced growth and no reproduction in *D. pulex* fed *Gleocapsa alpicola* and Lampert (1977) found reduced growth and low reproduction of *Daphnia* fed on *Synechococcus elongatus*. Since measured ingestion (Arnold 1971) and assimilation rates (Arnold 1971; Lampert 1977) were comparable to those of green algae (the green algae promoted better growth and reproduction in *Daphnia*), the investigators interpreted their results as due to toxic properties of the blue-greens. Nutritional inadequacy, however, cannot be discounted. The only study in which blue-greens were ingested and produced toxic effects (rapid mortality) showed that some strains of *Microcystis, Anabaena, Gleotrichia, Cylindrospermum,* and *Oscillatoria* were lethal to benthic ostracods (Mills and Wyatt 1974).

These properties (nutritional inadequacy, toxicity, poor manageability) of blue-greens can result in a greater expenditure of energy in filtration, manipulation, rejection, digestion, and detoxification than is gained from ingestion of the algae. This study was designed to assess separately the nutritional, handling, and toxic properties of algae that determine their quality as food for zooplankton. *Daphnia magna*, a natural herbivore found in ponds in Georgia (an extension of the range reported by Brooks 1957), was the test animal. Animals were provided with no food and low and high concentrations of cellular and filamentous forms of a known toxin-producing strain (Carmichael et al. 1975) of *Anabaena flos-aquae* (NRC-44-1, source W. Carmichael) and comparably sized cells of the green algae *Chlamydomonas reinhardi* (GMS⁻, source L. Provasoli).* The latter is commonly used as a high quality food source for *Daphnia*. *Anabaena flos-aquae* (NRC-44-1) produces anatoxin-a, an alkaloid similar in structure to cocaine and other uncharacterized toxins (Carmichael and Gorham 1978). Algal toxicity and nutritional adequacy were assessed by measuring growth rates and fitness parameters such as survivorship and fecundity, using life table methods. Growth, survivorship, and fecundity were expected to increase with increasing food concentration and to be low on unmanageable, undigestible, or nutritionally inadequate algae. The effect of toxicity was expected to be shown separately from nutritional quality if growth occurred but survivorship were lower than in animals fed nontoxic food or no food. Unmanageability and discrimination were assessed by measuring ingestion, filtering, and rejection rates.

We thank Wayne Carmichael for supplying us with an axenic culture of *A. flos-aquae* (NRC-44-1) of confirmed toxicity. John Hobbie provided facilities and John Helfrich counted our bacterial samples.

M. Pace, J. Gilbert, Y. Feig, and G. Rodgers reviewed the manuscript.

Materials and methods

Cultures and feeding suspensions—*Daphnia magna* was grown in aquaria with aged tapwater and fed *C. reinhardi* and *Ankistrodesmus falcatus*. Algae were batch cultured axenically in WC medium without silica and with glycylglycine buffer, biotin, B_{12}, and thiamine (Stein 1973). Algae were centrifuged, resuspended in aquarium water and supplied as food. All cultures were maintained in a 20°C walk-in environmental chamber with 16:8 LD "cool-white" flourescent lighting.

For the experiments, an axenic culture of a toxic strain of *A. flos-aquae* (NRC-44-1) was supplied by W. Carmichael. Cultures were grown axenically for 7-10 days in 4,000-ml Erlenmeyer flasks containing 2,000 ml of ASM-1 medium (Carmichael and Gorham 1974) with WC vitamins.

***Chlamydomonas reinhardi* Dangeard is occasionally misspelled *C. reinhardii*. The origin of this error is probably the index in Huber-Pestalozzi, G., 1961, Die Binnengewässer 16 (5):736. Transcriptional errors in culture collections have resulted in other misspellings, such as *C. reinhardti* and *C. reinhardtii* (see Starr, R. 1964. Jour. Botany 51 (9):1013-1044.

This culture method produced filaments in straight chains of 134 ± 28 cells filament^{-1}, with 4.0- X 7.6-μm vegetative cells (1.4 X 10^{-6} μg dry wt cell^{-1}). *Chlamydomonas reinhardi* (GMS$^-$ strain; source, L. Provasoli) was cultured as above with bubbling in WC medium with vitamins, producing 5.5- X 6.8-μm ovate single cells (8.0 X 10^{-5} μg dry wt cell^{-1}). Both algal growth media provide comparable nutrients. Cultures for feeding rate experiments were labeled with 20 μCi[^{14}C]NaHCO$_3$ per 100 cc of log growth phase culture 48 hours before each experiment.

Experimental suspensions of *C. reinhardi* were prepared by centrifugation (8,000 rpm for 20 min) and resuspension in 0.45-μm Millipore filtered, autoclaved, aged tapwater. Since algal cells with gas vacuoles are disrupted by centrifugation, *A. flos-aquae* was concentrated, washed, and resuspended by filtration with stirring (Carmichael and Gorham 1974). This removed the growth medium, buffers, and soluble algal products, shown in preliminary life table studies using toxic and nontoxic algae to reduce survivorship and fecundity (pers. obs.). Aliquots of *A. flos-aquae* suspensions were gently sonicated to produce primarily single (61%-1 celled, 25%-2 celled, 14%-3 celled) cell units. There was no detectable loss of label over a 6-hour period from cells due to sonication. Algal condition was checked visually and concentrations were quantified by Sedgwick-Rafter counts. Algal treatment concentrations were prepared by serial dilution.

Life table experiments. Life table experiments were conducted at 20°C in 3.5-5.0 μ Einsteins m^{-2} sec^{-1} "cool-white" flourescent lighting (16:8 LD) using improved standard procedures (Banta 1939; Anderson et al. 1937; Green 1956; Hall 1964; Parker 1966; Arnold 1971; Bottrell 1975). Equivalent-sized members of cohorts ≤24 hours old were isolated from a clone of *D. magna* and placed in shell vials containing 30 ml of the algal treatments. Algal treatments were 1 X 10^3, 1 X 10^4, and 1 X 10^5 cells cc^{-1} *C. reinhardi*; 1 X 10^3 and 1 X 10^5 cells cc^{-1} *A. flos-aquae* in filamentous form; 1 X 10^3 and 1 X 10^5 cells cc^{-1} *A. flos-aquae* as unicells; and 0.45-μm Millipore-filtered autoclaved aged tapwater without algae. Each treatment was run in two replicated experiments, each starting with 15-17 animals. No difference was found between replicates and these were pooled in subsequent presentations of the results.

Algal cell concentrations, pH, and O$_2$ in shell vials containing individual 2.5-mm *D. magna* were monitored at 24-hour intervals to determine the consistency of the experimental environment. Samples (20cc) of vial contents, algal cultures, and 0.45-μm Millipore-filtered, autoclaved water were also preserved in 2% formaldehyde and sent to J. Hobbie and J. Helfrich for bacterial counts (Hobbie et al. 1977). The results (Table 1) indicated that conditions remained constant over a 24-hour period except at the lowest food concentration. At 48 hours and longer intervals, all algal cell concentrations were significantly decreased, and a living bacterial flora of active orange rods developed. These bacteria are believed to survive *Daphnia* gut passage or to be gut flora. The initial bacterial cells were coccoid and were introduced with the water used to resuspend the algae. They probably do not represent a major food supplement for *D. magna* as they are much smaller than 1 μm, autoclaved, and are at concentrations well below those required for living bacteria to maintain, let alone grow, zooplankton (1 X 10^7 cc^{-1}: Tekuza 1974; see Porter et al. 1979). Therefore, 24 hours was used as the maximum interval between transfers. This is less than the intervals used (> 2 d) in previous life table studies (Hall 1964; Arnold 1971). A daily monitoring interval is also desirable because it is shorter than the duration of an instar under these experimental conditions.

Animals were transferred at the same time daily to clean vials with fresh feeding suspensions. Body length, shed carapace, egg number, young produced, and general condition were recorded. These data were used to calculate the following parameters—

Size at death: Body length (mm) from

Table 1. Environmental conditions over 24 hours in shell vials containing *Daphnia magna* and *Chlamydomonas reinhardi*.

	Time = 0 hours	Time = 24 hours	t-test
Cell concentration (cells cc^{-1})			
	$1.09\pm0.04\times10^5$ (26)*	$0.91\pm0.05\times10^5$ (26)	n.s.
	$1.22\pm0.15\times10^4$ (29)	$0.98\pm0.07\times10^4$ (29)	n.s.
	$1.35\pm0.10\times10^3$ (28)	$0.66\pm0.05\times10^3$ (28)	$p<0.01$
pH			
	7.50 ± 0.02 (10)	7.50 ± 0.02 (10)	n.s.
Oxygen (ppm)			
1×10^3 cells cc^{-1}	6.70 ± 0.05 (9)	6.72 ± 0.05 (9)	n.s.
1×10^4 cells cc^{-1}	6.70 ± 0.05 (10)	6.68 ± 0.06 (10)	n.s.
1×10^5 cells cc^{-1}	6.70 ± 0.05 (10)	6.68 ± 0.06 (10)	n.s.
Bacteria (cells cc^{-1})†			
	$6.80\pm2.30\times10^5$ (5)	6.00 ± 0.90 (5)	n.s.

*Values are mean ± SE (no. of vials).
†Bacteria are <1-µm inactive cells from 0.45-µm Millipore-filtered, aged autoclaved tapwater.

the anterior-most portion of the carapace to the base of the tail spine, dimension T in fig. 1 of Anderson et al (1973), at the last live measurement.

% Growth: Increase in body length (BL) during life, calculated as the $(BL_{death} - BL_{initial})/BL_{initial} \times 100$.

LT 50: Median age of death; time, in days, at which 50% of the experimental animals have died.

Average life span: Reported as the average number of days from emergence from the brood pouch until death; juvenile and adult life excluding embryonic period.

Number of instars: The number of carapace molts +1, considering the individual emerging from the brood pouch as the first instar.

Age at initial reproduction: Age in days at first appearance of egg brood pouch, i.e. attainment of reproductive maturity.

Mean generation time (T): Mean length of a generation for the experimental cohort calculated according to Poole (1974) as

$$T \approx \frac{\Sigma x 1_x m_x}{\Sigma 1_x m_x}$$

where x = day; 1_x = survivorship; m_x = age specific reproduction.

Net reproduction rate (R_0): The total number of female offspring produced per female in a single generation, or rate of population increase per generation, calculated (Poole 1974) as

$$R_0 = \sum_{x=0}^{n} 1_x m_x.$$

Intrinsic rate of increase (r): An instantaneous rate of increase, computed by iteration.

Differences between treatments were determined using Mann-Whitney U-tests. Slopes of survivorship curves were compared by ANCOVA on arcsine transformations of the probability data (Sokal and Rohlf 1969).

Feeding and filtering rates. Feeding and filtering rates were measured by using radiotracer techniques similar to those of Porter (1976) and Heisey and Porter (1977). Experimental water was aerated for

3–5 minutes to ensure oxygen saturation before the resuspension of algae (Heisey and Porter 1977). All experiments were run from 1200 to 1600 hours, at 20°C, under 3.5–5.0 μ Einsteins m^{-2} sec^{-1} "cool-white" fluorescent lighting.

Female *D. magna* were self-sorted by phototactic swimming through a series of different sized sieves. Animals of the desired size range (2.6–2.8 mm) were collected and acclimated for 1 hour to experimental conditions in unlabeled algal feeding suspensions. Feeding and filtering rates were obtained by transferring 10–12 acclimated animals to beakers containing 150 cc of radioactive labeled feeding suspension, stirring occasionally and allowing animals to feed for 20 minutes. Animals were then removed in 220-μm-mesh sieves, rinsed, and immobilized in boiling water. Immobilized *Daphnia* were measured and placed in scintillation vials (10 per vial). Three aliquots of each feeding suspension were filtered onto 0.45-μm Gelman glass-fiber filters and placed in scintillation vials. All algae and animal samples were digested for 24 hours with 1 cc of NCS tissue solubilizer (Amersham/Searle) at 55°C. Nine cubic centimeters of scintillation fluor (66 g PPO:82.5 mg POPOP:1.0 liter toluene) were added to cooled samples which were dark-adapted for 24 hours. These were counted in a Beckman LS-230 liquid scintillation counter and efficiencies were determined using the external standards method.

Ingestion and filtering rates were calculated according to Heisey and Porter (1977) and compared among treatments using a Mann-Whitney U-test for significance.

Rejection rates. Animals were attached by the head shield from a fine glass needle dipped in silicone grease. Animals were observed to remain alive and feeding in this position until the next molt, 2–3 days. For observation, animals were placed in 5×5×5-cm glass viewing chambers containing algal suspensions and allowed to acclimate for 20 minutes until consistent, rhythmic behavior patterns were maintained. Chambers were mounted on a Zeiss inverted microscope with a 63-mm Luminar lens as an objective and 10× eyepieces. This allowed a working distance of 6 cm and a 2 mm depth of field. This method (Porter et al. 1979) is an improvement over conventional techniques (Burns 1968; Webster and Peters 1978) in which animals are compressed, adhered to a surface, or allowed to swim free. Our procedure keeps the entire animal in the field of view for accurate, continuous observation while permitting freedom of movement of filtering and swimming appendages, abdominal claw, and carapace gape; and allows unimpeded flow of the filtering currents, water movements, and particles around the animal. Rejection rates were measured in a 20°C incubator under low intensity red light (Wratten filter No. 25) for 10-minute intervals using acclimated 2.5–2.7-mm animals. Suspensions of 10^2, 10^3, 10^4, 10^5, and 10^6 cells cc^{-1} of *C. reinhardi*, *A. flos-aquae* as filaments, and *A. flos-aquae* as cells were used. Individuals were exposed in random order to all cell concentrations. Significant differences between algal preparations were assessed using Mann-Whitney U-tests. Regressions were compared by ANCOVA on untransformed data.

Results and discussion

Growth parameters—Body size at death and percent growth show a significant decrease with decreasing cell concentrations of *C. reinhardi*. Values are greater for *C. reinhardii* than for *A. flos-aquae*; or for animals fed no food (Table 2). These parameters are higher for animals fed 1×10^5 cells cc^{-1} *A. flos-aquae* as cells or as filaments than for animals fed no food. However, animals fed no food have higher growth parameters than those fed 10^3 cells cc^{-1} *A. flos-aquae* as either cells or filaments. The observed growth rates represent increases in linear dimension, not necessarily in biomass, and in the case of the low rates for starved animals may only represent hydration after a molt. That animals have higher growth rates when fed *C. reinhardi* than equivalent or higher cell concentrations of *A. flos-aquae* or no food is also apparent

Table 2. Growth and suvivorship of *Daphnia magna* fed various cell concentrations of *Chlamydomonas reinhardi* (GMS⁻), *Anabaena flos-aquae* (toxic strain NRC–44-1) as filaments and as cells, and no food.

Treatment	Food concn (cells/cc)	Individuals	Size at death (mm)	% Growth	Instars	LT 50 (days)	Average life span (days)
No food							
1	0	33	1.00±0.02*	13.61± 1.07*	2.33±0.08*	3.71	4.68±0.33*
C. reinhardi							
2	1×10^3	13	1.30±0.14	53.69±13.10	5.26±1.03	7.00	13.42±2.88
3	1×10^4	22	2.92±0.11	209.09±13.12	15.45±1.08	45.50	42.86±3.76
4	1×10^5	18	3.66±0.30	356.00±36.80	13.00±1.59	27.50	33.17±5.22
A. flos-aquae filaments							
5	1×10^3	28	0.94±0.01	6.65±1.08	1.93±0.07	3.19	3.93±0.20
6	1×10^5	28	1.09±0.02	22.20±2.56	2.54±0.12	4.68	5.21±0.21
A. flos-aquae cells							
7	1×10^3	30	0.96±0.01	10.87±1.17	2.00±0.00	3.02	3.53±0.09
8	1×10^5	30	1.12±0.03	26.26±2.54	2.33±0.11	3.17	3.83±0.20

*Values are mean ± SE. All pair-wise differences among treatments are significant at least at the 0.05 level as determined by Mann-Whitney U-tests (2-tailed p), except: Size at death for treatments 2 vs. 6, 2 vs. 7, 2 vs. 8, 5 vs. 7, and 6 vs. 8; Percent growth: 1 vs. 7, 2 vs. 8; ($p<0.07$); Instars: 1 vs. 6, 1 vs. 8, 5 vs. 7, 6 vs. 8; Life span: 1 vs. 5, 5 vs. 7, 5 vs. 8.

from the body growth curves (Figs. 1 and 2). No significant differences are found between *A. flos-aquae* cells and filaments at the same concentrations, although the repeated growth patterns in Fig. 2 suggest that growth is better at high concentrations of cells. The number of instars is higher on *C. reinhardi* than on the other treatments but it is an insensitive indicator of differences among animals fed *A. flos-aquae* and no food as most animals in those groups molted only once.

High concentrations of *A. flos-aquae* as cells and as filaments promote higher growth than no food, but low concentrations of either form promote lower growth. This indicates that *A. flos-aquae* provides some nutrients for growth but less than *C. reinhardi*. At low concentrations it provides less than a threshold requirement and detrimental (i.e. toxic) properties result in lower growth than on no food at all.

Survivorship. Average lifespans (Table 2) are equivalent on 10^5 and 10^4 cells cc⁻¹ *C. reinhardi* and greater than on 10^3 cells cc⁻¹, no food, or on any *A. flos-aquae* treatment. A similar pattern is shown by the LT 50 data and survivorship curves (Figs. 3 and 4). Differences in lifespans among *A. flos-aquae* and no food treatments are less clear, due to the short time spans involved. Average lifespan is longer on the high concentration of *A. flos-aquae* in filamentous form than on the low concentration, on no food, or on either *A. flos-aquae* concentration in cellular form. No detectable difference in lifespan is found between no food and 10^3 cells cc⁻¹ *A. flos-aquae* as filaments, but lifespans are shorter on low and high concentrations of cellular *A. flos-aquae* than on no food.

The mortality patterns shown in the survivorship curves are more telling (Figs. 3 and 4). All animals fed *A. flos-aquae* died by day 6, those fed no food died by day 11, and those fed 10^3, 10^4, and 10^5 cells cc⁻¹ *C. reinhardi* died by days 34, 62, and 69 respectively. Survivorship curves show similar patterns of mortality throughout the lives of animals fed *C. reinhardi* and no food. However, high initial survivorship

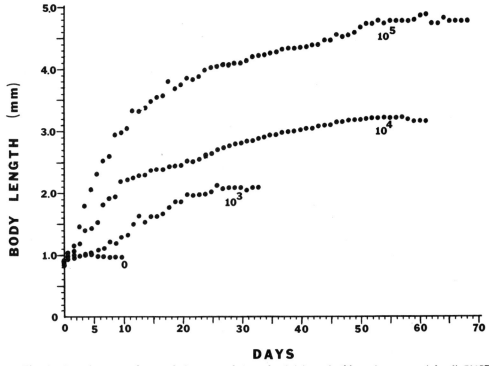

Fig. 1. Growth curves for *Daphnia magna* fed no food (0), and *Chlamydomonas reinhardi* GMS⁻ at concentrations of 1×10^3, 1×10^4, and 1×10^5 cells cc^{-1}.

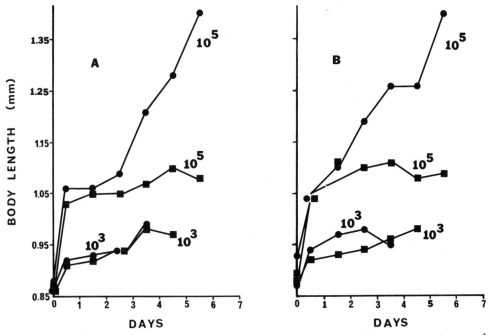

Fig. 2. Growth curves from replicate experiments (A and B) in which *Daphnia magna* were fed 1×10^3 and 1×10^5 cells cc^{-1} of toxic *Anabaena flos-aquae* NRC-44-1 as cells (●) and as filaments (■).

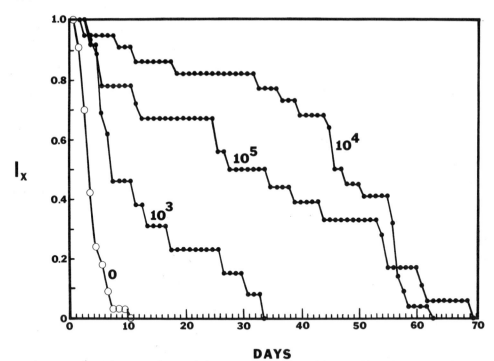

Fig. 3. Survivorship curves for *Daphnia magna* fed no food (0), and *Chlamydomonas reinhardi* (●) at concentrations of 1×10^3, 1×10^4, and 1×10^5 cells cc^{-1}.

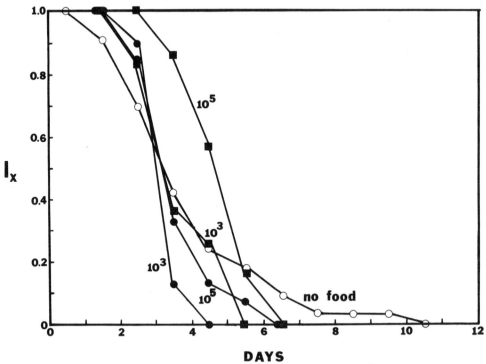

Fig. 4. Suvivorship curves for *Daphnia magna* fed no food (○) and 1×10^3 and 1×10^5 cells cc^{-1} of the toxin-producing strain of *Anabaena flos-aquae* NRC–44–1 in cellular (●) and filamentous (■) forms.

with abrupt mortality after day 2 occurs in animals fed *A. flos-aquae*. ANCOVA comparisons of angular transformed data (Sokal and Rohlf 1969) show that there are significant differences between the slope of the survivorship curve for animals fed no food and those for animals fed *A. flos-aquae* (all $p < 0.05$). The figure also suggests that mortality is more rapid for animals fed cells than filaments at each concentration. Shorter lifespans on 10^5 cells cc^{-1} of *A. flos-aquae* cells than on filaments or no food also suggest that cells are more readily ingested. This is confirmed by the ingestion rates reported later. These data indicate the *A. flos-aquae* toxin reaches a lethal dose by 48 hours of ingestion, a "slow death" effect (Carmichael and Gorham 1977), and that it is more readily available as cells than as filaments.

Fecundity. Reproduction occurred only in groups fed 10^4 and 10^5 cells cc^{-1} *C. reinhardi*. Initial age of reproduction, initial instar of reproduction and size at onset of reproduction were significantly lower on 10^5 than on 10^4 cells cc^{-1} (Table 3). The mean generation time, net reproduction rate, and intrinsic rate of increase, although not comparable statistically, show predicted differences. This indicates that *Daphnia* populations may be food-limited in nature where algal concentrations are often 10^4 cells cc^{-1} or less and are rarely of healthy log phase cells.

Ingestion and filtering rates—Ingestion rates (Table 4), measured either as cells ingested or as µg dry weight of algae ingested $animal^{-1} h^{-1}$ are equivalent for animals fed 10^4 and 10^5 *C. reinhardi* cells cc^{-1} and are significantly higher than for animals fed 10^3 cells cc^{-1}. This indicates that the two high cell concentrations of *C. reinhardi* are above the incipient limiting concentration for *D. magna* (in the experimental body-size range). Therefore, ingestion rates are maximal and independent of food concentration above 10^4 cells cc^{-1} of *C. reinhardi*. This is also illustrated as a highly significant reduction in the volume of feeding suspensions filtered clear of algae $animal^{-1} h^{-1}$ (filtering rate) at the highest cell concentration.

Ingestion rates and filtering rates for *A. flos-aquae* are significantly lower than those for equivalent cell concentrations of *C. reinhardi*. Ingestion rate differences between the two algae are even more pronounced when expressed in terms of µg dry weight $animal^{-1} h^{-1}$ because of the low dry weight per cell of the blue-green. *Anabaena flos-aquae* filaments are also ingested and filtered at significantly lower rates than equivalent cell concentrations of *A. flos-aquae* cells and any of the *C. reinhardi* concentrations. If animals are assessing food concentrations on the basis of cell volumes alone we would expect equivalent feeding rates on equivalent concentrations of the same-sized cells. If animals are assessing food concentrations on the basis of food content, as indicated by dry weight, we would expect higher feeding rates on *Anabaena* than on *Chlamydomonas*. Neither is the case. Our data show that *A. flos-aquae* cells are fed upon at lower rates than

Table 3. Reproduction parameters for *Daphnia magna* fed two concentrations of *Chlamydomonas reinhardi*. No reproduction was found in animals fed lower concentration of *C. reinhardi*, no food, or any concentrations of *Anabaena flos-aquae*.

| | *Chlamydomonas reinhardi* (cells cc^{-1}) | | | | |
	1×10^4			1×10^5	
Age of initial reproduction (days)	26.39±0.46	18	*	6.36±0.29	14
Instar at initial reproduction (instar No.)	12.44±0.28	18	*	6.00±0.15	14
Mean generation time T (days)	35.97	22	†	24.63	18
Net reproductive rate R_0 (♀/♀/gen.)	59.41	22	†	83.11	18
Intrinsic rate of increase r	0.20	22	†	0.29	18

*Significant differences were determined by Mann-Whitney U-tests, all $p<0.001$. Values are mean ± SE and number of individuals.
†Not testable. Values calculated according to Poole (1974).

Table 4. Ingestion and filtering rates of *Daphnia magna* fed various concentrations of *Chlamydomonas reinhardi* (GMS^{-1}) cells and a toxin-producing strain of *Anabaena flos-aquae* (NRC-44-1) in filamentous and cellular form. All concentrations on basis of cells cc^{-1}. Both algal cells are of equivalent size and shape.

Treatment	Food concn cells cc^{-1}	Replicates*	Ingestion rate† cells an^{-1} h^{-1} × 10^4	µg an^{-1} h^{-1}	Filtering rate† cc an^{-1} h^{-1}	Animal size† (mm)
C. reinhardi						
1	1×10^3	10	0.29±0.03‡	0.24±0.02‡	3.56±0.33‡	2.76±0.02
2	1×10^4	10	2.02±0.16	1.63±0.13	2.24±0.18	2.71±0.02
3	1×10^5	10	2.22±0.18	1.79±0.14	0.24±0.02	2.66±0.02
A. flos-aquae filaments						
4	1×10^4	15	0.16±0.01	0.0023±0.0001	0.063±0.004	2.87±0.16
A. flos-aquae cells						
5	1×10^4	14	0.44±0.01	0.0063±0.0002	0.163±0.012	2.64±0.03

*Number of replicates with 10 animals per replicate.
†Values are mean ± SE.
‡Significant differences exist among all treatments within a group ($p<0.001$), as determined by Mann-Whitney U-tests, except between ingestion rates at 1×10^4 and 1×10^5 cells cc^{-1} *C. reinhardi*.

equivalent-sized cells of *C. reinhardi* and are, therefore, being discriminated against on the basis of something other than size or shape—presumably taste. *Anabaena flos-aquae* cells packaged as filaments are ingested and filtered at the lowest rates, confirming reports that the filamentous form is relatively unavailable to zooplankton like *Daphnia* (Porter 1973, 1977; M. Lynch at this conf.; Webster and Peters 1978).

Rejection rates. Rejection rates increase with increasing cell concentration for all algae (Fig. 5) and the rate of increase (slope) is significantly higher on *A. flos-aquae* cells than on *C. reinhardi*. Since cells of both algae are of equivalent size and shape, animals are rejecting *Anabaena* on the basis of some other property—presumably taste. The rejection rate of *C. reinhardi* increases markedly above 10^4 cells cc^{-1}, the incipient limiting concentration. Below this level, rejection rates are equivalent to those in algae-free suspension (0 cells cc^{-1}) and probably represent a basal level of activity, other than food rejection, that involves flexion of the abdomen, such as preening, grooming, or aiding in peristalsis. Rejection rates on *A. flos-aquae* cells appear to be higher than rates on filaments at equivalent cell concentrations. However,

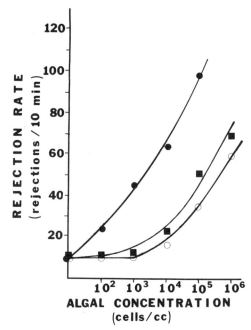

Fig. 5. Rejection rates of equivalent-sized *Daphnia magna* fed *Anabaena flos-aquae* (NRC-44-1) as cells (●) and as filaments (■) and *Chlamydomonas reinhardi* (GMS$^-$) cells (○).

since *A. flos-aquae* contain 134 ± 28 cells filament^{-1}, and encounter probability is directly related to particle concentration, filaments are encountered less frequently than cells although the same amount of algal material is encountered over time. A comparison of similar particle concentrations (i.e. 10^4, 10^5, and 10^6 cells cc^{-1} of filaments with 10^2, 10^3, and 10^4 cells cc^{-1} respectively of cells) shows no detectable difference in rejection rates.

Conclusion

Algal food quality is usually considered to be a function of nutrient content. However, other properties of algae, such as manageability and toxicity, can determine food quality. In this study, we find that *C. reinhardi* is a high quality food source for *D. magna*. Growth, survivorship, and reproduction increase with increasing concentrations of *C. reinhardi* and are greater than for animals fed no food. Animals appear to allocate energy in different ways as food concentrations increase. Below the incipient limiting food concentration, ingestion rates are low and filtering rates are high. Energy and nutrients gained go to maintenance, filtering, digestion, and growth. Above the incipient limiting concentration, ingestion rates are constant and filtering rates decrease significantly with increasing food concentration. Energy is allocated to maintenance, filtering, digestion, growth, and reproduction. As concentrations increase, less energy is expended in filtering and more is allocated to growth and reproduction.

In comparison, *A. flos-aquae* (NRC-44-1) is a low quality food source. This study shows that its food quality is related to low nutritive value, toxicity, and unmanageability in its naturally occurring filamentous form. *Anabaena flos-aquae* is of some nutritional value in that *Daphnia* growth rates increase with increasing algal concentration. *Anabaena flos-aquae* is, therefore, of low nutritional value. This may be because it has a significantly lower dry weight per cell than equivalent-sized cells of *C. reinhardi*.

Anabaena flos-aquae also produces toxins that are lethal to *Daphnia*. Toxic effects are apparent from the survivorship and feeding results. *Daphnia* have longer life spans on *C. reinhardi* than on *A. flos-aquae*. Animals have shorter life spans on *A. flos-aquae* cells than on no food. Survivorship curves show that there is an equal probability of mortality throughout life for animals fed *C. reinhardi* or no food, but there is an abrupt onset of mortality after 2 days in animals fed *A. flos-aquae*. This indicates that the toxin has the "slow death" effect (Carmichael and Gorham 1977) on its natural predator, while it has faster action when administered to organisms such as farm animals and other non-coevolved consumers. There is also a suggestion of earlier mortality in animals fed cells than those fed filaments. *Anabaena flos-aquae* toxin may also be detoxified or tolerated by animals that maintain a high food intake, since animals fed higher concentrations have less abrupt mortality patterns and higher growth rates than those fed low concentrations.

Anabaena flos-aquae may be discriminated against on the basis of taste. This is shown by the significantly lower ingestion and filtering rates and higher rejection rates on *A. flos-aquae* cells than on the same-sized cells of *C. reinhardi*. *Anabaena flos-aquae* filaments also present handling problems which result in lower ingestion and filtering rates than on cells. Since no differences in rejection rates are detected for equivalent particle concentrations of the two forms, the reduced ingestion of filaments may be due primarily to reduced accessibility, i.e. such as a reduced likelihood of entering the carapace or to an expenditure of time and energy in manipulation and orientation in the feeding apparatus of *Daphnia*. Easier accessibility resulting in higher ingestion rates of *A. flos-aquae* cells makes both nutrients and toxins more readily available and, therefore, growth rates are higher and survivorship is shorter on the cellular form than on the filaments.

When fed the toxic strain of *A. flos-aquae*, *Daphnia* expend energy in maintenance, filtering, digestion, manipulation of filaments, discrimination by rejection, and detoxification. Although *A. flos-aquae* supplies some nutrients, there is never sufficient energy for reproduction. This is probably due to toxins and inadequacy and not indigestibility, because previous studies show that blue-greens (including *Anabaena*) are assimilated at rates and efficiencies comparable to other algae that are good food sources (Arnold 1971; Lampert 1977). At low concentrations of *A. flos-aquae*, there is insufficient energy available for detoxification or tolerance; thus animals use body stores which result in growth rates and survivorships that are lower than on no food. At higher concentrations, the energy threshold for maintenance, digestion, and feeding is reached and animals show positive growth. Since cells are ingested more rapidly than filaments and with a smaller expenditure of energy in manipulation, they promote a higher initial growth rate yet lead to a shorter lifespan.

Toxicity of blue-green algae to their natural predators has been questioned in the literature. This study shows that a toxin-producing blue-green alga promotes growth while reducing survivorship when ingested as food by a natural planktonic herbivore. This study dissects the effect of poor nutritional quality from toxicity and shows that the toxin produced is lethal to zooplankton. Other studies using whole *A. flos-aquae* have had varied results (Arnold 1971; Mills and Wyatt 1974). These variations are due, in part, to differences among strains of blue-greens. Toxin production has been found to vary among different strain of *A. flos-aquae* (Carmichael and Gorham 1977, 1978). In particular, cultures of *A. flox-aquae* that were isolated from large, nonclonal colonies, such as strain NRC–44, have lost their toxicity over time, while clonal isolates from a single filament, such as NRC–44–1 used in this study, have maintained a constant rate of toxin production during years of culture (Carmichael and Gorham 1977). Potency loss occurred in the cultures used by Mills and Wyatt (Carmichael pers. comm.). This decrease, and the variety of toxins produced by each clone, indicates that toxins are produced at a cost to the alga and are not merely by-products or waste products of other metabolic processes. In laboratory cultures, nontoxin producing strains eventually outcompete and replace toxin-producing strains in nonclonal cultures. Selective pressures must favor the maintenance of toxin-producing strains in nature. We suggest that one such selection for their maintenance is their possession of an anti-herbivore device. Their presence may in turn select for tolerance or avoidance in zooplankton.

References

ANDERSON, B. G., H. LUMER, and L. J. ZUPANIC, JR. 1937. Growth and variability in *Daphnia pulex*. Biol. Bull. **73**: 444–463.

ARNOLD, D. E. 1971. Ingestion, assimilation, survival, and reproduction by *Daphnia pulex* fed seven species of blue-green algae. Limnol. Oceanogr. **16**: 906–920.

BANTA, A. M. 1939. Some studies on the physiology, genetics, and evolution of some cladocerans. Carnegie Inst. Wash. Pap. 39. 285 p.

BOTTRELL, H. H. 1975. Generation time, length of life, instar duration and frequency of moulting and their relationship to temperature in eight species of cladocera from the River Thames, Reading. Oecologia **19**: 129–140.

BROOKS, J. L. 1957. The systematics of North American *Daphnia*. Mem. Conn. Acad. Arts Sci. **8**: 1–180.

BURNS, C. W. 1968. Direct observations of mechanisms regulating feeding behavior of *Daphnia* in Lakewater. Int. Rev. Gesamten Hydrobiol. **53**: 83–100.

CARMICHAEL, W. W., D. F. BIGGS, and P. R. GORHAM. 1975. Toxicology and pharmacological action of *Anabaena flos-aquae* toxin. Science **187**: 542–544.

———, and P. R. GORHAM. 1974. An improved method for obtaining axenic clones of planktonic blue-green algae. J. Phycol. **10**: 238–240.

———, and ———. 1977. Factors influencing the toxicity and animal susceptibility of *Anabaena flos-aquae* (Cyanophyta) blooms. J. Phycol. **13**: 97–101.

———, and ———. 1978. Anatoxins from clones of *Anabaena flos-aquae* isolated from lakes in western Canada. Mitt. Int. Ver. Theor. Angew. Limnol. 21, p. 285–295.

CROWLEY, P. H. 1973. Filtering rate inhibition of *Dapnbia pulex* in Wintergreen Lake. Limnol. Oceanogr. **18**: 394–402.

FRYER, G. 1957. The food of some freshwater cyclopoid copepods and its ecological significance. J. Anim. Ecol. **26**: 263–286.

GENTILE, J. H. 1971. Blue-green and green algal toxins, p. 27–65. *In* S. Kadis et al. [eds.],

Microbial toxins, v. 7. Academic.

——, and T. B. MALONEY. 1969. Toxicity and environmental requirements of a strain of *Aphanizomenon flos-aquae* (L) Rolfs. Can. J. Microbiol. 15: 165–173.

GIBOR, A. 1956. Some ecological relationships between phytoplankton and zooplankton. Biol. Bull. 111: 230–234.

GORHAM, P. 1964. Toxic algae, p. 307–336. *In* D. F. Jackson [ed.], Algae and man. Plenum.

GREEN, J. 1956. Growth, size and reproduction in *Daphnia* (Crustacea: Cladocera). Proc. Zool. Soc. Lond. 126: 173–204.

HAERTEL, L. 1976. Nutrient limitation of algal standing crops in shallow prairie lakes. Ecology 57: 664–678.

HALL, D. J. 1964. An experimental approach to the dynamics of a natural population of *Daphnia galeata mendotae*. Ecology 45: 94–112.

HANEY, J. F. 1973. An in situ examination of the grazing activities of natural zooplankton communities. Arch. Hydrobiol. 72; 87–132.

HEISEY, D., and K. G. PORTER. 1977. The effect of ambient oxygen concentration on filtering and respiration rates of *Daphnia galeata mendotae* and *Daphnia magna*. Limnol. Oceanogr. 22: 839–845.

HOBBIE, J. E., R. J. DALEY, and S. JASPER. 1977. Use of Nuclepore filters for counting bacteria by fluorescence microscopy. Appl. Environ. Microbiol. 33: 1225–1228.

KEATING, K. I. 1978. Blue-green inhibition of diatom growth: transition from mesotrophic to eutrophic community structure. Science 199: 971–973.

LAMPERT, W. 1977. Studies on the carbon balance of *Daphnia pulex* as related to environmental conditions. I. Methodological problems of the use of ^{14}C for the measurement of carbon assimilation. Arch. Hydrobiol. Suppl. B 48, p. 287–309.

LEFEVRE, M. 1950. *Aphanizomenon gracile* Lem. Cyanophyte défevorable au zooplancton. Ann. Stn. Cent. Hydrobiol. Appl. 3: 205–208.

McMAHON, J. W., and F. H. RIGLER. 1963. Mechanisms regulating the feeding rate of *Daphnia magna* Straus. Can. Jour. Zool. 41: 321–332.

MILLS, D. H., and J. T. WYATT. 1974. Ostracod reactions to non-toxic and toxic algae. Oecologia 17: 171–177.

MOIKEHA, S. N., and G. W. CHU. 1971. Dermatitis-producing alga *Lyngkia majuscula* in Hawaii. II. Biological properties of the toxic factor. J. Phycol. 7: 8–13.

MORIARTY, D. J. W., J. P. E. C. DARLINGTON, I. G. DUNN, C. M. MORIARTY, and M. P. TEVLIN. 1973. Feeding and grazing in Lake George, Uganda. Proc. R. Soc. Lond. Ser. B 184: 299–319.

O'BRIEN, W. J., and F. DeNOYELLES, JR. 1973. Filtering rate of *Ceriodaphnia reticulata* in pond waters of varying phytoplankton concentrations. Am. Midl. Nat. 91: 509–512.

PARKER, R. A. 1966. The influence of photoperiod on reproduction and molting of *Daphnia scholderi* Sars. Physiol. Zool. 39: 266–279.

POOLE, R. W. 1974. An introduction to quantitative ecology. McGraw-Hill.

PORTER, K. G. 1973. Selective grazing and differential digestion of algae by zooplankton. Nature 244: 179–180.

——. 1975. Viable gut passage of gelatinous green algae ingested by *Daphnia*. Int. Ver. Theor. Angew. Limnol. Verh. 19: 2840–2850.

——. 1976. Enhancement of algal growth and productivity by grazing zooplankton. Science 192: 1332–1334.

——. 1977. The plant-animal interface in freshwater ecosystems. Am. Sci. 65: 159–170.

——, M. L. PACE, and J. F. BATTEY. 1979. Ciliate protozoans as links in freshwater food chains. Nature 277: 563–565.

POULET, S. A., and P. MARSOT. 1978. Chemosensory grazing by marine calanoid copepods (Arthropoda: Crustacea). Science 200: 1403–1405.

PROVASOLI, L., D. E. CONKLIN, and A. S. D'AGOSTINO. 1970. Factors inducing fertility in aseptic Crustacea. Helgol. Wiss. Meeresunters. 20: 443–454.

RICHMAN, S. 1958. The transformation of energy by *Daphnia pulex*. Ecol. Monogr. 28: 273–291.

SCHINDLER, D. W. 1968. Feeding, assimilation and respiration rates of *Daphnia magna* under various environmental conditions and their relation to production estimates. J. Anim. Ecol. 37: 369–385.

SCHINDLER, J. E. 1971. Food quality and zooplankton nutrition. J. Anim. Ecol. 40: 589–595.

SEALE, D. B., E. RODGERS, and M. E. BORASS. 1975. Effects of suspension-feeding frog larvae on limnological variables and community structure. Int. Ver. Theor. Angew. Limnol. Verh. 19: 3179–3184.

SOKAL, R. R., and F. J. ROHLF. 1969. Biometry. Freeman.

STANGENBERG, M. 1968. Toxic effects of *Microcystis aeruginosa* Kg. extracts on *Daphnia longispina* O. F. Müller and *Eucypris virens* Jurine. Hydrobiologia 32: 81–88.

STEIN, J. 1973. Phycological methods. Plenum.

TEKUZA, Y. 1974. An experimental study on the food chain among bacteria, *Paramecium* and *Daphnia*. Int. Rev. Gesamten Hydrobiol. 59: 31–37.

WEBSTER, K. E., and R. PETERS. 1978. Some size-dependent inhibitions of larger cladoceran filterers in filamentous suspension. Limnol. Oceanogr. 23: 1238–1245.

WIEGERT, R. G., and R. MITCHELL. 1973. Ecology of Yellowstone thermal effluent systems: Intersects of blue-green algae, grazing flies (*Paracoenia*, Ephydridae) and water mites (*Partnuiella*, Hydrachnellae). Hydrobiologia 41: 251–271.

26. Filtering Rates, Food Size Selection, and Feeding Rates in Cladocerans—Another Aspect of Interspecific Competition in Filter-Feeding Zooplankton

Z. Maciej Gliwicz

Abstract

As an alternate hypothesis to size-selective predation, another highly size-selective biotic factor is proposed, which, combined with unselective predation, would affect the structure of zooplankton communities. This hypothesis relies upon the abundant inedible net phytoplankton (large, mostly colonial or filamentous algae) which interfere to a higher degree with food collection (thus growth and fecundity) in larger cladocerans than in smaller cladoceran species. Larger cladocerans, because they ingest larger particle sizes, experience a reduction in efficiency of food collection under conditions of net phytoplankton blooms, while less affected smaller cladocerans do much better.

Although the Brooks and Dodson (1965) size-efficiency hypothesis has been convincingly criticized (Hall et al. 1976), there is still one aspect left which seems to be generally accepted. It is that the greater effectiveness of food collection in larger than in smaller filter-feeding species (one of the most important components of their competitive success) is not only due to the relatively higher filtering rates of larger-bodied species but also results from their ability to graze a wider size spectrum of food particles (the higher upper size limit of particles available). Not only the competitive aspect, but also the grazing effect upon the phytoplankton and the energy-material transfers along the food chains cannot be satisfactorily analyzed without detailed information on the size spectrum of grazed food particles. The latter aspect seems even more important than the filtering rate: assuming an even size distribution of particles in the environment (same numbers of particles of every size class) and their uniform spherical shape, it is easy to calculate that less than a twofold increase in the upper size limit of particles ingested would be enough for a 10-fold increase in feeding rate, while for the same increase we would need to increase the filtering rate 10 times if the upper size limit remained unaltered.

It is known that the effective filtering rate (the volume of water from which particles are not only collected but also ingested) may change widely up to its possible maximum, depending on food concentration and various aspects of an animal's physiological state. It is not known, however, how (and whether at all) the upper size limit of particles grazed changes, what the possible maximum size is, and how it is related to changes in the filtering rate. Might it be that a cladoceran feeding rate is regulated by changes in both filtering rate and upper size limit, as it certainly is in calanoid copepods?

Were the size spectrum of particles

Presentation of this paper at the conference was made possible by travel support from the National Science Foundation.

grazed constant, and were both the upper size limit and filtering rate known for each instar of important filter-feeding species, we could hopefully describe both the entire zooplankton feeding rate and the selective grazing effect of any zooplankton community upon the size distribution of phytoplankton. However, once we try to estimate these experimentally, our hope appears an obvious illusion.

The first obstacle that we encounter is a significantly different upper size limit of particles grazed (those collected in the filtering chamber) than of particles ingested (those inside the intestinum). Although this difference (Fig. 1) may be associated with the very "plastic nature" of the particles used in the experiment, it seems, however, that a different upper size limit should be considered depending on whether we are after the feeding rates (energy transfers and their efficiency) or after the grazing effect, since those collected yet undigested particles, especially when soft, must also be affected somehow by the filtering machinery of animals.

Another obstacle seems more difficult. It is the shape of the curve of filtering rate versus particle size. It never happens to be of type A but always of B or, more rarely, of type C (Fig. 2); which means that the grazing pressure of a cladoceran (cladoceran instar) is different on various algal sizes within the size spectrum of particles grazed (this also makes the feeding rate much more difficult to estimate). Such a shape (B) of the curve may be explained as a result of the feeding by many individuals (such estimations are mostly made on a large sample of animals—e. g. Gliwicz 1969, 1977; Bogdan and McNaught 1975; S. Richman at this conf.), some of which have had a higher and some of which have had a lower upper size limit of particles

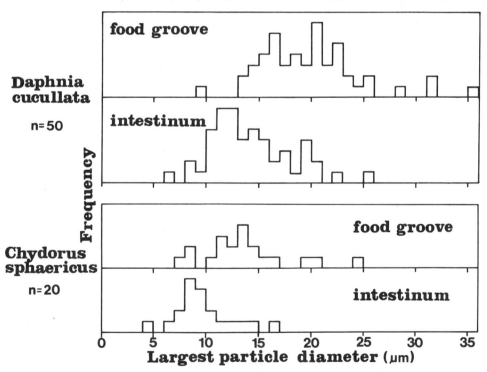

Fig. 1. Example illustrating differences in upper size limit of particles grazed (collected) and of particles ingested: distribution of animals in respect of sizes of largest particles (plastic beads) found in food groove and intestinum after 20 minutes in situ exposure of natural plankton to low concentrations of beads in eutrophic lake in summer 1973.

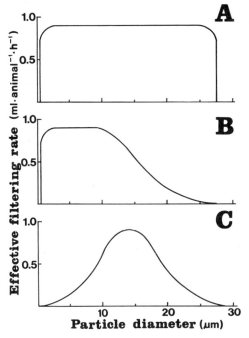

Fig. 2. Of three possible types of relation between filtering rate and size of food particles, type B is most frequently found in experimental measurements.

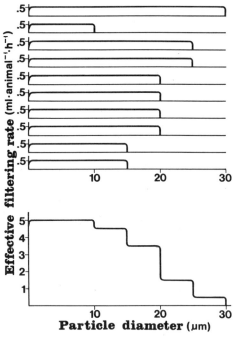

Fig. 3. Shape of curve B (Fig. 2) may result from feeding by many animals, some of which had higher (e.g. 30 μm) and some lower (e.g. 10 μm) upper size limits of particles collected and ingested. Assuming same filtering rate for all individuals and normal distribution of upper size limit for 10 individuals (top), we obtain B shape for overall feeding rate (bottom).

grazed (Fig. 3). Alternatively, the curve might also be explained, for a single cladoceran, as a result of either the animal's morphology or behavior (Fig. 4). The most convincing argument seems, however, to be on the basis of behavior. The curve may result from more frequent postabdominal rejections of larger than of smaller particles, which is certainly the case when blue-green filaments are considered (e. g. Burns 1968); it may also be a result of a change in the upper size limit with time (e. g. Fig. 4, b).

That the upper size limit of particles grazed is altered in time became evident from the "filtering rate versus particle size" curves obtained in various seasons of 1973 for some cladoceran species in a eutrophic lake (Gliwicz 1977). The most distinct changes were demonstrated in *Daphnia cucullata* Sars—a marked decrease in the mean upper size limit of particles grazed (much more reduced filtering rates of larger than of smaller particles) in summer, during net (50 μm) phytoplankton blooms dominated by peridinians and blue-greens (Fig. 5, left-hand side—July, August, September). The summer effective filtering rate was less dramatically decreased in *D. cucullata* than in those "spring" cladocerans that tended to keep their high upper size limit of the particles grazed unaltered. The latter had feeding rates, fecundity, and numbers much more reduced during the summer blooms and, consequently, were replaced by "summer" species. Those "summer" species (*Chydorus sphaericus* O. F. Müller and *Diaphanosoma brachyurum* Lievin) which had a much lower upper size limit of particles ingested, were most probably less affected by the presence of large algae that interfered with filtration. So, benefiting from the decrease in the

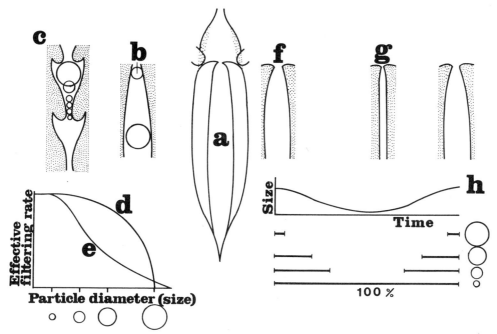

Fig. 4. Shape of curve B (Fig. 2) may also be explained on grounds of morphology or behavior of single animal. On the ventral side of *Daphnia* or *Bosmina*, we frequently notice that the gape between the edges of the carapace valves is not equally wide within its front part "a" where water with food suspension is sucked into filtering chamber. Since particles smaller than "b" may be brought into chamber through every part of the gape, while only larger ones pass through middle part of it, "d" type of curve would be expected; "e" type is possible if passage were rather more dependent on gape between edges of thoracic appendages, i.e. when water is pumped into chamber and if gape is, for instance, of "c" shape. However, a more realistic explanation is that either carapace gape or gape between appendages (probably both) alters chamber width in time. If there were regular carapace gape alterations between "f" and "g" dimensions, the smaller particles would be constantly (100% time) brought into filtering chamber and larger particles only during a limited time "h." Again, "e" type of curve would be expected.

"spring" species' numbers and from larger amounts of fine particulate material in the lake, *Chydorus* and *Diaphanosoma* demonstrated their highest fecundity during summer blooms.

The same set of data (Fig. 5, left-hand side) supplies us with an example of the C (Fig. 2) type of the "filtering rate versus particle size" curve. It may not be pure coincidence that this type was only once found experimentally in July, at the time of the highest abundance of hard-covered algal cells. The cells, mostly comprised of *Peridinium* sp. between 40–50 µm in diameter, are sufficiently small to be brought into a filtering chamber, but large and solid enough to make tight closure of the chamber difficult. As a result, particles <5 µm may possibly escape from the chamber during the abduction-adduction phase of filter feeding, when water is supposed to be pressed out through the setules. This leakage should also be considered as a possible cause of a drop in the effective filtering rate.

The summer decrease in the upper size limit of the particles grazed by *D. cucullata* (Fig. 5, left-hand side) was found to coincide nicely with the decrease in the proportion of the animals that had their carapace gape opened wider (Fig. 5, right-hand side). It seems, therefore, that *Daphnia* tends to close its carapace gape when there is abundant net phytoplankton in the lake which might obviously interfere with efficient food collection by the ani-

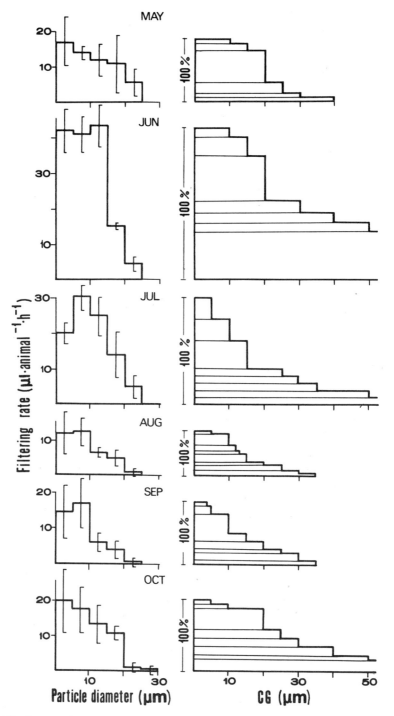

Fig. 5. Filtering rate (mean and standard deviation of sample) of *Daphnia cucullata* with respect to particles of various size ranges in different months of 1973 (left-hand side) and (right-hand side) percentage distribution of animals according to width of their carapace gape (CG) during same months in eutrophic lake. Filtering rate estimated from in situ 15-minute exposures of natural lake plankton to low concentrations of plastic beads; CG measured on fixed animals (details in Gliwicz 1977).

mals. In summer 1977, in vivo measurements of the width of the carapace gape were made in *D. cucullata* from lakes with little and with much net phytoplankton. The net phytoplankton in rich lakes was dominated by peridinians, blue-greens, colonial diatoms, and greens (details in Gliwicz and Siedlar, 1980). Whether we compared gape with the net phytoplankton fresh weight or with Carlson's (1977) "Trophic State Index," significant negative correlation was found—even stronger when only 20% of animals with narrowest gape were taken into consideration (Fig. 6).

That cladocerans narrow the gape between the edges of the carapace valves in the presence of dense net phytoplankton is also evident in other species. In 1977 gape narrowing was observed in *Daphnia hyalina pellucida* (Leydig) that were transferred from net phytoplankton-poor hypolimnetic water (where it exclusively remains during the summer blooms, perhaps not necessarily due to selective fish predation) into net phytoplankton-rich epilimnetic water which had been previously cooled down to the hypolimnetic temperature. In 1978, the phenomenon was also observed in *Simocephalus vetulus* (O. F. Müller) and *Ceriodaphnia quadrangula* O. F. Müller when both were transferred from net algae-free water of a small pond adjacent to the shore of a eutrophic lake, to the net algae-rich epilimnetic water of the lake.

Gape narrowing was also observed in the lab in *Daphnia magna* Straus (1.8–2.3 mm long) that were exposed to various concentrations of *Anabaena* filaments. Correlations were significant for animals with empty egg chambers and those bearing eggs or embryos (Fig. 7). Of interest is the pattern of *Daphnia* response to the introduction of blue-green filaments or of concentrated natural lake net phytoplankton into the medium. As soon as these interfering particles appear in the throughflow observation chamber where *Daphnia* were fastened to the end of a glass capillary tube, and as soon as the particles make physical contact with the animal's filtering machinery, the cladoceran makes frequent post-

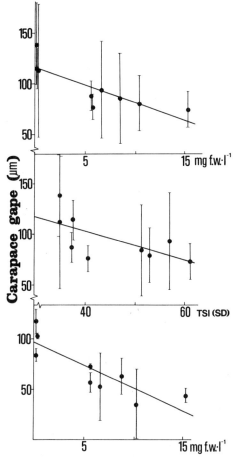

Fig. 6. Width of carapace gape (mean and standard deviation of sample) measured in live *Daphnia cucullata* in natural water from lakes with low and high net phytoplankton biomass expressed as fresh weight (top, $r = 0.71$, $p = 0.95$) or as Secchi disk-based Trophic State Index (middle, $r = -0.66$, $p = 0.95$). Same comparison shown for 20% of animals with narrowest carapace gape (bottom, $r = -0.85$, $p = .01$) demonstrates lakes in which daphnids tend to close gape more frequently (details in Gliwicz and Siedlar, in press).

abdominal movements to clean the food groove (Fig. 8, top). The movements are less and less frequent as the gape becomes narrower than previously in the absence of interfering particles. Eventually when the carapace gape is set at half the opening, the movements become almost as infrequent as they were before introduction of the interfering particles. Until the gape becomes

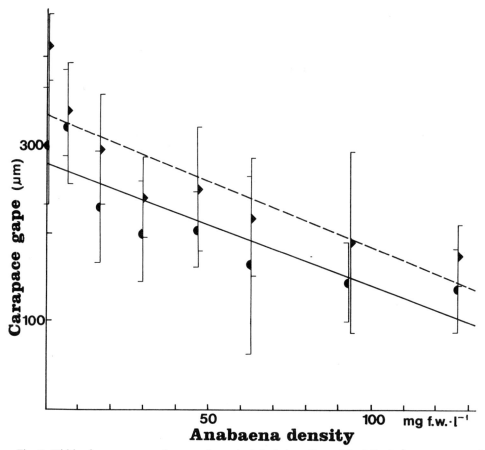

Fig. 7. Width of carapace gape (mean and standard deviation of sample) of *Daphnia magna* exposed for at least 30 minutes in various concentrations of *Anabaena* filaments. In both egg/embryo-bearing animals (dashed line, $r = -0.84$) and animals with empty egg chambers (solid line, $r = -0.88$), correlations are significant at $p = 0.99$ (details in Gliwicz and Siedlar, 1980).

reduced, only a very small part of the food collected could have been efficiently used. On the assumption that food collected requires 10 seconds to be pushed forward along the feeding groove and ingested, the percentage of time used for effective filtration (food collected and ingested) was calculated (e. g. 0% if there was not one pause longer than 10 seconds between subsequent movements of the postabdomen). This percentage is significantly decreased in the presence of interfering particles to become high again once the gape is narrowed (Fig. 8, bottom). It is not, however, much increased in those animals that retain a wider carapace gape.

Although one might expect that the chemical properties of blue-greens were responsible for the increased frequency of postabdominal rejections and, possibly, for the narrowing of the carapace gape, these reactions appeared to be induced simply by mechanical interference. While cellulose fibers of similar dimensions caused the same behavior in cladocerans, neither homogenized (ground) blue-greens nor extracts from concentrated log phase and dying cultures of *Anabaena* produced any effect.

So, summing up, the hungry cladoceran faces a dilemma when net phytoplankton blooms occur in the lake: it must either

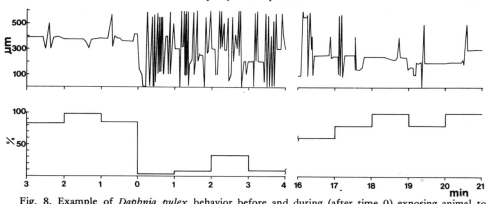

Fig. 8. Example of *Daphnia pulex* behavior before and during (after time 0) exposing animal to high density ($8 \cdot 10^4$ filaments $\cdot$ ml^{-1}) of *Anabaena* filaments. Changes in width of carapace gape (top) also reflect frequency of postabdominal movements. Percentage of time used for effective filtration (bottom) was calculated assuming collected particles need 10 seconds to be pushed forward and ingested (details in Gliwicz and Siedlar, in press).

narrow the carapace gape and not leave the edges of the thoracic appendages apart too much (to avoid interfering particles), or it must leave the carapace gape wide open so as not to lose larger food particles. In both cases, the feeding rate must drop: either because of the smaller amount of food entering the filtering chamber in each unit of water or as the result of a drop in the effective filtering rate (i. e. the volume of water from which food particles were collected and ingested).

This is also clearly seen in the relation between the filtering rate (of the smallest particles: 0–10 μm) and the maximum size of particles grazed by *D. cucullata*, *Bosmina coregoni* Baird, and *Diaphanosoma brachyurum* during the net phytoplankton blooms (peak in August) in the same eutrophic lake. While in May, June, and October, the largest particles are found in animals that have the highest filtering rate (highest chance of encountering larger particles which are less abundant in the environment), so in August the largest particles are found in animals having the lowest filtering rate (Fig. 9). In other words, those animals which also graze on larger particles (have their gapes opened apart) can efficiently filter neither larger nor smaller particles.

What should the conclusion be following this comment on the structure of zooplankton communities?

A filter-feeding cladoceran certainly does benefit from the high upper size limit of particles grazed, as has been pointed out by Brooks and Dodson (1965). It, however, benefits only while there are few large or filamentous algal forms that might interfere with food collection. The lower the upper size limit of particles grazed, the less a cladoceran is endangered by interfering algae, the less its feeding rate is decreased, and the less its growth and fecundity is halted. The species ability to change from the broader size spectrum of particles grazed (utilized in poor food conditions) into the narrower one, when both the smallest food particles and large inedible algae become abundant, seems to reflect its opportunistic strategy.

I am not quite certain whether all larger cladoceran species necessarily have their upper size limit of particles grazed higher, although, in general, this should be expected from field (Gliwicz 1977) and laboratory (Webster and Peters 1978) evidence. If the general relationship does hold, then abundant inedible net phytoplankton should be thought of as a highly size-selective factor co-responsible for the size structure of zooplankton communities. A shift in the zooplankton size distribution toward smaller forms could result as well from a combined effect of a nonselective predator and the inhibitory effects of filamentous algae, with the latter causing a greater decrease in fecundity of larger than of

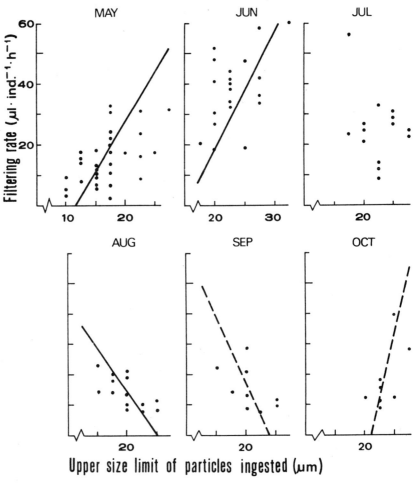

Fig. 9. Filtering rates for particles below 10-μm diameter vs. maximum size of particles ingested by *Daphnia cucullata* in various months of 1973 in eutrophic lake. Both x and y values obtained for individual animals from in situ 15-minute exposures of natural plankton to low concentrations of plastic beads. Solid regression lines—r significant at $p = 0.99$; dashed ones—correlation not significant.

smaller cladoceran species. Despite their general competitive priority, the larger species would become losers in competition games as soon as their fecundity dropped, exactly as predicted from the Gause's "winner becoming a loser" model (*see* Slobodkin 1963).

It seems reasonable enough to expect that quite a large part of all those examples of zooplankton size structure (also of those presented at this conf.), which are readily explained by selective predation (almost all is there: seasonal and long-term changes, oligo- and eutrophic lakes, tropical and temperate water, vertical and horizontal distribution, etc.) could, with no lesser confidence, be explained by another selective biotic or abiotic factor combined with nonselective predation. The very concept of selective predation is, no doubt, of vital importance for understanding the patterns of the structure of zoo- and phytoplankton communities. The trouble, however, is that it becomes attractive to follow the one-way road and to ignore the antichaotic forces of the negative feedback mechanisms in the plankton system. And this is perhaps why selective predation would no longer be sufficient in itself, when, as once we arrive at the highest predator trophic level—a dead end—we find no more predators to be used for further explanations.

References

BOGDAN, K. G., and D. C. McNAUGHT. 1975. Selective feeding by *Diaptomus* and *Daphnia*. Int. Ver. Theor. Angew. Limnol. Verh. 19: 2935-2942.

BROOKS, J. L., and S. DODSON. 1965. Predation, body size, and composition of the plankton. Science 150:28-35.

BURNS, C. W. 1968. Direct observations of mechanisms regulating feeding behaviour of *Daphnia* in lakewater. Int. Rev. Gesamten Hydrobiol. 53:83-100.

CARLSON, R. E. 1977. A trophic state index for lakes. Limnol. Oceanogr. 22:361-369.

GLIWICZ, Z. M. 1969. Studies on the feeding of pelagic zooplankton in lakes with varying trophy. Ekol. Pol. A 17:663-708.

———. 1977. Food size selection and seasonal succession of filter feeding zooplankton in an eutrophic lake. Ekol. Pol. 25:179-225.

———, and E. SIEDLAR. 1980. Food size limitation and algae interfering with food collection in *Daphnia*. Arch. Hydrobiol.

HALL, D. J., S. T. THRELKELD, C. W. BURNS, and P. H. CROWLEY. 1976. The size-efficiency hypothesis and the size structure of zooplankton communities. Annu. Rev. Ecol. Syst. 7:177-208.

SLOBODKIN, L. B. 1963. Growth and regulation of animal populations. Holt, Rinehart and Winston.

WEBSTER, K. E., and R. H. PETERS. 1978. Some size-dependent inhibitions of larger cladoceran filterers in filamentous suspension. Limnol. Oceanogr. 23:1238-1245.

27. Resource Characteristics Modifying Selective Grazing by Copepods

Donald C. McNaught, David Griesmer, and Michele Kennedy

Abstract

Algal resource characteristics, including the size, shape, and presence of a sheath, were examined for their modifying influence on selective grazing. Three green and two blue-green algae were grown in culture and fed in pairs to the crustacean plankton that occurred in Lake Huron in spring; animals were preconditioned before testing. Only two of seven forms were selective specialists, including adult *Diaptomus sicilis* and immature cyclopoid copepods. Three species of adult cyclopoids and one cladoceran were generalists in feeding. The selective grazers chose large and small plain cells over sheathed and spined algae. Certainly the presence of a sheath was overwhelmingly important in modifying resource utilization; sheathed algae exhibit increased fitness, as evidenced by their success in forming blooms.

An active controversy concerning the selection of phytoplankton foods by zooplankton currently revolves around the question of whether filtrators behave as though they had a fixed sieve, passively removing particles collected by filtration, or as active selectors. Originally Wilson (1973) suggested that calanoid copepods could track a biomass peak of food particles, removing those just larger than the modal size. More recently Richman et al. (1977) found that marine calanoids (*Acartia* and *Eurytemora*) first removed the largest cells and then actively "tracked" biomass peaks, taking large or small particles of the most abundant size. Frost (1977) simultaneously countered with the suggestion that *Calanus* indiscriminantly filtered food suspensions passively, capturing the larger particles more efficiently than the small. These current arguments do not negate selective feeding based on size alone, but debate the mechanisms and whether the process is active or passive.

Earlier Burns (1968) suggested that filtrators captured large particles in proportion to their own body size. However, some freshwater calanoids (*Diaptomus*) exhibited their highest filtration rates on small particles (McQueen 1970), suggesting long term adaptation to a certain resource size spectrum (McNaught 1975). Freshwater cladocerans also took very small particles, with *Daphnia* selecting those in the 8–16-μm range, and *Bosmina* those of a little smaller (3–14 μm) size (Gliwicz and Hillbricht-Ilkowska 1972). Obviously aquatic herbivores are selective feeders, and our assessment of the evidence of others suggests that calanoids are actively selective and highly specialized. In contrast, the cladocerans, which often feed in eutrophic environments, may be much more gener-

---

This paper is dedicated to Prof. A. D. Hasler in celebration of his contributions to aquatic ecology. It has been supported by U. S. EPA grant 803178, as a component of the southern Lake Huron study. Logistic support was provided by men of the RV *Simons*.

alized in their food habits, taking both large and small particles when these foods are presented in a properly designed experiment (Bogdan and McNaught 1975).

Factors other than size are basic to understanding the mechanisms of selective feeding. The interactions between size and other postulated factors have not been investigated. To the overemphasized factor of food size I would add its shape, whether it contains a sheath, its resistance to digestion, and its taste. Blue-green algae often have a sheath. Arnold (1971) has shown that *Daphnia pulex* rejected small, sheathed *Anacystis* and also that their assimilation of these blue-greens was lower than that upon other algae. These same blue-greens may simply not be digested and pass through the gut unharmed (Porter 1973). Thus factors other than size have been investigated, but the proper experiments to separate the independent effects of size, presence or absence of a sheath, shape, taste, and assimilability have not been performed. To test all factors simultaneously would require a complex experimental design and substantial resources. Thus, we will examine only the effects of size, shape, and the presence of a sheath on a resource's ingestability by the major herbivorous components of a natural community of zooplankton in southern Lake Huron.

Hypothesis and significance

Copepods and cladocerans found in southern Lake Huron have evolved in freshwater environments for over 12,000 years. The early day Laurentian Great Lakes were extremely oligotrophic, with phytoplankton populations dominated by small centric diatoms like *Cyclotella*, still prevalent in Lake Superior (Schelske et al. 1972). Calanoid copepods, which were the earliest inhabitants of these environments, may have fed upon such small diatoms, although both their feeding habits and food populations have evolved since (McNaught 1975). Thus the likelihood of size-selective grazing by extant forms is great. In more recent times, especially during the last 40 years, (McNaught and Scavia 1976) the cladocerans, which are typically warm-water inhabitants, have become abundant in the shoreward regions of Lake Huron and may constitute up to 85% of inshore populations. We have pictured them as generalists with regard to diet (they eat more blue-greens) and particularly with regard to size-selective feeding (they consume a broader range of sizes). Thus the cladocerans are marked generalists especially when contrasted with the highly specialized calanoids.

The purpose of these experiments was to define the mechanisms of selective feeding. The underlying factors are likely both quantifiable and interrelated. Size-selective feeding alone is too simple to explain the niche structure of the pelagic zooplankton; taste and shape of foods and other factors must be considered.

Understanding selective feeding is important to describing the behavior of herbivores at the organismic level, but it is imperative to define their combined impact on phytoplankton populations at the ecosystem level. Selective feeding also implies that certain potential food resources are ignored. Underutilized foods may grow unchecked into noxious nuisance blooms. In such systems selection on algal populations is likely based both on selective grazing (inhibition) and nutrient remineralization (stimulation) of algae by zooplankton.

Methods for laboratory studies of selective grazing

Seven common herbivorous crustaceans, including five adult stages (*Diaptomus sicilis*, *Cyclops bicuspidatus*, *Cyclops vernalis*, *Tropocyclops prasinus*, and *Eubosmina coregoni*) and two of their larval stages (copepod nauplii and cyclopoid copepodites), were fed mixtures containing 50% by volume of one of two blue-green algae (*Gloeocapsa* sp., *Anacystis nidulans*) and one of three green algae (*Scenedesmus quadricauda*, *Ankistrodesmus falcatus*, and *Pediastrum* sp.). Animals comprising the natural grazing assemblage were collected from waters of about 4°C. Before measuring ingestion with labeled algae, these herbivores were acclimated to the paired mixtures for

2 h. After feeding on the labeled mixture for 15 min, they were narcotized and killed; they were picked by species, digested in Protosol, and counted in a liquid scintillation counter. These experiments involved two treatments with replicates, one treatment in which the blue-green was tagged (50 µCi of ^{14}C) and the green was cold, and the other where the green algae was tagged (^{14}C) and the blue-green was cold. There were at least six internal replicates for each of these combinations. Thus 12 independent treatments resulted in 72 samples containing seven herbivores. Approximately two to three subsamples of 50 animals of each species were used; thus about 6×10^4 animals were used to provide data on selective feeding at the species/stage level. Differences in selectivity were detected by examining respective filtering rates, using the t-statistic for unpaired means and a two-level ANOVA. In addition, electivities were calculated according to Ivlev (1955).

Results

Selective cropping of paired resources. Zooplankton graze at varying rates on different food resources. The initial analysis of these experiments consisted of calculating zooplankton filtering rates (ml ind^{-1} h^{-1}) upon paired algal resources, wherein one of the six pairs was a sheathed blue-green and the other an unsheathed green. The sheathed blue-greens *A. nidulans* and *Gloeocapsa* sp. were paired with the unsheathed greens *S. quadricauda*, *A. falcatus*, and *Pediastrum* sp. Since the experiments were done in replicate the results were amenable to an unpaired t-test, since some samples yielded more subreplicates than others. Results showed whether zooplankton filtering rates were significantly higher on one algal species than another. However, these results did not indicate whether one food was selected or another rejected. This second analysis was accomplished with an electivity index. Those characters chiefly responsible (size and sheath) were further examined with a 2×2 factorial analysis.

Most crustacean zooplankters were not highly selective feeders when presented with paired resources made up of five common algae. This generalization is based on the lack of selection of food by five dominant species (adult stage) and two immature stages (nauplii and cyclopoid copepodites) common enough to analyze without low numbers a priori indicating insignificance. Three common adult cyclopoids (*C. bicuspidatus*, *C. vernalis*, and *T. prasinus*) were not selective on any of the six paired foods (Table 1). Likewise the early immature copepod nauplii and the cladoceran *E. coregoni* were not selective. Thus the predominant conclusion is that food selectivity is not a characteristic of these important herbivores; they took both sheathed and unsheathed blue-greens and greens of a variety of sizes. However, two dominant forms (adult *D. sicilis* and cyclopoid copepodites) were highly selective and deserve further analysis.

First we will analyze the original filtering rates of *Diaptomus* and the cyclopoid copepodites, as each showed very different ingestion of the paired resources. *Diaptomus* selected the small, sheathed *Gloeocapsa* over the large, plain *Pediastrum* (Table 2) with respective filtering rates of 0.087 vs. 0.043 ml ind^{-1} h^{-1}. Similarly the cyclopoid copepodites selected *Gloeocapsa* over *Scenedesmus* (filtering rates of 0.057 vs. 0.010 ml ind^{-1} h^{-1}). Thus the small sheathed blue-green was always selected. In contrast, the greens were always selected when paired with the large, sheathed blue-green *Anacystis*. *Diaptomus* selected the spined *Scenedesmus* over *Anacystis* (filtering rates of 0.021 vs. 0.009 ml ind^{-1} h^{-1}); as well as *Pediastrum* (filtering rates of 0.092 vs. 0.015 ml ind^{-1} h^{-1}). The large, sheathed *Anacystis* was never selected. Now we can determine whether these very different filtration rates were due to active selection or simply to ignorance of specific foods.

Electivity matrix of resource characteristics. It is desirable for better understanding of ecosystem function and later modeling to identify those resource characteristics selected by herbivorous zooplankton. Electivities on the five phytoplankton species

Table 1. Variations in filtering rate on green (G) vs. blue-green (BG) algae by common herbivores of Lake Huron. Significantly (NS—not significant) higher filtering rates on one member of pair indicated; (—) indicates lack of significant numbers of trials.

	Scenedesmus	Ankistrodesmus	Pediastrum
	Paired with small, sheathed blue-green Gloeocapsa		
Diaptomus sicilis (A)	NS	NS	BG*
Cyclops bicuspidatus (A)	NS	NS	NS
Cyclops vernalis (A)	NS	NS	NS
Tropocyclops prasinus (A)	NS	—	NS
Cyclopoid copepodites	BG*	NS	BG*
Copepod nauplii	NS	NS	NS
Eubosmina coregoni (A)	NS	—	NS
	Paired with large, sheathed blue-green Anacystis		
Diaptomus sicilis (A)	G†	NS	G*
Cyclops bicuspidatus (A)	NS	NS	NS
Cyclops vernalis (A)	—	—	—
Tropocyclops prasinus (A)	NS	NS	NS
Cyclopoid copepodites	NS	NS	G*
Copepod nauplii	NS	NS	NS
Eubosmina coregoni (A)	—	—	—

*$\alpha-0.05$.
†$\alpha-0.01$.

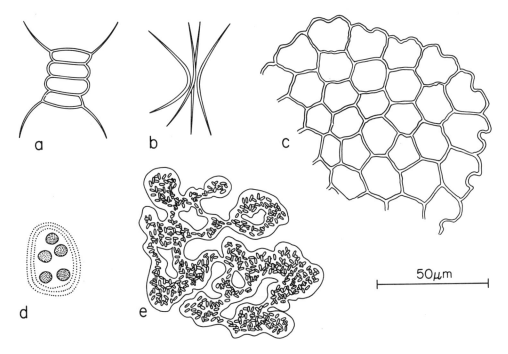

Fig. 1. Characteristics and relative sizes of the five resources (a = Scenedesmus, b = Ankistrodesmus, c = Pediastrum, d = Gloeocapsa, e = Anacystis).

presented in six pairs showed generally a selection for green algae when paired with a large blue-green and for a very small blue-green when paired with a variety of greens (Table 1). Thus neither the greens nor the blue-greens were selected consistently nor were the smaller species always preferred. Since the five resources had other characteristics than taxon and size, such properties deserve further attention.

Both blue-greens (*Gloeocapsa* and *Anacystis*) have a sheath or gelatinous matrix (Fig. 1). The species of *Scenedesmus (quadricauda)* was selected for its unusually long spines. *Ankistrodesmus*, usually found in fascicles of long, crescent-shaped cells, was considered unspined and clearly unsheathed. *Pediastrum*, a large colony of cells, was considered to have such short lobes that it was emarginate or unspined. Thus experimentally determined electivities on these various resources can be placed in a matrix, where size (large vs. small) can be paired with the presence of spines or the presence or absence of a sheath (Table 2). Ivlev's (1955) electivity index (E) was used, where:

$$E = (R-P)/(R+P)$$

where R is proportion of resources taken and P is proportion of total resources available.

The resource characteristics matrix (Tables 2, 3A) is useful in determining the degree to which size, sheath, or spines influence electivity. By summing the rows we obtain the mean electivity by size and by summing the columns the electivity for spined and sheathed cells. These electivities were then ranked (Tables 2, 3B). *Diaptomus sicilis*, a large, omnivorous calanoid copepod, selected small sheathed blue-green algae (*Gloeocapsa*) and large, plain green algae (*Pediastrum*) in proportions greater than

Table 2. Matrix of resource characteristics (A) and ranked food character preferences for *Diaptomus sicilis* (B) (E = electivity).

A.

	spined small	spined large	sheathed small	
small unspined and unsheathed	−0.20 / −0.02	−0.18 / −0.04	+0.04 / −0.02	$E = -0.03$ small, plain
large	−0.20 / −0.07	−0.62 / +0.20	+0.48 / −0.07	$E = 0.02$ large, plain
	$E = -0.20$ small, spined	$E = -0.40$ large, sheathed	$E = +0.26$ small, sheathed	

B. Preference by rank of food electivity

Characteristic	Order	E
small, sheathed	BG	+0.26
large, plain	G	+0.02
small, plain	G	−0.03
small, spined	G	−0.20
large, sheathed	BG	−0.40

offered (E = 0.26, 0.02). *Diaptomus sicilis* took small *Ankistrodesmus* and the small, sheathed blue-green (*Gloeocapsa*) in about the proportions present (E = -0.006 and -0.05). This numerous large herbivore rejected small spined greens (*Scenedesmus*, E = -0.2) and large sheathed blue-greens (*Anacystis*, E = -0.4). Selection ($E>0$) can be differentiated from simply ingesting what is available (E = 0) and ignorance or lack of utilization ($E<0$). But the difference between ignoring and actively rejecting a resource is not evident. Clearly *Diaptomus* (adults) reacted to food size as well as the presence or absence of sheaths and spines in feeding selectively.

The resource characteristics matrix (Table 3A) for cyclopoid copepodites was similar to that for *Diaptomus*. The highest electivity occurred when a large, plain cell (*Pediastrum*) was offered with a large, sheathed cell (*Anacystis*). In ranked order of preference, these numerous immature copepods selected large, plain and small, plain cells and ignored the spined and sheathed cells to differing degrees (Table 3B). Large, sheathed cells were taken less frequently than small, sheathed cells, and spined cells were taken the least. In fact, by number they constituted 50% of the food offered, but only 15.1% of that ingested when paired with *Gloeocapsa*. This analysis had one fault in that size and "sheathness" were not examined independently.

Effect of sheath and size on ingestion. A 2×2 factorial (parametric) design was used to discover whether the effect of the sheath or relative size was independently the most important factor influencing selection of food. Filtering rates for *Diaptomus* and cyclopoid copepodites on paired resources, previously shown to be significantly different (Table 4), were cast into a 2×2

Table 3. Matrix of resource characteristics (A) and ranked food character preferences for cyclopoid copepodites (B) (E = electivity).

A.

	spined small	spined large	sheathed small	
small unspined and unsheathed	-0.59 / +0.05	-0.11 / +0.03	-0.14 / +0.05	E = +0.04 small, plain
large	-0.10 / +0.05	-0.21 / 0.11	+0.22 / -0.10	E = +0.09 large, plain
	E = -0.35 small, spined	E = -0.16 large, sheathed	E = +0.04 small, sheathed	

B. Preference by rank of food electivity

Characteristic	Order	E
large, plain	G	+0.09
small, plain	G	+0.04
small, sheathed	BG	+0.04
large, sheathed	BG	-0.16
small, spined	G	-0.32

Table 4. Two-level ANOVA with unequal sample size summarizing effects of sheath and food size on filtering rates of *Diaptomus* and cyclopoid copepodites.

Source of variation	df	MS	F
Diaptomus			
Subgroups			
A (sheath)	1	0.43	71.0*
B (food size)	1	0.02	3.9
A × B (interaction)	1	0.03	4.8†
Within subgroups (error)	42	0.006	
Total	45		
Cyclopoid copepodites			
Subgroups			
A (sheath)	1	0.23	25.5*
B (food size)	1	0.03	2.9
A × B (interaction)	1	0.004	3.5
Within subgroups (error)	29		
Total	32		

* $\alpha = 0.001$.
† $\alpha = 0.05$.

factorial design to test the effect of size versus that of sheath. The presence or absence of a sheath was the overwhelming factor of importance determining whether a resource was ingested or ignored, both in the case of *Diaptomus* ($P < 0.001$) and cyclopoid copepodites ($P < 0.001$); it is noteworthy that the interaction was not significant (Table 4). Thus in modeling food electivity we can rank the sheath, and associated ramifications of taste and inhibitory substances combined with it, above size as a matter of critical importance. Two herbivores that were highly selective were inhibited by or ignored sheathed and spined forms and selected large and small plain resources. From the opposing viewpoint of resource fitness it is best to be large and sheathed, and this is precisely the direction in which blue-green algae which form blooms, like *Anacystis*, have evolved. *Diaptomus sicilis*, characteristic of oligotrophic environments with small-celled resources, unexpectedly did not select small cells but preferred larger resources during brief 15-min observations.

References

ARNOLD, D. E. 1971. Ingestion, assimilation, survival, and reproduction by *Daphnia pulex* fed seven species of blue-green algae. Limnol. Oceanogr. **16**: 906–920.

BOGDAN, K. G., and D. C. McNAUGHT. 1975. Selective feeding by *Diaptomus* and *Daphnia*. Int. Ver. Theor. Angew. Limnol. Verh. **19**: 2935–2942.

BURNS, C. 1968. The relationship between body size of filter-feeding Cladocera and the maximum size of particle ingested. Limnol. Oceanogr. **13**: 675–678.

FROST, B. W. 1977. Feeding behavior of *Calanus pacificus* in mixtures of food particles. Limnol. Oceanogr. **22**: 472–491.

GLIWICZ, M., and A. HILLBRICHT-ILKOWSKA. 1972. Efficiency of the utilization of nannoplankton primary production by communities of filter-feeding animals measured in situ. Int. Ver. Theor. Angew. Limnol. Verh. **18**: 197–203.

IVLEV, V. S. 1955. Experimental ecology and nutrition of fishes. Yale Univ. Press. [Original edition by Pishchemizdat.]

McNAUGHT, D. C. 1975. A hypothesis to explain the succession from calanoids to cladocerans during eutrophication. Int. Ver. Theor. Angew. Limnol. Verh. **19**: 724–731.

———, and D. SCAVIA. 1976. Application of a model of zooplankton composition to problems of fish introductions to the Great Lakes, p. 281–305. *In* R. P. Canale [ed.], Mathematical modelling of biochemical processes in aquatic ecosystems. Ann Arbor Sci.

McQUEEN, D. J. 1970. Grazing rates and food selection in *Diaptomus oregonensis* (Copepoda) from Marion Lake, British Columbia. J. Fish. Res. Bd. Can. **27**: 13–20.

PORTER, K. G. 1973. Selective grazing and differential digestion of algae by zooplankton. Nature **244**: 179–180.

RICHMAN, S., D. R. HEINLE, and R. HUFF. 1977. Grazing by adult estuarine calanoid copepods of the Chesapeake Bay. Mar. Biol. **42**: 69–84.

SCHELSKE, C. L., L. E. FELDT, M. S. SANTIAGO, and E. F. STOERMER. 1972. Nutrient enrichment and its effects on phytoplankton production and species composition in Lake Superior. Proc. 15th Conf. Great Lakes Res. **1972**: 149–165.

WILSON, D. S. 1973. Food size selection among copepods. Ecology **54**: 909–914.

28. *Aphanizomenon* Blooms: Alternate Control and Cultivation by *Daphnia pulex*

Michael Lynch

Abstract

Dense blooms of the blue-green alga, *Aphanizomenon flos-aquae*, are often associated with populations of the large filter-feeder, *Daphnia pulex*. These blooms are always characterized by colonies of "grass-blade" morphology. *Daphnia* may prevent the development of an *Aphanizomenon* bloom when the alga is unable to grow colonies large enough to avoid being grazed. However, once a grass-blade bloom is established, *Daphnia* may prolong the bloom's existence by removing potential algal competitors.

Recent proposals for improving water quality have included the suggestion that management for populations of large herbivores, such as *Daphnia*, will enhance water transparency by increasing the intensity of grazing on phytoplankton (Brooks 1969; Shapiro et al. 1975). However, while many algae may be depressed to very low levels by large herbivores, some phytoplankters directly benefit from such grazing activity (Porter 1977). One species in particular, *Aphanizomenon flos-aquae*, often reaches bloom proportions in the presence of dense *Daphnia pulex* populations. The morphology of this alga is unique; it grows from single cells (akinetes) into single filaments which may aggregate into large colonies (up to 3 by 30 mm) resembling small "grass-blades." All blooms of *Aphanizomenon* which I have seen co-occurring with *D. pulex* in Minnesota lakes and ponds have grass-blade morphology. Furthermore, grass-blade blooms rarely occur when *Daphnia* is not abundant.

The frequent association of *Aphanizomenon* with *D. pulex* is not restricted to Minnesota. Hrbáček (1964) studied the development of blue-green algal blooms in a variety of Czechoslovakian backwaters, fish ponds, and reservoirs. He found that while the intensity of *Aphanizomenon* blooms was related to the level of enrichment, their presence could not be explained in terms of nutrients alone. Rather, the blooms were generally associated with the presence of *Daphnia pulicaria* (a species nearly identical in morphology to *D. pulex*). The colonies had grass-blade morphology (J. Hrbáček pers. comm.). When *D. pulicaria* was removed by fish, *Aphanizomenon* was replaced by *Microcystis* (a colonial, gelatinous blue-green). Similar findings have been noted by Losos and Heteša (1973).

It appears that *Aphanizomenon* only maintains grass-blade morphology in the face of intense grazing pressure from large *Daphnia*. Many eutrophic lakes throughout North America have *Aphanizomenon*

This work was supported by NSF grant EMS 74–19490 to J. Shapiro. I am grateful to V. Smith and B. Monson for assistance and encouragement. I thank B. Forsberg, J. Hrbáček, J. Shapiro, and E. Swain for comments.

blooms in the absence of large *Daphnia*, but under these conditions the alga usually exists as single or small groups of filaments.

I here pose a hypothesis to explain the distribution and morphology of *A. flos-aquae* based on the abundance of large *Daphnia*. The data I present are drawn from studies conducted in Pleasant Pond (about 0.25 hectares, maximum depth = 2.5 m, and located about 10 km north of St. Paul, Minnesota) (Lynch and Shapiro in prep.). In 1975 the pond was without vertebrate predators and contained a large population of *D. pulex*. However, after the pond was divided by a polyethylene curtain in early 1976, a dense population of the zooplanktivorous fathead minnow (*Pimephales promelas*) developed in the southern half and removed all of the *Daphnia*; the north half remained without fish (Lynch 1979). Grass-blade blooms occurred in the pond at different times each year. While these blooms attained high densities even when *Daphnia* was abundant, they only commenced when *Daphnia* was scarce and the bottom waters were well oxygenated (Fig. 1).

The growth of an *Aphanizomenon* bloom in the presence of dense *Daphnia* populations may simply result from an expansion of individual colonies. It is unlikely that new colonies can develop from small cells or filaments when exposed to such intense grazing. However, a tempor-

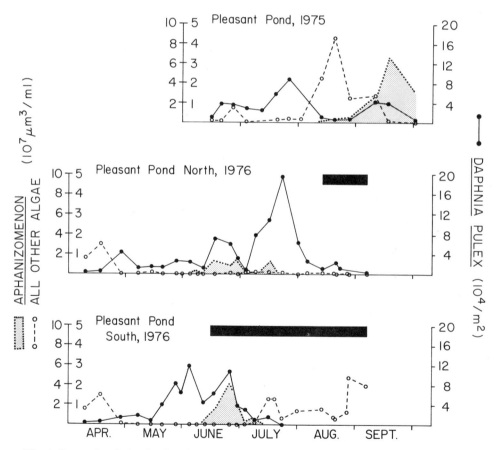

Fig. 1. Seasonal variation in abundance of *Aphanizomenon* "grass-blades" (stippled area), all other algae (dashed line), and *Daphnia pulex* (solid line). Solid bars indicate periods during which bottom waters were anoxic.

ary, spatial segregation from *Daphnia* could allow new *Aphanizomenon* colonies to reach a large enough size to avoid grazing upon exposure to *Daphnia*. The sediments may provide such a refuge. Since grass-blade blooms only commenced when the surface sediments were well oxygenated, it is possible that the blooms are initiated by an influx of new colonies which develop from akinetes deposited in the sediments.

Estimations of growth and loss rates for *Aphanizomenon* further implicate the importance of oxygenated bottom waters for the maintenance and development of a grass-blade bloom. Uptake of inorganic carbon was determined for samples incubated at a depth of 0.5 m for 24 h using track autoradiographic techniques (Knoechel and Kalff 1976). All incubations were done on cloudless days. Specific growth rates were calculated as

$$k = \ln \frac{\bar{x} \text{ cell C} + \bar{x} \text{ C fixed cell}^{-1} \text{ day}^{-1}}{\bar{x} \text{ cell C}}, \quad (1)$$

cell carbon being estimated from the cell carbon: cell volume regressions in Strathmann (1967). Specific loss rates (d) were calculated as the difference between the predicted growth rate (k) and the observed growth rate (r), where

$$r = \frac{\ln N_t - \ln N_o}{t}. \quad (2)$$

This analysis assumes that 24-h incubations give measurements close to net incorporation of carbon, i. e. growth; recent support for this contention can be found in Eppley and Sharp (1975) and Paerl and Mackenzie (1977). It also assumes that most cells are similar in size and composition.

Except for early May, estimates of *Aphanizomenon* loss rates were consistently negative in Pleasant Pond North (Table 1) indicating that the population was expanding at a higher rate than could be accounted for by its growth in the surface water. The bottom waters of Pleasant Pond North were well oxygenated throughout this period (Fig. 1). Loss rates were also negative in Pleasant Pond South early in the summer; however, when the bottom became anoxic in mid-June, the loss rates increased dramatically, becoming strongly positive. The regular appearance of these negative death rates, especially in Pleasant Pond North, suggests that they were not artifacts of horizontal or vertical transport of colonies. Rather, the results imply that a significant growth of *Aphanizomenon* occurs in the bottom waters when these are well oxygenated.

Consistent with this observation is the fact that artificial isolation of the water column from the sediments prevents the development and/or maintenance of a grass-blade bloom. Over a period of 2 years several experiments were done in Pleasant Pond in 1-m-diameter polyethylene bags suspended from the pond's surface and closed at the bottom (Lynch 1979). Grass-blade blooms never developed in any of these enclosures. Even when grass-blades were present at the outset of an experiment, they never persisted within an enclosure for more than 2 weeks. It is possible that in the absence of a suitable refuge *Aphanizomenon* was not capable of developing into large enough colonies to avoid grazing by the dense *Daphnia* (and *Ceriodaphnia*) populations present in many of these enclosures.

Aphanizomenon did become abundant as single filaments in several enclosures stocked with planktivorous fish (*Lepomis*) in 1975 (Lynch 1979). These fish altered the herbivore community from one dominated by *Daphnia* and *Ceriodaphnia* to one consisting primarily of *Bosmina* and rotifers. Analysis of the phytoplankton and zooplankton data suggests that the success of *Aphanizomenon* in fish enclosures may have resulted from the absence of large herbivores. There was a clear negative relation between the density of filamentous blue-greens (*Aphanizomenon* and *Anabaena*) and the abundance of *Daphnia* and *Ceriodaphnia* (Fig. 2). Furthermore, there appeared to be a threshold density of these herbivores below which filamentous blue-greens could escape predation.

Table 1. Specific growth (k) and loss (d) rates for *Aphanizomenon* (day^{-1}) in Pleasant Pond, 1976. Estimates derived from autoradiographic determinations (Knoechel and Kalff 1976).

		13 May	26 May	11 Jun	26 Jun	8 Jul	22 Jul	19 Aug
Pleasant Pond North	k	0.11	0.10	0.14	0.10	0.10	0.17	0.01
	d	0.33	−0.31	−0.12	−0.06	−0.26	−0.21	–
Pleasant Pond South	k	0.14	0.13	0.21	0.04	0.15		
	d	2.02	−1.59	−0.51	0.19*	0.73*		

*Anoxic bottom waters.

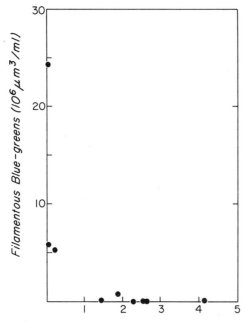

Fig. 2. Relation of abundance of filamentous blue-green algae to density of *Daphnia* + *Ceriodaphnia* in an enclosure experiment in Pleasant Pond (Lynch and Shapiro in prep.). Enclosures were stocked with different densities of a zooplanktivorous fish which selectively removed these herbivores. ●—Mean densities for last three sampling dates for different enclosures.

In terms of nutrient uptake, the formation of a large aggregation of filaments is probably detrimental—a tremendous amount of surface area is given up, particularly by filaments in the center of the colony. Since small filter-feeders are incapable of handling particles as large as those taken by large *Daphnia* (Burns 1968), there would be less selective pressure for *Aphanizomenon* to maintain large grass-blades in the absence of large *Daphnia*. This may explain the tendency for *Aphanizomenon* to form smaller colonies in communities dominated by small herbivores. However, even when "forced" to maintain grass-blade morphology, *Aphanizomenon* has a behavioral attribute which may offset the "cost of coloniality." The grass-blades are not constant with respect to size or shape. The individual filaments are constantly in motion, sliding past each other (Dodd 1954). Such behavior may serve to increase the exposed surface area of the individual filaments and may also increase the availability of nutrients by diminishing nutrient depletion in the immediate vicinity of the cells (Pasciak and Gavis 1974).

Finally, despite the fact that *Daphnia* may prevent the development of a grass-blade bloom under some circumstances, once a bloom is initiated it may actually provide a service to *Aphanizomenon* by removing potential algal competitors. At the peak of a bloom the water is always extremely transparent with the exception of the abundant grass-blades (Fig. 1). The grass-blades, being much larger than *Daphnia*, are not grazed.

From these observations, a hypothesis concerning the distribution of grass-blades can be made—

1. A grass-blade bloom will not develop when *Daphnia* is very abundant and the

bottom waters are anoxic; grass-blade development requires a temporary refuge from *Daphnia* grazing.

2. Once a grass-blade bloom is established, it will be maintained in the presence of dense *Daphnia* populations as long as there is a refuge for new *Aphanizomenon* colonies to develop in.

3. An established grass-blade bloom will not be maintained when *Daphnia* is reduced to low numbers.

This hypothesis is in accord with all aspects of the distribution of *Aphanizomenon* in Pleasant Pond except for the late June declines in 1976 when the bloom crashed in both sides of the pond after reaching high densities (Fig. 1). This crash did not appear to be a result of a shortage of phosphorus or nitrogen (Lynch and Shapiro in prep.). Since other algae did not increase during this decline, it is possible that some other micronutrient was limiting the growth of phytoplankton in the pond at that time. Soon after this crash, however, another grass-blade bloom developed in Pleasant Pond North. It commenced when *Daphnia* was rare and the bottom waters were well oxygenated, and continued to expand as *Daphnia* became very abundant. When *Daphnia* attained very high densities, the bloom crashed. A second bloom did not appear in Pleasant Pond South; instead, since the *Daphnia* was removed by fathead minnows, many other small species of algae increased dramatically. Since the curtain was removed in spring 1977, fathead minnows have continued to inhabit the pond at extremely high densities. Not only have *Daphnia* been completely eradicated, but grass-blades have not reappeared.

Further support for the hypothesis comes from a study I have done on a small enriched pond north of Minneapolis called Loch Loso. From late 1974 to mid-July 1975, the pond had a substantial population of *D. pulex* and the phytoplankton community was dominated by grass-blade *Aphanizomenon*. However, planktivorous fish, which were added to the pond in 1975, reproduced early that summer. Consequently all *Daphnia* was removed by mid-July; at that same time the *Aphanizomenon* bloom crashed and was replaced by other species of phytoplankton. In 1976 extremely dense populations of planktivorous fish developed; no *Daphnia* was noted, and grass-blades did not reappear. A complete fish-kill occurred during winter 1977, and in the summer large numbers of *D. pulex* reappeared. Once again a very dense bloom of grass-blades developed in the pond.

While my hypothesis is predictive in nature, it leaves several questions about the *Daphnia–Aphanizomenon* interaction unanswered. The mechanism regulating the clonal morphology of *Aphanizomenon* is not known. Grass-blades may be genetically distinct from their non-grass-blade counterparts. Alternatively, there might be a chemical or physical cue elicited by *Daphnia* which triggers the formation of grass-blades. In the absence of *Daphnia*, grass-blade colonies dissociate within 48 hours (B. Monson, V. Smith, and J. Shapiro, pers. comm.) Also in this regard, J. Almendinger and D. Tilman recently discovered that when exposed to grazing by *Daphnia*, *Nostoc*—a filamentous blue-green alga—forms dense aggregates of filaments.

The size which grass-blades must attain to acquire invulnerability to grazers is unclear, as is the mechanism of formation of new colonies within the water column. Dodd (1954) suggested that new colonies may be produced by the mechanical fragmentation of existing grass-blades. If this is true, then it is unclear why grass-blade blooms cannot establish when the sediments are anoxic; a grass-blade bloom could conceivably maintain itself in the absence of a refuge, if newly produced fragments were large enough to avoid grazers. Rose (1934) suggested that such fragmentation is not the usual method of reproduction.

Since other blue-green algae (*Oscillatoria, Gloeotrichia, Microcystis*) are known to develop into colonies on the sediment surface before entering the water column (Gerloff and Skoog 1954; Roelofs and Oglesby 1970; Gahler and Sanville 1971), this life-history strategy may be a common adaptation of blue-green algae against intense grazing by *Daphnia*. Thus, any efforts to "biologically control" the quantity of phytoplankton in lakes will have to consider the defensive mechanisms of these nuisance algae.

References

BROOKS, J. L. 1969. Eutrophication and changes in the composition of the zooplankton, p. 236-255. *In* Eutrophication: Causes, consequences, correctives. Publ. Natl. Acad. Sci. 1700.

BURNS, C. W. 1968. The relationship between body size of filter-feeding Cladocera and the maximum size of particle ingested. Limnol. Oceanogr. 13:675-678.

DODD, J. D. 1954. A note on the increase in flake size of *Aphanizomenon flos-aquae* (L) Ralfs. Proc. Iowa Acad. Sci. 60:117-118.

EPPLEY, R. W., and J. H. SHARP. 1975. Photosynthetic measurements in the central North Pacific: The dark loss of carbon in 24-h incubations. Limnol. Oceanogr. 20:981-987.

GAHLER, A. R., and W. D. SANVILLE. 1971. Characterization of lake sediment and evaluation of sediment-water nutrient interchange mechanisms in the upper Klamath Lake system. EPA Rep. Pac. NW Water Lab. Corvallis, Oregon. 45 p.

GERLOFF, G. C., and F. SKOOG. 1954. Cell contents of nitrogen and phosphorus as a measure of their availability for growth of *Microcystis aeruginosa*. Ecology 35:348-353.

HRBÁČEK, J. 1964. Contribution to the ecology of water-bloom-forming blue-green algae. *Aphanizomenon flos-aquae* and *Mycrocystis aeruginosa*. Int. Ver. Theor. Angew. Limnol. Verh. 15:837-846.

KNOECHEL, R., and J. KALFF. 1976. Track autoradiography: A method for the determination of phytoplankton species productivity. Limnol. Oceanogr. 21:590-595.

LOSOS, B., and J. HETEŠA. 1973. The effect of mineral fertilization and carp fry on the composition and dynamics of plankton. Hydrobiol. Stud. 3:173-217.

LYNCH, M. 1979. Predation, competition, and zooplankton community structure: An experimental study. Limnol. Oceanogr. 24: 253-272.

PAERL, H. W., and L. A. MacKENZIE. 1977. A comparative study of the diurnal carbon fixation patterns of nannoplankton and net plankton. Limnol. Oceanogr. 22:732-738.

PASCIAK, W. J., and J. GAVIS. 1974. Transport limitation of nutrient uptake in phytoplankton. Limnol. Oceanogr. 19:881-888.

PORTER, K. G. 1977. The plant-animal interface in freshwater ecosystems. Am. Sci. 65:159-170.

ROELOFS, T. D., and R. T. OGLESBY. 1970. Ecological observations on the planktonic cyanophyte *Gloeotrichia echinulata*. Limnol. Oceanogr. 15:224-229.

ROSE, E. T. 1934. Notes on the life history of *Aphanizomenon flos-aquae*. Univ. Iowa Stud. Nat. Hist. 16:129-142.

SHAPIRO, J., V. LAMARRA, and M. LYNCH. 1975. Biomanipulation: An ecosystem approach to lake restoration, p. 85-96. *In* P. L. Brezonik and J. L. Fox [eds.], Water quality management through biological control. Proc. Symp. Univ. Florida.

STRATHMANN, R. R. 1967. Estimating the organic carbon content of phytoplankton from cell volume or plasma volume. Limnol. Oceanogr. 12:411-418.

29. Zooplankton as Algal Chemostats: A Theoretical Perspective

Robert A. Armstrong

Abstract

A simple graphical method is developed for visualizing and analyzing the combined effects of herbivory and of competition for limiting nutrients on algal community structure. The model on which the graphical method is based is presented as a logical extension of chemostat theory and r-K theory; the qualitative predictions of the model are analyzed from this perspective. The graphical method should be useful for visualizing the quantitative implications of herbivore-algae interactions and for detecting subtle interactions among species in empirical investigations.

In an earlier paper (Armstrong 1979) I developed a graphical model of a system composed of one limiting nutrient, several producer (algal) species, and one herbivore (zooplankter) species. Here I develop this topic from a different perspective, a perspective which clearly shows the connections of this work to chemostat theory and to r-K theory (MacArthur and Wilson 1967; Pianka 1970). I then use this model to investigate theoretically the effects of nutrient enrichment and herbivory on the species composition of the algal community at equilibrium, and discuss the application of this model to empirical research.

Competition for a single limiting nutrient

Consider two algal species competing for a single limiting nutrient. Let the defining equations of algal growth be

$$\frac{dA_1}{dt} = A_1 \rho_1(S)$$

$$\frac{dA_2}{dt} = A_2 \rho_2(S) \tag{1}$$

where S is the concentration of the limiting nutrient (substrate), A_i is the density of algal species i, and $\rho_i(S)$ is the per capita growth rate of algal species i as a function of nutrient density. The curves $\rho_i(S)$ will have the general shapes shown in Fig. 1.

Since the competing algal species consume the limiting nutrient during growth,

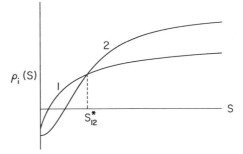

Fig. 1. Curves of algal per capita growth rates $\rho_i(S)$ vs. free nutrient concentration S. Species 1 is a superior competitor at low free nutrient densities; species 2 is superior at high free nutrient densities. Boundary between regions in which each is superior is at S_{12}^*. In simple competition for a single conservative resource, species 1 will replace species 2.

This paper is contribution 312 from the Program in Ecology and Evolution, SUNY, Stony Brook; partially supported by a University Award Fellowship from the Research Foundation of the State University of New York.

the density S of free nutrient must diminish as the algae grow. The simplest possible mathematical form of such diminution corresponds to the assumption that the limiting nutrient is "conservative" (Armstrong 1979). Under this assumption, the total amount of nutrient is fixed at some value S_T. At any given time, a portion S of the resource exists in a "free" state, capable of immediate use for algal growth, and a portion S_T-S is "bound" in the protoplasm of living organisms. The nutrient is assumed to regenerate instantaneously from the "bound" state to the "free" state upon the death of the individual by which it was bound. Lehman's (1979) observation of rapid nutrient regeneration in a zooplankton-alga system suggests that this assumption is a reasonable approximation to reality, even though not literally correct.

If we adopt the convention that algal densities A_i are to be measured as their nutrient equivalents (that is, if we agree that algal densities are to be measured in units of, say, μg P/liter), then the assumption that the resource is "conservative" can be expressed by the relationship

$$S_T = S(t) + A_1(t) + A_2(t). \qquad (2)$$

The system described by Eq. 1 and 2 is a system of two species and one limiting factor (Levin 1970). It can be shown that the species which can survive at the lowest concentration of free nutrient will eventually replace its competitor (Armstrong and McGehee 1980; see also Fig. 1). Alternatively, considering that the resource is conservative and that algal densities are measured as their nutrient equivalents, the algal species which maintains the highest density at equilibrium will replace its competitor (Fig. 2).

Density-independent mortality

If additional density-independent mortality is now added to the system, a la r–K theory (MacArthur and Wilson 1967; Pianka 1970), which competitor will prevail? We modify Eq. 1 and 2 to allow for this added mortality:

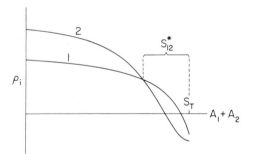

Fig. 2. Same algal growth curves as depicted in Fig. 1, but now plotted as functions of bound nutrient density A_1+A_2 instead of free nutrient density S, under assumption that resource is conservative. S_T (= $S+A_1+A_2$) is total nutrient (substrate) density.

$$\frac{dA_1}{dt} = A_1 [\rho_1(S) - \psi_1 \phi]$$
$$\frac{dA_2}{dt} = A_2 [\rho_2(S) - \psi_2 \phi] \qquad (3)$$
$$S = S_t - A_1 - A_2.$$

Here ψ_1 and ψ_2 are parameters which reflect the sensitivities of the two algal species to the added source of mortality; the variable ϕ sets the overall (constant) level of removal.

The behavior of model 3 parallels that of the previous model: that algal species which can furthest reduce the free nutrient concentration S at equilibrium, or, alternatively, that species which can maintain the largest standing crop at equilibrium, will eventually replace the other (Armstrong and McGehee, 1980).

We seek a method for plotting the curves $\rho_1(S)$ and $\rho_2(S)$ and the parameters ψ_1, ψ_2, and ϕ, such that the relative values of the species' equilibrium standing crops can be determined graphically for any given value of ϕ. Several methods are available. The method I find most convenient is to first plot the functions ρ_1/ψ_1 and ρ_1/ψ_2 against the total concentration of bound nutrient $S_T - S = A_1 + A_2$ (Fig. 3; see also Armstrong 1979). The level of density-independent mortality ϕ is then plotted on the same graph, and the equilibrium densities $A_1^*(\phi)$ and $A_2^*(\phi)$ are found (Fig. 3). That species with the largest value of $A_i^*(\phi)$ will replace the other at that

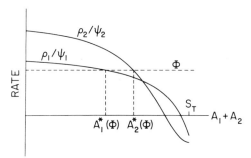

Fig. 3. Per capita algal growth rate curves ρ_1 and ρ_2, scaled for relative susceptibilities to density-independent mortality (ψ_i), and plotted on same graph as level of density-independent mortality ϕ. A_1^* (ϕ) and A_2^* (ϕ) are equilibrium densities (in monoculture) of algal species 1 and 2 at indicated mortality level ϕ, and measured as nutrient equivalents. At this level of density-independent mortality, alga 2 would eventually replace alga 1.

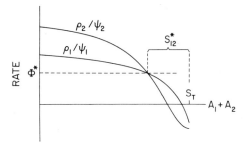

Fig. 4. For $\phi > \phi^*$, algal species 2 will replace alga 1 (since then A_2^* [ϕ] $> A_1^*$ [ϕ]); for $\phi < \phi^*$, algal species 1 will be superior. At critical removal rate ϕ^*, free nutrient density will be S_{12}^*. Value of S_{12}^* will be independent of total nutrient density S_T (cf. Figs. 1 and 2).

level of mortality. Note that if the level of mortality ϕ is too large, neither species will be able to survive.

A quantitative prediction, and tests with chemostats

Figure 3 immediately suggests the following prediction: There will exist a unique level of density-independent mortality ϕ^* such that for $\phi > \phi^*$ one algal species will be the victor and for $\phi < \phi^*$ the other species will win. The level ϕ^* will be independent of S_T, and at ϕ^* the free nutrient concentration will be some fixed value S_{12}^* (Fig. 4).

This last assertion follows from the fact that increasing the total nutrient level S_T by an amount δ translates the scaled productivity curves ρ_i / ψ_i a distance δ to the right but does not alter their shapes (Armstrong 1979). In particular, their point of intersection remains at the same height ϕ^* and the free nutrient concentration at this intersection remains at S_{12}^* (Fig. 4). The species which wins at $\phi > \phi^*$ is the "r"-species, since it can survive high rates of rarefaction (weighted, of course, by differential sensitivity to the agent of rarefaction). The species which wins at $\phi < \phi^*$ is the "K"-species, better adapted for competition at low nutrient levels.

If one were to perform a series of competition experiments at different levels of mortality ϕ and different total nutrient densities S_T, plotting the values of ϕ and S_T at which each alga is dominant, one would expect a graph such as in Fig. 5. Such a series of experiments can be performed with the use of chemostats. In a chemostat situation $\psi_1 = \psi_2 = 1$, and the level of mortality ϕ is the reciprocal of the residence time of the chemostat. The same conditions of global stability apply to chemostat models as to the simpler r-K models (Armstrong and McGehee in press).

Why would one want to perform such a series of experiments? Since the basic theoretical predictions of the model — that for $\phi > \phi^*$ species 2 will win, for $\phi < \phi^*$ species 1 will win, and that ϕ^* is independent of S_T — are independent of the exact functional forms of $\rho_1(S)$ and $\rho_2(S)$, such a plot cannot be used to discriminate among alternative models of algal species growth. Instead, the main use of such a series of experiments would be in detecting whether additional factors, such as allelopathy (Keating 1978), symbiosis, or the effects of a second limiting nutrient, must be considered. The value of the theoretical prediction of Fig. 5 is analogous to the value of the prediction of exponential growth in population size under constant environmental conditions: deviations from the prediction are easily detected, and signal

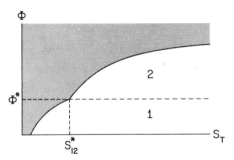

Fig. 5. Replacement diagram for algal species 1 and 2 in a parameter space of total nutrient density S_T and per capita removal rate ϕ. In stippled region, neither species can survive. Where conditions allow survival, alga 1 dominates at $\phi<\phi^*$ and alga 2 dominates at $\phi>\phi^*$.

that something "biologically interesting" is taking place. If data from a series of chemostat experiments are plotted in the same way as in Fig. 5, and the dividing line between the regions of competitive superiority is found to be other than a straight line at a constant ϕ, we can be sure that the situation involves more than simply competition for a single conservative nutrient.

Zooplankton as algal chemostats

The r–K model 3 can be modified to include herbivores as the source of mortality. The model becomes

$$\frac{dA_1}{dt} = A_1\,[\rho_1(S) - \psi_1\,Nb\,(A_1, A_2)]$$

$$\frac{dA_2}{dt} = A_2\,[\rho_2(S) - \psi_2\,Nb\,(A_1, A_2)]$$

$$S = S_T - A_1 - A_2. \qquad (4)$$

Here N is herbivore density (assumed to be fixed) and $b(A_1, A_2)$ is the per capita rate of consumption of algae. The quantity $b(A_1, A_2)$ can be interpreted as the "volume of water filtered" per individual, whereas the terms, $\psi_i b(A_1, A_2)$ can be interpreted as the "volume swept clear" per individual. Note that the consumption rates of the two algal species need not be equal. However, since ψ_1 and ψ_2 are fixed preferences, zooplankters are not allowed to "switch" from one algal species to another as the relative abundances of the algae change (Murdoch 1969, *see also* Wilson 1973; Frost 1977). The "no-switching" criterion can be interpreted as a constant ratio of volumes swept clear.

The model 4 is a two-species, two-limiting-factor model, and so admits of the possibility of coexistence of the algae or of multiple stable states. Before we can treat the model 4 in the same way we treated 3, we must reduce the system to a one-limiting-factor model by the restriction that

$$b(A_1, A_2) = b(A_1 + A_2). \qquad (5)$$

That is, the predator cannot saturate in an arbitrary manner with increasing prey density, but must saturate as a function of total algal density $A_1 + A_2$, measured, of course, in nutrient equivalents. Assumption 5 is made for ease of exposition. The data of Frost (1972) suggest that a more reasonable simplifying assumption may be $b(A_1, A_2) = b(\psi_1 A_1 + \psi_2 A_2)$, as used by Armstrong (1979). This latter assumption would allow the occurrence of coexistence or of alternative stable states. These phenomena are treated below.

We now proceed to graph the predation (herbivory) function $Nb\,(A_1 + A_2)$ on the same graph as the curves ρ_1/ψ_1 and ρ_2/ψ_2 Fig. 6). Again we seek a dividing line in parameter space between the region in which algal species 1 dominates and that in which species 2 dominates. In analogy with the r–K chemostat model (Figs. 3–5), the boundary will be the locus of points in parameter space at which the predation (herbivory)

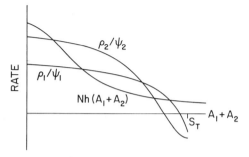

Fig. 6. Scaled alga per-capita growth curves ρ_1/ψ_1 and ρ_2/ψ_2 plotted on the same graph as an herbivory function $Nb\,(A_1 + A_2)$. Algal species 1 will come to dominate at this level of herbivory.

curve $Nb(A_1+A_2)$ passes through the intersection of the scaled algal growth curves ρ_1/ψ_1 and ρ_2/ψ_2. Analytically, the condition which must be satisfied is that

$$Nb(A_1^*+A_2^*) = (\rho/\psi)_{12}^* \qquad (6)$$

where $(\rho/\psi)_{12}^*$ is the value of the curves $\rho_1/\psi_1 = \rho_2/\psi_2$ at their point of intersection. If the value of free nutrient S at this intersection is denoted S_{12}^*, we can rewrite condition 6 in terms of $S_T - S_{12}^* = A_1^* - A_2^*$ as

$$N = (\rho/\psi)_{12}^*/b(S_T-S_{12}^*). \qquad (7)$$

A typical graph of condition 7 in the N-S_T parameter plane is shown in Fig. 7.

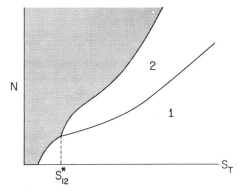

Fig. 7. Graph of regions in which algal species 1 and 2 dominate as functions of total nutrient density S_T and herbivore density N. In stippled region neither alga can survive. Dividing curve between region of dominance by alga 1 and region of dominance of alga 2 is given by $N = (g/\psi)_{12}^*/b(S_T-S_{12}^*)$, where $(g/\psi)_{12}^*$ and S_{12}^* are constants.

Actually, two additional restrictions are needed if condition 6 is to be interpreted as the boundary between two regions of algal dominance. First, the slope of the prediction curve $Nb(A_1+A_2)$ must be closer to zero than the slopes of either growth line at their common point of intersection. This condition assures the stability of the single-species equilibria on either side of the dividing line. Analytic conditions on the form of the function $b(A_1+A_2)$ are easily derived, but need not concern us here. Second, the densities of the algae at time $t = 0$ are assumed to be high enough that the herbivore cannot graze the algae to extinction before the predicted equilibrium is attained.

Again referring to Fig. 7, note that at large herbivore densities N and total nutrient densities S_T, the dividing curve between the two algae becomes a straight line. This phenomenon occurs because at large total nutrient densities S_T, total algal density (at the intersection of the growth curve) becomes large, and the herbivories' functional response curve saturates. That is, at large prey densities

$$(A_1+A_2) b (A_1+A_2) \sim H_s,$$

where H_s is the saturation rate of consumption. Therefore, at large values of S_T,

$$b(A_1^*+A_2^*) \sim H_s/(A_1^*+A_2^*)$$
$$= H_s/(S_T-S_{12}^*),$$

Substituting the above equation into (7) yields the relationship

$$N = (\rho/\psi)_{12}^* H_s^{-1}(S_T-S_{12}^*), \qquad (8)$$

which graphs as a straight line.

The regions occupied in Fig. 7 by algal species 1 and 2 are similar to those one would obtain from species growth curves such as in Fig. 6, using the same sort of decision criteria employed in the r-K chemostat section. Note that at any given value of S_T, alga 2 will be found at larger herbivore densities than will alga 1 (cf. Fig. 5). Finally, note that at very large values of N, the intensity of herbivory is so large that neither species of algae can survive (Fig. 7; see also Fig. 5).

Multiple species

The above methods can be extended in a straightforward manner to graphs of many algal species. Each species will occupy its own open-ended band sloping upward to the right on a plot of species in the $N - S_T$ parameter plane (Fig. 8). The dividing curves between species are scaled replicates of one another, since the shapes of all curves depend on the same function

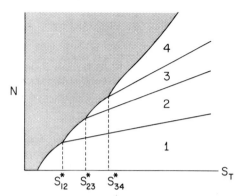

Fig. 8. Same type of graph as in Fig. 7, but displaying four species of algae, each superior in some section of parameter space. Species 1 is most "K-adapted"; species 2–4 are progressively more "r/ψ-adapted," as indicated by fact that boundaries between their regions of dominance occur at larger values of S_{ij}^* (see also Armstrong 1978). Boundaries between regions of dominance are here depicted as straight lines, as would be the case if $h(A_1+A_2)$ had Monod (or Michaelis-Menton) shape $h(A_1+A_2) = a/(b+A_1+A_2)$.

$h(S_T - S_{ij}^*)$; the curves differ only in the values of S_{ij}^* and of $(g/\psi)_{ij}^*$.

Discussion

Models describing the interaction of a single species of herbivorous zooplankter with several species of algal prey (Eq. 4) are similar in many respects to models describing algal competition in chemostats (Eq. 3). In particular, both models yield plots of the regions in parameter space in which different algal species dominate. The main difference between the chemostat plot (Fig. 5) and the herbivore plot (Figs. 7, 8) is that in the algal chemostat plot the dividing curve between regions of algal dominance is a horizontal line at constant flow rate ϕ^*, while in the alga-zooplankton plots the dividing curves slope upward to the right (Fig. 7). This difference reflects the facts that as the total nutrient concentration S_T increases, so does algal density, and that more herbivores are required to maintain the same per-unit-alga rate of algal removal when the algae are dense than when they are sparse. As a result, increasing the total nutrient concentration S_T without altering herbivore density N favors algal species which are better adapted to utilizing nutrients at low levels. (These species are the "K-adapted" species of the r-K continuum; see MacArthur and Wilson 1967; Pianka 1970). In contrast, lowering the total nutrient concentration S_T without changing the herbivore density N favors algal species with higher maximum growth rates r and lower susceptibilities to predation ϕ (see also Armstrong 1979).

Beyond this simple qualitative observation, of what use is this model? Or, in analogy with the empirical use of the r-K chemostat model, why would one perform a set of experiments with one herbivore species, several algal species, and one limiting nutrient and use the results to construct plots such as Figs. 7 and 8?

To answer this question, we must first discuss the circumstances under which construction of such a plot would probably not be desirable. First, one would not use such a plot to detect subleties in interactions between algal species, since these are much more easily detected by competition experiments in chemostats (a la Fig. 5). Second, one would not use such experiments to study an herbivore's preference among algae, or switching between algae as their relative abundances change, since both of these phenomena are more easily studied by more straightforward methods (e.g. Wilson 1973; Poulet and Marsot 1978; S. Richman and others at this conf.).

One would, on the other hand, use such a graph to represent the ecological significance of the above two classes of phenomena. For example, gelatinous green algae display a remarkable set of adaptations to a combination of predation pressure and low nutrient concentration (Porter 1976). Plotting the position of these algae in a diagram such as Fig. 8 would sharpen our awareness of the types of situations to which these algae are adapted.

As a second example, we know that selectivity, differential saturation of the predation curve $h(A_1, A_2)$, or switching behavior by the herbivores may allow coexistence or the occurrence of multiple stable states. Again, plotting real data in the manner of Fig. 8 will summarize the population consequences (for the algae) of these behaviors, since the tracts of single algal species in Fig. 8 will be supplemented by and/or replaced by tracts in which two or more algal species may coexist, or along which multiple stable states may occur.

Finally, plots of the form of Fig. 8 should be useful in detecting the existence of more complicated interactions between herbivores and algae, such as Porter's (1976) example of pseudo-mutualism. This type of interaction could not be detected in either an algal competition experiment or in a zooplankton feeding experiment, but could be detected by comparing the species occurrences predicted by a systems model based on feeding and competition experiments with those actually observed in a system containing all species. Undoubtedly many interactions of this sort remain to be discovered. This graphical method may serve as a valuable tool in finding them.

References

ARMSTRONG, R. A. 1979. Prey species replacement along a gradient of nutrient enrichment: a graphical approach. Ecology In press.

——, and M. E. GILPIN. 1977. Evolution in a time varying environment. Science 195: 591-592.

——, and R. McGEHEE. 1976. Coexistence of species competing for shared resources. Theor. Popul. Biol. 9: 317-328.

——, and ——. 1980. Competitive exclusion. Am. Nat. 115: 151-170.

FROST, B. W. 1972. Effects of size and concentration of particles on the feeding behavior of the marine planktonic copepod *Calanus pacificus*. Limnol. Oceanogr. 17: 805-815.

——. 1977. Feeding behavior of *Calanus pacificus* in mixtures of food particles. Limnol. Oceanogr. 22: 472-491.

KEATING, K. I. 1978. Blue-green algal inhibition of diatom growth: transition from mesotrophic to eutrophic community structure. Science 199: 971-973.

LEHMAN, J. T. 1980. Nutrient recycling as an interface between algae and grazers in freshwater communities (this symposium).

LEVIN, S. A. 1970. Community equilibria and stability, and an extension of the competitive exclusion principle. Am. Nat. 104: 413-423.

MacARTHUR, R. H., and E. O. WILSON. 1967. The theory of island biogeography. Princeton Univ. Press.

MURDOCH, W. W. 1969. Switching in general predators: experiments on predator specificity and stability of prey populations. Ecol. Monogr. 39: 335-354.

PIANKA, E. R. 1970. On r- and K-selection. Am. Nat. 104: 592-597.

PORTER, K. G. 1976. Enhancement of algal growth and productivity by grazing zooplankton. Science 192: 1332-1334.

POULET, S. A., and P. MARSOT. 1978. Chemosensory grazing by marine calanoid copepods (Arthropoda: Crustacea). Science 200: 1403-1405.

WILSON, D. S. 1973. Food size selection among copepods. Ecology 54: 909-914.

VI

Genetics, Demographics, and
Life Histories of Zooplankton

30. The Genetic Structure of Zooplankton Populations

Charles E. King

Abstract

From a genetic viewpoint, the most remarkable aspect of the groups comprising freshwater zooplankton is the diversity of their life history patterns. These range from obligate parthenogenesis in some rotifers to obligate sexual reproduction in copepods. Most rotifers and cladocerans fall between these extremes.

Zooplankton, perhaps more than most organisms, are subject to extensive temporal variation in their environment. Genetic studies of rotifers in particular have revealed considerable population differentiation through time; that is, the rotifers occupying a lake are subdivided into genetically distinct populations that succeed one another. Patterns of selection accompanying temporal subdivision of the environment are discussed for rotifers, and to a lesser extent for other zooplankton, in an attempt to examine the influence of sexual reproduction on population structure.

Calculations based on zooplankton population sizes and inferred mutation rates suggest that an enormous amount of genetic variation is produced by mutation, particularly in rotifers with their large population sizes and short generation times. Selection does not appear to produce as strong a temporal partitioning of the environment in cladocerans and copepods as in rotifers. In addition, mutation may be less effective in crustacean zooplankton than in rotifers as a generator of new genotypes capable of tracking environmental change. Finally, these data are considered in the context of current theories of the evolution of sexual reproduction. Thus sexual reproduction may have little significance to the adaptation of most rotifers except as a device for making resting eggs.

Limnologists have long been concerned with the ecological structure of zooplankton populations. The appearance of genetic research on field populations is quite new. As evident from the presentations at this symposium, knowledge in this area is rapidly expanding. Perhaps the major finding to date has been that the actual structure of rotifer and cladoceran populations is much more complex than that suggested by ecological studies. Where no recombination has been thought to occur, true sexual reproduction has been demonstrated. The opposite has also been shown—i.e. the production of pseudosexual eggs by the process of cryptoparthenogenesis. What appears to be a single well-defined population may in fact be a population that is rapidly changing in time, or even several populations succeeding each other as the lacustrine environment changes. Particularly in rotifers, it is becoming apparent that ex-

---

This research was supported by grants from the National Science Foundation. Although the views expressed herein are my own, both my conception of the problem and the final manuscript have benefited from critical discussion and review by Peter S. Dawson. I gratefully acknowledge this aid.

tremely rapid successions of populations may occur. These findings suggest that the "population" investigated in many limnological studies may be an artifact with closer affinities to griffins, unicorns, and mermaids than to the population as a biological unit.

Various aspects of genetic structure have recently been considered by King (1977a) for rotifers, and by Hebert (1978) for *Daphnia*. Here, I will briefly present some results from a recent study and then address three topics in a rather speculative fashion. These topics are the timing of sexual reproduction, the relative significance of recombination and mutation for different zooplankton populations, and the evolution of sexual reproduction. The majority of my remarks will be restricted to rotifers; however, other groups of zooplankton will be considered where feasible and appropriate.

Synopsis of Asplanchna girodi *study*

The predatory monogonont *Asplanchna girodi* is the only rotifer for which details of both genetic and ecological population structure are available. The study was initiated in Golf Course Pond (Tampa, Florida) during March 1976 and proceeded with plankton tows being taken at weekly, or more frequent, intervals until December 1977. About 150 amictic females were randomly removed from each field sample and used to initiate clones. Throughout the study frequent electrophoretic analyses (King 1977a) were made on the following enzymes: malic enzyme, malic dehydrogenase, glutamic dehydrogenase, esterase (both α- and β-specific), acetaldehyde oxidase, tetrazolium oxidase, phosphoglucose isomerase, glucose-6-phosphate dehydrogenase, and lactic dehydrogenase. Of these ten enzymes, only one, malic enzyme, was found to have more than one detectable allele in Golf Course Pond. (Variants have been found for PGI, 6-PGDH, and LDH in *A. girodi* from other central Florida ponds, and variants of α-esterase have been found in *A. girodi* from Oregon.) Thus the predominant impression obtained from these analyses is that *A. girodi* in Golf Course Pond has little genetic variation.

With a few exceptions, a minimum of 75 clones was used for electrophoretic determination of malic enzyme genotype frequencies. As reported in King (1977a), the population of *A. girodi* was composed exclusively of heterozygotes (alleles F and S) at this locus from March through the first week of May 1976. From the second week of May through August 1976 the population was composed entirely of homozygotes for another allele, VS. Thus a complete change in the genetic composition of *A. girodi* occurred in a 1-week period. That individuals taken before and after this change were from two different populations was further demonstrated by comparisons of their population dynamics. On the basis of life table experiments conducted under standard laboratory conditions, highly significant differences were found between the two populations in rates of increase (r), net reproductive rates (R_0), mean life expectancies (e_0), and equilibrium population sizes (K). Moreover, in spite of repeated attempts to make reciprocal crosses, F_1 progeny could not be obtained because males of one population displayed no mating behavior toward females of the other population. (Mating and subsequent resting egg production, however, occurred readily within each of the two populations.)

Asplanchna girodi was absent from Golf Course Pond for most of fall 1976, but returned in early December. At the time of its return, malic enzyme alleles F and S had respective frequencies of about 0.6 and 0.4 and the genotypes were in Hardy-Weinberg equilibrium. The equilibrium was maintained through February 1977 at which time the population disappeared. In order to study the anticipated fixation of heterozygotes (i.e. repeating the pattern found in the previous spring), daily or bidaily samples of the pond were taken from March through July 1977. The gene and genotype frequency analyses of these samples are given in Fig. 1. Not only did the heterozygotes fail to reach fixation,

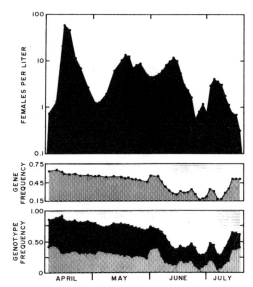

Fig. 1. Densities of *Asplanchna girodi* females (amictic plus mictic) in Golf Course Pond, 1977. Gene and genotype frequencies are for two alleles (F and S) of malic enzyme. Fast allele and fast homozygote are stippled, and heterozygote frequencies are given in shaded area of bottom panel.

but allele VS was not found once during the year 1977. Throughout most of spring 1977, alleles F and S were in Hardy-Weinberg equilibrium. However, during June and early July the frequency of allele F increased markedly. Whether this increase was due to selection on the malic enzyme locus directly or to selection at other loci is unknown. Experiments performed by T. W. Snell (pers. comm.) suggest that clones homozygous for the fast allele are much more resistant to toxic effects of blue-green algae (*Anabaena* sp. and *Lyngbia* sp.) than are either heterozygous clones or clones homozygous for the slow allele. These results could however be achieved by selection acting on linked loci. That is, since there is no recombination during parthenogenesis, linkage groups are not broken up. Selection acting on genes at one locus may therefore determine gene frequencies at linked loci. Moreover, during parthenogenesis the entire genome is effectively linked and so selection favoring either homozygotes or heterozygotes at a locus on one chromosome can produce fixation at a locus on a different chromosome. This latter process is sometimes referred to as "hitchhiking" or as "associative overdominance" (Angus 1978) if a heterozygote is fixed.

During the intensive sampling of spring 1977, each plankton sample was scored for numbers of amictic females, mictic females bearing resting eggs, and males. Whenever the total density of *A. girodi* exceeded 1.5 females per liter, males and usually females bearing resting eggs were found. By referring to the top panel of Fig. 1, we can see that bisexual reproduction started early in the population cycle and continued for most of the 4-month period.

Until recently, some workers have felt that rotifers lack true sexual reproduction (reviewed in King 1977a). However, by use of the malic enzyme system described above, it has been possible to document the existence of true sexual recombination in *A. girodi* (King 1977a; King and Snell 1977). And yet, as described above, we have found little genetic variation in this rotifer. The question thus arises: Given the apparent paucity of genetic variation in *A. girodi*, what function is served by sexual reproduction of this rotifer in Golf Course Pond? In other words: Why sex?

Timing of sexual reproduction in rotifers

Although most introductory biology texts state that sexual reproduction occurs at the end of the rotifer population cycle each year, research investigators have long recognized the fallacy of this description. Unfortunately, there are few good studies documenting the temporal pattern of sexual reproduction in monogonont rotifers. The following comments are based on data of Amren (1964) and Carlin (1943). While not the only studies presenting this information, they respectively represent one of the most thorough and one of the most extensive data sets available.

Amren (1964) studied the occurrence of *Keratella quadrata* and *Polyarthra dolichoptera* in a series of ponds on Spitsbergen Island, Norway. Figure 2 provides a summary of his results from one pond.

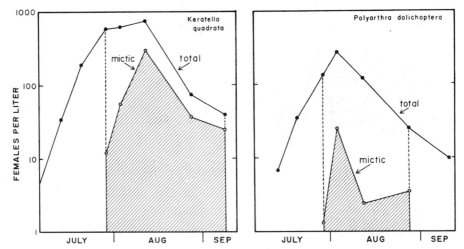

Fig. 2. Population composition of two rotifers in locality 1 on Spitsbergen Island (data from Amren 1964).

Note that individuals of both species initiated sexual reproduction at or slightly before the recorded population maximum. *Keratella quadrata* reproduced well into September, whereas no mictic females of *P. dolichoptera* were recorded after late August. Both populations can be typified as initiating sexual reproduction in the middle portion of the population cycle.

Carlin accumulated data on the timing of sexual reproduction in a number of rotifer species. His data do not include measures on the extent of sexual reproduction but do contain information on when males were found. Although considerable variation within species exists from year to year, three patterns can be drawn from Carlin's study. Individuals of some species, e.g. *Notholca caudata* (Fig. 3), tended to reproduce sexually in the early part of the population cycle, that is, while population size was still increasing. These species were relatively uncommon. A larger proportion of species followed the pattern of *K. quadrata* and *P. dolichoptera* depicted in Fig. 2 and initiated sexual reproduction near the population size maximum. *Polyarthra major* (Fig. 4) was typical of this second pattern. Third, the sexual reproduction of some species, such as *Polyarthra vulgaris* (Fig. 5), occurred toward the end of the population cycle. I will refer to these patterns as early cycle, midcycle, and late cycle.

Several implications of these patterns are explored in Fig. 6. This graphical model has three major assumptions. It is assumed that a population cycle is initiated when a cohort of resting eggs hatch. It is further assumed that the eggs giving rise to a single population hatch within a relatively short period and that they are genetically heterogeneous either because their parents were part of a heterogeneous population or because the eggs are derived from a number of different populations. (None of these assumptions has been investigated by limnologists; however, all three appear to be consistent with our present understanding of rotifer population structure.) As the population grows (top panel), the intensity of intrapopulation competition (via *r*-selection) also increases and the resultant selection (middle panel) leads to a decrease in variance among genotypes (bottom panel). If sexual reproduction occurs during the early part of this cycle, the resting eggs produced should represent a reasonably complete sample of the genetic variation of the original cohort.

As the population increases, variance among genotypes decreases. Consequently, if sexual reproduction occurs during the middle part of the population cycle, the progeny may have only a small portion of the original genetic variation. Two implications of this pattern deserve consideration. First, since high-density conditions are an indicator of successful environmental exploitation, density cues could well be used to indicate suitable conditions for sexual reproduction. Such cues have been found by Gilbert (1963, 1977) for *Brachionus calyciflorus* and by King and Snell (in prep.) for *A. girodi* in Golf Course Pond. Second, since resting eggs produced in

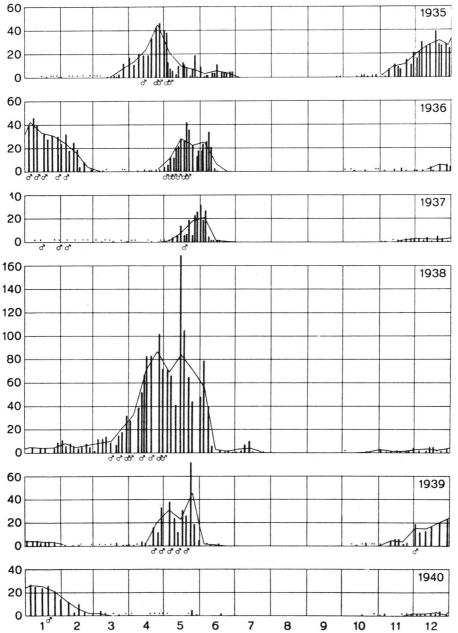

Fig. 3. Population sizes of *Notholca caudata* in the Motala River (from Carlin 1943); the symbol ♂ indicates presence of males. Abscissa is month, ordinate is no./liter.

early cycle are expected to have a higher genotypic variance than those produced in midcycle, the early cycle populations are expected to be less specialized in their habitat requirements than midcycle rotifers. Put differently, if two populations in a given lake are initiated at the same time, the population that starts sexual reproduction earlier in its growth cycle is expected to have the broader niche. However, one price of sexual reproduction is decreased potential for parthenogenetic

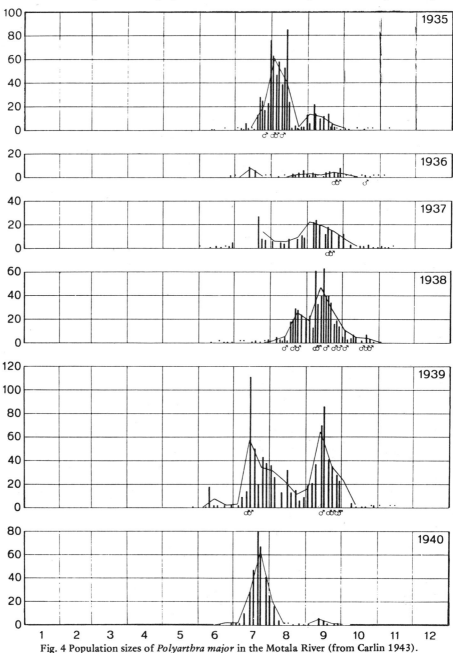

Fig. 4 Population sizes of *Polyarthra major* in the Motala River (from Carlin 1943).

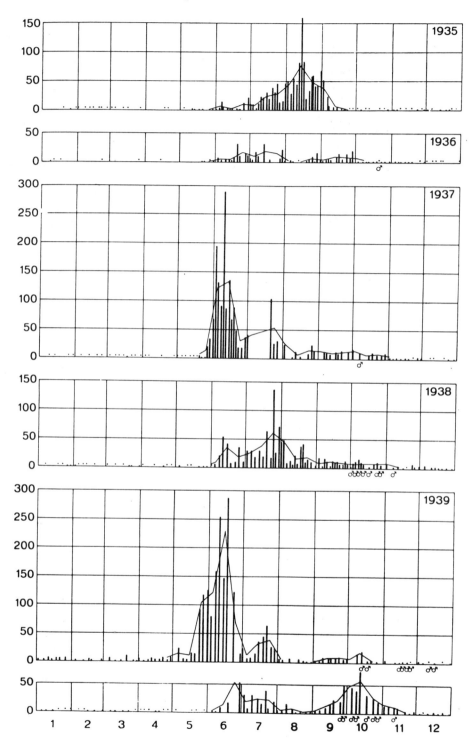

Fig. 5. Population sizes of *Polyarthra vulgaris* in the Motala River (from Carlin 1943).

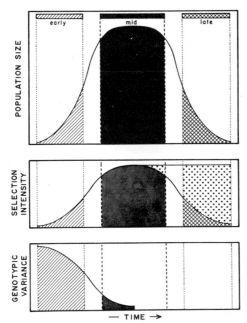

Fig. 6. Population growth cycle for a hypothetical monogonont rotifer. Middle panel indicates selection intensities derived from both density-dependent (bell-shaped curve) and density-independent (dotted area) sources. Bottom panel illustrates expected decay of variance among genotypes exposed to strong selection during parthenogenesis. Assumptions and further description given in text.

growth and hence decreased competitive ability. Thus the population with the broader niche may have the narrower period of occupancy.

The last group of monogonont rotifers is that defined by late cycle sexual reproduction. This group is exposed to a decreasing intensity of density-related selection (the cross-hatched area in the middle panel of Fig. 6), but as the habitat suitability decays they are subject to an increasing intensity of density-independent environmental displacement load (hatching in middle panel of Fig. 6). Environmental displacement load is a measure of the population's adaptation to a changing environment. A population exploiting an environment to which it is adapted will have minimal load. However, as the environmental state diverges from this "ideal" environment, the population will acquire environmental displacement load and have lower fitness. Since we have little, if any, idea about what makes a particular habitat suitable for a particular population, it is difficult to make statements about either the direction or intensity of selection during late cycle reproduction. We do know, however, that selection in any part of the growth cycle under parthenogenesis is more intense than a comparable regime of selection under sexual reproduction (King 1977b). The fixation of heterozygotes in the *A. girodi* population of Golf Course Pond, for example, was possible only because all reproduction was by parthenogenesis. Genetic variation may therefore be maintained under strong selection even if variation among genotypes is lost.

Environmental change accompanying population size reduction in late cycle raises an interesting question: What factors in such an environment can be used to induce mictic female production and subsequent recombination? Since population size is decreasing, it should not be anticipated that cues for sexual reproduction will be associated with correlates of fixed densities. The possibility exists that the population could respond to *decreases* in density, but this mechanism is, to my knowledge, without support. The possibility that the population cues on physical factors associated with habitat deterioration seems much more likely. Gilbert's (1968, 1977) demonstration that dietary tocopherol induces mictic female production in *Asplanchna sieboldi* and *A. brightwelli* is now well known. Although tocopherol response intensity is related to population density in the laboratory, there is a paucity of information on timing of sexual reproduction in nature. Because of the cryptic character of population succession in rotifers, simple correlation of sexual stages with population size (such as I did in citing results from Amren and Carlin) may obscure relevant events and processes. For these reasons it would be interesting to study the timing of sexual reproduction of *A. sieboldi* and *A. brightwelli* in a natural system in which patterns of population succession could be determined. It would also be very interesting to determine how sexual reproduction is initiated in *Euchlanis dilatata* since field data of both Carlin (1943) and King (1972) suggest that this is a late cycle rotifer.

It was suggested above that if sexual reproduction is delayed until midcycle, both variation among genotypes and niche breadth should be reduced relative to populations undergoing sexual reproduction in early cycle. This trend is expected to both continue and intensify for those rotifers reproducing sexually in late cycle. In this regard the rapid succession of *E. dilatata* populations reported by King (1972) may be a function more of the timing of reproduction than rapid environmental flux.

Recombination or mutation?

From the above discussion it is clear that sexual reproduction may serve different purposes for different rotifers. For those populations having recombination early in the growth cycle, sexual reproduction may result in the generation of a large number of novel genotypes. This function would appear to be greatly reduced in those populations having their sexual reproduction restricted to later portions of the growth cycle. Even if some genetic variation is maintained either directly or indirectly by selection, most variation among genotypes will be lost unless it is selectively neutral. The inevitable consequence of parthenogenesis in a stable environment is therefore loss of variation. However, even if populations are isolated units, there remains the possibility that mutation generates substantial amounts of variation. We might therefore change our focus to inquire: Which is of greater significance in the generation of variation among genotypes, recombination or mutation? Although the answer to this question remains for future research, it is appropriate at this time—particularly at a zooplankton (rather than rotifer) symposium—to view the problem in a broader taxonomic context.

The base data necessary to the following comments are: (1) bdelloid rotifers reproduce exclusively by parthenogenesis; (2) monogonont rotifers are predominant parthenogens and have distinct types of females (mictic and amictic) for sexual and parthenogenetic reproduction; (3) cladocerans, in general, are capable of combining both sexual and parthenogenetic reproduction in the same individual and appear to emphasize sexual reproduction more than rotifers; (4) copepods reproduce exclusively by sexual reproduction. With the felicitous addition of bdelloids to the category of "zooplankton," we can state that this group of limnetic organisms represents the entire spectrum of reliance on sexual reproduction. What is there in the biology of these groups that permits bdelloids, on one hand, to exist without sexual reproduction, and copepods, on the other hand, to exist without parthenogenesis? Further, why do monogonont rotifers and cladocerans appear to take two distinctly different intermediate positions?

Consider first the source of new genotypes. Since bdelloids reproduce by ameiotic diploid parthenogenesis and lack recombination, novel genotypes in a system closed to immigration result from the process of mutation. Although other sources—such as somatic crossing over—are in theory possible, their likelihood and effect would seem to be remote and insignificant. In contrast, recombination is an obvious source of new genotypes in copepods.

The relative significance of mutation and recombination is a function of population size. Consider a population with two alleles at a single locus, A and a. Suppose all individuals in the population are either AA or aa, but that the heterozygote has greater fitness than either homozygote. If population size is small, recombination is expected to produce a heterozygous individual more rapidly than mutation (of either A to a, or a to A), even if one of the two genotypes has a very low frequency. In a large population mutation may have increased significance simply because the probability of a particular allele recurring by mutation is an increasing function of population size. Note that I am assuming the pre-existence of genetic variation in this section. A similar argument is summarized later, but the pre-existence of variation is not assumed.

A graphical illustration of these relationships is presented in Fig. 7. It is assumed that all new genotypes are generated

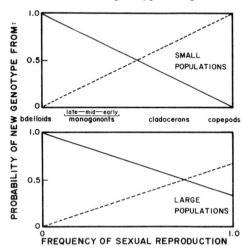

Fig. 7. Hypothetical significance of mutation and recombination in small and large populations of zooplankton groups varying in their frequency of sexual reproduction. Solid line—mutation; dashed line—recombination. Assumptions and further description given in text.

by either mutation or recombination and that the two processes are inversely related. From the discussion presented in the previous section, the late-, mid-, and early-cycle rotifers can be arrayed with the other groups as shown.

Since population size is of obvious importance to a discussion of the significance of mutation, estimates of the number of individuals per liter for a variety of planktonic species have been obtained from my study of Golf Course Pond, and Nauwerck's (1963) study of Lake Erken (Table 1). Population sizes are given in two forms: first for individuals/species/liter at time of maximum density, and second for the total number of individuals per species, per liter, per year. It must be emphasized that the second group of estimates may be quite rough since it is dependent upon both estimates of turnover times and interpolations of Nauwerck's graphs.

To estimate the number of mutations per locus, I used a mutation rate of 2×10^{-6} per locus. (The "2" is used since we are considering diploid loci.) For convenience, mutation numbers are given per 10^6 liter. The values in Table 1 are presented in Fig. 8 as a function of mean generation times estimated for the three groups. It is recognized that these generation time estimates are quite rough and do not recognize significant variation within groups. These values were chosen because they make the argument more conservative; i.e. rotifers probably tend to have shorter and cladocerans and copepods longer generation times than those given in Fig. 8. From the left panel of Fig. 8, it is apparent that population sizes of rotifers tend to be larger than those of cladocerans, which are, in turn, larger than those of copepods. However, note also that there is considerable overlap in peak population densities and, therefore, expected numbers of mutations.

If our view is broadened to encompass total individuals over the annual cycle, the graph in the right panel of Fig. 8 is generated and much of the similarity between groups disappears. Note particularly that there is no overlap between rotifers and cladocerans and that mean numbers of mutations/yr in the two groups are separated by an order of magnitude. From this result alone, the process of mutation may be expected to play a very different role in the three groups.

To obtain an estimate of the number of mutations per locus per year in the entire lake, we can multiply the last column of Table 1 by lake volume. *Keratella cochlearis* occurs in both lakes. In Golf Course Pond, a small shallow lake with a volume of 31.2×10^6 liter, this calculation suggests that there are more than 140,000 mutations at the average locus over the course of a year. For the same species in the much larger Lake Erken, the number of mutations exceeds one billion per locus. Even if only 1 of every

Table 1. Estimated densities of rotifers, cladocerans, and copepods in Lake Erken (Nauwerck 1963) and Golf Course Pond at population maxima (N_{max}/liter), and through the year (N_{tot}/liter·yr). Assumptions used in calculating mutations given in text.

	N_{max}/liter	Mutations per locus·10^6/liter	N_{tot}/liter·yr	Mutations per locus·yr·10^6 liter
I. Lake Erken (vol = 2 × 10^{11} liters)				
Rotifers				
Keratella cochlearis	55	110	3,000	6,000
K. hiemalis	165	330	4,000	8,000
Kellicottia longispina	50	100	2,000	4,000
Asplanchna priodonta	100	200	1,000	2,000
Polyarthra vulgaris	90	180	5,000	10,000
Filinia terminalis	140	280	2,000	4,000
Cladocerans				
Bosmina coregoni	18	36	40	80
Ceriodaphnia quadrangula	90	180	101	202
Diaphanosoma brachyurum	10	20	22	44
Daphnia longispina	12	24	39	78
Copepods				
Cyclops strenuus	6	12	8	16
Mesocyclops leuckarti	12	24	15	30
Eudiaptomus graciloides	78	150	150	300
II. Golf Course Pond (vol = 31.2 × 10^6 liters)				
Rotifers				
K. cochlearis	215	430	2,300	4,600
Brachionus havanaensis	58	116	395	790
Asplanchna girodi	94	186	435	870
Cladocerans				
Eubosmina tubicen	181	362	386	772
Daphnia ambigua	91	182	208	416
Copepods				
Diaptomus dorsalis	48	96	111	222
Cyclops ?	39	78	52	104
Mesocyclops edax	6	12	11	22

1,000 to 10,000 of these mutations confers a fitness advantage, the potential significance of this source of variation seems obvious.

Evolution of sexual reproduction

One of the recent debates in population biology has focused on attempts to explain the advantage of sexual reproduction. The initial problem was formulated by Fisher (1930) and by Muller (1932) who suggested that the major advantage of sexual reproduction is its ability to combine favorable mutations occurring in separate individuals. Muller's theory was taken up by Crow and Kimura (1965) and their brief paper both renewed interest and started the controversy. A review of these papers and subsequent developments is presented in Chapter 12 of Williams (1975).

Although a detailed consideration of "why sex?" is beyond the scope of this paper, it is appropriate to gather observations from the zooplankton relevant to this question. These observations will be discussed under four headings: Population

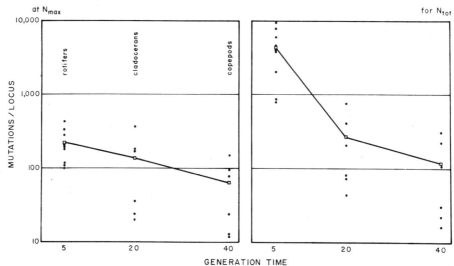

Fig. 8. Estimated numbers of mutations per diploid locus for rotifers, cladocerans, and copepods. ●—Based on values given in Table 1; □—(connected by lines) are averages of solid circles.

size, colonization, aphid-rotifer model, and resting eggs.

Population size. Muller's (1932) argument, as quantified by Crow and Kimura (1965), did not assume the pre-existence of genetic variation and so large population sizes were required for sex to acquire an advantage over mutation. Maynard-Smith (1971) suggested that this size should be an order of magnitude larger than the reciprocal of the rate of favorable mutation. In the preceding section I used 2×10^{-6} as the estimated mutation rate per diploid locus and 10^{-9} to 10^{-10} as the estimated rate of favorable mutations. Although these estimates are of highly uncertain validity, if they are within a few orders of magnitude of being correct they clearly require very large population sizes. The immediate supposition that these large numbers suggest why so many zooplankton species are parthenogenetic is clearly inconsistent with the observation that the phylogenetic progression from rotifers to higher organisms is accompanied by both more sex and smaller populations.

Colonization. Maynard-Smith (1971) suggested that the primary advantage of sex is to permit recombination of favorable genes from different sources during colonization. A situation is envisioned in which a new habitat is colonized by individuals from two different source populations. Even if each of the colonists is able to survive in the new habitat, a combination of their previously distinct genomes might well produce offspring better adapted than either parent.

A major difficulty with applying Maynard-Smith's suggestion to rotifers is that few species initiate sexual reproduction in the early part of the growth cycle, yet that is the portion in which sex should be concentrated in order to obtain recombinants between two different colonizers.

An argument based on colonization has aesthetic appeal for ephemeral populations and habitats that are recolonized each year. Most zooplankters and their lakes fit such a description. One further difficulty in assessing the applicability of this description is our lack of information on the source of resting eggs that give rise to rotifer populations. If a lake is occupied by three populations (A, B, and C) of a given species, it is conceivable that resting eggs from all three could hatch simultaneously and the eventual population would be determined by the outcome of competition. However, if the three populations are adapted to different parts of the annual cycle it is also conceivable, and probably more likely, that population A from the previous year will give rise to the new A, B to B, and C to C. Sex would not confer an advantage of the type envisioned by Maynard-Smith in such a system, although it would either in colonizing a previously unoccupied portion of the temporal spectrum or in colonizing an unoccupied lake.

As a final point, it is difficult to picture a better colonizer than a rotifer or cladoceran. Perhaps this is why so many species in these groups are so widely distributed. The thick-shelled, highly protected sexual egg of both groups, coupled with uniparental parthenogenesis, constitutes an ideal propagule. All that is lacking is some means of independent movement from one lake to another.

Aphid-rotifer model. This model was proposed by Williams and Mitton (1973) and expanded by Williams (1975). An analogy was drawn between sexual reproduction and a lottery: if a person were offered a number of copies of the same lottery ticket, or an equal number of different tickets, the correct choice would be obvious. If this analogy has currency, it resides in the value of sexually producing a large array of genotypes so that the likelihood of successful colonization is increased. However, it is not at all clear to me that the relevant biological payoff of reproduction is in enhancing future colonization. For instance, if a realtor offered a buyer a number of copies of a key for a house being purchased, or an equal number of keys to houses in some unknown distant city, the correct choice would be just as obvious as for William's analogy. That is, if the future environment is predictable, or if survival and successful reproduction in the present environment require offspring having uniform high fitness, then reproduction should be by parthenogenesis and not by sex.

It is difficult for me to be highly critical of the Aphid-Rotifer Model. However, I do have one fundamental reservation: it should obviously have been named the "Rotifer-aphid model."

Williams (1975) clearly recognizes that intense competition may occur between clones. Such competition is evident in the Golf Course Pond *A. girodi*, and has been recently demonstrated in rotifers under laboratory conditions by Snell (1977, 1978). Given intense interclonal competition, large population size may confer a competitive advantage. That is, if the environment is relatively constant, a large parthenogenetic population should outcompete a small sexual population. As pointed out earlier, parthenogenetic reproduction leads to both shorter generation times and larger offspring numbers than sexual reproduction. Additionally, with sexual reproduction a significant amount of recombinational load may be generated that is not associated with parthenogenesis.

A second source of advantages associated with the production of large numbers of genetically homogeneous eggs is in the area of Allee effects, i.e. the beneficial effects of high density on patterns of environmental exploitation. If for no other reason, mate location for rotifers is more likely in a high than in a low density population.

A third, and particularly relevant consideration is the genetic cost of sexual reproduction. An individual reproducing by diploid ameiotic parthenogenesis transmits all of her genes to each offspring. In contrast, progeny resulting from sexual reproduction have only half of the genetic complement of each parent. From the female parent's standpoint, sexual reproduction involves a 50% cost not associated with parthenogenesis. The advantage to sexual reproduction, as argued by Williams (1975), must be great enough to pay this 50% cost and still make a profit.

In this context, consider a rotifer population that has undergone strong selection during parthenogenesis. By the time maximum population size is reached, most of the competing genotypes will have been eliminated. The genetic variation remaining will be that which is neutral, that which has newly arisen through mutation, and that which is fixed in the polymorphic condition through either heterozygote advantage or associative overdominance. Given narrow niches, temporal restriction, and strong selection, the population is expected to be composed of a small number of genotypes. In this system the genetic correlation among mates will be very high and the process of sexual reproduction approaches the equivalent of inbreeding, if not parthenogenesis. Under these conditions, the 50% disadvantage of sex is greatly reduced and one must only look for a different type of advantage to explain its occurrence.

Resting eggs. Ruttner-Kolisko (1963) suggested that the major function served by bisexual reproduction was resting egg production rather than genetic recombination. The significance of the resting egg for colonization is obvious and needs no further mention. There have been a number of reports of pseudosexual resting eggs in both rotifers (*see* King 1977a) and cladocerans (*see* Hebert 1978). However, resting stages are generally produced by sexual rather than parthenogenetic reproduction. Bdelloid rotifers, which have no sexual reproduction, also have no resting eggs. Instead, survival during harsh periods is accomplished by cryptobiosis—an ability that is absent in both monogonont rotifers and cladocerans.

It should be clear from the discussion in this section that it is easier to relate the success of rotifers, and perhaps cladocerans also, more to their parthenogenetic reproduction than to their sexual reproduction. This of course in no way denies a role for recombination in sexual reproduction, nor has the information presented rendered any of the broader explanations untenable. What I have attempted to argue is that there may be no universal answer to this question. Further, at least for the present, a tenable hypothesis for rotifers is that sex exists because it is the way a rotifer makes resting eggs.

References

AMREN, H. 1964. Ecological and taxonomical studies on zooplankton from Spitsbergen. Zool. Bidr. Upps. 36:209–276.

ANGUS, R. A. 1978. *Daphnia* and the search for heterosis. Am. Nat. 112:955–956.

CARLIN, B. 1943. Die Planktonrotatorien des Motalaström. Medd. Lunds Univ. Limnol. Inst. 5. 255 p.

CROW, J. F., and M. KIMURA. 1965. Evolution in sexual and asexual populations. Am. Nat. 99:439–450.

FISHER, R. A. 1930. The genetical theory of natural selection. Clarendon.

GILBERT, J. J. 1963. Mictic female production in the rotifer *Brachionus calyciflorus*. J. Exp. Zool 153:113–124.

———. 1968. Dietary control of sexuality in the rotifer *Asplanchna brightwelli* Gosse. Physiol. Zool. 41:14–43.

———. 1977. Mictic female production in monogonont rotifers. Arch. Hydrobiol. Ergeb. Limnol. 8:142–155.

HEBERT, P.D.N. 1978. The population biology of *Daphnia* (Crustacea, Daphnidae). Biol. Rev. 53:387–426.

KING, C. E. 1972. Adaptation of rotifers to seasonal variation. Ecology 53:408–418.

———. 1977a. Genetics of reproduction, variation, and adaptation in rotifers. Arch. Hydrobiol. Ergeb. Limnol. 8:187–201.

———. 1977b. Effects of cyclical ameiotic parthenogenesis on gene frequencies and effective population size. Arch. Hydrobiol. Ergeb. Limnol. 8:207–211.

———, and T. W. SNELL. 1977. Sexual recombination in rotifers. Heredity 39:357–360.

MAYNARD-SMITH, J. 1971. What use is sex? J. Theor. Biol. 30:319–335.

MULLER, H. J. 1932. Some genetic aspects of sex. Am. Nat. 66:118–138.

NAUWERCK, A. 1963. Die Beziehungen zwischen Zooplankton und Phytoplankton im See Erken. Symb. Bot. Upps. 17. 163 p.

RUTTNER-KOLISKO, A. 1963. The interrelationships of the Rotatoria, p. 263–272. *In* E. C. Dougherty [ed.], The lower Metazoa. Univ. Calif. Press.

SNELL, T. W. 1977. Clonal selection: competition among clones. Arch. Hydrobiol. Ergeb. Limnol. 8: 202–204.

———. 1979. Competition and population structure in rotifers. Ecology 60: 494–502.

WILLIAMS, G. C. 1975. Sex and evolution. Monogr. Popul. Biol. 8:200p.

———, and J. B. MITTON. 1973. Why reproduce sexually? J. Theor. Biol. 39:545–554.

31. The Genetics of Cladocera

Paul D. N. Hebert

Abstract

Most cladocerans reproduce by cyclical parthenogenesis, although a few species are obligate parthenogens. Recombination does not occur during parthenogenesis, but sexual eggs are produced by normal meiosis. Selection experiments on clonal lines have been successful, possibly indicating high spontaneous mutation rates.

Genetic variation has been detected in cladoceran populations both by inbreeding studies and by electrophoresis. Visible polymorphisms have also been reported, but these probably involve unrecognized species complexes. Large gene frequency differences exist among populations in close proximity, indicating that populations are founded by few individuals and that migration between habitats is rare.

Many *Daphnia* populations are regularly refounded from sexual eggs. In such populations genotypic frequencies at enzyme loci are stable through time and usually approximate Hardy-Weinberg expectations. By contrast, populations of *Simocephalus*, with a similar life history, show heterozygote deficiencies and nonrandom associations of genotypes at different loci. These genotypic characteristics suggest that *Simocephalus* females are able to fertilize their own sexual eggs.

Some cladoceran populations reproduce by continued parthenogenesis and are subject to only sporadic sexual recruitment. In such populations genotypic frequencies commonly differ from Hardy-Weinberg expectations, usually because of a heterozygote excess. Genotypic frequencies at individual loci often seem to be shifted by natural selection, but the occurrence of nonrandom associations between genotypes at different loci indicates that selection is acting on gene complexes. Clonal variation persists in these populations, apparently because of ecological differences among clones.

The order Cladocera is a diverse taxon which includes 8 families, more than 50 genera, and probably nearly 1,000 species. Most of these species are able to reproduce both parthenogenetically and sexually. Parthenogenesis is, however, the usual mode of reproduction and populations often consist entirely of females. Parthenogenetic eggs are deposited into a brood pouch; embryogenesis occurs rapidly and the young are released as miniature versions of the adult. Parthenogenetic young are normally female, but in certain environments male eggs are produced. In addition to these two types of parthenogenetic egg, a female can produce sexual eggs which must be fertilized in order to develop. Species in the family Daphnidae release these sexual eggs into a structure called an ephippium, which is composed of the thickened membranes and cuticle of the brood pouch. The number of sexual eggs per ephippium is small; all species in the genus *Daphnia* release two eggs into each ephippium, while species of *Moina* and *Simocephalus* release a single egg. Among the Chydoridae, Macrothricidae, and Bosminidae the ephippium is not so well formed and a variable number of sexual eggs may be released. For instance, ephippia of *Eurycercus glacialis* contain from one to nearly fifty eggs (Kaiser 1959). The sexual eggs produced by species in the families Holopedidae, Leptodoridae, Polyphemidae, and Sididae are not enclosed in an ephippium and are released naked into the water.

Many cladocerans have become obligate parthenogens. These species continue to produce two sorts of eggs—parthenogenetic eggs which develop immediately and, in addition, parthenogenetic eggs which have a diapause capability. The latter eggs are embryologically similar to the sexual eggs produced by most species, but do not need to be fertilized to develop. In species belonging to taxa which normally form an ephippium, these diapausing eggs are enclosed within an ephippium.

Because there are so few animal groups which produce both sexual and asexual eggs, there has been a longstanding interest in cladoceran genetics. In the late 19th century Weismann initiated work on their cytogenetics—studies which have been nearly continuously in progress ever since. Unfortunately cladocerans have proven to be poor cytological material. Although in all studies to date it has been impossible to distinguish homologous pairs or to recognize chiasmata, chromosome counts have been made on a number of species in the family Daphnidae. All recent counts indicate that diploid chromosome numbers are close to twenty (Schrader 1926; Allan 1928; von Dehn 1948; Ojima 1958; Zaffagnini and Sabelli 1972). There has been no indication of polyploidy, although it might be expected to occur in cladocerans because of their ability to reproduce asexually.

Early cytological studies indicated that parthenogenetic eggs were formed by an essentially mitotic cell division. However, more recent work suggested that some meiotic behavior occurs during parthenogenetic oogenesis. Ojima (1958) observed a transitory synaptic concentration of chromosomes, while Bacci et al. (1961) noted pairing of chromosomes and subsequent deconjugation processes corresponding to the first division of normal meiosis. As the products of this division remained within the nuclear membrane, the ovum remained diploid. Bacci et al. (1961) proposed, however, that crossing over accompanied chromosome pairing and concluded that recombination can give rise to genetic variability within single parthenogenetic lines. Zaffagnini and Sabelli (1972) have argued that the chromosomal associations noted by Bacci et al (1961) resulted from preparative procedures. Their objection seems valid for genetic studies have confirmed the absence of recombination; the parthenogenetic offspring of females heterozygous for enzyme variants were themselves also heterozygous (Hebert and Ward 1972; Manning et al. 1978). Cytological studies have shown that chromosome numbers in males and females are similar, and that eggs which give rise to males are also produced mitotically. Cytogenetic study of *S. middendorffiana*, a species producing ephippial eggs not requiring fertilization, has shown that these eggs are also produced mitotically (Zaffagnini and Sabelli 1972). By contrast, in species which produce ephippial eggs requiring fertilization, both these eggs and the sperm are produced by normal meiosis (Ojima 1958; Zaffagnini and Sabelli 1972).

Aside from this work on cladoceran cytogenetics there has been continued interest in the factors which initiate the switch in cladoceran populations from asexual to sexual reproduction. Weismann (1893) felt that the switch was controlled by an internal clock—after a certain number of asexual generations a clone would reproduce sexually. This hypothesis was abandoned when it was shown that clones could be maintained through hundreds of generations of parthenogenesis in the laboratory (Agar 1914; Banta 1914). It is now recognized that environmental factors are of paramount importance in determining the switch from parthenogenetic to sexual reproduction. A plethora of factors have been implicated in determining the type of egg produced (Banta and Brown 1929; Banta 1939; Stross and Hill 1965). Parthenogenetic eggs developing into females are usually produced when food is abundant and cultures uncrowded. Crowding causes male eggs to be produced, while sexual eggs seem to be produced in response to a rapid deterioration in environmental suitability. Despite the large amount of work done on the topic, techniques for producing males and sexual females on demand are only moderately successful. This stands as one of the most significant problems to be solved if studies on cladoceran genetics are to become sophisticated.

Between 1910 and 1940 Banta and his associates did a tremendous amount of work on cladocerans. Banta's contribution to our understanding of sex determination is still widely appreciated, but his work on other aspects of cladoceran genetics is much less well known. Two of his experiments remain of particular interest. The first involved selection experiments on clones, designed to find out if it was possible to alter the expressivity of mutations (Banta 1939). Two mutations were studied—a gene causing sex integradation and a gene altering head shape. Breeding studies indicated that both these genes were dominant, but of variable expressivity. Starting from a single female heterozygous for one of the mutations, Banta selected among her offspring one female that showed little expression of the trait and another that showed maximum expression. These individuals were the founders of the low expression and high expression lines, which were then directionally selected for a number of generations. In nearly all the experiments there was a rapid response to selection—in the low line the phenotypic effect of the mutation was reduced, while in the high line its effect was enhanced (Fig. 1). Why Banta was able to obtain a response to selection in a clonal organism is not clear. Banta felt that the response was due to mutations occurring in the selected lines, but if this were indeed the case, then the frequency of mutation during parthenogenesis would need to be high. Determination of mutation rates at individual gene loci would be a worthwhile followup to this work.

Banta and his associates also carried out the first controlled breeding studies on cladocerans. In one experiment they were

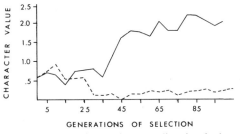

Figure 1. Results of divergent directional selection on expressivity of an excavated head mutation in *Daphnia*. Individuals with a character value of 0 show no expression of the mutation, while a character value of 3 indicates maximal excavation of head (data from Banta 1939).

able to show that clones derived by inbreeding a single parent clone were extremely variable in traits such as longevity, natality, and growth rates (Banta 1939). This observation provided the first convincing evidence that genetic variation was abundant in cladoceran populations, for the variation disclosed by inbreeding must have been derived from heterozygosity present in the parent clone. Unfortunately there was no way in which this line of study could be extended to provide gene frequency data, for the variable traits were under polygenic control. However, in the past decade electrophoresis has made it possible to study genetic variation at individual loci in cladoceran populations.

Before passing on to summarize the results of these studies, I want to consider some apparent cases of visible polymorphism for such variants seen initially to be analogous to the variation which has provided so much information about the genetic composition of lepidopteran and gastropod populations. These instances of polymorphism include forms of *Daphnia carinata* in India (O'Brien and Vinyard 1978), *D. lumholtzi* in Africa (Green 1967), *Ceriodaphnia cornuta* in Panama (Zaret 1969, 1972) and *Bosmina longirostris* in the western U.S.A. (Kerfoot 1975). Populations of these species have been found to include two or more morphologically distinct types—often one form has a helmet or spines while the other does not. In all cases other morphological and ecological differences between the morphs have been noted. For instance the forms of *C. cornuta* also differ in eye size, age to first reproduction, fecundity, and longevity (Zaret 1972). Such extensive differences are unlikely to be due to a single gene polymorphism, although the morphs could be under the control of supergenes, similar to those controlling the morphs of mimetic butterflies. However, there is no reason to postulate this until it can actually be shown that the coexisting forms regularly hybridize, and that a single female can produce offspring belonging to different forms. In none of the cases mentioned has this essential experiment been done. In the absence of such data it is more reasonable to treat the supposed forms as different species. Indeed genetic studies have now been done on two of the species. Populations of *Daphnia* in Australia, once all classified as *D. carinata*, are now recognized to belong to nine different species (Herbert 1977). These species are most readily distinguished by their differences in head shape; the characters normally used in species recognition are of little value. Species in the *D. carinata* complex often cohabit, but genetic analysis provided no evidence of introgression. Similarly it has now been shown that the two forms of *B. longirostris* represent different species which do not introgress (Manning et al. 1978). In conclusion, all cases of visible polymorphism so far noted in the Cladocera are likely to represent species complexes.

Although visible polymorphisms seem not to exist, electrophoresis has revealed the presence of variation at gene loci controlling enzyme phenotypes. In an analysis of 40 populations of *D. magna*, variation was found at 5 of the 15 loci studied (Herbert 1975; Young 1975). However none of the populations were polymorphic at more than three of these loci and six were completely monomorphic. On average, populations were polymorphic at 11.1% of their loci, a considerably lower frequency than the 30–50% values found in most other organisms which have been studied (Lewontin 1974). Even less variation was detected in populations of *D. carinata* (2 of 15 loci polymorphic); while in populations of the obligate parthenogen *D. cephelata* genetic variation was nearly absent. It seems then that local populations of *Daphnia* contain relatively little genetic variation. Whether this conclusion extends to other cladocerans is uncertain. In a study of *Simocephalus serrulatus* (Smith 1974; Smith and Reaser 1976) a much higher frequency of variation was recorded, but these data need to be treated with caution as no breeding studies were carried out to confirm interpretation of the allozyme patterns.

Regardless of its frequency, the presence of genetic variation makes it possible to obtain data on problems which are otherwise difficult or impossible to investigate. For instance, by looking at the similarity of gene frequencies among populations of a particular species one can obtain some idea about the amount of migration between populations. Such studies have now been carried out on several cladoceran species. In England, large differences in gene frequencies were observed between populations of *D. magna* only a few meters apart and the substitution of one allele for another was commonly noted among more distant populations (Hebert 1975). These allelic substitutions were almost certainly the result of founder effect rather than natural selection. Similar genetic differences have been observed among populations of *D. carinata* (Fig. 2), *D. cephalata*, *S. serrulatus*, and *B. longirostris* (Smith and Fraser 1976; Manning et al. 1978; Hebert in prep.). It seems justified to conclude that migration between cladoceran populations is limited and that most populations originate from a small number of ephippia. This genetic isolation of populations should also be reflected in morphological traits, although traditionally it has been argued that cladoceran species show little morphological variation over their large ranges (Brooks 1957; Hebert 1975). In most cases this conclusion has been based on the gestalt of the animals. It is true, for example, that specimens of *D. lumholtzi* from Africa, Asia, and Australia share many features, but it also seems likely that a careful study of meristic traits would reveal diagnostic

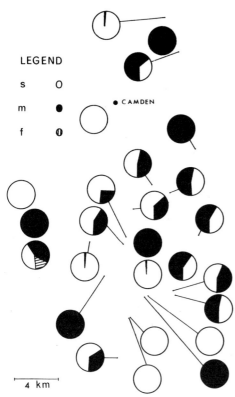

Fig. 2. Gene frequencies at leucine aminopeptidase-2 locus in natural populations of *Daphnia carinata* near Sydney, Australia.

differences not only among populations from the three areas, but also among individual populations in each area. Certainly workers studying cyclomorphosis have pointed out the marked variation in head shape among local populations of a particular species (Hutchinson 1967). While much of this variation may be related to environmental differences, it seems probable that at least some of the variation in head shape is genetically caused.

Gene frequency data have also made it possible to study the genetic effects of the variation in breeding systems which exists among cladocerans. For although most cladocerans are cyclical parthenogens, environmental conditions play a major role in determining the actual nature of their breeding system. In habitats which remain filled with water for only a portion of the year, populations survive the dry periods in the ephippial stage and are able to reproduce parthenogenetically for only a few generations at a time. In such intermittent populations there is a regular alternation of sexual and parthenogenetic reproduction, as long as the species involved produces its ephippial eggs sexually.

Genetic analysis has been carried out on intermittent populations of both *D. magna* (Hebert 1974a) and *D. carinata* (Hebert and Moran in prep.). Genotypic frequencies at polymorphic loci in these populations were normally in good agreement with Hardy-Weinberg expectations. This indicated that genotypes in the populations were mating randomly at the time of ephippial formation and that population sizes were large enough so that inbreeding was not significant. Genotypic frequencies were stable during annual cycles of parthenogenesis, indicating further that fitness differences between genotypes were not pronounced. In contrast to these results, work on intermittent populations of *S. serrulatus* revealed large deviations from Hardy-Weinberg expectations (Smith and Fraser 1976). In nearly all cases these deviations were due to the absence or extreme deficiency of heterozygotes. Positive assortative mating could explain the rarity of heterozygotes, but in these populations consistent associations between genotypes at different loci were also noted. For instance, in the Ranch Lake population there were only two common clones and each clone was homozygous for a different allele at four of the five loci studied. Such an association of homozygotes at different loci can only be explained if sexual reproduction is restricted. Yet these associations were noted immediately after the population had been re-established from ephippia. Two explanations can be suggested. Possibly two sibling species with distinct enzyme phenotypes were present in the habitat. The few heterozygotes observed would indicate that a limited amount of introgression was occurring between the species. This interpretation is weakened when the data on other populations of *S. serrulatus* are considered. In these populations heterozygote deficiencies were again noted, but associations between homozygotes at the four loci varied. These associations should have been stable if only two species were involved. A second explanation requires only that *S. serrulatus* females normally fertilize their own ephippial eggs, yet also retain the ability to cross-fertilize. This could be accomplished fairly simply. If the ephippial eggs undergo normal meiosis, then self-fertilization could be carried out by one of the polar bodies produced during oogenesis. The retention of a meiotic division would mean, moreover, that if males were present they could fertilize the egg. Such outcrossing would account for the retention of some heterozygosity in the populations. The apparent predominance of self-fertilization could indicate either that males were rare or that the sexual ova remain haploid for only a brief period.

Some cladoceran populations live in habitats that allow adults to survive throughout the year. Such permanent populations reproduce by continued parthenogenesis, for even during bouts of sexual reproduction some individuals in a population always reproduce asexually. There are undoubtedly differing degrees of population permanency. It seems likely that cladoceran populations in lakes have the longest histories of continued parthenogenesis, but unfortunately little work has been done on lake species. At present the only substantial data on permanent populations concern those of *D. magna* inhabiting small, but fairly permenent ponds. Although these populations probably do not have extremely long histories of parthenogenetic reproduction, there are striking differences between their genetic characteristics and those of intermittent populations of the same species.

Genotypic frequencies at polymorphic loci in these populations were rarely found to be in Hardy-Weinberg equilibrium (Hebert 1974b). In most instances these deviations were due to an excess of heterozygous individuals, and in some extreme cases populations consisted entirely of individuals heterozygous at one or more loci. Similar heterozygote excesses were noted in 3 of 5 *Bosmina* populations polymorphic at a phosphoglucose isomerase locus (Table 1).

Temporal analysis revealed that genotypic frequencies in permanent populations of *D. magna* were extremely unstable (Hebert 1974b, c; Hebert and Ward 1976). For instance, in the Audley End population the EST-1 MF heterozygote declined in frequency from 0.90 to 0.10 in less than a year (Hebert and Ward 1976). In this and many other similar cases population size remained large during the period of change in genotype frequencies, so random drift can be ruled out as the cause. The frequency changes must be due to selection and their magnitude implies that some genotypes leave 20-50% more offspring in each generation. In fact, by determining the brood size carried by an individual before electrophoresis, genotypic differences in egg production were documented (Hebert 1974b; Hebert and Ward 1976). Up to tenfold differences in parthenogenetic egg production were occasionally noted among genotypes and twofold differences were frequent. It is difficult to believe that such large differences in fitness result from allelic variation at a single locus, particularly as the same variants have no marked effect on fitness in intermittent populations. Additional studies have shown that in these permanent populations genotypes at different loci are nonrandomly associated (Hebert and Ward 1976; Young 1975). Such nonrandom associations mean that selection is not on variation at individual gene loci, but rather on whole gene complexes. The fitness differences associated via electrophoretic studies with allelic variation at individual loci are in fact due to genetic differences at many gene loci.

It is not yet clear why genotypic frequencies in permanent populations of *D. magna* change so dramatically. There is no obvious seasonal pattern to shifts in genotype frequency. They may be related to other environmental variations or perhaps they are triggered by the introduction of a new sexually produced genotype. Short-term studies of genetic variation at the phosphoglucose isomerase locus in *Bosmina*

Table 1. Genotypic frequencies at a phosphoglucose isomerase locus in five populations of *Bosmina longirostris*. Data from Manning et al. 1978 were reanalyzed by pooling samples from a particular habitat. Analysis restricted to samples with $n>15$ and with frequency of rare allele >0.05.

Population	n	Rare homozygote	Heterozygote	Common homozygote	Fit to X^2
Tub Lake	22	0	21	1	$p<0.001$
Chase Lake	60	2	35	23	$p<0.02$
Union Bay	414	6	62	346	$p>0.10$
Flowing Lake	19	0	16	3	$p<0.001$
Storrs' Pond	42	1	16	25	$p<0.80$

populations failed to reveal any drastic changes in genotypic frequencies, although some significant changes were observed (Manning et al. 1978). Before concluding that *Bosmina* populations are more genotypically stable than those of *D. magna*, it will be necessary to study a number of *Bosmina* populations for a longer period of time. Even in permanent populations of *D. magna*, genotypic frequencies at some loci were stable for up to a year.

The documentation of large changes in genotype frequencies within permanent populations raises the possibility that cyclomorphosis in cladocerans could be due, at least in part, to seasonal change in the frequencies of morphologically different clones. Certainly studies on cyclomorphosis have not been designed rigorously enough to exclude the possibility that genotypic variation plays a major role in the observed phenotypic changes. In fact Jacobs (1967) found that clones of *D. galeata* collected in summer produced larger crests than those from spring collections when both were cultured in the same environment. This observation suggests that selection pressures favored individuals producing large helmets during the summer months.

It has often been argued that clonal variation should be absent from isolated populations which reproduce parthenogenetically. Williams (1975) argued, for instance, that if the competitive exclusion principle were ever to apply it should certainly do so in these populations. Yet the work on *D. magna* has shown that selection does not act to eliminate all but the single fittest clone. Young (1975) was able to distinguish more than 30 different clones in one permanant population of *D. magna* and show further that this variation was maintained for nearly 2 years in the absence of any detectable sexual recruitment. There are two possible explanations for the maintenance of this diversity. Fitness differences among clones might be so small that competitive replacement is slow. This explanation is hard to reconcile with the major differences in fitness observed between clones. Their existence suggests that clonal coexistence is based instead on niche differentiation. It would be valuable to learn more about clonal coexistence in obligate parthenogens, for in these species there is no continual reintroduction of variation via sexual reproduction. Analysis to date has provided contradictory results. In populations of the Australian species *D. cephalata* genetic variation was detected, but individual populations always included only a single clone. In arctic Canada, populations of *D. middendorffiana* showed much more variation and a number of clones coexisted in many habitats.

References

AGAR, W. E. 1914. Reproduction in *Simocephalus vetulus*. J. Genetics **3**: 179-194.

ALLAN, E. 1928. A note on the chromosomes of *Moina macrocopa*. Science **67**: 18.

BACCI, G., C. COGNETTI, and A. M. VACCARI. 1961. Endomeiosis and sex determination in *Daphnia pulex*. Experientia **17**: 505-506.

BANTA, A. M. 1914. One hundred parthenogenetic generations of *Daphnia* without sexual forms. Proc. Soc. Exp. Biol. Med. 11: 180-182.

———. 1939. Studies on the physiology, genetics and evolution of some Cladocera. Carnegie Inst. Wash.

———, and L. A. BROWN. 1929. Control of sex in Cladocera. I. Crowding the mothers as a means of controlling male production. Physiol. Zool. 2: 80-92.

BROOKS, J. L. 1957. The systematics of North American *Daphnia*. Mem. Conn. Acad. Arts Sci. 13: 180 p.

DEHN, M. VON. 1948. Experimentelle Untersuchungen über den Generationswechsel der Cladoceren. Chromosoma 3: 167-193.

GREEN, J. 1967. The distribution and variation of *Daphnia lumholtzi* (Crustacea: Cladocera) in relation to fish predation in Lake Albert, East Africa. J. Zool. 151: 181-197.

HEBERT, P. D. N. 1974a. Enzyme variability in natural populations of *Daphnia magna*. III. Genotypic frequencies in intermittent populations. Genetics 77: 335-341.

———. 1974b. Enzyme variability in natural populations of *Daphnia magna*. II. Genotypic frequencies in permanent populations. Genetics 77: 323-334.

———. 1974c. Ecological differences between genotypes in a natural population of *Daphnia magna*. Heredity 33: 327-337.

———. 1975. Enzyme variability in natural populations of *Daphnia magna*. I. Population structure in East Anglia. Evolution 28: 546-556.

———. 1977. A revision of the taxonomy of the genus *Daphnia* (Crustacea, Daphnidae) in southeastern Australia. Aust. J. Zool. 25: 371-398.

———, and R. D. WARD. 1972. Inheritance during parthenogenesis in *Daphnia magna*. Genetics 71: 639-642.

———, and ———. 1976. Enzyme variability in natural populations of *Daphnia magna*. IV. Ecological differentiation and frequency changes of genotypes at Audley End. Heredity 36: 331-341.

HUTCHINSON, G. E. 1967. A treatise on limnology, v. 2. Wiley.

JACOBS, J. 1967. Untersuchungen zur Funktion und Evolution der Zyklomorphose bei *Daphnia*, mit besonderer Berücksichtigung der Selektion durch Fische. Arch. Hydrobiol. 62: 467-541.

KAISER, W. E. 1959. Biologiske og okologiske undersogelser over dafnierne *Eurycercus glacialis* Lilljeborg og *Eu. lamellatus* O. F. Müller. Flora Fauna 65: 17-34.

KERFOOT, W. C. 1975. The divergence of adjacent populations. Ecology 56: 1298-1313.

LEWONTIN, R. C. 1974. The genetic basis of evolutionary change. Columbia Univ. Press.

MANNING, B. J., W. C. KERFOOT, and E. M. BERGER. 1978. Phenotypes and genotypes in cladoceran populations. Evolution 32: 365-374.

O'BRIEN, W. J., and G. L. VINYARD. 1978. Polymorphism and predation: The effect of invertebrate predation on the distribution of two varieties of *Daphnia carinata* in South India ponds. Limnol Oceanogr. 23: 452-460.

OJIMA, Y. 1958. A cytological study on the development and maturation of the parthenogenetic and sexual eggs of *Daphnia pulex*. Kwansei Gakuin Univ. Annu. Stud. 6: 123-176.

SCHRADER, F. 1926. The cytology of pseudosexual eggs in a species of *Daphnia*. Z. Indukt. Abstamm. Vererbungsl. 40: 1-27.

SMITH, M.Y. 1974. A study of polymorphism in *Simocephalus serrulatus*. Ph. D. thesis, Univ. Cincinnati.

———, and A. FRASER. 1976. Polymorphism in a cyclic parthenogenetic species: *Simocephalus serrulatus*. Genetics 84: 631-637.

STROSS, R. G., and J. C. HILL. 1965. Diapause induction in *Daphnia* requires two stimuli. Science 150: 1462-1464.

WEISMANN, A. 1893. The germ-plasm: A theory of heredity. Walter Scott Ltd.

WILLIAMS, G. C. 1975. Sex and evolution. Princeton Univ. Press.

YOUNG, P. W. 1975. Enzyme polymorphisms and reproduction in *Daphnia magna*. Ph. D. thesis, Univ. Cambridge.

ZAFFAGNINI, F., and B. SABELLI. 1972. Karyologic observations on the maturation of the summer and winter eggs of *Daphnia pulex* and *Daphnia middendorffiana*. Chromosoma 36: 193-203.

ZARET, T. M. 1969. Predator-balanced polymorphism of *Ceriodaphnia cornuta* Sars. Limnol. Oceanogr. 14: 301-303.

———. 1972. Predator-prey interaction in a tropical lacustrine ecosystem. Ecology 53: 248-257.

32. An Analysis of the Precision of Birth and Death Rate Estimates for Egg-Bearing Zooplankters

William R. DeMott

Abstract

The sampling variability around egg ratio estimates of zooplankton birth and death rates is investigated using probability theory and computer simulation. The sampling variance of the egg ratio is shown to depend upon the sample size, the fraction of egg-bearing females in the population, and the mean and variance of the distribution of clutch sizes. From the simple assumptions that egg-bearers are randomly distributed within populations, that the sampling device collects an unbiased sample of the individuals it encounters, and that the duration of egg development is a fixed constant, formulas are derived for sampling confidence limits of the egg ratio and the instantaneous birth rate. Probability distributions generated by Monte Carlo simulation demonstrate that large sampling errors may occur even when these optimistic assumptions are met. Since the sampling variance of the death rate is equal to the sum of the sampling variances of the birth rate and the rate of population change, even modest precision in estimates of the instantaneous death rate requires intensive sampling.

Obtaining good estimates of birth and death rates for natural populations is a most difficult task. Edmondson (1960) provided a key to understanding zooplankton population dynamics when he showed that birth rates for animals which carry eggs until hatching can be estimated from the mean number of eggs per capita (the egg ratio) and the duration of egg development. In his method, births and deaths are assumed to be reasonably continuous, so that population change is described by the exponential growth equation

$$N_t = N_0 e^{rt} \tag{1}$$

where N_0 and N_t are the population size initially and t units of time later, r is the instantaneous rate of population change, and e is the base of natural logarithms. An estimate of the instantaneous death rate (d) is obtained from the difference of estimates of the instantaneous rates of birth (b) and population change (r)

$$d = b - r. \tag{2}$$

This egg ratio method and several variants have been applied extensively in studies of freshwater zooplankton population dynamics and in a large number of studies emphasizing zooplankton secondary productivity (*see* Edmondson 1974). The usual approach has been to relate changes in birth and death rates to temperature, food concentration, and the abundance of predators. Successful application of this technique to demographic problems depends, of course, on reasonably accurate and pre-

---

This work was supported in part by National Science Foundation grant DEB76-20238 to W. Charles Kerfoot. Dartmouth College provided computer time.

cise estimates of zooplankton birth and death rates. Accuracy requires unbiased formulas and reasonable underlying assumptions, while precision requires small sampling errors. Here I relate the precision of the egg ratio method estimates of birth and death rates to sample size, the distribution of eggs among females, and the coefficient of variation of estimates of population size. First, I derive an analytical formula for the variance of the egg ratio. Here samples from natural populations of five cladoceran species provide the basis for an analysis of the frequency distribution of cladoceran clutch sizes. Results of this analysis allow reasonable generalizations about the variance of egg ratio estimates from natural cladoceran populations. Finally, Monte Carlo computer simulation is used to generate sampling probability distributions for the instantaneous birth rate, the instantaneous rate of population change, and the instantaneous death rate.

I thank J. E. Frykman for answering questions on probability theory and suggesting the use of Monte Carlo simulation. W. C. Kerfoot, J. R. Litton, Jr., and R. I. Fletcher provided helpful criticism of earlier versions of this manuscript.

Variance of the egg ratio

The egg ratio can be defined as

$$\hat{\mathcal{E}} = pm \qquad (3)$$

where p is the egg-bearing fraction of the population and m is the mean number of eggs per egg-bearing female. To estimate the variance of the egg ratio we must account for variability in both the number of egg-bearing females counted and the number of eggs carried by egg-bearers. If we assume that egg-bearers are randomly distributed within the population and that the sampling device collects an unbiased sample of the individuals it encounters, we can obtain unbiased estimates of p and m from samples of a population. Given these assumptions, the number of egg-bearing females counted, R, has a binomial distribution, $b(n,p)$, with mean np and variance $np(1-p)$, where n is the total number of individuals counted

and p is the fraction of the population which is ovigerous. Let the distribution of T_i values, the number of eggs carried by the ith egg-bearing female, have mean m and variance σ^2.

Since

$$\hat{\mathcal{E}} = \frac{1}{n} \sum_1^R T_i$$

then

$$E(\hat{\mathcal{E}}) = \frac{E(R)}{n} \times E(T_i).$$

Let

$$S_R = \sum_1^R T_i,$$

and, given $E(T_i) = m$, $\text{Var}(T_i) = \sigma^2$, and $\text{Var}(R) = np(1-p)$. It is known that

$$E(S_R) = mE(R),$$

$$E(S_R^2) = \sigma^2 E(R) + m^2 E(R^2),$$

$$\text{Var}(S_R) = \sigma^2 E(R) + m^2 \text{Var}(R)$$

$$= \sigma^2(np) + m^2 \text{Var}(R).$$

So, finally

$$\text{Var}(\hat{\mathcal{E}}) = \frac{1}{n^2} \text{Var}(S_R)$$

$$= \frac{p}{n}[\sigma^2 + m^2(1-p)]. \qquad (4)$$

When females carry a maximum of only 1 egg, as often occurs in planktonic rotifer populations, the estimator of the variance of the egg ratio simplifies to

$$\text{Var}(\hat{\mathcal{E}}) = \frac{p}{n}(1-p). \qquad (5)$$

In practice, egg-bearers may show an overdispersed (more clumped than random) distribution within populations. When will this departure from our assumptions lead to serious underestimation of the variance of the egg ratio? A random distribution of egg-bearers can be expected when subsampling from a jar of plankton or sampling from well-mixed aquaria or beakers, as Marshall (1978) has recently done in a study of the effects of cadmium on *Daphnia* population dynamics. An estimate of the variance of the egg ratio calculated from

Eq. 4 should not differ significantly from the variance of the egg ratio between replicate samples from a single sampling station. If factors affecting births and deaths differ between sampling sites, Eq. 4 will underestimate the variance of the egg ratio. When Eq. 4 does seriously underestimate the between-sampling-site variance of the egg ratio, consideration should be given to calculating separate values of b and d for each sampling station or modifying the sampling program.

Sampling methods should be carefully designed to minimize systematic biases in estimates of the egg ratio. Since adult zooplankters often prefer greater depths than immature stages (e.g., see Zaret and Suffern 1976), samples should be taken from several depths, preferably from the entire water column. Individuals of a single species may differ in their ability to escape sampling devices. Strong swimmers, such as the cladoceran *Leptodora kindtii* and calanoid copepods, can avoid towed nets and small water bottles (Schindler 1969). Within a single species, fast-swimming adults should escape more frequently than juveniles, leading to systematic underestimation of the egg ratio. Rotifers may pass through netting unless very fine mesh (48-μm aperture or less) is used (Likens and Gilbert 1970). Since externally carried eggs increase effective body size, a disproportionately high percentage of egg-bearing rotifers will be found in samples collected with coarse nets.

Distribution of eggs among female Cladocera

The precision of egg ratio counts depends on the parameters p, m, and σ^2, which characterize the natural population, and the number of individuals counted, n. I have estimated p, m, and σ^2 for 30 populations including 5 cladoceran species, using samples collected from western Lake Erie during a 2-year study. The samples were taken by a team from the Center for Lake Erie Area Research using vertical tows, from bottom to surface of the water column, with a Wisconsin plankton net (mesh aperture 64 μm). Loss of eggs from cladoceran brood pouches was kept to a minimum by adding 40 g of sucrose per liter to a 4% Formalin preservative (Haney and Hall 1973). Estimates of p, m, and σ^2 were obtained by counting cladocerans and eggs until a minimum of 76 egg-bearing females from each population had been examined. The distribution of cladoceran clutch sizes is positively skewed since relatively scarce large females tend to carry more eggs ($p<0.05$ for 29 of 30 populations; g_1, statistic of skewness, <0). The coefficient of variation of mean clutch size (σ/m) is very large (mean C. V. = 0.51) and shows little variation between populations of a single species or between species (Fig. 1, correlation coefficient $r = 0.98$). While log transformation tends to normalize distributions of cladoceran clutch sizes (DeMott unpubl.), the large coefficient of variation means that precise estimates of both mean clutch size and the egg ratio require large samples of egg-bearers.

An estimator for the coefficient of variation of the egg ratio can be obtained by dividing the square root of the variance of the egg ratio (from Eq. 4) by the egg ratio (from Eq. 3):

$$V_{\hat{E}} = \sqrt{\frac{\frac{p}{n}[\sigma^2 + m^2(1-p)]}{pm}},$$

which is algebraically equivalent to

$$V_{\hat{E}} = \frac{\sqrt{\frac{\sigma^2}{m^2} + (1-p)}}{\sqrt{np}}. \quad (6)$$

If we consider σ/m a constant (see Fig. 1), the coefficient of variation of the egg ratio depends only on the sample size, n, and on the fraction of egg-bearers, p. The mean value of σ^2/m^2 for the data shown in Fig. 1 is 0.25. Note that $V_{\hat{E}}$ is inversely proportional to the square root of the expected number of egg-bearing females counted, np.

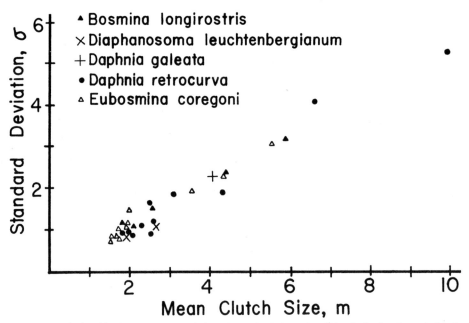

Fig. 1. Relationship between mean (m) and standard deviation (σ) of distribution of clutch sizes for Lake Erie cladoceran populations.

Confidence limits for egg ratio estimates

Given the derived variance of the egg ratio, we find that setting confidence limits requires some knowledge of the form of the distribution. According to the Central Limit Theorem, egg ratio estimates should approximate a normal distribution when random samples are moderately large, say 25 or greater, and the probability of samples with 0 egg-bearers is small, i.e. $np \geqslant 5$. When these conditions are met, confidence intervals for egg ratio estimates are given by the expression

$$pm \pm \sqrt{\text{Var}(\hat{\varepsilon})} \times t_{df} \qquad (7)$$

where $\text{Var}(\hat{\varepsilon})$ is calculated from Eq. 4 and Student's t has $n-1$ degrees of freedom. Monte Carlo simulation was used to confirm that egg ratio estimates approximate normal distributions and that the formulas for $\text{Var}(\hat{\varepsilon})$ and $V\hat{\varepsilon}$ give values which are approximated by random sampling. I chose three populations of *Daphnia retrocurva* representative of the range of values of p, m, and σ^2 for Lake Erie cladoceran populations: a spring population with a large mean clutch size ($p = 0.18$, $m = 9.93$, $\sigma^2 = 28.3$), a late spring population with a very low percentage of egg-bearing females and small mean clutch size ($p = 0.047$, $m = 2.51$, $\sigma^2 = 2.73$), and a summer population with a relatively high percentage of egg-bearers and small mean clutch size ($p = 0.28$, $m = 2.31$, $\sigma^2 = 1.25$). When p, m, and σ^2 from each of the three *Daphnia* populations were taken as fixed parameters, a computer generated sets of 5,000 samples of size $n = 25$, 100, and 300. Each individual had a probability p of being ovigerous and the number of eggs carried by each egg-bearer was randomly chosen from a log-normal distribution with the given mean m and the given variance σ^2. [Estimation of $\text{Var}(\hat{\varepsilon})$ and $V\hat{\varepsilon}$ requires unbiased estimates of the mean (m) and variance (σ^2) of the distribution of eggs among egg-bearing females, but does not depend on the form of the distribution.] The 95% confidence limits for computer generated samples were determined by counting 125 samples (2.5%) from each tail of distributions of 5,000 values of $\hat{\varepsilon}$. For $np >$

Table 1. Comparison of egg ratio statistics calculated from Eq. 3, 4, 6, and 7 with values generated by computer simulation. Values of p, m, and s are from Lake Erie *Daphnia retrocurva* populations sampled on 17 May 1976 (spring), 16 June 1976 (late spring), and 11 August 1975 (summer).

Sample size n	Egg ratio pm	Simulation mean egg ratio	Predicted $Var(\hat{E})$ Eq. 4	Simulation $Var(\hat{E})$	Predicted 95% C.I. Eq. 7	Simulation 95% C.I.	Predicted $V\hat{E}$ Eq. 6	Simulation $V\hat{E}$
				Spring				
25*	1.79	1.81	0.785	0.785	−0.039, 3.62	0.25, 3.7	0.496	0.490
100	1.79	1.78	0.196	0.198	0.913, 2.67	0.95, 2.7	0.248	0.250
300	1.79	1.79	0.0654	0.0668	1.29, 2.29	1.3, 2.3	0.143	0.144
				Late Spring				
25*	0.118	0.116	0.0161	0.0164	−0.146, 0.382	0.00, 0.44	1.09	1.10
100*	0.118	0.118	0.00411	0.00411	−0.008, 0.245	0.01, 0.27	0.543	0.544
300	0.118	0.118	0.00137	0.00134	0.0451, 0.191	0.05, 0.195	0.313	0.310
				Summer				
25	0.646	0.646	0.0568	0.0568	0.155, 1.14	0.20, 1.14	0.369	0.369
100	0.646	0.648	0.0142	0.0143	0.410, 0.882	0.41, 0.89	0.185	0.185
300	0.646	0.646	0.00474	0.00463	0.510, 0.782	0.51, 0.78	0.107	0.105

*$np \leq 5$.

5 there was close agreement between confidence intervals calculated from Eq. 7 and confidence intervals for sets of samples generated by the computer (Table 1). Good agreement was also found between values of $\hat{\varepsilon}$, Var($\hat{\varepsilon}$), and V$\hat{\varepsilon}$ calculated from Eq. 3, 4, and 6 and values of the mean, variance, and coefficient of variation of sets of 5,000 samples generated by computer simulation (Table 1).

Instantaneous birth rate

Edmondson (1960) calculated the instantaneous birth rate

$$b = \ln(\hat{\varepsilon}/D+1) \qquad (8)$$

where $\hat{\varepsilon}$ is the egg ratio, and D is the duration of egg development in days. This original formula for the instantaneous birth rate gives biased results, overestimating the value of b when egg development time is greater than 1.0 day (Edmondson 1968; Paloheimo 1974). Paloheimo (1974) has shown that an alternative form (eq. 11 in Edmondson 1968)

$$b = \ln(\hat{\varepsilon}+1)/D \qquad (9)$$

is the correct formula for calculating the instantaneous birth rate from egg ratio data. The egg ratio includes some moribund embryos as well as embryos carried by females that die before giving birth. Thus, the birth rate calculated from Eq. 9 is closer to a measure of the egg-laying rate than the live birth rate (Paloheimo 1974). Consideration of egg mortality and the egg age structure can be important in predicting zooplankton egg hatching rates in nature (Threlkeld pers. com.).

Sampling probability distributions for estimates of the instantaneous birth rate were generated from computer simulation by substituting a fixed egg development time and sets of 5,000 values of the egg ratio into the formula for b (Eq. 9). As before, the 95% confidence limits for samples generated by computer were determined by counting 125 samples from each tail of the distribution. If we assume that egg development time is a constant, confidence limits for the instantaneous

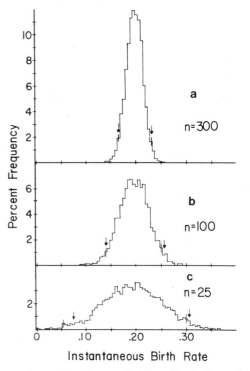

Fig. 2. Sampling probability distributions of instantaneous birth rate generated by Monte Carlo simulation using values of p, m, and σ^2 from summer *Daphnia* population, egg development time of 2.5 days, and three different sample sizes, n. Predicted 95% C. I. values (Eq. 7) shown by vertical bars are compared with simulation 95% C. I. values shown by arrows.

birth rate can be calculated by substituting the upper and lower confidence limits of the egg ratio from Eq. 7 into the formula for b (Eq. 9). Figure 2 shows three probability distributions for b generated by computer simulation. Values of p, m, and σ^2 from the summer *D. retrocurva* population and an egg development time of 2.5 days were held constant for three different values of sample size, n. Reasonably close agreement is shown between the 95% confidence intervals of samples generated by computer simulation (arrows) and the 95% confidence intervals determined by substituting the upper and lower confidence limits of the egg ratio into the formula for b (bars).

To estimate the variance of the instantaneous birth rate we must know the form of the distribution. Values of $\hat{\mathcal{E}}$ from natural populations are small, usually <1 and almost always <2. If we are given a distribution of small numbers, taking the logarithm of each number plus 1 has very little effect on the shape of the distribution. When estimates of the egg ratio have an approximately normal distribution, $b = \ln(\hat{\mathcal{E}} + 1)/D$ will also have an approximately normal distribution. Taking t_{df} and estimates of the upper and lower confidence limits of b (UCLb and LCLb) from Eq. 7, we can obtain an approximation of Var(b) from the expression

$$\text{Var}(b) = [(\text{UCL}b - \text{LCL}b)/2t_{df}]^2. \quad (10)$$

Values of Var(b) calculated from Eq. 10 differed from values for the variance of sets of b generated by Monte Carlo simulation by 0.1% to 6% for the three *Daphnia* populations when $np > 5$ and $n = 25, 100$, and 300.

Before proceeding further it is worthwhile to consider the implications of what we have learned about the sampling probability distributions of $\hat{\mathcal{E}}$ and b. For a given sample size, the differences in values of $V_{\hat{\mathcal{E}}}$ in Table 1 are primarily due to variation in the fraction of egg-bearing females, p. Small sample sizes of egg-bearing females lead to high relative variability in estimates of $\hat{\mathcal{E}}$. The coefficient of variation of the instantaneous birth rate, V_b, is affected by the expected number of egg-bearing females counted in a similar manner. When $\hat{\mathcal{E}}$ is very small, then $V_{\hat{\mathcal{E}}} \approx V_b$. Again, taking the logarithm of a number plus 1 has little effect on the shape of a distribution of small numbers. For the summer *Daphnia* population ($\hat{\mathcal{E}} = 0.65$) $V_b = 0.83 V_{\hat{\mathcal{E}}}$, while for the spring *Daphnia* population ($\hat{\mathcal{E}} = 1.8$) $V_b = 0.63 V_{\hat{\mathcal{E}}}$. To obtain a value of V_b of 0.1, a comfortably narrow margin for precise work, one must count about 240 *Daphnia* from the spring and summer populations and about 2,600 *Daphnia* from the late spring population. Counting about 60 *Daphnia* from the spring and summer samples and about 650 *Daphnia* from the late spring samples would result in a value of V_b of about 0.2. Estimates of b from samples as small as 25 are very imprecise even when p is relatively large (Fig. 2c).

Given the sensitivity of the various parameters in the formula for the variance of the egg ratio (Eq. 4), which studies of particular organisms yield the most precise estimates of the instantaneous birth rate? Species with small clutch sizes and a short period of immaturity should have a high mean percentage of egg-bearing females. Rotifers, for example, have very short generation times and many species usually carry but 1 egg. Since the mean value of $p \approx 0.3$ for the rotifer *Keratella cochlearis* (data from Edmondson 1960) on the average, we can calculate that $V_b \approx 0.2$ for counts of about 45 individuals and $V_b \approx 0.1$ for counts of 180 individuals. To obtain the same relative precision in estimating the birth rate of the cladoceran *Leptodora kindtii* (mean $p \approx 0.1$ calculated from data in Cummins et al. 1969), one would have to examine 250 or 1,000 individuals. For any species the estimates of $\hat{\mathcal{E}}$, and therefore b as well, will be imprecise when the percentage of egg-bearers is low, unless inordinately large samples are counted. George and Edwards (1974) increased sample size whenever the fraction of egg-bearers was low, providing an example of good sampling strategy.

Instantaneous death rate

One important reason for estimating the birth rate, b, is to allow calculation of the death rate, d, from the difference between the birth rate and the rate of population change, $d = b - r$. Since sampling errors in estimates of b and r are statistically independent, the sampling variance of d is equal to the sum of the sampling variances of b and r. If we assume unbiased samples and a constant rate of population change, the variance of r depends on the precision of two estimates of population size, N_o and N_t, and the length of time between sampling dates. Estimates of zooplankton numbers from samples of natural populations often reflect a patchy, overdispersed distribution of individuals

(Langeland and Rognerud 1974; Bottrell et al. 1976). According to Cassie (1971), counts of 100 individuals from plankton samples usually result in a coefficient of variation of population size, V_N, of about 0.25 to 0.45, although much higher values are not uncommon. Because of the overdispersed distribution of individuals, the best strategies for increasing the precision of estimates of zooplankton population size are to increase sample volume and the number of replicate samples.

Since each estimate of population size, N, is used twice in calculating r, large sampling errors in estimates of population size lead to negative correlation between successive estimates of r and therefore of d (e.g. see Cummins et al. 1969). Calculation of a running average death rate, d, should eliminate much of the scatter in death rate estimates due to imprecise estimates of population size.

Monte Carlo simulation was used to generate sets of 5,000 values of r for V_N = 0.10, 0.20, 0.30, and 0.40, and a 7-day sampling interval. Values of r were calculated by solving Eq. 1 for r, where N_o and N_t were randomly chosen from a normal distribution with mean and variance fixed to give the desired coefficient of variation of sample size, V_N. Given V_N, the variance of r is insensitive to the actual values of N_o and N_t. In this example t = 7 days, N_o = 50, and N_t = 100, giving a calculated value of r = 0.10. Monte Carlo simulation gave the following results: V_N = 0.10, Var(r) = 0.00042; V_N = 0.20, Var(r) = 0.0018; V_N = 0.30, Var(r) = 0.0056; and V_N = 0.40, Var(r) = 0.016. As V_N increases linearly Var(r) increases exponentially.

Sampling probability distributions of the instantaneous death rate were generated simply by subtracting randomly chosen values of r from randomly chosen values of b. Values of r for V_N = 0.10, 0.20, and 0.30 were subtracted from values of b for n = 100 individuals of the summer *D. retrocurva* population (Fig. 2b) giving three sets of 5,000 values of d (Fig. 3). In this set of examples Var(r) accounts for the greater part of the variance of

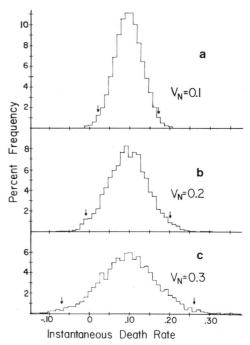

Fig. 3. Sampling probability distributions of instantaneous death rate generated by Monte Carlo simulation using formula $d = b-r$. Values of b are for summer *Daphnia* population (sample size = 100). Values of r were generated for three different coefficients of variation of population size, V_N. Arrows show simulation 95% C. I. values.

estimates of d until V_N is reduced to some value <0.20 but >0.10. Figure 3b shows that even "good" estimates of population size (V_N = 0.20) and counts of 100 individuals from a population with a high fraction of egg-bearers do not ensure precise estimates of d. The 95% confidence limits are about 0.0 to 0.2 per day, nearly the entire range of death rates which can be expected throughout the year for natural zooplankton populations. When V_N is as high as 0.30, estimates of d are very imprecise (Fig. 3c).

Conclusion

The sampling distributions presented in Figs. 2 and 3 show that large sampling errors may be associated with egg ratio method estimates of the instantaneous birth rate and the instantaneous death rate. Since we have assumed a random distribution of egg-bearers in the population, the variances of $\hat{e}$, b, and d determined by methods described in this paper are conservative estimates of the actual sampling variability which will be encountered in field studies. The accuracy of estimates of b and d, as affected by errors in estimates of the egg development time, sampling biases, and departures from constant exponential birth and death rates, has not been considered.

In nature, even given reasonably large samples, conclusions should always be based on trends in several estimates of b or d, and certainly not from single estimates. While the egg ratio method can be a key to understanding zooplankton population dynamics, these results emphasize that acceptable precision requires good sampling methods and counting large samples.

References

BOTTRELL, H. H., AND OTHERS. 1976. A review of some problems in zooplankton production studies. Norw. J. Zool. 24:419–456.

CASSIE, R. M. 1971. Sampling and statistics, p. 174–209. *In* W. T. Edmondson, and G. G. Winberg [eds.], Secondary productivity in fresh waters. IBP Handbook 17. Blackwell.

CUMMINS, K. W., AND OTHERS. 1969. Ecological energetics of a natural population of the predaceous zooplankter *Leptodora kindtii* Frocke (Cladocera). Oikos 20:189–223.

EDMONDSON, W. T. 1960. Reproductive rates of rotifers in natural populations. Mem. Ist. Ital. Idrobiol. 12:21–71.

——. 1968. A graphical model for evaluating the use of the egg ratio for measuring birth and death rates. Oecologia 1:1–37.

——. 1974. Secondary production. Mitt. Int. Ver. Theor. Angew. Limnol. 20, p. 229–272.

GEORGE, D. G., and R. W. EDWARDS. 1974. Population dynamics and production of *Daphnia hyalina* in a eutrophic reservoir. Freshwater Biol. 4:445–465.

HANEY, J. F., and D. J. HALL. 1973. Sugar-coated *Daphnia*: A preservation technique for Cladocera. Limnol. Oceanogr. 18:331–333.

LANGELAND, A., and S. ROGNERUD. 1974. Statistical analyses used in the comparison of three methods of freshwater zooplankton sampling. Arch. Hydrobiol. 73:403–410.

LIKENS, G. E., and J. J. GILBERT. 1970. Notes on quantitative sampling of natural populations of planktonic rotifers. Limnol. Oceanogr. 15:816–820.

MARSHALL, J. S. 1978. Population dynamics of *Daphnia galeata mendotae* as modified by chronic cadmium stress. J. Fish. Res. Bd. Can. 35:461–469.

PALOHEIMO, J. E. 1974. Calculation of instantaneous birth rate. Limnol. Oceanogr. 19: 692–694.

SCHINDLER, D. W. 1969. Two useful devices for vertical plankton water sampling. J. Fish. Res. Bd. Can. 26:1948–1955.

ZARET, T. M., and J. S. SUFFERN. 1976. Vertical migration in zooplankton as a predator avoidance mechanism. Limnol. Oceanogr. 21: 804–813.

33. Habitat Selection and Population Growth of Two Cladocerans in Seasonal Environments

Stephen T. Threlkeld

Abstract

Daphnia pulicaria has a cold-water distribution in midsummer in north temperate lakes while *Daphnia galeata mendotae* typically inhabits warmer waters during this period. In situ life table experiments were performed to assess the age-specific responses of these two species to the natural midsummer succession of phytoplankton and temperature in Wintergreen Lake (Michigan). Daily records of growth, survivorship, and reproduction were taken for cohorts of these two species incubated in the epilimnion and in the hypolimnion in cooled epilimnion water.

Reproduction by *D. pulicaria* was reduced in epilimnion incubations for a few days in midsummer at a time of high water temperatures and low algal densities. Comparison with hypolimnion incubations, where identical water was used, showed that food conditions alone were insufficient to account for the reduced reproduction, and experiments earlier and later in the summer indicated that high temperature alone was also insufficient to account for the reproductive slowdown. The increased duration of a molt (and reproductive) cycle in the colder waters relative to the length of time algal densities were reduced may have permitted *D. pulicaria* to continue reproduction in those incubations. *Daphnia galeata mendotae* was not adversely affected by these conditions, suggesting that this species has a more favorable metabolic balance at high temperatures under low food conditions than does *D. pulicaria*.

Comparison of life history characteristics of the two species in epilimnion and hypolimnion incubations revealed differences among characters known to be influenced by temperature (e.g. longevity and age of maturation). However, reproductive effort and size at maturation were not found to differ between depths of incubation for either species.

Although the habitat choices of these two species in midsummer are consistent with the importance of planktivory in their evolution, they are also consistent with maximizing population growth rate in a predictable fashion in the absence of predation.

Variation in the distribution and abundance of natural populations suggests mechanisms which regulate the ecological and evolutionary responses of species to their environments. When distribution and abundance patterns change seasonally, additionally powerful insights are often gained by assessing how biotic and abiotic factors potentially important in the regulation of these populations have also changed. The habitat in which an organism is most commonly found is usually considered its op-

---

Contribution No. 367 from the Kellogg Biological Station, Michigan State University.

timal habitat—the ultimate aim of any ecological study of distribution and abundance is to demonstrate how several environmental constraints have (either in ecological or evolutionary time) collectively accounted for the observed distribution patterns.

Here I consider possible reasons for the midsummer vertical distribution patterns of two limnetic *Daphnia* species in north temperate lakes. The two species, *Daphnia pulicaria* Forbes and *Daphnia galeata mendotae* Birge exhibit quite different vertical distribution patterns. *Daphnia pulicaria* typically inhabits the metalimnion and the cooler hypolimnion waters while *D. galeata mendotae* inhabits the warmer epilimnion regions (Birge 1895, 1898; Hall 1964; Ward and Robinson 1974; Tappa 1965; Haney and Hall 1975; Threlkeld 1977, 1979). Although this pattern is the basis for their designation as cold and warm water stenotherms, respectively (Hutchinson 1967), little experimental evidence has been available to evaluate whether temperature alone is sufficient to explain these patterns. Recent laboratory and theoretical work indicates that daphniid seasonal distribution may also be affected by food quality and quantity and by food-temperature interactions (Hrbáčková-Esslová 1962; Lynch 1977; Lampert 1977). Finally, differences in light penetration to the epilimnion and hypolimnion coupled with the difference in body size of these two species (*D. pulicaria* > *D. galeata mendotae*) may form the basis for an argument that the apparent habitat selection primarily results from the effects of visually orienting, size-selective predators.

In a recent analysis of the midsummer population dynamics of these two species in Wintergreen Lake (Michigan), Threlkeld (1979) concluded that aspects of food changes, water temperature, and size-selective predators were all contributory to the midsummer decline in population densities of both species. These analyses were based in large part on a series of in situ life table experiments where predators were excluded and where the age-specific responses of cohorts of these species to the natural succession of phytoplankton and temperature were assessed. My purpose here is to use a subset of these life table experiments to compare the life history characteristics of these two species as affected by food and temperature conditions. In particular, I will examine whether the midsummer habitat choices of these species can be explained on the basis of their responses to food and temperature conditions and without reference to the effects of size-selective predators.

This work was supported by NSF grant BMS 74-09013A01 to D. J. Hall and E. E. Werner. D. J. Hall, E. E. Werner, and M. Threlkeld provided encouragement and advice during the experimental period. M. Klug provided unpublished information on particulate organic carbon concentrations; D. Howard provided the July algal counts and assistance with the Coulter Counter analyses of algal samples. W. C. Johnson provided laboratory space at Wintergreen Lake. M. Lynch and D. Allan suggested using the field data to compare life history characteristics of these two *Daphnia* species, although I am responsible for the approach used.

Methods and materials

The experimental work described here was performed from 12 June to 12 September 1976, in Wintergreen Lake, a hypereutrophic lake situated on the Kellogg Bird Sanctuary of the Kellogg Biological Station, Michigan State University. Wintergreen Lake is relatively small (15 ha, max depth, 6.3 m) and stratifies thermally and chemically soon after spring overturn (Wetzel 1975; Threlkeld 1977). Deoxygenation of the hypolimnion at this time restricts the animal plankton to the upper 3.5 to 4.0 m. Aspects of the natural population dynamics of *D. pulicaria* and *D. galeata mendotae* in this lake have been described elsewhere (Threlkeld 1979); only the details of the in situ life table experimentation which help to measure the effects of temperature and food in the development of habitat choices by these two species in midsummer is described here.

Each life table experiment was started from individual female *Daphnia* isolated from the lake during this period and incubated at a centrally located deep station in the lake (station A: Manny 1972). The female was placed in a 75-ml clear-glass jar with screwcap and placed at 0.5 m depth. Such females and all other cohorts derived from them were incubated in lake water collected from 0.5 m and filtered through a 145 μm-mesh net to remove other zooplankton. When the females gave birth, they were removed and discarded and the life history characteristics (survivorship, growth, and reproduction) of their offspring followed until all members of the cohort were dead or until 12 September, when all remaining cohorts were terminated. During the life of the cohort members, newborn individuals were removed and used to start other cohorts or discarded. In addition to experiments incubated at 0.5 m, animals were also incubated in water from 0.5 m at a depth of 4.5 m after the freshly collected water had been cooled to temperatures at 4.5m. This particular manipulation provided a measure of the effect of temperature (where food conditions were identical) on the life history characteristics of these species. Both epilimnion (0.5 m) and hypolimnion (using 0.5 m water at 4.5 m) incubations were started on several dates during the summer to evaluate the effect of seasonal changes in natural food and temperature conditions on the age-specific responses of both *Daphnia* species.

Throughout the experimental period, animals were transferred daily by widebore pipette to clean jars and freshly collected water, and then returned to their in situ incubation depths within 2 hours of their removal from the lake. This daily transfer period permitted daily assessment of survivorship, reproduction, and growth of the members of each cohort. Survivorship was measured by counting the number of live animals during transfer to the clean jars; collection of dead individuals at the bottom of the previous day's jars provided a check on these daily survivorship counts. Reproduction was measured during the daily removal of newborn individuals (easily distinguished from adults by size). In addition, any dead newborn individuals were collected from the bottom of the jars and undeveloped eggs cast off with the molts were counted. Individual growth was assessed by measuring the molts (exclusive of tail spine) discarded during the previous day's incubations.

Information from these cohorts was combined according to cohort starting date and depth to give age-specific survivorship (l_x) and reproduction (m_x) schedules for each species. These schedules were then used to calculate the population (cohort specific) growth rate r (hereafter, cohort r), where $\Sigma l_x m_x e^{-rx} = 1.0$ (Lotka 1956). Conversion of body lengths to biomass was achieved by application of Burns' (1969) formula relating carapace length to dry weight. Reproductive effort was estimated as the percentage of adult net production (adult growth and reproduction) allocated to reproduction.

Comparisons of life history attributes of both species as measured in the epilimnion and hypolimnion incubations were based on all pairs of cohorts started on the same date in both the epilimnion and hypolimnion. Thus, only *D. pulicaria* cohorts started on 19, 21, 28, and 30 June, 5, 6, 12, 16, and 22 July, and 1 August (n = 10 pairs) were used for these comparisons, although additional cohorts were started on other dates in either the epilimnion or hypolimnion. Similarly, only the pairs of cohorts started on 5, 6, and 16 July and 1 August (n = 4 pairs) were used for analysis of *D. galeata mendotae* life history characteristics. *Daphnia pulicaria* was identified according to Hrbáček (1959) and Brandlova et al. (1972); *D. galeata mendotae* was identified according to Brooks (1957).

Estimates of the amount and quality

of food available to zooplankton in these experiments were based on measurements of chlorophyll a and pheopigments, particulate organic carbon, and algal size structure and taxonomic composition from water collected in the surface meter (0.5 m). Pigment analyses were performed on samples collected biweekly and filtered through Millipore filters (HA, 0.45 μm). Plant pigments were extracted in 90% aqueous acetone and absorbances read using a Hitachi Perkin-Elmer model 139 spectrophotometer. After acidification with 1 N HCl the samples were read again. Chlorophyll a and pheopigment concentrations were calculated from the absorbance equations of Wetzel and Westlake (1969). Samples were collected for particulate organic carbon analyses each week. Concentrations were determined using a Carlo Erba elemental analyzer model 1104 coupled with a CSI automatic digital integrator model 208. Algae, also collected each week, were preserved in Lugol's solution. Algal samples taken from 15 June to 3 August were analyzed for size structure using a model A Coulter Counter. In addition, July algal samples were allowed to settle and viewed with a Wild inverted microscope; identifications were made according to Prescott (1970).

Temperature measurements were made at 0.5 and 4.5 m each day throughout the experimental period with a YSI thermistor. This instrument, when combined with water flow from a cold tap, was also used to maintain in situ incubation jars at natural temperatures (± 1°C) during animal transfer procedures in the laboratory.

Results and discussion

Seasonal changes in food conditions included shifts in algal species composition and density, and fluctuations in chlorophyll a and particulate organic carbon. Figure 1 shows that all measures of phytoplankton biomass were lowest in mid- to late July and highest in early August. In June and from mid-August until the end of the experimental period, biomass estimates were of intermediate value. The rapid changes in biomass, chlorophyll a, and cell number in mid-July were accompanied by shifts in the species composition of the phytoplankton. The small algae *Dysmorphococcus* declined in numbers from early to late July, being replaced by the larger phytoplankton *Anabaena* and *Ceratium* (Fig. 1). *Volvox* was abundant in the water column in mid- and late July. In the period 10–27 July, densities for the water column averaged from 11.2 to 608 colonies per liter, as determined by counts from vertical net tows used for collection of zooplankton (Threlkeld unpubl.).

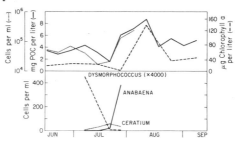

Fig. 1. Concentrations of particulate organic carbon (POC) and chlorophyll a per liter and phytoplankton cells per ml at 0.5 m in Wintergreen Lake, 1976. Cell densities of *Dysmorphococcus*, *Anabaena*, and *Ceratium* in July shown in lower panel.

Seasonal changes in temperature were greater at epilimnion than at hypolimnion depths (Fig. 2). Epilimnion temperatures ranged from 19.8°C (12 September) to 27.5°C (5 July), while hypolimnion temperatures ranged from 14.2°C (15 June) to 18.5°C (7 August). Epilimnion temperatures exceeded 25°C throughout much of mid-June, July, and a brief period in late August.

Daphnia pulicaria. All *D. pulicaria* cohorts started in the epilimnion between 12 June and 12 September 1976 had positive growth rates ($n = 17$, $\bar{r} = 0.21$). Similarly, population growth rates of hypolimnion cohorts were also positive throughout the experimental period ($n = 16$, $\bar{r} = 0.13$), although rates were generally lower than in epilimnion incubations. Epilimnion

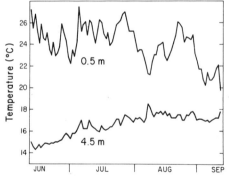

Fig. 2. Temperatures in epilimnion and hypolimnion incubations, 12 June–12 September 1976.

cohort growth rates were reduced in mid-July, although this reduction was not observed in hypolimnion incubations (Fig. 3). Comparison of cohort growth rates in June (22, 27, 28) and mid-July (12, 16, 19) showed that this reduction was in large part attributable to altered reproductive schedules rather than changed survivorship patterns (Table 1). Examination of the age-time distribution of reproduction for all epilimnion cohorts shows that the period from 16–21 July was associated with an age-independent reduction in reproduction (Fig. 3). In contrast, during the same period hypolimnion incubations did not show a similar reduction in reproduction although food conditions were identical.

Recovery of molts from the incubation jars provided specific information on one aspect of the observed reduction in reproduction in epilimnion incubations. It was found that production of eggs was reduced but not stopped during this period; the eggs that were produced were usually cast off in undeveloped condition with the molts. This abortion of eggs was not observed in hypolimnion incubations at the same time (Fig. 4).

The particular reasons for the observed abortion of eggs by *D. pulicaria* are somewhat unclear. Because identical food used in epilimnion and hypolimnion incubations

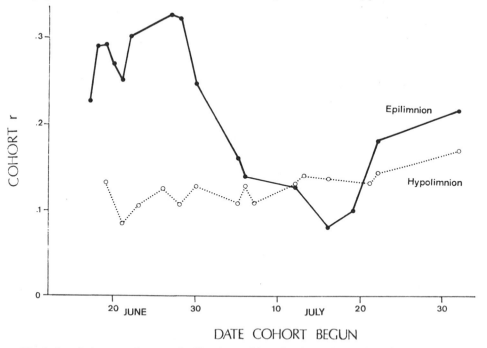

Fig. 3. Population growth rates of epilimnion and hypolimnion cohorts of *Daphnia pulicaria* (upper panel). Other panels (opposite page) show age-specific reproduction (m_x) of *D. pulicaria* in epilimnion (upper) and hypolimnion (lower) cohorts throughout experimental period. Vertical lines include 16–21 July 1976.

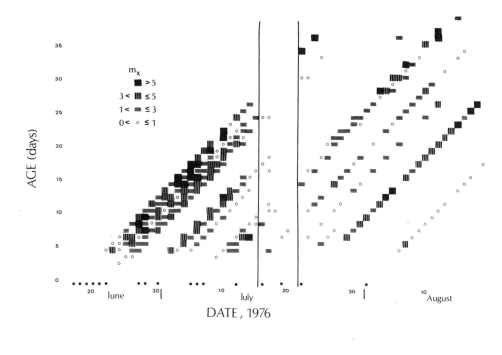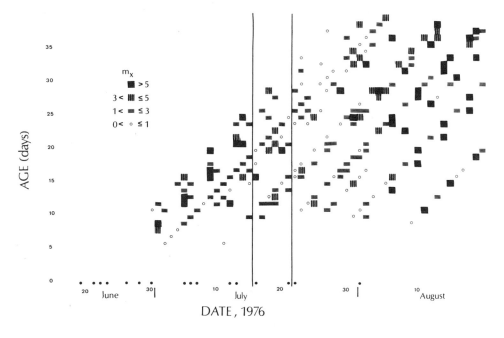

Table 1. Changes in *Daphnia pulicaria* epilimnion cohort growth rates, r, due to changes in age-specific reproduction (m_x) and survivorship (l_x) schedules from late June to mid-July 1976. Calculations made according to method of Keyfitz (1968, p. 189–193).

Cohort starting dates	r ($\bar{x} \pm$ SE)	Due to mortality change (%)	Due to reproduction change (%)	Total including interaction
22, 27, 28 Jun.	0.323 ± 0.01			
		−0.007 (3.4%)	−0.163 (74.5%)	−0.218
12, 16, 19 Jul.	0.105 ± 0.01			

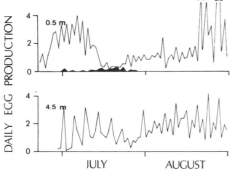

Fig. 4. Daily egg production (eggs adult female^{-1} day^{-1}) of *Daphnia pulicaria* in epilimnion (0.5 m) and hypolimnion (4.5 m) incubations. Solid areas indicate eggs that were aborted in undeveloped condition with molts.

resulted in different patterns of egg abortion, food alone is an insufficient explanation for this pattern. Similarly, because temperatures were equally high both earlier and later in the summer when no egg abortion was observed, temperature alone is also an insufficient explanation. However, food conditions did change during this period and it appears that an interaction of low food levels and high temperatures is the best explanation for the observed egg abortion. This appears to be one possible manifestation of recently proposed food-temperature interactions on the metabolism of *Daphnia* species (Lampert 1977; Lynch 1977), although the reasons for egg production under conditions which lead to egg abortion are unclear.

Although food and temperature changed considerably during the experimental period, major life history features of *D. pulicaria* incubated in the epilimnion and hypolimnion remained relatively unchanged during the period. Mean coefficients of variation for *D. pulicaria* life history attributes were 17.1 and 11.8% for epilimnion and hypolimnion cohorts. In contrast, population growth rates and recruitment rates were quite variable during the experimental period (C.V. = 28.0% and 71.5%).

Comparison of several life history characteristics of *D. pulicaria* between epilimnion and hypolimnion cohorts revealed few significant differences, mainly among characters known to be strongly influenced by temperature. Thus, the age at maturity was earlier and longevity shorter in epilimnion cohorts (Table 2). However, survival to maturity was lower in hypolimnion cohorts, although this did not produce a significant effect on longevity. In contrast, such plastic life history characteristics as size at maturity and reproductive effort (Stearns 1976, 1977) did not differ between epilimnion and hypolimnion cohorts.

Differences in population growth rate (r) and recruitment per female (R_o) of epilimnion and hypolimnion cohorts were not consistent with each other. Population growth rates were significantly greater in the epilimnion, but recruitment was greater in the hypolimnion (Table 2). This apparent inconsistency in population responses probably results from the stimulative effects of temperature on development rates and the deleterious effects of high temperatures on egg viability.

Daphnia galeata mendotae. All cohorts

Table 2. Comparison of life history characteristics of epilimnion and hypolimnion cohorts of *Daphnia pulicaria* and *Daphnia galeata mendotae*. Means, standard deviations, and probability statements (*t*-test) given.

	Epilimnion	Hypolimnion	Probability
Daphnia pulicaria (n = 10)			
Population growth rate, r (day^{-1})	0.21 (0.079)	0.13 (0.024)	<0.01
Recruitment, R_0	26.97 (27.78)	49.52 (19.80)	<0.05
Age at maturity (days)	5.3 (0.95)	11.0 (0.94)	<0.001
Size at maturity (mm)	1.06 (0.178)	1.12 (0.113)	0.2–0.4
Survival to maturity (l_x)	0.99 (0.03)	0.89 (0.11)	<0.01
Mean longevity (days)	20.12 (4.09)	38.30 (7.66)	<0.001
Reproductive effort (%)	65.11 (17.78)	72.50 (5.90)	0.1–0.2
Daphnia galeata mendotae (n = 4)			
Population growth rate, r (day^{-1})	0.20 (0.048)	0.12 (0.02)	<0.02
Recruitment, R_0	32.55 (22.62)	28.60 (16.44)	0.5–0.9
Age at maturity (days)	5.25 (0.5)	12.25 (0.96)	<0.001
Size at maturity (mm)	1.01 (0.11)	1.06 (0.06)	0.2–0.4
Survival to maturity (l_x)	0.97 (0.07)	0.89 (0.22)	0.4–0.5
Mean longevity (days)	23.77 (4.67)	35.87 (1.55)	<0.01
Reproductive effort (%)	68.16 (7.15)	68.46 (14.86)	0.9–1.0

of *D. galeata mendotae* showed positive population growth rates and, like *D. pulicaria*, tended to be greater in epilimnion (n = 13, $\bar{r}$ = 0.20) than in hypolimnion incubations (n = 5, $\bar{r}$ = 0.12) (Fig. 5). Epilimnion cohort growth rates did not show a consistent drop in mid-July as observed for *D. pulicaria*. Since epilimnion cohorts of *D. galeata mendotae* were not begun until early July, a comparison of changes in reproductive or survivorship schedules as a function of changes in food conditions is not possible for this species. However, an examination of age-specific reproduction patterns over time does not reveal any slowdown in mid-July or any apparent difference between epilimnion and hypolimnion incubations (Fig. 5). No abortion of eggs was observed for *D. galeata mendotae*.

Comparison of life history characteristics of epilimnion and hypolimnion cohorts of *D. galeata mendotae* revealed fewer differences than observed for *D. pulicaria*. Only age at maturity and longevity were significantly different between epilimnion and hypolimnion cohorts (Table 2), these both strongly influenced by temperature (Hall 1964; Bottrell et al. 1976). Survival to maturity, size at maturity, and reproductive effort did not differ between epilimnion and hypolimnion cohorts. Net recruitment per female did not differ, although population growth rate r was significantly greater in the epilimnion—a direct result of the shorter maturation times in those incubations.

General discussion

The relationship between habitat selection in a seasonal environment and population growth is complex. The decision to inhabit the epilimnion or hypolimnion necessarily involves tradeoffs in the rates of development and reproduction as well as sensitivity of the individual animals to rapid changes in food and temperature environments. Recent laboratory, field, and theoretical studies have emphasized the sensitivity of natural populations to

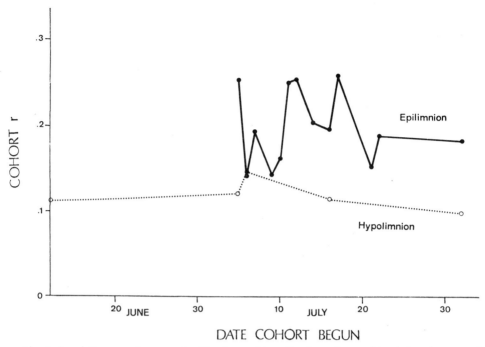

Fig. 5. Population growth rates of epilimnion and hypolimnion cohorts of *Daphnia galeata mendotae* (above). Other panels (opposite page) show age-specific reproduction (m_x) of *D. galeata mendotae* in epilimnion (above) and hypolimnion (lower) cohorts throughout experimental period. Vertical lines include 16–21 July 1976.

rapid changes in food conditions (e.g. Hall et al. 1976; Threlkeld 1976; Dagg 1977). The deleterious effects of such changes may be ameliorated by choosing a habitat where lower temperatures make the animals and population growth rates less sensitive because of lengthened development times.

It is now recognized that any number of selective pressures may result in identical or similar life history characteristics (Giesel 1976; Stearns 1977). Thus, it is worthwhile exploring how efficacious the habitat choices of these two *Daphnia* species are in maximizing population growth rates when variable food and temperature conditions are considered as the main selective constraints.

Of the two species studied here, only the habitat choice of *D. galeata mendotae* is clearly consistent with a strategy which maximizes population growth rate. Cohorts raised in the epilimnion exhibited the highest population growth rates and did not sacrifice juvenile survivorship or net recruitment per female. Age at maturity and longevity were reduced, but these results were expected from the direct effect which temperature has on these processes.

However, *D. pulicaria* does not appear to be distributed in a way which is advantageous to maximizing population growth. Instead, expected population growth rates for hypolimnion cohorts (where the species is usually found) were lower than found in the epilimnion (which the species seems to avoid). This is in spite of the reductions in epilimnion cohort growth rates during mid-July. Cohorts living in the hypolimnion do have greater recruitment per female (Table 2), although longer development times there make the population growth rates lower than in the epilimnion.

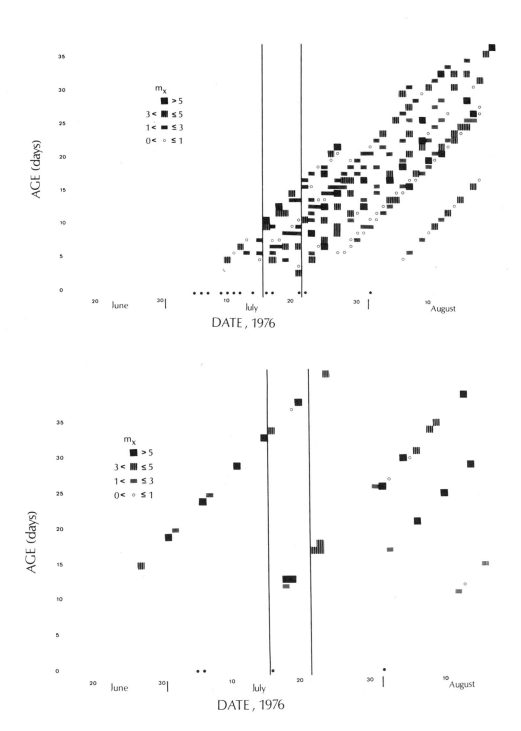

Thus, it is curious that *D. pulicaria* does not inhabit the epilimnion of Wintergreen Lake in midsummer or that of a variety of other lakes where temperatures exceed about 22°C (Threlkeld 1977; Haney and Hall 1975; Ward and Robinson 1974). It is possible that conditions which lead to the abortion of eggs and reduced population growth rate are common and prolonged in other lakes, thus preventing the animals from occupying the epilimnion. However, the hypereutrophic conditions of Wintergreen Lake would probably result in a greater frequency of such conditions there than in other lakes. Conversely, it is possible that conditions which lead to such reductions in population growth rate are sufficiently unpredictable that the more benign hypolimnion is consistently chosen as the preferred habitat. While it seems that colonization of the epilimnion would continue throughout the summer, this explanation appears to be the only one which considers the responses of *D. pulicaria* to food and temperature conditions without reference to predation. If size-selective predation is considered as an additional constraint on population distribution, then it is easier to understand the observed habitat choices of these two species.

Since size-selective predation is common, it is easy to suggest that these habitat choices merely reflect a response by the two species, according to their body sizes, which permits maximum population growth rate as a balance between predation intensity (which must vary at least with light levels and prey body size) and predator-free response to food and temperature conditions. Thus, the smaller-bodied *D. galeata mendotae* can choose an epilimnion habitat because the positive effect of epilimnion temperature on growth rate may be sufficient to offset the additional mortality which results from predation in the more well illuminated epilimnion. In contrast, *D. pulicaria* may tend to choose an environment (the hypolimnion) because, in spite of reduced population growth rates, light intensities are lower and predation intensity may be greatly reduced.

Although it is convenient to include size-selective predation in an explanation of these distribution patterns, it must also be noted that a simple choice by each species to live where maximal and predictable population growth rates can be achieved would also result in the same distribution. Thus, as Stearns (1977) pointed out, "Alternative explanation(s) of life history diversity. . . are possible and are rarely mutually exclusive. Multiple causation frequently operates on life histories." In the present study, constraints imposed by unpredictable food and temperature variation and by size-selective predators result in similar predictions of midsummer vertical distribution of these two *Daphnia* species.

References

BIRGE, E. A. 1895. Plankton studies on Lake Mendota. I. The vertical distribution of the pelagic Crustacea during July 1894. Trans. Wis. Acad. Sci. 10:421–484.
———. 1898. Plankton studies on Lake Mendota. II. The Crustacea of the plankton from July, 1894, to December, 1896. Trans. Wis. Acad. Sci. 11:274–451.
BOTTRELL, H. H., AND OTHERS. 1976. A review of some problems in zooplankton production studies. Norw. J. Zool. 24:419–456.
BRANDLOVA, J., Z. BRANDL, AND C. H. FERNANDO. 1972. The Cladocera of Ontario with remarks on some species and distribution. Can. J. Zool. 50:1373–1403.
BROOKS, J. L. 1957. The systematics of North American *Daphnia*. Mem. Conn. Acad. Arts Sci. 13:1–180.
BURNS, C. W. 1969. Relation between filtering rate, temperature and body size in four species of *Daphnia*. Limnol. Oceanogr. 14:693–700.
DAGG, M. J. 1977. Some effects of patchy food environments on copepods. Limnol. Oceanogr. 22:99–107.
GIESEL, J. T. 1976. Reproductive strategies as adaptations to life in temporally heterogeneous environments. Annu. Rev. Ecol. Syst. 7:57–79.
HALL, D. J. 1964. An experimental approach to the dynamics of a natural population of *Daphnia galeata mendotae*. Ecology 45:94–112.
———, S. T. THRELKELD, C. W. BURNS, AND P. H. CROWLEY. 1976. The size-efficiency hypothesis and the size structure of zooplank-

ton communities. Annu. Rev. Ecol. Syst. 7:177-208.

HANEY, J. F., AND D. J. HALL. 1975. Diel vertical migration and filterfeeding activities of *Daphnia*. Arch. Hydrobiol. 75:413-441.

HRBÁČEK, J. 1959. Über die angebliche Variabilitat von *Daphnia pulex* L. Zool. Anz. 162:116-126.

HRBÁČKOVÁ-ESSLOVÁ, M. 1962. Postembryonic development of cladocerans. I. *Daphnia pulex* group. Vestn. Cesk. Spol. Zool. 26: 212-233.

HUTCHINSON, G. E. 1967. A treatise on limnology, v. 2. Wiley.

KEYFITZ, N. 1968. Introduction to the mathematics of population. Addison-Wesley.

LAMPERT, W. 1977. Studies on the carbon balance of *Daphnia pulex* DeGeer as related to environmental conditions. IV. Determination of the "threshold" concentration as a factor controlling the abundance of zooplankton species. Arch. Hydrobiol. Suppl. 48, p. 361-368.

LOTKA, A. J. 1956. Elements of mathematical biology. Dover.

LYNCH, M. 1977. Fitness and optimal body size in zooplankton populations. Ecology 58:763-774.

MANNY, B. A. 1972. Seasonal changes in organic nitrogen content of net- and nanno-phytoplankton in two hardwater lakes. Arch. Hydrobiol. 71:103-123.

PRESCOTT, G. W. 1970. The freshwater algae. Brown.

STEARNS, S. C. 1976. Life-history tactics: A review of the ideas. Q. Rev. Biol. 51:3-47.

———. 1977. The evolution of life history traits: A critique of the theory and a review of the data. Annu. Rev. Ecol. Syst. 8:145-171.

TAPPA, D. W. 1965. The dynamics of the association of six limnetic species of *Daphnia* in Aziscoos Lake, Maine. Ecol. Monogr. 35: 395-423.

THRELKELD, S. T. 1976. Starvation and the size structure of zooplankton communities. Freshwater Biol. 6:489-496.

———. 1977. The midsummer dynamics of two *Daphnia* species in Wintergreen Lake, Michigan. Ph.D. thesis, Michigan State Univ.

———. 1979. The midsummer dynamics of two *Daphnia* species in Wintergreen Lake, Michigan. Ecology. 165-179.

WARD, F. J. AND G. G. C. ROBINSON. 1974. A review of research on the limnology of West Blue Lake, Manitoba. J. Fish. Res. Bd. Can. 31:977-1005.

WETZEL, R. G. 1975. Limnology. Saunders.

———, AND D. F. WESTLAKE. 1969. Periphyton, p. 33-40. *In* R. A. Vollenweider [ed.], Primary production in aquatic environments. IBP Handbook 12. Davis.

34. Seasonal Variation in the Sizes at Birth and at First Reproduction in Cladocera

David Culver

Abstract

Seven species of Cladocera from the Bay of Quinte, Lake Ontario, and Locust Point, Lake Erie, show a regular decline in size at first reproduction (SFR) associated with temperature fluctuations. Three of the species (*Eubosmina coregoni*, *Bosmina longirostris*, and *Chydorus sphaericus*) show major changes in neonate sizes. All species exhibit decreases in body size with increasing temperatures from spring to summer, then return to larger sizes with declining temperatures in the fall. Despite these pronounced changes in SFR, little overlap in neonate or early adult sizes occurred, as the size shifts occurred generally in tandem. In fact, successful appearances of new species fit into size voids, while decreases in abundance of species coincided with neonate or SFR overlaps.

Comparison of samples from within large enclosures without fish with those from the open waters of the Bay of Quinte shows similar trends in SFR and neonate size for both populations, implying that the decline in body size was not due directly to selective predation by fish. Because reduced SFR would increase vulnerability to invertebrate predators (copepods, *Asplanchna*, and *Leptodora*), invertebrate predation is unlikely to be the cause of this shift. I propose that a major component of the decline in SFR is an intrinsic change, caused ultimately by fish predation, beyond any effects of food. This change is stimulated proximately by rapid temperature change or some close correlate.

Hutchinson (1967) has summarized numerous examples of cyclomorphotic changes in the Cladocera; most of these studies emphasized differential growth of helmets, mucrones, etc. rather than overall changes in body size as it related to reproduction and population dynamics. Decline in body size through the spring to summer period has been reported for copepodites of *Calanus finmarchicus* collected in Vineyard Sound (Clarke and Zinn 1937). Similar patterns have been found for *Bosmina longirostris* in Frains Lake, Michigan (Kerfoot 1974, 1975a), and in eight lakes in Sweden (Stenson 1976). Similar results have been reported for *Daphnia rosea* (Dodson 1972). For cladocerans in the Bay of Quinte, Lake Ontario, and in Lake Erie (Culver and DeMott 1978), neonate length and size at maturity decline from spring to summer.

Whereas changes in helmet size and body proportions have been shown to involve predation defenses, variations in the SFR and neonate size affect reproductive output directly. Changes in the amount of energy invested in individual offspring through the seasons relative to the mother's own maintenance needs must affect fecundity.

This work was supported by grants from the Office of Water Resource Technology, U.S. Department of the Interior, and by The National Research Council of Canada, Queen's University, Environment Canada, and The Ohio State University. Computer calculations were performed with the aid of a grant from the Instruction and Research Computer Center, The Ohio State University.

The difficulty in aging cladocerans makes it desirable to perform productivity calculations based upon size classes, so it is essential to know at what stage a cladoceran becomes mature for this and other types of dynamics calculations. The decrease in size is also counter to predictions of the size efficiency hypothesis (Brooks and Dodson 1965; Hall et al. 1976), which suggests that in the absence of visual predators (e.g. fish) plankters with large body size will outcompete those that use similar resources but have small bodies. Accordingly, the purpose of this paper is to examine the seasonal variation in the size at first reproduction (SFR) and neonate size in the Bay of Quinte, Lake Ontario, to determine the benefits associated with this variation, and to establish whether the variation is a direct result of differential mortality of larger organisms, and if not, to postulate the ultimate and proximate causes of the changes in body size in these species.

I thank K. G. Lloyd, K. Mitchel, and J. Fletcher for helping make the thousands of measurements and enumerations required for this study, and S. Micol, R. Vaga, V. Ramos, and C. Scott for help in working the data into useful form. Computer programs were written primarily by E. W. Osman. The field site in Lake Ontario was maintained by the Canada Centre for Inland Waters; I thank D. R. S. Lean of that institution for providing support, both monetary and technical, throughout the field portions of this study.

Methods

Zooplankton samples were collected weekly from 24 April–9 October, 1974, in conjunction with a study of the effects of N and P loading on lake production. Samples came from either the open water of the Bay of Quinte, Lake Ontario (hereafter referred to as the bay), or from six 100,000-liter enclosures (25 m^2 surface area × 4 m deep), some of which were enriched with N and/or P. The effects of the enrichment on zooplankton will be reported elsewhere. Zooplankton samples were collected with a 0.5-m-tall Schindler-Patalas trap with a 65-μm net lowered twice to sample a 1-m stratum. Four samples thus represented the entire 4 m water column. Samples were preserved in sucrose and Formalin (Haney and Hall 1973) until counted and measured. Individual samples were subsampled quantitively and counted on a Bogorov tray. Cladocerans were measured to the nearest 10 μm using an ocular micrometer at 50× on a Wild M-5 dissecting microscope. Eggs or embryos in the brood pouch were counted and recorded along with the length of all ovigerous females.

The lengths reported herein were total body lengths, which were less satisfactory for indicating general size than lengths from the base of the antenna to the base of the tail spine (if any) for markedly cyclomorphotic forms (*Daphnia galeata mendotae* and *D. retrocurva*).

In this study, cladocerans were either from the enclosures (essentially no fish predation) or from the bay, where alewife (*Alosa pseudoharengus*) and fish larvae predation on zooplankton was considerable. Alewives occasionally jumped over the floats into the enclosures but were removed promptly by dipnet or gillnet. Samples from Locust Point, Lake Erie, were collected by W. R. DeMott using a vertical haul of a 64-μm net and preserved with buffered Formalin until counted and measured. This station was very similar in depth, exposure, and proximity of shore to the Bay of Quinte station and had many of the same species of Cladocera.

Temperatures were measured with a mercury thermometer for some dates, but more commonly with a continuously recording thermometer run by the Canada Centre for Inland Waters to record daily average temperatures for each enclosure and the bay.

Length-frequency distributions were constructed for all samples, whether collected inside enclosures or from the bay. The size at maturity, or more accurately the size at first reproduction (SFR), was determined from the 10th percentile on histograms for length among ovigerous individuals. This is assumed to be the size at which all individuals of that species and date reach maturity and ignores the few that reach maturity at abnormally small size. Neonate length was chosen as the 5th percentile to exclude those few individuals that were extruded from the brood pouch after preservation and thus did not undergo the swelling which occurs immediately after birth.

Percentiles have the additional advantage of giving SFR and neonate size for each date essentially independent of sample size. Since we worked with field populations, samples did not always have significant numbers of all species of interest, especially ovigerous ones. In many cases, the entire sample was examined for size-frequency measurements on rarer species, and often even then too few were found to construct reliable frequency diagrams. This was particularly true for the larger species during summer in the bay, where mortality from fish essentially excluded large species. The 5th and 10th percentiles were chosen arbitrarily, but there appears to be some justification for their continued use. Each size-frequency graph we drew involved 50 size classes, yet the size class containing the 10th percentile for ovigerous females often contained 8–20% of the animals observed, indicating that it was a good indication of the size at first reproduction.

Zooplankton dry weights were calculated from cladoceran lengths using regressions from Boucherle et al. (1979).

Temperature-dependent development times for eggs (which correspond to molting times in Cladocera), and for juveniles to adult stages, were taken from Hillbricht-Ilkowska and Patalas (1967) and Hall (1964).

Results

Cladoceran species occurring in the samples included *Alona affinis, Bosmina longirostris, Eubosmina coregoni, Ceriodaphnia lacustris, Chydorus sphaericus, Daphnia galeata mendotae, Daphnia pulicaria, Daphnia retrocurva, Diaphanosoma leuchtenbergianum, Holopedium gibberum,* and *Leptodora kindtii*. Of these, *A. affinis, D. pulicaria,* and *H. gibberum* were too rare to allow us to construct frequency distributions for length; the other eight species are considered in detail here.

Numerical abundance of the various species in the bay showed a pronounced seasonal pattern (Fig. 1) with some species (e.g. *C. lacustris, D. retrocurva,* and *D. leuchtenbergianum*) rare until midsummer while other species (*Eubosmina* and *Chydorus*) were present in varying abundance throughout the year. Abundance patterns inside the enclosures were different from those outside, with *D. galeata mendotae* common inside the enclosures throughout the year but scarce in the bay except for brief periods (Fig. 1).

Temperature in the Bay of Quinte was never stratified at the sample site. The maximum surface-bottom differential was 1°C; thus, vertical migration by zooplankton should not affect development times. Thermal patterns inside and outside the enclosures were similar. Temperature patterns show the effects of periods of fair weather which caused rapid temperature increases (Fig. 2). Stormy periods had greater turbulence, turbidity, and greater flushing rates for the bay.

SFR and neonate size showed a general pattern of decrease during summer followed by an increase in late fall (Figs. 3 and 4). The number of measurements available to construct the plotted SFR and neonate lengths for a given date, species, and station ranged from 14 to 300 with a mean of 73. We took 37,791 measurements. The patterns of SFR and neonate length for the two *Daphnia* species are less regular than those of the other species because total length measurements are sensitive to the development of helmets by these highly cyclomorphotic forms. The greater abundance of the large cladoceran forms in the enclosures is reflected in the greater number of observations plotted for these species (Fig. 4) as compared to the bay samples (Fig. 3). This is a direct effect of fish predation in the bay on the larger cladoceran species.

Eubosmina coregoni adults from Lake Erie (open squares marked "e", Fig. 3) also showed a seasonal decline in SFR, but at a much larger size than was observed for Lake Ontario. *Chydorus sphaericus* (open triangles marked "e", Fig. 3) showed an identical pattern. *Daphnia retrocurva*, plotted in top panel of Fig. 4 for convenience (open circles marked "e"), also showed the typical seasonal pattern, but was again much larger than the same species in Lake Ontario—approaching the SFR seen in the larger *D. galeata mendotae* from the Lake Ontario enclosures. It is surprising to find this difference in the SFR seasonal patterns of the two lakes since Lake Erie drains into Lake Ontario and the samples were collected from two similar nearshore stations.

Although differences between neonate and mature female lengths within a species are greater in spring than summer, this effect is less than it appears from Figs. 3 and 4. The neonates of a given date become mature at a later date (from 7 to 14 days, depending on temperature and species) when the SFR is typically lower than it was when they were born. If the weights instead of lengths at first reproduction and at birth are plotted against time, the effects are even more pronounced (Fig. 5).

Number of eggs produced per female was quite variable, depending upon female size and season for many species, except for *C. sphaericus* which never had more than two eggs. Representative plots for three species of Cladocera for spring and summer conditions (14.8 and 24.5°C; Fig. 6) demonstrate both differences in ranges of the size of reproductive females at these two periods and differences in fecundity (both maximum number of eggs per female and variance within a size class). Because eggs were found loose in samples, these egg counts must be considered minimal. Future studies should utilize the technique of Prepas (1978) to minimize this problem.

The tandem decrease among SFR and neonate lengths during the spring decline is quite striking (Figs. 3 and 4). Species which were rare in the early spring (*D. retrocurva*, *C. lacustris*, and *D. leuchtenbergianum*) all enter at SFRs and neonate sizes inter-

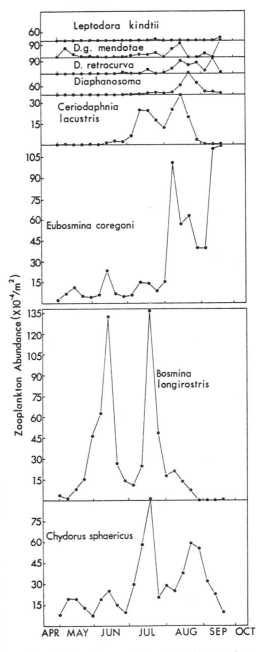

Fig. 1. Abundance of the eight major species of Cladocera in Bay of Quinte, 1974. Station depth, 4.0 m.

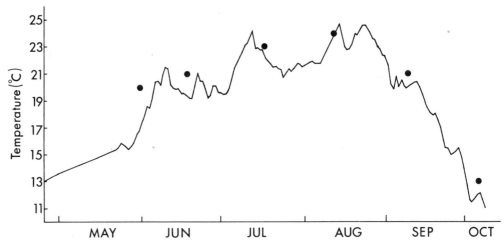

Fig. 2. Average daily temperature of Bay of Quinte, Lake Ontario, near Deseronto, Ontario, 1974. ●—Afternoon surface temperatures for the Locust Point station, Lake Erie, 1975; had these been continuous measurements, they probably would have corresponded closely to the Lake Ontario measurements.

mediate with those species already abundant and then vary with them. Upper size limits for these species are also discrete and decline seasonally along with SFR and neonate size (Fig. 7).

Comparing spring and summer frequency distributions (Fig. 8) demonstrates the narrow range of adult sizes represented in the bay plus the significant change in size ranges of *Eubosmina, Bosmina,* and *Chydorus*—the species which showed marked neonate declines. The three smallest species show no overlap between spring and summer size distributions.

Discussion

Although there have been few examinations of body size at maturity in field populations of Cladocera, work on single species by Kerfoot (1974, 1975a, b) and others mentioned in the introduction has indicated that the pattern seen in the bay is widespread. The data presented here indicate that the whole cladoceran community undergoes a coordinated shift in adult size and neonate sizes in certain species as the water warms in the summer and as it cools again in the fall. Questions raised by these patterns are many. What costs are associated with this shift and what benefits accrue to a species that changes size seasonally? Is the change caused by a direct factor such as food, physiological response to temperature, or differential mortality on one size range of organisms, or is it a cyclomorphotic change analogous to changes in helmet size, etc.? If it is a cyclomorphotic change, what is (are) the ultimate factor(s) responsible for this intrinsic characteristic in Cladocera, and what environmental cues initiate the growth response?

Fish predation is the most logical ultimate cause for a decline in SFR since fish show a pronounced seasonality in production of planktivorous fry and are predominately visual predators, favoring larger over smaller zooplankters. The effectiveness of these predators is demonstrated by the absence or extreme scarcity of the larger species of Cladocera in the bay in summer relative to the enclosures. This does not, however, explain why smaller species such as *Chydorus* follow the same patterns as do the larger ones. Shifts in SFR by those species exposed each year to fish predation may force concomitant shifts in the SFRs of smaller species to reduce competitive interaction.

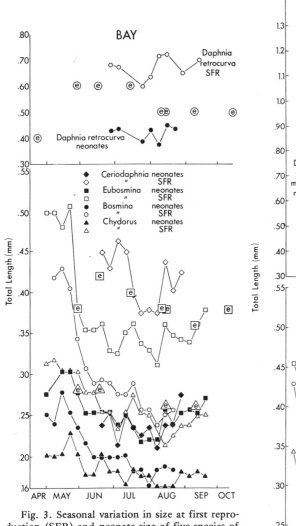

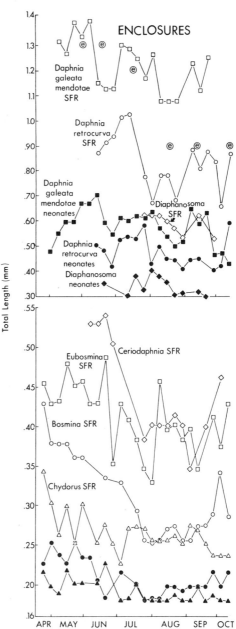

Fig. 3. Seasonal variation in size at first reproduction (SFR) and neonate size of five species of Cladocera from Bay of Quinte, Lake Ontario, and Locust Point, Lake Erie. Solid symbols—neonate sizes. Open symbols—size at first reproduction (SFR) as indicated by 10th percentile for length of ovigerous females. Larger species (*Daphnia retrocurva* and *D. galeata mendotae*) are rare in bay due to fish predation. Note that data from Lake Erie (marked "e") indicate that the organisms are born and mature at much larger sizes than those from Bay of Quinte for same time of year. Circles (marked "e") in upper panel are Lake Erie *D. retrocurva* neonates. SFR data for *D. retrocurva* from Lake Erie given in Fig. 4.

Fig. 4. Seasonal variation in size at first reproduction (SFR) and neonate size of seven species of Cladocera from enclosures in Bay of Quinte, Lake Ontario, and from Locust Point, Lake Erie. Symbols as in Fig. 3. Curves are not as smooth as those in Fig. 3 since these data represent samples lumped from three enclosures through mid-July, and then six enclosures through October. Some enclosures were enriched with N and/or P. Circles (marked "e") are Lake Erie *D. retrocurva* SFR.

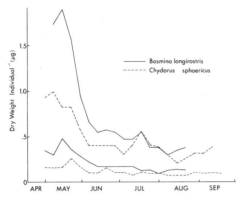

Fig. 5. Seasonal variation in dry weight of mature females and of neonates in the Bay of Quinte, 1974.

The proximate cause of the changes in SFR is probably one or more environmental cues. Examining the temperature pattern for the bay, one observes increases in four major bouts (Fig. 2). The first was in early May, the second in early June, the third in early July, and the fourth in mid-August. Declines in SFR (Fig. 3) occurred with each of these periods except that no samples were taken from the bay in April, so we must use values for SFR in the enclosures (Fig. 4) for comparison. SFR and neonate lengths began increasing as the water cooled, with the greatest changes occurring in late August and September, again inversely correlated with temperatures, with the greatest changes occurring when temperature rose or dropped rapidly.

My stations were unstratified, but thermal stratification has obviously had a great deal of input into the evolution of cyclomorphosis (Hutchinson, 1967). We are currently examining an excellent series of 10 monthly plankton samples from 30 offshore stations in Lake Erie taken by N. Watson (Canada Centre for Inland Waters) and will try to determine whether thermally stratified offshore zones show significant differences from those we observe in nearshore stations. Preliminary results show that there is a summer decline in SFR at these offshore stations.

Since most of the present study was carried out a posteriori with the exception of the comparison of fish and no-fish conditions, there are many questions left unanswered. The paucity of information on selective feeding preferences of freshwater copepods and rotifers makes it difficult to examine their role in the size-specific competition among herbivores in this system.

It is implicit in the pattern observed in SFR changes that they are a result of pronounced seasonality in predation, and thus would be unexpected in tropic zone lakes such as those currently being studied by M. Steinitz (pers. comm.) in the Galapagos Archipelago and in the Ecuadorian Andes. Any changes in SFR in those populations would be expected to be maintained by either stable polymorphisms caused by differential predation inshore–offshore (Zaret 1972) or related to environmental cues generated by such things as changes in rainfall patterns.

Ware (1975, 1977) pointed out that the size of fish eggs, and ultimately the fish larvae, also declines seasonally in response to increasing temperatures and that this results in the size of the fish larval forms being correlated with the decline in average size of zooplankters in the Gulf of St. Lawrence. Similar work should be done with freshwater larval fish to determine whether similar patterns have occurred in inland habitats. Recent migrants from estuarine systems (e.g. *Alosa pseudoharengus*) should be particularly fruitful organisms to study. Whatever the answers to these and other questions raised by these data, the interactions among size-selective predation, competition for food, and pronounced seasonality in the temperate zone have resulted in a complex pattern of cyclomorphic changes in the herbivorous Cladocera.

Cladoceran Size at Maturity 365

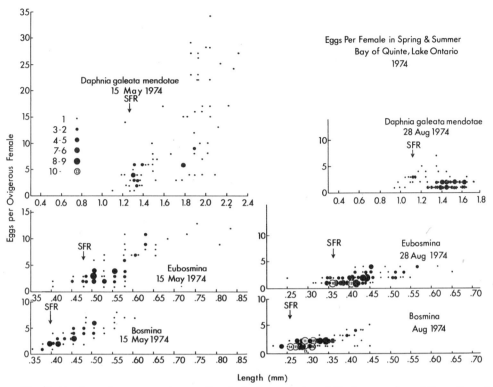

Fig. 6. Number of eggs per ovigerous female in three species of Cladocera collected from Bay of Quinte, for spring and summer 1974.

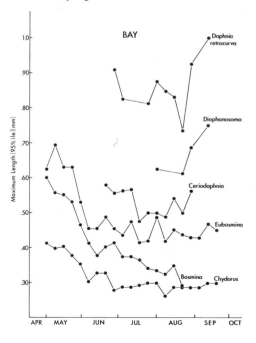

Fig. 7. Seasonal variation in maximal size of six species of Cladocera from Bay of Quinte, 1974. Plotted values represent length of individual that was at 95th percentile for length in samples collected from bay. Correct 95th percentile is somewhat smaller than this since measurements included all nonovigerous and ovigerous individuals measured. Rarer ovigerous organisms required a greater sample aliquot to achieve a useful sample size and, thus, are overrepresented in this measure. It is unlikely that pattern depicted will be altered by this bias.

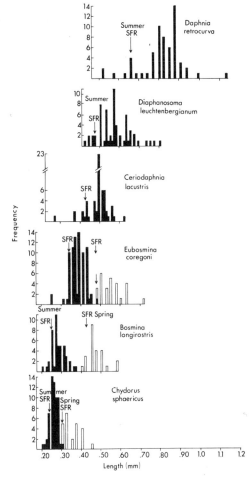

Fig. 8. Frequency diagrams for length of ovigerous cladocerans from Bay of Quinte. *Daphnia galeata mendotae* was too rare in the bay to be represented here. Spring samples came from 15 May, 1974, and summer samples from 28 August, 1974, except for *Bosmina longirostris*, which is represented here by a lumping of measurements from 14 and 21 August, 1974, due to its scarcity in the bay in late summer.

References

BOUCHERLE, M. M., D. A. CULVER, and J. W. FLETCHER. 1979. Dry weight estimates of biomass for twelve crustacean taxa from Lake Erie. J. Great Lakes Research. In press.

BROOKS, J. L., and S. I. DODSON. 1965. Predation, body size, and composition of plankton. Science **150**: 28–35.

CLARKE, G. L., and D. J. ZINN. 1937. Seasonal production of zooplankton off Woods Hole with special reference to *Calanus finmarchicus*. Biol. Bull. **73**: 464–487.

CULVER, D. A., and W. R. DeMOTT. 1978. Production of zooplankton at nearshore stations in Lakes Ontario and Erie. Int. Ver. Theor. Angew. Limnol. Verh. **20**:252–256.

DODSON, S. I. 1972. Mortality in a population of *Daphnia rosea*. Ecology **53**:1011–1023.

HAIRSTON, N. G., JR., and R. A. PASTOROK. 1975. Response of *Daphnia* population size and age structure to predation. Int. Ver. Theor. Angew. Limnol. Verh. **19**:2898–2905.

HALL, D. J. 1964. An experimental approach to the dynamics of a natural population of *Daphnia galeata mendotae*. Ecology **45**:94–111.

———, S. T. THRELKELD, C. W. BURNS, and P. H. CROWLEY. 1976. The size efficiency hypothesis and size structure of zooplankton communities. Annu. Rev. Ecol. Syst. **7**:177–208.

HANEY, J. F., and D. J. HALL. 1973. Sugar-coated *Daphnia*: A preservation technique for Cladocera. Limnol. Oceanogr. **18**:331–333.

HILLBRICHT-ILKOWSKA, A., and K. PATALAS. 1967. Methods of estimating production and biomass and some problems of quantitative calculation methods of zooplankton. Ekol. Pol. Ser. B. **8**:139–172. [Fish. Res. Bd. Can. Trans. Ser. 2781.]

HUTCHINSON, G. E. 1967. A treatise on limnology, v. 2. Wiley.

KERFOOT, W. C. 1974. Egg-size cycle of a cladoceran. Ecology **55**:1259–1270.

———. 1975a Seasonal changes of *Bosmina* (Crustacea, Cladocera) in Frains Lake, Michigan; laboratory observations of phenotypic changes induced by inorganic factors. Freshwater Biol. **5**:227–243.

———. 1975b. The divergence of adjacent populations. Ecology **56**:1298–1313.

PREPAS, E. 1978. Sugar-frosted *Daphnia*: An improved fixation technique for Cladocera. Limnol. Oceanogr. **23**:557–559.

STENSON, J. A. E. 1976. Significance of predator influence on composition of *Bosmina* spp. populations. Limnol. Oceanogr. **21**:814–822.

WARE, D. M. 1975. Relation between egg size, growth, and natural mortality of larval fish. J. Fish. Res. Bd. Can. **32**:2503–2512.

———. 1977. Spawning time and egg size of Atlantic mackerel, *Scomber scombrus*, in relation to the plankton. J. Fish. Res. Bd. Can. **34**:2308–2315.

ZARET, T. M. 1972. Predators, invisible prey, and the nature of polymorphism in the Cladocera (Class Crustacea). Limnol. Oceanogr. **17**:171–184.

35. Predation, Enrichment, and the Evolution of Cladoceran Life Histories: A Theoretical Approach

Michael Lynch

Abstract

Here I present a mechanistic theory for the evolution of life histories and utilize it to interpret life history variation in cladocerans; the theory is probably equally or more suitable for other zooplankton groups. Based on a knowledge of size-specific rates for feeding efficiency (a measure of the ability to harvest energy for growth and/or reproduction) and mortality, the model can be used to predict optimal sizes at birth and maturity, clutch size, and age at first reproduction.

The model predicts that when mortality increases with size, selection should be toward producing small offspring that mature at a small size, that when mortality decreases with increasing size, large sizes at birth and maturity and small clutch sizes should be selected for, that large offspring size but small adult size may be simultaneously favored when both large and small individuals are subjected to intense mortality, and that small-bodied genotypes and species should be increasingly favored as the nutritional status of the environment declines. All of these predictions seem consistent with what is presently known about cladocerans.

As in most organisms, the evolution of certain life-history traits in the Cladocera involves several tradeoffs. The rate at which one's genes enter the gene pool or at which a species comes to dominate a community depends not only on the number of offspring produced but also on the age at which they are produced. Since the ability to accumulate energy tends to increase with body size (Fig. 1), there is an energetic advantage to reproducing at a large size. However, postponing reproduction until attaining a large size also tends to reduce the rate of increase by increasing the age at maturity. The significance of this tradeoff can be reduced if increasing the size at maturity allows corresponding increases in the size of offspring produced. However, increasing the size of individual offspring necessarily reduces the number of them that can be produced (Fig. 2). In addition to these basic energetic constraints, the presence of predators in aquatic environments may modify the advantages of particular life histories by imposing high mortality rates on individuals of specific sizes. Thus, an interpretation of the adaptive significance of cladoceran life histories requires an evaluation of both the foraging ability and the vulnerability to predators of different-sized individuals.

I here present a theory which predicts both the optimal size at maturity and the optimal compromise between offspring

---

This work was supported by the Research Board, University of Illinois. I thank B. Bush, J. Lynch, and M. Shaffer for assistance with data analysis, and R. Armstrong and D. Allan for helpful comments.

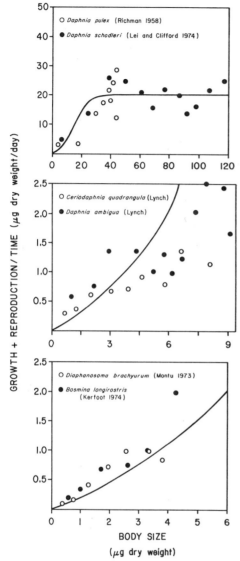

Fig. 1. Estimated feeding efficiency functions for six species of planktonic cladocerans (well-fed laboratory populations maintained at 20°C), assuming that body tissue and eggs are energetically equivalent on a dry weight basis (Lynch, in press). Solid lines represent Eq. 3 with $F_{max} = 20$ and $a = 0.2$ are are used throughout to represent feeding efficiency in environments of maximum nutritional value.

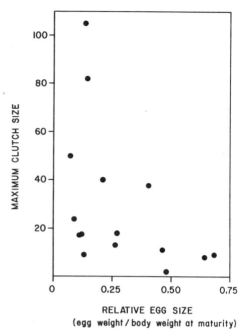

Fig. 2. Tradeoff between clutch size and relative egg size in cladocerans (data summarized in Lynch, in press).

size and number for organisms living in different environments. The simultaneous evolution of these two characters has not been previously considered in theoretical treatments of life histories (see Stearns 1976, 1977), despite the fact that an existing model (Gadgil and Bossert 1970) can be used to investigate the problem. The theory I present is most appropriate for examining intraspecific variability in life histories, but should also be valid for making interspecific comparisons of colonizing and competitive ability. Here I utilize the theory to examine the adaptive significance of cladoceran life histories; it may be equally or more suitable for copepods and rotifers, but the necessary background data for these groups are not available.

The Model

I start with two assumptions: that energy intake and mortality rates in cladocerans are primarily size- rather than age-dependent, and that the expenditure of energy on growth is negligible following the onset of maturity. If we assume that egg volume does not vary with age, the model then further assumes that the clutch size of an individual is constant throughout its life.

These conditions are frequently approximated by chydorids and large *Daphnia* species (Lynch, in press), but the adults of small planktonic species expend a considerable amount of their energy on growth (Fig. 3) and increase their clutch size throughout life. However, the fact that the rate of increase for cladocerans is most dependent on the age at first reproduction and the size of the first clutch (Lynch, in press) tends to minimize the effect which these restrictions have on the following theory. Thus, although the assumptions limit the precision of the model, they do not alter the qualitative results which I present below.

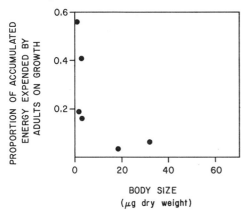

Fig. 3. Proportion of energy in excess of metabolic requirements expended on reproduction by adults of cladoceran species with different sizes at maturity (data summarized in Lynch, in press).

Under these conditions the stable age equation,

$$1 = \sum_{x=0}^{\infty} e^{-rx} l_x m_x,$$

can be rewritten as

$$1 = l_k m e^{-rk} \sum_{x=0}^{\infty} e^{-2rx} p^{2x} \quad (1)$$

where k is age at first reproduction (days), l_k is probability of survival to the age at first reproduction (a function of environment, offspring size, and size at maturity), m is number of offspring per clutch (a function of nutritional environment, offspring size, and size at maturity), p is day-to-day probability of adult survival (a function of environment and size at maturity), and r is rate of population increase (days^{-1}) given life table characteristics, k, l_k, m, and p. In the exponents to the right of the summation, 2 accounts for the fact that the time from egg laying to egg release is about 2 days at average summer temperature (20°C); egg development time does not vary significantly between species (Hebert 1978).

Since $e^{-2r}p^2 < 1$, the summation in Eq. 1 is approximated by a geometric progression,

$$\sum_{x=0}^{\infty} e^{-2rx} p^{2x} = \frac{1}{1 - e^{-2r} p^2},$$

and Eq. 1 simplifies to

$$e^{2r} - p^2 - l_k m e^{-r(k-2)} = 0. \quad (2)$$

If a functional relation exists between body size and the ability to gather energy for growth and/or reproduction (hereafter F, feeding efficiency), then the age at first reproduction (k) and the clutch size (m) can be determined given the sizes at birth (B_0) and maturity (B_{mat}). Existing information on this relation for well-fed laboratory populations of cladocerans at 20°C is presented in Fig. 1. Although the data cover species of very different sizes fed on a variety of algal types, they can be generally approximated by the same sigmoid function,

$$F = \frac{F_{max} + 1}{1 + F_{max} e^{-aB}} - 1 \quad (3)$$

where F is feeding efficiency, the rate of accumulation of biomass which can be used for growth and/or reproduction (μg dry wt/day); F_{max} is maximum attainable feeding efficiency; B is body size (μg dry wt); and a is a constant determining the initial slope of $F(B)$.

The growth rate of an individual of

size B is simply $F(B)$ until the size at maturity (B_{mat}) is attained and growth ceases. Allowing 2 days for the accumulation of energy for the first clutch and 2 additional days for egg development, and integrating Eq. 3 from the size at birth to the size at maturity, we find the age at first reproduction is

$$k = \frac{(B_{mat}-B_0) + (\frac{1+F_{max}}{a}) \ln(\frac{1-e^{-aB_{mat}}}{1-e^{-aB_0}})}{F_{max}} + 4. \quad (4)$$

The clutch size is determined by dividing the amount of energy that an adult accumulates in 2 days by the offspring size,

$$m = \frac{2}{B_0}\left[\frac{F_{max}+1}{1+F_{max}e^{-aB_{mat}}} - 1\right]. \quad (5)$$

Thus an increase in clutch size can only be accomplished at the expense of producing smaller individual offspring.

The probability of survival is also a function of body size so that adult survival

$$p = p(B)_{B_{mat}} \quad (6)$$

and the probability of surviving to release the first clutch

$$l_k = p^4 \prod_{t=0}^{k-5} p(B)_{B_t}. \quad (7)$$

The sufficiency of the model rests on its ability to predict the characteristics of cladoceran populations of known nutritional status and mortality rates. For well-fed populations at 20°C, $F(B)$ is approximated by setting $F_{max}=20$ and $a=0.2$ (Fig. 1). In Fig. 4A, actual ages at first reproduction are compared with those predicted by Eq. 4 for species of known offspring (egg weight) and adult size (data summarized in Lynch unpubl). The two estimates are significantly correlated ($p<0.01$), but the model consistently overestimates the actual age at first reproduction by about 3 days. This is largely a consequence of the model's tendency to under-

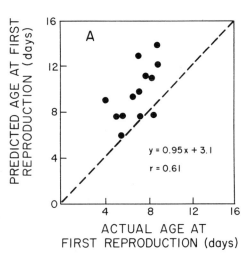
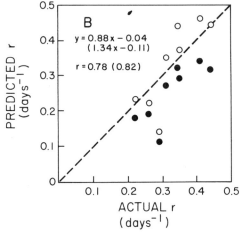

Fig. 4. A. Comparison of known ages at first reproduction and those predicted by Eq. 4 for cladocerans of known weights at birth and maturity. B. Comparison of known rates of increase and those predicted by Eq. 2 for species of known weights at birth and maturity; open circles and data in parentheses are for estimates with predicted dates at maturity reduced by 3 days as suggested in Fig. 4A (data summarized in Lynch, in press).

estimate the growth rates of very small individuals; the problem could be alleviated by setting the intercept of $F(B)$ larger than zero.

Assuming that mortality is negligible for well-nourished individuals until late in life, we can also use the model to predict laboratory population growth rates from a knowledge of offspring and adult size (using Eq. 4, 5, and 2). Predicted and actual growth rates (r) for well-fed laboratory populations are significantly correlated ($p<0.05$), although the model consistently underestimates r (Fig. 4B). By subtracting 3 days from the predicted age at first reproduction for these populations and recalculating r, the correlation between predicted and actual population growth rates is improved ($p<0.01$). The scatter in the relations in Fig. 4 may be primarily a consequence of the diversity of food types to which different populations were exposed.

By explicitly accounting for the relative energetic advantages of individuals of different sizes, the model provides a mechanistic basis for interpreting the evolution of life histories without resorting to the conceptual difficulties of reproductive effort. I now utilize the theory to examine the relative advantages of different combinations of offspring and adult size in various environments.

The consequences of size-selective predation

In the absence of size-selective predators and in an enriched environment (parameterized as in Fig. 1) the model predicts that species and genotypes which mature at a relatively large (but not the largest) size will be selected; their optimal offspring size will be intermediate (Fig. 5A). Under these conditions this strategy provides a favorable combination of a rapid attainment of maturity and moderate clutch size. On the other hand, small species in such environments will be best off producing the smallest offspring possible (here set at 0.5 µg dry weight) allowing a maximum clutch size; this is because for small species, which require only slight growth to attain maturity, reductions in offspring size produce only minor increases in the age at maturity.

In the following section I will show that the nutritional status of the environment greatly influences the relative advantages of particular life-history patterns. First, I wish to discuss the consequences of size-selective predation for the evolution of cladoceran life histories. These consequences are the same for all levels of food abundance.

A general assumption of many zooplankton ecologists is that environments with and without planktivorous vertebrates pose very different selective pressures for cladocerans (Dodson 1974; Kerfoot 1975; Hall et al. 1976; Lynch 1977). In environments containing visually feeding fish or salamanders, mortality rates are thought to be highest for large individuals and to decrease with decreasing size. According to the model, such environments tend to select not only for a small size at maturity but also for the production of large clutches of very small offspring (Fig. 5B). These results lend a theoretical basis to those studies which have noted a simultaneous reduction in offspring and adult size in species exposed to fish predation (Green 1967; Wells 1970; Warshaw 1972; Kerfoot 1974).

Environments from which vertebrates are absent pose a different pattern of mortality for cladocerans. In such cases, invertebrates, which tend to selectively prey upon small individuals, are the dominant predators (Dodson 1974; Kerfoot 1975; Lynch 1979). If predation intensity is simply a decreasing function of body size in these environments, then selection should be for a large size at maturity (Fig. 5C). All species (regardless of their size at maturity) should produce relatively large offspring, since this both increases the probability of survival to maturity and decreases the pre-reproductive period (Fig. 5C). Kerfoot (1974) has noted that, compared to the summer when fish predation is intense, during the winter when invertebrate predation may be of greater relative significance *Bosmina longirostris* increases both its size at maturity and its egg size. In vertebrate-free ponds, Dodson (1974) has found that *Daphnia middendorffiana* produces larger (and fewer)

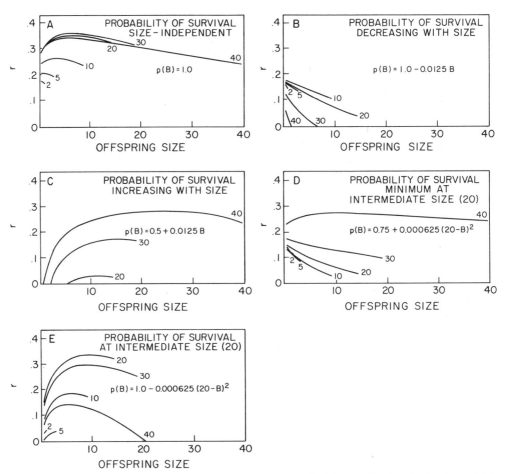

Fig. 5. A–E. Relative advantages of different combinations of adult and offspring size (μg dry weight) in enriched environments posing different patterns of size-selective mortality. Individual curves represent different sizes at maturity (2, 5, 20, 30, and 40 μg dry weight). For any combination of adult and offspring size, r was calculated using Eq. 2, 4, 5, 6, and 7. F_{max} = 20 μg dry weight/day and a = 0.2 as in Fig. 1. Increasing intensity of size-independent predation does not alter optimal strategies; qualitative effects of different mortality patterns do not vary with level of enrichment.

offspring when *Diaptomus* predation becomes intense; in these ponds, adult *D. middendorffiana* are too large to be consumed by *Diaptomus*, and adult size remains constant throughout the year.

The predation patterns existing in nature may actually be more complex than those presented above. In some cases mortality may be most intense for individuals of intermediate size; this would be characteristic of environments dominated by a predator, such as *Chaoborus* (Dodson (1974), having an upper and lower limit of prey sizes within the range of observed cladoceran sizes. The model predicts that such an environment provides an advantage to very large species (or genotypes) which also produce relatively large offspring (Fig. 5D); for these individuals adult mortality is minimal and offspring are capable of rapidly growing through the size range of intense mortality. Small species see this predation pattern simply as one of increasing mortality with size (as in Fig. 5B) and respond by producing a maximum number of small offspring.

Such a predation pattern may promote the maintenance of life-history polymorphisms in populations, since genotypes of initially different characteristics may evolve in opposite directions, depending on the relative ease with which offspring and adult size can be selected for. This pattern of mortality may also explain the different ways in which *Ceriodaphnia reticulata* and *Daphnia pulex* respond to *Chaoborus* predation (Lynch 1979). When *Chaoborus* became abundant in a vertebrate-free environment, *Ceriodaphnia* decreased its offspring size and size at maturity, while *Daphnia* increased both of these sizes. Such a pattern might arise if *Chaoborus* predation were most intense on the large instars of *Ceriodaphnia* and on the smallest *Daphnia* instars.

Many environments contain both vertebrate and invertebrate predators (for instance, a copepod and a fish), and under such circumstances the probability of mortality might actually be minimal for individuals of intermediate size. Under those circumstances, relative to the case in which mortality is size-independent, a reduction in the size at maturity to the size of minimum mortality may be selected for simultaneously with an increase in offspring size to allow rapid attainment of maturity (Fig. 5E). Such a pattern has been noted for *C. reticulata* which in the presence of both fish and *Cyclops* matures at a smaller size but produces larger eggs than in environments without fish and with fewer *Cyclops* (Lynch 1979).

The consequences of enrichment

While information on the relation of $F(B)$ to food abundance is scant, data for *D. pulex* (Richman 1958) and *Daphnia galeata mendotae* (Hall 1964) suggest that a decreased food supply reduces both F_{max} and the size at which it is attained. Such a relation can be approximated by keeping a in Eq. 3 constant and using F_{max} as an index of the nutritional status of the environment (Fig. 6).

Dilution of the food supply favors genotypes with a small size at maturity and small offspring size (Fig. 6). Large genotypes are selected against in dilute environments because their energetic advantage over smaller species is reduced while their pre-reproductive period is prolonged. The production of large clutches of small offspring is favored in nutritionally poor environments because it takes several days for even the largest possible juvenile to mature, and the additional time required for smaller juveniles to mature is offset by the larger numbers of them produced. This result is significant because it suggests that in some cases the optimal clutch size of individuals maturing at a particular size may actually increase with a decreasing food supply; the possibility of such a response implies that information on clutch size–body size relations may be an inadequate measure of the nutritional status of individuals, if egg size has not been accounted for.

The predictions of the model are in general agreement with what is known of the physiological response of cladocerans to a reduced food supply—a decrease in the size at maturity and frequently in the size at birth (Richman 1958; Hall 1964; Weglenska 1971; Vijverberg 1976). The results also provide a basis for interpreting the numerical dominance of small cladocerans in many oligotrophic lakes lacking vertebrates (Nilsson and Pejler 1973; Pope and Carter 1975). If small species have higher rates of increase than large ones at low food concentrations, then they cannot be competitively excluded by large species. In examining the interaction between natural populations of *D. pulex* and *C. reticulata*, I have discovered that while *Ceriodaphnia* is incapable of reducing the food supply to the extent that *Daphnia* can, it is less sensitive to resource depression than is *Daphnia* (Lynch 1978).

Interspecific variability in cladoceran life histories

This theory provides a framework for interpreting interspecific life-history pat-

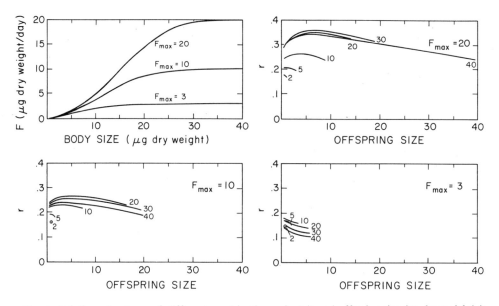

Fig. 6. Relative advantages of different combinations of adult and offspring size (μg dry weight) in environments of different nutritional value, where mortality is size-independent [$p(B) = 1.0$]. Individual curves represent different sizes at maturity (2, 5, 10, 20, 30, and 40 μg dry weight). For any combination of adult and offspring size, r was calculated using Eq. 2, 4, 5, 6, and 7. Curves are shown for $a = 0.2$ and $F_{max} = 20, 10$, and 3.

terns. Among the Cladocera, large species tend to be able to produce much larger clutches than can small species (Lynch unpubl.). This results for two reasons. First, large species produce eggs which are well within the expected size range for predator-free environments and much below the theoretical maximum size that they could produce, while small species produce eggs which are unexpectedly large and close to the theoretical maximum (Fig. 7). Second, the adults of small species allocate a greater proportion of their energy budget to growth than do larger species (Fig. 3 and Lynch, in press). Both of these characteristics of small species may be adaptations to reduce the impact of invertebrate predation.

Although large cladoceran species have not responded to the pressures of invertebrate predators by producing unexpectedly large juveniles, they do appear to mature at larger sizes than expected for predator-free environments. The theory predicts that in even the most enriched environments the optimal size at maturity will be less than the optimal foraging size (the size at which F_{max} is attained) in the absence of predators (Fig. 5A). The presence of a mortality pressure which decreases with increasing size may alter this result, favoring the postponement of maturity to a size close to or beyond that required to attain the maximum feeding efficiency. Many of the large *Daphnia* species appear to follow such a strategy, maximizing early growth to the optimal foraging size and then initiating reproduction (Lynch, in press.). Some large species, such as *Daphnia schodleri* (Fig. 1), may even grow well beyond the optimal foraging size.

By living in environments containing vertebrate predators, small species are forced to be suboptimal foragers. As they continue to grow after attaining maturity, they increase their feeding efficiency, but many of them apparently never attain an optimal foraging size (Fig. 1; and Lynch, in press.). This is a form of submergent behavior (Maiorana 1976).

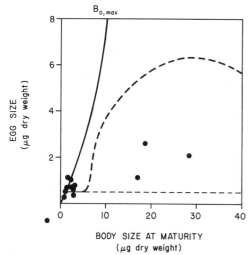

Fig. 7. Egg size vs. body size at maturity for cladocerans compared with theoretical maximum egg size and expected range for predator-free environments. Theoretical maximum egg sizes (represented by solid curve) for different sizes at maturity are determined with Eq. 5, setting $m = 1$, $F_{max} = 20$, and $a = 0.2$. Upper dashed line represents optimal egg sizes for different sizes at maturity in highly enriched environments (as determined in Fig. 1). In very dilute environments, optimal egg size for all sizes at maturity is minimal possible size (see Fig. 6); minimal possible egg size is assumed to be 0.5 µg dry weight and is represented by bottom dashed line (data summarized in Lynch, in press).

Discussion

The general agreement between the model's predictions and observations on life-history variation in cladocerans is very encouraging and suggests that it may be appropriate for examining more complex problems. There is no reason that the model cannot be modified to examine the consequences of continuous vs. curtailed growth following the onset of maturity or to investigate the problems associated with life in a variable or periodic environment. Perhaps of most significance, the theory provides a mechanistic framework for examining competition in age-structured populations.

This paper is exploratory in nature, and the shapes of the functions used for feeding efficiency (F) and probability of survival (p) should be considered speculative and at most approximate. The future application of the theory to examining life-history evolution in the Cladocera will require a more precise knowledge of these size-dependent relationships for populations living in different environments. Despite the fact that investigations of predation on zooplankton are now very numerous (see Hall et al. 1976), $p(B)$ has never been determined for a natural cladoceran population. Because of the logistical difficulties involved in marking and recapturing zooplankton, a direct estimate of this function is exceedingly difficult. This problem may be circumvented by comparing observed size-frequency distributions with those predicted on the basis of in situ growth and reproduction estimates, the difference being attributable to predators (Lynch in prep). The feeding efficiency functions can be directly obtained by measuring the rates at which individuals of different sizes incorporate energy into growth and reproduction; such measurements should also serve as an assay for the nutritional status of the environment.

References

DODSON, S. I. 1974. Zooplankton competition and predation: An experimental test of the size-efficiency hypothesis. Ecology 55:605–613.

GADGIL, M., and W. M. BOSSERT. 1970. Life historical consequences of natural selection. Am. Nat. 104:1–24.

GREEN, J. 1967. The distribution and variation of *Daphnia lumholtzi* (Crustacea: Cladocera) in relation to fish predation in Lake Albert, East Africa. J. Zool. Lond. 151:181–197.

HALL, D. J. 1964. An experimental approach to the dynamics of a natural population of *Daphnia galeata mendotae*. Ecology 45:94–112.

——, S. T. THRELKELD, C. W. BURNS, AND P. H. CROWLEY. 1976. The size-efficiency hypothesis and the size structure of zooplankton communities. Ann. Rev. Ecol. Syst. 7: 177–208.

HEBERT, P. D. N. 1978. The population biology of *Daphnia* (Crustacea, Daphnidae). Biol. Rev. 53:387–426.

KERFOOT, W. C. 1974. Egg-size cycle of a cladoceran. Ecology 55:1259–1270.

——. 1975. The divergence of adjacent populations. Ecology 56:1298–1313.

LEI, C., and H. F. CLIFFORD. 1974. Field and laboratory studies of *Daphnia schodleri* Sars from a winterkill lake of Alberta. Natl. Mus. Can. (Ottawa) Nat. Sci. Publ. Zool. 9:1-53.

LYNCH, M. 1977. Fitness and optimal body size in zooplankton populations. Ecology 58:763-774.

———. 1978. Complex interactions between natural coexploiters—*Daphnia* and *Ceriodaphnia*. Ecology 59:552-564.

———. 1979. Predation, competition, and zooplankton community structure: An experimental study. Limnol. Oceanogr. 24:253-272.

LYNCH, M. in press. The evolution of cladoceran life histories. Quart. Rev. Biol.

MAIORANA, V. C. 1976. Predation, submergent behavior, and tropical diversity. Evol. Theory 1:157-177.

MONTU, M. 1973. Crecimiento y desarrollo de algunas especies de cladóceras dulceacuícolas. I. *Diaphanosoma brachyurum* (Lievin, 1848). Physis (B) 32:51-59.

NILSSON, N., and B. PEJLER. 1973. On the relationship between fish fauna and zooplankton composition in north Swedish lakes. Rep. Inst. Freshwater Res. Drottningholm 53:51-77.

POPE, G. F., and J. C. H. CARTER. 1975. Crustacean plankton communities of the Matamek River system and their variation with predation. J. Fish. Res. Bd. Can. 32:2530-2535.

RICHMAN, S. 1958. The transformation of energy by *Daphnia pulex*. Ecol. Monogr. 28:273-291.

STEARNS, S. C. 1976. Life-history tactics: A review of the ideas. Q. Rev. Biol. 51:3-47.

———. 1977. The evolution of life-history traits: A critique of the theory and a review of the data. Ann. Rev. Ecol. Syst. 8:145-171.

VIJVERBERG, J. 1976. The effect of food quantity and quality on the growth, birth-rate and longevity of *Daphnia hyalina* Leydig. Hydrobiologia 51:99-108.

WARSHAW, S. J. 1972. Effects of alewives (*Alosa pseudoharengus*) on the zooplankton of Lake Wononskopomuc, Connecticut. Limnol. Oceanogr. 17:816-825.

WEGLENSKA, T. 1971. The influence of various concentrations of natural food on the development, fecundity and production of planktonic crustacean filtrators. Ekol. Pol. 19:427-471.

WELLS, L. 1970. Effects of alewife predation on zooplankton populations in Lake Michigan. Limnol. Oceanogr. 15:556-565.

36. Size-Selective Predation on Zooplankton

Barbara E. Taylor

Abstract

The effects of size-selective predation on Leslie matrix models of the life histories of five zooplankton species were investigated. The results suggest that exponential models, commonly used to analyze zooplankton population dynamics, may not describe adequately the effects of selective predation on zooplankton populations. These results are a consequence of various age-specific properties of the life histories of the prey populations. The relationships between the effects of selective predation and particular features of the life histories of the zooplankton are also considered.

A clear understanding of the role predation plays in structuring plankton communities depends on accurate evaluation of the effects of predators on the population dynamics of their prey. While many studies have shown that predation may have a strong effect on plankton communities (e.g. Hrbáček 1962; Brooks and Dodson 1965; Wells 1970; Warshaw 1972), the responses of zooplankton populations to predation have not been analyzed thoroughly.

The analysis of plankton population dynamics is often based on an exponential model of population growth (Edmondson and Winberg 1971). To make accurate predictions of population growth, the various assumptions of the model should be satisfied reasonably well. The extent to which these assumptions may be violated by zooplankton populations and the effects of these violations require further analysis (Edmondson 1974). Zooplankton populations exhibit age structure and variation in age-specific birth and death rates which may represent significant violations of the assumptions of these models.

Exponential models of population growth assume that each individual has an equal probability of dying and that each has an equal probability of reproducing during any period of time. When processes such as selective predation, which act differentially on the various individuals in a population, are considered, a simple exponential model may not be adequate to describe the dynamics of the population.

Using models of five zooplankton life histories and six regimes of size-selective predation, I will consider two questions. How does the life history or, more particularly, the age-specific variation in birth and death rates and size, affect the response of

---

I was supported by a National Science Foundation Graduate Fellowship and Department of Energy grant EV76S062225 to W. T. Edmondson while investigating this problem. The Graduate School of the University of Washington provided funds for computer use. W. T. Edmondson, R. A. Pastorok, T. M. Zaret, and T. W. Schoener contributed in various ways to this project. W. O. Tschumy and M. Slatkin were particularly helpful in discussing problems and criticizing the manuscript.

zooplankton populations to size-selective predation? What features of the life history determine that response?

The models

Life histories of five zooplankton species, three cladocerans and two copepods, were modeled using Leslie matrices (Leslie 1945). A Leslie matrix is a convenient way to describe mortality and fecundity in an age-structured population, and its properties have been thoroughly investigated. The species modeled were chosen principally on the basis of availability of the required data, but appear to be fairly typical of species found in plankton communities.

Life history parameters—Age-specific birth and death rates and sizes were estimated from data in the literature. The data used were from laboratory cultures given ample food and reared at temperatures at or near 20°C. The five species modeled were *Bosmina*, *Daphnia galeata*, *Daphnia pulex*, *Diaptomus*, and *Acanthocyclops*. Size and fecundity of *Bosmina* were estimated from data for *Bosmina longirostris* in Kerfoot (1974), of *D. galeata* from data in Hall (1964), of *D. pulex* from data in Frank et al. (1957) and Richman (1958). *Diaptomus* is a chimera constructed from size and fecundity data for *Diaptomus* (subgenus *Eudiaptomus*) *graciloides* from Weglenska (1971), size data for *Diaptomus* (subgenus *Leptodiaptomus*) *siciloides* from Comita and Tommerdahl (1960), and some of my unpublished measurements of *Diaptomus* (subgenus *Onychodiaptomus*) *hesperus* from Hall Lake, Washington. *Acanthocyclops* is based on growth and fecundity data for *Acanthocyclops viridis* in Smyly (1970), data on size and survivorship for the same species in Coker (1933), and some data in Auvray and Dussart (1966, 1967) on the growth of other cyclopoid species. Because little information about survivorship was found, rectangular (type I) survivorship curves similar to that for *D. pulex* were assumed for the other species. The time step of the matrix was 2.0 days for *Bosmina*, *Diaptomus*, and *Acanthocyclops* and 2.5 days for the two *Daphnia* species. These times were chosen to correspond to the duration of egg development, which is also the duration of the adult instar for the cladocerans. Table 1 summarizes characteristics of the life histories of the model species. Body length is given as total length exclusive of tail spine for the cladocerans and exclusive of caudal setae for the copepods.

Size-selective predation—Size-selective predation by both fish and invertebrate predators was modeled. Six different regimes of predation were tested on each zooplankton model. Preferences by the fish (fish 1 and 2) show definite lower, but not upper,

Table 1. Characteristics of the model zooplankton.

	Bosmina	Daphnia galeata	Daphnia pulex	Diaptomus	Acanthocyclops
Initial size (mm)	0.28	0.42	0.70	0.14	0.15
Size at first reproduction (mm)	0.39	1.08	1.78	1.54	1.75
Maximum size (mm)	0.63	2.26	2.70	1.54	1.75
Age at first reproduction (days)	6	7.5	10	26	36
Life span (days)	14.0	47.5	55.0	44.0	52.0
Size of first clutch (female eggs)	1.57	3.18	7.38	25.00	50.00
Average reproductive output (female eggs) R_0	11.22	269.80	199.26	62.42	279.32
Intrinsic growth rate of increase (per day) r	0.262	0.331	0.294	0.136	0.138
Generation time (days) T	9.23	16.91	18.01	30.40	40.81
Intrinsic growth rate of increase first clutch only (per day) r_1	0.058	0.149	0.195	0.119	0.105
Contribution of subsequent clutches to intrinsic growth rate (per day) $r - r_1$	0.204	0.182	0.099	0.017	0.033

limits of prey size within the range of prey considered. Preferences by the invertebrates (invertebrates 1 through 4) show definite lower and upper size limits. The functions used to calculate mortality due to predation for each of the predators are shown in Fig. 1. Mortality is given as q_x'', the fraction of the cohort alive at age x which dies by age $x + 1$ as a result of predation. The curve for fish 2 is based on an electivity curve for trout calculated from data in Galbraith (1967) by R. A. Pastorok. The curve for invertebrate 4 is based on an electivity curve for fourth instar *Chaoborus trivittatus* larvae from Pastorok (1978). The remaining curves were obtained by translating the original curves along the size axis. The heights of the curves were adjusted so that the maximum probability of death due to predation in one time step was equal to a probability of 0.5 in 2.5 days. Probability of death by predation was assumed to be independent of the probability of death by other causes. For each age class of each model, mortality due to predation was calculated as a function of the size specified for the animal at that age. This value was used to recalculate survivorship,

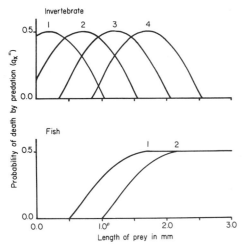

Fig. 1. Prey size preference curves for the predators. Probability of death by predation (q_x'') during a 2.5-day time step plotted against body length of prey. Curves for models with a 2-day time step are similar, except that q_x'' reaches a maximum value of 0.426 instead of 0.500.

$$p_x' = p_x p_x'' = p_x(1 - q_x'')$$

where p_x gives the fraction of the cohort surviving from age x to age $x + 1$, and $p_x = 1-q_x$. The prime designates a parameter including mortality due to both predation and natural of physiological causes; the double prime, a parameter including effects only of mortality due to predation. An unprimed parameter includes the effects only of physiological or natural mortality. The conventions are followed throughout this paper.

Analysis of the models—The population growth rate coefficient, stable age distribution, stable size distribution, and age-specific reproductive value were computed for each case. A total of 35 cases was considered. The behavior of each of the five species was analyzed under seven conditions: no predation, predation by each of the two fish, and predation by each of the four invertebrates. Two of the predators, fish 2 and invertebrate 4, had no effect on *Bosmina* because it was too small to be taken by them. The computations were performed by a CDC 6400 computer (Academic Computing Center, University of Washington). Library routines were used for eigenanalysis: International Mathematical and Statistical Libraries subroutine EIGRF, to calculate eigenvalues; and Boeing subroutine EIGCO1 to calculate eigenvectors. Age-specific reproductive value was obtained from the left principal eigenvector of the Leslie matrix (Pollard 1973).

Results

Some additional parameters describing the life histories of the model zooplankton species are also shown in Table 1. Population growth rate coefficient (r) for the population without predation and population growth rate coefficient based on only the first clutch (r_1) are given. The parameter b_i, where $b_i = r - r_1$, is a measure of the contribution of subsequent clutches to population growth. Expected number of offspring produced by an individual during its lifetime (R_0) and generation time (T) are also given. The copepods have lower population growth rate coefficients and longer generation times than the cladocerans, but the expected number of offspring produced is not always smaller for the copepods. The contribution of subsequent clutches to population growth is much higher for the cladocerans than for the copepods, suggesting that repeated reproduction is more important for the cladocerans.

Age-specific reproductive value (V_x) gives the expectation of future offspring, discounted according to the rate of growth of the population. Age-specific reproductive value, plotted as a function of body size, is shown in Fig. 2. Size at first reproduction

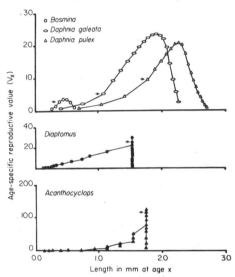

Fig. 2. Age-specific reproductive value as a function of body length. Age-specific reproductive value (V_x) plotted against size of animal at that age. Determinate growth pattern of the copepods results in a series of reproductive value points for maximum size. Arrow indicates reproductive value corresponding to age of first reproduction.

is marked with an arrow. Note that, because of their determinate growth pattern, the copepods show a series of reproductive values for the largest size class. For the copepods, the largest reproductive value occurs just before the age of first reproduction. That the age of largest reproductive value just precedes the age of first reproduction is an artifact of the discrete time formulation of the model. The first clutch is produced at the end of the time step preceding the one which begins with the age of first reproduction. For the cladocerans, reproductive value continues to increase after the age and size of first reproduction.

The difference between the population growth rate coefficients with and without predation ($r - r'$) was used as a measure of the effect of a predator on a prey species. This quantity may be regarded as an effective predation death rate coefficient, although some of the changes in the population growth rate coefficient will be explained by changes in the per capita birth rate rather than directly by changes in the death rate. The decrease in the population growth rate coefficient due to predation is shown in Fig. 3 for each case tested. The smallest prey species, *Bosmina*, is affected most by the predators with the smallest prey preferences. The larger *D. galeata* is affected more by predators with somewhat larger prey preferences. The largest species, *D. pulex*, is affected most by the fish and the invertebrates with the largest prey preferences. The two copepods, although fairly similar in size, are affected very differently by the predators. Predators preferring small prey have the greatest impact on *Diaptomus*, but predators preferring larger prey have the greatest impact on *Acanthocyclops*.

Importance of life history—The importance of the various age-specific properties, or the life histories, of the zooplankton in determining the response to predation was evaluated by examining the relationship between the predation rate and the resulting decrease in the population growth rate for the species. For the exponential model the relationship between these quantities is linear. If y is the decrease in growth rate of the population caused by predation and x is the rate of predation on the population, then

$$y = e^b x, \text{ where}$$
$$x = N_0 \left[e^{-dt} - e^{-(d + d'')t} \right],$$

N_0 is the initial size of the population, b and d are instantaneous birth and death rate coefficients for the population, and t is time.

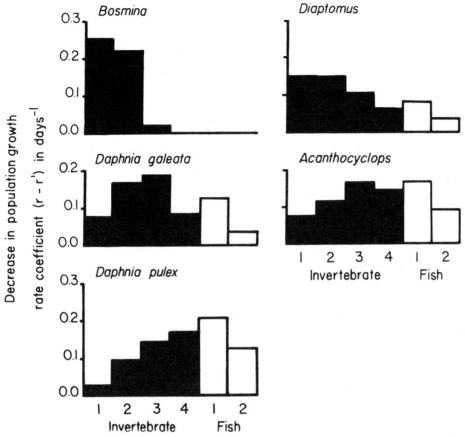

Fig. 3. Effect of predation for each predator and each prey species. Effect of predation given as decrease in population growth rate coefficient $(r - r')$ in units of days^{-1}.

If the age-specific properties of the members of the model populations are not important in determining the response to predation, this relationship will be satisfied. Deviations from this relationship will be caused by variations in e^b, which represents the per capita birth rate of the population.

Decrease in the growth rate of the population caused by predation is shown plotted against rate of predation for each of the model zooplankton species. Predation rate is calculated as the number of animals dying which would not die if predation were not acting on the population. The predation rate in number of animals dying per time step (d_x'') in each age class is

$$d_x'' = n_{x+1,t+1} - n_{x+1,t+1}$$
$$= n_{x,t}(p_x - p_x'),$$

and the total number dying (D'') is

$$D'' = \Sigma d_x''$$

where $n_{x,t}$ is the number of animals in age class x at time t. The stable age distribution for each regime of predation, normalized to give a total population of 1,000, was used to calculate this quantity. The decrease in population growth caused by predation was expressed in comparable units:

$$\Delta N_t'' = \Delta N_t - \Delta N_t' = N_t(e^r - e^{r'})$$

where $N_t = 1,000$ is the total population at time t.

Figure 4. shows that the relationship between decrease in population growth rate and rate of death due to predation by each predator is not linear for any of the model species,

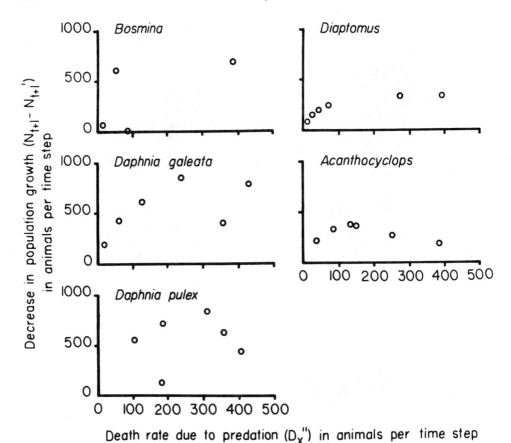

Fig. 4. Decrease in population growth due to predation plotted against predation rate. Both quantities calculated for a period of one time step (2 or 2.5 days) using an initial population size (N_t) of 1,000 animals. Decrease in population growth due to predation is difference between growth of populations without and with predation ($\Delta N_t - \Delta N_t'$). Predation rate is difference between number of individuals dying in populations with and without predation ($D_t - D_t'$).

and for four of the species it is not even strictly increasing. Therefore, simple exponential models will not describe accurately the effects of predation on these populations; their life histories are important.

Predictive value of life history traits— Because life history characteristics are important in determining the responses of a model population to predation, some feature of the life history may be sufficiently important to predict that response. Various life history traits were tested for their predictive values. They included: several indices of prey size, compared with prey preferences and effect of predation; the duration of the period of vulnerability to predation; and several measures of reproductive value of the prey taken by the predator.

Size—The relationships between zooplankton size, prey preference, and impact of predation were considered in two ways. First, given a prey species of a particular size, can the most effective predator be predicted on the basis of its prey size preferences? Second, given a predator with a particular selectivity, can the prey species on which it will have the greatest effect be predicted?

The relationship between two measures of prey size and the size preference of the most effective predator is shown in Fig. 5. The measures of prey size are size at first reproduction and maximum size. For the copepods, which do not grow after attaining

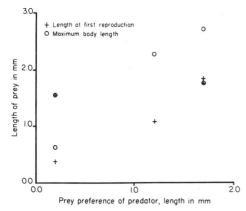

Fig. 5. Two indices of prey size plotted against prey size preference of predator with greatest effect on species. Predator effect is measured by decrease in population growth rate coefficient resulting from predation $(r-r')$. For fish predators, smallest prey size at which q_x'' attains its maximum value is used as preferred prey size.

maturity, these two values are equal. The most effective predator is defined to be the one which causes the greatest decrease in the population growth rate coefficient. Prey size preference is the size of prey for which the predator shows maximum preference or, in the case of the fish, the smallest prey for which it shows maximum preference. When all five prey species are considered, there is a positive relationship between maximum body length of the prey and the prey size of the most effective predator, but no relationship between size at first reproduction and prey size preference of the most effective predator. Both of these relationships become strongly positive when only the cladocerans are considered.

The predator preferring the largest prey has its greatest effect on the largest prey species, and the predator preferring the smallest prey has its greatest effect on the smallest prey species.

Duration of vulnerable period—The vulnerable period is the portion of the life span during which an animal may be taken by a predator. In these models vulnerability is determined by size, and the length of the vulnerable period by the size and growth rate of the animal. The effect of predation is shown plotted against the duration of the vulnerable period in Fig. 6. Duration of the vulnerable period is given in days. Effect of predation is given as the decrease in the population growth rate coefficient. There is no apparent relationship between these two quantities.

Reproductive value—Age-specific reproductive value indicates the relative effect on population growth of removing an individual of that age from the population. The removal of an individual of high reproductive value will have a larger effect than will removal of an individual of low reproductive value. To estimate the reproductive values of the prey taken by a particular predator, a measure of the similarity between size-specific predation rate and reproductive value curves was calculated. This measure is the difference between the sizes giving the maxima of the age-specific predation rate (d_x'') and reproductive value (V_x) curves. Effect of predation is shown plotted against this measure in Fig. 7. Effect of predation is given as the decrease in the population growth rate coefficient. No dependence of the effect of predation on this measure is apparent. The age-specific reproductive value of the size class receiving the maximum rate of predation and the age-specific reproductive value of the preferred prey size also showed no apparent relationship to the effect of predation. In all of these calculations, when a size class contained more than one age class, the reproductive value for the youngest age class was used.

Discussion

These results suggest that various age-specific properties of zooplankton populations are important in determining the responses of the populations to selective predation. If the age composition of the animals taken by a predator is identical to that of the prey population, the relationship between predation rate and the effect of predation on the population will be linear. However, selective predation has been extensively documented for planktivorous predators (e.g. Galbraith 1967; Brooks 1968; Werner and Hall 1974; Swift and Fedorenko 1975; Pastorok 1978). The models show that, when predation is selective, large de-

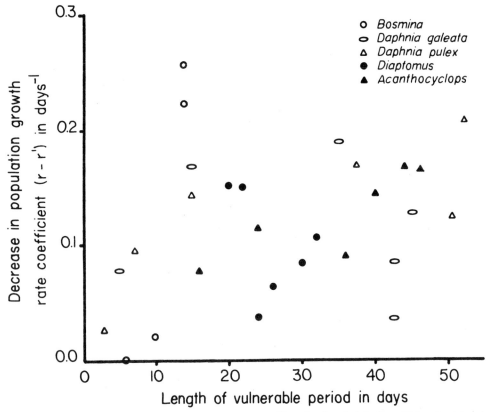

Fig. 6. Effect of predation plotted against a measure of length of period during which animal is vulnerable to predation. Effect of predation measured by decrease in population growth rate coefficient resulting from predation $(r - r')$. Length of vulnerable period given in days. For a given species, each point corresponds to one of six selective predation regimes, given in Fig. 1.

viations from this linear relationship may occur. Thus, simple exponential models, often used to analyze zooplankton population data, may not describe accurately either the effect of a particular level of predation on the growth of the prey population or the predation rate producing a particular decrease in the growth of the prey population. The results were obtained from model populations at a stable age distribution. Deviations from the stable age distribution may complicate further the behavior of real zooplankton populations.

Leslie matrix models with fixed coefficients were used because they provide a convenient format for describing an age-structured population and because the behavior of the population can be obtained analytically. A disadvantage of this type of model is the assumption that the age-specific probabilities of birth and death are constant and, thus, strictly density-independent. Because most or all natural populations are subject eventually to density-dependent processes, as well as to other sorts of environmental variation, these models may give short term, but not long term descriptions of the behavior of natural populations. The analysis assumes that the populations have attained a stable age distribution, a condition which may not be satisfied by natural populations. Because of these assumptions and the problems in estimating parameters for the models, the effects of the various types of predators on the prey species can

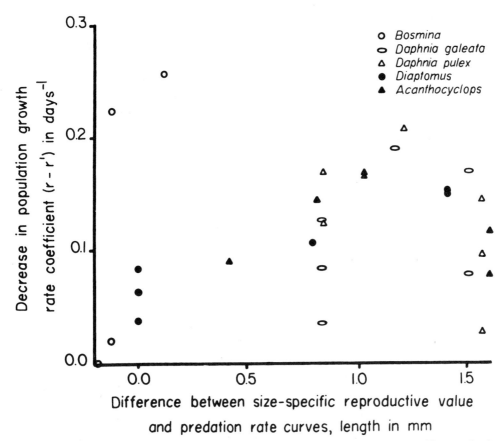

Fig. 7. Effect of predation plotted against a measure of similarity between size-specific reproductive value and predation death rate curves. Effect of predation measured by decrease in population growth rate coefficient resulting from predation $(r - r')$. Similarity between size-specific reproductive value and size-specific predation death rate curves is difference between sizes of animals at ages for which reproductive value and predation death rate reach their maxima. Age-specific reproductive value is V_x; age-specific predation death rate is $d_x'' = d_x' - d_x$. For a given species, each point corresponds to one of six selective predation regimes given in Fig. 1.

most satisfactorily be interpreted as an analysis of the sensitivity of the growth of the prey populations to predation.

Some particular features of the prey life histories provides bases for qualitative prediction of the effects of predation. Measures of size are of some use for ranking the effects of predation on the cladocerans. Only two copepod life histories were modeled, but their behaviors suggest that their responses to predation are less predictable. The unpredictability may result from the greater complexity and variability of their life histories.

None of the more quantitative measures of the effect of predation, predation rate, duration of the vulnerable period, and estimates of the reproductive value of the prey taken, was of much value in predicting the effect of a particular predator. Presumably, all of these factors are important, but their effects are confounding. Consequently, none of them, examined separately, is very useful.

Some qualitative explanations of the various responses are suggested by the reproductive value curves. A comparison of the size-specific reproductive values for the two copepods shows that *Diaptomus* has relatively larger reproductive values in the early stages. These greater reproductive values may explain the greater effect on *Diaptomus* of predators preferring small prey. The larger clutch of *Acanthocyclops* gives it extremely high reproductive values during the adult stage and reduces the relative reproductive value of the small stages. *Acanthocyclops* is much more strongly affected by predation on the larger stages.

Presumably, any process which reduces fecundity, such as competition for food, will alter the shape of the age- or size-specific reproductive value curve and, consequently, the effects of various predators. Reduced fecundity, which increases the relative reproductive value of early stages, would be expected to increase the effect of invertebrate predation on population growth. Other effects of competition for food may include delayed maturation, small size at maturity, and higher natural mortality. Smaller size at maturity may also increase the effect of predators preferring small prey, as may delayed maturation, which will increase the length of the vulnerable period. The effect of higher natural mortality on the reproductive value curve will depend on the age distribution of that mortality; but a uniform increase in mortality over all age classes will have no effect on the reproductive value curve or, consequently, on the response to various predators.

Conclusions

Life history appears to be very important in determining the response of zooplankton populations to predation. Exponential models may often be inadequate to describe accurately the effects of selective predation on natural populations. More accurate analysis of the effects of predation on zooplankton populations and communities will require more detailed life history models of the prey species as well as better descriptions of the behaviors and diets of their predators. Some features of the life histories may be used to make qualitative predictions of the responses of prey populations to predation.

References

AUVRAY C., and G. DUSSART. 1966. Role de quelques facteurs du milieu sur le développement post-embryonnaire des Cyclopides (Crustacés Copépodes). I. Généralites. Cas des *Eucyclops*. Bull. Soc. Zool. Fr. **91**: 477–491.

——, and ——. 1967. Roles de quelques facteurs du milieu sur le développement post-embryonnaire des Cyclopides (Crustacés Copépodes). II. Cas des *Cyclops* et influences des facteurs extérieurs. Bull. Soc. Zool. Fr. **92**: 11–22.

BROOKS, J. L. 1968. The effects of prey size selection by lake planktivores. Syst. Zool. **17**: 272–291.

——, and S. I. DODSON. 1965. Predation, body size, and composition of plankton. Science **150**: 28–35.

COKER, R. E. 1933. Influence of temperature on freshwater copepods (*Cyclops*). Int. Rev. Gesamten Hydrobiol. Hydrogr. **29**: 406–436.

COMITA, G. W., and D. M. TOMMERDAHL. 1960. The postembryonic developmental instars of *Diaptomus siciloides* Lilljeborg. J. Morphol. **107**: 297–355.

EDMONDSON, W. T. 1974. Secondary production. Mitt. Int. Ver. Theor. Angew. Limnol. 20 p. 229–272.

——, and G. G. WINBERG, [Eds.] 1971. Secondary productivity in fresh waters. IBP Handbook 17. Blackwell.

FRANK, P. W., C. D. BOLL, and R. W. KELLY. 1957. Vital statistics of laboratory cultures of *Daphnia pulex* de Geer as related to density. Physiol. Zool. **30**: 290–305.

GALBRAITH, M. G. 1967. Size-selective predation on *Daphnia* by rainbow trout and yellow perch. Trans. Am. Fish. Soc. **96**: 1–10.

HALL, D. J. 1964. An experimental approach to the dynamics of a natural population of *Daphnia galeata mendotae*. Ecology **45**: 94–112.

HRBÁČEK, J. 1962. Species composition and the amount of zooplankton in relation to the fish stock. Rozpr. Cesk. Akad. Ved Rada, Mat. Prir. Ved **72**(10): 116 p.

KERFOOT, W. C. 1974. Egg-size cycle of a cladoceran. Ecology **55**: 1259–1270.

LESLIE, P. H. 1945. On the use of matrices in certain population mathematics. Biometrika **33**: 183–212.

PASTOROK, R. A. 1978. Predation by *Chaoborus* larvae and its impact on the zooplankton community. Ph.D. thesis, Univ. Washington. 238 p.

POLLARD, J. H. 1973. Mathematical models for the growth of human populations. Cambridge Univ. Press.

RICHMAN, S. 1958. The transformation of energy of *Daphnia pulex*. Ecol. Monogr. **28**: 273–291.

SMYLY, W. J. P. 1970. Observation on rate of development, longevity and fecundity of *Acanthocyclops viridis* (Jurine) (Copepoda, Cyclopoida) in relation to type of prey. Crustaceana **18**: 21–35.

SWIFT, M. C., and A. Y. FEDORENKO. 1975. Some aspects of prey capture by *Chaoborus* larvae. Limnol. Oceanogr. **20**: 418–425.

WARSHAW, S. J. 1972. Effects of alewives (*Alosa pseudoharengus*) on the zooplankton of Lake Wononskopomuc, Connecticut. Limnol. Oceanogr. **17**: 816–825.

WEGLENSKA, T. 1971. The influence of various concentrations of natural food on the development, fecundity, and production of planktonic crustacean filtrators. Ecol. Pol. **19**: 428–471.

WELLS, L. 1970. Effects of alewife predation on zooplankton populations in Lake Michigan. Limnol. Oceanogr. **25**: 556–565.

WERNER, E. E., and D. J. HALL. 1974. Optimal foraging and the size selection of prey by the bluegill sunfish (*Lepomis macrochirus*). Ecology **55**: 1042–1052.

37. Some Aspects of Reproductive Variation among Freshwater Zooplankton

J. David Allan and Clyde E. Goulden

Abstract

Small, rapidly-reproducing freshwater zooplankton are excellent subjects for comparative life history studies. They show immense variation in reproductive traits such as generation time, fecundity and timing of reproduction, while small size and (in some taxa) ease of culturing allow controlled experimentation. This review considers demographic variation within and between major zooplankton groups, using information gathered from field and laboratory studies. We discuss the interrelationship of body size and demographic variation in determining the overall life history of an organism, and attempt to relate observed life histories to ecological and geographical distributions.

The small, rapidly reproducing freshwater zooplankton afford excellent opportunities for comparisons of reproductive and life history traits. Generation times range from a few days to one, sometimes two years, fecundity from perhaps a dozen to in excess of 10^3 offspring, and reproduction may be parthenogenetic or sexual, to contrast some obvious characteristics. Zooplankton are subject to extreme seasonal fluctuations of phytoplankton food abundance and composition (Porter 1977), as well as to strong size-selective predation (Brooks and Dodson 1965). The life histories of these organisms almost certainly mirror these seasonal events. In this review we discuss similarities and differences among zooplankton life histories. Ultimately, this information should help to answer two broad questions that motivate our approach. First, what are the observable interspecific differences between demographic variables and body size that result in a given life history pattern? Second, what is the relationship of life history patterns within and between major zooplankton groups and their geographic and ecological distributions?

There have been extensive descriptive studies of the life cycles of individual populations, but only recently has the adaptive significance of life histories been investigated from a theoretical or more general viewpoint (*see* Stearns 1976, 1977). For zooplankton, Heinrich (1962) made one of the first attempts at generalization by comparing the life histories of marine copepods with the pulse of phytoplankton in the oceans. Hutchinson (1967) compared the principal groups of zooplankton in terms of generation time, fecundity, voltinism, and feeding adaptations. He concluded that the zooplankton represent a continuum from small, opportunistic, rapidly reproducing forms (the rotifers); through the similar, but larger and slower growing cladocerans; to the comparatively long-lived copepods which are often annual, and occasionally biennial or even triennial. Allan (1976) drew on more recent studies

Supported by NSF grant DEB 76-20119.

and theoretical developments to extend this comparison.

We here consider demographic variables within and between major zooplankton groups. Over the past decade there has been considerable progress toward elucidating the relationship between body size and predation. But body size not only affects risk of predation, it also bears directly on the reproductive potential of the organism, as Brooks and Dodson (1965) observed and Hall et al. (1976) and Lynch (1977) subsequently discussed.

Our review is selective as dictated by the data available. Unfortunately there have been few demographic studies of copepods in the laboratory because copepods are difficult to culture. However, copepod life cycles are quite amenable to field analysis as the developmental stages are morphologically distinct and each generation appears as a cohort, often with little or no overlap in time. Thus, we have reviewed copepod life histories from data gathered from field studies. There is a substantial body of data on the energetics of growth for marine copepods reared in the laboratory, but those data are not relevant to this discussion. With two exceptions (Wesenberg-Lund 1908; Brooks 1946), cohorts are seldom distinguishable among the Cladocera, but these animals are easily cultured in the laboratory. Differences in demographic characters between populations, such as age of first reproduction, brood size as a function of body size, or brood size as altered by temperature and food concentration can be studied in the Cladocera; whereas we can only estimate these parameters for the copepods from field data. Finally, owing to the relative lack of information on rotifers, we attempt to indicate only their general relationship here.

We thank Linda Hornig for laboratory contributions and S. Dodson, D. Gill, D. Lonsdale, M. Reska, and G. Wyngaard for helpful comments on the manuscript.

Life cycles of calanoid copepods

Our knowledge of the life histories of copepods in freshwater is based almost entirely on studies of field populations. However, their complex life cycles and distinct reproductive pulses make it possible to characterize individual generations more distinctly than in either Cladocera or rotifers. We will review calanoid copepods only; cyclopoid copepod life histories have been discussed by Elgmork et al. (1978), Nilssen (1977), and Smyly (1973).

Calanoid copepod life histories may be characterized as univoltine or multivoltine, the latter consisting of two or more generations per year. In addition copepods are often acyclic, but those multivoltine populations that inhabit water bodies subject to complete anoxia, very low temperature, or drying are often monocyclic (sensu Hutchinson 1967).

Univoltine populations—Univoltine populations typically overwinter as late stage copepodites or as adults, less often as resting eggs. Reproduction begins in the late winter, and eggs are produced before ice melt or as the water begins to warm. The exact stimulant to egg formation and development is unknown. Nauplii hatch and swarm in the surface waters as the spring phytoplankton bloom develops. A typical seasonal pattern for a univoltine population is illustrated by *Mixodiaptomus laciniatus* (Fig. 1; data from Lago Maggiore: Ravera 1954).

In some univoltine populations, eggs may be produced in the autumn and the populations overwinter as nauplii that develop slowly but reach the copepodite stage in spring. This appears to be true for *Limnocalanus macrurus* in Georgian Bay, Lake Huron (Carter 1969), but is not a general pattern for the Great Lakes. Davis (1961) found this species breeding only in January and February in Lake Erie. Ekman (1907) and Gurney (1931) believed that *L. macrurus* only produced resting eggs but Roff (1972) has demonstrated that this is not true for *L. macrurus* populations from Lake Ontario and Resolute Lake, Cornwallis Island (northern Canada). Roff and Carter (1972) have found overwintering adults in Char Lake, Canada. In habitats

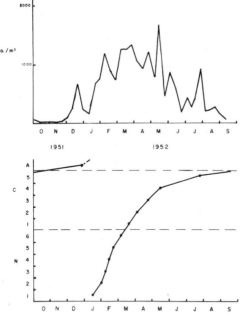

Fig. 1. Typical seasonal pattern for a univoltine copepod population, illustrated by *Mixodiaptomus laciniatus* in Lago Maggiore (data from Ravera 1954). Top panel gives total population size, bottom panel shows seasonal progression through six nauplial and six copepodid stages.

that dry or that are characterized by extreme cold, as is the habitat of *Limnocalanus johanseni* in Imikpuk, Alaska (Comita 1956), and *Hesperodiaptomus shoshone* at 3,400 m in Colorado (Maly 1973), adults may produce resting eggs in autumn and then die.

There are several marine copepods with life cycles similar to *L. macrurus*. These can be exemplified by *Calanus plumchrus* (Fulton 1973), which overwinters as a copepodite V or an adult and which begins breeding in January and February. At this time the adults are in deep oceanic waters; the eggs (which are buoyant) and the nauplii rise toward the surface water, arriving early in the spring pulse of phytoplankton. The nauplii grow rapidly and pass to copepodite IV while feeding on the temporarily abundant algae. The animals also accumulate large fat reserves. In June and July the late copepodites cease feeding and migrate to the deep water. They do not feed through the rest of the year but subsist on their fat reserves. *Calanus hyperboreus* (Conover and Corner 1968) has a similar life cycle but apparently does feed at a low rate as an adult. In most instances these adults die after breeding.

All univoltine populations studied (Table 1) have short breeding periods and do not have overlapping generations. This is in essence a semelparous rather than an iteroparous reproduction pattern (Cole 1954).

Multivoltine species—Most calanoid copepods probably have the potential to be multivoltine since all major taxonomic families of calanoids have multivoltine species (Table 2). In the Centropagidae, *Boeckella* species appear to be multivoltine (Bayly 1962). *Calanus finmarchicus* in Loch Striven (Marshall and Orr 1955) and elsewhere south of the Arctic Circle is bivoltine (Zenkevitch 1963). Most estuarine species, including *Acartia clausi* and *Acartia tonsa* (Deevey 1948; Conover 1956) and many diaptomids are multivoltine.

These populations generally have two or three generations per year. Some have four while the largest recorded number in freshwater habitats is five, but in marine species six or seven generations may be identified. In subtropical and tropical populations extreme multivoltinism may lead to complete overlap of generations (Heinrich 1962; Wyngaard pers. comm.).

As in univoltine populations, multivoltine populations also overwinter as late stage copepodite or adults. This is illustrated by *Leptodiaptomus minutus* (Fig. 2), which has one or more summer generations, the last of which forms the winter population. *Epischura nordenskiöldi* in Newfoundland (Davis 1972), a primarily predatory species (Strickler and Twombly 1975; Davis 1976) and *Epischura baikalensis* in Lake Baikal, primarily a herbivore (Kozhova 1956; Kozhov 1963), have two generations per year and have similar seasonal patterns to *L. minutus*. In Lago Maggiore, Italy, Ravera (1954) found new generations of *Eudiaptomus vulgaris* beginning in Jan-

Table 1. Species of calanoid copepods reported to be univoltine.

Centropagidae	
Limnocalanus johanseni	Imikpuk Lake, Alaska[1]
L. macrurus	Ekoln, Sweden[2]; Georgian Bay[3] and Cornwallis Island[4], Canada; Green Lake, Wisconsin[4]; Lake Erie, USA[5]
Calanidae	
Calanus finmarchicus	East Greenland[6]
Calanus plumchrus	Strait of Georgia, Canada[7]
Pseudocalanidae	
Senecella calanoides	Georgian Bay, Canada[3]
Temoridae	
Heterocope borealis	Bodensee, Germany[8]
H. saliens	Feldsee, Germany[9]; Lago di Monscera, Italy[10]
Diaptomidae	
Acanthodiaptomus denticornis	Lago di Monscera, Italy[10]
Arctodiaptomus bacillifer	Lago di Monscera, Italy[10]
Eudiaptomus graciloides	Esromsø, Denmark[11]
Leptodiaptomus ashlandi	Lake Washington, USA[12]
Mixodiaptomus laciniatus	Lago Maggiore, Italy[13]; Titisee, Germany[9]

1—Comita 1956; 2—Ekman 1907; 3—Carter 1969; 4—Roff and Carter 1972; 5—Marsh 1898; 6—Marshall and Orr 1955; 7—Fulton 1973; 8—Elster 1954; 9—Eichhorn 1957; 10—Bossone and Tonolli 1954; 11—Wesenberg-Lund 1904; 12—Comita and Anderson 1959; 13—Ravera 1954

uary, May, and August. Kuntze and von Klein studying *Eudiaptomus gracilis* populations in Schleinsee, Germany, in successive years found three or four generations each year. In Schleinsee and in Llyn Tegid in North Wales (Thomas 1961), this species reproduced in early spring (March–April), in late spring (May–June) and in autumn (October–November).

In Severson Lake, Minnesota, *Leptodiaptomus siciloides* has no winter generation (Comita 1972). The first generation hatches from winter resting eggs in April and May. Four subsequent summer generations develop in rapid succession, one each in June, July, August, and September. The last generation produces resting eggs and the adults then die. Severson Lake may become completely anoxic in winter (Schindler and Comita 1972) so that the winter resting egg would represent a means for *L. siciloides* to avoid extinction. In the marine species *Pseudocalanus minutus* in Loch Striven, the overwintering generation is followed by at least six summer generations (McLaren 1979), and *A. tonsa* has seven discrete summer generations off the Oregon coast (Miller et al. 1977).

Multivoltine species other than *L. siciloides* produce resting eggs and thus have monocyclic (sensu Hutchinson 1967) life cycles. *Acanthodiaptomus denticornis* from Germany (Eichhorn 1957) and *Skistodiaptomus oregonensis* in Teapot Lake in Ontario, Canada (Cooley 1971) are monocyclic. Teapot Lake is a small meromictic lake. The mixolimnion may become anoxic during the fall overturn, killing all zooplankton (Cooley 1971; Rigler and Cooley 1974). Many species that inhabit temporary water bodies, such as *Aglaodiaptomus clavipes* (Cole 1961) and *A. stagnalis* (Brewer 1964) may also be monocyclic but too little is known of the annual life cycles to be certain. *Acartia californiensis* in Oregon (Johnson unpubl.) and other marine cope-

Table 2. Species of calanoid copepods reported to be multivoltine

Centropagidae	
Limnocalanus grimaldii	Caspian Sea, USSR[1]
Boeckella propinqua	Mayor Island, New Zealand[2]
Calanidae	
Calanus finmarchicus	Loch Striven, Scotland[3]
Pseudocalanidae	
Pseudocalanus cf. *minutus*	Ogac Lake, Canada[4]
Temoridae	
Epischura baikalensis	Lake Baikal, USSR[5]
Epischura nordenskiöldi	Hogans Pond, Newfoundland[6]
Diaptomidae	
Acanthodiaptomus denticornis	Titisee, Germany[7]
Eudiaptomus gracilis	Schleinsee[8,9] and Bodensee, Germany[10]; Llyn Tegid, North Wales[11]; Greifensee and Vierwaldstättersee, Switzerland[12]
Eudiaptomus graciloides	Lake Erken, Sweden[13]
Eudiaptomus vulgaris	Lago Maggiore, Italy[14]
Leptodiaptomus ashlandi	Lake Erie[15] and Lake Michigan[16], USA
Leptodiaptomus minutus	Hogan's Pond, Newfoundland[6]; Lake Nipissing, Canada[17]
Leptodiaptomus sicilis	Lake Lenore, USA[18]
Leptodiaptomus siciloides	Lake Erie, USA[15]; Severnson Lake, Minnesota[19]
Acartiidae	
Acartia clausi	Tisbury Great Pond, Massachusetts[20]; Long Island Sound, USA[21]
Acartia tonsa	Tisbury Great Pond, Massachusetts[20]; Long Island Sound, USA[20]

1—Zenkevitch 1963; 2—Bayly 1962; 3—Marshall and Orr 1955; 4—McLaren 1965; 5—Kozhov 1963; 6—Davis 1972, 1976; 7—Eichhorn 1957; 8—Kuntze 1938; 9—von Klein 1938; 10—Elster 1954; 11—Thomas 1961; 12—Mittelholzer 1970; 13—Naumann cited in Hutchinson 1967; 14—Ravera 1954; 15—Davis 1961; 16—Wells 1960; 17—Langford 1938; 18—Edmondson et al. 1962; 19—Comita 1972; 20—Deevey 1948; 21—Conover 1956

pods that produce resting eggs (Hirota and Uno 1977) may also be monocyclic.

Seasonal abundance patterns—Calanoid copepods that are univoltine are most abundant in late winter and early spring (Fig. 1). In summer, late stage copepodites or adults retire to the cold hypolimnion and remain until the next breeding period. In *L. macrurus* the adults continue to feed and have a diel migration (Marsh 1898; Carter 1969), but in some *Calanus* species, including *C. plumchrus* (Fulton 1973), the copepodite V and adults remain at considerable depths and do not feed, apparently using the fat or wax stored in the oil sac.

Predaceous species such as *Heterocope borealis* (Elster 1954) and *Heterocope saliens* (Eichhorn 1957; Bossone and Tonolli 1954) are active predators in the epilimnion during July and August. The predator *E. nordenskiöldi* is bivoltine but is also

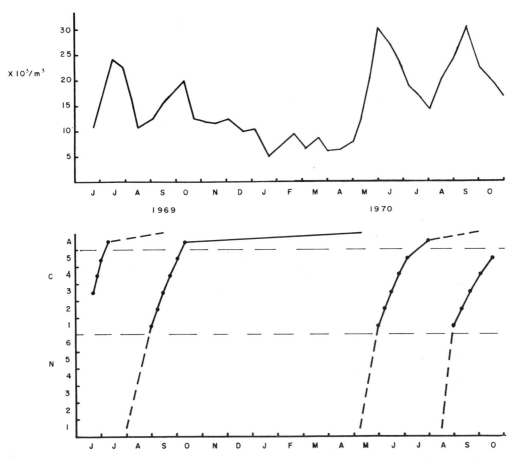

Fig. 2. Typical seasonal pattern for a multivoltine copepod population, illustrated by *Leptodiaptomus minutus* (data from Davis 1972). Top panel gives total population size, bottom panel shows seasonal progression through stages. This study observed one summer and one overwintering generation, but in some multivoltine species additional summer generations are observed (*see text*).

most active in midsummer (Davis 1972, 1976; Strickler and Twombly 1975).

Multivoltine populations also have peak abundance in spring as a result of large nauplii population produced by overwintering adults (Fig. 2). These populations decline rapidly but are replaced by one or more summer generations. When they co-occur, multivoltine species are more abundant than univoltine species (Ravera 1954; Davis 1961; Carter 1969).

Development times for calanoid copepods—Developmental rate of copepods is dependent on water temperature (Marshall and Orr 1955; McLaren 1974) but is not always a monotonic function. Geiling and Campbell (1972) found a plateau of developmental rates between 15° and 20°C in *Skistodiaptomus pallidus*. Landry (1975) did not see a similar plateau in *A. tonsa*.

Development times estimated from field studies of univoltine populations are given in Table 3 and for multivoltine populations in Table 4. The problems of estimating development times for copepods has been discussed by Rigler and Cooley (1974). We believe our use of estimates from the literature is not in serious error because our concerns are restricted to a comparison of relative rather than absolute development times. Development times do not

Table 3. Duration of developmental stages of some univoltine calanoid copepods, in days.

	Limnocalanus macrurus*	Limnocalanus johanseni†	Mixodiaptomus laciniatus‡	Heterocope borealis‡
NI	20		1.05	0.8
NII	22	≈4	1.65	2.6
NIII	14		2.95	2.9
NIV	15	2	3.55	3.0
NV	11	3.5	4.75	3.2
NVI	15	5	5.45	3.4
CI	11	3.5	7.95	3.8
CII	19	3	11.85	5.0
CIII	21	3	22.95	9.0
CIV	20	5	47.65	15.0
CV	50	16	99.55	18.6
Adult lifespan	ca. 65	ca. 120	ca. 150	ca. 60
(Body length)	(2–3 mm)	(2–3 mm)	(1–2 mm)	(3–5 mm)
NI–CV	218	45	210	67

*Carter 1969,   †Comita 1956   ‡Rigler and Cooley 1974.

Table 4. Duration of developmental stages of some multivoltine calanoid copepods, in days.

	Skistodiaptomus oregonensis*	Eudiaptomus gracilis†	Leptodiaptomus siciloides‡
NI	1.15		0.50
NII	1.77		0.76
NIII	1.92		1.22
NIV	2.23	17	1.05
NV	2.45		0.90
NVI	2.96		0.58
CI	3.44		2.07
CII	4.56		1.65
CIII	8.01	31	1.80
CIV	14.25		2.23
CV	16.71		3.36
Adult lifespan	—	—	13.50
(Body length)	(1.25–1.5 mm)		(1–1.3 mm)
NI–CV	60	48	16

*Rigler and Cooley 1974.   †von Klein 1938.   ‡Comita 1972.

appear to be related to body size, but are related to temperature and to the number of generations a population has per year. The late copepodite stages have the most variable development times. However, some of this variation must be environmentally induced. Laboratory cultures under constant conditions would result both in more rapid development and in more similar rates of passage through each stage (Miller et al. 1977; McLaren 1979). This could hardly be the case for many univoltine copepods, however, because they remain inactive at the same seasons that sympatric multivoltine species populations are actively growing.

The dependence of development time on temperature may be observed in the rates given for *L. macrurus* from Georgian Bay, Lake Huron (Carter 1969). This population produces eggs and nauplii in December. *Limnocalanus johanseni* nauplii hatch from resting eggs in July prior to complete ice melt in Imikpuk Lake, Alaska. Its development time is more comparable to naupliar growth rates of *M. laciniatus* and very similar to times of multivoltine copepods. *Mixodiaptomus laciniatus* rapidly passes through naupliar development and the first copepodite stage. Late copepodite stages develop very slowly. In Lago Maggiore, *M. laciniatus* and *E. vulgaris* both have generations beginning in January. Their developments are almost identical until the second copepodite stage. *Eudiaptomus vulgaris* matures by June while *M. laciniatus* is not mature until the following January. In *C. finmarchicus* which may be either univoltine or multivoltine, summer generations normally require about 30 days to reach the adult stage. In the arctic, and in winter generations, the developmental rate of stage IV and V copepodites is slowed considerably.

Spring and summer generations of multivoltine species require from 16 to 60 days to attain the adult stage. As expected, species with the most generations per year develop most rapidly. In Severson Lake, *L. siciloides* will reach the adult stage from nauplius I in 16 days during May and June (Comita 1972). *Eudiaptomus gracilis* and *S. oregonensis* with three generations per year develop in 48 and 60 days, respectively.

The length of life of adults of univoltine and multivoltine species also varies a great deal. Adults of univoltine populations that overwinter may live up to 200 days and die at the end of a brief reproductive period. If the population produces resting eggs the adults may live only a comparatively short time as in *L. johanseni*.

Adults of multivoltine populations may live 3–6 months in the winter, but spring and summer adults are short lived, surviving only a few weeks. The lifespan of adults appears to be inversely related to the number of summer generations. In a three generation per year population such as *E. vulgaris* in Lago Maggiore, summer adults may live 1 to at most 2 months (*see* Tonolli 1961: fig. 7). In Severson Lake, Comita (1972) estimated that adults of *L. siciloides* live only 11 to 17 days.

Conclusion—In contrast to Cladocera and rotifers, calanoid copepods have long maturation times. Though their lifetime fecundity may be high (Allan 1976), this cannot offset the effect of slow development in reducing the growth rate of the population. Copepods have a comparatively brief period of reproduction which, although it may involve several successive broods, marks the end of the lifespan. Generations follow one another in succession, without overlap, at least at temperate and high latitudes. This is in essence a semelparous reproductive pattern. Thus, the life history tactics of calanoid copepods are perhaps even more distinct from those of Cladocera and rotifers than indicated by Allan (1976).

Variation in reproductive traits

Theoretical work has indicated how a number of life history or demographic traits can influence the reproductive output of individuals and the per capita growth rate of populations (Cole 1954; Lewontin 1965; Stearns 1976). Among the most important of these are embryo or egg developmental rate, postembryonic maturation time, brood size (or number of eggs carried by a female at one time), and fecundity. Adult body size provides a convenient perspective from which to view this life history variation.

Maturation time—Maturation time, as Cole and Lewontin stressed, is extremely important to per capita growth rate. It is positively correlated with body size in the planktonic Cladocera: small species such as *Bosmina longirostris*, *Ceriodaphnia* spp.,

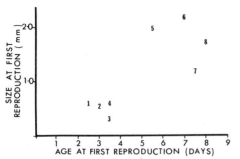

Fig. 3. A positive relationship exists between age and size of first reproduction. 1—*Moina micrura* (Murugan 1975); 2—*Scapholeberis kingi* (Murugan and Sivaramakrishnan 1976); 3—*Bosmina longirostris* (Kerfoot 1974); 4—*Ceriodaphnia reticulata* (Hall et al. 1970); 5—*Simocephalus acutirostratus* (Murugan and Sikaramakrishnan 1973); 6—*Daphnia magna* (Anderson 1932); 7—*D. galeata mendotae* (Hall 1964); 8—*D. pulex* (Richman 1958).

and *Moina micrura* become reproductive at an earlier age than do large species such as *Daphnia galeata mendotae* or *Daphnia pulex* (Fig. 3). These data point to the general trend but probably do not give the exact relationship, as various species included in the graph were raised at different temperatures. Egg developmental rates have been studied in detail by McLaren (1965) and Bottrell (1975) and will not be considered here. At least for the Cladocera which have been studied, embryonic maturation times are much more similar than postembryonic maturation times.

Brood size—Although temperature and food concentration have an impact on brood size in the zooplankton, there is an underlying relationship with body size that has a profound effect on per capita growth. The relationship between body size and brood size was first pointed out for copepods by Margalef (1953) and Ravera and Tonolli (1956), and for Cladocera by Green (1956). This can be seen in Fig. 4 for *Arctodiaptomus bacillifer* (data from Ravera and Tonolli 1956) and Fig. 5A and B for *B. longirostris* and *Daphnia longispina* (data from Green 1956).

In laboratory cultures of Cladocera (Fig. 6), the number of eggs produced per

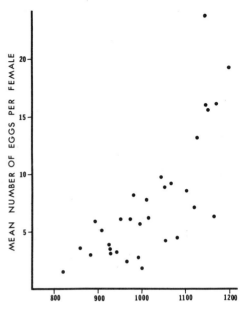

Fig. 4. A positive relationship exists between brood size and body size for calanoid copepods. Example shown is *Arctodiaptomus bacillifer* collected from the field (data of Ravera and Tonolli 1956).

brood clearly increases with age, levels off, and then declines. Since growth rate also decreases with age, size increments become smaller with each successive instar (Anderson 1932; Hall 1964; Richman 1958), presumably because more energy is channeled into egg production. Since body size must place an upper limit on clutch size, assuming adequate food concentration, it seems reasonable to conclude that the leveling off of brood size with age in the Cladocera is due to the slowing of growth.

However, in nature one rarely observes a decrease in clutch size at larger body sizes, as is implied by Fig. 6. If we assume that laboratory results properly describe the fundamental relationship between age and brood size, then we must ask, with Green (1956), why field data do not demonstrate decreased brood size in older, larger individuals (Fig. 5). Green suggested that in nature, adults rarely live to senescence. Given the characteristics of size-selective

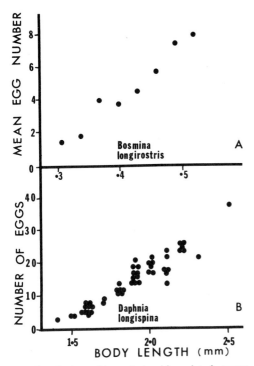

Fig. 5. A positive relationship exists between brood size and body size for cladocerans collected from field. Examples shown are (A) small *Bosmina longirostris* and (B) larger *Daphnia longispina* (data of Green 1956).

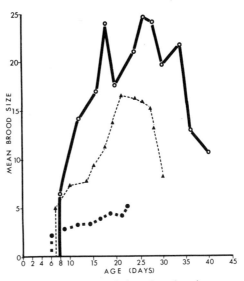

Fig. 6. Mean brood size plotted against mean age at which brood is hatched, for a large (*Daphnia pulex*, solid line), intermediate (*D. ambigua*, dashed line), and small (*Bosmina longirostris*, broken line) cladoceran. In addition to obvious differences in brood size, note that *B. longirostris* produces four broods in about the time that *D. ambigua* produces three, and *B. longirostris* produces five broods in about the time that *D. pulex* produces three. All at 20°C. *Daphnia pulex* from Richman (1958); *D. ambigua* and *B. longirostris* data from Goulden and Hornig (unpubl.).

predation by fish, we can easily support this conclusion. However, we also wish to raise a more fundamental concern regarding the use of field data. Because of the continuous nature of cladoceran reproduction, a plankton sample collected on one date may include representatives of at least three different overlapping generations. Individuals of each generation would have hatched and developed under different food and temperature regimes. Thus we caution against the use of field-derived data such as Fig. 4 for estimating the exact quantitative relationship between brood size and body size. Instead we suggest that unless it can be demonstrated that field data are based on individuals of a single generation, laboratory results should be used to quantify the relationship between brood size, body size, and age.

A positive correlation between mean maximum brood size and size at maturity appears to hold across species (Fig. 7) as well as within a species (Fig. 6). At one extreme *B. longirostris* may become reproductive at <300 μm, and rarely produce a clutch larger than six eggs. At the other extreme lie *D. pulex*, *Daphnia magna*, and *Simocephalus acutirostratus*, which mature at 1.7 to 2.1 mm long and routinely produce clutches of 22 to 27.

Egg size—It is at least plausible that an evolutionary choice can be made between producing many, small eggs or a few, large eggs. Burgis (1967) points out that *Ceriodaphnia pulchella*, compared to *Ceriodaphnia reticulata* of the same size, produces a lower total egg volume on the average, but more eggs. Clearly the reproductive output of *C. pulchella* is packaged in more, smaller units compared to its congener.

The positive relationship between body

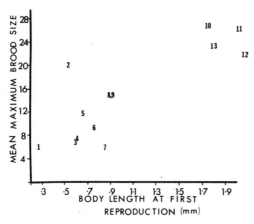

Fig. 7. Cladoceran species which become mature at a large size tend to achieve larger maximum brood sizes. Mean maximum brood, rather than single largest observation, was used. 1—*Bosmina longirostris* (Kerfoot 1974); 2—*Scapholeberis kingi* (Murugan and Sivaramakrishnan 1976); 3—*Moina micrura* (Murugan 1975); 4—*Ceriodaphnia pulchella* (Burgis 1967); 5—*C. quadrangula* (Allan unpubl.); 6—*C. reticulata* (Burgis 1967); 7—*C. megalops* (Burgis 1967); 8—*Daphnia ambigua* (Winner and Farrell 1976); 9—*D. parvula* (Winner and Farrell 1976); 10—*D. pulex* (Richman 1958; Winner and Farrell 1976); 11—*Simocephalus acutirostratus* (Murugan and Sivaramakrishnan 1973); 12—*D. magna* (Winner and Farrell 1976); 13—*D. schødleri* (Lei and Clifford 1974).

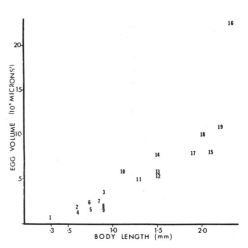

Fig. 8. A positive relationship exists between volume of a single parthenogenetic egg and size of a primaparous adult of that species. Where egg volume given as a range, midpoint of that range was used. 1—*Bosmina longirostris*, 2—*Ceriodaphnia quadrangula*, 3—*Daphnia ambigua* and *parvula* (Kerfoot 1974); 4—*Ceriodaphnia pulchella*, 5—*C. reticulata*, 6—*C. laticaudata*, 7—*C. megalops* (Burgis 1967); 8—*D. ambigua*, 9—*D. cucullata*, 10—*D. hyalina* var. *galeata*, 11—*D. curvirostris*, 12—*D. longispina*, 13—*D. obtusa*, 14—*D. pulex*, 15—*D. thomsoni*, 16—*D. magna* (Green 1956); 17—*Simocephalus vetulus*, 18—*Sida crystallina*, 19—*Eurycercus lamellatus* (Green 1966).

size and mean maximum brood size might imply a relatively constant egg size across species; in fact the volume of a single egg varies by more than an order of magnitude within the Cladocera (Green 1956). Egg volume correlates positively with the body length of the maternal female within a species (Green 1956; Kerfoot 1974), as well as with size of primaparous adult between species (Fig. 8; Green 1956, 1966; Kerfoot 1974). Egg size also correlates positively with maternal nutrition and inversely with temperature (Green 1966). Latitudinal differences exist with larger eggs found at higher latitudes; Green (1966) suggests that since larger eggs result in larger neonates, which in turn require fewer molts to become reproductive, larger eggs at higher latitudes may be an adaptation to compensate for the longer development time associated with colder climates.

Two additional factors affecting egg size deserve mention because they are both speculative and intriguing. Green (1966) suggested that lake inhabitants produce larger eggs than do pond inhabitants of the same species, as an adaptation to the lower food conditions of lakes as compared to ponds. Kerfoot (1974) argued that egg size varies seasonally as a response to size-selective predation pressure on neonates and adults. However, it is possible that temperature would produce the cycle Kerfoot observed by purely physiological mechanisms. That is, colder temperatures produce larger individuals. These larger individuals produce larger eggs to ensure that the growth interval from neonate to primiparous adult will stay constant—not to lessen invertebrate predation on neonates.

In comparing Figs. 7 and 8, it seems

that smaller cladocerans produce smaller eggs, quite likely owing to the limits of space in the brood chamber and to the fact that the growth interval to maturity would stay roughly constant. This allows smaller species to carry larger broods than would be possible if egg volume were constant throughout the order. However, the reduction in egg volume is not sufficient to obscure a positive correlation between body size and number of eggs carried. Quite possibly the evolutionary barrier to further reduction in egg size is the increased maturation time it would impose.

Curiously, one could argue that large species have reduced egg size for gains in fecundity, more so than have small species. The obvious cost of large size is the increased maturation time required to achieve it. Species which mature at small size also mature earlier (Fig. 3), and this compensates substantially for their relatively tiny egg output. Presumably, larger species could shorten their maturation time by producing fewer, larger eggs, but this does not occur.

Major differences between copepods and cladocerans

Iteroparity vs. semelparity—Cladocera and rotifers have the traits associated with high reproductive potential and rapid population growth that are collectively termed *r*-selected characters (Gill 1974). They mature rapidly, have a short generation time, and are iteroparous: females lay numerous successive clutches of eggs. Relative to the juvenile period, adults are long-lived with up to 80% of an individual's life spent in the reproductive stage. The result of a short maturation time coupled with iteroparous reproduction is a high reproductive potential and several successive generations which overlap. Overlapping generations are evident in both laboratory and field populations (Wesenberg-Lund 1908; Brooks 1946).

Copepods do not have the characteristics of *r*-selected populations. In contrast to cladocerans and rotifers, copepods are more semelparous (but not strictly so, as an individual often produces more than one brood). They mature much more slowly. Except for populations which overwinter as adults, most of the lifespan is spent as a juvenile and reproduction is usually for a comparatively brief period. Adult animals typically do not persist beyond the breeding period which produces the next brood (Thomas 1961; Comita 1972; Davis 1972). As a consequence, reproductive potential is low and populations in the field typically exhibit a cohort structure.

Tradeoffs among demographic variables— Because data from the laboratory concerning copepods are few, inferences must be drawn from field studies where environmental variation complicates the picture. Nevertheless a few comments are possible. Cladocerans and copepods differ in two important ways which may affect brood size-body size relationships. Adult copepods do not grow, whereas adult cladocerans continue to molt and increase their body size. In adult copepods all energy intake in excess of maintenance needs is channeled into egg production. Thus brood size-body size comparisons in copepods (Fig. 4) are between individuals only, whereas in cladocerans the relationship also holds within an individual over its lifetime (Fig. 6). A second major difference is that, in cladocerans, eggs are carried within the carapace; in copepods, eggs are carried external to the body. As a result, total brood size may be more strongly limited in the Cladocera than in the Copepoda, due to the volume of the brood chamber.

This comparison leads to a speculative interpretation of the advantage of indeterminant growth in the Cladocera. Because they carry their eggs internally, continued growth as adults allows continued increase in brood size. Presumably there is no basis for adults to channel any energy intake in excess of maintenance needs into further growth, unless a long-term reproductive benefit is achieved. The strong positive correlation between brood size and body size (Figs. 4 and 5) suggests a substantial long-term reproductive benefit of continued adult growth in the Cladocera; this would be especially true if the organism commonly experienced high food levels. An observation consistent with this argument is that the smallest cladocerans, such as *Bosmina*, which typically are most restricted in brood size (Fig. 7), show the greatest amount of growth after reaching reproductive maturity. However, it is also true that larger copepods produce more eggs (Marshall and Orr 1955). The gain in per capita growth rate obtained by increasing egg production is lessened as maturation times lengthen (*see below*), and this would favor indeterminant growth in cladocerans compared to copepods. It may also be relevant that cladocerans tend to be associated with more food-rich environments. In any event, all food intake by adult copepods, in excess of maintenance needs, is channeled into reproduction.

Another point of contrast concerns the relationship between body size and maturation time (Fig. 3). The positive relationship observed in cladocerans does not appear to hold for copepods. Tables 3 and 4 provide no evidence that larger species of copepods have longer development times. McLaren (1979) found that the longest development times occurred in the smallest two of the six marine species examined by Marshall in Loch Striven.

Sexual vs. parthenogenetic reproduction—Perhaps the most striking reproductive variable in zooplankton is the distribution of sex. In the copepods, only sexual reproduction is known. Among Cladocera, parthenogenetic reproduction is typical, with numerous generations produced in the absence of males and one, often synchronous, generation produced by sexual reproduction. Only *Daphnia middendorffiana* is known to forego sexual reproduction and produce resting eggs without fertilization, at the northern extent of its range (D. Frey unpubl.). In the monogonont rotifers, which includes all planktonic forms, reproduction is also facultatively parthenogenetic. Thus we find a clear contrast between copepods on the one hand and rotifers and cladocerans on the other in their mode of reproduction. This raises two questions. What are the advantages of each? Why do we not find parthenogenesis in copepods or obligate sexuality in cladocerans and planktonic rotifers?

Parthenogenesis seems to offer three advantages and one disadvantage. Individuals do not need to locate mates, lifetime reproduction is doubled, and successful genetic combinations are retained. However, new genetic combinations will not be as readily forthcoming.

Zooplankton populations may be quite rare for much of the year. Under these conditions, one might imagine that an advantage would accrue to individuals who needn't find mates to reproduce. This probably is an advantage to cladocerans and rotifers at the very beginning of a bloom. How much of a disadvantage to copepods is the necessity of finding mates? Sex pheromones are known in several marine species (Katona 1973; Griffiths and Frost 1976) and may occur in additional species as well. Furthermore, as discussed earlier, copepods typically show a cohort synchrony not discernible in populations of cladocerans and rotifers. At least in seasonally temperate environments, synchronous maturation of adults plus the release of sex pheromones must lessen the risk of not locating a mate. In temperate climate populations of cladocerans, sexual reproduction typically coincides with high population density, easing mate location.

A population in which all offspring are female will increase at the same rate as a population with twice the brood size and a 1:1 sex ratio. In other words, lifetime reproductive output, not intrinsic rate of increase, is doubled. The maximum reproductive benefit of doubling reproduction can be illustrated using the approximation formula

$$r \cong \frac{\log_e R_o}{T} ;$$

$R_o = \Sigma 1_x \times m_x$ = net reproductive rate: average number of female offspring produced by each female during her lifetime;

$T = \Sigma x \times 1_x \times m_x / \Sigma 1_x \times m_x$ = the mean generation time, which is average of time period over which progeny are produced;

1_x = probability of surviving to age x;
m_x = number of female offspring born to a female aged x during time interval $x \to x + 1$.

Doubling m_x has little effect on T, since m_x appears in both the numerator and the denominator. However, doubling m_x doubles R_o. Thus r increases by $(\log_e 2)/T$. As generation time lengthens, doubling R_o results in smaller and smaller gains in the intrinsic rate of increase (*see* Southwood 1976 for further discussion). Using the approximation of Allan (1976), one can determine that r_{max} at 20°C is 30% larger in rotifers and 11% larger in cladocerans than it would be if half the offspring were males. The smaller gain for cladocerans, relative to rotifers, is due to their longer generation time. Provided the survival of offspring is not reduced as a side effect of asexual reproduction, gains in r_{max} of 10–30% represent a substantial advantage.

The third advantage and one disadvantage center around the genetic implications of parthenogenetic reproduction. In cladocerans and in the monogonont rotifers, asexual reproduction is apomictic. The important consequence is that offspring are genetic replicas of their mothers. While it is often pointed out that sexual reproduction provides the benefit of (indeed, evolved for) new genetic combinations that may be more fit, especially if the environment is changing, others (Smith 1971; Williams 1975) have recently emphasized the cost of sexual reproduction. That is, once an individual has survived to reproductive maturity, thus conveying positive fitness at least to its parents, sexual recombination destroys that unique genotype and may lead most frequently to loss of fitness. Depending upon how closely the environment occupied by the offspring resembles that experienced by the parent, exact genetic duplication by apomictic parthenogenesis will be advantageous. Organisms which have shorter generation times will more closely meet this condition of environmental autocorrelation between parent and offspring. Thus it is not surprising that cladocerans and rotifers show parthenogenesis in contrast to copepods with their longer generation times, and it is also not surprising that when parthenogenesis becomes obligate it is in some rotifers.

Hebert's (1974) electrophoretic analysis of *D. magna* populations provides the most convincing evidence for this line of reasoning, as it clearly establishes that daughters are exact genetic replicas of mothers. During the course of a summer population increase, heterozygotes gradually increased in proportion. This is to be expected if heterozygotes have a fitness advantage, but it is difficult to observe in sexually reproducing populations because of the tendency for recombination to restore Hardy-Weinberg equilibrium.

In sum, it appears that parthenogenesis provides a clear gain in lifetime reproduction. The advantage in obviating the need to find mates is somewhat debatable, as sex pheromones and cohort synchrony evidently work for copepods. The genetic advantage or disadvantage depends on the temporal pattern of environmental change. A possible fourth advantage is that a population reproducing parthenogenetically can rebound more easily from heavy predation than a sexually reproducing population. Since commoness or rareness is a population rather than an individual attribute, however, it is difficult to construct a predator avoidance argument based on individual selection.

Environmental effects on reproductive patterns

The age of first reproduction, average brood size, and number of broods per lifetime are all affected by varying conditions of temperature and food. The role of temperature has received perhaps more attention, stemming initially from Deevey's (1960) observation that marine copepods collected during winter were larger than individuals collected during summer. McLaren (1963) discussed the strong inverse relation between temperature, development time, and body size, again using marine copepods. Raised at lower temperatures, zooplankton take longer to mature (both embryonic and postembryonic development are lengthened) and mature at a larger size which results in larger broods. The net effect of the tradeoff seems to be a reduction in r at lower temperatures (e.g. Hall 1964; Rippingale and Hodgkin 1974). McLaren (1974) has argued that vertical migration may provide a mechanism whereby the tradeoff between developmental rate and body size, which affects fecundity, can be minimized.

Food has received relatively less attention but also affects the same parameters. Weglenska (1971) demonstrated that reduced food ration decreases postembryonic developmental rate, body size at maturity, brood size, and number of broods per lifetime. Under laboratory conditions, Hall (1964) found that food concentration affected intrinsic growth rate in *D. galeata mendotae* principally via daily fecundity but to a lesser extent by delaying the onset of reproduction. However, he concluded that the effect of temperature was greater than that of food. By diluting or concentrating natural foods, Weglenska (1971) demonstrated a strong effect of food concentration on reproductive parameters. Postembryonic development was far more susceptible than embryonic development to changing food levels, while egg production per female lifetime varied 6–8-fold. Eggs/brood, broods/lifetime, and egg survival varied in the four cladocerans studied (*D. cucullata, D. longispina, Diaphanosoma brachyurum, Chydorus sphaericus*) and the one copepod (*Eudiaptomus graciloides*), except that the chydorid invariably produced a maximum of two eggs/brood.

The level at which food dilution significantly affects reproductive parameters may vary with the species and food conditions. *Eudiaptomus graciloides* appeared most susceptible to dilution of natural food. However, this food was mainly bacteria and consisted of only a few percent nannoplankton by weight; in all likelihood the copepod was least able to utilize bacteria, hence was most readily affected by dilution of the already rare nannoplankters. *Daphnia pulex* appears to require relatively high food concentrations for reproduction; Hrbáčková-Esslová (1963) was able to maintain it in situ in the Slapy Reservoir only by supplementing its food.

Weglenska argued that food and temperature are equally important determinants of such reproductive parameters as maturation time and egg production. It seems likely that during the spring phytoplankton pulse, increasing temperatures may have the greatest effect on developmental rates. In the summer, when edible food is less abundant (Porter 1976) while temperature changes little, developmental rates will be affected most by changing food conditions.

Intrinsic rate of natural increase, r_{max}

The intrinsic rate of natural increase, r, describes the per capita growth rate of an exponentially growing population. In the absence of crowding and under optimal physiological conditions, it tends to a maximum termed r_{max} (Andrewartha and Birch 1954). If food is plentiful and of high quality, the estimate of r should be the maximum for that temperature, hence identical to r_{max}. Typically, this was the aim of the investigator, and we take estimates of r to be attempts to estimate r_{max} for that temperature. However, as data will show (below), some efforts clearly resulted in low estimates, presumably due to inadequate nutrition. It is a useful statistic because it combines survival with fecundity and depends not just on total fecundity but also on its distribution over age. However, the careful and tedious laboratory cultivation required to estimate r has been done for only a handful of species. A shortcut which gives reasonable estimates was developed by Lewontin (1965) in a theoretical consideration of fecundity and development times, and used by Allan (1976) to provide a band of r_{max} values for rotifers, cladocerans, and copepods. Figures 9 and 10 reproduce Allan's (1976) estimates, superimposed on which are all values of r known to us where a complete life table was carried through.

Generally, with the possible exception of the copepods, it appears that the theoretical approximations (bands in Figs. 9 and 10) are close to the average for the group under consideration, but underestimate the range of variation.

Relatively little can be said about the copepods, as so few estimates of r are available, and most are for marine or estuarine species. Actual life table estimates, and thus the true range of r-values, are lower than Allan's estimates (Fig. 9). The maturation times used by Allan (1976) probably underestimate the actual time from egg to egg for typical freshwater species. In addition the model he used applies best to species undergoing continuous reproduction under constant conditions. The life cycles of copepods at temperate latitudes are sufficiently slow that the annual cycle of the environment entrains the life cycle of the animal; owing

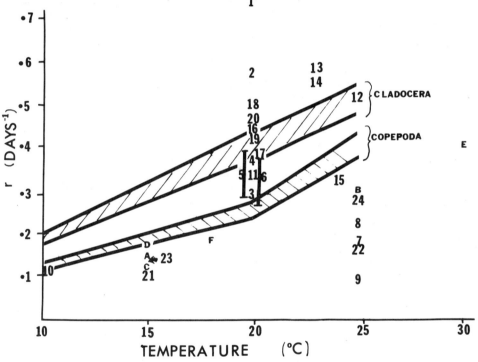

Fig. 9. Relationship of intrinsic rate of natural increase (r) to temperature in cladocerans and copepods. Bands are estimates of Allan (1976), while numbers or letters represent actual life table estimates. Cladocera: 1—*Daphnia pulex* fed *Ankistrodesmus*; 2—*D. pulex* fed yeast (Arnold 1971); 3—*D. pulex* (Frank et al. 1957); 4—*D. galeata mendotae*; 5—*D. ambigua*; 6—*Bosmina longirostris* (Goulden et al. 1978); 7—*Alona affinis*; 8—*A. guttata*; 9—*A. rustica* (Goulden and Berseth in Goulden et al. 1978); 10-12—*Daphnia galeata mendotae* (Hall 1964); 13—*Ceriodaphnia reticulata*; 14—*Simocephalus serrulatus* (Hall et al. 1970); 15—*Daphnia pulex* (Marshall 1962); 16—*D. magna* (Smith 1963); 17—*D. magna*; 18—*D. pulex*; 19—*D. parvula*; 20—*D. ambigua* (Winner and Farrell 1976); 21-22—*Pleuroxus denticulatus*; 23-24—*Chydorus sphaericus* (Keen 1967). Copepoda: A-B—*Gladioferens imparipes* (Rippingale and Hodgkin 1974); C—*Diaptomus pallidus*; D-E—*Acartia tonsa* (Heinle unpubl.); F—*Eurytemora affinis* (Daniels and Allan unpubl).

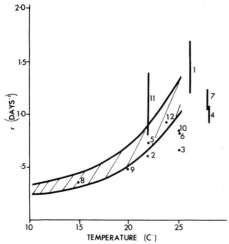

Fig. 10. Relationship of intrinsic rate of natural increase (r) to temperature in rotifers. Band is estimate of Allan (1976), while numbers represent actual life table estimates. 1—*Asplanchna sieboldi*, saccate, fed *Paramecium aurelia* or *Brachionus calyciflorus* (Gilbert 1977); 2-4—*A. sieboldi*, saccate, 5-7—*A. sieboldi*, campanulate (Gilbert 1976); 8-10—*Brachionus calyciflorus* (Halbach 1970); 11—*Euchlanis dilatata* (King 1965); 12—*B. calyciflorus* (Theilacker and McMaster 1971).

to the strong influence of temperature on development, this may reduce r substantially below its theoretical maximum.

Only three species of rotifers have been studied with respect to life table estimates of r (Fig. 10). Values may vary substantially based upon food used, and, for polymorphic forms, depending on the morph studied. Further improvement in culture techniques for rotifers and copepods clearly are needed before the true extent of natural variation in these groups will be revealed.

The cladocerans provide the most material for discussion as some 24 estimates utilizing 13 different species have been published (Fig. 9). The highest estimates of r, for *D. pulex* (points 1 and 2 in Fig. 9; Arnold 1971), if real, demonstrate that extremely high fecundity can produce correspondingly high per capita growth rates in a cladoceran that is larger and matures more slowly than most. We question the validity of these two estimates, however, as Arnold's highest value of r could be obtained only if an average *D. pulex* hatched (not just carried) 30 eggs per day, uninterrupted, for 100 days. This is so inconsistent with other data for *D. pulex* (e.g. Richman 1958) that we doubt Arnold's estimates. At the other extreme, Marshall (1962) evidently failed to achieve adequate culture conditions.

Of the estimates remaining, the chydorids *Alona* spp., *C. sphaericus* and *Pleuroxus denticulatus* clearly have extremely low values of r relative to more planktonic species. Certainly they would be unable to withstand the level of mortality common to planktonic species (20% per day or more—see Hall 1964; Dodson 1972; Allan 1973); presumably then, predation is less intense on these littoral and epibenthic species. The other planktonic cladocerans form a reasonably tight cluster. This might seem surprising considering the range in reproductive variation, from tiny *B. longirostris*, with 2 to 6 eggs per brood, up to *D. pulex* or *D. magna*, with several times the brood size and a longer life as well. In terms of population growth rate, the smaller brood size of small species (Fig. 7) is offset by the short maturation time also typical of small species (Fig. 3). With the data now available the conservative view would be a fairly close cancellation. While it is true that *B. longirostris* (maturing ca. 300 μm) has a lower r than *D. pulex* (maturing ca. 1.7 mm), *C. reticulata* (maturing ca. 600 μm) appears to have a higher r than either.

We conclude that, for a given temperature, the values of r for large and small planktonic cladocerans are remarkably close and not necessarily predictable from body size. Quite possibly, then, the crucial difference between species may lie less in their innate r_{max}, and more in how r responds to changing environmental conditions such as nutrition and temperature.

Reproductive patterns and the composition of the plankton

Two populations, initially at the same density, will diverge at the rate $e^{(r_1-r_2)t}$, where r_1 and r_2 are realized per capita growth rates of the two species and t is

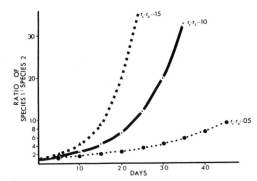

Fig. 11. Rate at which numbers of two species would diverge, given equal starting numbers and various differences in realized per capita growth rates. Note that for a difference in r of 0.05 day^{-1}, 28 days are required for faster growing species to attain 4 × the numbers of slower species, while this is achieved in 14 days and 7 days for r differences of 0.10 and 0.15.

time. Figure 11 portrays the rate of divergence in days, expressing the faster species as a multiple of the slower species, for differences in r of 0.05, 0.10, and 0.15 (days^{-1}). If one arbitrarily assumes that a fourfold difference in numbers would appear to the limnologist as a clear dominance of one species over another, then a difference in r_i of 0.05 day^{-1} probably is unimportant because it would take a month for a fourfold difference to be reached, and conditions rarely stay constant for that long in lakes and ponds. But for an r-difference of 0.10 or 0.15 day^{-1}, the fourfold dominance is reached in 2 weeks or 1 week respectively, with even more rapid divergence thereafter. While this is merely playing with numbers, it helps to give some feeling for the effect a "race in numbers" would have on zooplankton composition. It should be recognized that what is at issue here is differences in realized r. For example, two populations with r_{max} values of 0.40 and 0.25 day^{-1}, suffering respective mortalities of 0.15 and 0.10 day^{-1}, differ in realized r by 0.10 day^{-1}.

Given this perspective, the r_{max} difference between *D. pulex* and *B. longirostris* of 0.13 day^{-1} is sufficient to lead to rapid divergence, if neither predation nor competitive pressures act selectively against one species alone. However, as argued previously, this comparison lies toward the extreme of differences in r_{max} within the Cladocera.

What controls the size composition of the zooplankton remains a central question since Hrbáček et al. (1961) demonstrated that increased fish stock shifts the zooplankton composition toward smaller sizes. Brooks and Dodson (1965) provided further evidence of the effects of fish predation and argued that larger plankters predominate in the absence of fish due to their superior competitive ability. This second argument has received little support, and an alternate hypothesis has gained credibility from increasing evidence that smaller zooplankters are most susceptible to invertebrate predation. If invertebrate predators are of increased importance in the absence of vertebrates (fish) which normally control their numbers, then predation upon small zooplankton would increase, possibly resulting in the predominance of large forms (Dodson 1974).

Goulden et al. (1978) proposed a third hypothesis, that larger cladoceran species predominate in the absence of fish predation due to their greater reproductive rate, measured by r. Thus we have three apparently sufficient hypotheses for the predominance of larger zooplankton in the absence of vertebrate predation: greater competitive ability, lesser risk of invertebrate predation, and greater reproductive ability.

At present it appears that Goulden's r-hypothesis may apply in some specific comparisons. Within the Cladocera, *B. longirostris* is the smallest planktonic species and appears to have a lower r_{max} than most members of the genus *Daphnia*. It is reasonable to suppose that the higher r_{max} of *Daphnia* spp. contributes to their predominance over *Bosmina* in the absence of fish. Other small Cladocera in the Chydoridae also have much lower growth rates (Goulden et al. 1978; see Fig. 9). This may explain in part why such species rarely appear in the open water plankton. Finally, within the Daphniidae, the relationship between body size and r_{max} is weak at best and is unlikely to contribute greatly to observed patterns in size composition engendered by fish stock.

Two arguments seem important to us when relating patterns in reproductive rates to patterns in size composition of the plankton. The first is that per capita growth rates differ less than one would expect because size-related gains in fecundity are more or less exactly countered by size-related costs in slower maturation. As a result, the size variation among planktonic species of Cladocera can be viewed as a series of positions along the spectrum from low susceptibility to vertebrate predation (small size) to low susceptibility to invertebrate predation (large size) *yet achieving surprisingly similar growth rate (r_{max}) potential*.

The second argument relates realized r to environmental conditions. If species differ in their relationships of realized r to environmental conditions such as food (Hrbáčkova-Esslová 1963; Weglenska 1971; Goulden and Hornig unpubl.) and possibly temperature (Allan 1978), then the realized r for two species may differ far more than their r_{max}. One species may cease reproduction at a food level where another is quite capable of producing young. Hrbáček (1978) has suggested that r-specialists among the plankton are species such as *D. pulex* which have short generation times under high food levels but are unable to reproduce at low food levels. K-specialists such as *Daphnia pulicaria* mature more slowly at high food levels than do r-specialists such as *D. pulex*, but are able to reproduce at low food levels where *D. pulex* dies out. Thus response to food level is the critical variable in Hrbáček's r-K comparison, not the absolute value of r_{max}. Lampert's (1978) investigations of the food threshold at which reproduction ceases for various Cladocera should help to clarify this possibility.

Conclusions

The study of life history and reproductive variation in the zooplankton is very much in its infancy. Substantial differences exist between the Rotifera, Cladocera, and Copepoda, while the within-group variation also appears to have important implications for understanding patterns in the abundance of populations and the composition of species assemblages. On the basis of field analysis of life cycles and laboratory culture studies, we observe that rotifers and cladocerans are iteroparous with rapid maturation and spend a high proportion of their lifespan as reproductive adults. The result is a high reproductive potential and generations in the field which typically overlap. In contrast, copepods are more semelparous, with slower maturation times. Except for populations which overwinter as adults most of the lifespan is passed as a juvenile. Reproduction is for a comparatively brief period, and adult animals usually do not persist beyond the breeding period that produces the next brood. As a consequence, reproductive potential is low and populations in the field typically exhibit a cohort structure.

Life history variation can profitably be related to the species and the size composition of the zooplankton, both by considering the reproductive advantage that accrues to certain (possibly large) species compared to others, and by viewing reproductive variation as the consequence of size variation which places species along the spectrum from high invertebrate predation to high vertebrate predation. Because the relationship of r_{max} to body size is by no means simple and may be further complicated by differences between species in their response to changing food conditions—zooplankton researchers must view with considerable caution the prevailing idea (Pianka 1970) that r-strategists have high r_{max} and small body size compared to K-strategists. The ranking of life histories in terms of opportunism (Hutchinson 1967; Allan 1976): rotifers >

cladocerans > copepods corresponds to the seasonal timing of life cycles among these groups (Pejler 1962; Hurlbert et al. 1972); and is consistent with the strong association of cladocerans and rotifers with waters which are both productive and seasonally pulsed (Allan 1976). Thus we conclude that life cycles clearly relate to ecological distributions in both space and time.

Finally we note that zooplankton species are especially suitable for life history studies because of their small size, short generation time, feasibility for culturing, and immense variation in reproductive traits. Yet only the Cladocera have received even a modest amount of study and, as a result, our tentative generalizations rest too heavily on this one of the three important freshwater groups. We agree with Hutchinson (1967, p. 708): "A more comparative study of planktonic life histories is clearly needed in limnology."

References

ALLAN, J. D. 1973. Competition and the relative abundances of two cladocerans. Ecology 54:484-498.

———. 1976. Life history patterns in zooplankton. Am. Nat. 110:165-180.

———. 1978. The dynamics of a mixed population of *Daphnia*, and the associated cladoceran community. Freshwater Biol. 7:505-512.

ANDERSON, B. G. 1932. The number of preadult instars, growth, relative growth, and variation in *Daphnia magna*. Biol. Bull. 68:81-98.

ANDREWARTHA, H. G., and L. C. BIRCH. 1954. The distribution and abundance of animals. Univ. Chicago Press.

ARNOLD, D. E. 1971. Ingestion, assimilation, survival, and reproduction of *Daphnia pulex* fed seven species of blue-green algae. Limnol. Oceanogr. 16:906-921.

BAYLY, I. A. E. 1962. Ecological studies on New Zealand lacustrine zooplankton with special reference to *Boeckella propinque* Sars (Copepod: Calanoida). Aust. J. Mar. Freshwater Res. 13:143-197.

BOSSONE, A., and V. TONOLLI. 1954. Il problema della convivenza di *Arctodiaptomus bacillifer* (Koelb.) di *Acanthodiaptomus denticornis* (Wierz.) e di *Heterocope saliens* Lill. Mem. Ist. Ital. Idrobiol. 8:81-94.

BOTTRELL, H. H. 1975. The relation between temperatures and duration of egg development in some epiphytic Cladocera and Copepoda from the River Thames, Reading, with a discussion of temperature functions. Oecologia 18:63-84.

BREWER, R. H. 1964. The phenology of *Diaptomus stagnalis* (Copepoda: Calanoida): The development and the hatching of the egg stage. Physiol. Zool. 37:1-20.

BROOKS, J. L. 1946. Cyclomorphosis in *Daphnia*. I. An analysis of *D. retrocurva* and *D. galeata*. Ecol. Monogr. 16:409-447.

———, and S. I. DODSON. 1965. Predation, body size, and composition of plankton. Science 150:28-35.

BURGIS, M. J. 1967. A quantitative study of reproduction in some species of *Ceriodaphnia*. J. Anim. Ecol. 36:61-75.

CARTER, J. C. H. 1969. Life cycles of *Limnocalanus macrurus* and *Senecella calanoides*, and seasonal abundance and vertical distribution of various planktonic copepods in Parry Sound, Georgian Bay. J. Fish. Res. Bd. Can. 26:2543-2560.

COLE, G. A. 1961. Some calanoid copepods from Arizona with notes on congeneric occurrences of *Diaptomus* species. Limnol. Oceanogr. 6:432-442.

COLE, L. C. 1954. The population consequences of life history phenomena. Quant. Rev. Biol. 29:103-137.

COMITA, G. W. 1956. A study of a calanoid copepod population in an arctic lake. Ecology 37:576-591.

———. 1972. The seasonal zooplankton cycles, production and transformation of energy in Severson Lake, Minnesota. Arch. Hydrobiol. 70:14-66.

———, and G. C. ANDERSON. 1959. The seasonal development of a population of *Diaptomus ashlandi* March, and related phytoplankton cycles in Lake Washington. Limnol. Oceanogr. 4:37-52.

CONOVER, R. J. 1956. Oceanography of Long Island Sound. VI. Biology of *Acartia clausi* and *A. tonsa*. Bull. Bingham Oceanogr. Collect. 15:156-233.

———, and E. D. S. CORNER. 1968. Respiration and nitrogen excretion by some marine zooplankton in relation to their life cycles. J. Mar. Biol. Assoc. U. K. 48:49-75.

COOLEY, J. M. 1971. The effect of temperature on the development of resting eggs of *Diaptomus oregonensis* Lillj. (Copepoda: Calanoida). Limnol. Oceanogr. 16:921-926.

DAVIS, C. C. 1961. Breeding of calanoid copepods in Lake Erie. Int. Ver. Theor. Angew. Limnol. Verh. 14:933-942.

———. 1972. Plankton succession in a Newfoundland lake. Int. Rev. Gesamten Hydrobiol. 57:367-395.

———. 1976. Simultaneous quantitative comparison of planktonic crustacea in two Newfoundland

boreal lakes. Int. Rev. Gesamten Hydrobiol. 61:807-823.
DEEVEY, G. B. 1948. The zooplankton of Tisbury Great Pond. Bull. Bingham Oceanogr. Collect. 12:1-44.
———. 1960. Relative effects of temperature and food on seasonal variations in length of marine copepods in some eastern American and western European waters. Bull. Bingham Oceanogr. Collect. 17:54-85.
DODSON, S. I. 1972. Mortality in a population of *Daphnia rosea*. Ecology 53:1011-1023.
———. 1974. Adaptive change in response to size-selective predation: A new hypothesis of cyclomorphosis. Limnol. Oceanogr. 19:721-729.
EDMONDSON W. T., G. W. COMITA, and G. C. ANDERSON. 1962. Reproductive rate of copepods in nature and its relation to phytoplankton population. Ecology 43:625-634.
EICHHORN, R. 1957. Zur Populationsdynamik der calanoiden Copepoden in Titisee und Feldsee. Arch. Hydrobiol. Suppl. 24 (2), p. 186-246.
EKMAN, S. 1907. Über das Crustaceenplankton der Ekoln (Mälaren) and über verschiedene Kategorien von marinen Relikten in schwedisher Binnenseen. Zool. Stud. Festschr. Prof. T. Tulberg, Uppsala, p. 42-65.
ELGMORK, K., J. P. NILSSEN, T. BROCH, and R. ØVREVIK. 1978. Life cycle strategies in neighbouring populations of the copepod *Cyclops scutifer* Sars. Int. Ver. Theor. Angew. Limnol. Verh. 20: in press.
ELSTER, H. H. 1954. Über die Populationsdynamik von *Eudiaptomus gracilis* Sars und *Heterocope borealis* Fischer im Bodensee-Obersee. Arch. Hydrobiol. Suppl. 20, p. 546-614.
FRANK, P. W., C. D. BALL, and R. W. KELLY. 1957. Vital statistics of laboratory cultures of *Daphnia pulex* de Geer as related to density. Physiol. Zool. 30:287-305.
FULTON, J. 1973. Some aspects of the life history of *Calanus plumchrus* in the Strait of Georgia. J. Fish. Res. Bd. Can. 30:811-815.
GEILING, W. T., and R. S. CAMPBELL. 1972. The effect of temperature on the development rate of the major life stages of *Diaptomus pallidus* Herrick. Limnol. Oceanogr. 17:304-307.
GILBERT, J. J. 1976. Polymorphism in the rotifer *Asplanchna sieboldi*: biomass, growth and reproductive rate of the saccate and campanulate morphotypes. Ecology 57:542-551.
———. 1977. Effect of the non-tocopherol component of the diet on polymorphism, sexuality, biomass, and reproduction rate of the rotifer *Asplanchna sieboldi*. Arch. Hydrobiol. 80: 375-397.
GILL, D. E. 1974. Intrinsic rate of increase, saturation density, and competitive ability. II. The evolution of competitive ability. Am. Nat. 108:103-116.
GOULDEN, C. E., L. HORNIG, and C. WILSON. 1978. Why do large zooplankton species dominate? Int. Ver. Theor. Angew. Limnol. Verh. 20: 2457-2460.
GREEN, J. 1956. Growth, size and reproduction in *Daphnia* (Crustacea: Cladocera). Proc. Zool. Soc. Lond. 126:173-204.
———. 1966. Seasonal variation in egg production by Cladocera. J. Anim. Ecol. 35:77-104.
GRIFFITHS, A. M., and B. W. FROST. 1976. Chemical communication in marine planktonic copepods *Calanus pacificus* and *Pseudocalanus* sp. Crustaceana 30:1-8.
GURNEY, R. 1931. British fresh-water Copepoda, v. 1. The Ray Soc. Lond.
HALBACH, U. 1970. Einflus der Temperatur auf die Populations Dynamik des planktischen Rüdertieres *Brachionus calyciflorus* Palla. Oecologia 4:176-207.
HALL, D. J. 1964. An experimental approach to the dynamics of a natural population of *Daphnia galeata mendotae*. Ecology 45:94-112.
———. W. E. COOPER, and E. E. WERNER. 1970. An experimental approach to the production dynamics and structure of freshwater animal communities. Limnol. Oceanogr. 15:839-928.
———, S. T. THRELKELD, C. W. BURNS, and P. H. CROWLEY. 1976. The size-efficiency hypothesis and the size structure of zooplankton communities. Annu. Rev. Ecol. Syst. 7:177-208.
HEBERT, P. D. N. 1974. Enzyme variability in natural populations of *Daphnia magna*. I. Population structure in East Anglia. Evolution 28:546-556.
HEINRICH, A. K. 1962. The life histories of plankton animals and seasonal cycles of plankton communities in the oceans. J. Cons. Int. Explor. Mer 27:15-24.
HIROTA, R., and S. UNO. 1977. Seasonal abundance of the pelagic eggs of copepoda in the vicinity of Amakusa-Matsushima, Western Kyushu. Bull. Plankton Soc. Jpn. 24(2):77-84.
HRBÁČEK, J. 1978. Competition and predation in relation to species composition of freshwater zooplankton, mainly Cladocera, p. 307-353. *In* J. Cairns, Jr. [ed.], Aquatic microbial communities. Garland Ref. Library Sci. Technol. v. 15.
———, M. DROŘÁKOVA, V. KOŘÍNEK, and L. PROCHÁZKOVÁ. 1961. Demonstration of the effect of fish stock on the species composition of zooplankton and the intensity of metabolism of the whole plankton association. Int. Ver. Theor. Angew. Limnol. Verh. 14:192-195.
HRBÁČKOVÁ-ESSLOVÁ, M. 1963. The development of three species of *Daphnia* in surface water of Slapy reservoir. Int. Rev. Ges-

amten Hydrobiol. 48:325-333.

HURLBERT, S. H., M. S. MULLA, and H. R. WILSON. 1972. Effects of an organophosphorus insecticide on the phytoplankton, zooplankton, and insect populations of freshwater ponds. Ecol. Monogr. 42:269-299.

HUTCHINSON, G. E. 1967. A treatise on limnology, v. 2. Wiley.

KATONA, S. K. 1973. Evidence for sex pheromones in planktonic copepods. Limnol. Oceanogr. 18:574-583.

KEEN, R. E. 1967. Laboratory population studies of two species of Chydoridae (Cladocera: Crustacea). M.S. thesis, Mich. State Univ. 34 p.

KERFOOT, W. C. 1974. Egg-size cycle of a cladoceran. Ecology 55:1259-1270.

KING, C. E. 1965. Food, age, and the dynamics of a laboratory population of rotifers. Ecology 48:111-128.

KLEIN, H. VON. 1938. Limnologische Untersuchungen über das Crustaceenplankton des Schleinsees und zweier Kleingewässer. Int. Rev. Gesamten Hydrobiol. 37:176-233.

KOZHOV, M. 1963. Lake Baikal and its life. Monogr. Biol. V. 11. Junk.

KOZHOVA, O. M. 1956. On the biology of the *Epischura baicalensis* Sars of Lake Baikal. Izv. Biol.-Geogr. Inst. Irk. Univ. 16:92-120.

KUNTZE, H. 1938. Die Crustaceen des Schleinsees im Jahre 1935. Int. Rev. Gesamten Hydrobiol. 37:164-175.

LAMPERT, W. 1977. Studies on the carbon balance of *Daphnia pulex* de Geer as related to environmental conditions. IV. Determination of the "threshold" concentration as a factor controlling the abundance of zooplankton species. Arch. Hydrobiol. Suppl. 48, p. 361-368.

LANGFORD, R. R. 1938. Diurnal and seasonal changes in the distribution of limnetic crustacea of Lake Nipissing, Ontario. Univ. Toronto Stud. Biol. Ser. 45 (Publ. Ontario Fish. Lab. 56), 1-42.

LANDRY, M. R. 1975. Seasonal temperature effects and predicting development rates of marine copepod eggs. Limnol. Oceanogr. 20:434-440.

LEI, C. H., and H. F. CLIFFORD. 1974. Field and laboratory studies of *Daphnia schodleri* Sars from a winterkill lake of Alberta. Natl. Mus. Can. Zool. Publ. 9. 53 p.

LEWONTIN, R. C. 1965. Selection for colonizing ability. *In* H. G. Baker and G. L. Stebbins [eds.], The genetics of colonizing species. Academic.

LYNCH, M. 1977. Fitness and optimal body size in zooplankton populations. Ecology 58:763-774.

McLAREN, I. A. 1963. Effects of temperature on growth of zooplankton and the adaptive value of vertical migration. J. Fish. Res. Bd. Can. 20:685-727.

———. 1965. Some relationships between temperature and egg size, body size development rate, and fecundity, of the copepod *Pseudocalanus*. Limnol. Oceanogr. 10:528-538.

———. 1974. Demographic strategy of vertical migration by a marine copepod. Am. Nat. 108:91-102.

———. 1979. Generation lengths of some temperate marine copepods: Estimation, prediction and implications. J. Fish. Res. Bd. Can. 36: in press.

MALY, E. J. 1973. Density, size, and clutch of two high altitude diaptomid copepods. Limnol. Oceanogr. 18:840-848.

MARGALEF, R. 1953. Caracteres ligados a las magnitudes absolutas de los organismos y su significado sistemático y evolutivo. Publ. Inst. Biol. Appl. 12:111-121.

MARSH, C. D. 1898. On the limnetic crustacea of Green Lake. Trans. Wis. Acad. Sci. 11: 179-224.

MARSHALL, J. S. 1962. The effects of continuous gamma radiation on the intrinsic rate of increase of *Daphnia pulex*. Ecology 43: 598-607.

MARSHALL, S. M., and A. P. ORR. 1955. The biology of a marine copepod. [Reprint 1972] Springer.

MITTELHOLZER, Z. 1970. Populationsdynamik und Produktion des Zooplanktons inn Greifensee und im Vierwaldstättersee. Schweiz. Z. Hydrol. 32:90-149.

MILLER, C. B., S. K. JOHNSON, and D. H. HEINLE. 1977. Growth rates in the marine copepod genus *Acartia*. Limnol. Oceanogr. 22:325-335.

MURUGAN, N. 1975. Egg production, development and growth in *Moina micrura* Kurz (1874) (Cladocera: Moinidae). Freshwater Biol. 5:245-250.

———, and K. G. SIVARAMAKRISHNAN. 1973. The biology of *Simocephalus acutirostratus* King (Cladocera: Daphniidae)—Laboratory studies of life span, instar duration, egg production, growth and stages in embryonic development. Freshwater Biol. 3:77-83.

———, and ———. 1976. Laboratory studies on the longevity, instar duration, growth, reproduction and embryonic development in *Scapholeberis kingi* Sars (1903) (Cladocera: Daphniidae). Hydrobiologia 50:75-80.

NILSSEN, J. P. 1977. Cryptic predation and the demographic strategy of two limnetic cyclopoid copepods. Mem. Ist. Ital. Idrobiol. 34: 187-196.

PEJLER, B. 1962. The zooplankton of Ösbysjön, Djursholm. II. Further ecological aspects. Oikos 13:216-231.

PIANKA, E. R. 1970. On r- and K-selection. Am. Nat. 104:592-597.

PORTER, K. G. 1976. Enhancement of algal growth and productivity by grazing zooplank-

ton. Science 192:1332–1334.
———. 1977. The plant-animal interface in freshwater ecosystems. Am. Sci. 65:159–170.
RAVERA, O. 1954. La struttura demografica dei Copepodi del Lago Maggiore. Mem. Ist. Ital. Idrobiol. 8:109–150.
———, and V. TONOLLI. 1956. Body size and number of eggs in diaptomids, as related to water renewal in mountain lakes. Limnol. Oceanogr. 1:118–122.
RICHMAN, S. 1958. Transformation of energy by *Daphnia pulex*. Ecol. Monogr. 28:273–291.
RIGLER, F. H., and J. M. COOLEY. 1974. The use of field data to derive population statistics of multivoltine copepods. Limnol. Oceanogr. 19:636–655.
RIPPINGALE, R. J., and E. P. HODGKIN. 1974. Population growth of a copepod *Gladioferens imparipes* Thomson. Aust. J. Mar. Freshwater Res. 25:351–360.
ROFF, J. C. 1972. Aspects of the reproductive biology of the planktonic copepod *Limnocalanus macrurus* Sars, 1863. Crustaceana 22:155–160.
———, and J. C. H. CARTER. 1972. Life cycle and seasonal abundance of the copepod *Limnocalanus macrurus* Sars in a high arctic lake. Limnol. Oceanogr. 17:363–370.
SCHINDLER, D. W., and G. W. COMITA. 1972. The dependence of primary production upon physical and chemical factors in a small senescing lake, including the effects of complete winter oxygen depletion. Arch. Hydrobiol. 69:413–451.
SMITH, F. E. 1963. Population dynamics in *Daphnia magna* and a new model for population growth. Ecology 44:651–663.
SMITH, J. MAYNARD. 1971. What use is sex? J. Theor. Biol. 30:319–338.
SMYLY, W. J. P. 1973. Bionomics of *Cyclops strenuus abyssorum* Sars (Copepoda: Cyclopoida). Oecologia 11:163–186.
SOUTHWOOD, T. R. E. 1976. Bionomic strategies and population parameters, p. 26–48. *In* R. M. May [ed.], Theoretical ecology: Principles and applications. Saunders.

STEARNS, S. C. 1976. Life-history tactics: A review of the ideas. Q. Rev. Biol. 51:3–47.
———. 1977. The evolution of life-history traits: A critique of the theory and a review of the data. Annu. Rev. Ecol. Syst. 8:145–171.
STRICKLER, J. R., and S. TWOMBLY. 1975. Reynolds number, diapause, and predatory copepods. Int. Ver. Theor. Angew. Limnol. Verh. 19:2943–2950.
THEILACKER, G. H., and M. F. McMASTER. 1971. Mass culture of the rotifer *Brachionus plicatilis* and its evaluation as a food for larval anchovies. Mar. Biol. 10:183–188.
THOMAS, M. P. 1961. Some factors influencing the life history of *Diaptomus gracilis* Sars. Int. Ver. Theor. Angew. Limnol. Verh. 14:943–945.
TONOLLI, V. 1961. Studio sulla dinamica del popolamento di un copepoda (*Eudiaptomus vulgaris* Schmeid.). Mem. Ist. Ital. Idrobiol. 13:179–202.
WEGLENSKA, T. 1971. The influence of various concentrations of natural food on the development, fecundity and production of planktonic crustacean filtrators. Ekol. Pol. 19:427–473.
WELLS, L.-R. 1960. Seasonal abundance and vertical movements of planktonic crustacea in Lake Michigan. Fish. Bull. 60:343–369.
WESENBERG-LUND, C. 1904. Plankton investigations of the Danish lakes. Special part. Dan. Freshwater Biol. Lab. Op. 5.
———. 1908. Plankton investigations of the Danish lakes. General part. The Baltic freshwater plankton, its origin and variation. Dan. Freshwater Biol. Lab. Op. 5.
WILLIAMS, G. C. 1975. Sex and evolution. Princeton Univ. Press.
WINNER, R. W., and M. P. FARRELL. 1976. Acute and chronic toxicity of copper to four species of *Daphnia*. J. Fish. Res. Bd. Can. 33:1685–1691.
ZENKEVITCH, L. 1963. Biology of the seas of the U.S.S.R. Wiley.

38. Evolutionary Aspects of Diapause in Freshwater Copepods

Kåre Elgmork

Abstract

Diapause seems to be fundamentally identical in insects and cyclopoid copepods. Migration is intimately connected with diapause in both groups. A hypothetical evolutionary line may be traced from so-called active diapause, to oligopause and diapause. Varying environmental pressure in copepod populations in some shallow ponds indicates how the microevolution of diapause may have been initiated. The adaptive significance of diapause in copepods is that it serves as temporal escape from an unfavorable environment and a time of important events in the life cycle.

An important field of evolutionary ecology is the study of life history strategies, which includes diapause and migration. Life history patterns vary considerably in relation to latitude, altitude, climate, and various other abiotic and biotic factors. Detailed studies of the complex interaction between genes and environment in natural populations are, however, scarce, and this field of study is only at its starting phase. Some progress has been made in entomology, for, as recently summarized in Dingle (1978): "The evolution of phenological strategies should be an important and exciting subject for future research" (p. 52) and "modern genetic theory in diapause discussions may presage a new era of evolutionary thought" (p. xiv).

Diapause in freshwater copepods seems fundamentally identical with diapause in insects. This relation was first acknowledged by Cole (1953) and further elaborated by Elgmork (1959), Brewer (1964), and Elgmork and Nilssen (1978). The demonstration of regulation by photoperiod also strongly supports this view (Einsle 1964; Spindler 1969, 1970, 1971; Watson and Smallman 1971a, b). Diapause in freshwater copepods thus emerges as an integral part of arthropod diapause.

In this paper some evolutionary aspects of diapause in cyclopoid copepods are considered, mainly those based on studies in nature. In cyclopoids diapause occurs in the copepodid instars II–V and occasionally as adult females. Diapause in copepods has been reviewed by Elgmork (1967).

Macroevolution

Diapause in freshwater cyclopoids is now known to occur in a number of species, as summarized in Elgmork (1967:Table 1) and in Table 1 of this paper. Up to now diapause has most frequently been observed in the genus *Cyclops s. str.*, while more than one species with diapause is known in the genera *Diacyclops, Thermocyclops,* and *Mesocyclops*.

The diapause stage is most frequently observed in the copepod instar IV, which in many species seems to be the only diapause stage. Copepod instar V is the next most frequent stage, and diapause also occurs occasionally in other instars (Table 2).

I thank J. P. Nilssen and W. J. P. Smyly for critically reading the manuscript.

Table 1. Survey of diapause in freshwater cyclopoids established in the period 1966–1977.

No.	Species	Diapause stage, copepod instar	Encystment	Habitat	Main references
1	*Cyclops furcifer* Claus	IV	–	Shallow reservoir	Wierzbicka 1966, 1972a, b
2	*Diacyclops bicuspidatus* (Claus)	IV	–	Shallow reservoir	Wierzbicka 1966, 1972a, b
3	*Thermocyclops dybowskii* Lande	IV	–	Shallow tarns	Smyly 1967
4	*Diacyclops navus* Herrick	IV	–	Temporary ponds	Watson and Smallman 1971a, b
5	*Cyclops abyssorum* Sars	(IV) V	–	Limnetic in lakes	Smyly 1973; Nilssen and Elgmork 1977

Table 2. Distribution of diapause stages in freshwater cyclopoid copepods at present stage of knowledge.

Diapause stage	Frequency
Copepodid II	1
III	2
IV	15
V	6
Adult ♀	3

It is interesting to note that the diapause stage varies within closely related species, as e.g. within the genus *Cyclops*. In *C. strenuus, C. kolensis,* and *C. vicinus,* diapause occurs in nature in the copepodid instar IV, while the diapause stage in *C. abyssorum* is instar V. In *C. scutifer* diapause is found in copepodid stages II, III, IV, and V, with variable frequencies in natural populations (Elgmork 1962).

The manifestation of diapause in different instars within closely related species indicates that diapause has evolved independently in the different species after their taxonomical differentiation. The distribution of diapause stages within insect species indicates the same relation (Hoy 1978).

Let us then speculate about possible evolutionary steps to diapause. A distinction can be made between *quiescence* and *diapause* (Andrewartha 1952; Lees 1955; Elgmork 1959; Mansingh 1971). Quiescence is characterized as short-termed and irregular and directly imposed by adverse environmental factors. Quiescence is not fixed to a special ontogenetic instar and may be induced repeatedly in the same individuals. Diapause responds to predictable, cyclic changes in the environment that occur regularly in the seasonal cycle and manifests itself in definite ontogenetic stages.

It is natural to assume that diapause has evolved from the more "primitive" stage of quiescence (Mansingh 1971), but that the primitive phase has been preserved because it is highly adaptive under special conditions, e.g. being more flexible than diapause, which requires a longer preparatory phase. For example, a cyclopoid population that inhabits a temporary pond which dries up a week before the programmed diapause sets in would be doomed unless quiescence was an alternative.

In populations of cyclopoids, an intermediate form of restricted development has been demonstrated in lakes. It will be termed *active diapause*, for it indicates an arrested growth and development, but without dormancy in the sediments (Fig. 1) (cf. Coker 1933). Active diapause has been found in the same instars as the diapause stage (e.g. *M. leuckarti*: Einsle 1968; *C. abyssorum*: Einsle 1975; *C. scutifer*: Halvorsen and Elgmork 1976; Elgmork and colleagues unpubl.). The copepods in active diapause occur in the profundal zone, often close above the bottom. This type of arrest in development and reproduction, but not in activity, may be regarded as an evolutionary step to dormancy and inactivity.

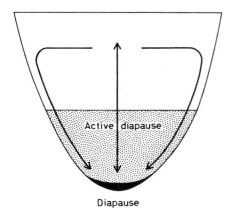 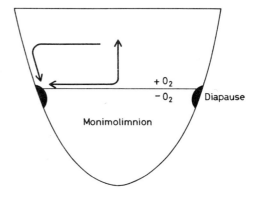

Fig. 1. Examples of selection of particular diapause sites in *Cyclops scutifer*. Arrows indicate hypothetical seasonal migration routes. (Right: redrawn from Elgmork and Langeland in press).

According to Mansingh (1971), a less pronounced form of dormancy may be classified as *oligopause*, i.e. one that represents an evolutionary step to diapause.

It is noteworthy that a relatively large number of cyclopoids that exhibit diapause are limnetic species which occur in the plankton of very large lakes, a phenomenon commented on by Elgmork (1967). Since then, diapause has been demonstrated in populations of *C. abyssorum* in several lakes of the English Lake District (Smyly 1973) and in a relatively large Norwegian lake (Nilssen and Elgmork 1977). Evolution has thus also favored diapause in populations that inhabit the limnetic zone of large lakes, where the adaptive value of diapause is less obvious.

In insect ecology and physiology seasonal migration seems to be intimately connected with diapause (Dingle 1978). One type of migration is to and from diapause sites. In freshwater cyclopoids the descent into the profundal zone and sediments for dormancy may be regarded as a migration (Fig. 1). The copepodid instars may remain either in active diapause or after a short time they may penetrate into the sediments. In certain cases it has been shown that relatively restricted areas of the lake bottom are chosen as diapause sites (Fig. 1), indicating a seasonal migration type of movement (Elgmork 1967, 1973, in prep.; Nilssen and Elgmork 1977). Orientation and orientation cues most likely are present, but are unknown.

When diapause ends, the copepodid instars migrate to the warmer, and more productive, euphotic zone of the lakes where further development and reproduction take place.

Microevolution

Sufficiently detailed studies are available to indicate some possible mechanisms of microevolution centering around diapause and its adaptive significance. However, so far no genetic analysis has been conducted.

One example may be drawn from the *C. strenuus* populations studied in four ponds and a lakelet by Elgmork (1955, 1959). A generalized scheme of the life histories of these populations is presented in Fig. 2.

Reproduction occurs in early spring and there is a rapid development of the new generation to copepodid, IV, which enters diapause in the mud in early summer. Diapause may last for about 10 months. Development is resumed in early spring and the adults die after reproduction.

In some years there was an irregular and partial termination of diapause in the autumn, sometimes leading to reproduction and the production of a new generation represented by a swarm of nauplii. With the onset of winter, oxygen was depleted under the ice and in the shallow ponds the entire population of nauplii died out during the anaerobic phase. The diapause stage in the mud was capable of passing the winter in

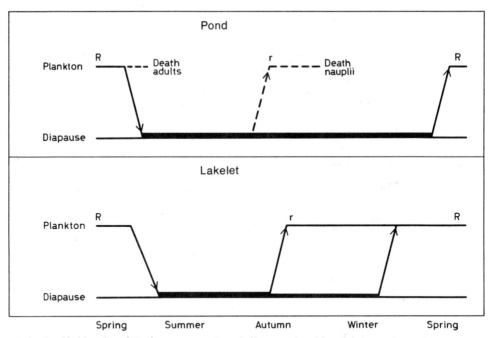

Fig. 2. Life histories of *Cyclops strenuus* in a shallow pond and in a lakelet (redrawn from Elgmork 1959): arrows down—initiation of diapause; arrows up—termination of diapause; dashed line—irregular, alternative development indicating unsuccessful evolutionary trials. R—Main reproduction periods; r—subordinate reproduction periods. Seasons defined as in Elgmork (1959).

anaerobiosis. In severe winters when some of the ponds became frozen to the bottom, the diapause stage was the only means of survival (Elgmork 1959: p. 101). There is thus a strong selection for the evolution and maintenance of diapause in these populations.

In certain years with a larger water volume, an autumnal period of reproduction may be more successful. Nauplii produced in autumn may reach the stage of large copepodids or adults and survive until the next spring. This was regularly the case in the lakelet (Bergstjern) situated in the same area as the much shallower ponds. The greater water volume of the lakelet never became totally depleted of oxygen. In both the ponds and the lakelet the diapause stages of *C. strenuus* were present in the sediments during the entire year. A fraction of the population was always present in diapause forming a "subterranean root system" (Fig. 2) with a high adaptive value in case of catastrophies at any time of the year.

Also in spring there were deviations from the main developmental pattern. A certain fraction of the population did not enter diapause but embarked on direct development to reproducing adults. Conditions for reproduction in summer were suboptimal as indicated by the smaller size of females and smaller clutch sizes. In many cases the adults died without reproducing. Again a selection favoring diapause was evident. In certain cases, however, reproduction may be successful, maintaining the genetical constitution for alternate developments.

A similar pattern of evolutionary variability emerges from the study of another freshwater planktonic cyclopoid, *C. scutifer*. A comprehensive study of life cycle strategies in this species is in progress (Elgmork 1962, 1965; Halvorsen and Elgmork 1976; Elgmork et al. 1978; Elgmork and colleagues in prep). This species shows a particularly flexible life history that extends from a simple annual cycle to complex life histories with life cycles of 2 and 3 years with and without diapause. Only a few examples with diapause are considered here.

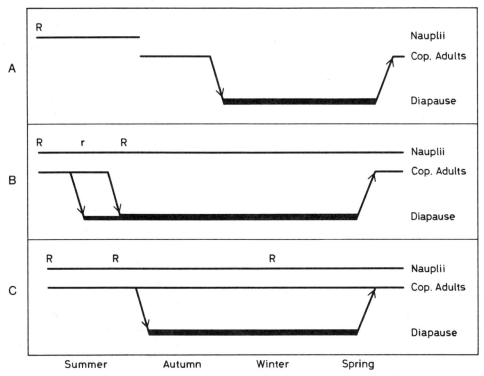

Fig. 3. Generalized scheme of life history patterns with diapause in Cyclops scutifer in three lakes in southern Norway. A—a small humic lake with oxygen depletion during winter in entire water volume (Elgmork unpubl.); B—population with 2- and 3-year life cycles and two diapause periods (Elgmork 1965); C—population with 1- and 2-year life cycles and winter reproduction (Elgmork and Langeland in press). Details as in Fig. 2.

Figure 3 presents three examples in which diapause has evolved as an important phase in the life histories. Figure 3A shows a simple annual cycle with diapause occurring in the first year of development. The winter is passed exclusively in diapause obviously as a response to a complete depletion of oxygen in the shallow (2.5 m deep), humic lakelet. Figures 3B and C represent complex life histories with fractioning of the populations and prolongation of parts of the population to 2 and 3 years. The diagrams indicate in situ distributions, disregarding the complexity of the cycles. In these examples diapause occurs together with developing strains of the populations in the plankton. The different life history strategies and the varying proportions of nauplii, copepodids, and adults in the plankton and diapause stages in the sediments are regulated by a complex interplay of such abiotic and biotic factors as lack of oxygen and food, competition, and predation. In the population in Fig. 3C, a period of reproduction occurs during the winter, coinciding with diapause in the mud. As egg-bearing females are relatively large and the egg sacs conspicuous, the predation pressure from fish must be very small (Elgmork and Langeland in press). Diapause as an escape from predation (cf. Nilssen 1977, Elgmork et al. 1978) can hardly be applied to this population. The lakelet in Fig. 3A is void of fishes and the presence of diapause in this habitat is best explained as a response to extreme abiotic environments.

Adaptive significance

The ultimate aim of the interaction of genes and environment is to optimize reproductive success. The adaptive significance of evolutionary alternatives in life history strategies may, however, be difficult to comprehend.

The adaptive significance of diapause in freshwater cyclopoids centers around two aspects, an escape in time from unfavorable environmental conditions and the timing of important events in the seasonal cycle. Both these principles are illustrated by the example in Fig. 2. The diapause stage is much more tolerant to extremes in the environment such as low temperature, lack of oxygen, etc. Under extreme conditions the diapause stage is the only means of survival in the locality. Diapause thus makes it possible for species to inhabit niches and regions which would otherwise be inaccessible. The diapause phase may be regarded as a buffer system in the life history, one that secures survival during unfavorable periods and catastrophies.

In the population of *C. strenuus* illustrated in Fig. 2, the diapause development during the winter leads to a very rapid development to adults and reproduction in early spring. By means of the diapause mechanism this species takes advantage of the phenological niche of early spring before all other members of the zooplankton community appear (Elgmork 1959, 1964). Generally diapause helps to time the reproductive phase at the optimal period of the year.

A further advantage of diapause may be an increase in the fecundity stimulated by the chilling effect during diapause (Elgmork 1964: p. 72).

With increasing water volume, survival value of diapause becomes less obvious. The effect of abiotic factors in large lakes is less severe than in the shallow localities discussed above. Biotic factors as food availability and community interactions such as competition and predation become more possible as selective forces. Observations of such mechanisms are at present practically non-existent. Nilssen (1977) presented some evidence that diapause in *C. abyssorum* may be regarded as an escape from predation by fish during the winter. Also in insects diapause is considered to minimize the risk of predation (Southwood 1978).

The study of the evolutionary ecology of freshwater cyclopoids is only in its starting phase. The results so far seem promising and this animal group appears well suited for studies of the evolutionary ecology of life histories because of its large number of well defined ontogenetical instars. It is fairly easy to obtain representative samples both of the population composition and of abiotic parameters. It should be emphasized that a life cycle study of a planktonic freshwater cyclopoid should include an investigation of the bottom sediment for a demonstration of a possible diapause—a stage so important in the life history of many cyclopoid copepods.

References

ANDREWARTHA, H. G. 1952. Diapause in relation to the ecology of insects. Biol. Rev. **27**: 50-107.

BREWER, R. H. 1964. The phenology of *Diaptomus stagnalis* (Copepods: Calanoida): the development and the hatching of the egg stage. Physiol. Zool. **37**: 1-20.

COKER, R. E. 1933. Arrêt du développement chez les copépodes. Bull. Biol. Fr. Belg. **67**: 276-287.

COLE, G. A. 1953. Notes on copepod encystment. Ecology **34**: 208-211.

DINGLE, H. [Ed.] 1978. Evolution of insect migration and diapause. Springer.

EINSLE, U. 1964. Larval Entwicklung von Cyclopoiden und Photoperiodik. Naturwissenschaften **14**: 345.

———. 1968. Die Gattung *Mesocyclops* im Bodensee. Arch. Hydrobiol. **64**: 131-169.

———. 1975. Revision der Gattung *Cyclops s. str.* speziell der *abyssorum*-Gruppe. Mem. Ist. Ital. Idrobiol. **32**: 57-219.

ELGMORK, K. 1955. A resting stage without encystment in the annual life cycle of the freshwater copepod *Cyclops strenuus strenuus*. Ecology **36**: 739-743.

———. 1959. Seasonal occurance of *Cyclops strenuus strenuus* in relation to environment in small water bodies in southern Norway. Folia Limnol. Scand. **11**: 196 p.

———. 1962. A bottom resting stage in the planktonic freshwater copepod *Cyclops scutifer* Sars. Oikos **13**: 306-310.

———. 1964. Dynamics of zooplankton communities in some small inundated ponds. Folia Limnol. Scand. **12**: 83 p.

———. 1965. A triennial copepod (Crustacea) in the temperate zone. Nature **205**: 413.

———. 1967. Ecological aspects of diapause in copepods. Proc. Symp. Crustacea **3**: 947-954.

———. 1973. Bottom resting stages of planktonic cyclopoid copepods in meromictic lakes. int. Ver. Theor. Angew. Limnol. Verh. 18: 1474-1478.

———, and A. LANGELAND. In press. *Cyclops scutifer* Sars—One and two year life cycles with diapause in the meromictic lake Blankvann. Arch. Hydrobiol.

———, and J. P. NILSSEN. 1978. Equivalence of copepod and insect diapause. Int. Ver. Theor. Angew. Limnol. Verh. 20:2511-2517.

———, ———, T. BROCH, and R. ØVREVIK. 1978. Life cycle strategies in neighbouring populations of the copepod *Cyclops scutifer* Sars. Int. Ver. Theor. Angew. Limnol. Verh. 20:2518-2523.

HALVORSEN, G., and K. ELGMORK. 1976. Vertical distribution and seasonal cycle of *Cyclops scutifer* Sars (Crustacea, Copepoda) in two oligotrophic lakes in southern Norway. Norw. J. Zool. 24: 143-160.

HOY, M. A. 1978. Variability in diapause attributes of insects and mites: Some evolutionary and practical implications. *In* H. Dingle [ed.] Evolution of insect migration and diapause. Springer.

LEES, A. D. 1955. The physiology of diapause in arthropods. Cambridge Univ. Press.

MANSINGH, A. 1971. Physiological classification of dormancies in insects. Can. Entomol. 103: 983-1009.

NILSSEN, J. P. 1977. Cryptic predation and the demographic strategies of two limnetic cyclopoid copepods. Mem. Ist. Ital. Idrobiol. 34: 187-196.

———, and K. ELGMORK. 1977. *Cyclops abyssorum*—Life cycle dynamics and habitat selection. Mem. Ist. Ital. Idrobiol. 34: 197-238.

SMYLY, W. J. P. 1967. A resting stage in *Cyclops dybowski* Lande (Crustacea: Copepoda). Naturalist 903: 125-126.

———. 1973. Bionomics of *Cyclops strenuus abyssorum* Sars (Copepoda: Cyclopoida). Oecologia 11: 163-186.

SPINDLER, K.-D. 1969. Untersuchungen zur Dormanz bei *Cyclops vicinus*. Naturwissenschaften 56: 93-94.

———. 1970. Die Bedeutung der Photoperiode für die Entwicklung von *Cyclops vicinus*. Zool. Anz. 33(suppl.): 190-195.

———. 1971. Dormanzauslösung und Dormanzcharakteristika beim Süsswassercopepoden *Cyclops vicinus*. Zool. Jahrb. Abt. Allg. Zool. Jb. Physiol. Tiere 76: 139-151.

SOUTHWOOD, T. R. E. 1978. Habitat, the templet for ecological strategies. J. Anim. Ecol. 46: 337-365.

WATSON, N. H. F., and B. N. SMALLMAN. 1971a. The role of photoperiod and temperature in the induction and termination of an arrested development in two species of freshwater cyclopoid copepods. Can. J. Zool. 49: 855-862.

———, and ———. 1971b. The physiology of diapause in *Diacyclops navus* Herrick (Crustacea, Copepoda). Can. J. Zool. 49: 1449-1454.

WIERZBICKA, M. 1966. Les résultats des recherches concernant l'état de repos (resting stage) des Cyclopoida. Int. Ver. Theor. Angew. Limnol. Verh. 16: 592-599.

———. 1972a. The metabolic products of copepodites of various Cyclopoida species during their resting stage. Pol. Arch. Hydrobiol. 19: 279-290.

———. 1972b. Distribution of Cyclopoida copepodids in the resting stage in the bottom sediments of astatic reservoirs. Pol. Arch. Hydrobiol. 19: 369-376.

39. When and How to Reproduce: A Dilemma for Limnetic Cyclopoid Copepods

Jens Petter Nilssen

Abstract

Reproductive strategies in limnetic cyclopoid copepods appear closely adapted to the predictability of the environment. A more predictable environment allows for a more exact timing of the amount of reproductive effort. A diapause phase may be beneficial for synchronization of emergence and reproduction and provides the diapausing animals with a predator refuge and an energy advantage at reproduction which may result in larger clutches than for animals inhabiting free waters during adverse seasons. Cyclopoid copepods without a diapause phase frequently fractionate the deme into different developmental lines and resultant length of life cycles. A common life cycle strategy in large oligotrophic lakes seems to be co-occurring annual and biennial lines with gene flow taking place between these two lines during midsummer. The tradeoff between longevity and reproduction has resulted in two different strategies: the larger sized, biennial fraction exhibits a "big-bang" strategy, producing large clutches within a restricted period of time when food is in abundance and competitors usually are absent. Conditions for the smaller sized, annual fraction are less predictable, resulting in production of smaller clutches extending over a greater part of the life period. Both the intervening diapause and the fractionation of the deme result in a finer adaptation to prevailing food abundance and predator activity in the immediate environment.

It is widely accepted that organisms assimilate a finite and limited amount of energy, and that they allocate it to maintenance, growth, reproduction, competition, and predator avoidance. Thus an organism that devotes a proportion of its total energy to avoidance of predation has correspondingly less available for the other compartments, i.e. there are tradeoffs among the state variables. However, an animal's ability to compete or to avoid predators is probably not a simple function of energy allocation, but morphological adaptation might also be important (cf. Allan 1976; Southwood 1976a). Tradeoffs between the different state variables over time represent the life history or life cycle strategy of that organism. Since evidence is accumulating that there is a real cost associated with reproducing (Stearns 1976), the reproductive strategies of organisms are important to elucidate. All such strategies may be thought of as compromises reflecting balances of selection pressures on parental survival, fecundity, survival of offspring, and other factors. My aim here is to investigate the reproductive strategies in some limnetic cyclopoid copepods, mainly those with seasonal growth arrest.

Reproduction in limnetic cyclopoid copepods

Limnetic cyclopoid copepods show obligate sexuality. Generally males appear prior to females and may also have a shorter life span (Elgmork 1959). Mating follows sexual

I thank Kåre Elgmork for critically reading this manuscript.

maturity, and the females may store sperm in a spermathecal sac, thus minimizing the need for subsequent mating (cf. Lucks 1937; Hopkins 1977). Fertilized eggs are carried in two egg sacs attached to either side of the genital segment that may be considered a parental care mechanism for increasing egg survival. After a specific, mainly temperature-dependent period (Smyly 1974; Munro 1974), the eggs hatch into a free-swimming larval stage, the nauplius. Six nauplier stages (Elgmork and Langeland 1970), five copepodid stages, and one adult stage are recognized.

Timing of the amount of reproductive effort

The timing of the amount of reproductive effort to coincide with the existence of favorable environments depends on a large number of phenological factors, among the most significant of which are food availability and competitor and predator avoidance for both offspring and the parent generation (Nilssen 1978a). The reproductive strategies are closely adapted to the predictability of the environment (cf. Cohen 1967, 1968; Nilssen 1978a). A more predictable environment allows for a more exact timing of reproduction. For instance in temperate, dimictic, oligotrophic lakes where environmental conditions can be predicted within relatively narrow limits, the main production of progeny usually coincides with the spring outburst of nannophytoplankton. Another peak in progeny production may coincide with the autumn nannoplankton peak (McLaren 1964; Halvorsen and Elgmork 1976; Nilssen 1978a).

In limnetic cyclopoids a period of growth arrest often intervenes leading to a prolongation of the life cycle (Fig. 1). The delayed maturity produced by a dormancy phase probably gives some reproductive advantage compared with the animals inhabiting the free waters (Fig. 1). The diapause fraction probably exhibits a higher survival rate, as they have a spatial refuge from fish predators by remaining in the sediments (Nilssen 1977). The extensive energy accumulation in the prediapause phase (cf. Elgmork and Nilssen 1978), the low metabolic cost of the diapause phase (Watson and Smallman 1971), and the observation of no growth whatsoever (Elgmork 1959), all assure that the cohort reviving from diapause has more energy available for reproduction compared with the cohort remaining in the plankton. Moreover, the energy accumulation during the predormancy period is readily mobilized and used in the first clutches after dormancy (cf. Vijverberg 1977), and would thus seem to be a way of spreading the energetic cost of reproduction over a physiologically manageable span of time. In addition, a diapause phase can be beneficial in resulting in a higher degree of synchronization of emergence and reproduction (Elgmork 1959; Nilssen and Elgmork 1977).

Often cyclopoid copepods that lack a diapause phase fractionate the deme into different lines with different development rates and resulting length of life cycles (Fig. 2; Halvorsen and Elgmork 1976; Elgmork et al. 1978). One line frequently shows a 1-year (annual) life cycle while the other line exhibits a biennial cycle, with gene flow occurring between these two lines during midsummer (cf. Axelson 1961). The fraction displaying delayed maturity (i.e. that which results in a biennial life cycle) exhibits some important demographic and bionomic advantages compared with the annual line: the delayed maturity increases the energy devoted to maintenance and growth, resulting in a larger size at maturity (cf. Axelson 1961; Halvorsen and Elgmork 1976). This development line consequently produces a larger clutch size than the annual line (cf. Halvorsen and Elgmork 1976).

A high degree of synchrony between adult males and females with respect to seasonal development appears to be necessary. The propinquity of individuals necessary for efficient reproduction could not be accomplished on the basis of randomly distributed development, except perhaps in populations of extremely high density and in seasonally stable environments. The proximate timing of reproductive effort may depend on a number of factors, the influences of which are largely unknown in

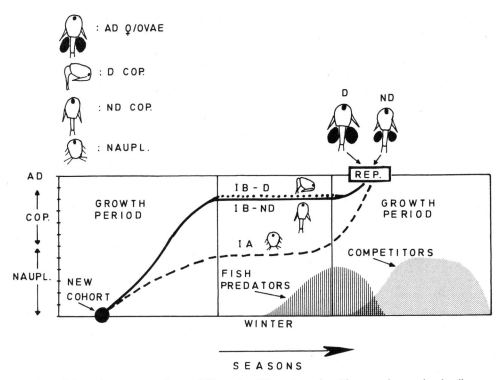

Fig. 1. Schematic summary of annual life cycle of limnetic cyclopoid copepods experiencing "coarse-grained" spatial environments. IA—Naupliar fraction during winter; IB—advanced copepodid fraction during winter; D—animals in diapause or egg-bearing females originating from diapausing individuals; ND—animals in a nondiapause phase or egg-bearing females originating from nondiapausing individuals; REP—reproduction period; AD—adults; COP—advanced copepodids; NAUPL—nauplii. Development line without diapause (—); development line with diapause (. . . . . .); alternative annual development line with nauplii during winter (- - - - -) (based on Smyly 1973a, b; Nilssen 1977, 1978a).

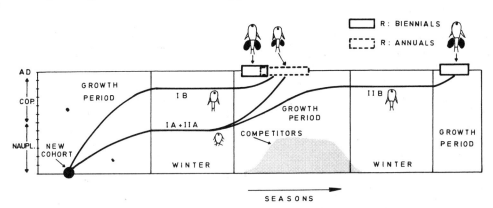

Fig. 2. Schematic summary of life cycle of limnetic cyclopoid copepods from large oligotrophic lakes with co-occurring annual and biennial lines without a diapause phase. R—Reproduction period; IA—Naupliar fraction of annual line; IIA—naupliar fraction of biennial line; IB—advanced copepodid fraction of annual line; IIB—advanced copepodid fraction of biennial line. Other details as in Fig. 1 (based on Axelson 1961; Halvorsen and Elgmork 1976; Nilssen 1978a, in prep.).

nature. Photoperiodism may frequently be important in the adaptation that promotes mating and reproduction for crustaceans and rotifers (cf. Segerstråle 1970, 1971; Clément and Pourriot 1972, 1974; Pourriot and Clément 1973). Under laboratory conditions Spindler (1969, 1971a, b) showed that photoperiodism was of decisive importance in timing the egg deposition, hatching, naupliar and copepodid development in *Cyclops vicinus*, but he was unable to assess its adaptive significance in nature.

Amount of reproductive effort

Clutch size—Clutch size in limnetic cyclopoid copepods generally increases with increasing food levels (Smyly 1973a, b). Clutch size seems to be highly dependent upon body size of mature females. In lakes where the biennial fraction greatly outnumbers the annual (*see Fig. 2*), this has resulted in a larger clutch size than if the annual fraction is more important (Halvorsen and Elgmork 1976). This works via the larger body sizes of the biennial females, since they as a rule far exceed those of the yearly fraction (cf. Axelson 1961; Halvorsen and Elgmork 1976).

In the majority of localities clutch size is considerably higher in late spring and early summer and decreases sharply after the first month of reproduction (Røen 1957). Thereafter clutch size is more or less constant with a possible secondary peak coinciding with the autumnal increase in nannoplankton (cf. Halvorsen and Elgmork 1976). The large clutch size in the early period of the reproductive season seems to depend mainly on the rapid mobilization of fat reserves to reproduction (cf. Vijverberg 1977). These reserves accumulated during the previous autumn production period (cf. Elgmork and Nilssen 1978). The large clutch size is also connected with very few competitors during spring (Nilssen 1978a). The differences in clutch size between seasons may also be explained in some populations by the various premature life histories of the females and the tradeoff strategy mentioned above: early in the season the biennial fraction, with larger body size, makes up the majority of the mature female population, while later in the season the annual fraction, with smaller female body size, dominates the female population.

Some largely eutrophic environments may produce polymorphic female populations during the spring outburst of algae in that an additional large morph occurs sympatrically with the perennial smaller form (Nilssen, unpubl.). The generally strong correlation between mature female body size and clutch size (cf. McLaren 1963, 1965, 1974; Walker 1970; Smyly 1973a), results in an increased clutch size at that time (Walker 1970).

In a polycyclic, polyphenic species, like *Cyclops strenuus* (Elgmork 1959; Einsle 1975), the first summer generation that resulted from the spring-reproducing cohort exhibits smaller female body size and correspondingly smaller clutch size than the females originating from diapausing individuals (Fig. 3). In contrast to the diapause fraction that profited from earlier gathered food reserves, in the short-lived summer generation (Einsle 1975), the foraging females themselves must gather material and energy for subsequent clutches.

In populations with strongly synchronized life histories, female body size and clutch size exhibit negligible seasonal variations (Nilssen and Elgmork 1977).

Numbers and frequency of clutches—The number of clutches generally depends on available food (cf. Marshall and Orr 1972) and premature history of the female population (Smyly 1970; Whitehouse and Lewis 1973). Cyclopoid copepods are probably iteroparous: stored sperm from one insemination is capable of fertilizing many broods (cf. Smyly 1970; Whitehouse and Lewis 1973; Hunt and Robertson 1977). The frequency of clutches may depend on water temperature, egg development time, and food availability for the reproducing females (cf. McLaren 1963, 1965; Corkett and McLaren 1969). More food available for the reproducing females generally results in more clutches (Smyly 1970), but the frequency of clutches with food in surplus depends on the temperature-dependent egg

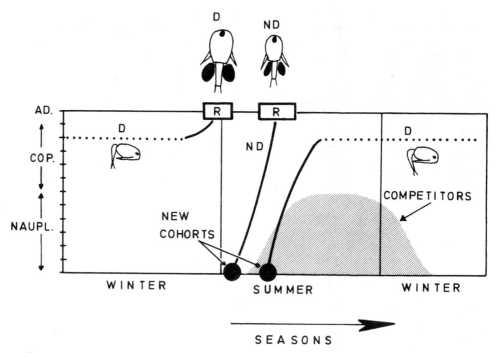

Fig. 3. Schematic summary of life cycle of free-living cyclopoid copepods displaying a biennial life cycle with one diapause phase. Details as in Fig. 1 (based on Elgmork 1959; Einsle 1975; Nilssen 1978a).

development time, which then poses a physiological bottleneck for the production of maximum numbers of offspring in a food surplus environment (cf. McLaren 1965; Corkett and McLaren 1969).

Generally, the length of the reproductive period depends on the predictability of the local environment. In large, predictable oligotrophic lakes the reproductive period is shorter than in unpredictable, eutrophic environments (e.g. Nilssen and Elgmork 1977).

In a number of lakes the tradeoff between longevity and reproduction has resulted in two different strategies: the biennial fraction exhibits a "big-bang" strategy (sensu Gadgil and Bossert 1970), reproducing when food is in abundance and when competitors usually are absent (see Fig. 2). The conditions for the annual fraction are less predictable, resulting in a reproductive effort extending over a greater part of the life period (Fig. 2). The 2-year fraction could theoretically have shortened its life cycle by reproducing the previous autumn, but every reproductive pattern should be determined by the relative importance of unsuccessful reproduction vs. adult mortality (cf. Schaffer 1974). Conditions for reproduction and growth of offspring are unfavorable during late autumn and early winter compared with spring; therefore the population uses the strategy of postponing reproduction until the subsequent spring.

Egg size—It is obvious that as the energy expended on individual offspring increases, the number of offspring that parents can produce decreases. While considerable interest has been devoted to clutch size and deposition of clutches in limnetic cyclopoid copepods, egg size studies are remarkably rare. Cyclopoid egg size probably does not reflect environmental conditions as closely as in the calanoid copepods (cf. Hutchinson 1951, 1967; Czeczuga 1959, 1960), since the latter usually are herbivorous members of the aquatic biota. It seems as though reproductive effort (clutch volume/body volume) is greater in more eutrophic environments (Smyly 1973a), thus being closer to r-strategists in the r–K continuum in such unpredictable environments (cf. Pianka 1970; Stearns 1976; Southwood 1976b). The body size of the reproducing female and the clutch volume in the majority of cases is greater in eutrophic lakes, but the egg size or the energy devoted to each egg as a parental care strategy may be greater under less eutrophic environments (cf. Czeczuga 1960). Despite the strongly positive correlation between female body size and clutch size, a constant egg size may be present even from females showing considerable size differences. In the dicyclic, polyphenic *C. strenuus*, the larger spring reproductive cohort and the smaller early summer reproductive cohort showed much the same egg size (cf. Elgmork and Halvorsen 1976), which suggests that a larger amount of parental care is devoted to the latter as a sort of K-strategy (cf. Pianka 1970) as a result of the considerable competition during summer (cf. Fig. 3).

Intra- and interdemic adaptations

Increasing food availability results in increasing clutch volume of cyclopoid copepods (Smyly 1973a), but clutch volume seems to a certain extent to be genetically determined. The addition of a similar amount of food to four adjacent populations of *Cyclops abyssorum* still resulted in largely different clutch volumes, with the individuals propagated from the more nutrient-rich environments showing the highest clutch volumes as in nature (Smyly 1973a). As a rule, large *C. abyssorum* appeared to invest more of the body mass in each clutch than smaller ones (Smyly 1973a), although copepods generally devote far less of their assimilated energy to reproduction than to maintenance (Comita 1964; Harris 1973). Generally clutch volume early in the reproductive season far exceeds that invested later (Røen 1957; Elgmork 1959), and the reproductive effort also seems to decrease (Comita 1964; Smyly 1973a). A small clutch need not imply a low reproductive effort however. Early in the season females have the large energetic reserves that accumulated during the previous growth season. Most of these reserves are probably rapidly mobilized and used in the first clutches, while the summer generation must collect material and the foraging females must devote energy to subsequent clutches (Fig. 3).

If the nutrient input to an aquatic ecosystem increases, this may lead to an imbalance in the biota and result in a less predictable environment (Natl. Acad. Sci. 1969; Nilssen 1978b). Generally, populations living in such environments reproduce many times, while females of populations from more predictable environments reproduce correspondingly fewer times (e. g. Nilssen and Elgmork 1977).

Under stable conditions, e. g. the tropics, the main developmental stages show an almost constant age structure throughout the year and reproduction is continuous (Burgis 1971). The cyclopoids show no diapause (Burgis pers. comm.), which probably would not be of selective advantage, since there is no reason to synchronize life cycle to any specific environmental factor in view of the constant predation patterns and food availability. Consequently reproduction in a temperate lake and a tropical lake differs considerably: in a temperate lake the timing of reproductive effort is important, egg volume is larger, female body volume is larger and clutch size per female is larger than in the tropics (Burgis and Walker 1972). The reproductive effort (egg volume/body volume) is much the same, however (Burgis and Walker 1972). One could expect reproductive effort to increase in the tropics in view of the larger number of biotic interactions, and a strategy selected under such an environment would be more of a K-strategy in the r–K continuum. The temperate lake used in this comparison was Loch Leven (Burgis and Walker 1972), a highly eutrophic lake, and the comparison involved a copepod with correspondingly large clutch volumes (Walker 1970), prolonged reproductive period (Walker 1970), and higher reproductive effort compared with oligotrophic lakes (cf. Smyly 1973a). In a predictable oligotrophic lake system the reproductive effort is most probably less than in the tropics (cf. Burgis and Walker 1972; Smyly 1973a).

Zooplankton are well equipped with chemical and mechanical receptors that are likely to detect seasonal changes in the environment (Strickler 1975) such as the spring outburst of algae providing substantial growth for the progeny. If in a changing environment, as in a temperate lake system, juvenile survival can be predicted by the parent, the life-history strategy should be to postpone or minimize reproduction until a favorable period is anticipated. The individuals then should expend a high amount of energy in reproductive effort, even if maximal reproductive effort would result in a short adult life span. The loss of reproductive potential resulting from reduced adult survival is outweighed by the selective advantage of early and maximal reproduction because juvenile survival is, in such a case, predictably high (cf. Giesel 1976). This strategy is commonly found in many "winter" species, i. e. *Cyclops strenuus* (cf. Elgmork 1959; Einsle 1975; Fig. 3), *C. vicinus* (cf. Einsle 1975), *Diacyclops thomasi* (cf. Nilssen 1978a), and in many populations of *C. scutifer* with annual and biennial development lines (cf. Halvorsen and Elgmork 1976; Elgmork et al. 1978; Fig. 2). In the case of *C. scutifer*, the biennial fraction reproduces first, exhibits larger female body size, and larger clutch volume per clutch in an environment usually absent of predators and competitors for offspring (Fig. 2). Adult survival is low—probably a mechanism to decrease cannibalism or as a result from predation by fish (Nilssen 1977). The annual fraction reproduces later, when more competitors and predators are present, have smaller body size, and smaller clutch volumes per clutch, but parent survival seems to be higher than in the biennial fraction. Consequently, the 2-year fraction often shows a higher fecundity than the annual fraction, and its total share in the progeny production of the local population may be comparable to the annual fraction in spite of its lower numbers (cf. Halvorsen and Elgmork 1976).

The adaptive significance of the splitting of the populations into fractions has recently been suggested by Nilssen (1977) to result from size-selective predation by fish. The share of the annual fraction in the total population increases with increasing fish predation, while decreasing fish predation may lead to a larger share of the biennial fraction. Therefore the reproductive strategy of the fractionating *C. scutifer* populations seems to be a compromise between size-selective predation and expectation of future offspring production. A more detailed account of this phenomenon is underway (Nilssen in prep.). Comparable results are given in McLaren (1966) for a marine carnivorous zooplankton. He reasoned that where generation length is set by marked seasonality of food supply, as in the arctic, high fecundity and associated large size and slow development may be selected for. A careful study of the benthic detrivore *Chironomus anthracinus* (Diptera: Chironomidae) by Jónasson (1965, 1970, 1972) deserves special mention since some of his findings parallel mine. The life cycle of this species also showed co-occurring annual and biennial lines, in which complex predation patterns determined the ratios between the different fractions.

References

ALLAN, J. D. 1976. Life history patterns in zooplankton. Am. Nat. **110**: 165–180.

AXELSON, J. 1961. On the dimorphism in *Cyclops scutifer* (Sars) and the cyclomorphosis in *Daphnia galeata* (Sars). Rep. Inst. Freshwater Res. Drottningholm **42**: 169–182.

BURGIS, M. J. 1971. The ecology and production of copepods, particularly *Thermocyclops hyalinus*, in the tropical Lake George, Uganda. Freshwater Biol. **1**: 169–192.

———, and A. F. WALKER. 1972. A preliminary comparison of the zooplankton in a tropical and a temperate lake (Lake George, Uganda and Loch Leven, Scotland). Int. Ver. Theor. Angew. Limnol. Verh. **18**: 647–655.

CLEMENT, P., and R. POURRIOT. 1972. Photopériodisme et cycle hététogonique chez certains Rotifères Monogonontes. I. Observations préliminaires chez *Notommata copeus*. Arch. Zool. Exp. Gen. **113**: 41–50.

———, and ———. 1974. Photopériodisme et cycle

hétérogonique chez Notommata copeus (Rotifère). III. Recherche du seuil minimal d'éclairement. Arch. Zool. Exp. Gen. 115: 641–650.

COHEN, D. 1967. Optimizing reproduction in a randomly varying environment when a correlation may exist between the conditions at the time a choice has to be made and the subsequent outcome. J. Theor. Biol. 16: 1–14.

——. 1968. A general model of optimal reproduction. J. Ecol. 56: 219–228.

COMITA, G. W. 1964. The energy budget of *Diaptomus siciloides*, Lilljeborg. Int. Ver. Theor. Angew. Limnol. Verh. 15: 646–653.

CORKETT, C. J., and I. A. McLAREN. 1969. Egg production and oil storage by the copepod *Pseudocalanus* in the laboratory. J. Exp. Mar. Biol. Ecol. 3: 90–105.

CZECZUGA, B. 1959. Oviposition in *Eudiaptomus glacilis* G. O. Sars and *E. graciloides* Lilljeborg (*Diaptomidae, Crustacea*) in relation to season and trophic level of lakes. Bull. Acad. Pol. Sci. Ser. Sci. Biol. 7: 227–230.

——. 1960. Zmiany płodności niektórych przedstawicieli zooplanktonu I. *Crustacea* jezior Rajgrodzkich. [in Polish, English summary]. Pol. Arch. Hydrobiol. 7: 61–93.

ENSLE, U. 1975. Revision der Gattung *Cyclops* s. str., speziell der *abyssorum*-Gruppe. Mem. Ist. Iatl. Idrobiol. 32: 57–219.

ELGMORK, K. 1959. Seasonal occurrence of *Cyclops strenuus strenuus* in relation to environment in small water bodies in southern Norway. Folia Limnol. Scand. 11: 1–196.

——, and G. HALVORSEN. 1976. Bodysize of free-living copepods. Oikos 27: 27–33.

——, and A. LANGELAND. 1970. The number of naupliar instars in Cyclopoida (Copepoda). Crustaceana 18: 277–282.

——, and J. P. Nilssen. 1978. Equivalence of copepod and insect diapause. Int. Ver. Theor. Angew. Limnol. Verh. 20: 2511–2571.

——, ——, T. BROCH, and R. ØVREVIK. 1978. Life cycle strategies in neighbouring populations of the copepod Cyclops scutifer Sars. Int. Ver. Theor. Angew. Limnol. Verh. 20: 2518–2523.

GADGIL, M., and W. H. BOSSERT. 1970. Life history consequences of natural selection. Am. Nat. 104: 1–24.

GIESEL, J. T. 1976. Reproductive strategies as adaptations to life in temporally heterogenous environments. Annu. Rev. Ecol. Syst. 7: 57–79.

HALVORSEN, G., and K. ELGMORK. 1976. Vertical distribution and seasonal cycle of *Cyclops scutifer* Sars (Crustacea, Copepoda) in two oligotrophic lakes in southern Norway. Norw. J. Zool. 24: 143–160.

HARRIS, R. P. 1973. Feeding, growth, reproduction and nitrogen utilization by the harpacticoid copepod, *Tigiopus brevicornis*. J. Mar. Biol. Assoc. U. K. 35: 785–800.

HOPKINS, C. C. E. 1977. The relationship between maternal body size and clutch size, development time and egg mortality in *Euchaeta norvegica* (Copepoda: Calanoida) from Loch Etive, Scotland. J. Mar. Biol. Assoc. U. K. 57: 723–733.

HUNT, G. W., and A. ROBERTSON. 1977. The effect of temperature on reproduction of *Cyclops vernalis* Fischer (Copepoda, Cyclo-Crustaceana 32: 169–177.

HUTCHINSON, G. E. 1951. Copepodology for the ornithologist. Ecology 32: 571–577.

——. 1967. A treatise in limnology, v. 2. Wiley.

JÓNASSON, P. M. 1965. Factors determining population size of *Chironomus anthracinus* in Lake Esrom. Mitt. Int. Ver. Theor. Angew. Limnol. 13 p. 139–162.

——. 1970. Population studies on *Chironomus anthracinus*. Proc. Adv. Study Inst. Dynam. Numbers Popul. (Oosterbeek, 1970), p. 220–231.

——. 1972. Ecology and production of the profundal benthos in relation to phytoplankton in Lake Esrom. Oikos 14(suppl.): 1–148.

LUCKS, R. 1937. Die Crustaceen und Rotatorien des Messinasees. Ber. Westpreussischen Bot. -Zool. Ver. 59: 59–101.

McLAREN, I. A. 1963. Effects of temperature on growth of zooplankton and the adaptive value of vertical migration. J. Fish. Res. Bd. Can. 20: 685–727.

——. 1964. Zooplankton of Lake Hazen, Ellesmere Island, and a nearby pond, with special reference to the copepod *Cyclops scutifer* Sars. Can. J. Zool. 42: 613–629.

——. 1965. Some relationships between temperatures and egg size, body size, development rate, and fecundity, of the copepod *Pseudocalanus*. Limnol. Oceanogr. 10: 528–538.

——. 1966. Adaptive significance of large size and long life of the chaetognath *Sagitta elegans* in the arctic. Ecology 47: 852–855.

——. 1974. Demographic strategy of vertical migration by a marine copepod. Am. Nat. 108: 91–102.

MARSHALL, S. M., and A. P. ORR. 1972. The biology of a marine copepod. 1st reprint. Springer.

MUNRO, I. G. 1974. The effect of temperature on the development of egg, nauplier and copepodite stages of two species of copepods, *Cyclops vicinus* Uljanin and *Eudiaptomus gracilis* Sars. Oecologia (Berl.) 16: 355–367.

NATIONAL ACADEMY OF SCIENCES. 1969. Eutrophication: Causes, consequences, correctives. Publ. 1700.

NILSSEN, J. P. 1977. Cryptic predation and the demographic strategy of two limnetic cyclopoid copepods. Mem. Ist. Ital. Idrobiol. 24: 187–195.

——. 1978a. On the evolution of life histories of limnetic cyclopoid copepods. Mem. Ist. Ital. Idrobiol. 36: 193–214.

―――. 1978b. Eutrophication, minute algae and inefficient grazers. Mem. Ist. Ital. Idrobiol. 36: 127-138.

―――, and K. ELGMORK. 1977. Cyclops abyssorum—life cycle dynamics and habitat selection. Mem. Ist. Ital. Idrobiol. 34: 197-238.

PIANKA, E. R. 1970. On r and K selection. Am. Nat. 104: 592-597.

POURRIOT, R., and P. CLÉMENT. 1973. Photopériodisme et cycle hétérogonique chez Notommata copeus (Rotifère Monogononte). II. Influence de la qualité de la lumière. Spectres d'action. Arch. Zool. Exp. Gen. 114: 277-300.

RØEN, U. 1957. Contributions to the biology of some Danish free-living freshwater copepods. Kgl. Dan. Vidensk. Selsk. Biol. Skr. 9(2): 1-101.

SCHAFFER, W. M. 1974. Optimal reproductive effort in fluctuating environments. Am. Nat. 108: 783-790.

SEGERSTRÅLE, S. G. 1970. Light control of the reproductive cycle of Pontoporeia affinis Lindström (Crustacea Amphipoda). J. Exp. Mar. Biol. Ecol. 5: 272-275.

―――. 1971. Light and gonad development in Pontoporeia affinis. Proc. European Mar. Biol. Symp. (4th), p. 573-581.

SMYLY, W. J. P. 1970. Observations on rate of development, longevity and fecundity of Acanthocyclops viridis (Jurine) (Copepoda, Cyclopoida) in relation to type of prey. Crustaceana 18: 21-36.

―――. 1973a. Clutch-size in the freshwater cyclopoid copepod, Cyclops strenuus abyssorum Sars in relation to thoracic volume and food. J. Nat. Hist. 7: 545-549.

―――. 1973b. Bionomics of Cyclops strenuus abyssorum Sars (Copepoda: Cyclopoida). Oecologia (Berl.) 11: 163-186.

―――. 1974. The effect of temperature on the development time of the eggs of three freshwater cyclopoid copepods from the English Lake District. Crustaceana 27: 278-284.

SOUTHWOOD, T. R. E. 1976a. Bionomic strategies and population parameters, p. 26-48. In R. M. May [ed.], Theoretical ecology—principles and applications. Blackwell.

―――. 1977b. Habitat, the templet for ecological strategies? J. Anim. Ecol. 46: 337-365.

SPINDLER, K.-D. 1969. Die Bedeutung der Photoperiode für die Entwicklung von Cyclops vicinus. Zool. Anz. 33(suppl.): 190-195.

―――. 1971a. Untersuchungen über den Einfluss äusserer Faktoren auf die Dauer der Embryonalentwicklung und den Häutungsrhythmus von Cyclops vicinus. Oecologia (Berl.) 7: 342-355.

―――. 1971b. Der Einfluss von Licht auf die Eiablage des Copepoden Cyclops vicinus. Z. Naturforsch. 26b: 953-955.

STEARNS, S. C. 1976. Life-history tactics: a review of the ideas. Q. Rev. Biol. 51: 3-47.

STRICKLER, J. R. 1975. Intra- and interspecific information flow among planktonic copepods: Receptors. Int. Ver. Theor. Angew. Limnol. Verh. 19: 2951-2958.

VIJVERBERG, J. 1977. Population structure, life histories and abundance of copepods in Tjeukemeer, the Netherlands. Freshwater Biol. 7: 579-597.

WALKER, A. F. 1970. The zooplankton of Loch Leven. M. Sc. thesis, Univ. Sterling. 123 p.

WATSON, N. H. F., and B. N. SMALLMAN. 1971. The physiology of diapause in Diacyclops navus Herrick (Crustacea, Copepoda). Can. J. Zool. 49: 1449-1454.

WHITEHOUSE, J. W., and B. G. LEWIS. 1973. Effect of diet and density on development, size and egg production in Cyclops abyssorum Sars., 1863 (Copepoda, Cyclopoida). Crustaceana 25: 225-236.

VII
Cyclomorphosis

40. Environmental Control of Cladoceran Cyclomorphosis via Target-Specific Growth Factors in the Animal

Jürgen Jacobs

Abstract

After a brief survey of the environmental mechanisms known to influence cyclomorphosis in Cladocera, new data are presented on the proximate determination of helmet allometry in *Daphnia*. It seems that cyclomorphogenic factors influence mitotic rates as well as growth rates of the epidermal helmet cells. When temperature is used to modify helmet allometry, mitotic rates are significantly more affected than cell growth rates. Yet, does the helmet react more sensitively to growth-controlling factors than does the rest of the body? If so, this would mean that cyclomorphosis of the helmet simply reflects the absolute growth rate of the animal. By applying several cyclomorphogenic factors in variable combinations, the growth rate of the helmet can be modified independent of the growth rate of the animal. Just about every possible combination of helmet and carapace growth rates can be produced. This falsifies the "sensitivity hypothesis" and proves that there are helmet-specific growth determinants. Experiments with immobilized animals show that these specific growth determinants do not act externally on the helmet. Since the helmet cells are not innervated, helmet-specific growth factors in the hemolymph must be responsible for cyclomorphosis in *Daphnia*. This fact supports the thesis of a selective advantage of daphnid cyclomorphosis.

Cyclomorphosis of rotifers and cladocerans has been investigated with respect to its causation by environmental factors, its genetics, and its adaptive significance and presumable evolutionary course. Most recent research has centered around the problem of adaptation especially in connection with selective predation (e.g. Brooks 1965; Brooks and Dodson 1965; Gilbert 1967, 1977; Hall 1967; Halbach 1971; Zaret 1972, 1975; Dodson 1974; Allan 1974; Kerfoot 1975a, 1977; Zaret and Kerfoot 1975; Jacobs 1966, 1977a, b; Lynch 1977). This contribution deals with the environmental control of cladoceran cyclomorphosis. The aim is to look closer at the unifying principle common to the various factors known to be influential.

Earlier experiments and field data have shown that temperature, food conditions, light, and turbulence modify relative growth rates of certain body parts of Cladocera (Ostwald 1904; Coker and Addlestone 1938; Brooks 1946, 1947; Hrbáček 1959; Hazelwood 1962, 1966; Jacobs 1961a, b, 1962, 1965, 1970; Egloff 1968; Kerfoot 1975b). In *Daphnia* the major changes involve the helmet blade of the head and the caudal spine. Less conspicuous features such as the rostrum and the antennal bristles may also be influenced. Warm, well fed, illuminated, and turbulated animals produce the largest helmets. Whether photoperiod has an influence independent of the sum of the received light quanta needs further clarification (Hazelwood 1962). Turbulence is virtually ineffective in the dark (Jacobs 1961a; Hazelwood 1966). Temperature and turbulence influence relative growth rates from the beginning of the second half of embryogenesis. Food becomes effective immediately after birth. All factors are influential as long as the animal is capable of growth (Jacobs 1961a). There is a compensatory mechanism, at least in daphnid helmet growth: the longer the helmet due to earlier influences, the slower it will grow (relative to the body) in later instars (and vice versa under otherwise identical conditions). This regulatory mechanism facilitates the establishment and maintenance of certain body proportions under given environmental circumstances, no matter what the earlier conditions may have been (Jacobs 1961a, b). Prenatal growth conditions may thus indirectly influence and even obscure postnatal effects (Jacobs 1970).

If several environmental factors influence the same phenotypical feature in the same fashion, the question arises what is common to the environmental factors or their action. An earlier interpretation of daphnid cyclomorphosis was that allometric growth of the helmet reflects absolute growth of the animals (Jacobs 1961b). If the helmet reacts more sensitively than the rest of the body to all factors that influence growth rates, then the helmet's relative growth rate would be the higher, the faster the animal grows. The temperature and starvation effects are easily explained by this hypothesis because short-helmeted, cold and/or starved animals definitely grow much slower than long-helmeted, warm and/or well fed animals. The same interpretation might also be extended to turbulence and light effects: It is plausible to assume that turbulated and illuminated animals have a higher metabolic rate than still and dark animals, and one might expect that higher metabolic rates would be correlated with higher growth rates. Therefore I wanted to test whether this hypothesis could be generalized for all cyclomorphogenic factors of the environment, thus yielding a unified causal explanation.

Animals and allometric measurements

The experimental animal was *Daphnia galeata mendotae* from Klamath Lake, Oregon (Fig. 1), a species which is easily kept in the laboratory with tapwater and

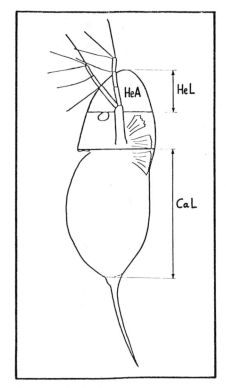

Fig. 1. *Daphnia galeata mendotae*, lateral view. CaL is carapace length, HeL is helmet length, HeA is helmet area.

green algae (*Selenastrum gracile*) or baker's yeast. All experiments were done with parthenogenic females. To measure the length of the carapace (as an index of body size) and the helmet, the animals were drawn in lateral view under the microscope with a camera lucida. A line was drawn between the rostral tip and the point between the insertion of the second and third antennal muscles on the dorsal side. A parallel line was drawn through the upper edge of the compound eye. The shortest distance from this line to the apex of the head was the helmet length. The area of the head above the line was the helmet area; the numbers of helmet cells (Table 1) refer to the epidermal cells in this area. The shortest distance from the baseline to the caudal end of the carapace (excluding the caudal spine) was the carapace length. The carapace area was assumed to vary approximately proportional

Table 1. *Daphnia galeata mendotae*. Allometric rates, b, based on measurements of neonates and adults at 5° and 24°C. CaL is carapace length, HeL is helmet length, HeA is helmet area (one side), HeCeNo is number of epidermal cells of helmet blade (one side), HeCeA is area of epidermal helmet cell. For some allometric rates CaL^2 was used as a measure proportional to carapace area.

Temp (°C)	Instar	CaL (μ)	HeL (μ)	HeA (μ²)	HeCeNo	HeCeA (μ²)	Allometric rates (b)				
							$\frac{HeL}{CaL}$	$\frac{HeA}{CaL^2}$	$\frac{HeCeNo}{CaL^2}$	$\frac{HeCeA}{CaL^2}$	$\frac{HeCeNo}{HeCeA}$
5°	Neon.	462	52	7,457	121	62	0.426	0.510	0.199	0.310	0.643
	Adults	1,269	80	20,918	181	116					
24°	Neon.	440	98	15,486	177	87	1.094	1.000	0.556	0.447	1.243
	Adults	1,312	324	137,835	596	231					
$b_{24°}/b_{5°}$		—	—	—	—	—	2.57	1.96	2.79	1.44	1.93

to the square of the carapace length. Allometric rates b were calculated according to the equation $y = ax^b$, where x and y are the phenotypic variables to be compared. b is the ratio between the absolute growth rates $\beta = dy/dt \cdot y$ and $\alpha = (dx/dt \cdot x)$; $b = 1$ indicates isometry, $b > 1$ positive allometry, $b < 1$ negative allometry. All experiments were carried out in temperature-controlled rooms (ca. ± 1°C).

Mitotic rate vs. cell growth rate

As a preliminary step it seemed useful to know how much the two possible determinants of tissue growth, viz. *mitotic rates* and *growth rates of the cells*, contribute to helmet cyclomorphosis. Simple counts and measurements on four neonate and adult helmets at 5° and 24°C showed that both phenomena were involved, but mitotic rates were more important (Table 1). The allometry of the helmet area relative to the carapace area at 24°C was about twice as high as that at 5°C ($b_{24}/b_5 = 1.96$), but the two responsible components differed markedly. The b_{24}/b_5 ratio of the allometries of cell *numbers* (which reflect mitotic rates) was 2.8 whereas the corresponding value for the allometries of cell *areas* (which reflect growth rates of the cells) was only 1.4. This means that mitosis is about twice as important as is cell growth. As a matter of fact, at 5°C the allometry of cell numbers vs. cell area was strongly negative; at 24°C it was positive (*see last column in Table 1*).

This finding suggested that variable helmet growth was something more than the simple reflection of a greater responsiveness of the helmet to environmental growth determinants. A quite specific and coordinated response of the helmet tissue appeared to be more likely. To get more information on this possibility, I made a number of experiments in which conflicting cyclomorphogenic factors were combined (for instance, turbulence with low food, or low light intensities with ample food). The idea was to try to uncouple absolute and relative growth, to make animals grow slowly and yet develop large helmets, and vice versa.

Absolute and relative growth rates

Six sets of experiments were carried out at 23°C. Each set consisted of two simultaneous experiments (a) and (b). In (a), growth-promoting conditions (rich food supply) were combined with "anticyclomorphogenic" factors (low light, individual maintenance in small beakers). In (b), bright light and turbulence in large vessels were combined with medium or low food conditions. In some of these experiments, the light was polarized in order to supply a special cue of orientation to the turbulated animals (Jander and Waterman 1960). Further details are given in Table 2, experimental sets 2-8. For comparison, two pairs of experiments were done at different temperatures (set 1) and food quantities (set 2). Except for set 2, food was controlled only in a qualitative way. As a measure of the nutritional state of the animals, the number

Table 2. Daphnia galeata mendotae. Pairs of experiments with opposing combinations of cyclomorphogenic environmental conditions. Each experiment lasted 5 days. Details given in text and Fig. 2.

Exp (set)	Growth Conditions					Growth Data			
	Illumination intensity (lux)	Mean temp (°C)	Food	Vol exp vessels (ml)	Anim/vessel	α (day^{-1})	β (day^{-1})	b^*	Mean eggs/indiv at end of exp
1 (a)	400	15	Selenastrum	1,000	40	0.090	0.066	0.729	0†
1 (b)	400	24		1,000	40	0.161*	0.156*	0.967	3.23
2 (a)	300	23.9	$5–10 \times 10^3$ cells/ml Selenastrum	500	40	0.081	0.060	0.734	0
2 (b)	300	23.9	$5–10 \times 10^5$ cells/ml	500	40	0.151*	0.160*	1.062	3.33
3 (a)	1,000	24.0	Baker's yeast	60	1	0.131	0.121	0.927	2.0
3 (b)	3,000‡	23.8		8,000§	60	0.134	0.143*	1.067	2.1
4 (a)	1,000	24.1	Selenastrum	60	1	0.113	0.098	0.870	1.04
4 (b)	12,000	23.8		8,000§	120	0.115	0.123*	1.075	0.19
5 (a)	150	23.8	Selenastrum	60	1	0.109	0.100	0.913	0.64
5 (b)	3,000‡	23.4		8,000§	60	0.103	0.120‖	1.167	0
6 (a)	300	23.2	Selenastrum	60	1	0.149	0.121	0.807	4.72
6 (b)	3,000‡	22.9		8,000§	60	0.106*	0.108‖	1.016	0.35
7 (a)	5	23.9	Selenastrum	60	1	0.118	0.107	0.912	1.16
7 (b)	12,000	24.0		8,000§	110	0.104	0.179*	1.713	0.15
8 (a)	200	24.2	Selenastrum	60	1	0.091	0.649	0.533	0.58
8 (b)	12,000	23.7		8,000§	125	0.056*	0.069*	1.238	0

* Differences between experiments (a) and (b) significant with $p < 0.005$.
† Animals not yet mature.
‡ Polarized.
§ Turbulence by bubbling air or by circulation.
‖ Differences between experiment (a) and (b) significant with $p < 0.05$.

of eggs per animal was counted at the end of each experiment. Each pair of experiments was started by placing neonates from the same stock culture in both experimental arrangements at the same time. A third group of neonates was measured to establish average initial body proportions. Each experiment lasted 5 days.

From initial and final body proportions (see Fig. 1), the absolute growth rate α of the carapace length CaL (α = dCaL/dt·CaL) as a measure of the overall growth rate of the animal, the absolute growth rate β of the helmet length HeL (β = dHeL/dt·HeL), and the allometric rate b of the helmet ($b = \beta/\alpha$) were calculated for each individual.

Figure 2 shows the results of these experiments. In sets 1 and 2 (temperature and food as environmental variables), there was a positive correlation between α and b: the faster the animals grew, the higher was the growth rate of the helmet relative to the rest of the body. This would be in accordance with the "sensitivity hypothesis." Sets 3–8, however, disprove this hypothesis as a general principle: The specific as well as the relative growth rates of the helmets were always greater in the large vessel (b) than in the individual cultures (a), but the growth rates of the animals as measured by α varied independently. All kinds of combinations between absolute and relative

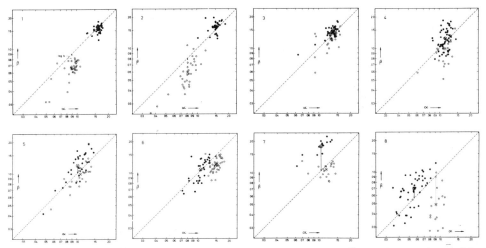

Fig. 2. Correlation between specific growth rates of helmet length (β) and carapace length (α) of *Daphnia galeata mendotae* in eight pairs of experiments under various environmental conditions (*see Table 2*). Open circles—experiment (a); solid circles—experiment (b). Circles with crosses are corresponding mean values. Double log plot. Vertical distance of a value from line of isometric growth ($\alpha = \beta$, dashed line) is log of rate of allometry $b = \beta/\alpha$. Above line of isometry, allometry is positive; below, it is negative.

growth were found. In sets 3, 4, 5, and 7, α was almost identical in both experiments, in sets 6 and 8, α was significantly smaller in the large vessel than in the individual cultures.

Thus, the growth rate of the helmet is obviously independent of the animals' general growth rate. There must then be a helmet-specific agent that stimulates growth and mitosis of the epidermal cells of the helmet independent of the cells of the rest of the body.

External vs. internal control

The most logical conclusion would seem to be that there are specific "helmet-factors" within the animals, the production of which is somehow initiated and/or regulated under appropriate environmental conditions. However, thus far the possibility cannot be excluded that the environment acts *externally* on the helmet. Although it might seem unlikely, all available data would be in accordance with the interpretation that mechanical external stimulation triggers helmet tissue growth: Warm, illuminated, well fed, and turbulated animals are more active than cold, dark, starved, and "still" animals, and therefore may experience stronger mechanical forces on the thin helmet blade.

In order to exclude this perhaps remote possibility, I manipulated experimentally the mechanical stress on the helmet at different temperatures: I compared the allometry of helmet length between the first and second instar of six groups: (1) free-swimming animals at 15°C; (2) animals "planted" at 15°C with the caudal spines in a plasticene bed so that they were immobilized but could move their 2nd antennae (Fig. 3a); (3) as (2) but with the antennae amputated so that the animals could not create any substantial water currents (Fig. 3b); (4)–(6) as (1)–(3) but at 23°C. The experiments were started with newly hatched neonates (not more than 2 hours after birth) and ended when the animals had molted into the second instar. All animals were kept individually and measured before and after the experiment. The planted animals had no molting problems, the old integument remained attached to the plasticene (Fig. 3). Antennal activities were recorded several times in free-swimming as well as "planted" animals. There were no appreciable differences. Details are given in Table 3. No doubt, the helmets of cold,

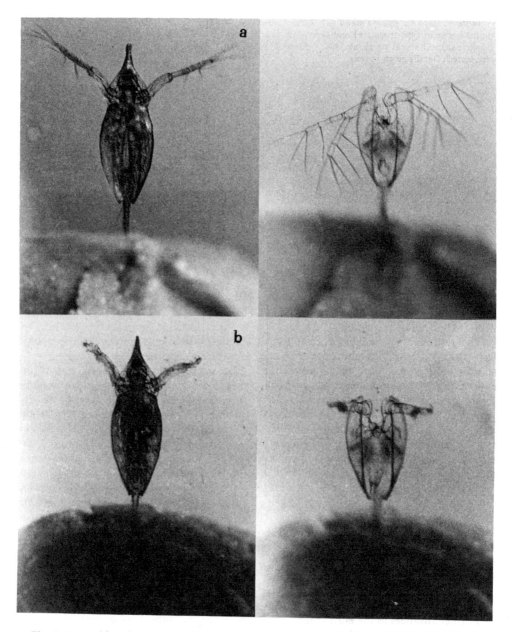

Fig. 3. Intact (a) and amputated (b) neonates of *Daphnia galeata mendotae*, "planted" with caudal spines in a plasticene bed. Pictures on right show situation after molt into second instar: cast cuticle remains attached to plasticene, animal is released into water.

Table 3. *Daphnia galeata mendotae*. Mean values ± SE of allometric rates b (helmet length vs. carapace length) between first (neonate) and second instars, in six groups of animals kept under various conditions of external mechanical stress, at 15°C (groups 1-3) and 23°C (groups 4-6). In each group, 20 individuals were tested. Details given in text.

Temp. (°C)	Group	Allometric rate b ±SE	antennal beats/min ± SE (No. measurements)
15	1: Free-swimming	1.43 ± 0.09	115 ± 4 (132)
15	2: Attached, antennae intact	1.34 ± 0.09	140 ± 5 (152)
15	3: Attached, antennae amputated	1.38 ± 0.11	139 ± 7 (144)
23	4: Free-swimming	1.76 ± 0.09	114 ± 4 (98)
23	5: Attached, antennae intact	1.84 ± 0.07	119 ± 6 (110)
23	6: Attached, antennae amputated	2.01 ± 0.13	104 ± 5 (78)

free-swimming animals must have experienced more mechanical stress than the planted and amputated warm animals. Since the experiments were done in small glass vessels (glass cylinders, 5 cm diam., ca. 30 ml) with illumination from below, the free-swimming animals often had contact with the bottom of the vessel.

The results were straightforward: There were no significant differences between the allometric rates at a given temperature, but each group of warm animals had a significantly higher allometric rate than each group of cold animals. On the average, $b_{23} = 1.35\ b_{15}$. My conclusion is that the specific control of helmet growth is not mediated externally. Otherwise, the free individuals should have grown higher helmets than the attached individuals. Since there is no innervation of the epidermal cells of the helmet, the target-specific growth factors must be borne in the hemolymph. Thus the criteria for one or more growth hormones seem to be fulfilled.

Discussion

Further advances in the elucidation of the environmental control mechanisms of cladoceran cyclomorphosis will probably come from the fields of neurobiology and developmental physiology. Since there is no single receptor mechanism for all effective environmental factors, the first common member in the chain of causation must lie within the animals. Most environmental factors affect the animal via sense organs. It may therefore be worthwhile to look for a seasonal cycle of neurosecretion, or to try to alter the activity of neurosecretory cells (for instance, in the esophogeal ganglia) by cyclomorphogenic factors. Another promising approach would be short-term tissue cultures of helmet epidermis with culture medium containing extracts of low- or high-helmeted forms, using radioisotopes to detect small differences of cell growth. This technique would also be feasible to find out whether the negative relation between helmet size and subsequent allometry (*see above*) lies within the helmet cells themselves.

Concerning the question of the adaptive significance of cyclomorphosis, the presence of a target-specific internal mechanism to regulate helmet growth may be regarded as strong evidence for a specific adaptive function of the helmet. Why else should it have evolved? If all other evidence is added— conspicuousness of the phenomenon, convergent evolution in several species, negative feedback between helmet size and helmet growth, experiments on selective predation— the conclusion seems inevitable that there must be a strong fitness component associated with cyclomorphosis. Thus far the only unquestionable demonstration of the function of protuberances as adaptive features appear to be supplied by the rotifer *Brachionus calyciflorus* which makes protective spines in the presence of the predatory rotifer *Asplanchna* (De Beauchamp 1952; Gilbert 1967; Halbach 1971). Similar functional explanations have been offered for cladoceran cyclomorphosis, invoking effects on the visibility, the swimming capacity, or the obstruction of the predator's handling techniques (Brooks 1965; Jacobs 1967; Dodson 1974; Kerfoot 1975a). But there is no example in which the exact nature of such mechanisms has been fully analyzed. In the case of *Brachionus*, the most effective causative factor is also the adapted-to factor, viz. the predator *Asplanchna*. In cladocerans all evidence points toward a disjunction between immediate causation and ultimate adaptation. Such a separation is only understandable if both factors or groups of factors were correlated in a predictable way during the course of evolution. As a matter of fact, the coupling seems to be very loose. Temperature, light, food, and turbulence exhibit appreciable independent variability in the natural environment. Furthermore all four parameters are not strictly correlated with predator activity. Of course, long-term evolution works with averages. It might therefore be useful to measure the extent of long-term correlation between predation, allometry, and cyclomorphogenic factors in various habitats, and to check whether those factors that show the highest correlation with the predation pressure are also the most effective ones with respect to allometry.

In conclusion it can be stated that much work has been done but the final causes and consequences of cladoceran cyclomorphosis, and their interconnections, are still on the list of most wanted information in limnological research.

References

ALLAN, J. D. 1974. Balancing predation and competition in cladocerans. Ecology **55**: 622-629.
BROOKS, J. L. 1946. Cyclomorphosis in *Daphnia*. I. Analysis of *D. retrocurva* and *D. galeata*. Ecol. Monogr. **16**: 409-447.
———. 1947. Turbulence as an environmental determinant of relative growth in *Daphnia*. Proc. Natl. Acad. Sci. **33**: 141-148.
———. 1965. Predation and relative helmet size in cyclomorphic *Daphnia*. Proc. Natl. Acad. Sci. **53**: 119-126.
———, and S. I. DODSON. 1965. Predation, body size, and composition of plankton. Science **150**: 28-35.
COKER, R. E., and H. H. ADDLESTON. 1938. Influence of temperature on cyclomorphosis in *Daphnia longispina*. J. Elisha Michell Sci. Soc. **54**: 45-75.
De BEAUCHAMP, P. 1952. Variation chez les Rotifères du genre *Brachionus*. C. R. Acad. Sci. Paris **235**: 1355-1356.
DODSON, S. I. 1974. Adaptive changes in plankton morphology in response to size-selective predation: A new hypothesis of cyclomorphosis. Limnol. Oceanogr. **19**: 721-729.
EGLOFF, D. A. 1968. The relative growth and seasonal variation of several cyclomorphic structures of *Daphnia catawba* Coke in natural populations. Arch. Hydrobiol. **65**: 325-359.
GILBERT, J. J. 1967. *Asplanchna* and posterolateral spine production in *Brachionus calyciflorus*. Arch. Hydrobiol. **64**: 1-62.
———. 1977. Defenses of males against cannibalism in the rotifer *Asplanchna*: Size, shape and failure to elicit tactile feeding responses. Ecology **58**: 1128-1135.
HALBACH, U. 1971. Zum Adaptivwert der zyklomorphen Dornenbildung von *Brachionus calyciflorus* Pallas (Rotatoria). I. Räuber-Beute-Beziehung in Kurzzeit-Versuchen. Oecologia **6**: 267-288.
HALL, D. J. 1967. The distribution and variation of *Daphnia lumholtzi* (Crustacea, Cladocera) in relation to fish predation in Lake Albert, East Africa. J. Zool. **151**: 181-197.
HAZELWOOD, D. H. 1962. Temperature and photoperiod effects on cyclomorphosis in *Daphnia*. Limnol. Oceanogr. **7**: 230-232.
———. 1966. Illumination and turbulence effects on relative growth in *Daphnia*. Limnol. Oceanogr. **11**: 212-216.
HRBAČEK, J. 1959. Circulation of water as a main factor influencing the development of helmets in *Daphnia cucullata* Sars. Hydrobiologia **13**: 170-185.
JACOBS, J. 1961a. Cyclomorphosis in *Daphnia galeata mendotae*, a case of environmentally controlled allometry. Arch. Hydrobiol. **58**: 7-71.
———. 1961b. On the regulation mechanism of environmentally controlled allometry (heterauxesis) in cyclomorphic *Daphnia*. Physiol. Zool. **34**: 202-216.
———. 1962. Light and turbulence as co-determinants of relative growth rates in cyclomorphic *Daphnia*. Int. Rev. Gesamten Hydrobiol. **47**: 146-156.
———. 1965. Control of tissue growth in cyclomorphic *Daphnia*. Naturwissenschaften **52**: 92-93.
———. 1966. Jährliche Zyklen des Adaptivwertes und ökologische Einnishung bei Daphnien. Verh. Dtsch. Zool. Ges. **30**: 290-296.
———. 1967. Untersuchungen zur Funktion und Evolution der Zyklomorphose bei *Daphnia*, mit besonderer Berücksichtigung der Selektion durch Fische. Arch. Hydrobiol. **62**: 467-541.

——. 1970. Multiple Determination der Zyklomorphose durch Umweltfaktoren. Eine Regressionsanalyse an *Daphnia galeata* Sars im Freiland. Oecologia 5: 96–126.

——; 1977a, b. Coexistence in similar zooplankton species by differential adaptation to reproduction and escape in an environment with fluctuating food and enemy densities. I. A model. II. Field data analysis of *Daphnia*. Oecologia 29: 233–247; 30: 313–329.

JANDER, R., and T. H. WATERMAN. 1960. Sensory discrimination between polarized light and light intensity patterns by arthropods. J. Cell Comp. Physiol. 56: 137–159.

KERFOOT, W. C. 1975a. The divergence of adjacent populations. Ecology 56: 1298–1313.

——. 1975b. Seasonal changes of *Bosmina* (Crustacea, Cladocera) in Frains Lake, Michigan: Laboratory observations of phenotypic changes induced by inorganic factors. Freshwater Biol. 5: 227–243.

——. 1977. Competition in cladoceran communities: The cost of evolving defenses against copepod predation. Ecology 58: 303–313.

LYNCH, M. 1977. Zooplankton competition and plankton community structure. Limnol. Oceanogr. 22: 775–777.

OSTWALD, W. 1904. Experimentelle Untersuchungen über den Saisonpolymorphismus bei Daphniden. Arch. Entwicklungsmech. Org. (Wilhelm Roux) 18: 415–451.

ZARET, T. M. 1972. Predator's invisible prey, and the nature of polymorphism in the Cladocera (Class Crustacea). Limnol. Oceanogr. 17: 171–184.

——. 1975. Strategies for existence of zooplankton prey in homogeneous environments. Int. Ver. Theor. Angew. Limnol. Verh. 19: 1484–1489.

——, and W. C. KERFOOT. 1975. Fish predation on *Bosmina longirostris*: Body size selection versus visibility selection. Ecology 56: 232–237.

41. Seasonal Changes in Size at Maturity in Small Pond *Daphnia*

Donald J. Brambilla

Abstract

Midsummer declines in carapace length at maturity in cladocerans may be adaptations to seasonal variation or the result of growth limitation through environmental deterioration. I report on a study of variation in size at maturity in two small pond populations of *Daphnia pulex*. In one pond, animals are subject to seasonal changes in the intensity of visual predation while in the other visual predators are never present. Fluctuations in body size occur in both populations. In the first pond these fluctuations are caused by a developmental polymorphism. Decreased size at maturity is hypothesized to be an adaptation to visual predation. In the second pond, food limitation is the major cause of size reductions. An adaptive mechanism need not be postulated to explain size variation in this case.

Midsummer declines in carapace lengths at first reproduction are commonly observed in temperate cladoceran populations (Berg 1931; Green 1966; Brooks 1965; Dodson 1974a; Kerfoot 1974; Daborn et al. 1978; Brambilla unpubl.). In some populations reduced size at maturity is associated with the development of cephalic helmets, crests or spikes, while in others size varies seasonally without any appreciable change in head shape. The former pattern, known as cyclomorphosis, is confined to epilimnetic lake populations (Brooks 1964; Hutchinson 1967, p. 947) while the latter is typical of hypolimnetic and pond populations (Brooks 1964; Dodson 1974a; Daborn et al. 1978; Brambilla unpubl.).

Size change has been extensively studied only in populations which display cyclomorphotic changes in head shape. Changes in body size and head shape are thought to be parts of intricate adaptations to a seasonally varying environment. This idea stems from the demonstration that both variations are caused by developmental polymorphisms cued by the same environmental variables (Coker and Addlestone 1938; Brooks 1947; Jacobs 1961, 1962) and the belief that such a complex ontogenetic pattern could not arise by chance alone. The roles of changes in head shape are still debated (Brooks 1965; Dodson 1974b; Hebert 1978), but reductions in carapace length at maturity are generally thought to be adaptations to seasonally intense fish predation. Fish search for prey visually and selectively remove the largest or most visible prey (Brooks and Dodson 1965; Brooks 1968; Zaret 1972; Zaret and Kerfoot 1975). Reductions in

---

This study was supported by two grants-in-aid from Sigma Xi. I thank my doctoral committee (K. G. Porter, D. W. Tinkle, P. Kilham, G. F. Estabrook, and B. S. Low) for advice, encouragement, and criticism of my dissertation from which this paper is derived. F. C. Evans permitted use of the E. S. George Reserve and W. S. Benninghoff permitted use of the Matthaei Botanical Gardens. I am grateful to A. Unal, and B. and A. Brouillet for field assistance and E. L. King for laboratory assistance.

size will reduce visibility, which should be adaptive when such predators are present.

Populations with constant head shape have not been as thoroughly studied, and size changes in them could be the result of growth limitation through environmental deterioration rather than adaptive responses. There has been no demonstration of a developmental polymorphism or other compelling evidence of an adaptive mechanism. Size may be reduced in the presence of visual predators, but this is not sufficient reason to conclude that predators are the ultimate cause of this variation because size changes also occur in populations that are never subject to visual predation (Daborn et al. 1978). Food supplies are usually at a minimum in midsummer, which is when visual predators are most abundant. Food limitation can reduce size at maturity (Banta 1939; Green 1956, 1966; Hrbáčková-Esslová 1963; Hall 1964) and may be sufficient to explain the variations observed in hypolimnetic and pond populations.

Temperature variation has also been implicated in changes in size at maturity. In cyclomorphotic species temperature acts as a cue for size change (Hutchinson 1967). In noncyclomorphotic species high temperature may also be correlated with small size at maturity (MacArthur and Baillie 1929) but in this case the role of temperature as a cue for change, a form of physiological stress or a correlate of environmental variation is not clear.

Here I report on a study of variation in carapace lengths at first reproduction in two small pond populations of *Daphnia pulex*. In one pond, changes in the intensity of visual predation occur seasonally while in the other visual predators are absent. Fluctuations in body size occur in both populations. In the first pond these fluctuations are caused by a developmental polymorphism and are hypothesized to be an adaptation to visual predation while in the second food limitation is the major cause of decreased size at maturity.

Materials and methods

The two ponds were deliberately chosen for this study because of their differences in predation patterns. WWL Pond is a small vernal pond located on the E. S. George Reserve, Pinkney, Michigan. It is formed from rain and melting snow in the spring and usually lasts for 2–4 months before drying. At maximum extent it has an area of about 150 m^2 and a maximum depth of about 0.5–0.75 m. Both figures are variable depending on yearly changes in precipitation. Several species of the salamander *Ambystoma* breed in the pond shortly after thaw, and their larvae may be visual predators on *Daphnia* (Dodson and Dodson 1971; Sprules 1972). Thus in spring, WWL *Daphnia* should be free from visual predation, during the period before salamander eggs hatch. In summer visual predation on *Daphnia* should be intense. Two dipteran larvae, *Mochlonyx* and *Chaoborus*, also occur in this pond and may prey on *Daphnia*.

Rash Pond is a permanent pond in an abandoned gravel pit at the Matthaei Botanical Gardens, Ann Arbor, Michigan. At maximum extent in the spring the pond is about 1,000 m^2 in area and about 2 m deep. There are no salamander populations in the area and no fish in the pond. The pond contains *Chaoborus* larvae and damselfly and dragonfly nymphs, all of which may prey on *Daphnia*.

I used several methods for sampling *Daphnia* and predator populations. In 1976 in WWL Pond I took triplicate net tows of known length at 5–7-day intervals, using a 20-cm-diameter, 64 μm-mesh plankton net. At the end of the season, when the pond was too shallow for tows, I continued triplicate sampling by pouring 3 liters through the net with a small beaker. In 1978 the pond was never deep enough for tows, so triplicate collections were made by pouring 8–16 liters through the net with a 1-gallon widemouth jar. Again, at the end of the season a small beaker was used for sampling. Rash Pond was studied only in 1978, using triplicate diagonal tows at 5–8-day intervals. All samples were preserved in sucrose and Formalin to prevent egg loss (Haney and Hall 1973).

Salamander larvae were sampled by Bailey's (1952) triple catch method.

In the laboratory each plankton sample was subsampled three times. *Daphnia, Mochlonyx*, and *Chaoborus* were counted, dipteran head capsule widths were measured, *Daphnia* were scored for the presence of eggs or ephippia, and adult carapace lengths and egg numbers were measured. Eggs in early cleavage stages were dissected from the brood chambers and lengths and widths measured so that egg volumes could be estimated. Eggs were assumed to approximate an ellipsoid for volume calculations.

Salamander larvae prey size preferences were measured by in situ feeding experiments. Gut analysis of a few larvae taken from the pond showed that they do indeed prey upon *Daphnia*. However gut contents provide little information about the role of salamander larvae in *Daphnia* size variation. If *Daphnia* are maturing at small sizes when salamanders are present, guts will not contain large prey and we will not be able to assess the relative fitnesses of large and small sizes at maturity under salamander predation. Fortunately GR Pond, another vernal pond near WWL Pond, lacks salamander populations and contains large *D. pulex* adults. This extended size range was offered to *Ambystoma* larvae in an attempt to demonstrate relative predation on different body sizes and, by implication, exclusion of large forms.

I set up four plastic pens in WWL Pond, each having an open surface of about 1.0 m^2 and containing about 300 liters of pond water. Plankton was collected from GR Pond by repeated net tows, mixed, divided into four aliquots, and dumped into the bags. Samples were collected by pouring 10 liters through a net with a 1.0-liter beaker. Salamander larvae were then added to the bags in densities of 1, 5, 10, and 20 per bag. After 24 hours I again sampled the plankton in all four bags. All *Daphnia* present in both sets of samples were counted and all carapace lengths were measured. The animals were then divided into four size classes and per capita rates of change were calculated for each size class at each predator density. Regressions of rate of change on predator density were constructed for each size class and used to compare relative rates of predation.

Feeding preferences of *Mochlonyx* and *Chaoborus* larvae were estimated by comparing size distributions of prey in guts to size distributions in the ponds. This technique suffers from the same limitation as the use of gut contents for measuring feeding preferences of salamander larvae. However other studies indicate that these predators prefer small *Daphnia* as prey and may avoid adults. Gut contents were examined to confirm or deny this with the intention of proceeding to feeding experiments if significant predation on adults was found. The *Daphnia* in the guts were not sufficiently intact to permit direct measurement. Postabdominal claws were measured and used to estimate carapace lengths from regressions of carapace length on claw length. Regressions were constructed using data derived from animals sampled from the pond on the day the predators were collected. Carapace length and claw length are highly correlated in all samples ($r > 0.90$). *Chaoborus* feeding preferences were measured in Rash Pond and GR Pond, and *Mochlonyx* feeding preferences were measured in WWL Pond. Feeding preferences of *Chaoborus* in GR Pond and WWL Pond should be very similar as the ponds are similar in community structure and the *Chaoborus* larvae in both are probably derived each year by migration from the same permanent ponds.

For all predators prey size preferences were calculated using Ivlev's (1961) electivity index:

$$E_i = (Y_i - P_i)/(Y_i + P_i)$$

where Y_i is the proportion of the *i*th size class in the gut and P_i is the proportion of this class in the pond. Values can range from -1.0 to 1.0, with 0 indicating no preference, positive values indicating preferred prey, and negative values indicating avoidance of that size class in the diet.

I used two methods to assess the extent

of food limitation in *Daphnia*. First phytoplankton samples were collected at regular intervals. A 250-ml sample of middepth water was collected every 5–8 days and preserved in Lugol's solution. The samples were counted on an inverted microscope and population densities and cell sizes of all species present were recorded (Vollenweider 1969). In addition, when zooplankton samples showed that size at maturity was decreasing in the ponds, a phytoplankton enrichment experiment was carried out to see if size at maturity would increase with elevated food. I placed 10 randomly selected subadult *Daphnia* in each of six 1-gallon widemouth jars. The mouths of the jars were covered with 102 μm nylon mesh and the jars were placed in the pond. Every 2 days laboratory raised *Chlamydomonas* was added to three of the jars in quantities sufficient to raise phytoplankton densities by 10^5 cells/ml. The other three jars served as controls. After 7–14 days, depending on the experimental run, the jars were removed from the pond, the *Daphnia* were counted, carapace lengths were measured, eggs or embryos were counted and, where possible, eggs were dissected out and lengths and widths measured for egg volume calculations.

The impact of food limitation on Rash Pond animals was also measured in the laboratory. Laboratory cultures were established in pond water by removing animals from the pond in April, when they were maturing at large sizes, and in June, when size at maturity was reduced (Fig. 1). These cultures are referred to as Rash Early and Rash Late, respectively. Newborn individuals from each culture were collected, isolated, and raised under two food levels that approximated maximum and minimum phytoplankton densities observed in the pond. The animals were held in shell vials containing 30 ml of Whatman glass-fiber-filtered pond water and either 10^3 or 10^5 cells/ml *Chlamydomonas*. They were transferred to fresh vials of water and food daily. When the first clutch was deposited in the brood chambers, the following data were recorded: carapace lengths, days to maturity, egg number and, where possible, egg lengths and widths for volume calculations. Carapace length at maturity, egg volume and age at maturity were all tested against food level and time of collection from the pond by analysis of variance corrected for unequal subgroup sizes (Snedecor and Cochran 1967).

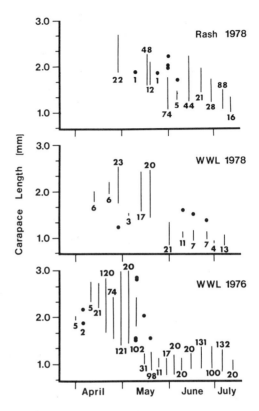

Fig. 1. Ranges of carapace lengths of mature females, excluding those carrying ephippia. Dots indicate single females.

An estimate of total reproductive and growth potential of *Daphnia* was also made by raising individuals in the laboratory with excess food. For this I used animals from the two Rash Pond cultures and from two WWL Pond cultures which were also established when size at maturity in the pond was large (WWL Early) and small (WWL Late) (Fig. 1). Newborn animals were isolated in shell vials containing 30 ml of filtered pond water and held in an incubator at 20°C with a L:D cycle of

16:8 hours. Each day they were fed *Chlamydomonas* in quantities sufficient to raise cell densities to over 10^5 cells/ml. When animals reached the primiparous instar I recorded ages and carapace lengths. Ages and sizes were also recorded at subsequent molts. Periodically a few randomly selected animals were killed so that egg numbers and egg volumes could be measured. Five eggs were measured from each individual and a mean volume was calculated.

If size variation is an adaptation to predation then it is possible that changes in size at maturity are cued by some environmental parameter, correlated reliably with predator abundance, in much the same way that helmet growth in cyclomorphotic populations is cued by temperature, turbulence, or daylength (Coker and Addlestone 1938; Brooks 1947, 1966; Jacobs 1961, 1962, 1967). Preliminary data showed that daughters of animals removed from WWL Pond, when size at maturity was small, matured at a larger size in the laboratory than did their counterparts in the field. I investigated the possibility of environmental cuing of size changes by placing animals from laboratory cultures in pens in the pond and comparing their development to that of animals held in the laboratory and to that of animals in the pond. The pens were made by cutting the sides and bottoms from 2-quart freezer containers and covering the holes with 102 µm nylon mesh. With the tops snapped in place the pens float just below the surface of the pond. Six newborn *Daphnia* were placed in each pen. The pens were left in the pond for varying lengths of time so effects on more than one generation could be examined. This procedure will not identify the cue, if one exists, but should indicate whether a cued response does exist.

Results

Population growth patterns in the two ponds are very similar (Fig. 2). *Daphnia* hatch from ephippia shortly after thaw and population densities are initially low. This is followed by rapid exponential growth which terminates in a midseason period of ephippia production. Populations then decline briefly followed by a period of renewed growth. In WWL Pond population growth terminates as the pond dries. In both 1976 and 1978, when the pond had shrunk to a small puddle, the surface was covered with floating ephippia. Thus there is probably a second period of ephippia production as this pond is drying. Rash Pond is permanent and the *Daphnia* population there may persist into the fall. However the study was terminated in mid-July.

Adult size distributions also follow similar patterns in the two ponds (Fig. 1). The exephippio generation matures at carapace lengths of 1.66 mm or more and may eventually grow to 2.66 mm. Their daughters are mature at 1.5–1.66 mm and may or may not reach the same maximum size as the first generation. In WWL Pond in 1976, there are two sample days on which all mature females were > 2.2 mm. These samples do not represent an increase in size at maturity as the females that are present are all exephippios which are beyond the primiparous instar.

There is an abrupt midseason change in the size distributions of mature females in WWL Pond, coinciding with the first period of ephippia production. During the second part of the season animals are mature at 0.8–1.0 mm and, aside from an occasional individual, never achieve sizes > 1.33 mm. Thus there is virtually no overlap in adult size distributions between early and late season *Daphnia* in this pond.

In Rash Pond there is also an abrupt decline in size at maturity in late May and early June. However the pattern of size variation is different in the two ponds. Size at maturity is variable in June and July in Rash Pond and is often > 1.33 mm. Minimum mature sizes are only rarely as small as those found in WWL Pond. Also the decrease in size at maturity in Rash Pond does not coincide with a drop in the maximum size attained by mature females. In the samples taken in June and July animals still reached sizes > 1.5 mm.

Laboratory results—Animals raised in the laboratory show a remarkable degree of sim-

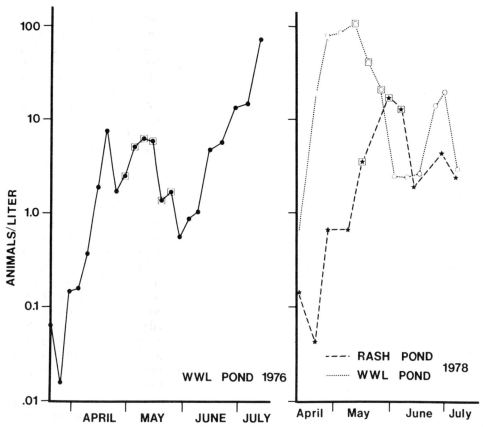

Fig. 2. Population growth of *Daphnia pulex* in WWL and Rash Ponds. Points enclosed by squares are days on which some adults were carrying ephippia.

ilarity compared to animals collected at different times in the ponds (Table 1). Early and late season *Daphnia* from WWL Pond have similar sizes at maturity ($t=1.4$, df-25, $p>0.15$) and egg volumes ($t=0.6$, df=24, $p>0.50$). Mean ages at which the first clutch is deposited in the brood chamber are also similar ($t=1.0$, df=25, $p>0.30$) as are mean ages at which the first clutch is born ($t=0.2$, df=17, $p>0.50$), indicating that egg development times are the same. The two Rash Pond groups also have similar sizes at maturity ($t=1.3$, df=15, $p>0.15$), egg volumes ($t=.6$, df=8, $p>0.60$), ages at maturity ($t=0.6$, df=15, $p>0.50$), and ages at birth of the first clutch ($t=1.6$, $p>0.20$, Mann-Whitney U-test) when raised under similar conditions in the laboratory.

The changes in size at maturity seen in the field can be explained by food limitation, developmental polymorphism, seasonal succession of genotypes, or midseason replacement by a smaller sibling species. The disappearance of size differences in laboratory cultures argues against the latter two possibilities but is consistent with both of the first two. Food limitation or other forms of environmental deterioration could take place in the field but resulting size changes would disappear when both groups of animals are raised under the same conditions in the laboratory. Field differences arising from developmental polymorphisms could also disappear under uniform laboratory conditions. Thus it is not possible at this point to determine the mechanism of size variation in either pond.

Predators on Daphnia—In WWL Pond there are two phases to the seasonal distribution of predators (Fig. 3). *Mochlonyx*

Table 1. Results of laboratory culture of *Daphnia pulex* from WWL and Rash Ponds. Numbers of females used for each determination in parentheses. Early and Late defined in text.

	A*	B†	C‡	D§
WWL Early	1.45 (13)	6.15 (13)	8.33 (6)	0.0033 (12)
WWL Late	1.52 (14)	6.00 (14)	8.33 (6)	0.0032 (14)
Rash Early	1.71 (8)	6.87 (8)	10.50 (4)	0.0044 (7)
Rash Late	1.64 (9)	7.11 (9)	10.00 (3)	0.0041 (3)

*Mean carapace length at maturity in mm.
†Time from birth to appearance of first clutch in brood chamber in days.
‡Time from birth to birth of first clutch in days.
§Mean egg volume in mm^3.

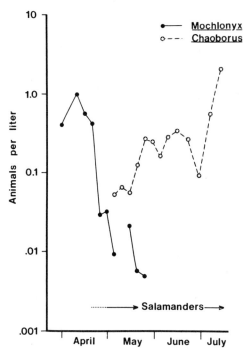

Fig. 3. Densities of predators in WWL Pond, 1976. Densities of *Mochlonyx* and *Chaoborus* larvae are pooled 3rd and 4th instars only. Salamander line indicates time of occurrence only and not density.

larvae are present in the early months but absent during the middle and later parts of the season. *Chaoborus* and *Ambystoma* larvae are present only during the middle and late parts of the season. First instar *Chaoborus* appear about 1 month after spring thaw. They molt at about weekly intervals so fourth instar individuals appear about 3 weeks later. All 4 instars are present for the rest of the summer. *Ambystoma* larvae hatch from eggs about 6 weeks after thaw and remain in the pond for the rest of the summer, metamorphosing to adults as the pond dries. The salamander line in Fig. 3 indicates time of occurrence only and not density. In 1975, during a pilot study, I found that there were 636±600 larvae present in the pond in June while in 1976 in mid-June there were 349±252 larvae in the pond. The 1976 figure represents about 5 larvae/m^2.

The predator distribution pattern in Rash Pond is different from that in WWL Pond. *Chaoborus* overwinter as 4th instar larvae and are present as potential predators when *Daphnia* hatch from ephippia in early spring (Fig. 4). Shortly thereafter larvae disappear from the pond as they transform to pupae. During May most of the population is composed of 1st and 2nd instar individuals, but by the end of May a substantial population of 3rd and 4th instar larvae is again present (Fig. 4, bottom panel). Larvae again disappear from the pond in early July. Just before this disappearance the population is composed mainly of late instar individuals and it is likely that their disappearance results from a second round of pupation. Large damselfly and dragonfly nymphs are also present throughout the season in this pond but gut analyses indicate that, while they feed heavily upon clad-

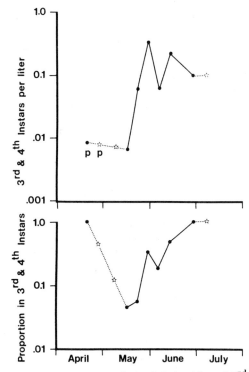

Fig. 4. Upper panel: Pooled densities of 3rd and 4th instar *Chaoborus* larvae in Rash Pond, 1978. Lower panel: Proportion of *Chaoborus* larvae which are in 3rd and 4th instars. P—samples with pupae; ☆—samples with no larvae.

ocerans and other organisms found in the littoral zone, they rarely feed upon *D. pulex*. Therefore these nymphs are excluded from the rest of this study.

Abystoma *predation*—Gut contents of a few salamander larvae indicate that even small individuals prey upon *Daphnia* (Table 2). Twenty *Ambystoma* guts were examined over 5 sample days—2 during a pilot study in 1975 and 3 in 1976. The salamanders had a snout-vent length range of 0.6–2.2 cm. There were *Daphnia* in 16 of the guts. Salamanders without *Daphnia* had snout-vent lengths of 0.6–1.2 cm while those with *Daphnia* had snout-vent lengths of 1.1–2.2 cm. Larvae in this pond and in others metamorphose at 3.5–8.2 cm, depending upon species, location, and year to year variations in the environment (Uzzell 1962; Wilbur 1971; Dodson and Dodson 1971) but even the largest larvae are predators on zooplankton (Dodson and Dodson 1971). Thus *Ambystoma* predation on *Daphnia* in WWL Pond starts shortly after these predators hatch from eggs and continues for the rest of the season. Also the onset of salamander predation coincides roughly with the midseason decline in *Daphnia* size at maturity.

The salamander predation experiment shows that *Ambystoma* larvae prefer the largest available *Daphnia* as prey (Fig. 5, Table 3). Regressions for the two middle size classes have negative slopes while regressions for the largest and smallest size classes have slopes that are not significant. The general negative relationship between body size and removal rate, and the negative slopes, indicates that predation intensity increases with prey body size and salamander density. For the largest size class predation intensity is high and independent of density while for the smallest *Daphnia* predation intensity is low and not significantly dependent on density. Bartlett's test of homogeneity of variances

Table 2. Size distributions of salamander larvae with and without *Daphnia pulex* in their guts. N—Number; SVL—snout-vent length.

	N	SVL range (cm)	N With Daphnia	SVL range with Daphnia (cm)	SVL range without Daphnia (cm)
4 Jun 75	4	1.2–1.6	4	1.2–1.6	—
10 Jun 75	2	1.6–1.9	2	1.6–1.9	—
30 Apr 76	5	0.6–1.2	2	1.2	0.6–1.0
9 May 76	5	1.1–1.6	4	1.1–1.6	1.2
19 May 76	4	1.7–2.2	4	1.7–2.2	—
Σ	20	0.6–2.2	16	1.1–2.2	0.6–1.2

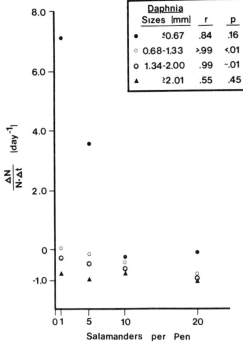

Fig. 5. Per capita rates of change of *Daphnia* size classes as functions of salamander larva density in salamander predation experiment in WWL Pond.

shows that variances are significantly heterogeneous ($\chi^2=45.4$, df=3, $p<0.005$), so the regressions cannot be compared by analysis of covariance. However at all salamander densities, the per capita rates of change of *Daphnia* are ordered according to size class in such a way that the smallest size class has the most positive rate and the largest has the most negative rate. Friedman's method for randomized blocks (Sokal and Rohlf 1969) shows this order to be highly significant ($\chi^2=12$, df=3, $p<0.01$). Thus at any given salamander density, predation falls first and most heavily on the largest *Daphnia*. As this size class is depleted, preference switches to the next largest size class. This suggests that *Ambystoma* larvae in WWL Pond have definite feeding preferences for the largest available prey.

Mochlonyx and Chaoborus predation – Electivities for these two predators are presented in Fig. 6 along with the values from the salamander predation bag containing five predators. I chose the latter because it approximates natural densities for 1975 and 1976 more closely than the other bags. In contrast to the salamanders, the dipterans show a strong preference for the smallest *D. pulex* in a prey assemblage. For *Mochlonyx*, electivities are positive for 1st and 2nd instar *Daphnia* while for *Chaoborus* electivities are positive for the first three instars but higher for the 2nd and 3rd than for the 1st. This difference arises from seasonal changes in size distributions in prey populations and not from a difference in size preferences on the part of the predators. Both species have upper size limits of preferred prey of about 0.7 mm. Thus dipteran predation falls most heavily on young *Daphnia* and has little or no impact on adults.

Food supplies and feeding – Eucaryote phytoplankton in WWL and Rash Ponds is dominated by flagellates in terms of cell numbers and total biomass in all samples. Flagellates are never <90% of the

Table 3. Regressions of per capita rate of change of *Daphnia* on salamander larva density for the four *Daphnia* size classes in salamander predation experiment. p is significance of slope of each regression.

Size class	Size range (mm)	Slope	Intercept	r^2	p
1	0–0.67	−0.36	5.82	0.708	0.15
2	0.68–1.33	−0.04	0.05	0.992	0.004
3	1.34–2.00	−0.04	−0.27	0.977	0.01
4	2.00	−0.01	−0.87	0.297	0.45

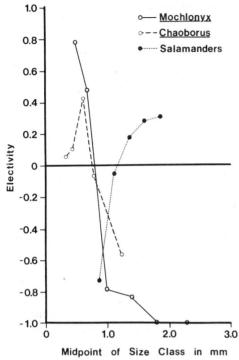

Fig. 6. Ivlev's (1961) electivity index plotted against midpoint of *Daphnia* size class for which it was calculated. *Mochlonyx* values are for pooled 3rd and 4th instar larvae from WWL Pond, *Chaoborus* values are for pooled 3rd and 4th instar larvae from GR Pond, and salamander larva values are for the fine-salamander pen in salamander predation experiment.

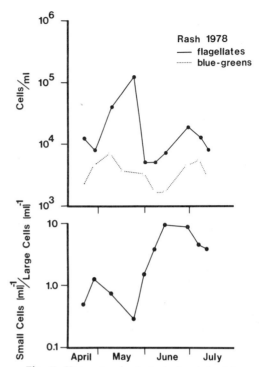

Fig. 7. Upper panel—phytoplankton densities in Rash Pond, 1978. Lower panel—ratio of density of small flagellates (4 μm) to density of large flagellates (6 μm).

eucaryote cells counted so the trends in their numbers and size distributions are used to illustrate trends in *Daphnia* food supplies. This simplification actually serves to minimize food supply changes through time because other cell types, such as diatoms, are most abundant when flagellates are at their highest densities and largest mean cell volumes. By excluding these other cell types, peak food levels are underestimated by a greater amount than minimum food levels. However neither level is underestimated sufficiently to interfere with this analysis.

Flagellates occur primarily in two groups: species with cell diameters or lengths <4 μm and species with sizes >6 μm. Very few cells fall between the two ranges. Aside from dinoflagellates, which are never abundant, the larger category has an upper limit of about 12-14 μm. The two size categories differ in mean diameter by a factor of roughly 4-6 and hence in volume by a factor of 56-216, depending upon the pond and sample date. This disparity in cell sizes means that both cell numbers and relative volumes must be considered in discussing food supplies for *Daphnia*.

In Rash Pond there is a peak of flagellate abundance in late spring (Fig. 7). At this time large cells are more abundant than small cells (Fig. 7, bottom panel). An abrupt drop in total flagellates coincides with a decrease in mean cell size. This suggests a severe decrease in food supplies for *Daphnia* in terms of available food volume. This decrease occurs at about the same time as the decline in *Daphnia* size at maturity, suggesting that food limitation may have a major impact on the Rash Pond population.

The phytoplankton enrichment experiments show that food limitation is important in this case. Per capita growth rates and mean sizes of 1st and 2nd instar animals are significantly greater in enriched jars than in controls (Fig. 8, Table 4). Animals in enriched jars carry larger eggs than animals present in the pond at the same time. Unfortunately none of the mature females in control jars were carrying eggs in an early enough stage of development for measurement. However if 1st instar sizes are smaller in controls than in enriched jars then egg volumes in controls should also be smaller and thus more similar to eggs carried by animals in the pond than to eggs carried by females in enriched jars. This suggests that food limitation results in smaller size at birth in Rash Pond.

The disparity between 2nd instar sizes is greater than the difference between 1st instar sizes. Thus growth rates per instar are also food limited. Reductions in size at birth and in growth rate may result in lower size at maturity. Minimum size at maturity is about 1.33 mm in enriched jars but only about 1.00 mm in the pond at the same time. The minimum size of egg-bearing females in controls is also about 1.33 mm. However these animals are probably not in the primiparous instar.

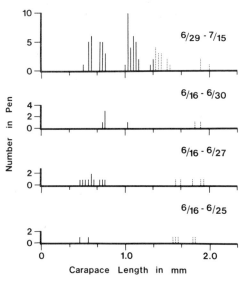

Fig. 8. Size-frequency distributions for *Daphnia* in phytoplankton enrichment experiment in Rash Pond, 3-12 July 1978. Distributions constructed by pooling either the three enriched or control jars. Dotted lines—egg-bearing females.

I base this assertion on the number of offspring present in the controls when the experiment was terminated. There were 58 young *Daphnia* in the three jars, or about 4.83 offspring per mature female. Mean clutch size is 2.55 but was probably somewhat smaller when these adults were younger. This suggests that an average adult had produced about two clutches in the controls by the time the experiment ended. Thus it was not until the 3rd mature

Table 4. Results of phytoplankton enrichment experiment carried out in Rash Pond, 3-12 July 1978. Numbers of animals (or jars for bottom line) used for each determination given in parentheses.

	Enriched	Controls	Pond 11 Jul 78	t	p
First instar					
Carapace lengths (mm)	0.52 (48)	0.50 (24)	–	3.7	<0.001
Second instar					
Carapace lengths (mm)	0.67 (76)	0.62 (31)	–	4.8	<0.001
Egg volumes (mm^3)	0.0033 (14)	–	0.0028 (12)	2.5	∼0.017
Per capita rate of					
change (per day)	1.15 (3)	0.20 (3)	–	4.4	∼0.012

instar that control females reached the size of primiparous females in the enriched jars.

While enrichment was designed to elevate phytoplankton to peak natural densities, size at maturity and egg size did not achieve the high levels observed in the laboratory or earlier in the year in the pond. There are two possible reasons for this. First, I observed an accumulation of green material on the bottoms of the enriched jars, indicating that the *Chlamydomonas* were rapidly settling out of suspension. Therefore it is likely that phytoplankton densities in the enriched jars were substantially less than peak natural densities. Although more food was available in enriched than in control jars, some amount of limitation may still have taken place. It is also possible that body size is limited by temperature as in MacArthur and Baillie's (1929) laboratory study. Minimum size of mature females is highly correlated with temperature in Rash Pond ($r = -0.712$, df=8, $p=0.01$). However there are several sample days on which temperature was close to the value used in the laboratory experiments reported in Table 1 (Fig. 9). In those samples, size at maturity was large when food was abundant and small when food was scarce. This is weak evidence, but does suggest that food supply has a greater impact on body size than does temperature.

The laboratory feeding experiment further supports the idea that food supply is responsible for size variation in Rash Pond (Table 5). Analyses of variance show significant main effects of food level on carapace length at maturity, mean egg volume, and age at which the first clutch is deposited in the brood chamber (Table 6). The main effects of time of collection from the pond and the interaction terms are not significant. At high food levels carapace

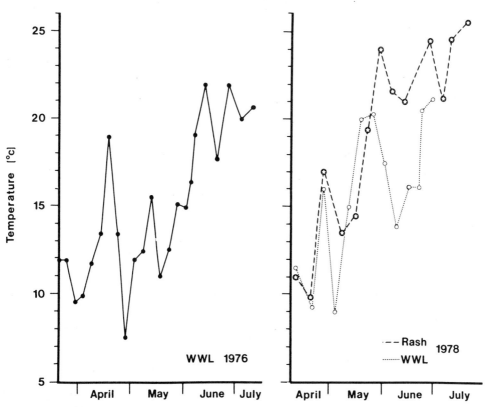

Fig. 9. Middepth temperatures in WWL and Rash Ponds taken at about midday.

Table 5. Results of Rash Pond laboratory feeding experiment. 10^3 and 10^5 are food levels as cells/ml. Numbers of females used for each determination given in parentheses. Early and late defined in text.

	A*		B†		C‡	
	10^3	10^5	10^3	10^5	10^3	10^5
Rash early	1.32 (8)	1.70 (9)	11.3 (8)	6.7 (9)	0.0050 (5)	0.0066 (9)
Rash late	1.34 (9)	1.70 (9)	12.4 (8)	6.2 (9)	0.0046 (9)	0.0070 (9)

*Mean carapace length at maturity, in mm.
†Age at which first clutch appears in brood chamber, in days.
‡Mean egg volume, in mm^3.

Table 6. Analyses of variance for Rash Pond laboratory feeding experiment. Culture—early and late as defined in text.

Source	df	M.S.	F	p
Carapace length at maturity				
Food level	1	1.1772	141.83	<0.001
Culture	1	0.0019	0.23	>0.50
Food level × culture	1	0.0023	0.28	>0.50
Error	31	0.0083		
Age at which first clutch appears in brood chamber				
Food level	1	230.90	43.66	<0.001
Culture	1	0.05	0.01	>0.75
Food level × culture	1	2.56	0.48	>0.50
Error	31	5.28		
Egg volumes				
Food level	1	3.20×10^{-5}	58.8	<0.001
Culture	1	8.07×10^{-8}	0.15	>0.50
Food level × culture	1	1.27×10^{-6}	2.31	~0.15
Error	28	5.48×10^{-7}		

lengths at maturity are similar to early season values in the pond while at low food levels carapace lengths are similar to late season values. This suggests that most or all of the size variation in Rash Pond can be attributed to food supply.

The low food level portions of this experiment were run at cell densities that are substantially lower than late season densities observed in the pond. However, the *Chlamydomonas* used in the field and laboratory experiments have cell diameters of 6–8 μm and thus would be categorized as large flagellates in the phytoplankton counts described above. The large size of these cells, relative to late season cell sizes, compensates for the difference in phytoplankton densities. Thus the low food level in the laboratory accurately represents minimum food levels in the pond.

WWL Pond has a series of peaks in flagellate abundance during the season (Fig. 10). Peaks in April and June are dominated by large cells while the peak in May is dominated by small cells (Fig. 10, bottom panel). This indicates a peak

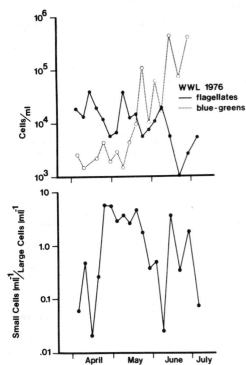

Fig. 10. Upper panel—phytoplankton densities in WWL Pond, 1976. Lower panel—ratio of density of small flagellates (4 μm) to density of large flagellates (6 μm).

in food volume in April, a low point in May, and a peak again in June. Blue-green algae are abundant during the latter part of the season but their cells are generally about 1.0 μm in diameter and thus they represent a small fraction of the total phytoplankton volume present.

If food supply determines size at maturity in WWL Pond, there should be a decrease in body size in midseason and an increase as food supplies increase again in June. However, while the transition to small size does occur during the low point in food abundance there is no subsequent increase in size. Therefore size at maturity in this population does not seem to be controlled by food supply over the range of abundances examined here.

Phytoplankton enrichment experiments did not work in WWL Pond. The experiments were tried twice in 1976 and once in 1978; in all three attempts all animals died within a few days. The experiments worked well in Rash Pond using the same equipment and food supplies so it is unlikely that the heavy mortality in WWL Pond is the result of a flaw in experimental design. Also the type of food used produced good survival and reproduction in the laboratory and occurs naturally in WWL Pond. Apparently there is something wrong with enclosing *Daphnia* in 1-gallon widemouth jars in WWL Pond, but the nature of the problem is not clear.

The pen experiments suggest that food limitation is not the reason for size variation in WWL Pond (Fig. 11, Table 7). Size at maturity is smaller in the pens than in the laboratory results shown in Table 1 ($t=3.6$, df=25, $p<0.01$). However size at maturity in the pens is significantly >1.0 mm, which is about the minimum size of late season mature females in the pond ($t=7.2$, df=12, $p<0.001$). The size differences between penned and pond animals are not simply a density effect, as densities in the pens are equal to or greater than densities in the pond at the same time. Egg volumes in the pens are smaller than egg volumes in the laboratory ($t=4.1$, df=20, $p=0.01$) and early season egg volumes in the pond ($t=6.0$, df=27, $p<0.001$) but substantially larger than egg volumes of females present in the pond when the experiment was run ($t=10.7$, df=16, $p<0.001$).

In the Rash Pond laboratory feeding experiment and in other studies (Banta 1939; Green 1956, 1966; Hrbáčková-Esslová 1963; Hall 1964), reduced food levels have led to declines in egg volume and size at maturity. Thus it is likely that the differences between penned animals and laboratory and early season pond animals can be attributed to reduced late season food supplies. This does not mean that late season size reductions in the pond can be attributed to food limitation. Minimum size of late season females in the pond does not fluctuate with phytoplankton volume. In addition, if food limitation is responsible for size variation then penned animals should be similar in size to animals

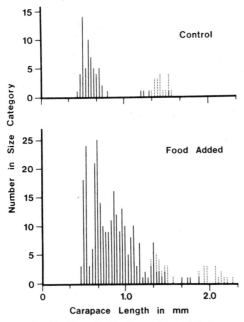

Fig. 11. Size distributions of *Daphnia* in pens in WWL Pond. Dates indicate dates on which pens were put into and removed from pond so that *Daphnia* could be counted and measured. Dotted lines indicate females carrying eggs or ephippia.

in the pond at the same time. This was the case in Rash Pond and in the phytoplankton enrichment controls, but was not true in WWL Pond.

Discussion

This study asks two related questions: are body size variations in small pond *Daphnia* adaptive and, if so, what is the nature of the adaptation? Williams (1966) noted that variation could often be explained on grounds which did not require adaptive arguments and that natural selection should only be invoked to explain variation when absolutely necessary. For small pond *Daphnia* this means that the observation of changes in size at maturity is not sufficient to require an adaptive explanation. Decreased size may also be the result of environmental deterioration and, unless smaller animals have higher fitness under such conditions, we need not resort to natural selection for an explanation. Here I discuss the nature of size variation in each pond in turn.

In Rash Pond, food limitation is clearly the major force affecting size at maturity. When food available in the pond was supplemented with large quantities of high quality food, there was an increase in size at birth, growth per instar, and size at maturity. These parameters did not reach laboratory or early season pond values but the discrepancies can be explained by incomplete elimination of food limitation in the experiments. In the laboratory, food limitation caused reductions in size at maturity and egg size and increases in age at maturity. These results were the same for animals taken from cultures established from the pond when size at maturity was large and when size at maturity was small. This suggests that size variation in Rash Pond is not the result of a develop-

Table 7. Carapace lengths at maturity and mean egg volumes for WWL Pond animals held in mesh pens in pond, raised in laboratory (Table 1), and found in pond. Numbers of animals used for each determination given in parentheses. Carapace length at maturity for penned animals determined from pen removed on 15 July 1978 by assuming all egg-bearers <1.66 mm are primiparous.

	Pens	Lab	Pond 1978	
			11 Apr–19 May	16 Jun–15 Jul
Carapace length at maturity, in mm	1.43 (13)	1.52 (14)	—	—
Egg volume, in mm^3	0.0020 (8)	0.0031 (14)	0.0028 (21)	0.0012 (10)

mental polymorphism or seasonal succession of genotypes or sibling species.

It is possible that animals with smaller body sizes are more fit when food is limited. Such a situation would clearly warrant an adaptive explanation. However the response of cyclomorphotic species to seasonal variation suggests that this is unlikely. In these species, carapace lengths at maturity are generally largest in winter and early spring, which is when food tends to be most limited in temperate lakes (Hutchinson 1967 for review). This is true for large and medium size cyclomorphotic *Daphnia* species and for small cladocerans such as *Bosmina* (Kerfoot 1974). Thus decreased size in Rash Pond can be explained by the physiological limitations imposed by environmental deterioration and there is no need to suggest that natural selection is involved.

Size variation in WWL Pond clearly requires an adaptive explanation. Late season animals mature at very small sizes in the pond but at much larger sizes when raised in the laboratory. In terms of size at birth and size and age at maturity, laboratory-raised late season *Daphnia* are very similar to early season animals from both the pond and laboratory. When late season animals are returned to the pond, as they were in the pen experiments, they do not revert to the typical late season form. Failure to revert to small size at maturity occurs at natural densities and at a time when small forms are present in the pond. Thus it is unlikely that small size results from either density dependence or food limitation. Temperature effects also cannot explain size variation in this case. The laboratory experiments were run at late season temperatures but did not result in late season reduced size distributions. In addition, animals in the pens are subject to physical conditions that are very similar to those in the pond but penned *Daphnia* continue to mature at a large size while their counterparts in the pond are maturing at a small size.

If size is not limited by food or physical conditions then some other explanation must be sought. Reductions in size at maturity can be explained by a developmental polymorphism cued by an as yet unidentified environmental variation. Removal of animals from the pond interrupts the cue, resulting in similarity to early season forms in laboratory experiments. Apparently the cuing mechanism cannot be restored by returning animals to the pond in mesh pens, indicating that either the cue is complex or there is something different about living in the pond and living in an enclosure.

It is difficult to conceive of a developmental polymorphism that is not an adaptive response. However the ultimate cause for this response is not yet clear. A likely candidate is the seasonal variation in size-selective predation imposed by salamander larvae. Predation on *Daphnia* starts shortly after salamander larvae hatch from eggs. The salamander predation experiment clearly suggests that at natural predator and prey densities, predation rates on large size classes are so high that large *Daphnia* could be eliminated from the pond in a few days. Thus when *Ambystoma* predation is intense, a prey organism that matures at a small size should have higher fitness than one maturing at a larger size. In addition, the time required for elimination of large size classes by predation is far shorter than the time required for competitive interactions (Sprules 1972; Dodson 1974a), so it is unlikely that changes in competition are responsible for size variation. On the other hand, larger individuals carry more eggs and should have a reproductive advantage when salamander larvae are absent. Therefore when size-selective predation on large forms is low, as in early spring, *Daphnia* matures at a relatively large size and when this predation is intense, as in late spring and summer, *Daphnia* switches to maturing at smaller sizes.

The evolution of a developmental polymorphism depends on the assumption that either prey can detect the presence of a predator and respond very rapidly or that the timing of predator abundance is very predictably correlated with changes in some other environmental variable to which the prey can respond. *Brachionus calyciflorus* can sense the presence of the predatory rotifer *Asplanchna* and produces offspring that are spined, making them less susceptible to predation (Gilbert 1967). The extremely short generation time of the prey rotifer is certainly conducive to the evolution of this type of response. In *Daphnia* the time from birth to maturity is probably too long for such a direct response to work. In WWL Pond, animals take slightly more than 8 days from birth to the time they drop their first clutch (Table 1). The salamander feeding experiment suggests that predation rates are so high that it is unlikely that large *Daphnia* could survive long enough to produce offspring which mature at a smaller size when cued by the presence of salamanders. However *Ambystoma* larvae do display a reasonably predictable seasonal pattern of abundance, so a response to an associated cue is certainly possible. Body size changes in this pond may then be analogous to helmet development in lake populations in that physical parameters cue an adaptive response to predation. It is unfortunate that the data do not permit the identification of a possible cue. This subject certainly merits further investigation.

References

BAILEY, N. J. T. 1952. Improvements in the interpretation of recapture data. J. Anim. Ecol. 21:120–127.

BANTA, A. M. 1939. Studies on the physiology, genetics, and evolution of some Cladocera. Carnegie Inst. Wash. Dep. Genetics Pap. 39. 285 p.

BERG, K. 1931. Studies on the genus *Daphnia* O. F. Müller with especial reference to the mode of reproduction. Vidensk. Medd. Dan. Nat. Foren. 92:1–222.

BROOKS, J. L. 1947. Turbulence as an environmental determinant of relative growth in *Daphnia*. Proc. Natl. Acad. Sci. 33:141–148.

———. 1964. The relationship between the vertical distribution and seasonal variation of the limnetic species of *Daphnia*. Int. Ver. Theor. Angew. Limnol. Verh. 15:684–690.

———. 1965. Predation and relative helmet size in cyclomorphic *Daphnia*. Proc. Natl. Acad. Sci. 53:119–126.

———. 1966. Cyclomorphosis, turbulence, and overwintering in *Daphnia*. Int. Ver. Theor. Angew. Limnol. Verh. 16:1653–1659.

———. 1968. The effects of prey size selection by lake planktivores. Syst. Zool. 17:273–291.

———, and S. I. DODSON. 1965. Predation, body size, and composition of plankton. Science 150:28–35.

COKER, R. E., and H. H. ADDLESTONE. 1938. Influence of temperature on cyclomorphosis of *Daphnia*. J. Elisha Mitchell Sci. Soc. 54: 45–75.

DABORN, G. R., J. A. HAYWARD, and T. E. QUINNEY. 1978. Studies on *Daphnia pulex* Leydig in sewage oxidation ponds. Can. J. Zool. 56:1392–1401.

DODSON, S. I. 1974a. Zooplankton competition and predation: an experimental test of the size efficiency hypothesis. Ecology 55: 605–613.

———. 1974b. Adaptive change in plankton morphology in response to size-selective predation: A new hypothesis of cyclomorphosis. Limnol. Oceanogr. 19:721–729.

———, and V. E. DODSON. 1971. The diet of *Ambystoma tigrinum* larvae from western Colorado. Copeia 1971:614–624.

GILBERT, J. J. 1967. *Asplanchna* and posterolateral spine production in *Brachionus calyciflorus*. Arch. Hydrobiol. 64:1–62.

GREEN, J. 1956. Growth, size and reproduction in *Daphnia* (Crustacea: Cladocera). Proc. Zool. Soc. Lond. 126:173–204.

———. 1966. Seasonal variation in egg production of Cladocera. J. Anim. Ecol. 35:77–104.

HALL, D. J. 1964. An experimental approach to the dynamics of a natural population of *Daphnia galeata mendotae*. Ecology 45:94–112.

HANEY, J. F., and D. J. HALL. 1973. Sugar-coated *Daphnia*: A preservation technique for Cladocera. Limnol. Oceanogr. 18:331–332.

HEBERT, P. D. N. 1978. Cyclomorphosis in natural populations of *Daphnia cephalata* King. Freshwater Biol. 8:79–90.

HRBÁČKOVÁ-ESSLOVÁ, M. 1963. The development of three species of *Daphnia* in the surface water of the Slapy Reservoir. Int. Rev. Gesamten Hydrobiol. 48:325–333.

HUTCHINSON, G. E. 1967. A treatise on limnology, v. 2. Wiley.

IVLEV, V. S. 1961. Experimental ecology of the feeding of fishes. Yale Univ. Press.

JACOBS, J. 1961. On the regulation mechanism of environmentally controlled allometry (heterauxesis) in cyclomorphic *Daphnia*. Physiol. Zool. 34:202–216.

———. 1962. Light and turbulence as co-determinants of relative growth rates in cyclomorphic *Daphnia*. Int. Rev. Gesamten Hydrobiol. 47:146–156.

———. 1967. Temperature, food, turbulence as natural determinants of cyclomorphosis in *Daphnia*. Naturwissenschaften 54:207.

KERFOOT, W. C. 1974. Egg-size cycle of a cladoceran. Ecology 55:1259–1270.

MacARTHUR, J. W., and W. H. T. BAILLIE. 1929. Metabolic activity and duration of life I. Influence of temperature on longevity in *Daphnia magna*. J. Exp. Zool. 53:221–242.

SNEDECOR, G. W., and W. G. COCHRAN. 1967. Statistical methods. Iowa State Univ. Press.

SOKAL, R. R., and F. J. ROHLF. 1969. Biometry. Freeman.

SPRULES, W. G. 1972. Effects of size-selective predation and food competition on high altitude zooplankton communities. Ecology 53:375–386.

UZZELL, T. M., Jr. 1962. Morphology and biology of the salamanders of the *Ambystoma jeffersonnianum* complex. Ph.D. thesis, Univ. Michigan. 170 p.

VOLLENWEIDER, R. A. [ed.]. 1969. Measuring primary productivity in aquatic environments. IBP Handbook 12. Blackwell.

WILBUR, H. M. 1971. Competition, predation and the structure of the *Ambystoma-Rana sylvatica* community. Ph.D. thesis Univ. Michigan. 142 p.

WILLIAMS, G. C. 1966. Adaptation and natural selection. Princeton Univ. Press.

ZARET, T. M. 1972. Predator-prey interaction in a tropical lacustrine ecosystem. Ecology 53:248–257.

——, and W. C. KERFOOT. 1975. Fish predation on *Bosmina longirostris*: body-size selection versus visibility selection. Ecology 56:232–237.

42. The Genetic Component of Cyclomorphosis in *Bosmina*

Robert W. Black

Abstract

The morphology of an inshore Lake Washington population of *Bosmina* was examined over several seasons. Four of the seven morphological characters examined showed seasonally related changes in average size; this cyclomorphosis was unrelated to the age structure of the population but was due in part to changes in the relative abundance of two reproductively isolated populations. Re-evaluation of the data, after taking the species heterogeneity into account, showed that only one of the two species was cyclomorphic and that cyclomorphosis in this species had a strong genetic component.

Cyclomorphosis, the seasonal change in morphology within a plankton population, has been documented in dinoflagellates, rotifers, copepods, and cladocerans (*see* Hutchinson 1967). The ultimate cause of cyclomorphosis has been much discussed in the literature (Hutchinson 1967; Dodson 1974; Kerfoot 1974, 1977b; Hebert 1978), but is not yet clearly understood; it may well be related to patterns of selective predation by temporally abundant predators (Brooks 1965; Dodson 1974). Laboratory studies by Brooks (1947), Jacobs (1961), and Kerfoot (1972, 1975a) on cyclomorphic Cladocera have shown that cyclomorphosis probably includes the differential morphological development of one or more genotypes in seasonally distinct lake environments. Although Jacobs and Kerfoot both suggested that cyclomorphosis may also include a genetic component, the findings of these researchers on the phenotypic plasticity of single genotypes are so dramatic that many biologists have disregarded the possible importance of a genetic component and believe that cyclomorphosis is solely the effects of environmental conditions on body shape (Hebert 1978). I have documented and analyzed patterns of morphological change in a *Bosmina* population in Lake Washington and have determined that cyclomorphosis in this cladoceran includes a strong genetic component.

Background and methods

This study was initiated in spring 1975. At that time, it was commonly thought that *Bosmina longirostris* was a widespread and phenotypically variable species. Not only had Kerfoot (1972, 1974, 1975a) documented cyclomorphosis in a *B. longirostris* population in Frains Lake, Michigan, but Kerfoot (1975b) and Herbst (1962) had collectively reported the occurrence of

---

This research was supported by NSF grant No. 031-1141A to L. B. Slobodkin. I thank L. B. Slobodkin and W. T. Edmondson for their invaluable help and support throughout this study. The research presented here benefited from discussions with R. A. Pastorok, W. C. Kerfoot, T. Zaret, and C. King and from the technical assistance of C. Peterson and advice of A. Litt. I appreciate the help and support of A. Sauer throughout the data analysis and preparation of this manuscript.

several distinct morphs of this species in North America and Europe. Kerfoot, in fact, documented the simultaneous occurrence of two such morphs in several locations in Lake Washington.

In an attempt to determine if cyclomorphosis occurred in Lake Washington *Bosmina*, I examined the morphology of animals collected every 1-2 weeks from an inshore study site. The study site was a small, shallow arm of Union Bay, Lake Washington, and was bounded by land on three sides and by fixed, floating logs and vegetation on the fourth. Although the margins of the site contained scattered stands of emergent and floating aquatic plants, all sampling was done in the center of the site which was 2m deep and free of rooted aquatics. The zooplankton community at the study site was typical of the inshore sites studied by Kerfoot (1975*b*, 1977*a*) and Kerfoot and Peterson (in press) and commonly included the following genera: *Asplanchna, Bosmina, Ceriodaphnia, Cyclops, Daphnia, Diaphanosoma, Diaptomus, Epischura,* and *Polyphemus.*

Quantitative and qualitative plankton samples were taken regularly at the study site. Quantitative samples were taken through the repeated use of a 2-liter van Dorn bottle. Water samples obtained in this way were strained through No. 20 plankton netting and the organisms collected were preserved in alcohol. Large qualitative plankton samples were collected with long hauls of a No. 20 net; half of each sample was preserved in alcohol, the other half was returned alive to the laboratory.

About 65 *Bosmina* were randomly chosen from certain samples and mounted in glycerin on a slide; these animals were taken from quantitative samples during most of 1975 and from qualitative samples from then on. Each mounted *Bosmina* was then examined and the following characters measured or scored (*see Fig. 1*): total length, carapace length, carapace height, antennule length, number of antennule segments, mucro length, and number of mucro sutures. All measurements were done with an ocular micrometer in a Bausch and Lomb compound microscope (430X).

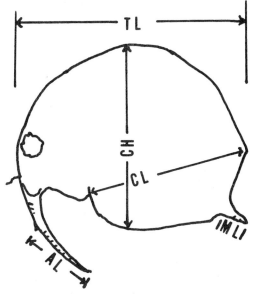

Fig. 1. Characters measured in *Bosmina longirostris* showing alignment of total length (TL), carapace length (CL), carapace height (CH), antennule length (AL), and mucro length (ML) measurements.

To determine if the two morphs Kerfoot reported were genetically distinct and not merely the extremes of phenotype that can be shown by a single genotype, I began cloning *Bosmina* in fall 1975. Assuming that a given genotype will always produce the same phenotype or range of phenotypes when acclimated to and raised under a single set of environmental conditions, I kept the cloning medium, food type and concentrations, temperature, and photoperiod constant and examined clones for differences in the morphological characters listed above.

Each clone was initiated with a single female that was chosen at random from a live sample. Cloning was done in shell vials that contained 10 ml of a modified Provasoli medium and at least 10^4 cells/ml of both *Chloridella minuta* and *Bracteacoccus minor*. The vials were stored in a constant temperature box set at 15°C and a long day photoperiod (14:10, L:D). The clones were examined for survival and reproduction every 2 or 3 days and received additions of at least 10^4 cells/ml of both algal species

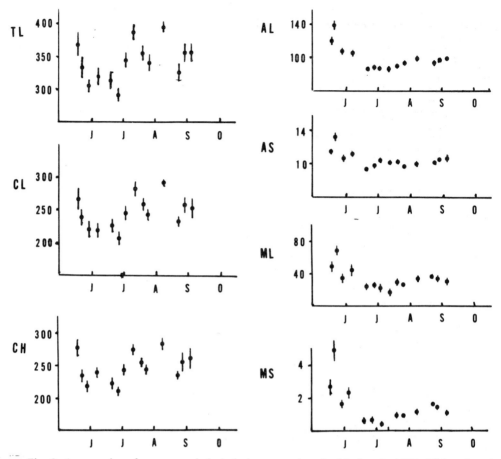

Fig. 2. Average size of seven morphological characters plotted with date in 1975. All length and height measurements are in micron units; number of antennule segments (AS) and the number of mucro sutures (MS) are counts. Vertical bars = ± 1 SE.

at these times. The clones were maintained in this manner for at least three generations (about 24 days).

During the cloning period, the shed carapaces and the corpses of all unsuccessful clones were collected as the clones failed and were mounted in glycerin on a slide. At the end of the cloning period all successful clones were terminated and the surviving animals and shed carapaces preserved in a similar manner. All of the preserved material was then scored and measured as described above.

Patterns of morphological change in 1975

Figure 2 presents plots of average size versus the sampling date for the seven morphological characters in 1975. Two patterns of change are apparent: three characters (total length, carapace length, and carapace height) fluctuate irregularly throughout the year, while the other four show a smooth seasonal transition from large spring to smaller summer sizes. Table 1 summarizes Kruskal-Wallis tests which compared the average size of a character across sampling dates and showed that every character examined showed significant changes in size during 1975.

The changes in average total length, carapace length, and average carapace height during 1975 reflect changes in the age structure of the population. Figure 3 shows that each of these characters increase

Table 1. Summary of Kruskal-Wallis tests (H) comparing average character size over 16 sampling dates in 1975.

Character	df	H	Significance
Total length	15	95.55	$p < 0.005$
Carapace length	15	95.52	$p < 0.005$
Carapace height	15	91.95	$p < 0.005$
Antennule length	15	157.92	$p < 0.005$
Antennule segments	15	116.26	$p < 0.005$
Mucro length	15	193.25	$p < 0.005$
Mucro sutures	15	176.70	$p < 0.005$

Table 2. Summary of Kendall rank correlations (r) between mean character size and proportion of population that is of reproductive size.

Character	No. of samples	r	Significance
Total length	16	0.80	$p < 0.001$
Carapace length	16	0.80	$p < 0.001$
Carapace height	16	0.74	$p < 0.001$
Antennule length	16	−0.05	ns
Antennule segments	16	0.04	ns
Mucro length	16	−0.08	ns
Mucro sutures	16	−0.17	ns

in size with age. Changes in the frequency of young or old animals in the population could thus produce changes in the average size of these characters. That such changes in the age structure of the population occur is shown in Fig. 4 which plots the proportion of the population that is reproductively mature against date. Figure 5 shows that the strong correlations exist between the proportion of mature individuals in the population and the average size of each character, and Table 2 shows that these correlations are statistically significant.

The observed changes in average antennule and mucro length, the average number of antennule segments and the average number of mucro sutures are not affected by changes in the age structure of the population. Figure 6 shows that none of these characters increase in size with age, so changes in the age structure of the population cannot influence them. This is clearly demonstrated by Fig. 7 which shows no correlation between the size of these characters and the proportion of the population that is of reproductive size.

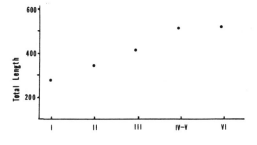
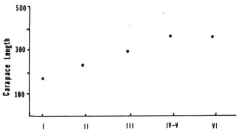
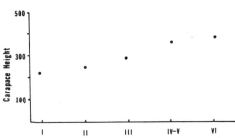

Fig. 3. Average size of three characters plotted with age (Instar). (Animals of given ages provided by W. C. Kerfoot).

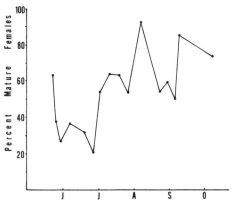

Fig. 4. Proportion of population that is reproductively mature plotted during summer 1975.

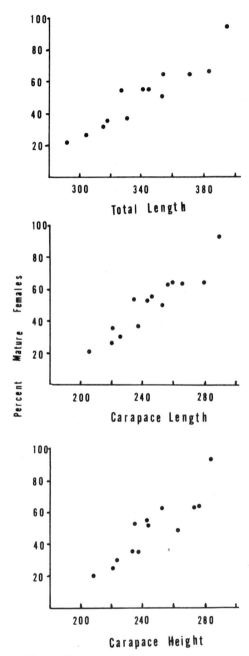

Fig. 5. Proportion of population that is of reproductive size plotted with average size of three characters.

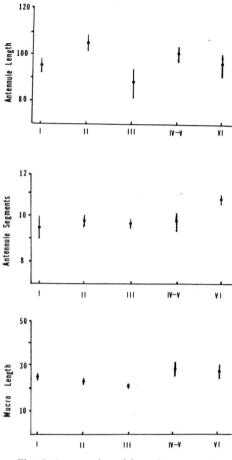

Fig. 6. Average size of four characters plotted with age (Instar). (Animals of given ages provided by W. C. Kerfoot).

The two morphs of *Bosmina* present in Lake Washington differed primarily in the sizes of these latter four characters. Although Kerfoot (pers. comm.) has found the number of mucro sutures to be the only consistent distinguishing character between the two morphs, Table 3 shows that in certain samples at least, the longer featured individuals (long-featured genotype, cf. Manning et al. 1978: i.e. with mucro sutures present) had much longer antennules and mucrones, both of which contained greater numbers of segments and sutures than did the short-featured morph. Figure 8 shows that large changes in the relative abundance of the two morphs occurred in 1975. It seemed likely then that changes in the relative abundance of the two morphs would produce changes in the average sizes of these characters. Figure 9 and Table 4

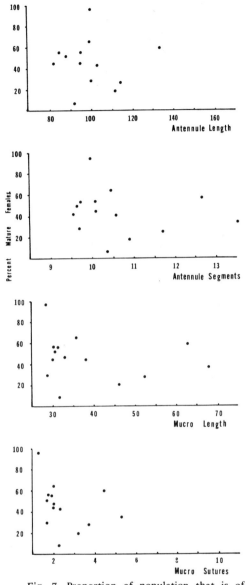

Fig. 7. Proportion of population that is of reproductive size plotted with average size of four characters.

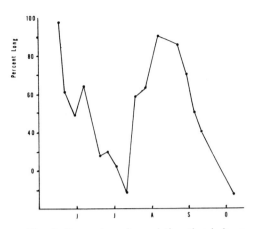

Fig. 8. Proportion of population that is long-featured morph plotted with date.

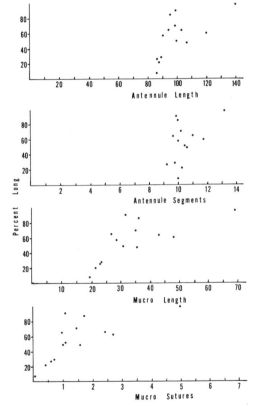

Fig. 9. Proportion of population that is long-featured morph plotted with average size of four characters.

show that this was apparently the case for three of these characters; for average antennule and mucro length and the number of mucro sutures all showed significant correlations with the proportion of long-featured morphs in the population.

Table 5 shows the average antennule and mucro length, the average number of antennule segments and the average number of mucro sutures of clones initiated with

Table 3. Morphology of long- and short-featured morph individuals (all animals collected 20 May 1975).

Individual	Total length	Carapace		Antennule		Mucro	
		length	height	length	segments	length	sutures
Long-featured morph							
5	307.4	220.4	220.4	159.5	13	78.3	6
7	391.5	258.1	261.0	145.0	14	84.1	5
11	365.4	278.4	261.0	153.7	15	75.4	6
13	362.5	261.0	269.7	98.6	11	49.3	4
Short-featured morph							
10	319.0	232.0	237.8	87.0	9	14.5	0
14	348.0	252.3	263.9	104.4	10	17.4	0
16	461.1	342.2	348.0	84.1	10	20.3	0
21	342.2	252.3	243.6	98.6	10	20.3	0

Note: all length and height measurements in micron units.

long- and short-featured *Bosmina* in fall 1975. Examination of this table shows that long-featured clones have consistently longer antennules and mucrones which contain a greater number of segments and sutures respectively than do short-featured morph clones. This indicates that genetic differences exist between morphs with respect to these characters. This demonstration initially suggested that cyclomorphosis in *Bosmina* includes a genetic component. Since I knew that the morphs were genetically different and had already documented the occurrence and effects of changes in the relative abundance of each morph on the average size of cyclomorphic characters, it seemed obvious that changes in the size of these characters directly reflected the occurrence of genetic change in the population.

Table 4. Summary of Kendall rank correlations (r) between average character size and proportion of long-featured morph individuals in population.

Character	No. of samples	r	Significance
Total length	16	−0.05	ns
Carapace length	16	−0.07	ns
Carapace height	16	−0.07	ns
Antennule length	16	0.45	$p < 0.02$
Antennule segments	16	0.10	ns
Mucro length	16	0.67	$p < 0.001$
Mucro sutures	16	0.69	$p < 0.001$

The magnitude of genetic differences

In 1976 Manning et al. (1978) used electrophoretic techniques to show that the

Table 5. Average character size of long- and short-featured morph clones isolated from a single live haul (September 1975).

Clone	Antennule						Mucro					
	length			segments			length			sutures		
	n	$\bar{x}$	SE	n	$\bar{x}$	SE	n	$\bar{x}$	SE	n	$\bar{x}$	SE
Long-featured morph												
6	2	98.6	14.5	2	12.0	2.0	5	39.4	2.8	5	2.0	0.0
7	6	95.7	2.4	6	11.3	0.3	6	41.6	2.3	5	2.4	0.2
Short-featured morph												
1	7	75.1	5.3	7	11.0	0.2	6	16.4	1.6	6	0.0	0.0
2	3	88.0	7.0	3	10.7	0.3	4	18.1	0.7	4	0.0	0.0
3	3	77.3	1.0	3	10.3	0.3	4	19.6	2.1	4	0.0	0.0
4	6	91.8	6.2	5	11.0	0.3	6	21.8	0.6	6	0.0	0.0
5	2	79.8	7.2	2	11.5	0.5	2	18.8	1.4	2	0.0	0.0
8	3	105.4	6.4	3	10.6	0.3	2	18.8	1.4	2	0.0	0.0
9	6	87.0	3.0	6	10.8	0.4	5	20.3	1.6	5	0.0	0.0

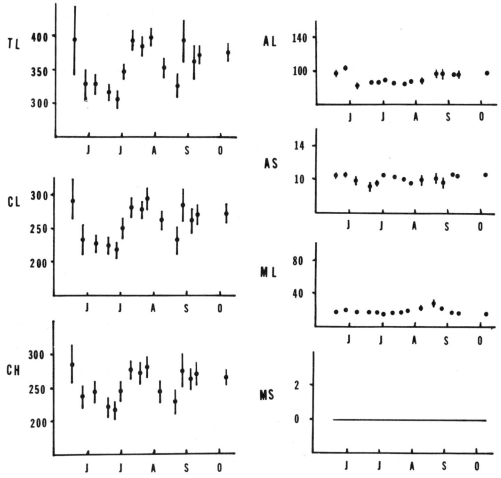

Fig. 10. Average size of seven characters in short-featured species plotted with date.

two *Bosmina* morphs present in Lake Washington were not two morphs of a single species, but were distinct species. This discovery further complicated the investigation of seasonal changes. It meant that data sets of the two species should be analyzed separately. Furthermore, the notion that the cyclomorphosis was produced by changes in the relative abundance of just two morphs was perhaps oversimplistic. In fact, a more detailed description was necessary to reveal cyclomorphotic patterns within the two reproductively isolated populations.

Figures 10 and 11 plot average size versus the sampling date for the seven morphological characters of the long- and short-featured species after the original data set had been separated into the proper species sets. These figures show that the patterns of morphological change described earlier for total length, carapace length, and carapace height are also found in each species when examined alone. Figures 12 and 13 and Table 6 show that similar, significant correlations exist between the age structure of each species population and the average size of these three characters, as was shown previously.

The two species show markedly different patterns of change in average antennule and mucro length and the average numbers of segments or sutures found in each. Figures 10 and 11 show that the short-featured

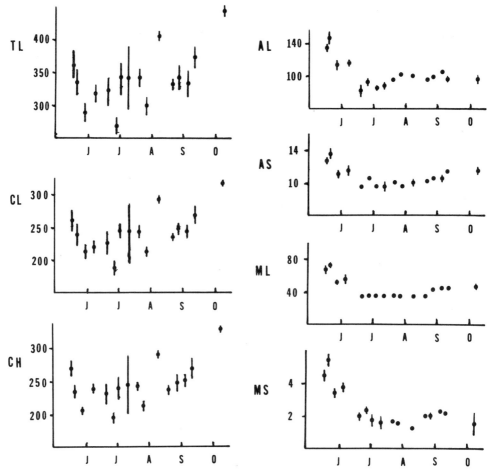

Fig. 11. Average size of seven characters in long-featured species plotted with date.

species exhibits no seasonally related changes in the average size of these characters, while the long-featured species shows marked cyclomorphosis. Figures 14 and 15 show that no correlations exist between the average size of these characters and the age structure of the population. The initial averaging of character size across the two species thus masked the presence of cyclomorphosis in one *Bosmina* species and its absence in the second species. What was interpreted as cyclomorphosis in a single species was then not just due to the changing frequencies of two morphologically distinct species, but was due to the occurrence of cyclomorphosis in one of the species.

A common problem?

Mistaking morphologically similar, yet distinct species for morphs or variants of a single species is probably a common occurrence. Just as Manning et al. (1978) showed what is conventionally thought to be a single species (*B. longirostris*) to actually be a complex of two species, Hebert (1977) used electrophoretic techniques to distinguish between several similar Australian *Daphnia* species that were once thought to be variants of a single species. Furthermore, Hebert (1978) has suggested that what was once thought to be a cyclomorphosis in *Daphnia carinata* is actually the seasonal

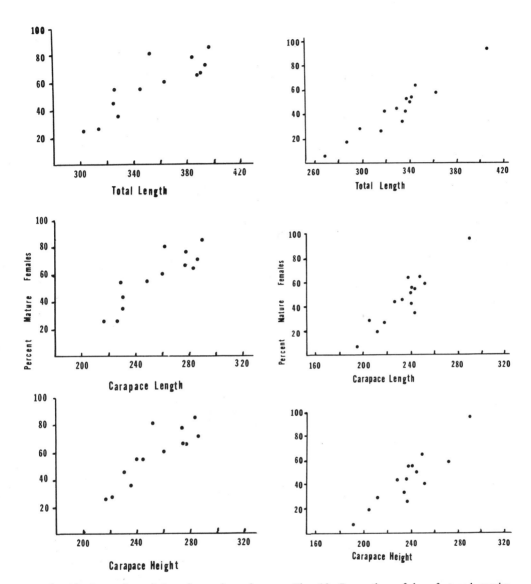

Fig. 12. Proportion of short-featured species population that is of reproductive size plotted with average size of three characters.

Fig. 13. Proportion of long-featured species population that is of reproductive size plotted with average size of three characters.

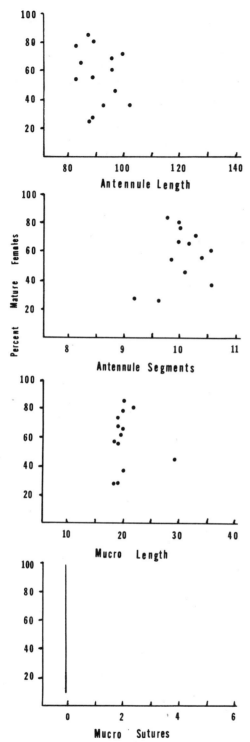

Fig. 14. Proportion of short-featured species population that is of reproductive size plotted with average size of four characters.

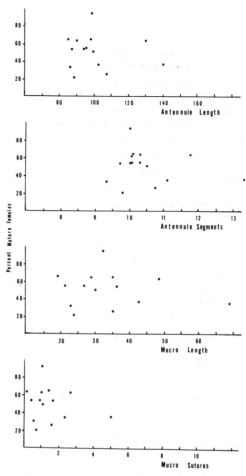

Fig. 15. Proportion of long-featured species population that is of reproductive size plotted with average size of four characters.

succession of several of these similar species. Since limnologists are just recently turning to electrophoresis as a tool to help distinguish between similar species, it is likely that upon examination other species that are now thought to be phenotypically quite variable will be shown to be complexes of similar species. Similarly, reports of cyclomorphosis that are based solely on the measurement of field animals and which do not include supporting laboratory or electrophoretic studies may be found to have mistakenly lumped similar species together and to have failed to document cyclomorphosis.

Clonal differences within a species

Table 7 presents the average size of the

four cyclomorphic characters in nine clones of the long-featured species that were isolated from a single live haul in 1976. An examination of this table reveals that significant differences in these characters exist between clones of this species, even when the clones were isolated from the same sample and raised under identical conditions. This indicates not only that genetic differences occur between clones with respect to certain morphological characters, but that the field population was genetically rather diverse with respect to these characters at the time the clones were isolated.

Table 8 presents a summary of the average size of cyclomorphic characters of five sets of long-featured clones that were isolated in spring 1976 and 1977, summer 1976, and fall 1975. Note that the average character size across clones is large in spring and smaller in summer and fall. This pattern is the same as is found in field samples within a year for this species and suggests that the seasonal differences in the morphology of field animals reflects underlying genetic differences between the animals and the occurrence of morphologically related genetic changes in the population.

Speculation

I suggest that just as the two species of Lake Washington *Bosmina* have been shown to change in relative abundance during the year, so do several genetically distinct clones of a single species change in frequency. These changes in frequency may be due to random events, but are more probably due to selective events that involve the clones' reproductive capacities (*see* King

Table 6. Summary of Kendall rank correlations (r) between average character size of a species and proportion of population that is of reproductive size.

Character	No. of samples	r	Significance
Long-featured species			
Total length	16	0.88	$p < 0.001$
Carapace length	16	0.87	$p < 0.001$
Carapace height	16	0.73	$p < 0.001$
Antennule length	16	−0.08	ns
Antennule segments	16	0.02	ns
Mucro length	16	−0.05	ns
Mucro sutures	16	−0.39	$p < 0.05$
Short-featured species			
Total length	15	0.68	$p < 0.001$
Carapace length	15	0.66	$p < 0.001$
Carapace height	15	0.62	$p < 0.01$
Antennule length	15	−0.05	ns
Antennule segments	15	0.09	ns
Mucro length	15	0.04	ns
Mucro sutures	15	0.00	ns

Table 7. Average character size of long-featured species clones isolated from a single live haul (July 1976).

| | Antennule | | | | | | Mucro | | | | | |
| | length | | | segments | | | length | | | sutures | | |
Clone	n	$\bar{x}$	SE	n	$\bar{x}$	SE	n	$\bar{x}$	SE	n	$\bar{x}$	SE
1-3	14	104.8	2.6	14	11.4	0.3	13	39.9	0.7	13	2.0	0.3
2-1	7	106.5	5.4	7	12.6	0.5	8	42.4	1.8	8	1.9	0.2
3-7	14	95.9	1.6	14	10.7	0.2	15	28.8	0.9	14	1.2	0.2
4-1	8	87.7	4.3	8	10.9	0.4	9	25.4	1.5	9	1.2	0.2
4-3	8	99.7	3.6	8	10.9	0.6	13	36.1	1.4	13	1.8	0.2
4-7	7	101.1	4.8	7	10.7	0.6	9	46.1	1.3	9	2.4	0.2
9-2	4	90.6	3.8	4	10.0	0.0	4	23.2	3.6	4	0.8	0.2
9-7	13	96.4	2.0	12	11.1	0.3	12	28.8	0.9	12	1.2	0.1
11-2	6	97.6	3.1	6	10.3	0.2	11	32.4	2.8	11	1.9	0.3
	$F=3.19$			$F=2.78$			$F=21.99$			$F=4.83$		
	$p < 0.005$			$p < 0.025$			$p < 0.001$			$p < 0.001$		

Note: length measurements are in micron units. Symbols defined as follows: n—number of measurements made for a given clone; $\bar{x}$—average size of a character for a given clone; SE—standard error of each average value. Results of an analysis of variance which compared sizes of these characters across the nine clones are also presented.

Table 8. Average size of cyclomorphic characters in five sets of long-featured clones isolated between fall 1975 and spring 1977.

	Antennule				Mucro			
	length		segments		length		sutures	
Date of isolation	a	$\bar{x}$	a	$\bar{x}$	a	$\bar{x}$	a	$\bar{x}$
Spring								
1976	8	109.0	8	11.4	8	47.1	8	2.5
1977	5	112.4	5	11.6	5	50.0	5	2.5
Summer								
1976—July	9	97.8	9	11.0	9	33.7	9	1.6
1976—August	11	104.2	11	10.4	11	31.4	11	1.5
Fall								
1975	2	97.2	2	11.7	2	40.5	2	2.2

Note: length measurements are in micron units. Symbols defined as follows: a—number of clones examined; $\bar{x}$—average size of a character across a set of clones.

1972; Hebert 1974; Kerfoot 1977a) or their ability to escape predation. If the clones that change in frequency are morphologically distinct, then changes in the average morphology of field animals must necessarily occur. An example of such change is cyclomorphosis in the long-featured *Bosmina* species, which involves changes in the frequency of clones that have been shown to be morphologically distinct when grown under identical conditions.

Conclusions

1. The analysis of morphological change in *B. longirostris* in Lake Washington in a traditional manner may conceal intriguing phenomena; for example, what was once considered to be the cyclomorphosis of a single *Bosmina* species has been shown to be in part a two-species phenomenon.

2. Cyclomorphosis occurs in only one of the two *Bosmina* species present in Lake Washington.

3. Cyclomorphosis in the long-featured *Bosmina* species occurs in four characters and is unrelated to changes in the age structure of the population.

4. Morphological differences in cyclomorphic characters occur between clones of the long-featured species.

5. Changes in the frequency of morphologically distinct clones probably produces the observed cyclomorphic changes in the long-featured species.

References

BROOKS, J. L. 1947. Turbulence as an environmental determinant of relative growth in *Daphnia*. Proc. Natl. Acad. Sci. 33: 141–148.
———. 1965. Predation and relative helmet size in cyclomorphic *Daphnia*. Proc. Natl. Acad. Sci. 53: 119–126.
DODSON, S. I. 1974. Adaptive change in plankton morphology in response to size selective predation: A new hypothesis of cyclomorphosis. Limnol. Oceanogr. 19: 721–729.
HEBERT, P. D. N. 1974. Ecological differences between genotypes in a natural population of *Daphnia magna*. Heredity 33: 327–337.
———. 1977. A revision of the taxonomy of the genus *Daphnia* in southeastern Australia. Aust. J. Zool. 25: 371–398.
———. 1978. The adaptive significance of cyclomorphosis in *Daphnia*: more possibilities. Freshwater Biol. in press.
HERBST, H. Y. 1962. Blattfusskrebse. Kosmos, Stuttgart.
HUTCHINSON, G. E. 1967. A treatise on limnology, v. 2. Wiley.
JACOBS, J. 1961. Cyclomorphosis in *Daphnia galeata mendotae* Birge, a case of environmentally controlled allometry. Arch. Hydrobiol. 58: 7–71.
KERFOOT, W. C. 1972. Cyclomorphism of the genus *Bosmina* in Frains Lake, Michigan. Ph.D. Thesis, Univ. Michigan.
———. 1974. Egg-size cycle of a cladoceran. Ecology 55: 1259–1270.
———. 1975a. Seasonal changes of *Bosmina* in Frains Lake, Michigan: Laboratory observations of phenotypic changes induced by inorganic factors. Freshwater Biol. 5: 227–253.
———. 1975b. The divergence of adjacent populations. Ecology 56: 1298–1313.
———. 1977a. Competition in cladoceran communities: The cost of evolving defenses against

copepod predation. Ecology **58**: 303–313.

———. 1977*b*. Implications of copepod predation. Limnol. Oceanogr. **22**: 316–325.

———, and C. PETERSON. In press. Predatory copepods and *Bosmina*: replacement cycles and further influences of predation upon prey reproduction. Ecology.

KING, C. 1972. Adaptation of rotifers to seasonal variation. Ecology **53**: 408–418.

MANNING, J., W. C. KERFOOT, and E. M. BERGER. 1978. Phenotypes and genotypes in cladoceran populations. Evolution **32**: 365–374.

43. Perspectives on Cyclomorphosis: Separation of Phenotypes and Genotypes

W. Charles Kerfoot

Abstract

Apparent morphological cycles in supposed "species" may involve seasonal species successions (taxocene succession among sibling species), clonal replacements (substitutions among parthenogenetic lineages that belong to the same biological species), or phenotypic flexibility within parthenogenetic lineages (cyclomorphosis). While the three different patterns may ultimately reflect the magnitude of external factors (e.g. seasonally varying predation), cyclomorphosis is an interesting developmental adaptation to a predictably fluctuating environment. It apparently evolves mainly in cyclical parthenogens where the lifespans of genotypes greatly exceed the lifespans of phenotypes (individuals). The prevalence of cyclomorphosis within the Rotifera and Cladocera, and its virtual absence in the Copepoda, can be closely associated with reproductive modes and prey defenses.

The term "cyclomorphosis" was proposed by Lauterborn (1904) to describe the seasonal, and hence cyclical, changes in body form found in successive generations of many planktonic organisms. These changes affect such different aspects of the phenotype as body size, tail spine length, head ornamentation (helmets or pointed spikes), and pigmentation, and involve such phylogenetically diverse groups as dinoflagellates, protists, rotifers, and cladocerans (for a modern review of specific cases, *see* Hutchinson 1967, p. 810-954). Implicit in most early studies was the assumption that these phenotypic cycles involved phenotypic plasticity within a single population (but for a discussion of genetic differences *see* Jacobs 1961) and that the changes were intimately related to changes in the physical properties (temperature, oxygen concentration, turbulence, viscosity) of the individual's immediate environment (e.g. *see* Wesenberg-Lund 1926).

More recently, refinements in the study of cyclomorphosis have featured the division of causal mechanisms into two broad categories, "proximate" and "ultimate"; a distinct, and useful, tendency to restrict the use of the term "cyclomorphosis" to describe only the phenotypic (developmental) modifications which occur among clonal (genetically identical) lineages; the development of electrophoretic techniques, the running of clonal experiments, and the use of mating studies to clarify the genetic component in seasonal cycles; and the identification of several ultimate agents.

It now seems certain that many of the cyclomorphotic adjustments, at least in the more spectacular cases, are ultimately related to predation. Phenotypic responses can be associated with the pattern of preda-

---

This research was supported by NSF grant DEB 76-20238.

tion: either predators that continually inhabit the littoral or pelagic environments of large lakes and that select for spatially stable littoral "races," or predators that fluctuate seasonally to produce cyclomorphotic sequences in their prey (de-Beauchamp 1952; Pourriot 1964; Jacobs 1965; Brooks 1965, 1967; Gilbert 1966; Gilbert and Waage 1967; Halbach 1971; Halbach and Jacobs 1971; Zaret 1972a; Kerfoot 1972, 1974, 1975a b; Dodson 1974a). The problems of spatially stable "races" and of cyclomorphotic sequences must be intimately connected, for the slight, but distinctive, features that characterize "races" and cyclomorphotic phenotypes within both the rotifers and the Cladocera are similar; only the pattern is different (spatial versus seasonal). Within the Cladocera, the morphological adjustments affect spine lengths (Halbach and Jacobs 1971; Zaret 1972a; Kerfoot 1975b), body pigmentation and eye-pigment diameter (Green 1971; Zaret 1972a, b; Zaret and Kerfoot 1975), head size and shape (Green 1967; Dodson 1974a), egg size and exoskeleton thickness (Kerfoot 1974, 1977b; Dodson 1974a, b; Jacobs 1977, 1978).

An important task of modern research is the investigation of specific feeding patterns in vertebrate and invertebrate predators (e.g. fishes, mysids, *Leptodora*, salamanders, midge larvae, predatory copepods, and *Asplanchna*) which favor the evolution of particular morphological defenses. For example, in the littoral, adult fish or their fry may draw heavily upon zooplankton during early spring (Hrbáček 1962; Hrbáček and associates 1960, 1961, 1965; Brooks and Dodson 1965; Brooks 1968; Nilson and Pejler 1973; Hall et al. 1970; Zaret 1972a; Zaret and Kerfoot 1975; Werner and Hall 1977; Kerfoot 1975b; Kerfoot and Peterson 1979). The visual, particulate (Janssen 1976) foraging of most fry would favor small size or increased inconspicuousness of zooplankton (Hrbáček and Hrbáčeková-Esslová 1960; Brooks and Dodson 1965; Brooks 1968; Zaret and Kerfoot 1975; O'Brien et al. 1979). In contrast, in the open-water regions of many holarctic lakes, predatory copepods which sense prey primarily by means of mechanoreceptors (Strickler and Bal 1973; Strickler 1975; Kerfoot 1977a, 1978) may draw heavily from small or fragile-bodied zooplankton populations (Kerfoot 1975b, 1977a; Dodson 1975; Lane et al. 1976; Confer and Cooley 1977; Brandl and Fernando 1978). Prey have evolved a variety of evasive tactics and morphological defenses to reduce the risk from grasping predators, and these responses are implicated in cyclomorphoses (Kerfoot 1972, 1974, 1975a, b, 1977a, b; Dodson 1974a; O'Brien and Vinyard 1978; O'Brien et al. 1979). Finally, the most conclusive evidence for cause and effect comes from studies on rotifers. Here Pourriot (1964) and Gilbert (1966, 1967), working with the observations of de Beauchamp (1952), have identified a factor produced by the predaceous rotifer *Asplanchna* which stimulates the production of spines in several prey, especially in the rotifer *Brachionus calyciflorus*. These experiments have since been verified and expanded to include other predator-prey pairs (Halbach 1969, 1971; Halbach and Jacobs 1971; Pourriot 1974; Green and Lan 1974).

Lest anyone suppose that the study of cyclomorphosis is now a closed field, there is still much that needs to be done on the description, interpretation, and dynamics of seasonal changes. In this paper, in keeping with the spirit of "perspectives," I will discuss a broad array of subjects: the origin of cyclomorphosis, multivariate techniques for measuring phenotypic cycles, population dynamics that underlie cycles in small and large lakes, and reasons for the lack of cyclomorphosis in co-occurring taxa. Throughout this discourse, however, I will continue to stress one important theme: that the separation and independent consideration of phenotypes and genotypes gives considerable insight into the origin of cyclomorphosis and serves to clarify the population dynamics that underlie seasonal cycles.

The origin of cyclomorphosis

Cyclomorphosis and parthenogenesis have occasionally been associated in a causal sense, but the relationship has always been vaguely and poorly defined (Hartmann 1917; Hutchinson 1967, p. 810 and 856). To fully appreciate how cyclomorphosis functions as an adaptation and how it might relate to parthenogenesis, we must acknowledge the ex post facto nature of natural selection, recognize the anticipatory character of the developmental response, and distinguish between the *longevity of phenotypes* (individuals) *and of genotypes*. Cyclomorphosis is an adaptation to a predictably fluctuating environment. It is intriguing because successive generations respond, rather than single long-lived individuals.

In a fluctuating environment, because the evolutionary process is by natural selection, as pointed out by Levins (1968), "there is always the danger that by the time the adaptation has taken place the environment has changed." For example, during the winter, those genotypes that are best adapted to winter conditions survive in greater numbers and leave more offspring for the next generation. The final generations are best adapted to winter conditions, but must develop during the summer. Likewise, those generations that survive best and leave the most offspring during summer must face winter. The species is "always lagging behind the environment, doing the right thing for the previous situation."

Under this selective regime, as illustrated in Fig. 1, natural selection is apt to favor *genotypes* which could developmentally adjust to a seasonal sequence of environments, acquiring sensitivity to cues which would

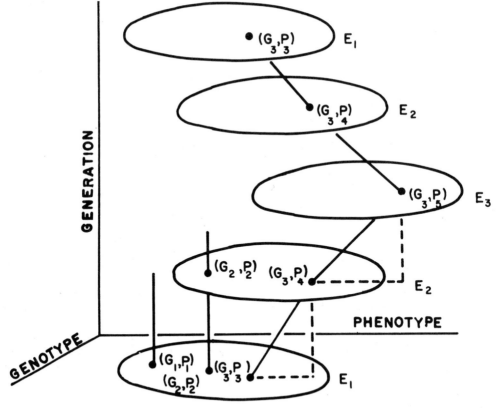

Fig. 1. Circumstances favoring evolution of flexible phenotypes in successive generations of clonal lineages. Three genotypes (G_1, G_2, G_3) are followed through a cyclical sequence of environments (E_1, E_2, E_3), where circles enclose fit phenotypes (P_i) in each environment. In this example, genotypes (G_1, G_2) that produce inflexible phenotypes are eliminated, while a genotype (G_3) that produces a flexible phenotype survives cyclical environmental changes.

anticipate changes and produce various phenotypes in successive generations. This adaptation is found in long-lived species, such as birds and mammals, that moult appropriate coat colors during successive seasons. In these vertebrates, the longevity of individuals and of genotypes is similar, i.e. one generation. In cyclomorphosis, however, apomictic parthenogenesis produces successive generations that are identical in genotype, although individual phenotypes (lifespans) are short. The seasonal fitness of an entire lineage, i.e. the success of the genotype, depends upon the combined fitness of the component phenotypes. The adaptation involves the production of different phenotypes in successive generations. The longevity of genotypes is crucial (and presumably could extend for prolonged periods, perhaps for years) rather than the longevity of a single individual per se. The circumstances of the adaptation are seriously disrupted or severely modified by sexual reproduction which destroys genotypes, decreases parent-offspring resemblance, and introduces individually unpredictable variation.

While focusing attention away from the origin of cyclomorphosis as such, the theoretical separation between proximate cues (the stimuli and reactions which govern the time, place, and magnitude of morphological responses) and ultimate advantages (the selective benefits of the morphological responses) is both an important and convenient convention. Natural selection has favored sensitivity to proximate stimuli because these cues most reliably announce environments that influence individual fitness. In cladocerans and rotifers, proximate cues seem to fall into two very different classes. One class is comprised of physical factors (e.g. temperature, turbulence, and light: Brooks 1946, 1947; Jacobs 1961, 1962, 1967) which, in various combinations, tell the organism the appropriate time or place to change its morphology. Another quite different sort of cue is a substance (metabolic by-product or pheromonal substance), associated with a predator, which prey have acquired sensitivity to and which stimulates the development of protective defenses (e.g. *Asplanchna* and *Brachionus*: Gilbert 1966, 1967; Halbach 1969, 1971). This kind of very specialized response may evolve where predation comes primarily from one predator.

In a cyclomorphism related to predation, both kinds of cues (physical combinations and specific chemical stimuli) serve the same purpose, i.e. to indicate to the organism the probability of its encountering a particularly disadvantageous situation. Most of these sophisticated stimulus-response systems are the end-product of considerable evolution, are anticipatory reactions, and depend upon a predictable (contingent: Lee pers. commun.) association between proximate cues and ultimate (adaptive) benefit.

Seasonal cycles: Cyclomorphosis and succession

In describing the nature of population responses in lakes, we encounter two major problems: one is semantic and the other biological. Seasonal sequences in many lakes entail not only simple morphological adjustments within coexisting lineages, but may involve a variety of population responses at several levels in the community. These responses may include regular (seasonal) developmental (phenotypic) responses within lineages (i.e. cyclomorphosis in its strictest sense), clonal replacements (within a biological species), or species succession (taxocene succession). If we restrict the term "cyclomorphosis" to describe the developmental (phenotypic) adjustments within individual lineages of parthenogenetically produced generations, which I believe we should, we are left with no corresponding formal terms for regular clonal replacements, or species successions, in lakes. Toward a more coherent semantics, I suggest use of the above terms (cyclomorphosis, clonal replacement, taxocene succession).

In the following pages, my aim is to consider the description of regular seasonal cycles and to discuss the magnitude of such cycles in the context of underlying population responses.

Description of cyclomorphosis in small

ponds or lakes—In small ponds or lakes, prey populations are often caught in a seasonal progression of predators, without any major spatial refuge. This type of selective regime would certainly promote the regular phenotypic flexibility of parthenogenetic lineages, or a regular clonal succession. For the moment, though, let us bypass the matter of population dynamics and consider only the problem of adequately describing the phenotypic cycles.

The description of seasonal cycles in the size or shape of organisms is a deceptively difficult subject, handled by a variety of methods in the past. Traditionally, most studies of cyclomorphism have treated spectacular changes in single features, and yet have recognized (e.g. Jacobs 1961, 1967) that several features of the phenotype are usually changing simultaneously. Significant seasonal changes are depicted by plotting the sampling means (with accompanying standard errors, standard deviations, and ranges) against sampling dates (time) for several months or even years. Cyclomorphosis is then confirmed by demonstrating oscillating patterns, where the period of oscillation is 1 year (Fig. 2, upper panel). The plot, then, is one of variable (character state) against time (character versus time space). The cyclomorphosis is judged similar from year to year if the phase (timing) and magnitude of the oscillations are similar. Recent refinements on this theme have used the concept of heterauxesis (Brooks 1947; Jacobs 1961; Hutchinson 1967) to characterize the growth rates of features relative to some standard dimension (usually a measure of body length). In this way, characters which show regular shifts in allometric growth patterns can be analyzed simply and portrayed in meaningful ways.

If more than one feature of the phenotype shows significant cycles, the features (variables) either can be treated independently, i.e. each one plotted against time and checked for yearly oscillations, or can be handled simultaneously. While not customarily done, because of its complexity, simultaneous consideration of several

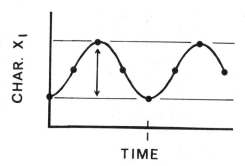

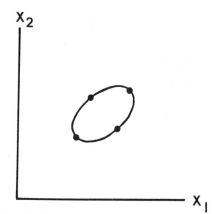

Fig. 2. Two ways to plot cyclomorphotic changes. Upper panel—variable against time; lower panel—variable against variable.

features (variables) seems important for three reasons: it allows full use of information found in covariance patterns among the features; since fitness is a property of the entire phenotype, multivariance description facilitates accurate calculation of phenotypic fitness; and since phenotypic adjustments may involve compromises to one or more opposing internal or external selective pressures, accurate assessment requires the simultaneous consideration of more than one aspect of the phenotype.

Multivariate descriptions of seasonal changes also potentially provide a more precise, inclusive, and formal definition of seasonal cycles. If we characterize each individual in a population sample by a set of n measurements, e.g. (carapace length, mucro length . . . antennule length), a set of vectors $(x_{ij}; j = 1, n)$ represents the value of the ith variable (feature or character) on the jth individual. A set of monthly samples can be summarized, then, by a set of mean vectors ($\bar{x}_{ik}$;$k = 1,N$) for N monthly samples, and a corresponding set of $n \times n$ covariance matrices V_k, derived from the variation of individuals about the character means. For the moment, let us consider the circumstances where we select characters that are ontogenetically independent of each other (correlation coefficient = 0.00). In multivariate space, the variables (the character means, $\bar{x}_{ik}$) can be plotted against each other (character $\times$ character space). For example, Fig. 2 (lower panel) illustrates a plot of two hypothetical variables in two-dimensional character space. In a cyclomorphotic pattern, the means for each sampling period should trace out a loop or, in the unusually unique case where fluctuations between all variables are perfectly correlated, a line. If the cyclomorphosis is indeed regular and repeatable, the monthly means should describe a closed loop, i.e. one that begins again where the last year's points left off. In multidimensional space (n-dimensional space), the loop would be characterized by direction (vector coordinates) and magnitude. In this way, a cyclomorphosis would be defined as a phenotypic cycle within a lineage which traces a closed loop in n-dimensional space, where one course about the loop lasts a single year. In theory, if a cyclomorphosis were identical from year to year, the corresponding yearly loops would be identical in magnitude and direction. Similarity of points along the loop could be measured from a single reference point (Fig. 3), e.g. using the Euclidean metric

$$D_{(k=1,N)} = [\sum_{i=1}^{n} (\bar{x}_{il} - \bar{x}_{ik})^2]^{1/2} \quad (1)$$

to measure the distance between monthly samples. Alternatively, distance could also be measured around the loop, using a circumferential technique.

In practice, there are usually conventions for reducing scaling effects, and techniques for treating correlated characters

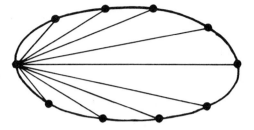

Fig. 3. Linear (Euclidean) measurement of distance between initial reference point and nine different successive points along cyclomorphotic loop in n-dimensional variable space.

(Sneath and Sokal 1973). For example, even with uncorrelated characters, certain variables may show greater fluctuations because they are larger in absolute size. To correct for scaling effects and to compare the relative magnitude of cycles, fluctuations in variables can be standardized by dividing differences between means by the within-population standard deviation of the respective variables. Compensation for use of moderately to highly correlated characters within samples requires other, much more complicated, procedures and frequently involves use of factor or principal components analysis to summarize redundant patterns (Sneath and Sokal 1973; Searle 1966).

In one of the most widely used descriptive techniques, that of Principal Component Analysis, the original axes (x_{ij}-coordinates that describe monthly means) can be rigidly rotated to form a new set of coordinates (y_{ij}-coordinates) for each point. Individual variation around the estimated means can be shown by plotting the original individual values in terms of the new system of coordinates or by drawing equal frequency ellipses around the means (Jolicoeur and Mosimann 1960). Empirically, principal components analysis often distinguishes size from shape variation (Jolicoeur and Mosimann 1960; Boyce 1965, 1969) and this feature may be of some use in cyclomorphotic studies. For example, consider the hypothetical bivariate swarm of points plotted in Fig. 4. If x_1 and x_2 represent the mean length and height, respectively, of a series of monthly samples, then the points along the cluster describe populations

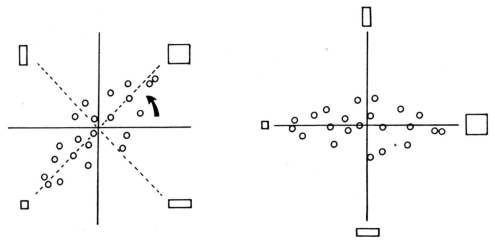

Fig. 4. Principal components ordination using cosine transformation. Original axes (*x*-coordinates) are rigidly rotated to form new set of axes (*y*-coordinates) that, in this case, separate size from shape variation.

that are large in the first quadrant, tall in the second, small in the third, and squat in the fourth. If we subject the original axes to a rigid rotation through an angle θ to form the new orthogonal (perpendicular) axes, y_1 and y_2, we find that the new axes now describe the points in terms of linear combinations of the original variables,

$$y_{1j} = x_{1j} \cos \theta + x_{2j} \sin \theta$$

$$y_{2j} = x_{2j} \sin \theta + x_{2j} \cos \theta \quad (2)$$

(Searle 1966; Pielou 1969). When viewed along the new system of coordinates, increases in body dimensions (small to large) are often separated from changes in body proportions (tall to squat). Transformations of this kind can be important in summarizing trends among large numbers of variables measured on individuals from single samples (within-sample variation) or applied to summarize trends among monthly samples (between-sample variation). Care must be taken, however, not to confuse description with underlying population responses, although multivariate description may aid population analysis by illuminating subtle heterogeneities.

A case study—The small cladoceran *Bosmina longirostris* occurs year-round in Frains Lake, Washtenaw County, Michigan, where it undergoes a yearly cyclomorphosis. Frains Lake is a small (6.5 ha), eutrophic lake distinguished by a single symmetrical basin (10 m deep at its center), a rapidly deoxygenated summer hypolimnion, and luxuriant littoral vegetation. Detailed aspects of its biology have been described elsewhere (Davis 1968; Davis et al. 1972; Kerfoot 1972, 1974, 1975a; Roth 1971; Storch 1971; Allan 1977) and will not be repeated here. The Frains Lake population of *Bosmina* was assigned to genus and species using the criteria of Goulden and Frey (1963) and Deevey and Deevey (1971).

In 2 years of sampling this species in Frains Lake, I found males only twice and each time (October 1969 and November 1968) they amounted to < 2% of the total samples. Based upon these observations, I suspect that sexual reproduction is limited to the late fall season in Frains Lake, occurs in only a small proportion of lineages, and that apomictic parthenogenesis occurs throughout most of the year. Thus the population of *Bosmina* in Frains Lake corresponds closely to the theoretical situation discussed earlier (i.e. in Fig. 1).

Although I took samples from other stations, primarily to check on the amount of spatial heterogeneity (which proved slight: Kerfoot 1972), I mainly took monthly samples from the central basin of the lake through a period of 2 years, from September 1967 through October 1969.

All individuals came from vertical net hauls with a No. 10 mesh Nitex net (125-μm aperture), preserved in 5% Formalin, mounted on glass slides, and measured at 400X under a Wild Heerbrugg stereomicroscope using a calibrated ocular micrometer. A conscious effort was made to measure 50 individuals from each sample, although sampling and scoring problems often lowered the totals somewhat.

Nine features (variables) were measured on *Bosmina*: 1) carapace length: the distance from the top of the posterior fissure to the anteroventral margin of the carapace; 2) carapace height: the distance from the ventral margin of the carapace to the highest opposing point on the dorsal margin; 3) mucro length: the length of the "tail" measured from the base, taken as the ventral extension of the fissure margin; 4 and 5) left and right mucro sutures: the number of sutures or ridges along the ventral margin of the mucro; 6) facet length: the length of the facet rim formed along the lateral margin of the head shield above the insertion of the antennule; 7 and 8) left and right antennule segments: the number of segments, not counting the very small tip segments (terminal ones disregarded if less than half the length of an ordinary segment), counted on the antennule from a toothlike process (midway down the antennule) to its distal tip; segments are outlined by circular sutures which girdle the antennule to its tip; and 9) antennule length: the length of the antennule as measured from the base of the "tooth" to its distal tip. Presumably the sutures and segments correspond to cell boundaries and cells, respectively (Hutchinson 1967, p. 852). In addition, a 10th character, egg length, was measured on parthenogenetic eggs from mature adults.

In order to clarify the counting and measuring techniques, the various measurements are shown in Fig. 5, upper panel, while representative counts are given in Fig. 5, lower panel, for antennules and mucrones on summer (July) and winter (January) individuals.

The nine basic characters were examined for significant seasonal trends. Table 1 lists the monthly means, standard deviations, and ranges for each variable. Variables were also checked for correlations within samples (x_{ij}) and between sample means ($\bar{x}_{ik}$). The resulting 22 monthly within-sample correlation matrices are too cumbersome to include here, but can be found in Kerfoot (1972); Table 2 presents two examples, one from a summer sample, the other from a winter sample. The correlations between sample means are presented in Table 3. Comparison of the two kinds of correlations reveals consistently high correlations within and between samples for certain characters expected to exhibit ontogenetic relationships (carapace length, carapace height, facet length) and most left-right pairs (left mucro sutures, right mucro sutures; left antennule segments, right antennule segments). The first set of characters were expected to have high correlations, since body growth should affect all centrally located parts. Likewise, the correlations between left and right pairs were expected, since animals are expected to be symmetrical. On the other hand, correlations within samples between either mucro length or antennule length and the other length characters were always consistently low, although these correlations did show seasonal trends (Fig. 6). All features, however, showed strong seasonal correlations (Table 3). Thus while mucro and antennule lengths show almost no relationship to only weak relationships within samples, the seasonal association between these variables, and between the principal nine characters and egg length, was consistently strong. Features were generally short in summer, long in winter.

The cyclomorphosis of *B. longirostris* in Frains Lake is described in Fig. 7. The variation in feature lengths can now be combined with information already available from previous studies on egg sizes (Kerfoot 1974, 1975a). In the summer, *Bosmina* carry small eggs which give rise to small first instar young that have short features. In late fall, small individuals carry large eggs which develop into large

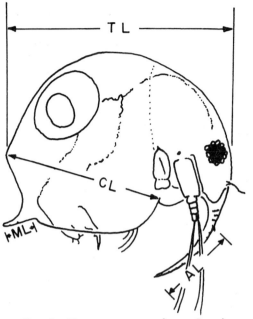
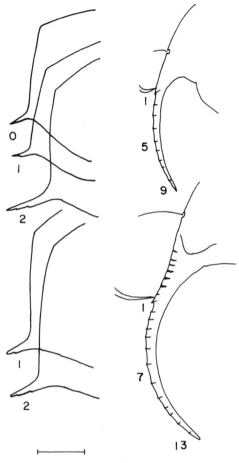

Fig. 5. Characters measured on *Bosmina*. Upper panel—adult *Bosmina*, showing location of length measurements (TL, total length; CL, carapace length; ML, mucro length; AL, antennule length); lower panel—meristic characters, showing representative counts on mucrones (left) and antennules (right). Scale bar on lower panel is 50 μm (from Kerfoot 1975a).

first instar young with longer features and thicker exoskeletons. The cycle reverses itself in early spring as large winter females carry many small eggs that develop into small, short-featured first instar young. The seasonal cycle of mucro sutures and antennule segments is given in Fig. 8. Both cycles were significant, although the seasonal variation was not great.

All the original characters were examined for significant cyclomorphotic (seasonal) patterns. The original set of ten variables was reduced to five by eliminating those characters which were greatly influenced by either age structure or seasonal growth patterns (e.g. carapace length, carapace height, and facet length) and by combining bilateral pairs (e.g. left and right mucro sutures and antennule segments). Of the remaining five variables, the monthly means of four changed by more than one standard deviation during the two years of sampling:

egg length, 2.6; mucro sutures, 1.4; antennule segments, 2.7; and rescaled antennule length, 4.4. When the variables were plotted against each other in bivariate plots or in a principal components plot of all variables, their means described the loops expected of cyclomorphotic patterns (Fig. 9, left panels; the jaggedness is caused largely by stochastic errors associated with sampling, whereas translational differences probably represent real differences).

To measure the loops in n-dimensional space, I standardized each variable by dividing by the average standard deviation, thus rescaling variables into a common scale. Measurement of the distances between the first sample (September 1967) and subsequent samples, using the Euclidean

Table 1. Characters measured on Bosmina from plankton samples (means, SD in parentheses, ranges below) All lengths in microns.

Date collected	N	Carapace Length	Carapace Height	Mucro Length	Mucro Sutures (l)	Mucro Sutures (r)	Facet length	Antennule Segments (l)	Antennule Segments (r)	Antennule Length
10 Sept 67	67	234.1(43.1) 201.2–319.5	228.7(44.0) 153.9–319.5	31.7(8.3) 16.6–52.1	0.82(0.82) 0–3	0.76(0.80) 0–2	31.7(5.4) 18.9–42.6	10.31(0.74) 9–12	10.24(0.99) 8–12	105.6(8.0) 87.6–127.8
22 Oct 67	40	304.9(58.9) 208.3–390.6	310.1(70.1) 198.8–402.4	42.6(10.7) 23.7–66.3	1.45(0.85) 0–3	1.48(0.85) 0–2	41.7(8.3) 29.4–54.4	11.32(1.25) 9–14	11.30(1.07) 9–13	143.0(14.2) 89.9–170.4
21 Dec 67	44	274.6(48.3) 227.2–411.9	260.6(61.3) 189.4–437.9	39.8(10.2) 23.7–54.4	1.84(0.61) 0–3	1.64(0.65) 0–3	36.0(7.8) 26.0–59.2	11.91(1.24) 9–15	11.95(1.18) 9–14	160.2(14.7) 134.9–194.1
17 Feb 68	41	342.7(60.1) 239.1–433.2	346.3(65.1) 224.9–428.4	40.7(8.0) 26.0–56.8	1.83(0.59) 0–3	1.83(0.59) 0–3	49.0(8.0) 33.1–66.3	11.93(1.13) 10–15	11.93(1.47) 9–15	163.8(14.2) 127.8–189.4
1 Mar 68	41	362.6(55.6) 243.8–435.5	373.7(63.2) 236.7–459.2	43.17.8	1.90(0.58) 1–3	1.88(0.46) 1–3	48.3(8.0) 30.8–61.5	12.46(1.21) 10–15	12.41(1.26) 9–15	171.6(12.3) 144.4–196.5
1 Apr 68	39	344.6(67.0) 224.9–459.2	370.9(87.6) 217.8–565.7	42.6(7.3) 30.8–59.2	1.77(0.54) 1–3	1.77(0.54) 1–3	49.7(10.9) 28.4–68.6	11.67(2.00) 7–15	12.03(1.53) 8–15	170.0(15.6) 142.0–201.2
29 Apr 68	62	317.9(61.3) 203.6–440.3	347.2(84.5) 203.6–523.1	40.5(7.8) 28.4–63.9	0.90(0.78) 0–2	0.81(0.88) 0–3	46.4(9.9) 28.4–68.6	11.98(1.31) 9–16	11.94(1.34) 9–15	154.8(12.5) 130.2–191.7
31 May 68	45	281.0(54.2) 189.4–383.5	277.6(56.6) 184.6–383.5	32.9(7.6) 21.3–49.7	0.42(0.69) 0–2	0.42(0.69) 0–2	36.9(7.3) 26.0–56.8	10.47(1.06) 8–13	10.76(1.07) 8–13	121.7(11.8) 101.8–153.9
1 Jul 68	41	262.3(53.5) 175.2–359.8	250.9(50.7) 177.5–336.1	38.6(9.9) 21.3–61.5	1.05(0.95) 0–3	1.02(1.04) 0–3	33.1(8.5) 16.6–47.3	10.32(0.99) 8–12	10.41(1.12) 7–13	119.8(9.9) 104.1–142.0
28 Jul 68	43	246.9(40.2) 161.0–314.8	251.4(46.2) 158.6–336.1	33.6(10.4) 14.2–61.5	0.84(0.97) 0–3	0.53(0.80) 0–2	32.7(5.7) 21.3–42.6	9.44(1.14) 7–12	9.70(1.60) 6–12	100.8(10.2) 87.3–125.5
7 Sept 68	40	265.8(52.8) 168.1–355.1	270.5(60.4) 165.7–390.6	39.3(11.4) 18.9–61.5	1.28(1.04) 0–4	1.25(1.01) 0–3	34.6(7.8) 21.3–49.7	10.27(0.93) 8–12	10.17(1.11) 8–13	119.3(10.7) 87.6–144.4
28 Sept 68	47	265.3(45.2) 189.4–374.0	251.1(49.5) 177.5–362.2	35.0(9.5) 23.7–59.2	0.76(0.81) 0–3	0.74(0.77) 0–3	35.7(7.8) 21.3–54.4	10.43(0.93) 8–12	10.38(1.01) 8–12	114.1(10.4) 92.3–137.3
9 Nov 68	44	319.5(45.9) 227.2–390.6	332.6(60.1) 208.3–423.7	39.3(5.7) 28.4–52.1	1.48(0.76) 0–3	1.39(0.75) 0–3	43.1(6.4) 26.0–54.4	11.52(0.88) 10–13	11.52(0.88) 10–13	160.7(11.1) 144.4–189.4
20 Jan 69	42	285.7(43.6) 224.9–362.2	282.9(55.6) 222.5–414.2	35.5(5.2) 26.0–45.0	1.60(0.59) 0–3	1.38(0.70) 0–3	40.0(6.2) 28.4–54.4	12.05(1.13) 9–15	12.33(0.93) 10–14	169.0(13.0) 139.7–208.3
8 Feb 69	25	297.8(47.3) 236.7–392.9	302.5(58.0) 227.2–423.7	35.7(5.0) 28.4–47.3	1.64(0.64) 0–3	1.56(0.65) 0–3	40.9(5.9) 28.4–54.1	11.76(1.54) 8–16	12.08(1.22) 9–15	167.8(15.9) 132.6–217.8
19 Apr 69	43	323.1(54.1) 201.2–426.1	352.7(61.8) 213.0–471.0	38.3(6.2) 26.0–54.4	0.93(0.77) 0–2	0.91(0.81) 0–3	46.6(6.6) 30.8–59.2	11.63(1.38) 10–14	11.72(1.22) 9–14	151.5(14.7) 125.5–191.7
1 May 69	38	296.3(72.0) 191.7–430.8	316.0(86.2) 196.5–454.5	36.5(5.2) 23.7–45.0	0.84(0.79) 0–2	0.50(0.56) 0–2	42.8(11.8) 26.0–68.6	11.45(0.98) 10–13	11.58(1.24) 9–14	148.6(12.11) 125.5–177.5
21 May 69	43	290.7(57.5) 172.8–400.0	289.2(65.6) 168.1–419.0	34.3(5.9) 23.7–52.1	0.84(0.65) 0–2	0.86(0.74) 0–2	40.0(9.5) 23.7–59.2	11.47(1.18) 8–14	11.40(1.26) 9–14	131.1(12.8) 106.5–165.7

Table 1 (continued)

Date	n									
26 June 69	39	283.3(42.4) 191.7–376.4	293.7(42.1) 220.1–362.2	35.0(8.8) 7.1–49.7	0.77(0.78) 0–2	0.69(0.83) 0–2	36.4(7.6) 21.3–54.4	11.15(1.20) 7–13	10.77(0.87) 9–12	120.5(10.2) 99.4–142.0
21 Jul 69	33	243.3(33.4) 175.2–312.4	237.9(38.6) 170.4–319.5	29.6(6.6) 18.9–45.0	0.45(0.71) 0–2	0.30(0.59) 0–2	32.9(5.9) 21.3–45.0	10.45(1.09) 9–13	10.42(1.37) 7–14	106.0(9.2) 78.1–123.0
7 Aug. 69	44	260.8(33.6) 172.8–319.5	245.2(29.6) 156.2–295.9	32.7(5.0) 23.7–45.0	0.77(0.77) 0–2	0.68(0.80) 0–2	29.4(4.5) 21.3–40.2	10.27(0.92) 8–12	10.18(0.92) 8–12	106.5(7.8) 87.6–127.8
9 Oct. 69	40	280.3(42.8) 203.6–397.7	279.8(39.1) 198.8–402.4	33.6(6.9) 26.0–66.3	0.50(0.60) 0–2	0.38(0.63) 0–2	39.1(6.6) 28.4–54.4	10.15(1.08) 8–13	10.25(1.08) 8–12	119.1(9.9) 99.4–137.3

Table 2. Within-sample correlations between characters (Frains Lake, Michigan: upper right—20 Jan, 1969; lower left—28 Jul, 1968).

Variable	1	2	3	4	5	6	7	8	9
1. Carapace length	1	0.965*	0.462*	0.253	0.150	0.890*	0.030	−0.004	0.595*
2. Carapace height	0.961*	1	0.522*	0.231	0.071	0.893*	−0.009	0.009	0.623*
3. Mucro length	0.536*	0.569*	1	−0.034	0.156	0.477*	0.297	0.199	0.475*
4. Mucro segments (1)	0.117	0.133	0.644*	1	0.446*	0.323	−0.007	−0.194	0.026
5. Mucro segments (r)	0.410*	0.413*	0.632*	0.728*	1	0.242	0.225	−0.088	0.208
6. Facet length	0.900*	0.905*	0.533*	0.059	0.363	1	0.011	−0.020	0.532*
7. Antennule segments (1)	−0.076	−0.137	−0.238	−0.105	−0.109	−0.144	1	0.335	0.307
8. Antennule segments (r)	0.226	0.189	0.136	0.152	0.282	0.215	0.525*	1	0.112
9. Antennule length	−0.191	−0.288	0.051	0.265	0.132	−0.157	0.115	0.315	1

* 0.05 confidence level or better.

Table 3. Correlations between characters for 18 monthly samples.

Variable	1	2	3	4	5	6	7	8
1. Carapace length	1.000							
2. Carapace height	0.974*	1.000						
3. Mucro length	0.756*	0.733*	1.000					
4. Mucro segments (1)	0.632*	0.547†	0.742*	1.000				
5. Mucro segments (r)	0.664*	0.562†	0.767*	0.976*	1.000			
6. Facet length	0.946*	0.962*	0.689*	0.562†	0.573†	1.000		
7. Antennule segments (1)	0.817*	0.779*	0.607*	0.699*	0.686*	0.836*	1.000	
8. Antennule segments (r)	0.795*	0.763*	0.565†	0.695*	0.667*	0.826*	0.986*	1.000
9. Antennule length	0.823*	0.794*	0.673*	0.799*	0.765*	0.833*	0.950*	0.968*

* 0.01 confidence level.
† 0.05 confidence level.

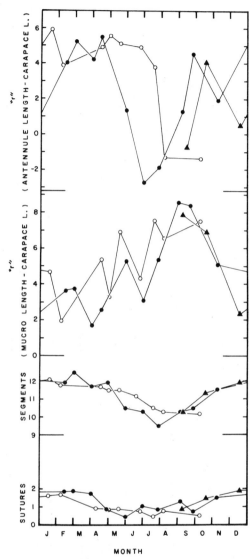

Fig. 6. Seasonal variation in within-sample correlations (Pearson product-moment) between antennule length and carapace length (upper), mucro length and carapace length (middle; y-axis in both upper and middle portions calibrated in decimals, i.e. $2 = 2 \times 10^{-1}$). Seasonal variation in mean segment and suture counts given in bottom portion of graph.

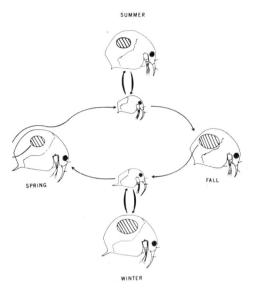

Fig. 7. Cyclomorphosis of *Bosmina longirostris* in Frains Lake, Michigan.

distance measure given by Eq. 1, gave the magnitudes shown in Fig. 9, right panel. Distances varied from nearly zero to slightly over four equivalent standard deviations, as the phenotypic transformations progressed through two cycles. Since winter individuals cloned in the laboratory could regress down to, and even beyond, summer feature lengths (Kerfoot 1975a), the cycles were presumed to represent progressive developmental modifications within clonal lineages (i.e. cyclomorphosis).

Complex cycles in large lakes: Union Bay, Lake Washington—During the past 7 years, my colleagues and I have investigated the interrelationships between two guilds of predators (visually foraging fishes and grasping copepods) and "morphs" of *B. longirostris* in Union Bay, Lake Washington. In Union Bay, a shallow bay off the western shore of Lake Washington, *Bosmina* populations demonstrate a marked spatial pattern that shows consistent seasonal transformations. At the beginning of each spring in the nearshore areas, a variety of clones emerge from ephippia or are born from a few overwintering females. These clones differ in several features, namely the length of body projections (antennules and mucrones), the size of eggs carried in the brood chamber, and the transparency and thickness of the exoskeleton (Kerfoot 1977b). Certain of these lines dominate the nearshore environment, then expand to meet the resident

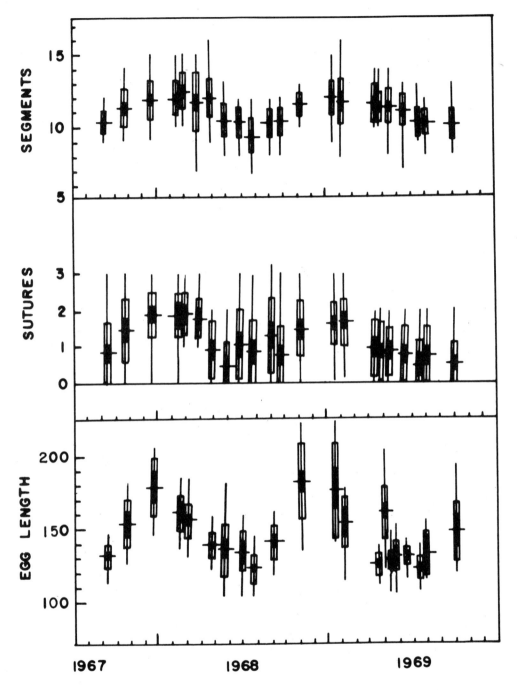

Fig. 8. Cyclomorphosis of specific features (sample mean, 95% confidence limits of mean, 1 SD about mean, and sample range given by horizontal line, two vertical bars, and vertical line, respectively).

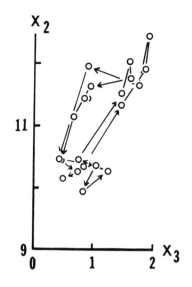

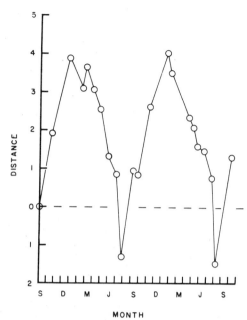

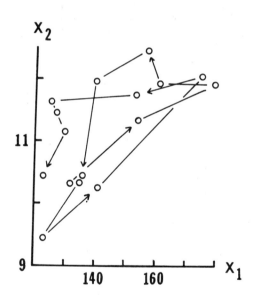

Fig. 9. Cyclomorphotic loops and sample similarities. Left panels—bivariate plots of variables (antennule segments against mucro sutures, antennule segments against egg length); right panel—phenotypic distance between samples in n-dimensional space, using Euclidean measure.

population of long-featured *Bosmina* in the offshore environment. Increasing densities of predators superimpose upon this basic pattern to rearrange areas of dominance and to sharpen boundaries between the various "morph" populations.

Long-featured "morphs" are heavily favored in the offshore regions, where predation comes primarily from predatory copepods (*Epischura nevadensis, Cyclops vernalis*, and *C. bicuspidatus*: Kerfoot 1975a, b, 1977a, 1978; Kerfoot and Peterson 1979). These grasping (raptorial) predators sense prey movements by means of mechanoreceptors on their first antennae (Strickler and Bal 1973), then capture with a quick leap or directed swoop (Strickler 1975; Strickler and Twombly 1975; Kerfoot 1975a, 1978). The long antennules and mucrones of the long-featured phenotype protect *Bosmina* against damage during an attack. The fused antennules protect the delicate swimming appendages (antennae), which are folded into a lateral groove behind the curved antennules, while the stout mucrones protect the vulnerable ventral area (Kerfoot 1977a, 1978).

In the nearshore areas of Lake Washington embayments, and especially along the lily-pad margins of Union Bay, populations of *Bosmina* show considerable spatial variation of phenotypes. Early spring transects show open-water regions of Union Bay dominated by long-featured "morphs," the outer fringes of the lily-pad margin dominated by intermediate "morphs", and the inner region of the lily-pad margin dominated by short-featured "morphs" (Kerfoot 1975b; Kerfoot and Peterson 1979). During the year, there is a consistent seasonal cycle at all stations along the margin, and that seasonal cycle contrasts strongly with the apparent morphological stability of the central Lake Washington *Bosmina* populations.

In Union Bay, there is a close association between the distribution of visual and grasping predators and the dominance of particular *Bosmina* "morphs." Visually hunting fish fry (predominately those of the three-spined stickleback, *Gasterosteus aculeatus*) hide among the lily pads. In early spring, these fry selectively remove the large, predatory copepods (especially *Epischura*) that prey upon *Bosmina*, the large herbivorous competitors of *Bosmina* (e.g. *Daphnia* and *Diaphanosoma*), and the more conspicuous, long-spined morphs. Short-featured, transparent morphs are favored in the nearshore environment, not only because they are less conspicuous, but also because they produce more eggs, as less yolk is required for defenses in the first instar young (Kerfoot 1977b; Kerfoot and Peterson 1979, in press; Kerfoot and Pastorok 1978). In situ competition experiments, run inside suspended gallon-jar enclosures, showed that short-featured "morphs" survive well in all regions of Lake Washington and could out-reproduce or out-compete longer-featured "morphs" at most stations (Kerfoot 1977b).

Although the direction of phenotypic changes in Union Bay *Bosmina* appears superficially similar in timing and sense, but not in magnitude, to those transformations observed in Frains Lake, a close inspection at many stations reveals the dynamics to be quite different. First of all, the seasonal pattern has a strong spatial component. Many morphological "morphs" or "varieties" are simultaneously present at individual stations throughout the year, but fluctuate greatly in frequency. Secondly, the magnitude of the phenotypic fluctuations is much greater than in Frains Lake. For example, Fig. 10, upper panels, illustrates the spatial and seasonal patterns of mean feature lengths (using antennule segments and mucro sutures as indicators of lengths) at four sampling stations in Union Bay and Lake Washington. Central stations in Lake Washington, as represented by station MP (Madison Park) in the center of Lake Washington, show little variation in space and time. All individuals at these stations are long-featured, and the mean lengths show a monotonous consistency throughout the year, with only a slight *increase* during summer months. In strong contrast, populations of *Bosmina* in Union Bay exhibit clinal variation in space as, early in the spring, long-featured "morphs" dominate offshore regions while intermediate- and short-featured "morphs" dominate inshore regions. During early June to July, the mean feature lengths at all stations along this cline *decline* dramatically. By late summer, *Bosmina* populations within the lily-pad margin (stations N, H) become very homogeneous and dominated by short-featured "morphs." Careful inspection of Fig. 10, left panels, however, reveals that during periods of transition the variance about the means increases greatly. Replacements are occurring, rather than the smooth morphological transitions characteristic of cyclomorphosis. Fig. 10, right panel, shows that during the summer, Union Bay stations generally varied by $< 3°C$ on any single date.

Clonal replacement was suspected from the onset of these investigations, not only because all phenotypes would co-occur simultaneously throughout the year, but also because clonal lines established in the laboratory maintained phenotypic differences when raised in identical environments (Kerfoot 1975b). Moreover, bivariate plots of mucro and antennule lengths, done for large sample sizes, disclosed that individuals often fell into three modal groups (Fig. 11). These modal groups are evident in frequency diagrams done during transitional periods

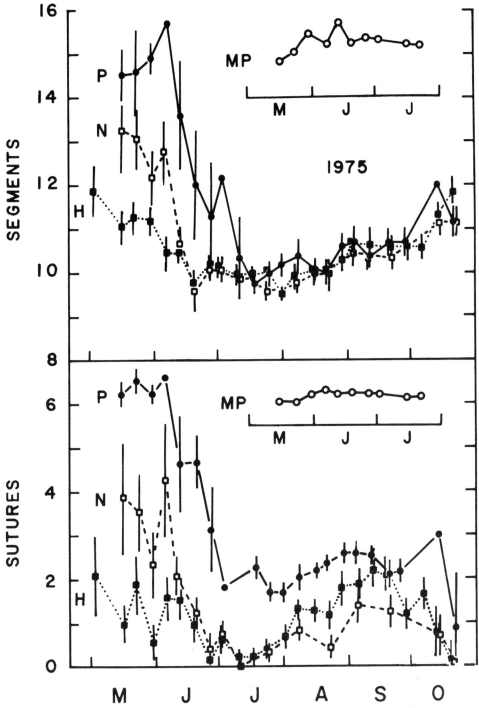

Fig. 10. Changes in feature lengths at four stations in Lake Washington and associated water temperatures (1975 data). Upper panels—seasonal variation in antennule segments and mucro sutures at four stations (MP, Madison Park, shifted to right to eliminate confusion, but vertical scale and position same; P, N, and H are shallow Union Bay stations; vertical bars = 2 SE); right panel—water temperatures at the four stations (MP—○, temperature taken as 0-10-m average; P—□; N—■; H—●; all Union Bay temperatures taken at 0.5-m depth).

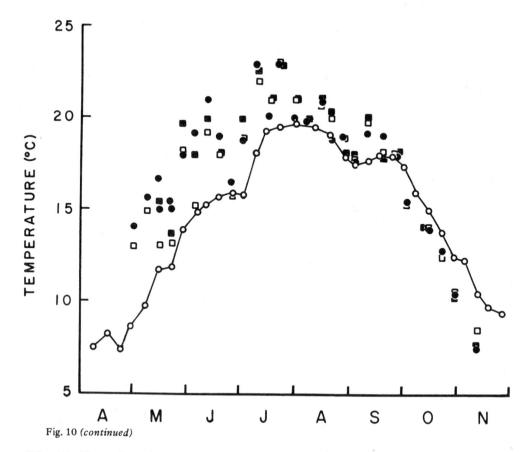

Fig. 10 *(continued)*

(Fig. 12). The early spring decline in feature lengths is not a smooth decline, but rather the successive replacement of long-featured and intermediate clones by short-featured clones.

Electrophoresis of long-featured and intermediate "morphs" (offshore "intermediates") revealed that these populations were generically isolated from short-featured "morphs." Not only was there no evidence of hybridization between these two groups, but each group contained between 3–5 unique alleles at the PGI locus (Manning et al. 1978).

Reciprocal translocation experiments—To investigate whether cyclomorphosis was occurring within the broad pattern of clonal replacement, we approached the problem with two different designs. One, reported in detail by R. Black (at this Conf.), consisted of extending and refining initial cloning experiments. A single population of *Bosmina* (station G) was repeatedly sampled during the year, individuals transferred from nature to the laboratory to establish clonal lines, and the resulting clones kept in the same environment for three generations. In this way, the morphological changes of lineages could be compared between clones and through time. These experiments conclusively demonstrated fine-scale clonal replacements, and additionally suggested slight cyclomorphosis within intermediate-length "morphs."

The second approach consisted of moving individuals from one location to another within the natural habitat. In these reciprocal transfers, inshore populations were placed in clear gallon jars suspended at offshore stations and offshore populations were suspended at inshore stations. *Bosmina* individuals were sampled at one site, then

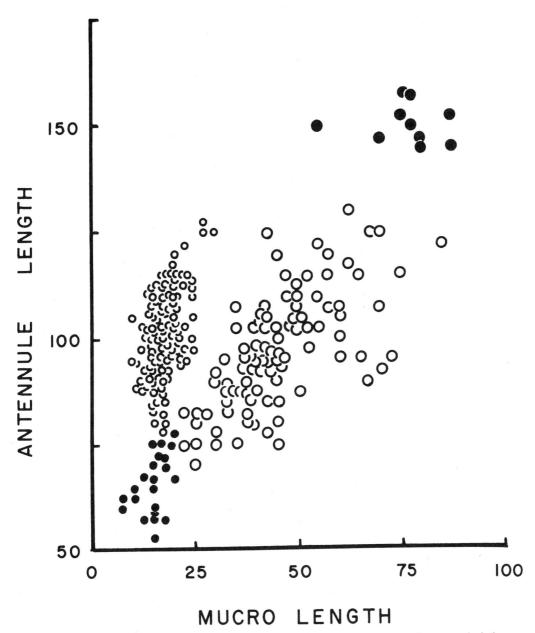

Fig. 11. Clonal nature of phenotypic variation in *Bosmina*. Three most common phenotypes include short-featured (o), intermediate-featured (O), and long-featured phenotypes (●). *Bosmina* with short, hooked antennules (•) are also found in this sample (these phenotypes carry the short-featured genotype, and hooked morphotype can be induced in crowded cultures: Manning et al. 1978). All from inshore station H, 6 June 1975.

transferred to a second site where they were placed in water from the new site, the gallon jar covered by No. 20 mesh Nitex netting, and left suspended at 1–2-m depth for 1–2 weeks. Control jars were treated similarly, but left at the original sampling site (details described in Kerfoot 1977b). Since the filtered water placed within the jar was

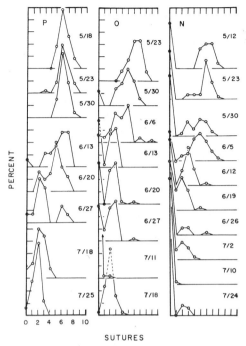

Fig. 12. Frequency distributions for mucro suture counts at three Union Bay stations, 12 May–25 July 1975. Station P is farthest from shore, and station N is closest (sample sizes, 50–100 indiv.).

from the final site, and since water continued to circulate (though at a reduced rate) throughout the experiment, the enclosed populations were exposed to many of the physical, chemical, and food factors peculiar to the new site. Of course, turbulence within the jars was presumably substantially decreased from natural values.

The results of the 1975 transfers are given in Table 4. When transferred to the offshore environment (Webster Pt. lighthouse near the mouth of Union Bay), the inshore population sample of Bosmina was composed of the typical midsummer assemblage: mostly short-featured (0 mucro sutures) with a few intermediates (1–3 sutures). After 1–2 weeks of suspension in the offshore environment, there were no consistent or major differences between the reference sample (original population introduced into bottles), inshore controls (inshore), and offshore experimentals (offshore). However, there was a slight indication that intermediates had increased in suture counts, as their peak counts became more easily distinguishable from those of the short-featured "morph." The same experiment run on long-featured "morphs" transferred to an inshore station (station W) showed evidence for slight phenotypic regression at all locations, including within the control jars. We must conclude, then, that short-featured clones are largely noncyclomorphotic, but that intermediates and long-featured "morphs" are capable of some phenotypic plasticity. The amount, or rate, of change for both intermediates and long-featured "morphs" is trifling compared to the major morphological shifts during the clonal replacement sequence in early spring. Thus the cyclical morphological patterns observed in Union Bay consist basically of a seasonal sequence of species and clonal replacements, where two of the subpopulations (intermediates and long-featured modes) exhibit limited cyclomorphosis.

Geographic patterns: Regional comparisons between lakes—Bosmina populations have been sampled regionally in Michigan, Washington, and New England lakes. Common to all surveys is the discovery of considerable geographic variation in feature lengths of Bosmina. Often adjacent water bodies contain short-featured and long-featured populations and, less often but not uncommonly, bimodal or multimodal frequency distributions for sutures and antennules, presumably indicating the presence of one or more subpopulations. For example, Fig. 13 gives the frequency distributions of mucro and antennule counts for several ponds and lakes in Vermont and New Hampshire. In New England, as elsewhere, short-featured Bosmina forms are characteristic of small, shallow ponds, intermediate-length forms are characteristic of moderate-sized oligotrophic lakes, and long-featured Bosmina forms are characteristic of large lakes (e.g. Lake Champlain). Sometimes single samples contain multimodal distributions (e.g. Storrs Pond), and electrophoresis has confirmed that short-featured populations and intermediates are as reproductively isolated in

Table 4. Results of transfer experiments (N = No. of *Bosmina* scored; reference sample represents sample from animals introduced into jars; 20 individuals initially introduced into each jar).

Date	Collection site	Incubation site	N	Sutures									Segments										
				0	1	2	3	4	5	6	7	8	9	10	11	12	13	14	15	16	17	18
8 June	Inshore	Reference	55	78	13	2	7					9	42	35	11	4						
25 June	,,	Inshore	50	74	10	6	8	2			2	6	40	32	12	8						
,,	,,	,,	50	70	12	14	4				2	6	46	30	17							
,,	,,	,,	20	95	0	0	5						12	53	29	6						
,,	,,	,,	92	49	14	21	15	1				7	37	32	22	3						
2 July	,,	,,	121	72	7	12	9					6	21	36	38	6						
,,	,,	,,	54	76	7	2	13	2				4	35	31	27	4						
,,	,,	,,	26	85	4	8	4			4		4	40	28	12	12						
,,	,,	,,	44	75	7	11	7						30	43	23	5						
,,	,,	,,	55	64	15	18	2						31	40	20	7						
25 June	,,	Offshore	50	96	0	0	2	2				2	24	42	16	2	0	4				
,,	,,	,,	12	58	17	8	17					8	25	42	25							
,,	,,	,,	11	100									18	27	55							
,,	,,	,,	42	88	0	5	7					10	45	33	7	5						
2 July	,,	,,	53	75	0	11	13					9	32	34	19	6						
,,	,,	,,	58	93	2	3	2				2	7	34	47	7	3						
,,	,,	,,	80	96	1	1	1	1			1	9	31	38	14	6						
,,	,,	,,	44	80	9	7	5					5	45	32	16	2						
8 June	Offshore	Reference	50	3		12	44			40	4				6	8	18	16	22	14	10	6
25 June	,,	Offshore	13			15	77			8					8	0	8	31	31	23		
,,	,,	,,	25		4	20	52			24					12	16	16	20	12	24		
,,	,,	,,	5			40	40			20						20	40	20				
,,	,,	,,	14			21	57			21			20	7		7	20	7	14	21	7	
,,	,,	,,	6		17	33	50							7			7	36	7	67		
15 June (totals)			63		3	22	57			17			2	2	8	10	19	13	21	25	2		
2 July	Offshore	Offshore	55	26	22	55	20	4					2	2	15	24	22	15	22	2	2		
,,	,,	,,	23		35	35	4																
,,	,,	,,	18	11	22	61	6									19	25	6	19				
25 June	Offshore	Inshore	12			17	42	33				8			12	17	25	8	42	25	8		
,,	,,	,,	5			20	80								20	20	20	40	20				
2 July (totals)		Inshore	16	6	6	56	38									6	6	31	31	13	6		

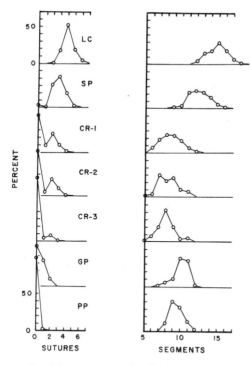

Fig. 13. Frequency distributions for mucro sutures and antennule segments, single plankton samples from several Vermont and New Hampshire sites (LC—Lake Champlain, 14 July 1977; SP—Storr's Pond, south basin, 18 May 1978; CR-1—Connecticut River backwater pond, 1 mile N of Hanover Bridge, 16 July 1978; CR-2—backwater pond, 4 miles N of Hanover Bridge, 16 July 1978; CR-3—backwater pond, 6.8 miles N of Hanover Bridge, 16 July 1978; GP—Golf Course Pond, Hanover, 4 August 1978; PP—Pond View Pond, Rte. 10, 10 July 1978).

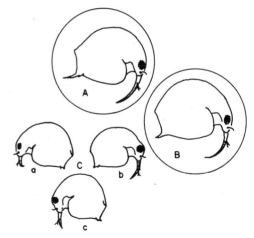

Fig. 14. Varieties of European *Bosmina longirostris* recognized by European limnologists, arranged in groups corresponding to phenotypic categories in Lake Washington studies. A—Long-featured phenotypes, *B. l.* var. *similis*; B—intermediate-featured phenotypes, *B. l.* var. *pellucida*; C—short-featured phenotypes, which include *B. l.* var. *curvirostris* (a), *B. l.* var. *typica* (b), and *B. l.* var. *brevicornis* (c). Sketches of types follow Herbst (1962).

these ponds as they are in Lake Washington. In fact, although 3,000 miles apart, the electrophoretic patterns of New Hampshire short-featured *Bosmina* are very similar to morphologically identical populations in Lake Washington and surrounding lakes (Manning et al. 1978). When we treat short-featured, intermediate, and long-featured modes, we seem to be dealing with separate, undescribed species within the genus *Bosmina*. This situation, however, is not without historical precedent, for the European taxonomy of *B. longirostris* recognizes several "forms" (Fig. 14); three of which correspond, respectively, to short-, intermediate-, and long-featured "morphs": *Bosmina longirostris* forma *typica*, *pellucida*, and *similis* (Lilljeborg 1900; Herbst 1962). While more electrophoretic information on many more loci and cross-continental comparisons seem necessary before a final decision can be made, these "forms" will probably turn out to be biological species, reproductively isolated from each other.

If, however, we ignore the source of these samples and plot individuals solely by their mucro and antennule counts, we can observe that over the geographic range of *Bosmina*, there is a close correspondence between the lengths of these two features. While there is often little within-sample correlation, between lakes we find that short-mucroned populations usually have short antennules, long-mucroned populations usually have long antennules (Fig. 15). This kind of regional pattern seems indicative of strong natural selection along the short-long feature axis. Associated with this axis, we also find a gradient of injury and a correspondence with *Epischura*, much as in Lake Washing-

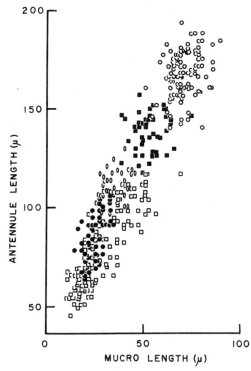

Fig. 15. Geographic association between antennule and mucro lengths (□—CR series; ●—PP sample; 0—GP sample; ■—SP sample; o—LC sample).

ton (Table 5). Long-featured populations have higher incidences of injuries and are almost always found associated with *Epischura*. In New England, however, the species is *Epischura lacustris*, rather than *E. nevadensis* as found in the Pacific northwest.

Discussion

If the population dynamics of variation seem to differ so greatly between small and large lakes, what elements tie such disparate cycles together? I believe the connection is one of magnitude, i.e. the magnitude of the selective forces (the intensity of selection) that acts along the short-featured/long-featured axis. If we compare the absolute magnitude of fluctuations in Frains Lake, Lake Washington, and between extensive geographic samples we find the least amount of change in the small lake and the greatest, as we might expect, between the geographic samples. While the changes at some Lake Washington stations are spectacular (Lake Washington was purposely selected for intensive study because of such dramatic spatial and temporal variations), the average amount of seasonal change is only moderate

Table 5. Dominant phenotpyes, predatory copepods, and incidence of regenerated mucrones or antennules.

Site	Dominant Bosmina phenotypes	Largest common predatory copepod	% Regenerated Mucrones	% Regenerated Antennules
Frains Lake				
summer	short-featured	*Orthocyclops*	0–3.9 (mean 0.9)	0–7.8 (mean 1.9)
winter	intermediate	*Cyclops bicuspidatus* *Mesocyclops* (Apr)	2.1–5.0 (mean 3.4)	2.0–17.5 (mean 9.4)
Lake Washington, MP station				
summer	long-featured	*Epischura nevadensis*	2–35	13–69
winter	long-featured	*Epischura* (rare in Mar and Apr)	0–22	4–24
Pond View Pond				
July	short-featured	*Cyclops bicuspidatus*	0	1.0
Storrs Pond				
south end	short-featured	*Cyclops bicuspidatus*	0	0
center	intermediate, long-featured	*Epischura lacustris*	0–4	6–10
Lake Champlain				
July 77	long-featured	*Epischura lacustris*	4–6	8–14

at most Union Bay stations (Kerfoot 1975*b*). The morphological changes, whether underlain by cyclomorphosis, clonal succession, or species succession are reflecting population and species responses to selective regimes of progressively greater magnitude.

Beyond the realm of magnitude, there seems a legitimate question of pattern. Small ponds or lakes can be small enough, especially in situations where mixing is extensive, i.e. where the surface is exposed to wind action and the littoral zone is small, that clonal lineages or clonemates can be dispersed throughout the lake during a single season. In this kind of environment, there is no refuge from predation, and successive lineages must pass through a seasonal progression of predators, some with quite different biases in their foraging. This environment would promote the phenotypic flexibility of lineages. In contrast, the extensive littoral environment of large lake bays, as in Lake Washington, provides spatial refugia for clones. While the balance shifts markedly between short-, intermediate-, and long-featured clones, none of these clones is really eliminated from the littoral, each of these modal phenotypes can be found somewhere at all times of the year. We might surmise that the magnitude of seasonal shifts outstrips the physiological capabilities of single populations to adapt or adjust by phenotypic changes (both to the rate and magnitude of early spring shifts) and that the presence of spatial refugia allows certain lineages the opportunity of passing through the seasons without stringent selection for phenotypic change.

Having attempted to embrace so many aspects of cyclomorphosis and seasonal changes, I have left until last one of the most difficult and most glaring exceptions to most generalizations: the lack of cyclomorphosis in free-living copepods. As I alluded to earlier in discussing what I believe is the origin of cyclomorphosis, I contend that the answer to this exception lies partly in the mode of reproduction characteristic of almost all copepods, i.e. the bisexual habit of reproduction which results in recombination during each generation and which hence destroys genotypes. In these organisms, the lifespans (so to speak) of genotypes are short, female-offspring resemblance is reduced, and hence the fundamental conditions that favor phenotypic flexibility are greatly altered. In addition to this primary reason, I believe that the defensive tactics of copepods place them under a radically different selective regime than that imposed on rotifers and cladocerans.

During their phylogenetic evolution, copepods have adapted to aquatic existence in a fundamentally different way than cladocerans or rotifers. In copepods, the thoracic legs are modified for locomotion, while those of cladocerans are modified for filter feeding. Nauplii, copepodites, and adult stages are capable of swift evasive maneuvers ("escape reactions"), during which they are able to reach speeds around 10–35 cm/sec (Strickler 1975; Strickler and Twombly 1975). In contrast, cladocerans use a single pair of appendages, the second antennae, as their sole means of locomotion through the water and are generally capable of only feeble to moderate escape velocities (Kerfoot 1978; Lehman 1977). Rotifers generally move even more slowly through water, propelled by ciliary waves, although a few show evasive "escape reactions" (e.g. *Polyarthra*). Thus the three taxa are predisposed to use fundamentally different defenses against predatory copepods that are encountered in open waters: copepods have evolved *precontact* defenses, while most rotifers and cladocerans have evolved predominantly *postcontact* defenses (Fig. 16). The consequences of these avenues of adaptation are important to body shape, reproduction, and life history.

Because copepods (calanoid and cyclopoid adult and copepodite stages; many cyclopoid nauplii) escape before contact with a predator, there has been no strong selection for variety in body shape. The epidermis is not strongly chitinized, the body is flexible for rapid evasive movement, and a variety of structures project beyond the body covering. Eggs can be

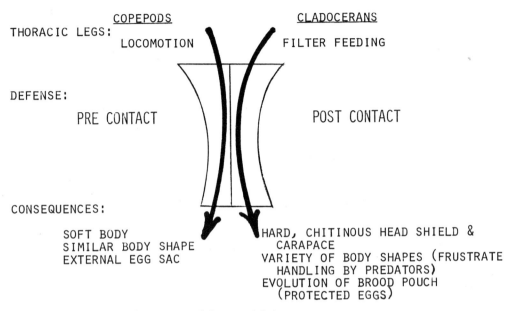

Fig. 16. Precontact and postcontact defenses and their consequences.

carried in external sacs without much danger that those eggs will be stripped off and devoured by an attacking invertebrate predator. The highly successful evasive tactics of immature and adult copepods seem a prerequisite to the prolonged development (delayed maturity) characteristic of these animals. For example, compare Confer and Cooley's (1977) estimates of instar mortality for copepods with corresponding estimates from cladocerans (Hall 1964; Kerfoot 1975b).

In contrast, most cladocerans are less likely to avoid the swift attack of predatory copepods (*Heterocope, Epischura*, or large-bodied cyclopoids) or the phantom midge larvae (*Chaoborus*) and have evolved morphological defenses that deter predators once contact has taken place. These *postcontact* defenses include the evolution of a hard, chitinized head shield and carapace, elaboration of spines and helmets to frustrate handling, and evolution of a brood pouch to protect eggs and embryos (which would be stripped and consumed by attacking copepods, if carried externally). Naupliar development is shortened and compressed within a single adult instar. As if to provide the exceptions that prove the rule, the genus *Diaphanosoma*, which has developed greatly enlarged second antennae and an effective "escape reaction," is characterized by frail morphology and little cyclomorphosis.

In closing, while I acknowledge that many of the above suggestions are conjectural, my intent has been to stimulate consideration of phenomena that are all too often casually accepted. If I have provoked some readers to question, rather than to accept blindly, the origins of natural patterns, then I have achieved my desired goal.

References

ALLAN, J. D. 1977. An analysis of seasonal dynamics of a mixed population of *Daphnia*, and the associated cladoceran community. Freshwater Biol. 7: 505–512.

BOYCE, A. J. 1965. The methods of quantitative

taxonomy with special reference to functional analysis. Ph.D. thesis, Oxford Univ. 190 p.
———. 1969. Mapping diversity: a comparative study of some numerical methods, p. 1-31. *In* A. J. Cole [ed.], Numerical taxonomy. Academic.
BRANDL, Z., and C. H. FERNANDO. 1978. Prey selection by the cyclopoid copepods *Mesocyclops edax* and *Cyclops vicinus*. Int. Ver. Theor. Angew. Limnol. Verh. **20**: 2505-2510.
BROOKS, J. L. 1946. Cyclomorphosis in *Daphnia*. I. An analysis of *D. retrocurva* and *D. galeata*. Ecol. Monogr. **16**: 409-447.
———. 1947. Turbulence as an environmental determinant of relative growth in *Daphnia*. Proc. Natl. Acad. Sci. **16**: 409-447.
———. 1965. Predation and relative helmet size in cyclomorphotic *Daphnia*. Proc. Natl. Acad. Sci. **53**: 119-126.
———. 1967. Cyclomorphosis, turbulence and overwintering in *Daphnia*. Int. Ver. Theor. Angew. Limnol. Verh. **16**: 1653-1659.
———. 1968. The effects of prey size selection by lake planktivores. Syst. Zool. **17**: 272-291.
———, and S. DODSON. 1965. Predation, body size, and composition of the plankton. Science **150**: 28-35.
CONFER, J. L., and J. M. COOLEY. 1977. Copepod instar survival and predation by zooplankton. J. Fish. Res. Bd. Can. **34**: 703-706.
DAVIS, M. B. 1968. Pollen grains in lake sediments: redeposition caused by seasonal water circulation. Science **162**: 796-799.
———, L. B. BRUBAKER, and J. M. BEISWINGER. 1972. Pollen grains in lake sediments: pollen percentages in surface sediments from southern Michigan. Quat. Res. **1**: 450-467.
De BEAUCHAMP, P. 1952. Un facteur de la variabilité chez les rotifères du genre *Brachionus*. C. R. Acad. Sci. Paris **234**: 573-575.
DEEVEY, E. S., and G. B. DEEVEY. 1971. The American species of *Eubosmina* Seligo (Crustacea, Cladocera). Limnol. Oceanogr. **16**: 201-218.
DODSON, S. I. 1974*a*. Adaptive change in plankton morphology in response to size-selective predation: A new hypothesis of cyclomorphosis. Limnol. Oceanogr. **19**: 721-729.
———. 1974*b*. Zooplankton competition and predation: An experimental test of the size-efficiency hypothesis. Ecology **55**; 605-613.
———. 1975. Predation rates of zooplankton in arctic ponds. Limnol. Oceanogr. **20**: 426-433.
GILBERT, J. J. 1966. Rotifer ecology and embryological induction. Science **151**: 1234-1237.
———. 1967. *Asplanchna* and posterolateral spine production in *Brachionus calyciflorus*. Arch. Hydrobiol. **64**; 1-62.
———, and J. K. WAAGE. 1967. *Asplanchna*, *Asplanchna*-substance, postero-lateral spine length variation in the rotifer *Brachionus calyciflorus* in a natural environment. Ecology **48**: 1027-1031.

GOULDEN, C., and D. FREY. 1963. The occurrence and significance of lateral head pores in the genus *Bosmina* (Cladocera). Int. Rev. Gesamten Hydrobiol. **48**: 513-522.
GREEN, J. 1967. The distribution and variation of *Daphnia lumholtzi* (Crustacea: Cladocera) in relation to fish predation in Lake Albert, East Africa. J. Zool. Lond. **151**: 181-197.
———. 1971. Associations of Cladocera in the zooplankton of the lake sources of the White Nile. J. Zool. Lond. **165**: 373-414.
———, and O. B. LAN. 1974. *Asplanchna* and the spines of *Brachionus calyciflorus* in two Javanese sewage ponds. Freshwater Biol. **4**: 223-226.
HALBACH, U. 1969. Räuber und ihre Beute: der Anpassungswert von Dornen bei Rädertieren. Naturwissenschaften **56**: 142-143.
———. 1971. Zum Adaptivwert der zyklomorphen Dornenbildung von *Brachionus calyciflorus* Pallas (Rotatoria). I. Räuber-Beute-Beziehung in Kurzeit-Versuchen. Oecologia **6**: 267-288.
———, and J. JACOBS. 1971. Seasonal selection as a factor in rotifer cyclomorphosis. Naturwissenschaften **57**: 1-2.
HALL, D. J. 1964. An experimental approach to the dynamics of a natural population of *Daphnia galeata mendotae*. Ecology **45**: 94-111.
———, W. E. COOPER, and E. E. WERNER. 1970. An experimental study of the regulation of freshwater invertebrate populations. Limnol. Oceanogr. **15**: 839-928.
HARTMANN, O. 1917. Über die temporale Variation bei Copepoden (Cyclops, Diaptomus) und ihre Beziehung zu der bei Cladoceren. Z. Indukt. Abstammungs. Vererbungsl. **18**: 22-43.
HERBST, H. V. 1962. Blattfusskrebse. Kosmos, Stuttgart.
HRBÁČEK, J. 1962. Species composition and the amount of zooplankton in relation to the fish stock. Rozpr. Cesk. Akad. Ved Rada Mat. Prir. Ved **72**: 1-116.
———, M. DVOŘÁKOVÁ, V. KOŘÍNEK, and L. PROCHÁZKOVÁ. 1961. Demonstration of the effect of the fish stock on the species composition of zooplankton and the intensity of metabolism of the whole plankton association. Int. Ver. Theor. Angew. Limnol. Verh. **14**: 192-195.
———, and M. HRBÁČEKOVÁ-ESSLOVÁ. 1960. Fish stock as a protective agent in the occurrence of slow developing dwarf species of strains of the genus *Daphnia*. Int. Rev. Gesamten Hydrobiol. **45**: 355-358.
———, and M. NOVOTNÁ-DVOŘÁKOVÁ. 1965. Plankton of four backwaters related to their size and fish stock. Rozpr. Cesk. Akad. Ved Rada Mat. Prir. Ved **75**: 3-10.
HUTCHINSON, G. E. 1967. A treatise on limnology, v. 2. Wiley.
JACOBS, J. 1961. Cyclomorphosis in *Daphnia*

galeata mendotae, a case of environmentally controlled allometry. Arch Hydrobiol. **58**: 7–71.

——. 1962. Light and turbulence as co-determinants of relative growth rates in cyclomorphotic *Daphnia*. Int. Rev. Gesamten Hydrobiol. **47**: 146–156.

——. 1965. Control of tissue growth in cyclomorphotic *Daphnia*. Naturwissenshaften **52**: 92–93.

——. 1967. Untersuchungen zur Funktion und Evolution der Zyklomorphose bei *Daphnia*, mit besonderer Berücksichtigung der Selektion durch Fische. Arch. Hydrobiol. **62**: 467–541.

——. 1977. Coexistence in similar zooplankton species by differential adaptation to reproduction and escape in an environment with fluctuating food and enemy densities. II. Field data analysis of *Daphnia*. Oecologia **30**:313–329.

——. 1978. Coexistence of similar zooplankton species by differential adaptation to reproduction and escape in an environment with fluctuating food and enemy densities. III. Laboratory experiments. Oecologia **35**: 35–54.

JANSSEN, J. 1976. Feeding modes and prey size selection in the alewife (*Alosa pseudoharengus*). J. Fish. Res. Bd. Can. **33**: 1972–1975.

JOLICOEUR, P., and J. E. MOSIMANN. 1960. Size and shape variation in the painted turtle. A principal component analysis. Growth **24**: 339–354.

KERFOOT, W. C. 1972. Cyclomorphosis of the genus *Bosmina* in Frains Lake, Michigan. Ph.D. thesis, Univ. Michigan.

——. 1974. Egg-size cycle of a cladoceran. Ecology **55**: 1259–1270.

——. 1975*a*. Seasonal changes of *Bosmina* (Crustacea, Cladocera) in Frains Lake, Michigan: laboratory observations of phenotypic changes induced by inorganic factors. Freshwater Biol. **5**: 227–243.

——. 1975*b*. The divergence of adjacent populations. Ecology **56**: 1298–1313.

——. 1977*a*. Implications of copepod predation. Limnol. Oceanogr. **22**: 316–325.

——. 1977*b*. Competition in cladoceran communities: The cost of evolving defenses against copepod predation. Ecology **58**: 303–313.

——. 1978. Combat between predatory copepods and their prey: *Cyclops, Epischura*, and *Bosmina*. Limnol. Oceanogr. **23**: 1089–1102.

——, and R. A. PASTOROK. 1978. Survival versus competition: evolutionary compromises and diversity in the zooplankton. Int. Ver. Theor. Angew. Limnol. Verh. **20**: 362–374.

——, and C. PETERSON. 1979. Ecological interactions and evolutionary arguments: Investigations with predatory copepods and *Bosmina*. *In* pp. 159–196 *in* Halbach and Jacobs [ed.], Population Ecology. Fortschr. Zool. 25, 2/3. Gustav Fischer.

——, and ——. In press. Predatory copepods and *Bosmina*: Replacement cycles and further influences of predation upon prey reproduction. Ecology.

LANE, P., M. KLUG, and L. LOUDEN. 1976. Measuring invertebrate predation in situ on zooplankton assemblages. Trans. Am. Microsc. Soc. **95**: 143–155.

LAUTERBORN, R. 1904. Die cyklische oder temporale Variation von *Anuraea cochlearis*. Part II. Verh. Nat. -Med. Ver. Heidelb. **7**: 529–621.

LEHMAN, J. T. 1977. On calculating drag characteristics for decelerating zooplankton. Limnol. Oceanogr. **22**: 170–172.

LEVINS, R. 1968. Evolution in changing environments. Princeton Monogr. Princeton Univ. Press.

LILLJEBORG, W. 1900. Cladocera Sueciae. (Grundlegende Arbeit über die Cladoceren-Systematik; Sehr Eingehende Beschreibungen, vortreffliche Abbildungen.). Nova Acta regiae societatis scient. Upsaliensis, Ser 3, 19:1–701.

MANNING, B. J., W. C. KERFOOT, and E. M. BERGER. 1978. Phenotypes and genotypes in cladoceran populations. Evolution **32**: 365–374.

NILSSON, N., and B. PEJLER. 1973. On the relation between fish fauna and zooplankton composition in north Swedish lakes. Fish. Bd. Sweden Rep. **53**: 51–77.

O'BRIEN, W. J., D. KETTLE, and H. RIESSEN. 1979. Helmets and invisible armor: Structures reducing predation from tactile and visual planktivores. Ecology: **60**:287–294.

——, and G. L. VINYARD. 1978. Polymorphism and predation: The effect of invertebrate predation on the distribution of two *Daphnia carinata* varieties in South India ponds. Limnol. Oceanogr. **23**: 452–460.

PIELOU, E. C. 1969. An introduction to mathematical ecology. Wiley-Interscience.

POURRIOT, R. 1964. Étude experimentale de variations morphologiques chez certaines espèces de rotifères. Bull. Soc. Zool. Fr. **89**: 555–561.

——. 1974. Relations prédateur-proie chez les rotifères: influence du prédateur (*Asplanchna brightwelli*) sur la morphologie de la proie (*Brachionus bidentata*). Ann. Hydrobiol. **5**: 43–55.

ROTH, J. C. 1971. The food of *Chaoborus*, a plankton predator, in a southern Michigan Lake. Ph.D. thesis, Univ. Michigan. 94 p.

SEARLE, S. R. 1966. Matrix algebra for the biological sciences. Wiley.

SNEATH, P. H. A., and R. R. SOKAL. 1973. Numerical taxonomy. Freeman.

STORCH, T. A. 1971. Production of extracellular dissolved organic carbon by phytoplankton and its utilization by bacteria. Ph.D. thesis, Univ. Michigan. 130 p.

STRICKLER, J. R. 1975. Swimming of planktonic

Cyclops species (Copepoda, Crustacea): Pattern, movements and their control, p. 599–613. *In* T. Y.-T. Wu et al. [eds.], Swimming and flying in nature. Plenum.

——, and A. K. BAL. 1973. Setae of the first antennae of the copepod *Cyclops scutifer* (Sars): Their structure and importance. Proc. Natl. Acad. Sci. **70**: 2656–2659.

——, and S. TWOMBLY. 1975. Reynolds number, diapause, and predatory copepods. Int. Ver. Theor. Angew. Limnol. Verh. **19**: 2943–2950.

WERNER, E. E., and D. J. HALL. 1977. Competition and habitat shift in two sunfishes (Centrarchidae). Ecology **58**: 869–876.

WESENBERG-LUND, C. 1926. Contributions to the biology and morphology of the genus *Daphnia*. Kg. Dan. Vidensk. Selsk. Skr. Nat. -Mat. Afd. (Ser. 8) **11**(2): 92–250.

ZARET, T. 1972*a*. Predator-prey interaction in a tropical lacustrine ecosystem. Ecology **53**: 48–57.

——. 1972*b*. Predators, invisible prey, and the nature of polymorphism in the Cladocera (class Crustacea). Limnol. Oceanogr. **17**: 171–184.

——, and W. C. KERFOOT. 1975. Fish predation on *Bosmina longirostris*: Body-size selection versus visibility selection. Ecology **56**: 232–237.

44. Dimorphic *Daphnia longiremis:* Predation and Competitive Interactions between the Two Morphs

W. John O'Brien, Dean Kettle, Howard Riessen,
David Schmidt, and David Wright

Abstract

A dimorphic cladoceran *Daphnia longiremis*, is commonly found in arctic Alaskan lakes with fish, and the exuberant, or *cephala*, form is often found where the invertebrate predator *Heterocope septentrionalis* also occurs. In a series of experiments designed to determine the relative susceptibility of the two morphs of *D. longiremis* to predation by *Heterocope* and by fish, the smaller *typica* form was found to be much more heavily preyed upon by *Heterocope* than would be predicted by the difference in size. Little difference was found between the morphs in susceptibility to fish predation. A life table experiment showed the exuberant morph to increase at a rate 1.5 times that of the *typica* form, and this result was confirmed by *in situ* population growth experiments. Thus it appears that the helmet of *D. longiremis cephala* is an adaptation that effectively reduces invertebrate predation while neither increasing vulnerability to fish nor adversely affecting growth and reproduction.

Limnologists have long been interested in the phenomenon of cyclomorphosis (Zacharias 1894), and especially in cladocerans (*see* Hutchinson 1967 for a full review). In several tropical species, two or more morphs are continually present throughout the year (Zaret 1972; Green 1967); these are more accurately termed dimorphic rather than cyclomorphotic, but the similarity of the exuberant structures developed suggests that the selective forces behind such structures may be similar. *Daphnia longiremis*, a cold-water stenothermic species of the holarctic limnoplankton, is also a dimorph whose forms occur at the same time but in different lakes (Riessen and O'Brien 1980). As with the tropical dimorphs, the exuberant structures which develop in *D. longiremis* closely resemble those of temperate cyclomorphotic species.

Research on cyclomorphosis has focused on both the proximate triggering mechanisms and the ultimate selective advantages conferred by the existence of different morphological types. Early work on the adaptive value of exuberant anterior and dorsal structures, often termed helmets, of certain daphnids centered around the idea that these structures increased buoyancy (*see* Hutchinson 1967).

Much more recently predation, either by small planktivorous fish (Jacobs 1965; Brooks 1965; Zaret 1972) or invertebrates (Dodson 1974; Kerfoot 1977; O'Brien and Vinyard 1978), has been suggested as the selective force molding these interesting structures.

This work was supported by NSF funds OPP75-12949, DPP76-80652, and DPP77-22994.

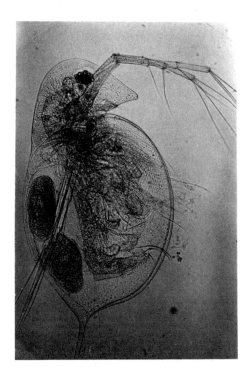

Fig. 1. Body morphology of nonhelmeted *Daphnia longiremis* f. *typica* (left) and anterior body morphology of helmeted *Daphnia longiremis* f. *cephala* (above).

The present work focuses on the advantages of helmet development in *Daphnia longiremis* f. *typica* (Riessen and O'Brien 1980) and this form's competitive interactions with the exuberant morph *Daphnia longiremis* f. *cephala*.

We thank all those at the Toolik Lake research facility who helped make this work possible and enjoyable. Special thanks go to J. Haney and C. Buchanan who contributed a number of ideas and techniques useful in this project. We thank M. O'Brien for editorial help and B. Archinal for office help.

Description of the two morphological forms—Riessen and O'Brien (1980) have described the morphology of the two forms of *Daphnia longiremis*, forma *typica* and forma *cephala*. The *typica* form is exactly as described by Brooks (1957), but the *cephala* form is significantly larger (up to 2.4 mm in total length) and can attain quite large helmet development (Fig. 1).

Distribution of the species and morphs—Brooks (1957) thought the large helmet of *D. longiremis* forma *cephala* (Riessen and

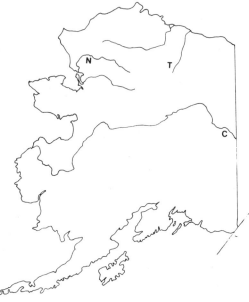

Fig. 2. Map of Alaska (excluding Southeast) showing lakes where *Daphnia longiremis* was collected. N—Noatak River valley, 43 lakes and ponds sampled. C—Charley River area along Yukon River, 8 lakes and ponds sampled. T—Toolik Lake area, 25 lakes and ponds sampled.

O'Brien (1980) to be rare. However, in qualitative collections of zooplankton throughout arctic and subarctic Alaska we have consistently found this large morph. The principal sites sampled include lakes and ponds throughout the Noatak River Valley (O'Brien et al. 1975; O'Brien 1975), the Charley River region (O'Brien and Huggins 1976), and the Toolik Lake area (Fig. 2;

O'Brien et al. (1979). Of the 76 lakes and ponds sampled in these three regions 20 contained sizable populations of *D. longiremis*. In all 20 lakes planktivorous fish were either known (9 lakes) or strongly suspected to be present (11 lakes). (A lake deeper than 6 m with a flowing outflow during midsummer almost always will contain at least arctic grayling.) In 49 lakes or ponds where fish were certainly absent, *D. longiremis* was never found, while only 7 lakes thought likely to contain planktivorous fish lacked *D. longiremis* at the time of sampling. Thus there is a strong correlation between the presence of planktivorous fish and *D. longiremis*.

Even more interesting is the distribution of the two morphs with respect to the predaceous copepod *Heterocope septentrionalis*. Out of the 20 collections of *D. longiremis*, 6 contain individuals which Riessen and O'Brien (1980) term *Daphnia longiremis* forma *cephala*. These 6 also contain *Heterocope*. Of the 14 lakes containing *D. longiremis* forma *typica*, none contains *Heterocope* (Table 1). This striking correlation suggested that the *Heterocope* might be responsible for the observed distribution and a series of experiments were designed to illustrate such an interaction.

Table 1. Distribution of the two morphs of *Daphnia longiremis* with respect to *Heterocope septentrionalis*.

Heterocope septentrionalis	Daphnia longiremis	
	f. typica	f. cephala
Present	0	6
Absent	14	0

Methods

The experimental phase of this research was conducted in the Toolik Lake region of arctic Alaska during the summers of 1976 through 1978 as part of the NSF Research on Arctic Tundra Environments and Arctic Lake Process Studies programs. Toolik Lake is located 68°37′ N, 149°35′W, about 200 km south of Prudhoe Bay along the oil pipeline haul road.

Predation—Experiments were conducted to determine the relative susceptibility of the two morphs of *D. longiremis* to predation by *H. septentrionalis* and by planktivorous fish.

The invertebrate predator *H. septentrionalis* was collected from a shallow pond near the laboratory, and its feeding on a wide range of *Daphnia pulex* and a few sizes of both *D. longiremis* forma *typica* and *D. longiremis* forma *cephala* was measured. Most of the methods employed have been described elsewhere (O'Brien et al. 1979; O'Brien and Schmidt 1979) and will be only briefly covered here.

The predators were maintained in the laboratory for several days and starved for a day immediately preceding an experiment. Freshly collected prey were sized using an ocular micrometer with a Wild M-5 dissection microscope, and 10 to 20 (exact numbers recorded) were placed in 300-ml long cylindrical Plexiglas chambers with No. 25 plankton netting affixed to each end. In each series, one chamber served as a control; to each of the others we added 4 to 8 (generally 6) *H. septentrionalis*. The *Heterocope* averaged 2.0 mm in cephalothorax length.

The cylinders with the animals enclosed were submerged in a small aquarium which was placed in a nearby pond to provide some degree of thermal stability in the absence of sophisticated temperature control equipment. The top of the aquarium was covered with black plastic to prevent direct exposure to sunlight under which *D. longiremis* would clump at the bottom of the cylinders. Considerable diffuse light entered the aquarium from the sides and all zooplankton appeared to disperse throughout the chambers. After about 12 h (exact times recorded) the aquarium was retrieved and the numbers of prey remaining were counted using the M-5 microscope.

The impact of *Heterocope* predation was considered in terms of the number of prey consumed. The rate of decrease of the prey population is expressed as the product of an instantaneous feeding rate K (feeding rate coefficient: Dodson 1975):

$$dP/dt = KPX$$

where the density of the prey is P per liter and the density of the predator is X per liter. The integral of this equation is used to solve for K and is a simple negative exponential equation (Dodson 1975). The dimensions of K are liters per day.

We have determined the vulnerability of a wide variety of arctic zooplankton to predation from small lake trout by measuring the reactive distance of trout to each type of prey (Kettle and O'Brien 1978). We report here only the comparison of one size of *D. longiremis* forma *typica* and two sizes of *D. longiremis* forma *cephala* to a linear regression of a variety of sizes of lake-dwelling *Daphnia middendorffiana* that are much more translucent than those typical of arctic ponds (Kettle and O'Brien 1978).

The lake trout used in these experiments were 9 to 11 cm in total length and were maintained in a Plexiglas aquarium (30 × 30 × 120 cm) which also served as the experimental reactive distance chamber. Procedures used to measure the reactive distance are thoroughly described in Kettle and O'Brien (1978).

Because most arctic zooplankton species populations, including *D. longiremis*, develop as discrete cohorts, direct comparisons of predation rates between sizes and species are difficult or impossible to make. We therefore established "standard" regression lines showing the relationship between size and vulnerability to predation for species available in a variety of sizes. For invertebrate predation we used *D. pulex*, which occurs in the Toolik area only in shallow ponds that lack *Heterocope*; to determine typical reactive distance to lake daphnids we used lake-dwelling *D. middendorffiana* from Toolik Lake, which are much more translucent than usual arctic pond populations (Kettle and O'Brien 1978). These regression lines are then used as predictors of the vulnerability of species at particular sizes to determine if the two morphs of *D. longiremis* are more or less susceptible to invertebrate or fish predation than a "standard" prey. A 95% confidence interval is developed for each standard linear regression; observed means from other species and the associated confidence intervals can then be compared to these lines. When the confidence intervals at the same size overlap, we consider the observed mean to be essentially the same as that predicted by the line. When the two confidence intervals do not overlap, we consider the mean feeding rate coefficient or reactive distance to be different from that predicted by the line.

Competition—Experiments were also conducted to determine whether either morph of *D. longiremis* has a marked advantage in the absence of predation.

In early July 1977, a life table experiment was set up to study the comparative survivorship and fecundity of *D. longiremis* forma *typica* and *D. longiremis* forma *cephala*. Thirty embryo-carrying forma *cephala* females from Toolik Lake and 30 forma *typica* females from a nearby lake, N–2, were collected and isolated into separate 300-ml Plexiglas cylinders with No. 25 plankton netting affixed to each end. These cylinders were kept in a 18-liter aquarium filled with equal parts of Toolik and N–2 lake water, and the aquarium was placed in a 50-liter water bath maintained in a small unheated laboratory. Two days later 40 individuals born of N–2 females and 40 born of Toolik Lake females were placed in 150-ml experimental cylinders, 20 to a cylinder, and the four cylinders were returned to the aquarium. The water in the aquarium was aerated and changed every 4 days at which time a small amount of concentrated phytoplankton which passed through a 44-μm plankton net was added. All four cylinders were checked for survivorship and production of young every 2 days. Total chlorophyll *a* in the water varied from 5 to 10 μg Chl *a*/liter and was always slightly higher (up to 150%) than Toolik Lake or N–2. Water temperature generally ranged between 10° and 20°C, as compared to maximum epilimnetic temperatures in Toolik and N–2 of about 18°C in mid-July. The experiment was terminated on 4 August as the camp was beginning to be shut down.

In summer 1978 on 30 June an in situ

experiment was performed in which 20 egg-carrying forma *cephala* females isolated from Toolik Lake and 20 forma *typica* females from N-2 were placed into one of three 1.3-liter Plexiglas cylinders. Two cylinders of N-2 animals and one of Toolik animals were placed in a plastic-coated wire basket and sunk to 3 m in Toolik Lake, and two cylinders ot Toolik animals and one of N-2 animals were similarly sunk to 3 m in N-2; The contained populations were left to grow undisturbed until 26 July 1978 when the contents of all the cylinders were preserved in a mixture of 4% Formalin and 50% ethyl alcohol. Subsequently the population was counted and helmet development estimated as given in Riessen and O'Brien (1980).

Results and discussion

Heterocope *predation*—As has been shown by Dodson (1975) for *Heterocope* and by many others for other invertebrate predators, these animals tend to feed much more heavily upon small crustacean zooplankton prey than upon larger prey. As shown by the solid line in Fig. 3 and as detailed in O'Brien et al. (1979) there is almost an order of magnitude decline in feeding rate coefficient when *Heterocope* are offered 2-mm versus 1-mm *Daphnia pulex*. Given the strong correlation between the occurrence of *D. longiremis* forma *cephala* and that of *Heterocope*, it might be expected the exuberant form would actually be preyed upon less than the nonhelmeted morph; as Fig. 3 shows, this is indeed the case. The *typica* form is preyed upon at the same rate as similar-sized *D. pulex*, whereas in 10 out of 14 trials the *cephala* form suffered no mortality when exposed to *Heterocope* and in the 4 other trials the mortality is much lower than that of similar-sized *D. pulex* or *D. longiremis* forma *typica*. Most notable is that the death rate is much lower than can be attributed to increased size due to helmet development.

The actual mechanism by which *D. longiremis* forma *cephala* is able to reduce mortality from *Heterocope* is not known. Dodson (1974) suggested that exuberant structures may be more difficult for tactile

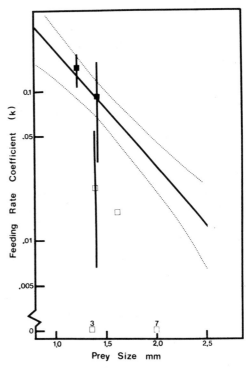

Fig. 3. Predation rate of *Heterocope septentrionalis* on *D. pulex*, *D. longiremis* f. *typica*, and *D. longiremis* f. *cephala*. Solid line—linear regression of 26 observations of *H. septentrionalis* feeding on groups of different-sized *D. pulex*; dotted line—95% confidence interval around this line; ■—observations of *Hetercope* feeding on *D. longiremis* f. *typica*; □—observations of *Heterocope* feeding on *D. longiremis* f. *cephala*; solid bars—95% confidence intervals around these means. Numbers "3" and "7" indicate number of zero feeding rate observations on *D. longiremis* f. *cephala*.

predators to handle. The greater body width to body length ratio and longer tail spines of the exuberant morph (O'Brien et al. 1979) would certainly support this contention, but we have not been able to observe enough predator-prey interactions to confirm such a handling difference. O'Brien and Vinyard (1978) showed that an extremely helmeted morph of a tropical daphnid, *D. carinata* forma *cephalata*, reduced predation by an invertebrate predator (a notonectid) by evading attack rather than escaping from grasp. While such a mechanism cannot be ruled out without observational data, during routine handling of

D. longiremis forma *cephala* we have not seen any of the adroit evasion skills that were very obvious with *D. carinata* forma *cephalata*. Kerfoot (1977) has pointed out that certain varieties of *B. longirostris* with elongated antennules are less susceptible to predation from *Epischura nevadensis*, at least partly because their elongated and enlarged antennules protect the swimming antennae once an individual *Bosmina* is under attack. In several scanning electron micrographs we have noted that the antennae of *D. longiremis* forma *cephala* are lodged within a groove formed by the great exaggeration of the rostrum and anterior portion of the body. This groove may provide some protection. Even without knowing the precise mechanism, however, the *Heterocope* predation experiments (Fig. 3) and the strong correlation between the occurrence of *D. longiremis* forma *cephala* and that of *Heterocope* (Table 1) leave little doubt that these exuberant structures reduce predation from this invertebrate, tactile predator.

Planktivorous fish predation—In the Toolik area, *D. longiremis* is found only in those lakes having planktivorous fish. In this situation selective pressure from an invertebrate tactile predator which feeds heavily upon small crustacean zooplankton cannot be met simply by increasing body size. Such an increase in size and thereby in visibility would probably increase the susceptibility of *D. longiremis* to fish predation through increasing the probability of being located. Recent research with fresh-water planktivorous fish has shown that not only are larger zooplankton fed upon more frequently by fish (Hrbáček 1962; Brooks and Dodson 1965) but that the reason this is so is that the fish are better able to locate large-sized zooplankton than small ones (Werner and Hall 1974; Confer and Blades 1975; Vinyard and O'Brien 1976).

Such was certainly the case with such facultative planktivores as lake trout and arctic grayling. The reactive distance of small lake trout to a wide variety of arctic zooplankton species has been documented elsewhere (Kettle and O'Brien 1978); its reactive distance to lake-dwelling *D. midden-*

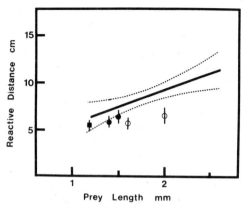

Fig. 4. Reactive distance of small (9–11-cm SL) lake trout to *Daphnia middendorffiana*, *D. longiremis* f. *typica*, and *D. longiremis* f. *cephala*. Solid line—reactive distance to different-sized lake-dwelling *D. middendorffiana*; dotted line—95% confidence interval around this line; ■—reactive distance to 1.2-mm *D. longiremis* f. *typica*; ○—reactive distance to 2.0-mm and 1.6-mm *D. longiremis* f. *cephala*; ●—same observations but plotted as core body size of *D. longiremis* f. *cephala* (1.5 for 2.0-mm animals and 1.35 for 1.6-mm animals); solid bars—95% confidence intervals around these means.

dorffiana is shown in Fig. 4. Clearly, large daphnids can be located at greater distances. If we assume that planktivorous fish search a roughly spherical volume of which the reactive distance is the radius, a 2.5-mm *D. middendorffiana* is almost 10 times more likely to be located than is a 1.0-mm *D. middendorffiana*.

The reactive distance of lake trout to 1.2-mm *D. longiremis* forma *typica* is not significantly different from the reactive distance to 1.2-mm lake *D. middendorffiana* (Fig. 4). However, small lake trout appear to have more difficulty locating the helmeted morph. The reactive distance to 2-mm *D. longiremis* forma *cephala* is 70% that to 2-mm *D. middendorffiana*, and, as shown in Fig. 4, the reactive distances to the helmeted morph are significantly lower than that to either lake dwelling *D. middendorffiana* or *D. longiremis* forma *typica*. However, when just the core body size of *D. longiremis* f. *cephala* (the length from the compound eye spot to the end of the postabdomen) is considered, these adjusted points fall within the predicted reactive distance of lake trout to typical lake daphnids (Fig. 4). Recent work with arctic grayling feeding on these two morphs showed the same relationship (Schmidt pers. comm.). These findings for the helmeted form of *D. longiremis* support the contention of Brooks (1965) that the helmets of cyclomorphotic *Daphnia* may be body elaborations not easily seen by fish. However, the results of *Heterocope* feeding experiments clearly also support Dodson's (1974) hypothesis that the helmet effectively reduces invertebrate predation beyond what would be predicted from size alone. As we never saw *D. longiremis* forma *cephala* evade a planktivorous fish suction attack, helmet development does not appear to allow evasion from fairly large fish, as might be inferred from Jacobs (1965).

Competition—Kerfoot and Pastorok (1978) have recently suggested that the energetic cost to exuberant morphs or species which elaborate various defensive structures may be such that their competitive abilities and indeed population growth rates may suffer. To determine the relative competitive capabilities of *D. longiremis* forma *typica* and *D. longiremis* forma *cephala* we performed a carefully controlled laboratory life table experiment with the results shown in Fig. 5 and 6. Contrary to our expectations and the prediction of Kerfoot and Pastorok (1978), the exuberant morph of *D. longiremis* demonstrated greater survivorship, earlier age at first reproduction, and considerably greater fecundity than *D. longiremis* forma *typica*. The overall rates of increase for the *typica* morph were 0.059 and 0.059 and for the *cephala* morph 0.084 and 0.091, almost 1.5 times higher. If we rule out physical interference, which is quite unlikely under the extremely low field densities of these animals (1 animal per 1–10 liters), these experiments predict *D. longiremis* forma *cephala* the better competitor and even suggest it should exclude the nonhelmeted form in any direct competition.

Perhaps the higher rate of growth of the *cephala* morph should have been expected as it is considerably larger than the *typica* morph. *Cephala* adults consistently grow longer than 2 mm (Riessen and O'Brien in press) and develop broods of 8 to 10 eggs, whereas the *typica* adults rarely reach 1.3 mm or develop broods of more than 5 to 7 eggs.

The in situ population growth experiments confirmed the findings of the life table experiments. Whether incubated in Lake N-2 or Toolik Lake the *cephala* morph always attained the greater average

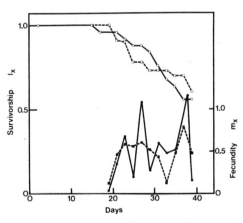

Fig. 5. Life table results for *D. longiremis* f. *typica*. Lower two lines represent fecundity values, upper two lines represent suvivorship values. Solid line—one replicate; dotted line—second replicate.

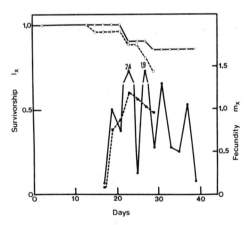

Fig. 6. As Fig. 5, but for *D. longiremis* f. *cephala*.

Table 2. In situ population growth of *D. longiremus* f. *typica* and *D. longiremis* f. *cephala* grown in both Toolik Lake in which *D. longiremis* f. *cephala* occurs and Lake N-2 in which *D. longiremis* f. *typica* occurs. Densities are those resulting from 20 egg-bearing females placed in each cylinder. *r*-value is daily rate of growth which would yield such an increase in density corrected for initial 20 animals.

D. longiremis f. cephala from Toolik Lake		*D. longiremis* f. typica from Lake N-2	
Density	r value	Density	r value
Incubated in Toolik Lake			
585	0.130	91	0.058
		462	0.121
Incubated in Lake N-2			
301	0.104	126	0.071
280	0.102		
$\bar{X}$ 388	0.112	226	0.083

population density (Table 2). The population densities of the *cephala* morph in cylinders incubated in Lake N-2 were more than twice those of the *typica* morph. While there was more variability in the Toolik Lake-incubated cylinders, the average increase in density of the *cephala* morph was still better than 150% that of the *typica* morph. The average population increase occurring in cylinders placed in both lakes was 0.112 for the *cephala* populations compared to 0.083 for the *typica* populations. These rates of increase are greater than found in the life table experiments because the cylinders were initially stocked with adult animals already containing eggs. If we assume that at the temperatures in both lakes the egg-carrying females in the cylinders would begin releasing young in 5 days, then recomputed *r* values from the life table (using age 5 days in the beginning of the reproductive phase in the life table) are fairly similar to the rates of increase observed in the cylinders. For the *typica* form the life table *r* values become 0.125 and 0.126—only a little higher than the observed mean *r* value from the cylinders (Table 2). The *cephala* life table values become 0.185 and 0.170—quite a bit higher than the *r* values computed from growth in the cylinders. Of course population densities generally were high in these cylinders and may have begun reducing the *cephala* birth rates.

An important result of these experiments was that no cylinder showed a negative rate of population increase. Thus both morphs could increase their density in both lakes when isolated from mortality due to predation. It therefore appears that although *D. longiremis* forma *cephala* may be the better competitor, such factors as competition for food resources are not important in regulating the distribution of these two morphs, at least during the summer months.

General discussion

Daphnia longiremis f. *cephala* thus appears to be specially adapted to reduce invertebrate predation from *Heterocope* without increasing visible body size which would increase vulnerability to planktivorous fish. Surprisingly this morph is also the more fecund and would seem to have such a competitive advantage that one wonders why lakes containing only planktivorous fish should contain *typica* rather than the more competitive *cephala*. Unfortunately we can only speculate, but surely the answer, in part, lies with the considerably different size of the two morphs. The *cephala* morph adults can and frequently do become 2–2.4 mm long with even their core body size being 1.5 to 2 mm long whereas the *typica* morph populations rarely attain a size of even 1.3 mm (Riessen and O'Brien in press). Then taking an intermediate adult of each morph, say 1.8 mm for *cephala* and 1.0 mm for *typica* the vulnerability to planktivorous fish can be calculated. The regression from Fig. 4 suggests that a 1.0-mm daphnid would have a reactive distance of 4.5 mm whereas a 1.8-mm lake daphnid would have a reactive distance of 8.3 mm. If we assume planktivorous fish search a spherical volume with the fish's eye at the center of the sphere and the radius being the reactive distance, then a 1.8-mm adult *D. longiremis* f. *cephala* would be more than six times more vulnerable to planktivorous fish than an adult *D. longiremis* f. *typica* of 1.0-mm size. Such an argument is based on situations where planktivory is fairly intense and one may wonder what would occur at lesser levels of fish predation. It seems likely that at intense levels of planktivory the *typica* morph has the advantage due to its smaller size and that this intense planktivory also acts to exclude *Heterocope* since it is quite large—lake adults having a cephalothorax length > 2 mm (Kettle and O'Brien 1978). When planktivory is lessened, then the *typica* morph would have less of an advantage so that the competitive superiority of the *cephala* morph would come into play and quite likely *Heterocope* would become present and abundant.

References

BROOKS, J. L. 1957. The systematics of North American *Daphnia*. Mem. Conn. Acad. Arts Sci. **13**: 180 p.
———. 1965. Predation and relative helmet size in cyclomorphic *Daphnia*. Proc. Natl. Acad. Sci. **53**: 119-126.
———, and S. I. DODSON. 1965. Predation, body size and composition of plankton. Science **150**: 28-35.
CONFER, J. L., and P. I. BLADES. 1975. Omnivorous zooplankton and planktivorous fish. Limnol. Oceanogr. **20**: 571-579.
DODSON, S. I. 1974. Adaptive change in plankton morphology in response to size-selective predation: A new hypothesis of cyclomorphisms. Limnol. Oceanogr. **19**: 721-729.
———. 1975. Predation rates of zooplankton in arctic ponds. Limnol. Oceanogr. **20**: 426-433.
GREEN, J. 1967. The distribution and variation of *Daphnia lumholtzi* (Crustacea: Cladocera) in relation to fish predation in Lake Albert, East Africa. Proc. Zool. Soc. Lond. **151**: 181-197.
HRBÁČEK, J. 1962. Species composition and the amount of zooplankton in relation to the fish stock. Rozpr. Cesk. Akad. Ved Rada. Mat. Prir. Ved **72**(10): 116p.
HUTCHINSON, G. E. 1967. A treatise on limnology, v. 2. Wiley.
JACOBS, J. 1965. Significance of morphology and physiology of *Daphnia* for its survival in predator-prey experiments. Naturwissenschaften **52**: 141.
KERFOOT, W. C. 1977. Implications of copepod predation. Limnol. Oceanogr. **22**: 316-325.
———, and R. A. PASTOROK. 1978. Survival versus competition: evolutionary compromises and diversity in the zooplankton. Int. Ver. Theor. Angew. Limnol. Verh. **20**: 362-374.
KETTLE, D., and W. J. O'BRIEN. 1978. Vulnerability of arctic zooplankton species to predation by small lake trout. J. Fish. Res. Bd. Can. **35**: 1495-1500.
O'BRIEN, W. J. 1975. Some aspects of the limnology of the ponds and lakes of the Noatak drainage basin, Alaska. Int. Ver. Theor. Angew. Limnol. Verh. **19**: 472-479.
———, and D. HUGGINS. 1976. A limnological investigation of the lakes and streams of the Charley River area, p. 282-325. *In* S. B. Young [ed.], The environment of the Yukon-Charley River Area, Alaska. Contrib. Center Northern Stud. 2.
———, ———, and F. DeNOYELLES, JR. 1975. Primary productivity and nutrient limiting factors in lakes and ponds of the Noatak River Valley, Alaska. Arch. Hydrobiol. **75**(2): 263-275.
———, D. KETTLE, and H. RIESSEN. 1979. Helmets and invisible armor: Structures reducing predation from tactile and visual planktivores. Ecology **60**: 287-294.
———, and D. SCHMIDT. 1979. Arctic *Bosmina* morphology and copepod predation. Limnol. Oceanogr. **24**: 564-568.
———, and G. L. VINYARD. 1978. Polymorphism and predation: The effect of invertebrate predation on the distribution of two *Daphnia carinata* varieties in South India ponds. Limnol. Oceanogr. **23**: 452-460.
———, C. BUCHANAN, and J. F. HANEY. 1979. Arctic zooplankton community structure: Exceptions to some general rules. Arctic **32**: 237-247.
RIESSEN, H. P., and W. J. O'BRIEN. 1980. Re-evaluation of the taxonomy of *Daphnia longiremis* Sars 1862 (Crustacea, Cladocera): Description of a new morph from Alaska. Crustaceana. **38**: 1-11.
VINYARD, G. L., and W. J. O'BRIEN. 1976. Effect of light intensity and turbidity on the reactive distance of bluegill sunfish. J. Fish. Res. Bd. Can. **33**: 2845-2849.
WERNER, E. E., and D. J. HALL. 1974. Optimal foraging and size-selection of prey by bluegill sunfish (*Lepomis macrochirus*). Ecology **55**: 1042-1052.
ZACHARIAS. O. 1894. Beobachtungen am Plankton des grossen Plonersees. Forschungsher. Biol. Stn. Plon **2**: 91-137.
ZARET, T. M. 1972. Predator-prey interaction in a tropical lacustrine ecosystem. Ecology **53**: 234-257.

VIII

Predation,
Prey Vulnerability,
Zooplankton Composition

45. Variation among Zooplankton Predators: The Potential of *Asplanchna, Mesocyclops,* and *Cyclops* to Attack, Capture, and Eat Various Rotifer Prey

Craig E. Williamson and John J. Gilbert

Abstract

Observation of different freshwater invertebrate predators reveals the presence of a diversity of attack, capture, and ingestion behaviors. The rotifer prey in turn may have a wide variety of morphological and behavioral defense mechanisms. These complex interactions are largely overlooked by the "black-box" approach of enclosure experiments.

Reduced ingestion rates of the invertebrate predators may result from problems in handling prey at different levels of the encounter, attack, capture, and ingestion sequence. Direct observation is needed for a better understanding of predator-prey interactions and of the potential influence of different invertebrate predators on the population dynamics of rotifer communities.

Results of the interactions investigated suggest that: *Cyclops bicuspidatus* cannot be an important predator of *Asplanchna*; *A. girodi* can only rarely capture *Polyarthra vulgaris*; *A. girodi* is likely to be a more efficient predator of *Keratella cochlearis* than *Mesocyclops edax*; *M. edax* could be an important predator of *P. vulgaris*; *M. edax* readily preys on *A. girodi* and so may greatly reduce predation by this rotifer on *K. cochlearis*.

Predators of freshwater zooplankton can be divided into two broad categories, invertebrate predators and vertebrate predators. Vertebrate predators are generally larger than their invertebrate counterparts and selectively consume larger or more visible prey, while invertebrate predators consume smaller, more easily grasped organisms (Brooks and Dodson 1965; Dodson 1970; Dodson and Dodson 1971; Sprules 1972; Zaret 1972; Dodson 1974a; Confer and Blades 1975; Kerfoot 1975; Swift and Fedorenko 1975; Zaret and Kerfoot 1975). However, combining all the invertebrate predators into a single category is overly simplistic and obscures the great variety of prey-capture behavior displayed by this group.

Considerable emphasis has been placed on the impact of invertebrate predation on zooplankton communities. Most of these experiments have been of an enclosure or exclosure design where the impact of predation is assessed by observing initial versus final densities of prey under variously manipulated predator densities (McQueen 1969; Dodson 1970, 1974b, 1975; Sprules 1972; Fedorenko 1975; Confer and Blades 1975; Brandl and Fernando 1975, 1978; Gophen 1977; Kerfoot 1977a). Due to their inherent "blackbox" design, however, enclosure experiments largely ignore the complex series of behavioral interactions which determine the ultimate fate of predator and prey following an encounter. An evaluation of these interactions is critical to an understanding of the effect of invertebrate predators on zooplankton communities.

There are very few previous studies

which have employed methods of direct observation to quantify predator-prey interactions. Kerfoot (1977b, 1978) has examined copepod predation on *Bosmina*, while Brandl and Fernando (1978) and Gilbert and Williamson (1979) have looked at the dynamics of predation by copepods and rotifers. Aside from these latter works, most studies on invertebrate predation have emphasized crustacean zooplankton as both predator and prey and have ignored the rotifers. This is true in spite of the fact that rotifers often form a substantial component of the zooplankton community, both in terms of their numerical density and biomass.

When a predator encounters a prey organism, a complex series of behavioral interactions may follow. These interactions can be broken into several sequential components including encounter, attack, capture, and ingestion. Each component can in turn be expressed as a probability of the preceding component (Fig. 1). Numerous direct observations of predator-prey interactions are necessary to establish these probabilities. When used in conjunction with data from enclosure or exclosure experiments, these probabilities and the accompanying behavioral observations can contribute greatly to understanding the role of predator-prey interactions in determining zooplankton community structure.

In this study we use predator-prey interactions involving rotifers in an attempt to demonstrate the importance of quantifying the behavior of invertebrate predators and their prey. Our results suggest that different outcomes are expected for predators in the same major taxon as well as for predators from two different taxa (e.g. a rotifer and a copepod). Moreover, within the same taxon, different outcomes are also expected for two similarly sized species, and for males and females of the same species.

We thank Thomas M. Frost and W. Charles Kerfoot for reading and improving the manuscript and A. B. Farrell for permission to study Star Lake.

Methods

In the present experiments the behavior of two species of predatory copepods, *Mesocyclops edax* and *Cyclops bicuspidatus*, and one species of predatory rotifer, *Asplanchna girodi*, was examined. The four species of rotifers used as prey were *Asplanchna girodi*, *Keratella cochlearis*, *Polyarthra vulgaris*, and *Brachionus calyciflorus*. The lengths of these organisms are given in Table 1. All organisms except *Brachionus* and some *Asplanchna* were collected from Star Lake, Norwich, Vermont. All *Brachionus* and some *Asplanchna* were obtained from laboratory clones (*see* Gilbert 1975; Gilbert and Litton 1978). Experimental techniques involved placing one to three dozen prey in a depression dish with 1 ml of filtered (Reeve Angel 934 AH glass-fiber paper) Star Lake water. The predator was then gently added, and the numbers of subsequent interactions

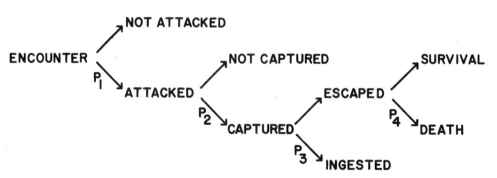

Fig. 1. Predator-prey diagram (P_1–P_4 represent associated probabilities).

Table 1. Body lengths of predator and prey organisms.

Organism	Sex	Length (μm)	Comments
Mesocyclops edax	♀	1,500	excludes caudal setae
M. edax	♂	970	
Cyclops bicuspidatus	♀	1,400	excludes caudal setae
C. bicuspidatus	♂	1,000	
Asplanchna girodi	♀	900	Star Lake adults
A. girodi	♀	600	laboratory clone adults
A. girodi	♀	300–400	laboratory clone juveniles
Keratella cochlearis	♀	180	lorica length excluding caudal spine
Brachionus calyciflorus	♀	170	clone S4, lorica length
Polyarthra vulgaris	♀	140	body length

with prey (encounters, attacks, captures, and ingestions) were scored.

For *Asplanchna* predators, an encounter was defined as such when the prey was directly contacted by the predator's corona. An attack consisted of directed movements of the head of the predator, often accompanied by movement of the jaws. Attacks were not scored for *Asplanchna-Polyarthra* interactions because the escape response of *Polyarthra* usually occurred too quickly for *Asplanchna* to elicit a clear attack response. For *Cyclops* predators, an encounter was defined as prey contact with any part of the body of the copepod (excluding the caudal setae). *Cyclops* was an extremely adept predator and generally responded to even the slightest prey contact over most parts of its body. Occasionally *Cyclops* responded to a prey before any direct contact was made. The reaction distance in these cases was always much less than 0.5 mm. *Mesocyclops* was much less responsive and never initiated an attack unless it directly contacted its prey. Furthermore, the prey contact had to be almost exclusively with the feeding chamber of the predator in order for an attack response to be elicited. Thus for *Mesocyclops* an encounter was defined as prey contact with any part of the second antennae, with the long setae on the fourth segment of the first antennae, or with parts of the first antennae more proximal to the mouth. Attack for both copepods was defined as a directed lunge toward the prey accompanied by grasping movements of the mouthparts.

The first encounter-attack was discounted for each *Cyclops* as it was found to be highly variable in comparison with subsequent probabilities. This was most likely a handling artifact. *Mesocyclops* did not seem to be similarly affected. From one to several captures and ingestions were scored for each predator, depending on the experiment. *Asplanchna* predators were starved for 5 to 9 hours, while copepod predators were starved for 5 to 9 days. Shorter starvation periods resulted in very low attack probabilities.

All behavioral observations were made with a Wild M-5 stereomicroscope. In order to minimize abnormal behavior caused by temperature changes, most experiments were performed at a temperature close to that of Star Lake at the time the animals were collected. When cloned *Asplanchna* were used as predators, experiments were conducted at room temperature. For observations at lower temperatures, a Plexiglas experimental chamber was constructed and used in conjunction with a Brinkmann MGW Lauda RC20 circulating water bath; in this way, a constant temperature in the depression dish was maintained throughout the course of an experiment.

Results and discussion

Rotifers and copepods utilize different mechanisms to propel themselves and thus exhibit radically different swimming behaviors. This was clearly evident in our experiments. *Asplanchna* swam about the experimental chamber in a smooth gliding motion which was occasionally interrupted by an abrupt change in direction. Both copepod predators swam in a jerky, "hop and sink" manner previously described by Strickler (1975) and Kerfoot (1978). The swimming behavior of *Mesocyclops* consisted of short rests between prolonged hop and sink sequences. *Cyclops*, on the other hand, spent more time at rest, either on the bottom of the dish or ventral-side up in an inclined position at the surface. When it did swim, *Cyclops* moved in a modified hop and sink manner which involved coordinated movement of its antennules, swimming legs, and feeding appendages. The actual movements of all the appendages were slower than those of *Mesocyclops*. The forward motion of *Cyclops* was directed more horizontally rather than at an elevated angle, as occurred in the hop and sink mode of *Mesocyclops*. The ability of a copepod predator to respond to prey contact appeared unchanged regardless of its activity state. The one exception was the lack of responses observed at moderate to high swimming velocities. This may be caused by a decrease in the sensitivity of the predator's receptors due to the increase in water disturbance around a fast-moving predator (Gerritsen and Strickler 1977).

The *Asplanchna* and copepod predators also exhibited substantially different attack behaviors. Attacks by both *Mesocyclops* and especially *Cyclops* included high accelerations, rapid grasping movements of the feeding appendages, and often some impressive gymnastics. Those by *Asplanchna* were somewhat slower and involved primarily orientation of the head and grasping with the jaws. These differences in attack mechanisms led to very different capture and handling abilities for the two types of predators. This can be clearly seen by the ability of *Mesocyclops* to pounce on and capture the rotifer *Polyarthra* in spite of the latter's ability, after being contacted by a predator, to instantaneously propel itself great distances by means of its paddle-like appendages. At the same time, *Polyarthra*'s escape response was so effective against *Asplanchna* that the latter usually did not even have a chance to attack (Table 2). The relative effectiveness of these two predators is reversed when they are preying on the rotifer *Keratella* (Table 2). However, the crucial factors here are the handling and ingestion of the prey rather than the capture of the prey. *Mesocyclops* was often unable to penetrate the hard lorica of *Keratella* and rejected the prey unharmed, while *Asplanchna* could swallow the prey whole (Gilbert and Williamson 1979).

It seems logical that representatives from two major taxa, such as Rotifera and Copepoda, would possess different prey capture and handling abilities. Similarly, it comes as no surprise that larger *Bosmina* instars are more readily attacked by a larger cyclopoid copepod than by a smaller one (Kerfoot 1977b). However, what might not be expected is that two similarly sized cyclopoids may possess differences in their preferences toward large versus small prey. Adult females of *Mesocyclops* were more responsive to *Asplanchna* prey than to the smaller *Brachionus*, while adult female *Cyclops* were more likely to attack and capture *Brachionus* than *Asplanchna* (Table 3). The explanations involved here are most likely differences in the size or orientation of the grasping mouthparts of the two predators, and a lower response threshold for *Cyclops*. *Cyclops* had little or no success in capturing *Asplanchna* due to the prey's ability to retract its corona and expose only smooth turgid cuticle to the attacking predator. Because of this, *Cyclops* was unable to grasp even the smaller juveniles of *Asplanchna*. Kerfoot (1977b) noted a similar decrease in attack and capture ability with increasing prey size for *Cyclops*

Table 2. *Mesocyclops* and *Asplanchna* predation on the rotifers *Polyarthra* and *Keratella* at 23° C.

Predator		Prey		Probabilities	
Genus	no. tested		no. attacks	Capture after attack	Ingestion after capture
♀ *Mesocyclops* (adults)	20	*Polyarthra*	39	0.49	1.00
	25	*Keratella*	24	0.92	0.23
♀ *Asplanchna* (laboratory clone adults)	31	*Polyarthra*	*	*	1.00
	29	*Keratella*	46	0.89	0.73

* Although attack data are unavailable here, probability of capture after encounter ($N = 346$ encounters) for this interaction was very low (0.01).

Table 3. Predation by *Mesocyclops* and *Cyclops* on laboratory clones of *Asplanchna* and *Brachionus*.

Female predator		Prey			Probabilities			
Genus	no. tested		Exp temp (°C)	no. encounters	Attack after encounter	Capture after attack	Ingestion after capture	Ingestion after encounter
Mesocyclops	19	*Brachionus*	10	389	0.11	1.00	1.00	1.1×10^{-1}
Mesocyclops	18	*Asplanchna* (adults)	10	140	0.30	0.78	1.00	2.3×10^{-1}
Cyclops	10	*Brachionus*	0	130	0.86	0.87	0.96	7.2×10^{-1}
Cyclops	9	*Asplanchna* (juveniles)	20	95	0.69	0.04	0.11	3.0×10^{-3}
Cyclops	5	*Asplanchna* (adults and juveniles)	0	147	0.38	0.00	0.00	0.00

bicuspidatus predation on *Bosmina*. *Mesocyclops* did not have trouble grasping either rotifer prey species. However, it reacted less often to contact with *Brachionus* than with *Asplanchna*, presumably due to the weaker stimulus created by the smaller size of *Brachionus*.

Differences in prey capture and handling ability were apparent even between two sexes of the same species of copepod. Adult females of *Mesocyclops* were fully capable of capturing and ingesting adult *Asplanchna*, while the smaller adult males of *Mesocyclops* were unable to do so (Table 4). This is apparently a size-related phenomenon, as males were successful in capturing and ingesting juvenile *Asplanchna*.

Starved male and female cyclopoids may also exhibit different appetites. In *Cyclops*, males and females exhibited differences in the quantity of *Brachionus* consumed. Starved adult male *Cyclops* allowed to feed to repletion consumed an average of 4 or 5 *Brachionus*, while

Table 4. Predation by male and female *Mesocyclops* on *Asplanchna* at ~20° C.

Mesocyclops		Asplanchna		Probabilities		
Sex	no. tested	Size and origin	no. attacked	Capture after attack	Ingestion after capture	Ingestion after attack
Females	24	Star Lake adults (900 μm)	86	0.26	0.77	0.20
Males	13	Star Lake adults (900 μm)	67	0.01	0.00	0.00
Males	12	Laboratory clone juveniles (300–400 μm)	27	0.41	1.00	0.41

Table 5. Comparison of appetites of starved male and female *Cyclops* permitted to feed until satiation on *Brachionus*. Juvenile *Brachionus* and partial ingestions of adult *Brachionus* scored as half an individual when determining quantities consumed.

Cyclops	No. Brachionus consumed per Cyclops (N = 5 Cyclops)		Partial: Total ingestions (N = 10 Cyclops)
	Mean	Range	
Male	4.4	3.5–6.0	0.42
Female	11.6	10.0–16.0	0.05

adult female *Cyclops* under the same conditions consumed an average of 11 or 12 *Brachionus* (Table 5). In addition, male *Cyclops* often ingested only part of a captured *Brachionus* (Table 5). Anderson (1970) found that the ingestion rates of females of *Diaptomus nevadensis* and prairie populations of *D. arcticus* (Calanoida) were about twice those of males, while no appreciable differences in ingestion rates were observed between males and females of *D. shoshone* and mountain populations of *D. arcticus*. Anderson (1970) also observed partial ingestion of prey by male *Diaptomus*.

Previous experiments by others indicate that even different populations of the same or similar species may exhibit different prey preferences. For example, *Cyclops bicuspidatus thomasi* is reportedly unable to prey on copepodites of *Diaptomus oregonensis* or *D. hesperus* in one system (McQueen 1969), while it preys extensively on copepodites of *D. tyrelli* and *D. sicilis* in another system (Anderson 1970). In addition, enclosure experiments have shown that *Mesocyclops leuckarti* prefers larger cladoceran prey (Monakov and Sorokin 1959) while its congener, *M. edax*, prefers the smaller copepodites of *Diaptomus* (Confer 1971).

Conclusions

It is apparent that important differences

in prey capture and handling abilities exist between broad taxa of invertebrates, between genera within the same major taxon, between males and females of the same species, and even between different populations of the same species. These differences are often, but not exclusively, due to size differences between predators. As Kerfoot (1978) has stressed, the varying efficiencies in prey-handling ability are often attributable to elaborate prey defense mechanisms rather than to differences between predator attack and capture mechanisms alone. There seems to be a delicate balance between specific predator-prey associations such that a prey's adaptation against one type of invertebrate predator is not necessarily effective against other invertebrate predators. This is clearly the case for the rotifer *Keratella*, whose hard lorica is quite effective in preventing copepod but not *Asplanchna* predation. On the other hand, if a predator evolves sophisticated attack and capture mechanisms which enable it to capture a particularly agile prey, it may sacrifice its ability to ingest prey protected by various morphological adaptations. This seems to be the case for *Mesocyclops* predation on *Polyarthra* and *Keratella*.

Several precautions must be observed when interpreting data collected on predator-prey interactions under artificial conditions. First, the effect that an artificial surface and its corresponding boundary layer may have on an isolated predator is uncertain. However, there were no observable differences in the behavioral responses of *Asplanchna* or copepods swimming near the surfaces of the experimental vessel versus through the interior of the vessel, except for a change in swimming direction when the surface of the vessel was encountered. It might be preferable to employ larger experimental vessels, but then encounters between predator and prey become more difficult to observe without jarring the vessel and the organisms. Second, because all of the prey offered at one time were of the same species, no information is available on how behavioral probabilities might be altered in multispecific prey assemblages. Third, in the above experiments, predators were starved and prey were abundant. These conditions cause maximal predator response and relatively high encounter rates. The resulting behavioral probabilities are thus more useful in a comparative rather than an absolute context. Thus, it is valid to say that *Cyclops* is more reactive than *Mesocyclops* when small rotifers are encountered, but caution must be used in extrapolating the numerical probabilities to natural situations.

Even though our results on predator-prey interactions were conducted under quite artificial conditions, they permit us to make several, specific conclusions regarding the potential of the predators tested to prey on the various rotifers tested. Since *C. bicuspidatus* could not grasp and eat the *A. girodi* it attacked, this predator probably never effectively preys on *Asplanchna* under natural conditions. Similarly, since *A. girodi* could only rarely catch *P. vulgaris*, it is likely that this would also be the case in nature. Since *A. girodi* was much more likely to ingest captured *K. cochlearis* than *M. edax* was, the former predator may be expected to have a greater impact on natural populations of this prey rotifer than the latter predator. *Mesocyclops edax*, however, could easily capture and eat *P. vulgaris* and therefore might well influence the survivorship of this rotifer in the field. Finally, since *M. edax* could effectively capture and ingest *A. girodi*, it may reduce the effect of this rotifer on *K. cochlearis*. Several of these conclusions were validated in experimental laboratory communities (Gilbert and Williamson 1979), and in 1 liter laboratory enclosure experiments in which *C. bicuspidatus* was offered natural assemblages of rotifers (Magnien and Williamson, unpubl. data).

Determination of numerical probabilities, such as we have made in the present study, may prove useful in refining, interpreting, and extending mathematical models, such as the one of Gerritsen and Strickler (1977), which attempt to predict encounter rates on the basis of the encounter radius of the predator and the velocities of both predator and prey. Our observations support those of Gerritsen and Strickler which indicate that the "encounter radius" of a copepod predator is a complex concept and not a simple function of predator size. We found that many factors contribute to the size of this "encounter radius," including the size, speed, and manner of locomotion of the prey. This is particularly well demonstrated by the fact that *C. bicuspidatus* responds to cladoceran prey at a distance of several millimeters (Kerfoot 1978), but responds to rotifer prey only at distances of <0.5 mm, and usually only after direct contact. This difference may be due to the ability of the copepod to react to greater levels of water disturbance caused by the swimming behavior of cladocerans versus rotifers. In this respect, the slower, smoother swimming behavior of rotifers may be viewed as a fortuitous trait that reduces the frequency of encounters with a predator.

The partial ingestion of rotifers by certain copepods, as reported here for *Cyclops*, necessitates careful interpretation of gut content analyses, especially when studies are quantitative in nature. The rotifer components of a copepod's diet are often analyzed by identification of the trophi of the prey. In the case of partial ingestions of the prey, the trophi may be rejected and the number of rotifers consumed by the copepod vastly underestimated.

In certain cases, predation may also be highly specific on different life stages of the same prey. In the laboratory, when fed prey at lake densities, *C. bicuspidatus thomasi* does not prey on copepodites of *D. oregonensis* or *D. hesperus*, but readily consumes their nauplii (McQueen 1969). Such varied responses of the same or similar predator species indicate that much remains to be understood regarding the dynamics of invertebrate predator-prey interactions. Direct observation and quantification of behavioral probabilities are critically needed to understand the nature of each level of interaction following an encounter. Elucidation of the finer details of these interactions will require the use of high-speed cinematography.

References

ANDERSON, R. S. 1970. Predator-prey relationships and predation rates for crustacean zooplankters from some lakes in western Canada. Can. J. Zool. 48:1229-1240.

BRANDL, Z., and C. H. FERNANDO. 1975. Food consumption and utilization in two freshwater cyclopoid copepods (*Mesocyclops edax* and *Cyclops vicinus*). Int. Rev. Gesamten Hydrobiol. 60:471-494.

——, and C. H. FERNANDO. 1978. Prey selection by the cyclopoid copepods *Mesocyclops edax* and *Cyclops vicinus*. Int. Ver. Theor. Angew. Limnol. Verh. 20:2505-2510.

BROOKS, J. L., and S. L. DODSON. 1965. Predation, body size, and composition of plankton. Science 150:28-35.

CONFER, J. L. 1971. Intrazooplankton predation by *Mesocyclops edax* at natural prey densities. Limnol. Oceanogr. 16:663-666.

——, and P. I. BLADES. 1975. Omnivorous zooplankton and planktivorous fish. Limnol. Oceanogr. 20:571-579.

DODSON, S. I. 1970. Complementary feeding niches sustained by size-selective predation. Limnol. Oceanogr. 15:131-137.

——. 1974a. Adaptive change in plankton morphology in response to size-selective predation: A new hypothesis of cyclomorphosis. Limnol. Oceanogr. 19:721-729.

——. 1974b. Zooplankton competition and predation; an experimental test of the size-efficiency hypothesis. Ecology 55:605-613.

——. 1975. Predation rates of zooplankton in arctic ponds. Limnol. Oceanogr. 20:426-433.

——, and V. E. DODSON. 1971. The diet of *Ambystoma tigrinum* larvae from western Colorado. Copeia 1974 (4):614-624.

FEDORENKO, A. Y. 1975. Feeding characteristics and predation impact of *Chaoborus* (Diptera, Chaoboridae) larvae in a small lake. Limnol. Oceanogr. 20:250-258.

GERRITSEN, J. and J. R. STRICKLER. 1977. Encounter probabilities and community structure in zooplankton: a mathematical model. J. Fish. Res. Bd. Can. 34:73-82.

GILBERT, J. J. 1975. Polymorphism and sexuality in the rotifer *Asplanchna*, with special reference to the effects of prey-type and clonal variation. Arch. Hydrobiol. 75:442-483.

——, and J. R. LITTON, JR. 1978. Sexual reproduction in the rotifer *Asplanchna girodi*: effects of tocopherol and population density. J. Exp. Zool. 204:113-122.

——, and C. E. WILLIAMSON. 1979. Predator-prey behavior and its effect on rotifer survival in associations of *Mesocyclops edax, Asplanchna girodi, Polyarthra vulgaris*, and *Keratella cochlearis*. Oecologia 37:13-22.

GOPHEN, M. 1977. Food and feeding habits of *Mesocyclops leuckarti* (Claus) in Lake Kinneret (Israel). Freshwater Biol. 7:513-518.

KERFOOT, W. C. 1975. The divergence of adjacent populations. Ecology 56:1298-1313.

———. 1977a. Competition in cladoceran communities: the cost of evolving defenses against copepod predation. Ecology 58:303-313.

———. 1977b. Implications of copepod predation. Limnol. Oceanogr. 22:316-325.

———. 1978. Combat between predatory copepods and their prey: *Cyclops, Epischura*, and *Bosmina*. Limnol. Oceanogr. 23:1089-1102.

McQUEEN, D. J. 1969. Reduction of zooplankton standing stocks by predaceous *Cyclops bicuspidatus thomasi* in Marion Lake, British Columbia. J. Fish. Res. Bd. Can. 26:1605-1618.

MONAKOV, A. V. and Y. I. SOROKIN. 1959. Experimental studies of the carnivorous feeding of *Cyclops* by means of an isotope method. Dokl. Akad. Nauk SSSR 125(1):201-204 [Russian]; 319-321 [English].

SPRULES, W. G. 1972. Effects of size-selective predation and food competition on high altitude zooplankton communities. Ecology 53:375-386.

STRICKLER, J. R. 1975. Swimming of planktonic *Cyclops* species (Copepoda, Crustacea): Pattern, movements and their control, p. 599-613. *In* T. Y. -T. Wu et al. [eds.], Swimming and flying in nature, v. 2. Plenum.

SWIFT, M. C., and A. Y. FEDORENKO. 1975. Some aspects of prey capture by *Chaoborus* larvae. Limnol. Oceanogr. 20:418-425.

ZARET, T. M. 1972. Predator-prey interaction in a tropical lacustrine ecosystem. Ecology 53:248-257.

———, and W. C. KERFOOT. 1975. Fish predation on *Bosmina longirostris*: body size selection versus visibility selection. Ecology 56:232-237.

46. The Predatory Feeding of Copepodid Stages III to Adult *Mesocyclops leuckarti* (Claus)

C. D. Jamieson

Abstract

The impact of predation by *Mesocyclops leuckarti* on other zooplankton species was studied experimentally. Predatory behavior when attacking a variety of species is described. Male and female *M. leuckarti* differed in their method of attacking prey species: males chased and then attacked a prey species whereas females would not chase, but instead, pounced on nearby prey.

The predation rate of adult male and female and copepodid V *M. leuckarti* typically increased linearly with increasing prey density. At high densities the predation rate of adult females was very variable. Copepodid III and IV had a curvilinear relationship between predation rate and increasing prey density. These stages were not able to kill large and/or mature individuals. All stages of *M. leuckarti* showed preferences for small zooplankton without hard exoskeletons. The swimming motion of the prey species also affected the ease of capture. Temperature only affected predation rate slightly. The predation rates of stages III–IV were reduced in the presence of abundant phytoplankton. Various aspects of the impact of predation by *M. leuckarti* on a species in a zooplankton community are discussed.

Predatory cyclopoid copepods are recognized as important components of aquatic food chains, but much of our present knowledge about their diets is qualitative. Fryer (1957a, b) reported that the food of carnivorous cyclopoid copepods includes nauplii and copepodids of their own and of other species, cladocerans, dipteran larvae, ostracods, rotifers, and indeterminate mush. *Mesocyclops leuckarti* is a very common predatory cyclopoid copepod but little is known about what, or how much, it kills and eats. Here I examine the predatory feeding of *M. leuckarti* under laboratory conditions to provide quantitative data to help elucidate interactions in the zooplankton community.

Methods

Experimental animals—Live animals (*M. leuckarti* and prey species) for experiments were collected from the following lakes near Hamilton, New Zealand: Lake Waahi, Lake D, Lake Cameron, Lake Rotoroa, two Melville dams, and the ponds on the university campus. Collecting was usually done once or twice weekly depending on the numbers of animals being used in experiments. Because the size of *M. leuckarti* collected from different habitats and at different times of the year could vary, the animals were measured after collection and only those populations in which the length of the adult female cephalothorax was 1.2

I particularly thank M. A. Chapman and J. D. Green for their advice and encouragement throughout the couse of this study.

mm were used for experiments. The animals were measured using a stereoscopic microscope with a micrometer eyepiece at a magnification of 80×.

Feeding observations—Starved *M. leuckarti* copepodids were introduced to a variety of prey species (*Bosmina meridionalis* Sars, *Chydorus sphaericus* Muller, *Cereiodaphnia dubia* Richard, *Calamoecia lucasi* Brady, calanoid nauplii, copepodid stages I–III *M. leuckarti*) in a 5-ml beaker containing 2–3 ml of filtered river water. Observations were made of the anmals' reaction to the various prey species with a stereomicroscope.

Predation experiments—The following general methods was used for most predation experiments.

To determine predation rate the prey and usually 10 predators were counted and pipetted from a petri dish under the stereomicroscope into a 1-liter opaque Perspex tube, with bolting silk of 100-µm mesh size across the bottom, standing in a 1-liter beaker containing filtered river water.

Accurate pipetting and recovery of animals was essential because of low predation rates. The animals were counted twice, firstly when being sucked up into the pipette and secondly when in the pipette (by holding the pipette near a light source, which was an easy and accurate method of checking numbers) before being placed in the experimental chamber. A plankton sorter as described by Furnass and Findley (1975) was tried initially, however, it did not provide the accuracy required.

Before each experiment *M. leuckarti* was conditioned for 24 hours at the same prey concentration and temperature used in the experiment. Copepodids had a 24-hour period of starvation before actual conditioning. This reduced the chance of moulting occurring during an experiment without affecting the predation rate.

During experiments prey were replaced every 6 hours to maintain the prey concentration (referred to hereafter as replacement). Preliminary experiments showed that if this was not done predation rates were much reduced. Because of the discontinuous nature of cyclopoid feeding it was necessary to estimate the numbers of prey to be replaced during the course of an experiment. A variation of ±10% from the set density in the numbers of prey was regarded as acceptable, except when low concentrations (5/liter, 10/liter) were used. Then the level of acceptable variation was ±20%. If the counts of prey remaining at the end of an experiment exceeded these ranges the results were discarded. After developing familiarity with the predation rates to be expected, it became possible to estimate very accurately the necessary replacement rate.

After the 24-hour experimental period the predators and prey were transferred into a petri dish by squirting distilled water through the bolting silk base of the chamber. By holding the chamber up to the light a check could be made to ensure that all animals had been transferred. The predators were immediately removed from the petri dish, counted, and placed in another experimental chamber with the same prey concentration. Checks showed that these techniques for counting, transferring, and recovery of the predators were 100% reliable.

The handling of *M. leuckarti* adult females and copepodids was facilitated by anaesthetizing them with carbonated water. Male *M. leuckarti* could not be treated in this way as they either died immediately or later during the experiment. At the end of an experiment the prey remaining were killed with Formalin and transferred to a squared Perspex counting tray mounted on the moving stage of a stereomicroscope. A magnification of 32.5× was used for counting and the numbers were recorded on a Clay-Adams tally counter. The prey animals were counted at least twice or until the same result was obtained twice in succession. All prey animals were checked for attack marks and if these were present they were scored as having been preyed upon.

A modification was made to the method when experiments with nauplii were carried out. Because of their size it was impossible to check whether all animals had been removed from the experimental chamber. Experiments with nauplii were carried out in 1-liter beakers. At the conclusion of the experiment the water in the beaker was filtered through a small circular sieve (1-cm radius) of 70-μ mesh size. The beaker was rinsed with distilled water and the rinsings were also filtered. The nauplii were then transferred to a petri dish by squirting with distilled water. The mesh area was small enough that one could be sure all animals had been transferred. Checks showed that no animals were lost with this method.

Preliminary trials—A comparison of daytime and nightime predation rates (Table 1), made in a series of 10-hour experiments, indicated that there was no preferred time for killing (Student's t-test).

Table 1. Number of copepodid stages I–III *C. lucasi* killed per day vs. per night per adult female *M. leuckarti*.

Trial No.	Predation rate	
	0830–1830	2030–0630
1	2.6	2.1
2	2.4	1.8
3	2.16	2.4
4	2.6	2.6
5	2.1	1.9
Mean	2.37	2.16
SD	0.24	0.34
$t = 1.3, n = 8, P = 0.28$ n.s.		

To test the effects of prior feeding on the experimental results, the predation rates of adult female *M. leuckarti* were compared when they had been kept for 24 hours before the experiment in experimental conditions, when they were taken directly from an aquarium stocked with mixed prey, and when they had been starved for 24 hours before the experiment. Groups of 10 *M. leuckarti* were allowed to feed in a 1-liter chamber for 24 hours on adult *C. lucasi* which were kept by replacement at a density of 20 per liter. The results (Table 2) showed that there was a statistically significant difference (Duncan's multiple range test, $P < 0.05$) between predation rates of starved animals and the other two pre-experimental treatments. The results from animals kept in the same conditions as the experimental were less variable.

As Anraku (1964) found that the size of the experimental container used in studies on the predation of the marine copepod

Table 2. Numbers of adult *C. lucasi* killed per day per adult female *M. leuckarti* subjected to different pre-experimental conditioning regimes.

Trial No.	No. killed		
	Conditioned	Aquarium	Starved
1	0.43	0.16	1.0
2	0.38	0.14	0.58
3	0.5	0.29	1.0
4	0.43	0.14	1.55
5	0.38	0.38	1.0
Mean	0.42	0.22	1.08
SD	0.05	0.11	0.26

Tortanus discaudatus affected the results, trials were carried out with *M. leuckarti* to determine whether they also were affected. Groups of 10 adult female *M. leuckarti* and adult *C. lucasi* at a density of 50/liter were placed in 250-ml, 500-ml, 750-ml, or 1,000-ml beakers. The experiment was run for 24 hours at 20°C with replacement of prey. The results showed that the predation rate increased with the size of the container up to a beaker size of 750 ml and then remained constant (Fig. 1). On the basis of these results it was decided that experiments be conducted in 1-liter chambers.

Brandl and Fernando (1974) found that *Acanthocyclops vernalis* had a higher feeding rate on the first experimental day than on later days. This was thought to be a result of handling. A preliminary experiment using adult female *M. leuckarti* and adult

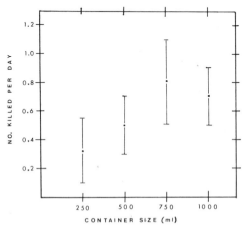

Fig. 1. Effect of container size on predation rate. ●—mean; bars—range of results.

C. lucasi at a concentration 20/liter with replacement showed that there was no statistically significant difference between the predation rates of *M. leuckarti* handled on three successive days and those that were handled once in a period of either 2 or 3 days (Table 3).

Table 3. Number of prey killed per day per adult female *M. leuckarti* under different handling regimes.

Trial No.	Handled Each day			Handled once in 2 days	Handled once in 3 days
	1	2	3		
1	0.3	0.4	0.4	0.4	0.6
2	0.5	0.4	0.4	0.4	0.4
3	0.5	0.5	0.5	0.5	0.5
4	0.5	0.4	0.6	0.6	0.6
5	0.4	0.6	0.3	0.6	0.3
Mean	0.44	0.46	0.48	0.46	0.48
SD	0.09	0.09	0.13	0.09	0.13

The discontinuity of the feeding of cyclopoid copepods and their cannibalistic tendencies are well known (Confer 1971; Brandl and Fernando 1974). Cannibalism by *M. leuckarti* occurred only once in the whole series of experiments and thus was never a problem. Trials were carried out to determine the number of predators necessary to negate the effect of discontinuous feeding in experiments and to see if any predator interaction took place which would affect the predation rate. Trials were carried out using groups of 5, 10, or 20 adult female *M. leuckarti* with 20 adult *C. lucasi* per liter as prey for a period of 24 hours. The results (Table 4) showed that with higher numbers of predators the predation rate was less variable. There did not appear to be any interactions between predators that affected the predation rate (i.e. the rates were not significantly different). On the basis of these results it was decided that 10 predators per experimental chamber would be sufficient to give repeatable results. However, at times only eight or nine predators were used. If one or two animals died during an experiment they were assumed not to have killed any prey animals, but if more than two deaths occurred the experimental results were discarded.

Table 4. Predation rate of adult female *M. leuckarti* at different predator densities.

Trial No.	Predator density		
	5	10	20
	Predation rate		
1	0.0	0.5	0.45
2	0.4	0.4	0.5
3	0.2	0.5	0.3
4	0.4	0.3	0.4
5	0.4	0.6	0.4
6	0.5	0.4	0.3
Mean	0.31	0.45	0.39
SD	0.18	0.10	0.08
Duncan's multiple range test, $P = 0.05$			

Results

Feeding behavior—The feeding methods of cyclopoid copepods have been described by a number of authors (Dzyuban 1937; Rylov 1948; Fryer 1957a, b; Monakov 1959; Anraku 1963; Confer 1971; Brandl and Fernando 1974; Strickler and Twombly 1975; Kerfoot 1977). Fryer gave a detailed account of the positions and roles of mouthparts in the feeding of *Macrocyclops albidus* and, as he found that *Acanthocyclops viridis* had no recognizable differences from it in its feeding mechanism, he suggested that the account would hold for all carnivorous cyclopoid species. Most investigators include only observations of the attack by a predator on a prey species and the result of that encounter. However, more recently, many of the mechanisms and forces involved in the recognition and capture of prey species have been elucidated (Strickler and Bal 1973; Strickler 1975a, b; Strickler and Twombly 1975).

Mesocyclops leuckarti adult females captured moving prey that were nearby (within 0–0.75 mm) by pouncing. Other attacks were made when the prey species had itself contacted *M. leuckarti*. Most of the successful attacks by females were made when they were in a resting position on the bottom of the container or sinking passively after an upward movement caused by a flick of the abdomen. When female *M. leuckarti* swam around the container prey species usually attempted to avoid them either by moving quickly in another direction (e.g. *C. lucasi*, calanoid nauplii) or by ceasing to swim and remaining motionless (e.g. *C. dubia, B. meridionalis, C. sphaericus*). On occasions, prey species, particularly *C. dubia*, collided with a female *M. leuckarti* when she was motionless or less frequently when she was swimming; usually she would attempt to capture the cladoceran.

In contrast, male *M. leuckarti* did not wait and pounce on prey but, on locating a potential victim, would follow them for short distances. When close enough, the male would attack from underneath the animal if *C. dubia* was the prey and from behind with other species. This attack was essentially the same as that described by Kerfoot (1977) for *Cyclops vernalis* when attacking *Bosmina*. Avoidance reactions by prey were not as pronounced for males. It was common for a male to grasp another male with its antennae and also to grasp the larger *M. leuckarti* copepodid stages (as when attempting to mate), although no attempt was made to consume them. However, stages I–III *M. leuckarti* were followed and captured from behind with the mouthparts and consumed. The chasing method of attack was not often used by males to capture *C. lucasi* because these prey usually avoided swimming males near them. Captures of the latter prey occurred if a male swam close enough and the *C. lucasi* failed to move far enough away when the male made a quick change of direction and pounced.

The ability of cyclopoids to hold on to their prey has been remarked upon by observers, and many of the previously cited studies have found that a cyclopoid holding on to prey can be pipetted into another container without losing its hold. Both male and female *M. leuckarti* varied in their ability to hold different prey species. Nauplii when caught were generally consumed in a few seconds, whilst swimming, and were never dropped. *Calamoecia lucasi* were also held tightly enough so I could transfer a *M. leuckarti* with a pipette without the animal dropping its prey. However most attempts to transfer a *M. leuckarti* which was holding a *C. dubia* failed. *Mesocyclops leuckarti* usually did not swim with larger prey. Interestingly, a *M. leuckarti* feeding on a *C. dubia* would drop it if bumped by another *C. dubia* and would attempt to capture the second animal. During one observation period two adult female *M. leuckarti* killed five *C. dubia* in 30 minutes but consumed only small portions of two of them.

Response to increasing prey density—The functional responses of *M. leuckarti* to increasing prey density are shown in Figs. 2–6. A stepdown predictive analysis was used to find the most significant relationship between prey density and predation rate.

Adult female *M. leuckarti* killed the greatest number of any particular prey species and also the widest range of prey species offered. The relationship between their predation rate and the density of prey was always linear, except with *C. sphaericus*, where a higher density was required before predation was significant. A feature of these results was that when the overall numbers of a particular species of prey killed per day were low (over the range of concentrations tested) the variability in results was greater than in experiments with prey species where a higher predation rate was obtained (compare Fig. 2 A with 2 G).

Adult male *M. leuckarti* could not kill as wide a range of prey species or as many individuals as adult females could. However, the males also showed a linear relationship between predation rate and prey density.

The range of prey species able to be killed and the predation rates of female stage V copepodids were similar to those found for adult males. Only female stage V copepodids were used, as male stage V copepodids tend to pass through this stage very quickly. The similarity in size of stage V copepodids (cephalothorax length = 1.05 ± 0.1 mm) and male *M. leuckarti* (cephalothorax length 0.85 ± 0.05 mm) probably accounts for this similarity in behavior.

Stage IV and stage III copepodids had the lowest predation rates and were unable to kill any large prey species offered. In contrast to the older stages they had a curvilinear relationship between predation rate and increasing prey density. However, whereas with calanoid nauplii the relationship

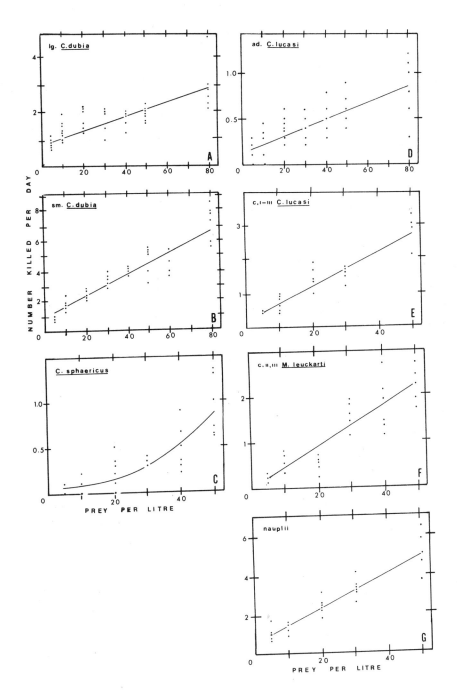

Fig. 2. Predation rate (K) of copepodid stages III-IV *Mesocyclops leuckarti* as a function of prey density (d). Results obtained for each prey density are shown and equations for fitted regression lines are given. (* $P = 0.001$.)

Adult females. A—Prey: large *Ceriodaphnia dubia* (body length 1.15-1.35 mm), $K = 0.809 + 0.026d$, $F = 132.097^*$, $R^2 = 0.677$. B—Prey: small *Ceriodaphnia dubia* (body length 0.55-0.65 mm), $K = 0.967 + 0.071d$, $F = 344.781^*$, $R^2 = 0.854$. C— Prey: *Chydorus sphaericus*, $K = 0.099 - 0.007d + 0.0004d^2$, $F = 50.546^*$, $R^2 = 0.697$. D—Prey: adult *Calameocia lucasi*, $K = 0.126 + 0.009d$, $F = 82.429^*$, $R^2 = 0.596$. E—Prey: copepodids I-III *Calamoecia lucasi*, $K = 0.210 + 0.050d$, $F = 220.156^*$, $R^2 = 0.884$. F—Prey: copepodids II-III *Mesocyclops leuckarti*, $K = 0.013 + 0.045d$, $F = 186.987^*$, $R^2 = 0.813$. G—Prey: calanoid nauplii, $K = 0.450 + 0.089d$, $F = 190.154^*$, $R^2 = 0.864$.

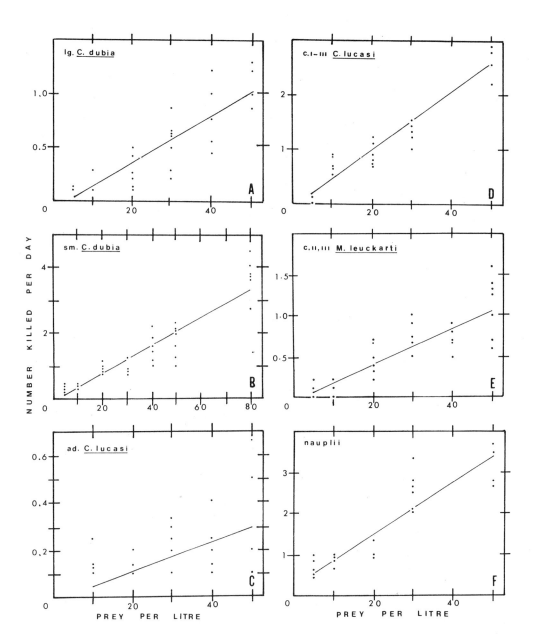

Fig. 3. As Fig. 2. Adult males. A—Prey: large *Ceriodaphnia dubia*, $K = 0.007 + 0.014d + 0.0016d^2$, $F = 95.536^*$, $R^2 = 0.820$. B—Prey: small *Ceriodaphnia dubia*, $K = -0.131 + 0.043d$, $F = 327.414^*$, $R^2 = 0.874$. C—Prey: adult *Calamoecia lucasi*, $K = -0.017 + 0.055d$, $F = 24.778^*$, $R^2 = 0.360$. D—Prey: copepodids I–III *Calamoecia lucasi*, $K = -0.113 + 0.055d$, $F = 350.291^*$, $R^2 = 0.926$. E—Prey: copepodids II–III *Mesocyclops leuckarti*, $K = 0.058 + 0.023d$, $F = 117.893^*$, $R^2 = 0.724$. F—Prey: calanoid nauplii, $K = 0.223 + 0.065d$, $F = 169.777^*$, $R^2 = 0.867$.

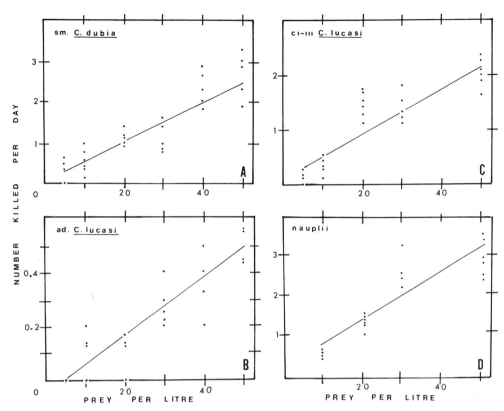

Fig. 4. As Fig. 2. Copepodid V. A—Prey: small *Ceriodaphnia dubia*, $K = 0.043 + 0.050d$, $F = 171.142^*$, $R^2 = 0.811$. B—Prey: adult *Calamoecia lucasi*, $K = -0.061 + 0.011d$, $F = 139.894^*$, $R^2 = 0.809$. C—Prey: copepodid I–III *Calamoecia lucasi*, $K = 0.115 + 0.041d$ $F = 107.468^*$, $R^2 = 0.799$. D—Prey: calanoid nauplii, $K = 0.247 + 0.057d$, $F = 64.708^*$, $R^2 = 0.746$.

was convex, with small *C. dubia* and stage I–III *C. lucasi* it was concave. That is, with nauplii there was a gradual increase in the predation rate up to a maximum, but for both *C. lucasi* and *C. dubia* a higher prey density was required before a significant number (only results including two or less zero values) of prey were killed. Thus it would appear that stage III and IV copepodids are less efficient predators than are the older stages, requiring a greater amount of time for locating, capturing, and eating a prey item.

It should be noted that, in spite of pre-experimental conditioning, an unavoidable feature of experiments with stage III and IV copepodids was that at times moulting occurred. If no more than 40% of the predators moulted during an experiment the results were included. I assumed that this was realistic because developmental rate experiments (Jamieson in prep.) with various prey species offered in excess indicated that the number of prey killed on the day that moulting occurred was similar to the numbers prior to moulting and because moulting would also be a feature of the environmental situation.

All stages had the highest predation rates when fed small and/or immature prey species, whereas the lowest predation rates were in most cases with larger or mature prey items. However, there is one exception to this, since with all stages of *M. leuckarti* a greater number of small *C. dubia* were killed at high densities than were nauplii. This appears to be a contradictory result as nauplii were preferred (*see*

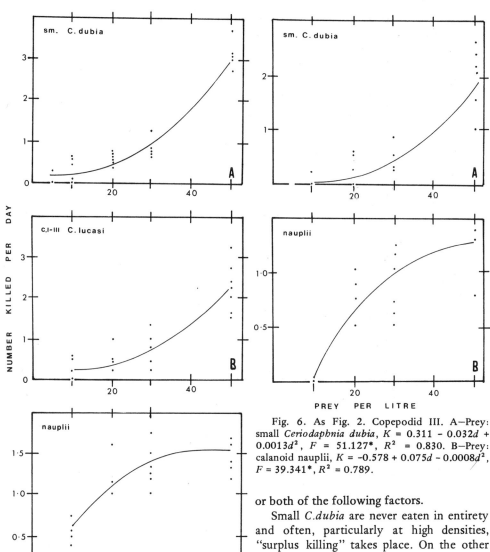

Fig. 5. As Fig. 2. Copepodid IV. A—Prey: small *Ceriodaphnia dubia*, $K = 0.265 - 0.022d + 0.0016d^2$, $F = 250.620^*$, $R^2 = 0.949$. B—Prey: copepodid stages I–III *Calamoecia lucasi*, $K = 0.194 - 0.013d + 0.0006d^2$, $F = 46.549^*$, $R^2 = 0.809$. C—Prey: calanoid nauplii, $K = -0.140 + 0.084d - 0.0010d^2$, $F = 22.040^*$, $R^2 = 0.688$.

Fig. 6. As Fig. 2. Copepodid III. A—Prey: small *Ceriodaphnia dubia*, $K = 0.311 - 0.032d + 0.0013d^2$, $F = 51.127^*$, $R^2 = 0.830$. B—Prey: calanoid nauplii, $K = -0.578 + 0.075d - 0.0008d^2$, $F = 39.341^*$, $R^2 = 0.789$.

preference experiments) to small *C. dubia* and were also considerably smaller. This apparent contradiction could be due to one or both of the following factors.

Small *C. dubia* are never eaten in entirety and often, particularly at high densities, "surplus killing" takes place. On the other hand nauplii are totally consumed, and in all the experiments with them as prey only once was a partly consumed nauplius found. Furthermore, it is probable that there is a greater loss of body parts and fluids when feeding on *C. dubia*, which is usually shredded, than on nauplii which are consumed whole. Brandl and Fernando (1975) found that the assimilation efficiencies of *Mesocyclops edax* were much higher for nauplii and protozoan prey than for larger crustacean prey. This corresponded to their finding that whenever the prey was torn to pieces most of the loss between the killed and assimilated biomass took place before the food entered the mouth of the predator.

The reason why a greater number of *C. dubia* are killed only at high densities is probably related to the type of swimming movement of both prey species and the ability of *M. leuckarti* to capture them. Calanoid nauplii glide in the water column and often appear to be virtually at rest, whereas *C. dubia* moves with a jerky continuous beating of the second antennae. At a low density of *C. dubia* there would be less chance of capture by *M. leuckarti* and once captured little likelihood of *M. leuckarti* being disturbed whilst feeding. At a high density the chances of capturing and killing *C. dubia* are increased but so is the likelihood of being disturbed by another *C. dubia* whilst feeding. Observations showed that *M. leuckarti* would release a captured *C. dubia* if bumped into by another, which it would then attempt to capture. Thus, more would be killed but less of each would be consumed. With nauplii the amount of each consumed would remain constant over the range of densities, since it would be unlikely that any disturbance by nauplii would take place.

Table 5 summarizes the range of prey species able to be killed by each stage and shows the lowest density (per liter) of each prey species at which significant predation occurs.

Effect of high densities—The range of prey densities initially tested included those commonly found in New Zealand lakes (Chapman et al. 1975). In an attempt to determine whether the predation rate of older copepodid stages continued to increase with density, some further experiments were carried out with small *C. dubia* (Fig. 3B).

Holling (1959) has described three types of functional response to increasing prey density. All responses have a leveling off point at a high density where the interplay of the four components of the response of a predator to prey density (rate of successful search, the time over which predators are exposed to prey, the time spent handling prey, and hunger) will be such that the maximum predation rate has been reached.

Table 5. Prey species killed by various stages of *M. leuckarti*. Significant predation occurs at prey densities down to 5/liter unless otherwise indicated. Large *C. dubia* body length, 1.15–1.35 mm; small *C. dubia* body length, 0.55–0.65 mm.

Prey species killed	♀	♂	V	IV	III
Large *C. dubia*	*	*10			
Small *C. dubia*	*	*	*	*10	*20
C. sphaericus	*20				
Adult *C. lucasi*	*	*40	*30		
Stg. I, II, III *C. lucasi*	*	*	*	*20	*40
Stg. II, III *M. leuckarti*	*	*10			
Nauplii	*	*	*	*	*20

*$P = 0.05$.

However, the results I obtained at high prey densities did not indicate such a point. Dodson (1975) also had difficulty finding predation rate saturation densities for the predatory copepods he used. *Heterocope septentrionalis* reached this point at an initial prey density of 455/liter with copepod nauplii but he was unable to find the saturation density of *Acanthocyclops vernalis*.

A feature of predation rates at the high densities tested was the extreme variability in results, e.g. with small *C. dubia* as prey at a density of 80/liter the mean predation rate was $4.92 \pm SD = 1.03$. I decided to attempt to reduce this variability and to see whether particularly high predation rates obtained for some experimental predator groups would be maintained. Table 6 shows the results from an experiment using the same five groups of predators on five successive days.

In all groups there were marked variations in the predation rate from day to day and there were no consistent patterns of change from day to day between the groups. Furthermore, at this high density a far greater number of animals were only partially consumed, whereas this was not common at low densities. Thus, although predation rate varies it is possible that consumption rate could be much less variable. It would seem probable that interference by other *C. dubia* as previously explained could have been the

Table 6. Number of small *C. dubia*, density 80/liter killed by adult female *M. leuckarti* on consecutive days.

Trial No.	Number killed				
	Day 1	Day 2	Day 3	Day 4	Day 5
1	6.4	4.2	6.2	7.7	5.2
2	4.7	5.3	6.2	7.5	4.9
3	4.6	6.5	5.3	10.0	5.8
4	4.0	7.8	6.4	4.1	5.3
5	7.8	3.5	6.2	6.6	5.7
Mean	4.92	5.95	6.02	7.32	5.3
SD	1.03	1.55	0.49	2.44	0.37

Table 7. Mean numbers of small *C. dubia* killed by adult female *M. leuckarti* at various temperatures.

Density of prey/liter	Number killed			
	25°C	20°C	15°C	10°C
50	3.55*	4.7	3.82	3.9
		----*	----	----
40	3.1	3.8	3.4	3.2
30	3.05	3.41	2.98	2.77
20	2.77*	2.63*	1.98	2.44
	----	----	----*	----
10	1.8*	1.6	1.2	1.4
	----	----	----*	----
5	1.16*	0.8	0.53	0.56

*Significantly different at P = 0.01 from groupings indicated by dashes. (----) Groupings from Duncan's multiple range test.

cause of this feature. Less variability and a saturation point may have been found if very small prey (e.g. nauplii) had been used, where interference would not have taken place.

Brandl and Fernando (1974) obtained wide variation in the number of *Ceriodaphnia reticulata* consumed by *A. vernalis* and suggested a need for long term experiments over 3 to 5 days. They used prey densities ranging between 800 and 1,250 per liter (my calculation from their data) where, if it is assumed that *A. vernalis* is similar to *M. leuckarti*, one would have to expect such variability.

As the younger stages of *M. leuckarti* have a type 2 response (Holling 1959) with nauplii as prey it is assumed that the older stages would also behave in this way. Very high prey densities were not tested because they are unrealistic and of little relevance to the environmental situation.

The effect of temperature and density on predation rate—A series of experiments, using adult female *M. leuckarti* and 5 to 50 small *C. dubia* as prey, was carried out at 10°, 15°, 20°, and 25°C to assess the effect of temperature on predation rate at different densities. I preconditioned *M. leuckarti* for at least 5 days at the experimental temperature and density. The results are shown in Fig. 7; they suggest that there was little difference in predation rates at each density at the different temperatures. The results of a statistical analysis (Duncan's multiple range test) are shown in Table 7.

Two trends emerged from the results. Firstly, the predation rates at 15°C and

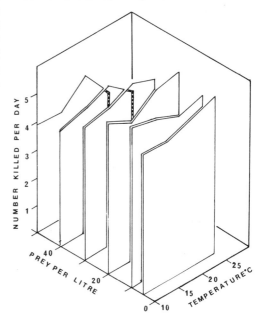

Fig. 7. Effect of temperature and density on predation rate of adult female *M. leuckarti* when offered small *C. dubia* as prey.

10°C did not differ significantly, except at a density of 20/liter where the mean result at 15°C appears to be unusually low. Secondly, the highest predation rates at low densities occurred at 25°C whereas at high densities they occurred at 20°C. This change became evident at a density of 30/liter. At a density of 40/liter the predation rate at 25°C was lower (though not significantly so) than that found at all other temperatures, but it was significantly lower at a density of 50/liter.

Thus it would appear that the predation rates of *M. leuckarti* are not very dependent on temperature, although significant differences do occur at 25°C and 20°C at low and high densities respectively. This apparent anomaly can perhaps be understood if it is considered in relation to the interference hypothesis proposed earlier. At low densities the faster swimming of both predator and prey at 25°C (Strickler 1975a) would presumably result in more predator-prey interactions than at 20°C; and because of low prey numbers interference by prey would be infrequent. With increasing prey density, predator-prey interactions also increase as do interference effects. However, with faster moving prey at 25°C the interference becomes so frequent that *M. leuckarti* would have less time in which to make a successful capture, whereas at 20°C interference would be less frequent and there would be less likelihood of disturbance in the time required to capture and kill a prey item.

It is interesting to note that workers investigating the effect of temperature on cyclopoid predation rates, using only one density in each case, have found both an increase with increasing temperature and an increase up to a maximum with a decrease at a higher temperature. Brandl and Fernando (1975) working with *Mesocyclops edax* found an increase in its predation rate on *Ceriodaphnia quadrangula* with increasing temperatures from 8°C to 28°C. Monakov (1959) also found a dependence of predation rate on temperatures varying from 8°C to 27°C with *A. vernalis*, but maximum consumption was found at 20°C with a decrease at 27°C. Unfortunately, the actual densities used in each trial were not given.

The effect of algal density—As all previous experiments were done using filtered river water, which was regarded as being equivalent to oligotrophic conditions, a series of experiments with stages III–VI *M. leuckarti* using copepodid stage I–III *C. lucasi* as prey were done with water collected from a eutrophic lake (L. Koutu, Cambridge) to determine whether high concentrations of algae affected predation. This water resembled a green soup and had 2.5×10^6 algal colonies per ml (genera listed in Table 8).

The results are shown in Fig. 8 and Table 9. A regression line was calculated for adult females using a stepdown predictive analysis (Wilson 1976). The relationship between predation rate and prey density remained linear in high algal densities. Because of insignificant predation rates by other stages at low densities in high algal densities, insufficient data were obtained to determine if the type of relationship between predation rate and prey density would be the same as that found without algae present.

Table 8. Algal composition of eutrophic lake water.

Algae	% Composition
Microcystis sp.	89
Trachelomonas sp.	6
Staurastrum sp.	
Melosira sp.	
Cryptomonas sp.	5
Closterium sp.	

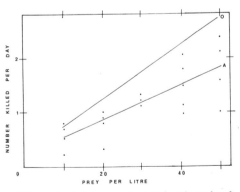

Fig. 8. Predation rate of adult female *M. leuckarti* when offered copepodids I–III *C. lucasi* in algal-rich water (A) and in algal-free water (O). ●—Experimental results with high algal densities. Line for algal-free water was extracted from Fig. 2 E.

Table 9. Number of stage I–III C. lucasi killed by various stages of M. leuckarti in presence (A) and absence (0) of algae.

Predator Conditions	♀		♂		C.V		C. IV		C. III	
	A	O	A	O	A	O	A	O	A	O
50/liter										
1	1.0	3.0	2.0	3.0	1.55	2.1	0.44	1.0	0.33	1.3
2	2.1	2.9	2.7	3.0	0.9	2.0	0.4	0.8	0.22	1.2
3	1.56	2.1	1.81	2.23	0.22	2.37	0.25	1.2	0.5	1.4
4	2.37	2.88	1.16	2.9	1.25	2.25	0.44	1.38	0.3	1.0
5	2.1	3.33	1.9	2.77	1.0	1.63	0.3	1.6	0.2	1.0
6		2.1	1.5	2.6	1.22	1.87	0.5	0.8	0.22	
7		3.0								
Mean	1.82	2.75	1.84	2.75	1.02	2.03	0.38	1.13	0.29	1.18
SD	0.24	0.17	0.21	0.12	0.18	0.11	0.03	0.11	0.04	0.03
30/liter										
1	1.2	1.5	0.55	1.0	0.3	1.5				
2	1.22	1.5	0.77	1.5	0.6	1.77				
3	1.11	1.6	0.63	1.3	0.25	1.1				
4	1.1	1.2	0.33	1.5	0.33	1.22				
5	1.3	1.7	0.4	1.4	0.4	1.22				
6	1.33	1.77	0.5	1.2	0.38	1.33				
Mean	1.21	1.54	0.53	1.3	0.37	1.3				
SD	0.03	0.08	0.06	0.08	0.04	0.09				

With all stages lower numbers were killed in high algal densities. This decrease in the numbers killed showed a progressive increase from mature M. leuckarti to stage III copepodids. At a prey density of 50/liter for adult males and females the number killed per day was about 75%, stage V 50%, stage IV 33%, and stage III 25% of that in the absence of algae. A greater prey density was also necessary for the younger stages before significant predation occurred.

Thus a higher concentration of algae in the environment will reduce the predation rate of M. leuckarti. The effect is more marked in the younger stages than in mature individuals. Presumably the degree to which predation rate decreases would depend on the algal concentration.

Preferences, switching, and training—To establish the manner in which M. leuckarti might affect the structure of its prey populations it is also necessary to determine the sorts of prey that it is likely to take when faced with a choice. Predators might feed on different prey species in simple proportion to the relative abundance of the prey in the environment, but could show more complicated behavior by "switching," concentrating their attacks on the most abundant species available to them. Murdoch (1969) has suggested that the most likely mechanism for switching is that the predator becomes "trained" to the most abundant species which is then disproportionately attacked. I did preference, switching, and training experiments to explore these effects.

When offered a choice of small C. dubia and of stages II–III copepodites, M. leuckarti females preferred the former ($c = 2.53$) whilst males showed no significant overall difference between the species, with the results being very variable. Observations showed that the method of attack used by males, that of pursuing their prey, was more successful for the capture of fast-swimming cyclopoid copepodids than was the female's pause and pounce attack. When pursued by males, C. dubia often gave an avoidance reaction in which, by closing the valves tightly, it became invulnerable to capture. Thus both species present problems for the male to capture (one because of its fast swimming and maneuverability, the other because of its avoidance behavior) and therefore the number of each killed would be variable. The females' method of attack is well adapted for the capture of *C. dubia*.

The problems encountered by males in capturing *C. dubia* are further shown by the next preference experiment. There was no significant difference between the predation rates on small *C. dubia* and younger stages of *C. lucasi* ($c = 2.92$). As the swimming movements of *C. lucasi* would not make it as difficult to capture as young *M. leuckarti*, the preference is therefore to be expected.

All stages preferred calanoid nauplii to *C. lucasi* copepodids with the preference values becoming larger as the predator got smaller (female $c = 2.02$, C.III $c = \infty$.

It is assumed that if "a" is preferred to "b" and "b" is preferred to "c" then "a" will also be preferred to "c." Thus, of the prey items offered, the smallest calanoid nauplii are considered to be the most preferred by *M. leuckarti*.

Preferences: Experiments were carried out to determine if, when two prey species were equally abundant, *M. leuckarti* would attack and kill more of one than the other, that is, if it showed a "preference." It should be emphasized, however, that this term is used in a very general sense since preferences are influenced by a complex of factors, including the ability of *M. leuckarti* to physically capture and eat a particular animal, as well as the size, swimming movements, and general behavior of the prey.

Equal numbers of two prey species were supplied to groups of *M. leuckarti* following the same general procedure used in the previous experiments. Replacement of both prey species was carried out relative to the numbers of each being killed, which were determined in an initial trial experiment. The results are shown in Table 10. The preference value c was calculated (Murdoch 1969) as:

$$c = \frac{\text{No. prey species ``a'' in diet}}{\text{No. prey species ``b'' in diet}}$$

when both were equally abundant in the environment.

A frequently noticed feature of cyclopoid predation is a preference for smaller prey species (Confer 1971; Brandl and Fernando 1974; Dodson 1975; Hall et al. 1976). This was clearly shown by *M. leuckarti*: the younger stages were unable to kill large *Ceriodaphnia*. Preference experiments with small and large *C. dubia* gave values of $c = 2.34$ with adult females and $c = \infty$ with males, in favor of the smaller prey. Younger stages also had high preference values for the smaller prey of different species.

A further preference experiment was done with adult females when calanoid nauplii and cyclopoid nauplii were the prey species offered. A preference value of $c = 3.0$ was found in favor of calanoid nauplii. Calanoid and cyclopoid nauplii are very similar in size, shape, and presumably palatability, thus the obvious preference for calanoid nauplii is thought to be a result of differences in their swimming movements, since calanoid nauplii glide through the water with occasional jerky jumps, but cyclopoid nauplii have a quick darting motion at all times. As the adult female is the most able predator and should have the least difficulty in attacking cyclopoid nauplii it may be assumed that a similar or greater preference would be shown by other stages.

Switching: Preference values were established by measuring the ratio of the numbers of two prey species attacked by *M. leuckarti* when given equal numbers of each prey. "Switching," when the most abundant prey is killed to a greater extent, has been shown to generally occur when the preference at equal prey densities is weak ($c \approx 1$) and when there is a large amount of heterogeneity in results (Murdoch 1969; Murdoch and Marks 1973; Murdoch et al. 1975). A weak preference is defined as being close to 1.0 and always < 3. A strong preference is > 3.

An experiment was done to test for switching behavior of adult female *M. leuckarti* using small and large *C. dubia* where the preference value for small *C. dubia* was $c = 2.18$. Prey were offered in the following proportions: 1:9, 1:4, 2:3, 1:1, 3:2, 4:1, 9:1, with a total prey density of 50/liter. The expected proportions of the two prey species to be consumed were

Table 10. Results of preference (c) experiments. Means, standard errors (SE), and the P and t values for statistically significant differences (Student's t-test).

	No. prey offered to M. leuckarti*										
	a. 25 small C. dubia b. 25 large C. dubia				a. 20 small C. dubia b. 20 stg. II–III M. leuckarti				a. 20 stg. I–III C. lucasi b. 20 small C. dubia			
	♀		♂		♀		♂		♀		♂	
No. killed In trial	a	b	a	b	a	b	a	b	a	b	a	b
1	2.0	0.7	1.25	0.0	2.1	1.0	0.4	1.2	2.55	1.77	1.6	1.0
2	2.0	0.5	2.0	0.0	1.8	0.9	0.55	0.44	1.66	1.9	1.3	0.3
3	2.1	0.53	1.3	0.0	2.6	0.8	1.0	0.22	1.9	1.9	1.77	0.5
4	1.9	1.2	1.86	0.0	1.66	0.22	0.62	1.75	1.7	1.2	1.4	0.0
5	2.1	1.0	1.8	0.0	1.6	0.5	1.0	0.2	1.85	1.13	1.88	0.6
6	2.4	0.9	1.6	0.0	1.7	1.2	0.62	0.25	1.8	1.44	2.0	0.88
7	1.6	1.4			2.33	0.9	0.66	0.55			2.9	1.1
8							1.3	0.86				
Mean	2.01	0.86	1.63	0.0	1.97	0.78	0.76	0.68	1.9	1.5	1.84	0.63
SE	0.09	0.14	0.12	0.0	0.14	0.12	0.10	0.19	0.13	0.14	0.20	0.15
t	7.98		13.58		6.82		0.37		2.09		4.84	
P	<0.01		<0.01		<0.01		N.S		<0.05		<0.01	
c	2.34		∞		2.53		1.12		1.12		2.92	

	a. 15 calanoid nauplii b. 15 stg. I–III C. lucasi										a. 15 calanoid nauplii b. 15 cyclopoid nauplii	
	♀		♂		C V		C IV		C III		♀	
	a	b	a	b	a	b	a	b	a	b	a	b
1	2.2	1.4	1.5	0.3	1.6	0.0	1.22	0.22	1.0	0.0	2.0	0.88
2	1.8	0.8	1.14	0.14	0.8	0.2	1.0	0.27	0.7	0.0	1.7	0.3
3	2.0	1.0	1.0	0.33	1.1	0.2	1.0	0.0	0.75	0.0	2.33	0.66
4	2.1	0.88	1.7	0.4	1.5	0.1	1.1	0.1	0.88	0.0	2.33	0.5
5	2.1	1.2	1.33	0.66	1.33	0.11	1.1	0.22	0.8	0.0	2.5	1.25
6	2.3	0.9	1.5	0.4	1.4	0.2					2.25	0.75
Mean	2.08	1.03	1.36	0.37	1.28	0.14	1.08	0.16	0.82	0.0	2.1	0.7
SE	0.07	0.09	0.10	0.06	0.11	0.03	0.04	0.05	0.05	0.0	0.11	0.13
t	9.21		8.49		10.0		14.37		16.4		8.22	
P	<0.01		<0.01		<0.01		<0.01		<0.01		<0.01	
c	2.02		3.68		9.14		6.75		∞		3.0	

* Indicated by stage.

calculated from the equation given by Murdoch (1969):

$y = 100\ cx/c100 - x + cx$,

y = percentage prey in diet,
x = percentage prey in environment,
c = preference value (as defined earlier).

Fig. 9 summarizes the results obtained with the different prey combinations and shows that switching did not occur. The proportions of small and large C. dubia in the diet were similar to the expected proportions. Adult female M. leuckarti kills its prey as a simple function of its relative abundance in the environment.

Training: The explanation proposed by Murdoch (1969) for switching is that the predator becomes trained to one species which is then disproportionately attacked. Thus, although switching did not occur it was of interest to further investigate whether M. leuckarti can be "trained," as it is still possible that with a lower preference value between two species switching could occur. Furthermore, if a training situation existed in the natural environment, that is an abundance of one species, this might affect the

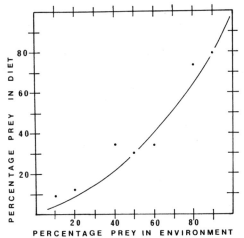

Fig. 9. Percentage of large *C. dubia* in diet (●) as a function of relative abundance in environment of adult female *M. leuckarti*. Expected proportions of large *C. dubia* in diet with no switching shown by solid line which was calculated with $c = 2.34$ (small *C. dubia* was alternate prey).

Table 11. Numbers of stage I–III *C. lucasi* (density 20/liter) killed by adult female *M. leuckarti* after training.

Trial No.	No. killed	
	Trained on small Ceriodaphnia	Control
1	2.66	1.4
2	2.4	1.9
3	2.5	1.66
4	2.22	1.33
5	1.8	1.0
6	1.8	1.7
Mean	2.23	1.33
SD	0.36	0.45
$t = 14.40, P < 0.01$		

predation rate on other species which later became abundant.

Groups of adult female *M. leuckarti* were given small *C. dubia* as a prey species at a density of 20/liter for 5 days. On day 6 they were given stage I–III *C. lucasi* at the same density for 5 days. Control groups were kept for 5 days with the same density of *C. lucasi* (Table 11).

If training on one prey species occurred, a decrease in the predation rate of another species would be expected. The number of *C. lucasi* killed by *M. leuckarti* after training on small *C. dubia* was significantly greater on the first day of the experiment than the number killed in controls, indicating that training had not taken place. On the following day the number of *C. lucasi* killed was comparable with that when normal conditioning had taken place.

The greater number of *C. lucasi* killed after training may have been a result of *M. leuckarti* not being as sated by small *C. dubia*. Such a dissimilarity in satiation could be a function of one or more of the following: less food being available as a result of less complete consumption; less food being available because of differences in prey size; *C. dubia* not being as adequate a food as *C. lucasi*.

Thus adult female *M. leuckarti* did not show preferences in prey species but could not be trained and did not show switching behavior. *Mesocyclops leuckarti* takes prey simply as a function of its relative abundance in the environment. As all stages showed similar preferences, it is assumed that similar responses to switching and training experiments would also be found.

Discussion

Quantitative aspects of cyclopoid predation have been investigated frequently (Anderson 1970; Brandl and Fernando 1974, 1975; Confer 1971; Dodson 1975; Monakov 1959; McQueen 1969; Stepanova 1970). Unfortunately it seems that many of these workers have neglected to consider factors resulting from experimental design and technique which might affect predation rates.

Firstly, it appears that none of the studies cited above assessed the effect of container size on predation rate and most used vessels that contained < 200 ml of water. In this study the predation rate of *M. leuckarti* was reduced when the volume of the experimental container was < 750 ml.

Secondly, few attempts have been made to control the pre-experimental feeding history and often unrealistic treatments have been imposed. For example, Monakov (1959) starved his animals for 48 hours preceding the experiment and Anderson (1970) starved his for 12 hours. However, Brandl and Fernando (1974) did note that pre-experimental feeding history affected predation rates of *A. viridis* and suggested that cyclopoid copepods should be fed with exactly the same kind of prey before any quantitative measurement. I found that if *M. leuckarti* had the same pre-experimental as experimental conditions, I obtained more reliable results. When *M. leuckarti* was starved, the predation rate was higher.

Anraku (1964) found that the physiological condition of two marine predatory copepods, *Tortanus discaudatus* and *Centropages typicus*, affected predation rate. More animals were killed in November than January, when environmental temperatures were colder (but not experimental temperatures). Although I did not investigate this effect as such with *M. leuckarti*, I found that results from experiments conducted 3 months later than the initial experiment did not differ.

The discontinuous feeding of cyclopoid copepods has created difficulties in experimentation and appears to have caused some workers to use unnaturally high prey densities, while others have increased the duration of the experiment. Increasing the duration of an experiment with prey replacement did not affect the predation rate of *M. leuckarti*. However, as most of the cited workers failed to replace prey which were killed, it seems likely that fewer prey would have been killed per day later in an experiment than initially. Increasing the duration of the experiment without replacing prey and using unnaturally high densities imposes limitations on the extrapolation of results to environmental situations. It would seem better to use larger numbers of predators, as there does not appear to be any interaction between them that affects predation rate, to obtain reliable results.

The predation rates of all stages of *M. leuckarti* were density-dependent. This has also been found for other cyclopoid copepods despite the above-mentioned errors in experimentation. Confer (1971) found that adult female *Mesocyclops edax* killed 0.07 to 0.6 *Diaptomus floridanus* copepodids per day at densities from 10–30 per liter. Adult female *M. leuckarti* killed from 0.66 to 1.55 stage I–III *C. lucasi* copepodids over the same density range. The predation rate of *M. edax* may be less than it would have been had experiments been conducted in larger containers, had the pre-experimental history been controlled, and had killed prey been replaced.

The importance of using naturally occurring prey species is apparent from my results. Most workers have found that cyclopoid copepods prefer small prey species to large but this is not the only factor that determines the suitability of prey. *Mesocyclops leuckarti* could not kill *Bosmina meridionalis* even though very small animals were offered. Smyly (1970) also observed that the size of the prey is less important than its nature. The results of my study suggest that there are three factors which determine the suitability of prey.

1. The nature of the prey—the body of the prey must be soft enough for *M. leuckarti* to grip whilst eating. For example *C. sphaericus* was difficult for *M. leuckarti* to capture because its hard exoskeleton and distinctly rounded valves prevented grasping with the first maxillae. Only adult female *M. leuckarti* which have the largest, and presumable strongest, mouthparts were able to capture very small *C. sphaericus*. *Ceriodaphnia dubia*, which unlike *C. sphaericus* has valves that gape widely, could be caught by stages III–IV but only if attacked between the valves. Furthermore, the exoskeleton of *C. dubia* was never consumed, whereas the softer exoskeleton of species such as *C. lucasi* was often consumed.

2. The swimming motion of the prey species also affects the ease of capture. This is perhaps best shown by the preference of adult female *M. leuckarti* for calanoid nauplii over cyclopoid nauplii. The two types of nauplii are of similar size and shape and apparently only vary in their swimming motion. Strickler and Twombly (1975) suggested that the capture of prey is a dynamic interaction and have shown that the speed of the prey is important in capture. However, velocity is not the only important parameter of motion. Certain methods of attack are more successful for capturing a species that swims in a particular fashion.

For example female *M. leuckarti* showed a preference for *C. dubia* over cyclopoid copepodids whereas male *M. leuckarti* did not have a preference. Observations of methods of attack provided the explanation for this difference. Predation by female *M. leuckarti* is of a pause and pounce type, while males usually chase their prey. Thus, fast-moving young cyclopoid copepodids would be more difficult for females to capture than for males.

3. As stated previously, small prey are preferred to large. *Mesocyclops leuckarti* killed many more small *C. dubia* than large when they were equally abundant. Younger stages of *M. leuckarti* being smaller were not able to capture large prey species at all. Thus the size of prey species available decreases with a decrease in size of predator. Presumably there is also a lower limit to prey size but this was not tested in these experiments.

The predation rate of *M. leuckarti* was only slightly affected by temperature. However, development rates of zooplankton are strongly temperature-dependent; the rate of increase of a population is much slower at colder temperatures. Therefore, it would seem probable that a prey population might be exposed to greater predation pressure at lower temperatures than at higher temperatures, especially as *M. leuckarti* prefers small immature prey species which would be present for a longer time. This is a good strategy for *M. leuckarti* since most herbivorous zooplankton have much higher calorific values in winter due to lipid storage, e.g. *C. lucasi* (Green 1976).

The predation rate of *M. leuckarti* also depends on the abundance of phytoplankton. *M. leuckarti* killed fewer prey in high algal densities, presumably because the algae provided an additional food source. Generally if there is an increase in phytoplankton abundance, herbivorous zooplankters also increase, decreasing the percentage mortality upon a prey population by predation.

Mesocyclops leuckarti could not be trained and did not show switching behavior, rather it killed suitable prey in simple proportion to the relative abundance of the prey in the environment. Thus it would appear that *M. leuckarti* is opportunistic in its feeding habit and would exploit suitable prey populations as they became available. Also since it is an omnivore it should be able to maintain populations in the absence of animal prey and would thus be able to inflict considerable mortality on a developing prey population.

The results have considerable relevance to the size-efficiency hypothesis proposed by Brooks and Dodson (1965) to account for the frequently observed inverse relationship between the abundance of small- and of large-bodied herbivorous zooplankton in freshwater lakes. They stated that, as large herbivorous zooplankters are more efficient at collecting particulate matter they will competitively eliminate small planktonic herbivores in the absence of predation. But when predation by vertebrates is intense, size-dependent predation will eliminate large forms allowing small zooplankters (rotifers and small cladocerans) to become dominant. More recently, it has been suggested (Hall et al. 1976) that vertebrate predation restricts the maximum adult body size of zooplankton populations that can increase and invertebrate predation may restrict the minimum size that can increase and both may aid the decline of zooplankton populations limited by food. Kerfoot (1977) has since suggested that in the absence of fishes the evolutionary consequences of invertebrate predation results in selection for larger prey organisms. Although *M. leuckarti* prefers small zooplankters, many of these are immature stages of larger-bodied zooplankters. These include large cladocerans and calanoid copepods while many of the small cladocerans, e.g. *B. meridionalis* and *C. sphaericus*, cannot be captured with ease. This is obviously a good strategy for *M. leuckarti*: to concentrate on the smaller forms not preyed upon by fish and avoid competition. (Unfortunately there is no qualitative and quantitative information on the suitability of rotifers as prey, but presumably this would depend upon their shape and structure, swimming behavior, and size).

Thus the effect of predation by *M. leuckarti* would not be the same as envisaged in the size-efficiency hypothesis, that is a shift in the body size of the planktonic community, or as Kerfoot suggested. Instead, it would affect populations directly by reducing their numbers. Those populations affected will depend upon the composition of the zooplankton community.

The impact of predation on a species in a community will depend upon the density, age structure, and rate of development of the *M. leuckarti* population, the diversity of the zooplankton community, the density and size/age structure of a prey population, and the algal concentration. Thus in some communities, where there is a large constant *M. leuckarti* population, predation will affect other zooplankton species. Where the *M. leuckarti* population is variable, predation will have an effect only for short periods and is unlikely to be sustained. In others, where numbers of both *M. leuckarti* and prey populations are low, predation would not contribute significantly to mortality.

References

ANDERSON, R. S. 1970. Predator-prey relationships and predation rates for crustacean zooplankters from some lakes in west Canada. Can. J. Zool. 48: 1229-1240.
ANRAKU, M. 1963. Feeding habits of planktonic copepods. Inform. Bull. Plankton Soc. Jpn. 9: 10-34.
———. 1964. Some technical problems encountered in quantitative studies of grazing and predation by marine planktonic copepods. J. Oceanogr. Soc. Jpn. 20(5): 19-29.
BRANDL, Z., and C. H. FERNANDO. 1974. Feeding of the copepod *Acanthocyclops vernalis* on the cladoceran *Ceriodaphnia reticulata* under laboratory conditions. Can. J. Zool. 52: 99-105.
———, and ———. 1975. Food consumption and utilization in two freshwater cyclopoid copepods (*Mesocyclops edax* and *Cyclops vicinus*). Int. Rev. Gesamten Hydrobiol. 60: 471-494.
BROOKS, J. L., and S. I. DODSON. 1965. Predation, body size and composition of plankton. Science 15: 28-35.
CHAPMAN, M. A., GREEN, J. D., and V. H. JOLLY. 1975. Zooplankton, p. 209-230. In V. H. Jolly and J. M. Brown [eds.], New Zealand lakes.
CONFER, J. L. 1971. Intrazooplankton predation by *Mesocyclops edax* at natural prey densities. Limnol. Oceanogr. 16: 663-665.
DODSON, S. I. 1975. Predation rates of zooplankton in arctic ponds. Limnol. Oceanogr. 20: 426-433.
DZYUBAN, N. A. 1937. On the nutrition of some cyclopoidea (Crustaceae). Dokl. Akad. Nauk SSSR 17(6): 319-322.
FRYER, G. 1957a. The food of some freshwater cyclopoid copepods and its ecological significance. J. Anim. Ecol. 26: 263-286.
———. 1957b. The feeding mechanism of some freshwater cyclopoid copepods. Proc. Zool. Soc. Lond. 129: 1-25.
FURNASS, T. I., and W. C. FINDLEY, 1975. An improved sorting device for zooplankton. Limnol. Oceanogr. 20: 295-297.
GREEN, J. D. 1976. Population dynamics and production of the calanoid copepod *Calamoecia lucasi* in a northern New Zealand lake. Arch. Hydrobiol. 50: 313-400.
HALL, D. J., S. T. THRELKELD, C. W. BURNS, P. H. CROWLEY. 1976. The size-efficiency hypothesis and the size-structure of zooplankton communities. Annu. Rev. Ecol. Syst. 7: 177-208.
HOLLING, C. S. 1959. The components of predation as revealed by a study of small mammal predation on the European pine sawfly. Can. Entomol. 91: 293-320.
KERFOOT, W. C. 1977. Implications of copepod predation. Limnol. Oceanogr. 22: 316-325.
MCQUEEN, D. J. 1969. Reduction of zooplankton standing stocks by predaceous *Cyclops bicuspidatus* Thomasii in Marion Lake, British Columbia. J. Fish. Res. Bd. Can. 27: 13-20.
MONAKOV, A. V. 1959. The predatory feeding of *Acanthocyclops viridis* Jur. (Copepoda, Cyclopoidea). Freshwater Biol. Assoc. Transl. (N.S.) 12.
MURDOCH, W. W. 1969. Switching in general predators: Experiments on predator specificity and stability of prey populations. Ecol. Monogr. 39: 335-354.
———, S. AVERY, and M. E. B. SMYTH. 1975. Switching in predatory fish. Ecology 56: 1094-1105.
———, and J. R. MARKS. 1973. Predation by coccinellid beetles, experiments on switching. Ecology 54: 160-167.
RYLOV, M. 1948. Cyclopoidea of fresh waters. Fauna SSSR 3(3): 309p.
SMYLY, W. J. P. 1970. Observations on rate of development, longevity and fecundity of *Acanthocyclops viridis* Jurine (Copepoda Cyclopoida) in relation to type of prey. Crustaceana Leiden 18: 21-36.
STEPANOVA, L. A. 1970. Rations of *Mesocyclops leuckarti* Claus and *Leptodora kindtii* Fouke

populations in Lake Ilmen. Hydrobiol. J. **8**: 70-72.

STRICKLER, J. R. 1975a. Swimming of planktonic *Cyclops* species (Copepoda, Crustacea): Pattern, movements and their control, p. 599-613. *In* T. Y. - T. W. et al. (eds.), Swimming and flying in nature, v. 2. Plenum.

——. 1975b. Intra and interspecific information flow among planktonic copepods: Receptors. Int. Ver. Theor. Angew. Limnol. Verh. **19**: 2950-2958.

——, and A. K. BAL. 1973. Setae of the first antennae of the copepod *Cyclops scutifer* Sars: Their structure and importance. Proc. Natl. Acad. Sci. **70**: 2656-2659.

——, and S. TWOMBLY. 1975. Reynolds number, diapause and predatory copepods. Int. Ver. Theor. Angew. Limnol. Verh. **19**: 2942-2949.

WILSON, J. B. 1976. TEDDYBEAR statistical program. Tech. Rep. T5. Bot. Dep., Univ. Otago. 85 p.

47. Selection of Prey by *Chaoborus* Larvae: A Review and New Evidence for Behavioral Flexibility

Robert A. Pastorok

Abstract

A review of the literature on *Chaoborus* feeding indicates that the composition of the larval diet, its selective effect on the zooplankton community, and the diel cycle of foraging activity vary between lakes as well as seasonally within a single lake. An examination of the crop contents of fourth instar *C. trivittatus* in Lake McDonald (Washington State), however, reveals a pattern of size selection that is relatively stable over time. Electivity values are always positive for medium-sized *Daphnia* because these prey encounter *Chaoborus* more frequently than do smaller prey, yet they are more vulnerable to attack and capture than are larger prey. Thus, this aspect of the diet can be explained entirely on the basis of differences in prey availability that are temporally constant.

Although *Chaoborus* is usually regarded as an opportunistic predator, the results of this study show that preferential feeding behavior may outweigh the effects of differences in the vulnerability of prey. Prestarved larvae prefer *Diaptomus* over *Daphnia* at high food levels but are nondiscriminatory when food is scarce. Satiated larvae prefer copepods regardless of food concentration. Whereas the functional response of *Chaoborus* to the density of a single species of prey resembles Holling's type-2 curve, the presence of an alternative food results in an atypical curve for the inferior prey. In this case, the rate of feeding increases with prey density at first, but then reaches a peak and declines with further elevations of food abundance. Thus, spatial and temporal variation in the diet of *Chaoborus* may be the result of behavioral plasticity exhibited by the predators, differences in the availability of prey, or a combination of both factors.

The predatory larvae of *Chaoborus* are a significant component of the limnetic food web in many lakes. Parma (1971) has reviewed the general dietary habits and predators of *Chaoborus*. Recently, predatory zooplankters have been implicated as powerful agents in the selection of morphological, behavioral, and reproductive characteristics of prey species (Dodson 1974a, b; Kerfoot 1975, 1977; O'Brien and Vinyard 1978). My purpose here is to summarize the information available on the feeding behavior of *Chaoborus*, giving special attention to selective predation, and to present new evidence for behavioral adjustment of dietary habits to ambient resource conditions.

I thank W. T. Edmondson for continuous support and advice during the development of this paper. J. R. Strickler aided in filming the attack behavior of *Chaoborus*. Discussions with W. C. Kerfoot and T. M. Zaret were instrumental in forming my thoughts on zooplankton ecology.

Materials and methods

The experimental organisms were *Chaoborus trivittatus* (fourth instar), *Daphnia*

Supported by ERDA contract AT (45-1)-2225-T23 to W. T. Edmondson.

pulicaria, and *Diaptomus franciscanus* collected fresh from Lake McDonald (Washington). Details of methods outlined here are given by Pastorok (1978).

For analysis of larval diet in the lake, a coarse-mesh (1,090 μm) net was used to collect only the largest *Chaoborus* (mostly instar IV and some instar III). The sample was narcotized immediately and preserved in Formalin later, following the method of Swift and Fedorenko (1973). These techniques minimized the problems of net feeding and loss of crop contents during preservation.

Fourth instar *C. trivittatus* were taken from narcotized samples, and their crop contents were carefully removed and spread on a slide for examination under 120× power. The abundance of various prey items was determined by counting whole animals (rotifers, *Chaoborus* larvae), pairs of mandibles (copepods), or postabdominal claws (*Daphnia*). Each postabdominal claw of *Daphnia* was measured to the nearest 0.004 mm. Total body length of the prey was estimated from claw size using a regression equation derived from measurements of intact organisms ($r > 0.95$).

The relative abundance of major prey in the lake was determined by counting a subsample of at least two Clarke-Bumpus samples (Nos. 10 and 20 nets). The size distribution of *Daphnia* in the lake was determined by measuring the total length of every individual in the subsample to the nearest 0.03 mm.

The degree of selection for a given prey was measured by an electivity coefficient recommended by Jacobs (1974):

$$D = \frac{r - p}{r + p - 2rp}$$

where r is the proportion of the given prey in the diet, and p is the proportion of that prey in the environment.

The strike efficiency on various sizes of *Daphnia* was determined by offering several *Daphnia* of a given size to individual *Chaoborus* that had been starved for 1–2 days, and scoring number of attacks and number of successful ingestions. All observations were made at room temperature in a well lighted area. The maximum attack range of *Chaoborus* was noted during these experiments and verified by high speed (100 fr/s) photography.

To test for selection of prey species, I placed two or three *Chaoborus* larvae in a gallon jar containing 2–3 liters of lake water and a mixture of *Daphnia* and *Diaptomus* (1:1) at a known density. The total length of prey was between 1.4 mm and 1.6 mm, the size range most acceptable to the predators. Since mean lengths of the two species were equal, strike efficiencies were also the same (Swift and Fedorenko 1975). The jars were placed in a darkened environmental chamber at 15°C. Larvae were allowed to feed for an alloted time, 10–75 h, depending on the density being tested; then, the remaining prey were removed and counted. Because *Chaoborus* ate only 15–30% of the available prey, food was not replenished during these experiments.

The effect of larval hunger state on the selection of prey was tested by examining three treatment groups: "Satiated"—larvae that were offered saturation densities of prey (either *Diaptomus* or *Daphnia*) for 3 days before the experiment; "lake"—larvae that were collected from the lake just before sunset and transferred to the experimental jars within 5 h (<10% of these individuals had food visible in their crops); "starved"—larvae that were held without food for 3 days before the experiment.

Sensory abilities, search behavior, and prey capture

The larva of *Chaoborus*, being specialized for existence as a planktonic predator, has a complex morphology and behavior, atypical among dipteran larvae as a whole (*see* Leydig 1851; Akehurst 1922; Schremmer 1950; Duhr 1955*a, b*). Each larva has an extensive array of mechanoreceptors to monitor water displacements and vibrations; simple and plumose setae adorn the entire surface of the body while chordotonal organs lie just inside the integument. In adult insects, chordotonal organs are known to be receptors of vibrations through the substrate or through the air (Autrum 1964). Although *Chaoborus* does have functional compound eyes, most workers agree that prey is located primarily by mechanoreception of hydrodynamic signals generated by the prey's swimming and feeding movements (Harper 1907; Duhr 1955b; Horridge 1966). For example, many species of *Chaoborus* feed at night when eyes would be useless. The exact role of vision in prey capture by species that feed in the daytime is unknown, but based on the following evidence, I assume it is secondary to the tactile sense even in day-feeding animals:

1) *Chaoborus* feeds at the same rate in the dark as in the light (Duhr 1955b);
2) when a fine glass probe vibrating with a small amplitude is brought near to a *Chaoborus*, the larva responds with attack and grasping motions over distances up to 5 mm (Harper 1907; Horridge 1966; pers. obs.);
3) when a prey item, stunned by an unsuccessful attack, falls to the bottom of an experimental vessel and remains motionless, it is not pursued by the nearby larva; as soon as the prey resumes locomotion the larva attacks;
4) a moving prey can be perfectly visible, yet if it broadcasts no mechanical stimuli (e.g. if the prey is confined within a glass tube), it will not be attacked by a nearby larva.

With respect to their tactile habits, *Chaoborus* larvae resemble the other carnivorous invertebrates found in the plankton of lakes. The mechanoreceptors and search behavior of the predatory copepods *Cyclops* and *Epischura* are well known also (Fryer 1957a; Strickler and Bal 1973; Strickler 1975; Strickler and Twombly 1975). Because of the feeding mechanisms of these invertebrates, Kerfoot (1975, 1977) and Zaret (1975, 1978) refer to them collectively as "blind" or primarily nonvisual, grasping predators.

Some zooplankters find food or mates by "tracking" chemical trails left by the prey or members of the opposite sex as they move through the water (Kittredge et al. 1974; Hamner and Hamner 1977). In *Chaoborus*, distance chemoreception appears to be limited, prey are never tracked, and information about the taste of food items is conveyed by contact chemoreceptors located on the labrum (Duhr 1955a; also observations 2 and 3 above).

Harper (1907) and Duhr (1955b) described the general aspects of searching and prey capture by *Chaoborus*. While hunting prey, *Chaoborus* adopts a characteristic "ambush strategy." With the aid of buoyancy organs in the fore and aft portions of the body, the larva is able to maintain its position or drift slowly through the water column without causing turbulence in the surrounding medium (cf. Bardenfleth and Ege 1916). The larva initiates an attack only when the prey swims within immediate range, usually inside a radius of about a third the larva's body length (Fig. 1). The attack itself is stereotypic and rapid, being completed within a few hundreths of a second (pers. obs. from high speed films—100 fr/s). In contrast to other motor behaviors, the posterior end of the body remains stationary during an attack motion; only the anterior portion (head plus thorax) moves toward the prey as the larva bends in the middle. If a prey approaches from the rear, *Chaoborus* will sometimes rotate up to 180° before a strike is made.

During an attack, the prehensile antennae, which articulate at the tip of the larval head capsule, reach out and, with the aid of long distal setae, sweep the prey toward the mouth. The mandibles, characterized by long pointed teeth, swing laterally, and a mandibular fan, composed of a dozen or more long curved spines, opens on each side in front of the teeth. The antennae, mandibles, and mandibular fans form a trap basket (Fangkorb: Schrem-

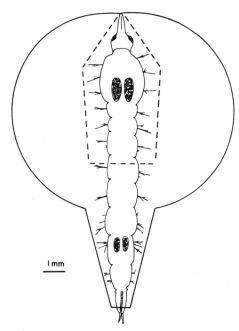

Fig. 1. Attack range of a fourth instar *Chaoborus trivittatus*. Solid line—maximum; dashed line—minimum (cf. Pastorok 1978).

mer 1950) in which the prey is held. Food is stuffed into the mouth by the combined action of the labrum and alternating movements of the mandibles. Prey are ingested whole, although large items may be crushed before being swallowed. Exceptionally large prey, such as mosquito larvae or other *Chaoborus*, may be only partly swallowed so that some portion of the food hangs out of the mouth (Meinert 1886 [cited in Berg 1937]). A detailed functional anatomy of the larval mouthparts is given by Schremmer (1950), Cook (1956), and Pucat (1965).

After prey are swallowed, they enter a muscular pharynx, often referred to as the crop, where mastication and preliminary digestion takes place (Montshadsky 1945; Gersch 1952; Duhr 1955a). Only liquid portions of the food accompanied by digestive enzymes pass into the midgut. The undigested remains of prey, primarily the sclerotin and chitin of the cuticle, are expelled through the mouth; residence time in the crop varies from 3-12 hours depending on size of the ration, species of prey, and water temperature (Montshadsky 1945; Gersch 1952; Goldspink and Scott 1971; Fedorenko 1973; pers. obs.). Additional prey may be eaten and digested before indigestible parts are expelled. After a large meal, a digestive pause may take place before searching is resumed. Otherwise, digestion and other feeding activities occur simultaneously.

A diel cycle in feeding activity occurs in many *Chaoborus* populations, with a greater number of larvae obtaining food sometime during the night (Roth 1971; Fedorenko 1973, 1975a; Green et al. 1973; Swift 1976). This pattern is undoubtedly related to the diel vertical migration of the larvae; migration behavior is reviewed by Goldspink and Scott (1971), Parma (1971), and Swift (1976). As Berg (1937) and Roth (1971) suggested, daytime feeding is unlikely in species that migrate to deep waters where prey are scarce. Berg (1937), however, could find no definite feeding rhythm in the population at Esrom Lake, probably because of difficulties in distinguishing "fresh" food from prey that were captured hours before (cf. Fedorenko 1973, 1975a; Swift 1976). After counting the larvae with "fresh" food throughout a daily cycle, Fedorenko (1973, 1975a) concluded that peaks in feeding activity at dawn and dusk on *Diaptomus kenai* and at midnight on *Bosmina longirostris* were the result of a greater likelihood of predator-prey encounters associated with greater overlap in their spatial distributions brought about by nocturnal migration. It is unlikely that the feeding rhythm itself is governed by endogenous factors since the larvae will feed at all times of the day and night if prey are available (Swift 1976).

In contrast, Goldspink and Scott (1971) and Sikorowa (1973) found that the proportion of *Chaoborus* with food in the crop remained relatively constant throughout the day and night. Fedorenko (1973) suggested that these larvae might have been planktonic near the mud surface during the daytime in lakes where food was equally abundant at all depths. In some lakes, significant feeding may occur during the benthic phase of the migration cycle to mask the rhythm of planktonic feeding (see below).

General dietary habits

Although the diet of *Chaoborus* encompasses potentially all planktonic animals, the crustaceans and rotifers form the bulk of the food (Sikorowa 1968, 1973; Parma 1971 and references therein; Fedorenko 1973, 1975a, b; Lewis 1977). In some situations, larvae may feed during the benthic phase of the migration cycle, consuming chironomids, oligochaetes, harpacticoid copepods, and other organisms found in or upon the surface of the mud (Sachse and Wohlgemuth 1916; Alverdes 1926; Tubb and Dorris 1965). Jonasson (1955) attributed the disappearance of oligochaetes in stored samples of benthic mud to predation by *Chaoborus*. Parma (1971) and Swuste et al. (1973) directly observed predation on *Tubifex tubifex* and *Limnodrilus hoffmeisteri* by *C. flavicans* larvae when the animals were buried in a transparent agar. In another series of experiments using glass tubes filled with water overlying a muddy substrate, these investigators discovered that predation on zooplankton (*Daphnia*) by *Chaoborus* was diminished when benthic prey were available in large numbers. In a particular lake, the importance of benthic feeding will depend on the density and mobility of both predator and prey (Parma 1971). Parma noted that the horizontal movements of *Chaoborus* in muddy substrates are limited to only a few centimeters. Consequently, the density and mobility of the prey would have to be quite high before benthic forms comprised a significant portion of the diet and feeding in the planktonic phase was inhibited.

Although *Chaoborus* is primarily a carnivore, algae are sometimes found in the foregut in sufficient quantity to make it unlikely that it is derived from herbivorous prey (Herms 1937; Deonier 1943; Green et al. 1973; Pollard 1976). Pollard (1976) concluded that *Chaoborus* might switch to feeding on algae whenever animal prey become scarce. In the crops of fourth instar *C. nyblaei*, he found the algae *Trachylomonas*, *Phacus*, *Glenodinium*, *Scendesmus*, *Pediastrum*, *Pandorina*, and *Synedra* (also see Sikorowa 1968). Parma (1971) and Swuste et al. (1973) actually raised *Chaoborus* from eggs to pupae in the presence of filamentous algae without zooplankton. Perhaps the mandibular fans, which are used to trap animal prey, are also used to filter large algae (or protozoa) from the water.

Several studies indicate that composition of the larval diet varies seasonally, depending at least in part on the numerical availability of different foods and the amount of overlap in the spatial distributions of predator and prey. Sikorowa (1968) found that *Chaoborus* ate mainly copepods in late autumn, winter, and spring, whereas the consumption of Cladocera and larger organisms (e.g. insect larvae) increased during the summer. Over a period of only 5 months, Fedorenko (1973, 1975a) observed considerable variation in the diets of several instars of *C. trivittatus* and *C. americanus*. In this case, the spatial availability of prey as well as their relative abundance clearly influenced what foods were eaten. Pollard (1976) found that the fourth instar larvae shifted from a copepod diet to one composed primarily of algae as winter progressed and copepods became less abundant. When rotifers became dominant in the spring, they were found among the crop contents in increasing numbers.

Selective feeding

Prey selection is defined here as any difference between the distribution of prey species or prey sizes in the environment and the composition of the diet. Most studies of predation fail to distinguish between active selection, in which certain prey types are avoided or rejected by the predator, and passive selection, in which some foods are overrepresented in the diet simply because they are more susceptible to encounter, capture, and ingestion (cf. Gilbert 1976). I will reserve the use of the term "preference" to specify cases of active selection where, presumably, the "preferred" food is high in either nutrients or energy content and the predator therefore behaves as an optimal forager (cf. Schoener 1971; Pyke et al. 1977).

Size-selection—Table 1 summarizes the experimental information on size-selective predation by *Chaoborus* larvae. This table includes only those studies in which the various kinds of prey were simultaneously available to the predator; McQueen (1969) has shown that generalizing about prey selection from the results of single species feeding experiments can lead to erroneous conclusions. Experiments in which the prey types were from different genera were also excluded; in these cases, which are discussed below, selection may be influenced by body shape and behavior as well as body size.

In general, third and fourth instar *Chaoborus* select prey in the 0.4–1.9 mm size range. (Table 1). In this respect, the larvae are similar to other large invertebrate predators such as *Leptodora, Diaptomus,* and *Epischura* (Dodson 1974a; Confer and Blades 1975). The juvenile stages of these crustaceans, however, may be herbivorous (Moshiri 1968; Cummins et al. 1969; Maly and Maly 1974). Observations on the food of first and second instar *Chaoborus* in nature show that they consume mainly small animals such as rotifers, nauplii, small copepodites, and protozoa (Deonier 1943; Sikorowa 1968; Fedorenko 1973, 1975a; Lewis 1977). This pattern of size selection is similar to the feeding of adult and subadult states of cyclopoid copepods (cf. Fryer 1957b; Brandl and Fernando 1974, 1975; Kerfoot 1977).

The upper limit to the size of food ingested by a particular instar is undoubtedly governed by the larva's ability to manipulate and swallow large prey (Deonier 1943; Main 1953; Swift and Fedorenko 1975; Pastorok 1978; Fig. 2). Fedorenko (1975a) concluded that fourth instars had the most diverse diet of any larval stage partly because they were able to handle nearly all prey sizes. Elsewhere, I have shown that older instars consume fewer small prey

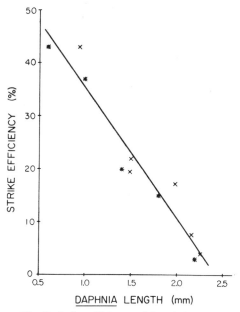

Fig. 2. Strike efficiency of fourth instar *Chaoborus trivittatus* on various sizes of Daphnia. *—Data from Swift and Fedorenko (1975); ×—this study.

than expected on the basis of plankton composition because the per capita encounter rate between *Chaoborus* and these small, slow-moving items is so low (Pastorok 1978). In Lake McDonald, predation by *Chaoborus* falls heaviest on prereproductive instars of *Daphnia pulicaria*; most reproductive instars of this large daphnid species are beyond the range of prey sizes which *Chaoborus* can handle (Fig. 3).

Selection of prey species—After summarizing his own work and reviewing the literature on the dietary habits of *Chaoborus*, Lewis (1977) gave the following rank order of selectivity for some common food items: *Bosmina* > copepod copepodids and adults, *Diaphanosoma* > rotifers > nauplii, *Daphnia*. Furthermore, calanoid copepods are selected over cyclopoids (Anderson and Raasveldt 1974), and *Daphnia* are taken more than *Holopedium* (Allan 1973).

Lewis (1977) observed a diel cycle in electivity which he attributed to a change in larval behavior. Apparently, *Chaoborus* larvae of Lake Lanao exercise their prefer-

Table 1. Size selection by Chaoborus.

Chaoborus spp (Predator)	Prey sp.	Prey size (mm total length) Available	Positively selected	Source	Notes
C. americanus	Daphnia sp.	0.4–1.7	0.7–1.5	Dodson 1974a, fig. 5, 7	Electivity coefficients for five prey size classes determined by laboratory experiments
C. americanus instar IV	Daphnia pulicaria D. rosea	Large Small	Small	Anderson and Raasveldt 1974, table 6	41 large tested, 33 small Daphnia, 20 larvae eaten; no size data for Diaptomus sicilis, also available during experiments
C. americanus 9.5 mm	Daphnia pulex D. rosea	2.2 1.5	1.5	Sprules 1972, table 3	Laboratory experiments
	Diaptomus shoshone D. coloradensis	2.4 1.3	2.4		
C. flavicans instar IV	D. pulex	0.67–2.24	0.67–1.31	Dodson 1970	Electivity coefficients for seven prey-size classes determined by laboratory experiments
C. flavicans	Copepods	Nauplii Post-naupliar stages	Nauplii	Kajak and Ranke-Rybicka 1970, table 1, fig. 1	Field experiments at natural prey densities in Lake Mikolajskie; selection was slight and probably not significant
C. nyblaei and C. punctipennis	Holopedium gibberum	Various sizes	"Small"	Allan 1973, table 4	Predation within field enclosures usually resulted in a significant increase in mean size of remaining prey as compared to controls
	Daphnia parvula	Various sizes	"Small"		

Predator	Prey	Prey size range (mm)	Predator/prey size ratio	Reference	Notes
C. punctipennis instar IV (7.58–9.19-mm total length)	*D. parvula* and *D. ambigua*	0.4–0.6 0.7–0.9 1.0–1.2 1.3–1.5	0.7–0.9 3 cases 1.0–1.2 1 case 1.3–1.5 1 case	Roth 1971	^{14}C-tracer field experiments
	Ceriodaphnia reticulata	0.3–0.4 0.4–0.5 0.6–0.7	0.4–0.5 4 cases 0.6–0.7 2 cases		
	Bosmina longirostris	0.2–0.3 0.4–0.6	0.2–0.3 1 case 0.4–0.6 6 cases		
	Mixed cyclopoid copepods	Nauplii: 0.1–0.2 Copepodites: 0.4–0.6 0.7–0.8 1.0–1.2	0.4–0.8 11 cases 1.0–1.2 2 cases		
C. trivittatus? instar IV	*D. rosea*	Various sizes	<1.5	Main 1953	Qualitative laboratory observations
C. trivittatus instar IV (= his *C. nyblaei*)	*D. pulex*	0.67–2.24	0.84–1.31	Dodson 1970, table 7	*See above* Dodson 1970
C. trivittatus instar IV	*D. pulicaria*	0.5–2.6	1.1–1.9	Pastorok 1978	Data from field observations and laboratory experiments
C. trivittatus instar IV	*Diaptomus franciscanus*	0.5–1.6	1.0–1.6	Pastorok (unpubl.)	Field data

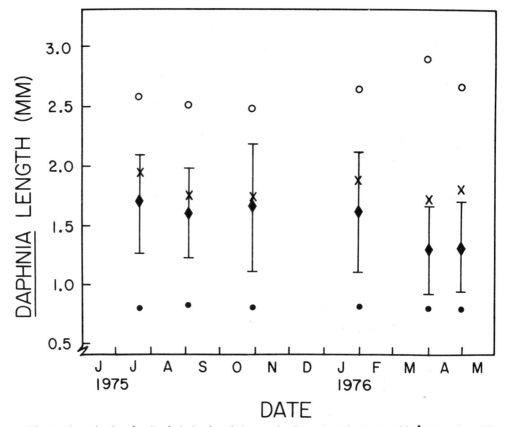

Fig. 3. Size selection for *Daphnia* by fourth instar *Chaoborus* in Lake McDonald. ♦—Prey size with highest electivity (D) value; vertical bar—range of positive electivity; ○, ●—size of largest and smallest *Daphnia* in plankton sample; ×—size of primiparous instar.

ence for *Bosmina* and *Diaphanosoma* only at night when the larvae are in the upper waters where food is abundant. To avoid predation by visually oriented fishes, *Chaoborus* spends the day at lower depths where food is scarce, forcing the larvae to feed opportunistically.

Sikorowa (1968) observed seasonal changes in electivity (*see above*), but copepods ranked high throughout most of the year. Sikorowa also found that the electivity index for a given prey varied between lakes. In contrast to Berg's (1937) suggestion, however, the composition of the diet was not correlated with habitat type.

Despite spatial and temporal variation in electivity indices, the copepodids and adults of diaptomid species consistently rate as one of the most important foods of *Chaoborus* larvae in nature (Deonier 1943; Main 1953; Sikorowa 1968, 1973; Dodson 1970; Kajak and Ranke-Rybicka 1970; Goldspink and Scott 1971). In feeding experiments, *Chaoborus* selects *Diaptomus* over *Daphnia* (Deonier 1943; Main 1953; Sprules 1972; Swuste et al. 1973; Anderson and Raasveldt 1974). Main (1953) and Roth (1971) suggested that copepods are easily swallowed by *Chaoborus* because of their torpedolike shape, whereas a somewhat round cladoceran such as *Daphnia* might be harder to handle. Swift and Fedorenko (1975) confirmed this hypothesis, but they also observed that the overall strike efficiency of larvae for these two prey types was the same as long as prey of equivalent total length were compared (also see Pastorok 1978). Because the copepods have a highly developed escape response, they are harder to catch initially than are cladocerans, but once they are caught, they are more likely to be swallowed.

Roth (1971) speculated that copepods might be easier to locate than *Daphnia* and hence ingested more frequently. However, the results of Swuste et al. (1973) and Pastorok (1978) indicate that *Chaoborus* actually encounters *Daphnia* more readily on a per capita basis. According to these workers the observed selection is due to larval behavior, not the differential vulnerability of prey species; that is, *Chaoborus* rejects *Daphnia* when *Diaptomus* is available. This preferential behavior is expressed only when food is abundant and predators are satiated after each feeding bout (Fig. 4). Larvae that are starved before a feeding experiment are initially nonselective (Pastorak 1978), but within a short time ($\sim$3 h), the predators exhibit a preference for copepods at densities of 40 prey/liter and above. Larvae that are satiated before the experiments initially select copepods regardless of food concentration (Fig. 4). However, if these predators are exposed to low prey densities for 4 days or longer, they relax their food preferences and become nonselective (Pastorok 1978).

Larvae taken directly from Lake McDonald at sunset exhibit a slight bias toward copepod prey (Fig. 4). Since their hunger level is probably between that of the starved and the satiated groups (Pastorok 1978), selectivity is inversely related to hunger. The larval diet in Lake McDonald, however, consists mainly of *Daphnia*, not *Diaptomus*, because *Chaoborus* encounters the former more readily and prey densities are low (< 20/liter) causing the larvae to be relatively opportunistic (Table 2; Pastorok 1978).

Parma (1969) and Pastorok (1978) have shown that *Chaoborus* grows faster and pupates sooner when fed on a diet of copepods than when fed *Daphnia* alone. Thus, the food value of *Diaptomus* for the larvae appears to be higher than the food value of *Daphnia*; this is probably related

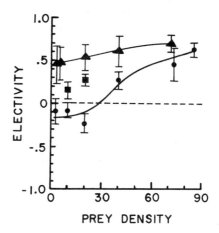

Fig. 4. Electivity (D) of fourth instar *Chaoborus* for *Diaptomus* when *Daphnia* is alternative prey vs. prey density, No./ liter). ▲—Satiated larvae; ■—lake larvae; ●—starved larvae; vertical bar, ± SE.

Table 2. Electivity (D) of fourth instar *Chaoborus trivittatus* for *Daphnia* and *Diaptomus* in Lake McDonald.

	Time	Daphnia	Diaptomus
1975			
22 Jul	2176	0.97	−0.97
3 Sep	2080	0.71	−0.67
28 Oct	1929	0.90	−0.84
1976			
29 Jan	2205	1.00	−1.00
1 Apr	2009	0.66	−0.88
30 Apr	2120	0.11	0.20
1 May	0245	0.46	0.16

to the higher caloric content and digestibility of the former (Pastorok 1978).

The ability to choose only the most valuable foods under the appropriate conditions (e.g. high prey densities, low hunger states of the predator) depends on discriminatory powers of the predator's sensory system. Perhaps, as Strickler (1975) suggests, *Chaoborus* distinguishes copepods from cladocerans on the basis of the species-specific patterns of hydrodynamic disturbances generated by the swimming prey. In some cases, *Chaoborus* may recognize prey by their taste, shape, or hardness. Swuste et al. (1973) noted that *C. flavicans* rejects ostracods after handling them for a short time; these prey are notorious for having tough (probably indigestible) exoskeletons.

An organism's ability to alter its foraging mode in accordance with its level of hunger depends on the existence of an internal monitor of the degree of satiation or some correlate of it. Dethier and Gelperin (1967) and Dethier (1969) discussed the neurological mechanisms underlying the regulation of food intake by the blowfly *Phormia regina*; proprioceptors located in the foregut and the abdominal wall provide negative feedbacks (I in Fig. 5) that inhibit the feeding response when the crop and gut are full. Figure 5 is a modification of their model to explain selective feeding by *Chaoborus* on two kinds of prey, *Diaptomus* and *Daphnia*. Suppose that hungry *Chaoborus* enter a dense patch of prey at the beginning of a daily feeding period (say at sunset) when the crop and midgut are empty. The degree of sensory stimulation (S) resulting from encounters with either the preferred prey (*Diaptomus* = S_c) or the alternate prey (*Daphnia* = S_d) would be high and approximately the same ($S_c = S_d$ at time 0 in Fig. 5). Peripheral sensors would adapt to stimulation from a preferred prey, at a slower rate than that for the alternate prey so the curve S_c falls less steeply than the curve S_d. As the gut fills, inhibitory feedback (I) increases. The net arousal for the copepod ($E_c = S_c - I$) is positive for a longer time than net arousal for the daphnid ($E_d = S_d - I$). Feeding occurs only when $E > 0$, so feeding on the copepod occurs more frequently than on the cladoceran (Fig. 5). Note that even when prey densities are constantly high, stimulatory and inhibitory levels fluctuate around their mean values and some alternate prey are always eaten (Figs. 4, 5). Holling (1965) proposed a similar explanation involving fluctuating hunger levels to explain the mixed diet (i.e. partial preferences) of vertebrate predators.

Differential adaptation to stimuli from two prey species is not the only way in which selection can occur. Integration of impulses in segmental ganglia or in the central nervous system may determine a species-specific feeding threshold. In this case, the zero line in Fig. 5 would differ among prey species.

In Holling's (1965, 1966) model of predation by the mantid *Hierodula crassa*, hunger was measured in terms of fullness of the gut and various predatory behaviors (including search, pursuit, strike, and ingestion) occurred at their respective hunger thresholds. For example, searching behavior and pursuit might occur at a lower state of arousal ($E < 0$ in Fig. 5) than that necessary to release ingestion behavior.

To account for "wasteful killing" by damselfly naiads, Johnson et al. (1975) proposed a two-compartment model of hunger involving control thresholds in the

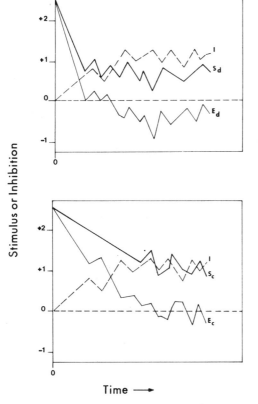

Fig. 5. A model of one possible neural mechanism underlying selective feeding. Symbols explained in text; vertical scale in arbitrary units.

foregut and the midgut. Thus, as the work of Dethier and Gelperin (1967) suggests, the I-curve in Fig. 5 could be decomposed into two components, each one corresponding to the respective inhibitory impulse which would originate in a region of the gut as it becomes full.

Feeding rates of Chaoborus *and impact on the zooplankton community*

Table 3 summarizes the available information on the maximum rate of feeding by *Chaoborus* larvae. Except in the case of Roth (1971), who used a radiotracer method, all predation rates were estimated from feeding trials in which the number of prey eaten was calculated as the difference between the number of individuals available at the start of the trial and the number remaining at the end. Results obtained by this approach agree with estimates of feeding rates which are based on the number of prey found in the crop and known digestion times (Fedorenko 1973). The values for food intake as a percentage of body weight were taken from Lewis (1977). It should be emphasized that these are maximum rates, and many of the criticisms of Hall et al. (1976), concerning appropriate exposure times of predator and prey, unnaturally high prey densities, and failure to control the feeding history of predators before the experiment, apply to these studies. Values from Kajak and Ranke-Rybicka (1970) are based on 10 h of nightime feeding, and values from Roth (1971) represent a sunset to sunrise feeding period at various times of year. All other predation rates assume 24 h of feeding. Nevertheless, one cannot conclude that the daily feeding rate would be lower if the time of exposure of predator and prey is <24 h; after going without food for a short time, many arthropods exhibit "compensatory feeding" in which the intake of food is elevated due to the storage of peritrophic membrane (cf. Johnson et al. 1975).

In general, old larvae eat more prey of a given size than do young ones, and maximum feeding rate increases as prey size decreases (Table 3). Much of the variation in these data is probably due to temperature effects and differences in prey density (cf. Fedorenko 1973, 1975a,b). Lewis (1977) concluded that food intake as a percentage of body weight was greater in younger larvae. When compared with the food intake of other predatory invertebrates, the values for *Chaoborus* are rather low. For example, *Leptodora* ingests 30–50% of its body weight per day (Karabin 1974), and *Mesocyclops* may consume over 100% of its weight in a single day (Brandl and Fernando 1975). The values for other invertebrates range from 15% to 168% (Hall et al. 1976).

At a given temperature, the feeding rate of *Chaoborus* on a single prey species is related to prey density by a type-2 functional response curve (Fedorenko 1975b). More complex curves are possible when alternative foods are available to the predator. For example, Fig. 5 shows the feeding rates of fourth instar larvae on *Daphnia* and *Diaptomus* when the two species of prey are presented simultaneously (data are mean number of prey eaten per day per predator for larvae starved 3 days before the experiment; cf. Fig. 4). *Chaoborus* maintains maximum consumption rates on its preferred prey (*Diaptomus*) and a typical type-2 curve is obtained for this species. The response to an alternative food (*Daphnia*), however, reflects a more complex feeding behavior usually associated with vertebrate predators. The alternative food is partially dropped from the diet when preferred items are abundant, resulting in a bell-shaped functional response (Fig. 6; cf. fig. 6.6: Holling 1965). In Holling's (1965) model, the attack rate of a vertebrate predator on distasteful prey decreases at high prey densities because the learned association between the stimulus from the prey and its distastefulness becomes stronger as contacts become more frequent. I believe an alternative explanation exists for the behavior of *Chaoborus*. *Daphnia* need not be unpalatable to the larvae; rather, they have a lower food value than *Diaptomus*, hence *Chaoborus* specializes on the "better" prey when resources are abundant (Pastorok 1978).

Table 3. Feeding rates of *Chaoborus* larvae.

Species—Instar	Prey	Prey abundance (No./liter)	Daily food intake		Reference
			No. prey	% Larval Body wt.	
C. americanus—IV	*Cyclops b. thomasi*, *Diaptomus sicilis*	154	1.26	1.7	Anderson and Raasveldt 1974
	Daphnia pulicaria *D. rosea, D. sicilis*	154	1.53		
C. americanus—II	Copepod nauplii	160	20.0		
III	*Diaptomus tyrelli*	33–66	8.0		Fedorenko 1975*b*
IV	*Diaptomus kenai*	10	3.5		
	D. tyrelli	40	18.0		
C. borealis—IV	*Aedes* larvae, *Daphnia*	73	0.4		James and Smith 1958
	Aedes larvae	37	0.4		
	Aedes larvae	73	0.6		
Eckstein 1—III	*Tropodiaptomus gigantoviger*		1.4	2.8	Lewis 1977
IV	*T. gigantoviger*		3.7	2.1	
C. flavicans—IV	*Daphnia*	99	4.4	0.8	Dodson 1970
C. flavicans—IV	Copepods, Cladocerans	8	8.8	3.6	Kajak and Ranke-Rybicka 1970
	Copepods, Cladocerans	44	8.0	12.5	
Chaoborus sp.—IV	*Diaptomus shoshone*, *D. coloradensis*	20	3.2		
	D. rosea, *D. coloradensis*	50	3.3		
	Daphnia pulex *D. shoshone*	25	0.9	6.3	Sprules 1972
	D. pulex, *D. coloradensis*	38	5.2		
	D. pulex, D. rosea	25	1.8		
C. trivittatus (= *C. nyblaei*)—IV	*Daphnia*	57	3.9	0.6	Dodson 1970
C. trivittatus—II	*Diaptomus tyrelli*	5	2.2	13.2	
III	*D. tyrelli*	5	6.5	6.7	
	D. kenai	2.5	2.5		
	Diaphanosoma brachyurum	70	28.0		Fedorenko 1975*b*
IV[y]	*D. kenai*	9	3.5		
IV[o]	*D. tyrelli*	80	18.0	8.4	
	D. kenai	5	4.0		

The predatory impact of *Chaoborus* will depend on larval density, prey density, and the supply of resources for the prey as well as individual predation rates. Table 4 gives the percentage of the prey population removed per day by *Chaoborus* as determined by various investigators. In many cases, *Chaoborus* exerts a significant in-

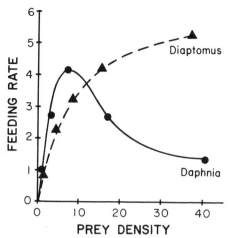

Fig. 6. Differential functional response to two species of prey by fourth instar *Chaoborus trivittatus* (mean No. prey pred.$^{-1}$ day^{-1} vs. No. prey/liter for each prey species).

Table 4. Impact of *Chaoborus* predation on prey populations.

Prey	% removed per day	Reference
Daphnia parvula	18.0 (12–23)	Allan 1973
Daphnia rosea	5.5 (1–12)	Dodson 1972
Diaphanosoma brachyurum	8.0 (1–19)	"
Diaptomus kenai	9.0 (5–28)	Fedorenko 1975b
Diaptomus tyrelli	3.0 (0–10)	"
Copepod nauplii	2.0 (0–5)	"
Copepoda, Cladocera	7.0	Kajak and Ranke-Rybicka 1970
Copepoda, Cladocera	13.0	"
Copepoda, Cladocera	10.0	Kajak and Rybak cited in Hillbricht-Ilkowska et al. 1975
Bosmina longirostris	3.5	Roth 1971
Ceriodaphnia reticulata	1.3	"
Daphnia ambigua and *D. parvula*	0.6	"
Tropocyclops prasinus	1.9	"
Thermocyclops nauplii	1.6	Lewis, in press
Thermocyclops copepodids	15.6	"
Thermocyclops adults	5.0	"
Tropodiaptomus nauplii	18.5	"
Tropodiaptomus copepodids	9.6	"
Tropodiaptomus adults	8.0	"
Diaphanosoma	20.0	"
Moina	26.0	"
Bosmina	16.0	"

fluence on the population abundance of its prey. Smyly (1976) observed up to a fourfold increase in the abundance of *Diaptomus gracilis* when *Chaoborus* larvae were excluded from large experimental tubes. In the absence of *Chaoborus*, *Daphnia rosea* populations may double their abundance within a month (Neill unpubl.). Allan (1973) found that *Chaoborus* predation determined the relative abundance of two competing species of Cladocera. By preying on *Daphnia*, *Chaoborus* prevented the population from increasing at the expense of an inferior competitor, *Holopedium*. The latter was protected from larval predation by its large gelatinous sheath.

After experimental removal of *Chaoborus* from their lakes, Smyly (1976) and Neill (pers. comm.) documented a shift in dominance relations among the zooplankton, but they observed no species invasions or extinctions. On the other hand, Lynch (1977) found that *Chaoborus* excluded *Ceriodaphnia* from experimental vessels, and Von Ende (1975) has shown that *C. americanus* causes the extinction of other *Chaoborus* species in the lakes where it occurs. *Chaoborus* is probably responsible for the absence of large copepods from alpine ponds (Sprules 1972; Anderson and Raasveldt 1974). Juveniles of large, slowly growing species such as *Diaptomus shoshone*, *D. nevadensis*, and *Epischura nevadensis* spend a long time within the range of prey sizes which is highly vulnerable to predation by *Chaoborus* (Table 1). In contrast, large, rapidly growing cladocerans like *Daphnia pulex*, *D. pulicaria*, and *Simocephalus vetulus* may coexist with *Chaoborus* by exploiting a strategy of "escape in size" (Fig. 3; cf. Dodson 1974a; Paine 1976; Kerfoot and Pastorok 1978; Zaret 1978).

References

AKEHURST, S. C. 1922. Larva of *Chaoborus crystallinus* (DeGeer) (*Corethra plumicornis* F.). J. R. Microsc. Soc. **15**:341–372.

ALLAN, J. D. 1973. Competition and the relative abundances of two Cladocerans. Ecology **54**:484–498.

ALVERDES, F. 1926. *Corethra*–und Ephemeriden Larven nach Unterbrechung ihrer Bauchganglienkette. Z. Vgl. Physiol. **3**:558–594

ANDERSON, R. S., and L. G. RAASVELDT. 1974. *Gammarus* and *Chaoborus* predation. Can. Wildl. Serv. Occas. Pap. 18.

AUTRUM, H. 1964. Anatomy and physiology of sound receptors in invertebrates. *In* R. G. Busnel [ed.], Acoustic behaviour of animals. Elsevier.

BARDENFLETH, K. S., and R. EGE. 1916. On the anatomy and physiology of the airsacs of the larva of *Corethra plumicornis*. Vidensk. Medd. Dan. Naturhist. Foren. Kbh. **67**:25–42.

BERG, K. 1937. Contributions to the biology of *Corethra* Meigen (*Chaoborus* Lichtenstein). Biol. Medd. **13**:1–101.

BRANDL, Z., and C. H. FERNANDO. 1974. Feeding of the copepod *Acanthocyclops vernalis* on the cladoceran *Ceriodaphnia reticulata* under laboratory conditions. Can. J. Zool. **52**:99–105.

——, and ——. 1975. Food consumption and utilization in two freshwater cyclopoid copepods (*Mesocyclops edax* and *Cyclops vicinus*). Int. Rev. Gesamten Hydrobiol. **60**:471–494.

CONFER, J. L., and P. I. BLADES. 1975. Omnivorous zooplankton and planktivorous fish. Limnol. Oceanogr. **20**:571–579.

COOK, E. G., 1956. The neritic Chaoborinae (Diptera: Culicidae). Tech. Bull. Minn. Agr. Exp. Stn. 218. 102 p.

CUMMINS, K. W., and others. 1969. Ecological energetics of a natural population of the predaceous zooplankter *Leptodora kindtii* Focke (Cladocera). Oikos **20**:189–223.

DEONIER, C. C. 1943. The biology of the immature stages of the Clear Lake gnat (Diptera, Culicidae). Ann. Entomol. Soc. Am. **36**: 383–388.

DETHIER, V. G. 1969. Feeding behavior of the blowfly. Adv. Stud. Behav. **2**:112–266.

——, and A. GELPERIN. 1967. Hyperphagia in the blowfly. J. Exp. Biol. **47**:191–200.

DODSON, S. I. 1970. Complementary feeding niches sustained by size-selective predation. Limnol. Oceanogr. **15**:131–137.

——. 1972. Mortality in a population of *Daphnia rosea*. Ecology **53**:1011–1023.

——. 1974a. Zooplankton competition and predation: An experimental test of the size-efficiency hypothesis. Ecology **55**:605–613.

——. 1974b. Adaptive change in plankton morphology in response to size-selective predation: A new hypothesis of cyclomorphosis. Limnol. Oceanogr. **19**:721–729.

DUHR, B. 1955a. Über den chemischen Sinn, die Darmperistaltik und die Bildung der peritrophischen Membran der *Corethra* Larve (*Chaoborus crystallinus* De Geer). Zool. Jahrb. Abt. Allg. Zool. Physiol. Tiere **65**:315–333.

——. 1955b. Über Bewegung, Orientierung und Beutefang der *Corethra* Larve (*Chaoborus crystallinus* De Geer). Zool. Jahrb. Abt. Allg. Zool. Physiol. Tiere **65**:387–429.

FEDORENKO, A. Y. 1973. Predation interactions between zooplankton and two species of *Chaoborus* (Diptera, Chaoboridae) in a small coastal lake. M. S. thesis, Univ. British Columbia. 123 p.

——. 1975a. Instar and species-specific diets in two species of *Chaoborus*. Limnol. Oceanogr. **20**:238–249.

——. 1975b. Feeding characteristics and predation impact of *Chaoborus* (Diptera, Chaoboridae) larvae in a small lake. Limnol. Oceanogr. **20**:250–259.

FRYER, G. 1957a. The feeding mechanism of some freshwater cyclopoid copepods. Proc. Zool. Soc. Lond. **129**:1–25.

——. 1957b. The food of some freshwater cyclopoid copepods and its ecological significance. J. Anim. Ecol. **26**:263–286.

GERSCH, M. 1952. Experimentelle Untersuchen über den Verdauungstraktus der Larve von *Chaoborus* (*Corethra*), Z. Vgl. Physiol. **34**: 346–369.

GILBERT, J. J. 1976. Selective cannibalism in the rotifer *Asplanchna sieboldi*: Contact recognition of morphotype and clone. Proc. Natl. Acad. Sci. **73**:3233–3237.

GOLDSPINK, C. R., and D. B. SCOTT. 1971. Vertical migration of *Chaoborus flavicans* in a Scottish loch. Freshwater Biol. **1**:411–421.

GREEN, J., S. A. CORBET, and E. BETNEY. 1973. Ecological studies on crater lakes in West Cameroon. The blood of endemic cichlids in Barombi Mbo in relation to stratification and their feeding habits. J. Zool. Proc. Zool Soc. Lond. **170**:299–308.

HALL, D. J., S. T. THRELKELD, C. W. BURNS, and P. H. CROWLEY. 1976. The size-efficiency hypothesis and the size structure of zooplankton communities. Annu. Rev. Ecol. Syst. **7**:177–208.

HAMNER, P., and W. M. HAMNER. 1977. Chemosensory tracking of scent trails by the planktonic shrimp *Acetes sibogae australis*. Science **195**:886–888.

HARPER, E. H. 1907. The behavior of the phantom larvae of *Corethra plumicornis* Fabricius. J. Comp. Neurol. **17**:435–456.

HERMS, W. B. 1937. The Clear Lake gnat. Bull. Calif. Agr. Exp. Sta. 607. 22p.

HILLBRICHT-ILKOWSKA, A., Z. KAJAK, J.

EJSMONT-KARABIN, A. KARABIN, and J. RYBAK. 1975. Ecosystem of the Mikolajski Lake. The utilization of the consumers production by invertebrate predators in pelagic and profundal zones. Pol. Arch. Hydrobiol. 22:53-64.

HOLLING, C. S., 1965. The functional response of predators to prey density and its role in mimicry and population regulation. Mem. Entomol. Soc. Can. 45:1-60.

———. 1966. The functional response of invertebrate predators to prey density. Mem. Entomol. Soc. Can. 48:1-86.

HORRIDGE, G. A. 1966. Some recently discovered underwater vibration receptors in invertebrates, p. 395-405. In H. Barnes [ed.], Some contemporary studies in marine science. Allen and Unwin.

JACOBS, J. 1974. Quantitative measurement of food selection—a modification of the forage ratio and Ivlev's electivity index. Oecologia 14:413-417.

JAMES, H. G., and B. C. SMITH. 1958. Observations on three species of *Chaoborus* Licht (Diptera: Culicidae) at Churchill, Manitoba. Mosq. News 18:242-248.

JOHNSON, D. M., B. G. AKRE, and P. H. CROWLEY. 1975. Modeling arthropod predation: Wasteful killing by damselfly naiads. Ecology 56:1081-1094.

JONASSON, P. M. 1955. The efficiency of sieving techniques for sampling freshwater bottom fauna. Oikos 6:183-207.

KAJAK, Z., and B. RANKE-RYBICKA. 1970. Feeding and production efficiency of *Chaoborus flavicans* Meigen (Diptera, Culicidae) larvae in eutrophic and dystrophic lakes. Pol. Arch. Hydrobiol. 17:225-232.

KARABIN, A. 1974. Studies on the predatory role of the cladoceran, *Leptodora kindtii* (Focke), in secondary production of two lakes with different trophy. Ekol. Pol. 22:295-310.

KERFOOT, W. C. 1975. The divergence of adjacent populations. Ecology 56:1298-1313.

———. 1977. Implications of copepod predation. Limnol. Oceanogr. 22:316-326.

———, and R. A. PASTOROK. 1978. Survival versus competition: evolutionary compromises and diversity in the zooplankton. Int. Ver. Theor. Angew. Limnol. Verh. 20:362-374.

KITTREDGE, J. S., F. T. TAKAHASHI, J. LINDSEY, and R. LASKER. 1974. Chemical signals in the sea: Marine allelochemics and evolution. Fish. Bull. 72:1-11.

LEWIS, W. M. 1977. Feeding selectivity of a tropical *Chaoborus* population. Freshwater Biol. 7:311-325.

———. in press. A zooplankton community analysis. Springer-Verlag.

LEYDIG, F. 1851. Anatomisches und Histologisches über die Larve von *Corethra plumicornis*. Z. Wiss. Zool. 3:435-451.

LYNCH, M. R. 1977. Predation, competition, and zooplankton community structure. Ph.D. thesis, Univ. Minnesota, Minneapolis. 185 p.

McQUEEN, D. J. 1969. Reduction of zooplankton standing stocks by predacious *Cyclops bicuspidatus thomasi* in Marion Lake, British Columbia. J. Fish. Res. Bd. Can. 26:1605-1618.

MAIN, R. A. 1953. A limnological study of *Chaoborus* (Diptera) in Hall Lake, Washington. M.S. thesis, Univ. Washington, Seattle. 106 p.

MALY, E. J., and M. P. MALY. 1974. Dietary differences between two co-occurring calanoid copepod species. Oecologia 17:325-333.

MONTSHADSKY, A. S. 1945. On the mechanism of digestion in the larvae of *Chaoborus* (Diptera, Culicidae). Zool. Zh. 24:90-99. [In Russian, English summary.]

MOSHIRI, G. A., 1968. Energetics of zooplankton *Leptodora kindtii* Focke (Crustacea: Cladocera) and selected prey. Limnol. Oceanogr. 14:475-484.

O'BRIEN, W. J., and G. L. VINYARD. 1978. Polymorphism and predation: The effect of invertebrate predation on the distribution of two varieties of *Daphnia carinata* in South India ponds. Limnol. Oceanogr. 23:452-460.

PAINE, R. T., 1976. Size-limited predation: An observational and experimental approach with the *Mytilus-Pisaster* interaction. Ecology 57:858-874.

PARMA, S. 1969. The life cycle of *Chaoborus crystallinus* (DeGeer) (Diptera, Chaoboridae) in a Dutch pond. Int. Ver. Theor. Angew. Limnol. Verh. 17:888-894.

———. 1971. *Chaoborus flavicans* (Meigen) (Diptera, Chaoboridae): An autecological study. Ph.D. thesis, Groningen. 128 p.

PASTOROK, R. A. 1978. Predation by *Chaoborus* larvae and its impact on the zooplankton community. Ph.D. thesis, Univ. Washington, Seattle. 238 p.

POLLARD, J. E. 1976. The life cycle and trophic relationships of *Chaoborus nyblaei* in a small coastal pond. M.A. thesis, Humboldt State Univ. 66 p.

PUCAT, A. M. 1965. The functional morphology of the mouthparts of some mosquito larvae. Quaest. Entomol. 1:41-86.

PYKE, G. H., H. R. PULLIAM, and E. L. CHARNOV. 1977. Optimal foraging: A selective review of theory and tests. Rev. Biol. 52:137-154.

ROTH, J. C. 1971. The food of *Chaoborus*, a plankton predator, in a southern Michigan lake. Ph.D. thesis, Univ. Michigan, Ann Arbor. 94 p.

SACHSE, R., and R. WOHLGEMUTH. 1916. Die Nahrung der fische die Teichwirtschaft wichtigen niederen Tiere. Allg. Fisch.-Zig. 41:50-56.

SCHOENER, T. W. 1971. Theory of feeding

strategies. Annu. Rev. Ecol. Syst. 2:369–404.

SCHREMMER, F. 1950. Zur Morphologie und funktionellen Anatomie des Larvenkopfes von *Chaoborus* (*Corethra* auct.) *obscuripes* V. D. Wulp (Diptera, Chaoboridae). Oesterr. Zool. Z. 2:471–516.

SIKOROWA, A. 1968. Resistance of *Chaoborus* Licht. larvae to lack of food. Ekol. Pol. Ser. A 16:243–251.

——. 1973. Morphology, biology, and ecology of species belonging to the genus *Chaoborus* Lichtenstein (Diptera, Chaoboridae) occurring in Poland [In Polish, English Summary]. Zesz. Nauk. Akad. Roln.-Tech. Olsztynie Ochr. Rybactuo Srodladowe 1:1–121.

SMYLY, W. J. P. 1976. Some effects of enclosure on the zooplankton in a small lake. Freshwater Biol. 6:241–251.

SPRULES, W. G. 1972. Effects of size-selective predation and food competition on high altitude zooplankton communities. Ecology 53:375–386.

STRICKLER, J. R. 1975. Intra- and interspecific information flow among planktonic copepods: Receptors. Int. Ver. Theor. Angew. Limnol. Verh. 19:2951–2958.

——, and A. K. BAL. 1973. Setae of the first antennae of the copepod *Cyclops scutifer* (Sars): Their structure and importance. Proc. Natl. Acad. Sci. 70:2656–2659.

——, and S. TWOMBLY. 1975. Reynolds number, diapause and predatory copepods. Int. Ver. Theor. Angew. Limnol. Verh. 19:2943–2950.

SWIFT, M. C. 1976. Energetics of vertical migration in *Chaoborus trivittatus* larvae. Ecology 57:900–914.

——, and A. Y. FEDORENKO. 1973. A rapid method for the analysis of the crop contents of *Chaoborus* larvae. Limnol. Oceanogr. 18: 795–798.

——, and A. Y. FEDORENKO. 1975. Some aspects of prey capture by *Chaoborus* larvae. Limnol. Oceanogr. 20:418–426.

SWUSTE, H. F. J., R. CREMER, and S. PARMA. 1973. Selective predation by larvae of *Chaoborus flavicans* (Diptera, Chaoboridae). Int. Ver. Theor. Angew. Limnol. Verh. 18:1559–1563.

TUBB, R. A., and T. C. DORRIS. 1965. Herbivorous insect populations in oil refinery effluent holding pond series. Limnol. Oceanogr. 10: 121–134.

VON ENDE, C. 1975. Organization of bog lake zooplankton communities: Factors affecting the distribution of four *Chaoborus* species (Diptera, Chaoboridae). Ph.D. thesis, Univ. Notre Dame. 107 p.

ZARET, T.M. 1975. Strategies for existence of zooplankton prey in homogeneous environments. Int. Ver. Theor. Angew. Limnol. Verh. 19:1484–1489.

——. 1978. A predation model of zooplankton community structure. Int. Ver. Theor. Angew. Limnol. Verh. 20:2996–2500.

48. The Effects of an Introduced Invertebrate Predator and Food Resource Variation on Zooplankton Dynamics in an Ultraoligotrophic Lake

Stephen T. Threlkeld, James T. Rybock,
Mark D. Morgan, Carol L. Folt, and Charles R. Goldman

Abstract

The establishment of *Mysis relicta* in Lake Tahoe, California-Nevada, between 1969 and 1971 contributed to the disappearance of the cladocerans *Daphnia pulicaria*, *Daphnia rosea*, and *Bosmina longirostris* from the pelagic zooplankton in 1970–1971. Analyses of *Daphnia* population dynamics revealed that increased mortality rates were at least partially responsible for their disappearance, although reductions in birth rates were also observed. The loss of cladocerans was sustained until late 1974 when, during a decline of *Mysis*, *Bosmina* reappeared for about 6 months. Examination of *Mysis* stomachs from this period revealed the presence of *Bosmina* remains; laboratory studies and analyses of *Mysis* fecal pellets from 1970 indicate that *Mysis* readily consumes *Daphnia* and *Bosmina*. Currently, cladocerans exist only in Emerald Bay and in Tahoe Keys, a bay and a harbor of Lake Tahoe, respectively, which are well isolated from the main body of the lake and which are also more productive than the lake.

Since the decline of cladocerans, mean annual zooplankton densities have generally been inversely related to *Mysis* density, suggesting the importance of *Mysis* predation in the regulation of noncladoceran zooplankton (*Epischura nevadensis*, *Diaptomus tyrelli*, and *Kellicottia longispina*) population dynamics. Examination of *Mysis* stomach contents recently collected from the lake reveals remains of all of the noncladoceran zooplankton.

The impact of vertebrate and invertebrate predation serves as a focal point in many recent studies of zooplankton distribution and abundance. Less attention has been given to the role of food resource variation in zooplankton dynamics, primarily because food resources are not often thought to limit zooplankton populations or, for the duration of most zooplankton studies, that food resource variation is low and of minor consequence. However, in long term studies of zooplankton populations, it is probably unsafe to assume that food resource variation is negligible, especially where natural or cultural eutrophication adds additional complexity to any assessment of food resources.

In Lake Tahoe, California-Nevada, our attention has recently been focused on the impact of the introduced predator *Mysis relicta* on the zooplankton of this ultraoligotrophic lake. We here present four examples spanning the data record of the last 11 years and various trophic conditions where *Mysis* appears to have had an important effect on the zooplankton. We review some previously published information but emphasize current studies and analyses in

---

This research was supported from 1967–1971 by EPA grant 16010 DBU and from 1972–1978 by NSF grants GI-22, 74-22675, and DEB 76-11095.

the evaluation of the importance of *Mysis* predation and food resource variation to the dynamics of Lake Tahoe zooplankton. Although our main approach has been spatial and temporal analyses of zooplankton and *Mysis* distribution and abundance, this information has been supplemented with the results of analyses of predator food habits and population birth rate dynamics and field and laboratory experimentation on the feeding, survival, and reproduction of resident zooplankton. We thank R. C. Richards, H. Morgan, E. de Amezaga, and B. Yung for technical assistance.

Methods and materials

Since August 1967, zooplankton and phytoplankton have been collected at an index station along the west shore of Lake Tahoe (z = 200 m) (Fig. 1), a location which provides representative samples of both whole lake phytoplankton (Goldman and Armstrong 1969) and zooplankton (Richerson 1969; Roth and Goldman in prep.).

Fig. 1. Bathymetric map of Lake Tahoe slowing index station and other sampling locations discussed in text. Contour interval is 50 m. Shaded area indicates areas with water depth < 100 m.

In addition to these collections, made each week in summer and about every 10 days in winter, regular measurements of primary productivity, temperature, and light have also been taken at this station (Goldman 1974). Recent modifications in zooplankton sampling design and techniques will be discussed here. Prior to June 1977, zooplankton were collected with a Clarke-Bumpus sampler equipped with a 80-μm-mesh net. Since that time, additional collections have been made with a 0.75-m-diameter net with 80-μm-mesh net and equipped with a TSK flowmeter for net tow calibration. As will be discussed later, the increased volume filtered by this larger net has been useful in evaluating the abundance of rare zooplankton species important to this study.

Zooplankton sampling in 1977–1978 also includes transects in Emerald Bay and Tahoe Keys, two productive embayments of Lake Tahoe (Fig. 1). Zooplankton have been collected regularly since June 1977 at a central station in Emerald Bay, since October 1977 at three stations in the bay, and at one station just off its mouth with a 0.75-m net identical to the one used for index station sampling. In all cases, vertical net tows were made from the bottom to the surface, being about 20 m in length at all stations except the central Emerald Bay station, which has a water depth of about 60 m. In Tahoe Keys, six stations have been sampled since March 1978 at about 10-day intervals with a 30-cm-diameter, 80-μm-mesh net towed vertically from the bottom to the surface ($\leqslant$ 3 m). Six synoptic stations (3 at 10–20-m and 3 at 100-m depth) off the south Lake Tahoe shelf have been sampled each month since May 1978 with the 0.75-m net. All zooplankton samples have been enumerated to species, sex, and number of eggs carried per female in each population. The entire volume of the synoptic station samples have been examined for cladoceran zooplankton, although index, Emerald Bay, and Tahoe Keys zooplankton samples are usually split (with a Folson plankton splitter) to permit counting of from 100–500 individuals.

Since the introduction of *M. relicta* in

Lake Tahoe in 1963 (Linn and Frantz 1965), four different methods have been used to evaluate changes in its abundance. Since 1968, and at approximately yearly intervals thereafter, daytime benthic trawls have been made with an epibenthic sled (Linn and Frantz 1965) at various locations around the lake at depths from 15 to 100 m. The size structure of the *Mysis* population was determined by measurements of total length (tip of rostrum to tip of telson) on aliquots of at least 100 animals collected in these trawls. Creel censuses by the Nevada Department of Fish and Game have provided information since 1968 on the frequency of angler-caught lake trout which were found to consume *Mysis*, and from 1972 to the present have provided seasonal averages of the weight of *Mysis* consumed by these fish. In November 1975, nighttime vertical net tows with a 0.67-m-diameter mouth, 500-μm-mesh net were made at various locations around the lake (*see* Morgan et al. 1978). Since October 1976, these collections have been made with paired 0.75-m nets mounted on a Bongo frame. The Emerald Bay and South Lake synoptic stations which were sampled for smaller zooplankton with 0.75-m nets were also sampled for *Mysis* with these Bongo nets.

Grazing rates by adult (1.0–2.0 mm) *Daphnia pulicaria* on a variety of algal species representative of those in Lake Tahoe were determined using both ^{14}C and cell count methods (for review, *see* Rigler 1971). Algal genera used and their maximum linear dimensions were: *Stichococcus* (5 μm), *Chlorella* (5 μm), *Choricycstis* (5 μm), *Ocystis* (10 μm), *Ankistrodesmus* (25 μm), and *Navicula* (10–15 μm). All experiments were performed in the dark at $10°C$ and terminated after a 15-minute feeding period.

Food habits of *M. relicta* and *Epischura nevadensis* were determined by stomach analyses of lake-collected specimens. Immediately upon collection (with vertical net tows, Bongo tows, or Tucker horizontal tows), the animals were preserved in 5% Formalin and returned to the lab. There, specimens of *Mysis* and *Epischura* were rinsed and sorted: their stomach contents were then dissected out. Prey parts (mandibles, postabdominal claws, mucrones, etc.) were identified by comparison to dissected parts from known prey species. In addition, fecal pellets conforming in size and shape to *M. relicta* fecal pellets were removed from 1970 zooplankton samples and treated in the same manner as stomachs removed by microdissection techniques.

Preliminary life table experiments were performed from 1 May to 29 June 1978 with specimens of *D. pulicaria* and *D. ambigua*. *Daphnia pulicaria* was obtained from laboratory cultures established from collections made at Indian Creek Reservoir (Alpine Co., California) in October 1977. *Daphnia ambigua* was obtained from Tahoe Keys (Lake Tahoe) immediately before experimental use. Individual gravid females of each species were isolated in 150-ml clear glass jars and suspended in surface lake water obtained along the west shore of the lake (about 1 km north of the index station in water about 3 m deep: *see* Fig. 1). Survivors were counted at about 5-day intervals when the lake water was also changed.

Results and discussion

Dynamics of Mysis relicta *in Lake Tahoe*—*Mysis relicta* Loven, first introduced in 1963 as forage for the lake trout *Salvelinus namaycush* (Linn and Frantz 1965), was first collected in epibenthic trawls and found in lake trout stomachs in 1968 and 1969. Numbers of *Mysis* collected per 15-minute sled trawl increased dramatically from August 1969 ($\bar{x} \pm 2$ SE = 408 ± 399) to December 1971 (145,000 ± 106,926). During the same period, the frequency of 38–50-cm lake trout consuming *Mysis* increased from 1.4% to 60%. Current life history studies of *M. relicta* in Lake Tahoe and Emerald Bay (M. Morgan unpubl. data) show that recruitment of juveniles only occurs between March and June, thus limiting the possible times of increase between August 1969 and December 1971 to either the spring of 1970 or 1971. The size distribution of *Mysis* collected with the epibenthic trawls in December 1971 was unimodal with a mean of 15.69 mm (± 2 SE = 0.34). In order to achieve this size distribution by late 1971, it is most likely that the numerical increase observed between 1969 and 1971 occurred in spring 1970. This would require growth by newborn *Mysis* (3–4 mm) to a length of about 10–11 mm in the first year, a rate well within the range observed currently in Lake Tahoe and Emerald Bay and in other lakes (Threlkeld et al. in press). However, if the observed increase in *Mysis* was due to recruitment in early 1971, then an unreasonably high growth rate by first year *Mysis* would be required. Thus, establishment of high densities of *M. relicta* in Lake Tahoe (on the order of 250–300 m^{-2}: see Morgan et al. 1978) probably resulted from excellent recruitment to its population in spring 1970.

From 1971 to 1974, both the numbers of *Mysis* obtained with the epibenthic trawls and the weight of *Mysis* consumed by lake trout have decreased (Morgan et al. 1978) (Fig. 2.). In early 1975, *Mysis* densities apparently increased again (Fig. 2), but since that time, *Mysis* densities have again decreased and are currently (July 1978) at about 27 m^{-2}.

Morgan et al. (1978) showed that *M. relicta* is widely distributed throughout Lake Tahoe, although it is relatively scarce in shallow areas (< 20 m) during times of thermal stratification. These areas make up < 5% of the surface area of the whole lake, making any effects of *Mysis* potentially lakewide. Vertical migration by *Mysis* usually includes the entire water column, except during thermal stratification (July – September) when the upper 15 m are avoided. During this time, *Mysis* is most concentrated at night from about 15 to 100 m, while nighttime distributions in other seasons include the entire upper 100 m. During the day, *Mysis* is found on the bottom in water shallower than 200 m and in the pelagic zone in waters deeper than 200 m (Morgan et al. 1978).

The disappearance of cladocerans—The

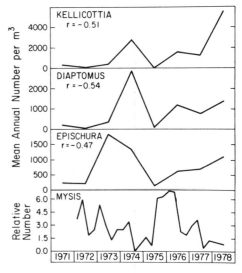

Fig. 2. Mean annual densities of *Kellicottia longispina*, *Diaptomus tyrelli*, and *Epischura nevadensis* for the years 1971–1977. Data from 1978 based on first 6 months of year. Correlation coefficients between mean annual zooplankton densities and those of *Mysis* shown for each species. Relative *Mysis relicta* densities based on mean weight consumed per 38–50-cm lake trout per 3-month period also shown.

most dramatic correlate of the establishment of *M. relicta* in Lake Tahoe in 1970 was the disappearance of the cladocerans *Daphnia pulicaria*, *D. rosea*, and *Bosmina longiostris* from the plankton (Goldman 1974; Richards et al. 1975). Analysis by Goldman et al. (1979) of the population dynamics of the pooled *Daphnia* species revealed major changes in population parameters from the start of sampling in 1967 until the elimination of these species in 1970. Table 1 shows that during the *Daphnia* population decline in 1967–1968 and 1968–1969, birth and death rate statistics were similar. However, the population decline in 1970, which led to the disappearance of *Daphnia* from Lake Tahoe, began much earlier in the year (4 June 1970), with similar birth rates but much higher (about 23%) death rates than during previous winters. The similarity between the birth rates during this period and the periods of population decline in 1967–1969 is especially important because they were as low as winter birth rates but

Table 1. Comparison of birth and death rate statistics for *Daphnia* from 1967-1970 during periods of sustained population decline and during periods chronologically equivalent to time period of 1970 population decline.

Time period	r	b	d
21 Sep 1967–8 Feb 1968	−0.031	0.004	0.035
1 Dec 1968–15 Mar 1969	−0.029	0.005	0.034
4 Jun 1970–22 Dec 1970	−0.039	0.007	0.046
5 Jun 1968–22 Dec 1968	*	0.015	0.019
7 Jun 1969–19 Jan 1970	*	0.013	0.006

*Periods 5 June–22 December 1968 and 7 June 1969–19 January 1970 included both population increases and decreases, making calculations of mean sustained rates of population change meaningless.

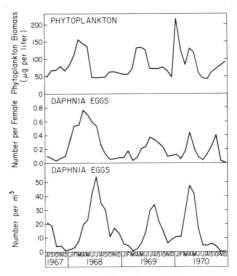

Fig. 3. Mean monthly values of phytoplankton biomass (μg liter^{-1}), eggs per female *Daphnia*, and *Daphnia* eggs per m^3 for period 1967–1970.

occurred much earlier in the year in 1970. The second half of Table 1 shows that when periods of time chronologically equivalent to the June to December 1970 period are compared, birth rates are about half those observed in earlier years. At the same time, death rates were substantially higher than in earlier years. Although the same analyses were not performed for *Bosmina*, 1970 was also notable in that *Bosmina* failed to exhibit its regular autumnal peak in abundance; by May 1971, *Bosmina* had also virtually disappeared from the regular plankton catches (Goldman 1974; Richards et al. 1975).

The changes in *Daphnia* birth rates from 1968 to 1970 are hypothesized to be related to fluctuations in phytoplankton biomass available to the daphnids. Results of the *Daphnia* grazing experiments show that *Daphnia* was capable of using a wide variety of algal sizes and types as food; only the feeding experiment with *Choricystis* (5 μm) failed to indicate significant feeding by *Daphnia*. However, without additional information on the consumption of Lake Tahoe algae by cladocerans, we can safely examine *Daphnia* reproduction only in relation to total phytoplankton biomass. Figure 3 shows that the monthly mean number of eggs per female *Daphnia* and per m^3 follows the peaks of phytoplankton biomass. Peaks in individual reproduction followed phytoplankton biomass by about 1 month ($r = 0.36$, $P < 0.05$), while increases in eggs m^{-3} followed more slowly (with a 3-month lag, $r = 0.64$, $P < 0.01$). Although various hypotheses are under investigation to account for this annual variation in phytoplankton biomass we are not yet able to explain it satisfactorily. We note, however, that such variations in phytoplankton are not uncommon in other lakes where long records have been kept (cf. Hutchinson 1967; Wetzel 1975). A decline of this type in Lake Tahoe phytoplankton appears to be partly responsible for the 30% increase in the rate of *Daphnia* population decline observed in 1970 (Table 1).

The increase in death rates of *Daphnia* from 1968 and 1969 to 1970 has been most closely linked with increases in the abundance of *M. relicta*. Immediately following recruitment to the *Mysis* population in early 1970, *Daphnia* began its final decline from the lake. The vertical distribution of *Daphnia*, *Bosmina*, and *Mysis*, as well as the food habits of *Mysis* make the assignment of the increased death rates to *Mysis* especially plausible. Richerson (1969), on the basis of nine diel series conducted in 1968, found that *Daphnia* and *Bosmina* were most concentrated in the hypolimnion of Lake Tahoe, and in particular from about 150 to 30 m in depth. This pattern varied little with season or on a diel basis. Thus, the nighttime vertical distributions of *Mysis* and these cladocerans were completely overlapping. That *Mysis* was consuming *Daphnia* and *Bosmina* during this period was confirmed by analyses of fecal material taken from 1970 zooplankton samples. Microscopic examination of 35 *Mysis* fecal pellets collected between June and December 1970 revealed the presence of mandibles and postabdominal claws of *Daphnia* and mucrones and mandibles of *Bosmina*. Such predation on cladocerans is not surprising in view of the observed voracity of *Mysis* in laboratory feeding experiments where these cladocerans are offered as prey (Cooper and Goldman in press; Lasenby and Langford 1973). In addition, since virtually the entire *Daphnia* population was exposed to *Mysis* predation both horizontally and vertically, the observed increase in death rates and the overall impact of the *Daphnia* population is not surprising.

The relative importance of predation by *Mysis* and the observed changes in egg production by *Daphnia* to its decline in 1970 is uncertain. Most questionable is whether annual variation in egg production by *Daphnia* was a necessary ingredient to the observed decrease in population densities in 1970. That *Daphnia* was largely a victim of increased death rates in 1970 has been suggested by Threlkeld (1979) on the basis of analyses of the age distribution of eggs being carried by the population at the time of the 1970 population decline. Theoretical expectations for declining populations where mortality is of little importance (relative to declining rates of egg production) are for an egg age distribution dominated by older eggs (Edmondson 1968; Threlkeld 1979). However, analyses of the egg age distribution of *Daphnia* in 1970 showed that younger stages dominated, in direct contrast to this expectation but as expected for a population declining because of increased predation rates.

Although low rates of egg production may have been only incidental to the decline of *Daphnia* from Lake Tahoe, the possibility of re-establishment necessarily depends on the magnitude of population birth rates. Although *Daphnia* had virtually disappeared by the end of 1970, it has been collected sporadically at the index station in 1971 (6 December: 1 specimen), in 1972 (12 July: 1 specimen), and in 1973 (1 October: 7 specimens). Similarly, although *Bosmina* was last abundant in early 1971, it was also found at the index station in 1973 (1 October: 3 specimens) and from 8 October 1974 to 9 April 1975 (*see next section*). Thus, it appears that a very rare residual population of cladocerans persisted in the lake through 1974. Because of the large volume of the lake (156 km^3), rapid recolonization would almost have to depend on recruitment from some residual population. If we assume that population densities are at least three orders of magnitude less than their minimum levels in 1967–1971 (about 5 m^{-3}), the importance of the low temperatures of the hypolimnion habitats where these animals lived and the low rates of egg production become very important to chances of recolonization. Based on minimum rates of increase for *Daphnia* in 1967–1970 ($r = 0.005$), we find an increase of three orders of magnitude would take about 3.8 years. But even at the maximum observed rates of population growth in 1967–1970 ($r = 0.018$), the same increase would still take about 1 year. Likewise, a three-orders-of-magnitude increase in *Bosmina* densities at average rates of population increase observed between 1967 and 1970 ($r = 0.049$) would take about 140 days. Thus, very low productivity and cold hypolimnion temperatures combine to make these animals and their seasonal abundance patterns very sensitive to the introduced predator *Mysis relicta*.

Recent zooplankton changes in Lake Tahoe—Since the disappearance of *Daphnia* and *Bosmina* in 1970–1971, populations of the remaining zooplankton species of Lake Tahoe (*Diaptomus tyrelli*, *E. nevadensis*, and *Kellicottia longispina*) have fluctuated

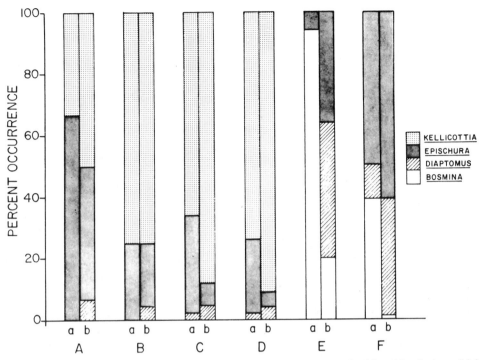

Fig. 4. Frequency of occurrence of prey items in *Mysis relicta* stomachs (a) and in plankton (b) in Lake Tahoe (A–D) and in Emerald Bay (E–F). Dates and total number of prey items recovered from stomach analyses (in parentheses) are as follows: A—8 November 1977 (226); B—9 December 1977 (177); C—22 January 1978 (169); D—14 February 1978 (232); E—29 November 1977 (50); F—13 February 1978 (101).

dramatically, but in general their densities have been inversely related to estimates of *Mysis* density (Fig. 2). Analyses of *Mysis* stomach contents show extensive use of these prey items (Fig. 4). Thus, control of zooplankton dynamics in Lake Tahoe by *Mysis* predation may be even broader in scope than of cladoceran prey species alone.

In contrast to *Mysis* predation, annual changes in primary production as a measure of potential food production rates do not provide a satisfactory explanation for the observed zooplankton fluctuations in these years. The regression of mean annual zooplankton densities against mean annual primary productivity for 1971–1977 is not significantly different from zero ($r = 0.078$, $P > .5$), although the general increase in primary productivity from 1958 to the present (Goldman 1974) may have permitted the increased maxima in zooplankton populations which have been observed.

Zooplankton dynamics in 1974 and 1975 are particularly interesting for the information they provide on relationships between *M. relicta* dynamics and those of their cladoceran and noncladoceran prey. Zooplankton densities were several orders of magnitude greater in 1974 than in 1975 (Fig. 5). Both copepod species reached very high densities in early summer, with both populations then showing a decrease in density (Goldman et al. 1979). These decreases were marked by the elimination of adults and followed by declining numbers of copepodids and nauplii. Although this pattern is similar to the annual dynamics of both *Epischura* and *Diaptomus* in other years, the copepod population declines in 1974 were different in two respects: the severity of the declines, spanning about two orders of magnitude in density over a 2-month period, and the almost total loss of adults as the first indication of the decline.

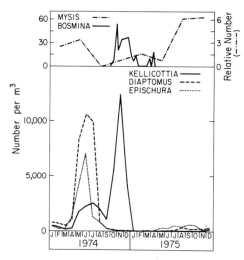

Fig. 5. Densities of dominant zooplankton species in Lake Tahoe, 1974–1975. *Diaptomus* and *Epischura* numbers include all stages. Relative numbers of *Mysis relicta* based on seasonal averages of weight consumed by 38–50-cm lake trout.

After the decline of these two copepod populations, the *Kellicottia* population increased to higher densities than had been observed for this species in Lake Tahoe at any other time. It was also at this time that *B. longirostris* began a sustained reappearance lasting about 6 months, this being the first time in 3 years that *Bosmina* was collected in any appreciable numbers. In its reappearance, *Bosmina* conformed exactly to the timing of its seasonal peaks in abundance in the years 1967–1970. Population numbers declined in the winter of 1974–1975, and no specimens were collected after 9 April 1975.

It was during this period that *M. relicta* was at its lowest density since its establishment in 1970. Examination of *Mysis* stomachs from those collected with the epibenthic trawl in December 1974 showed that *Mysis* consumed *Bosmina* during this period. However, even though *Mysis* consumed *Bosmina*, the reduced *Mysis* densities may have permitted *Bosmina* to exhibit its normal autumnal population increase. The reappearance and subsequent decline of *Bosmina* were contained within a period of relatively uniformly low *Mysis* abundance; thus, while low *Mysis* abundances may have permitted *Bosmina* to reappear in 1974–1975 (in contrast to the winters of 1971–1972, 1972–1973, and 1973–1974), the timing of the reappearance in 1974 and the decline in 1975 is very similar to non-*Mysis* years and should probably be interpreted as the result of typical autumn and winter conditions (e.g. temperature, food).

The year 1975 was unique in the failure of the noncladoceran zooplankton to reestablish high population densities similar to those in 1974. This appears to be related to increased population densities of *Mysis* during spring 1975 which may have effectively limited the development of these zooplankton populations. With reduced densities of *Mysis* from 1976 to the present, zooplankton densities have again increased, thus permitting the possible repetition of this zooplankton-*Mysis* cycle at some future time.

Emerald Bay: Where Bosmina *and* Mysis *coexist*—Emerald Bay is an embayment of Lake Tahoe about 2.5 km long and 0.77 km wide and having a maximum depth of about 60 m. It is virtually separated from the much deeper main body of Lake Tahoe by a shallow sill at its mouth of from 2–4-m depth. Boat traffic in and out of the bay is heavy in summer, and prevailing southwest winds may result in the mixing of surface water of Emerald Bay and Lake Tahoe across the sill. Emerald Bay is one of two places in Lake Tahoe where *Bosmina* occurs and the only area where *Bosmina* and *Mysis* coexist. The bay is generally more fertile than the main body of the lake as revealed by several measures of primary and secondary productivity in Table 2.

Zooplankton collections made at the central deep station (z = 60 m) in Emerald Bay show that *Bosmina* was most abundant seasonally in late fall and early winter 1977–1978, as was recorded for Lake Tahoe in the years 1967–1970 and 1974–1975 (Fig. 6). The pattern of seasonal occurrence in Emerald Bay is noteworthy in two respects. First, the increase in *Bosmina* population densities observed in 1977 came at a

Table 2. Comparison of several measures of primary and secondary productivity in Lake Tahoe and Emerald Bay.

	Lake Tahoe	Emerald Bay
Primary productivity* (mg C m^{-2} day^{-1})	34.1	65.3
Secchi depth (m)†	30	14
Eggs per gravid ♀ *Diaptomus* ‡	10.4	15.6
Eggs per ♀ *Bosmina* §	0.12	0.74
Eggs per ♀ *Kellicottia* ‡	0.18	0.30
Eggs per ♀ *Mysis* (>13.5 mm) ‡	1.59	11.66

* Based on comparisons of measurements made on 15 dates between 1968 and 1970.

† Based on measurements made in October and November 1977.

‡ Based on samples collected from October 1977 to May 1978.

§ Based on samples taken in Lake Tahoe in 1974-1975 and in Emerald Bay in 1977-1978.

time when *Mysis* was increasing, and not decreasing as observed in other *Bosmina* studies in Lake Tahoe. In addition, *Mysis* densities (150-200m^{-2}) were greater than or equal to densities observed in Lake Tahoe in years when *Bosmina* was absent (1971-1973 and 1975-1978). Second, the increase in *Bosmina* densities in 1977 coincides with the seasonal minima of *Epischura*, which is known to be a predator of *Bosmina* in other lakes (e.g. Kerfoot 1975). Thus, *Bosmina* population densities in Emerald Bay appear to be regulated by different factors than operate in the main body of Lake Tahoe. Examination of the spatial distribution of *Bosmina* in October 1977 in Emerald Bay appears to confirm that differences exist between the apparent control of *Bosmina* in Lake Tahoe and in Emerald Bay. Here, *Bosmina* was found to be most abundant at the southwest end of the bay where both *Epischura* and *Mysis* were rare, and virtually absent at the northeast end where *Epischura* was abundant but *Mysis* was rare (Fig. 7). *Bosmina* was rarely collected at a shallow station (z = 20 m) off the mouth of Emerald Bay, even though *Mysis* has also been rare at this location throughout the sampling program.

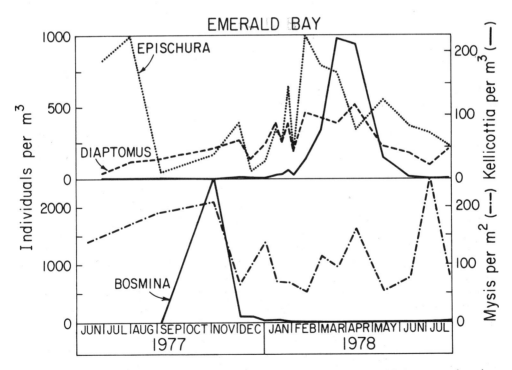

Fig. 6. Densities of dominant zooplankton species in Emerald Bay, 1977-1978, at central station (z = 60 m).

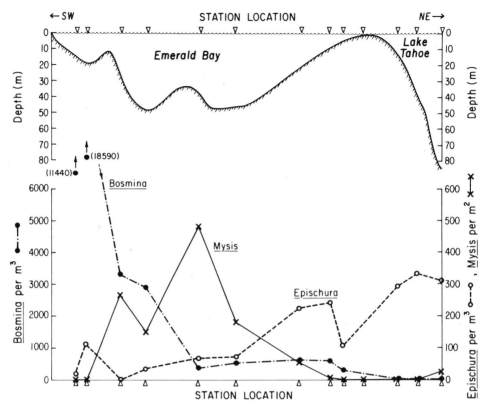

Fig. 7. Densities of dominant zooplankton species along a transect in Emerald Bay, 31 October 1977. Cross-sectional map of Emerald Bay showing station locations also given.

In spite of these obvious relationships among *Epischura*, *Bosmina*, and *Mysis*, we do not feel that *Bosmina* distribution and abundance is satisfactorily explained by *Epischura* distribution alone. Analyses of stomach contents of *Epischura* ($n = 7$ groups totalling 140 individuals) and *Mysis* ($n = 20$ groups totalling 353 individuals) collected from October 1977 through February 1978 where *Bosmina* was also present reveal that *Mysis*, but not *Epischura*, feed on *Bosmina* in Emerald Bay. The lack of *Bosmina* body parts in *Epischura* stomach contents does not appear to result from differences between *Mysis* and *Epischura* feeding behavior because *Epischura* ($n = 20$), fed on *Bosmina* in laboratory cultures, was found to contain numerous *Bosmina* body parts. Thus, an analysis of *Bosmina* distribution and abundance without reference to *Epischura* must be attempted.

The lack of *Bosmina* at the shallow station off the mouth of Emerald Bay may result from lower amounts of food resources available there, this being suggested by reduced particulate organic carbon concentrations at this station (July 1978: 0.12 mg C liter^{-1}) compared to stations within the bay (0.30 mg C liter^{-1}) and by the lower egg per female ratios found for Lake Tahoe *Bosmina* than found in Emerald Bay (Table 2). However, the spatial distribution and seasonal occurrence of *Bosmina* within Emerald Bay is difficult to explain on the basis of food resource variation. Analyses of chlorophyll flourescence (29 November 1977: CV = 28.8%) and particulate organic carbon (24 July 1978: CV = 3.8%) indicate little resource variation among the three regularly sampled stations within Emerald Bay. However, variations in the seasonal distribution of *Mysis* may help to account for the spatial and temporal distribution of *Bosmina*. These appear to be related to the bathymetry of the basin, as well as to effects of thermal stratification on vertical distribution of *Mysis*. Summer thermal stratification appears to limit *Mysis* to areas of the bay deeper than about 20 m, thus effectively eliminating this predator from both the northeast and southwest ends of the bay. Following reproduction by *Mysis* in the spring, *Mysis* juveniles have been found to populate the northeast end of the bay more readily than the southwest end, probably as a result of the distribution of deep sill barriers to juvenile dispersal (Fig. 7). From May to July 1978, densities at the northeast end averaged 6.5 times those found at the southwest end ($\pm$ 2 SE = 2.7). As thermal stratification occurs, migration of the population to the deep portion occurs, leaving the two shallow ends open for recolonization by *Bosmina*. However, because *Bosmina* has been effectively eliminated from the northeast end by spring by predation and possibly by diffusion losses into Lake Tahoe, the main source of a population increase in the fall may be at the southwest end of the bay (from 30 January to 3 July 1978, mean *Bosmina* densities at the northeast end were 0.83 m^{-3}, while at the southwest end mean densities were 4.34 m^{-3}). Any residual effect of *Epischura* predation, if taking the form of prey injury or partial consumption of prey, simply reinforces this pattern of distribution and autumnal colonization of Emerald Bay by *Bosmina*. The overall increase in population density in the fall may emanate from this source throughout the bay, being regulated in part by the distribution of the *Mysis* population.

Tahoe Keys: Leakage of Bosmina *and* Daphnia *into the lake and their fate*—Two marina basins (Tahoe Keys) at the south end of Lake Tahoe provide another opportunity to evaluate the relationship of cladoceran dynamics to the distribution of *Mysis* and variations in resource productivity. Beginning in March 1978, zooplankton samples have been collected along two transects of three stations each originating in the two marinas of Tahoe Keys and extending about 75 m offshore. These transects differ significantly from the Emerald Bay transects in being over uniformly shallow water ($\leqslant$ 3 m) which effectively prevents *Mysis* from being present nearby. As such, the distribution of zooplankton along these transects provides information on the effect of differences in productivity without the complicating effect of predation by *M. relicta*. Limited studies indicate that the marina basins are considerably more productive (25 July 1978: 178.9 mg C m^{-3} day^{-1}) than either Lake Tahoe (25 July 1978: 1.99 mg C m^{-3} day^{-1}) or Emerald Bay (28 July 1978: 10.0 mg C m^{-3} day^{-1}), and consistently (March - July 1978) have Secchi disk readings of $<$ 3 m (compared to 10–15 m in Emerald Bay and from 25–35 m in Lake Tahoe).

Since March 1978, both *D. ambigua* and *B. longirostris* have been found at stations both inside and outside these marinas, although their densities and those of their eggs are obviously much lower in Lake Tahoe than in the marina centers (Fig. 8). Mean densities over the sampling period reflect the degree of physical exposure of the two marinas and their connection to the lake. The eastern marina is connected by a long, protected channel to the lake. In the lake off this marina, mean *Daphnia* and *Bosmina* densities are usually 0.5% of those found in the marina center. Offshore from the western marina, which has a mouth about twice as wide as does the eastern marina and with a less-protected channel half as long, mean densities of *Bosmina* and *Daphnia* are from 3–4% of those in the marina center. Dispersal of these cladocerans into the lake is probably passive and considerable dilution occurs at the lake mouth. However, *Daphnia* and *Bosmina* are not found to reach the edge of the south lake shelf (0.35 km offshore) where *Mysis* is found. *Mysis* collected at these stations did not include cladoceran prey remains. Thus, the question of the fate of the introduced cladocerans must be considered without respect to predation activities of *Mysis*.

Enclosure experiments (1 May – 29 June 1978) along the west shore of the lake (north of the index station in surface water) showed that *D. ambigua* females were unable to replace themselves ($r = -0.40$; $R_0 = 0.091$), although *D. pulicaria* obtained from Indian Creek Reservoir could do so ($r = 0.005$, $R_0 = 1.37$) during the same period. *Daphnia ambigua* from Tahoe Keys may not serve as an adequate source of recolonization because this species may be unable to live in the lake, even when predators are excluded. Similar experiments have not been performed with *B. longirostris*, but it seems likely from our Emerald Bay experience that reproduction would be much greater in Tahoe Keys than in the main body of the lake.

Another explanation for the inability of *Daphnia* and *Bosmina* to colonize the entire shelf area of the south end of the lake includes predation by *E. nevadensis*. However, as in the Emerald Bay studies, adult *Epischura* collected from stations 3 and 6 (Fig. 8) had not fed on either *Daphnia* or *Bosmina*, although laboratory experimentation showed that *Epischura* could take these prey and that body parts could then be recovered in stomach analyses.

General discussion

The spatial and temporal dynamics of zooplankton in Lake Tahoe appear to be most satisfactorily explained by variations in resource productivity and the distribution and abundance of the introduced predator *M. relicta*. Two important questions are raised by these spatial and temporal relationships of zooplankton and mysids. First, what is the relative importance of *Mysis* predation and food resource variation to the development of zooplankton populations? Second, how important are physical factors such as diffusion, bathymetry, or lake size to the distribution and abundance of zooplankton when an omnivore such as *M. relicta* is introduced into a system? Our studies provide some insight into the relative importance of such biotic and abiotic factors in ultraoligotrophic lakes perturbed by the introduction of a predator species.

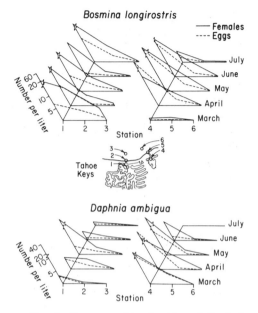

Fig. 8. Mean monthly densities of *Daphnia ambigua* and *Bosmina longirostris* females and eggs at six stations along two transects originating in two Tahoe Keys marinas, March–July 1978.

The dramatic decline of cladocerans in 1970–1971, as well as the limited reappearance by *Bosmina* in 1974–1975, from Lake Tahoe appears most closely linked causally with variation in the abundance of *M. relicta*, shown by stomach analyses and laboratory feeding studies to be particularly voracious on these prey. However, the inability of cladocerans to repeatedly colonize the open waters of Lake Tahoe appears to be partially due to very low rates of egg production in these infertile waters, as well as the lack of any spatial or temporal refuge from the predation activities of *Mysis*, a relationship resulting directly from the bathymetric character of the lake (very steep sides exposing virtually the entire population to *Mysis* predation activity). The very low productivity also appears responsible for the limited colonization abilities of cladocerans being presently introduced from Tahoe Keys and Emerald Bay. The large volume of the lake makes any successful cladoceran recolonization dependent on reduced *Mysis* densities for several years, a condition which has simply not been met since 1970.

The introduction of *M. relicta* to other lakes has often resulted in changed abundance of cladoceran zooplankton and has even resulted in shifts in species composition (reviewed by Gosho 1975). A particularly interesting situation has been observed in Pend Oreille, Idaho, where a warm water daphnid, *D. galeata mendotae*, has almost completely replaced the cold water form, *D. thorata*, which was present before *Mysis* became established there (Rieman 1977). The elimination of *D. thorata*, if due to *Mysis* predation, is very similar to circumstances in Lake Tahoe. However, the ability of a new species to colonize the epilimnion would necessarily depend on sufficient resources and time, two features not present in Lake Tahoe. The current situation in Emerald Bay appears more similar to that in other lakes where cladocerans coexist with introduced or native *Mysis*. Bathymetric features allow a sufficiently large refuge for *Bosmina* to persist, while the bay's higher productivity permits rapid recolonization each year. Even the high densities of *Mysis* found in Emerald Bay at this time are insufficient to prevent the seasonal increase of *Bosmina* populations, a pattern which must result from a positive balance of *Bosmina* recruitment over mortality within the bay as a whole. In Lake Tahoe the more widespread distribution of *Mysis* (over more of the basin), coupled with its lower productivity, makes cladoceran recolonization especially difficult, even though *Mysis* densities are also lower.

The distribution of cladocerans surviving in many lakes with *Mysis* confirms these patterns: usually the cladocerans are epilimnetic and occur when *Mysis* is restricted to deeper, cooler waters by summer thermal stratification (Rieman 1977; Watson and Carpenter 1974). In addition, the lakes generally have sufficient spatial refugia (resulting from more gentle bathymetry than in Lake Tahoe) or high productivity permitting rapid recolonization when *Mysis* vertical distribution becomes restricted.

The response of the remaining zooplankton in Lake Tahoe to variations in *Mysis* abundance is also not surprising considering that *Mysis* readily consumes all of these prey species. What is of interest is the eventual outcome of the predator-prey oscillations in Lake Tahoe. Which species, if any, will become the next casualty of *Mysis* predation? And what will be the magnitude of any future oscillations in zooplankton and *Mysis* abundance? As with cladocerans, the relatively low productivity of Lake Tahoe will make these prey species especially vulnerable to rapid changes in *Mysis* abundance or distribution, as seen in the reduced development of zooplankton in 1971 and 1975. Perhaps most important to the dynamics of the two copepod species is that their distributions include a vertical refuge from *Mysis* predation in midsummer, when thermal stratification limits *Mysis* to the hypolimnion and concentrates *Epischura* (all stages) and *Diaptomus* (adults and copepodids) in the epilimnion. The ability of *Kellicottia* (which is concentrated in the hypolimnion) to withstand *Mysis* predation may stem entirely from its rapid population growth rates and reduced selection by *Mysis* (Fig. 4).

Thus, the physical character of Lake Tahoe and its embayments contribute to the observed distributions of *Mysis* and its zooplankton prey. The sensitivity of these zooplankton prey to the predatory activities of *Mysis* is accentuated by the very low productivity of Lake Tahoe and the almost complete lack of spatial and temporal refuges (except in Emerald Bay and Tahoe Keys) from this predator.

References

COOPER, S. D., and C. R. GOLDMAN. In press. *Mysis relicta* predation on zooplankton. Limnol. Oceanogr.

EDMONDSON, W. T. 1968. A graphical model for evaluating the use of the egg ratio for measuring birth and death rates. Oecologia 1: 1–37.

GOLDMAN, C. R. 1974. Eutrophication of Lake Tahoe emphasizing water quality. EPA 660/3-74-034. U.S. GPO, Washington, D. C.

———, and R. ARMSTRONG. 1969. Primary production studies in Lake Tahoe, California–

Nevada. Int. Ver. Theor. Angew. Limnol. Verh. 17: 49-71.

——, M. D. MORGAN, S. T. THRELKELD, and N. ANGELI. 1979. A population dynamics analysis of the cladoceran disappearance from Lake Tahoe, California-Nevada. Limnol. Oceanogr. 24: 289-297.

GOSHO, M. E. 1975. The introduction of *Mysis relicta* in freshwater lakes. Univ. Wash. Coll. Fish. Circ. 75-2. 66 p.

HUTCHINSON, G. E. 1967 A treatise on limnology, r. 2. Wiley.

KERFOOT, W. C. 1975. The divergence of adjacent populations. Ecology 56: 1298-1313.

LASENBY, D. C., and R. R. LANGFORD. 1973. Feeding and assimilation of *Mysis relicta*. Limnol. Oceanogr. 18: 280-285.

LINN, J. D., and T. C. FRANTZ. 1965. Introduction of the opossum shrimp (*Mysis relicta* Loven) into California and Nevada. Calif. Fish Game 51: 48-51.

MORGAN, M. D., S. T. THRELKELD, and C. R. GOLDMAN. 1978. Impact of the introduction of kokanee (*Oncorhynchus nerka*) and opossum shrimp (*Mysis relicta*) on an alpine lake. J. Fish. Res. Bd. Can. 35: 1572-1579.

RICHARDS, R. C., C. R. GOLDMAN, T. C. FRANTZ, and R. WICKWIRE. 1975. Where have all the *Daphnia* gone? The decline of a major cladoceran in Lake Tahoe, California-Nevada. Int. Ver. Theor. Angew. Limnol. Verh. 19: 835-842.

RICHERSON, P. J. 1969. Community ecology of the Lake Tahoe Plankton. Ph.D. thesis, Univ. california, Davis. 111p.

RIEMAN, B. E. 1977. Limnological studies in Lake Pend Oreille. Idaho Dep. Fish Game: Lake Reservoir Invest. Job Rep. F-53-R-12. Job IV-d.

RIGLER, F. H. 1971. Zooplankton, p. 228-255. *In* W. T. Edmondson and G. G. Winberg [eds.], Secondary productivity in fresh waters. IBP Handbook 17. Blackwell.

THRELKELD, S. T. 1979. Estimating cladoceran birth rates: The importance of egg mortality and the egg age distribution. Limnol. Oceanogr. 24(4): 601-612.

——, M. D. MORGAN, J. T. RYBOCK, and C. R. GOLDMAN. In press. Zooplankton community structure in Lake Tahoe. *In* C. R. Goldman and R. L. Leonard [eds.], Lake Tahoe: A research monograph.

WATSON, N. H. F., and G. F. CARPENTER. 1974. Seasonal abundance of crustacean zooplankton and net plankton biomass of Lakes Huron, Erie and Ontario. J. Fish. Res. Bd. Can. 31: 309-317.

WETZEL, R. G. 1975. Limnology. Saunders.

49. Odonate "Hide and Seek": Habitat-Specific Rules?

Dan M. Johnson and Philip H. Crowley

Abstract

Odonata, dominant invertebrate predators in the littoral zone of many lakes, face the familiar problem of avoiding predators (especially fish) and of capturing prey (Cladocera, midge larvae, and others). Niche space is apparently partitioned by odonates primarily along a habitat (microhabitat) axis, though seasonality and food may also be important. Fish predators appear to restrict odonates to one of two species-specific life-styles: a sluggish, cryptic, smaller, tactile, slow-growing type; and an active, larger, visual, rapidly growing type. Coexistence of Odonata and benthic Cladocera depends on apparent shifts in predator and prey behavior with prey densities, stabilizing properties of the numerical, developmental, and functional responses of these predators to their prey, and the complexity of littoral structure.

Odonate naiads are among the most abundant organisms in fresh-water littoral communities (Ball and Hayne 1952; Gerking 1962; Macan 1964), often attaining biomass standing crops considerably greater than their benthic prey (Benke 1976). Analyses of gut contents and fecal pellets show that odonate naiads have catholic diets (Chutter 1961; Pritchard 1964; Ross 1967; Benke 1972; Kime 1974); the diet of damselfly naiads (suborder Zygoptera) has usually been found to include cladocerans, which dominate numerical analyses, and midge larvae, which contribute most of the biomass (Macan 1964; Fischer 1966; Lawton 1970; Pearlstone 1971; Martin unpubl.). Unlike "typical invertebrate predators" in recent discussions of the size structure of zooplankton communities (Hall et al. 1976; Lynch 1977), odonates should influence *all* size classes of their zooplankton prey—even the large prey that are relatively safe from *Chaoborus* (Dodson 1974) and cyclopoid copepods (Kerfoot 1977).

There are only a few experimental studies of the influence of odonates on zooplankton populations. For example, Hall et al. (1970) altered the densities of invertebrate predators (mainly Odonata and Hemiptera) in experimental ponds and found an inverse relation between the abundance of these predators and the abundance of benthos and zooplankton. (They also detected a shift toward smaller species of zooplankton in "high predator" ponds, though it is unclear to what extent the odonates might be responsible for this result.) Keen (1973) suggested that high mortality rates of chydorid populations in Lawrence Lake might be attributable, in part, to odonate naiads present in the same habitat. And Johnson (1973) credited *Ischnura verticalis* with imposing heavy, although intermittent, mortality on *Simocephalus* populations in experimental pools. Subsequent analyses of his data (Johnson et al. 1975; Crowley 1975a) suggest that the naiads may have exerted temporary "control" over prey at low prey densities. We here consider some problems and "strategies" for coexistence—with other odonates, with predaceous fish, and with their zooplankton prey.

We thank C. Coney for help with the sampling at Bays Mountain, A. Wilson for suggestions on an earlier draft of the manuscript, and K. L. Johnson for help with the typing. This research was supported by NSF grants DEB 78-17518 (to D.M.J.) and DEB 78-02832 (to P.H.C.), a research assistantship from the East Tennessee State University Research Development Committee (to DMJ), and a University of Kentucky Faculty Summer Research Fellowship (to P.H.C.).

Niche partitioning by odonates

Many species of odonates often coexist in one body of water (Bick and Bick 1958; White 1963; Kormondy and Gower 1965; Benke and Benke 1975; Voshell and Simmons 1978). Our own collections at Bays Mountain Park (Sullivan Co., Tenn.) contain more than forty species. Such diversity among members of one order has provoked considerable interest in niche partitioning among odonate naiads. Following Schoener (1974), we focus briefly on time (seasonal segregation), food, and habitat as possible niche axes.

Seasonal segregation of life histories may reduce overlap among species coexisting in similar habitats (Corbet 1962; Kormondy and Gower 1965; Kime 1974; Sawchyn and Gillott 1975; Benke and Benke 1975; Ingram 1976; Ingram and Jenner 1976). Since larger size classes of naiads can eat larger size classes of prey (Chutter 1961; Kime 1974; Thompson 1975), diet overlap may be reduced whenever two odonates are concentrated in different size classes at any particular time (Hutchinson 1959). But many odonate populations have such broad size-frequency distributions that even species whose modal size classes are quite different may overlap considerably. Benke (1978) demonstrated the existence of significant competitive and predatory interactions among odonate species with quite different life histories.

Most studies of odonate diet suggest that naiads attack any prey large enough to detect and small enough to handle. When comparison of diet and prey availability implies selective predation, the results can usually be explained by microhabitat specificity (Chutter 1961) or differential detectability (Ross 1967; Martin unpubl.). Thus it seems likely that the influence of odonates on littoral zooplankton communities depends on the microdistribution of potential prey (Quade 1969, 1973; Lang 1970; Whiteside 1974) and on prey behavior—in the context of the naiads' own behavior and morphology.

Habitat segregation is clearly of primary importance: a remarkable adaptive radiation within odonate groups has generated morphological and behavioral specializations for living in particular microhabitats (Wright 1943; Sherk 1977; Nestler 1978). Corbet (1962) discerned three dominant themes underlying this evolutionary process: modifications of body form and behavior to make naiads less conspicuous to fish; modifications of antennae, eyes, and extensible labium to help naiads detect and capture the prey characteristic of the habitat; and modifications of rectal gills or caudal lamellae for obtaining sufficient oxygen.

Our sweep-net collections from ten different stations indicate the patterns of habitat segregation at Bays Mountain Park (Table 1). One sampling station is in the fishless 0.58-ha Ecology Pond; the others are located in five distinct littoral habitats in Bays Mountain Lake, an 15-ha eutrophic lake containing bluegill sunfish (*Lepomis macrochirus*) and large-mouth bass (*Micropterus salmoides*). The habitats sampled range from horizontally layered leaf litter (detritus) to closely packed vertical columns of emergent rushes (*Eleocharis*). Four habitat types are represented by two sampling stations each, one station at each end of the lake. We emphasize the following points about these data—

1. Though we have collected more than 40 species as adults or naiads in this locality, there are only a few that are abundant in the habitats sampled. The others may be transient, rare, or living in other habitats.

Table 1. Total numbers of naiads collected from dominant odonate taxa in nine monthly sweep-net samples (July-November 1977, March-June 1978) at 10 sampling stations in Bays Mountain Park, Sullivan County, Tennessee.

	Lake (bluegill and large-mouth bass)					Pond (no fish)
	Detritus	Vallisneria and Najas (submerged)	Scirpus (submerged and emergent)	Eleocharis (emergent)	Nelumbo (floating-leaved)	Detritus and Najas
Zygoptera						
Argia fumipennis violacea	3,7	0,0	0,0	2,9	4	0
Enallagma aspersum	0,0	0,0	0,0	0,0	0	138
Enallagma signatum	18,24	44,11	6,2	1,1	0	0
Enallagma traviatum	60,78	45,58	31,25	10,2	12	1
Ischnura verticalis (and I. posita)	8,0	4,0	15,0	39,9	11	3
Anisoptera						
Celithemis spp.	15,17	13,9	17,5	8,2	7	0
Tetragoneuria cynosura	59,35	7,14	3,4	8,1	9	2
Libellula spp.	8,0	2,2	20,8	1,3	9	6
Plathemis lydia	0,0	0,0	1,0	0,0	0	19

2. Among the most abundant species, two (*Enallagma aspersum* and *Plathemis lydia*) are found almost exclusively in Ecology Pond. Other similarly distributed species not included in Table 1 are *Lestes eurinus* and *Anax junius*, both known to be active, visual predators.

3. *Argia fumipennis violacea* is found only in habitats that have a significant amount of detritus. We suspect it may be exploiting microhabitats that contain relatively low oxygen concentrations.

4. *Ischnura verticalis* (and *I. posita*) generally abound in the emergent rushes (cf. Walker 1953), but the two rush-sampling stations differ markedly. We attribute this difference to the intensity of predation by small bluegill; at the site with more *Ischnura*, the *Eleocharis* is more densely packed and extends farther from shore.

5. Two congeneric species (*Enallagma traviatum* and *E. signatum*) are both particularly abundant in the detritus and submerged macrophytes (cf. Garman 1927). We have evidence for some seasonal segregation of these species (*E. signatum* tends to emerge later), but their size classes overlap considerably throughout the year. Niche partitioning by these two species needs further study.

6. The most abundant dragonfly species in our collections is *Tetrogoneuria cynosura*. It is found predominately in detritus habitats, appears to have a 2-year life cycle in this lake (as in Kormondy 1959), and emerges synchronously early in the spring (see Lutz and Jenner 1964).

Fish as keystone predators

There is reason to believe that predaceous fish assume the role of "keystone predators" (Paine 1966), determining the species composition and abundance in odonate communities (Kime 1974; Nestler 1978). As relatively large invertebrates, odonate naiads are attractive food for size-selective fish (Hall et al. 1970; Werner and Hall 1976, 1977; Macan 1977). The risks associated with exposure to fish predation may largely determine which search strategies by odonates can succeed in particular habitats. Odonate species can be separated into two groups according to their method of searching and its implications for growth (Corbet 1962; Richard 1970; Sherk 1977; Nestler 1978).

Cryptic "sit-and-wait" odonates that rely primarily on tactile detection of prey. They can tolerate long periods without food, and they exhibit relatively slow growth and long generation times. This strategy seems relatively compatible with the presence of fish predators.

Actively searching odonates that rely more heavily on vision and live in habitats where fish predation is less intense (dense weed-beds and shallow water) or absent. They often achieve very rapid growth resulting in short generation times, larger final size, or both.

A limited amount of information suggests that the ratio of dragonflies (Anisoptera) to damselflies (Zygoptera) in a community may depend on the presence or absence of fish considered to be weed-bed specialists. Dick's Pond (Benke and Benke 1975) supports a large population of the dollar sunfish *Lepomis marginatus*, while Lake Isaqueena (Nestler 1978, pers. comm.) contains the green sunfish *L. cyanella*. The latter has been shown to specialize on the fauna of littoral vegetation, especially damselfly naiads (Werner and Hall 1976, 1977), and the former may be considered its ecological equivalent (E.E. Werner pers. comm.). The odonate communities of both lakes are dominated by dragonflies and contain relatively small numbers of damselflies. On the other hand, the odonate communities in Hodson's Tarn (Macan 1964, 1966, 1977) and in Bays Mountain Lake (Table 2 and personal observations) are dominated by damselflies. The fish in these cases are relatively generalized predators: trout *Salma trutta* in Hodson's Tarn and the bluegill sunfish *L. macrochirus* and large-mouth bass *M. salmoides* in Bays Mountain Lake. Corroborating this relation between particular fish species and the dominance of the different odonate suborders in other communities would raise some intriguing questions: Do damselfly naiads compete with small size classes of dragonflies for food or space? Are damselflies outcompeting dragonflies in the absence of fish? How do small dragonfly naiads avoid being eliminated by such specialized fish?

Coexistence of Odonata and Cladocera

At this point we address some questions about the coexistence of odonates and Cladocera in the littoral zone. Could odonates selectively eliminate cladocerans that exhibit certain types of behavior? What prevents odonates from decimating the littoral benthic cladocera? Can odonates regulate the densities of their prey?

Table 2. Disc equation parameters for experiments with various damselfly naiads preying on cladocerans: a—successful attack rate, liters per naiad per hour; b—handling time, naiad-hours per prey; $1/ab$—half-saturation constant, prey per liter. Parameters apply to curves for eating rate per naiad, except parenthetical values for *I. ramburii* from a curve for killing rate per naiad.

	Daphnia			Simocephalus		
	a	b	$1/ab$	a	b	$1/ab$
Ischnura elegans (Lawton et al. 1974)	1.66	0.03	20.1	0.50	0.06	33.3
Ischnura ramburii (Akre 1976)	0.31 (0.42)	0.11 (0.06)	29.3 (39.7)	0.09	0.08	138.9
Anomalagrion hastatum (Akre 1976)	0.45	0.29	7.7	0.11	0.36	25.3
Lestes disjunctus (Johnson and Tschumy in prep.)	0.65	0.26	5.9			

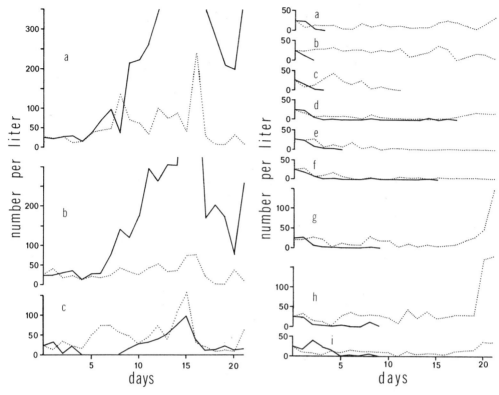

Fig. 1. *Daphnia* (solid line) and *Simocephalus* (dotted line) population densities in replicated 2-liter beakers with algae renewed daily. No predators present. Peak densities reached offscale are (a) 1,190 liter^{-1} and (b) 590 liter^{-1}.

Fig. 2. *Daphnia* (solid line) and *Simocephalus* (dotted line) population densities in 2-liter beakers with algae renewed daily. Experiments with last instar damselfly naiads (*Ischnura ramburii*) present: a, b, and c—four naiads per beaker; d, e, and f—two naiads per beaker; g, h, and i—one naiad per beaker.

Many odonates have adopted a sit-and-wait searching mode in order to avoid attracting the attention of fish. It seems likely that predation by such odonates would fall most heavily on those cladoceran prey species that are most active. Furthermore, Pritchard (1965) has shown that the types of motion exhibited by prey is also important—a jerky motion is more easily detected than a gliding motion. The two Cladocera most often used in laboratory studies provide an ideal comparison of these behavioral characteristics: *Daphnia* moves continuously with a jerky motion, while *Simocephalus* spends much of its time attached to weeds or substrate and swims more smoothly. *Daphnia* has proved the better competitor when the two species are cultured together in structurally simple containers (Frank 1952; and pers. obs.).

Figures 1 and 2 show the result of a simple experiment conducted by an introductory biology class at Rice University (spring semester 1972). *Daphnia magna* and *Simocephalus vetulus* populations were established in 2-liter beakers, and a fresh supply of algae from a mixed culture diluted to a standard density (16 Klett colorimeter units) was provided each day. Populations were censused daily by either subsampling or direct counts. *Daphnia* proved to be dominant in control populations (Figure 1), though *Simocephalus* also attained reasonably high densities. But

Daphnia was eliminated very rapidly from all beakers with damselfly naiads present, while *Simocephalus* persisted at low densities in all but one case. At least under these experimental conditions, *Simocephalus* behavior lowers the effect of odonate predation.

The answers to the second and third questions above depend largely on the responses of odonate predation to changes in prey density: the numerical response, the developmental response, and the functional response.

Odonata respond reproductively to prey densities only after a time lag much longer than cladoceran generation times, and many other factors (e.g. alternate prey, temperature, fish) may mask any significant coupling of cladoceran densities at one time with odonate densities later. Yet, a strong numerical response through predator mortality (starvation) at low prey densities could relieve predation pressure and tend to dampen large fluctuations in prey density (Crawley 1975). When free-swimming Cladocera are abundant, fish generalists like bluegill may concentrate on them, allowing more odonate naiads to survive and also consume benthic Cladocera (Hall et al. 1970). Naiads may be less active at high prey densities (*see below*), further reducing their vulnerability to fish predators. A greater frequency of molting with more food available would partly ameliorate the accumulation of parasites and fungi within instars. In contrast, all of these potential sources of mortality should assure lower naiad densities when alternate prey densities are low.

A closely related phenomenon is the developmental response (Murdoch 1971): Greater prey availability allows naiads to reach larger instars more quickly, thus, they can kill and consume prey much more effectively (Thompson 1975). Though this response tends to dampen density fluctuations by prey, the time lags inherent in the development process must sharply limit its effectiveness. Regulation by developmental response may be further complicated by behavioral changes associated with development within each instar (Johnson et al. 1975: fig. 2).

The functional response of odonate naiads to cladoceran prey has received considerable recent scrutiny in the laboratory (Lawton et al. 1974; Akre 1976; Akre and Johnson 1979; Johnson and Tschumy in prep.), suggesting that naiads with a single prey show a decelerating increase in predation rate with prey density to an asymptote at high densities (Holling's type 2 curve, generally represented by his "disc equation": Holling 1959 a,b). The parameters of the disc equation, a (effective volume within which all prey are successfully attacked, per predator per unit time) and b (predator-time spent processing each prey and thus unavailable for searching), indicate the approximate range over which predation rate is sensitive to prey density via the half-saturation constant $1/ab$. Values of $a, b,$ and $1/ab$ obtained from experimental data are presented in Table 2. We suspect that the half-saturation constant, particularly when measured for several species under comparable conditions, may be a useful comparative indicator of the capabilities and strategies of odonate predators (*see* Crowley 1975b).

If alternative prey are present and hunger is allowed to vary during the experiment, predator behavior may change with prey density, and the disc equation cannot strictly apply. Lawton et al. (1974), Akre (1976), and Akre and Johnson (1979) have found changes in the attack parameter a with prey densities. In the latter two papers, the change in a is attributed to a shift between search modes—from slow walking at low densities of *Daphnia* and high densities of *Simocephalus,* to sit-and-wait at higher *Daphnia* and low *Simocephalus* densities (*see Appendix*); the shift also assures a sharp rise in mortality to dampen any prey density increase crossing the mode boundary (Fig. 3). Another kind of naiad behavior that can augment the functional response of odonates to prey at higher densities is "wasteful killing" (Johnson et al. 1975). We found that naiads of *Ischnura ramburii* can sometimes kill many more cladoceran prey than they consume when prey are abundant; wasteful killing is much less frequent at low prey densities.

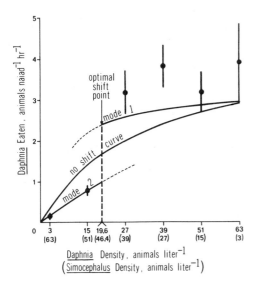

Fig. 3. Number of *Daphnia* eaten per *Anomalagrion hastatum* naiad per hour vs. the densities of *Daphnia* and *Simocephalus* in complementary density experiments (Akre 1976). ●—means of five replicate experiments, bars each extend 1 SE from mean. Following analysis in appendix, we calculate prey densities at which an optimally foraging naiad should shift between search modes 1 and 2; H_1^* and H_2^* are densities of prey species 1 and 2 along boundary between the two search modes. With "no-shift" two-prey type 2 functional response curve parameterized independently for comparison, we fit a_1' and a_2' to appropriate extreme pairs of data points (3:63, 15:51 and 51:15, 63:3, respectively) and find an "optimal shift point" in qualitative agreement with data pattern. Parameter values for Eq. 1 are: $a_1 = 0.450$, $a_2 = 0.110$, $a_1' = 0.189$, $a_2' = 0$ liters naiad^{-1} hr^{-1}; $b_1 = b_2 = 0.305$ naiad-hr prey^{-1}.

A final point here about the functional response: Crowley (1975a) has argued that predators with type 2 functional responses in the laboratory may develop type 3 (sigmoid) responses in structurally complex environments like the littoral zone of lakes. If prey preferentially occupy safer places of finite capacity, then prey become more vulnerable to predators at higher densities (see Benke 1978). Similarly, some Cladocera (e.g. *Simocephalus*) may become more active at higher densities, increasing the risk of encountering ambush predators (Crowley 1975a; cf. flies and spiders in Haynes and Sisojevic 1966).

The extent and structural complexity of littoral zones also suggest that such systems should be functionally subdivided into intensively interacting local subpopulations that exchange a few dispersing naiads and Cladocera. This kind of "cellular" structure can stabilize a system that, if completely homogeneous, would be unstable (Caswell 1978). Furthermore, the high ecological viscosity, or fine-scale structure, associated with dense weed-beds and detritus should help stabilize intracellular (local) naiad-cladoceran interactions (see Crowley 1978).

In conclusion, we attempt to bring our observations about Odonata and Cladocera in the littoral zone into sharper focus by presenting the list of testable hypotheses in Table 3.

Table 3. Testable hypotheses about coexistence within odonate communities and persistence of odonate-cladoceran predator-prey interactions.

1. Fish are keystone predators in odonate communities.
2. Habitat (microhabitat) is dominant odonate niche axis.
3. Odonates are nonselective within range of prey they perceive in a given microhabitat.
4. Odonata-Cladocera coexistence in littoral zone is promoted by:
   a. Strong numerical response to prey density by odonates through mortality imposed by fish, other odonates, and cannibalism.
   b. Significant developmental response to prey density by odonates.
   c. Supra-proportional (density-dependent) functional response to prey density by odonates brought about by
      (i) presence of relative refuges in structurally complex habitats,
      (ii) wasteful killing behavior, and
      (iii) shifting among search modes.
   d. Tendency for structurally complex habitats to become functionally subdivided or "cellular."

Appendix: Optimal foraging in multiple search modes

Increasing the proportion of a particular prey in the diet by a predator with increases in the relative density of the prey is called "switching" (Murdoch 1969). Since switching is defined only for changes in *relative* prey density, the concept cannot properly be applied when changes in absolute (but *not* relative) density alter proportions of prey in the diet. Also, switching has been used primarily, if not exclusively, to describe cases in which the diet (and the behavior that determines it) changes smoothly with relative prey densities; mathematically, the attack parameter a is a continuous, monotonic function of relative densities (e.g. Murdoch and Oaten 1975).

To allow for a more general relation between prey densities and diet and to emphasize discrete rather than continuous changes in diet with densities, we focus here on discrete modes of predator behavior (*see* Marten 1973; Krebs 1973; Akre 1976; Murdoch 1977; Akre and Johnson 1979). These behavioral modes influence the attack parameter by specifying how and where a predator searches for prey. We believe this approach to be particularly applicable to insect predators (for odonates *see* Corbet 1962), some birds (Royama 1971), fish (Murdoch et al. 1975), and many other general predators.

In this appendix, we use a general two-prey functional response model to determine the mode of predator searching behavior that maximizes prey consumption as a function of prey densities (the "optimal search mode"; cf. Marten 1973). We then compare the result to functional response data for *Anomalagrion hastatum* naiads preying on *Daphnia* and *Simocephalus* at complementary densities (from Akre 1976; Akre and Johnson 1979).

To keep the argument simple and general, consider a predator that encounters and consumes two species of prey randomly; that is, either prey are distributed randomly or predators search for them at random. Suppose the predator spends b_1 and b_2 time units precessing each prey found of species 1 and 2, respectively, regardless of the search mode; this "handling time" includes all prey processing time not available for search-pursuit, capture, and digestive pause. Further, suppose there exist two distinct search modes, each with characteristic rates of successful attack by the predator for each prey: In mode 1, species 1 is attacked at a_1 and species 2 at a_2' volume units per predator per time; in mode 2, species 1 is attacked at a_1' and species 2 at a_2 volume units per predator per time. If $E_1(H_1,H_2)$ represents the consumption rate of prey 1 in search mode 1 per predator as a function of prey densities 1 and 2, $E_1'(H_1,H_2)$ is consumption rate of prey 1 in search mode 2, and so on, then

$$E_1 = -\frac{dH_1}{dt} = \frac{a_1 H_1}{1 + a_1 b_1 H_1 + a_2' b_2 H_2},$$

$$E_1' = -\frac{dH_1}{dt} = \frac{a_1' H_1}{1 + a_1' b_1 H_1 + a_2 b_2 H_2},$$

$$E_2 = -\frac{dH_2}{dt} = \frac{a_2 H_2}{1 + a_1' b_1 H_1 + a_2 b_2 H_2},$$

$$E_2' = -\frac{dH_2}{dt} = \frac{a_2' H_2}{1 + a_1 b_1 H_1 + a_2' b_2 H_2}. \quad (1)$$

These equations are a differential-equation form of the two-prey type 2 functional response derived by Murdoch (1973) and Lawton et al. (1974), here presented for both prey in each of the two different searching modes.

The maximal total consumption rate per predator, $\max(E_1 + E_2', E_1' + E_2)$, is achieved when the predator searches in the appropriate mode at given prey densities. The boundary separating the regions of prey-density space in which the alternative modes are optimal is found by setting $E_1 + E_2' = E_1' + E_2$ and solving for the prey densities. the resulting relationship is depicted in Fig. 4.

This analysis can readily be performed using Akre and Johnson's data on the damselfly-cladoceran interaction cited above. Results of that study suggested a shift in search modes between density ratios (*Daphnia: Simocephalus*) 15:51 and 27:39 (Fig. 3); we wondered if predators were behaving optimally with respect to the alternative

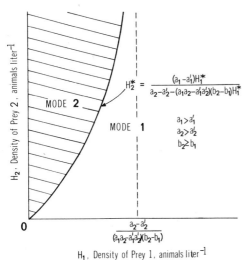

Fig. 4. General relation between discrete search modes and densities of alternative prey. H_1^* and H_2^* are densities of prey species 1 and 2 along boundary between two search modes. Notice that unless handling times of both prey species are the same (as in the special case of Figs. 3 and 5), boundary between modes is curved, approaching an asymptote at some positive finite density of prey handled more quickly. Also note that by this scheme it is quite possible for a predator to shift searching modes even if relative densities of prey remain constant; concept of "switching," as originally formulated, does not encompass this possibility (see Murdoch 1969).

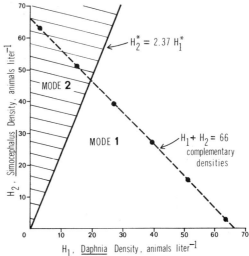

Fig. 5. Relation between search modes and prey densities for *Anomalagrion hastatum* preying on *Daphnia* and *Simocephalus*. Complementary density combinations (●) have been examined experimentally (Akre 1976). Optimal shift point at density combination 19.6:46.4 (*Daphnia:Simocephalus*) found from analysis presented here agrees with apparent shift between 15:51 and 27:39 illustrated in Fig. 3. Parameter values as in Fig. 3.

search modes. The parameters a_1, a_2, b_1, and b_2 were estimated from experiments without alternative prey present (i.e. independent of data in Fig. 3); lacking suitable independent estimates of a_1' and a_2', we fit these to data for the density extremes at which the appropriate search modes were most likely in effect—3:63 and 15:51 (mode 2, a_1') and 51:15 and 63:3 (mode 1, a_2'). Then the equations 1 were formulated; the mode 1, mode 2, and no-shift ($E_1 + E_2$) functional response curves were drawn; the mode boundary line was found; and the optimal shift point was determined (see Figs. 3 and 5). Since the calculated shift point (19.6:46.4) falls between the densities bounding the apparent shift in Fig. 3, the results are consistent with optimal foraging in two alternative search modes.

References

AKRE, B. G. 1976. The effect of alternate prey presence on the functional response by larval damselflies to changes in cladoceran prey density, M. A. thesis, Rice University, Houston, Texas. 132 p.

———, and D. M. JOHNSON. 1979. Switching and sigmoid functional response curves by damselfly naiads with alternate prey available. J. Animal Ecol. 48: 703-720.

BALL, R. C., and D. W. HAYNE. 1952. Effects of the removal of the fish population on the fish-food organisms of a lake. Ecology 33: 41-48.

BENKE, A. C. 1972. An experimental field study on the ecology of coexisting larval odonates. Ph.D. thesis, Univ. Georgia, Athens, 112 p.

———. 1976. Dragonfly production and prey turnover. Ecology 57: 915-927.

———. 1978. Interactions among coexisting predators—a field experiment with dragonfly larvae. J. Anim. Ecol. 47: 335-350.

———. and S. S. BENKE. 1975. Comparative dynamics and life histories of coexisting dragonfly populations. Ecology 56: 302-317.

BICK, G. H., and J. C. BICK. 1958. The ecology of the Odonata at a small creek in southern Oklahoma. J. Tenn. Acad. Sci. 33: 240-251.

CASWELL, H. 1978. Predator-mediated coexistence: a non-equilibrium model. Am. Nat. 112: 127-154.

CHUTTER, F. M. 1961. Certain aspects of the morphology and ecology of the nymphs of several species of *Pseudagrion* Selys (Odonata). Arch. Hydrobiol. 57: 430-463.

CORBET, P. 1962. A biology of dragonflies. Witherby.

CRAWLEY, M. J. 1975. The numerical responses of insect predators to changes in prey density. J. Anim. Ecol. 44: 877-892.

CROWLEY, P. H. 1975a. Spatial heterogeneity and the stability of a predator-prey link. Ph. D. thesis, Michigan State Univ. 96 p.

———. 1975b. Natural selection and the Michaelis constant. J. Theor. Biol. 50: 461-475.

———. 1978. Effective size and the persistence of eco-systems. Oecologia 35: 185-195.

DODSON, S. I. 1974. Adaptive change in plankton morphology in response to size-selective predation: A new hypothesis of cyclomorphosis. Limnol. Oceanogr. 19: 721-729.

FISCHER, Z. 1966. Food selection and energy transformation in larvae of *Lestes sponsa* (Odonata) in astatic waters. Int. Ver. Theor. Angew. Limnol. Verh. 16: 600-603.

FRANK, P. W. 1952. A laboratory study of intraspecies and interspecies competition in *Daphnia pulicaria* (Forbes) and *Simocephalus vetulus* (O. F. Müller). Physiol. Zool. 25: 173-204.

GARMAN, P. 1927. Guide to the insects of Connecticut. Conn. State Geol. Nat. Hist. Surv. Bull. 39. 331 p.

GERKING, S. D. 1962. Production and food utilization in a population of bluegill sunfish. Ecol. Monogr. 32: 31-78.

HALL, D. J., W. E. COOPER, and E. E. WERNER. 1970. An experimental approach to the production dynamics and structure of freshwater animal communities. Limnol. Oceanogr. 15: 839-928.

———, S. T. THRELKELD, C. W. BURNS, and P. H. CROWLEY. 1976. The size-efficiency hypothesis and the size structure of zooplankton communities. Annu. Rev. Ecol. Syst. 7: 177-208.

HAYNES, D., and P. SISOJEVIC. 1966. Predatory behavior of *Philodromus rufus* Walckenaer (Araneae: Thomisadae). Can. Entomol. 98: 113-133.

HOLLING, C. S. 1959a. The components of predation as revealed by a study of small mammal predation of the European pine sawfly. Can. Entomol. 91: 293-320.

———. 1959b. Some characteristics of simple types of predation and parasitism. Can. Entomol. 91: 385-398.

HUTCHINSON, G. E. 1959. Homage to Santa Rosalia or Why are there so many kinds of animals? Am. Nat. 80: 440-457.

INGRAM, B. R. 1976. Life histories of three species of Lestidae in North Carolina (Odonata: Zygoptera). Odonatologica 5: 231-244.

———. and C. E. JENNER. 1976. Life histories of *Enallagma hageni* (Walsh) and *E. aspersum* (Hagen) (Zygoptera: Coenagrionidae). Odonatologica 5: 331-345.

JOHNSON, D. M. 1973. Predation by damselfly naiads on cladoceran populations: fluctuating intensity. Ecology 54: 251-268.

———, B. H. AKRE, and P. H. CROWLEY. 1975. Modeling arthropod predation: wasteful killing by damselfly naiads. Ecology 56: 1081-1093.

KEEN, R. 1973. A probabilistic approach to the dynamics of natural populations of the Chydoridae (Cladocera, Crustacea). Ecology 54: 524-534.

KERFOOT, W. C. 1977. Implications of copepod predation. Limnol. Oceanogr. 22: 316-325.

KIME, J. 1974. Ecological relationships among three species of aeshnid dragonfly larvae (Odonata: Aeshnidae). Ph.D. thesis. Univ. Washington, Seattle. 142 p.

KORMONDY, E. J. 1959. The systematics of *Tetragoneuria* based on ecological, life history, and morphological evidence (Odonata: Corduliidae). Misc. Publ. Univ. Mich. 107. 79 p.

———, and J. L. GOWER. 1965. Life history variations in an association of Odonata. Ecology 46: 882-886.

KREBS, J. R. 1973. Behavioral aspects of predation, p. 73-111. *In* Perspectives in ethology. P. P. G. Bateson and P. H. Klopfer [eds.], Plenum.

LANG, K. 1970. Distribution and dispersion of the Cladocera of Lake West Okoboji, Iowa. Ph. D. thesis, Univ. Iowa. 184 p.

LAWTON, J. H. 1970. Feeding and food energy assimilation in larvae of the damselfly *Pyrrhosoma nymphula* (Sulzer) (Odonata: Zygoptera). J. Anim. Ecol. 39: 669-689.

———, J. R. BEDDINGTON, and R. BONSER. 1974. Switching in invertebrate predators, p. 141-158. *In* M. B. Usher and M. H. Williamson [eds.], Ecological stability. Halsted.

LUTZ, P. E., and C. E. JENNER. 1964. Life-history and photoperiodic responses of nymphs of *Tetragoneuria cynosura* (Say). Biol. Bull. 127: 304-316.

LYNCH, M. 1977. Fitness and optimal body size in zooplankton populations. Ecology 58: 763-774.

MACAN, T. T. 1964. The Odonata of a moorland fishpond. Int. Rev. Gesamten. Hydrobiol. 49: 325-360.

———. 1966. The influence of predation on the fauna of a moorland fishpond. Arch. Hydrobiol. 61: 432-452.

———. 1977. The influence of predation on the composition of fresh-water animal communities. Biol. Rev. 52: 45-70.

MARTEN, G. G. 1973. An optimization equation

for predation. Ecology 54: 92-101.

MURDOCH, W. W. 1969. Switching in general predators: experiments on predator specificity and stability of prey populations. Ecol. Monogr. 39: 335-354.

———. 1971. The developmental response of predators to changes in prey density. Ecology 52: 132-137.

———. 1973. The functional response of predators. J. Appl. Ecol. 10: 335-342.

———. 1977. Stabilizing effects of spatial heterogeneity in predator-prey systems. Theor. Pop. Biol. 11: 252-273.

———, S. L. AVERY, and M. E. B. SMYTH. 1975. Switching in predatory fish. Ecology 56: 1094-1105.

———, and A. OATEN. 1975. Predation and population stability. Adv. Ecol. Res. 9: 1-131. Academic.

NESTLER, J. M. 1978. Resource allocation by 3 co-occurring genera of Anisoptera (Odonata) nymphs from Lake Isaqueena, South Carolina. Bull. Ecol. Soc. Am. 59: 56.

PAINE, R. T. 1966. Food web complexity and species diversity. Am. Nat. 100: 65-75.

PEARLSTONE, P. 1971. Observations of a population of damselfly larvae in Marion Lake, p. 64-66. In Marion Lake Project, IBP Can. Rep. 1970-1971.

PRITCHARD, G. 1964. The prey of dragonfly larvae (Odonata: Anisoptera) in ponds in northern Alberta. Can. J. Zool. 42: 785-800.

———. 1965. Prey capture by dragonfly larvae (Odonata: Anisoptera). Can. J. Zool. 43: 271-289.

QUADE, H. W. 1969. Cladoceran faunas associated with aquatic macrophytes in some lakes in northwestern Minnesota. Ecology 50: 170-179.

———. 1973. The abundance and distribution of littoral Cladocera as related to sediments and plants. Ph. D. thesis, Indiana Univ. 176 p.

RICHARD, G. 1970. New aspects of the regulation of predatory behavior of Odonata nymphs, p. 435-451. In L. R. Aronson et al. [eds.], Development and evolution of behavior. W. H. Freeman.

ROSS, Q. E. 1967. The effect of different naiad and prey densities on the feeding behavior of Anax junius (Drury) naiads. M. S. thesis, Cornell Univ.

ROYAMA, T. 1971. A comparative study of models for predation and parasitism. Res. Pop. Ecol. Kyoto 1 (suppl.): 1-91.

SAWCHYN, W. W., and C. GILLOTT. 1975. The biology of two related species of coenagrionid dragonflies (Odonata: Zygoptera) in western Canada. Can. Entomol. 107: 119-128.

SCHOENER, T. W. 1974. Resource partitioning in ecological communities. Science 185: 27-39.

SHERK, T. E. 1977. Development of the compound eyes of dragonflies (Odonata). I. Larval compound eyes. J. Exp. Zool. 201: 391-416.

THOMPSON, D. J. 1975. Towards a predator-prey model incorporating age structure: the effects of predator and prey size on the predation of Daphnia magna by Ischnura elegans. J. Anim. Ecol. 44: 907-916.

VOSHELL, J. R., and G. M. SIMMONS. 1978. The Odonata of a new reservoir in the southeastern United States. Odonatologica 7: 67-76.

WALKER, E. M. 1953. The Odonata of Canada and Alaska, v. 1. Toronto Univ. Press.

WERNER, E. E., and D. J. HALL. 1976. Niche shifts in sunfishes: experimental evidence and significance. Science 191: 404-406.

———, and ———. 1977. Competition and habitat shift in two sunfishes (Centrarchidae). Ecology 58: 869-876.

WHITE, H. B. 1963. Seasonal distribution and abundance of Odonata at a large pond in central Pennsylvania. Proc. North Central Branch, Entomol. Soc. Am. 43: 120-130.

WHITESIDE, M. C. 1974. Chydorid (Cladocera) ecology: Seasonal patterns and abundance of populations in Elk Lake, Minnesota. Ecology 55: 538-550.

WRIGHT, M. 1943. The effect of certain ecological factors on dragonfly nymphs. J. Tenn. Acad. Sci. 18: 172-196.

50. Alewives (*Alosa pseudoharengus*) and Ciscoes (*Coregonus artedii*) as Selective and Non-Selective Planktivores

John Janssen

Abstract

Alewives and ciscoes have both selective and nonselective methods for feeding on zooplankton. Factors that favor the nonselective feeding mode include large fish size, small prey size, high prey density, and low light intensity. Alewives and ciscoes can feed in the dark and this feeding is not selective. The advantages of feeding in the dark are considered along with some field evidence that it does occur. Possible effects of nonselective planktivores and fish with both selective and nonselective modes of feeding on zooplankton communities are also considered.

The work of Brooks and Dodson (1965) on the effects of the alewife (*Alosa pseudoharengus*) and the blueback herring (*A. aestivalis*) on the zooplankton communities of some Connecticut lakes generated much interest in the role size-selective predation by fish and other vertebrates plays in structuring zooplankton communities. The effect that nonselective fish or fish with both selective and nonselective methods of feeding on zooplankton might have on zooplankton communities, or that such dichotomous feeding modes might exist, has been largely ignored.

The evidence that a fish is size selective has generally been of three types.

1. Indirect evidence based on the effect of an introduced fish on a lake that formerly did not have the fish or based on comparisons of lakes that have fish with those that do not. Examples include the work of Brooks and Dodson (1965), Wells (1970), Hutchinson (1971), and Warshaw (1972) for the alewife, Nilsson and Pejler (1973) for *Coregonus* in Swedish lakes, and Hrbáček (1972), Galbraith (1967), Hall et al. (1970), Reif and Tappa (1966), and others for other species.

2. Direct evidence based on comparing the stomach contents of the fish with the composition of the zooplankton. Examples include the reports of Hutchinson (1971), Warshaw (1972), Rhodes and McComish (1975) for alewives, Brooks' (1968, 1969) use of the data of Berg and Grimaldi (1966) and Tonolli (1962) for *Coregonus*, and Galbraith (1967) for yellow perch (*Perca flavescens*) and rainbow trout (*Salmo gairdneri*). This method is subject to O'Brien and Vinyard's (1974) criticism that, because fish and zooplankton form patches both vertically and horizontally, and fish concentrations and zooplankton concentrations need not overlap, what a plankton net collects may not be what is available to the fish. The method is also subject to Gannon's

---

I thank Bill McNown for helping with the cisco experiments, the University of Wisconsin-Milwaukee's Center for Great Lakes Studies for providing space for much of this work, and W. J. O'Brien and R. Drenner for reading earlier drafts of this paper.

(1976) criticism that different zooplankters digest at different rates thus biasing counts of prey in stomachs.

3. Laboratory observations of feeding with subsequent analysis of fish stomach contents or zooplankton remaining in the test chamber. Examples of such work include Brooks (1968) for young of the year alewives, Ivlev (1961), and Werner and Hall (1974).

Nonselective feeding modes, i.e. modes in which the fish does not orient to individual prey before taking them, have been described for a variety of marine fishes. Examples include the menhaden (*Brevoortia tyrannus*) (which seems to be an obligate filter feeder: Durbin and Durbin 1975), northern anchovy (*Engraulis mordax*: Leong and O'Connell 1969; O'Connell 1972), Pacific mackerel (*Scomber japonicus*: O'Connell and Zweifel 1972), and another scombrid, *Rastrelliger kanagurta* (Colin 1976).

Here I consider two freshwater fishes, the alewife (a marine invader of freshwaters) and the cisco (*Coregonus artedii*, a native freshwater fish), which have both selective and nonselective modes for feeding on zooplankton.

Feeding modes of alewives and ciscoes

Janssen (1976, 1978a) described modes of feeding on zooplankton for alewives and ciscoes. These are briefly summarized in Table 1. Both alewife and cisco have particulate and gulping feeding modes; filtering was found only in the alewife. Both species also have a modified particulate feeding mode which I term "darting" in which the fish quickly rushes at the prey while sucking. This mode is used for agile prey such as *Mysis relicta* and calanoid copepods. In a laboratory experiment under lighted conditions and with high prey density (254 *Daphnia*/liter), Janssen (1976) showed that alewives larger than 178-mm total length (TL) fed by filtering and were not selective. Alewives 124–152-mm TL fed by gulping and were size selective, while alewives < 114-mm TL were particulate feeders and were more size selective than gulpers. Both filterers and gulpers took more than one prey per feeding movement (mouth opening and closing). Close observation of particulate feeders in aquaria indicates they take only one prey per feeding movement. Conditions favoring filtering in alewives and gulping in alewives and ciscoes include large fish size, small prey size, and high prey density. The exact thresholds for various modes are not known and seem likely to be affected by experience (Janssen 1978a). Prey size and density determine whether northern anchovies and Pacific mackerel filter feed or particulate feed (Leong and O'Connell 1969; O'Connell 1972; O'Connell and Zweifel 1972).

The prey density used in the above experiment with alewives is high, but zooplankton densities near 254/liter do occur in alewife-containing waters: Roth and Stewart (1973) reported densities of *Bosmina longirostris* up to 180/liter for southeastern Lake Michigan; Gannon (1974); found *B. longirostris* in concentrations up to 230/liter in southern Green Bay; and Boucherle and Frederick (1976) found *Daphnia galeata mendotae* and *D. retrocurva* in swarms of 10,000 to 20,000/liter at Put-In-Bay Harbor, Lake Erie. Concentrations of prey in patches are likely very important for filter-feeding fish.

As the gulping and filtering modes could potentially be used for feeding in the dark, experiments were conducted to determine if alewives and ciscoes could so feed.

Table 1. Description of feeding modes of alewives and ciscoes (adapted from Janssen 1976, 1978a).

Particulate feeding: Fish visually orients to each prey, capturing it by suction while swimming.

Gulping: Fish constantly opens and closes mouth, pumping in water about 2–3 times per second while swimming.

Filter-feeding mode: Alewife only

Filtering: Fish swims with mouth held agape while beating hard with its tail. A filtering bout lasts about 0.5–2 seconds and is followed by mouth closing and then another filtering bout.

Materials and methods

Ciscoes were collected in late November 1975 from Pallette Lake (Vilas County, Wisconsin). Total lengths of collected fish were about 200–250 mm. Fish were reared in a circular tank (2.44-m diam., 0.45 m deep, 2.1 m^3) with flowing water and fed frozen and dried food until 18 April 1976. From 19–26 April, fish were fed pond-collected *Daphnia pulex* at night with the laboratory lights out and the tank completely covered by two overlapping sheets of plywood. Eight experiments were run from 28 April until 12 May, four with the tank covered as described above and four (also at night) with a fluorescent light on above the uncovered tank. Fish were allowed to feed for 15 minutes after introduction of *Daphnia*. After the feeding the lights were turned on and the tank uncovered (dark experiments) and one or (usually) two fish quickly captured and anesthetized (MS-222: ethyl *m*-aminobenzoate methanesulfonate). Stomach contents were removed and preserved in 10% Formalin. Ten 1–2-liter water samples were taken from scattered locations in the tank, the *Daphnia* anesthetized, screened on Nitex, rinsed, and preserved in 10% Formalin. Remaining fish were left undisturbed until the next experiment. *Daphnia* from stomachs and samples were later counted and about 200 from the combined water sample and each fish (unless fish consumed < 200) measured using an ocular micrometer in a compound microscope.

About 25 alewives of various sizes were captured on 31 May 1976 from Milwaukee Harbor by dip-netting and reared in a tank identical to that of the ciscoes. Feeding was as follows: from 31 May–1 June fish were fed *Daphnia* in an uncovered tank during the day, from 2–8 June fish were fed in the evening about 30 minutes before dark with the laboratory lights off, and from 9–16 June they were fed only after dark, with the lights off, in a covered tank. On 18 June I ran an experiment using all the surviving alewives (15, ranging from 92 to 185 mm). Test procedure was the same as for the dark experiments for ciscoes except 1-liter samples were taken and 100 *Daphnia* from stomachs (if present) and the combined water samples were measured.

Results

Prey size in fish stomachs and water samples were compared statistically, using a Chi-square contingency table with six *Daphnia* size categories (corresponding to divisions in the ocular micrometer) for alewives and the median test (a type of Chi-square test) for ciscoes (Conover 1971). A level of $P < 0.05$ was considered significant.

Results for the ciscoes are given in Table 2. All eight fish from the light experiments were significantly size selective while none was of those fed in the dark. Note that some feeding occurred in the dark at a prey density of only 15.7/liter.

Results for the alewives are given in Table 3. The prey density was 139/liter. Note that all but the smallest alewife had eaten *Daphnia* and that only the 145-mm fish had a significant Chi-square value. However, for 14 fish at a significance level of 0.05 about $0.05 \times 14 = 0.7$ significant differences are expected by chance alone. Thus for both alewives and ciscoes, feeding in the dark is possible and the feeding is not selective. These results also demonstrate that small alewives can feed non-selectively, as was found previously for large alewives feeding in a lighted tank (Janssen 1976).

Discussion

Given that alewives and ciscoes will feed in the dark in the laboratory, is there reason to believe they will in lakes? There is a potential selective advantage for feeding in the dark. Both the alewife and the cisco are schooling fish subject to predation by larger fish. In the laboratory their schools disperse when they feed, however, and in Lake Michigan echosounding recordings indicate that the schools disperse at night (Brandt 1976; pers. obs.). Eggers (1976) argued that schooling interferes with feeding as individual visual fields overlap and prey density will decrease toward the trailing part of

Table 2. Results of study of ciscoes feeding in dark and light. Statistical test: median (Conover 1971). NS—not significant.

	Fish size TL (mm)	Prey conc No./liter	No. prey eaten	Daphnia median size (mm)		χ^2	Signif.
				Fish	Sample		
Dark							
6 May 76	220	15.7	20	1.48	1.40	0.16	NS
	231	15.7	33	1.70	1.40	2.67	NS
4 May 76	252	48.0	341	1.47	1.44	1.62	NS
	223	48.0	308	1.40	1.44	0.76	NS
10 May 76	217	89.8	258	1.26	1.37	1.70	NS
	214	89.8	60	1.35	1.37	0.38	NS
28 Apr 76	213	139	1,203	1.32	1.32	0.0	NS
Light							
2 May 76	250	21.3	314	2.18	1.28	126.02	0.001
	226	21.3	290	2.25	1.28	96.29	0.001
12 May 76	242	43.3	105	2.08	1.00	84.90	0.001
	207	43.3	322	2.05	1.00	78.35	0.001
30 Apr 76	246	46.4	1,566	1.67	1.35	39.85	0.001
	215	46.4	833	2.01	1.35	81.58	0.001
8 May 76	225	390.2	1,150	1.47	1.16	7.76	0.01
	200	390.2	908	2.14	1.16	160	0.001

the school. Theoretical (Cushing and Harden-Jones 1968) and experimental (Neill and Cullen 1974) work indicates that schooling functions to reduce predation by visual predators. Feeding movements may also attract predators. Fish predators tend to select prey from schools that move differently than other members of the school (Hobson 1968). Hobson (1963) noted that anchovetas (*Cetengraulus mysticetus*) were preferentially eaten by gafftopsail pompano (*Trachinotus rhodopus*) when the anchovetas were schooling with flatiron herring (*Harengula thrissina*). Hobson suggested that the flash produced by the anchoveta's gill covers when it made feeding movements attracted the predators. Thus, if energetically feasible, planktivory at night might be an effective means of obtaining food while reducing predation.

There is some evidence from the field that alewives and ciscoes do feed mostly at night during some times of the year. On 20 September 1976, I observed a school of small alewives (50–70-mm TL) feeding near a Chicago breakwater (Janssen 1978b). Feeding began at about sunset and continued at least until it was too dark to see. Initially most fish were feeding particularly, but, by the time I could barely see, nearly all were filtering. My laboratory study indicates that they could continue to feed past dark. I had no means to capture any fish, but I used an abandoned jar and waded in among the fish to take some water samples. Prey were mostly *Cyclops bicuspidatus thomasi* and mostly concentrated in the top few centimeters of water. From a preserved sample I estimated that there were ca. 4,000/liter.

Gately (1978), in a study of the alewives of Echo Lake in Maine, found that alewives 92–179-mm TL (mean = 142 mm) had their stomachs fullest at night during July and August and fed mainly on Cladocera and *Chaoborus* at night. He also noted that the fish migrated vertically at night. Unfortunately he had no plankton samples.

Table 3. Results of study of alewives feeding in dark. Statistical test: Chi-square contingency table with 5 df (Conover 1971). NS—not significant. Adapted from Janssen (1978b).

	Fish size TL (mm)	No. prey eaten	Median Daphnia size (mm)	χ^2	Signif.
Large alewives	185	676	1.2	2.74	NS
	181	2,215	1.1	4.52	NS
	178	829	1.1	4.13	NS
	172	1,764	1.1	5.12	NS
	170	2,154	1.1	2.15	NS
	170	2,615	1.1	1.10	NS
	152	2,091	1.0	4.86	NS
	145	1,161	1.1	11.54	$P<0.05$
Small alewives					
	106	91	1.2	6.12	NS
	105	75	1.2	4.73	NS
	102	66	1.1	7.94	NS
	96	484	1.0	8.56	NS
	95	287	1.1	2.35	NS
	95	398	1.1	5.10	NS
	92	0	—	—	—
	Sample 278/20 liters		1.2		

Emery (1973) made some limited underwater observations on ciscoes feeding in some Ontario lakes. He characterized them as night planktivores. Engel (1972, 1976) found that ciscoes in Pallette Lake were along the bottom in the thermocline during the day and moved offshore along the thermocline at night to feed. When comparing the stomach contents with the zooplankton he found no evidence of size selectiveness. Another fish, small coho salmon (*Oncorhynchus kisutch*), was size-selective.

What effect might a fish with both selective and nonselective feeding modes have on zooplankton communities? If both modes are used, the overall effect is that fish are size-selective although the intensity of bias is diluted. The same conclusion can be made about a population in which some individuals are selective while others are not, or a fish community in which some species are selective while others are not.

What of the possible situation for a lake in which all fish are not size selective? If the fish feed in a particular stratum, then zooplankton species and sizes that are found in that stratum will experience the most predation. If one species of zooplankton tends to form patches while others do not, and the fish tend to concentrate filtering in patches, then that zooplankter will also experience more predation. While the small alewives I observed filtering in Lake Michigan were almost certainly not size selective, they were not sampling the lake randomly.

If the total impact of the fish is completely random as regards zooplankton species and size, then some of the effects attributed to size-selective predation might also be found. One effect is a shift to smaller body size (Brooks and Dodson 1965; Brooks 1968, 1969; Wells 1970; Hutchinson 1971; Warshaw 1972; and many others). Organisms such as Cladocera have indeterminant growth; older individuals are generally larger even when mature. Continuous nonselective predation should shorten the average life expectancy of a cladoceran. However, because the stable age distribution is also dependent on the population growth rate, the average age and size need not change (B. Taylor pers. comm.). Early maturation time should be selected for, however, with the result that some energy is devoted to reproduction at an earlier age, thus slowing growth. One would also expect species with short generation times to be favored. Hall et al. (1976) noted that, although there are few studies with the appropriate information, the limited data suggest that generation time and body length at maturity are positively correlated. Finally, Hall et al. (1970) and Neill (1975) found that size-selective predation allowed more species to coexist in the zooplankton community. Slobodkin (1964) found that intense nonselective artificial predation on laboratory populations of *Chlorohydra viridissima* and *Hydra littoralis* allowed coexistence, whereas without predation, the *Chlorohydra* eliminated *Hydra*. Dayton (1971) found that logs crashing into the rock intertidal zone created clear patches that subsequently had higher diversity than patches that had not been cleared. The logs might be considered as analogous to a nonselective predator.

In conclusion, at least some freshwater fishes do have selective and nonselective feeding modes. The relative importance of the two food gathering modes needs more study. Such fish may have effects on zooplankton communities similar to those attributed to size-selective fish planktivores.

References

BERG, A., and E. GRIMALDI. 1966. Ecological relationships between planktophagic fish species in the Lago Maggiore. Int. Ver. Theor. Angew. Limnol. Verh. **16**: 1065–1073.

BOUCHERLE, M. M., and V. R. FREDERICK. 1976. *Daphnia* swarms in the harbor at Put-In-Bay. Ohio J. Sci. **56**: 90–91.

BRANDT, S. 1976. Acoustic determination of fish distribution and abundance in Lake Michigan with special reference to temperature. M. S. thesis, Univ. Wisconsin.

BROOKS, J. L. 1968. The effects of prey-size selection by lake planktivores. Syst. Zool. **17**: 272–291.

———. 1969. Eutrophication and changes in the composition of zooplankton, p. 236–255. *In* Eutrophication: Causes, consequences, correctives. Natl. Acad. Sci. Publ. 1700.

———, and S. L. DODSON. 1965. Predation, body size, and composition of plankton. Science. **150**: 28–35.

COLIN, P. L. 1976. Filter feeding and predation on the eggs of *Thallasoma* sp. by the scombrid fish *Rastrelliger kanagurta*. Copeia **1976**: 596–597.

CONOVER, W. J. 1971. Practical nonparametric statistics. Wiley.

CUSHING, D. H., and R. R. HARDEN-JONES. 1968. Why do fish school? Nature **218**: 918–920.

DAYTON, P. K. 1971. Competition, disturbance, and community organization: The provision and subsequent utilization of space in a rocky intertidal community. Ecol. Monogr. **41**: 351–388.

DURBIN, A. G., and E. G. DURBIN. 1975. Grazing rates of the Atlantic menhaden *Brevoortia tyrannus* as a function of particle size and concentration. Mar. Biol. **33**: 265–277.

EGGERS, D. M. 1976. Theoretical effect of schooling by planktivorous fish predators on the rate of prey consumption. J. Fish. Res. Bd. Can. **33**: 1964–1971.

EMERY, A. R. 1973. Preliminary comparisons of day and night habits of freshwater fish in Ontario lakes. J. Fish. Res. Bd. Can. **30**: 761–774.

ENGEL, S. S. 1972. Utilization of food and space by cisco, yellow perch, and introduced coho salmon, with notes on other species, in Pallette Lake, Wisconsin. Ph. D. thesis, Univ. Wisconsin. 289 p.

———. 1976. Food habits and prey selection of coho salmon (*Oncorhynchus kisutch*) and cisco (*Coregonus artedii*) in Pallette Lake, Wisconsin. Trans. Am. Fish. Soc. **105**: 607–614.

GALBRAITH, M. G. 1967. Size-selective predation on *Daphnia* by rainbow trout and yellow perch. Trans. Am. Fish. Soc. **96**: 1–10.

GANNON, J. E. 1974. The crustacean zooplankton of Green Bay, Lake Michigan. Proc. 17th Conf. Great Lakes Res. **1974**: 28–51.

———. 1976. The effects of differential digestion rates of zooplankton by alewife, *Alosa pseudoharengus*, on determinations of selective feeding. Trans. Am. Fish. Soc. **105**: 89–95.

GATELY, G. F. 1978. Competition for food between landlocked smelt (*Osmerus mordax*) and landlocked alewives (*Alosa pseudoharengus*) in Echo Lake, Maine. M. S. thesis, Univ. Maine, Orono. 94 p.

HALL, D. J., W. E. COOPER, and E. E. WERNER. 1970. An experimental approach to the production dynamics and structure of freshwater animal communities. Limnol. Oceanogr. **15**: 839–928.

———, S. T. THRELKELD, C. W. BURNS, and P. H. CROWLEY. 1976. The size-efficiency hypothesis and the size structure of zooplankton communities. Annu. Rev. Ecol. Syst. **7**: 177–208.

HOBSON, E. S. 1963. Selective feeding by the gafftopsail pompano *Trachinotus rhodopus* (Gill) in mixed schools of herring and anchovies in the Gulf of California. Copeia **1963**: 593-596.

———. 1968. Predatory behavior of some shore fishes in the Gulf of California. Res. Rep. U. S. Fish. Wildl. Serv. 73. 92 p.

HRBÁČEK, J. 1962. Species composition and the amount of zooplankton in relation to the fish stock. Rozpr. Cesk. Akod. Ved, Roda Mat. Priv. Ved. **72**: 1-116.

HUTCHINSON, B. P. 1971. The effect of fish predation on the zooplankton of ten Adirondack Lakes, with particular reference to the alewife, *Alosa pseudoharengus*. Trans. Am. Fish. Soc. **100**: 325-335.

IVLEV, V. S. 1961. Experimental ecology of the feeding of fishes. Yale Univ. Press.

JANSSEN, J. 1976. Feeding modes and prey size selection in the alewife (*Alosa pseudoharengus*). J. Fish. Res. Bd. Can. **33**: 1972-1975.

———. 1978a. Feeding behavior repertoire of the alewife, *Alosa pseudoharengus*, and the ciscoes *Coregonus hovi* and *C. artedii*. J. Fish. Res. Bd. Can. **35**: 249-253.

———. 1978b. Will alewives feed in the dark? Environ. Biol. Fish. **3**: 239-240.

LEONG, R. H. J., and S. P. O'CONNELL. 1969. A laboratory study of particulate and filter feeding of the northern anchovy (*Engraulis mordax*). J. Fish. Res. Bd. Can. **26**: 557-582.

NEILL, S. R. ST. J., and J. M. CULLEN. 1974. Experiments on whether schooling by their prey affects the hunting behavior of cephalopods and fish predators. J. Zool. Lond. **172**: 549-569.

NEILL, W. E. 1975. Experimental studies on microcrustacean competition, community composition and efficiency of resource utilization. Ecology **56**: 809-826.

NILSSON, N., and B. PEJLER. 1973. On the relation between fish fauna and zooplankton composition in north Swedish lakes. Rep. Inst. Freshwater Res. Drottningholm **53**: 51-77.

O'BRIEN, W. J., and G. L. VINYARD. 1974. Comment on the use of Ivlev's electivity index with planktivorous fish. J. Fish. Res. Bd. Can. **31**: 1427-1429.

O'CONNELL, C. P. 1972. The interrelation of biting and filtering in the feeding activity of the northern anchovy (*Engraulis mordax*). J. Fish. Res. Bd. Can. **29**: 285-293.

———, and J. R. Zweifel. 1972. A laboratory study of particulate and filter feeding of the Pacific mackerel, *Scomber japonicus*. Fish. Bull. **70**: 973-981.

REIF, C. B., and D. W. TAPPA. 1966. Selective predation: Smelt and cladocerans in Harvey's Lake. Limnol. Oceanogr. **11**: 437-438.

RHODES, R. J., and T. S. McCOMISH. 1975. Observations on the adult alewife's food habits (Pices: Clupeidae: *Alosa pseudoharengus*) in Indiana's waters of Lake Michigan in 1970. Ohio J. Sci. **75**: 50-55.

ROTH, J. C., and J. A. STEWART. 1973. Nearshore zooplankton of southeastern Lake Michigan, 1972. Proc. 16th Conf. Great Lakes Res. **1973**: 132-142.

SLOBODKIN, L. B. 1964. Experimental populations of Hydrida. J. Anim. Ecol. 33 suppl.: 133-148.

TONOLLI, V. 1962. L'attuale situazione del populomento planctonico del Lago Maggiore. Mem. Ist. Ital. Idrobiol. **15**: 81-134.

WARSHAW, S. J. 1972. Effects of alewives (*Alosa pseudoharengus*) on the zooplankton of Lake Wononskopomuc, Connecticut. Limnol. Oceanogr. **17**: 816-825.

WELLS, L. 1970. Effects of alewife predation on zooplankton populations in Lake Michigan. Limnol. Oceanogr. **15**: 556-565.

WERNER, E. E., and D. J. HALL. 1974. Optimal foraging and the size selection of prey by the bluegill sunfish (*Lepomis macrochirus*). Ecology **55**: 1042-1052.

51. The Roles of Zooplankter Escape Ability and Fish Size Selectivity in the Selective Feeding and Impact of Planktivorous Fish

Ray W. Drenner and Steven R. McComas

Abstract

Zooplankter escape abilities determined the feeding selectivity of filter-feeding gizzard shad (*Dorosoma cepedianum*) and filter-feeding Mississippi silversides (*Menidia audens*). Size selectivity was the primary determinant of the feeding selectivity of silversides which were feeding using a combination of particulate and filter feeding. Because the planktivore community of Lake Texoma is composed of both filter and particulate feeders, its selective impact on zooplankton is a function of both zooplankter escape ability and size selectivity.

It was not until Hrbáček (1962) and Brooks and Dodson (1965) demonstrated that the addition of fish to a lake resulted in a shift to smaller-bodied zooplankton species that limnologists recognized that fish predation can influence zooplankton community structure. Because of these studies, most research on planktivore feeding has focused on the fish's active selection of large-bodied prey (Werner and Hall 1974; Confer and Blades 1975; O'Brien et al. 1976). However, fish size selectivity is not the only factor governing selective feeding. Zooplankters have species-specific abilities to escape fish attacks which can result in fish having a passive feeding selectivity for zooplankter prey with poor escape abilities (Singarajah 1969, 1975; Starostka and Applegate 1970; Strickler 1975; Confer and Blades 1975; Janssen 1976; Drenner et al. 1978). We here present the initial results of a study of the relative roles of fish size selectivity and zooplankter escape ability in influencing the selective impact of planktivore feeding on the zooplankton community of Lake Texoma, Oklahoma.

Methods

Patten et al. (1975) found that shad and minnow-like fish comprised 87% of the planktivore biomass of a cove at Lake Texoma. In a series of laboratory experiments, we determined the feeding selectivities of dominant representatives of these two fish types, adult gizzard shad (*Dorosoma cepedianum*) and Mississippi silversides (*Menidia audens*), respectively. Shad (mean S.L. 180 mm) and silversides (mean S.L. 56 mm) were seined or electrofished from Lake Texoma and acclimated to 85-liter pools and 15-liter aquariums, respectively. Feeding experiments were conducted in the light (11–23 Einst/m^2/sec)

Financial support provided by Texas Christian University Research Foundation grant B7783. We thank L. Hill and B. Kimmel for their aid, the University of Oklahoma Biological Station for research facilities, and the Environmental Protection Agency for use of the EPA mobile research laboratory. We acknowledge the editorial assistance of G. Kroh and the technical assistance of K. Brown, C. Cahill, and B. Looney.

and in the dark (0 Einst/m^2/sec). An experiment began when a mixture of zooplankton was added to the fish containers. Initial species densities ranged from 18.4 to 4,100 organisms/liter. Shad and silverside feeding trials lasted 0.5 and 5 hours, respectively. Fish feeding rates were determined by monitoring changes in zooplankton densities. Zooplankters were sampled by lowering a clear Plexiglas tube, 5.8 cm in diameter, onto a rubber stopper lying on the pool bottom. The tube, which now contained a column of water, was removed from the pool and its contents preserved. Feeding rate constants were computed using Dodson's (1975) equation: feeding rate constant $(k) = -\ln (P_T/P_I)/(XT)$. P_I and P_T are initial and final zooplankter densities per liter, X is the density of predators per liter, and T is time in hours. The dimensions of k are liters per hour. Feeding rate constants per gram of fish body weight (k/g) were then calculated by dividing k by the mean biomass of individual gizzard shad (105.7 g) and Mississippi silversides (1.4 g).

The zooplankton used in the experiments were freshly captured from Lake Texoma or a neighboring farm pond. The farm pond had to be used to supply *Daphnia* spp. and *Diaphanosoma brachyurum*, whose population decreased in Lake Texoma in early spring. Cladoceran and copepod body length was measured as the distance from the anterior-most edge of the head to the base of the tail spine or the end of the thorax, respectively. Mean zooplankter sizes were: copepod nauplii (200 µm), cyclopoid copepods (mostly adult copepodites, 305 µm), *Ceriodaphnia reticulata* (381 µm), *D. brachyurum* (672 µm), *Daphnia* spp. (701 µm), and calanoid copepods (mostly adult *Diaptomus* spp., 717 µm).

Drenner et al. (1978) measured zooplankter escape abilities by determining the relative capture probabilities of a simulated fish suction for all zooplankter types used in these experiments, except *D. brachyurum*. In this study, we used the same suction simulator to determine *D. brachyurum*'s capture probability. The simulator was a siphon system which inhaled water into a tube (Drenner 1977), mimicking the flow of water into the rounded mouths of suction feeding fishes (Alexander 1967a). In these simulations 10 ml of water was inhaled in 0.4 sec into a tube of 1.0-cm I. D.

Results and discussion

Zooplankter capture experiments—The capture frequencies of zooplankters in front of the fish suction simulator's intake tube are shown in Fig. 1. We defined 100% capture probability as the area under the capture curve for escapeless bubbles and heat-killed zooplankters. We then computed the relative capture probabilities of a live zooplankter species by dividing the area under the zooplankter's capture frequency curve by the area under the curve for bubbles and heat-killed zooplankters. *Diaphanosoma brachyurum*'s capture probability (0.49) was less than the other cladocerans *C. reticulata* (0.96) and *Daphnia galeata* (0.92) but greater than the capture probabilities of adult cyclopoid copepods (mostly *Cyclops* spp. and *Mesocyclops* spp.) (0.28) and adult *Diaptomus pallidus* (0.07).

Fish feeding experiments—The predator-prey interaction between fish and zooplankters can be broken into a sequence of six primary events: search, encounter, pursuit, attack, capture, and filtration. Each event has a probability associated with it such that the probability of a prey being eaten is equal to a multiple of the six event probabilities. When a predator encounters more than one prey at a time, such as in the high prey densities used in our experiments, only the pursuit, attack, capture, and filtration events can determine feeding selectivity.

Large adult shad filter-fed in the light and the dark using a rapid sequence of suctions to capture zooplankters. These suctions were not visually directed at individual prey items. Therefore, when shad are filter feeding they encounter, pursue, and attack zooplankter prey in proportion to their densities. Measurements of shad gillraker spaces showed that large shad should filter

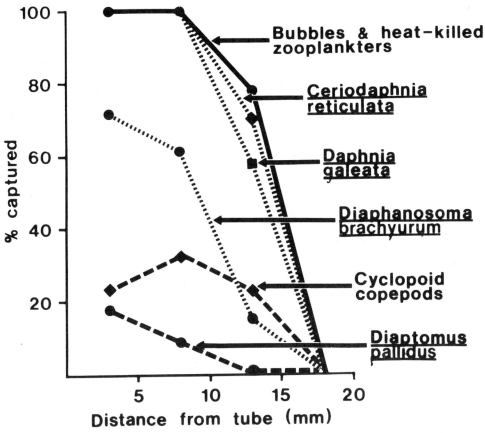

Fig. 1. Percentage of particles and organisms captured by siphon system vs. distance from tube. Points plotted at distances 3, 8, 13, and 18 mm represent averages of capture success within intervals of 1–5 mm, 6–10 mm, 11–15 mm, and 16–20 mm. Part of figure modified from Drenner et al. (1978).

particles ≥ 100 μm with 100% filtering efficiency (Drenner 1977). Therefore, capture probability should be the only event controlling the feeding selectivity of gizzard shad for zooplankter prey above 100 μm in size.

In our feeding experiments, shad feeding rate constants increased as a linear function of capture probability (Fig. 2). These differential feeding rates would result in shad having a passive feeding selectivity for cladocerans and nauplii prey. Smith (1971) found cladocerans and nauplii but no adult copepods in shad stomachs although copepods were present in the lake studied. Cramer and Marzolf (1970) suspected zooplankter escape ability as a factor when they found that gizzard shad stomachs had a lower proportion of *Diaptomus* than lake samples.

To confirm that our results were caused by zooplankter evasion of the shad's suction intake, we determined shad feeding rate constants for shad feeding on heat-killed zooplankters that could not escape. Heat-killed zooplankters are all captured with a 100% relative capture probability regardless of body form (Drenner et al. 1978). In our experiments, shad had about equal feeding rate constants for the heat-killed zooplankter species (Fig. 2). The regression lines for feeding rate constants on heat-killed and live zooplankters intersect at a capture probability of 1.00 where heat-killed zooplankters have the same capture probability as live zooplankters,

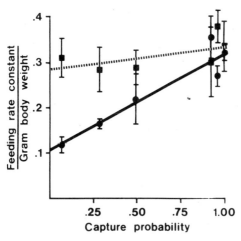
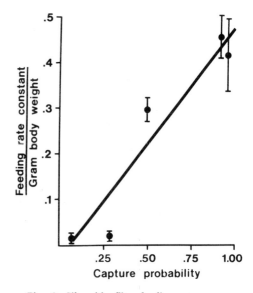

Fig. 2. Shad feeding rate constants for zooplankter prey species vs. capture probabilities of simulated suction. Solid line is least-squares linear regression line for live zooplankters ($Y = 0.219X + 0.106$, $r^2 = 0.89$). Dashed line is regression line for heat-killed zooplankters plotted at live capture probability ($Y = 0.053X + 0.283$, $r^2 = 0.36$). Points for live and heat-killed zooplankters are means of 8 (4 light and 4 dark) and 3 (light) feeding trials, respectively. From left to right, points represent feeding rate constants for adult calanoids, adult cyclopoids, *Diaphanosoma*, *Daphnia*, *Ceriodaphnia*, and nauplii. 100% capture probability used for copepod nauplii obtained from Rosenthal (1969). Bars represent ±1 SE.

Fig. 3. Silverside filter-feeding rate constants for zooplankter prey species vs. capture probabilities of simulated suction. Line is least-squares regression line ($Y = 0.520X - 0.037$, $r^2 = 0.90$). Points are means of 4 (dark) feeding trials. From left to right, points represent feeding rate constants for adult calanoids, adult cyclopoids, *Diaphanosoma*, *Daphnia*, and *Ceriodaphnia*. Bars represent ±1 SE.

confirming that zooplankter escape is the factor controlling shad feeding rate and feeding selectivity.

In the dark experiments, silversides could not visually encounter, pursue, and attack prey and therefore captured prey by filter feeding. Preliminary measurement of the spaces between the gillrakers show that silversides would filter particles > 200 μm in diameter with 100% efficiency. The filter-feeding rate constants of silversides on prey larger than 200 μm (nauplii were excluded) increased as a linear function of capture probability (Fig. 3), again showing that zooplankter escape determined the selectivity of filter feeders.

Silversides fed on zooplankters in the light primarily by visually pursuing, attacking, and capturing individual prey items. The particulate feeding was accompanied by filter feeding. This particulate/filter feeding was size selective for large-bodied zooplankters with feeding rate constants increasing as a linear function of prey body size (Fig. 4). However, zooplankter escape ability still influenced feeding rates. Aquarium observations of particulate-feeding silversides showed that silversides had lower probabilities of capturing adult cyclopoids (0.53) and adult calanoids (0.73) than of capturing *Diaphanosoma brachyurum* (0.78) or *Daphnia* spp. (0.98). Also, the filter feeding that accompanied the particulate feeding supplemented feeding on cladocerans only, since silversides were not able to filter feed on the evasive copepods (Fig. 3). These results show that

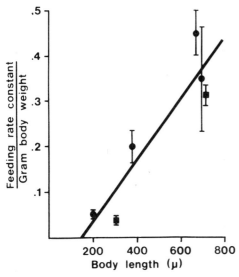

Fig. 4. Silverside particulate/filter-feeding rate constants for zooplankter prey species vs. zooplankter body length. Line is least-squares regression line ($Y = 0.001X - 0.100$, $r^2 = 0.84$). From left to right, circles represent feeding rate constants for nauplii, *Ceriodaphnia*, *Diaphanosoma*, and *Daphnia* and squares represent adult cyclopoids and adult calanoids. Bars represent ±1 SE.

the primary determinant of the feeding selectivity of particulate/filter feeders, e.g. silversides, feeding in the light is their selection of large-bodied prey. Zooplankter escape ability is then a secondary determinant, controlling the feeding selectivity on prey of similar size.

The probability of a particulate-feeding silverside capturing adult copepods (that it had attacked) was much higher than the capture probabilities of our simulated suction. Unlike shad, atherinids such as silversides have protrusible mouths (Alexander 1967b) which can be used to position the suction closer to the prey. As Fig. 1 shows, the closer the suction is to the prey, the higher the capture probability. Visual observation confirms that silversides use a more intense suction for adult copepods than cladocerans. Confer and Blades (1975) also found that particulate-feeding sunfish could recognize copepod prey and alter their feeding mechanics to increase capture success. Like silversides, the sunfish also had higher capture probabilities than our simulated suction, capturing *Daphnia* spp. on 100% of its attacks, *Diaptomus sicilis* on 79%, and *Diaptomus ashlandi* on 39% (Confer and Blades 1975).

Impact of fish feeding on zooplankton community structure—To calculate the feeding impact of the planktivore community on the zooplankter species, we multiplied mean shad and silverside feeding rate constants (k/g) for light and dark experiments by the fish's respective lake biomass, 0.391 and 0.541 g/m^2 (Patten et al. 1975), giving an estimate of feeding rate constants per m^2 (k/m^2). The total relative impact of fish feeding on each zooplankter species was then determined by adding shad and silversides k/m^2 together. The total k/m^2 values for the zooplankter species are shown in Table 1. These calculations show

Table 1. Total of shad and silverside feeding rate constants (k) per m^2. Capture probabilities obtained with simulator. Body length is mean body length of zooplankters in experiment.

	total k/m^2	capture probability	body length (μm)
Cyclopoids	0.08	low (0.28)	small (305)
Calanoids	0.14	low (0.07)	large (718)
Nauplii	0.15	high (1.0)	small (200)
Ceriodaphnia	0.27	high (0.96)	medium (381)
Diaphanosoma	0.29	medium (0.49)	large (672)
Daphnia	0.36	high (0.92)	large (701)

that zooplankters that are small and able to evade capture such as adult cyclopoid copepods would experience the lowest relative predation rates. Adult calanoid copepods, nauplii, *Ceriodaphnia*, and *Diaphanosoma* would sustain only intermediate predation by having either reduced body size or good escape ability. *Daphnia* spp., which were both large and easily captured, would be subject to both fish size selectivity and high capture success and therefore would experience the greatest relative predation pressure. The heavy predation on large cladocerans relative to copepods may account for the relative numerical dominance of copepods in fish-containing lakes (Brooks 1968). Because the escape ability of copepods can offset fish size selectivity, copepods are often the largest zooplankton species in lakes containing planktivorous fish (Brooks and Dodson 1965; Hutchinson 1971).

The estimated high predation on *Daphnia* spp. and *D. brachyurum* may account for their population decrease in Lake Texoma following the spring fish spawn. After their decline in the lake, we obtained these large cladocerans from a farm pond. In this way, our experiments could cover the full zooplankton complement. Other investigations of lakes that already contain planktivorous fish should be careful not to miss the real selective impact of the planktivore community. For example, field analysis of fish stomach contents may actually be studies of the fish's selection from an array of ecological leftovers. If the fish are influencing the zooplankton community structure, the only zooplankter species available are the prey which remain after the fish have eliminated their most selected prey.

Conclusion

Fish feeding mechanisms determine the relative roles of planktivore size selectivity and zooplankter escape ability in influencing fish feeding selectivity. Zooplankter escape ability determines the selectivity of filter feeders. Size selectivity by fish is the dominant factor controlling the feeding selectivity of particulate feeders. Zooplankter escape is a secondary factor.

Like Lake Texoma, many lakes and reservoirs have fish communities composed of filter feeders and particulate feeders. Filter-feeding shad often account for 50% of the fish biomass in the reservoirs of the central U.S. (Jenkins 1967). Therefore, it is necessary that zooplankter escape as well as size selectivity by fish be considered when assessing the influence of fish on the zooplankton community structure. Such studies should find that fish feeding not only shifts the community toward smaller-bodied zooplankton but that fish feeding results in a decline in the populations of the most easily captured zooplankton prey species.

References

ALEXANDER, R. McN. 1967a. Functional design in fishes. Hutchinson.
———. 1967b. Mechanisms of the jaws of some atheriniform fish. J. Zool. Lond. 151:233-255.
BROOKS, J. L. 1968. The effects of prey size selection by lake planktivores. Syst. Zool. 17:273-291.
———, and S. I. DODSON. 1965. Predation, body size, and composition of plankton. Science 150:28-35.
CONFER, J. L., and P. I. BLADES. 1975. Omnivorous zooplankton and planktivorous fish. Limnol. Oceanogr. 20:571-579.
CRAMER, J. D., and G. R. MARZOLF. 1970. Selective predation on zooplankton by gizzard shad. Trans. Am. Fish. Soc. 99:320-332.
DODSON, S. I. 1975. Predation rates of zooplankton in arctic ponds. Limnol. Oceanogr. 20:426-433.
DRENNER, R. W. 1977. The feeding mechanics of the gizzard shad (*Dorosoma cepedianum*). Ph.D. thesis, Univ. Kansas, Lawrence.
———, J. R., STRICKLER, and W. J. O'BRIEN. 1978. Capture probability: the role of zooplankter escape in the selective feeding of planktivorous fish. J. Fish. Res. Bd. Can. 35:1370-1373.
HRBÁČEK, J. 1962. Species composition and the amount of the zooplankton in relation to the fish stock. Rozpr. Cesk. Akad. Ved Rada Mat. Prir. Ved 72:1-114.
HUTCHINSON, B. P. 1971. The effect of fish predation on the zooplankton of ten Adirondack lakes, with particular references to the alewife, *Alosa pseudoharengus*. Trans. Am. Fish. Soc. 100:325-335.
JANSSEN, J. 1976. Selectivity of an artificial filter feeder and suction feeders on calanoid copepods. Am. Midl. Nat. 95:491-493.
JENKINS, R. M. 1967. The influence of some environmental factors on standing crop and harvest of fishes in U.S. reservoirs, p. 298-321. *In* Reservoir fisheries resources. Symp. Am. Fish. Soc.
O'BRIEN, W. J., N. A. SLADE, and G. L. VINYARD. 1976. Apparent size as the determinant of prey selection by bluegill sunfish (*Lepomis macrochirus*). Ecology 57:1304-1310.
PATTEN, B. C., D. A. Egloff, and T. H. RICHARDSON. 1975. Total ecosystem model for a cove in Lake Texoma, p. 205-421. *In* B. C. Patten [ed.], Systems analysis and simulation in ecology, v. 3. Academic.
ROSENTHAL, H. 1969. Untersuchungen über

das Beutefangverhalten bei Larven des Herings *Clupea harengus*. Mar. Biol. (Berl.) 3:208-221.

SINGARAJAH, K. V. 1969. Escape reactions of zooplankton: the avoidance of a pursuing siphon tube. J. Exp. Mar. Biol. Ecol. 3:171-178.

———. 1975. Escape reactions of zooplankton: effects of light and turbulence. J. Mar. Biol. Assoc. U.K. 55:627-639.

SMITH, A. D. 1971. Some aspects of trophic relations of gizzard shad, *Dorosoma cepedianum*. Ph.D. thesis, Virginia Polytech. Inst., Blacksburg.

STAROSTKA, V. J., and R. L. APPLEGATE. 1970. Food selectivity of bigmouth buffalo, *Ictiobus cyprincellus*, in Lake Poinsett, South Dakota. Trans. Am. Fish. Soc. 99:571-576.

STRICKLER, J. R., 1975. Intra- and interspecific information flow among planktonic copepods: receptors. Int. Ver. Theor. Angew. Limnol. Verh. 19:2951-2958.

WERNER, E. E., and D. J. HALL. 1974. Optimal foraging and the size selection of prey by bluegill sunfish (*Lepomis macrochirus*). Ecology 55:1042-1052.

52. The Effect of Prey Motion on Planktivore Choice

Thomas M. Zaret

Abstract

This paper investigates fish predation on the three Bosminidae of Gatun Lake, Panama, and its consequences for community species composition in the lake limnetic zones. A comparison of zooplankton collections before and after predation by the atherinid *Melaniris chagresi* for three dates indicates a single example of significant decrease in prey mean body-size and three significant decreases in prey eye-pigmentation diameter for *Bosminopsis deitersi* and *Eubosmina tubicen*. The frequency and magnitude of these changes were considerably less than those shown previously for the third sympatric species, *Bosmina longirostris*. Calculations of fish electivity indices show that fish select *Bosmina* preferentially over the other two Bosminidae, even though *Bosmina* has relatively the smallest mean body-size and smallest eye-pigmentation diameter of the three species. These results conflict with previous conclusions that prey body-size and prey visibility are sufficient to explain planktivore selection of prey items. It is suggested that effects of prey motion can most reasonably explain these data, and one possible model is proposed in which planktivores use a searching image based on prey motion. This model is discussed in light of previous studies on searching images in fishes.

There is considerable current interest in the mechanisms by which plankton-feeding fishes, here referred to as planktivores, effect the great degree of prey selectivity for which they are noted. Following the seminal paper of Brooks and Dodson (1965), which indicated that these fishes were removing preferentially the largest prey forms, research on planktivore feeding has proceeded in two general directions. First, several works have extended the concept of body-size predation initially proposed by Brooks and Dodson. These contributions have developed different models to explain fish selection by considering some combination of the fish's visual field, prey susceptibility based on body-size, and encounter probability of the predator and different types of prey (Werner and Hall 1974; Confer and Blades 1975; O'Brien et al. 1976; Eggers 1977). A second approach has explored other prey characteristics which can help explain predator selectivity (for review *see* Zaret 1975). To date these include prey visibility (e.g. Zaret 1972; Zaret and Kerfoot, 1975; Mellors 1975; Hairston 1977) and prey escape responses, including vertical migration patterns (Zaret and Suffern 1976) and direct escape abilities (Szlauer 1965, 1968). This direction is now undergoing considerable diversification and other aspects of predator-prey interaction soon may be uncovered which bear on planktivore selectivity.

A previous paper (Zaret and Kerfoot

Financial support from NSF grant GB 33396 to T. M. Zaret.

1975) examined the effects in Gatun Lake, Panamá, of the planktivorous fish *Melaniris chagresi* on one of its prey items, the cladoceran *Bosmina longirostris* (O. F. Müller). In that paper, we concluded that the predator selected *Bosmina* individuals according to the size of the heavily pigmented compound eye. This was called by the general term "visibility predation" since the predator chose items according to the amount of visible pigmentation in the prey item, and we distinguished it from body-size predation. This interpretation, emphasizing total pigmentation as a key to predator electivity, does not conflict with the observation that planktivores will feed more heavily on the largest of otherwise identical zooplankters as found by various workers (Hrbáček 1962; Brooks 1968; Werner and Hall 1974). The choice of larger individuals by the fish occurs because: first, there is a within-species direct correlation between eye size and body size so that larger individual crustaceans possess larger eyes; and second, the animal carapace itself does contribute some degree of pigmentation which is seen by the planktivores. In any mixed prey species assemblage, however, it is not always the largest species for which the planktivore has the greatest electivity (e.g. *see* Ivlev 1961; Brooks 1968), but more often prey with the largest total pigmentation.

Bosmina longirostris is one of three Bosminidae in Gatun Lake. Also present are *Bosminopsis deitersi* Richards, a seasonally abundant species, and *Eubosmina tubicen* (Brehm), a relatively rare species in this lake. Although these species represent three different genera, they are extremely similar, both in general appearance and size (maximum total length 300 ± 50 microns). In fact, *Bosmina* and *Eubosmina* are very difficult to distinguish with less than 400X magnification (*see* Goulden and Frey 1963). *Bosminopsis* can be separated readily from the other two by the way its antennules diverge at their distal end and by the thin mucrones (*see* Fig. 1).

This contribution focuses interest on a neglected prey characteristic, prey motion, whose importance for fishes has been recognized for decades, yet for which relatively little research effort has been expended. In this paper, I present a predation analysis of three Bosminidae in Gatun Lake and suggest that the data on prey selection by the planktivore *Melaniris* can only be explained by considering prey motion as a necessary component of the predation process. The approach consists of: an intensive examination of *Eubosmina* and *Bosminopsis* populations in natural situations before and after feeding by *Melaniris*; and an analysis of the electivities of the fish for the three Bosminidae by comparing fish stomach contents with zooplankton collections. All values for *Eubosmina* are presented, although in several cases its rareness makes any calculations meaningless due to small sample size.

I thank especially W. C. Kerfoot for his help in this study, including assistance with the statistics, editorial help, and the contribution of Fig. 1. I also thank W. T. Edmondson and R. T. Paine for their helpful criticism and support, and the Smithsonian Tropical Research Institute for use of research facilities.

Methods

The zooplankton which formed the basis for the study of Bosminidae were collected from an area called Slothia Island Station, where large schools of *Melaniris* were abundant, and consisted of two kinds of zooplankton hauls: morning samples taken before the feeding schools had arrived from their nocturnal resting area; and afternoon samples taken in the same location after the fish had fed on the zooplankton community for 3–4 hours. Collections were made with a No. 25 (64 µm mesh aperture) nylon net. There are no strong vertical migration patterns among the Cladocera of the lake, and the eye pigmentation itself shows < 5% average pigment migration with changing light intensities (unpubl. laboratory data). Further, samples were only used for dates when there was no significant horizontal water movement (i.e. windless days). Thus, any differences in animals

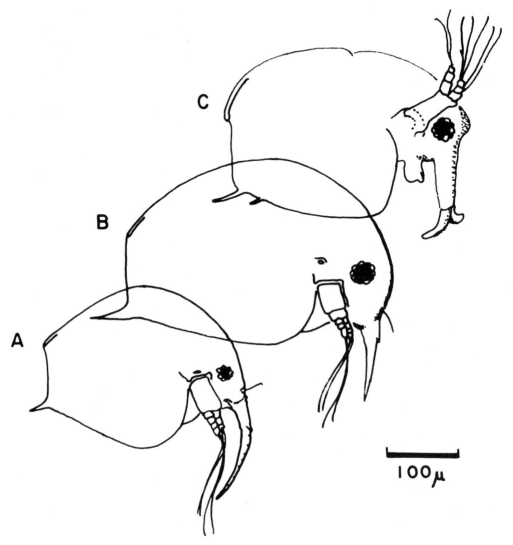

Fig. 1. Camera lucida drawings of the three Bosminidae: A. *Bosmina longirostris*; B. *Eubosmina tubicen*; C. *Bosminopsis deitersi*.

from morning versus afternoon collections should be due solely to the actions of the predators.

There are three dates, providing three replicates of this before-and-after natural field experiment. For each of the Formalin-preserved collections an attempt was made to remove at least 40 individuals of each species in order of encounter. These animals were mounted on slides in glycerin and measured under 400X magnification. Measurement included total body length and eye-pigmentation diameter along the main body axis, total length being from the base of the mucro to the top of the head, anterior to the eye (*see* Zaret and Kerfoot 1975: fig. 1). The term eye-pigmentation diameter is used rather than eye diameter since often the former value is actually smaller than the latter. Data were tested for differences over the day for body length and eye-pigmentation diameter. Comparisons of the change in relative proportion of each Bosminidae species over the course of the day came from the same plankton collections.

The electivity coefficients (E: Ivlev 1961) for *Melaniris* on the three prey items were calculated from the entire zooplankton assemblage, not only Bosminidae. For this reason the percentages do not total 100%. Collections were made with a No. 25 plankton net from an area called the Buoy Tow Sample Area by towing over a distance of about 100 m to minimize sampling errors from possible horizontal gradients in zooplankton distribution. These samples, preserved with Formalin, were later subsampled and counted under a dissecting microscope for plankton species composition. *Eubosmina* can be distinguished from *Bosmina* with a dissecting scope by observing differences in the shape and the angle formed by the mucrones, and the amount of space between the area of the compound eye and the edge of the carapace. The plankton tows were surface hauls taken about 1 hour after sunset and were followed immediately by fishnet tows to capture *Melaniris* individuals which had just been feeding on the zooplankton and were now hovering almost motionless in the surface waters where they could be captured easily. In the Buoy Tow Sample Area, *Melaniris* individuals feed on zooplankton only from about 1700 hours to sunset (see Zaret 1971).

From fish stomach contents *Bosminopsis* could be identified under the dissecting scope without problem; the distinction between *Bosmina* and *Eubosmina* was more difficult. To determine the *Melaniris* electivity for *Eubosmina* versus *Bosmina*, we tried to count at least 100 individuals of these two species per zooplankton haul, and 50 from fish stomachs, although this was not always possible. This latter count was made by removing individuals from fish stomach collections, mounting in glycerin on slides, and counting under 400X magnification. This method was necessary to ensure accurate identification since normally identifying features were often obscured after ingestion by *Melaniris*.

The formula for E, Ivlev's electivity, is

$$\frac{r_i - P_i}{r_i + P_i}$$

where r_i is the ratio or relative proportion which species i constitutes in the fish's stomach, and p_i is the proportion that species i constitutes in the zooplankton assemblage, including all lake species. The electivity value ranges from -1 (avoidance) to 0 (item taken randomly, i.e. no selection), to $+1$ (item always selected). To get as accurate as possible an electivity value, I analyzed paired zooplankton and fish collections from nine different dates over an entire year. These values derived from 76 fish stomachs containing over 4,700 Bosminidae which were identified and counted.

Slothia Island collections

Results—The Slothia Island samples compare the morning and afternoon populations of *Bosminopsis* and *Eubosmina*. By measuring body size and eye-pigmentation diameter of all individuals, we can determine whether there are significant population changes in these parameters over the day. The values are presented in Table 1, along with other pertinent data. For *Bosminopsis*, both the mean eye-pigmentation diameter and mean body size show a decrease on all dates, but of these there are only three examples of significant change (Student's-t). Body size is significantly reduced on 5 May ($p < 0.001$), and eye-pigmentation diameter on 1 April (< 0.001), and 5 May (< 0.01). For *Eubosmina* its rareness results in small sample sizes and, hence, difficulty in establishing the significance of changes. For the three dates *Eubosmina* shows no significant change in mean body size and a significant decrease in eye-pigmentation diameter only on 5 May ($p < 0.01$).

A test of correlation of eye-pigmentation diameter on body size examined the relationship between these two parameters over the day. *Bosminopsis* main-

Table 1. Slothia Island Station. Data from *Bosminopsis* and *Eubosmina* population. Measurements (in microns) of body-size (x) and eye-pigmentation diameters (y), with correlation coefficient (r) and significance of correlative coefficient (p).

Date (1969)	Time	n	$\bar{x}$	$\bar{y}$	S_x	S_y	S_{yx}	r	p	t
Bosminopsis										
9 Apr	A.M.	50	279	22.6	30.1	2.60	1.89	0.679	<0.001	
	P.M.	27	274	19.6	35.4	3.33	2.51	0.639	<0.001	0.348
2 May	A.M.	57	291	21.0	40.3	3.32	2.04	0.786	<0.001	
	P.M.	42	288	20.2	33.0	2.67	2.08	0.620	<0.001	2.09*
5 May	A.M.	48	280	22.4	30.5	2.65	2.14	0.576	<0.001	
	P.M.	45	248	22.1	31.8	2.95	2.56	0.478	<0.001	0.870
Eubosmina										
9 Apr	A.M.	17	322	26.0	50.6	4.15	1.97	0.873	<0.001	
	P.M.	16	330	26.0	41.7	3.76	2.75	0.655	<0.01	2.40*
2 May	A.M.	13	327	23.8	64.8	4.28	2.86	0.720	<0.02	
	P.M.	10	324	22.1	43.6	2.58	1.61	0.753	<0.01	0.235
5 May	A.M.	8	327	26.7	58.0	5.20	2.25	0.887	<0.01	
	P.M.	4	260	21.4	21.6	2.45	1.95	0.436	>0.10	1.04

*Significant at $p < 0.05$.

tained a highly significant correlation of body size to eye-pigmentation diameter on all dates, p always < 0.001, although all dates also showed a reduction in the correlation coefficient (r) between morning and afternoon samples. The *Eubosmina* values varied, due mostly to small sample sizes, showing a decreased correlation on 9 April and 5 May and an increased correlation on 2 May. A test for level of significance for the observed change in the correlation coefficient (r) over the day was performed on the r values using a "z transformation" (Snedecor and Cochran 1956), which approximately normalizes the sampling distribution. This permits use of a t-test on the daily difference in "z" values divided by the standard error. Only two dates showed a significant change in the correlation coefficient over the day, *Bosminopsis* on 5 May and *Eubosmina* on 9 April.

Conclusions—The results of 9 April and 5 May, which showed three significant decreases in the population mean for eye-pigmentation diameter, and only one reduction in mean body size, suggest that fish are selecting prey according to visibility, removing the largest-eyed *Bosminopsis* and *Eubosmina* on these dates. Even though predation by *Melaniris* at the Slothia Island station can reduce the Bosminidae population mean body size and eye-pigmentation diameter over the course of 1 day, the animals encountered in this area several days later may have means similar to the initial values. These later Bosminidae have two sources of origin. First, although *Melaniris* feeding activities are confined to the top 1 m of water, the depth at the Slothia Island station is a maximum of 16 m, and reproduction is occurring at all depths. The daily mixing of Gatun Lake waters allows the reintroduction to the surface water of Bosminidae from the rest of the water column. Second, there is a great deal of horizontal water mixing due to the almost daily (usually early afternoon) winds on Gatun Lake, and animals from other areas of the lake can enter via passive mixing to the few and distinctly defined breeding areas where *Melaniris* predation is so intense. On some dates, *Melaniris* predation reduced cladoceran populations in the top 1 m of water by 100% over the day (unpubl. data). This is not surprising in view of the very strong selection according to visibility previously demonstrated for *Melaniris* feeding on *Bosmina* (Zaret and Kerfoot 1975). In the present study, despite the apparent indication of visibility selection, the magnitude of change is much less than that for the previous study of *Bosmina*. The test of correlation, which showed a decrease over the course of the day in five of six cases, remained relatively high, at least for *Bosminopsis*, on all dates even after fish had fed for several hours. The conclusion is that, although visibility selection by the predator is the dominating selective factor, there is relatively little lasting effect from predation on the population for either body size or eye-pigmentation. These somewhat surprising results lead to two possible explanations: predators are taking prey items but not solely based on body size or eye-pigmentation diameter; little selection for these two species is occurring.

Support for the latter explanation—that *Melaniris* selection is relatively lower for these two species—comes from data examining the diurnal change in the relative proportion of each species. These values, comparing percentage composition of each of the three Bosminidae species for the morning and afternoon collections for these three dates, are presented in Table 2. These changes are highly significant for two of these three dates (χ^2-tests of homogeneity for total Bosminidae changes give values of 10.8, 47.9, and 1.4 for 9 April, 2 May, and 5 May, respectively). Values for the first two dates are significant ($p < 0.005$ on both dates), while that for the third is not. On all dates, decreases in the *Bosmina* population coincide with increases in the relative abundance of *Bosminopsis*. *Eubosmina* shows a decrease on two dates and no change on the third, but again the interpretation is complicated by small sample sizes.

Buoy Tow collections

These data suggest that fish predation falls heavily on *Bosmina* relative to the other two Bosminidae. To determine more exact relative values, Table 3 presents an electivity coefficient (E: Ivlev 1961) for *Melaniris* on the three prey species. This value is calculated from the entire zooplankton assemblage, not only Bosminidae. These collections came from the Buoy Tow Sample Area (*see methods*) where *Melaniris* feeds daily for a few hours before sunset.

In examining Table 3, it should be remembered that many factors are involved in affecting the electivity of a fish predator, including predator density, prey density, and predator degree of satiation (*see* Ivlev 1961 for a complete discussion). Also, seasonal variation in these factors can produce a great variance in electivity. (For instance on 24 September the electivities for all three Bosminidae are very high, indicating that fish were taking almost all individuals encountered of these three species. This is probably because on this date there was a very low density of

Table 2. Slothia Island Station. Change over day in relative abundance of three Bosminidae species, with levels of significance for daily change in total Bosminidae population.

Date (1969)	Time	N	Bosmina %	Bosminopsis %	Eubosmina %	χ^2	p
9 Apr	A.M.	200	37.0	46.0	17.0	10.8	<0.005
	P.M.	100	22.0	66.0	12.0	—	
2 May	A.M.	100	30.0	52.0	9.0	—	
	P.M.	100	5.0	95.0	0.0	47.9	<0.005
5 May	A.M.	200	38.0	53.5	8.5	—	
	P.M.	35	31.4	54.3	14.3	1.4	>0.50

Table 3. Ivlev's electivity coefficients (E) of Melaniris on Bosmina and Bosminopsis, ranging from −1 (avoidance) to 0 (random feeding) to +1 (complete selection for that item). Included are number of fish stomachs opened, and absolute numbers of prey and percentages of each species in total plankton and fish stomachs. Eubosmina numbers were too low for meaningful electivity calculations for most dates. Electivities based on total zooplankton assemblage, including non-Bosminidae, so that percentages do not sum to 100%.

Date (1969)	No. fish	Bosmina				Bosminopsis				Eubosmina		
		plankton		stomach	E	plankton		stomach	E	plankton	stomach	E
25 Jan	8	18	2%	447 45%	0.91	366	49%	401 40%	−0.10	2 0%	1 0%	—
14 Feb	5	43	4%	357 55%	0.86	899	80%	250 39%	−0.34	12 1%	0 0%	—
26 Mar	9	2	1%	92 47%	0.96	49	19%	50 26%	0.16	0 0%	0 0%	—
15 Apr	8	12	2%	104 24%	0.85	222	43%	203 47%	0.04	1 0%	4 1%	—
15 May	10	24	6%	446 38%	0.73	30	7%	38 3%	−0.40	0 0%	0 0%	—
4 Jul	10	27	9%	687 37%	0.61	39	12%	198 11%	−0.04	5 2%	0 0%	
13 Aug	10	124	2%	636 74%	0.95	50	8%	95 11%	0.16	20 3%	12 1%	−0.50
24 Sep	7	46	6%	108 36%	0.71	24	3%	66 22%	0.76	6 1%	23 9%	0.80
26 Nov	9	50	11%	324 49%	0.63	18	4%	215 33%	0.78	1 0%	0 0%	—
Totals	76	346		3,201		1,697		1,516		47	40	
				Average E 0.80					0.11			

prey per individual predator.) The relative scarcity of Eubosmina is clearly reflected in the great variation of values for E. To reduce the likelihood of erroneous assumptions due to seasonal variations, Table 3 presents an average E for each species.

Results—In general, the data show a striking consistency over the year. Melaniris highly prefers Bosmina over Bosminopsis and Eubosmina on seven of the nine dates, the exceptions being 24 September when all three species had about the same values, and 26 November when Bosminopsis had the highest value. The average electivity values for each species indicate: Bosmina with E = 0.80, meaning very high selection by Melaniris; Bosminopsis with E = 0.11, meaning only slight selection; and Eubosmina basically uninterpretable because of low numbers. These data substantiate the fact that Melaniris predation falls most heavily on Bosmina with Bosminopsis suffering relatively little effects.

Discussion

There thus appears to be an inconsistency in the theory of planktivore selection developed up to this point. If Melaniris selects prey items strictly according to visibility, which is supported by the field collection comparisons as well as previous investigations, why should Bosmina, the smallest-eyed and smallest-bodied of these three Bosminidae (see Table 1 here and table 1 of Zaret and Kerfoot 1975), be taken at a rate substantially greater than the other two? These results conflict also with studies suggesting that the largest or largest-appearing prey individual will be most liable to predation (Brooks 1968; Werner and Hall 1974; O'Brien et al. 1976; Eggers 1977).

One possible explanation is that some other aspect of prey "attractability" is operating, such as the component of prey motion. The remainder of this paper develops this idea by showing that the importance of prey motion has support in the literature by placing the component of motion in a model that provides a logical and meaningful interpretation of these data from Gatun Lake.

The importance of motion—There are two distinct ways by which the motion of zooplankton prey can significantly affect the rate of predation. First, there is a direct swimming escape response by animals that is well recognized by planktologists, as seen in studies of zooplankton "net avoidance" (*see* Fleminger and Clutter 1965; Szlauer 1965, 1968). Second, there is the general swimming locomotion of individual plankton species which can produce varying degrees of conspicuousness (*see* Brooks 1968).

The literature indicates that motion detection is important to fishes. It has even been demonstrated in some fish species (such as the common goldfish, *Carassius auratus*) that the fish responds not only to motion per se, but actually possesses receptors that distinguish fast speeds from slow speeds and the fish uses these in different responses (Ingle 1968). However, the absolute visibility of the perceiving fish still depends on the contrast relationship between the object and its background (Hemmings 1966; Hester 1968) and the visual pigment of the fish's eyes are adapted to maximize this contrast according to the light regime in its natural environment (Lythgoe 1966). Thus, if an object is visible at 10 m but not at 10.1 m, no amount of motion will make that object visible to the fish at 10.1 m. However, once the object is within the visible range, motion may increase its conspicuousness, affecting the probability that the prey item will be noticed by the fish. From this discussion it seems reasonable to assume that once the prey items are within their visible range, different motion components of prey can produce different responses in the predatory fishes.

There are several field and laboratory studies that document the role of prey motion in predator electivity. Lindström (1955) recognized the importance of this motion to fish predators when he observed in the laboratory that the fry of char, *Salmo alpinus*, eat moving pike fry, but lose interest when the pike are motionless. In a study of the three-spined stickleback, *Gasterosteus aculeatus*, found in streams along the coast of Washington State, McPhail (1969) concluded that motion was one of the key components for its natural predator, the endemic western mudminnow *Novumbra hubbsi*. Finally, Ware (1973) showed that rainbow trout, *Salmo gairdneri*, could locate moving prey more successfully than stationary ones with otherwise identical properties. These studies support the conclusion that prey species in motion are most vulnerable to fish predation and suggest that selection will tend to favor the existence of those prey species that move least. While it may be impossible for a zooplankter to remain motionless for long periods of time, if only to oppose its passive sinking speed, it appears that the evolved methods of locomotion such as that of *Bosminopsis* reduce the general motion component, at least to the human eye, and this could result in a lowered rate of predation by fishes.

I will suggest a strictly theoretical framework which could provide an explanation for the data on the predator feeding pattern observed in these Bosminidae studies. First, an initial detection by the fish of individual prey items with the greatest motion component (this being a species-specific characteristic). Second, once species are ranked in some hierarchy of choice based on motion, selection within each species proceeds according to the individual's total pigmentation. This explains why within each Bosminidae species those individuals with the greatest total pigmentation are selected first, but why in the mixed species assemblage more individuals of *Bosmina* may be removed even though other Bosminidae are more visible.

Motion selection mechanism—It is possible that increased conspicuousness in certain prey types may enable the fish to form what Tinbergen (1960) referred to initially as a "searching image," and which has been modified subsequently (because of ambiguities with this original term) to refer specifically to changes in feeding behavior related to vision or visual clues (Dawkins 1971). It is not my purpose here to discuss the relative merits of this concept. For this, an excellent review and discussion can be found in Krebs (1973). It is clear that the phenomenon is real; what is still being debated is to what extent a searching image is used by various predators (Krebs 1973; Paulson 1973). Here, I suggest the possibility that planktivores can facilitate prey-capture success by learning to recognize prey items according to their motion. There is an accumulating amount of empirical evidence from experiments which have shown that experience can influence vertebrate prey-capture success (e.g. de Ruiter 1952; Holling 1959; Beukema 1968; Croze 1970). Also, the searching image concept in fishes has been suggested (Popham 1941, 1942) and recently tested and substantiated by Beukema (1968) with *Gasterosteus* and by Ware (1971) in a series of carefully controlled laboratory experiments with *Salmo*. Although there are some semantic criticisms of the latter two studies (Krebs 1973), the main point is that experience plays an important role and, as Ware (1972) concluded, the development of this type of response based on experience might act as a positive feedback to improve the efficiency of vertebrate predators, whether we call this a searching image or not.

Final conclusions

Planktivores such as *Melaniris* may be able to distinguish prey forms according to motion, leading to species searching image formation. Then from among all individuals of this preferred-motion prey population, some individuals are more visible than others (namely, those having the greatest amount of pigmentation) and will be eaten first. This model, incorporating an interaction between motion and visibility components, yields one possible explanation for the otherwise inexplicable data for the Gatun Lake Bosminidae. The concept of a searching image in planktivores may also help to explain one of the seemingly incredible abilities of planktivores: to select unerringly one prey type in the presence of other almost identical ones, and to consume 100% of this single form from among a mixed species assemblage (e.g. Green 1967). It is apparent that a model of planktivore choice based on prey body size or prey visibility can explain fish predation on individuals within single-species prey population, but is inadequate to explain data sets from mixed species assemblages. It is likely that future studies on prey motion will be able to provide a better understanding of the mechanism of feeding by selective freshwater planktivores.

References

BEUKEMA, J. J. 1968. Predation by the three-spined stickleback (*Gasterosteus aculeatus* L.): the influence of hunger and experience. Behaviour 31:1-126.
BROOKS, J. L. 1968. The effects of prey size selection by lake planktivores. Syst. Zool. 17:272-291.
——, and S. I. DODSON. 1965. Predation, body size, and composition of plankton. Science 150:28-35.
CONFER, J. L., and P. I. BLADES. 1975. Omnivorous zooplankton and planktivorous fish. Limnol. Oceanogr. 20:571-579.
CROZE, H. 1970. Searching image in carrion crows. Z. Tierpsychol. 5 (suppl.):86 p.
DAWKINS, M. 1971. Perceptual changes in chicks: another look at the "search image" concept. Anim. Behavior 19:556-574.
EGGERS, D. M. 1977. The nature of prey selection by planktivorous fish. Ecology 58:46-59.
FLEMINGER, A., and R. I. CLUTTER. 1965. Avoidance of towed nets by zooplankton. Limnol. Oceanogr. 10:96-104.
GOULDEN, C. E., and D. G. FREY. 1963. The occurrence and significance of lateral head pores in the genus *Bosmina* (Cladocera). Int. Rev. Gesamten Hydrobiol. 48:513-527.
GREEN, J. 1967. The distribution and variation of *Daphnia lumboltzii* (Crustacea: Cladocera) in relation to fish predation in Lake Albert, East Africa. J. Zool. 151:181-197.
HAIRSTON, N. G., JR. 1977. The adaptive significance of carotenoid pigmentation in *Diaptomus* (Copepoda). Ph.D. thesis, Univ. Washington.
HEMMINGS, C. C. 1966. Factors influencing the

visibility of objects underwater, p. 359–374. In R. Bainbridge et al. [eds.], Light as an ecological factor. Brit. Ecol. Soc. Symp. 6.

HESTER, F. J. 1968. Visual contrast thresholds of the goldfish (*Carassius auratus*). Vision Res. 8:1315–1336.

HOLLING, C. S. 1959. The components of predation, as revealed by a study of small mammal predation of the European pine sawfly. Can. Entomol. 91:293–332.

HRBÁČEK, J. 1962. Species composition and the amount of zooplankton in relation to the fish stock. Rozpr. Cesk. Akad. Ved Rada Mat. Prir. Ved 72 (10): 116 p.

INGLE, D. 1968. Spatial dimensions of vision in fish, p. 51–59. In D. Ingle [ed.], The central nervous system and fish behavior. Univ. Chicago Press.

IVLEV, V. S. 1961. Experimental ecology of the feeding of fishes. Yale Univ. Press.

KREBS, J. R. 1973. Behavioral aspects of predation, p. 73–111. In P. P. G. Bateson and P. H. Klopfer, Perspectives in ethology. Plenum Press.

LINDSTRÖM, T. 1955. On the relation fish-size-food size. Inst. Freshwater Res. Drottningholm 36, p. 133–147.

LYTHGOE, J. N. 1966. Visual pigments and underwater vision, p. 375–391. In R. Bainbridge et al. [eds.], Light as an ecological factor. Brit. Ecol. Soc. Symp. 6.

McPHAIL, J. D. 1969. Predation and the evolution of a stickleback (*Gasterosteus*). J. Fish. Res. Bd. Can. 26:3183–3208.

MELLORS, W. K. 1975. Selective predation of ephippial *Daphnia* and the resistance of ephippial eggs to digestion. Ecology 56:974–980.

O'BRIEN, W. J., N. A. SLADE, and G. L. VINYARD. 1976. Apparent size as the determinant of prey selection by bluegill sunfish (*Lepomis macrochirus*). Ecology 57:1304–1310.

PAULSON, D. R. 1973. Predator polymorphism and apostatic selection. Evolution 27:269–277.

POPHAM, E. J. 1941. The variation in the color of certain species of *Arctocorisa* (Hemiptera: Corixidae) and its significance. Proc. Zool. Soc. Lond. Ser. A 111:135–172.

———. 1942. Further experimental studies on the selective action of predators. Proc. Zool. Soc. Lond. Ser. A 112:105–117.

RUITER, L. DE. 1952. Some experiments on the camouflage of stick caterpillars. Behaviour 4:222–232.

SNEDECOR, G. W., and W. C. COCHRAN. 1956. Statistical methods. Iowa State Univ. Press.

SZLAUER, L. 1965. The refuge ability of plankton animals before plankton-eating animals. Pol. Arch. Hydrobiol. 13:89–95.

———. 1968. Investigations upon ability in plankton crustacea to escape the net. Pol. Arch. Hydrobiol. 15:79–86.

TINBERGEN, L. 1960. The natural control of insects in pinewoods. I. Factors influencing the intensity of predation by songbirds. Arch. Neerl. Zool. 13:265–343.

WARE, D. M. 1971. Predation by rainbow trout (*Salmo gairdneri*): the effect of experience. J. Fish. Res. Bd. Can. 28:1847–1852.

———. 1972. Predation by rainbow trout (*Salmo gairdneri*): the influence of hunger, prey density, and prey size. J. Fish. Res. Bd. Can. 29:1193–1201.

———. 1973. Risk of epibenthic prey to predation by rainbow trout (*Salmo gairdneri*). J. Fish. Res. Bd. Can. 30:787–797.

WERNER, E. E., and D. J. HALL. 1974. Optimal foraging and the size selection of prey by the bluegill sunfish (*Lepomis macrochirus*). Ecology 55:1042–1052.

ZARET, T. M. 1971. The distribution, diet, and feeding habits of the atherinid fish *Melaniris chagresi* in Gatun Lake, Panama Canal Zone. Copeia 1971:341–343.

———. 1972. Predators, invisible prey, and the nature of polymorphism in the Cladocera (Class Crustacea). Limnol. Oceanogr. 17:171–184.

———. 1975. Strategies for existence of zooplankton prey in homogeneous environments. Int. Ver. Theor. Angew. Limnol. Verh. 19:1484–1487.

———, and W. C. KERFOOT. 1975. Fish predation on *Bosmina longirostris*: visibility selection versus body-size selection. Ecology 56:232–237.

———, and J. S. SUFFERN. 1976. Vertical migration in zooplankton as a predator avoidance mechanism. Limnol. Oceanogr. 21:804–813.

53. Selective Predation by Zooplankton and the Response of Cladoceran Eyes to Light

John L. Confer, Gregory Applegate, and Christine A. Evanik

Abstract

The diameter of the pigmented portion of the eye of three species of Cladocera was found to contract in light and dilate in the dark. Laboratory studies with dark-adapted *Daphnia magna* showed a decrease of 22% in the pigment diameter after 8 h of light. Field studies from ponds without fish showed a maximum decrease of 18% for *Bosmina longirostris* about 6 h after sunrise and a maximum decrease of 24% for *Daphnia pulex* about 16 h after sunrise. The field samples for *Bosmina* and for *D. pulex* both showed dilation of the eyes during the night.

The daytime decrease in *Bosmina* eye size from ponds without fish was as large as a previously observed decrease that has been ascribed to selective predation by fish.

The diameter of the pigmented portion of the eye of Cladocera has been related to predation by fish. For instance, convincing evidence demonstrates that *Melaniris* will prey on the morph of *Ceriodaphnia cornuta* with a large eye rather than the morph with a small eye (Zaret 1969). It has been suggested that large seasonal changes in the eye pigment diameter of cyclomorphic *Daphnia* are related to seasonal changes in the intensity of fish predation, smaller eyes occurring during the season when fish are more abundant (Zaret 1972). Furthermore, it has been suggested that the eye pigment diameter of *Bosmina longirostris* decreases during daylight hours due to selective removal of the large-eyed individuals by planktivorous fish (Zaret and Kerfoot 1975).

With regard to the diurnal variation in eye size, we wish to propose a simpler explanation. Our data show that light alone can cause diurnal variations in the eye pigmentation diameter of several Cladocera. Both laboratory experiments and studies of zooplankton in ponds without fish have shown contraction of eye pigment diameter in light and dilation in dark. Since this response to light is as large as the diurnal change previously reported for *Bosmina*, we see no reason to invoke selective predation by fish as an additional explanation.

Our laboratory studies were done primarily with *Daphnia magna*. Twenty *Daphnia* of 3.2 to 3.6 mm as measured by a filar micrometer at 12X were used in all treatments. The 20 animals were placed in individual beakers, acclimated to either dark or light for at least 15 h, and then either exposed to varying durations of light or dark. Animals were preserved with a sugar-enriched 8% Formalin solution (Haney and Hall 1973). The animal lengths were then remeasured and those that had molted were discarded. The maximum width of the eye

This work was supported by NSF-URP grant SM176-83107.

pigmentation was measured on the remaining animals at 50X for all *Daphnia* and at 200X for all *Bosmina* experiments.

The shape of the pigmented area of a cladoceran eye is quite irregular and sometimes quite variable. Some of the measured eyes were much more than 2 SD beyond the mean diameter for any treatment. In the following statistical analyses these values have been discarded. However, all analyses were also done using all the values, and these analyses are available on request.

Field samples were collected from three of the experimental ponds maintained by Cornell University. These ponds were virtually free of fish due to a winter kill after a record-breaking snowfall. No fish were ever seen in the three sampled ponds even though the senior author snorkled in one pond and we repeatedly searched all shorelines. Thus, the changes in pigment diameter reported for our field samples cannot be ascribed to fish predation. Field samples were all collected with a plankton tow net of No. 10 mesh size. Tows were drawn across the pond width with the net slightly below the surface.

Our results are as follows: Preliminary studies demonstrated that both *D. magna* and *Daphnia pulex* exposed to light had smaller eyes than when exposed to dark. We then wondered about the time necessary for the pigment diameter to change by its maximum amount. This rate of change was examined using *D. magna* during our laboratory studies. Figure 1 shows the rate of change for dark-conditioned *Daphnia*. Beakers with animals sorted for uniform size were removed from the dark, placed on white paper, and exposed to two 40-W light bulbs.

The maximum change in the mean values was from 0.17 mm at zero h to 0.14 mm at 8 h—a 22% decrease. Tukey's test for the Honestly Significant Difference at the 95% confidence level was applied to the data. The pigment diameters at zero and 1 h were significantly different from those at 5 and 8 h. While the means show a decrease in diameter throughout the sampling duration, the eyes reached their statistically minimum size after 3 h of light.

Figure 2 shows the rate of dilation of *D. magna* pigment diameter. The maximum change in the mean diameter was 14% from 0.15 mm at zero h to 0.17 mm at 11 h. The zero-h sample is statistically different from the samples at 5, 8, and 11 h. The 11-h sample is different from the zero- and 2-h sample, but the 5-h sample is not distinguishable from 8 or 11 h.

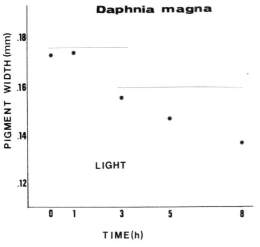

Fig. 1. Decrease in the pigment width of *Daphnia magna* eyes preconditioned to at least 12 h of dark. The lines above the points indicate groups of means that are statistically indistinguishable by Tukey's test for the Honestly Significant Difference.

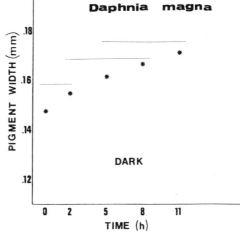

Fig. 2. Increase in the pigment width of *Daphnia magna* eyes preconditioned to at least 12 h of dark (see Fig. 1 for explanation of lines).

Field samples were collected over a 24-h period on three occasions. All pigment diameters were determined on animals sorted for a body size range of 10%. One pond with *D. pulex* and *B. longirostris* was sampled on a heavily overcast day. Another *B. longirostris* pond was sampled on a sunny day, and a third pond with *D. pulex* was sampled on another sunny day.

Figure 3 shows the change in *Bosmina* pigment diameter on the overcast day when the total irradiation was 104 ly (Langleys). The irradiation was measured by a United States weather station about 5 miles from the ponds. The *Bosmina* eyes shrank by a maximum of 14% from zero h to 16 h. Tukey's test showed that the sample from zero h was significantly different from all other daytime samples. None of the other samples were statistically distinguishable even though there is a trend for decrease in diameter over the entire daylight period. Eye shrinkage during the day was matched by dilation during the night. The 0500-hours sample of the second day was statistically indistinguishable from the 0500-hours sample of the first day, but both were different from all others.

A second pond with *Bosmina* was sampled on a day when the irradiation was 586 ly (Fig. 4). These samples showed a maximum decrease of 18% at 6 h. This compares to a maximum decrease of 14% on the overcast day. The mean diameters of the 12-h sample (1700-hours) and the 16-h sample (2100-hours) are larger than that of the 6-h sample. The increase in pigment diameter by 2100 was statistically different from the 6 h sample (1100 hours).

We have no certain explanation for this unique afternoon increase. However, if someone wishes to study this phenomenon further, the following tentative explanation should be considered.

We observed practically no animals in the midday samples collected in the normal manner. Consequently, we pulled the net slower in order to collect animals at greater depth. Possibly, these peculiar results are due to a relationship among light intensity, depth, and pigment diameter.

The *Bosmina* eyes dilated during the night as they had during the previous *Bosmina* sampling so that the 0500-hours samples of the first and second days were statistically indistinguishable.

Daphnia pulex was collected along with *Bosmina* on the heavily overcast day when total irradiation was 104 ly. Eye pigment diameters of *D. pulex* measured from samples taken at 0500, 1700, and 0100

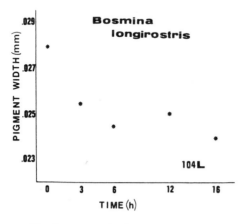

Fig. 3. Decrease in the pigment width of *Bosmina longirostris* eyes on a heavily overcast day with irradiation about 104 ly. The zero-h sample was collected at 5 AM (see Fig. 1 for explanation of lines).

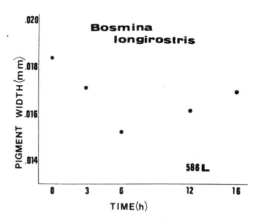

Fig. 4. The decrease in the pigment width of the eye of *Bosmina longirostris* during a day when irradiation was 586 ly. The zero-h sample was collected at 5 AM (see Fig. 1 for explanation of lines).

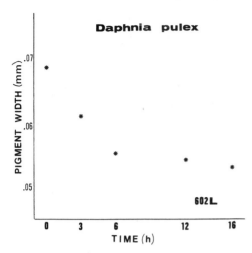

Fig. 5. Decrease in pigment width of *Daphnia pulex* eyes on a day with irradiation of 602 ly. The zero-h sample was collected at 5 AM (see Fig. 1 for explanation of lines).

hours of the second day showed no change. Thus, the same low light intensity that did induce a response in *Bosmina* had no effect on *D. pulex*.

Daphnia pulex was sampled from another pond on a bright day with 602 ly of irradiation (Fig. 5). Significant changes did occur on this date. A maximum decrease of 24% occurred from zero h (0500) to 16 h (2100). Tukey's test showed that the zero-h sample was different from all others. Although the mean pigment diameters decreased throughout the daylight sampling, none of the samples from 6 h (1100) through 16 h are statistically distinguishable. Dilation of the eyes occurred during the night so that the sample at 0500 hours of the second day was not statistically distinguishable from the sample at 0500 of the first day.

Collectively these results show that all three of the tested cladoceran species have contraction/dilation of eye pigment. *Daphnia magna* and *D. pulex* were tested in the laboratory where there is little possibility of variables other than light having an effect. The field samples of *D. pulex* and *B. longirostris* also showed dilation when collected from ponds without fish where light is almost certainly the controlling variable. A daytime decrease in the eye pigment diameter of *B. longirostris* has been reported before (Zaret and Kerfoot 1975). In this study, samples from Gatun Lake were collected in the "early morning" and compared to samples collected "usually 4–6 h later." This sampling routine matches the times when our data show the largest response to light. The magnitude of decrease observed on three dates in Gatun Lake ranged from 3–18%. The largest decrease that we observed for *B. longirostris* was also a decrease of 18%, which occurred over 6 h. Thus, light-induced contractions of the pigment diameter of the eye observed with *B. longirostris* from ponds without fish accounts entirely for the decrease for *B. longirostris* observed in Gatun Lake.

References

HANEY, J. F. and D. J. HALL. 1973. Sugar-coated *Daphnia*: A preservation technique for Cladocera. Limnol. Oceanogr. 18: 331–333.

ZARET, T. M. 1969. Predation-balanced polymorphism of *Ceriodaphnia cornuta* Sars. Limnol. Oceanogr. 14: 301–303.

———. 1972. Predators, invisible prey, and the nature of polymorphism in the Cladocera (class Crustacea). Limnol. Oceanogr. 17: 171–184.

———, and W. C. KERFOOT. 1975. Fish predation on *Bosmina longirostris*: Body-size selection versus visibility selection. Ecology 56: 232–237.

Addendum

There are three strong reasons that we see for disbelieving that visually selective predation on large-eyed *Bosmina* causes the diurnal decrease in mean eye size reported for Gatun Lake (Zaret and Kerfoot 1975). The first involved the maximum visual acuity of fish. The second derives from the lack of a decrease in variance for the eye size during the day which should occur according to the postulates of the visual selective predation hypothesis. The third reason is that light-induced reduction in eye size explains the data for both the eye size and the variance as observed in Gatun Lake.

Visual acuity—The maximum diurnal change was observed on 5 May 1969 when the mean eye size decreased from 0.0222 to 0.0182 mm. The mean was reduced by more than twice the standard deviation

(1.88μm) of the morning population. (Zaret has informed me that the values for the standard deviation of the eye sizes in their table 2 [Zaret and Kerfoot 1975] are too small by a factor of 10.) This change and postulated mechanism requires more than selectively preying on the extremely big eyes. If only the extremely large eyes were selectively removed, the mean could not shift to a size class more than 2 SD below the mean. The postulated mechanism requires that the fish prey selectively on the larger eyes within the range of 0.0182–.0222 mm. Thus, the visual hypothesis requires that the fish selectively prey on animals whose eyes differ by as little as 0.002–0.003 mm. Measurements of fish reaction distance (Ware 1973; Werner and Hall 1974; Confer and Blades 1975; O'Brien et al. 1976) suggest that 3-mm opaque particles (Ware 1973) or opaque *D. magna* can be detected from 25 cm to as much as 45 cm by different species of fish. In proportion, a 0.003-mm difference in eye diameter would increase reaction distance by at most 0.045 cm.

Blaxter (1970) summarized data on fish visual acuity from nine studies involving about 40 species. He reported that "particularly low values (for angle of resolution) of 2'–4'" (low angle of resolution means high acuity) were found for some fish. With this acuity, a 0.003-mm difference would add 0.52 cm to the detection distance. These "particularly low values" were obtained by optical tests of the lenses and histological analyses of the density of cones in the part of the retina with greatest visual acuity. A zooplankton might be anywhere within the fish's three-dimensional visual field. Thus, the 0.003-mm difference in eye size is likely to result in a difference in strike distance which is much less than 0.5 cm and approaching 0.045 cm. It is most unlikely that the postulated, highly selective predation would result from increments in detection distance of between 0.5 and 0.05 cm.

Variance—The second argument is based on the irreconcilable lack of a decrease in the standard deviation for eye size during the day. Zaret and Kerfoot (1975) postulated selective removal of large eyes so that only those more than 2 SD smaller than the mean remained on, e.g., 5 May 1969. If this selective removal were true, then the "population range would be truncated" (Zaret and Kerfoot 1975). In fact, the range would be immensely reduced, resulting in an immense decrease in the population variance. For all three pair of morning/afternoon samples, the standard deviation increases during the day: 1.82/2.16, 2.40/2.64, 1.88/2.73. The lack of a large decrease in variance during the day is, we believe, definitive evidence that selective removal of large-eyed individuals is not the appropriate explanation.

Alternative—As we have just described, the maximum decrease of 18% in the eye diameter observed by Zaret and Kerfoot (1975) is exactly equal to the maximum percent change in eyes of *B. longirostris* we observed in a pond without fish. Furthermore, the variance resulting from this light-induced reduction in eye size is completely compatible with the Gatun Lake data. During the day that we observed an 18% decrease in eye size, the standard deviation did not change: 1.61 (0 h), 1.98 (3 h), 1.38 (6 h), 1.51 (12 h), 1.69 (16 h). Thus, in all respects light-induced changes explain all the Gatun Lake data on *Bosmina* eye size.

References

BLAXTER, J. H. S. 1970. Light; animals; fishes, p. 213–320. *In* O. Kinne [ed.], Marine ecology, v. 1, part 1.

CONFER, J. L., and P. I. BLADES. 1975. Omnivorous zooplankton and planktivorous fish. Limnol. Oceanogr. 20: 571–579.

O'BRIEN, W., N. A. SLADE, and G. I. VINYARD. 1976. Apparent size as the determinant of prey selection by bluegill sunfish (*Lepomis macrochirus*). Ecology 57: 1304–1310.

WARE, D. M. 1973. Risk of epibenthic prey to predation by rainbow trout (*Salmo gairdneri*). J. Fish. Res. Bd. Can. 30: 787–797.

WERNER, E. E., and D. J. HALL. 1974. Optimal foraging and the size selection of prey by the bluegill sunfish (*Lepomis macrochirus*). Ecology 55: 1042–1052.

ZARET, T. M., and W. C. KERFOOT. 1975. Fish predation on *Bosmina longirostris*: body-size selection versus visibility selection. Ecology 56: 232–237.

Commentary: Transparency, Body Size, and Prey Conspicuousness

W. Charles Kerfoot

Before addressing the questions raised by Confer (at this conf.) let me briefly summarize the central issues discussed in Zaret and Kerfoot (1975) and elaborate on each point.

1. Pigmentation, and not body size as such, is a more appropriate index of prey conspicuousness to visual predators. The evolution of increased transparency is one of the most important and widespread protective adaptations against visually feeding predators (Greze 1963). The resulting crypsis provides an evolutionary solution to the problem of concealment against a continually changing background. While body pigmentation is usually highly correlated with general body size, several notable large-bodied prey survive in pelagic regions because of their extreme transparency (e.g. *Chaoborus, Leptodora, Daphnia galeata mendotae*).

2. In species that are already highly transparent, eye pigmentation may be an important component of general conspicuousness to visual predators. The eye pigment region is usually the most optically dense area of the entire body. While the function of the eye is not known, its role in orientation is strongly suspected (Siebeck 1968, at this conf.).

3. Intense visual predation by fishes in very clear tropical lakes explains the occurrence of different eye-size "morphs" (not morphs in the sense of different phenotypes within a single brood, but genetically different clones) of *Daphnia, Ceriodaphnia,* and *Bosmina*. Despite its small size, even the small-bodied cladoceran *Bosmina longirostris* shows spatial and temporal differences in eye size associated with fish predation in Gatun Lake, Panama.

Although the effects of visual predation are found even in clear arctic lakes (Kettle and O'Brien 1978), responses would be especially pronounced in tropical latitudes. Under tropical climates, fish activity is year-round, waters are brightly illuminated, and prey generation times are short. For example, before the peacock bass (*Cichla oscellaris*) entered Gatun Lake in 1970, adult planktivorous fishes (principally the silversides minnow, *Melaniris chagresi,* but also *Astyanax ruberrimus* and three other characins), their fry, and the fry of several other herbivorous, insectivorous, and piscivorous species drew heavily from the zooplankton found along the littoral margin of the lake (Zaret and Paine 1973). During this time, water temperatures were nearly constant and high (28°C min, 30°C max), illumination was bright (midday subsurface incident radiation, I_0, ranged between 2-6 kcal/m^2/min; the mean percentage of daylight hours that experienced direct sunlight ranged between 40% in the wet season to 75% in the dry season), and light penetration was deep (Secchi disk reading varied between 5.3-9.0 m; 1% light penetrated to between 15 and

20 m; all 1969–1970 values summarized in Gliwicz 1976).

Under these conditions, fish predation can be especially pronounced. Although there seems little doubt that visual selection by fishes is responsible for the evolution of highly transparent, small-eyespot forms of *Daphnia* in Africa (the small-eyed and helmeted *D. lumholzi* in Lake Albert: Green 1971), and of *Ceriodaphnia* in both the Old World and New World tropics (the small-eyed and horned *C. cornuta*: Green 1971; Zaret 1972a), the extension of similar reasoning to explain comparable variation between clones and species of *Bosmina* is questioned by Confer (at this conf.) largely because *Bosmina* is so small (generally 200–600 μm long) that differences in its eye-pigment diameters approach the theoretical visual threshold for detection by certain fishes and because the eye pigment of many cladocerans undergoes a diurnal contraction (day) and dilation (night) cycle within the simple eyespot.

Unfortunately, while the investigations of Zaret and Kerfoot (1975) treated both spatial and temporal variation between different *eye-sized* clones, only pigment diameters were reported, and thus the data could include pigment expansion-dilation effects. To clarify the difference between these variables, I will distinguish between measurements of pigment diameter (PD) and of eye diameter (ED: *see insert in Fig. 3*).

In Gatun Lake, *Bosmina* populations show considerable spatial variation in pigment diameters. There is also abundant evidence of long-term temporal changes associated with the drastic reduction of *Melaniris* predation after 1970. For example, Fig. 1 compares samples of *Bosmina* taken in 1974 (these selected samples were all taken within a few days of each other, and around midday, 1200–1600 hours, to minimize effects of diurnal pigment cycles) with a representative Slothia sample (high *Melaniris* predation: Zaret and Kerfoot 1975). As this figure shows, *Bosmina* from the Trinidad arm of Gatun Lake, near the mouth of the Rio Trinidad in a region of reduced water transparency (Secchi disk reading, 1.5 m; high phytoplankton density and suspended silt load), are generally large-bodied, large-eyed forms, although a few small-eyed individuals are present in the sample (Fig. 2, arrow). *Bosmina* from the

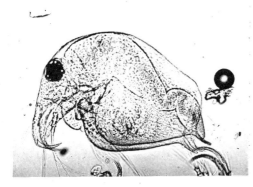
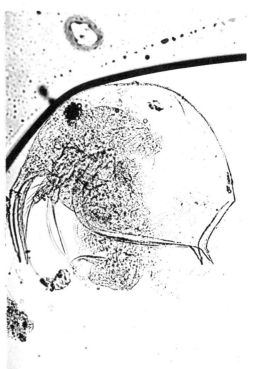

Fig. 1. Large- and small-eyed *Bosmina longirostris* (large-eyed clone from Star Lake, Vermont; small-eyed from *Hydrilla* margin, Frijoles Bay, Gatun Lake).

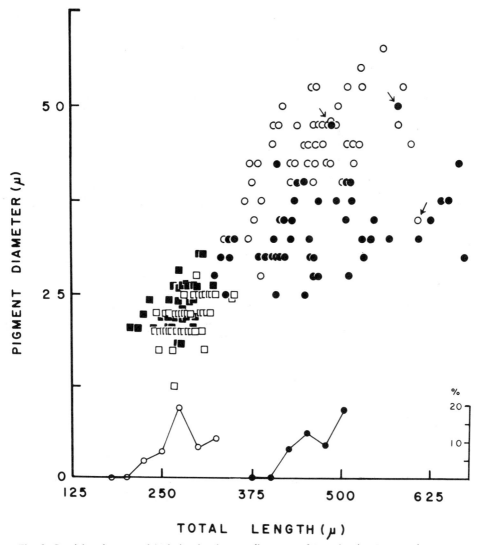

Fig. 2. Spatial and temporal variation in pigment diameters of *Bosmina* (○—Gatun Lake near mouth of Rio Trinidad, 20 July 1974; ●—Panama Canal channel, 25 July 1974; ■—off *Hydrilla* margin, Frijoles Bay, 25 July 1974; □—off *hydrilla* margin along Slothia Island Bay, 9 April 1969). Frequencies of egg- or embryo-carrying adults plotted along bottom axis (connected open circle, combined 1969 Slothia samples; connected solid circles, 1974 open-water Frijoles Bay samples).

center of the Panama Canal channel, midway between Barro Colorado Island and the mainland, are large-bodied but generally small-eyed as adults (here too there is appreciable variance in pigment diameters and a few large-eyed individuals: *see arrows* in Figure 2. The final two samples, both from the upper meter of water just off the outer margin of the *Hydrilla* beds in Frijoles Bay, Barro Colorado Island (solid squares, 1974; open squares, 1969), come from areas of *Melaniris* feeding. Both of these latter samples contain very small-bodied individuals. In a sense, the 1969 and 1974 samples form a continuum in space and time. In turbid waters, *Bosmina* suffers little mortality from visual predators, and thus can grow as large, and possess as large an eyespot, as

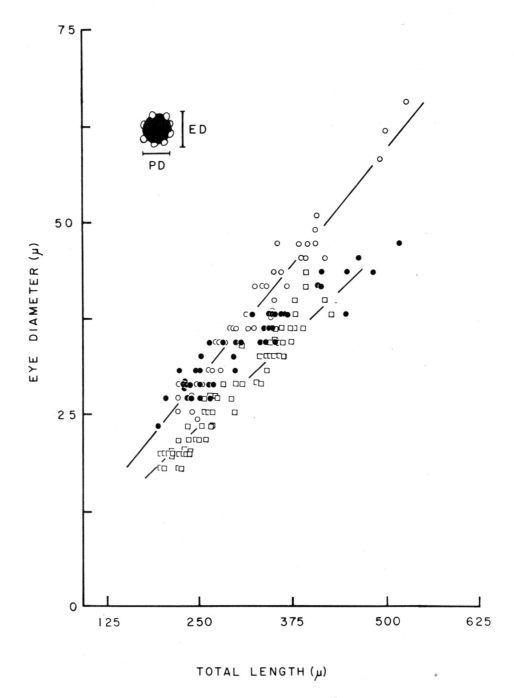

Fig. 3. Eye diameter variation in laboratory clones ○—Star Lake, Vermont; ●, □—two different clones from Frijoles Bay, Gatun Lake. Lines show course of growth if eye diameter remained 12% (upper line) or 9% (lower line) of total body length. Insert shows measurements of eye (ED) and pigment (PD) diameters.

its northern temperate counterparts. In the Panama Canal channel, within the clear, open waters of Gatun Lake (although the surface area is very large, 430 km^2, the mean depth is only 12.5 m; maximum depth, 29 m) away from the shallow coves of Barro Colorado Island, there is some planktivore activity, but not nearly as much as in places along the *Hydrilla* margin, where *Melaniris* concentrates its feeding schools. Even off the *Hydrilla* margin, there appears to be a major difference between *Bosmina* samples taken in 1969 and 1974. Although both samples appear truncated with respect to body length, the 1974 samples are composed chiefly of juveniles (1st and 2nd instars), while the 1969 samples contain dwarf forms: *Bosmina* that reach reproductive maturity at <250 μm and that produce substantially smaller-sized eggs and neonates (egg size, 109.9μm±2.0 SE, N = 6; neonates, 186–216μm, 204.5μm±1.8 SE, N = 27; Fig. 2 shows differences in maturity as percentages of egg- or embryo-carrying adults in 1974 and 1969 samples). Before 1970, the higher densities and greater spatial distribution of *Melaniris* in Frijoles Bay (Zaret and Paine 1973) is clearly associated with much smaller-bodied and more slowly-growing clones. Against this background information, the importance of eye pigment diameter relative to body length, per se, no longer seems mysterious. Before 1970, in regions of heavy *Melaniris* predation, the expression of body length was so restricted that eye pigment variation could contribute a major component to general conspicuousness.

As alluded to earlier, relative pigment diameter (PD/TL) is highly variable in Gatun Lake samples of *Bosmina* and related to relative eye diameter (ED/TL). To clarify the underlying causes of this variation, we shipped zooplankton samples from Frijoles Bay to Dartmouth College, removed *Bosmina*, and established clonal cultures. These clones were grown in 50-ml vials provided with mixed Provasoli's medium, fed *Chlamydomonas* every 4 days, and maintained in an environmental chamber under controlled conditions (Lab/line;

Table 1. Regressions for eye diameter on total body length of *Bosmina longirostris* (N = sample size, r = correlation coef., Y = eye diameter, X = total length).

Clone source	N	Regression equation	SE(B)	r
Star Lake	42	Y = .122X − 0.005	.008	.923
Frijoles Bay (clone A)	43	Y = .067X + 0.728	.004	.944
Frijoles Bay (clone 3)	66	Y = .104X − 0.140	.004	.953

constant illumination, temperature, 20°C). *Bosmina longirostris* from a local, small and highly eutrophic lake (Star Lake, Vermont) was cloned in similar manner. As an interesting footnote, during the 10 years of intensive sampling in Gatun Lake, we (T. Zaret and I) never saw any sexual activity among any cladoceran species (presumably the result of low densities, 3–20 total cladocerans/liter)—an important fact which may contribute to the clonal diversity in the lake. However, under culture conditions, many of the well-fed and yet crowded lines showed male production, and ephippial females could be induced in starved, high-density, crowded cultures.

Comparison of relative eye sizes among the clonal lines confirmed the eye-size basis of pigment diameter variation in Gatun Lake. Moreover, eye growth provided evidence for a simple mechanism, suggested earlier (Zaret and Kerfoot 1975), that contributes to eye-size variation between clones and that lowers correlations between pigment diameter and total length. While Star Lake *Bosmina* have relatively large eyes (ca. 12% total length) that grow in direct proportion to body length (isauxetic growth), Frijoles Bay clones either have relatively small eyes (ca. 9–10% total length) throughout life, or, if large-eyed at birth, some clones have eyes that grow more slowly than body length (bradyauxetic growth). As expected from their restricted genetic variance and culture conditions (uniform environment), variances of eye diameter around regression lines are reduced, as compared to field samples (Fig. 3; Table 1 gives regressions). Thus the spatial and temporal variation of pigment diameter in Gatun Lake involves real differences in the eye diameter of clonal lineages, not just expansion-dilation cycles of pigment within the eye. Not only are small-eyed lineages present, but large-eyed young often show allometric growth of eyes. The bradyauxetic developmental response is understandable if eyespot pigmentation contributes to general conspicuousness, but difficult to understand otherwise.

Direct evidence for the enhanced conspicuousness of *Bosmina* with enlarged pigment diameters came from fish-choice experiments. Our methods are briefly summarized below.

Following Milinski (1977), a 20-liter aquarium (sand bottom) was divided in half with a plastic partion which included a slip door (11 × 6 cm.). Half contained plants and was aerated. The other half contained a row of three 20-ml glass vials (2.5 cm diameter). Two non-glare 75-W bulbs hung above the aquarium, directly over the opposite ends. The whole assemblage was placed in a walk-in environment chamber (Module-lab, Hudson Bay Co., set at constant temperature, 20°C). Surface illumination during experiments (measured with a Gossen Lunasix Exposure meter) was ca. 910 lux, while the illumination through the sides of the tank was ca. 280 lux. Millimeter rulers placed along the bottom of the aquarium provided distance calibration.

Preliminary to measurements of reactive distance and vial choice, the fish *Lepomis* fry, ca. 2.7–cm. snout-fork length) was trained to move through the opened slip door to hunt for *Daphnia pulex*. After initial training on *Daphnia*, *Bosmina* was offered in the open water. At first the fish appeared confused by the change of prey type, but within a few hours it adjusted to spotting and individually picking out *Bosmina*. These steps completed, we set up the experiment.

For the experiment, a trained bluegill fry (*Lepomis macrochirus*) was starved for 48 hours before choice tests. The prey—Frijoles Bay clones of *Bosmina*—received two kinds of treatment. A single population of clonemates was split into two groups. Both groups were placed in Provasoli's medium for 1 hr before the experiment. Of the two groups, one was given a suspension of India ink particles 15 min before the experiment, allowed to ingest the particles, and then washed in Provasoli's medium. Thus one group of otherwise identical organisms had evacuated guts, while the other group had "super-eyespot" pigment areas, following Zaret (1972b). Comparably sized adults from each of the two treated groups were then placed individually, or in groups of 10, into one of the three vials. Before the experiment, the bluegill was placed in the aerated end of the aquarium and allowed to acclimate. When the door was lifted, it would then enter the experimental end of the aquarium, move forward first without perceiving the prey, then would suddenly orient toward one of the three vials, approach, and attempt to "bite" at a swimming *Bosmina*. The distance at which the fish first oriented toward the vial was recorded as its "reactive distance" (RD), following Confer and Blades (1975). After a "bite", the fish was returned to its original chamber, new *Bosmina* placed in the vials, and the vials rearranged in random order. Two levels of response are discussed here: when presented with two empty vials and one that contained *Bosmina*, correct choice of the vial containing *Bosmina*; and, given a reaction, the difference in RD between normal and "super-eyespot" *Bosmina*.

The results of these experiments are summarized in Table 2. Of the 77 responses to groups of *Bosmina*, in all cases the fish selected the vial that contained *Bosmina*. In 53 instances out of the 77, the fish attempted to "bite" at an individual *Bosmina*. Reaction distance (RD) values for the evacuated-gut group averaged 3.5 cm, while RD values for the "super-eyespot" group averaged 6.4 cm (differences between the two groups were highly significant, $P = 0.001$:

Table 2. Reactive distances on evacuated and "super-eyespot" Bosmina.

Treatment	Group Size	No. of trials	Reactive distance mean ± SE (SD) cm	t-value	Significance
Evacuated	10	29	3.52±0.28 (1.53)	6.05 (75 df)	$P \ll 0.001$
Super-eyespot	10	48	6.93±0.31 (2.17)		
Evacuated	1	28	3.93±0.35 (1.84)	2.82 (54 df)	$P < 0.01$
Super-eyespot	1	28	5.50±0.44 (2.30)		

Student's t-test). Because there were 10 *Bosmina* in each vial, the large reactive distances could result from "group" effects, i.e. enhanced effect due to the presence of more than one individual in a swarm (Milinski 1977). However, trials run on individual *Bosmina* gave comparable results (Table 2). Out of 59 responses, however, the fish chose incorrectly three times, once biting at an air bubble. Because the vials were 2.5 cm deep, the RD values here are obviously minimal estimates, with only the relative differences being crucial. Yet they are not unreasonable estimates, since Kettle and O'Brien (1978) report RD values between 5.07–6.19 cm for arctic *Bosmina*, using lake trout fingerlings (*Salvelinus namaycush*, 10 cm long).

Thus both the circumstantial evidence from zooplankton samples and the experimental fry choices (by a nonspecialized, facultative planktivore) support the view that, in tropical *Bosmina*, transparency is great enough to make artificially increased eye pigmentation contribute to over-all prey conspicuousness.

While the increased transparency of zooplankton species subject to visually feeding fishes is strongly supported by recent studies on reactive distances (Kettle and O'Brien 1978), certain criticisms by Confer (Confer et. al. 1978; at this conf.) require comment. For instance, appropriate tests of Zaret's "eye pigmentation" hypothesis should be run on highly transparent zooplankton, not on "translucent" *Daphnia pulex* (Confer et. al. 1978). Secondly, the selective "reaction" or response of populations to the selective foraging bias of fishes should not be confused with the predators ability to discriminate differences. *Melaniris* in Gatun Lake are preying on zooplankton that have pigment diameters between 17–50 μm while the *differences* (the response) before and after predation may move the mean pigment diameter 2–4 μm. Finally, there is no reason why the variance in eye pigment diameter need be "truncated" by selective predation (Falconer 1960), although movement of the mean by two standard deviations would assume selective removal of around 95% of the individuals from the original (early dawn) population. Whereas mortality rates this high are uncommon in general within Gatun Lake and Frijoles Bay, they are not unusual in the restricted areas where *Melaniris* feeds. Samples of zooplankton taken immediately off the *Hydrilla* margins, in areas of *Melaniris* schooling, contain ratios of dead:live *Bosmina* which often vary between 23–170:1. The dead *Bosmina* are flattened, partially decomposed, and yet articulated (rostra and carapace attached) - closely resembling ones which have passed through the guts of laboratory fishes.

Since the diurnal pigment diameter difference reported in Zaret and Kerfoot (1975) could have included pigment dilation/contraction cycles, we have begun to investigate eye morphology and reaction to light. Despite its rather simple appearance, the eyespot contains numerous internal complexities that suggest sophisticated move-

Table 3. Eyespot responses of Star Lake *Daphnia pulicaria* and *Bosmina longirostris* to a dark-light shift (all lengths in mm; all measurements made at 500X).

Size Class	Control			1 Hr			3 Hr			5 Hr			8 Hr		
	N	$\bar{x}$	SD	N	$\bar{x}$	SD	N	$\bar{x}$	SD	N	$\bar{x}$	SD	N	$\bar{x}$	SD
Daphnia					Eye diameter/Total length										
1.0	44	0.101	0.011	9	0.096	0.008	15	0.097	0.010	14	0.105	0.009	15	0.110	0.012
1.1–1.5	35	0.092	0.010	15	0.092	0.008	25	0.096	0.013	30	0.098	0.011	29	0.092	0.009
1.5	12	0.096	0.013	19	0.093	0.005	30	0.094	0.011	27	0.092	0.010	27	0.087	0.008
all	91	0.097	0.011	43	0.093	0.007	70	0.095	0.012	71	0.097	0.011	71	0.094	0.012
					Pigment diameter/Total length*										
1.0		0.091	0.010		0.087	0.007		0.087	0.009		0.094	0.006		0.097	0.010
1.1–1.5		0.084	0.009		0.083	0.008		0.087	0.013		0.088	0.010		0.081	0.008
1.5		0.087	0.010		0.084	0.006		0.084	0.009		0.083	0.009		0.077	0.008
all		0.087	0.010		0.084	0.007		0.086	0.011		0.087	0.010		0.083	0.011
Bosmina					Eye diameter/Total length										
0.200–0.299	87	0.139	0.010	58	0.135	0.012	75	0.138	0.018	72	0.138	0.013	74	0.137	0.011
0.300–0.399	24	0.127	0.011	24	0.131	0.013	16	0.123	0.025	21	0.134	0.017	19	0.124	0.010
0.400–0.499	14	0.129	0.013	9	0.137	0.013	7	0.126	0.005	8	0.132	0.006	8	0.128	0.009
all	125	0.135	0.012	91	0.134	0.013	98	0.135	0.020	101	0.137	0.013	101	0.134	0.012
					Pigment diameter/Total length*										
0.200–0.299		0.122	0.010		0.119	0.011		0.118	0.018		0.119	0.011		0.115	0.011
0.300–0.399		0.112	0.012		0.113	0.015		0.111	0.015		0.114	0.017		0.104	0.007
0.400–0.499		0.114	0.013		0.119	0.010		0.107	0.006		0.124	0.037		0.108	0.007
all		0.119	0.012		0.117	0.012		0.116	0.017		0.118	0.016		0.113	0.011

*same sample sizes as above

ments (lateral and interior muscles individually attached to each lens, lens cup attachments to a central core; Kerfoot and Zelazny, unpubl.). hence the importance of distinguishing between changes in eye diameter (lens position) from changes in pigment diameter (centrally pigmented portion of eye). For organisms, we used *Bosmina longirostris* and *Daphnia pulicaria* from Star Lake, Vermont. The experimental protocol was similar to that used by Confer et al. (at this conf.), i.e. 500 ml beaker of animals placed in the dark (at 20°C) for 15–23 hours and then transferred to brightly lighted conditions (40 W bulb, ca. 1400 lux). Animals left under dark conditions served as controls, while animals removed from illuminated beakers at various times (1, 3, 5, and 8 hrs after initiation of illumination) served as experimentals. All cladocerans were immediately preserved in 8% Formalin and measured under 500X. As Table 3 shows, while there is some indication of decrease under high light levels, no significant differences in eye diameter (ED) or pigment diameter (PD) occurred in either *Daphnia* or *Bosmina*. While these results are disappointing because they give no indication which components are involved in contraction, since contraction did not occur, they may indicate that smaller species of cladocerans do not undergo contraction cycles as pronounced as in *Daphnia pulex*.

Despite the objections raised by Confer et al. (at this conf.), I believe I have shown that selective predation and not physiological accommodation is responsible for the eye size differences seen in Gatun Lake *Bosmina*. The fact that the size and development of the entire eye, not just the pigmented area, shows such marked spatial variation, and that both field and laboratory show little response in PD/ED measurements, makes the situation in these *Bosmina* comparable to previously described variation in tropical *Daphnia lumholzi* and *Ceriodaphnia cornuta* (Green 1971; Zaret 1972a).

References

CONFER, J. L., and P. BLADES. 1975. Omnivorous zooplankton and planktivorous fish. Limnol. Oceanogr. 20: 571–579.

———, G. L. HOWICK, M. H. CORZETTE, S. L. KRAMER, S. FITZGIBBON, and R. LANDESBERG. 1978. Visual predation by planktivores. Oikos 31: 27–37.

FALCONER, D. 1960. Introduction to quantitative genetics. Ronald Press.

GLIWICZ, M. 1976. Plankton photosynthetic activity and its regulation in two Neotropical man-made lakes. Pol. Arch. Hydrobiol. 23: 61–93.

GREEN, J. 1971. Associations of Cladocera in the zooplankton of the lake sources of the White Nile. J. Zool. Lond. 165: 373–414.

GREZE, V. N. 1963. The determination of transparency among planktonic organisms and its protective significance. Dokl. Biol. Sci. (Engl. Transl. Dokl. Akad. Nauk. SSSR Ser. Biol.) 151 (2): 956–958.

KETTLE, D., and W. J. O'BRIEN. 1978. Vulnerability of arctic zooplankton species to predation by small lake trout (*Salvelinus namaycush*). J. Fish. Res. Bd. Can. 35: 1495–1500.

MILINSKI, M. 1977. Do all members of a swarm suffer the same predation? Z. Tierpsychol. 45: 373–388.

SIEBECK, O. 1968. "Uferflucht" und optische Orientierung pelagischer Crustaceen. Arch. Hydrobiol. Suppl. 35, p. 1–118.

ZARET, T. M. 1972*a*. Predator-prey interactions in a tropical lacustrine ecosystem. Ecology 53: 248–257.

———. 1972*b*. Predators, invisible prey, and the nature of polymorphism in the Cladocera (class Crustacea). Limnol. Oceanogr. 17: 171–184.

———, and W. C. KERFOOT. 1975. Fish predation on *Bosmina longirostris*: body-size selection versus visibility selection. Ecology 56: 232–237.

———, and R. T. PAINE. 1973. Species introduction in a tropical lake. Science 182: 449–455.

54. Predation Pressure from Fish on Two *Chaoborus* Species as Related to Their Visibility

Jan A. E. Stenson

Abstract

Larvae of *Chaoborus obscuripes* and *C. flavicans* differ with respect to pigmentation and behavior. These factors are significant for a visually dependent predator. *Chaoborus obscuripes*, which has more pigment and the least marked diel migration pattern may hence be more susceptible to predation. This suggestion is supported by results from a field study where *C. obscuripes* appeared after an experimental reduction of the fish population and a feeding experiment where fish preyed significantly more on *C. obscuripes*. The significantly higher survival of *C. flavicans* when exposed to fish is also due to its ability to penetrate into the sediment.

Chaoborus larvae are common components of zooplankton in lakes and ponds, and several species often coexist in the same body of water. The physical and chemical properties of the water probably have no major influence on the distribution and abundance of larvae. Instead these factors are probably controlled by a series of biotic factors among which food densities and predation from fish seem to be important (e.g. Stahl 1966; Saether 1972; Pope et al. 1973; Stenson 1976).

Previous knowledge about the way in which planktivorous fish select prey organisms makes it possible to predict the probable effects of fish predation on the abundance and distribution of *Chaoborus* larvae. Species with morphological and/or behavioral characteristics which make them easier to detect by eye are likely to be subjected to a greater predation pressure, as fish rely mainly on visual stimuli for detection of prey. It may therefore be difficult for those easily detectable larvae to coexist with fish.

I here present some experiments in which the validity of the above hypothesis has been checked. The experiment and observations were made in Lilla Stockelidsvatten in southwest Sweden.

Field experiment

The main questions were the following: does an experimental elimination of the fish population result in an increase in the number of larvae, or a changed species composition, and, are there any significant morphological and/or behavioral differences between the already existing species and any additional species that may turn up. A more detailed description of this experiment is given elsewhere (Stenson, 1978).

Results

The only *Chaoborus* species present together with fish was *C. flavicans* (Meig.).

This investigation was financially supported by grants from the National Swedish Environment Protection Board, the National Research Council of Sweden, and the Royal Fishery Board of Sweden.

The majority of the larvae was of the first two instars, and only very few individuals of the older instars were found in the samples taken before the fish elimination.

After the fish removal there was a marked increase in the total abundance of *Chaoborus* (Fig. 1) and also an increase of the older instars. The most noticeable change, however, was the appearance of a second species, *C. obscuripes* (V. D. Wulp), that was never before recorded in this lake. These changes after the elimination of fish, i. e. the increase in number and the appearance of a new species, indicate that fish predation may be one key factor in limiting larval population.

Because visual stimuli are perhaps the most important factors in the selection of prey by fish, both the pigmentation and behavior of the two species must be considered in seeking an explanation for the presumed differential predation.

Pigmentation—It seems more appropriate to estimate differences in pigmentation rather than differences in size since the two species are of about the same length. The eye size is shown in Fig. 2, according to which *C. obscuripes* has significantly larger eyes than *C. flavicans*. Apart from having a larger eye, the head region of *C. obscuripes* is also darker than that of *C. flavicans* because of heavier pigmentation on the mandibles. Finally, the body of *C. obscuripes* is a more yellowish brown (and hence darker) than are the other species.

Behavior—The diel migration of larvae in *C. flavicans* is well documented. The third and especially the fourth instar larvae undergo a marked diel vertical migration with the greatest abundance in the surface layers during night-time (e. g. Teraguchi and Northcote 1966; Goldspink and Scott 1971). *Chaoborus obscuripes* on the other hand does not show this migration pattern, and larvae of all instars are present in the upper layers in daytime (unpubl.). This species is consequently found to a greater

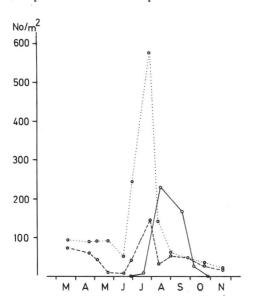

Fig. 1. Abundance of *Chaoborus* larvae in Lilla Stockelidsvatten (midday samples) from May to November. When fish were present (1973) only *C. flavicans* was present—solid line. After fish had been experimentally removed, *Chaoborus* spp. occurred in significantly higher numbers (1976)—dotted line. *Chaoborus obscuripes* was not recorded when fish were present, but when lake was devoid of fish (1976) it was fairly abundant—dashed line.

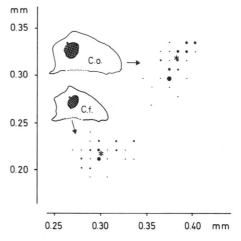

Fig. 2. Eye size of fourth instar larvae of *C. flavicans* and *C. obscuripes*. Eye length on horizontal axis and eye width on vertical. Mean eye length: *C. obscuripes*, 0.38±0.003 mm; *C. flavicans*, 0.30±0.003 mm; mean eye width: *C. obscuripes*, 0.32±0.004 mm; *C. flavicans*, 0.22±0.002 mm (mean ± 95% C. I.). Asterisks show position of mean values. Five different point sizes denote 1-5 specimens. Figure shows also outline of head capsules of fourth instar larvae of *C. obscuripes* (above) and *C. flavicans* (below).

degree than *C. flavicans* in the same part of the water as fish and is hence exposed to fish during the day, when vision is important in the predatory behavior of the fish. To summarize: the differences in visibility (pigmentation) and behavior (diel migration pattern) between the two species imply that the more conspicuous *C. obscuripes* will be more easily discovered by fish.

Laboratory experiment

To test how relevant these differences in fact are, I carried out the following two tank experiments (A and B). Following the experimental procedures shown in Fig. 3, I offered fourth instar larvae of the two *Chaoborus* species to fish. A more detailed description of these experiments is given elsewhere (Stenson in prep.).

A. One larva of each species is presented to a fish (Fig. 3A). The results (Table 1) show a significant preference for *C. obscuripes* ($p < 0.01$; χ^2-test). My interpretation of this result is twofold: *C. obscuripes* is preferred because of pigmentation or preferred because of a more conspicuous locomotion pattern. To discriminate between these two possible answers, the same experiment was repeated but with larvae of *C. flavicans* which were vital-stained (Bismarck brown) to approximate the body color of *C. obscuripes*. The most pigmented (stained) larva was preferred in every experimental set-up ($p < 0.01$; χ^2-test), but there was no difference between *C. obscuripes* and stained *C. flavicans*.

B. When fish were placed into a tank where five larvae of each species had acclimatized for 10 minutes, only those larvae which had entered the sediment survived the experimental period (Table 2). Survival was significantly higher for *C. flavicans* ($p < 0.001$; χ^2-test).

Discussion

Chaoborus obscuripes has in the literature been reported mostly from shallow nutrient-poor or meso- and polyhumic ponds (e.g. Parma 1969; Nillsen 1974). There are probably no direct connections between these abiotic lake characteristics and the occurrence of this species. These small shallow ponds are very often without fish, a fact which, when considering the results of this study, probably causes the positive correlation between this lake type and the

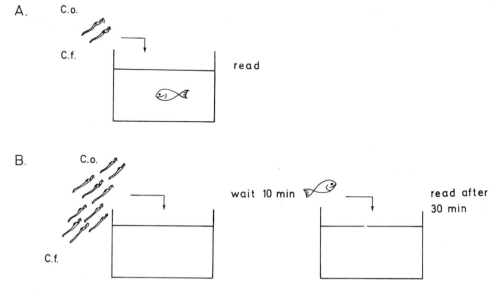

Fig. 3. Laboratory experiment procedures. A. Input of one *C. obscuripes* and one *C. flavicans* larvae (instar IV) to a tank with a fish. Which species will be eaten first? B. Input of five larvae of each species (instar IV). How many larvae of each species survive 30 minutes of exposure to fish after 10 minutes of acclimatization.

Table 1. Results of experiments: one *Chaoborus* larva of each species (*C. o.*—*Chaoborus obscuripes*; *C. f.*—*Chaoborus flavicans*) offered to fish simultaneously. Numbers represent occasions on which respective *Chaoborus* species was eaten first. Two fish species were used, blue goramy *(Trichogaster trichopterus)* and perch *(Perca fluviatilis)*.

Fish species	C.o.	C.f.	χ^2	C.f. stained	C.f.	χ^2	C.o.	C.f. stained	χ^2
T. trichopterus	27	8	10.31 $p<0.01$	25	9	7.53 $p<0.01$	21	23	0.12
P. fluviatilis	47	22	9.06 $p<0.01$	29	7	13.44 $p<0.01$	18	16	0.09

Table 2. Number of larvae of each species of *Chaoborus* remaining after 30 minutes exposure to fish *(Trichogaster trichopterus)*.

| | Experiment No. | | | | | | | | | χ^2 |
	1	2	3	4	5	6	7	8	9	
C. obscuripes	0	0	1	2	0	0	2	0	0	449.14
C. flavicans	2	3	2	4	3	3	5	2	3	$p<0.001$

presence of *C. obscuripes*. The conclusion, that fish are responsible, is recently supported in many west Swedish lakes. In these, the fish population is declining due to acidification and *C. obscuripes* is expanding its range.

If this species is adapted to lakes without fish predation pressure, it has no need for an energy-wasting "predator-avoiding" migration to the benthic or the hypolimnic zone at the end of the night. *Chaoborus obscuripes* can spend a considerably longer time than the vertically migrating *C. flavicans* in the upper layers where prey organisms are found, an adaptive advantage in the absence of significant numbers of predators. There is no need for *C. obscuripes* to reduce its visibility in an environment that lacks significant predators. Although all *Chaoborus* species probably rely mainly on tactile stimuli when they prey, "day-feeders" may also incorporate visual components. Therefore large eyes may constitute more of an adaptive advantage to a species such as *C. obscuripes*, which is active partly in the daytime, than to a species which is active exclusively at night.

Final remarks

One question which may arise from the discussion above is to what extent these two sympatric species compete. Since it is impossible to discuss competition in terms of niche utilization and niche overlap, with respect to all limiting factors, we can restrict the discussion to the feeding niche. Fedorenko (1975) studied the diet of larval instars of two *Chaoborus* species. Although overlap in prey selection existed, there were significant interspecific differences in the diet. These differences are, according to Fedorenko, mainly a function of the spatial distribution of predators and prey organisms and not a function of existing differences in morphology e. g. mouth diameter. Fourth instar larvae of the species with the largest mouth diameter, however, were more successful in eating large copepods.

The significant difference in mouth size between *C. flavicans* and *C. obscuripes* found in this study (*C. flavicans*, 0.63 ± 0.025 mm; *C. obscuripes*, 0.79 ± 0.019 mm; mean mouth diameter of fourth instar larvae ± 95% C. I.) may hence have ecological consequences. *Chaoborus obscuripes* may be at an advantage in an environment with decreasing fish predation and where the mean size of zooplankton increases and small cladocerans are replaced by larger copepods. However, it is difficult to conclude from such data, without adequate experiments, that two species compete.

References

FEDORENKO, A. Y. 1975. Instar and species-specific diets in two species of *Chaoborus*. Limnol. Oceanogr. **20**: 238-249.

GOLDSPINK, C. R., and D. B. C. SCOTT. 1971. Vertical migration of *Chaoborus flavicans* in a Scottish loch. Freshwater Biol. **1**: 411-421.

NILLSEN, J. P. 1974. On the ecology and distribution of the Norwegian larvae of *Chaoborus* (Diptera, Chaoboridae). Norsk Entomol. Tidskr. **21**: 37-44.

PARMA, S. 1969. Notes on the larval taxonomy, ecology and distribution of the Dutch *Chaoborus* species (Diptera, Chaoboridae). Beaufortia **17**: 21-50.

POPE, G. F., J. C. H. CARTER, and G. POWER. 1973. The influence of fish on the distribution of *Chaoborus* spp. and density of larvae in the Matamek river system, Québec. Trans. Am. Fish. Soc. **102**: 707-714.

SAETHER, O. A. 1972. Chaoboridae. Binnengewaesser **26**: 257-280.

STAHL, J. B. 1966. Coexistence in *Chaoborus* and its ecological significance. Invest. Indiana Lakes Streams **7**: 99-113.

STENSON, J. A. E. 1976. Significance of predator influence on composition of *Bosmina* spp. populations. Limnol. Oceanogr. **21**: 814-822.

——. 1978. Differential predation by fish on two species of Chaoborus (Diptera, Chaoboridae). Oikos.

TERAGUCHI, M., and T. G. NORTHCOTE. 1966. Vertical distribution and migration of *Chaoborus flavicans* larvae in Corbett Lake, British Columbia. Limnol. Oceanogr. **11**: 164-176.

IX

Spatial and Temporal Aspects of Community Structure:
Micro- and Macrogeographic Patterns

55. Evidence for Stable Zooplankton Community Structure Gradients Maintained by Predation

William M. Lewis, Jr.

Abstract

Zooplankton were sampled on transects across Lake Lanao, a large tropical lake, on five different dates representing the full range of annual conditions. The sampling stations ranged in depth between 25 and 80 m. Regression analysis showed that abundance is linearly related to depth for almost all species and developmental stages and that the relation is quite stable through time. Herbivorous cyclopoid nauplii, copepodids, and adults showed a strong positive relation between abundance per unit area and depth, as did carnivorous *Chaoborus* larvae of all instars. All other herbivores, including calanoids, cladocerans, and a rotifer species, showed negative depth-abundance relationships, and thus decreased in areal abundance toward midlake. The gradients ranged in intensity but were especially strong for cladocerans. The hypothesis is formulated that herbivore abundance gradients are created and maintained by predation through *Chaoborus*, the dominant primary carnivore. The feeding rates, abundance-depth relation, and feeding selectivity of all four *Chaoborus* instars were used to calculate potential predation losses as a percentage of the stock of each prey type, assuming an average community structure. These losses were then regressed against depth and proved to have significant slope, thus yielding predation gradients for individual prey species. The predation gradients were compared to abundance gradients and showed a significant negative relationship. The evidence for maintenance of abundance gradients by predation is thus very strong. This mechanism is apparently responsible for maintenance of an unexpected amount of pattern in the zooplankton community structure of the limnetic zone, and for spatial diversification of the zooplankton.

It is now well established that competing species partition environmental resources, and that this partitioning is accomplished in large part by dietary, spatial, or temporal segregation (e.g. MacArthur 1958; Schoener 1974, 1975; Cody 1974; Roughgarden 1974). Mechanisms and strategic bases for separation of many kinds of organisms remain to be worked out, however. The major focus thus far has been on vertebrates and higher plants (Whittaker 1967), although a few comprehensive studies of invertebrate groups are also available (e.g. Green 1971; Lane 1975).

Mechanisms for partitioning of resources among plankton species have been of special interest since Hutchinson (1961) first pointed out the heuristic value of analyzing resource partitioning in a community which occupies an unstructured habitat. For phytoplankton, the mechanisms most likely to facilitate resource partitioning include temporal separa-

---

This work was supported by NSF grants BG41293 to D. G. Frey and BMS 75-03102, DEB 7604300 to the author. The Mindanao State University and the Ford Foundation generously furnished facilities for the fieldwork.

tions based on varying optima and tolerances for resource supply and attrition factors (e.g. Hutchinson 1967; Allen and Koonce 1973; Grenny et al. 1973; Titman 1976; Lewis 1977a), and spatial separation based on ephemeral horizontal spatial patchiness and possible vertical spatial variation as well (Margalef 1958; Richerson et al. 1970; Sandusky and Horne 1978). The relative importance of these mechanisms and their interaction have not yet been well established. For the zooplankton, which are capable of extensive vertical movement, there is evidence that vertical spatial separations interacting with temporal separations facilitate resource partitioning (Sandercock 1967; Miracle 1974; Lane 1975; Makarewicz and Likens 1975). Horizontal spatial patchiness has also been documented (e.g. Hutchinson 1967; Wiebe 1970; Mullin and Brooks 1976; Lewis 1978a), but its relative importance in separation of zooplankton species is not obvious at present.

The present study gives evidence for stable horizontal spatial gradients in the relative abundances of euplanktonic zooplankton species in a large lake. Horizontal spatial variation can be subdivided into a fixed component and an ephemeral component. In terms of an analysis of variance model, the fixed component is a main effect caused by differing average conditions between stations, whereas the ephemeral component is a space-time interaction resulting from patchiness of changing character caused by moving water masses. The separation of these components in Lake Lanao has been discussed elsewhere (Lewis 1978a). The separation technique showed that in Lake Lanao ephemeral horizontal spatial variation generally exceeds horizontal spatial variation in magnitude. Fixed horizontal variation will be the focus of the present discussion, however, because fixed variation proves to be of special interest in connection with zooplankton community structure in Lake Lanao. The relation of the fixed component of variation to other components is shown schematically in Fig. 1.

Fixed horizontal variation in a plankton population must be based on fixed features of the planktonic habitat. Two general kinds of fixed features seem to be worthy

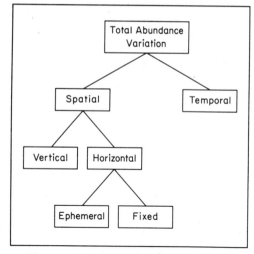

Fig. 1. Schematic organization of components of variance for plankton abundance in lakes.

of consideration: physicochemical gradients created and maintained by a point source of heterogeneity, such as a river, and depth. In the open ocean, both of these sources of variation are ordinarily absent (depth is functionally infinite), hence one would expect fixed spatial variation to be nil except over regions so large that climatic differences come into play. In lakes, persistent horizontal physicochemical gradients emanating from point sources are frequently not well developed, particularly if the flushing rate of the lake is low. Depth, however, provides a universal fixed gradient in lakes. In fact the depth gradient creates two major communities in lakes, the littoral and the limnetic. It is well known that these two communities differ in composition, complexity, and function from each other. The present study considers only the plankton community in the limnetic zone. The depth gradient extends across the limnetic zone, but crosses no obvious plankton ecozones, hence it is common to visualize the plankton zone as having essentially no stable community structure gradients. The qualitative composition of plankton is in fact typically uniform over the limnetic zone if ephemeral patchiness is averaged through time. The question to be tested here is whether the depth gradient in the limnetic zone is in fact associated with subtle but important quantitative changes in the relative abundance of species. This line of inquiry is a logical extension of recent studies which have shown that horizontal gradients in morphology of species populations can be quite striking in some cases (Green 1967; Zaret 1972a; Kerfoot 1975). The same kinds of forces which maintain polymorphisms in species might also maintain community structure gradients, even in the largely unstructured limnetic zone.

Data collection and study area

The study of spatial variation was conducted on Lake Lanao, Philippines. Lanao is ideal for the study of fixed spatial variation associated with depth because it is large (360 km^2), contains a substantial depth gradient (maximum depth, 112m; mean, 60 m), and lacks any detectable chemical gradients arising from point sources (flushing time, 6.5 yr). The physicochemical features and plankton biology of the lake have been thoroughly described (Frey 1969; Lewis 1973, 1974, 1977b). Herbivorous zooplankton in the limnetic zone include one cyclopoid copepod species, one calanoid copepod species, three cladoceran species, and seven rotifer species (Table 1). The present analysis excludes the six least abundant rotifer species, as these did not provide a firm statistical basis for analysis. The only planktonic carnivore is *Chaoborus*, which is represented by large populations of a single unnamed species (Eckstein Form 1; Lewis 1975).

Heterogeneity studies were made on five different dates spanning the full range of annual conditions (Table 2). On each date, samples were taken at eight stations, 2.5 km apart along two transects representing most of the lake inside the 10-m contour. The station depths ranged from 25 m to 80 m. Duplicate tows were taken at each station on each date with a calibrated metered net. The tows were later corrected for filtration efficiency using the meter readings. The transect was run in reverse order on alternate dates. A detailed comparison of metered net samples with Schindler-Patalas trap samples using a 35-μm net showed that los-

ses through the net meshes for species discussed here were negligible (Lewis 1979). Samples were preserved in the field with Lugol's solution and counted later with a dissecting scope.

Analytical methods

For each species or developmental stage, the relation between abundance per unit surface area and depth was tested with a linear regression analysis. Potential compli-

Table 1. List of euplanktonic zooplankton in Lake Lanao and mean abundances (entire water column, 0–45 m) at main sampling station (N = 52 weeks).

Species/Stage	Mean abundance indiv./liter
Copepoda	
Thermocyclops hyalinus	
Adult	10.88
Copepodid	25.97
Nauplius	88.87
Tropodiaptomus gigantoviger	
Adult	2.13
Copepodid	4.04
Nauplius	3.56
Cladocera	
Diaphanosoma (modigliani/sarsi)	4.79
Moina micrura	0.43
Bosmina fatalis	1.50
Rotifera	
Conochiloides dossuarius	17.48
Hexarthra intermedia	4.27
Polyarthra vulgaris	0.81
Keratella procurva	0.68
Keratella cochlearis	3.96
Trichocerca brachyurum	0.26
Tetramastix opoliensis	4.33
Diptera	
Chaoborus (Eckstein 1)	0.160

Table 2. Conditions in Lake Lanao on the five sampling dates (data for main station in 45 m of water).

Date (1971)	Net production mg C/m^2/day	Mixed layer (m)	O_2 at 35 m mg/liter	Zooplankton indiv./liter
25 Feb	1,190	60	4.18	130
15 Apr	1,440	25	5.08	280
13 May	770	15	4.15	190
10 Jun	1,940	15	2.50	150
8 Jul	2,530	30	3.00	310

Table 3. Relation of water depth to mean population size per unit area. Column 1 gives correlation coefficient derived from model I regression of abundance (No./m^2) and depth. Column 2 gives probability associated with regression slope. Column 3 gives numbers of individuals at a station 25 m deep, as determined by regression analysis. Column 4 gives expected slope of regression on assumption that each added meter of depth adds 1/25 of number at a station 25 m deep. Column 5 gives observed slope. Column 6 shows observed slope as a percentage of expected.

Species/Stage	(1, 2) Abundance vs. depth r p		(3) Thousands indiv./m^2 @ 25 m	(4) Expected slope thousands indiv./m^2/m depth	(5) Observed slope	(6) Obs./Exp. × 100(%)
Herbivores						
Cyclopoids—*Thermocyclops hyalinus*						
Nauplii	0.24	0.02	3,888	155.5	37.0	23.8
Copepodids	0.52	0.00	959	38.3	24.1	62.8
Adult ♀♀	0.45	0.00	112	4.5	5.63	125.9
Calanoids—*Tropodiaptomus gigantoviger*						
Nauplii	-0.19	0.05	336	13.5	-2.46	-18.2
Copepodids	0.09	0.21	—	—	—	0.0
Adult ♀♀	-0.33	0.00	91	3.63	-0.83	-23.0
Cladocera						
*Diaphanosoma modigliani**	-0.26	0.01	355	13.8	-3.14	-22.7
Moina micrura	-0.31	0.00	34	1.36	-0.42	-31.9
Bosmina fatalis	-0.17	0.07	109	4.34	-1.22	-28.1
Rotifers						
Conochiloides dossuarius	-0.15	0.09	1,080	43.2	-6.81	-15.8
Predators						
Chaoborus						
Instar 1	0.37	0.00	0.161	0.0064	0.0081	125.8
Instar 2	0.42	0.00	0.150	0.0060	0.0036	60.0
Instar 3	0.33	0.00	0.128	0.0051	0.0035	68.0
Instar 4	0.54	0.00	0.000	0.0000	0.0302	∞

* Includes some *D. sarsi*.

cations were checked as follows. 1. Nonlinearity. A depth-abundance relation might exist in nonlinear form, so the analyses were repeated after semilog and log transformations. The transformations did not substantially change the results. The simple linear model was therefore retained.

2. Time-space interactions. As the analysis extended over five different dates, it would have been possible for depth-abundance relations to be time-dependent. Multiple regression controlling for time showed that this was not the case, so the simple regression was retained.

Results

Table 3 lists the species and stages included in the study and the results of regression analysis for each. The first column in the table gives the Pearson product-moment correlation coefficient abundance vs. depth as derived from the regression slope (model I regression: Sokal and Rohlf 1969) and the second column indicates the significance level of the regression slope. The depth-abundance relation is highly significant for most species and stages. No relation could be conclusively demonstrated for calanoid copepodids, and this group will be considered to show no relationship (it is quite likely that β error in the analysis is responsible for the failure of the demonstration here, in view of the demonstrated relationship in both younger and older stages). The relations for *Bosmina* and *Conochiloides* are borderline but will be retained for further analysis.

The relations in Table 3 leave much variance unaccounted for, even though they are for the most part significant statistically. This is because the ephemeral component of spatial variance (Fig. 1), which typically accounts for a major portion of total spatial variance (Lewis 1978a) is intentionally excluded from the analysis and thus appears as part of the error variance.

One important aspect of the regressions is that the cyclopoid groups all have positive slopes, indicating an increase in numbers per m^2 with increasing depth, while all other herbivore species show negative or zero slopes, indicating the opposite trend or no trend. The predator *Chaoborus* shows a trend similar to the cyclopoids.

Column 3 of Table 3 gives the average numbers of organisms at a station 25 m deep. These were computed from the regression lines (X = depth, Y = indiv./m^2) and compare closely in all cases with the mean numbers of organisms actually observed at the shallowest stations (25 m).

The fourth column in Table 3 is the expected slope of the depth-abundance relation. Expectation is based on the assumption that each meter of depth beyond the 25-m station will contribute 1/25 of the abundance per unit area at the 25-m station. Expectation here provides a null hypothesis and not a prediction of reality, as will become apparent. Since the euphotic zone falls within the top 25 m (1% light = 9–18 m, depending on date), the expected slope basically assumes that a 1-m layer from the nonproducing zone of the lake is equivalent in its support capacity to a 1-m layer from the producing zone.

The observed slopes of depth vs. abundance, as derived from the regression analysis, are given in column 5 of Table 3. As might be anticipated, they are always lower than the expected slopes of column 4. The negative values for all species except *Thermocyclops* are counter-intuitive, however, as they suggest that addition of 1 m to the nonproducing zone detracts substantially from the support capacity of the producing zone (support capacity refers to the sum of growth and mortality control factors).

The last column in Table 3 shows the ratio of observed to expected slopes expressed as a percentage, which facilitates comparisons between species and stages. There are four possible results, all of which are shown in the table.

1. Percentage > 100% (cyclopoid adults, *Chaoborus* instars 1 and 4). This indicates that the addition of a meter of depth to the nonproducing zone increases the support capacity of the water column for this particular species or stage by more than an equal amount of depth in the producing zone. The remarkable avoidance of shallower water by large *Chaoborus* (zero abundance at 25 m) has been documented elsewhere (Lewis 1975) and appears to result from downslope movement necessitated by fish predation or by elimination of large individuals due to fish predation.

2. Percentage > 0 and < 100% (cyclopoid nauplii, copepodids, *Chaoborus* instars 2 and 3). Abundance per unit surface increases with addition of depth below 25 m, but only by a fraction of the amount expected from a straightforward extrapolation of numbers at the 25-m station.

3. Percentage 0% (calanoid copepodids). Adding depth below 25 m is inconsequential to the support capacity of the water column.

4. Percentage < 0% (all cladocerans, rotifers, two stages of calanoids). Adding depth below 25 m reduces the support capacity of the upper 25 m.

Discussion

The analysis shows that a fixed and temporally stable horizontal spatial variation in the limnetic zone is statistically detectable for almost all species and developmental stages, that the fixed spatial variation can be expressed in terms of linear relationships between abundance per unit surface area and depth, and that the nature of the fixed spatial variation is quite different between the coexisting

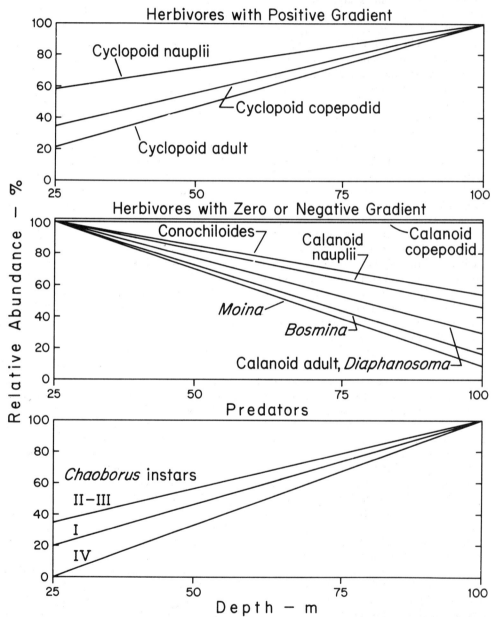

Fig. 2. Abundance gradients with depth in Lake Lanao shown as percentage of maximum abundance.

species. Figure 2 illustrates the spatial gradients that were detected in the analysis. In order to facilitate comparison, the abundance of each species in Fig. 2 is expressed as a percentage of its abundance at the depth where abundance is maximum. The changes shown in Fig. 2 take the form of clinal gradations in relative abundance of species rather than sharp transitions in abundance.

Simple ecological explanations for the herbivore gradients in Fig. 2 could be based on corresponding gradients of either the phytoplankton food resource or predation. Extensive studies of primary production and phytoplankton composition and distribution (Lewis 1974, 1978b) show that no comparable fixed gradients exist in the food resource. Thus predation is the most likely mechanism by which the gradients are maintained.

Chaoborus is by far the most important zooplankton predator in Lanao. The fish fauna is largely endemic and does not include efficient first-order planktivores of great abundance in the pelagic zone. The feeding habits of the Chaoborus population are known in considerable detail (Lewis 1977b, 1979), so it is possible to convert the Chaoborus abundance gradients of Fig. 2 to gradients of attrition for all prey species. This was done as follows.

1. The average dietary composition was obtained for each instar from a previous study (Lewis 1977b). This dietary composition reflects considerable feeding selectivity. Cladocerans and copepod copepodid/adults are major components of the diet. The cladocerans in particular are eaten in amounts far exceeding their relative abundance in the plankton.

2. Feeding rates were obtained for each Chaoborus instar by cohort analysis of the Chaoborus abundance data over an 18-month interval and approximations of growth efficiency (Lewis 1979). The food intake computed on this basis varies between 33 and 90% of body wt day^{-1}, depending on the age of the predator. These estimates were independently confirmed by measurements of herbivore loss rates.

3. The total potential loss of each food type to predation in terms of weight per unit surface area was calculated as follows:

$$L_{k,z} = \sum_{i=1}^{4} I_i W_{i,z} P_{i,k}$$

where $L_{k,z}$ is the predation loss of prey type k at a station in water of depth z to Chaoborus of all instars (mg/m^2/day, assuming initial community structure is as given in Table 1), I_i is the food intake rate of Chaoborus of instar i (fraction of body weight per day), $W_{i,z}$ is the weight per unit area at a station of depth z of Chaoborus instar i (mg/m^2), and $P_{i,k}$ is the proportion of prey type k in the diet of Chaoborus of instar i.

4. The predation loss of each prey type was then expressed as a percentage of the total stock of prey available (again assuming initial community structure as given in Table 1) to obtain relative loss:

$$RL_{k,z} = (L_{k,z}/B_k) N 100$$

where $L_{k,z}$ is the absolute daily loss as defined above, $RL_{k,z}$ is the relative daily loss of prey type k at a station of depth z (%/day), and B_k is the stock of prey type k (mg/m^2).

5. The values of $RL_{k,z}$ were then tested for linear regression with z. All relationships were highly significant (Fig. 3a). A regression line was obtained for each prey type and these are shown in Fig. 3a.

It should be noted that these computations leading to Fig. 3a are not intended to represent actual loss rates, but rather potential rates based on an average community structure as set forth in Table 1. In actuality the community structure responds to these predation gradients and thus the relative abundances are adjusted away from the average according to predator abundance and composition of a particular site. Figure 3a merely shows how this adjustment could be maintained by predation. In a situation approaching equilibrium, predation pressure on certain species and stages in excess of

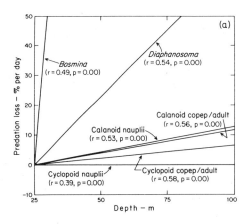

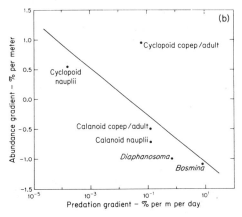

Fig. 3. a. Predation gradients with depth in Lake Lanao computed for a fixed community structure of prey (average annual abundance of all species at central sampling station in 45 m of water as shown in Table 1). Loss expressed as percentage of population per day eaten by all instars of *Chaoborus*. b. Relation of predation gradients to abundance gradients in Lake Lanao herbivores.

their growth capacity will cause drastic decline in abundance of these forms until the predator reduces its selectivity for them (Lewis 1977b).

The analysis leading to Fig. 3a omits *Moina* and *Conochiloides*. *Moina* is such a small proportion of the available prey that a predation analysis on the species is not possible, even though the abundance data are sound. *Conochiloides* lacks distinctive skeletal parts resistant to maceration in the crop of *Chaoborus* so that predation data based on crop analysis are not comparable to the data for other species. In addition, the copepodid and adult stages are combined in Fig. 3a because they could not be distinguished in the crop analysis.

Figure 3a shows that considerable predation gradients exist in Lake Lanao and that the severity of the gradients varies greatly between species and stages. There is an overall increase in the severity of predation toward deep water, coincident with the increase in numbers of all *Chaoborus* instars toward deep water (Fig. 2). Because of the selective feeding on cladocerans, especially in larger *Chaoborus* instars, the gradients are very steep for *Diaphanosoma* and *Bosmina*. Nauplii of both copepods, by contrast, are seldom eaten by *Chaoborus* and thus experience very limited gradients.

If predation gradients (Fig. 3a) establish and maintain abundance gradients (Fig. 2), then the two should be quantitatively related. The relation will not be linear because the effect of a given predation gradient on a species will depend on the average production rate of the species and all other sources of mortality besides *Chaoborus* predation. The relationship is shown in Fig. 3b in a semilog format, which condenses the great range of predation gradients. It is clear that the predation and abundance gradients are related. The correlation of points shown in Fig. 3b is significant at $P = 0.05$ and accounts for 51% of the total variance in gradients ($r = 0.77$, $r^2 = 0.51$). The circumstantial evidence for maintenance of the abundance gradients by predation is therefore very strong.

Figure 3b has two general implications for the organization of the plankton community. First it is evident that a physically and chemically unstructured plankton environment can support an unexpected degree of stable pattern in community structure. While several cases of fixed horizontal pattern in a single species based on predation have been documented (Green 1967; Zaret 1972a; Kerfoot 1975). The existence of predation-maintained fixed patterns in the structure of entire limnetic zooplankton assemblages has apparently not been demonstrated.

It is presently clear that predation accounts for many differences in herbivore composition between lakes (e. g. Hrbáček et al. 1961; Brooks and Dodson 1965; Brooks 1968; Dodson 1970; Zaret 1972b). The Lanao data suggest that predation plays a similar role within lakes, except that the influence is more subtle and affects the relative abundance of species rather than the species composition.

A second implication of the analysis is that predation helps maintain complexity in the lake zooplankton. While some generalized arguments and experimental evidence support this notion for some communities (Paine 1966), no demonstration has ever been given for lake zooplankton. The fixed predation gradients in Lake Lanao could easily account for the persistence of species that would be eliminated if predation were uniform or randomly variable over the lake. It thus seems possible that the gradients enhance the overall biotic diversity of herbivores.

If predation gradients can reasonably account for herbivore abundance gradients, then a second step toward the mechanistic explanation for fixed patterns in community structure would involve an analysis of the factors controlling the distribution of primary carnivores. Such an explanation would involve fish predation on *Chaoborus*. The behavior and distribution of *Chaoborus* both suggest that deep water offers refuge from predation, largely because of the daily retreat of larvae to water of low oxygen content (Lewis 1975, 1977b). Organization of the zooplankton thus in large part seems to be determined in a stepwise manner from the top trophic levels. Although one should not overlook the certainty that other mechanisms will affect the other components of variation (Fig. 1), the separation of variance components seems to simplify mechanistic approaches to community structure.

References

ALLEN, T. F., and J. F. KOONCE. 1973. Multivariate approaches to algal strategems and tactics in systems analysis of phytoplankton. Ecology 54: 1234-1246.
BROOKS, J. L. 1968. The effects of prey size selection by lake planktivores. Syst. Zool. 17: 272-291.
———, and S. I. DODSON. 1965. Predation, body size, and composition of the plankton. Science 150: 28-35.
CODY, M. L. 1974. Competition and the structure of bird communities. Princeton Univ. Press.
DODSON, S. I. 1970. Complementary feeding niches sustained by size-selective predation. Limnol. Oceanogr. 15: 131-137.
FREY, D. G. 1969. A limnological reconnaissance of Lake Lanao, Philippines. Int. Ver. Theor. Angew. Limnol. Verh. 17: 1090-1102.
GREEN, J. 1967. The distribution and variation in *Daphnia lumholtzi* (Crustacea:Cladocera) in relation to fish predation in Lake Albert, East Africa. J. Zool. Lond. 151: 181-197.
GREEN, R. H. 1971. A multivariate statistical approach to the Hutchinsonian niche: bivalve mulluscs of central Canada. Ecology 52: 225-229.
GRENNY, W. J., D. A. BELLA, and H. C. CURL. 1973. A theoretical approach to interspecific competition in phytoplankton communities. Am. Nat. 107: 405-425.
HRBÁČEK, J., M. DVOŘÁKOVA, V. KOŘÍNEK, and L. PROCHÁZKOVÁ. 1961. Demonstration of the effect of fish stock on the species composition of zooplankton and the intensity of metabolism of the whole plankton association. Int. Ver. Theor. Angew. Limnol. Verh. 14: 192-195.
HUTCHINSON, G. E. 1961. The paradox of the plankton. Am. Nat. 95: 137-146.
———. 1967. A treatise on limnology, v. 2. Wiley.
KERFOOT, W. C. 1975. The divergence of adjacent populations. Ecology 56: 1298-1313.
LANE, P. A. 1975. The dynamics of aquatic systems: a comparative study of the structure of four zooplankton communities. Ecol. Monogr. 45: 307-336.
LEWIS, W. M., JR. 1973. The thermal regime of Lake Lanao (Philippines) and its theoretical implications for tropical lakes. Limnol. Oceanogr. 18: 200-217.
———. 1974. Primary production in the plankton community of a tropical lake. Ecol. Monogr. 44: 377-409.
———. 1975. Distribution and feeding habits of a tropical *Chaoborus* population. Int. Ver. Theor. Angew. Limnol. Verh. 19: 3106-3119.
———. 1977a. Ecological significance of the shapes of abundance-frequency distribution for coexisting phytoplankton species. Ecology 58: 850-859.
———. 1977b. Feeding selectivity of a tropical *Chaoborus* population. Freshwater Biol. 7: 311-325.
———. 1978a. Comparison of spatial and temporal heterogeneity in lake plankton by means of variance components. Ecology 59: 666-671.
———. 1978b. Spatial distribution of the phytoplankton in a tropical lake. Int. Rev. Gesamten Hydrobiol. 63: 619-635.

———. 1979. A zooplankton community analysis. Springer-Verlag, New York.

MacARTHUR, R. H. 1958. Population ecology of some warblers of northeastern coniferous forests. Ecology 39: 599–619.

MAKAREWICZ, J. C., and G. E. LIKENS. 1975. Niche analysis of a zooplankton community. Science 190: 1000–1003.

MARGALEF, R. 1958. Temporal succession and spatial heterogeneity in natural phytoplankton, p. 323–349. In A. Buzzati-Traverso [ed.], Perspectives in marine biology. Univ. Calif. Press.

MIRACLE, M. R. 1974. Niche structure in freshwater zooplankton: a principle components approach. Ecology 55: 1306–1316.

MULLIN, M. M., and E. R. BROOKS. 1976. Some consequences of distributional heterogeneity of phytoplankton and zooplankton. Limnol. Oceanog. 21: 784–796.

PAINE, R. T. 1966. Food web conplexity and species diversity. Am. Nat. 100: 65–76.

RICHERSON, P., R. ARMSTRONG, and C. R. GOLDMAN. 1970. Contemporaneous disequilibrium, a new hypothesis to explain the "paradox of the plankton". Proc. Natl. Acad. Sci. 67: 1710–1714.

ROUGHGARDEN, J. D. 1974. Species packing and the competition function with illustrations from coral reef fish. Theor. Pop. Biol. 5: 163–186.

SANDERCOCK, G. A. 1967. Coexistence of calanoid copepods. Limnol. Oceanogr. 12: 97–112.

SANDUSKY, J. C., and A. J. HORNE. 1978. A pattern analysis of Clear Lake phytoplankton. Limnol. Oceanogr. 23: 636–648.

SCHOENER, T. 1974. Resource partitioning in ecological communities. Science 185: 27–39.

———. 1975. Presence and absence of habitat shift in some widespread lizard species. Ecol. Monogr. 45: 233–258 . . . .

SOKAL, R. R., and F. J. ROHLF. 1969. Biometry. Freeman.

TITMAN, D. 1976. Ecological competition between algae: experimental confirmation of resource-based competition theory. Science 192: 463–465.

WHITTAKER, R. H. 1967. Gradient analysis of vegetation. Biol. Rev. 42: 207–264.

WIEBE, P. H. 1970. Small-scale distribution in oceanic zooplankton. Limnol. Oceanogr. 15: 205–217.

ZARET, T. M. 1972a. Predator-prey interaction in a tropical lacustrine ecosystem. Ecology 53: 248–257.

———. 1972b. Predators, invisible prey, and the nature of polymorphism in the Cladocera (Class Crustacea). Limnol. Oceanogr. 17: 171–184.

56. Relationships between Trout and Invertebrate Species as Predators and the Structure of the Crustacean and Rotiferan Plankton in Mountain Lakes

R. Stewart Anderson

Abstract

The structure of the zooplankton communities of 320 lakes in the Canadian Rocky Mountains was examined to determine the dominant invertebrate predators in lakes where trout species were present, the dominant and subdominant invertebrate predators where trout species were not present, the dominant nonpredaceous crustacean species, and the presence or absence of abundant rotifer populations in all lakes. Differences were associated with the species of trout present, e.g. cutthroat (*Salmo clarki*) lakes most closely resembled the fishless lakes. The presence or absence of species of *Hesperodiaptomus* is undoubtedly the greatest factor determining the occurrence and abundance of rotifers in the lakes surveyed. *Gammarus, Chaoborus*, and cyclopoid copepods probably exert little control over rotifer numbers in these lakes. There was no clear evidence that competition from cladocerans was an important factor controlling rotifer numbers.

There have been many recent studies of the effects of the introduction of various fish species on the structure of the zooplankton communities in lakes (e.g. Brooks and Dodson 1965; Hillbricht-Ilkowska 1964; Hrbáček and Novotná-Dvořáková 1965; Stenson 1972; Wells 1970), but most of these studies have involved very few lakes. Not many studies of this type have involved salmonids (e.g. Galbraith 1967, 1975; Nilsson and Pejler 1973; Walters and Vincent 1973) and these studies did not include a consideration of the rotifers. There are also few surveys of fish-zooplankton relationships in large numbers of relatively uniform lakes which have had fairly stable communities for several years (e.g. Galbraith 1975—28 lakes; Nilsson and Pejler 1973—65 lakes). There has been a recent proliferation of theories and hypotheses concerning the possible roles of competition, predation, and combinations of these, on zooplankton community structure. However, there is still a need for surveys of uniform and stable groups of lakes, for the proof of the theoretical pudding still lies in the eating.

This survey summarizes data from 320 small mountain lakes in the Canadian Rocky Mountains in the vicinity of the Continental Divide (Fig. 1). Most of these lakes had no fish originally. Those that did

---

I thank the many people who helped me with the field studies, the most frequent helpers being S. Anderson, D. Cadman, A. Colbeck, D. Donald, R. Green, J. Hill, J. Kilistoff, D. Krochak, D. Mayhood, D. Mudry, and B. Smiley. I especially thank G. Thompson and D. Radford of the Alberta Fish and Wildlife Division for providing samples and information on lakes outside the parks. I am also grateful to the Resource Conservation Section of Parks Canada, Western Region, and to G. W. Scotter of the Canadian Wildlife Service for providing help and encouragement in this study. Most of this study was conducted in conjunction with ongoing limnological studies done for Parks Canada by the Canadian Wildlife Service and funded by Parks Canada.

contained populations of cutthroat trout (*Salmo clarki*) or Dolly Varden char (*Salvelinus malma*), two of the three "trout" species native to the area. A very few of the lakes under consideration may have contained minnows, suckers, or whitefish. In recent years, many of these lakes have been stocked regularly or occasionally with various "trout" species, mainly brook trout (*Salvelinus fontinalis*), cutthroat trout, rainbow trout (*Salmo gairdneri*), or golden trout (*Salmo aguabonita*). Detailed or survey studies concerning most of these lakes have been published (e.g. Anderson 1971, 1972, 1974, 1975, 1977; Anderson and Raasveldt 1974; Bajkov 1927, 1929; Rawson 1942, 1953). A summary of the distribution and general location of the lakes surveyed follows.

Fig. 1. Map showing general location of lakes included in this study (hatched region).

Mountain National Parks of western Canada
- Banff   70
- Glacier   2
- Jasper   107
- Kootenay   7
- Mount Revelstoke   5
- Waterton Lakes   40
- Yoho   30

Vicinity of these National Parks
- Province of Alberta   56
- Province of B. C.   3

Total No. of lakes   320

The material in this survey has been drawn from samples and data collected over the past 14 years. Most material has come from data gathered for the preparation of resource conservation and management reports prepared by the Canadian Wildlife Service for Parks Canada. Data and samples for many lakes in the vicinity of the mountain National Parks were provided by the Alberta Fish and Wildlife Division, and much of the information for these lakes has been used in the preparation of survey and management reports compiled for or by the division.

Methods

Criteria for selecting the lakes for this study—The basic criterion for selection was that "trout" were present or, in fishless lakes, that stocked trout could conceivably survive the winter in at least some years. Whether a fish population (real or postulated) could maintain its numbers by natural propagation was not a criterion for either group. The "lakes" should be greater than 500 m^2 in surface area, have a maximum depth > 1 m, and have a bottom at least 80% free of macrophytes or debris (e.g. submerged dead trees). Virtually all lakes with established fish populations had these features. When it was not possible to categorize a lake because of conflicting or inadequate data, it was omitted from consideration. For example, lakes which could be expected to have fish (i.e. fish known to be in the watershed and no natural impediments to their movement) but for which no catch or observational data existed were omitted. Lakes where the fish community was known or believed to be comprised of over 15% whitefish, suckers, chub, or dace were omitted. Pike lakes and lakes with established lake trout populations were also omitted, because their communities were usually too complex to permit valid comparison to the majority of the lakes. Rainbow trout lakes and the few golden trout lakes were considered as one group. In cases where there were recent introductions or other drastic changes which could affect the zooplankton composition, lakes were only included where conditions had been fairly stable for at least 5 years.

Invertebrate predator and prey criteria—Dominance in crustacean plankters was numerical. Dominance in invertebrate predators was determined somewhat subjectively, utilizing counts from benthic and planktonic collections, observational data, and in some cases a consideration of body size relative to numbers present. It was usually rather obvious which species were dominant. Dominance in fish communities was determined from catch data (where available) or from stocking or creel census records.

Dominant diaptomid predators were almost entirely *Hesperodiaptomus arcticus* and *H. shoshone*. Dominant cyclopoid predators were almost always *Diacyclops bicuspidatus thomasi*; a few were *Acanthocyclops vernalis*, and *Macrocyclops albidus* dominated once. Crustacean "grazers" included known or suspected filter feeders and particulate feeders: large cladocerans were mainly species of *Daphnia* (> 1.5 mm, head-plus-valve); small cladocerans included *Ceriodaphnia* spp., *Bosmina longirostris*, *Holopedium gibberum*, and a few populations of small *Daphnia* and *Chydorus sphaericus* (< 1.5 mm, head-plus-valve); small diaptomid copepods included *Acanthodiaptomus denticornis*, *Aglaodiaptomus* spp., and *Leptodiaptomus* spp. (< 1.5 mm, body length excl. urosome); cyclopoid copepod "grazers" included mainly *Eucyclops agilis*, although a few populations of the nauplii and early copepodids of other species were believed to best fit this category.

Results

The general structure of the zooplankton in the lakes with fish is summarized in Table 1. The subdominant predators, if there were any, are listed in column 2. Concerning dominant crustacean "grazers," only one was counted per lake. In very few cases was there any question as to whether the dominance should be ascribed to one group rather than another. Occasionally the numbers of small cladocerans and small diaptomids were very close, in which case dominance was assigned to the species group which was most abundant when counts for all available samplings were considered. There was an obvious correlation between the presence of brook trout or rainbow trout and the occurrence of cyclopoids as the dominant invertebrate predator. Abundant *Hesperodiaptomus* spp. were uncommon with either brook trout or rainbow trout. In the 11 brook trout lakes (‡—Table 1) where large diaptomids (*Hesperodiaptomus* spp.) coexisted with the fish, the fish populations were almost all sparse. Furthermore, most of these lakes are very deep, and many receive glacial runoff and are rather silty. Most of the rainbow-*Hesperodiaptomus* co-occurrences involved sparse fish populations also. In cutthroat trout lakes, it was common for *Hesperodiaptomus* spp. and trout to occur together. Large cladocerans (*Daphnia* spp.) were almost as common in the fish lakes as in the fishless lakes (Table 2), although the pigmented *Daphnia middendorffiana* was seldom found at all in the fish lakes.

In the lakes with fish (Table 1), rotifers were usually rare or absent if *Hesperodiaptomus* was present, but they were usually very abundant when cyclopoid numbers were high. There was no apparent relationship between the occurrence of abundant cladocerans and the presence or absence of rotifers. When rotifers did occur (rarely) in moderate or large numbers with *Hesperodiaptomus* spp., they were usually *Kellicottia longispina* or *Conochilus unicornis*, a feature previously reported (Anderson 1977). *Kellicottia* was the most frequent rotifer to occur at low numbers if *Hesperodiaptomus* was present. Anostraca were never found in the presence of fish, although they did occur in a few large, permanent lakes without fish and without abundant *Gammarus* or *Chaoborus* predators (Table 2).

Most fishless lakes of this study (Table 2) were dominated by *Hesperodiaptomus* spp. The two greatest exceptions were the *Chaoborus* lakes, where *Hesperodiaptomus* was rare if present at all, and cyclopoid

Table 1. Analysis of zooplankton structure for four groups of lakes (total = 177) where fish were believed to be principal predators on large zooplankters and invertebrate predators. Numbers are numbers of lakes falling into each category.

Dom. vertebrate pred. on zoopl.*	Subdom. pred. on zoopl.†	Dom. large Cladocera	small Cladocera	crustacean "grazers" small Diaptomus	cyclopoids	none or very sparse	Rotifers abundant	sparse or absent	Anostraca
Brook trout	A−1	1	0	0	0	0	1	0	0
(Salvelinus	B−1	1	0	0	0	0	1	0	0
fontinalis)	C−11‡	1	0	4	2	4	2	9	0
74	D−52	11	20	15	0	6	47	5	0
(11)	E−6	0	0	6	0	0	6	0	0
	F−3	0	0	3	0	0	2	1	0
Rainbow and	A−0	0	0	0	0	0	0	0	0
golden trout	B−0	0	0	0	0	0	0	0	0
(Salmo	C−6	1	0	4	1	0	2	4	0
gairdneri and	D−39	14	9	16	0	0	35	4	0
S. aguabonita)	E−4	0	0	3	0	1	2	2	0
50 (7)	F−1	0	0	1	0	0	1	0	0
Cutthroat	A−1	0	1	0	0	0	1	0	0
trout	B−0	0	0	0	0	0	0	0	0
(Salmo	C−29	8	1	8	6	6	4	25	0
clarki)	D−10	3	3	1	0	3	8	2	0
47	E−7	0	1	5	0	1	2	5	0
(4)	F−0	0	0	0	0	0	0	0	0
Dolly	A−0	0	0	0	0	0	0	0	0
Varden	B−0	0	0	0	0	0	0	0	0
(Salvelinus	C−2	1	0	1	0	0	1	1	0
malma)	D−4	3	0	1	0	0	1	3	0
6	E−0	0	0	0	0	0	0	0	0
(3)	F−0	0	0	0	0	0	0	0	0

*Where fish populations known to be sparse or in low density, No. of lakes in parentheses; other number represents total no. of lakes.
†For categories, see Table 2; "F" = second salmonid from first column.
‡Special remarks on this group given in text.

lakes, a few of which at one time have had fish populations. *Hesperodiaptomus* numbers were seldom very high when *Gammarus* was abundant. Rotifers rarely occurred when *Hesperodiaptomus* was common to abundant, and when they did, it was usually *Kellicottia* or *Conochilus* as noted above. It was most common for no crustacean "grazers" or only *Daphnia* to occur with *Hesperodiaptomus*. Small cladocerans were only about half as common in fishless lakes as in lakes with fish, and nearly half of these in fishless lakes were *Holopedium gibberum* (†, ‡—Table 2), a species which seems to be able to resist predation by *Hesperodiaptomus* or *Chaoborus*.

Lakes with neither fish nor arthropod predators (Table 2) seemed to have a 50/50 chance of having fairly abundant small crustaceans or rotifers. These lakes were generally very oligotrophic.

Discussion

The composition of the zooplankton in *Chaoborus* and *Gammarus* lakes suggests a similar pattern of predation on the zooplankton as is apparent for brook trout and rainbow trout. In the presence of the two invertebrate predators, there were no anostracans, few large diaptomids, and few small diaptomids, although cyclopoids were rather common. These data are in

Table 2. Analysis of zooplankton structure for five groups of lakes (total = 143) where no fish are present. Dominant predators on zooplankton, if present, were invertebrates. Numbers are numbers of lakes falling into each category.

Dom. invert. pred. on zoopl.	Subdom. pred. on zoopl.*	Dom. crustacean "grazers"					Rotifers		Anostraca
		large Cladocera	small Cladocera	small Diaptomus	cyclopoids	none or very sparse	abundant	sparse or absent	
A. Gammarus lacustris 14	B—1	0	0	0	1	0	1	0	0
	C—3	2	1	0	0	0	1	2	0
	D—10	3	2	4	0	1	8	2	0
	E—0	0	0	0	0	0	0	0	0
B. Chaoborus spp. 14	A—0	0	0	0	0	0	0	0	0
	C—0	0	0	0	0	0	0	0	0
	D—10	5	3†	1	0	1	10	0	0
	E—4	2	1	1	0	0	3	1	0
C. Hesperodiaptomus spp. 86	A—4	1	0	3	0	0	2§	2	0
	B—2	1	0	1	0	0	1	1	0
	D—15	4	1†	1	0	9	0	15	1
	E—65	27	4‡	16	1	17	3§	62‖	9
D. Cyclopoid copepods 18	A—1	0	0	1	0	0	1	0	0
	B—2	1	0	1	0	0	2	0	0
	C—1	0	0	1	0	0	1	0	1
	E—14**	2	5	4	0	3	12	2	1
E. No known vert. or arthropod predators 11	none 11	0	0	6	0	5	6	5	0

*For categories, see col. one.
†All were *Holopedium gibberum*.
‡Half were *H. gibberum*.
§Dominated by *Kellicottia* or *Conochilus*.
‖Sparse rotifers often only *Kellicottia*.
**Includes one *Polyphemus pediculus* subdominant which was rather sparse.

agreement with experimental and field studies reported earlier (Anderson and Raasveldt 1974) and with my results of laboratory studies of *Chaoborus* predation (R. S. Anderson at this conf.). The data in the present study also suggest that *Chaoborus* and *Gammarus* predation on rotifers is light, although Comita (1972) and Hillbricht-Ilkowska et al. (1975) suggest that *Chaoborus* predation on rotifers can be significant.

In these lakes, it was rare for rotifers to occur with *Hesperodiaptomus* species. Earlier feeding studies indicated that these copepods were capable of heavy predation on rotifers of many species (Anderson 1970), although some species with spines (e.g. *K. longispina*) or colonial forms (e.g. *C. unicornis*) seem to be able to avoid predation (Anderson 1977). Small cladocerans and cyclopoid copepods did not commonly occur in large numbers with *Hesperodiaptomus*. The latter are known to feed voraciously on both groups (Anderson 1970; Dodson 1974).

The invertebrate communities in lakes dominated by cyclopoids as the only predator resembled the invertebrate communities of brook trout lakes. Predaceous cyclopoids will readily eat the nauplii and early copepodids of *Hesperodiaptomus* and are therefore capable of excluding this diaptomid from lakes where the cyclopoids are well established (Anderson 1970). Such a case could occur following the stocking and eventual disappearance of trout in a lake. Follow-up studies in such lakes are needed to test the permanence of this sort of species shift. Cyclopoids seem to have no suppressing effect on rotifer populations in these lakes, although Hillbricht-Ilkowska et al. (1975) indicate that rotifers are sometimes subject to heavy predation by cyclopoids. In the mountain lakes surveyed here, rotifer populations actually seem to be enhanced by the presence of cyclopoids.

The invertebrate communities in cutthroat trout lakes most closely resemble pristine fishless lakes. This is interesting in view of the fact that this trout species was the most commonly occurring native trout species in the waters of the east slopes of the Canadian Rockies. Some remarkable populations of native cutthroat trout occurred in some of these watersheds before the manipulations of recent years (e.g. Spray River watershed: Miller and Mac Donald 1950). The late-spring spawning activity of cutthroat trout in mountain lakes may be the feature that permits the coexistence of the trout species and some of its food organisms. We have often observed (Anderson and Donald unpubl.) that the larvae and emerging adults of Plecoptera, Ephemeroptera, and Trichoptera were often abundant in cutthroat lakes at the time of spring spawning (when fish do not eat; sometimes as late as July or early August), whereas they were rare or absent in nearby brook trout lakes at the same time of year. Similar behavior may permit *Hesperodiaptomus* to coexist with cutthroat.

In summary, it is probable that *Hesperodiaptomus* species are an important control over rotifer numbers in these mountain lakes, and that cyclopoid copepods do not suppress rotifer numbers and may, in fact, enhance their numbers. This study provided no good evidence for competition from cladocerans as a control over rotifers, although a number of other studies show or suggest a strong competitive relationship between the two groups (Daborn et al. 1978; Dumont 1972; Hrbáček and Novotná-Dvořáková 1965). In these lakes, it is likely that the introduction of fish has enhanced the development of rotifer populations indirectly by suppressing the natural predators on rotifers rather than by adding something to the food supply which would favor rotifers (cf. Hillbricht-Ilkowska 1964).

Conclusions

1. The invertebrate communities in pristine lakes in the study area are likely to be dominated by species of *Hesperodiaptomus*.

2. The abundance of rotifers in these lakes seems to be inversely correlated with the abundance of species of *Hesperodiaptomus*.

3. The presence of fish alone is not responsible for increases in rotifer numbers in these lakes.

4. The occurrence of abundant rotifers is more closely related to the presence of cyclopoids than to fish.

5. Of the trout species commonly stocked in small lakes in the general study area, cutthroat trout, followed by Dolly Varden, seem to have the least impact on natural zooplankton communities.

6. Brook trout, followed by rainbow trout, seem to have the greatest impact on the natural communities.

References

ANDERSON, R. S. 1970. Predator-prey relationships and predation rates for crustacean zooplankters from some lakes in western Canada. Can. J. Zool. **48**:1229–1240.

———. 1971. Crustacean plankton of 146 alpine and subalpine lakes and ponds in western Canada. J. Fish. Res. Bd. Can. **28**:311–321.

———. 1972. Zooplankton composition and change in an alpine lake. Int. Ver. Theor. Angew. Limnol. Verh. **18**:264–268.

———. 1974. Crustacean plankton communities of 340 lakes and ponds in and near the National Parks of the Canadian Rocky Mountains. J. Fish. Res. Bd. Can. **31**:855–869.

———. 1975. An assessment of sport-fish production potential in two small alpine waters in Alberta, Canada. Symp. Biol. Hung. **15**:205–214.

———. 1977. Rotifer populations in mountain lakes relative to fish and species of copepods present. Arch. Hydrobiol. Ergeb. Limnol. **8**:130–134.

———, and L. G. RAASVELDT. 1974. *Gammarus* predation and the possible effects of *Gammarus* and *Chaoborus* feeding on the zooplankton composition in some small lakes and ponds in western Canada. Can. Wildl. Serv. Occas. Pap. 18, p. 1–23.

BAJKOV, A. 1927. Reports of the Jasper Park investigations, 1925–26. I. Fishes. Contrib. Can. Biol. Fish. N.S. 3:379–404.

———. 1929. Reports of the Jasper Park investigations, 1925–26. VII. A study of the plankton. Contrib. Can. Biol. Fish. N.S. 4:345–396.

BROOKS, J. L., and S. I. DODSON. 1965. Predation, body size, and composition of plankton. Science 150:28–35.

COMITA, G. W. 1972. The seasonal zooplankton cycles, production, and transformation of energy in Severson Lake, Minnesota. Arch. Hydrobiol. 70:14–66.

DABORN, G. R., J. A. HAYWARD, and T. E. QUINNEY. 1978. Studies on *Daphnia pulex* Leydig in sewage oxidation ponds. Can. J. Zool. 56:1392–1401.

DODSON, S. I. 1974. Zooplankton competition and predation: an experimental test of the size-efficiency hypothesis. Ecology 55:605–613.

DUMONT, H. J. 1972. A competition-based approach to the reverse vertical migration in zooplankton and its implications, chiefly based on a study of the interactions of the rotifer *Asplanchna priodonta* (Gosse) with several Crustacea Entomostraca. Int. Rev. Gesamten Hydrobiol. 57:1–38.

GALBRAITH, M. G., JR. 1967. Size-selective predation on *Daphnia* by rainbow trout and yellow perch. Trans. Am. Fish. Soc. 96:1–10.

———. 1975. The use of large *Daphnia* as indices of fishing quality for rainbow trout in small lakes. Int. Ver. Theor. Angew. Limnol. Verh. 19:2485–2492.

HILLBRICHT-ILKOWSKA, A. 1964. The influence of the fish population on the biocenosis of a pond, using Rotifera fauna as an illustration. Ekol. Pol. Ser. A 12:453–503.

———, Z. KAJAK, J. EJSMONT-KARABIN, A. KARABIN, and J. RYBAK. 1975. Ecosystem of the Mikojajshie Lake. The utilization of the consumers production by invertebrate predators in pelagic and profundal zones. Pol. Arch. Hydrobiol. 22:53–64.

HRBÁČEK, J., and M. NOVOTNÁ-DVOŘÁKOVÁ. 1965. Plankton of four backwaters related to their size and fish stock. Rozpr. Cesk. Akad. Ved Rada Mat. Prir. Ved 75:3–65.

MILLER, R. B., and W. H. MacDONALD. 1950. Preliminary biological surveys of Alberta watersheds, 1947–1949. Alberta Dep. Lands Forests Rep. 139 p.

NILSSON, N. -A., and B. PEJLER. 1973. On the relation between fish fauna and zooplankton composition in north Swedish lakes. Rep. Inst. Freshwater Res. Drottningholm 53:51–77.

RAWSON, D. S. 1942. A comparison of some large alpine lakes in western Canada. Ecology 23:143–161.

———. 1953. The limnology of Amethyst Lake, a high alpine type near Jasper, Alberta. Can. J. Zool. 31:193–210.

STENSON, J. A. E. 1972. Fish predation effects on the species composition of the zooplankton community in eight small forest lakes. Rep. Inst. Freshwater Res. Drottningholm 52:132–148.

WALTERS, C. J., and R. E. VINCENT. 1973. Potential productivity of an alpine lake as indicated by removal and reintroduction of fish. Trans. Am. Fish. Soc. 102:675–697.

WELLS, L. 1970. Effects of alewife predation on zooplankton populations in Lake Michigan. Limnol. Oceanogr. 15:556–565.

57. Zoogeographic Patterns in the Size Structure of Zooplankton Communities, with Possible Applications to Lake Ecosystem Modeling and Management

W. Gary Sprules

Abstract

A principal components ordination based on the taxonomic structure of zooplankton communities of lakes from Colorado and two areas of Ontario led to the regional grouping of lakes expected when there are major taxonomic differences among faunas. However a similar analysis based on the more functional characterization of these communities by the size and feeding ecology of the zooplankters also led to distinctive regional differences, implying that these Colorado and Ontario lakes differ in fundamental ecological respects. For a second set of Ontario lakes, this functional characterization of zooplankton communities led to the separation of shallow, turbid lakes dominated by small species from deeper, clearer lakes dominated by larger species as well as a separation of lakes on the basis of surface pH. Evidence is presented that patterns in the size-feeding ecology structure of limnetic zooplankton communities reflect major patterns in energy flow through lake ecosystems. The size of herbivorous species is an indication of whether primary production is consumed directly, or indirectly through detrital pathways, the relative abundance of predaceous species is related to the efficiency of secondary production, and the size of zooplankters in general affects the feeding efficiency of planktivorous fish. Some recommendations are made for the use of this technique of classification in lake ecosystem modeling and management.

Many investigations (Armitage and Davis 1967; Carter 1971; Patalas 1971; Pennak 1957; Sprules 1977) have identified clear relations between the species composition of limnetic zooplankton communities and limnological characteristics of the lakes in which they are found. Not only does this lead to increased understanding of the ecology of individual species and species associations, it also provides a technique for lake classification which is useful in the design of whole lake experiments (Patalas 1971, and other papers in the same journal issue) and possibly in lake management (Sprules 1977). However this taxonomic approach to community structure may not be the most useful, for it is becoming increasingly apparent that energy flow within aquatic ecosystems is more a function of the size of organisms than it is of their taxonomic status (Kajak and Hillbricht-Ilkowska 1972; Kerr 1971, 1974a; Sheldon et al. 1977). In limnetic freshwater ecosystems, interactions and energy flow between primary and secondary producers, between herbivorous and predaceous zooplankters, and between zooplankters and fish planktivores are strongly dependent upon particle size distributions at each of these levels (Brooks and Dodson 1965; Hall et al. 1976; Hillbricht-Ilkowska 1977). Because

---

This work was supported by the National Research Council of Canada and the Canadian National Sportsmen's Fund.

it reflects these dynamic processes, a size-feeding ecology characterization of zooplankton communities would thus seem to be a more fruitful basis than a taxonomic one upon which to compare and classify lakes within regions and between geographic areas. To my knowledge no one has explored this extension of Brooks and Dodson's (1965) original observations, although much has been learned of size-dependent competition and predation in zooplankton communities (Hall et al. 1976). This more functional approach to lake classification will be particularly useful for lake management and modeling purposes since zooplankton communities of different size structures channel energy from primary producers to fish planktivores along different paths (Hillbricht-Ilkowska 1977).

My purpose here is twofold—first to illustrate the use of size-feeding ecology classifications of zooplankters in identifying lake types within and between geographic regions, and second, to present an analysis of the relations between the size structure of zooplankton communities and energy flow through them.

I thank K. Patalas for helpful discussions and G. Stirling and R. Newhook for processing zooplankton samples. The Lake Ecosystem Working Group (LEWG) at the University of Toronto provided data as well as the opportunity to refine these ideas through interaction with stimulating colleagues.

Methods

The analysis of patterns in zooplankton community structure in lakes from different geographic regions is based on published data. Patalas (1964) presented data on the proportionate numerical abundance of crustacean zooplankton species from 52 lakes near Boulder, Colorado, ranging in altitude from 1,515–3,620 m above sea level and similar data are available (Patalas 1971) for zooplankton communities from 45 lakes in granite basins in the Experimental Lakes Area near Kenora, Ontario. Zooplankton data for 19 lakes in granite basins in the Gull Lake drainage of the Haliburton region, southern Ontario, were analyzed in another context (Sprules 1977) and are available from me. Details of zooplankton sampling and enumeration techniques, which are comparable for all three regions, are given in the papers cited.

Principal components analysis (Morrison 1967) was used to display the relationships among lakes from the three regions according first to a taxonomic characterization of zooplankton communities, and second to a size-feeding ecology characterization. Prior to both analyses species occurring in fewer than 4% of the lakes in any particular region were eliminated. For the taxonomic analysis, this left a raw data matrix of the proportionate numerical abundance of 43 crustacean zooplankton species in 119 lakes. The ordination was based upon the species variance-covariance matrix derived from this matrix, and species correlations with the principal components were used to identify those species that were the most important discriminators among lake groups.

For the size-feeding ecology analysis, the 43 zooplankton species were regrouped into eight classes: four sizes of herbivorous species and four sizes of carnivores. Individual species were classed as primarily herbivorous or carnivorous according to my own experience and published data on feeding ecology (e.g. Fryer 1957; Hutchinson 1967; Kajak and Hillbricht-Ilkowska 1972; Monakov 1976). Adult body sizes for the herbivorous species were taken as the midpoint of ranges published by Edmondson (1959); individual species were assigned to one of four size classes (Table 1) which were chosen so that each class had roughly comparable numbers of species. The same procedure was applied to the carnivorous species (Table 1) but, because they tend to be larger, the actual size classes differed from the herbivorous ones. This procedure resulted in a raw data matrix of eight size-feeding ecology classes of zooplankters in 119 lakes. This matrix was analyzed as described above for the taxonomic classification.

A modification of this approach was applied to part of a large data set from 36

Table 1. Occurrence of species in the Colorado, Haliburton, and Kenora lakes and species correlations with components from taxonomic ordination.

Species	Lakes in which species occurs (%)			Correlation with principal component	
	Colorado	Haliburton	Kenora	1	2
Herbivore 1					
Bosmina longirostris (O.F. Müller)	43	95	92	-0.05	-0.04
Chydorus sphaericus (O.F. Müller)	39	21	16	0.08	0.08
Ceriodaphnia lacustris Birge	4	26	12	-0.11	0.01
Ceriodaphnia quadrangula (O.F. Müller)	35	0	0	-0.10	0.17
Daphnia ambigua Scourfield	24	0	0	0.05	0.07
Herbivore 2					
Ceriodaphnia reticulata (Jurine)	6	0	0	-0.26†	0.21†
Daphnia longiremis Sars	0	68	4	0.15	-0.03
Daphnia retrocurva Forbes	0	58	37	0.10	-0.12
Diaphanosoma leuchtenbergianum Fischer	0	0	45	0.01	-0.23†
Diaphanosoma brachyurum (Liévin)	18	100	39	-0.06	-0.11
Diaptomus minutus Lilljeborg	0	68	71	0.29†	-0.39‡
Diaptomus nudus Marsh	24	0	0	-0.08	0.22†
Diaptomus oregonensis Lilljeborg	0	95	45	-0.11	-0.16
Diaptomus siciloides Lilljeborg	27	0	0	-0.16	0.18
Eucyclops agilis (Koch)	6	0	0	-0.06	0.04
Herbivore 3					
Daphnia catawba Coker	0	11	18	0.10	-0.02
Daphnia dubia Herrick	0	79	0	0.12	-0.04
Daphnia galeata mendotae X *thorata* (Forbes)	0	0	8	0.03	-0.09
Daphnia laevis Birge	8	0	0	0.09	0.02
Daphnia rosea Sars	33	0	0	-0.24†	0.59†
Diaptomus coloradensis Marsh	12	0	0	-0.05	0.11
Diaptomus sicilis Forbes	0	20	21	0.03	-0.07
Eucyclops speratus (Lilljeborg)	8	16	0	0.08	0.04
Herbivore 4					
Daphnia galeata Sars *mendotae* Birge	6	89	41	0.09	0.0
Daphnia pulex Leydig	4	0	12	-0.07	0.05
Daphnia pulex carina Kiser	4	0	0	0.05	0.04
Daphnia schødleri Sars	10	0	12	-0.12	0.10
Holopedium gibberum Zaddach	8	95	71	-0.01	-0.05
Sida crystallina (O.F. Muller)	0	68	0	0.0	0.0
Carnivore 1					
Tropocyclops prasinus mexicanus Kiefer	0	74	82	-0.37‡	-0.74‡
Polyphemus pediculus (L)	0	63	0	0.0	0.0
Carnivore 2					
Cyclops bicuspidatus thomasi Forbes	51	89	63	0.05‡	0.06
Cyclops vernalis Fischer	33	58	29	-0.35‡	0.36‡
Mesocyclops edax (Forbes)	12	95	77	-0.15	-0.18
Orthocyclops modestus (Herrick)	4	32	22	-0.08	0.0
Carnivore 3					
Cyclops scutifer Sars	0	16	0	-0.01	-0.06
Diaptomus leptopus Forbes	20	0	24	-0.17	0.05
Epischura lacustris Forbes	0	84	53	0.04	-0.19 §
Macrocylcops albidus (Jurine)	6	0	0	-0.06	0.04
Carnivore 4					
Diaptomus shoshone Forbes	6	0	0	-0.06	0.03
Leptodora kindtii (Focke)	0	100	24	0.0	0.0
Limnocalanus macrurus Sars	0	0	14	0.0	0.0
Senecella calanoides Juday	0	21	12	0.06	0.0

central Ontario lakes currently under study by the Lake Ecosystem Working Group at the University of Toronto. These lakes (LEWG lakes) were chosen to provide a variety of basin morphometries and bedrock types ranging from the Kawartha lakes in the south to Sudbury in the north. Each lake was visited once between 28 June and 6 July 1978 when a bottom-to-surface tow-net sample of zooplankton was taken with a Wisconsin-style plankton net with upper and lower openings of 30- and 50- cm diameter, respectively, and 110-μm nylon mesh. The net was fitted with a digital flowmeter (Dycus and Wade 1977) so that the volume of water filtered could be determined. Samples were preserved in a 3% buffered Formalin and 4% sucrose mixture. Data on surface pH, surface conductivity, Secchi depth, and total number of fish species were also collected. Morphometric data were obtained from the Ontario Ministries of Natural Resources and Environment.

In the laboratory a subsample of at least 5% of the total zooplankton sample was processed as follows. All zooplankton ranging from the smallest copepod nauplii and rotifers to the largest Crustacea and *Chaoborus* sp. larvae were counted. A microscope viewing screen (Wild Leitz) in combination with a rotating, circular counting chamber was used to class each individual encountered as either carnivorous or herbivorous; it was then placed into a predetermined size class using a scaled plastic "ruler" and tallied. The carnivores comprised principally *Chaoborus* larvae, *Leptodora*, *Polyphemus*, *Mysis*, *Asplanchna*, all stages of cyclopoid copepods except nauplii, and adults of the large calanoid copepods such as *Epischura*, *Limnocalanus*, and *Senecella*. All the remaining cladocerans, rotifers, and small calanoids were classed as herbivores. Size classes were chosen so that the total range of sizes for each feeding type was divided into five roughly equal millimeter intervals.

Herbivores

1	2	3	4	5
<0.3	0.3–0.49	0.5–0.84	0.85–1.19	>1.2

Carnivores

1	2	3	4	5
<0.5	0.5–0.89	0.9–1.19	1.2–1.49	>1.5

This led to a raw data matrix of the density (numbers per liter) of 10 size-feeding ecology classes of zooplankton in 36 lakes. After a natural logarithm transformation, these data were statistically treated as described above. As an additional aid in identifying lake groups, an unweighted pair group cluster analysis of the transformed data using arithmetic averages (Sneath and Sokal 1973) with taxonomic distance as a measure of interlake similarity was superimposed on the ordination. To provide a limnological interpretation of the resulting patterns, I calculated correlations between the limnological variables (all but pH transformed by natural logarithm) and the principal components.

All statistical analyses were done at the University of Toronto Computing Centre using two statistical packages, NTSYS (Rohlf et al. 1971) and SPSS (Nie et al. 1975).

Results and discussion

Colorado, Haliburton, Kenora comparisons—The first three principal components from an analysis of the taxonomic structure

---

Table 1 *(continued)*

†Probability *(P)* of obtaining observed product-moment correlation if true correlation is zero (two-tailed text) < 0.01.
‡$P < 0.01$.
§$P < 0.05$.

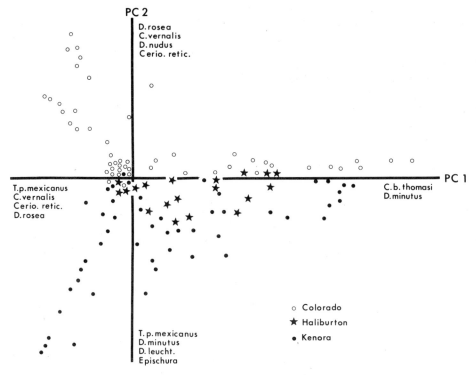

Fig. 1. Ordination of Colorado, Haliburton, and Kenora lakes based on taxonomic characterization of zooplankton communities. PC1 and PC2—first and second principal components. Species having significant correlations with the components are shown.

of the zooplankton communities in the Colorado, Kenora, and Haliburton lakes account for cumulative totals of 19.0, 32.0, and 43.8% of the variation in zooplankton community structure among lakes. The most important variation, along the first component (Fig. 1), is from lakes at the right-hand end characterized by high relative abundances of *Cyclops bicuspidatus thomasi* and *Diaptomus minutus* (Table 1) versus those toward the left-hand end characterized by lower relative abundances of these species and higher abundances of *Tropocyclops prasinus mexicanus*, *Cyclops vernalis*, *Daphnia rosea*, and *Ceriodaphnia reticulata*. More interesting, the Colorado lakes are almost completely separated from the others along the second component (Fig. 1). This is to be expected for the zooplankton fauna of the Colorado lakes is quite different from that of the Haliburton and Kenora lakes (Table 1). According to the

Table 2. Correlations of zooplankton size-feeding ecology classes with components from functional ordination of the Colorado, Haliburton, and Kenora lakes.

Size-feeding ecology class*	Correlation with principal component	
	1	2
Herbivore	−0.10	−0.19†
1	−0.10	−0.19†
2	0.37‡	0.77‡
3	0.46‡	−0.75‡
4	−0.03	0.07
Carnivore		
1	0.31‡	0.24§
2	−0.96‡	−0.01
3	0.20†	−0.06
4	0.08	−0.07

*1–4 designates size from smallest to largest.
†Probability (P) of obtaining observed product-moment correlation if true correlation is zero (two-tailed test) < 0.05.
‡P < 0.001.
§P < 0.01.

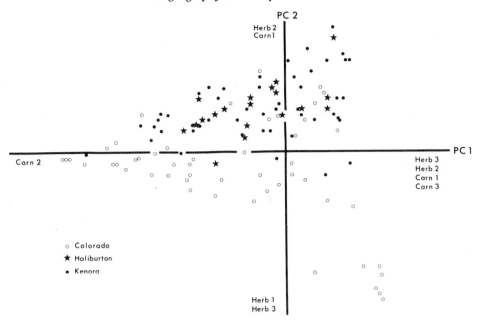

Fig. 2. Ordination of Colorado, Haliburton, and Kenora lakes based on functional characterization of zooplankton communities. PC1 and PC2 as in Fig. 1. Zooplankton classes having significant correlations with the components are shown.

species correlations with the second component (Table 1), the Colorado lakes, which generally fall on the upper half of the ordination, are characterized by greater relative abundances of *D. rosea, C. vernalis, Diaptomus nudus,* and *C. reticulata* in contrast to the Kenora lakes in particular which have greater relative abundances of *T. p. mexicanus, D. minutus, Diaphanosoma leuchtenbergianum,* and *Epischura lacustris*. The central position of the Haliburton lakes on the second component indicates that the composition of these communities is intermediate between those of Colorado and Kenora. Thus the ordination based on the taxonomic structure of zooplankton communities separates the lakes reasonably well by geographic region as would be expected when corresponding faunas differ.

Of greater interest is the ordination based on the functional classification of zooplankters (Fig. 2) into size-feeding ecology groups (Table 1). In this case the first three components account for cumulative totals of 30.3, 51.4, and 69.7% of the variation in community structure among lakes. More variance is accounted for in this ordination than the last, probably because the communities are characterized by 8 rather than 43 variables. The first principal component indicates that the most important variation in community structure is from lakes toward the right end of the component, which are characterized by relatively low abundances of carnivore 2 and higher abundances of herbivore 2, herbivore 3, carnivore 1, carnivore 3, to lakes toward the left end, which show the reverse trend (Fig. 2, Table 2).

The Colorado lakes are again separated along the second component from the Haliburton and Kenora lakes, although the pattern is less pronounced than for the taxonomic analysis. This is of particular interest for it indicates that there are quite distinct regional differences in size-feeding ecology or *functional* aspects of the zooplankton communities. According to species correlations with the second component (Table 2), the Colorado lakes, which fall generally in the lower half of the ordination, have high relative abundances of large and small herbivores (herbivore 1, 3) and lower relative abundances of medium herbivores and small carnivores (herbivore 2, carnivore 1). The Haliburton and Kenora lakes, which fall primarily on the upper half of this component, show the reverse trend. According to current theory (Hall et al. 1976) the greater importance of large herbivores in Colorado lakes in contrast to that of small carnivores in Ontario lakes could indicate that food-competitive interactions are predominant in structuring zooplankton communities of the former lakes whereas predation is more predominant in the latter. Whatever the exact interpretation, the fact that this approach can identify regional differences in functional aspects of community structure is of interest. In challenging ecologists to determine the basic causes of these differences, the approach will lead to increased understanding of ecosystem dynamics, especially because variations in the size-feeding ecology structure of zooplankton communities reflect variations in ecological interactions and energy flow within lakes (see below). This is not to completely denigrate the taxonomic approach described above, for the insight it provides into regional changes in species associations, particularly those involving widely distributed species such as *Bosmina longirostris* and *C. b. thomasi* are of ecological interest. Nevertheless the use of zoogeographic comparisons as a tool (MacArthur 1972) in ecological studies of zooplankton communities is more likely to be fruitful if it is based on functional characterizations of the animals.

LEWG lakes—While a size-feeding ecology approach to the structure of zooplankton communities seems promising, there are two drawbacks to simply rearranging existing taxonomic data on zooplankton species abundances. One is that adult sizes must usually be obtained from general taxonomic works which report only ranges and the other is that developmental stages within species are rarely distinguished. A more exact characterization of the size structure would be to directly assign each animal encountered in a sample of the community to a size class. Large and small species as well as developmental stages within a species can thus be differentiated. Explicit recognition of the changing ecological role of animals as they grow in size and sometimes switch from herbivores to carnivores is clearly of importance. As outlined in the methods section, this more exact sizing was applied to zooplankton samples from 36 lakes in central Ontario.

The first three principal components from an analysis of these data accounted for cumulative totals of 50.1, 66.9, and 79.4% of the variation in community structure among lakes. As the raw data for this analysis are absolute abundances rather than the relative abundances of the previous section, the first component (Fig. 3) is simply an abundance axis—lakes toward the right end of the component (group A in Fig. 3) have low densities of animals whereas those at the other end (group B) have higher densities. This is indicated by the significant, positive correlations of all but the last category of zooplankter with the first component (Table 3). None of the measured limnological characteristics of the lakes correlate with this pattern (Table 4). Within each of the major groups those lakes falling toward the positive end of the second component are characterized by high relative

Table 3. As Table 2 for LEWG lakes.

Size-feeding ecology class	Correlation with component		
	1	2	3
Herbivore			
1	0.78*	0.26	0.53*
2	0.72*	0.38†	−0.47‡
3	0.80*	−0.18	−0.32
4	0.63*	−0.55*	0.28
5	0.51‡	−0.46‡	0.25
Carnivore			
1	0.46‡	0.68*	0.07
2	0.86*	−0.23	−0.17
3	0.41†	−0.60*	−0.06
4	0.36†	−0.48‡	0.16
5	0.08	0.08	0.23

*Probability (P) of obtaining observed product-moment correlation if true correlation is zero (two-tailed test) < 0.001.
†$P < 0.05$.
‡$P < 0.01$.

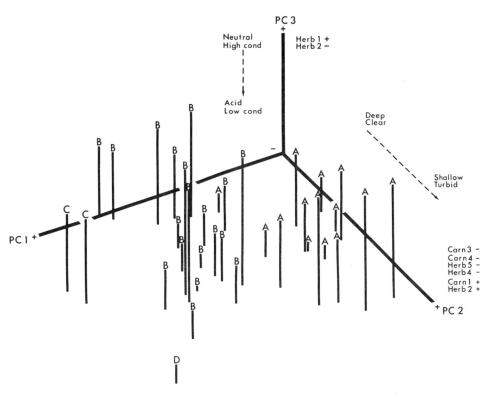

Fig. 3. Ordination of LEWG lakes. Zooplankton classes having significant positive (+) or negative (−) correlations with the components are shown. Limnological interpretation of components from Table 4. Letters (A, B, etc.) designate lake groups identified by cluster analysis. Positive and negative ends of components are identified. PC1, PC2, PC3—first, second, and third components.

Table 4. Correlations of limnological variables with components from ordination of LEWG lakes.

Limnological variables	Correlation with component		
	1	2	3
Maximum depth (m)	−0.12	−0.64*	−0.33
Mean depth (m)	−0.20	−0.59*	−0.33
Surface area (ha)	−0.10	0.10	0.09
Secchi disk transparency (m)	0.0	−0.52*	−0.34
Surface conductivity (μmhos cm^{-2} sec^{-1})	0.07	0.13	0.53*
Number of fish species	−0.15	0.19	−0.01
Surface pH	−0.02	0.06	0.60*

*Probability of obtaining observed product-moment correlation if true correlation is zero (two-tailed test) < 0.001.

abundances of carnivore 1, herbivore 2 and low relative abundances of carnivore 3, carnivore 4, herbivore 4, herbivore 5, whereas those toward the negative end show the reverse trend. Thus, along this component, lakes with generally large zooplankters are differentiated from those with smaller ones. Associated with this is a gradient from deep, clear lakes to shallow, turbid ones—i.e. lakes with large species tend to be deep and clear whereas those with small species tend to be shallow and turbid (Table 4). Despite the lack of significant correlation with number of fish species, one is tempted to conclude that vertebrate predation pressures in the shallow lakes are higher than those in the deep lakes and thus eliminate large zooplankters (Hall et al. 1976). The third component, accounting for only 16% of the variance among lakes, separates lakes toward its positive end (tall stalks in Fig. 3), which have high relative abundances of herbivore 1 and lower abundances of herbivore 2, from those at its negative end (short stalks), which show the opposite trend. In this case the former lakes tend to have water of comparatively high conductivity with pH close to neutral whereas the latter tend to be more acidic with water of lower conductivity (Table 4). With the recent (Dillon et al. 1978) documentation of early stages in the industrial acidification of central Ontario lakes, the relationship between pH and size-feeding ecology structure of zooplankton communities is particularly interesting.

The clear patterns in the relationships among lakes that emerge from this analysis based on direct size measurements of limnetic zooplankters with no reference to taxonomy suggest that the technique is promising and should be investigated further. Possible applications to lake ecosystem modeling and management are presented following the discussion on energy flow through zooplankton communities of differing size structure.

Size-feeding ecology structure and energy flow—While students of patterns in the size structure of zooplankton communities have, for some time, recognized underlying causal mechanisms such as size-selective vertebrate (Galbraith 1967; Hall et al. 1970) and invertebrate (Dodson 1974; Fedorenko 1975; McQueen 1969) predation and selective availability to herbivores of only certain algal sizes and types (Gliwicz 1969a; Porter 1977), the implications of these interactions to energy flow through pelagic ecosystems are only beginning to be considered (Kerr 1974b; Hillbricht-Ilkowska 1977). Much information about this energy flow can be deduced from analyzing the size structure of zooplankton communities.

In a series of interesting studies (Gliwicz 1969a, b, 1975; Gliwicz and Hillbricht-Ilkowska 1972; Hillbricht-Ilkowska et al. 1966, 1972; Hillbricht-Ilkowska and Spodniewska 1969; Hillbricht-Ilkowska and Weglénska 1970), a distinction has been made between those lakes in which energy flows from primary to secondary producers via a detrital food chain and those in which it flows via a grazing food chain. In the former lakes, more typically very productive, the secondary production is predominantly by small species or "microfiltrators" (e. g. *Bosmina, Ceriodaphnia, Chydorus,* small rotifers such as *Keratella* and *Filinia,* copepod nauplii, certain protozoans) rather than larger ones. In these lakes most phytoplankton are about $10\mu m$ or larger (colonial blue-greens and diatoms, dinoflagellates) and thus are unavailable to the microfiltrators which are restricted to particles from one to several micrometers. Bacteria and detritus constitute the bulk of particles of this size so that microfiltrators ingest on average 5–15% algae, 10–20% detritus, and 70–80% bacteria. Only 33% of the net algal production is consumed (Table 5); these small herbivores rely much more heavily on the detrital and bacterial products of primary production. Because much energy is lost through bacterial respiration as phytoplankton biomass is decomposed, only about 8% of the net algal production appears as net secondary production in these lakes (Table 5).

By contrast, in less productive lakes, herbivorous zooplankters comprise predominantly "macrofiltrators" (copepodites and adult calanoid copepods, *Diaphanosoma, Daphnia,* some larger rotifers), which specialize on particles from roughly $10-20\mu m$. Since edible nannoplankton of this size tend to predominate in these lakes, the macrofiltrators directly consume, on the average, 62% of the net algal production (diet is up to 50% algae, up to 30% bacteria). With the extra energy loss through bacterial metabolism being less important in these lakes, 17% of the net algal production, or roughly twice as much as in productive lakes, ends as net secondary production (Table 5).

Of course the absolute value of primary production in lakes of different trophy will vary, generally being higher in more eutrophic lakes, but the fact remains that the

Table 5. Means and ranges of various trophic efficiencies in pelagic ecosystems.

Ratio	Detritus food chain	Grazing food chain
Net carnivore production / Net herbivore production	0.236* (< 0.10 - > 0.30)	0.176 (< 0.10 - > 0.30)
Predator consumption / Net herbivore production	0.66† (0.20 - 1.0)	
Net herbivore production / Net primary production	0.08 (0.01 - < 0.10)	0.17 (0.10 - 0.20)
Herbivore consumption / Net primary production	0.33 (0.05 - 1.0)	0.62 (0.05 - 1.0)

*All data from Hillbricht-Ilkowska's (1977) summary of a large number of water bodies.
†All lakes combined (no distinction between food chains).

size structure of the zooplankton community is directly related to the efficiency of energy transfer from algae to secondary producers. At one extreme, in lakes which tend to have a preponderance of very small herbivores (e. g. shallow, turbid LEWG lakes, *see above*), energy is more likely to flow indirectly from phytoplankton to zooplankton through bacteria and detritus; whereas at the other extreme, in lakes which have more large herbivores (deep, clear LEWG lakes), energy flows more directly from phytoplankton to zooplankton. As Hillbricht-Ilkowska (1977) summarized "...the size structure of phytoplankton ... together with the composition of the consumers characterized by different size range of selected food particles, are more decisive for the efficiency of food utilization than the absolute concentration of food suspension, or the intensity of primary production."

The total production of the zooplankton community (herbivores plus carnivores) is also dependent on its size structure. Other things being equal, production will be higher in a community comprised principally of herbivorous species than in one in which there is the additional trophic level of invertebrate predators with attendant respiratory and other losses of about 80% (Table 5). In two comparable lakes in the USSR, Svyatoye and Svetloyar, in which the ratio of herbivorous to predaceous zooplankter biomass is 5:1 and 38:1 respectively, the corresponding net productions of total zooplankton are 5.77 and 16.12 kcal m^{-2} season^{-1} (Petrova et al. 1975). These differences in secondary production are important when one considers the energy available to planktivorous fish. Soviet workers have defined "effective zooplankton production," i. e. that available to fish, as the sum of the net production of the predaceous and nonpredaceous zooplankters less that production of nonpredaceous species which is consumed by the predators (predator ration). For example, in Red Lake, (USSR) production during the 1964 growing season by herbivorous zooplankters was 88.4 kcal m^{-2} (Table 6). Of this, 31 kcal m^{-2} were consumed by predaceous species whose resultant net production was 9.9 kcal m^{-2}, leaving 67.3 kcal m^{-2} available to fish. This is only 75% of the 88.4 kcal m^{-2} which would have been available to fish had invertebrate predation on zooplankton been unimportant. For a series of Soviet lakes (Table 6) this fraction varies from 0.30 to 0.93 with most values clustering around 0.75-0.80. This ratio will depend upon assimilation efficiencies of predaceous zooplankters and the degree to which they obtain energy exclusively from herbivorous zooplankters (Krivoe Lake: Table 6) but in general it appears as if invertebrate predators effectively reduce by 20-25%

Table 6. Components of limnetic zooplankton production in various USSR lakes.

Water body*	Herbivorous zooplankton production (H)	Predaceous zooplankton		"Effective" zooplankton production (H+P-R)	Units ($k\,cal\cdot m^{-2}$)	$\frac{H+P-R}{H}$	Source
		production (P)	ration† (R)				
Unfertilized pond	107.0	11.0	37.5	80.5	per season‡	0.75	Lyakhanovich 1973
Fertilized pond	227.0	28.0	84.0	171.0	per season	0.75	Lyakhanovich 1973
Red Lake	88.4	9.9	31.0	67.3	per season	0.76	Andronikova et al. 1972
Lake Baikal	80.5	8.1	38.2	50.4	per year	0.63	Moskalenko and Votinsev 1972
Lake Naroch	55.4	19.5	29.2	45.7	per season	0.82	Winberg et al. 1972
Lake Myastro	116.7	44.6	66.8	94.5	per season	0.81	Winberg et al. 1972
Lake Batorin	138.0	53.8	80.6	111.2	per season	0.81	Winberg et al. 1972
Kiev Reservoir	199.0	6.9	21.6	184.3	per season	0.93	Gak et al. 1972
Rybinsk Reservoir	78.0	6.5	18.5	66.0	per season	0.85	Sorokin 1972
Krugloe Lake	11.0	2.0	4.6	8.4	per year	0.76	Alimov and Winberg 1972
Krivoe Lake	17.0	2.6	14.5	5.1	per year	0.30§	Alimov and Winberg 1972

*All in USSR.
†Total consumption by predaceous zooplankters.
‡Season is generally the open water period, May–October.
§Ration is high, probably because consumption includes much more than just herbivorous zooplankton.

the zooplankton production available to fish.

Thus in zooplankton communities in which predaceous species, particularly large ones, are abundant the efficiency of energy transfer out of the zooplankton will be lower than in those communities with few predators and primarily herbivorous species. This is precisely the kind of pattern in community structure that is reflected by the size-feeding ecology classification of zooplankton communities presented here. Incidentally, because of this effect of invertebrate predators on energy transfer out of the zooplankton community, the recent controversy over whether small species of zooplankton are eliminated by food competition with large herbivores or by predation by large carnivores (Dodson 1974; Lynch 1977) assumes considerable practical importance.

While the efficiency with which energy is transferred out of the zooplankton community is an inverse function of the abundance of predaceous zooplankters, the feeding efficiency and growth of fish planktivores is directly related to the size of their food particles (Galbraith 1975; Hall et al. 1970; Kerr 1971, 1974b; Kerr and Ryder 1977). Kerr (1971) has provided a theoretical framework for increased growth efficiency of fish with increased size of particle ingested, which basically considers the energy expenditure required to capture a prey item versus the energy content of the ingested item. He provides field data to support this for the piscivorous lake trout and argues that by similar processes the selection of large zooplankters by planktivorous fish should lead to increased growth efficiency. Werner (1974) supports this by noting that, for small fish, a relatively small shift away from the optimal diet generates a considerable increase in effort expended in food gathering. Kerr's predictions are indirectly supported by Galbraith (1975), who showed that the rainbow trout quality of a lake (catch per unit effort) is greatest in lakes with abundant large daphnids, and more directly by Northcote (1972) who documented increased size, weight, and growth rate of rainbow trout and kokanee coincident with their increased predation on introduced *Mysis relicta*, a large predaceous zooplankter. After its introduction to Lake Tahoe, California-Nevada, *M. relicta* appeared with increased frequency in kokanee and lake trout stomachs (Richards et al. 1975) with resultant increased growth of these species (Morgan et al., 1978).

These data provide good evidence that planktivorous fish can feed and grow more efficiently when feeding on large zooplankters than on small ones. If we recall that invertebrate predators decrease the total production of zooplankton, it is clear that, at the very least, fish planktivore production is a complex tradeoff between the amount of energy flowing from zooplankton and the feeding efficiency of the fish. Less energy may be available to fish when invertebrate predators are abundant but, if these predators are large, the fish can utilize the decreased production more efficiently. Again, the important point is that these processes are clearly related to the size and feeding ecology of limnetic zooplankters—variations in which can be of predictive value.

General discussion

To a certain extent the zooplankton community can be considered a pivotal component of freshwater pelagic ecosystems. On the one hand, phytoplankton-zooplankton relationships determine the nature of energy flow from the primary to the secondary producers and, on the other hand, the nature of the zooplankton community governs, to a large extent, energy flow to planktivorous fish (and thence to top carnivores). For this reason analyses of patterns in the structure of zooplankton communities alone can provide useful information about the nature of lake ecosystems. It has been argued here that such an analysis should be based on the functional, size-feeding ecology characterization of the zooplankton association, for this reflects dynamic processes in the ecosystem much more closely than does a simple taxonomic characterization. Comparisons of lakes from disparate zoogeographic or faunal regions on this basis can help determine the extent to which ecosystem dynamics vary from region to region and thus lead to new questions about lake ecology. For instance the possibility that predaceous zooplankters differed in relative importance in Colorado and Ontario lakes emerged from comparing these regions. Certainly lakes from these regions showed distinct patterns in the structure of their zooplankton communities, suggesting their ecosystems operate on different functional bases.

Because this approach provides clues to the nature of energy flow through the algal and fish components of lake ecosystems—the two components of greatest social importance—there are possibilities for its use in other contexts. A first step in modeling energy flow through lakes should be to determine whether there are major "types" of lakes with distinct pathways of energy flow. Grouping lakes by the size-feeding ecology structure of their zooplankton communities could be a first step. On the basis of what has been argued here, this could lead to separation of lakes on the basis of a) the relative abundance of predaceous species with implications for efficiency of energy transfer out of the zooplankton community, b) the relative abundance of small herbivorous species with implications about the direct or indirect utilization of phytoplankton production, or c) the relative abundance of large species with implications for planktivorous fish feeding and growth efficiency. The degree to which different models, or at least major submodels, for these various lake types were necessary could be explored on the basis of this analysis. These models are of use in lake ecosystem management, but direct use of patterns in community structure could also be made. Thus lakes with high abundances of large species are likely to provide the best growth potential for introduced fish species that are planktivorous for part of their life. This is basically an extension of Galbraith's (1975) recommendations. Similarly, one should expect blooms of large "nuisance" algae more in lakes dominated by small species of herbivorous zooplankton than in those with larger species. Of course this lake classification approach is no substitute for detailed data on production within various ecosystem components, but the main point is that it can provide a preliminary identification of lake types that are likely to differ in critical aspects of energy flow. Furthermore, the expertise required for functional zooplankton community characterization is considerably less than that necessary for a complete taxonomic characterization.

In developing this approach, some key research problems which should be pursued by zooplankton ecologists have emerged. At an early stage it became clear that little is known about the feeding ecology of many zooplankton species and, more important, about changes in feeding mode through life. As it is quite likely that many individuals are omnivorous, the necessity for refining the feeding ecology characterization used here becomes clear. More data on the amount of secondary production consumed by invertebrate predators in lakes are required. This is especially important when considered along with competitive interactions among herbivorous zooplankters, for clearly a community in which competition dominates will have a higher "effective" production than will one in which predator-prey interactions dominate. And, of course, no study of energy flow to fish is reasonable without detailed data on feeding and growth efficiency in relation to particle size ingested. For instance, what happens to planktivorous fish production after the well established (Hall et al. 1976) elimination of the preferred large prey? Presumably it decreases considerably. If so, are there techniques for maintaining the critical abundance of these large prey in managed lakes?

Finally, there is the intriguing possibility that the size structure of zooplankton communities changes in response to socially important stresses such as acid, thermal, and nutrient. There was an indication that variations in surface pH are affecting the communities in the LEWG lakes. It is worth determining whether changes in size structure can provide an early warning about deteriorating lake quality.

References

ALIMOV, A. F., and G. G. WINBERG. 1972. Biological productivity of two northern lakes. Int. Ver. Theor. Angew. Limnol. Verh. **18**: 65–70.

ANDRONIKOVA, I. N., V. G. DRABKOVA, K. N. KUZMENKO, N. F. MICHAILOVA, and E. A. STRAVINSKAYA. 1972. Biological productivity of the main communities of the Red Lake, p. 57–72. *In* Z. Kajak and A. Hillbricht-Ilkowska [eds.], Productivity problems of freshwaters. PWN Pol. Sci. Publ., Warsaw.

ARMITAGE, K. B., and M. DAVIS. 1967. Population structure of some pond microcrustacea. Hydrobiologia **29**: 205–225.

BROOKS, J. L., and S. I. DODSON. 1965. Predation, body size, and composition of plankton. Science **150**: 28–35.

CARTER, J. C. 1971. Distribution and abundance of planktonic Crustacea in ponds near Georgian Bay (Ontario, Canada) in relation to hydrography and water chemistry. Arch. Hydrobiol. **68**: 204–231.

DILLON, P. J., AND OTHERS. 1978. Acidic precipitation in south-central Ontario: recent observations. J. Fish. Res. Bd. Can. **35**: 809–815.

DODSON, S. I. 1974. Zooplankton competition and predation: an experimental test of the size-efficiency hypotheses. Ecology **55**: 605–613.

DYCUS, D. L., and D. C. WADE. 1977. A quantitative-qualitative zooplankton sampling method. J. Tenn. Acad. Sci. **52**: 2–5.

EDMONDSON, W. T. [ed.] 1959. Freshwater biology, 2nd. ed. Wiley.

FEDORENKO, A. Y. 1975. Feeding characteristics and predation impact of *Chaoborus* (Diptera, Chaoboridae) larvae in a small lake. Limnol. Oceanogr. **20**: 250–258.

FRYER, G. 1957. The food of some freshwater cyclopoid copepods and its ecological significance. J. Anim. Ecol. 26: 263-286.

GAK, D. Z., AND OTHERS. 1972. Productivity of aquatic organism communities of different trophic levels in Kiev Reservoir, p. 447-455. In Z. Kajak and A. Hillbricht-Ilkowska [eds], Productivity problems of freshwaters. PWN Pol. Sci. Publ., Warsaw.

GALBRAITH, M. G. 1967. Size-selective predation on Daphnia by rainbow trout and yellow perch. Trans. Am. Fish. Soc. 96: 1-10.

———. 1975. The use of large Daphnia as indices of fishing quality for rainbow trout in small lakes. Int. Ver. Theor. Angew. Limnol. Verh. 19: 2485-2492.

GLIWICZ, Z. M. 1969a. The share of algae, bacteria and trypton in the food of the pelagic zooplankton of lakes with various trophic characteristics. Bull. Acad. Pol. Sci. Cl. 2 17: 159-165.

———. 1969b. Studies on the feeding of pelagic zooplankton in lakes with varying trophy. Ekol. Pol. Ser. A. 17: 663-708.

———. 1975. Effect of zooplankton grazing on photosynthetic activity and composition of phytoplankton. Int. Ver. Theor. Angew. Limnol. Verh. 19: 1490-1497.

———, and A. HILLBRICHT-ILKOWSKA. 1972. Efficiency of the utilization of nannoplankton primary production by communities of filter feeding animals measured in situ. Int. Ver. Theor. Angew. Limnol. Verh. 18: 197-203.

HALL, D. J., W. E. COOPER, and E. E. WERNER. 1970. An experimental approach to the production dynamics and structure of freshwater animal communities. Limnol. Oceanogr. 15: 839-928.

———, S. T. THRELKELD, C. W. BURNS, and P. H. CROWLEY. 1976. The size-efficiency hypothesis and the size-structure of zooplankton communities. Annu. Rev. Ecol. Syst. 7: 177-208.

HILLBRICHT-ILKOWSKA, A. 1977. Trophic relations and energy flow in pelagic plankton. Pol. Ecol. Stud. 3: 3-98.

———, Z. GLIWICZ, and I. SPODNIEWSKA. 1966. Zooplankton production and some trophic dependences in the pelagic zone of two Masurian lakes. Int. Ver. Theor. Angew. Limnol. Verh. 16: 432-440.

———, and I. SPODNIEWSKA. 1969. Comparison of the primary production of phytoplankton in three lakes of different trophic type. Ekol. Pol. Ser. A 17: 241-261.

———, and T. WEGLÉNSKA. 1970. Some relations between production and zooplankton structure of two lakes of a varying trophy. Pol. Arch. Hydrobiol. 17: 233-240.

———, ———, T. WEGLÉNSKA, and A. KARABIN. 1972. The seasonal variation of some ecological efficiencies and production rates in the plankton community of several Polish lakes of different trophy, p. 111-129. In Z. Kajak and A. Hillbricht-Ilkowska [eds.] , Productivity problems of freshwaters. PWN Pol. Sci. Publ., Warsaw.

HUTCHINSON, G. E. 1967. A treatise on limnology, v. 2. Wiley.

KAJAK, Z., and A. HILLBRICHT-ILKOWSKA [Eds.] 1972. Productivity problems of freshwaters. PWN Pol. Sci. Publ., Warsaw.

KERR, S. R. 1971. A simulation model of lake trout growth. J. Fish. Res. Bd. Can. 28: 815-819.

———. 1974a. Theory of size distribution in ecological communities. J. Fish. Res. Bd. Can. 31: 1859-1862.

———. 1974b. Structural analysis of aquatic communities. Proc. Int. Congr. Ecol. 1st. 1974: 69-74.

———, and R. A. RYDER. 1977. Niche theory and percid community structure. J. Fish. Res. Bd. Can. 34: 1952-1958.

LYAKHANOVICH, V. P. 1973. The biotic matter circulation and energy flow in the ecosystems of fish ponds. Int. Ver. Theor. Angew. Limnol. Verh. 18: 1809-1817.

LYNCH, M. 1977. Zooplankton competition and plankton community structure. Limnol. Oceanogr. 22: 775-777.

MacARTHUR, R. H. 1972. Geographical ecology. Patterns in the distribution of species. Harper and Row.

McQUEEN, D. J. 1969. Reduction of zooplankton standing stocks by predaceous Cyclops bicuspidatus thomasi in Marion Lake, British Columbia. J. Fish. Res. Bd. Can. 26: 1605-1618.

MONAKOV, A. B. V. 1976. Nutrition and trophic relations of freshwater copepods. Nauka, Leningrad.

MORGAN, M. D., S. T. THRELKELD, C. R. GOLDMAN. 1978. Impact of the introduction of kokanee [Oncorhynchus nerka] and opossum shrimp [Mysis relicta] on a sub-alpine lake. J. Fish. Res. Bd. Can. 35: 1572-1579.

MORRISON, D. F. 1967. Multivariate statistical methods. McGraw-Hill.

MOSKALENKO, B. K., and K. K. VOTINSEV. 1972. Biological productivity and balance of organic substance and energy in Lake Baikal, p. 207-226. In Z. Kajak and A. Hillbricht-Ilkowska [eds.], Productivity problems of freshwaters. PWN Pol. Sci. Publ., Warsaw.

NIE, H. H., C. H. HULL, J. G. JENKINS, K. STEINBRENNER, and D. H. BENT. 1975. SPSS: statistical package for the social sciences, 2nd. ed. McGraw-Hill.

NORTHCOTE, T. G. 1972. Some effects of mysid introduction and nutrient enrichment on a large oligotrophic lake and its salmonids. Int. Ver. Theor. Angew. Limnol. Verh. 18: 1096-1106.

PATALAS, K. 1964. The crustacean plankton

communities in 52 lakes of different altitudinal zones of northern Colorado. Int. Ver. Theor. Angew. Limnol. Verh. 15: 719-726.

———. 1971. Crustacean plankton communities in forty-five lakes in the Experimental Lakes Area, northwestern Ontario. J. Fish. Res. Bd. Can. 28: 231-244.

PENNAK, R. W. 1957. Species composition of limnetic zooplankton communities. Limnol. Oceanogr. 2: 222-232.

PETROVA, M. A., T. S. YELAGINA, V. K. SPIRIDONOV, and T. A. FILTAKINA. 1975. Production of planktonic crustaceans of two secondary oligotrophic lakes. Gidrobiol. Zh. 11: 61-65.

PORTER, K. G. 1977. The plant-animal interface in freshwater ecosystems. Am. Sci. 65: 159-170.

RICHARDS, R. C., C. R. GOLDMAN, T. C. FRANTZ, and R. WICKWIRE. 1975. Where have all the Daphnia gone? The decline of a major cladoceran in Lake Tahoe, California-Nevada. Int. Ver. Theor. Angew. Limnol. Verh. 19: 843-849.

ROHLF, F. J., J. KISHPAUGH, and D. KIRK. 1971. NT-SYS. Numerical taxonomy system of multivariate statistical programmes. Tech. Rep., SUNY, Stony Brook.

SHELDON, R. W., W. H. SUTCLIFFE, JR., and M. A. PARANJAPE. 1977. Structure of pelagic food chain and relationship between plankton and fish production. J. Fish. Res. Bd. Can. 34: 2344-2353.

SNEATH, P. H. A., and R. R. SOKAL. 1973. Numerical taxonomy: the principles and practice of numerical classification. Freeman.

SOROKIN, J. I. 1972. Biological productivity of the Rybinsk reservoir, p. 493-504. *In* Z. Kajak and A. Hillbricht-Ilkowska [eds.], Productivity problems of freshwaters. PWN Pol. Sci. Publ., Warsaw.

SPRULES, W. G. 1977. Crustacean zooplankton communities as indicators of limnological conditions: an approach using principal component analysis. J. Fish. Res. Bd. Can. 34: 962-975.

WERNER, E. E. 1974. The fish size, prey size, handling time relation in several sunfishes and some implications. J. Fish. Res. Bd. Can. 31: 1531-1536.

WINBERG, G. G., AND OTHERS. 1972. Biological productivity of different types of lakes, p. 383-404. *In* Z. Kajak and A. Hillbricht-Ilkowska [eds.], Productivity problems of freshwaters. PWN Pl. Sci. Publ., Warsaw.

58. Chydorid Cladoceran Assemblages from Subtropical Florida

Thomas L. Crisman

Abstract

The distribution of chydorid cladocerans from 52 Florida lakes was analyzed as a function of 13 physical and chemical parameters. Three major lake types were identified that represented a transitional series of increasing alkalinity, conductivity, and phosphorus concentration. The distribution of each lake type corresponded closely with the areal extent of major geologic units. Chydorid assemblages were reconstructed from subfossil remains isolated from surface sediments, and groups of associations for all species were constructed by R-mode principal components analysis. Although 13 factors were identified, only the first 2 were considered valid. The relationship between each chydorid factor and individual lacustrine parameters was determined by Pearson correlation of factor scores with each physical and chemical parameter. The contribution of the present investigation to understanding the biogeography of North American chydorids is discussed.

In the past, littoral cladocerans have been ignored by most limnetic zooplankton investigators, yet paleolimnologists have long recognized the potential of littoral cladocerans (predominately Chydoridae) for reconstructing lake trophic histories (Frey 1955, 1960; Goulden 1964, 1966; Harmsworth 1968; Crisman 1976) due to both the abundance and ease of specific identification of subfossil remains and the fact that subfossil assemblages are representative of the living assemblage that produced them (Mueller 1964). Although recent investigations have been able to correlate changes in subfossil cladoceran assemblages with historical evidence for trophic state oscillation (Bradbury and Megard 1972; Birks et al. 1976), precise interpretations of past changes in cladoceran assemblages have been hindered by a lack of detailed data on the ecology of individual species.

Data on the geographical distribution and ecology of littoral cladocerans are incomplete. Investigations on both the distribution of individual species with respect to lake trophic state and possible geographical variations in intraspecific responses to trophic state are vitally needed.

At present, much of our understanding of littoral cladoceran ecology is based on investigations of the northern European fauna. The investigation of Whiteside (1970) on subfossil chydorid assemblages isolated from surface sediments of 77 Danish lakes represents the first quantitative analysis of littoral cladoceran ecology. Whiteside divided his study lakes into three groups (clear-water lakes, ponds and bogs, polluted clear-water lakes) on the basis of seven

Contribution No. 186 of the Limnological Research Center, University of Minnesota.

physical-chemical parameters. The relative abundances of the 22 most common chydorid species were then used as variables in a multiple discriminant analysis of the three lake groups in order to define major species groupings. In addition, the distribution of each chydorid species was correlated with individual lake parameters. More recently, Beales (1976) correlated the distribution of individual chydorid species from 59 English lakes with the same seven physical-chemical parameters utilized by Whiteside (1970), but unlike the latter, used R-mode principal components analysis to define major species groupings and the controlling parameters of each group.

These investigations, in addition to recent work in Czechoslovakia (Sládečková and Sládeček 1977), have provided valuable information on the ecology of European chydorids, but interpretation of North American assemblages based on European data must be approached cautiously. European species designations have been applied to morphologically similar forms in North America without detailed taxonomic analysis. Taxonomic uncertainties, in addition to possible geographic differences in intraspecific responses to various lacustrine parameters, demonstrate the need for a detailed ecological analysis of North American chydorids.

Two quantitative analyses of chydorid distributions within North American lake types are of note. DeCosta (1964) studied the latitudinal distribution of chydorid species in the Mississippi River Valley by examining chydorid assemblages from 45 lakes along a transect from northern Minnesota to Louisiana. Three species groupings (northern, southern, eurytopic) were defined statistically by the distribution of 15 chydorid species which DeCosta attributed to differential climatic responses of individual species.

Synerholm (1974) related the distribution of subfossil remains of chydorid species isolated from surface sediments of 32 Minnesota lakes to changes in lake conductivity. She divided the lakes into three arbitrary groups on the basis of increasing conductivity, and by means of discriminant analysis correctly assigned 81% of the lakes to the proper conductivity group on the basis of the percent composition of four chydorid species: *Chydorus sphaericus, Acroperus harpae, Leydigia leydigi*, and *Alona intermedia*.

The investigation reported here is intended as a baseline survey of subtropical chydorid assemblages in Florida with particular emphasis on the geographic distribution of individual species within the state as a reflection of lake chemistry and trophic state. All chydorid assemblages have been reconstructed from subfossil remains isolated from surface sediments. Subfossil assemblages represent an integration of the living chydorid assemblage for the entire lake basin and thereby minimize complications arising from patchy distributions of individual species within a single basin. It is believed that these data will contribute to an understanding of the magnitude of inter- and intracontinental variations in the response of individual chydorid species to physical and chemical parameters.

I thank G. Jacobson for assistance in the field, J. Shapiro for providing facilities for chemical analyses, D. G. Frey for assistance with chydorid taxonomy, and H. E. Wright, Jr., for guidance and financial support. Special thanks to E. S. Deevey, Jr., for supplying extensive field equipment while I was collecting samples in Florida. I also thank S. O. and S. P. Crisman for their support and assistance in preparing the final manuscript. This investigation was completed while I was a postdoctoral research fellow at the Limnological Research Center of the University of Minnesota as part of an NSF grant awarded to H. E. Wright, Jr.

Methods

The investigation is based on analyses of surface sediment and water samples collected during 1976 from 52 lakes throughout peninsular Florida (Fig. 1). Lakes were selected such that all major natural vegetation and surficial geologic

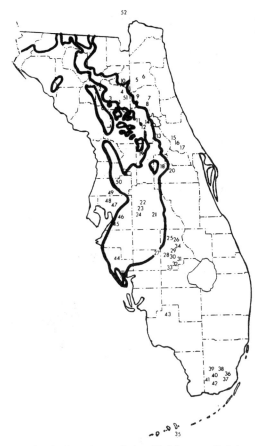

Fig. 1. Locations of the 52 lakes studied in peninsular Florida. Darkly outlined area represents areal extent of phosphate deposits.

units within the study area would be sampled. All chemical and physical analyses were performed on raw water samples collected at a depth of 1 m according to techniques outlined in *Standard Methods* (Am. Public Health Assoc. 1975). Physical and chemical parameters routinely analyzed included: calcium, magnesium, sodium, potassium, iron, chloride, silica, sulfate, total phosphorus, Kjeldahl nitrogen, specific conductivity, alkalinity, and Secchi disk transparency.

Surface sediments were collected from the deepest area of each lake by means of a Hongve sediment sampler (Hongve 1972). Sediments collected by this method were assumed to represent the past 5-10 years of deposition. Chydorid remains were isolated from sediments by deflocculating about 1 cc of sample in hot 10% KOH for 30 min followed by treatment with hot 10% HCl to remove carbonates. The sediment residue was then filtered through a 37-μm-mesh phosphor-bronze screen with distilled water, centrifuged, decanted, and suspended in tertiary butyl alcohol (TBA). Slides were prepared by applying aliquots of this mixture onto a mounting medium of heated silicone oil and adding a glass cover slip following evaporation of the TBA.

Slides were counted at a magnification of 400 diameters, with difficult identifications made under oil immersion. About 200 chydorid remains were counted from each lake, and both fragmented and whole remains were utilized. Two fragmented half-remains have been equated with one whole remain. The population size of each species was estimated from individual chydorid remains in an attempt to account for interspecific differences in exoskeletal fragmentation. The maximum numbers of individuals represented by each fragmented and whole-remain category for each species were calculated, and the largest resulting estimate of individuals from a single remain category was selected as the population estimate for that species. Each chydorid species was then expressed as a percent of the total chydorid assemblage for a given lake.

Lake chemistry

The distribution of four select physical and chemical parameters in peninsular Florida lakes is shown in Fig. 2. Three major lake groups have been defined based on lacustrine physical-chemical parameters, and they conform closely with variations in surficial geology of the state. The first lake group (I) is located principally in central Florida east of Tampa Bay. These lakes are characterized by high values for phosphorus (>100 μg/liter) and conductivity (>150 μmhos/liter), intermediate alkalinity values (<100 mg $CaCO_3$/liter), and low Secchi disk transparency (<2 m). The distribution of Group I lakes (Fig. 2)

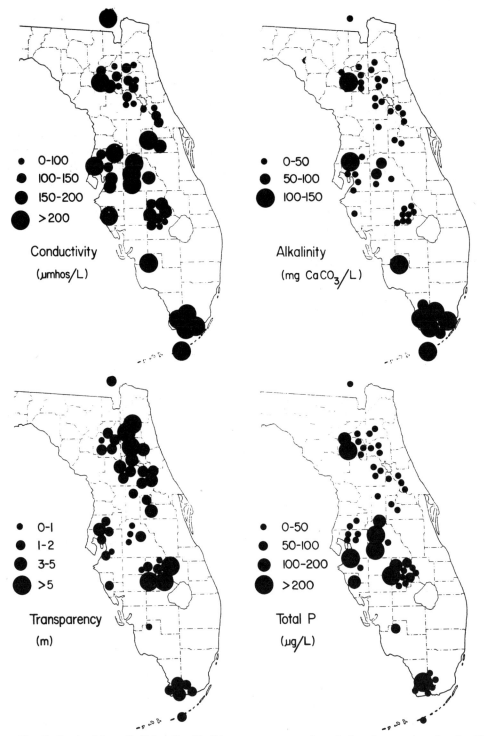

Fig. 2. Conductivity, alkalinity, Secchi disk transparency, and total phosphorus values for the 52 lakes studied.

corresponds closely with the extent of Miocene phosphate deposits of the Hawthorn and Bone Valley Formations (Fig. 1).

The lakes of Group II are located principally in eastern Florida from Clay County in the north to Highlands County in the south and thereby form a fringe adjacent to the eastern and southern extent of the phosphatic sand deposits. Group II lakes are located in highly leached Pleistocene and Recent sand deposits and are characterized by low phosphorus (< 50 μg/liter), and alkalinity values (< 50 mg $CaCO_3$/liter), high Secchi disk transparency (usually > 5 m), and low to intermediate conductivity (< 200 μmhos/liter).

Group III lakes are located in southern Florida, south of Lake Okeechobee, and are generally characterized by low to intermediate phosphorus values (< 200 μg/liter), intermediate to high transparency (> 2 m), and extremely high conductivity (> 200 μmhos/liter) and alkalinity (> 100 mg $CaCO_3$/liter). The chemistry of these lakes is largely a reflection of the fact that these basins are located in calcareous deposits, the Miocene age Tamiami Formation and the Pleistocene age Miami Oolite. Greater similarity is displayed between Group I and III lakes than either group displays separately with Group II lakes. The principal distinguishing characteristic between Group I and III lakes is the presence of higher alkalinity and much lower phosphorus values in the latter group.

The above discussion is intended as a general summary of Florida lake chemistry. The detailed quantitative investigations upon which the above remarks were extracted will be published elsewhere.

Chydorid distributions in Florida

Several chydorid species display a preference for either hard-water (Groups I and III) or soft-water (Group II). The distribution of the four species displaying the greatest preference for hard-water lakes is shown in Fig. 3. These include *Alona setulosa, Chydorus sphaericus, Euryalona orientalis*, and *Leydigia* spp. (*L. acanthocercoides, L. leydigi*). Of these, *C. sphaericus* has the broadest distribution, being present, if not dominant, in all three lake types. The fauna of most hard-water lakes is dominated by *C. sphaericus* with *L. acanthocercoides* and *L. leydigi* as the major subdominants.

Alona affinis, Alonella globulosa, Alonella hamulata, and *Chydorus piger* are largely restricted to the soft-water lakes of Group II (Fig. 4). At least ten other species including *Acroperus harpae, Alona guttata, Monospilus dispar*, and *Rhynchotalona falcata* tend to favor soft-water lakes, but their distribution is not as restricted as those species depicted in Fig. 4. As mentioned earlier, *C. sphaericus* is often dominant in Group II lakes, but *A. harpae* and *A. affinis* are major subdominants.

In general, chydorid faunas of soft-water lakes in Florida are characterized by greater species richness and diversity (Fig. 5) than hard-water lakes. Maximal Shannon-Wiener diversity (> 2.0) is displayed by chydorid faunas of the soft-water lakes of Group II with lower values (< 2.0) recorded from hard-water lakes of Groups I and III. A similar trend of decreasing faunal diversity with increasing alkalinity and conductivity was reported by Whiteside (1970) in Denmark and Synerholm (1974) in Minnesota.

While distribution patterns for individual species provide useful qualitative data, the response of each species to individual physical and chemical parameters is not quantified. To this end, R-mode principal components analysis was used to separate the 36 chydorid species isolated from Florida lake sediments into groups of association (Fig. 6). Thirteen factors were identified, with 40% of the observed variance explained by the first three factors.

Acroperus harpae, Alona affinis, Alonella globulosa, Chydorus piger, Alona rustica, and *Chydorus faviformis* display strong positive loadings (> 0.2) on factor 1, while *C. sphaericus, Alona circumfimbriata, Euryalona orientalis*, and *Dunhevedia crassa* display negative loadings. Eighteen addi-

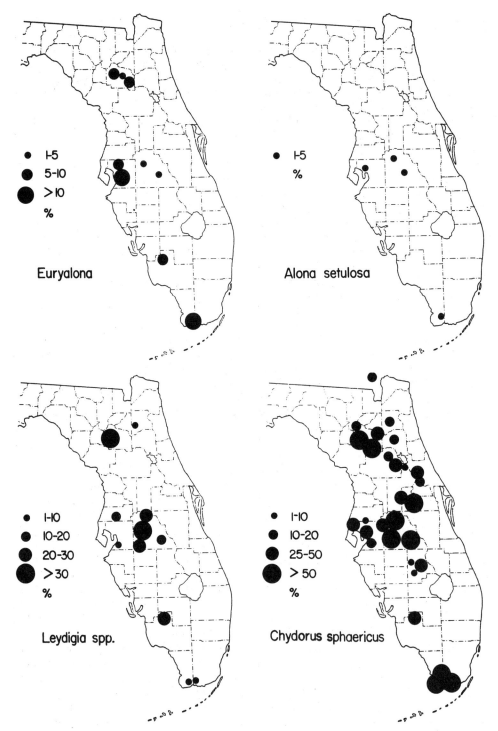

Fig. 3. Distribution of four chydorid species characteristically found in hard-water lakes expressed as percent of total chydorid fauna of each basin.

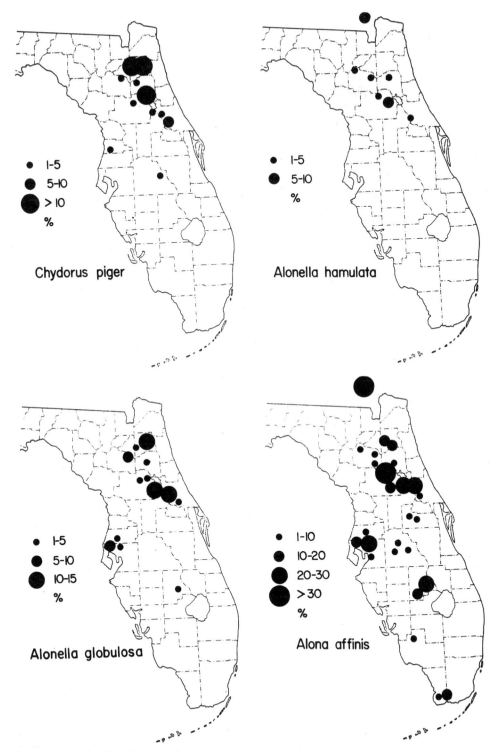

Fig. 4. As Fig. 3, but in soft-water lakes.

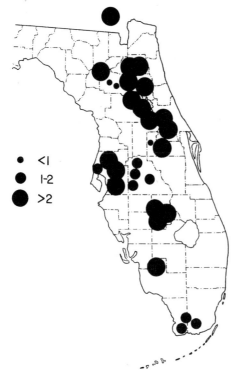

Fig. 5. Shannon-Wiener diversity indices for chydorid assemblages throughout peninsular Florida.

R-mode principal components analysis. The degree of association of the chydorid assemblage of any lake with each factor could then be calculated based on the percentage composition of the chydorid assemblage and the previously determined factor loadings for each species. The faunal assemblage of each basin was therefore treated as a single composite species whose relation to each chydorid factor grouping could be expressed as a factor score. Correlation of factor scores with the water chemistry of each basin then provided a measure of the importance of individual parameters in controlling chydorid factor groups. The relationship of chydorid factors 1 and 2 with individual lacustrine parameters is shown in Fig. 7.

Chydorid factor 1 is negatively correlated with calcium, sulfate, phosphorus, magnesium, and conductivity, with other parameters displaying weaker control. In effect, factor 1 divides the chydorid fauna of Florida (Fig. 6) into soft-water (positive loadings) and hard-water (negative loadings) assemblages.

Factor 2 is most strongly controlled by positive correlations with chloride, sodium, alkalinity, potassium, and conductivity, and negative correlation with transparency. Comparison of the factor loadings and geographic distributions of individual chydorid species indicates that factor 2 represents a subdivision of the soft-water fauna characteristic of Group II lakes according to tolerance for increasing conductivity and alkalinity. Species limited to more dilute waters are represented by negative loadings on factor 2, while those soft-water taxa able to tolerate a wider range of conductivity and alkalinity display positive loadings on factor 2.

A similar species replacement series with increasing conductivity and alkalinity has been noted for European (Whiteside 1970; Beales 1976) and North American (Synerholm 1974) faunas. Synerholm (1974) constructed species groupings in Minnesota based on lake conductivity, with high conductivity lakes characterized by *C. sphaericus*, *A. circumfimbriata*, and

tional species were more weakly associated with factor 1, having loadings between −0.2 and +0.2.

Factor 2 is controlled by positive loadings for *Alona quadrangularis*, *Eurycercus* spp., *Alonella exigua*, *Disparalona rostrata*, *Alona costata*, *C. faviformis*, and *A. hamulata*, and by negative loadings for *R. falcata*, *Graptoleberis testudinaria*, *A. rustica*, and *Alona verrucosa*. With the exception of *A. rustica* and *C. faviformis*, none of these species were strongly associated with factor 1.

Although 13 factors were identified by discriminant analysis, only the first 2 are considered in detail. The other 11 factors are controlled by species with such limited distributions and abundance that any interpretations would be extremely speculative.

The degree of association of individual chydorid species with each factor (species grouping) was determined previously by

Florida Chydorid Cladocerans

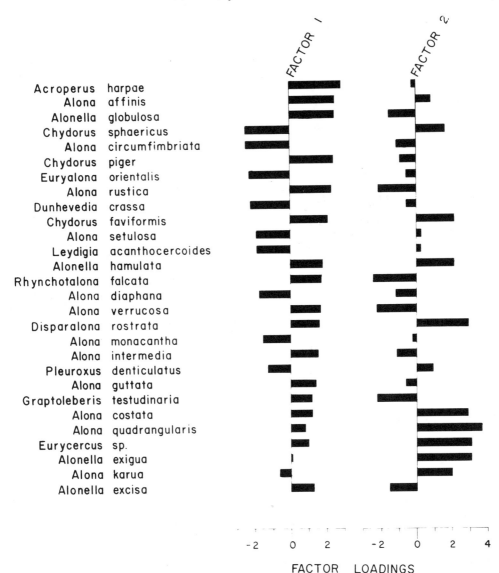

Fig. 6. Factor loadings on first two factors calculated from R-mode principal components analysis for 28 most common chydorid species.

Pleuroxus spp. and low conductivity lakes by 13 species that are also common in Florida soft-water lakes.

No attempt was made in the present investigation to define the distribution of each chydorid species as a function of a fixed range for each lacustrine parameter. To do so would deny the interaction of individual parameters and assume that the entire potential distribution of each species was defined by the 52 lakes that were analyzed. While the species groupings that have been based on lacustrine physical and chemical parameters provide a valid interpretation of chydorid distributions in Florida, it must be kept in mind that chydorid species may be responding to lacustrine chemistry indirectly rather than directly. The species composition and abundance of aquatic macrophytes and

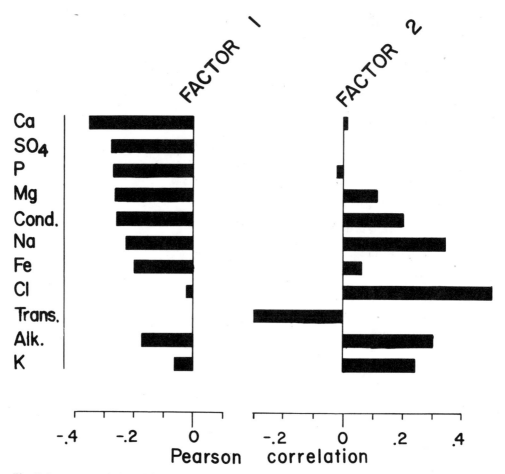

Fig. 7. Pearson correlation of factor scores for factors 1 and 2 with individual lacustrine parameters.

sessile algae, the major substrate and food of chydorids, respectively, are directly under the control of lacustrine physical and chemical parameters. In addition, population levels of potential vertebrate and invertebrate predators on chydorids are often directly related to the extent of littoral macrophyte cover and sediment composition for protection and reproduction. The present investigation, then, has examined the final product (chydorid distributions) as a function of the ultimate controlling factor (lacustrine chemistry) without considering the intermediate steps in the process (biotic interactions).

Latitudinal distributions in North America

DeCosta (1964) studied the latitudinal distribution of chydorid species in the Mississippi River Valley by examining chydorid assemblages from 45 lakes along a transect from northern Minnesota to Louisiana. Based on the distribution of 15 indicator species, DeCosta (1964) statistically defined three species groupings (northern, southern, curytopic) for the Mississippi River Valley (Table 1). The eurytopic assemblage displayed the widest distribution and was dominant in the transition zone between northern and southern faunas (35.80°N–39.25°N). North of 39.25°N and south of 35.80°N the northern and southern assemblages were dominant respectively. DeCosta noted that the zone of northern-southern faunal transition was coincident with an

Table. 1. Latitudinal assemblages of chydorids constructed from Mississippi River Valley transect of DeCosta (1964).

Northern species
Acroperus harpae
Alona quadrangularis
Alonella excisa
Eurycercus lamellatus
Graptoleberis testudinaria
Southern species
Alona karua
Alonella hamulata
Chydorus spp.
Euryalona occidentalis
Leydigia acanthocercoides
Eurytopic species
Alona rectangula
Camptocercus rectirostris
Chydorus globosus
Leydigia leydigi
Pleuroxus denticulatus

area transitional between dimictic and warm monomictic lakes. He therefore reasoned that faunal assemblages were geographically limited by macroclimatic conditions and thermal regimes of individual lakes.

Examination of chydorid assemblages from Florida revealed a number of inconsistencies with DeCosta's (1964) scheme. All five of the species classified by DeCosta as northern, in addition to possibly *Alona affinis*, *A. guttata*, *A. intermedia*, *A. rustica*, *Alonella exigua*, *Chydorus faviformis*, *C. piger*, *C. sphaericus*, *Disparalona acutirostris*, *Disparalona rostrata*, and *Rhynchotalona falcata*, are equally common in similar lake types in both the northern and southern United States. With the exception of *Alona rectangula*, which is most likely a misnomer, the eurytopic category thus may be expanded to include several species commonly classified as northern elements. No species have been deleted from DeCosta's southern category, but a number of species including *Alona diaphana*, *A. monacantha*, *A. verrucosa*, and *Alonella globulosa* may be added to this category.

The present data suggest that when similar lake types are compared in both the north and south, species once thought to be predominantly northern elements will be more widespread than previously noted. Although most chydorid species are widely distributed in eastern North America, it is suggested that the southern category is characterized by greater species richness than the northern category. Finally, these data suggest that any future biogeographic surveys must consider the importance of lake type in controlling chydorid distributions.

References

AMERICAN PUBLIC HEALTH ASSOCIATION. 1975. Standard methods for the examination of water and wastewater.

BEALES, P. W. 1976. Paleolimnological studies of a Shropshire mere. Ph.D. thesis, Cambridge Univ.

BIRKS, H. H., M. C. WHITESIDE, D. M. STARK, and R. C. BRIGHT. 1976. Recent paleolimnology of three lakes in northwestern Minnesota. Quat. Res. 6:249-272.

BRADBURY, J. P., and R. O. MEGARD. 1972. Stratigraphic record of pollution in Shagawa Lake, northeastern Minnesota. Bull. Geol. Soc. Am. 83:2639-2648.

CRISMAN, T. L. 1976. North Pond Massachusetts: Postglacial variations in lacustrine productivity as a reflection of changing watershed-lake interactions. Ph.D. thesis, Indiana Univ., Bloomington.

DeCOSTA, J. 1964. Latitudinal distribution of chydorid cladocera in the Mississippi Valley, based on their remains in surficial lake sediments. Invest. Indiana Lakes Streams 2:65-101.

FREY, D. G. 1955. Langsee: a history of meromixis. Mem. Ist. Ital. Idrobiol. 8 (Suppl.):141-164.

———. 1960. The ecological significance of cladoceran remains in lake sediments. Ecology 41:684-699.

GOULDEN, C. E. 1964. The history of the cladoceran fauna of Esthwaite Water (England) and its limnological significance. Arch. Hydrobiol. 60:1-52.

———. 1966. La Aguada de Santa Ana Vieja: An interpretive study of the cladoceran microfossils. Arch. Hydrobiol. 62:373-404.

HARMSWORTH, R. V. 1968. The developmental history of Blelham Tarn (England) as shown by animal microfossils, with special reference to the cladocera. Ecol. Monogr. 38:223-241.

HONGVE, D. 1972. En bunnhenter som er lett a lage. Fauna 25:281-283.

MUELLER, W. P. 1964. The distribution of cladoceran remains in surficial sediments from three northern Indiana lakes. Invest. Indiana Lakes Streams 6:1-63.

SLÁDEČKOVÁ, A., and V. SLÁDEČEK. 1977. Periphyton as indicator of the reservoir water quality. II. Pseudoperiphyton. Arch. Hydrobiol. Ergeb. Limnol. 9:177-191.

SYNERHOLM, C. C. 1974. The chydorid cladocera from surface lake sediments in Minnesota and North Dakota. M.S. thesis, Univ. Minnesota, Minneapolis.

WHITESIDE, M. C. 1970. Danish chydorid cladocera: modern ecology and core studies. Ecol. Monogr. 40:79-118.

59. Structure of Zooplankton Communities in the Peten Lake District, Guatemala

Edward S. Deevey, Jr., Georgiana B. Deevey, and Mark Brenner

Abstract

Cypria petenensis, the only known New World pelagic ostracod, is endemic to the Peten lake district. Outside Lake Peten Itza, the type locality, it is now known from eight smaller lakes of similar chemistry, but is scarce or absent in waters rich in $MgSO_4$. *Botryococcus* dominates the phytoplankton, but Lakes Yaxha and Sacnab, where year-round studies were conducted, also contain blue-green algae and much inorganic seston. Most zooplankton populations were minimal in August 1973, when rains were delayed, lake levels were low, and primary productivity was maximal. *Cypria* is a regular and abundant zooplankter, reproducing all year except perhaps at the end of dry season. After copepods (*Diaptomus dorsalis, Mesocyclops inversus, M. edax, Tropocyclops prasinus mexicanus*), ostracods are numerically dominant throughout the year. *Eubosmina tubicen* is the only important cladoceran; *Daphnia* is absent and *Bosminopsis, Ceriodaphnia*, chydorid and macrothricid Cladocera, and rotifers are notably scarce. At least three kinds of planktivorous fishes are present and are probably responsible for the small size and low diversity of the zooplankton. Day-night studies in Lakes Yaxha and Quexil show adult *Cypria* (> 0.5 mm) of both sexes to be migratory, avoiding upper waters except at night; juveniles are pelagic in daylight as well as at night. As *Chaoborus* is also migratory, the predators avoided by migration are presumably the fishes. *Cypria* populations regularly show seven distinct size-classes, including adults. The smallest class, containing instars I and II, can be recruited and pass into the next size-class at intervals of a few days, but proof that diurnal samples are biased by migration invalidates estimates of population growth and mortality. Historical (stratigraphic) data suggest that this community has been stable even in seriously impacted lakes since the 16th century, when the southern Maya Lowlands were depopulated and tropical forest was restored.

For analysis of the impact of an urban civilization on tropical forest ecosystems, two similar and adjacent closed lakes were chosen for limnological and historical study. As stated in the research proposal, the objectives were "to search the sedimentary record for evidence of Maya land use; to learn enough of the special limnology, botany, and archaeology to evaluate the record of airborne and waterborne substances; and to evaluate the basin-wide budgets of water, carbon, and major cations and anions at the present time and as they were modified by Maya exploitation of the environment." The two lakes are situated at 17°04'N lat., 89°17-18'W long., altitude ca. 183 m, at the east end of the Peten lake chain in northern Guatemala, in the heavily forested and nearly uninhabited heart of the Maya Lowlands.

Lake Yaxha (area 7.4 km², z_{max} 27 m), heavily urbanized in Classic and Postclassic Maya time, supported Preclassic and Classic nonurban populations about twice as large as those of unurbanized Lake Sacnab (area 3.9 km², z_{max} 13 m); the Sacnab subbasin contained no Postclassic occupation (Rice 1976, 1978). Deforestation of the lake district, first demonstrated at Lake Petenxil (Tsukada 1966; Cowgill et al. 1966), began with the appearance of settled agriculturists about 1000 B. C. With urbanization, lacustrine siltation and nutrient loading increased exponentially, in proportion to population densities, but while post-Maya nutrient inputs continue today at the high rates attained in Classic time, siltation rates diminished after the Classic maxima (Deevey et al. 1979). In the only long sedimentary sequence yet available, that from Lake Quexil, 2 km from Lake Petenxil, Mayan disturbance is seen palynologically and sedimentologically as a 2500-yr interruption, ca. 1000 B. C.-A. D. 1550, in a 9000-yr-long fossil record otherwise dominated by dense tropical forest (Vaughan 1976, 1978, 1979; Deevey 1978). Modern disturbance is mainly confined to the immediate vicinity of the large, central Lake Peten Itza, including Petenxil and Quexil, and to the savanna districts west and south of the lake chain.

Although water and plankton samples have been taken at most of the major lakes, detailed limnological observations were made throughout 1973-1974 only on Yaxha and Sacnab. Zooplankton samples from Quexil (area 2.1 km^2, z_{max} 32 m, altitude 110 m), taken (by Schindler trap, as were most others) in 1978, are also reported here. The emphasis in all our work having been on historical interpretation, little or no information has been obtained on fishes, molluscs, or macrobenthos, which do not occur as fossils in deep-water sediments. Diatoms and rotifers are scarce in these lakes and have not been given close study either in plankton or in sediments. *Botryococcus*, the dominant phytoplankter, is a common fossil, but sediment-trap and short-core studies (Deevey et al. 1977) prove its fossilization (like that of *Eubosmina* and *Cypria*) to be nonquantitative. Animal microfossils in older sediments, showing the communities to have been stable at least since the 17th century, are being reported by Brenner. Hydrological and hydrochemical data are being published elsewhere (Deevey et al. 1980).

We thank the cognizant Guatemalan government agencies, IGN, IDAEH, and FYDEP, and many friends in Guatemala for numerous courtesies. Most of the 1973-1974 plankton samples were collected by John Cooley, Habib Yezdani, and Hague Vaughan. Fieldwork in 1978 was assisted by Samuel Garrett-Jones and M. J. Flannery. A list of Peten fishes, including those of Yaxha and Sacnab, collected in 1973 by Reeve Bailey and Donn Rosen, was provided in a private communication by Bailey. Our work in Guatemala has been supported by the National Science Foundation (BMS 72-01859 and, currently, DEB 77-06629).

Regional limnology

The central Peten is a low-lying, steeply dissected karst plateau developed on Cretaceous and Tertiary limestones, locally dolomitized and interbedded with evaporites (Vinson 1962). All surface drainage is interior, although some tributaries of Rios Usumacinta and Mopan may have underground sources in the central region. The climate is warm and humid, with rainfall averaging 1,600 mm, only 5-10% of which falls in the 5 months (January-May) of the dry season. Relief is related both to solution of carbonate and sulfate rocks and to an E-W-striking system of *en echelon* faults. The chain of E-W elongate, closed-lake basins, 100 km long from Lake Perdida on the west to Lake Sacnab on the east, occupies the principal graben system. All lakes that have been mapped, except Macanche, show pronounced N-S asymmetry, with gently shelving south shores and deep trenches aligned at the feet of fault scarps on the north shores. The largest lake, Lake Peten Itza, has not been mapped, but is probably also the deepest, reaching at least 60 m in the trench. Next deepest is Macanche, with z_{max} = 57 m. Sinkholes, with water depths > 20 m, are common both within and outside the lake basins. Other topographic lows are occupied by swampy *bajos*, some of which may have held shallow lakes in Maya times (Cowgill and Hutchinson

1963; Harrison 1977). Bathymetric maps of Yaxha and Sacnab have been published (Deevey et al. 1977); maps of Quexil, Macanche, Salpeten, and Petenxil have been completed (Deevey et al. in prep.).

Chemically, most Peten lake and stream waters show $CaCO_3$ hardness with t.d.s. < 250 ppm and are unremarkably fresh. Higher salinities, up to 4,810 ppm (Salpeten) are traceable to high $SO_4^=$, to high Mg^{++}, or in one case (Macanche) to both. Lake Monifata I, moderately saline (t.d.s. 2,050 ppm), exceptionally rich in $SO_4^=$ but poor in Mg^{++}, is a deep sinkhole just outside the shoreline of shallow, dilute Lake Petenxil. Similarly wide hydrochemical variations were reported from many of the same waters by Brezonik and Fox (1974), who did not visit Yaxha or Sacnab. We presume them to reflect independent and very localized occurrences of anhydrite or gypsum and of dolomite.

Lakes Yaxha, Sacnab, and Quexil, whose zooplankton has been sampled repeatedly, are very similar chemically and in most other limnological features. The important points of difference emerge only in the paleolimnological record. Oxygen curves are clinograde, but no other chemical stratification has been seen, and complete exhaustion of O_2 in deep water is rare and temporary. Thermal stratification (observed only during daylight) is regular at all seasons, but probably breaks down at night. Here as elsewhere in the humid tropics, diel air-temperature variation is often wider than seasonal variation of mean temperature; the polymixia we infer from chemical evidence of mixing would be expected where very dilute waters, with vertical thermal gradients ranging between ca. 24° and ca. 28°C, are regularly subject to nocturnal cooling. The suggestion of Brezonik and Fox (1974), that Quexil may be meromictic, is not confirmed by our data, but meromixis may occur in saline sinkhole lakes, e. g. Lake Paxcaman (which we have not studied) and Lake Monifata I.

Phytoplankton productivity, as measured by dark-and-light bottle experiments, is only moderately high in these lakes. Five experiments in Yaxha, 1973–1974, gave a mean of 251.6 ± 121.6 mg C m^{-2} day^{-1}, while three experiments in Sacnab gave an identical mean, 251.7 ± 110.0 mg C. A single 1978 experiment in Quexil gave a similar result, 198 mg C m^{-2} day^{-1}. These inadequate measurements, 5 years apart, do not contradict our firm impression that Quexil is somewhat more productive than Yaxha and Sacnab. Quexil is believed not to have been urbanized in Classic Maya time, and the layer of montmorillonitic "Maya silt," where penetrated by single borings in the three lakes, is thinner and more deeply buried by post-Maya sediments in Quexil. Secchi disk transparencies are low, 90–140 cm, in all three lakes, but the seston that darkens Yaxha and Sacnab, as caught in sediment traps (Deevey et al. 1977) is apparently resuspended silt (loss on ignition 20–40%), whereas Quexil's seston is mainly dead phytoplankton and forms fresh sediment with 50–60% loss on ignition.

Today, as far as our data show, Yaxha and Sacnab differ from each other no more than any two lakes of similar water type but different size and depth. Quexil, though possibly less impacted by Maya disturbance, resembles both. Sacnab phytoplankton is slightly richer in blue-green algae than Yaxha phytoplankton, and siltier tow and trap samples make zooplankton counting more difficult in Sacnab, but zooplankton communities of all three lakes are remarkably similar in composition and structure. Moreover, although our stratigraphic data are incomplete and made quantitatively uncertain by diagenesis, the fossilizing (mainly ostracod and cladoceran) components of the zooplankton have been similar in the three lakes and stable over at least four centuries of post-Maya time.

Zoogeography

Perspective on zooplankton composition is given by the checklist of entomostracans in 10 Peten lakes (Table 1). Of 32 taxa recorded, as many as 23 were found in limited sampling of the largest lake, Peten Itza. Lake Yaxha, by far the most intensively and frequently sampled, yielded 18 taxa. The

Table 1. Entomostracan zooplankton, in ten lakes of the Department of El Peten, Guatemala

	Peten Itza	Yaxha	Sacnab	Quexil	Macanche	Monifata I	Salpeten	Sacpuy	Oquevix	Petenxil
COPEPODA										
Diaptomus dampfi	+	–	–	–	–	–	–	–	–	–
D. dorsalis	+	+	+	+	+	+	+	+	+	+
Calanoid sp.	+	–	–	–	–	–	–	–	–	–
Cyclops varicans rubellus	–	+	+	–	–	–	–	+	–	+
Eucyclops agilis	+	+	+	–	+	–	–	–	–	–
Mesocyclops edax	+	+	+	+	–	+	–	–	–	+
M. hyalinus	–	–	–	+	–	+	–	+	?	+
M. inversus	+	+	+	+	+	+	+	–	+	+
M. leuckarti	–	–	–	–	–	–	–	–	–	–
Tropocyclops prasinus mexicanus	+	+	+	+	+	+	+	+	–	+
parasitic cyclopoid	+	+	–	–	–	–	+	–	–	+
Nitocra spinipes	+	+	+	+	+	+	–	+	+	+
OSTRACODA										
Cypria petenensis	+	+	+	+	+	+	–	+	+	+
CLADOCERA										
Diaphanosoma sp.	–	+	–	–	–	–	–	–	–	–
Ceriodaphnia cf. *pulchella*	+	+	–	–	+	–	–	–	–	–
C. cf. *rigaudi*	–	–	–	–	+	–	–	–	+	–
Daphnia sp.	–	–	(+)	–	–	–	–	–	–	–
Simocephalus serrulatus	+	+	–	–	–	–	–	–	–	–
Eubosmina tubicen	+	+	+	+	+	–	–	–	+	+
Bosminopsis deitersi	+	–	–	+	+	–	–	–	–	–
Ilyocryptus sordidus	+	–	+	–	–	–	–	–	–	–
I. spinifer	–	–	+	–	–	–	–	–	–	+
Macrotrix cf. *rosea*	+	+	+	–	–	–	–	–	+	+
Alona affinis	+	–	–	–	–	–	–	–	–	+
A. cf. *circumfimbriata*	+	–	+	–	–	–	–	–	–	+
A. verrucosa	+	–	–	+	–	–	–	–	+	–
Alonella cf. *excisa*	+	–	–	+	–	–	–	–	–	+
C. hybridus group	–	+	–	+	–	–	–	–	–	+
C. pubescens	+	+	–	–	–	–	–	+	+	+
Leydigia sp.	+	+	–	–	–	–	–	–	–	–
Pseudochydorus globosus	+	–	–	–	–	–	–	–	–	–

other fresh water lakes (Sacnab, Quexil, and the very shallow Petenxil and Oquevix, sampled only once each) produced 11 to 15 entomostracans; thus, counting the shallow, dilute Lake Sacpuy, with 7 taxa in a single sample, the modal entomostracan diversity in fresh water lakes is 11–13 taxa. Even if raised to 18 by more frequent sampling and closer attention to chydorid cladocerans, as in Yaxha, the diversity seems very low. Even fewer kinds are found in the sulfate-rich saline lakes Salpeten and Monifata I. Results from $MgSO_4$-rich Macanche are difficult to interpret; and several Clarke-Bumpus hauls yielded only 8 entomostracans, virtually all those expected in open water, but *Cypria petenensis* was absent from nearly all samples.

Generalizations drawn from uneven sampling in 10 lakes of rather different chemistry are necessarily tentative. The nuclear entomostracan community of open waters consists of six species common to 7 or more of the 10 lakes. In order of frequency of occurrence they are *Diaptomus dorsalis* (all 10 lakes), *C. petenensis* (9), *Tropocyclops prasinus mexicanus* (9), *Mesocyclops inversus* (8), *M. edax* (7 or 8), and *Eubosmina tubicen* (7). Two chydorids, *Alona cf. circumfimbriata* and a problematical *Chydorus* of the *barroisi* group, will probably be found in more lakes, but do not belong to open water. *Chydorus sphaericus* was expected in lakes with blue-green algal plankton, but has not been encountered; it is represented in seven lakes by *C. pubescens*.

The pelagic ostracod *C. petenensis* is not only endemic to the Peten (though no longer to Lake Peten Itza, the type locality: Brehm 1932, 1939; Ferguson et al. 1964), but nearly unique in the world's lakes. Another *Cypria*, *C. javanus*, has a pelagic subspecies, *C. j. pelagica*, in Java, Bali, and Ceylon (Klie 1933; Apstein 1907, as *Cypria purpurascens*). *Eubosmina tubicen*, though first seen in Lake Peten Itza (Brehm 1939), is common in the American tropics and ranges north to Nova Scotia (Deevey and Deevey 1971). Of 32 taxa in Table 1, only *Diaptomus dampfi* remains apparently endemic to Lake Peten Itza (Brehm 1939).

Some endemism is known among Peten fishes, but the best-known (generic) case, the cichlid *Petenia splendida*, is now reported (Puleston 1977) from the Rio Hondo drainage in Belize. Except for species of *Diaptomus*, few entomostracans of open-water plankton have narrow ranges anywhere. What is surprising about the Peten community is not the banality of the crustacean fauna (other than *Cypria*), but the absence of many widespread species that might be expected in tropical and subtropical lakes situated on well-known avian flyways. A single dead *Daphnia*, possibly a contaminant, was encountered in Lake Sacnab. *Ceriodaphnia*, *Bosminopsis*, and macrothricid and chydorid Cladocera are rare; *Scapholeberis* is absent. Cyclopoid and diaptomid copepods are abundant, but their specific diversity is low. No "archaic" or troglobiont Malacostraca (or fishes or amphibians), such as might be expected in deep lakes in a karst region, are known in the Peten.

For individual, relatively small lakes, which lack the habitat diversity of lakes as large as Peten Itza, we consider the low specific diversity (and small body size) of the zooplankton community to reflect the presence of at least three species of planktivorous fishes: the clupeid *Dorosoma petenense*, the characin *Astyanax fasciatus*, and an unnamed atherinid, *Melaniris* sp. (R. M. Bailey pers. comm.). More generally, as rotifers and even diatoms appear to be scarce and of low diversity, we suspect the shortness of our list reflects not only the abundance and variety of fishes, but the prevalence of inedible blue-green algae and *Botryococcus*.

Low phytoplankton diversity may eventually find a chemical explanation. Guatemalan highland lakes are notably rich in diatoms (Deevey 1957), perhaps because of soda-rich volcanic rocks that do not occur in the Peten. We also speculate, but cannot yet prove, that prevalence of blue-green algae is another example of Mayan environmental impact, the familiar result of forest clearance, intensive agriculture, and excessive nutrient loading.

Seasonal variation

Arithmetic mean plankton concentrations in Lake Yaxha, 1973–1974, computed from vertical Schindler-trap series, are shown for five selected dates in Fig. 1. Several other series, taken in Lake Yaxha in February and March 1973, in the 24-m sinkhole at its SW corner in February, April, and August, and in Lake Sacnab in all 1973–1974 periods except August, gave closely comparable results. Seasonal variation of community structure is unspectacular, but the August minima for blue-green algae, *Diaptomus, Cypria,* and *Eubosmina* are statistically valid. A seasonal pattern is not evident in the Sacnab data, as the August zooplankton series was missed in that lake. Heavy rains were delayed until August in 1973, the driest year of record at Flores, lake levels were low, and Cl^- and HCO_3^- concentrations were slightly higher than in February or November. Dark-and-light bottle experiments also showed maximal productivities in both lakes in August, but statistical significance cannot be claimed for these figures.

Vertical migration

Depth distributions of the dominant zooplankters are remarkably uniform and constant over the seasons in Yaxha and Sacnab. Most animals avoid the upper 1–2 m during daylight, but no species avoids O_2-deficient depths, where *Chaoborus* lurks. In the usual arrangement, modal populations of *Diaptomus* lie below those of *Eubosmina*, with *Cypria* predominating above (Fig. 2). The largest of the three cyclopoids, *Mesocyclops edax*, is too scarce to show a significant depth preference over *M. inversus*. Cyclopoid and diaptomid nauplii are not separated in the counts.

Nocturnal distributions, also shown in

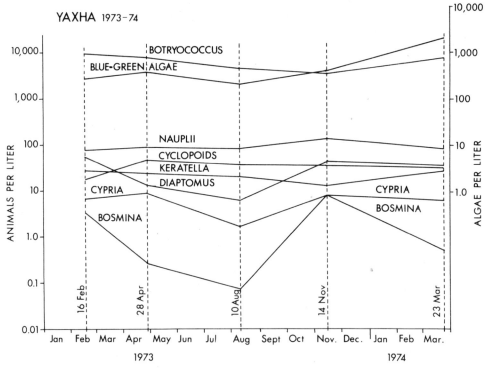

Fig. 1. Arithmetic mean standing crops of the principal plankters of Lake Yaxha, measured in five 10-liter Schindler-trap series between February 1973 and March 1974. "*Bosmina*" is *Eubosmina tubicen*; *Diaptomus* is *D. dorsalis*.

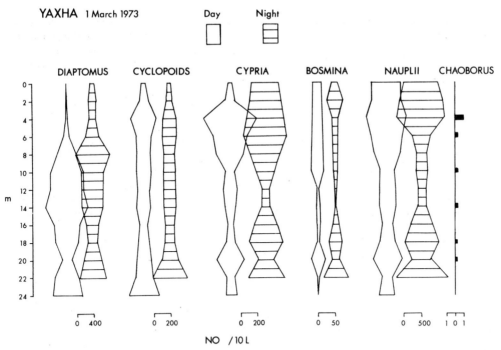

Fig. 2. Diurnal and nocturnal depth distributions of principal zooplankters in Lake Yaxha, 1 March 1973. Both series of 30-liter Schindler-trap samples were taken by J. Cooley and H. Vaughan.

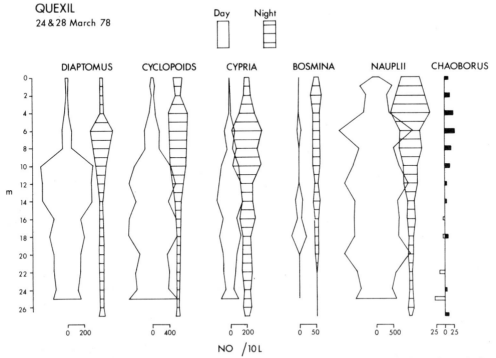

Fig. 3. Diurnal (24 March) and nocturnal (28 March 1978) depth distributions of principal zooplankters in Lake Quexil.

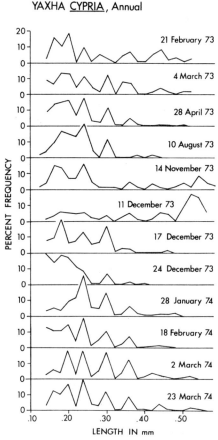

Fig. 4. Size-frequency distributions of *Cypria petenensis* populations in Lake Yaxha, on selected dates between February 1973 and March 1974.

Fig. 2, were examined only once in Lake Yaxha, on 1 March 1973. Vertical migration of all the principal zooplankters is not obvious in this figure but was convincingly shown in Lake Quexil on two dates in March 1978. One of the two sets of Quexil data is graphed in Fig. 3. The most conspicuous migration in both lakes, that of adult *Cypria*, is not shown separately in either figure. In fact, as Fig. 4 shows, significant numbers of adult ostracods were taken in Yaxha's daytime plankton only twice: November and December 1973. *Cypria* size-frequency distributions on ca. 40 other dates in Lake Yaxha and ca. 20 dates in Lake Sacnab showed juveniles only. Sacnab was not studied at night, but all nocturnal hauls on three lakes (Yaxha, Quexil, Peten Itza) contained adult ostracods in large numbers.

Cypria adults must spend the daylight hours on the lake bottom, but they have not been found either in washed Ekman samples or in muddy Schindler-trap samples immediately above the bottom. Their density in nighttime plankton is so low, 1 to 2 per cm^2 of lake surface, that special methods will be needed to find them in the mud. As *Chaoborus* shows similar migratory behavior (Figs. 2, 3), the predators avoided by adult *Cypria* are presumed to be the fishes.

Ostracod populations

Like other ostracods, *C. petenensis* has seven juvenile instars. Adult females (instar VIII) carry 2-10 eggs in a brood chamber until they hatch as instar I; the sex ratio is close to unity in all of our few large samples of adults. Size-frequency distributions in Lake Yaxha are shown for selected dates in Figs. 4 and 5. Closely similar results were obtained for the Sacnab population, a single example being shown in Fig. 5. Six distinct size-classes of juveniles are regularly present, adults, when captured, making the 7th. The smallest size-class usually contains instars I and II and is sometimes fused with instars III and IV.

Analysis of the age structure of a natural animal population is rarely easy. Data like those shown for *Cypria* could not be obtained in routine sampling for copepods or cladocerans. Where reproduction is continuous and population size is constant throughout the year, changing age structure should give invaluable information about recruitment, individual and population growth, and mortality.

Recruitment of the Yaxha *Cypria* population is apparent between December 1973 and January 1974 (Fig. 4.). In daily samples during the week of 2 March 1973 (Fig. 5), both recruitment of the smallest size-class and its passage into the next larger size-class appear as clearly as if a single experimental cohort were being observed.

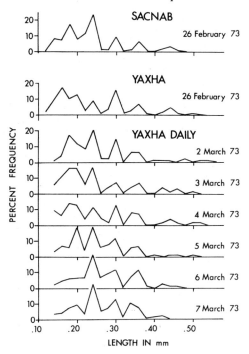

Fig. 5. Size-frequency distributions of *Cypria petenensis* populations in Lake Sacnab, 26 February 1973, in Lake Yaxha on same date, and at daily intervals in Lake Yaxha in week of 2 March 1973.

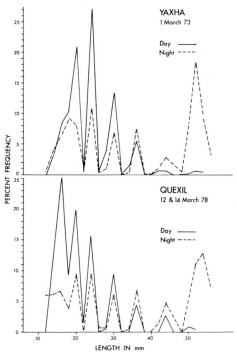

Fig. 6. Diurnal and nocturnal size-frequency distributions of *Cypria petenensis* populations in Lake Yaxha, 1 March 1973, and in Lake Quexil, 12-14 March 1978. Adults are present only at night.

Identical results, not shown, were obtained during the same week of 1974.

This evidence of rapid recruitment and growth, over intervals of a few days, accords with our intuition and may be acceptable. Unfortunately, although many size-frequency histograms appear also to provide estimates of growth and age-specific mortality, the dynamics of this population cannot be inferred from such data. Their defects become obvious in Fig. 6, which proves daytime samples to be severely biased by absence of adults, both in Yaxha and in Quexil. Until unbiased samples of the entire population can be obtained, e. g. by routine nocturnal studies, precise estimates of mortality, growth, and (probably) recruitment remain frustratingly beyond reach.

References

APSTEIN, C. 1907. Das Plancton im Colombo- See auf Ceylon. Zool. Jahrb. Abt. Syst. **25**: 201-244.

BREHM, V. 1932. Notizen zur Süsswasserfauna Guatemalas und Mexikos. Zool. Anz. **99** (3-4): 63-66.

———. 1939. La fauna microscopica del Lago Petén, Guatemala. Esc. Nac. Cienc. Biol. An. **1**: 173-204.

BREZONIK, P. L., and J. L. FOX. 1974. The limnology of selected Guatemalan lakes. Hydrobiologia **45**: 467-487.

COWGILL, U. M., and G. E. HUTCHINSON. 1963. El Bajo de Santa Fé. Trans. Am. Phil. Soc. **53**: 1-5.

———, ———, and others. 1966. The history of Laguna de Petenxil, a small lake in northern Guatemala. Mem. Conn. Acad. Arts. Sci. **17**: 126 p.

DEEVEY, E. S. 1957. Limnologic studies in Middle America with a chapter on Aztec limnology. Trans. Conn. Acad. Arts. Sci. **39**: 213-328.

———. 1978. Holocene forests and Maya disturbance near Lake Quexil, Peten, Guatemala. Pol. Arch. Hydrobiol. **25**: 117-128.

———, and G. B. DEEVEY. 1971. The American

species of *Eubosmina* Seligo (Crustacea, Cladocera). Limnol. Oceanogr. 16: 201-218.
——; D. S. RICE; P. M. RICE; H. H. VAUGHAN; M. BRENNER; and M. S. FLANNERY, 1979. Mayan urbanism: impact on a tropical karst environment. Science 206: 298-306.
——, H. H. VAUGHAN, and G. B. DEEVEY. 1977. Lakes Yaxha and Sacnab, Peten, Guatemala: planktonic fossils and sediment focusing, p. 189-196. *In* H. L. COLTERMAN [ed.], Interactions between sediments and fresh water. Junk, The Hague, and PUDOC, Wageningen.
——, M. BRENNER; M. S. FLANNERY; and G. H. YEZDANI. 1980. Lakes Yaxha and Sacnab, Peten, Guatemala: limnology and hydrology. Arch. Hydrobiol., Suppl., 57: 419-460.
FERGUSON, E. JR., G. E. HUTCHINSON, and C. E. GOULDEN. 1964. *Cypria petenensis*, a new name for the ostracod *Cypria pelagica* Brehm 1932. Postilla Peabody Mus. Nat. Hist. 80. 4 p.
HARRISON, P. D. 1977. The rise of the *bajos* and the fall of the Maya, p. 469-508. *In* N. Hammond [ed.], Social process in Maya prehistory. Academic.
KLIE, W. 1933. Die Ostracoden der Deutschen Limnologischen Sunda-Expedition. Arch. Hydrobiol. Suppl. 11, p. 447-502.
PULESTON, D. E. 1977. The art and archaeology of hydraulic agriculture in the Maya Lowlands, p. 449-467. *In* N. Hammond [ed.], Social process in Maya prehistory. Academic.
RICE, D. S. 1976. Middle Preclassic Maya settlement in the central Maya Lowlands. J. Field Archaeol. 3: 425-445.
——. 1978. Population growth and subsistence alternatives in a tropical lacustrine environment, p. 35-61. *In* P. D. Harrison and B. L. Turner [eds.], Pre-Hispanic Maya agriculture. Univ. New Mexico Press.
TSUKADA, M. 1966. The pollen sequence. Mem. Conn. Acad. Arts Sci. 17: 63-66.
VAUGHAN, H. H. 1976. Prehistoric disturbance: the area of Flores, Peten, Guatemela [abstr.]. Bull. Ecol. Soc. Am. 57 (1): 9.
——. 1978. An absolute pollen diagram from Lake Quexil, Peten, Guatemala [abstr.]. Bull. Ecol. Soc. Am. 59 (2): 97.
——. 1979. Prehistoric disturbance of vegetation in the area of Lake Yaxha, Peten, Guatemala. Ph.D. thesis, University of Florida.
VINSON, G. L. 1962. Upper Cretaceous and Tertiary stratigraphy of Guatemala. Bull. Am. Assoc. Pet. Geol. 46: 425-456.

60. Systematic Problems and Zoogeography in Cyclopoids

U. Einsle

Abstract

The problems of zoogeographic studies on crustaceans can be demonstrated especially well with the distribution of the genus *Cyclops* s. str. in western Europe. Depending upon which key characters are used, quite different geographic ranges could be drawn.

Because *Cyclops* populations have both high local and temporal variability, the basis for all zoogeographic efforts must be a clear systematic concept with a definite criterion for the level of species. This has been found in the chromatin diminution.

Several examples demonstrate that the modifications in the external appearance of *Cyclops* forms are variable enough that a zoogeographical statement can be given only after a cytological test.

Zoogeographic studies in a traditional sense, i.e. the investigations on the distribution of species over the continents, are dependent upon the reliability of the species definition. Especially within groups living in the plankton of lakes, the local and temporal variation often crosses the definition of described species and subspecies. For example, two distribution maps of *Cyclops abyssorum* could be drawn for Europe in the course of 1 year. If the samples are taken in the spring, the result could be quite different from samples taken in autumn, during which the effect of temporal variation is most significant.

Fortunately, the phenomenon of chromatin diminution allows a new basis for the taxonomy of the genus *Cyclops*. Using this criterion (Fig. 1), we can distinguish at least four groups. Since some species with clear morphological markings belong to one and the same diminution type, they can be distinguished without difficulty only by using their special chromatin criteria.

Based upon these clearly distinguishable differences, it was possible to arrange a classification of the several previously described subspecies or races; here the taxon term "race" may only indicate the existence of morphological discontinuities in the species without implying any genetic judgment. According to E. Mayr, a subspecies describes a collection of populations which has reached a certain autonomy in its genetical substance; these characteristics are not at the level of a species. They can be associated with ecological or morphological particularities.

Especially in *C. abyssorum* the possibilities of local and temporal modifications, and a beginning speciation, are obvious. The limits between these two stages, i.e. subspecies to species level, seem so

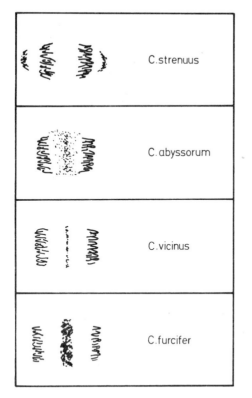

Fig. 1. Representative examples of chromatin-diminution types.

fluid and yet so difficult to confirm. Even in experiments, the fluctuations of general morphology are so surprising that one can, at best, only measure and record the differences, without assessing genetic relationships.

One can presume that the type of *C. abyssorum* described by Sars represents in a way a certain "average form," from which the several phenotypes in a lake could be derived or could revert to under certain conditions. In an exaggerated sense, one could say that—at least potentially—the entire spectrum of *abyssorum* phenotypes could be found in the temporal variation of a single population (Fig. 2).

These statements can be demonstrated by several examples.

1. A good illustration of the variability exhibited by an *abyssorum* type was seen this year in the Lake of Constance. Its *abyssorum* population is normally split into two ecotypes, *C. a. praealpinus* and *C. a. bodanus*. The two ecotypes can be distinguished by differences in body length, some of their external proportions, and their different biotopes in the lake (*C. a. praealpinus* lives in the pelagic zone, while *C. a. bodanus* inhabits the profundal region: Fig. 3).

Normally at the end of April or at the beginning of May, there appears a generation of *C. a. praealpinus* at low density (animals per m^3), but with a strikingly long body length. Individuals often nearly reach the proportions of small *bodanus* types.

During the extremely cold spring months of 1978 the warming of waters were so retarded that only by the middle of May had surface temperatures reached about 10°C. In 1978, the large spring generation appeared at the beginning of June. These animals, however, were so long that, even with a morphometric analysis, *praealpinus* and *bodanus* could no longer be separated.

We might suppose that the body length and the general appearance of the benthic *bodanus* is caused by the lower water temperatures at depth in the lake. If by chance the population living in the pelagic zone moves into regions of similar temperature, the animals would be modified to resemble the *bodanus* type.

2. *Cyclops abyssorum tatricus*. The *tatricus* type of *C. abyssorum* is mainly marked by its broad second thoracomer with the "lobus" and by its occurrence in high alpine lakes (over ca. 1,800 m). The chromatin diminution is typical for *C. abyssorum* (Fig. 4).

Breeding experiments carried out in the laboratory at Konstanz (400 m) showed that, during the first generation following breeding, individuals lost the "lobus" feature and exhibited all the characteristics of *C. abyssorum* found in the original type described by Sars. Crossing experiments with *C. abyssorum* from the Lake of Constance were successful, and the next generation proved to be fertile (Fig. 5).

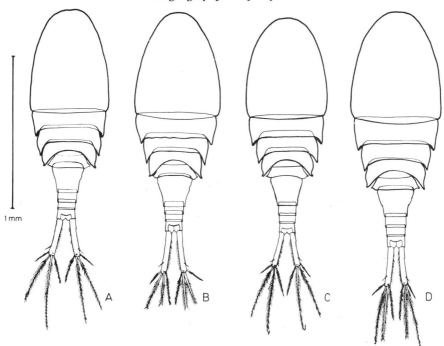

Fig. 2. *Cyclops abyssorum*, scandinavian forms: A) Aikesträsk Farö/Gotland, 21 September 1969 (Arnemo coll.); B) Stensfjord, summer 1972 (Elgmork coll.); C) Lønavatn, 14 July 1969 (Zool. Mus. Oslo); D) Langsvatn, 4 July 1970 (Zool. Mus. Oslo).

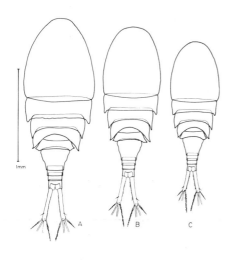

Fig. 3. *Cyclops abyssorum*, Bodensee-Obersee, 10 July 1973: A) *bodanus* form; B) "Intermediate"; C) *praealpinus* form (Einsle coll.).

The occurrence of *C. a. tatricus* is not tightly correlated with altitude. There are numerous populations with intermediate features in comparable regions. The temporal variation has not been well studied, for in many cases the determination of local types seems in error.

However, it seems evident that there must be some external factors which determine the morphological development of this *Cyclops* type in high mountain lakes. One can only suppose that the annual course of temperatures and—perhaps mainly—solar radiation could effect these phenotypes.

3. As a further example of the kinds of difficulties encountered in zoogeographic studies, one ought to mention the *C. abyssorum* populations in Italy. In the alpine lakes of Italy, there are intermediates between a strong *tatricus* type and a "northern *abyssorum*" type.

The populations in larger north Italian lakes belong to the *praealpinus* type, re-

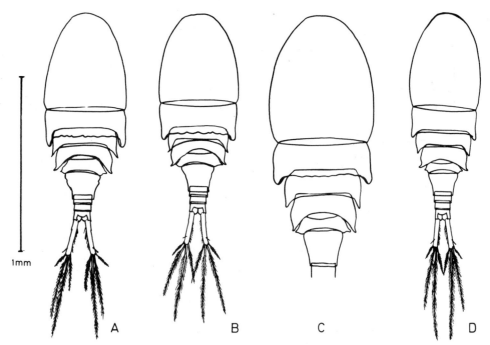

Fig. 4. *Cyclops abyssorum, tatricus* form: A) Kalbelesee, 4 October 1968 (Amann coll.); B) Drachensee, 12 September 1959 (Pechlaner coll.); C) Stabbio (Piora), 16 August 1949 (Ambühl coll.); D) Königsee, 22 May 1974 (Einsle coll.).

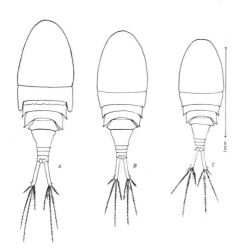

Fig. 5. *Cyclops abyssorum* f. *tatricus*, Kalbelesee: A) typical female; B) first generation offspring; C) bastard from *tatricus* (♀) X *praealpinus* (♂) cross.

sembling those found in lakes north of the Alps. In Lago Maggiore, one can observe this type splitting into two different, sympatric pelagic races (*C. a.* forma *novarensis*).

Another group of *C. abyssorum* is found in the lakes of the Apennines, a mountain range in middle Italy. These populations correspond most closely to types from the islands of Corsica, Sardinia, Elba, and Capraia. While the chromatin diminution was studied in a population from Lago di Chiusi, seasonal variation and diapause behavior are unknown at the moment (Fig. 6).

In the lakes near the vicinity of Rome live smaller *abyssorum* types. They can be distinguished from the animals which occur in the Apennin and the islands, and thus remain recognized as a different taxon: *C. a.* forma *laevis*.

Therefore, in a synopsis of Italian populations one cannot clearly separate *abyssorum* types and hence one cannot draw a distributional map for types. There are

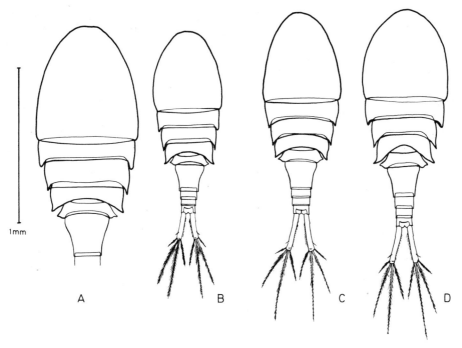

Fig. 6. *Cyclops abyssorum, divulsus* forms from Italy: A) Pozze Rio Vento (Lazio), 20 October 1968 (Stella coll.); B) Pozze Rio Vento (Lazio), 20 October 1968 (Stella coll.); C) Lago di Chiusi, 2 April 1968 (Einsle coll.); D) Lago di Paterno, February 1965 (Viganò coll.).

some tendencies, more or less obviously dependent on altitude, character of the lake, water temperature, and other factors, toward geographic differentiation, and yet the entire geographic pattern does not fit conveniently into regional subunits.

Summary

All investigations on the distributions of *C. abyssorum* and, in general, on most planktonic crustaceans, can be fit into only the broad concept of a species. A detailed assignment of populations to regional subspecific categories can result only in confusion, especially when little is known about ecological details.

In spite of these reservations, such an attempt to describe regional differentiation could be of great interest, as it might demonstrate rapid speciation in a group of freshwater crustaceans. This kind of attempt would be beneficial only if it incorporates ecological information, for the following reasons:

The body length and various morphological proportions seem partially controlled by temperature (molting frequency, absolute length).

The elevation of a water body above sea level can also influence aspects of the phenotype (e.g. the *abyssorum* type *C. a. tatricus*).

The degree of diapause behavior can vary considerably. In the *divulsus* group one finds a real dormant stage, whereas in the *praealpinus* group one finds only a cryptic retardation in development.

The intensity of diapause, together with external conditions (e.g. periods of overturn, extremely high water temperatures, degree of solar radiation) will influence the intensity of morphological variation with time (temporal).

Long-term changes in the character of biotopes (e.g. eutrophication) appear to cause modifications in the appearance of animals.

At least in the genus *Cyclops* there remains the possibility that several species can be defined by a set of reliable criteria, which are not influenced by external environmental conditions. These criteria involve the phenomenon of chromatin diminution. Observed phenotypic discontinuities are generally less reliable indicators of species, since they relate less to genetic differences and often indicate only the influence of environmental conditions. Therefore, a zoogeographical study of *Cyclops* populations ought to be a serious investigation of both taxonomic criteria and ecological facts.

61. Zooplankton and the Science of Biogeography: The Example of Africa

Henri Jean Dumont

Abstract

A zooplankton-based zoogeography makes sense if its objectives and constraints are well specified. Unlike the situation in vertebrates, the formative geological events that marked the end of the Mesozoic were of little consequence to the distribution of higher order zooplankton taxa. The Tertiary left traces in some old lakes where speciation has proceeded to the generic level, and in Australia and South America, where endemic genera of Cladocera evolved. Calanoid copepods evolved endemic genera in all continents. The transition between the Tertiary and the Quaternary, marked by the onset of glaciations in the northern hemisphere, allowed allopatric speciation in the Old and the New world. The most detailed information is, however, obtained from a consideration of Pleistocene geology and present zooplankton distributions. Little speciation occurred during this short period, but numerous patterns of distribution took shape, as exemplified by Africa. Here, the Quaternary was tectonically uneventful, but important changes in climate took place. *Daphnia* is at present absent from the aseasonal equatorial forest and Guinea savannah; deserts are characterized by large species living in astatic water bodies, while the south, the eastern plateau, the Maghreb, and a wedge extending from the Nile sources to the Senegal River hold three species that are particularly sensitive to climatic change. *Bosmina longirostris* behaves exactly like these. *Chydorus* is absent from the Sahara, and replaced by Aloninae and *Pleuroxus aduncus*. Some *Alona* species indicate recent climatic changes in northwest Africa (vicariant subspecies), while others are witness to the long and uninterrupted existence of evergreen forests along the Gulf of Guinea (endemic species). In general, zooplankton react promptly to climatic change, but relict populations outside the main area of distribution may persist for longer times. Hypotheses based on zooplankton distributions should, in most cases, fall in the time range covered by paleolimnological research and can thus be confirmed or refuted.

Ball (1975) writes: "groups which are known to be distributed by passive dispersal make poor subjects for biogeographical enquiry . . . and should not form part of general hypotheses." The internal contradiction in this statement is that a hypothesis is not general if a number of constituent categories are excluded a priori, and the challenge is to prove that the plankton, which by virtue of its drought-resistant stages is ideally suited to passive dispersal, is still valuable to biogeography.

It appears that a large number of zooplankton species are cosmopolitan; at least, it is not possible today to prove that they are not. If we keep in mind that

---

This work was supported by the Fund for Collective Fundamental Research, Belgium (grant 2.0009.75).

it is immensely more difficult to prove that something is not than that something is, it is fortunate that evidence is now accumulating that many zooplankters occupy restricted ranges (Rotifers: Pejler 1977; Cladocera: Wagler 1936; Brooks 1957; Goulden 1968; Smirnov 1971 and older refs. therein; Frey 1961, 1965; Copepoda Calanoida: Tollinger 1912).

The objectives of the present paper are to discuss plankton biogeography relative to general biogeography, to define objectives and an appropriate time-scale, and to apply this to the African continent.

I am indebted to J. Den Hengst (Holland) for samples from Moçambique. Some samples from lowland localities in Zaïre were originally collected by the late Prof. P. Van Oye. E. Lamoot (Gent) kindly let me use his data and great experience of Ivory Coast plankton. Dr. B. Hickel (Plön) made a short series of samples from Tibesti available to me (leg. P. Quézel), and L. Chelazzi (Firenze) provided material from Somalia. Prof. T. Monad (Paris) sent samples from Jeisen. Dr. G. Fryer (Ambleside) and Dr. J. Green (London) kindly communicated unpublished data on Lake Bangweulu area and Nigeria, respectively.

Time, lifetime of freshwater biotopes, and the availability of space and hyperspace

In discussing plankton biogeography relative to "classical" biogeography, scale effects have to be taken into account. Classical biogeography tends to measure time on the time-scale of the Mesozoic, when the major formative events of the present world took place. Many of its "centers of origin," "tracks," and "barriers" trace back to Cretaceous times. These, by lack of a fossil record and lack of continuity of biotopes, are irrelevant to planktologists. Only fully untestable hypotheses can be made. The oldest lakes of the world were formed around mid-Tertiary. One (L. Baikal) has an endemic genus of Cladocera. By analogy, we could assume that some endemic genera in Australia and South America are of the same age, hence that the isolation of these continents had then become sufficient to hinder passive dispersal. Again, this is vague and uncertain. Our factual background begins to grow solid only with the onset of the glaciations marking the transition between the Tertiary and the Pleistocene. We can be rather confident that considerable speciation took place since that time, as exemplified by the freshwater plankton of North America and Eurasia, which has numerous congeneric but different species. Limnology deals, as a rule, with environments that came into existence during the Upper Pleistocene. These environments are too young to have permitted endemic speciation. The time-scale of zooplankton biogeography is thus from the end of the Tertiary through the Pleistocene.

Importantly, tertiary lakes can be envisioned as areas of the world that have been ecologically stable or, at least, have not lost their identity since the end of the Cenozoic. Such areas can be predicted to be identifiable from endemic elements in their zooplankton (e.g. African evergreen forests). Further, all endemic elements in the plankton of old lakes pertain to the littoral or the benthos, while their limnetic faunas are dull and poor in species. This confirms rather than contradicts modern ecological theory: lakes, however ancient, are subject to regular water renewal, and are therefore continuously immigrated into and lose elements from their limnetic community. Endemism is therefore an improbable process in the pelagic zone of a single lake, but not along its edges and at its bottom. Further, it has been postulated (Hutchinson 1964) that the number of limnetic species in modern lakes exceeds the number of niches available, but that fluctuating environmental conditions make the elimination of supernumerous species a slow process. Ancient lakes, then, are places which should be near equilibrium.

Endemics of old lakes tend to spread little outside their biotope. Since no two lakes are identical, endemism in a single spot supposes specialization to unique conditions, and therefore, dispersal may be prevented by absence of a niche.

This is a point which I consider of great importance in balancing passive dispersal. However ingenious the dispersing strategies of the zooplankton (resting stages, parthenogenetic reproduction), no dispersal is possible if no adequate niche is encountered. A species cannot conquer space if adequate hyperspace is not available. Tautological as this may appear, it is often ignored in biogeography.

As is well known, niche boundaries are set by a number of environmental variables (climatic and edaphic factors) and biological variables (competition, predation). Ecology, and especially the branch dealing with community structure, tends to give paramount importance to the second category. In plankton biogeography, conversely, we often have no choice but to relate distribution patterns to changes in environmental conditions. However, in some cases, and postulating that the ecology of the past is that of today, we can understand patterns best from species interactions. I shall now exemplify those views, taking the African continent as an example.

The African climate during the Upper Pleistocene

Tectonically, the Upper Pleistocene was a period of relative calm, but climatological changes were important. Moreau (1963) calls attention to a reversal in importance between lowland and montane biomes during the Würm-Wisconsin. At the height of this glaciation, roughly during 50000–18000 B.P., the average temperature of the African continent was lowered by about 5°C. This means that a temperate and seasonal climate extended from the cape to the eastern Mediterrancean, sending a westward wedge from Ethiopia to the Camerouns. The Sahara was bridged or compressed by pluvials on several occasions. The most recent one, corresponding to the Würm, can be dissected into stadials which were of two types: cold winter rains of Mediterranean origin, and warm summer rains of equatorial origin. Both did not necessarily coincide, but they corresponded to advances of Palaearctic, viz. Ethiopian, faunas into the present desert. A broad girdle of fossil dunes extending well into the present savannah show that periods of southward expansion of the desert also occurred. I have argued (Dumont 1978) that the most recent hyperarid spell on the Sahel was from 5500–3000 B.P. and coincided in time with the Neolithic pluvial over the Sahara, i.e. all climatic zones had shifted south in belts parallel to the equator. On such occasions, the rainforest of west Africa was reduced and dissected, but it never did disappear completely. This evergreen forest is thus one of the oldest features of the African lowland biome, since the Congo cuvette was arid and covered by sand blown in from the Kalahari for long periods during the mid-Pleistocene, at least. In east Africa, there is again evidence of several pluvial epochs. This part of the continent, uplifted and broken during the Tertiary, presents a mosaic of climates and faunas that probably has not been fundamentally altered during the Pleistocene. Southern Africa, finally, witnessed climatic fluctuations, but the changes were far less important than elsewhere. Moreover, it so happens that large parts of southern Africa (Angola, Moçambique) are poorly known from a limnological point of view, and form the weaker parts of the present discussion. The island of Madagascar is not dealt with here.

The nature of the biogeographical evidence

Two main handicaps should be mentioned. First is the state of the art in the taxonomy of the groups involved. A well-known aphorism says that we hold of the distributional record only a taxonomic version, and that no taxonomy is ever final. Students of plankton groups will not contradict this, especially as some pioneer workers have left us with a rather confused situation. However, refinements are continuously being introduced and, at the genus level, we are beginning to face a stable situation, but new species will need to be described by at least another generation of limnologists. Second is the number

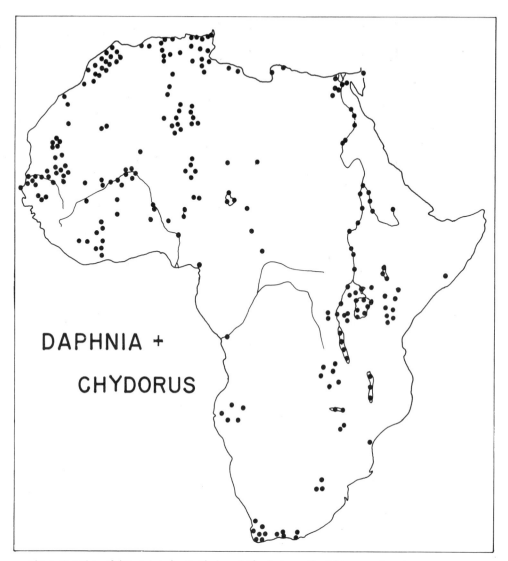

Fig. 1. Overview of data points for *Daphnia* and *Chydorus* on the African continent.

of data points and their distribution over the area under study (Fig. 1). No absolute rule exists. Figure 1 shows that not all parts of Africa have received equal attention. Relatively well-studied areas are the east African great lakes, the Nile area, the Maghreb countries, the Cape Province in south Africa, and the areas covered by the Institut Français d'Afrique Noire in west and northwest Africa. Recently, Lake Chad has been thoroughly surveyed by ORSTOM. Conspicuous blank areas, conversely, are the former Portuguese colonies (Guinea, Angola, Moçambique), and the lowland part of the Congo Valley. About a third of the points on Fig. 1 are furthermore unreported, based on unstudied collections made available to me, or on collections made by me and my associates in a geographically systematic manner in the Sahara over the past 4 years.

Not only does the absolute number of the data points matter, but also the detail in which a given area has been investigated.

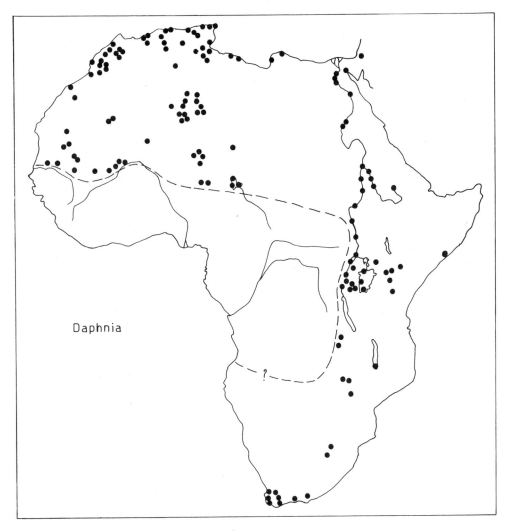

Fig. 2. Distribution of the genus *Daphnia* in Africa (dashed line represents margin of geographic range).

The Maghreb is known from a year-round study by Gauthier (1928); a summary of long teamwork on Lake Chad is given by Carmouze et al. (1972); west Africa is comparatively well known through studies by Green (1961) on the Sokoto River, Nigeria, and by Gauthier (1951) on Senegal. I also rely heavily on several years of uninterrupted observations of various types of aquatic biotopes in the Ivory Coast by E. Lamoot (unpubl.). Studies on the Nile area, including its source lakes have recently been summarized by Rzóska (1976). A particularly rich source of information is the unpublished thesis by Moghraby (1972) on the Blue Nile. From South Africa, we have reliable data only from Cape Province and the environs of Pretoria (*see* Harding 1961).

The patterns

Cladocera—About ten species of *Daphnia* (Fig. 2) are known to occur in Africa. None has ever been found in the evergreen rainforest zone and adjacent guinean sa-

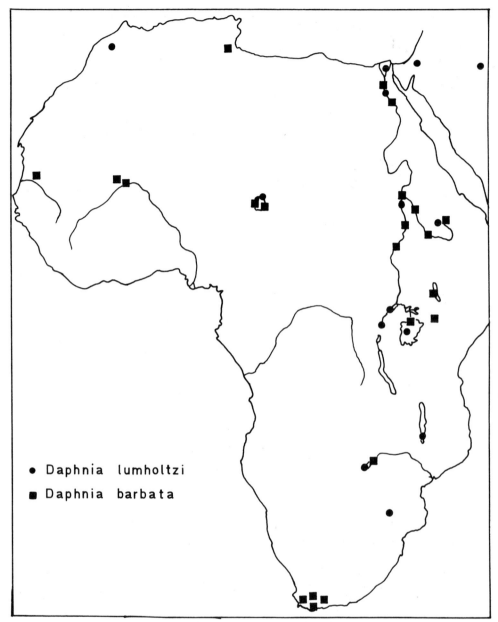

Fig. 3. Distribution of *D. lumholtzi* and *D. barbata*. Third species, *D. longispina*, shows same pattern but is not indicated on figure.

vannah. Both zones have no or poorly defined seasons. At the same latitude, the east African plateau has distinct seasons, and several *Daphnia* spp. occur here. The ecology behind this pattern is intriguing and completely unstudied. The production of resting eggs in *Daphnia* is controlled by variation in environmental factors, such as temperature and photoperiod (Stross and Hill 1968), and possibly, the too constant tropical environment is unfit for *Daphnia* populations in the long run.

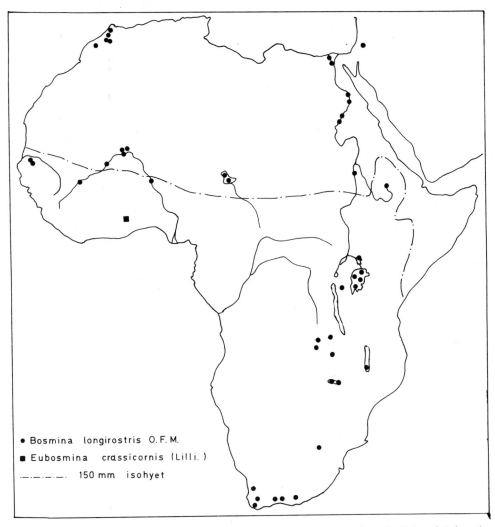

Fig. 4. Distribution of *Bosmina longirostris* in Africa; single occurrence in Lake Volta of *Eubosmina crassicornis* also indicated (an isohyet is a line of equal precipitation).

Different species have different distributions. Three, *D. longispina* O.F.M., *D. lumholtzi* Sars (Fig. 3), and *D. barbata* Weltner (Fig. 3) occur throughout the eastern half of the continent, have isolated populations in the Maghreb and send a westward wedge from the Nile sources to Lake Chad, the Niger, and the Senegal River. These species do not occur in the rivers themselves, but in lakes north of their basins, flooded during monsoon only, and situated in subdesert country. These lakes show strong seasonal fluctuations in physical and chemical conditions (best studied in Lake Chad: Carmouze et al. 1972), and the rate of reproduction of phyto- and zooplankton is strikingly seasonal too. Further north, in the Sahara, no *Daphnia* are found in permanent waters. However, true permanent waters in the Sahara always have dense populations of planktivorous fish (*Barbus* spp.), and their zooplankton consists of small chydorids and rotifers only. In astatic and semipermanent lakelets, *D. similis* and *D. magna* are found, but there is little doubt that

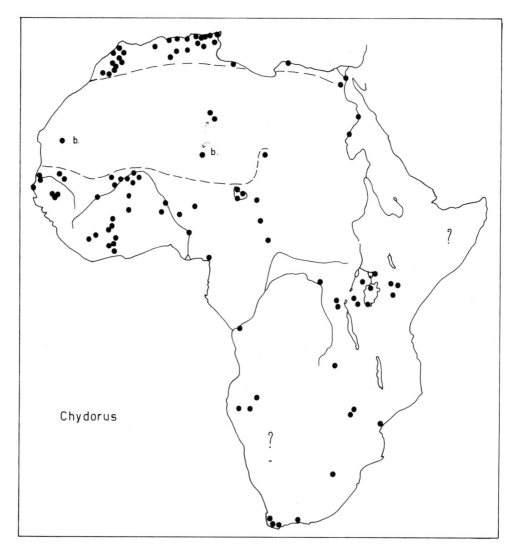

Fig. 5. Distribution of genus *Chydorus* in Africa (b = *C. barroisi*).

during every pluvial the three species discussed earlier (the "trio") formed a continuum between the great Sahel rivers and the Maghreb. The region bordering the Nile River, in which all three still occur today, resembles the environment during a pluvial epoch: *D. similis* and *D. magna* occur in its valley too, but only in marginal pools. Repeated southward expansions of the Sahara, on the other hand, also occurred during the Pleistocene (*see below*), permitting a similar movement of the *Daphnia* trio. The present populations along the major west African rivers might be called two-way relicts. They, finally, represent an interesting analog to fish distribution: the Nile, Logoni-Chari, Niger, and Senegal Rivers have almost identical fish faunas, due to communications between these respective river basins during the Pleistocene (*see* Beadle 1974).

The distribution of *Bosmina longirostris* O.F.M. (Fig. 4) duplicates that of the *Daphnia* trio. The ecological background

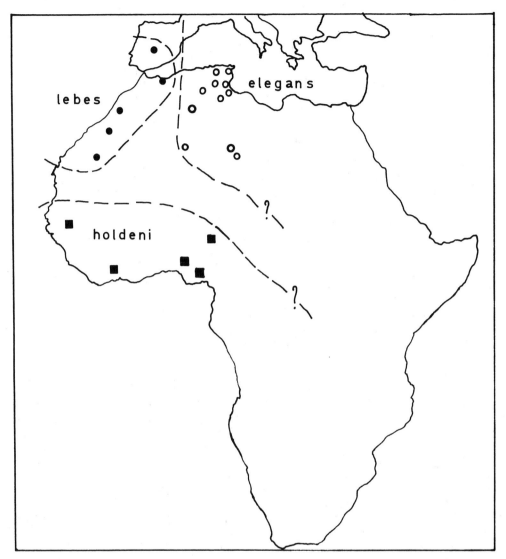

Fig. 6. Distribution of *Alona e. elegans*, *A. e. lebes*, and *Alona holdeni* as examples of restricted distributions of chydorids in north and west Africa.

must thus be analogous too, but needs to be investigated. In manmade Lake Volta, Kořínek (1970) found *Eubosmina crassicornis* Lilljeborg. This is a puzzling occurrence, possibly an introduction. The third African representative of the Bosminidae is *Bosminopsis deitersi* Richard. It is found in the equatorial and subequatorial zones of all continents. Such tropicopolitan occurrences are widespread among Cladocera, but it remains to be seen whether they are species or superspecies, i.e. groups of closely related but different species.

It is not yet clear how many species of *Chydorus* (Fig. 5) occur in Africa, but there are at least five and possibly more than ten. The majority of published records refers to *C. sphaericus* (O.F.M.). This species is widely considered to be the most ubiquitous cladoceran in exis-

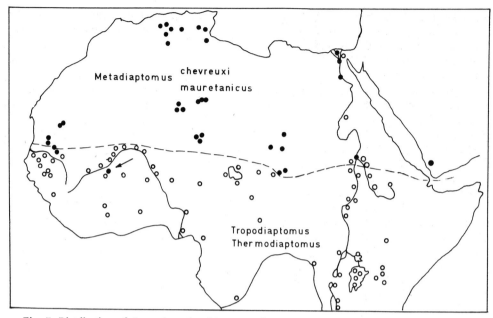

Fig. 7. Distribution of the major calanoid genera of northern half of Africa. Arrow indicates site of a relict population of *M. mauretanicus* in Dogonland, Mali.

tence. Remarkably, the distribution pattern of the genus in Africa demonstrates regional patterns: *Chydorus* is widely distributed on this continent, but is not found in the Sahara. Possibly, the same goes for Kalahari-Namib, but no data for this area are available. The Nile, again, behaves as a "pluvial" Sahara: *Chydorus* is found at several points all along this river. There are four exceptions to the Sahara rule along its southern border. Three pertain to *C. barroisi* Richard, a taxon with tropicopolitan distribution and questionable affinity with the genus *Chydorus* (D. G. Frey pers. comm.). The third record pertains to an animal close to *C. sphaericus*, collected in the Tibesti Mountains (leg. P. Quézel). Strikingly, the space normally taken by *Chydorus* is filled by various Aloninae, above all *Oxyurella tenuicaudis* and *Pleuroxus aduncus* (Jurine).

The absence of *Chydorus* from the Sahara is an excellent argument against the sacrosanct effectiveness of passive dispersal. The central mountain axis of the Sahara is an important pathway for migrating birds, and waterfowl frequently halt on gueltas (mountain lakelets) to feed and rest. Introductions of resting stages of zooplankton, including *Chydorus*, must thus regularly occur, yet remain unsuccessful. In my mind, this indicates that at least one fundamental element of the *Chydorus* niche is not realized in the Sahara, although I am unable to specify what this "factor" might be.

Some distribution patterns within *Alona* spp. (Fig. 6) again relate to late Pleistocene climatic changes. *Alona elegans* Kurz occurs in two distinct subspecies on the continent: *lebes* is found between central Spain and northern Mauretania, fringing the Atlantic Ocean; *elegans* s. str. is found in the central Maghreb and advances deep into the Sahara, possibly even further south. Spain and Atlantic north Africa acted as an important faunal refuge for Paleoarctic species at the height of glaciations. Many of these species were limited by the Atlas Mountains. The distribution of *lebes* nicely conforms to this.

The west African coastal area, on the other hand, which is possibly the only zone where the evergreen forest subsisted throughout the Pleistocene, can be expected to have evolved endemic species among Cladocera. Several chydorids were described from here by Brehm (1934), but in such a superficial way that most of them have (perhaps prematurely?) been synonymized with known species (Smirnov 1971). The first well-characterized endemic from this general area was made known by Green (1952): *Alona boldeni*. Later finds have shown that this species might range over the whole of west Africa. In addition I know of one, possibly two, more undescribed *Alona* from the Ivory Coast, confirming the same line of thinking. Secondary (during the Late Pleistocene) extension of these species into the Congo basin arid zone, after the Pleistocene had come to an end, may logically be expected.

Copepoda (Fig. 7)—Little can, at present, be said about the cyclopoids, since quite a few genera need revision. Many species are littoral, and thus considerable endemism should be expected, particularly in the genus *Eucyclops*, which has about ten endemic species in Lake Tanganyika alone. Among other things, there is now reason to suppose that *Eucyclops serrulatus* Fisher (= *agilis* Koch), a species considered to be cosmopolitan, occurs neither in Africa nor in America.

In north Africa, relict populations of various Cyclopoids occur disjunctly from their stock populations. The northern *Cyclops strenuus* Fisher has been found in the central Saharan mountains (Ahaggar and Tassili-n-Ajjer), whereas equatorial species have remained established in the Maghreb steppes (*Afrocyclops gibsoni*, *Thermocyclops oblongatus* Sars).

Interesting coherent patterns appear, above all, in the calanoids and this is also true on a world-wide scale. In fact, the ranges of most calanoids are so limited that this produces a new problem. Many species, having resting stages, could indeed be expected to disperse over wide areas, but this evidently does not happen (Fig. 7). In the northern half of the African continent, there is a sharp separation between the tropical genera, *Thermodiaptomus*

and *Tropodiaptomus*, and the Saharan representatives of the genera, *Metadiaptomus*. The borderline coincides with the limit of southward extent of *Daphnia* and *Bosmina*. Locally, and especially in small ponds, as in the inundation ponds along the Nile, both groups may co-occur. As soon, however, as the pelagic space of a lake becomes large, *Metadiaptomus* is replaced by *Thermo-* and *Tropodiaptomus*. This is exemplified in the Tagant hills of Mauretania, where *Metadiaptomus* lives in the smaller ponds, and *Thermo-* and *Tropodiaptomus* in the larger lakes. Among other things, a latitudinal shift of climatic belts implies an expansion or shrinking of lakes, thus a displacement of the "front" between the calanoids. However, relict populations out of range may survive for a variable period of time. Thus, a population of *Metadiaptomus* discovered on the plateau of Bandiagara, Mali, was used as evidence for a former southward displacement of the Sahara, and, on archaeological grounds, was assigned an age of not more than 3000 years (Dumont 1978). As among Cladocera, the Nile again behaves like a pluvial Sahara: *Tropodiaptomus* and *Thermodiaptomus* reach as far north as Cairo.

Rotifers—Of all planktonic groups, the taxonomy of rotifers is in the poorest state of development, and chronological questions in this group are consequently numerous. The number of undescribed or ill-described species probably equals the number of known species. Specifically, Africa has been only fragmentarily explored, often uncritically. Therefore, little can be added to Pejler's (1977) discussion: among Brachionidae (the only family with a relatively stable taxonomy), four taxa now appear to be restricted to the Ethiopian region, but no distribution patterns can yet be given.

Conclusion

The examples of clearly defined ranges in African plankton species and groups demonstrate that passive dispersal, however important, is successful only if suitable niches are ultimately encountered. Evidently, if in a given freshwater biotope all niches are occupied, colonization by passive dispersers will not normally take place. However, if niche boundaries change, as in the case of changing climates, rapid community changes can be expected, even on a continental scale. This prompt reaction of the zooplankton to changing conditions on the one hand and the great probability of leaving relict populations, reflecting former conditions, on the other hand, make this community a valuable tool in evaluating Pleistocene geological events. A strong interaction between plankton biogeography and paleolimnology will doubtlessly be beneficial to both disciplines.

It is to be expected that we shall get a better insight into distributional patterns as our taxonomies improve. To paraphrase Hutchinson (1967), nature has split species often much further than taxonomists. What we now discuss as distribution patterns of species might well, in many cases, refer to units of higher taxonomic rank.

References

BALL, I. R. 1975. Nature and formulation of biogeographical hypotheses. Syst. Zool. 24: 407–430.

BEADLE, L. C. 1974. The inland waters of tropical Africa. Longman.

BREHM, V. 1934. Cladocera; *In* Voyage de Ch. Allualud et P. A. Chappuis en Afrique Occidentale Francaise. Arch. Hydrobiol. 26: 50–90.

BROOKS, J. L. 1957. The systematics of North American *Daphnia*. Mem. Conn. Acad. Arts Sci. 13:180 p.

CARMOUZE, J. P., and OTHERS. 1972. Contribution à la connaissance du Bassin Tchadien. Sommaire. Grandes zones écologiques du Lac Tchad. Cah. O.R.S.T.O.M. Hydrobiol. 6:103–169.

DUMONT. H. J. 1978. Neolithic hyperarid period preceded the present climate of the Central Sahel. Nature 274:356-358.

FREY, D. G. 1961. Differentiation of *Alonella acutirostris* (Birge) and *Alonella rostrata* (Koch, 1841). Trans. Am. Microsc. Soc. 80: 129–140.

———. 1965. Differentiation of *Alona costata* Sars from two related species (Cladocera, Chydoridae). Crustaceana 8:159–173.

GAUTHIER, H. 1928. Recherches sur la faune des eaux continentales de l'Algérie et de la Tunisie. Thesis D.S., Paris. 420 p.

———. 1951. Contribution à l'étude de la faune des eaux douces au Sénégal (Entomostracés). Minerva, Alger. [Privately Publ.] 169 p.

GOULDEN, C. E. 1968. The systematics and evolution of the Moinidae. Trans. Am. Phil. Soc. N.S. 58:101 p.

GREEN, J. 1961. Zooplankton of the river Sokoto. The Crustacea. Proc. Zool. Soc. Lond. 138:415–453.

HARDING, J. P. 1961. Some South African Cladocera collected by Dr. A. D. Harrison. Ann. S. Afr. Mus. 46:35–46.

HUTCHINSON, G. E. 1964. The lacustrine microcosm reconsidered. Am. Sci. 52:334–341.

———. 1967. A treatise on limnology, v. 2. Wiley.

KOŘÍNEK, V. 1970. Comparative study of head pores in the genus *Bosmina* Baird (Crustacea, Cladocera). Vist. Co. spel. Zool., 35:275–296.

MOGHRABY, A. I. 1972. The zooplankton of the Blue Nile. Ph.D. thesis, Univ. Khartoum.

MOREAU, R. E. 1963. Vicissitudes of the African biomes in the late Pleistocene. Proc. Zool. Soc. Lond. 141:395–421.

PEJLER, B. 1977. On the global distribution of the family Brachionidae (Rotatoria). Arch. Hydrobiol. 53 (Suppl.):255–306.

RZÓSKA, J. 1976. The Nile, biology of an ancient river. Junk. 417 p.

SMIRNOV, N. N. 1971. Chydoridae fauny mira. *In* Fauna of the USSR. Crustacea, v. 1, No. 2. Nauka, Leningrad. 531 p.

STROSS, R. G. and J. C. HILL. 1968. Photoperiod control of winter diapause in the freshwater crustacean, *Daphnia*. Biol. Bull. 134:176–198.

TOLLINGER, M. A. 1912. Die geographische Verbreitung der Diaptomiden. Zool. Jahrb. Syst. 30:1–302.

WAGLER, E. 1936. Die Systematik und geografische Verbreitung des Genus Daphnia O. F. Müller mit besonderer Berücksichtigung der südafrikanischen Arten. Arch. Hydrobiol. 30:505–556.

62. Species Richness and Area in Galapagos and Andean Lakes: Equilibrium Phytoplankton Communities and a Paradox of the Zooplankton

Paul Colinvaux and Miriam Steinitz

Abstract

We have studied a series of equatorial lakes from the point of view of the theory of island biogeography. All the lakes occupy closed basins and are thus isolated as islands of the oceans are isolated. A preliminary examination of species richness in a suite of Galapagos lakes, that included both fresh and salt water lakes, produced a remarkable correlation between species richness of phytoplankton and lake area. Arguments were marshalled to show how this correlation might be spurious, but the proposition that a causal connection between species richness and area in equatorial closed lakes might exist in some suites of lakes was tested. A suite of freshwater lakes from the Galapagos and from mainland Ecuador was sampled for species richness. A strong correlation between species richness and area was found. It is argued that the numbers of phytoplankton species might be sensitive not to the area of open water per se but to the area of diverse habitat at the periphery. A model is put forward which explains number of phytoplankton species coexisting in the open waters of lakes as being determined by the rate of competitive exclusion in the open water, the rate of colonization from the shallows, and the rate at which the shallows themselves receive immigrants. The model is consistent both with our data and with the theory of island biogeography. In addition, we find an anomalously low number of zooplankton species, suggesting that an explanation for this is not at present available. It is "the paradox of the zooplankton" that there are few species compared to the phytoplankton which supports them.

We report an analysis of species richness in a suite of closed basin lakes in the Andes and the Galapagos. We undertook this research in the belief that closed basins offer special advantages to limnology. Closed basins may be particularly satisfactory for recording a history of the lake ecosystem in the sediments, which was our original reason for studying Galapagos lakes (Colinvaux 1969, 1972). But closed basins also offer advantages for biogeographic studies in that they may properly be likened to the islands discussed by MacArthur and Wilson (1967). We began this research by using the collections of Galapagos lake biota made during our reconnaissance of the Galapagos lakes (Colinvaux 1968) to construct species area curves for the phytoplankton of a variety of Galapagos lakes. We found a correlation which, although open to different interpretations, encouraged us to make further investigations of the possibilities of closed equatorial lakes as biogeographic islands. To this end we chose a suite of lakes straddling the equator in the Andes of Ecuador and sampled their biota. At

Contribution No. 18 of the Charles Darwin Foundation for the Galapagos Islands.

the same time, we made more collections on the Galapagos. The data from these collections reveal strong correlations between species richness and area for phytoplankton, which we discuss in terms of the equilibrium theory of island biogeography and the "paradox of the plankton" (Hutchinson 1961). At the same time, we find that the number of species of grazing plankters in these lakes is so low that we suggest that this constitutes a "paradox of the zooplankton." This work is part of a continuing program of research in the Galapagos Islands and the Andes supported by NSF grants DEB77-19267, GB-4654, GB-7154, and GB-29065X. A crucial part of the fieldwork was supported by a grant from The Ohio State University Development Fund and by a grant from Sigma Xi. We thank Mr. and Mrs. H. Steinitz for the use of their home as a base for operations in Ecuador. R. Kannan, H. Calles, F. Arcos, C. and M. Leck, and J. Gordillo helped with many problems in the field operations. D. Culver and J. Downhower provided a continual flow of useful advice. The hospitality and encouragement of R. Patrick and C. Reimer of the Philadelphia Academy of Sciences gave us confidence in diatom taxonomy.

The preliminary observation on Galapagos lakes

This investigation began with the plotting of a species-area relationship of net phytoplankton collected during a limnological reconnaissance of the Galapagos Islands (Fig. 1). The lakes named in Fig. 1 are those described in Colinvaux (1968) and represent nearly the entire suite of Galapagos lakes other than small (ca. 10 m) ponds. Only two of the lakes are of freshwater (El Junco and Tortoise Pond). One lake has juvenile volcanic water (Fernandina). The remainder are all coastal lakes containing various concentrations of seawater. It does not seem reasonable that lakes so different in water chemistry should have their numbers of phytoplankton species solely determined by some correlate of lake morphometry and size as would be

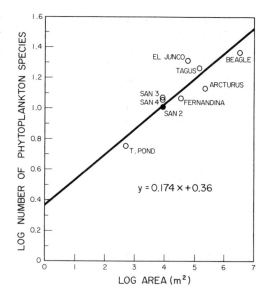

Fig. 1. Regression line for log area of a suite of Galapagos lakes with log number of phytoplankton species. Data of this figure are from collections made in Galapagos limnological reconnaissance (Colinvaux 1968) and include lakes differing very widely in salinity. Correlation coefficient (r) = 0.88.

represented by area, and yet we find a correlation coefficient of 0.888 (Fig. 1) suggesting just that.

Correlations, even "good" ones, can be spurious and it is not hard to discover ways in which this data set, or our treatment of it in Fig. 1, could have led us astray. Removing the largest lake (Beagle, a saline lake) and the smallest (Tortoise Pond, freshwater) would make the correlation disappear. The one freshwater lake other than a pond (El Junco) is the most "scattered" point in the data set. The species number in the collections from the three shallowest lakes (Tortoise Pond, El Junco, Santiago Lagoons) is sensitive to the inclusion of benthic diatoms in the net plankton sample so that the species number in these lakes might change with different collecting efforts more than might the species number in the deeper lakes. Three data points are made up of a suite of lagoons on Santiago that are so close together (separated by narrow spits a few tens of meters across) that they might reasonably be collapsed into a single sample. Our taxonomy of phytoplankton in these widely different fresh and salt water lakes was imprecise. Arguments like these can readily be marshalled to show that our correlation might be meaningless.

And yet arguments can also be advanced to suggest that some underlying correlations of species richness with lake area might exist, even if rather different slopes should define the correlations of fresh and salt water lakes. It is to be remembered that these lakes all occupy closed basins and are thus possible islands in the sense of MacArthur and Wilson (1967), so that a relationship between species richness and surface area is plausible. Then, some models to explain the paradox of the plankton (Hutchinson 1961, 1967) rely on the use of the bottom mud to provide some parameters of niche space even for open water species, and available mud is a likely correlate of lake area.

Other lakes have been examined with the equilibrium theory of island biogeography in mind (Hutchinson 1967) with inconclusive results, but this may well be because of lake data used in the analysis. The tropical lakes studied by Hutchinson (Ruttner 1952) were all very large by the standards of our lakes (in the tens of kilometers range) and they were all occupying open basins fed by the streams of very considerable catchments. These large lakes, therefore, were more comparable to land continents than to the islands of MacArthur and Wilson.

There is a literature applying the MacArthur and Wilson theory to very small (glass jars or even pitcher plants) "lakes," often showing very striking correlations of the MacArthur and Wilson type (Maguire 1963, 1971a, b; Cairns et al. 1969; Dickson and Cairns 1972; Patrick 1967). Keddy (1976) discussed the number of *Lemna* species as a function of lake area. Otherwise we know of no attempt to examine the species richness of natural lakes in closed basins designed to see if the number of species in any part of the lake biota is determined by those obvious correlates of the limnic habitat, size, and morphometry of the lake. Accordingly we have followed the promptings of Fig. 1 and sampled a suite of equatorial lakes occupying closed basins to test the proposition that a relationship between species richness and area might exist. We chose only freshwater lakes, avoiding both the saline Galapagos lakes and volcanic lakes containing juvenile water of unusual chemistry. Because there are very few suitable freshwater lakes on the Galapagos Islands, we have gone to mainland Ecuador for 6 of the 11 lakes comprising our data set.

Methods: Field collections

The Galapagos lakes were sampled once: in July 1977. The Andean lakes were sampled three times (except as noted in lake descriptions): in December 1975, July or August 1977, and March 1978. The Andean samples span what seasonality there is in the local climate, the wet and the dry (Fig. 2). Samples for plankton analyses were taken in three ways from each lake: with a Schindler trap in open water near the lake center at a series of depths; preserved as whole water samples taken with a 1-liter Van Dorn sampler; and by net tows (No. 20 net, silk aperture 76 μm), from open water in the middle along a lake radius to the shore. The net was not allowed to enter emergent vegetation but otherwise included plankton

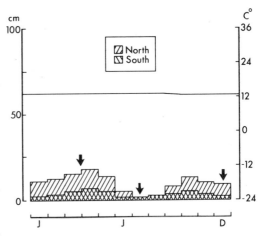

Fig. 2. Average monthly temperature and precipitation for stations located near northern and southern Andean lakes included in this study. Arrows point to months in which lakes were sampled (adapted from Ferdon 1950).

Table 1. Physical parameters and number of species in Galapagos and Andean lakes. Lakes arranged in order of area and thus appear in same order in Fig. 3.

Lake	Depth (m)	Area (m^2)	Altitude (m)	Phytoplankton				Zooplankton			
				Cyano- phyta	Chloro- phyta	Bacil- laro- phyta	Eugle- nophyta & Chry- sophyta	Clado- cera	Cope- poda	Roti- fera	Ostra- coda
Cuicocha	132	$6,570 \times 10^3$	3,068	11	24	39	5	3	2	3	0
Chicapan	48	$6,202 \times 10^3$	2,661	12	23	36	3	3	1	4	1
Yaguar- cocha	9	$2,300 \times 10^3$	2,210	7	20	53	4	2	1	3	1
Yambo	23	238×10^3	2,600	10	13	36	2	1	1	2	1
El Junco	6	57×10^3	700	5	18	51	2	2	2	3	0
Cunro	5	12.2×10^4	2,800	10	22	41	2	2	1	6	0
Limpio- pungo	0.65	10×10^4	3,888	4	10	33	1	1	1	1	1
Tortoise Pond 2	1	284	550	3	5	21	3	2	1	0	0
Frigate Bird Pond	0.5	65	50	5	6	8	2	2	1	0	1
Tortoise Pond 1	0.16	41	550	3	2	19	0	0	1	1	0
Posa Floreana	1.5	8	200	0	2	11	1	0	1	1	1

of the shallows in the catch. All plankton collections were preserved with Lugol's medium. A complete supporting limnological survey was made of each lake, including chemical analyses of the open water done on fresh samples by standard methods (Am. Public Health Assoc. 1971). These data constitute the first survey available for most of the suite of lakes and will be published elsewhere.

Methods: Laboratory and taxonomic procedure

Collections were examined with a Leitz Diavert inverted microscope. All identifications were made under oil-immersion using a 90X APO N. A. 1.32 Leitz objective with 10X eyepieces and Koehler illumination. The data necessary for the analysis described here do not include relative abundance, but abundance data are being collected using standard settling chambers and aliquots of the Schindler trap and Van Dorn bottle samples. Since more species were present in the net plankton samples than the others (presumably a function of the much larger volume of water sampled), the species lists were prepared from the radial net tows.

The species lists of all algae except diatoms were prepared as follows. Suspensions were allowed to settle in standard chambers until the density of the deposit was considered optimum for efficient searching and counting, essentially a density that provided algae to every oil-immersion field of view without serious overlap of specimens. Zooplankters had been removed from the suspensions with a looped needle operated under a dissecting microscope. The entire bottom surface of each settling chamber was searched, first with a 10X objective to identify the larger algae, and then with the oil-immersion objective, a procedure that occupied a minimum of three laboratory days per preparation. The total number of species found in this search constitutes the species richness of phytoplankton less diatoms as listed in Table 1.

Identification of algae other than diatoms was achieved using standard keys, particularly the following: Prescott 1951; Smith 1950; Ralfs 1848; Cooke 1887. It was possible to identify about 90% of the algae using the above keys but the results were checked against specialist monographs and regional reviews when these were available, a procedure that was to give us assurance of both the applicability of these keys to the equatorial algae and the ubiquity of most phytoplankton species (Drouet 1937, 1938; Foerster 1969, 1974; Hegewald et al. 1978; Bourrelly 1966, 1968, 1970, 1972; Thienemann 1938; West and West 1904–1923). As one example, we keyed out *Scenedesmus intermedius* Chod. in Prescott and compared our specimen with that described in the recent paper on Peruvian phytoplankton by Hegewald et al. (1978), finding that Prescott could provide us with all the taxonomic data except the strain name "Hegewald." We believe our taxonomy to be accurate, certainly with no confusions that would affect our estimates of the numbers of species present in the samples.

We treated the diatoms separately. The procedure was to take rich suspensions from the net plankton samples, boil them for about 20 min in strong nitric acid, wash the suspension with distilled water to adjust the pH to neutral, and make preparations from the residue. Subsamples of extract were added to a microscope slide until the resulting preparation had a proper density for efficient work and the sample was then mounted in Hyrax. The entire area of the cover slip was then searched with the oil-immersion objective to obtain a list of all species present. The following works were used as an aid to diatom taxonomy: Cleve-Euler 1968; Patrick and Reimer 1966, 1975; Van Heurck 1896; Cleve 1893–1896; Zanon 1927–1928. In addition, one of us (M. S.) took the entire collection of diatom preparations to the department of Limnology of the Academy of Natural Sciences at Philadelphia and spent a month checking the taxonomy against the reference collections there and with the counsel of R. Patrick and C. Reimer. Patrick (1970) had already worked on the diatom collections of our earlier Galapagos reconnaissance (Colinvaux 1968). No serious problems of diatom taxonomy arose and we believe our species lists to be both complete and accurate.

We have also investigated the species richness of the zooplankton in our collections. For this purpose we used both the net tows and the Schindler trap samples, thus compiling our species lists from samples taken at a representative series of depths. Since many of our lakes tended to be polymictic and we sampled widely, both in vertical and horizontal planes, it is unlikely that we missed any concentrations of zooplankters. We can assign generic names to all specimens with confidence and we have experienced no difficulty in separating species as clear taxonomic entities in each individual lake. We have, therefore, confidence in our separate estimates of species richness as numbers. However, on the advice of C. Goulden and D. Frey, we are refraining from giving species attributes to most of the zooplankters. This must await revision of the taxonomy of the tropical zooplankton now being undertaken by these and other workers. Zooplankton species richness is included in Table 1.

Brief description of the lakes

There are 11 lakes in our sample: 5 from the Galapagos Islands and 6 from the Andes of Ecuador. All the lakes are within 1°7′ of the Equator. Except for El Junco, the Galapagos lakes can more properly be called ponds and comprise the smaller end of our size range. The Andean lakes are all from the inter-Andean plateau or above, between 2,000 and 3,900 m. All the lakes occupy closed basins. Little or nothing has previously been published about most of these lakes, except for brief references in geography and geology books on Ecuador (Wolf 1934; Sauer 1971; Terán 1966, 1975). In Table 1 we give the area, a depth measurement taken in the deeper part, and the elevation of each. What follows is a very brief description of each lake, arranged in order of area beginning with the smallest.

Posa—Isla Floreana, Galapagos. An artificially excavated seepage in the western interior of the island. The pond is on the flank of the island volcano, in the wetter part though not in the permanent stratus cloud of the summit. Cattle are kept out of the pond by barbed wire but the land about has been farmed for some decades.

Tortoise Pond 1.—Isla Santa Cruz, Galapagos. Shallow pond in the wet forest of the south side of the island, in land set aside as a reserve for the tortoises which use the pond as a wallow. Only 16 cm deep when sampled and almost completely dry in June 1966.

Frigate Bird Pond.—Isla Santa Cruz, Galapagos. A pond of freshwater near the northern coast of the island close to the Canal de Itabaca. This is in the dry zone and yet the water chemistry is comparable to that of the tortoise ponds of the well-watered zone high on the other side of the island. It seems possible that the pond is spring-fed from the island water table and may thus be permanent. The depth over most of its extent was 60 cm when it was sampled. This was in an exceptionally dry period, encouraging the suggestion that the pond is permanent. In contrast to the other shallow lakes in our series, Frigate Bird Pond had no aquatic macrophytes and no large or filamentous algae, and hence few places for epiphytes to attach.

Tortoise Pond 2.—Isla Santa Cruz, Galapagos. 1.5 km south of Tortoise Pond 1. This is the pond included in the original analysis of Fig. 1. Maximum depth in both 1966 and 1977 was about 1 m. If this pond ever dries completely, it would be only after an exceptional drought. Used by tortoises as a wallow and well-known to National Park Service officials as a center for tortoises. It supports an almost closed stand of *Azolla microphylla*.

Limpiopungo.—Shallow lake in a basin formed by glacial morraines at the base of the Cotopaxi volcano, Cotopaxi National Park. At 3,888 m this is the highest lake in our suite and was sampled only in March 1978.

Cunro.—Imbabura Province, near hamlet of Cunro. Maare lake is very similar in appearance to El Junco (Colinvaux 1968). In a well-watered, fertile countryside apparently farmed and settled for centuries. Sediments were cored to the lake basin and found to be 5.5 m thick. It was sampled twice: July 1977 and March 1978.

El Junco.—Isla San Cristobal, Galapagos. Maare lake in the rim of the main volcanic peak of the island. Within the clouds for most of the year. Radiocarbon dates on sediment cores show that the lake has held water continuously for 10,000 years (Colinvaux 1968, 1969, 1972; Colinvaux and Schofield 1976a, b; Schofield and Colinvaux 1969).

Yambo.—Cotopaxi Province, near Latacunga. Apparently occupying a graben in the floor of a rift. Long, narrow, and steepsided. In a semidesert region of scrub trees. Sediments were cored to the lake basin under 20 m water and found to be 3 m thick (mentioned in Whitton 1968).

Yaguarcocha.—Imbabura Province, just north of the city of Ibarra. The lake occupies the lowest part of a broad, fertile valley between mountains. It is now surrounded by an automobile race track.

Chicapan (San Pablo).—Imbabura Province, near Otavalo. Occupies a broad depression or valley floor in mountainous country. Inlet streams drain rich farmland and there is sewage and phosphatic detergent entering the lake from settlements.

Cuicocha.—Imbabura Province close to the town of Cotacachi. A deep maare lake with two islands formed by tufa cones. Inlet streams are fed by melting snows on Cotacachi volcano and the catchment has no significant settlement or farming (figured in Sauer 1971).

Results: The species–area correlations

The species richness for each lake is given in Table 1 as number of species in each major algal or zooplanktonic taxon.

Table 2. Correlation coefficients (r) and slopes (z) of regression lines of log of area of Galapagos and Andean lakes with log of number of species of Chlorophyta, Cynaophyta, Bacillarophyta, and total phytoplankton. All r values are significant to 99% level. Highest correlation is obtained between area and number of diatoms species in shallow lakes (those <10m deep), when Galapagos Frigate Bird Pond, which has no macrophytes or filamentous algae to provide attachment for epiphytes, is excluded.

Variable	r	z
Total phytoplankton, 11 lakes	0.92	0.1179
Chlorophyta, 11 lakes	0.90	0.27
Cyanophyta, 11 lakes	0.82	0.12
Bacillarophyta, 11 lakes	0.83	0.104
Bacillarophyta, 8 shallow lakes	0.896	0.147
Bacillarophyta, 7 shallow lakes Frigate Bird Pond excluded	0.954	0.126

The complete species lists for the algae are available and will be published elsewhere. In Table 2 we give the correlation coefficients (r) for log area and log number of species of algae in the collections from each lake and the slopes of the regression lines (z). In every instance the correlation coefficients are significant at the 99% confidence level. Figure 3 plots the correlation of total phytoplankton in all lakes and Fig. 4 the diatoms alone for the eight lakes we have designated as "shallow"(<10 m). The number of species of zooplankton, both herbivores and carnivores, was always small and correlations with area are neither possible nor apparent.

In Table 2 and in Fig. 4, we have subdivided the data set in two ways designed to reveal any special influence on the correlations of diatoms and shallowness of lake basins. We do this because a large proportion of our diatom taxa appears to be epiphytic or semibenthic, being associated particularly with rooted aquatics, filamentous blue-green algae, or *Azolla*. This

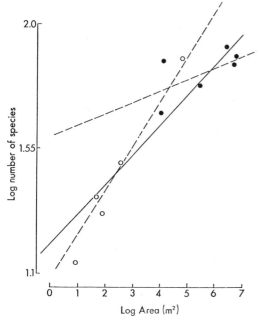

Fig. 3. Regression lines for log area of Galapagos and Andean freshwater lakes and log total number of phytoplankton species. Solid line is for all 11 lakes studied. Dashed lines are for Galapagos freshwater lakes (☆) and Andean lakes (●) separated. Slope of regression line for Galapagos lakes alone (z = 0.18) is higher than that for Andean lakes alone (z = 0.043), suggesting an isolation effect for Galapagos lakes. Lakes can be identified from their relative areas (listed in order in Table 1).

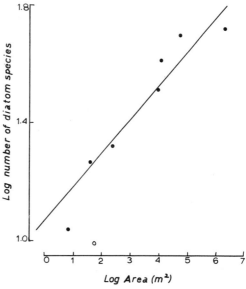

Fig. 4. Regression line for area of shallow Galapagos and Andean lakes (< 10 m deep) and number of diatom species. Correlation coefficient r = 0.95. ○–Galapagos Frigate Bird Pond that has no macrophytes of filamentous algae and not included in calculated regression.

is not to say that these diatom species are not a usual constituent of the phytoplankton of open waters in these lakes for we believe that they are always present. These forms get into the open water when they are stirred by waves or are torn adrift. But we expected that habitats for these diatoms might be more abundant in shallow lakes. The contribution of these numerous epiphytic or semibenthic species then might well be responsible for a considerable part of the very striking general correlation between algal species richness and area. We defined "shallow" as meaning all lakes with a maximum depth of 10 m. Eight lakes are in this class. One of the eight, however, is the Frigate Bird Pond which was bare of rooted aquatics, filamentous algae, or other substrates for epiphytic diatoms, and had accordingly a low number of diatom species. The species–area plot for the Frigate Bird Pond is drawn on Fig. 4 as an open circle and is not included in the calculation of the regression line. This procedure should place the highest possible emphasis on the contribution of diatoms of the shallows to the species–area regressions. The results (Table 2) show that there is some substance to our suspicion that diatom species numbers should reflect area of available bottom, in that the highest correlation coefficients are between diatoms species and area of shallow lakes. Removing diatoms from the species sum of either set of lakes (total or shallow alone), however, still leaves the regressions of remaining algae with area significant at the 99% confidence level.

It will be noticed that these regressions are not open to the same orders of doubt that could be offered to the original correlation of species and area in the very variable array of Galapagos lakes in Fig. 1. None of this latter set of lakes has chemistry so aberrant that it was occupied by a unique assemblage of algae, and in fact there is very considerable overlap in the species lists. We took care that a sufficient range of lake sizes would be represented so that the regressions would not be sensitive to selective removal or additions of lakes to the list. And we believe the taxonomic effort to have been both thorough and consistent.

We are, of course, aware that a species list of limnoplankters is always dependent on the thoroughness of the search, because there will always be some species too rare to be discovered by any degree of effort. We notice, however, that our species lists are already large by the standards of tropical limnology and that our totals (Table 1) are comparable to published totals for much larger tropical lakes (Lewis 1974, 1978; Ruttner 1952). Indeed we remark that our diatom lists tend to be much larger than for some of these lakes, a fact that we attribute to the techniques of our search. Most diatoms cannot be found or identified unless concentrated and cleaned in boiling nitric acid, so that they are missed in examination of whole plankton samples. And yet it is still true that consistency is probably even more important for investigations of comparative species richness than is intensity of effort, at least beyond some threshold. We believe that the methods of search that we have described above assure that our taxonomic effort has been both thorough and consistent from lake to lake.

The objection may be made that our use of net samples means that we miss some of the nannoplankters, perhaps particularly flagellates, from our lists. We provided a direct guard against this in three lakes (El Junco, Yambo, Cunro) by examining the whole water samples collected with the Van Dorn sampler, finding no species that were not already included in our net samples. It is true that nannoplankters escape from nets, even when these become clogged at the end of a tow, so that net counts do not yield relative abundance data. But large numbers of flagellates and other nannoplankters are present in our net samples all the same. We think it at least as likely that we collect representatives of all flagellates and other nannoplankters living in a lake by sweeping a very large volume with a net that allows some individuals to escape than by sweeping clean a single liter. We also note the objection that our net tows were of surface water only. For the most part our lakes were polymictic with little sign of thermal stratification, suggesting that algae ought to be stirred through the surface parts of the water column. When examining the Schindler trap samples of deeper water for zooplankton we kept watch for algae not in our net samples but discovered none.

In studies of species richness in temperate lakes it is necessary to sample frequently throughout the year so that changing species populations of the annual phytoplankton succession are included. This is because the relative abundance changes with the seasons in temperate lakes to such an extent that the common species of one season may be either too rare to be discovered in another season, or may actually have left the open water for the mud. This is much less of a problem in tropical lakes where there is little seasonality, and Lewis (1978) found that both species richness and relative abundance did not change very significantly from month to month in Lake Lanao. We guarded against the possibility of missing some ephemeral species in our lists by sampling as many of our lakes as was practicable at different times of the year. Most of the Andean lakes were sampled at extremes of the two climatic "seasons" experienced locally, the wet and the dry (Fig. 2). These seasons are not pronounced by the standards of temperate seasons. We counted the algae in the collections for both seasons from lakes Cuicocha, Yaguarcocha, Chicapan, Yambo, and Cunro, finding that there was little increase in total species richness when the lists from the different seasons were added. Comparisons of the species lists for the two Galapagos lakes sampled in 1966 and 1977 (El Junco and Tortoise Pond 2) also revealed no differences that could not be accounted for by the different methods of collection and preservation used. We conclude, therefore, that the increase in species richness which would result from a program of round-the-year sampling in these lakes is probably small.

Zooplankters were immensely abundant in all the lakes, so much so that a major task in making the algal counts was the prior removal of zooplankton with a hooked needle. But, although abundant, the species lists were always small (Table 1). These lists of zooplankton species were compiled from the net samples, from Schindler trap samples at representative depths, and from the different seasonal collections. As with the algae, it seems unlikely that more frequent sampling would increase the lists significantly.

The results of this analysis of species richness in this suite of equatorial lakes occupying closed basins may be summarized as follows.

Species richness of phytoplankton is strongly correlated with lake area.

There appears to be a particularly strong correlation between the area of shallow lakes and the species richness of diatoms, many of which are epiphytic or benthic.

The species–area relationship holds good for small lakes which have dried up within the last few decades and for permanent lakes known to have been in existence for 10,000 years.

The species richness of zooplankton in all the lakes is an order of magnitude less than the species richness of phytoplankton.

What can area mean to phytoplankton?

Even the smallest of our lakes contains an immense volume of water compared to the needs of phytoplankters existing at populations which may reach densities of $10^6 \, ml^{-1}$. One might expect, therefore, that any lake with an area of more than a few hundred square meters would effectively be infinitely large for the purpose of population dynamics of phytoplankters and that lake area would be irrelevant to the determination of diversity or species richness. In what ways, then, could area influence population histories? We suggest that area is important in that it reflects the mud surface available at the periphery and that this mud surface influences population histories, and the outcome of competition, because it provides a refugium from which the open water can be colonized continuously.

In this suggestion we follow the reasoning of Hutchinson (1961) in his original explanation for the paradox of the plankton when he postulated that coexistence was predicated on being provided with periodic opportunities to escape the open water competitions by descending to the mud. Hutchinson was arguing for conditions in temperate lakes with seasonal histories. We suggest that the mechanisms might just as well apply to all lakes: that the provision of fresh immigrants from shallows or mud might prolong competition in the open water indefinitely so that exclusion would be prevented.

Our results that the highest correlations are found between the areas of shallow lakes and the species richness of diatoms, although far from conclusive, are in keeping with this suggestion. Many diatoms are known to have needs for the substrates provided in shallows, either on mud itself or on the larger plants associated with shallows and mud. If the area of mud can have an effect on the number of species, it should be particularly noticeable in this taxon. The data show that it is.

If we shift the argument, thus, from the area of the water to the area of the mud, it might still seem that we are faced with a problem of scale. Like the volume of water, the area of mud in a lake might seem nearly infinite in terms of the needs of species populations of microscopic algae. But the problem of scale might not be so insuperable when cast in terms of mud because there can be great local variability in the qualities of the mud surface in providing microhabitats. There is some variability from lake to lake in our suite (Table 1), in spite of the significant species–area regression. We suggest that this variability reflects factors like shoreline variability, depth gradients, rooted aquatic plants, and local mud physiography.

We suggest, therefore, that the number of species of phytoplankters may reflect the number of ways in which the mud surface can be shared between niches. We would expect this variety of opportunities provided by mud to increase with lake area. It would be interesting to see an analysis of species richness in a larger suite of closed lakes where shoreline and mud physiography were individually scored.

On lakes as islands

A strong correlation of species number with area is a prediction of the equilibrium theory of island biogeography (MacArthur and Wilson 1967). The number of species on each island is predicted to be a function of the rate of immigration and the rate of local extinction. Area enters into both sides of the equilibrium. It influences immigration by determining the size of the target set to dispersing migrants, and hence the rate of immigration. But area probably has a larger effect on local extinction, since large islands should permit larger populations, hence lowering extinction rates until population sizes are reduced by competition between growing numbers of species. Tests of the theory by constructing species–area curves for islands of varying remoteness have mostly used bird data, but there is a growing literature applying the model to other organisms and in various special circumstances (Simberloff 1974; Brown 1971; Barbour and Brown 1974; Schoener 1969; Diamond 1973). A historical test of the model against the pollen record for one Galapagos Island suggests that it is less satisfactory for the species richness of vegetation. Perhaps this is because invasion of established vegetation with a complete cover is very difficult for opportunist (*r*-selected) species which are adapted to island dispersal (Colinvaux and Schofield 1976a, b). This is a problem that migrant, microscopic algae would not encounter.

We argue above that our lakes are fully the equivalent of oceanic islands from the point of view of the model: they are closed basins of water separated from other similar habitats by relatively large expanses of inhospitable land. The demonstration that species richness of algae in the lakes is strongly correlated with area is in general keeping with the predictions of the model. If this is to be taken as evidence that species number has been determined by the mechanisms of MacArthur and Wilson then it would follow that interspecific competition determines species number.

One of the striking bits of evidence put

forward by MacArthur and Wilson in support of the theory was that extremely remote islands had fewer bird species per unit area than islands close to continents. This showed up as a steeper slope to the regression line relating species number to area in sets of remote islands, and MacArthur and Wilson suggested that slopes of these regressions might prove an indicator of the remoteness of a site from the source of propagules. This result is seen as a consequence of the fact that the rate of immigration is expected to fall with distance to be covered by the migrants. In Fig. 3. we have identified the plots of Galapagos and Andean lakes separately. The separate samples of lakes become small, although the separate regressions are still significant. The slopes derived from these small data sets differ as would be expected by MacArthur and Wilson.

Phytoplankters may be expected to be very efficient dispersers. Evidence for this is found in the fact that the phytoplankton flora of remote places is similar, there appearing to be a global species-pool for phytoplankton. The distinctly different slopes of the regressions for Galapagos and Andes lakes illustrated in Fig. 3 must be evaluated with this fact in mind because "failure to reach the island" cannot be used as an argument to explain the absence of taxa.

We suggested in the last section that lake area is important to population dynamics and competition between phytoplankters because it reflects the area of mud at the periphery. The outcome of competition in the open water is seen as depending on the rate of colonization of the open water from the mud at the periphery. In a like way, it may be that competition for niche space at the periphery is dependent on colonization of the mud. This can come about in two ways: by returning colonists from the open water or by the arrival of immigrants through long-distance transport (Fig. 5). We suggest, therefore, that the steeper slope of the regression for Galapagos lakes reflects the relatively slow rate of invasion by migrants from outside, even though arrival is essentially continuous. Mainland lakes have their competitions pushed in the direction of more coexistence because of the higher rates of immigration. A conclusion so interesting warrants testing with a series of lakes more numerous than in our sample.

It is worth emphasizing that our results are for equatorial lakes where there is little seasonal change of a kind that might influence the outcome of phytoplankton competition. The processes described in Fig. 5 should be in progress in temperate or seasonal lakes as well, but it could be argued that successional phenomena would obstruct the outcome of competitions so that the rates of local extinctions would be slower. This is an echo of the argument which concludes that temperate lakes should have a greater species richness than tropical lakes if the counts for all seasons of the year are summed. We suggest that a repeat of our work with sets of closed basins on islands and continents in temperate latitudes might serve to test the extent to which competitive exclusions run their course to extinction in temperate lakes.

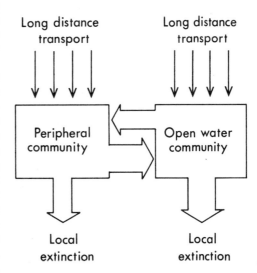

Fig. 5. Scheme describing flux of species richness through a closed-basin (island) lake. Different slopes of regressions in Fig. 3 suggest that competitive interactions among phytoplankton might be sensitive to rate of arrival of fresh propagules by long-distance transport.

The saline Galapagos lakes revisited

The discovery of the species–area relationship in the Andean and Galapagos freshwater lakes prompts a re-examination of the original correlation shown in Fig. 1. It now becomes more plausible that there could be a causal connection underlying the correlation between species richness and area in even this chemically diverse group of lakes. All the lakes meet the first requirement of the analysis that they be closed basins. All except one of the lakes (Fernandina) meet the unspoken requirement that the lakes have a chemistry which is tolerable to a very large species-pool of algae. The Galapagos saline lakes met this requirement in that their basic ionic composition is that of seawater (Colinvaux 1968), making it possible that they might be accessible to the algae of parts of the ocean system. We note particularly that lagoons with fluctuating salinities are an abundant and presumably ancient equatorial habitat which might have promoted species populations or ecotypes of algae suited to varying salinities like those of the Galapagos lakes.

"Saline" or "high conductance" lakes of the Fernandina class are another matter in that their ionic composition is rarely encountered, with the result that very few organisms seem adapted to live in them. Their water, high in magnesium sulfate, is partially derived from juvenile volcanic water and they are, like Fernandina itself, typically to be found occupying the floors of calderas. We sampled one such caldera lake in the Andes (Quilotoa) according to our usual sampling schedule and found its total biota to consist of seven species of algae. The diatoms of this collection are of species found only in such aberrant water (Patrick and Reimer 1966, 1975). Fernandina, therefore, should be excluded from the analysis of Fig. 1.

Apart from lakes of peculiar ionic composition, however, it seems not unreasonable that the mechanisms of competition at the periphery, which we suggest are operating in the freshwater lakes, are operating in the saline lakes also. It may be of interest to resample the Galapagos lakes, together with similar closed basins elsewhere, to see to what extent the correlation of species richness with area holds good when applied to mixed suites of lakes with admixtures of seawater as well as freshwater.

The paradox of zooplankton

In all our lakes the number of zooplankton species is an order of magnitude less than the number of species of algae which support them. The typical pattern is one copepod, usually a cyclopoid, a few cladocerans, a few rotifers, and sometimes a single ostracod (Table 1). In some lakes the single cyclopoid species is joined by a single harpacticoid species, but this is found only in the shallows. In El Junco, the third copepod is a calanoid, and, although we have not completed our counts of relative abundance, it appears to be about two orders of magnitude more common than the cyclopoid. The commonest copepods are immensely abundant in all the lakes. The commonest cladocerans are usually much less abundant than the commonest copepod. It is probably true to say that a search in the shallows, particularly on the rooted aquatic plants, would reveal a number of species of chydorids not in our list, but our data show, nevertheless, that the species richness of the zooplankton of the open water is very low.

Low species richness may well be a general condition of tropical lakes (Lewis 1978; Ruttner 1952; Deevey at this conf.), and even in temperate lakes the species richness of zooplankton is very much lower than that of phytoplankton. We suggest that a paradox of some interest is involved in the low species richness of zooplankters.

The idea of a cropping principle is now widely accepted in ecology. Predation and herbivory are considered to be diversity-inducing mechanisms, whether in offshore marine communities (Paine 1969), pastures in Wales (Harper 1969), or in the insect-plant systems of tropical rain forests (Whittaker 1969). It is necessary to ask why the great diversity of the algae in a lake does not support a corresponding array of herbivores to graze them. One herbivorous copepod and a few crustacean and rotifer species, intuitively, do not seem a proper response to an array of 50 or more choices of plant food.

For herbivory to induce diversity, or for many herbivores to collect to crop a diverse array of plants, it is necessary that herbivores be able to select different kinds of plants from among those available. Ever since "Santa Rosalia" (Hutchinson 1959) it has been a matter of question whether filter feeders like most zooplankters could in fact distinguish between phytoplankters. But Porter (1977) and Strickler (at this conf.) now provide strong evidence that considerable discrimination is, in fact, both possible and usual. It may not be unrealistic to compare the discriminatory powers of copepods for their plant food with those of ungulates cropping in the great array of species in a pasture. A superficial view of a cow likens the animal to a mowing machine that ingests everything soft and green, but real cows and their relatives discriminate closely, with the consequence that many species of ungulates get their living from the mixture of plants in prairie or savannah. The species richness of both plants and herbivores in grasslands is high. But in lakes the plants are species-rich, the herbivores species-poor.

It is tempting to suspect a nebulous connection between low species richness of zooplankton and the parthenogenetic habit of many of them, postulating that the mechanism of speciation has been slowed in these systems. We would then argue that herbivore variety might even be provided by an array of clones so that species become superfluous. This argument, however, may be met with the objection that rather similar processes ought to supress algal species richness which they do not do. Indeed, algae ought to serve as a model for species richness in zooplankters. Even though algae may have specialized needs for nutrients or choose between wavelengths of light, they are largely indiscriminant feeders. They all "filter" the sun. And, like the zooplankters, they persist for long intervals as asexual clones. But, far from serving as a model for zooplankter variety, the algae have evolved an order of magnitude more coexisting species than the herbivores that feed on them. Copepods, of course, are not parthenogenetic and yet we often find but one species. One possibility that deserves exploring is that the relative numbers of plant and animal limnoplankters results from their different antiquities. The species of smaller algae are thought to be very ancient, certainly as old as the early Tertiary or late Mesozoic, but zooplankter taxa may be much younger. If speciation is slow in taxa of parthenogenetic habit, then the paucity of zooplankton species might be a function of the comparative youth of the major taxa.

It might seem that the number of species of zooplankters in open water ought to be set by processes of gain and loss of species from the shallows as is postulated for the phytoplankton (Fig. 5). Like the algae, the zooplankters have clear affinities with ancestral or supporting populations living a largely benthic life in the shallows. But the process of invading the open water does not lead to a diverse array of herbivore species. The "paradox of the phytoplankton" is that there seems to be too many species of them. We suggest it is just as interesting to reflect on the "paradox of the zooplankton," the fact that there seem to be too few species of these.

References

AMERICAN PUBLIC HEALTH ASSOCIATION. 1971. Standard methods for the examination of water and wastewater, 13th ed.
BARBOUR, C., and J. BROWN. 1974. Fish species diversity in lakes. Am. Nat. **108**: 473–479.
BOURRELLY, P. 1966. Les algues d'eau douce. Initiation à la Systématique I. Les Algues Vertes. Boubée & Cie.
———. 1968. Les algues d'eau douce. Initiation à la Systématique II. Les algues jaunes et brunes. Boubée & Cie.
———. 1970. Les algues d'eau douce III. Algues bleues et rouges. Boubée &. Cie.
———. 1972. Les algues d'eau douce. I. Les Algues Vertes, 2nd. ed. Boubée & Cie.
BROWN, J. 1971. Mammals on mountain tops: non-equilibrium insular biogeography. Am. Nat. **105**: 467–478.
CAIRNS, J., M. L. DAHLBERG, K. L. DICKSON, N. SMITH, and W. T. WALLER. 1969. The relationship of freshwater protozoan communi-

ties to the MacArthur-Wilson equilibrium model. Am. Nat. 103: 439-454.

CLEVE, P. T. 1893-1896. Les Diatomées de l'Equateur. Le Diatomiste V. 2. M. J. Tempère, Paris.

CLEVE-EULER, A. 1968. Die Diatommeen von Schweden und Finnland. V. 2-5. J. Cramer, New York.

COLINVAUX, P. A. 1968. Reconnaissance and chemistry of the lakes and bogs of the Galapagos Islands. Nature 219: 550-594.

———. 1969. Paleolimnological investigations in the Galapagos Archipelago. Mitt. Int. Ver. Theor. Angew. Limnol. 17, p. 126-130.

———. 1972. Climate and the Galapagos Archipelago. Nature 240: 17-20.

———, and E. K. SCHOFIELD. 1967a. Historical ecology in the Galapagos Islands. I. A. holocene pollen record from Isla San Cristobal. J. Ecol. 64: 989-1012.

———, and ———. 1976b. Historical ecology in the Galapagos Islands. II. A holocene spore record from El Junco lake, Isla San Cristobal. J. Ecol. 64: 1013-1026.

COOKE, M. C. 1887. British desmids. Suppl. to British fresh-water algae. Williams & Norgate.

DIAMOND, J. M. 1973. Distributional ecology of New Guinea birds. Science 179: 759-769.

DICKSON, K. L., and J. CAIRNS, JR. 1972. The relationship of fresh-water macroinvertebrate communities collected by floating artificial substrates to the MacArthur-Wilson equilibrium model. Am. Midl. Nat. 88: 68-75.

DROUET, F. 1937. The Brazilian Myxophyceae. I. Am. J. Bot. 24: 598-608.

———. 1938. The Brazilian Myxophyceae II. Am. J. Bot. 25: 657-666.

FERDON, E. N., JR. 1950. Studies in Ecuadorian geography. Monogr. School Am. Res., Santa Fe, New Mexico.

FOERSTER, K. VON. 1969. Amazonische Desmidieen. Amazoniana 2: 5-232.

———. 1974. Amazonische Desmidieen. 2. Teil: Areal Maues-Abacaxis. Amazoniana 5: 135-242.

HARPER, J. L. 1969. The role of predation in vegetational diversity. Brookhaven Symp. Biol. 22: 48-62.

HEGEWALD, E., A. ALDAVE, and E. SCHNEPF. 1978. Phytoplankton of La Laguna, Huanuco (Peru). Arch. Hydrobiol. 82: 207-215.

HUTCHINSON, G. E. 1959. Homage to Santa Rosalia or Why are there so many kinds of animals? Am. Nat. 93: 145-159.

———. 1961. The paradox of the plankton. Am. Nat. 95: 137-146.

———. 1967. A treatise on limnology, v. 2. Wiley.

KEDDY, P. A. 1976. Lakes as islands—distributional ecology of two aquatic plants, *Lemma minor* L. and *Lemma trisulca* L. Ecology 57: 353-359.

LEWIS, W. M., JR. 1974. Primary production in the plankton community of a tropical lake. Ecol. Monogr. 44: 377-409.

———. 1978. A compositional, phytogeographical and elementary structural analysis of the phytoplankton in a tropical lake: Lake Lanao, Philippines. J. Ecol. 66: 213-226.

MacARTHUR, R. H., and E. O. WILSON. 1967. The theory of island biogeography. Princeton Univ. Press.

MAGUIRE, B., JR. 1963. The passive dispersal of small aquatic organisms and their colonization of isolated bodies of water. Ecol. Monogr. 33: 161-185.

———. 1971a. Community structure of protozoans and algae with particular emphasis on recently colonized bodies of water. *In* J. Cairns, Jr. [ed.], Structure and function of fresh-water microbial communities. Am. Microsc. Soc.

———. 1971b. Phytotelmata: biota and community structure determination in plant-held waters. Annu. Rev. Ecol. Syst. 2: 439-464.

PAINE, R. T. 1969. The *Pisaster-Tegula* interaction: prey patches, predator food preference, and intertidal community structure. Ecology 50: 950-961.

PATRICK, R. 1967. The effect of invasion rate, species pool, and size of area on the structure of the diatom community. Proc. Natl. Acad. Sci. 58: 1335-1342.

———. 1970. The diatom flora of some lakes of the Galapagos islands. Nova Hedwigia Z. 31: 495-510.

———, and C. REIMER. 1966. The diatoms of the United States. I. Monogr. Acad. Nat. Sci. Phila. 13.

———, and ———. 1975. The diatoms of the United States. II-Part 1. Monogr. Acad. Nat. Sci. Phila. 13.

PORTER, K. G. 1977. The plant-animal interface in freshwater ecosystems. Am. Sci. 65: 159-170.

PRESCOTT, G. W. 1951. Algae of the western Great Lakes area (exclusive of desmids and diatoms). Cranbrook Inst. Sci. Bull. 31.

RALFS, J. 1848. The British Desmidieae. Reeve, Benham and Reeve, London.

RUTTNER, F. 1952. Planktonstudien der Deutschen Limnologischen Sunda-Expedition. Arch. Hydrobiol. 21: 1-274.

SAUER, W. 1971. Geologie Von Ecuador. Gebrüder Borntraeger, Berlin.

SCHOENER, T. 1969. Size patterns in West Indian *Anolis* lizards: I. Size and species diversity. Syst. Zool. 18: 386-401.

SCHOFIELD, E. K., and P. A. COLINVAUX. 1969. Fossil *Azolla* from the Galapagos Islands. Bull. Torrey Bot. Club 96: 623-628.

SIMBERLOFF, D. A. 1974. Equilibrium theory of island biogeography and ecology. Annu. Rev. Ecol. Syst. 5: 161-182.

SMITH, G. M. 1950. The fresh-water algae of the United States, 2nd ed. McGraw-Hill.

THIENEMANN, A. 1938. Das Phytoplankton des Suesswassers. Die Binnengewaesser 16.

TERÁN, F. 1966. Geografía del Ecuador. Edit. Colón, Quito, Ecuador.

———. 1975. Nuestras Lagunas Andinas. Geografía e Historia. Casa Cultura Ecuatoriana.

VAN HEURCK, H. 1896. A treatise on the Diatomaceae [transl. W. E. Baxter]. William Wesley and Son, London.

WEST, W. Y., and G. S. WEST. 1904-1923. A monograph of the British Desmidiaceae. I-V. Lam. R. Soc., London.

WHITTAKER, R. H. 1969. Evolution of diversity in plant communities. Brookhaven Symp. Biol. **22**: 178-260.

WHITTON, B. A. 1968. Phytoplankton from Ecuadorian lakes. Nova Hedwigia Z. **16**: 267-268.

WOLF, T. 1934. Geography and geology of Ecuador. Grand & Troy Ltd., Toronto.

ZANON, V. 1927-1928. Diatomeas del rio Napo (Ecuador, S. A.) Contributo al conocimiento de la flora. Atti Accad. Pont. Nuovi Lincei **81**.

X

Community Structure:
Experimental and Theoretical Approaches

63. Breaking the Bottleneck: Interactions of Invertebrate Predators and Nutrients in Oligotrophic Lakes

William E. Neill and Adrienne Peacock

Abstract

The dynamics and limits of invertebrate predation upon planktonic prey species in an oligotrophic lake were explored through a graded series of experimental disturbances of the predator-prey interactions in 10 m^3 and 17 m^3 in situ enclosures. Manipulating densities of *Chaoborus* spp. and *Cyclops bicuspidatus thomasi* simultaneously with food supplies to grazer species (via differential fertilization of identical plankton assemblages) revealed significant quantitative and qualitative changes in the roles of invertebrate predators in this community's organization. At low food levels, juvenile predators suffered such high springtime mortality that few predators remained to have significant effects upon grazer prey by midsummer. Moderate food levels enhanced grazer abundance more than predator abundance, and prey became proportionally less vulnerable to predation, despite higher predation rates per predator. At highest food levels, springtime juvenile predator survivorship was markedly increased, thereby reducing summertime prey abundances below those of unenhanced control enclosures and of similarly fertilized enclosures lacking all predators. Under these high productivity conditions, *Cyclops* predation greatly reduces nearly all species' abundances, but is unable to eliminate any, whereas *Chaoborus* predation suddenly eliminates nearly all prey species when larvae molt to fourth instar. These results indicate that this plankton assemblage is surprisingly resistant to moderate changes in available food and predator abundances, and invertebrate predators are relatively unimportant to most species. Very substantial changes in grazer food supply (5–10X) will, however, favor 100–1,000X increases of predators, and significant demographic effects on prey, including elimination of species, result.

Unlike communities composed of long-lived plants or animals, "structure" in zooplankton communities is dynamic relative to the time frame of the investigator's interests. Unfortunately, it is becoming clear that even dynamic plankton communities must be routinely monitored for many years before anything like a full range of system behaviors can be observed and described (e.g. W. T. Edmondson at this conf.). Further, the very nature of rare events or "atypical" constellations of factors predisposes such descriptive studies to a limited range of testable questions about system organization. We feel that one of the strongest alternative methods for probing suspected organizational relationships among ecosystem components is to perturb them experimentally and to track their responses. Observations on

---

Supported by research grants from the University of British Columbia and the People and Government of Canada through the National Research Council.

the motion of a perturbation wave through a community can provide detailed information on the kinds and strengths of relationships and may be very useful in detecting conditions establishing alternative, locally stable states. For example, a graded series of disturbances, carefully chosen to perturb a particular relationship, may expose the limits (boundary conditions) at which the relationship begins to break down. We here report on the use of this method to examine the relative importances of invertebrate predation and food supply in organizing the relative abundances of crustacean species in an oligotrophic bog lake.

It has proven surprisingly difficult to demonstrate that invertebrate predation has significant mortality effects on prey that are not balanced by reproduction. Even the archetypal studies of Hall (1964) and DeBernardi (1974) are subject to criticism because of their probable miscalculation of prey birth rates, thereby overestimating subsequent juvenile mortality from invertebrate predators (Threlkeld unpubl.). Clearly, the mere presence of probable antipredator adaptations or even numerous prey individuals found in predator gut tracts are insufficient indicators as well. In fact, in any lake the prey species able to coexist with predators are among the least likely to be strongly affected by predation pressures.

There are two major limits on invertebrate predator effectiveness as controllers of prey community demography, despite apparently wide tastes in prey. First, many seem to require relatively high densities of prey to have significant predation rates per predator, and prey are usually least vulnerable demographically when very abundant. Secondly, many invertebrate predators seem to have a poor potential for rapid numerical adjustment to increases in prey abundance. They are usually slow recruiters due to lengthy development periods and/or vulnerable stages in the life history. Predator histories with such "bottlenecks" in development may offer an important mechanism of coexistence of predator and prey species within a given lake, yet be very sensitive to changes in conditions that alter development (e.g. food quantity, quality, and temperature).

Oligotrophic lakes offer an excellent opportunity to examine the importance of developmental bottlenecks of invertebrate predators on community structure and dynamics, since such bottlenecks are most likely to be present in these environments. By experimentally altering food supplies and community composition, it is possible to evaluate the effect of removing the bottleneck in the predators' life history while simultaneously observing the responses of the prey.

We here present preliminary results of some experimental studies on the impact of invertebrate predators when lake fertility is altered. Two main concerns are examined: the constraints on invertebrate predators in oligotrophic lakes, and the responses of low nutrient systems to experimentally enhanced fertility.

Materials and methods

Study area—The study was conducted in one of the montane lakes in the Research Forest of the University of British Columbia. Although these lakes have been described in detail elsewhere (Efford 1967; Northcote and Clarotto 1975; Neill 1978), a summary of relevant information follows.

In general these lakes are small, oxygen-rich, nutrient-poor, and deep. Particulate organic matter concentrations (particles < 30 μm in diameter or "grazable seston") are low. Before recent fish introductions (Northcote and Clarotto 1975; Northcote et al. 1978) about half the lakes were fishless. Gwendoline Lake was chosen for this study, as the only significant predators in the plankton are invertebrate. No fish had ever been recorded in the lake before the experiment.

All lakes contain a similar assemblage of crustaceans with differences in some copepod species between fish and fishless lakes. Gwendoline Lake differs from nearby Placid Lake by a single species of calanoid (Table 1). *Cyclops bicuspidatus thomasi*

Table 1. Zooplankton species assemblage in Gwendoline and Placid Lakes in the U.B.C. Research Forest. Although relative densities of species in these lakes may differ, only a single species of calanoid copepod is replaced: *Diaptomus oregonensis* in Placid Lake by *D. leptopus* in Gwendoline Lake.

Zooplankton species*	Gwendoline	Placid
Diaptomus kenai	A	A
Diaptomus oregonensis	—	A
Diaptomus leptopus	A	—
Cyclops bicuspidatus thomasi	R	A
Tropocyclops prasinus prasinus	A	R
Macrocyclops albidus	R	R
Eucyclops agilis	R	R
Daphnia rosea	A	A
Holopedium gibberum	A	A
Diaphanosoma brachyurum	A	A
Bosmina longirostris	A	A
Ceriodaphnia quadrangula	R	R
Polyphemus pediculus	A	A
Chydorus spp.	R	R

*Key to symbols: A—abundant; R—rare; (—)—absent.

is abundant only in Placid where the chaoborid species is *Chaoborus flavicans*. Conversely *Tropocyclops prasinus prasinus* is abundant in Gwendoline where only *C. trivittatus* and *C. americanus* are found. Both of these species of chaoborids remain in the plankton during the day while *C. flavicans* in Placid seeks refuge in the benthos during daylight hours.

Experimental methods—To examine the effect of nutrient addition and invertebrate predator impact on the zooplankton community, we carried out large in situ experiments from mid-May to late September over several years. *Chaoborus* manipulations were conducted in 12 enclosures constructed after a design modified from Kořínek (1971) and described by Neill (1978). These tubular enclosures, holding about 17 m³ were made of 4 mil, clear polyethylene plastic, 1.5 m wide and 10 m deep. They were sealed at the bottom and suspended from a wooden and polystyrene frame floating on the surface. Plastic screening was used to prevent egg laying in the *Chaoborus*-free enclosures. The enclosures used in the *Cyclops* study were similar but were 6 m deep and held 10 m³ of water. Bags were filled by pumping lake water through a 54-μm-mesh plankton net that removed all crustaceans but permitted grazable seston to pass through. *Cyclops* collected in Placid Lake and separated from the other crustaceans by a series of sieves was added to the *Cyclops*-supplemented enclosures at Placid Lake densities. Natural lake densities of Gwendoline Lake crustaceans were added to all enclosures from pooled zooplankton samples obtained from depth-stratified horizontal tows with a Clark-Bumpus sampler.

Six experimental distrubances (each with two replicates) were produced in the *Chaoborus* experiments: three food levels and two *Chaoborus* conditions in all combinations. To alter food levels, we used phosphate-nitrate fertilizer (ratio of 1:10) (NaH_2PO_4 and KNO_3) at final phosphate concentrations of 0, 3, and 16 μg liter^{-1} ("low" fertilizers), 80 μg liter^{-1} ("medium" fertilizer), and 400 and 2,000 μg liter^{-1} ("high" fertilizers). Fertilizers were added only once in mid-May, although partial mixing of hypolimnetic and epilimnetic waters in bags was carried out via occasional bubbling during July and August. *Chaoborus* adults were permitted to lay eggs in six enclosures representing all fertilizer levels, while all *Chaoborus* were excluded from six others. Because relatively few eggs were laid in some enclosures, first instar *Chaoborus* larvae were supplemented in mid-July to approach lake densities. Abundances of *Chaoborus* IV larvae were thus the result of differing survival rates at the several fertilizer levels.

Four experimental disturbances in factorial design were produced in the *Cyclops* study: two enclosures contained "high" fertilizer levels (500 μg liter^{-1} PO_4) while two received no fertilizer. The same fertilizer cocktail used in the above *Chaoborus* experiments was used here. To one enclosure in each of the fertilizer conditions, adult *C. bicuspidatus thomasi* from Placid Lake were supplemented to match Placid Lake *Cyclops* densities. *Chaoborus* was excluded from all four conditions. This experiment thus examined *Cyclops* predation under conditions of high and low nutrients and high and low *Cyclops* abundance.

Zooplankton sampling was done weekly in these enclosures and the lake with an electric bilge pump. Each 100-liter sample was filtered through a 54-μm plankton net and preserved in sugar-Formalin. Samples were examined in toto under 25× magnification for species' abundances and reproductive conditions. Temperature and O_2 profiles were monitored via occasional sampling. Water samples were collected to determine ash-free dry weights of particles passing a 30-μm mesh (grazable seston) and for microscopic examination of algal size composition and category (green, blue-green, colonial, etc.).

Life histories—The two major plankton invertebrate predators in the Research Forest lakes are *Chaoborus* larvae and the cyclopoid copepod *C. bicuspidatus thomasi*. The species of *Chaoborus* and abundance of *Cyclops* varies with the lake as described. In Gwendoline, there are two species of *Chaoborus*, *C. americanus* and *C. trivittatus*. The former is univoltine while the latter may take 2 years to mature. Emergence and egg laying occur in late spring to early summer. By August most larvae have matured to the fourth and final instar.

Cyclops is a rare species in Gwendoline but abundant in nearby Lake Placid. *Chaoborus* copepodites IV and V appear in the plankton in late April to early May and mature to adults by mid-May. Eggs develop and by the end of June only a few old adults are left, the main population being represented by naupliar forms. By August the population is dominated by copepodites IV and V, some of which diapause although many remain in the plankton. Adults are reproductive again in late August and early September.

The prey species, following a similar pattern, begin reproduction in the spring. Calanoid copepods produce no more than two generations during the year while the cladocerans are all multivoltine, producing parthenogenic young through the spring and summer period and sexual eggs in the late fall.

Results

Chaoborus × fertility experiments—The following community responses to fertilization in the absence and presence of *Chaoborus* were observed (Fig. 1) during August.

There was an enormous demographic response in enclosures lacking *Chaoborus* (Fig. 1a, b, c) mainly by *r*-strategist cladocerans, particularly chydorids. Rare species in the lake, such as *Cyclops* and other cyclopoids, increased in number significantly, while the abundance of calanoids did not increase, even under high nutrient conditions. In spite of these demographic responses to fertilization, even the high nutrient enclosure is still recognizable as the same community. Although we intentionally attempted to create this condition by carefully choosing our fertilizers to minimize changes other than increased food concentration, the response may be more general. W. T. Edmondson's comments on the lack of differential species responses among zooplankton to nutrient perturbations in Lake Washington are instructive.

Relative to the situation without *Chaoborus*, there was little expansion of animal density with increasing fertilizer load when *Chaoborus* was present (Fig. 1d, e, f). These predators effectively and generally dampened growth. At low fertilizer levels (0, 3, and 16 μg liter^{-1} PO_4), the enclosures with *Chaoborus* showed only a small predator effect, mainly permitting rare species to become more common. Perhaps this small predatory restriction of population growth reduced the demand for scarce resources, allowing rare species to increase slightly. There was almost no detectable additional impact of *Chaoborus* at medium fertilizer (80 μg liter^{-1} PO_4) over that at low fertilizer. By contrast there was an enormous *Chaoborus* impact at high nutrient levels (400 μg liter^{-1} PO_4) with virtual elimination of all species except *Chydorus*. Consequently there was a major

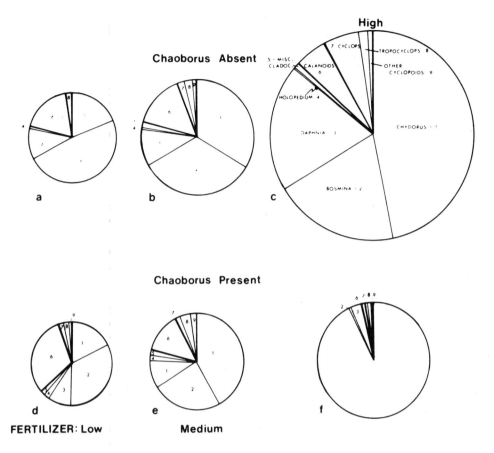

Fig. 1. Crustacean community responses by late summer to fertilizer and *Chaoborus* manipulations. Size of pie diagrams and segments reflect average abundances of crustaceans sampled during last 2 weeks of August in experimental enclosures: a—1,801; b—2,768; c—9,781; d—1,533; e—2,253; f—3,145 individuals per 100-liter sample. Key to fertilizer designations: low—0, 3, and 16 μg liter^{-1}; medium—80 μg liter^{-1}; high—400 and 2,000 μg liter^{-1} PO$_4$.

change in the organizational importance of *Chaoborus* in the community.

Constraints on Chaoborus *and the interactions with fertility*—The mechanism of the differential success of *Chaoborus* under high fertility can be explained by examining *Chaoborus* survivorship under different nutrient conditions (Fig. 2). At low fertility (lake and low fertilizer enclosures), there was enormous instar I and II mortality. This juvenile mortality decreased significantly at medium fertilizer conditions. Consequently early *Chaoborus* mortality appears to be food regulated in Gwendoline Lake. Crop contents indicated that instar I was feeding primarily on nauplii and rotifers. The rotifer populations were extremely low in the lake and similar densities were found in the low fertilizer enclosures (Fig. 3). Although rotifer abundance increased at medium fertilizer levels, it was not until high nutrient conditions were created that rotifers became numerous. This bloom occurred just as *Chaoborus* was in instar I and had a dramatic impact on juvenile survivorship (Fig. 2). As a result there were many more instar IV *Chaoborus* in the plankton by late summer and their impact was catastrophic upon virtually all species except *Chydorus* (Fig. 1). Fer-

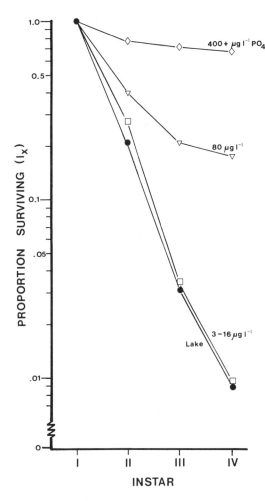

Fig. 2. *Chaoborus trivittatus* survivorship by instar in experimental enclosures receiving different fertilization treatments. Points are conservative estimates based on peak instar I abundances because egg abundance was unreliably censused.

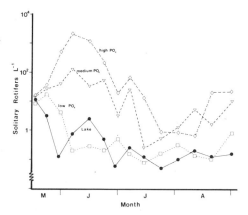

Fig. 3. Seasonal patterns of solitary rotifer densities in experimental enclosures receiving different fertilizer treatments.

tilization removed the bottleneck in the development of *Chaoborus* and changed the numerical response to one in which the predator was capable of overriding prey responses to high nutrients.

Cyclops × *fertility experiments*—Like *Chaoborus*, *Cyclops* is an invertebrate predator apparently capable of significant demographic impact on its prey (McQueen 1969) although this copepod is much smaller in size than *Chaoborus* larvae. As *Cyclops* is rare in Gwendoline Lake, adults were transferred from nearby Placid Lake to examine the impact of this predator upon community responses to fertilization. *Chaoborus* was eliminated from all these enclosures.

There was a major increase in animal density (about 2 times) in the fertilized enclosure without *Cyclops* (Fig. 4). *Tropocyclops* made the most dramatic increase, in contrast to the species' response in the *Chaoborus* experiments above. Because the enclosures were filled with the May assemblage of animals and *Tropocyclops* does not appear in the plankton until mid-July, *Tropocyclops* was added to all enclosures in the *Cyclops* experiments to equal summertime lake densities. As *Tropocyclops* was not supplemented in the *Chaoborus* experiments, they remained rare. Improved survivorship of *Tropocyclops* in the absence of *Cyclops* was consistent with earlier results of *Cyclops* removal experiments in nearby Placid Lake and consequently of interest here.

As in the *Chaoborus* experiments, calanoid copepods were unable to increase densities in the high nutrient environment over the lake nutrient condition. *Chydorus* increased while *Bosmina* decreased in relative density. This behavior echoed that in the *Chaoborus* experiments (Fig. 1). Similarly, rare species in Gwendoline Lake (such as *Cyclops*) became countable due to natural

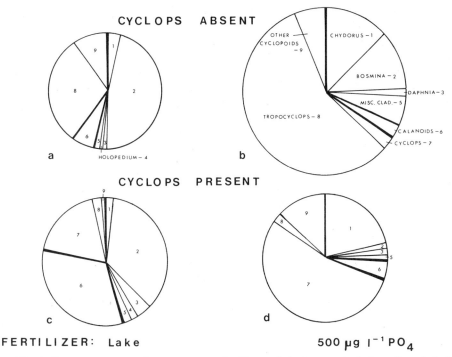

Fig. 4. Crustacean community responses to fertilizer and *Cyclops* manipulations. Size of pie diagrams and segments reflect average abundances of crustaceans sampled during last 2 weeks of August in experimental enclosures: a—1,736; b—4,035; c—2,099; d—1,911 individuals per 100-liter sample. Fertilizer designations: lake levels in a, c; 500 μg liter^{-1} PO$_4$ added to b, d.

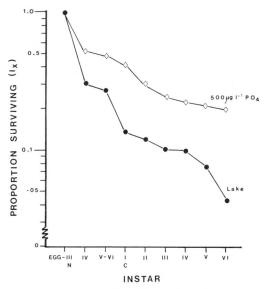

Fig. 5. *Cyclops bicuspidatus thomasi* survivorship by instars in experimental enclosures with fertilizer added and lake level nutrients. Field data used to calculate duration of each instar.

recruitment after fertilization. Again, community organization did not change drastically after fertilization. The same community is recognizable regardless of nutrient condition.

The only major effects at low fertility when *Cyclops* was present were the decrease of *Tropocyclops* and an increase in calanoid copepods. In the absence of *Chaoborus*, *Cyclops* had no difficulty surviving and recruiting in the Gwendoline Lake environment. Carnivory by *Cyclops* may also have released some food resources for the herbivores, including the calanoids.

Without fertilization *Cyclops* increased slowly. At lake nutrient levels this species may take many years to become a significant element of the community. Survivorship curves (Fig. 5) indicate heaviest mortality occurred in the egg-naupliar III stages and naupliar V to copepodite I stages. Survivorship in these early stages has been shown to be food-dependent (Rigler and Cooley 1974). In Gwendoline Lake, *Cyclops* survivorship suggested the early stages are sensitive to increased fertility as mortality was significantly reduced in the high nutrient enclosure (Fig. 5).

Although there was a large decrease in density of the total community when *Cyclops* survived at higher densities in the fertilized enclosure, there was no drastic change in community organization. *Bosmina*, *Tropocyclops*, and the calanoids all decreased. As the calanoids were unable to increase their numbers in high nutrient conditions relative to the rest of the community, their decrease cannot be attributed solely to an increase in *Cyclops*. However, *Tropocyclops* and *Bosmina* decreases do appear to be connected to enhanced survival of the predator.

Constraints on Cyclops *and the interaction with fertility*—Heavy naupliar mortalities were alleviated with increasing fertility. Although survivorship was enhanced, and appeared to lower the total numerical response of the community to nutrient enrichment, the increased *Cyclops* survivorship was insufficient to overwhelm the prey. We conclude that the prey community in oligotrophic Gwendoline Lake is relatively well buffered against variation in *Cyclops* numbers produced by fluctuations in food resources. During the herbivorous stages, *Cyclops* is experiencing the same constraints as the prey community. Increases in nauplia survival parallel increased prey survival, thereby buffering the community against sudden changes in predator numbers.

Discussion

In the oligotrophic lake community studied here, experimental perturbations of food levels revealed that invertebrate predators have relatively little effect on community organization because of poor early survival. In the case of *Chaoborus*, additional nutrients had little effect until sufficient resources were available to stimulate rotifer populations. Unlike cladocerans, these smaller *r*-strategists seemed unable to sequester resources when food supplies were low and hence could not respond quickly to slightly increased nutrients. Unpublished evidence from species removal experiments indicates that competition between cladocerans and rotifers may generally reduce *Chaoborus* success in our oligotrophic lakes. Only at relatively high fertilizer levels was rotifer growth response sufficiently dramatic to provide *Chaoborus* instar I and II with the means to push past the bottleneck created by heavy juvenile mortality. The results were catastrophic, indicating a new configuration of community organization. It is tempting to invoke such an explanation for sudden changes in seasonal plankton assemblages characteristic of many eutrophic ponds.

Within the enhanced food range tested, *Cyclops* is not capable of such an impact even with greatly improved survivorship under high nutrient conditions. Consequently this predator probably cannot be considered "keystone" in Paine's sense, whereas *Chaoborus* could be a potential organizing agent if juvenile survival is good. Because *Cyclops* is herbivorous in the nauplia stage, it utilizes a food resource similar to the grazer community and nutrient enrichment affects survivorship of both predator and prey in similar manners. As *Cyclops* enters the predatory instars, its diet is probably restricted to smaller members of the prey assemblage, unlike *Chaoborus* III and IV which can and do eat all available zooplankters.

The influence of *Chaoborus* on community organization cannot be assessed independently of its survivorship, the grazer food available in the lake, and the temporal effects of its life history in that lake. This assessment is complicated by the fact that a predator's influences on community composition may have a historical context (species which cannot coexist are absent or rare). Response rates of some species in oligotrophic systems may be so slow without enhancement of fertility that even the effect of predator removal may not be apparent for a long

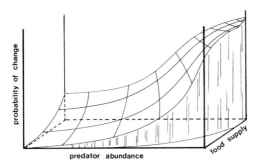

Fig. 6. Apparent response surface to food/predator perturbations of Gwendoline Lake zooplankton community.

time relative to the researchers' interest.

Results of these large manipulations of the plankton indicate that the natural community in our study lake is very resistant to disturbances by predators and nutrient pulses of this kind. Although other combinations of nutrients might have different effects, this fertilizer mix, applied over a very broad range of concentration, did little but increase crustacean densities. Figure 6 summarizes our current view of the response surface for food/predator disturbances in our oligotrophic lakes. We feel that the probability of change in community composition and dynamics with disturbance varies in a complex way with food supply to the grazers and predator abundance. Depending upon the combination of algal and predator conditions produced by a disturbance, the likelihood of current community organization to persist declines at sharply different rates. Unfortunately, it may be some time before we can predict the direction of that change.

Conclusions

Examining the responses to graded perturbations of invertebrate predators and food supply in an oligotrophic lake led to the following conclusions—

1. In the nutrient-poor lake, invertebrate planktonic predators do not have major compositional or demographic effects on the resident lake community—especially on the dominants.

2. By slightly reducing the total abundance of dominant species, invertebrate predators permit the persistence of a few rare species by reducing competition for scarce resources.

3. With small increases in food resources, survivorship of nearly all species is enhanced. The relative impact of *Chaoborus* decreases in consequence until some "threshold" is reached at which rotifer survival is increased. With high rotifer densities *Chaoborus* survival is high and then it has a great effect on community dynamics and composition.

By contrast, *Cyclops* probably never has this keystone impact in nature, due to a life-history strategy with slower and more restricted functional and numerical responses to improved conditions relative to that of the prey.

Hence predicting dynamical and compositional effects of invertebrate predation depends upon information about the relative survivorship responses of predator and prey species.

References

DeBERNARDI, R. 1974. The dynamics of a population of *Daphnia hyalina* Leydig in Lago Maggiore, northern Italy. Mem. Ist. Ital. Idrobiol. 31:221-243.

EFFORD, I. E. 1967. Temporal and spatial differences in phytoplankton productivity in Marion Lake, British Columbia. J. Fish. Res. Bd. Can. 24:2283-2307.

HALL, D. J. 1964. An experimental approach to the dynamics of a natural population of *Daphnia galeata mendotae*. Ecology 45:94-112.

KOŘÍNEK, V. 1971. Direct determination of zooplankton reproduction, 170-172. *In* W. T. Edmondson and G. G. Winberg [eds.], Secondary productivity in fresh waters. IBP Handbook 17. Blackwell.

McQUEEN, D. J. 1969. Reduction of zooplankton standing stocks by predaceous *Cyclops bicuspidatus thomasi* in Marion Lake, B.C. J. Fish, Res. Bd. Can. 26:1605-1618.

NEILL, W. E. 1978. Experimental studies on factors limiting colonization by *Daphnia pulex* Leydig of coastal montane lakes in British Columbia. Can. J. Zool. 56: 2498-2507.

NORTHCOTE, T. G., and R. CLAROTTO. 1975. Limnetic macrozooplankton and fish predation in some coastal British Columbia lakes. Int. Ver. Theor. Angew. Limnol. Verh. 19: 2378-2393.

———, C. W. WALTERS, and J. M. B. HUME. 1978. Initial impacts of experimental fish introduction on the macrozooplankton of small oligotrophic lakes. Int. Ver. Theor. Angew. Limnol. Verh. **20**: 2003–2012.

RIGLER, F. H., and J. M. COOLEY. 1974. The use of field data to derive statistics of multivoltine copepods. Limnol. Oceanogr. **19**: 636–655.

64. Foundations for Evaluating Community Interactions: The Use of Enclosures to Investigate Coexistence of *Daphnia* and *Bosmina*

W. Charles Kerfoot and William R. DeMott

Abstract

In the open-water (pelagic) regions of ponds and lakes, the minute size of plants (phytoplankton) and the relatively structureless form of the physical environment combine to make herbivorous zooplankton especially vulnerable to visual and grasping predators. Since many intuitive arguments, based largely on the unpredictability and short duration of individual phytoplankton "blooms," argue against resource specialization, recent conceptual models of zooplankton community structure have incorporated increasingly finer details of predation to explain herbivore coexistence. However, more formal models of simple predator-prey interactions generally require additional assumptions to allow the coexistence of morphologically diverse herbivores. In our experiments, these additional factors appear to relate to resource heterogeneity. Although *Daphnia* can severely eliminate or depress *Bosmina* when both species share a single algal resource under laboratory conditions, natural populations of these two genera do not seriously depress each other when released from fish predation. Moreover, simple enclosure experiments failed to document strong competition in nature, and the two genera responded differently to artificially created specific "phytoplankton blooms." Hence the coexistence of *Bosmina* and *Daphnia* in natural waters depends, at least in part, either on utilization of, or preference for, different resources.

Before considering the particulars of *Daphnia-Bosmina* coexistence in natural waters, a brief discussion of the general biotic environment seems appropriate to place these organisms in their proper setting. In the open-water aquatic (pelagic) and terrestrial ecosystems, plant-animal interactions operate in many fundamentally different ways. In terrestrial communities, plant-animal interactions are exceedingly complex and intertwined, while, at least with regard to large-scale structural aspects, in pelagic communities they are relatively simple. For example, in terrestrial communities plants often either form a carpet upon or a canopy above the soil layer. Their relatively long life spans make them a major structural component of the environment, as they provide prey with shelter and refuge from environmental conditions or predators. Terrestrial plants serve not only as a platform for life, but also indirectly or directly supply nutrients and food for higher trophic levels in complex ways by either losing (through defoliation; activity of leaf, bark, or root parasites; action of sap-sucking insects) or offering (evolution of disperser interactions) only parts of their complete bodies. The evolution of terrestrial plants and animals is so interrelated with intimate mutualisms and coevolution (Gilbert and Raven 1975) that many basic theories of animal diversity depend upon plant spatial heterogeneity (MacArthur and MacArthur 1961; MacArthur 1965), while

---

This research was supported by NSF grant DEB76-20238. Ron Slade, Dean Kellogg, and Julie Kuo Syang DeMott assisted. The members of the Lake Mitchell Trout Club and Jerry Ketch, caretaker, provided cooperation and permission to perform experiments in their lake.

many basic theories of plant diversity depend upon animal diversity (Janzen 1970; Harper 1977).

In contrast to the terrestrial community, the plants of the open-water environment, i. e. the phytoplankton, are characterized by minute size (most are between 1–25 μm in greatest diameter), low biomass, rapid turnover (short generation time), and erratic and uncorrelated fluctuations in time and space (Allen and Koonce 1973; Lewis 1977). In studies of herbivorous zooplankton, because individual phytoplankton species show pulses of relatively short duration and unpredictable timing, most workers have argued against resource specialization and for opportunistic feeding (Hutchinson 1967; Levinton 1972). Customarily zooplankton have been considered, if they specialized at all, to differ mainly in the mean sizes of particles ingested, primarily as a mechanical reflection of body size (e. g. Burns 1968; Gliwicz 1969). The reliance upon a generalized, opportunistic feeding mode has underlain developing conceptual models of community structure during the past 10 years and has serious implications for more formal models of community structure.

Conceptual models of zooplankton community structure—Before 1960, the general problem of zooplankton coexistence and diversity, if recognized at all, focused upon the influence of physical and chemical variables upon individual species or small groups of species, although a few studies explored simple competitive interactions (Frank 1952, 1957). Biotic interactions were thought to be dominated by competition, that is, if the environment ever approached equilibrium conditions (Hutchinson 1953, 1961). From these early studies came a rather elegant picture of "niche" structure, as different-sized zooplankton, competing in an exploitative manner, were presumed to superimpose themselves upon the resource spectra provided by the phytoplankton against a background of chemical and physical gradients (summarized in Hutchinson 1967, p. 231–233).

However, since the pioneering work of Hrbáček and his associates in Czechoslovakian lakes and backwater ponds (Hrbáček 1962; Hrbáček and Novotná-Dvořáková 1965), predation has played a progressively greater role in conceptual models of community interactions. Three major hypotheses of zooplankton community structure are shown in Fig. 1. While all these hypotheses consider questions of coexistence in open waters, the hypotheses more aptly document the gradual clarification of community interactions (in the face of accumulating insights and experimental evidence) rather than offering discrete, alternative, and competing viewpoints. Significantly, in each model the zooplankton are presumed to compete for a common resource pool (the bacteria, phytoplankton, and detritus of the open water), with the exception that larger zooplankton can ingest larger particles.

Historically, each of the hypotheses incorporates progessively more information about the kinds, and selective biases, of aquatic predators. The publication of the popular "size-efficiency" hypothesis (Brooks and Dodson 1965) focused attention upon the size structure of zooplankton communities and the effects of fish predation. Noticing that the zooplankton communities of many New England lakes and ponds were dominated primarily by small-bodied species (rotifers, *Bosmina, Tropocyclops*), while others were dominated by large-bodied species (*Daphnia, Mesocyclops, Epischura*), Brooks and Dodson postulated that the differences resulted from a dynamic balance between fish predation and zooplankton competition. Fishes, they concluded, by selectively removing the larger-bodied species, forced communities toward small body size, while competition forced them back toward large body size. In their scheme, competitive superiority was envisioned as a simple function of herbivore body size, based upon the presumably increased filtering efficiency of larger-bodied species. The theory has five points:

1) Planktonic herbivores all compete for the fine particulate matter 1–15 μm long of open waters.
2) Larger zooplankters compete more

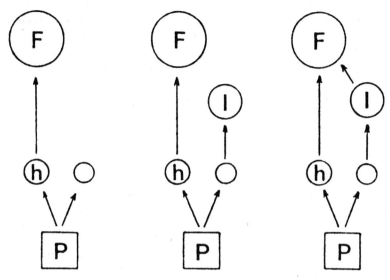

Fig. 1. Three conceptual models of zooplankton community structure: The Brooks-Dodson size-efficiency hypothesis, the Dodson complementary predation hypothesis, and the modified keystone predator hypothesis (P—small particulate matter, including bacteria, phytoplankton, and detritus; h—herbivores, principally rotifers, cladocerans, and copepods; I—carnivorous or omnivorous invertebrates; F—fishes or other large visual predators). Arrows designate direction of energy flow.

efficiently and can also take larger particles.

3) When predation is of low intensity, the small planktonic herbivores will be competitively eliminated by large forms (dominance of large Cladocera and calanoid copepods).

4) When predation is intense, size-dependent predation will eliminate the large forms, allowing the small zooplankters (rotifers, small Cladocera) that escape predation to become the dominants.

5) When predation is moderately intense, it will, by falling more heavily upon the larger species, keep the populations of these more effective herbivores sufficiently low to prevent elimination of slightly smaller competitors.

In 1970, against a background of accumulating evidence for effects of invertebrate predators, Dodson (1970) proposed the "complementary predation" hypothesis. The complementary predation hypothesis, like the size-efficiency hypothesis, implied that larger zooplankton were competitively superior to smaller-bodied species, but introduced invertebrate predation and incorporated a delicate balance between the selective biases of visual (fishes or salamanders) and grasping (invertebrate) predators. Visually feeding predators, Dodson envisioned, removed larger zooplankton which would otherwise monopolize resources, and thus allowed smaller-bodied species to penetrate the community and to increase in abundance. Invertebrate predators, such as *Chaoborus*, could then enter the community to feed upon the smaller species. Subsequently these intermediate-sized grasping predators increased in abundance and, by virtue of their size-selectivity against smaller-bodied species, maintained the advantage of large-bodied species. Thus each group of predators (visually foraging vertebrates and grasping invertebrates) ensured resources for the other in a delicately balanced, complementary, manner (Fig. 1).

More modern conceptual models of aquatic community interactions have recognized important interactions between predators and incorporated schemes which more directly address diversity, e. g. the "keystone predator" hypothesis of Paine (Paine 1966, 1969; first applied to marine intertidal communities). Recent embellishments continue to emphasize the role of predation in promoting and increasing diversity but differ from earlier versions in several important particulars (Dodson 1974; Neill 1975; Kerfoot 1975). The more current versions, here collectively termed the "modified keystone" hypothesis, recognize 1) that visually feeding fishes prey heavily upon larger-bodied grasping invertebrate predators, and hence indirectly determine the abundance and effect of grasping predators (Kerfoot 1975); and 2) that the biased selective nature of visual and grasping predators can favor the evolution of defenses in prey which, while ensuring survival, detract from competitive ability (Zaret 1972a; Kerfoot 1977; Kerfoot and Pastorok 1978; Jacobs 1978). This counterevolution is extremely important because it ensures that the competitive dominant will remain the most susceptible to predation; and that the visibility, and not the body size, per se, of zooplankton is the crucial measure of conspicuousness (Zaret 1972a, b; Zaret and Kerfoot 1975; Kettle and O'Brien 1978). While few can argue that these models provide a more accurate picture of general zooplankton interactions, there are disquieting implications for diversity that emerge from more formal models of coexistence based solely upon simple interactions between prey and predators.

Formal models of predator-induced coexistence—Over the past few years, the numbers of papers examining the question of predator-induced prey coexistence have continued to grow. One of the simpler forms of Lotka-Volterra models involve statements of the general form:

$$\frac{dN_1}{dt} = r_1 N_1 \left(\frac{K_1 - N_1 - aN_2}{K_1}\right)$$

$$-f_1(N_1, N_2, P) \qquad (1)$$

$$\frac{dN_2}{dt} = r_2 N_2 \left(\frac{K_2 - N_2 - \beta N_1}{K_2}\right)$$

$$-f_2(N_1, N_2, P) \qquad (2)$$

where the function $f_i(N_1, N_2, P)$ expresses the losses to prey populations (N_1, N_2) due to predation from a single predator (P). Here the usual predator response term [$dP/dt = g(N_1, N_2, P)$] is dropped, uncoupling the numerical fluctuations of predator population size from those of the prey. Independence of predator population change and prey removal rates in specific paired cases is not unreasonable where predator densities are regulated by higher trophic interactions other than those directly considered, or where the predator may have a wide array of other prey species at its disposal (Jacobs 1977a).

While the Lotka-Volterra logistic equations are arbitrarily chosen because of their simplicity and historical value, they are sufficiently complex to reveal important ways in which predation alters the outcome of competition between two species and thus effects the chances of coexistence. The specific Lotka-Volterra interactions used here are

$$\frac{dN_1}{dt} = r_1 N_1 \left(\frac{K_1 - N_1 - aN_2}{K_1}\right)$$

$$- m_1 N_1 P \qquad (3)$$

$$\frac{dN_2}{dt} = r_2 N_2 \left(\frac{K_2 - N_2 - \beta N_1}{K_2}\right)$$

$$- m_2 N_2 P \qquad (4)$$

where modified logistic equations (Gause and Witt 1935; Slobodkin 1961; Emlen 1973) describe the interactions between two competing prey (one susceptible to predation, N_1; the other resistant, N_2) and a selective predator (P). Here N_i, K_i, r_i, and m_i describe population densities (variable), saturation values in the absence of predation (constant), intrinsic rates of natural increase (constant), and specific removal rates (constant). At equilibrium,

$$r_1 (K_1 - N_1 - aN_2) - m_1 K_1 P = 0 \quad (5)$$

$$r_2 (K_2 - N_2 - \beta N_1) - m_2 K_2 P = 0 \quad (6)$$

that is, following Emlen's (1973) approximate solutions,

$$\overset{*}{N_1} = \frac{r_2 K_1 [r_1 - m_1 P] - r_1 K_2 a [r_2 - m_2 P]}{(1 - \beta a) r_1 r_2} \quad (7)$$

$$\overset{*}{N_2} = \frac{r_1 K_2 [r_2 - m_2 P] - r_2 K_1 \beta [r_1 - m_1 P]}{(1 - \beta a) r_1 r_2} \quad (8)$$

The two prey coexist when $\overset{*}{N_1} > 0$, $\overset{*}{N_2} > 0$, i.e. when

$$\frac{K_1}{aK_2} > \frac{(r_2 - m_2 P) r_1}{(r_1 - m_1 P) r_2} > \frac{\beta K_1}{K_2}. \quad (9)$$

In the absence of predation ($m_1 = m_2 = 0$), species 2 (the resistant species) is excluded if $\beta K_1/K_2 > 1$, i. e. if protection incurs a cost in terms of competitive ability. Predation allows coexistence of the two prey species when

$$\frac{K_1}{aK_2} > \frac{(r_2 - m_2 P) r_1}{(r_1 - m_1 P) r_2} > \frac{\beta K_1}{K_2} > 1. \quad (10)$$

However, if P and m_1/m_2 are too high, the susceptible prey species is excluded.

In order to visualize the situations described by this model, we have plotted the isoplanes for the two prey species in Fig. 2. The intersections of the isoplanes upon the N_1, N_2 plane (the base) give the familiar isoclines of Gausian competition ($P = 0$). Intersections of the same planes on the vertical axis (P) correspond to points where the removal rate by predation (m_iP)

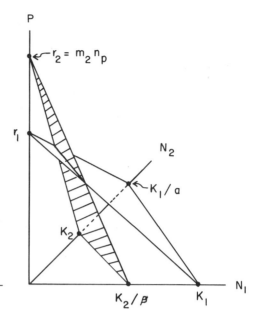

Fig. 2. Equilibrium isoplanes for a two prey-one predator interaction. Here, in "indifferent equilibrium" case (saturation isoclines on the N_1, N_2 basal plane are parallel), the two planes intersect along a line at only one predator density.

balances the intrinsic rates of increase (r_i). Beyond this point, the prey can no longer maintain populations in the face of removal rates (predation).

As Eq. 9 and 10 suggest, the outcomes of interactions are very sensitive to losses in K, a, and r associated with the evolution of resistance. If values are nearly identical, as before the origin of genetic differences in the same stem stock, the two isoplanes will overlap completely. Under these circumstances, coexistence is unstable because the outcome is subject to random events, although exclusion may proceed at a slow rate depending upon exact conditions (Parrish and Saila 1970; Jacobs 1977a; Caswell 1978; Vance 1978). If the isoplanes are distinct, yet parallel (i. e. in the improbable case where $K_2 = K_1$, $a = 1/\beta$), they cross only at a single point along the P axis (Fig. 2). This configuration, described by Jacobs (1977a) as a case of "indifferent equilibrium" is inherently unstable in the presence of fluctuating predator densities and is very sensitive to changes in r_i, K_i, a, and β. If, however, the competitive ability (a) and carrying capacity (K_2) of the resistant prey are influenced unequally by the evolution of resistance, the declination of the isoplanes will change, and the N_2 isoplane will rotate clockwise or counterclockwise with respect to the stationary N_1 isoplane. If the rotation is clockwise (Fig. 3a), the two isoplanes will intersect along a line which spans a range of predator densities, but which is inherently an unstable equilibrium, i. e. where only one prey persists with the predator, depending upon initial concentrations. If the rotation is counterclockwise, the two isoplanes will again intersect along a line which spans some range of predator densities, but here the equilibrium is stable (Fig. 3b), i. e. both prey coexist with the predator. Hence if the evolution of resistance more seriously affects the prey's ability to utilize resources (variety of food, amount of food) than ability to suppress its competitors, then a stable balance can be maintained within a range of predator densities. While the models are very simple, one disquieting observation is that the range of predator densities over which stable conditions prevail is very sensitive to the divergence of the two isoplanes. This range is broadest when only slight differences in the declination of the isoplanes, and in the absolute differences of parameters, exist initially. Major differences in r_2, K_2, m_2, and a lead to very restricted ranges. Thus two very important conclusions come from these relationships. First, the conditions which allow stable coexistence between two species, under the influence of predation, contain a rather restricted, yet important, set of assumptions. Secondly, as opposed to conclusions based upon purely competitive interactions, coexistence is more likely, the more initially similar the species (Kerfoot and Peterson 1979). Simpler and more complicated forms of Eq. 1, using coupled predator response equations, even more strongly emphasize the restricted set of

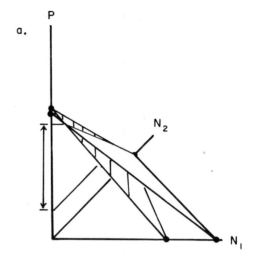

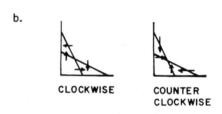

Fig. 3. Rotation of resistant-prey isoplane as saturation values (K) and competitive abilities (β) are unequally affected by evolution of resistance. Note that the two planes intersect over a broad range of predator densities in Fig. 3a. In Fig. 3b, a horizontal cross section (at a single value of P) near middle of zone of isoplane overlap distinguishes between unstable and stable configurations.

conditions for coexistence of both prey and predator, even though limit-cycle behavior emerges as an alternative (Cramer and May 1971; May 1973; Yodzis 1976, 1977; Caswell 1978; Vance 1978). For this reason, and to provide an alternative mechanism, many workers have advised renewed focus on functional responses of predators (Caswell 1978; Vance 1978), or the application of nonequilibrium models (Caswell 1978).

Thus the dilemma: while predators can stabilize species associations, selective removal appears to be more effective in stabilizing closely related species or clones, and less effective upon distantly related pairs. The two herbivorous cladocerans, *Bosmina* and *Daphnia*, which played a major role in the formulation of Brooks and Dodson's size-efficiency hypothesis, are very different in both size and morphology, and yet they very commonly coexist in ponds and lakes throughout the world. In the predator-based conceptual models of community interactions, while selective predation can certainly shift community composition and create spatial heterogeneity, the coexistence of *Bosmina* and *Daphnia* would seem unlikely without recourse to alternative contributing factors (such as partial non-overlap of resources). To investigate the coexistence of *Daphnia* and *Bosmina*, we have examined the outcomes of competition under both laboratory and field conditions, using artificially restricted, natural, and manipulated resource spectra.

Methods

As mentioned above, to examine the coexistence of *Bosmina* and *Daphnia*, we decided to test for competition in the laboratory and extensively in a small New England lake. The results would contrast the outcome of competition for a single, shared and limiting resource in the laboratory with foraging upon a varied resource spectrum in nature. In laboratory 500-ml glass beakers, *Daphnia pulex*, *Bosmina longirostris*, and *Chydorus sphaericus* competed for a single algal species (*Chlamydomonas reinhardi*, Indiana collection No. 90, 3–7 μm diameter, mode 5 μm, while in Lake Mitchell *Daphnia* and *Bosmina* competed for a variety of naturally occuring bacteria, phytoplankton, and detritus. Field investigations to study the coexistence of *B. longirostris* and *Daphnia* spp. involved use of both large (12,000 liter) and small (3.78 liter) enclosures with natural food or with natural food supplemented with algae or bacteria.

In the laboratory, cladocerans were placed in 500-ml glass beakers filled with 400 ml of Provasoli's mixed medium (D'Agostino and Provasoli 1970), kept at 15°C on a 16H:8H light-dark cycle (Lab-Line environmental chamber, Lab-Line Instr., Inc.) and fed *Chlamydomonas* (10^5 cells/ml) every 4 days. The algae were raised on modified Sager and Granik medium (Sager and Granik 1953) in Petri dishes (agar substrate), resuspended in Provasoli's medium, and diluted to appropriate concentrations. While the Petri dish plating technique allowed a simple check for bacterial contamination in phytoplankton cultures, no attempt was made to run the actual competition experiments under axenic conditions. Several weeks before initiating competition, single-species cultures were started in 500-ml beakers, their medium changed every 4 days (old medium siphoned off and replaced with new medium, algae added), until cladocerans became quite abundant. To initiate competition, a selected number of cultures were split in half, 200 ml of medium that contained one species placed together with 200 ml of medium that contained the second species. Unsplit cultures served as controls. In this procedure, the *Daphnia* × *Chydorus* experiments began with 15 cultures for each species, then continued with 12 single-species controls (actually 12 *Daphnia* controls and 11 *Chydorus* controls, since one *Chydorus* control was accidentally lost) and 6 competition beakers (mixtures of *Daphnia* and *Chydorus*). In the *Daphnia* × *Bosmina* experiments, the sequence began with 8 cultures for each species and continued, after mixing, with 5 single-species controls and 6 competition beakers. In all experiments, both controls and competition beakers were destructively sampled at weekly intervals. Cultures were stirred, 1/20th the volume filtered through No. 20 mesh Nitex netting (75μm opening); all cladocerans were preserved in 10% buffered Formalin and counted.

Enclosure experiments were placed in Lake Mitchell, near Sharon, Vermont; a small, shallow and mesotrophic lake (11.2 ha; Z max = 5 m; chlorophyll *a* range 1.1–15.5 μg/liter). Maintained as a private trout club, Lake Mitchell was selected because the lake is protected (shielded from strong winds), less prone to vandalism (on private land; a caretaker could view the enclosures daily), only weakly stratified during summer, and mesotrophic (hence less likely to cause periphyton build up on enclosure sides). Additionally, fish predation does not appear to be great. Fishes include only rainbow trout (*Salmo gairdneri*; stocked as fingerlings and adults), brook trout (*Salvelinus frontinalis*; stocked as adults), and an unidentified minnow; carefully kept records document the numbers of trout stocked and caught each year. Dominant zooplankters include the herbivorous cladocerans *Daphnia pulicaria* [*D. pulex* according to J. L. Brooks (1957); but more recent workers assign populations of this phenotype to *D. pulicaria*], *D. rosea*, and *Bosmina longirostris* (intermediate form), and the omnivorous copepod *Mesocyclops edax*. A few other zooplankters occasionally exceed 1 indiv./liter (*Ceriodaphnia reticulata, Diaphanosoma brachyurum, Tropocyclops prasinus*, and *Chydorus sphaericus*).

Large-scale experiments during 1977 and 1978 tested whether *Daphnia* would replace *Bosmina* under near-natural conditions in the absence and presence of fish predation. In these experiments, natural assemblages of zooplankton were enclosed in large, rectangular (2 × 2 × 3 m) plastic bags (0.2-mm-thick polyethylene) supported by wooden frames. The framework consisted of pine beams (2 × 4 inch and 2 × 2 inch), bolted together and reinforced by crossbeams. Polyethylene sheeting was stretched over the sides and bottom, sealed, and stapled to the frame. Strips of fiberglass screening, placed between the plastic sheeting and staples, at sites of attachment, prevented the plastic from ripping during transport, installment, and during periods of wave action. Each side of the bag also contained a 200-cm^2 window (No. 20 mesh Nitex netting) to permit limited circulation between enclosure and surrounding lake water. The enclosures were constructed at Lake Mitchell, placed on rowboats, and ferried into position. There they were lowered on their sides until completely full, then tilted upright. Ropes drawn through eyelets at the top and bottom corners were secured to cinderblocks on the lake floor, anchoring the enclosures in water about 3.5 m deep.

To test the effects of various levels of fish predation in 1977, we introduced rainbow trout fingerlings (*S. gairdneri*) into two enclosures shortly after repositioning (three 5-cm fingerlings into enclosure "B"; twelve into "C"). The third enclosure ("A") served as a control (actually an important exclosure, since it excluded naturally occuring rainbow trout from foraging upon the enclosed zooplankton). Zooplankton, water transparency, temperature, chlorophyll *a*, and net phytoplankton were regularly monitored within all enclosures and compared with lake values. Zooplankton samples were taken with a 25-liter Schindler trap from three different depths (0–0.7 m, 1–1.7 m, 2–2.7 m) and combined into one composite sample. Each composite sample thus contained 75 liters of water sampled from surface to near-bottom strata. On each sampling date, four composite samples were taken from each enclosure, two each from opposing sides. Lake zooplankton assemblages were sampled in a similar manner, in the vicinity of the enclosure site and from three other stations ("lake" densities). Zooplankton samples were preserved in a Formalin-sucrose mixture (5% Formalin with 40 g/liter sucrose) which prevents "ballooning" and egg-loss (Haney and Hall 1973).

Since species other than *Daphnia* and *Bosmina* were present in enclosures (e. g. *Diaphanosoma brachyurum, Ceriodaphnia reticulata, Mesocyclops edax, Tropocyclops prasinus, Chydorus sphaericus*), albeit scarce, and conceivably could have exerted some unknown influence upon competition dynamics, we decided to conduct small-enclosure experiments using only *Daphnia* and *Bosmina* concurrently with the large-scale enclosure experiments. These enclosures followed the design described by Kerfoot (Kerfoot 1977; Kerfoot and Pastorok 1978), i. e. glass jars (1 gallon) covered by Nitex netting, (70 μm in 1977, 48 μm in 1978), placed in sets of three within wire baskets, and suspended at about 1–2m depth beside the large-scale enclosures. Before conducting these experiments, and in order to answer some questions about food exchange through the netting, we conducted dye experiments.

Gallon (3.78 liter) glass jar containers, suspended in wire baskets and covered by

Nitex netting, have yielded reliable results in oligotrophic to mesotrophic lakes (Kerfoot 1977). Yet because these vessels are often used as modified settling chambers, termed "sediment traps" (Davis 1967), there is some justifiable concern about the "naturalness" of the environment within the enclosure. Aside from the fact that they enclose a rather small volume of water (adequate for natural densities of most herbivores, e. g. *Daphnia* and *Bosmina*, but confining for normal dispersion of many predators, e. g. *Epischura* or *Leptodora*), previous to our experiments we have known little about the amount of water and phytoplankton exchange between the lake and suspended jars. For example, rapid flushing of resources, i. e. a high renewal, relative to grazing, rate would preclude competition since the enclosed grazers could have little or no effect on their resources (Lynch 1977). On the other hand, some exchange is desirable to minimize "enclosure effects." To examine exchange, we added methylene blue dye to a series of jars covered by various mesh sizes of Nitex netting and to a control jar (plastic lid screwed on, impermeable to exchange). Otherwise the jars were treated identical to routine competition experiments. A spectrometer was used to measure relative changes in dye concentrations and from these data we calculated water exchange rates. The test results (Fig. 4) demonstrated significant mixing between lake and bottles, including effects of mesh size and weather conditions on the rate of dilution. Concentration values suggested that water exchange was strongly affected by wind, as wave action produced a pumping action, bobbing the bottles up and down through the water column. If we assume a simple exponential dilution of dye, the mean water exchange rates were 24% per day with 48-μm-mesh netting, 26% per day with 70-μm netting, and 148% per day with 280-μm netting. Using introduced dye, we routinely monitored exchange rates during the competiton experiments. Over the course of our experiments, mean exchange rates ranged from 11–16% per day; a reasonable compromise between rates too

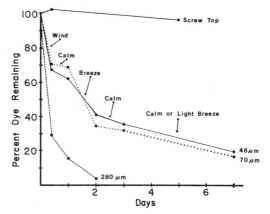

Fig. 4. Decline in concentration of methylene blue dye within suspended 4-liter jars (top line is control, others give dilutions within jars covered by different-sized meshes). Experiments run during early June, 1978.

high for grazers to depress resources and rates which allow no exchange. Since periphyton and detritus blocked only about 10–20% of the total netting aperture area at the termination of competition experiments, there appeared to be a continuous exchange of water and small food particles between the lake and experimental jars during these experiments.

In the small-enclosure experiments, enclosed Cladocera were given different treatments. Always, control jars received 40 *Bosmina* or 20 *Daphnia* each, while competition jars received a mixture of 40 *Bosmina* and 20 *Daphnia*. Jars were left suspended for 2½–4 weeks. In treated experiments, both control and competition jars were enriched with either *Chlamydomonas* (4–21 July 1978 = 12,500 cells/ml every 3 days; 2–28 September 1978 = 6,250 cells/ml every 3 days) and/or with *Aerobacter* (0.25 μg per ml dry wt every 3 days; 2–28 September 1978). The supplemental food regime provided the opportunity to directly test general food limitation of *Bosmina* and *Daphnia* populations, as well as effects of specific resource pulses (phytoplankton "blooms") under near-natural concentrations. Water temperature and chlorophyll *a* concentrations were monitored during each experiment and, at the end of each interval, zooplankton were preserved and water samples taken for phytoplankton analysis (Lugol's samples; cell counts, volumes, and size-frequency spectra using Particle Data Electrozone/Celloscope model 112LTHN/ADC, 48-μm aperture, with added computer interface programs). All zooplankton contaminants were also preserved, identified, and their concentrations determined. In addition, we calculated several demographic parameters for lake and enclosure populations. The number of eggs per reproductive adult (mean brood sizes, egg ratio for reproductively mature females) was determined directly under a dissecting microscope (Zeiss 1-B, 60X), while egg ratios and birth rates were calculated for samples according to the methods described by Edmondson (1968) and Paloheimo (1974).

Results and discussion

In laboratory beaker experiments, when both species were fed the same resource (*Chlamydomonas*), *Bosmina* was severely depressed and eventually eliminated in all beakers (Fig. 5). Although all instars of *Daphnia* and *Bosmina* ingested *Chlamydomonas*, *Daphnia* populations reached much higher standing biomass totals than *Bosmina*, suggesting that they can utilize this resource more effectively. Shortly after initiation of competition, *Bosmina* controls plateaued at ca. 2.9×10^3 individuals, while *Daphnia* maintained control levels around 2.0×10^3 individuals. In the competition beakers, following the halving of density during mixing, *Daphnia* quickly rebounded to control levels, while *Bosmina* declined rapidly to extinction within 21 days.

In contrast to the *Daphnia* × *Bosmina* experiments, when *Daphnia* competed against a small zooplankter (*Chydorus*) that foraged mainly on the bottom and sides of beakers, both species persisted for extended periods of time. In these experiments, 15 days after initiation of competition, *Daphnia* in control beakers plateaued at ca. 2.0×10^3 individuals, while *Chydorus* in control beakers gradually increased from 1.6×10^3 to 3.6×10^3 individuals. The continued increase of *Chydorus* within control beakers was associated with behavioral changes (movement

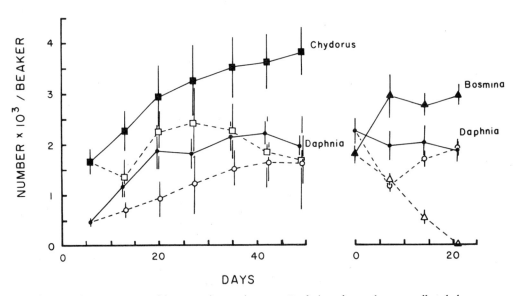

Fig. 5. Laboratory competition experiments between *Daphnia pulex* and two smaller cladocerans: *Chydorus sphaericus* and *Bosmina longirostris* (*Chydorus*: ■—controls; □—competition; *Daphnia*: ●—controls; ○—competition; *Bosmina*: ▲—controls; △—competition). Vertical lines give 95% C. I. for means, days represent number of days since initiation of competition. All experiments run at 15°C.

from sides to interior: Kerfoot and Pastorok 1978). Although *Chydorus* densities began to decline in competition vessels after 25 days, both species persisted for up to and beyond 45 days. These and other experiments (Kerfoot and Pastorok 1978; Neill pers. commun.) suggest that when *Daphnia* and *Bosmina* share a single algal resource under laboratory conditions, *Daphnia* can quickly depress or eliminate *Bosmina*.

In contrast, the large-scale (12,000 liter) enclosure experiments run in Lake Mitchell tested whether *Daphnia* would replace *Bosmina* under near-natural conditions on a varied resource spectrum. Only the results of the 1977 large enclosures are discussed here. Replicated (6 large enclosures) experiments run during summer and fall of 1978 verified the 1977 results, and increase our confidence in their generality (DeMott manus.). As described above, fish were introduced into two enclosures (rainbow trout fingerlings, *S. gairdneri*, ca. 5 cm long; 3 in enclosure "B", 12 in enclosure "C"), while the third enclosure ("A") served as a control. Two important points need re-emphasis here. The control enclosure ("A") actually served as an exclosure experiment, since the bag effectively excluded any naturally occurring trout fingerlings from foraging on the enclosed zooplankton community; hence if the "size-efficiency" hypothesis applies, *Daphnia* would be expected to suppress and eventually eliminate *Bosmina*. In the experimental enclosures, where fish suppress *Daphnia*, *Bosmina* and other small species should increase in density as they feed upon the surplus resources.

Table 1 documents that the large enclosures verified the selective effects of fish predation, as *Daphnia* declined in "B" and became virtually extinct in "C." The declining *Daphnia* populations in these enclosures were characterized by skewed age distributions (loss of large adults) and a decrease in the smallest observed size for primiparous individuals (from 1.30 to 1.15 mm in total length). The large enclosure experiments, however, did not confirm expectations about *Bosmina* response. *Daphnia* did not replace *Bosmina* in the control enclosure ("A"), nor did *Bosmina* increase in fish enclosures any more than in the control enclosure. Absolute densities of *Bosmina* increased within all plastic bags, yet the increases were similar and apparently independent of *Daphnia* concentration. Fluctuations of egg ratios and birth rates for *Daphnia* and *Bosmina* were similar between control ("A") and lake populations (Fig. 6), reinforcing our confidence in the results (Table 1). The extremely low birth rates observed at the beginning of the experiment suggested that lake populations were severely limited by food resources and that the sharp decline of *Daphnia* in both the lake and the fishless control enclosure ("A") was due to starvation. A high flushing rate, due to exceptionally heavy rain during the last week in September and the first week in October, may explain the decline in all lake populations relative to control enclosure populations during the later half of the experiment. While *Bosmina* did not show a noticeable enrichment within predation enclosures, *Tropocyclops* dramatically increased with *Daphnia* removal (Table 1). Adults of *M. edax*, a potential predator of *Tropocyclops*, were most abundant at intermediate levels of fish predation (encl. "B").

When placed together in small-enclosure experiments (3.78-liter bottles) for 17–26 days, *Bosmina* and *Daphnia* showed weak, yet variable competitive interactions with no instances of competitive exclusion (Table 2). During four intervals, involving numerous bottles, *Daphnia* showed significant depression in competition vessels (under conditions of unenriched, natural resources) only during the 16 June–3 July 1978, and 2–28 September trials, although densities were generally lower in competition jars than in controls. Likewise, *Bosmina* showed significant depression in competition jars only during the 2–28 September 1978 interval, although densities again were generally less in competition jars than in controls.

When the natural phytoplankton, bacteria, and detritus were supplemented with specific, laboratory-grown algae or bacteria, zooplankton responses to the

Table 1. Results of enclosure experiments (all estimates based on sample size of 4). Results of large-scale enclosures given as numbers/liter±1 SE. Lake temperature (°C) given for 2-m depth; surface and bottom temperatures varied by barely more than 1°C during time interval of experiments.

	Date	Lake	A0	B3	C12	Lake temp.
Daphnia	9 Sep	17.2±1.9	12.6±1.8	8.6±1.8	2.7±1.0	18.9
	12 Sep	20.5±2.8	13.1±0.6	7.9±1.5	0.8±0.5	17.8
	16 Sep	21.8±4.1	7.5±1.1	4.4±1.2	0.3±0.1	17.0
	20 Sep	5.2±0.6	5.9±1.1	1.7±0.3	0.1±0.0	15.9
	27 Sep	3.0±1.0	2.8±0.5	0.2±0.1	0.0±0.0	13.1
	4 Oct	3.5±1.0	3.5±0.8	0.0±0.0	–	12.1
	11 Oct	2.4±0.8	2.4±0.5	0.0±0.0	–	10.0
	18 Oct	0.6±0.2	0.6±0.1	0.0±0.0	–	7.9
	25 Oct	0.8±0.5	0.5±0.3	0.0±0.0	–	7.2
Bosmina	9 Sep	16.0±1.9	19.6±3.0	47.7±7.6	12.3±4.6	
	12 Sep	64.7±5.5	28.1±9.3	44.5±10.3	16.4±2.7	
	16 Sep	55.4±16.9	25.0±8.7	39.2±7.5	12.2±0.7	
	20 Sep	31.6±8.5	62.6±18.1	44.5±4.5	23.1±2.9	
	27 Sep	33.4±6.7	46.4±6.4	54.2±11.5	67.2±10.5	
	4 Oct	24.2±4.0	83.9±7.3	53.6±3.2	54.5±9.7	
	11 Oct	12.9±6.7	58.3±6.2	59.8±8.7	57.2±6.2	
	18 Oct	8.7±3.6	86.7±12.7	116.4±21.4	118.7±16.8	
	25 Oct	43.8±13.8	31.8±8.7	91.5±20.7	54.5±10.6	
Tropocyclops	9 Sep	1.7±0.2	1.7±0.3	1.7±0.4	1.4±0.2	
	12 Sep	1.8±0.4	1.7±0.5	1.5±0.1	0.9±0.3	
	16 Sep	2.3±0.5	1.7±0.1	1.6±0.3	1.1±0.2	
	20 Sep	1.7±0.2	2.3±0.3	3.1±0.2	3.7±0.2	
	27 Sep	3.4±0.5	3.2±0.5	5.5±0.8	9.5±1.2	
	4 Oct	1.1±0.3	1.4±0.3	6.3±0.7	8.9±1.4	
	11 Oct	0.7±0.2	0.8±0.2	4.2±0.8	9.6±1.9	
	18 Oct	0.2±0.0	0.4±0.1	4.5±0.7	9.8±1.4	
	25 Oct	0.3±0.1	0.1±0.1	3.6±1.0	5.2±1.0	

food pulse were both obvious and interesting. Enrichment with *Chlamydomonas* dramatically increased *Daphnia* densities 20–30X during the 4–17 July experiments, while half that concentration stimulated a 2–3 fold increase during the 2–28 September experiments (Table 2). Enrichment with comparable biomass of *Aerobacter* during 2–28 September also more than doubled *Daphnia* concentrations over unenriched controls. Interestingly, enrichment with *Chlamydomonas* did not significantly alter *Bosmina* densities during the 4–21 July experiments, but did dramatically increase densities 5–8X during the 2–28 September experiments (Table 2). Simultaneous enrichment with both *Chlamydomonas* and *Aerobacter* increased both *Bosmina* and *Daphnia* densities, but only slightly above results when *Chlamydomonas* was used alone. While no competition was significantly demonstrated during the 4–21 July enrichment with *Chlamydomonas*, competition effects were significant during all 2–28 September enrichment experiments. Although neither species could eliminate the other, *Daphnia* consistently had more of a suppressing effect on *Bos-*

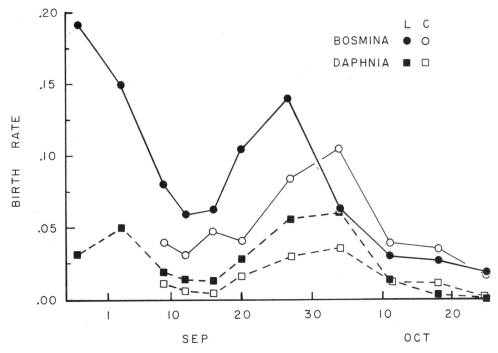

Fig. 6. Birth rates (b) for *Bosmina* and *Daphnia* in Lake Mitchell (L) and in the large enclosures (C).

mina under algae enrichment, while *Bosmina* suppressed *Daphnia* more strongly under bacterial enrichment.

Although zooplankton contaminants often pose a problem in long-duration in situ experiments, the occurrence of contaminants during our Lake Mitchell study was sporadic, their densities were usually low, and in no instance did contaminant species appear to effect either *Bosmina* or *Daphnia*. Rotifer and copepod densities were <1 per liter except in the 2–28 September experiments, when *Tropocyclops* reached densities of 3/liter in unenriched jars and 24/liter in jars enriched with *Chlamydomonas*. *Chydorus sphaericus* was present as a contaminant during all experiments, but only reached high densities (16–65/liter) during the midsummer experiments (in three of the *Chlamydomonas*-enriched jars: one *Bosmina* control, one *Daphnia* control, and one competition jar). Neither *Bosmina* nor *Daphnia* were affected by the presence of *Chydorus*, a finding consistent with the results of our *Daphnia-Chydorus-Bosmina* laboratory competition experiments (Fig. 5). Due to the presence of *Bosmina*, results from two *Daphnia* control jars were discarded (16 June–3 July experiment: 27 *Bosmina*, 20 *Daphnia*; 4–21 July experiment: 5 *Bosmina*, 6 *Daphnia*).

Certain aspects of the enrichment experiments warrant additional comment. During the 2–28 September experiments, both *Bosmina* and *Daphnia* responded to additions of *Chlamydomonas* and/or bacteria, and exhibited significant competitive interactions. In contrast, no competition was uncovered during the 4–21 July *Chlamydomonas*-enrichment experiments—a surprising result considering the magnitude of *Daphnia* response to enrichment (20–30X increase). Data obtained by particle counter techniques, taken 3 days after the last addition of *Chlamydomonas*, confirm that the high densities of *Daphnia* within enriched control and competition jars had depressed the total volume of fine parti-

Table 2. Results of *Bosmina-Daphnia* competition in natural and enriched small enclosures (Asterisks after species name indicate significant enrichment effects, where $*p < 0.05$, $**p < 0.01$, and $***p < 0.001$; other asterisks indicate significant density depression between competition and control jars).

Date	Incubation period (days)	Temp(°C)	Water exchange (% day)	Chl a (Lake) (μg/liter)	Treatment	Species	N	Controls	N	Competition
16 Sep–9 Oct 1977	23	16.8–10.2	NR	5.4–15.5	Natural algae	*Bosmina*	2	25.9±2.4	6	19.7±2.9 $p<0.30$
						Daphnia	1	14.5	6	20.2±2.0 $p<0.40$
16 Jun–3 Jul 1978	18	16.4–21.2	15.6 (T=4.4)†	1.4–2.7	Natural algae	*Bosmina*	3	35.1±7.4	3	22.4±5.9 $p<0.30$
						Daphnia	2	8.8±0.9	3	1.5±0.3 $p<0.01**$
4–21 Jul 1978	17	20.5–22.8	11.2 (T=6.2)	1.1–3.6	Natural algae	*Bosmina*	3	2.1±0.6	3	1.9±0.6 $p<0.70$
						Daphnia	2	3.2±0.5	3	2.3±0.3 $p<0.30$
					Chlamydomonas 12,500 cell/ml Every 3 days	*Bosmina*	3	2.1±0.2	3	3.2±1.2 $p<0.50$
						*Daphnia****	3	84.4±5.2	3	66.5±8.2 $p<0.20$
2–28 Sep 1978	26	19.5–14.0	NR	4.3–8.1	Natural algae	*Bosmina*	3	33.4±4.9	3	8.5±2.2 $p<0.01**$
						Daphnia	3	16.1±1.5	3	12.1±0.8 $p<0.05*$
					Chlamydomonas 6,250 cells/ml Every 3 days	*Bosmina***	3	167.2±16.5	3	66.0±9.1 $p<0.01**$
						*Daphnia****	3	49.5±3.1	3	26.8±6.7 $p<0.05*$
					Bacteria (0.25μg per ml) Every 3 days	*Bosmina***	3	85.0±25.9	3	78.8±26.2 $p<0.80$
						*Daphnia**	3	35.2±4.7	3	2.5±1.1 $p<0.01**$
					Chlamydomonas & bacteria Every 3 days	*Bosmina***	3	173.7±30.1	3	72.9±9.1 $p<0.05*$
						*Daphnia****	3	54.3±6.5	3	33.1±3.2 $p<0.05*$

† T = time (in days) for 50% exchange.

Table 3. Particle counter (electronic) measurements of total particulate volume (between 1.3–19-μm equivalent spherical diameter) given as mean ($\times 10^{-4}$ μm^3/ml) ± SE.

Source	N	Volume
Lake Mitchell (3 Jul)	1	16.6
Lake Mitchell (11 Jul)	2	22.6±0.2
Lake Mitchell (21 Jul)	2	21.2±2.6
Unenriched jars	6	14.0±1.7
Enriched *Bosmina* control	3	20.9±4.6
Enriched *Daphnia* control and competition	4	8.4±0.2

culate matter (Table 3). The volume of fine particulates found in enriched *Bosmina* control jars, where *Bosmina* showed no response to enrichment, and in unenriched jars was similar to values for Lake Mitchell. Hence the enclosure results not only document release of *Daphnia* from natural conditions of food limitation within Lake Mitchell, but also demonstrate density-dependent depression of phytoplankton following increased herbivore abundance. The failure of *Bosmina* to respond to the artificially created *Chlamydomonas* "bloom" and to show little evidence of suppression in competition jars with *Daphnia*, strongly suggest that (at the time of the experiments) these small herbivores were reacting to a set of resources not shared with *Daphnia*.

Conclusions

While congeners (i. e. species within the genus *Bosmina* or *Daphnia*) can show dramatic replacements during field competition experiments (Kerfoot 1977; Kerfoot and Pastorok 1978; Jacobs 1977b; DeMott manus.), suggesting that morphologically similar pairs of species do compete intensively in nature for shared resources, the results reported here suggest that competitive interactions between *Bosmina* and *Daphnia* in nature are relatively weak. If forced to share a single, limiting resource under artificial and restricted laboratory conditions, *Daphnia* can rapidly outcompete *Bosmina*. However, if both species are allowed to utilize the natural

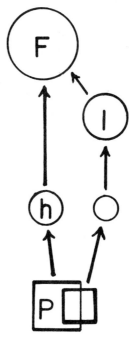

Fig. 7. Incorporation of particulate heterogeneity into conceptual models of zooplankton coexistence. Symbols as in Fig. 1.

resource spectra of lakes, competition is difficult to demonstrate, coexistence is prolonged, and the two species often show independent fluctuations in demographic variables and densities. The complete absence of *Bosmina* response during one experimental series to *Chlamydomonas* enrichment and the enhanced ability of *Bosmina* to suppress *Daphnia* in the presence of *Aerobacter* strongly suggest that *Bosmina* may utilize alternative resources (such as certain bacteria) in preference to certain algal species.

The individualistic responses of *Bosmina* and *Daphnia*, we believe, strongly suggest that the resource spectrum of bacteria, algae, and detritus in Lake Mitchell is diverse enough, and the foraging of both genera specialized enough, that both species can coexist in the absence of appreciable predation from either fishes or predatory invertebrates. Thus, conceptual models of zooplankton community interactions must incorporate resource diversity as a factor which stabilizes coexistence between zooplankters and which hence promotes zooplankton diversity (Fig. 7).

That zooplankton possess the sensory and behavioral equipment to be choosy about their diets certainly appears confirmed by recent disclosures on feeding habits (Friedman and Strickler 1975; Poulet and Marsot 1978; Strickler at this conf.). Although cladocerans may not be as selective as copepods in their choice and manipulation of foods, there are arguments for a certain degree of resource specialization (Gliwicz 1969, 1977; Lynch 1978).

References

ALLEN, T., and J. F. KOONCE. 1973. Multivariate approaches to algal strategems and tactics in systems analysis of phytoplankton. Ecology 54: 1234–1246.

BROOKS, J. L. 1957. The systematics of North American *Daphnia*. Mem. Conn. Acad. Arts Sci. 13: 1–180.

———, and S. I. DODSON. 1965. Predation, body size, and composition of plankton. Science 150: 28–35.

BURNS, C. W. 1968. The relationship between body size of filter-feeding Cladocera and the maximum size of particle ingested. Limnol. Oceanogr. 14: 675–678.

CASWELL, H. 1978. Predator-mediated coexistence: a nonequilibrium model. Am. Nat. 112: 127–154.

CRAMER, N. F., and R. M. MAY. 1971. Interspecific competition, predation, and species diversity: a comment. J. Theor. Biol. 34: 289–293.

D'AGOSTINO, A., and L. PROVASOLI. 1970. Dixenic culture of *Daphnia magna* Straus. Biol. Bull. 139: 485–494.

DAVIS, M. B. 1967. Pollen deposition in lakes as measured by sediment traps. Geol. Soc. Am. Bull. 78: 849–858.

DODSON, S. I. 1970. Complementary feeding niches sustained by size-selective predation. Limnol. Oceanogr. 15: 131–137.

———. 1974. Zooplankton competition and predation: An experimental test of the size-efficiency hypothesis. Ecology 55: 605–613.

EDMONDSON, W. T. 1968. A graphical model for evaluating the use of the egg ratio for measuring birth and death rates. Oecologia 1: 1–37.

EMLEN, J. M. 1973. Ecology: An evolutionary approach. Addison-Wesley.

FRANK, P. W. 1952. A laboratory study of intraspecies and interspecies competition in *Daphnia pulicaria* (Forbes) and *Simocephalus vetulus* (O. F. Müller). Physiol. Zool. 25: 173–204.

———. 1957. Coactions in laboratory populations of two species of *Daphnia*. Ecology 38: 510–519.

FRIEDMAN, M. M., and J. R. STRICKLER. 1975. Chemoreceptors and feeding in calanoid copepods (Arthropoda: Crustacea). Proc. Natl. Acad. Sci. 72: 4185–4188.

GAUSE, G. F., and A. A. WITT. 1935. Behavior of mixed populations and the problems of natural selection. Am. Nat. 69: 596–609.

GILBERT, L. E., and P. H. RAVEN. [eds.]. 1975. Coevolution of animals and plants. Univ. Texas Press, Austin.

GLIWICZ, Z. M. 1969. Studies on the feeding of pelagic zooplankton in lakes with varying trophy. Ekol. Pol. 17: 663–708.

———. 1977. Food size selection and seasonal succession of filter feeding zooplankton in an eutrophic lake. Ekol. Pol. 25: 179–225.

HANEY, J. F., and D. J. HALL. 1973. Sugar-coated *Daphnia*: A preservation technique for Cladocera. Limnol. Oceanogr. 18: 331–333.

HARPER, J. L. 1977. Population biology of plants. Academic Press.

HRBAČEK, J. 1962. Species composition and the amount of zooplankton in relation to the fish stock. Rozpr. Cesk. Akad. Ved Rada Mat. Prir. Ved 72: 1–116.

———, and M. NOVOTNA-DVORAKOVA. 1965. Plankton of four backwaters related to their size and fish stock. Rozpr. Cesk Akad. Ved Rada Mat. Prir. Ved 75: 3–10.

HUTCHINSON, G. E. 1953. The concept of pattern in ecology. Proc. Acad. Nat. Sci. Phila. 105: 1–12.

———. 1961. The paradox of the plankton. Am. Nat. 95: 137–146.

———. 1967. A treatise on limnology, v. 2. Wiley.

JACOBS, J. 1977a. Coexistence of similar zooplankton species by differential adaptation to reproduction and escape, in an environment with fluctuating food and enemy densities. I. A model. Oecologia 19: 233–247.

———. 1977b. Coexistence of similar zooplankton species by differential adaptation to reproduction and escape, in an environment with fluctuating food and enemy densities. II. Field data analysis of *Daphnia*. Oecologia 30: 313–329.

———. 1978. Coexistence of similar zooplankton species by differential adaptation to reproduction and escape, in an environment with fluctuating food and enemy densities. III. Laboratory experiments. Oecologia 35: 35–54.

JANZEN, D. H. 1970. Herbivores and the number of tree species in tropical forests. Am. Nat. 104: 501–528.

KERFOOT, W. C. 1975. The divergence of adjacent populations. Ecology 56: 1298–1313.

———. 1977. Competition in cladoceran communities: The cost of evolving defenses against copepod predation. Ecology **58**: 303–313.

——, and R. A. PASTOROK. 1978. Survival versus competition: Evolutionary compromises and diversity in the zooplankton. Int. Ver. Theor. Angew.Limnol. Verh. **20**: 362–374.

——, and C. PETERSON. 1979. Ecological interactions and evolutionary arguments: Investigations with predatory copepods and *Bosmina*, pp. 159–196. *In* Population Ecology (U. Halbach and J. Jacobs, ed.). Fortschr. Zool. **25**, 2/3: 1–409. Gustav Fischer.

KETTLE, D., and W. J. O'BRIEN. 1978. Vulnerability of arctic zooplankton species to predation by small lake trout (*Salvelinus namaycush*). J. Fish. Res. Bd. Can. **35**: 1495–1500.

LEVINTON, J. 1972. Stability and trophic structure in deposit-feeding and suspension-feeding communities. Am. Nat. **106**: 472–486.

LEWIS, W. M. 1977. Ecological significance of the shapes of abundance-frequency distributions for coexisting phytoplankton species. Ecology **58**: 850–859.

LYNCH, M. 1977. Zooplankton competition and plankton community structure. Limnol. Oceanogr. **22**: 775–777.

——. 1978. Complex interactions between natural coexploiters—*Daphnia* and *Ceriodaphnia*. Ecology **59**: 552–564.

MacARTHUR, R. H. 1965. Patterns of species diversity. Biol. Rev. **40**: 510–533.

——, and J. MacARTHUR. 1961. On bird species diversity. Ecology **42**: 594–598.

MAY, R. 1973. Stability and complexity in model ecosystems. Princeton Univ. Press.

NEILL, W. E. 1975. Experimental studies of microcrustacean competition, community composition and efficiency of resource utilization. Ecology. **56**: 809–826.

PAINE, R. T. 1966. Food web complexity and species diversity. Am. Nat. **100**: 65–75.

——. 1969. The *Pisaster-Teguia* interaction: Prey patches, predator food preference, and intertidal community structure. Ecology **50**: 950–961.

PALOHEIMO, J. E. 1974. Calculation of instantaneous birth rate. Limnol. Oceanogr. **19**: 692–694.

PARRISH, J. D., and S. B. SAILA. 1970. Interspecific competition, predation and species diversity. J. Theor. Biol. **27**: 207–220.

POULET, S. A., and P. MARSOT. 1978. Chemosensory grazing by marine calanoid copepods (Arthropoda: Crustacea). Science **200**: 1403–1405.

SAGER, R., and S. GRANIK. 1953. Nutritional studies with *Chlamydomonas reinhardti*. Ann. N. Y. Acad. Sci. **56**: 831–838.

SLOBODKIN, L. B. 1961. The growth and regulation of animal populations. Holt, Rinehart and Winston.

VANCE, R. R. 1978. Predation and resource partitioning in one predator-two prey model communities. Am. Nat. **112**: 797–813.

YODZIS, P. 1976. The effects of harvesting on competitive systems. Bull. Math. Biol. **38**: 97–109.

——. 1977. Harvesting and limiting similarity. Am. Nat. **111**: 833–843.

ZARET, T. 1972a. Predator-prey interactions in a tropical lacustrine ecosystem. Ecology **53**: 248–257.

——. 1972b. Predators, invisible prey, and the nature of polymorphism in the Cladocera (class Crustacea). Limnol. Oceanogr. **17**: 171–184.

——, and W. C. KERFOOT. 1975. Fish predation on *Bosmina longirostris*: Body-size selection versus visibility selection. Ecology **56**: 232–237.

65. The Inadequacy of Body Size as an Indicator of Niches in the Zooplankton

B. W. Frost

Abstract

Theoretical analyses of competitive interactions among filter-feeding zooplankters frequently utilize body size as a basis for scaling rates of physiological processes. Although metabolic rate seems to be adequately represented by a function which depends on body size and is independent of species, other processes, such as maximum filtration rate, ingestion rate, development rate and growth rate, may follow size-related functions which vary among species. This is exemplified by results of physiological studies on the marine filter-feeding copepods *Calanus pacificus* and *Pseudocalanus* sp. The effects on zooplankton species composition of unpredictable interspecific variation in size-related physiological processes is demonstrated with simulations of the dynamics of competing populations of *C. pacificus* and *Pseudocalanus* sp.

Stimulated by the pioneering study of Brooks and Dodson (1965), several recent theoretical analyses of potential competitive interactions among filter-feeding zooplankters have utilized body size as a means of scaling rates of physiological processes for animals of different sizes (Allan 1974; Hall et al. 1976; Lynch 1977). Because these studies all involve time-independent analytical approaches they are limited to suggesting potential competitive advantage, in terms of fitness or rate of population increase, for one particular size of planktonic filter feeder under a defined, unchanging set of environmental conditions. Further, they necessarily ignore the fact that species of filter feeders experience substantial changes in body size during development and that competition may be most intense among different sizes or developmental stages of species (Neill 1975). An equally serious limitation of the aforementioned theoretical studies concerns the validity of the central assumption that body size may indeed define the rate of physiological processes among different sizes and species of filter feeders.

It is well known that respiration rate R of planktonic animals can be described by a power function of body weight

$$R = aW^b$$

where a varies with species and temperature, but b tends toward 0.75 for all species. A similar relationship seems to describe maximum assimilation rate A as a function of body size for planktonic animals (e.g. Dagg 1976; Lampert 1977)

$$A = cW^d.$$

Recognizing that respiration and assimilation are the major gain and loss components of material or energy budgets, we then find, following Steele (1974), the

---

Contribution 1063 from the Department of Oceanography, University of Washington, Seattle 98195. Research supported by National Science Foundation grant OCE 74–22640.

dependence of maximum individual growth rate G_m on body size can be described as

$$G_m = cW^d - aW^b. \quad (1)$$

Growth rate as a function of body size therefore depends on the pattern of variation of both the weight coefficients a and c, and the values of the exponents b and d in the power functions. Both Allan (1974) and Lynch (1977) assumed that a, c, b, and d were constants and that b and d had different values. If b and d were indeed different their relative magnitudes would certainly be critical in determining size-dependent patterns of growth rate. Yet, if anything, the experimental data suggest great similarity of b and d (e.g. Dagg 1976; Lampert 1977) and in this paper they are assumed equal. Since for the moment a and c may be assumed constant for a single species under specific environmental conditions,

$$G_m = fW^b. \quad (2)$$

If food concentration is below that permitting G_m, then individual growth rate depends upon filtration rate and filtration efficiency which may both also be functions of body size (Hall et al. 1976; Lynch 1977).

In previous theoretical studies it was assumed that the weight coefficients, such as a and c in Eq. 1, are either constants or vary with temperature, but do not vary within species or interspecifically with body size. It is my intent here to illustrate a case in which the coefficients do vary unpredictably among species and to demonstrate the effect of such variation on potential competitive interactions.

Background

The zooplankton in mid- to high-latitude oceans, particularly in the northern hemisphere, tends to be dominated in terms of biomass by filter-feeding copepods of the genus *Calanus*. In the North Atlantic Ocean, which has been most intensively studied, *Calanus finmarchicus* frequently constitutes more than 90% of the biomass of filter-feeding copepods during the growing season (e.g. Bainbridge and Forsyth 1972; Cushing and Vucetic 1963; Sherman and Perkins 1971; Wiborg 1954). The remarkable feature of this pattern of dominance is that *C. finmarchicus* is a very large-sized species of copepod (adult about 100 µg C body weight) whereas the other numerically important genera of filter feeders which co-occur with *C. finmarchicus* (*Acartia, Paracalanus, Pseudocalanus, Temora*) are much smaller ($\lesssim$ 20 µg C body weight as adults). A similar case of dominance in the zooplankton is evident in waters over continental shelves in temperate-boreal waters of the North Pacific Ocean, although the species are different. For example, *Calanus pacificus* and *Calanus marshallae* are the North Pacific analogues of the closely related *C. finmarchicus* (Frost 1974). In the open subarctic Pacific, species of large-sized filter feeders also dominate the zooplankton for reasons touched upon later.

The analysis presented below is illustrated with data on two North Pacific species of filter-feeding copepods, *C. pacificus* and *Pseudocalanus* sp. (Fig. 1). The latter species closely resembles, but seems to be distinct from, the North Atlantic *P. elongatus* (Boeck). *Calanus pacificus* and *Pseudocalanus* sp. co-occur and are abundant in Puget Sound, Washington, and have been the subjects of numerous studies on feeding, development, growth, and reproduction.

Methods

Some factors affecting the relative abundance of *C. pacificus* and *Pseudocalanus* sp. were investigated by means of a modified form of Landry's (1976) simulation model of temperate ocean plankton dynamics. The multiple cohort reproduction version of the model was used. One modification was a more realistic description of reproduction and survival of eggs and the first two nauplier stages. Specifically, the factor X, used by Landry (1976) to convert stored reproductive products to recruited stage III nauplii, was deleted and eggs and the first two nauplier stages were included in each daily

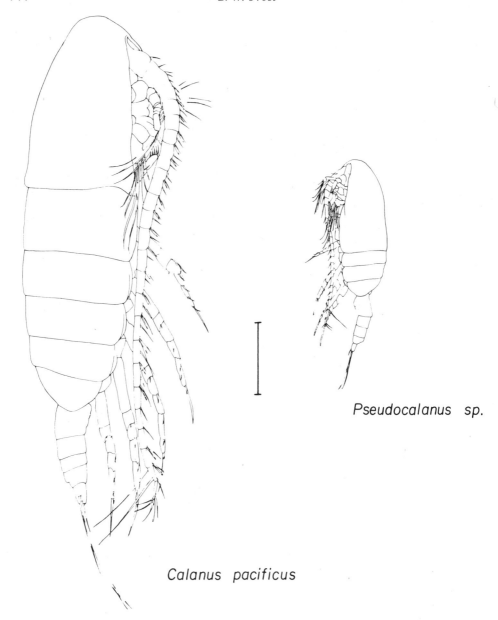

Fig. 1. Adult females (lateral view) of *Calanus pacificus* and *Pseudocalanus* sp. Scale bar—0.5 mm.

estimate of population size and subjected during their development to the population predation rate. The prediction of the model with this modification (Fig. 2) differed relatively slightly from that of Landry's (cf. his fig. 2B, P1 = 75), the main departure being larger population size of the copepod because eggs and the first two naupliar stages were included. The model explicitly describes changes in population structure so that population growth of the copepod was realized as a series of rela-

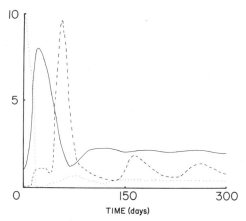

Fig. 2. Simulated dynamics of plankton in a temperate ocean. Dotted line—dissolved plant nutrient (as μg-atoms N liter^{-1}); solid line—phytoplankton concentration (as μg C liter^{-1} × 10^{-2}); dashed line—grazing zooplankton assumed to be a copepod similar to *Calanus finmarchicus* (as no. of individuals m^{-2} × 2 · 10^{-5}).

tively distinct cohorts (Fig. 3). Notice that initial conditions of the model favor a spring bloom of phytoplankton.

The model was further modified to represent simultaneously the population dynamics of two species of filter-feeding copepods, similar to *C. pacificus* and *Pseudocalanus* sp. Population growth of each species was simulated by a submodel identical to that used for population growth of *Calanus* in Fig. 2. However, several parameters, those specifying maximum growth and metabolic rates, shape of the ingestion curve, and predation rate could be varied independently in the two species. Initial conditions for the model were equal population biomasses, consisting wholly of adults, of the two species and a low phytoplankton concentration (15 μg C liter^{-1}). To avoid large weight losses in adults (which would be heavily biased against *Pseudocalanus* because of its smaller size) early in the simulation when the phytoplankton concentration was low, I took weight loss of adults to be negligible if assimilation was less than respiration. To emphasize the role of pure competition for phytoplankton between the species, the description of predation was simple and of the purely density-dependent type. Population sizes of the two species were summed each day to determine the coefficient for density-dependent predation rate. This necessarily

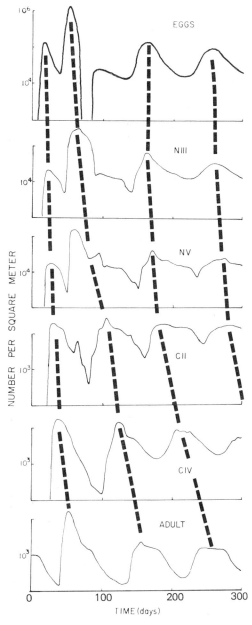

Fig. 3. Abundance of selected developmental stages of copepod population represented in simulation in Fig. 2. Dashed lines connect apparent cohorts.

introduced a bias against the rarer species, the effect of which was subsequently investigated as described later. Four runs of the simulation model are presented (Table 1) and the rationale for specific differences between runs is discussed in detail below.

Experimental data, some currently unpublished, were used to establish values of parameters specifying rates of feeding, development, growth, and reproduction of the two species. Body weights of eggs and adults were determined in the laboratory (Table 2). Optimal development times and growth rates for both species were determined in laboratory cultures provided with excess food (Frost unpubl.; Vidal 1978). Feeding rates of adult females of both species were determined under identical laboratory conditions using one species of diatom as food.

Results

Figure 4a depicts the simulated dynamics of *Calanus* and *Pseudocalanus* assuming that growth in both species is described by a single weight-dependent function of the form of Eq. 1. That is, in terms of Eq. 1, the coefficients c and a (C, E and F combined, in Table 1) are identical for the two species. Further, feeding rate as a function of food concentration (specified by the parameters D and $P1$ in Table 1) is assumed identical for the two species. *Pseudocalanus* quickly dominates and *Calanus* ultimately goes extinct. Steele (1974) noted the same

Table 1. Values of parameters and initial conditions for four runs of a simulation model of plankton including two species of copepods. Subscripted parameters refer to the copepod species (c—*Calanus*; p—*Pseudocalanus*). Dashes indicate values unchanged from Run 1. Units for nitrate are carbon equivalents. Units of A, C, E, and F are μg C μg C^{-1} day^{-1}; B, D, NIT, NO, $PHYT$, $P1$ are μg C liter^{-1}; G, U, and V are fraction day^{-1}; H is numbers m^{-2}.

	Definition	Run 1	Run 2	Run 3	Run 4
A	Maximum uptake of nitrate by phytoplankton	0.2	—	—	—
B	Nitrate concentration for $A/2$	96.0	—	—	—
C_c	Maximum ingestion of phytoplankton by copepod	1.23	—	—	—
C_p	"	1.23	0.97	0.97	0.97
D_c	Phytoplankton concentration at $C \times W^{0.7}/2$	100.0	—	—	—
D_p	"	100.0	—	50.0	50.0
E_c	Maximum copepod respiration dependent on ingestion	0.3	—	—	—
E_p	"	0.3	—	—	—
F_c	Copepod respiration independent of ingestion	0.2	—	—	—
F_p	"	0.2	—	—	—
G	Maximum predation mortality of copepod	0.1	—	—	—
H_c	Copepod abundance for $G/2$	1×10^5	—	—	—
H_p	"	1×10^5	—	—	2×10^5
U	Fraction of copepod excretion in upper layer	0.4	—	—	—
V	Fraction of upper layer replaced by mixing	0.01	—	—	—
NIT	Initial nitrate concentration	756.0	—	—	—
NO	Nitrate concentration in lower layer	756.0	—	—	—
$PHYT$	Initial phytoplankton concentration	15.0	—	—	—
$P1_c$	Copepod feeding threshold	40.0	—	—	—
$P1_p$	"	40.0	—	20.0	20.0

Table 2. Weights (μg C) of eggs and adult females used in simulation model. Weights of adult females are values obtained for laboratory cultures maintained at 12°C with optimal food. Weight of egg of *C. pacificus* directly measured; that of *Pseudocalanus* sp. estimated from measurements of egg diameter (120 μm) and diameter : weight relationship for *C. pacificus* (see also McLaren 1969; Paffenhöfer and Harris 1976).

	Weight (μg C)	
	egg	adult ♀
Pseudocalanus sp.	0.07	8.0
Calanus pacificus	0.25	100.0

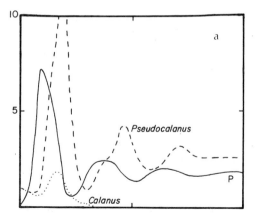

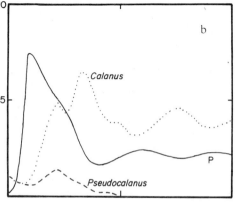

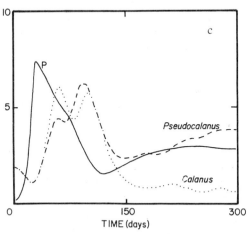

Fig. 4. Simulated dynamics of plankton containing two species of grazing copepods similar to *Calanus pacificus* and *Pseudocalanus* sp. Phytoplankton concentration (P, μg C liter^{-1} × 10^{-2}) is plotted together with population biomasses of the two grazer species (g C m^{-2} × 0.5 in a and b; g C m^{-2} in c). a—simulation Run 1; b—simulation Run 2; c—simulation Run 3.

result with a similar model and suggested that *Pseudocalanus* excludes *Calanus* because it matures faster. This is evident in Fig. 4a; *Pseudocalanus* begins increasing much sooner than *Calanus*. However, as noted earlier, in the ocean *Calanus* species dominate the zooplankton, so the model should be scrutinized for possible flaws in the description of population growth of the copepods.

In the model, dependence of individual growth rate on body size is described by Eq. 1 for optimal conditions. Following Steele (1974), we assume $b = d = 0.7$ so that in terms of Eq. 1 and 2 optimal individual growth rate is

$$G_m = cW^{0.7} - aW^{0.7} = fW^{0.7}. \quad (3)$$

In Run 1 a single value of f was used for both species and the value was derived from estimated growth rate of *Calanus*. Noting that

$$G_m = dW/dt = fW^{0.7},$$

then integrating with respect to time, we obtain

$$W_t = (0.3ft + W_o^{0.3})^{3.33} \quad (4)$$

where W_t is the weight at time t and W_o is the weight of the egg. Equation 4 was fitted to data for *C. pacificus* and describes individual growth reasonably well (Fig. 5). The sigmoid shape of the growth curve is grossly represented in the model by two

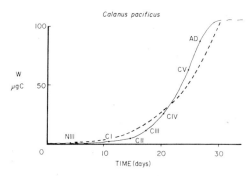

Fig. 5. Optimal growth curve, in terms of body carbon W, for *Calanus pacificus* at 12° C. Solid line and dots—observed optimal growth (Vidal 1978); dashed line—fit of Eq. 4 using data in Table 2, median development time of 31 days, and no growth for first 4.2 days after fertilization of egg.

Table 3. Optimal development times (days from fertilization of egg until molt to adult stage) for *Calanus pacificus* and *Pseudocalanus* sp. Unpublished data based on laboratory cultures of both species maintained with excess food.

(°C)	Pseudocalanus sp.	Calanus pacificus
10	34	36
12	26	26
15.5	21	19

restrictions: that a newly recruited nauplius III is the same weight as the egg and that later stages of copepods grow to a maximum size at maturity (100 µg C in the case of *C. pacificus*), with any additional growth channeled into reproductive products.

For *C. pacificus* at 12°C the median development time (defined as the time from fertilization of the egg until the molt to adult stage) is about 31 days (Vidal 1978), giving from Eq. 4 a value of 0.427 for the coefficient f. If this value is applied to *Pseudocalanus*, where $W_0 = 0.07$ and $W_t = 8.0$ µg C (Table 2), then *Pseudocalanus* would mature in only 15.2 days. The growth curves for *C. pacificus* and *Pseudocalanus*, with $f = 0.427$ for both, are shown in Fig. 6. Because with $f = 0.427$, *Pseudocalanus* would reach maturity much earlier than *Calanus*, it would have the advantage of early reproduction and, as illustrated in simulation Run 1 (Fig. 4a), would quickly swamp *Calanus*.

Faster maturation in *Pseudocalanus*, while perhaps intuitively expected for the smaller species (e.g. Hall et al. 1976), is not supported by results of laboratory studies of development rates. Indeed, *C. pacificus* and *Pseudocalanus* sp. have remarkably similar development times (Table 3). These results agree with those of other

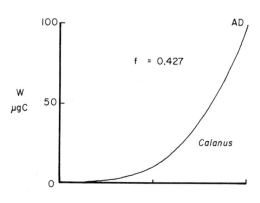

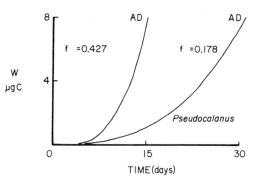

Fig. 6. Modeled optimal growth of *Calanus pacificus* and *Pseudocalanus* sp. (AD—adult stage). Curve with $f = 0.427$ was used in all simulations for *Calanus* and in simulation Run 1 for *Pseudocalanus*. Curve with $f = 0.178$ for *Pseudocalanus* was used in simulation Runs 2–4.

studies on the same or closely related species (e.g. Mullin and Brooks 1970; Paffenhöffer and Harris 1976). Noting from Table 2 that during development

Pseudocalanus increases in weight by about 114 times whereas the factor for *C. pacificus* is 400, we find the two species must have very different growth rates because they take about the same time to mature.

The relatively slower growth rate of *Pseudocalanus* can be introduced into the model by specifying for *Pseudocalanus* a median development time of 31 days (i.e. the same as for *Calanus*) and solving for f in Eq. 4. The corrected optimal growth curve, when $f = 0.178$, is shown in Fig. 6 (lower panel). To incorporate the corrected f in the simulation model we note from Eq. 1 that under optimal conditions

$$G_m = fW^b = (eC - E - F)W^b$$

where e is assimilation efficiency (assumed 70%) and C, E, and F are optimal specific rates of ingestion, feeding-dependent respiration, and resting respiration. Thus f could be corrected either by decreasing C or increasing E and F. Available data on feeding rates and metabolism of *Calanus* and *Pseudocalanus* do not suggest a particular choice; it is assumed that maximum specific ingestion rate differs between the two species. Therefore, simplifying the equation and substituting values from Table 1 we find that

$$f = 0.7C - 0.5$$

and the required adjustment of the ingestion relationship for *Pseudocalanus* is shown in Fig. 7b.

Simulation Run 2 was made with the corrected value of f for *Pseudocalanus*. In this run, even though the two species mature at the same time, *Calanus* dominates and *Pseudocalanus* eventually disappears (Fig. 4b). The superiority of *Calanus* is due to its much higher potential reproductive rate. Optimal reproductive rate is equivalent to optimal growth rate in the adult stage (Eq. 2) divided by the weight of the egg. Thus, optimal reproductive rates are

$$\text{Calanus: } \frac{0.427 (100)^{0.7}}{0.25} = 42.9$$

$$\text{eggs } \female^{-1} \text{day}^{-1};$$

$$\text{Pseudocalanus: } \frac{0.178 (8)^{0.7}}{0.07} = 10.9$$

$$\text{eggs } \female^{-1} \text{day}^{-1}.$$

These rates are reasonably close to my unpublished observations and other observations for the same or closely related species (Paffenhöfer 1970; Corkett and McLaren 1969). The point to be drawn from comparisons of Fig. 4a and b is that a difference in specific growth rate between two species, which is totally unpredictable from body size, leads to radically different predicted structure in the zooplankton.

Calanus and *Pseudocalanus* coexist in the ocean, which could be due to a variety of reasons, but one potentially important aspect of the competitive interaction, also dependent on body size relationships, has not yet been addressed. In the model copepods compete for phytoplankton, and the relationship between ingestion rate I, food concentration P, and body size is given by

$$I = \frac{C(P - P1)}{D + (P - P1)} \times W^b \quad (5)$$

where $P1$ is a food concentration below which ingestion ceases and D is the food concentration at which ingestion rate is half the maximum rate. In Run 1 parameters of Eq. 5 were identical for the two species (Fig. 7a), whereas in Run 2 the parameter C took different values for the two species (Fig. 7b). Results of laboratory feeding experiments with adult females of *C. pacificus* and *Pseudocalanus* sp. indicate that the other parameters, D and $P1$, also differ significantly between the two species (Fig. 8). To approximate the differences, and noting the variation in the data, I took the values of D and $P1$ for *Pseudocalanus* to be half those for *Calanus* (Fig. 7c). The difference is manifested physically by much smaller

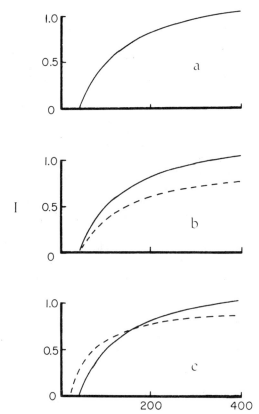

Fig. 7. Specific ingestion rate I (μg C μg C^{-1} day^{-1}) as a function of food concentration P (μg C liter^{-1}). a—Relationship used for both species in simulation Run 1 and for *Calanus* in all succeeding simulations; b—ingestion curve for *Pseudocalanus* (dashed line) adjusted to give lower optimal specific growth rate (simulation Run 2); c—ingestion curve for *Pseudocalanus* (dashed line) adjusted to give more efficient feeding at low food concentration (simulation Runs 3 and 4).

intersetule spacing of the maxillary filter in *Pseudocalanus* than in *C. pacificus* (Frost unpubl.). With these parameter changes incorporated in simulation Run 3 the model suggests long-term coexistence of the two species with *Pseudocalanus* dominating (Fig. 4c).

In addition to competitive interactions, patterns of predation mortality can affect the relative abundance of species of filter-feeding zooplankton. Thus far in the model, predation has been described solely as a function of population size of the species combined. Following Landry (1976), I took the predation rate in simulation Runs 1–3 (Fig. 4) as

$$\text{predation rate} = \frac{GZ}{H + Z} \qquad (6)$$

where G is the maximum predation rate (0.1 in all simulations), Z the combined population sizes of the two species, and H the combined population size at which predation rate is 0.5. As noted in the methods, this procedure for computing predation tends to bias against the rarer species.

In view of their very different body sizes at similar developmental stages, it is possible that *Calanus* and *Pseudocalanus* are exposed to different types and intensities of predation. A simple way to investigate the consequences of this in the model is to assume different predators on each species and calculate predation rate of

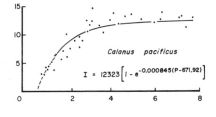

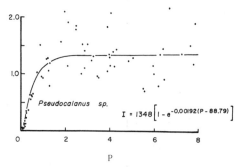

Fig. 8. Ingestion rate I (cells ingested copepod^{-1} h^{-1} × 10^{-3}) as a function of food concentration P (cells ml^{-1} × 10^{-3}) for adult females of *Calanus pacificus* and *Pseudocalanus* sp. feeding on unicells (about 11-μm cell diameter) of *Thalassiosira fluviatilis*. Data for *C. pacificus* from Frost (1972); data for *Pseudocalanus* obtained under identical experimental conditions (Evans and Frost unpubl.). Equations are for Ivlev curves (Parsons et al. 1967) fitted to data by least-squares nonlinear regression.

each species independently of the other. This was done in Run 4 where, also, to account for the different population sizes attained by the two species in Runs 1 and 2, different values of the parameter H were used. The values of H were 1×10^5 for *Calanus* and 2×10^5 for *Pseudocalanus* (Table 1), which in simulation Run 4 generally gives somewhat higher predation rate for *Pseudocalanus* than for *Calanus*. As a result, the model predicts coexistence of the two species, but dominance by *Calanus* both during and after the simulated spring bloom (Fig. 9).

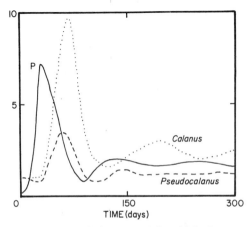

Fig. 9. Simulation Run 4 in which the two copepod species are subjected to different predation rates. Phytoplankton concentration P (μg C liter^{-1} $\times$ 10^{-2}) plotted together with population biomasses of the two copepod species (g C m^{-2} $\times$ 0.5).

The results of the four simulations (Figs. 4 and 9) are sufficient to demonstrate the potential range of predicted outcomes of interspecies interactions for analyses which assume particular types of size-dependent physiological processes. Similar shifts in relative abundance of species occur in more complicated models (e.g. that of Steele and Frost 1977; Frost unpubl.), although size-dependent food utilization assures coexistence of species.

Elaboration of the model to incorporate other intraspecific and interspecific variations in physiological processes would only serve to strengthen the conclusions and in most instances would require biological knowledge not yet available. For example, simplifications implicit in Eq. 1 and 5 will be unrepresentative if coefficients and exponents vary with age or stage of development, such as between the nauplian and the copepodid stages. Similarly, interspecific variations in the time between molt to the adult stage and production of the first brood could strongly affect the outcome of simulated competition between species. Finally, intraspecific and interspecific variations in behavior could promote alterations in relative abundance of species: species of *Calanus*, including *C. pacificus*, are diel vertical migrators in the late copopodid stages, whereas most species of *Pseudocalanus* seem not to migrate. The importance of these and other factors in influencing the relative abundance of species will be appreciated only after further experimental investigation of behavior, physiology, and growth in planktonic animals.

Discussion

Results of the simulation model indicate that in a zooplankton assemblage the relative abundance of species may depend heavily on biological characteristics of species that are unique and unpredictable from consideration of body size alone. The relatively high growth rate of *C. pacificus*, which may be shared by closely related species such as *C. finmarchicus* (Marshall and Orr 1955), is one striking example. Indeed, this feature alone may very well explain the dominance of the zooplankton by species of *Calanus* in the mid to high latitudes of the North Atlantic Ocean and in waters over continental shelves in the temperate-boreal North Pacific Ocean.

Because of the great importance of *Calanus* species in marine food webs of the northern hemisphere oceans, it is of interest to inquire whether the growth pattern of *Calanus* or that of *Pseudocalanus* is more unusual. Arbitrarily choosing studies for which reliable data are available on growth and development, temperate-boreal filter-feeding calanoid copepods representing a

50-fold range of adult weight have remarkably similar development times (Table 4a). For the same species, growth rate seems to increase with increasing adult size; this is indicated by the ratio of weights of the adult and egg for species with similar development times (Table 4b). More data of this sort are needed for other temperate-boreal species, but even with limited information *Temora longicornis* seems exceptional in that it has very small eggs and a high growth rate. *Paracalanus parvus* (intermediate in size between *Acartia* and *Pseudocalanus*) also has tiny eggs and relatively high optimal growth rates (D. Checkley pers. comm.). Therefore, neither body size in general nor adult size in particular is likely to be a good predictor of maximum growth rate or development in species of filter-feeding copepods, unless perhaps the species are very closely related. Other factors, such as size-selective predation, may promote adaptive patterns in growth and development which are the inverse of patterns expected from variations in body size (e.g. Kerfoot 1977).

Similarly, the relatively more efficient feeding by *Pseudocalanus* at low food concentrations, although perhaps intuitively expected on the basis of body size, is also species-specific and unpredictable (Nival and Nival 1973). For example, in open waters of the subarctic Pacific Ocean the zooplankton is dominated by two species of filter-feeding calanoid copepods, *Neocalanus plumchrus* and *Neocalanus cristatus*, which, although attaining very large body size (about 200 and 700 µg C, respectively, in copepodid stage V), have maxillary filters which are both much larger in area and have much smaller intersetule spacing relative to body size than *Pseudocalanus* (Heinrich 1963). Steele and Frost (1977) mentioned some consequences of this for the zooplankton assemblage in the subarctic Pacific. Egloff and Palmer (1971) described a case in which dependence of filter area on body size differs for congeneric species of cladocerans.

Indeed, for each of the variables listed by Hall et al. (1976: table 1) as dependent

Table 4. a—Optimal development time (days from fertilization of egg until molt to adult stage). b—Weight (µg carbon) of eggs and adults in four temperate-boreal species of filter-feeding calanoid copepods. Data compiled from Mullin and Brooks (1970), Paffenhöfer (1970), Harris and Paffenhöfer (1976), Paffenhöfer and Harris (1976), Landry (1978), and results of unpublished studies.

	a		
	Temp (°C)		
	10	12	15
Acartia clausi	31	25	19
Pseudocalanus sp.	34	27	21
Temora longicornis	—	25	—
Calanus pacificus	36	26	18–20
	b		
	Egg	Adult	Adult:Egg
Acartia clausi	0.035	2.2	63
Pseudocalanus sp.	0.1	10.0	100
Temora longicornis	0.04	16.0	400
Calanus pacificus	0.25	100.0	400

on body size and affecting population growth rates, it is likely that unpredictable variations exist at least interspecifically and possibly even intraspecifically. Thus, understanding the size composition and relative abundance of species in natural zooplankton assemblages will require considerable specific biological knowledge of the component species.

References

ALLAN, J. D. 1974. Balancing predation and competition in cladocerans. Ecology 55: 622–629.

BAINBRIDGE, V., and D. C. T. FORSYTH. 1972. An ecological survey of a Scottish herring fishery. V. Bull. Mar. Ecol. 8:21–57.

BROOKS, J. L., and S. I. DODSON. 1965. Predation, body size, and composition of plankton. Science 150:28–35.

CORKETT, C. J., and I. A. McLAREN. 1969. Egg production and oil storage by the copepod *Pseudocalanus* in the laboratory. J. Exp. Mar. Biol. Ecol. 3:90–105.

CUSHING, D. H., and T. VUCETIC. 1963. Studies on a *Calanus* patch. III. J. Mar. Biol. Assoc. U.K. 43:349–371.

DAGG, M. J. 1976. Complete carbon and nitrogen budgets for the carnivorous amphipod, *Calliopius laeviusculus* (Krøyer). Int. Rev. Gesamten Hydrobiol. 61:297–357.

EGLOFF, D. A., and D. S. Palmer. 1971. Size relations of the filtering area of two *Daphnia* species. Limnol. Oceanogr. 16:900–905.

FROST, B. W. 1972. Effects of size and concentration of food particles on the feeding behavior of the marine planktonic copepod *Calanus pacificus*. Limnol. Oceanogr. 17: 805-815.

——. 1974. *Calanus marshallae*, a new species of calanoid copepod closely allied to the sibling species *C. finmarchicus* and *C. glacialis*. Mar. Biol. 26:77-99.

HALL, D. J., S. T. THRELKELD, C. W. BURNS, and P. H. CROWLEY. 1976. The size-efficiency hypothesis and the size structure of zooplankton communities. Annu. Rev. Ecol. Syst. 7: 177-208.

HARRIS, R. P., and G. A. PAFFENHÖFER. 1976. Feeding, growth and reproduction of the marine planktonic copepod *Temora longicornis* Müller. J. Mar. Biol. Assoc. U.K. 56: 675-690.

HEINRICH, A. K. 1963. On the filtering ability of copepods in the boreal and the tropical regions of the Pacific [In Russian], Tr. Inst. Okeanol. Akad. Nauk SSSR 71:60-71.

KERFOOT, W. C. 1977. Competition in cladoceran communities: The cost of evolving defenses against copepod predation. Ecology 58:303-313.

LAMPERT, W. 1977. Studies on the carbon balance of *Daphnia pulex* as related to environmental conditions. I, II. Arch. Hydrobiol. Suppl. 48 (3/4), p. 287-335.

LANDRY, M. R. 1976. The structure of marine ecosystems: an alternative. Mar. Biol. 35:1-7.

——. 1978. Population dynamics and production of a planktonic marine copepod, *Acartia clausi*, in a small temperate lagoon on San Juan Island, Washington. Int. Rev. Gesamten Hydrobiol. 63:77-119.

LYNCH, M. 1977. Fitness and optimal body size in zooplankton populations. Ecology 58: 763-774.

McLAREN, I. A. 1969. Population and production ecology of zooplankton in Ogac Lake, a landlocked fiord on Baffin Island. J. Fish. Res. Bd. Can. 26:1485-1559.

MARSHALL, S. M., and A. P. ORR. 1955. The biology of a marine copepod. Oliver and Boyd.

MULLIN, M. M., and E. R. BROOKS. 1970. The effect of concentration of food on body weight, cumulative ingestion, and rate of growth of the marine copepod *Calanus helgolandicus*. Limnol. Oceanogr. 15:748-755.

NEILL, W. 1975. Experimental studies of microcrustacean competition, community composition and efficiency of resource utilization. Ecology 56:809-826.

NIVAL, P., and S. NIVAL. 1973. Efficacité de filtration des copépodes planctoniques. Ann. Inst. Oceanogr. 49:135-144.

PAFFENHÖFER, G. A. 1970. Cultivation of *Calanus helgolandicus* under controlled conditions. Helgol. Wiss. Meeresunters. 20:346-359.

——, and R. P. HARRIS. 1976. Feeding, growth and reproduction of the marine planktonic copepod *Pseudocalanus elongatus* Boeck. J. Mar. Biol. Assoc. U.K. 56:327-344.

PARSONS, T. R., R. J. LeBRASSEUR, and J. D. FULTON. 1967. Some observations on the dependence of zooplankton grazing on the cell size and concentration of phytoplankton blooms. J. Oceanogr. Soc. Jpn. 23: 10-17.

SHERMAN, K., and H. C. PERKINS. 1971. Seasonal variation in the food of juvenile herring in coastal waters of Maine. Trans. Am. Fish. Soc. 100:121-124.

STEELE, J. H. 1974. The structure of marine ecosystems. Harvard Univ. Press.

——, and B. W. FROST. 1977. The structure of plankton communities. Phil. Trans. R. Soc. Lond. 280:485-534.

VIDAL, J. 1978. Effects of phytoplankton concentration, temperature, and body size on rates of physiological processes and production efficiency of the marine planktonic copepods, *Calanus pacificus* Brodsky and *Pseudocalanus* sp. Ph.D. thesis, Univ. Washington. 207 p.

WIBORG, K. F. 1954. Investigations on zooplankton in coastal and offshore waters of western and northwestern Norway. Fiskeridir. Skr. Ser. Havunders. 11 (1):1-246.

66. Dynamic Energy-Flow Model of the Particle Size Distribution in Pelagic Ecosystems

William Silvert and Trevor Platt

Abstract

Energy flow through a pelagic ecosystem is described by a nonlinear, time-dependent, and spatially inhomogeneous master equation. It is assumed that within broad functional groupings the behavior of organisms depends only on size.

One way to study the general properties of ecosystems is to analyze how the organisms are distributed by size. This approach has proved to be informative in the pelagic marine ecosystem, both observationally (Sheldon and Parsons 1967; Sheldon et al. 1972) and theoretically (Kerr 1974; Platt and Denman 1977, 1978; Silvert and Platt 1978).

The theoretical work published to date has dealt only with simplified linear interactions between organisms in different size ranges, and with one exception (Silvert and Platt 1978) has been restricted to the steady state case. In this paper we present a general nonlinear description of energy flow in which the interparticle dynamics are analyzed in detail and spatial heterogeneity is admitted. Although the analysis is still anchored firmly on particle size as the crucial variable, we allow a limited degree of distinction between functional groups of organisms.

We first define and derive a master equation for energy flow in ecosystems based on the amount of energy within discrete compartments. The principal forms of energy exchange within an ecosystem are discussed and used to derive a continuous form of the master equation. Finally we analyze several models of increasing complexity and look at the role of nonlinearity and feedback in determining the dynamic response of an ecosystem to perturbation.

Master equation for energy flow

In any ecosystem there is a continual flow and dissipation of energy. Since energy is a conserved quantity, we can follow it through the system as it is changed from solar radiation to chemical and mechanical energy, and ultimately is converted to heat. Alternatively, we can use the rough equivalency between energy and biomass to use mass or carbon atoms to trace the flow of energy. The choice depends in large part on the particular system under consideration and the experimental techniques which are used, so we will use mass and energy interchangeably in our discussion.

In order to characterize the flow of energy through an entire system, a high degree of aggregation is desirable in describing organisms. We shall treat two particles as equivalent if their ecological roles are similar, and if they are of the same size and at the same location. More precisely, we define compartments such that all particles of a given type which fall within a specified size range and lie within a given volume of space are identified with the same compartment and treated as identical. The number of different types that can be specified is arbitrary but should be kept small. In previous work (Platt and Denman 1977, 1978; Silvert and Platt 1978) we treated all particles of the same size as equivalent, but a more reasonable approach is to deal with three categories: plants, animals, and detritus. The size of the compartments in terms of particle size and volume can be either finite or infinitesimal; the latter choice leads to a continuous formulation, which will be discussed later.

Let b_i be the energy content of all the particles in the ith compartment, which we shall assume is represented by their combined biomass. The principle of conservation of energy is expressed by the equation

$$\frac{db_i}{dt} = S_i = \sum_j (T_{ij} - T_{ji}) \quad (1)$$

where S_i is the net rate at which energy flows into compartment i and is expressed in terms of the transfer rates T_{ij}, which represents the rate at which energy flows into compartment i from compartment j. It is assumed that all energy sources and sinks are represented by compartments, including the sun and sediments as well as other compartments of living organisms. We shall refer to this equation and its generalizations as master equations, following the usage of Pauli (1928).

The structure of the principal transfer terms is discussed in the following sections, followed by the derivation of a continuous form of the master equation. Some of the processes which transfer energy are basically the same in both formulations, such as reproduction and predation, while other processes that involve the exchange of energy across the interface between two adjacent cells are quite different in the continuous formulation; these include advection, diffusion, and growth.

On summing the master equation over all compartments we obtain the general expression of energy conservation

$$\frac{d}{dt} \sum_i b_i = 0. \quad (2)$$

However, since we need to represent the sources and sinks in the environment by compartments, this equation is of limited use. We shall adopt the convention that compartment 0 represents the environment. If, for example, we treat the entire ecosystem as one compartment so that b_1 is the total biomass, then Eq. 1 reduces to

$$\frac{db_1}{dt} = T_{10} - T_{01} \quad (3)$$

which states that the rate of change of the biomass is equal to the rate at which energy enters the system minus the rate at which it is lost and is a more useful expression of the principle of conservation of energy.

Reproduction

Unfortunately, there are a number of unresolved problems in the treatment of reproductive flows. Most obvious of these is that reproduction in most organisms follows a seasonal cycle, and the size of the gonad at time of spawning depends not on the current energy flow, but rather on the integrated ration over the entire spawning cycle. The bioenergetics of the egg stage are also difficult to deal with, since eggs are eaten and respire but are not consumers, and, in cases where eggs are laid in clumps, they may attract predators that normally would not consume such small prey. In many species there is also a strong link between spawning and mortality that should be taken into account.

A linear model of reproduction, in cases where density-dependence does not have to be taken into account, gives rise to transfer terms of the form

$$T_{ij}^{(\text{rep})} = f_{ij} b_j \quad (4)$$

where f_{ij} represents the fecundity of particles in compartment j which produce eggs, etc., in compartment i. We can also use Eq. 4 to represent the effect of weight loss or mortality by the parent stock which causes an additional transfer of energy to a different compartment. Fortunately, although these terms are difficult to evaluate, their effect on the dynamics of the model is less than that of nonlinear terms, such as predation. In addition, the magnitude of the reproductive terms averaged over time (as opposed to their instantaneous values during spawning) may make it possible to approximate or even ignore them without greatly affecting the validity of the results. This is only true of energy-flow models, and not of models based on particle number, since an accurate treatment of reproduction is clearly essential if numbers are to be calculated correctly. The distinction between bioenergetic models and those based on particle numbers is discussed later in connection with the von Foerster equation and its applicability to ecosystem modeling.

Predation

We shall use predation in a general sense to categorize all processes whereby particles are consumed by other particles, including grazing on phytoplankton and bacterial decomposition. It is an extremely important mechanism for energy transfer in ecosystems, and it is also the cause of many of the interesting dynamic effects that arise because of nonlinearity and feedback. We have therefore attempted to build a simple but realistic model of the predation process, based mainly on generalization of the disc equation (Holling 1959; Rashevsky 1959; Murdoch 1973).

The rate at which energy flows from compartment j to compartment i because of predation defines $T_{ij}^{(\text{pred})}$. However, during a unit time interval the predator will spend a fraction t_H handling its prey and a fraction $t_F = 1 - t_H$ foraging. The rate at which the predators encounter and take prey is proportional to t_F the fraction of time spent foraging, times the prey abundance b_j and times the abundance of predators b_i. We can therefore write the transfer rate in the form

$$T_{ij}^{(\text{pred})} = \pi_{ij} b_i b_j t_F \quad (5)$$

where we call π_{ij} the predation coefficient. The handling time is proportional to the total amount of prey taken per unit predator biomass, so we can write it as

$$t_H = \sum_k b_{ik} \pi_{ik} b_k t_F \quad (6)$$

where b_{ik} is handling time for a unit mass of prey k ingested by predator i. Solving Eq. 5 and 6 for the predation rates we obtain

$$T_{ij}^{(\text{pred})} = \frac{\pi_{ij} b_i b_j}{1 + \sum_k b_{ik} \pi_{ik} b_k} . \quad (7)$$

The flow of energy from j to i due to predation is proportional to the biomass of predators, b_i, but depends on the abundances of all available prey. For example, if the prey in some third compartment k increase, then the predator will take more of this prey and will therefore spend more time handling its food. Because this reduces t_F, the predation on compartment i will decrease. The generalized disc equation described here thus appears to incorporate a degree of predator switching, although there is no true functional response to changes in prey abundance in the model (Murdoch 1973).

There are two special cases of Eq. 7 that are of interest. First of all, if we consider predation on one prey compartment only, then

$$T_{ij}^{(\text{pred})} = \frac{\pi_{ij} b_i b_j}{1 + b_{ij} \pi_{ij} b_i} \quad (8)$$

which is equivalent to the usual form of the disc equation. Secondly, if the prey concentrations b_k or the handling times b_{ik} are small we obtain

$$T_{ij}^{(\text{pred})} = \pi_{ij} b_i b_j \quad (9)$$

which is the type of quadratic interaction typically used in Lotka-Volterra models of interacting populations. We shall refer to the case described by Eq. 9 as a Lotka-Volterra interaction, even though the term is commonly used in a more restrictive sense (Goel et al. 1971).

The simplest model of predation is one in which only one of the π_{ij} for a given value of i is nonzero, i. e. the compartments are chosen large enough so that all the prey fall within one of them. In this case Eq. 8 applies, or, if the prey density is low, we can use Eq. 9 with just a single j for each i. We know of only one published calculation of the predation coefficients for this case, that carried out by Ware (1978) for several species of pelagic fish. Ware obtained a power-law relationship of the form $\pi_{ij} \sim w_i^{-\delta}$ where δ falls in the range 0.16 to 0.24. For comparison, Fenchel (1974) concluded that the specific growth rate scales as $w^{-0.28}$. If we assume that the growth efficiency is constant, this implies that the abundance of prey, b_j, scales with predator size as $w_i^{-\epsilon}$ where ϵ falls in the range 0.04 to 0.12. Using Ware's value of $(0.07)^3$ for the prey-predator size ratio gives an abundance ratio of prey to predators of 1.4 to 2.6, which is consistent with observation (Sheldon et al. 1972).

The gross energy input to compartment i from predation, $\Sigma_j T_{ij}^{(\text{pred})}$, does not reflect all the energy transfers involved. Some of the energy in the prey may be lost in the process of foraging and consumption and in the handling of food, some is lost through the inefficiency of assimilation, and some goes into metabolic costs (specific dynamic action and respiration) and does not contribute to the biomass within compartment i. The amount of energy that is ultimately converted to biomass is a complicated function of the ration, which is difficult to incorporate in a minimally structured model of the type described here.

Paloheimo and Dickie (1966) have shown that growth efficiency is a decreasing function of ration, which means that respiration and other metabolic costs rise faster than ration does. However, since growth and hence growth efficiency must fall to zero at some finite subsistence ration, there must be a low-ration regime for which the growth efficiency increases with ration (Kerr 1971). For the present we shall assume that the growth efficiency K_1 is constant and has the same value for all organisms, so that the specific growth rate of particles in compartment i is

$$G_i = K_1 \sum_j T_{ij}^{(\text{pred})} / b_i$$

$$= \frac{K_1 \sum_j \pi_{ij} b_j}{1 + \sum_j h_{ij} \pi_{ij} b_j}. \quad (10)$$

Growth and advection: continuous formulation

So far we have considered the rate of energy transfer due to reproduction and predation, which are mechanisms for exchanging energy between compartments corresponding to quite different particle sizes. The processes of growth, advection, and diffusion are qualitatively different in that they exchange energy by the transport of particles across the interface between adjacent compartments. These processes are best dealt with by developing a continuous model.

We define a biomass distribution function $\beta_k(w, \mathbf{x})$ such that the biomass within any compartment encompassing particles of type k in the size range $[w_1, w_2]$ and volume V is

$$b = \int_{w_1}^{w_2} dw \int dV \beta_k(w, \mathbf{x}). \quad (11)$$

The rates at which b changes due to growth and advection are

$$\left(\frac{db}{dt}\right)_{\text{growth}} = \int dV \left[\beta_k(w_1, \mathbf{x}) g_k(w_1) - \beta_k(w_2, \mathbf{x}) g_k(w_2)\right] \quad (12)$$

where $g_k(w) = dw/dt$ is the growth rate for particles of type k, and

$$\left(\frac{db}{dt}\right)_{\text{adv}} = -\int \beta_k(w, \mathbf{x}) \mathbf{v}(\mathbf{x}) \cdot d\mathbf{s} \quad (13)$$

which is the integral of the advective flow across the surface of the compartment. These can be converted to volume integrals by the use of Green's theorem, and since the result obtains for any compartment, we obtain

$$\frac{D\beta}{Dt} \equiv \frac{\partial \beta}{\partial t} + \frac{\partial}{\partial w}(\beta g) + \text{div}(\beta v) = \sigma \quad (14)$$

where we define the flow derivative $D\beta/Dt$ by the above equation and the source term σ represents the net energy flow due to all causes other than growth and advection, e. g. reproduction, predation, and diffusion. Equation 14 is the continuous analog of Eq. 1, and we shall therefore refer to it as the continuous bioenergetic master equation.

The master equation is not unique, and we can derive a number of alternate forms by transformations of various kinds. In particular, by defining a particle number density function $n_k(w, \mathbf{x})$ we can derive the master equation

$$\frac{Dn}{Dt} = \frac{\sigma - ng}{w} \quad (15)$$

which is a generalization of the von Foerster equation (Sinko and Streifer 1967; Streifer 1974; Levin 1976). However, the particle number density is a very sensitive function of particle size, and particle number (unlike energy) is not a conserved quantity; therefore it is very difficult to work with particle number as a primary variable in general ecosystem modeling and we shall confine our attention to the distribution of biomass rather than particles.

Diffusion

In addition to advective flows, there is a net transport of particles through a heterogeneous system by the process of diffusion. This contributes a term of the form (Levin 1976)

$$\sigma_{\text{diff}} = \text{div}(K \text{ grad } \beta) \quad (16)$$

to the master equation where K is the diffusion parameter and may be a tensor function in nonisotropic systems. It scales as L^2/T where L is the range of diffusive processes and T is the corresponding time, and experimentally L and T are related by $L \sim T^{1.17}$ (Okubo 1971). A reasonable time scale T would be the lifetime of the particle, but there does not appear to exist a meaningful set of data to indicate how lifetime scales with particle size; we have therefore used the generation time which according to Bonner (1965, p. 16-17) and to Fenchel (1974) scales as $w^{-0.28}$. The diffusive range of a particle over this time then scales as $L \sim (w^{0.28})^{1.17} = w^{0.33}$, so the range is proportional to the linear dimensions of the particles. Consequently the diffusion parameter scales as $K \sim L^2/T \sim w^{0.38}$

We find it somewhat surprising that the product of the two exponents 1.17 for the diffusive range of ocean circulation and 0.28 for the lifetime of living organisms should turn out to be almost exactly 1/3, and it appears that the proportionality between the range of organisms and their linear dimension may be a coincidence.

Linear form of the model

If the growth term g is independent of the biomass density function β and the source term σ is linear in β, the master equation is homogeneous and linear in β and thus relatively easy to solve. For a homogeneous system it becomes

$$\frac{D\beta}{Dt} = \frac{\partial \beta}{\partial t} + \frac{\partial}{\partial w}(\beta g) = -\mu \beta. \quad (17)$$

The static form of this equation was first derived and solved by Platt and Denman (1977, 1978), and the above dynamic model was developed by Silvert and Platt (1978). In the static case, if g and $w\mu$ (which have the same dimensions) scale as w^γ, then the biomass distribution function β scales as $w^{-(\gamma + w\mu/g)}$. Perturbations of the distribution propagate up the distribution at a rate given by g, which is a direct consequence of the linearity of the model. Since there is no feedback in the linear model, it cannot reflect density-dependent effects and predator-prey cycles. However, it does appear to illustrate some of the characteristics of pelagic ecosystems which are close to a stationary state.

Quadratic form of the model

To simplify discussion of the nonlinear properties of the model, we consider a limiting form in which all terms but the time derivative are of second order in β. We ignore reproduction, which is a linear flow term, and assume that all mortality is due to predation so that the mortality for particles in compartment i is

$$Z_i = \sum_j T_{ji}^{(\text{pred})}/b_i = \sum_j \pi_{ji} b_j \quad (18)$$

(we also assume that there is no saturation in the feeding process so Eq. 9 is valid). The master equation for a homogeneous system can be written in the form

$$\frac{\partial \beta}{\partial t} = -\frac{\partial}{\partial w}(\beta g) + \left(\frac{g}{w} - Z\right)\beta$$

$$= -w\frac{\partial}{\partial w}(\beta G) - Z\beta \quad (19)$$

where $G = g/w$ is the specific growth rate. The left-hand side of this equation is linear in β, while the right-hand side is a homogeneous second order function of β. If β is a stationary solution of the equation, $\partial \beta/\partial t = 0$, then $a\beta$ is also a stationary solution where a is an arbitrary constant. Consequently the equilibrium form of the biomass distribution is independent of the density of the system in this approximation. However the relaxation time for perturbations of the system, $\tau = (\partial \ln \beta/\partial t)^{-1}$, scales as $1/a$. For example, if we double the value of β everywhere then $\partial \beta/\partial t$ increases by a factor of four for a comparable perturbation, so the time it takes the system to respond is halved. This occurs because the increased density of the system with all other quantities (e.g. the predation coefficients) kept constant means a doubling of the turnover rates.

This does not mean that a plankton community which becomes enriched through eutrophication will necessarily exhibit a spring bloom two or three times a year because of the speeding up of the time scale for the system. The actual dynamics of an ecosystem depend on how the natural time scale of the system relates to the frequency of the external shocks it gets from interaction with its environment, and a major change in the dynamic response of the system is most likely to arise when an external periodic perturbation matches one of its characteristic internal cycles (Silvert and Smith in prep.). Thus a change in the response time of the system may not affect the speed of system processes when these involve an external driving force, but may instead show up as changes in the magnitudes of these processes.

In order to find a stationary solution for $\beta(w)$ we shall make the assumption suggested by Kerr (1974) of a fixed prey-predator size ratio q, so that the ration and growth of particles of size w are proportional to $\beta(qw)$ and their mortality is proportional to $\beta(w/q)$. We can write the equations for growth and mortality in the form

$$G(w) = K_1 q \rho(w) \beta(qw) \quad (20)$$

and

$$Z(w) = \rho(w/q)\beta(w/q) \quad (21)$$

by analogy to Eq. 9 and 10. It can be verified that these expressions satisfy the conservation law implied by Eq. 2, namely

$$K_1 \sum_i \sum_j [T_{ij}^{(\text{pred})} - T_{ji}^{(\text{pred})}]$$

$$= \int G(w)\beta(w)dw - K_1 \int Z(w)\beta(w)dw$$

$$= 0. \quad (22)$$

If we define the function $\phi(\ln w) = \beta(w)G(w)$ we find that the stationary solutions of Eq. 19 are equivalent to solutions of the first-order difference equation

$$\phi'(\ln w) + \phi(\ln w - \ln q)/K_1 q = 0 \quad (23)$$

and depend on the value of $\phi(w)$ over some finite size range, such as the interval $[qw_o, w_o]$ (J. Murray pers. comm.). One possible solution is $\phi(\ln w) \sim e^{-k \ln w} = w^{-k}$ where k must satisfy the condition $K_1 k = q^{k-1}$. If the factor $\rho(w)$ scales as w^Δ this implies that $\beta(w) \sim w^{-(\Delta+k)/2}$ and growth $G(w) \sim w^{(\Delta-k)/2}$.

If predation scaled in a perfectly isometric fashion (e. g. if fish had large cilia instead of gill rakers), then we could assume that the range of particle sizes acceptable as prey scales as w and that the ability to identify and capture prey is the same for all organisms, so that $\rho(w)$ is proportional to the appropriate predation coefficient π_{ij} times w. This gives values of $\Delta = 1 - \delta$ in the range 0.76 to 0.84, based on Ware's data (Ware 1978). Using a value of 0.2 for K_1 and $q = (0.07)^3$ we get $k = 1.18$ and thus β scales as $w^{-\epsilon}$ with ϵ in the range 0.97–1.01. The corresponding exponent for specific growth is $(\Delta - k)/2 = -0.17$ to -0.21, which is somewhat less in magnitude than the commonly accepted value of -0.28 (Fenchel 1974) but is consistent with some of the values obtained by Ware (1978).

These results agree with evidence that the biomass per logarithmic size interval, $w\beta(w)$, is constant (Sheldon et al. 1972, 1977), although if we carry this to its logical conclusion a value of $\epsilon \leqslant 1$ implies that the total biomass of the system is infinite; obviously the abundance of infinitely large animals is determined by factors other than bioenergetics (Sheldon and Kerr 1972).

We based these conclusions on the assumption of isometric predation, but it is likely that organisms which actively hunt for their prey have larger prey-predator size ratios (closer to unity) than filter feeders; Ursin's analysis of feeding behavior in cod and dab indicates that this is the case, the optimal value of q being 0.006 for cod and 0.001 for dab (Ursin 1973). A larger value of q leads to a more rapid decrease in particle density with size; if we use Ursin's value of $q = 10^{-3}$ we get $k = 1.21$, which increases ϵ by about 1%, so the difference is small. Similarly, an increase in growth efficiency from 0.20 to 0.25 would decrease ϵ by about 1%. Thus the relative constancy of $w\beta(w)$ seems to be a fairly robust consequence of the model in relation to changes in prey-predator size ratio or growth efficiency.

The inclusion of linear loss terms in the model is a major mathematical complication and we have not yet been able to obtain an analytic solution. Qualitatively we can conclude that such terms, which correspond to processes like standard metabolism and forms of mortality other than predation, are most important at low densities and cause the biomass distribution function $\beta(w)$ to fall off more rapidly than predicted by the homogeneous quadratic model. In many pelagic systems predation is the dominant form of energy exchange, but it is interesting to note that in deep benthic low-energy systems the particle size distribution falls off more rapidly than in shallower systems where the energy density is higher (Thiel 1975). These higher-energy benthic systems appear to have the same particle-size distribution as that we predict for pelagic systems (Ursin 1973).

Feedback structure and dynamic behavior

Although the general time-dependent solution of the master equation is not known, the qualitative nature of the dynamic effects implied by the model can be investigated by looking at the role of feedback. In the time-dependent linear model in which growth and loss are fixed functions of w independent of the biomass distribution, there is no feedback and any perturbation of the system propagates up the biomass scale at a rate given by the growth term (Silvert and Platt 1978). In the more general formulation developed here there are two principal feedback mechanisms, reproduction and predation; reproduction contributes terms which are linear in β, while predation gives rise to quadratic terms.

To illustrate these terms, the response of the model to a step-function increase in phytoplankton biomass is shown in Fig. 1. The two dashed lines represent the station-

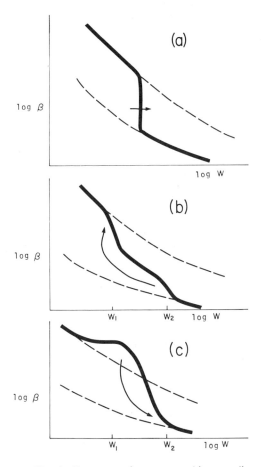

Fig. 1. Response of ecosystem biomass distribution to a sudden increase in phytoplankton standing stock. a—Linear model with no feedback. b—Linear model with feedback through reproduction, as indicated by arrow. c—Nonlinear model with predation flows, as indicated by arrow.

ary biomass distribution curves corresponding to two different total biomass densities, and the solid line represents the biomass distribution at some intermediate time. Figure 1a shows the intermediate behavior in the absence of feedback (Silvert and Platt 1978). The increase in energy propagates up the distribution like a front with no disturbance on either side of it. When reproduction is taken into account as shown in Fig. 1b, so that there is feedback but the equation is still linear, there is still basically a front, but now it is spread out. Because the reproductive flow represented by the arrow has not yet reached its equilibrium level, the biomass at w_1 is less than its limiting value. In concrete terms this means that copepods and other invertebrate zooplankters equilibrate faster than ichthyoplankton. Finally the effect of nonlinear feedback through predation is shown in Fig. 1c. Because the predation flow is in the opposite direction from the reproductive flow, the effect is reversed; the biomass at w_1, instead of being depressed by the low reproductive flow from w_2, is enhanced and overshoots its equilibrium value because of reduced grazing pressure. This gives rise to a wave-like instability that may take a long time to die out. Whether these oscillations actually do die out probably depends on the magnitude of the transfer terms and is not known at the present time; preliminary computer simulations seem to indicate that the damping times are very large, if in fact they are finite at all (Silvert and Smith in prep.).

The response of the system to a periodic driving force, such as the diurnal or seasonal variations in light intensity, is greatly complicated by the fact that the delays in the feedback loops vary from hours to decades. A basic principle of feedback analysis is that large resonant effects can be expected when the period of a disturbance matches the time it takes energy to flow around a feedback loop, and this condition is bound to be met at some part of the biomass distribution (Silvert and Smith in prep.). In effect there exists a group of particles whose prey are small enough so that they can track the perturbation, but whose predators are too big to track it. During the part of the cycle when prey are abundant, this group of particles can increase without suffering increased mortality, since their predators cannot respond in time. Their relative response is greater than that of smaller particles. This gives rise to a drastic increase in predation on their prey, so that the number of smaller particles is destabilized. The energy wave which started off propagating up the particle size distribution is partially reflected and interferes with itself. The process is similar to the breaking of waves, where the energy reflected from a sloping beach due to the reduced speed of waves in shallow water gives rise to a dramatic localized instability.

It has long been known that many marine organisms respond to environmental variability, an effect normally ascribed to direct environmental impact on these species, as through larval survival. This model suggests that there may also be a fundamental bioenergetic mechanism such that small oscillations in the food supply can drive large resonant swings in population.

Concluding remarks

We have attempted to show how a master equation might be constructed to describe the energy flux through the pelagic ecosystem. For a complete description, one component of this master equation describes the forcing of the system through primary production by insolation, i. e. the input of photosynthetically produced material at the small-particle end of the biomass distribution. The forcing equation need not use organism size as the independent variable, although there is reason to believe that this would not be inappropriate (e. g. Banse 1976).

Elsewhere in the spectrum, organism size is the most attractive choice for independent variable, excluding models based on trophic levels or species-by-species descriptions. The growth function can be related to size with a high degree of confidence (Fenchel 1974). The other leading term in the master equation—predation—can be written in terms of the relative sizes of the interacting organisms with a confidence commensurate with our faith in size selectivity of feeding. What is known about the structure of the food web then appears as a transition matrix whose elements express the probability that an organism of a given size will ingest any organism of another size. In the extreme simplification of the trophic level model with constant size differential between levels, this transition matrix reduces to a single number (Kerr 1974).

In specific applications of the formalism developed here, individual terms in the master equation may be written in as much detail as can be justified by knowledge of the system in question. It is also worth pointing out that the formalism could be applied to discrete and limited parts of the size distribution: its use is by no means restricted to whole ecosystem representation. For example, the grazing on phytoplankton by copepods, recently modeled by Steele and Frost (1977), could be handled very well in terms of organism size through the master equation, particularly since much of the experimental evidence on this subsystem includes the size-dependence of the feeding processes.

References

BANSE, K. 1976. Rates of growth, respiration and photosynthesis of unicellular algae as related to cell size—A review. J. Phycol. **12**: 135–140.

BONNER, J. T. 1965. Size and cycle. Princeton Univ. Press.

FENCHEL, T. 1974. Intrinsic rate of natural increase: The relationship with body size. Oecologia **14**: 317–326.

GOEL, N. S., S. C. MAITRA, and E. W. MONTROLL. 1971. On the Volterra and other nonlinear models of interacting populations. Rev. Mod. Phys. **43**(2, part): 231–276.

HOLLING, C. S. 1959. Some characteristics of simple types of predation and parisitism. Can. Entomol. **91**: 385–398.

KERR, S. R. 1971. Analysis of laboratory experiments on growth efficiency of fishes. J. Fish. Res. Bd. Can. **28**: 801–808.

———. 1974. Theory of size distribution in ecological communities. J. Fish. Res. Bd. Can. **31**: 1859–1862.

LEVIN, S. A. 1976. Population dynamic models in heterogeneous environments. Annu. Rev. Ecol. Syst. **7**: 287–310.

MURDOCH, W. W. 1973. The functional response of predators. J. Appl. Ecol. **10**: 335–342.

OKUBO, A. 1971. Oceanic diffusion diagrams. Deep-Sea Res. **18**: 789–802.

PALOHEIMO, J. E., and L. M. DICKIE. 1966. Food and growth of fishes. III. Relations among food, body size, and growth efficiency. J. Fish. Res. Bd. Can. **23**: 1209–1248.

PAULI, W. 1928. Über das H-Theorem vom Anwachsen der Entropie vom Standpunkt der neuen Quantenmechanik, p. 30–45. *In* Probleme der modernen Physik (Arnold Sommerfeld zum 60 Geburtstage, gewidmet von seinen Schülern). Hirzel, Leipzig.

PLATT, T., and K. DENMAN. 1977. Organization in the pelagic ecosystem. Helgol. Wiss. Meeresunters. **30**: 575–581.

———, and ———. 1978. The structure of pelagic marine ecosystems. Rapp. P.-V. Reun. Cons. Int. Explor. Mer **173**: 60–65.

RASHEVSKY, N. 1959. Some remarks on the

mathematical theory of nutrition of fishes. Bull. Math. Biophys. **21**: 161–183.

SHELDON, R. W., and S. R. KERR. 1972. The population density of monsters in Loch Ness. Limnol. Oceanogr. **17**: 796–798.

———, and T. R. PARSONS. 1967. A continuous size spectrum for particulate matter in the sea. J. Fish. Res. Bd. Can. **24**: 909–915.

———, A. PRAKASH, and W. H. SUTCLIFFE, JR. 1972. The size distribution of particles in the ocean. Limnol. Oceanogr. **17**: 327–340.

———, W. H. SUTCLIFFE, JR., and M. A. PARANJAPE. 1977. The structure of the pelagic food chain and the relationship between plankton and fish production. J. Fish. Res. Bd. Can. **34**: 2344–2353.

SILVERT, W., and T. PLATT. 1978. Energy flux in the pelagic ecosystem: A time-dependent equation. Limnol. Oceanogr. **23**: 813–816.

SINKO, J. W., and W. STREIFER. 1967. A new model for age-size structure of a population. Ecology **48**: 910–918.

STEELE, J. H., and B. W. FROST. 1977. The structure of plankton communities. Phil. Trans. R. Soc. Lond. **280**: 485–534.

STREIFER, W. 1974. Realistic models in population ecology. Adv. Ecol. Res. **8**: 199–266.

THIEL, H. 1975. The size structure of the deep-sea benthos. Int. Rev. Gesamten Hydrobiol. **60**: 575–606.

URSIN, E. 1973. On the prey size preferences of cod and dab. Medd. Dan. Fisk. Havunders. **7**: 85–98.

WARE, D. M. 1978. Bioenergetics of pelagic fish: Theoretical change in swimming speed and ration with body size. J. Fish. Res. Bd. Can. **35**: 220–228.

Appendix. The Program

Monday, August 21

Aspects of Behavior and Physiology

J. Rudi Strickler, Chairman

9:00–9:05. Official Welcome, W. Charles Kerfoot, Dartmouth College; Greetings to Third Special Symposium, George W. Saunders, President, American Society of Limnology and Oceanography, June 1977–1978

Session I. *Fluid Dynamics and Swimming Behavior: Life in a Viscous Medium*

9:05–9:25. Visual observation of live zooplankters. J. Rudi Strickler, University of Ottawa, Ottawa, Ontario, Canada

9:25–9:55. The animal and its viscous environment. R. Zaret, The Johns Hopkins University, Baltimore, Maryland

9:55–10:10. Swimming movements and energetics of *Cyclops*. J. Rudi Strickler and M. Alcaraz, University of Ottawa, Ottawa, Ontario, Canada

10:15–10:20. Discussion

10:20–10:30. Coffee Break

Session II. *Orientation in Space and Time*

10:30–11:00. Vertical and horizontal orientation of zooplankton. H. Otto Siebeck, Zoologisches Institut der Universität München, W. Germany

11:00–11:25. Model of zooplankton vertical migration. L. Giguère, Simon Frazier University, Burnaby, British Columbia, Canada

11:25–11:50. Vertical migrations of arctic zooplankton. C. L. Buchanan, University of New Hampshire, Durham, New Hampshire

11:50–12:00. Discussion

12:00–1:30. Lunch

1:30–1:50. Vertical migration of water mites. H. Riessen, University of Ottawa, Ottawa, Canada

1:50–2:10. Experimental studies on vertical migration. R. Bohrer, Dalhousie University, Halifax, Nova Scotia, Canada

2:10–2:30. Time and space considerations in vertical migration. D. Meyers, University of Wisconsin, Madison, Wisconsin

2:30–2:40.	Coffee Break

Session III. Pigments and Zooplankton Behavior

2:40–3:10.	Panel Discussion. Pigments and zooplankton behavior: Do recent experiments shed light on old controversies? J. Ringelberg, Universiteit van Amsterdam, Netherlands (moderator). Participants: J. Gerritsen; N. Hairston, Jr.; W. Lewis; J. Ringelberg; O. Siebeck

Session IV. Diapause and Cyclomorphosis

3:10–3:35.	Evolutionary aspects of diapause in freshwater copepods. Kaare Elgmork, University of Oslo, Norway
3:35–4:00.	Panel Discussion. Kaare Elgmork, moderator. Participants: U. Einsle, J. Jacobs, J. P. Nilssen, R. Stross
4:00–4:10.	Coffee Break

Session V. Encounters of the First Kind: Social Interactions

4:10–4:40.	Adaptive responses to encounter problems. J. Gerritsen, The Johns Hopkins University, Baltimore, Maryland
4:40–5:10.	Ritualized mating behavior in copepods. P. Blades, Harbor Branch Foundation, Ft. Pierce, Florida
5:10–5:30.	Discussion

Tuesday, August 22

Zooplankton Food-Gathering

Session I. Rotifer Feeding and Population Dynamics.
J. J. Gilbert, Dartmouth College, Section Chairman

9:00–9:35.	Laboratory studies of suspension feeding in rotifers: A review with some new observations. P. L. Starkweather, Dartmouth College, Hanover, New Hampshire
9:35–10:00.	Population dynamics of a suspension feeding rotifer in continuous culture. M. E. Boraas, Pennsylvania State University, University Park, Pennsylvania
10:00–10:30.	Feeding behavior of *Asplanchna*. J. J. Gilbert, Dartmouth College, Hanover, New Hampshire
10:30–11:00.	Break

Session II. Copepod and Cladocera Feeding Dynamics.
S. Richman, Lawrence University, Section Chairman

11:00–11:20.	Grazing interactions in the freshwater environment. S. Richman, Lawrence University, Appleton, Wisconsin
11:20–11:30.	Discussion
11:30–11:50.	Grazing interactions in the marine environment. P. L. Donaghay, Oregon State University, Corvallis, Oregon
11:50–12:00.	Discussion
12:00–1:00.	Lunch
1:00–1:20.	Resource chatacteristics modifying selective grazing. D. C. McNaught, State University of New York, Albany, New York

1:20–1:30.	Discussion
1:30–1:50.	Chemosensory grazing and food gathering by marine copepods. S. A. Poulet and P. Marsot, INRS-Oceanologie, Quebec, Canada
1:50–2:00.	Discussion
2:00–2:20.	A model of the copepod filter-feeding mechanism based on electron microscopic studies: small vs. large particle feeding, food processing, and morphologic indicators of ecological and evolutionary relationships. M. M. Friedman, University of Pennsylvania, Philadelphia, Pennsylvania
2:20–2:30.	Discussion
2:30–2:50.	On the importance of "threshold" food concentrations for filter-feeding zooplankton. Winfried Lampert, J. W. Goethe-Universität, Frankfurt am Main, West Germany
2:50–3:00.	Break
3:00–4:30.	Round Table Discussion: Zooplankton feeding processes. S. Richman, Moderator. Participants: C. M. Boyd, R. J. Conover, B. W. Frost, M. Gliwicz, D. R. Heinle, W. Lampert, K. G. Porter
4:30–4:40.	Break

Session III. Food Gathering and Community Structure

4:40–5:05.	The inadequacy of body size as an indicator of niches in the zooplankton. B. W. Frost, University of Washington, Seattle, Washington
5:05–5:30.	Dynamic energy-flow model of the particle size distribution in pelagic ecosystems. W. Silvert and T. C. Platt, Bedford Institute of Oceanography, Dartmouth, Nova Scotia, Canada

Wednesday, August 23

The Genetics and Life Histories of Zooplankton

J. David Allan, Section Chairman

9:00–9:30.	The genetic structure of zooplankton populations. Charles King, Oregon State University, Corvallis, Oregon
9:30–10:00.	Studies on the genetics of cladocerans. P. Hebert, University of Windsor, Windsor, Ontario, Canada
10:00–10:30.	Geography of gamogenesis among the cladocera. D. Frey, Indiana University, Bloomington, Indiana
10:30–10:45.	Break
10:45–11:00.	An analysis of the precision of birth and death rate estimates for egg-bearing zooplankters. W. R. DeMott, Dartmouth College, Hanover, New Hampshire
11:00–12:15.	Round Table Discussion: Genetics and mating systems in zooplankton. P. Hebert, moderator. Participants: L. Lee, P. Hebert, C. King, B. Black, E. Berger
12:15–1:45.	Lunch
1:45–2:15.	Reproductive patterns in zooplankton. D. Allan, University of Maryland, College Park, Maryland, and C. Goulden, Academy of Natural Sciences, Philadelphia, Pennsylvania

2:15–2:45.	Habitat selection, resource allocation, and population growth of cladocerans in seasonal environments. S. Threlkeld, Tahoe Research Group, University of California, Davis, California
2:45–3:15.	Life history variation in *Chaoborus*. C. von Ende, Northern Illinois University, DeKalb, Illinois
3:15–3:40.	Break
3:40–4:10.	The evolution of cladoceran life histories. M. Lynch, University of Illinois, Urbana, Illinois
4:10–4:35.	When and how to reproduce: A dilemma for limnetic cyclopoid copepods. J. P. Nilssen, University of Oslo, Norway
4:35–5:00.	Structure of zooplankton communities in the Peten lake district, Guatemala. E. S. Deevey, G. B. Deevey, M. Brenner, Florida State Museum, University of Florida, Gainesville, Florida
5:00–6:00.	Life history variation and seasonality in zooplankton. J. David Allan, moderator. Participants: J. D. Allan, W. DeMott, M. Lynch, W. C. Kerfoot, W. Lampert, S. Threlkeld, J. Jacobs, J. Gilbert, J. P. Nilssen

Thursday, August 24

Populations and Communities

Session I. Cyclomorphosis
W. C. Kerfoot and J. O'Brien, Section Chairmen

8:30–8:55.	Environmental determination of cladoceran cyclomorphosis. J. Jacobs, Zoologisches Institut der Universität München, West Germany
8:55–9:15.	Cyclomorphosis and population dynamics of *Bosmina*. R. Black, SUNY at Stony Brook, New York
9:15–9:35.	Perspectives on cyclomorphosis: Phenotypes and genotypes. W. C. Kerfoot, Dartmouth College, Hanover, New Hampshire
9:35–9:55.	Seasonal size changes in small pond *Daphnia pulex* populations. D. Brambilla, University of Michigan, Ann Arbor, Michigan
9:55–10:15.	The proximate selective factors in the cyclomorphosis of *Daphnia longiremis*. J. O'Brien, University of Kansas, Lawrence, Kansas
10:15–10:30.	Discussion

Session II. The Reasons for Vertical Migration

10:30–11:45.	Formal Presentations: N. Hairston, T. Zaret, J. O'Brien, P. Lane, D. McNaught
11:45–12-00.	Round Table Discussion: Ultimate causes of vertical migration and patchiness. T. Zaret (moderator), University of Washington, and J. O'Brien, section chairmen. Participants: B. Frost, H. Dumont
12:00–1:00.	Lunch

Session III. Zoogeography and Distribution of Zooplankton
K. Porter, University of Georgia, Section Chairwoman

1:00–1:20.	Patterns of zooplankton distributions in Africa. H. Dumont, Rijksuniversiteit Gent, Gent, Belgium

1:20–1:40.	Species richness and area in Galapagos and Andean lakes: Equilibrial phytoplankton communities and a paradox of the zooplankton. P. Colinvaux and M. Steinitz, Ohio State University, Columbus, Ohio
1:40–2:00.	Population maintenance mechanisms of copepods in the Oregon coastal zone. C. B. Miller, Oregon State University, Corvallis, Oregon
2:00–2:20.	Zoogeographic patterns in the size structure of zooplankton communities with possible applications to lake ecosystem modeling and management. W. G. Sprules, Erindale College, University of Toronto, Ontario, Canada
2:20–2:30.	Notes on the distribution of Central American zooplankton. T. Zaret, University of Washington, Seattle, Washington

Session IV. Plant-Herbivore Interactions
K. Porter, Section Chairwoman

2:30–2:55.	Toxicity, manageability, and nutritional quality of the blue-green alga *Anabaena flos-aquae*. K. Porter and J. Orcutt, University of Georgia, Athens, Georgia
2:55–3:20.	Nutrient recycling as an interface between algae and grazers in freshwater communities. J. Lehman, University of Michigan, Ann Arbor, Michigan
3:20–3:45.	Zooplankton as algal chemostats: A theoretical perspective. Robert Armstrong, SUNY at Stony Brook, New York
3:45–4:10.	Comments on zooplankton-phytoplankton interactions. M. Gliwicz, University of Warsaw, Warsaw, Poland
4:10–4:20.	Recent changes in the phytoplankton of Lake Washington. W. T. Edmondson, University of Washington, Seattle, Washington
4:20–4:30.	*Aphanizomenon* blooms: Alternative control and cultivation by *Daphnia pulex*. M. Lynch, University of Illinois, Urbana, Illinois
4:30–4:45.	Coffee Break

Session V. The Ecology of The Littoral Region: An Ignored Area
Dewey Meyers, University of Wisconsin, Madison, Wisconsin, Section Chairman

5:05–5:25.	Odonate "hide and seek": Habitat-specific rules. D. M. Johnson, East Tennessee State University, Johnson City, Tennessee, and P. H. Crowley, University of Kentucky, Kentucky
5:25–5:50.	(Florida lakes). T. Crisman, University of Florida, Gainesville, Florida
5:50–6:15.	Round Table Discussion: Littoral zone ecology. Dewey Meyers, moderator. Participants: M. Whiteside, G. Fryer, D. Frey, D. Phoenix, C. Goulden, T. Crisman

Friday, August 25

Community Structure

Session I. Zooplankton Competition, Invertebrate Predation, and Community Structure
W. C. Kerfoot, Section Chairman

8:30–8:50.	Experimental studies on the predatory feeding of copepodid stages III-IV of *Mesocyclops leuckarti* Claus. C. Jamieson, University of Otago, Dunedin, New Zealand

8:50–9:10.	Size selection by predaceous zooplankton. J. L. Confer, Ithaca College, New York
9:10–9:30.	Preferential feeding by *Chaoborus* larvae: Laboratory and field observations. R. Pastorok, University of Washington, Seattle, Washington
9:30–9:55.	Direct observation of predator-prey interactions between zooplankters. C. E. Williamson and J. J. Gilbert, Dartmouth College, Hanover, New Hampshire
9:55–10:20.	Breaking the bottleneck: Interactions of invertebrate predators and nutrients in oligotrophic lakes. W. E. Neill and A. Peacock, University of British Columbia, Vancouver, Canada
10:20–10:45.	Enclosure experiments: Evaluating the importance of competition and predation. W. C. Kerfoot and W. DeMott, Dartmouth College, Hanover, New Hampshire
10:45–11:10.	Evidence for predictable spatial patterns in the structure of a lake zooplankton assemblage. W. M. Lewis, Jr., University of Colorado, Boulder, Colorado
11:10–11:35.	The role of invertebrate predation in maintaining the diversity of herbivores. P. A. Lane, Dalhousie University, Halifax, Nova Scotia
11:35–12:00.	Discussion
12:00–1:00.	Lunch

Session II. Vertebrate Predation and Community Structure
J. O'Brien, University of Kansas, Section Chairman

1:00–1:05.	Introduction. J. O'Brien
1:05–1:30.	The effect of prey motion on planktivore choice. T. Zaret, University of Washington, Seattle, Washington
1:30–1:55.	Alewives (*Alosa pseudoharengus*) and ciscoes (*Coregonus artedii*) as selective and nonselective predators. J. Janssen, Biology Department, Loyola University, Chicago, Illinois
1:55–2:20.	The selective impact of a planktivorous fish community: Roles of zooplankton escape ability and planktivore size selectivity. R. Drenner, Texas Christian University, Texas
2:20–2:45.	Relationships between trout and invertebrate species as predators and the structure of the crustacean and rotiferan plankton in mountain lakes. R. S. Anderson, University of Calgary, Calgary, Alberta, Canada
2:45–3:00.	Break
3:00–5:00.	Round Table Discussion: Predation versus competition in the structuring of zooplankton communities. D. Hall, Michigan State University, East Lansing, Michigan, moderator. Participants: J. Jacobs, T. Zaret, G. Vinyard, J. O'Brien, D. Hall, M. Gliwicz, J. P. Nilssen, Jan A. E. Stenson

Evening Sessions

Tuesday Night, August 22

Barclay Room
(Regular Meeting Room)

Stoneman Room

7:30–7:55. Changes in feeding ability associated with metamorphosis in the ctenophore *Mnemiopsis leidyi*. D. J. Lonsdale, Chesapeake Biological Lab., University of Maryland, Solomons, Maryland.

Film Session: Fluid Dynamics, Copepod filter-feeding

Thursday Night, August 24

Barclay Room

Stoneman Room: Informal gatherings

7:30–7:55. *Mysis relicta* in Lake Tahoe: Some effects of an introduced species on the zooplankton of an ultraoligotrophic lake. S. T. Threlkeld, J. T. Rybock, M. D. Morgan, and C. R. Goldman, Institute of Ecology and Division of Environmental Studies, University of California, Davis, California

7:55–8:20. Biogeography of zooplanktonic communities of Spain. M. R. Miracle, Universidad de Barcelona, Barcelona, Spain

8:20–8:45. Life history calculations: The effects of selective predation upon copepods and cladocerans. B. Taylor, University of Washington, Seattle, Washington

8:45–9-05. Long-term laboratory studies of *Chaoborus* predation on crustacean plankton species. R. S. Anderson, University of Calgary, Alberta, Canada

9:05–9:30. Viscosity changes and swimming efficiency of *Daphnia*. M. Mort, Dartmouth College, Hanover, New Hampshire

Index of Species

Acanthocyclops (see also *Cyclops*), 378, 380–382, 384–386
   A. viridis, 378, 521, 534
   A. vernalis (see also *Cyclops vernalis*)
      distribution, 637
      feeding, 520, 527–529
Acanthodiaptomus denticornis
   distribution, 637
   feeding, 54
   pigmentation, 91–93, 95–96, 99
   reproduction, 391–392
   swimming, 54, 59
Acartia
   feeding, 185, 188, 190–191, 196, 203, 215–216, 234, 239, 292, 743, 752
   A. clausi
      feeding, 201, 206, 209–215, 231, 234–239, 247
      nutrient regeneration, 259
      reproduction, 390, 392, 752
      vertical migration, 120
   A. tonsa, 14
      feeding, 190, 192, 194, 196
      nutrient regeneration, 259
      reproduction, 390–393, 403
Acercus, 24
   A. torris, 25
Acetes sibogae australis, 213
Acroperus harpae, 658, 661, 665, 667
Aetideus divergens, 207–208
Afrocyclops gibsoni, 695
Aglaodiaptomus (see also *Diaptomus*), 637
   A. californiensis, 391
   A. clavipes, 391
   A. stagnalis, 391
Alona, 404, 685, 694–695
   A. affinis, 360
      distribution, 661, 663, 665, 667, 672
      reproduction, 403

   A. circumfimbriata, 661, 664–665, 672–673
   A. costata, 664–665
   A. diaphana, 665, 667
   A. elegans, 694
   A. elegans elegans, 693–694
   A. elegans lebes, 693–694
   A. guttata, 403, 661, 665, 667
   A. holdeni, 693, 695
   A. intermedia, 658, 665, 667
   A. karua, 665, 667
   A. monocantha, 665, 667
   A. quadrangularis, 664, 667
   A. rectangula, 667
   A. rustica, 403, 661, 664, 665, 667
   A. setulosa, 661–662, 665
   A. verrucosa, 664–665, 667, 672
Alonella excisa, 80, 665, 667, 672
   A. exigua, 664–665, 667
   A. globulosa, 661, 663, 665, 667
   A. hamulata, 661, 663–665, 667
Anax junius, 571
Anomalagrion hastatum, 572, 575–577
Anuraeopsis fissa, 167
Arctodiaptomus bacillifer, 391, 396
Argia fumipennis violacea, 571
Arrenurus, 21
Asplanchna, 457
   as prey, 59, 509, 515
   feeding, 54, 158–171, 358, 645
   induced polymorphism, 67, 167–168, 435–436, 454, 471, 473
   nutrient regeneration, 259
   A. brightwelli, 155–160, 163–164, 166–167, 322
   A. girodi
      as prey, 509–515
      feeding, 161–164, 166–170, 509–515
      genetics, 316–317, 319, 322, 325, 327
   A. herricki, 159–160, 166

Asplanchna (continued)
  A. intermedia, 158–167
  A. priodonta
    feeding, 159–160, 164, 166–167, 170
    genetics, 325
    vertical migration, 120
  A. sieboldi
    feeding, 54, 158–166, 168–170
    reproduction, 322, 404
    swimming, 54
  A. silvestri, 166

Boeckella, 390
  B. propinqua, 392
Bosmina
  as prey, 19, 26, 378–382, 384–385, 512, 522, 543, 546, 551
  competition, 725–726, 731–733, 735–740
  cyclomorphosis, 456–457
  distribution, 720–721
  feeding, 285, 292, 301, 650
  genetics, 335
  reproduction, 365, 378–382, 384–385, 405
  swimming, 7–9, 12, 16–17
  B. coregoni (see also *Eubosmina coregoni*)
    feeding, 289
    genetics, 325
  B. fatalis, 627–630, 632
  B. longirostris, 13, 581
    as prey, 143, 145, 483, 502, 541, 545, 551, 555, 558–567, 594–597, 599–601, 604–616
    competition, 731–734
    cyclomorphosis, 456–468, 476–490
    distribution, 637, 644, 648, 685, 691–692, 717
    feeding, 54
    genetics, 331–332, 334–335, 395–398, 456–468
    nutrient regeneration, 252
    reproduction, 143, 340, 358, 360–364, 366, 368, 371, 378, 403–405
    swimming, 19, 54
    vertical migration, 71, 138, 140–141, 143, 145
    var. *brevicornis*, 490
    var. *curvirostris*, 490
    var. *pellucida*, 490
    var. *similis*, 490
    var. *typica*, 490
  B. meridionalis, 519, 522, 534–535
Bosminopsis, 669, 673
  B. deitersi, 594–601, 672, 693
Brachionus
  as prey, 158, 163, 169
  Asplanchna-induced polymorphism, 67, 436, 473
  feeding, 153, 156
  B. bidentata, 168

B. calyciflorus
  as prey, 160, 166–170, 510–514
  Asplanchna-induced polymorphism, 168, 435, 454, 471
  culture, 173–182
  feeding, 151–156
  reproduction, 319, 404
B. havanaensis, 325
B. urceolaris sericus, 168

Calamoecia lucasi, 519–527, 529–535
Calanus
  competition, 742–752
  feeding, 191, 203, 207–208, 215, 292
  nutrient regeneration, 258, 392
  reproduction, 50
  C. finmarchicus
    feeding, 187, 201–202, 209, 212–213, 743
    nutrient regeneration, 259
    reproduction, 358, 390–392, 395, 751
    vertical migration, 111, 115–116
  C. helgolandicus, 120, 206, 242, 259–260
  C. hyperboreus, 191, 202, 206, 259, 390
  C. marshallae, 239, 743
  C. pacificus
    feeding, 199, 202–204, 207, 234, 742–744, 747–752
    reproduction, 47, 747–752
  C. plumchrus, 390–392
Camptocercus rectirostris, 667
Candacia bradyi, 206
Celithenis, 571
Centropages typicus, 48–50, 534
Ceriodaphnia, 457
  as prey, 301–302, 551, 609–610
  distribution, 637, 669, 673
  feeding, 650
  reproduction, 395
  C. cornuta, 331–332, 604, 610
  C. dubia, 519, 522–528, 530–535
  C. lacustris, 360–361, 363, 365, 368, 644
  C. laticaudata, 398
  C. megalops, 398
  C. pulchella, 397–398, 672
  C. quadrangula, 80, 287, 325, 368, 398, 529, 644, 717
    as prey, 529
    distribution, 644, 717
    feeding, 287
    genetics, 325, 368
    reproduction, 398
    vertical migration, 80
  C. reticulata, 732
    as prey, 373, 528, 545, 551, 588–591
    distribution, 644, 646–647

Index of Species

reproduction, 396–398, 403–404
vertical migration, 80, 82–84
C. rigaudi, 672
Chaoborus
as prey, 583, 609, 618–619
distribution, 635, 637–639, 669, 717–723
feeding, 54, 55, 60, 109, 122–128, 372–373, 439–440, 444–447, 493, 538–543, 546–551, 569, 621, 625, 629–633, 635, 637–639, 645, 715, 717–723, 727
swimming, 54
vertical migration, 106, 122–128, 135, 674–676
C. americanus
distribution, 717–718
feeding, 542, 544, 550–551
vertical migration, 106, 108
C. borealis, 550
C. crystallinus, 123
C. flavicans
as prey, 618–622
distribution, 717
feeding, 542, 544, 547, 550
vertical migration, 106–108
C. nyblaei, 542, 544–545, 550
C. obscuripes, 618–622
C. punctipennis, 87, 544–545
C. trivittatus
distribution, 717–718
feeding, 122, 124, 379, 538, 542, 545, 547, 550–551
vertical migration, 106, 108, 122, 127
Chironomus anthracinus, 424
Chromogaster, 167
Chydorus
as prey, 19
distribution, 667, 685, 688, 692–694, 717, 719–720
feeding, 650
swimming, 12
C. barroisi, 673, 692, 694
C. faviformis, 661, 664–665, 667
C. globosus, 667
C. hybridus, 672
C. piger, 661, 663, 665, 667
C. pubescens, 672–673
C. sphaericus
as prey, 20, 519, 522–523, 527, 534–535
competition, 731–732, 734–735, 737
distribution, 637, 644, 658, 661–662, 664–665, 667, 673, 693–694
feeding, 283–285
reproduction, 358, 360–365, 402–404
vertical migration, 80–88
Codonella, 164
Conochiloides dossuarius, 627–629, 632

Conochilus unicornis, 637–639
Cyclops 252, 457 (see also *Acanthocyclops*, *Diacyclops*)
as prey, 55, 588
diapause, 411–412
distribution, 679, 681, 684
feeding, 17, 19, 26, 515, 540, 717–723
genetics, 325
reproduction, 60, 373
swimming, 17, 52, 55–56, 59
C. abyssorum, 412–413, 416, 423, 679–683
f. *bodanus*, 680–681
f. *divulsus*, 683
f. *laevis*, 682
f. *novarensis*, 682
f. *prealpinus*, 54, 59, 680–683
f. *tatricus*, 680–683
C. bicuspidatus (see also *Diacyclops bicuspidatus*), 19, 293–295, 483, 491, 509–516
C. bicuspidatus thomasi (see also *Diacyclops bicuspidatus thomasi*)
as prey, 550, 583
distribution, 644, 646, 648
feeding, 54, 514, 516, 715–718, 721
swimming, 54
C. furcifer, 412, 680
C. kolensis, 412
C. vernalis (see also *Acanthocyclops vernalis*)
distribution, 644, 646–647
feeding, 293–295, 483, 522
C. scutifer, 186
diapause, 412–415
distribution, 644
feeding, 54
reproduction, 48, 424
swimming, 17–19, 52, 54, 56–60
vertical migration, 71
C. strenuus
diapause, 412–414, 416
distribution, 680, 695
genetics, 325
reproduction, 421, 423–424
C. varicans rubellus, 672
C. vicinus, 412, 421, 680
C. vicinus lobosus, 54, 59
Cypria, 670, 673
C. javanus, 673
C. javanus pelagica, 673
C. petenensis, 669, 672–673, 674–676
C. purpurascens, 673

Daphnia, 451
as prey, 17, 21, 59, 123, 439–440, 453–454, 538, 542–544, 546–551, 555, 559–560, 573–577, 581, 588, 590–592, 604–605, 609–610

competition, 484, 725–726, 731–733, 735, 740
cyclomorphosis, 429, 452–454, 503
distribution, 637–638, 650, 669, 672–673, 685, 688–691
feeding, 234, 242, 264, 269–271, 279, 285, 292, 299–304, 373, 559
genetics, 316, 331–332, 464
nutrient regeneration, 252, 256, 260
pigmentation, 92, 109
reproduction, 329, 338, 352, 369, 374, 405, 558–560
swimming, 12, 56
vertical migration, 65, 76–77, 131

D. ambigua
as prey, 545, 551, 557, 565–566
distribution, 644
feeding, 368
genetics, 325
reproduction, 398, 403

D. barbata, 690–692

D. carinata, 331–333, 464
f. *cephalata*, 501

D. catawba, 644

D. cephalata, 332, 335

D. cucullata, 283–287, 289–290, 398, 402

D. curvirostris, 398

D. dubia, 644

D. galeata
as prey, 380–382, 384–385, 588–589
cyclomorphosis, 335
pigmentation, 92
reproduction, 340, 378

D. galeata gracilis, 94

D. galeata mendotae, 346–356, 567
as prey, 581, 609
cyclomorphosis, 430–436
distribution, 644
feeding, 54, 89, 373
reproduction, 143, 346–356, 359–361, 363, 366, 396, 403
swimming, 54, 59
vertical migration, 143–145

D. hyalina f. *galeata*, 398

D. hyalina pellucida, 287

D. laevis, 644

D. longiremis
distribution, 644
f. *cephala*
as prey, 497–505
competition with f. *typica*, 497–505
vertical migration, 71
f. *typica*
as prey, 497–505
competition with f. *cephala*, 497–505
vertical migration, 74–75

D. longispina
distribution, 690–692
genetics, 325
pigmentation, 91
reproduction, 266, 396–398, 402
vertical migration, 77

D. lumholtzi, 331–332, 610, 690–692

D. magna
as prey, 573–574, 604–605, 608
distribution, 691–692
feeding, 54, 88, 268
genetics, 332–335, 401
nutrient regeneration, 259, 270–279, 288
reproduction, 396–398, 403–404
swimming, 28, 54

D. middendorffiana
as prey, 371–372, 500, 502
distribution, 637
genetics, 330, 335
reproduction, 330, 371–372, 400
vertical migration, 73–74

D. obtusa, 398

D. parvula, 139–140, 398, 403, 544–545, 551

D. pulex
as prey, 23, 123, 373, 380–382, 384–385, 438–454, 499–501, 544–545, 550–551, 582, 604–607, 614–615
competition, 731–732, 734
distribution, 644
feeding, 54, 87, 264–266, 289, 293, 299–300, 303, 368, 373
nutrient regeneration, 259
reproduction, 264–266, 378, 396–398, 402–406, 438–454
swimming, 8, 15, 19, 54
vertical migration, 74–75, 77

D. pulex carina, 644

D. pulex obtusa, 92, 94

D. pulicaria, 360, 732
as prey, 109, 539, 543–545, 550–551, 555, 557–558, 616
feeding, 299, 557
reproduction, 346–356, 406, 566

D. retrocurva
as prey, 581
distribution, 644
reproduction, 340–344, 359–361, 363, 365

D. rosea, 732
as prey, 544–545, 550–551, 555, 558
distribution, 644, 646–647, 717
feeding, 265–266
nutrient regeneration, 259
reproduction, 265–266, 358

D. schodleri, 368, 374, 398, 644

D. similis, 691–692
D. thomsoni, 398
D. thorata, 567, 644
D. tibetana, 109
Diacyclops (see also *Cyclops*), 411
   D. bicuspidatus (see also *Cyclops bicuspidatus*), 412
   D. bicuspidatus thomasi (see also *Cyclops bicuspidatus thomasi*)
      feeding, 637
   D. navus, 412
   D. thomasi, 424
Diaphanosoma, 457
   as prey, 493, 543, 546, 551
   distribution, 650, 672
   feeding, 54, 484, 650
   swimming, 12, 54
   D. brachyurum, 732
      as prey, 550–551, 588–592
      distribution, 644, 717
      feeding, 284–285, 289, 368
      genetics, 325
      reproduction, 402
   D. leuchtenbergianum
      distribution, 644, 646–647
      feeding, 54
      reproduction, 340, 360–361, 363, 365
      swimming, 54
   D. modigliani, 627–628, 630, 632
   D. sarsi, 627–628, 630, 632
Diaptomus (see also *Aglaodiaptomus, Hesperodiaptomus, Leptodiaptomus*), 457, 673
   as prey, 122, 378, 380–382, 384–385, 457, 538, 546–549, 588–589
   feeding, 185–186, 188, 190–191, 194, 247, 292, 372, 514, 543
   nutrient regeneration, 252, 256
   pigmentation, 99, 106
   swimming, 12
   vertical migration, 71, 141, 143, 145
   D. arcticus (see also *Hesperodiaptomus arcticus*), 514
   D. ashlandi (see also *Leptodiaptomus ashlandi*)
      as prey, 591
      feeding, 54, 190, 192–196, 219–220, 228–232
      swimming, 54, 59
   D. coloradensis, 544, 550, 644
   D. dampfi, 672–673
   D. dorsalis, 325, 669, 672–675
   D. floridanus, 534
   D. franciscanus, 19, 106, 108–109, 539, 545
   D. gracilis (see also *Eudiaptomus gracilis*), 551
   D. hesperus (see also *Onychodiaptomus hesperus*), 106, 108, 514, 516
   D. kenai
      as prey, 55, 106, 108–109, 122–127, 541, 550–551
      distribution, 717
      feeding, 54
      pigmentation, 93, 98, 106
      swimming, 54
      vertical migration, 106, 108–109
   D. minutus (see also *Leptodiaptomus minutus*), 231, 644, 646–647
   D. leptopus, 644, 717
   D. nevadensis, 92–96, 98–107, 514, 551
   D. nudus, 644, 646–647
   D. oregonensis (see also *Skistodiaptomus oregonensis*)
      as prey, 514, 516
      distribution, 644, 717
      feeding, 190–196, 219–221, 223, 227–232
   D. pallidus (see also *Skistodiaptomus pallidus*)
      as prey, 588–589
      feeding, 186–189, 192–196
      reproduction, 403
      vertical migration, 140–141
   D. shoshone (see also *Hesperodiaptomus shoshone*), 514, 544, 550–551, 644
   D. sicilis (see also *Leptodiaptomus sicilis*)
      as prey, 514, 544, 550, 591
      distribution, 644
      feeding, 292–298
      pigmentation, 94, 98–107
      vertical migration, 98–107
   D. siciloides (see also *Leptodiaptomus siciloides*)
      distribution, 644
      feeding, 190, 193–196, 219–221, 223–232
   D. superbus, 96
   D. tyrelli
      as prey, 106, 108–109, 514, 550–551, 555, 558, 560–563, 567
      feeding, 54
      swimming, 54
Diplodontus, 21, 25
Disparalona acutirostris, 667
   D. rostrata, 664–665, 667
Dunhevedia crassa, 661, 665

Enallagma aspersum, 571
   E. signatum, 571
   E. triviatum, 571
Eucalanus, 196
   E. crassus, 241, 243–244, 247
Epischura, 457
   distribution, 645–646, 726
   feeding, 490–491, 493, 540, 543, 563

nutrient regeneration, 256
swimming, 15, 243, 733
E. baikalensis, 390-391
E. lacustris, 54, 491, 644, 647
E. nevadensis, 252
  as prey, 551, 555, 558, 560-561
  distribution, 491
  feeding, 54, 483, 502, 557
  swimming, 54, 59
E. nordenskioldi, 14, 390-392
Eubosmina, 670, 674
  E. coregoni (see also *Bosmina coregoni*)
    feeding, 293-295
    reproduction, 340, 358, 360-363, 365
  E. crassicornis, 691, 693
  E. tubicen
    as prey, 594-600
    distribution, 669
    genetics, 325
    vertical migration, 672-675
Euchlanis dilatata, 323, 404
Eucyclops, 695
  E. agilis, 637, 644, 672, 695, 717
  E. serrulatus, 29, 695
  E. speratus, 644
Eudiaptomus gracilis (see also *Diaptomus gracilis*), 394-395
  E. vulgaris, 390, 392, 395
Euryalona, 662
  E. occidentalis, 667
  E. orientalis, 661, 665
Eurycercus, 664
  E. glacialis, 330
  E. lamellatus, 398, 667
Eurytemora, 190, 206, 215, 239, 292
  E. affinis
    feeding, 190, 192, 194, 196, 219-220, 228-230, 232
    reproduction, 47-50, 503
  E. herdmani, 201, 209-212, 214-215
Eylais, 21, 25

Filinia, 164, 650
  F. mystacina, 165
  F. terminalis, 325

Gammarus, 635, 637-639
  G. lacustris, 639
Gladioferens imparipes, 403
Graptoleberis testudinaria, 664-665, 667

Hesperodiaptomus (see also *Diaptomus*), 635, 637-640
  H. arcticus (see also *Diaptomus arcticus*), 637
  H. shoshone (see also *Diaptomus shoshone*), 390, 637

Hexarthra intermedia, 627
Heterocope, 493, 497, 501, 504-505
  H. borealis, 391-392, 394
  H. saliens, 391-392
  H. septentrionalis
    feeding, 497, 499, 501-503, 527
    vertical migration, 71, 74-75
Hexarthra mira, 164
Holopedium, 543, 551
  H. gibberum, 360
    as prey, 544
    distribution, 637-639, 644, 717
    vertical migration, 71
Hydrachna, 21-25
Hydryphantes, 21, 25
Hygrobates, 21

Ilyocryptus sordidus, 672
  I. spinifer, 672
Ischnura elegans, 572
  I. posita, 571
  I. ramburii, 572-574
  I. verticalis, 569, 571

Kellicottia longispina
  as prey, 555, 558, 560-563, 567
  distribution, 637-639
  genetics, 325
Keratella, 167-170, 515, 650, 674
  K. cochlearis
    as prey, 167-170, 509-513, 515
    distribution, 627
    genetics, 324-325
    reproduction, 343
  K. procurva, 627
  K. quadrata, 167, 317-318
  K. hiemalis, 325

Labidocera, 50, 243
  L. aestiva, 39-50
  L. jollae, 206
  L. trispinosa, 50
Lecane, 166
Lepadella, 166
Leptodiaptomus (see also *Diaptomus*), 637
  L. ashlandi (see also *Diaptomus ashlandi*)
    reproduction, 391
  L. minutus (see also *Diaptomus minutus*), 390, 392-393
  L. sicilis, (see also *Diaptomus sicilis*)
    reproduction, 392
  L. siciloides (see also *Diaptomus siciloides*), 378
    reproduction, 391-392, 394-395
Leptodora
  as prey, 609, 645

feeding, 358, 471, 543, 549, 733
  swimming, 7
  L. kindtii, 360-361
    distribution, 644
    reproduction, 343
    swimming, 339
Lestes disjunctus, 572
  L. eurinus, 571
Leydigia, 661-662, 672
  L. acanthocercoides, 661, 665, 667
  L. leydigi, 658, 661, 667
Libellula, 571
Limnocalanus, 645
  L. grimaldii, 392
  L. johanseni, 390-391, 394-395
  L. macrurus, 389-392, 394-395, 644
Limnochares, 21, 24
  L. americana, 25

Macrocyclops albidus, 29, 521, 637, 644, 717
Macrothrix rosea, 672
Mesocyclops, 411, 491, 515, 549, 588, 726
  as prey, 588
  diapause, 411
  distribution, 726
  feeding, 491, 515, 549
  M. edax, 732
    distribution, 644, 669, 672-674
    feeding, 509-515, 526, 529, 534
    genetics, 325
    vertical migration, 140
  M. inversus, 669, 672-674
  M. leuckarti, 325, 412, 514, 518-536, 672
Metadiaptomus, 695
  M. chevreuxi, 694
  M. mauretanicus, 694
Metridia pacifica, 120
Mixodiaptomus laciniatus, 28, 29, 31, 35, 37, 389-391, 394-395
Mochlonyx, 439, 440, 443-444, 446-447
Moina, 329, 551
  M. micrura, 396, 398, 627-628, 630, 632
Monospilus dispar, 661
Mysis, 645
  M. relicta, 555-567, 581, 653

Neocalanus cristatus, 752
  N. plumchrus, 752
Nitocra sinipes, 672
Notholca caudata, 318-319

Oithona similis, 201
Onychodiaptomus hesperus (see also *Diaptomus hesperus*), 378
Orthocyclops, 491
  O. modestus, 644

Oxyurella tenuicaudis, 694

Pachycheles pubescens, 204
Paracalanus, 743
  P. parvus, 752
Piona, 21, 25, 129, 135
  P. carnea, 135
  P. constricta, 15, 130, 135
  P. limnetica, 129
  P. rotunda, 129
Plathemis lydia, 571
Pleuroxus, 665
  P. aduncus, 685, 694
  P. denticulatus, 403, 404, 665, 667
Polyarthra, 11, 167, 169, 492, 515
  P. dolichoptera, 167, 317-318
  P. euryptera, 169
  P. major, 318, 320
  P. remata, 167
  P. vulgaris
    as prey, 167, 169-170, 509-513, 515
    distribution, 627
    genetics, 325
    reproduction, 318, 321
Polyphemus, 20, 54, 457, 645
  P. pediculus, 639, 644, 717
Pompholyx sulcata, 167
Pontellopsis villosa, 50
Pseudocalanus
  competition, 742-752
  feeding, 206-208, 215-216
  reproduction, 47
  P. elongatus, 743
  P. minutus
    feeding, 199-201, 203-204, 208
    reproduction, 391-392
    vertical migration, 111, 113-114, 117-119
Pseudochydorus globosus, 80-88, 672
Pseudodiaptomus, 247

Rhynchotalona falcata, 661, 664-665, 667

Sapphirina angusta, 206
Scapholeberis, 673
  S. kingi, 396, 398
Senecella, 645
  S. calanoides, 391, 644
Sida crystallina, 398, 644
Simocephalus, 329, 569, 572-577
  S. acutirostratus, 396-398
  S. exspinosus, 77
  S. serrulatus, 332-334, 403, 672
  S. vetulus, 287, 398, 551, 573-574
Skistodiaptomus oregonensis (see also *Diaptomus oregonensis*), 14
  reproduction, 391, 394-395

S. *pallidus* (see also *Diaptomus pallidus*)
  reproduction, 393
Synchaeta grandis, 167
  S. kitina, 167
  S. oblonga, 167
  S. pectinata, 167

Temora, 206, 215–216, 247, 743
  T. longicornis
    development, 752
    feeding, 201
    vertical migration, 111, 113, 115, 117–119
  T. turbinata, 50
Tetragoneuria cynosura, 571
Thalassiosira fluviatilis, 111–112
Thermocyclops, 411, 551
  T. dybowskii, 412
  T. hyalinus, 259, 627–629
  T. oblongatus, 695

Thermodiaptomus, 694–695
Tortanus discaudatus
  feeding, 206, 520, 534
  vertical migration, 111, 115–117
Trichocerca, 167
  T. brachyurum, 627
  T. pusilla, 167
  T. similis, 167
Tropocyclops, 720–722, 726, 735–737
  T. prasinus, 293–295, 551, 732
  T. prasinus prasinus, 717
  T. prasinus mexicanus
    distribution, 644, 646–647, 669, 672–673
Tropodiaptomus, 551, 694–695
  T. gigantoviger, 550, 627–628

Unionicola, 21, 129–130, 135
  U. crassipes, 130, 133, 135

General Index

Active space, 246
Advection of energy flow, 757–758
Aggregation of bluegreens in presence of *Daphnia*, 303
Akinesis, 17, 19–20
Algae in diet of:
  Chaoborus, 542
  cladocerans, 75, 81, 91, 265–266, 268–280, 282–289, 292–304, 349, 356, 432, 441–443, 447–453, 457–458, 500, 557, 559, 573, 731, 733–740
  copepods, 112, 201–204, 219–232, 235–239, 242, 245–248, 292–298, 529–530, 535, 750
  rotifers, 151–157, 159–161, 166–167, 173–182, 317
  Zooplankton in general, 251, 258, 261, 267, 282, 299, 305–311, 389–390, 650–654, 673, 697, 708–709, 715, 723, 725–727, 745–752, 762
Algal succession, natural, 299–304, 349
Allocation of energy (*see* Energy)
Allometry, 429–436, 474; of *Daphnia* helmets, 429–436
Alternative stable states, 716
Ambush in mating, 60
Ambush predators, 54, 60, 540, 575
Aphid-rotifer model, 326–328
Aposematic selection, 20
Apparent-size hypothesis, 142
Arctic zooplankton, 69–78, 497–505
Assimilation, efficiency of, 749; assimilation rates, 265, 724–743, 749
Attack behavior of:
  Chaoborus, 123, 540–541
  cyclopoid copepods, 17, 512, 521–522, 534–535
  rotifers, 159, 512
  see also Predation, Predator-prey interactions
Avoidance of:
  approaching predators, 12, 60, 167, 418, 501–502, 530–531, 546 (*see also* Escape reaction)

  lake outlets, 36
  light, 98
  shad filter-feeding, 588–590
  shores (*see* Shore avoidance, Uferflucht)
  suction, 588–590
  walls, 8, 515

"Big-Bang" strategy, 418, 422
"Black-Box" design of enclosure experiments, 509
Blasius problem, 7
Biomass distributions for entire pelagic ecosystem, 757–762
Blue-green algae:
  digestion and assimilation of, 269, 293–298
  effects upon zooplankton densities, 268–280
  food quality of, 269, 293–298
  grass-blade morphology of, 299–304
  in fish enclosures, 301–302
  inhibition of zooplankton grazing, 267, 268–280, 282–291
  reduction of toxicity by bacteria, 269
  rejection of: by *Asplanchna*, 161; by Cladocerans, 269, 278–279, 287–289
  resistance to, 317
Body shape, 10–19, 26
  of odonates in relation to fish predation, 570, 572
  of zooplankton in relation to:
    predation in general, 436, 456, 470–471, 492–493, 535–536
    by *Chaoborus*, 123, 546
    by Copepods, 12, 16–19, 26, 483, 492–493, 497–505, 512, 515, 534
    by Rotifers, 162–171, 435, 454, 471
    by fishes, 438, 497–505
  mating, 39–50
  streamlining, 11, 16
  as distinguished from size, 475–476
Boltzmann's constant, 205

Bottlenecks:
  physiological, 422
  population, 715–716, 722
Brightness distribution of light, 29–38
  in littoral and pelagic zones, 30–37
  orienting vertical migration, 29, 35
  symmetry under water, 29–38
Buccal funnel of rotifers, 153–156; rejection of food from, 156

Campanulate morphotype of *Asplanchna*, 159–166
Cannibalism, 161–165, 424, 521
Chemical trails, 540 (*see also* Pheromones)
Chemically induced developmental changes:
  stimulated in rotifers by *Asplanchna*-substance, 167–168, by dietary tocopheral, 322
  induced in algae by *Daphnia*-substance, 303
Chemostat cultivation: of algae, 173–176; of rotifers, 173–176; theoretical aspects of, 305–311
Chemotropotaxis, 214
Chordotonal organs, 539–540
Chromatin diminution, 679
Cingulum, 153
Circadian rhythms, 52
Cladoceran:
  distribution, 637–638, 644, 646–648, 650, 661, 663, 665, 667, 669, 672–685, 688–694, 720–721
  feeding, 54, 87–89, 234, 242, 264–266, 268–271, 279, 284–285, 287, 289, 292–295, 299–304, 358, 368, 373, 471, 484, 543, 549, 557, 559, 650, 725–740 (*see also* Feeding behavior)
  genetics, 316, 325, 330–335, 368, 395–398, 401, 456–468
  reproduction, 193, 264–266, 329–330, 338, 340–344, 346–356, 358–366, 368–369, 371–372, 374, 378, 396–398, 400, 402–406, 438, 454, 558–560, 566
  swimming, 7, 12, 19, 28, 54, 59, 339
  vertical migration, 65, 71, 73–77, 80–88, 131, 138, 140–141, 143–145
Clonal lines, 335, 457–458, 461–462, 466–468, 481
Coexistence, general, 308–311, 335, 570–571, 575, 585, 716, 728–731, 740
  of clonal lines, 335
  of competing species:
    by niche partitioning, 570–571
    list of possible causes, 575
    theoretical conditions for, 308–311, 570, 728–731
  of predators and prey, 716
Community structure, 170–171, 305–311, 625–633, 635–640, 669, 714–763

comparison of terrestrial and pelagic, 725–726
conceptual models of, 726–728
effects of algae upon, 282–291
effects of predation upon, by:
  Chaoborus, 625–633, 715, 718–719
  Cyclops, 715, 719–723
  rotifers, 170–171
energy flow through, 742–762
formal mathematical models of, 728–731
incorporation of food heterogeneity into schemes of, 739–740
of algae, 305–311
of Guatemalan lakes, 669
of zooplankton:
  as described by principal components ordination techniques, 642–650
  as described by relative abundances, 635–640:
    with fish absent, 639
    with fish present, 638
review of major hypotheses relating to role of predation, 726–728
simulation models of, 742–752
(*see also* Geographic considerations of community structure)
Comparative morphology of calanoid copepod mouthparts, 19, 187–188, 190–196
Compensation for current, orientation of body as, 36
Compensatory feeding, 549
Competition:
  between *Bosmina* and *Daphnia*, 725–740
    in laboratory beakers, 725, 731, 734–735, 739
    in natural lakes, 725, 731–740
  between *Calanus* and *Pseudocalanus*, simulated, 742–752
  between clones, 327, 335, 484
  between *Daphnia* and *Simocephalus*, 573–574
  between rotifers and cladocerans, 640
  related to shifts in body size of species, 362
  structuring communities, 648
Compound eyes, of *Bosmina*, 595–597, 609–616; of *Chaoborus*, 540, 619
Concept of "race" in copepod taxonomy, 679
Continuous culture techniques (chemostats), 173–182; for algae, 173; for algal-rotifer interactions, 173
Copepod:
  distribution, 491, 637–639, 644–649, 669, 672–674, 679–681, 684, 695, 717, 726
  (*see also* Geographic considerations of community structure)
  feeding, 11, 17, 19, 26, 54, 185–196, 198–216, 219–232, 237–239, 241–248, 292–298, 372, 483–484, 790–791, 793, 797, 499, 501–503, 509–516, 518–536, 540, 549, 557,

637, 650, 715-723, 742-752 (*see also*
  Feeding behavior, Predation, Predator-
  prey interactions)
 genetics, 325, 679-680
 mating, 39-50, 60-61
 orientation, 28-37
 reproduction, 47-50, 60, 358, 373, 390-395,
  403, 421, 423-424, 503, 747-752
 swimming, 8, 11-19, 52-61, 83, 96, 243, 512,
  573, 733 (*see also* Swimming behavior)
 vertical migration, 28, 35, 37, 71, 74-75,
  98-109, 110-120, 138-146
Copulatory position for calanoid copepods, 46
Coronal region, of rotifers, 12, 153, 158-160,
 164. Movements, 159; of *Brachionus*, 12
 153; of *Asplanchna*, 158-159
 receptors, 159-160
Cost of reproduction, 418
Counter-evolution of prey, 728
Cruciform morphotype of *Asplanchna*, 161-166
Crypsis, 572, 609
Cryptobiosis, 328
Cryptoparthenogenesis, 315
Cyclomorphosis, 77, 333, 335, 358, 362, 364,
 428-505
 as an adaptation, 472-473, 492
 described by closed loops, 474-475, 483
 distinction between size and shape components
  of 475-476
 in cladocerans:
  Bosmina, 358, 362, 453, 456-468, 470-471,
   476-483, 486, 488, 492
  Daphnia, 77, 335, 429, 438, 453-454, 497
  of rostrum and antennal bristles, 429
  head shape, 333, 335, 358, 429-436, 438,
   497
  spine lengths, 358, 429, 438, 471
  in dinoflagellates, 456; and protists,
   470; and rotifers, 429, 435-436, 454,
   471, 473
 in relation to swimming behavior, 77
 involving size reductions, 358, 362, 364,
  439, 452-454, 477-478, 481
 multivariate descriptions of, 474-483
 proximate and ultimate causes, 364, 429-436,
  452-454, 470, 473, 497
 food, 429-436, 439
 abiotic:
  illumination, 429-436, 473; polarized,
   431
  photoperiod, 429, 442
  temperature, 429-436, 442, 473
 biotic:
  predation, 358-359, 429, 456, 468,
   470-474, 484, 490-493, 497-505;
   invertebrate, 358, 435-436, 483, 490-

  493, 497, 499-505; vertebrate, 358-
  359 362, 436, 438-439, 497-500,
  502-505
 related to vertical migration, 77
 scaling effects, 475
 ways of describing, 474

Defenses:
 of algae, against grazing, 293, 303-304
 of zooplankton:
  general, 11-17, 67, 358, 429, 468, 470-471,
   473, 492-493, 497-505, 509, 591-592,
   594
  against predatory copepods, 16-19, 483-484,
   490-491, 497-505, 512, 515, 521-522,
   534-535 (*see also* Predator-prey interactions)
  against predatory insect larvae, 547
  against predatory rotifers, 158, 162, 164-
   169, 471, 473
  against visually feeding vertebrates, 10, 16,
   19-26, 497-505, 591-592 594, 609
  against suction attack of fishes, 591-592
  primary and secondary, 16
Deme, 418
Demographics:
 general, 271-272, 274-277, 337-410
 in situ life table experiments, 346-356
 responses of:
  cladocerans, 143, 271-277, 279, 339-344
   346-356, 358-366, 367-375, 377-386,
   388-389, 395-407
  copepods, 143, 378, 380-386, 388-396,
   399-400, 402-404, 406, 411-416, 418-
   424
  rotifers, 343, 399-404, 406
 techniques for estimating parameters, 271-
  272, 337-345, 378-380
 traits:
  age-specific responses, 346, 351, 355, 357,
   377-386, 390, 393-394
  birth rates, 342-343
  brood (clutch) sizes, 339-341, 352, 365,
   378, 395-398, 418-419, 421-424
  death rates, 343-344
  delayed maturity, 369
  fecundity, 143, 268, 277, 284, 289-290,
   351, 355, 388, 395
  intrinsic rates of increase, 272, 277, 378,
   402-404
  population growth rate (*see* Population
   dynamics)
  reproductive value, 380, 383, 385-386
  size at maturity, 352-353, 358-380, 382-
   383, 386, 395-400, 402, 406, 439-454
  See also Life history traits

Detection of prey (*see* Sensory structures)
Diapause, 411–416, 418–420, 422–423, 683–684, 718
  active diapause, 411–412
  and quiescence, 412
  oligopause, 411, 413
  photoperiod regulation, 411
  relation to oxygen depletion, 413–416
  relation to predation, 415–416, 418–419, 424
Diffusion of energy flow, 758
Disc equation parameters:
  for damselfly naiads preying on cladocerans, 572
  generalization of 756
Diversity, 661–664, 673–674, 697–709
  of algae, 697–709
  of chydorids in Florida lakes, 664
  of zooplankton, reduced in tropical lakes, 673, 708–709
    suspected associations with:
      predatory fishes, 673
      blue-green algae, 673–674
Dormancy phase, 419, 683 (*see also* Diapause, Resting eggs)

Echo location of walls, 8
Effects of grazing:
  upon nutrient budgets, 251
  upon potential algal competitors, 261
Efficiency of energy transfer, 651–654
Efficient predators, 52
Egg ratio, estimates of birth and death rates, 337–345, 558–559
Eggs:
  abortion of, 350–353, 356
  cladoceran:
    male eggs, 331
    parthenogenetic, 264–267, 271, 277, 327, 329–330, 334, 337–343, 348, 350, 352–353, 356, 360–361, 365, 368, 371, 373, 375, 395–399, 402–404, 422–423, 440, 471–478, 481–484
    pseudosexual, 328
    resting, 327–328, 329–331, 334, 400
  copepod, 389, 395–396, 399, 403, 415, 419–423; resting eggs, 389–392, 395
  developmental times, 360, 369 395
  rotifer:
    parthenogenetic, 169–170
    pseudosexual, 315
    resting eggs, 315, 319, 326–328
  size of, 368, 371, 373, 375, 397–399, 422–423, 471, 477–478, 481–484
  volume of, 398, 441–444, 448
Electivity index:
  of Ivlev, 167, 294–297, 440, 447, 544, 594, 597, 599–600
  of Jacobs, 539
  as applied to
    Chaoborus feeding on various prey, 544–547
    Chaoborus, *Mochlonyx*, and salamanders feeding on *Daphnia*, 447
    microcrustaceans feeding on algae, 294–297
    rotifers feeding on algae, 167
Encounter:
  models, 53, 515–516
    for maximization of, 52
    for minimization of, 52
  probability of, 52–53, 510
  radius, 52, 516, 541
Endemic genera and species, 686, 695
  relation to benthic, littoral, and pelagic habitat, 686
Endogenous rhythms, 65, 77, 80
  of grazing, 91
  in relation to carotenoid pigmentation content, 91–92
Energy, allocation of, 122–128, 367–371, 375, 399, 418–419
Energy flow through ecosystems, 642, 650–654, 754–762
  as disrupted by perturbations, 754–762
  as related to nature of interactions, 650–654
  as related to size structure, 642, 650–654
  efficiency of, 651–654
  fraction available to fish, 651–652
  relation to growth of fish, 652–653
Ephippia, 329, 333, 440, 442, 481
"Escape in size" strategy, 551
Escape reaction, 12, 60, 167, 169, 186, 492–493, 511–512, 515, 521–522, 531, 547, 587–592, 594, 601
  of cladocerans, 19, 493
    Daphnia, 19
    Diaphanosoma, 493, 589–592
  of copepods, 12, 18, 60, 186, 492, 521–522, 531, 547
    Cyclops, 18
    Diaptomus, 547
    nauplii, 531
  of rotifers, 12, 167, 169, 492, 511–512, 515
    Polyarthra, 167, 511
  reaction to:
    filter-feeding fish, 590–592
    lowered glass tube, 594
    towed nets, 601
    suction, 588–590
Exoskeleton thickness, 471, 481
Expanded feeding envelope, 185, 191
Eye pigment diameter, 471, 594–599, 609–616, 619

diel dilation-contraction cycles of, 595, 604–608, 610, 615–616

Feedbacks in satiation response of predators, 548
Feedbacks within particle-size distributions of pelagic ecosystems, 754, 760–762
Feeding behavior:
  of Cladocerans, herbivorous:
    general, 11, 54, 87–89, 234, 242, 264–280, 282–295, 299–304, 368, 373, 557, 559, 650, 725–740
    energy to move water, 242
    filtering rates, 286, 289–290
    food rejection, 268, 273, 278–279
    gape of carapace during, 285–289
    inhibition of feeding by blue-greens, 264, 268–280, 282–289
    shape of algae, effects upon feeding, 293
    taste, 268, 278–279
    threshold concentrations of algae for filter-feeding, 264–267, 277, 279
    toxicity of algae, 268–280
  of cladocerans, omnivorous or predatory, 358, 471, 543, 549, 733
  of copepods, herbivorous calanoid (*see also* Filter-feeding mechanisms):
    general, 11, 185–196, 198–216, 219–232, 234–239, 241–248, 292–298, 650–654, 708–709, 725–727, 742–752
    active and passive selection, 185, 209, 220, 216, 228, 231–232, 235–239, 246–248, 292–298
    chemosensory grazing, 199, 209–216, 236–239, 220, 231, 242–247
    coordination problems, 204, 246
    direct observations of feeding, 244–248, of food transfer, 205, 246–247
    effects of previous feeding history, 236–237, 239
    filtering currents, 242–247
    hypothesized transfer of particles by mouthparts, 203–209, 234–239
    impaction feeding, 198, 205–209, 213, 215–216, 220, 232, 234
    list of mechanisms for removing food particles, 205, 232
    mouthpart oscillation rates, 190, 204, 215, 244
    niche separation based on, 232
    on natural particle distributions, 200–201, 219–232
    particle production by grazing, 231
    particle size selectivity hypothesis (*see* Filter-feeding mechanisms)
    raptorial feeding, 198–199, 203, 206–209, 215–216
    reconstruction of, from SEM studies, 194
    rejection of inert particles, 236–239
    release of nutrients, 251–261
    shape of algae, effects upon feeding, 220, 228, 231, 246
    structure of second maxillae, 193, 203
    switching of behavioral mode, 199, 203, 209 220, 308
  of omnivorous or predatory calanoid copepods (*see* Predator-prey interactions)
  of cyclopoid copepods, 10, 16–19, 26, 203, 206, 292–295, 297–298, 509–516, 518–536, 549, 716–723 (*see also* Predator-prey interactions)
    active and passive selection, 292–295, 297–298, 509–516, 518–536
    effects of previous feeding history, 532–533, 535
    shape of algae, effects upon feeding, 294–298
    switching behavior, 530–533, 535
  of rotifers
    herbivorous, 11, 152–157, 173–182
      density-dependent, functional response (monospecific food), 153
      in mixed cultures, 152, 155–156
    omnivorous, 158–172, 358, 509–515 (*see also* Predator-prey interactions)
      active and passive selection, 165–166
      cannibalism (*see* Cannibalism)
      chemical inhibition of, 165
      chemical stimulation of, 159–161, 166, 213
Feeding efficiency functions:
  Chaoborus-Diaptomus, 124
  cladocerans-algae, 282, 368
Filter-feeding mechanisms, 185, 188–196, 198–216
  fluid velocity and particle capture, 205, 208
  list of mechanisms, 205, 232
  preliminary observations on, in cladocerans, 282–291
    particle capture, 283–289
    particle rejection, 268, 273, 278–279, 284–287 (*see also* Blue-greens, Inhibition hypothesis)
  specific hypotheses of, for copepods:
    "biomass peak," 199–203, 220, 228–232
    "fixed-sieve," 185, 188–190, 200, 203, 207, 292
    "impaction," 194, 205–209, 215–216, 232
    "particle feeding efficiency," 199–201, 234
    "particle size selectivity," 199–201, 203–204, 228–230, 232, 261, 292
Fluid flow, 7
Fluid inertia, 3; relation to propulsion mechanisms, 3, 7, 247

Fluid mechanics, 4–7
Form (*see* Shape)
Functional responses (predator-prey):
   general, 522–528, 538, 549–551, 574–577, 723, 730, 756
   Chaoborus, 538, 549–551, 723
   cyclopoid copepods, 522–528
   odonate larvae, 569, 574–577
   (*see also* Predation)

Gene hitchhiking, 317, 334
Genetics:
   general, 315–336
   aphid-rotifer model, 327–328
   associative overdominance, 317
   assortative mating, 333
   average heterozygosity, 332
   chromosome numbers, 330
   colonization, 326–327
   cryptobiosis, 328
   cryptoparthenogenesis, 315
   cytogenetics, 330
   electrophoretic analysis of, 316–317, 332–335, 462–464, 466, 470
   embryogenesis, 329
   environmental displacement load, 322
   enzyme polymorphisms, 316–317, 330, 462–464, 466, 470
   evolution of sexual reproduction, 315, 317–328
   gene frequencies, 316–317, 332–335
   gene linkage, 317
   genetic variance, 318–320, 323, 326, 332
   Hardy-Weinberg equilibrium, 316–317, 329, 333–334, 401
      attained, 316–317, 333
      departures from, 316–317, 329
         in *Asplanchna*, 316–317
         in *Daphnia*, 334
         in *Bosmina*, 334–335
         in *Simocephalus*, 333–334
   heterozygote advantage, fixation, or predominance, 316–317, 329
   hitchhiking, 317, 334
   intensity of selection hypothesized, 322
   meiosis, 329
   migration frequency, 332
   mutation rates, 315, 323–328, 331
   of systematic groups:
      cladocerans, 77, 315, 323–335, 401, 462–464, 466, 470
      copepods, 315, 323–326, 679–680, 684
      rotifers, 315–328
   overdominance, 317, 328, 334
   parthenogenesis (*see* Parthenogenesis)
   pseudosexual eggs, 315, 328, 330
   recombination, 316, 323–328
   relation to population size, 323–328
   resting eggs (*see* Eggs)
   selection experiments, 331
   sexual reproduction (*see* Sexual reproduction)
   timing of sexual reproduction, 317–323
Geographic considerations of community structure, 625–711
   in tropical lakes:
      Lake Gatun, Panama, 594–601, 609–616
      Lake Lanao, Phillipines, 625–633
      Peten Lake chain, Guatemala, 669–677
   in Canadian rocky mountain lakes, 635–640
   in Colorado mountain and Ontario lakes, 642–654
   in African lakes, 685–696
   in Florida lakes, 657–667
   in Galapagos and Andean lakes, 697–709
   low species richness and diversity of zooplankton in tropical lakes, 673, 698, 700, 703, 705, 708–709
      absence of *Daphnia* in equatorial forests, 669, 673, 685, 689–692
      in relation to fish predation, 673, 691–692
      in relation to prevalence of inedible blue-green algae, 673
      in relation to seasonality of climate, 689–690
   ordination by taxonomic criteria, 645–647
   ordination by size and trophic criteria:
      using general sizes, 647–648
      using full developmental stages, 648–650
   associated with invertebrate and fish predation, 625–633, 635–640, 649–650, 652–654
   (*see also* Community structure)
Geographic distribution:
   arguments for using cladocerans, copepods, and rotifers in zoogeographic studies, 685–696
   difficulties with use of zooplankton in zoogeographic studies, 679–684, 685
   impact of geological events and climate upon, 685–687, 694–695
   in relation to physical and chemical parameters, 657–667, 673, 698, 708
   of algae in Galapagos and Andean lakes, 697–709
   of individual species (*see* Index of Species)
   of littoral cladocerans in:
      Danish lakes, 657–658, 664
      English lakes, 658
      Florida lakes, 657, 667
      Minnesota lakes, 658, 664, 666–667
      Mississippi river valley lakes, 658
   of pelagic entomostracans in Guatemalan lakes, 669, 671–677
   of zooplankton in mountainous trout lakes, 635–640

dominance of *Hesperodiaptomus*, 637, 640
inverse correlation between rotifers and *Hesperodiaptomus*, 637, 640
of zooplankton in African lakes, 685–696
taxonomic uncertainties on identifications, 627, 658, 687, 699
Geotaxis, negative, 28, 35
Gillraker spaces in silversides, 590
"Grass-blade" morphology, 299; of *Aphanizomonon* in association with *D. pulex*, 299–304
Growth, 143, 264–265, 268–271, 273–275, 280, 367–375, 388–407, 429
  as effected by blue-greens, 273–275
  body and helmet growth in *Daphnia*, 429–436
  of *Chaoborus*, 122–128
  of cladocerans, 264–265, 268, 273–275, 367–375, 388–389, 399–400, 405–407
    Bosmina, 368
    Ceriodaphnia, 368
    Daphnia, 265, 273–275, 368, 373–374, 431, 438–454
  of copepods
    freshwater, 143, 389–395, 401–402
      Heterocope, 394
      Leptodiaptomus, 393
      Limnocalanus, 394
      Mixodiaptomus, 390, 394
    marine, 389, 742–752
      Calanus, 748
      Pseudocalanus, 748
  simulated, 143, 742–752
Growth efficiency, 757

Hardy-Weinberg equilibrium, 316–317, 329, 333–334, 401 (*see also* Genetics)
Hyperspace, 687

Illumination, brightness distribution under water, 29–31 (*see also* Light intensity)
Indifferent equilibrium, 729–730
Inhibition hypothesis, 284–291
Isometric predation, assumption of, 760
Ivlev curves, 750

Keystone predator hypothesis, 728; modified version of, 727–728
Keystone predators, 571, 575, 722–723
Kinematic reversibility, 7
Kinetic theory of gases applied to movement, 53

Leslie matrix models of life history traits, 377–386
Life cycles of copepods, 411–416, 418–424
Life histories, 367–375, 377–386, 388–407, 411–416, 418–424

Life history tables for:
  Asplanchna, 316
  Acanthocyclops, 378
  Bosmina, 378
  Daphnia magna, 277
  Daphnia galeata, 353, 378
  Daphnia longiremis forma *cephala*, 500, 503–504
  Daphnia longiremis forma *typica*, 500, 503–504
  Daphnia pulex, 378
  Daphnia pulicaria, 353
  Diaptomus, 378
  heteroparous pattern, 390, 399, 406, 421
  multivoltine populations, 389–395, 718
  semelparous pattern, 390, 399, 406
  univoltine populations, 389–391, 394, 718
  See also Demographics
Life history traits, age-specific:
  general, 367–375, 377–386, 388–407, 744
  brood size of
    Arctodiaptomus bacillifer, 396
    Bosmina longirostris, 143, 365, 397
    Daphnia, 266
      D. ambigua, 397
      D. galeata, 365
      D. longispina, 397
      D. pulex, 397
    Eubosmina, 365
  fecundity:
    Acanthocyclops, 378
    Bosmina longirostris, 143, 378
    Daphnia galeata, 143, 355, 378
    D. magna, 143
    D. pulex, 265, 378
    D. pulicaria, 351–355
    D. rosea, 265
    Diaptomus
  age at sexual maturity:
    Acanthocyclops, 378
    Bosmina longirostris, 378
    Daphnia galeata, 353, 378
    D. magna, 277
    D. pulex, 378
    D. pulicaria, 353
    Diaptomus, 378
  reproductive value:
    Acanthocyclops, 380
    Bosmina longisrostris, 380
    Daphnia galeata, 380
    D. pulex, 380
    Diaptomus, 380
  survivorship:
    general, 367–375, 377–386
    specific:
      Chaoborus trivittatus in enclosures, 719

Life history traits, (*continued*)
   Cyclops bicuspidatus in enclosures, 721
   Daphnia pulex and *D. galeata* exposed to ultraviolet radiation, 94
   D. magna fed *Chlamydomonas* or blue-greens, 274, 276–277
   D. galeata and *D. pulicaria* in situ experiments, 353
   Diaptomus nevadensis exposed to blue light, 95–96
   (*see also* Demography, Life history tables, Reproductive characteristics)
Light intensity:
   as a factor in cyclomorphosis, 432
   as incident seasonal radiation, 70
   in temperate lakes, 101–105, 132, 141
   isopleths of, at various depths in a lake, 71, 141
   under arctic conditions, 69–70
Light refraction, creating "window" (Snell's window) underwater, 29
Littoral crustaceans:
   competition of, 569, 572–573
   copepods, 29, 521, 637, 644, 717
   defensive tactics of, 19–20
   distribution of chydorid cladocerans, 657–668, 672–673, 685, 688, 692–694, 717, 719–720
   feeding of, 287, 650
   genetics, 77, 329, 332–334
   locomotion of, 12, 80
   orientation and movement, 29, 36–37
   reproduction of, 358–365, 396–398, 402–404
   vertical migration of, 80–89
Locomotion, 11 (*see also* Swimming behavior)
Loricae of rotifers, 166
Lotka-Volterra models, 728–730, 756–757

Master equations, 754–755, 758, 760, 762
Mating behavior:
   chemical advertisement of receptivity, 48 (*see also* Pheromones)
   female reinforcement of male behavior, 48
   mate recognition, 39, 48
   mate-specific swimming behavior, 48, 52
   of calanoid copepods, 39–51
   of cladocerans, 421
   of cyclopoid copepods, 52, 60–61
   of rotifers, 61, 421
   stereotyped behavior in relation to hybridization, 49
Maxwell-Boltzmann velocity distribution, 53
Metabolic costs of cladocerans, 265
Migration:
   horizontal, 28–37 (*see also* Shore avoidance, Uferflucht)
   related to dispersal, 685–696, 697–709
   of algae, 697–709
   of zooplankton, 685–687, 695–696
   scheme of transport (long-distance), 707
   toward the littoral region, 37
   transverse to current, 36
   vertical (*see* Vertical migration)
Mimicry, 21
Mouthparts of herbivorous copepods, 185, 187–196, 203–209, 215–216, 232, 234–236, 239, 242–247

Navier-Stokes equations, 5
Newton's second law of mechanics, 4–6
Niche space, 569, 699
Niches, 687, 742
Nonequilibrium models, 730
Numerical responses of predators to prey density, 569, 574, 716, 719
Nutritional adequacy of algae, 268–280

Ocellus: movement in *Mixodiaptomus* and *Daphnia*, 29; hypothetical, minimal pigment cups, 33–35
Optimal foraging, 543; with multiple search modes, 575–557
Orientation, 28–38
   in relation to lake outlets, 36
   optical, 28–38, 65, 72–73
   preferred body, 28
   primary and secondary, 28
   of body in brightness field, 29–31, 33–34, 65–67, 72
   of dead animals, 28
   of littoral and pelagic organisms, 29, 36–37
Overdominance, associative, 317, 334

Palatability, 10, 19–26, 531; of algae, 268–280, 293, 298; of water mites to fish and invertebrates, 21–23
Paradox: of the plankton, 698, 706; of the zooplankton, 698
Parthenogenesis, 315, 323–328, 329–400, 476, 687, 709
   advantage of, 327–328, 400–401
   cyclical, 323, 329, 388
   facultative, 400
   obligate, 315, 323, 329, 400, 476
   (*see also* Cryptoparthenogenesis)
Particle selectivity curve, 235
Particle size distributions, 175–177, 198–216, 219–232, 234–239, 650–654, 726, 734, 739, 754–762
   for algae:
      Chlamydomonas reinhardi, 266–227
      Cosmarium, 227
      Pediastrum, 227

for algae under grazing from:
  Acartia, 238
  Brachionus, 176
  Diaptomus, 226–227
  Pseudocalanus, 200
expected results of feeding, 284
manufactured microcapsules (simulated food particles) grazed by *Acartia* and *Eurytemora*, 211–212
maximum limitations, 285, 290
natural particles under grazing
  by the calanoid copepods:
    Acartia, 201
    Diaptomus, 224–225, 229
    Eurytemora, 201, 229–230
    Oithona, 201
    Pseudocalanus, 200–201
    Temora, 201
  by the cladocerans:
    Chydorus, 283
    Daphnia, 283
Partitioning of resources, 570–571, 625–626.
  (*see also* Competition)
Passive dispersal, 685–687, 694–696, 707
Patchiness: of algae, 626; of zooplankton, 626
Perturbations and perturbation waves, 715–716
Pharynx of rotifers, 159
Pheromones of copepods, 47, 48, 53, 60, 400–401
Photoprotection: against ultraviolet radiation, 92–94, 98–99; against blue-light radiation, 92–95, 98–109
Phytoplankton species richness, 697; area-species relationships, 698, 703
Pigmentation:
  geographic distribution of pigmented forms, 92, 99, 106, 109
  nature of pigments, 91–92, 98–99
  photoprotection function, 91–96, 98–109
  relation to visual predation, 471, 595–602, 609–616, 618–622
  seasonal variation in, 99–103
  warning coloration, 20
Plankton sorter, 519
Polymorphisms, 95–96, 161–168, 316–317, 331–335, 421, 453, 457, 460–464, 466, 481–493, 497–505
  body size, 421, 453
  color, 95–96
  dimorphism, 497
  electrophoretic, 316–317, 332–335, 462–464, 466, 486, 488, 490
  in:
    Asplanchna, 161–165
    Bosmina, 457, 460–464, 481–493
    Brachionus, 167, 168

Daphnia, 497–505
Filinia, 168
use of term polymorphism, 331–332, 609
visual, 331–352
(*see also* Cyclomorphosis, Genetics)
Population dynamics
  Growth rates (population), 143–146, 173–182, 268, 277, 299–304, 305–311, 337–357, 364–374, 377–386, 402–404, 555, 558–560, 728–730, 743–752, 760–762
  logistic models of, 182, 728–730
  predator-prey oscillations, 179–182, 567
  steady state, 178–182
  (*see also* Demographics, Life history traits, life tables, Reproductive characteristics)
Predation:
  breakdown of predator-prey interaction into various probabilistic components, 510, 588
  density-dependent influences, 521–528, 538, 574–577, 716, 723
  effects upon diversity, 715–723, 726–740
  energetic considerations of, 53, 59, 122–128
  general features of population responses:
    developmental, 569, 574
    functional, 522–528, 538, 574–575, 723, 756
    numerical, 569, 574, 716, 723
  interference between predators, 521, 527–528, 529
  mechanisms of fish predation, 581–585
  modifying influences:
    algae density (as alternative food), 529–530, 535
    alternative prey, 551
    conditioning, 519–520, 533–534
    containers, 515, 520, 533
    experience or training, 23–26, 532–533, 535, 602
    pauses (discontinuity) in feeding, 521, 534
    prey motion, 594–602
    satiation, 533, 539, 548, 574
    temperature, 528–529, 535
    review of hypotheses, 726–728
    switching behavior, 530–533, 535, 756
  (*see also* Attack behavior, Cyclomorphosis, Defenses, Wasteful killing)
  (*see also, for systematic treatment of predation,* Predator-prey interactions)
Predator-prey interactions:
  invertebrate predators
    copepods, 10–19, 39, 52–61, 104–106, 471, 483, 490–493, 497, 499–505, 509–516, 518–536, 557, 563–566, 635, 637–640
    calanoid copepods, 98, 104–106, 497–505, 514, 557, 563–566, 635, 637–640

Predator-prey interactions, *(continued)*
    Diaptomus articus—D. sicilis, 514
    Diaptomus nevadensis—D. sicilis, 98, 104–106, 514
    Epischura lacustris—Bosmina longirostris, 491
    Hesperodiaptomus—rotifers, 635, 637–640
    Heterocope septentrionalis—Daphnia longiremis, 497–505
  cyclopoid copepods, 16–19, 26, 509–516, 518–536, 549, 717–723
    Cyclops, in general, on various prey: *Bosmina,* 17–19, 26; *Chydorus,* 19–20; *Daphnia,* 17–19
    Cyclops bicuspidatus on various prey: *Asplanchna,* 509, 512–516; *Diaptomus,* 514, 516; *Tropocyclops,* 716–723
    Mesocyclops, in general, on various prey: calanoid nauplii, 519, 522–532, 534; its own immature stages, 522–523, 530, 532, 534; microcrustaceans, 518–537; rotifers, 509–516
    Mesocyclops edax, on various prey: *Asplanchna girodi,* 509, 511–516; *Ceriodaphnia quadrangulata,* 529; *Diaptomus floridanus,* 534; *Keratella cochlearis,* 509, 512–516; *Polyarthra vulgaris,* 509, 512–516
    Mesocyclops leukarti, on various prey: *Bosmina meridionalis,* 519, 534–535; *Calamoecia lucasi,* 519, 521–527, 529–534; *Ceriodaphnia dubia,* 519, 522–535; *Chydorus sphaericus,* 519, 522–527, 534–535
  larvae (predatory) of insects, 106–109, 122–128, 439–440, 446–447, 471, 538–554, 569–577, 715, 718–723, 727
    midge larvae (*Chaoborus*), 106–109, 122–128, 439–440, 446–447, 471, 538–551, 621–622, 625, 627–633, 715, 718–723, 727
    Chaoborus, in general on various items: algae, 542
      benthic fauna, 542
      summary of items, 544–545, 550–551
    Chaoborus, on various prey:
      Chaoborus—Daphnia pulex, 439–440, 446–447, 471
      Chaoborus americanus and *C. trivittatus— Diaptomus kenai,* 106, 108–109, 122–128, 541
      Chaoborus flavicans—Diaptomus sicilis, 106–107
      Chaoborus trivittatus, on various prey: *Bosmina longirostris,* 541; *Diaptomus hesperus,* 106, 108; *Diaptomus kenai,* 106, 108, 541; zooplankton, 715, 718–723
      Chaoborus trivittatus and *C. flavicans* on various prey: *Daphnia pulicaria,* 109, 538–539, 543–551; *Diaptomus franciscanus,* 106, 108–109, 538–539, 547, 549, 551
  Mochlonyx larvae, 439–440, 446–447
  odonate larvae, 569–577
    Anomalagrion hastatum—Simocephalus, 573–576
    Ischnura verticalis—Simocephalus, 569
  mysids, 555–568
    Mysis relicta, on various prey: *Bosmina longirostris,* 555, 559–567; *Daphnia,* 555, 559–560; *Diaptomus tyrelli,* 555; *Kellicottia longispina,* 555
  rotifer (*Asplanchna*), 158–171
    Asplanchna girodi on various prey: *Keratella cochlearis,* 509; *Polyarthra vulgaris,* 509
  vertebrate predators
    fishes, in general, upon various prey: *Chaoborus,* 609–616, 618–622, 629, 633; microcrustaceans, 10–11, 16, 88, 138–146, 300–301, 347, 356, 358–359, 363–364, 415, 571, 484, 497, 499–500, 502–505, 509, 580–585, 587–592, 609–616, 629, 631, 635–646, 649, 651–654, 669, 676, 725–728, 732, 735–736, 739–740
    odonates, 570–575
    water mites, 10, 19–26, 135
    in specific interactions:
      Alosa pseudoharengus—Daphnia pulex, 581–584
      Coregonus artedii—Daphnia pulex, 581–583
      Dorosoma cepedianum—zooplankton, 587–588, 590–592
      Lepomis macrochirus—Bosmina, 614–615
      Melaniris chagresi—Bosminidae, 594–600, 609–616
      Menida audens—zooplankton, 587–588, 590–592
      Perca fluviatius and *Trichogaster trichopterus—Chaoborus,* 621
    salamanders (*Ambystoma*), 439–440, 445–446, 471
Predictability of the environment, 418, 423, 473
Principal components ordination:
  of community composition, based upon: taxonomic criteria, 642–650; body size and trophic criteria, 642–650; physical and chemical characteristics of lakes, 661–666
  of morphology for separation of size and shape

components in studies of cyclomorphosis, 475–476
Propagation of perturbations of pelagic biomass structure, 754, 760–762
Proprioceptors, related to satiation of predators, 548
Protective coloration of water mites, 10, 19–26
Pseudosexual eggs, 315, 328, 330 (*see also* Genetics)
Pseudotrochal "screens," 152, 154–156
Pseudotrochus, 153–154; of cirri, 154

Receptors (*see also* sensory structures)
Remineralization of nutrients, 251
Replacement cycles, definition of, 473
  cryptic succession in rotifers, 316, 322
  in cladocerans, 443
    in *Daphnia*, 464
    in *Bosmina*, 464, 466–468, 470, 484–488
Reproductive characteristics:
  age at first reproduction, 277, 353, 367, 369–371, 373, 383, 395–300
  as influenced by blue-greens or fouling phytoplankton, 268–280, 284–285, 290
  birth rates, 337, 342–345, 377, 379–380, 559
  brood size, 143, 265–267, 330, 334, 338–340, 343, 361, 365, 367–371, 378, 396–406, 414, 442
  fecundity, 143, 265–268, 277, 284, 289–290, 351, 355, 378, 397, 418, 559
  intrinsic rate of natural increase (max.), 402–404
Reproductive value, 380
  (*see also* Demographics, Eggs, life histories, Life history tables, Life history traits)
Reproductive effort, 371, 418–419, 421
Reproductive structures of the calanoid copepod *Labidocera*, 40–47
  coupling plate, 40, 42, 46
  fifth leg of males, 41
  gonopore, 42, 44
  pit-pores, 42, 45, 47
  spermatophore, 42–43
Resource heterogeneity, 725
Response surface, 723
Resting eggs, 330, 687, 690 (*see also* Eggs)
Reynolds number, 3, 6, 7, 11–12
  definition of, 6
  relation to streamlining, 12–16
  uses and implications of, 6, 11–12, 12–16
r-K theory, 305–311, 318, 399, 406, 423, 706, 718, 722
  analogy in nutrient-algae-herbivore system, 305–311
  as applied to:
    algae, 706
    cladocerans, 399, 406

copepods, 399, 423
rotifers, 318, 399
Rotifer
  defenses against *Asplanchna* predation, 158–172; against *Asplanchna* cannibalism, 164–165
  distribution, 627, 637–639
  feeding, 54, 151–171, 358, 509–515 (*see also* Feeding behavior)
  food specific behaviors, 151, 153–155 (*see also* Feeding behavior)
  genetics, 316–317, 319, 322, 325, 327
  reproduction, 319, 321–322, 342, 404
  swimming, 54
  swimming behavior, 11–12, 61
  vertical migration, 120

Saccate morphotype of *Asplanchna*, 159–162, 165–166
Search behavior,
  of fishes, 16, 142, 580–581, 588–591, 594–595, 600–602
  of predatory copepods, 16–20, 60
    Cyclops, 17–20, 60, 540
    Epischura, 14–15, 540
    Mesocyclops, 522
  of odonates, 573, 576–577
Searching image, 602
Secretory structures:
  anal pit pores, 42, 45
  labial pores, 10, 187, 192, 196
Sensory structures:
  general, 10, 39, 209, 424
  aesthetase hairs, 47
  on various groups
    cladocerans, 12–13
    copepods, 12, 14, 39, 47–48, 185–190, 209
      chemoreceptors, 12, 14, 39, 47–48, 186–189, 209, 215–216, 245
      mechanoreceptors, 12, 14, 39, 186–189, 209, 215, 244
        used in food-gathering: on second maxillae, 193, 209; on mouthparts in general, 188–189, 215–216, 245
      receptors (unspecified)
        used in mating, 42, 45, 209
        found on gonopore, 44
        on male chela, 45, 47
    midge larvae (*Chaoborus*), 539–546 (*see also* Chordotonal organs)
    rotifers, 159–160
Sequence of behavioral components in predator-prey interactions, 510, 527, 548, 588
Sexual reproduction
  general, 315–328, 329–334
  evolution of, 325–328
  induction of, 319, 322–330

Sexual reproduction, (*continued*)
  in cladocerans, 315, 323-328, 329-334,
  in copepods, 315, 323-326, 400, 418-424
  in rotifers, 317-328
  self-fertilization, 334
  timing of, 317-323, 419-424
Shape:
  general body form, 10-19, 26, 474
    of various groups, general body form, 231, 243, 246-247, 268-270
      algae, 220, 292-298, 299-304
      cladocerans, 11, 13, 16, 596
      copepods, 11-12, 14-19, 679-684, 744
      rotifers, 11
      midge larvae, 539-541
    body projections, 429-436, 438, 454, 456-468, 470-471, 477-493, 497-505 (*see also* Cyclomorphosis)
    of copepod mouthparts, 185-190, 192-196, 206-209, 239, 246
    relationship to:
      feeding, 285, 543-547
      mating, 39-50
      predator handling of prey, 16, 123, 161-172, 268, 299-304, 331-332, 358, 428-505, 509, 543-547
      Reynolds number, 11-16
Shore avoidance, 28, 31, 55, 59 (*see also* Uferflucht)
Sibling species, 470
Size distributions (*see* Particle size distributions)
Size-efficiency hypothesis, 282, 289, 359, 405, 535-536, 726-727
Size-related physiological processes, models of, 368-370, 742-752
Size-selective feeding:
  general, 293, 371-375, 377-386, 388, 404-405, 512-513, 515, 527, 531, 534-535, 543, 650
  by: *Chaoborus*, 372-373 538, 542-546
    cladocerans:
      herbivorous, 283, 293
      carnivorous, 543 549
    copepods:
      herbivorous, 198, 293
      carnivorous, 512-513, 515, 543
    rotifers, 176
Spatial heterogeneity
  algae, 631
  clonal, 476, 483
  components
    depth-abundance relationship, 627-632
    ephemeral component, 626, 629
    fixed component, 626-633
    limnetic, 626-633
    littoral, 626
  tropical cladocerans, 625-633

Sperm, storage of in copepods, 419
Spermathecal sac, 419
Spermatophore, 39
  coupler, 40, 42, 46
  discharge of, 47
  morphology of, 40, 42
  multiple attachments of, 48
  preparation for placement of, 45, 46
  range of variation of, 50
  transfer of, 43, 47
Stable states (*see* Alternative stable states)
Starvation, effects upon:
  Asplanchna, 160, 162
  helmet growth, 430
  later food ingestion, 264, 267
  mortality in presence of light, 93
  nutrient regeneration, 254
  vertical migration, 74-75
Stokes approximation, 6
Stretch receptors in stomach, hypothesized for:
  Asplanchna, 160; *Chaoborus*, 548-549
Strike efficiency of *Chaoborus*, 543, 546
Stroking, 47
Strouhal number, definition, 6; uses, 6, 8-9
Structure, chemical, of carotenoids, 92
Submergent behavior, 374
Surplus killing, 526-528 (*see also* Wasteful killing)
Survivorship (*see* Demographics, Life history traits)
Suspension feeders (*see* Filter-feeding)
Swimming behavior, 7, 11-19, 52-66, 73
  avoidance reaction (*see* Avoidance, Escape reaction)
  diel cycles of swimming activity, 73-74, 77
  energetic costs of, 53, 59
  flexibility of, 12
  in relation to mating, 47-50, 60-61
  of confined animals, 8-9
  of various organisms:
    cladocerans, 8, 11-16, 73
    copepods, 12
      calanoids
        gliding, 14
        turning, 15
      cyclopoids, 12, 17-18, 59, 153, 158
    ostracods, 8
    rotifers, 11-12, 61
    water mites, 15
Swimming speeds:
  average speeds, 12, 52-59
  comparison between herbivorous and predatory cladocerans, 54-55
  during vertical migration, 33
  escape velocities, 19
  horizontal and vertical components, 55
  instantaneous speeds, 12, 19

of blue and red morphs (copepod), 96
of various organisms:
  Chaoborus, 54–55
  cladocerans, 8–9, 56, 59, 83
  copepods, 8, 54
    calanoids, 54
    cyclopoids, 54, 56–58
  ostracods, 8
  rotifers, 54, 56, 58, 158
Swimming wakes (*see* Wakes)
Switch, from asexual to sexual reproduction, discussion of factors which initiate, 330–331
Switching by predatory copepods, 530–533, 535, 736 (*see also* Predation)
Synchronization of emergence and reproduction, 419, 421
System organization, 715

Taste of prey, 547
Threshold concentrations, 264–267, 277; definition, 264–265
Toxicity:
  of algae as food,
    for cladocerans, 268
    for blue-greens, 264, 268, 317
Transparency, as a general adaptation to visual predators, 609, 728
Trophi, 159
Trophic status in relation to swimming speeds, 54

Uferflucht, 28, 35, 37 (*see also* Shore avoidance)

Velocity: 3-dimensional definition, 4; instantaneous (*see* Swimming speeds)
Vertical migration:
  connection with shore avoidance, 28
  definition of, 65
  list of causes for, 66
    proximate causes, 65, 71, 111
      definition of, 65
      directional stimuli, 65–67, 72–73, 80
      lag times, 82–83
      releasing stimuli, 65–67, 71–72, 74–76, 80–89, 119
    ultimate causes, 65, 138–139
      definition of, 65
  in relation to,
    avoidance of predation, 66, 88–89, 104–106, 135, 138–146, 621, 633, 669, 676
    reduction of competition, 66, 88, 111, 120
    food-gathering (niche differentiation), 66, 74–76, 111–120, 123, 136, 138
    metabolic efficiency, 66, 122–128, 135–136, 138, 402
    "optimal" light intensity (preferendum hypothesis), 66, 70–73
    photo damage, 98, 106–109
  modifying influences, 67, 76–78, 130
    pH, 130
    light, 69, 80–84, 129, 132–133
    temperature, 74–76, 129, 130, 133
    age, 74–76
    oxygen concentration, 80, 85–89, 129–130, 133
    genotypic constitution, 66–67, 77
  reverse migrations, 72, 87, 136
Viscosity, 3, 5–7
  definition, 5
  effects upon movement, 7–9
  implications for animal size and shape, 11, 16, 242–248
Visual acuity of fishes, 141
Visual observations, 10
Visual pigments, 76
von Foerster equation, 756, 758

Wakes, of prey, 14–16, 18, 547
Wasteful killing, 548–547 (*see also* Surplus killing)
Water mites:
  distastefulness of, 10, 19–26, 135
  feeding, 136
  general coloration, 20–26
  life history stages, 129
  predators upon, 21–23, 135
  swimming of, 15
  vertical migration of, 129–136

Zeitgeber, 73

Library of Congress Cataloging in Publication Data

Main entry under title:

Evolution and ecology of zooplankton communities.

    Papers presented at a symposium at Dartmouth College, August 20–25, 1978, sponsored by the American Society of Limnology and Oceanography.
    Includes bibliographies and indexes.
    1. Zooplankton—Ecology—Congresses.
2. Zooplankton—Evolution—Congresses. I. Kerfoot, W. Charles, 1944– II. American Society of Limnology and Oceanography.
QL143.E96   592   80-50491
ISBN 0-87451-180-1